Florencio Zaragoza Dörwald

Lead Optimization for Medicinal Chemists

Related Titles

Smith, D. A., Allerton, C., Kalgutkar, A. S., van de Waterbeemd, H., Walker, D. K.

Pharmacokinetics and Metabolism in Drug Design

2012
ISBN: 978-3-527-32954-0

Gad, S. C. (ed.)

Development of Therapeutic Agents Handbook

2012
ISBN: 978-0-471-21385-7

Tsaioun, K., Kates, S. A. (eds.)

ADMET for Medicinal Chemists
A Practical Guide

2011
ISBN: 978-0-470-48407-4

Sotriffer, C. (ed.)

Virtual Screening
Principles, Challenges, and Practical Guidelines

2011
ISBN: 978-3-527-32636-5

Curry, S. H., Whelpton, R.

Drug Disposition and Pharmacokinetics
From Principles to Applications

2011
ISBN: 978-0-470-68446-7

Rankovic, Z., Morphy, R.

Lead Generation Approaches in Drug Discovery

2010
ISBN: 978-0-470-25761-6

Han, C., Davis, C. B., Wang, B. (eds.)

Evaluation of Drug Candidates for Preclinical Development
Pharmacokinetics, Metabolism, Pharmaceutics, and Toxicology

2010
ISBN: 978-0-470-04491-9

Faller, B., Urban, L. (eds.)

Hit and Lead Profiling
Identification and Optimization of Drug-like Molecules

2009
ISBN: 978-3-527-32331-9

Florencio Zaragoza Dörwald

Lead Optimization for Medicinal Chemists

Pharmacokinetic Properties of Functional Groups and Organic Compounds

WILEY-VCH

WILEY-VCH Verlag GmbH & Co. KGaA

The Author

Dr. Florencio Zaragoza Dörwald
Lonza AG
Rottenstrasse 6
3930 Visp
Switzerland

Cover illustration:
Mobile in the style of Alexander Calder.

Library of Congress Card No.: applied for

British Library Cataloguing-in-Publication Data
A catalogue record for this book is available from the British Library.

Bibliographic information published by the Deutsche Nationalbibliothek
The Deutsche Nationalbibliothek lists this publication in the Deutsche Nationalbibliografie; detailed bibliographic data are available on the Internet at <http://dnb.d-nb.de>.

Cover Adam-Design, Weinheim

Typesetting: Laserwords Private Limited, Chennai, India

Printing and Binding: Printed and bound by CPI Group (UK) Ltd, Croydon, CR0 4YY

Printed on acid-free paper

Print ISBN: 978-3-527-33226-7
ePDF ISBN: 978-3-527-64566-4
oBook ISBN: 978-3-527-64564-0
ePub ISBN: 978-3-527-64565-7
Mobi ISBN: 978-3-527-64567-1

C9783527332267_270924

Contents

Introduction

The aim of this book is to provide medicinal chemists with a guide to the effect of structural modifications and functional groups on the pharmacokinetic (PK) properties of compounds, illustrated by a full collection of PK data. How does the plasma half-life and oral bioavailability change on small structural modifications? Which structural elements are allowed for a drug, and which are interchangeable, without significantly altering the PK properties? How can the oral bioavailability and half-life of a lead be increased? Questions of this type, critical to every small molecule drug development program, are addressed herein.

The tables contain most drugs for which human PK data have been reported. Some compounds in clinical development or with animal PK data only have also been included. The compounds listed include currently available drugs, withdrawn drugs, and compounds that failed during clinical trials. Compounds are arranged by keeping structurally closely related drugs together, to show variations of PK properties on minimal structural modification. This kind of information should be particularly valuable for medicinal chemists in their endeavor to improve the PK of their leads.

PK data are no natural constants inherent to a compound, such as its melting point or density, but depend on the sex, age, and condition of the patient, his diet, the dose of the drug, and the hour at which the measurement is performed. In a typical study with 5–10 healthy volunteers, half-life and oral bioavailability may show strong interindividual variations. Even for a single patient, PK properties of a drug will change with each new dosing. Thus, the establishment of a true PK SAR (structure activity relationship) for drugs would require PK data obtained from huge groups of people. Understandably, such information is not available.

The accuracy of compound quantization in plasma and thus the quality of PK data has increased dramatically in recent years. Moreover, a more complete PK profile is requested today by the regulatory agencies than in the past. Therefore, older PK data is usually less accurate and less complete than more recent data.

Because PK is highly variable between individuals, even high-quality data must be used with caution by physicians. Also, medicinal chemists should keep in mind that (i) reported PK values are always inaccurate and (ii) vary strongly between patients. Nevertheless, short of anything better, the values reported herein should give chemists an estimate of the PK properties of compound classes and a knowledge of how structural modifications will probably affect PK.

The PK data and other information presented herein was collected from various sources:

1) Books: Goodman and Gillman's *The Pharmacological Basis of Therapeutics,* 9th edn, Hardman, J. G.; Limbird, L. E. eds., McGraw-Hill, 1996. Forth, W.; Henschler, D.; Rummel, W. *Pharmakologie und Toxikologie*. Mannheim: B.I. Wissenschaftsverlag, 1987. Jack, D. B. *Handbook of Clinical Pharmacokinetic Data*. Macmillan Publishers Ltd, 1992. Bochner, F.; Carruthers, G.; Kampmann, J.; Steiner, J. *Handbook of Clinical Pharmacology*, 2nd edn, Boston, Toronto: Little, Brown & Co, 1983;
2) Peer-reviewed publications, for example, Obach, R. S.; Lombardo, F.; Waters, N. J. Trend analysis of a database of intravenous pharmacokinetic parameters in humans for 670 drug compounds. *Drug Metab. Dispos.*, 2008, **36**, 1385–1405; Smith, D. A.; Jones, B. C.; Walker, D. K. Design of drugs involving the concepts and theories of drug metabolism and pharmacokinetics. *Medicinal Research Reviews*, 1996, **16**, 243–266;
3) Public online databases: EMA-filings (www.ema.europa.eu), the swiss Arzneimittel Kompendium (www.kompendium.ch), the drug bank (www.drugbank.ca), banque des donnêes automatisêe sur les medicaménts (www.biam2.org), clinical drug use (www.clinicaldruguse.com), PharmGKB (www.pharmgkb.org), medsafe (www.medsafe.govt.nz).

On minor discrepancies between sources, the mean value was calculated. In case of major inconsistencies, preference was given to newer data. Clearance (CL) is reported as given in the literature; its consistency with the reported values of $t_{1/2}$ and V was not verified.

The values for polar surface area (PSA) and log P were calculated with ACD/Labs (Advanced Chemistry Development), Version 11.02 (uncharged compounds), or with Molinspiration Cheminformatics (www.molinspiration.com) (charged compounds).

The historical background of many drugs and therapeutic strategies was gathered from the books cited above, from peer-reviewed articles, and from the excellent assays of Walter Sneader, Professor at the Department of Pharmaceutical Sciences at the University of Strathclyde in Glasgow, Scotland.

Glossary and Abbreviations

ACE	angiotensin converting enzyme
ADME(T)	absorbtion, distribution, metabolism, excretion, (toxicity)
ADP	adenosine diphosphate
AMPA	α-amino-3-hydroxy-5-methyl-4-isoxazolepropionic acid receptor, a glutamate receptor
ATP	adenosine triphosphate
AUC	area under the curve; area under a plot of the concentration of the drug in plasma as function of time
AUMC	area under the first moment of the concentration–time curve
bbb	blood–brain barrier
BCL	B-cell lymphoma
BCNU	*N*,*N*′-bis(2-chloroethyl)nitrosourea, carmustine
bp	boiling point
CDK	cyclin-dependent kinases
CETP	cholesteryl ester transfer protein
CCK	cholecystokinin
CETP	cholesteryl ester transfer protein
CGRP	calcitonin-gene-related peptide
CL	clearance in ml min^{-1} kg^{-1} = dose/AUC; amount of plasma that is freed of the drug in a given time; $CL = \ln 2 \times (V/t_{1/2})$
CNDAC	1-(2-C-cyano-2-deoxy-β-D-arabinopentofuranosyl)cytosine
CNS	central nervous system
COMT	catechol-O-methyltransferase
COX	cyclooxygenase, enzyme involved in the conversion of arachidonic acid into prostaglandines
CPIB	*p*-chlorophenoxyisobutyric acid
CYP	cytochrome *P*450, group of hepatic enzymes that mediate oxidative metabolism
DMPK	drug metabolism and pharmacokinetics
DPPIV	dipeptidylpeptidase IV
EMA	European Medicines Agency
F	oral bioavailability; AUC (oral dose)/AUC (iv dose)
FDA	Food and Drug Administration
GABA	γ-aminobutyric acid, central neurotransmitter
GI	gastrointestinal

GLP-1	glucagon-like peptide 1
GnRH	gonadotropin-releasing hormone, also called LHRH
GPCR	G-protein coupled receptor
HCV	hepatitis C virus
hERG	human ether-à-go-go related gene
HIV	human immunodeficiency virus
HMG CoA reductase	3-hydroxy-3-methylglutaryl coenzyme A reductase, enzyme involved in the biosynthesis of cholesterol
5-HT	5-hydroxytryptamine, serotonin
im	intramuscular
ip	intraperitoneal
iv	intravenous
LH	luteinizing hormone
LHRH	luteinizing-hormone-releasing hormone, also called *GnRH*
log*P*	log of partition coefficient of drug in a two-phase system octanol/water
Δlog*P*	log *P* (octanol/water) − log *P* (cyclohexane/water)
MAO	monoamine oxidase
MAP	mitogen-activated protein
MEK	mitogen-activated protein kinase kinase
mp	melting point
MRT	mean residence time = AUMC/AUC
Mwt	molecular weight
nd	not determined
NF-κB	nuclear factor kappa-light-chain-enhancer of activated B cells
NK	neurokinin
NMDA	*N*-methyl D-aspartic acid
NSAID	nonsteroidal antiinflammatory drug
PARP	poly(ADP-ribose)polymerase
pb	extent to which drug is bound to plasma proteins
PDE	phosphodiesterase
P-gp	P-glycoprotein, a 170 kDa protein encoded by the multidrug resistance gene 1, which pumps drugs out of the resistant cells
PK	pharmacokinetic(s)
po	per os, oral
PSA	polar surface area
PPAR-γ	peroxisome-proliferator-activated receptor γ
QT	interval time for both the ventricular depolarization and repolarization
R&D	research and development
SAR	structure activity relationship
sc	subcutaneous
SSRI	selective serotonin reuptake inhibitor
$t_{1/2}$	plasma half-life of unchanged drug after oral administration to humans
ur	fraction of orally dosed drug being excreted unchanged in urine
V	volume of distribution = MRT × CL
VEGF	vascular endothelial growth factor

Part I
Introduction

1
The Drug Discovery Process

Drugs are compounds that interact selectively with certain proteins in the human body and, thereby, suppress or activate biochemical pathways or signal transmission. Although the structures of modern drugs hardly allow to guess their origins, these were mostly natural products, discovered empirically, and used for centuries [1]. While synthesizing and evaluating new structural analogs of known hormones or natural drugs, new therapeutic applications often emerged. A fruitful starting point for the development of new drugs has in fact often been an old drug [2]. Illustrative examples of "drug evolution" are shown in Scheme 1.1.

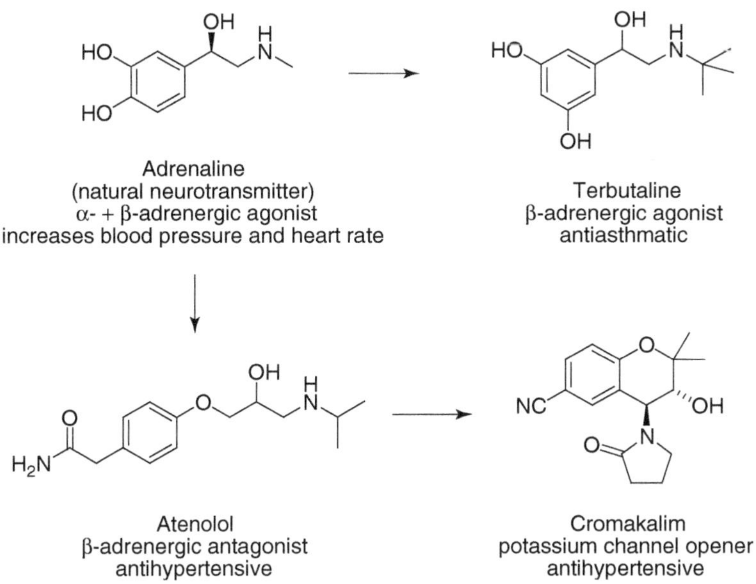

Scheme 1.1 Examples of the development of new drugs from old drugs.

Lead Optimization for Medicinal Chemists: Pharmacokinetic Properties of Functional Groups and Organic Compounds, First Edition. Florencio Zaragoza Dörwald.

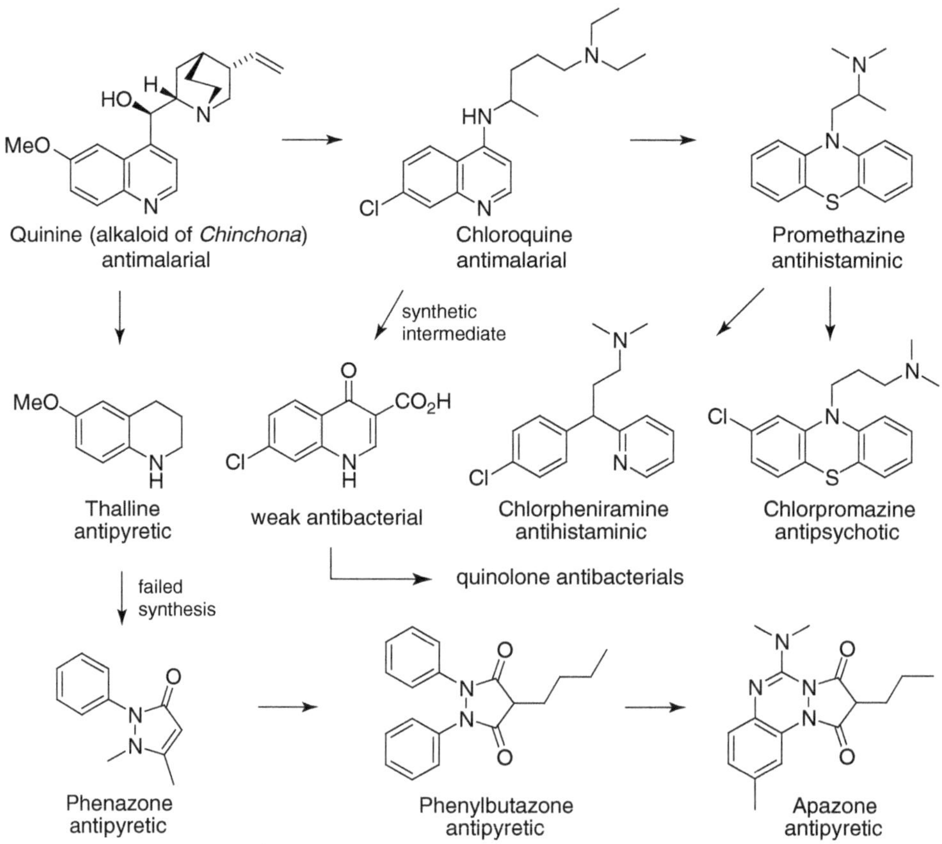

Scheme 1.1 *(Continued)*.

To initiate a drug discovery program with hormones, natural products, or old drugs as leads has a number of advantages: the biochemical concept is already proven (the compound "works"), the target is "druggable," and, importantly, the lead structure has acceptable or at least promising PK/ADME (pharmacokinetics/absorption, distribution, metabolism, and excretion) properties; otherwise, it would not work.

High-throughput screening of compound collections only rarely provides leads for new drugs [3]. Notable exceptions include the dihydropyridine calcium channel blockers, benzodiazepines, and sulfonamide antibacterials. These important drug classes resulted from testing drug-unrelated chemicals.

New proteins are constantly being discovered, and many of them are potential targets for therapeutic intervention. It must be kept in mind, though, that only few drugs have been successfully developed from scratch, starting with a biochemical hypothesis. The odds for succeeding are higher with a lead that works *in vivo* or by optimizing or exploiting a side effect of an old drug.

Today, the development of a new drug usually comprises the following stages:

1) **Discovery**
   A target protein is selected, for example a receptor or an enzyme. If no lead structure is available, high-throughput screening of suitable, leadlike compounds may yield some weak ligands (hits). Systematic structural modification of these, supported by *in vitro* assays, may provide a lead, that is, a compound with an unambiguous dose–response relationship at the target protein. Further structural modifications aim at improving potency at the target, selectivity (low affinity to other proteins), water solubility, pharmacokinetics (PK, oral bioavailability, half-life, CNS penetration, etc.), therapeutic index (LD_{50}/ED_{50}), and general ADMET properties (absorption, distribution, metabolism, excretion, and toxicity). In addition to *in vitro* assays, the medicinal chemist will need guidance by more time- and compound-consuming *in vivo* pharmacology.

2) **Preclinical development**
   Once a development candidate has been chosen, the following steps will be initiated: chemical scale-up of the selected compound, formulation, stability studies, more detailed metabolic, toxicological, and PK studies (only in animals, typically rodents, dogs, pigs, or primates).

3) **Clinical development**

Phase I	Healthy volunteers; determination of the suitable dose for humans
Phase II	First studies with a small group of patients
Phase III	Extended clinical trials

Thus, the fateful selection of a development candidate, which will either fail or succeed during the expensive preclinical or clinical development, is already taken in the discovery phase. Success in drug development is, therefore, primarily dependent on the medicinal chemist, on his ability to design and prepare the compound with the desired biological properties. No matter how hardworking and talented the members of the preclinical and clinical development team are, if the medicinal chemist has not delivered the right compound, the whole project will fail. And the later the recognition of the failure, the larger the costs.

Thus, pharmaceutical companies should allocate significant resources to the hiring and training of their medicinal chemists.

1.1
Pharmacokinetics–Structure Relationship

During the discovery phase of a new drug, two different, mutually independent sets of properties of the compound must be optimized: (i) potency and selectivity at the target protein and (ii) ADME and toxicity. The most critical ADME/PK parameters are as follows:

- **Plasma half-life ($t_{1/2}$)**: The time required for the plasma concentration of a drug to drop by 50%. A constant half-life means that the rate of elimination of a drug is a linear function of its concentration (first order kinetics). This is never exactly the case, and $t_{1/2}$ will usually increase as the concentration of the drug declines.

- **Oral bioavailability (*F*)**: The fraction of a drug that reaches systemic circulation after oral dosing. Oral bioavailability is determined by dividing the area under the curve (AUC) for an oral dose by the AUC of the same dose given intravenously. A low *F* means that either the drug is not absorbed from the gastrointestinal (GI) tract or that it undergoes extensive first-pass metabolism in the liver.

- **Plasma protein binding (pb)**: Hydrophobic compounds will bind unspecifically to any hydrophobic site of a protein. For this reason, high-throughput screening often yields hydrophobic hits (which are difficult to optimize and should be abandoned). Plasma proteins, such as albumin or α-glycoproteins, may also bind to drugs and thereby reduce their free fraction in plasma, their renal excretion, their ability to cross membranes (also the blood–brain barrier (bbb)), and their interaction with other proteins (metabolizing enzymes, the target protein). Binding to plasma proteins also prevents highly insoluble compounds from precipitating upon iv dosing and helps to distribute such drugs throughout the body.
  The half-life of peptides may be increased by preventing their renal excretion through enhanced binding to albumin. This can be achieved by acylating the peptide with fatty acids or other plasma protein binding compounds.
  Plasma protein binding is usually determined by equilibrium dialysis or ultrafiltration. Both techniques exploit the ability of certain membranes to be permeable to small molecules but not to proteins or protein-bound small molecules. The clinical relevance of plasma protein binding has been questioned [4].

- **Volume of distribution (*V*)**: Amount of drug in the body divided by its plasma concentration. The volume of distribution is the volume of solvent in which the dose would have to be dissolved to reach the observed plasma concentration. Compounds with small volumes of distribution (i.e., high plasma concentration) are often hydrophilic or negatively charged molecules that do not diffuse effectively into muscle and adipose tissue. Compounds strongly bound to plasma proteins will show small volumes of distribution as well. Hydrophobic and/or positively charged molecules, however, readily dissolve in fat and interact strongly with the negatively charged cell surfaces (phospholipids) and, often, have large volumes of distribution (i.e., low plasma concentrations). Experimental and computational methods have been developed to estimate the volume of distribution in humans [5].

The typical volumes of body fluids are as follows:

Plasma	0.04–0.06 $l\,kg^{-1}$
Blood	0.07 $l\,kg^{-1}$
Extracellular fluid	0.15 $l\,kg^{-1}$
Total body water	0.5 $l\,kg^{-1}$

The "ideal" volume of distribution depends on the targeted half-life and pharmacological activity. For antibiotics or antivirals directed toward intracellular pathogens, a high tissue distribution (large volume of distribution) would be desirable. For short-acting anesthetics or antiarrhythmics or for compounds with a low safety margin, smaller volumes of distribution may enable a better control of drug plasma levels.

- **Clearance (CL)**: The rate at which plasma is freed of drug, the remainder of the drug diluting into the freed volume. If a constant concentration C of a drug is to be attained, the infusion rate must be CL $\times C$. CL is related to other PK parameters: $\text{CL} = \text{dose}/\text{AUC} = \ln 2 \times (V/t_{1/2})$. The liver blood flow in humans is about 25 ml $min^{-1}\,kg^{-1}$.

With the aid of high-throughput *in vitro* assays, potency and selectivity at a given target protein can often be rapidly attained. The most promising strategy for ligand optimization is to start with a small, hydrophilic (i.e., leadlike, *not* druglike) compound identified by screening leadlike compounds at high concentrations (leads are usually smaller, more hydrophilic, and less complex than drugs and show lower affinity to proteins). Then, potency and selectivity is enhanced by systematically introducing lipophilic groups of different shapes at various positions of the hit [6]. Numerous examples of such optimizations have been reported; three examples are shown below (Scheme 1.2 [6c, 7], see also Chapter 72; discovery of losartan).

Unfortunately, the structure–activity relationship (SAR) resulting from such examples is of little use for medicinal chemists, as these only hold for specific target proteins and lack general applicability.

Far fewer examples have been reported of the optimization of PK and ADME because this requires large amounts of compound and tedious, expensive *in vivo* assays. Because few compounds reach the clinic, human data are scarce.

The ADME properties of xenobiotics are always determined by their interaction with the same set of proteins. Therefore, the study of PK–structure relationships should provide the medicinal chemist with valuable general guidelines for the optimization of leads. Structural features that strongly modulate ADME will always do so, no matter what the target protein of the drug might be.

Therefore, a thorough understanding of ADME– and PK–structure relationships should greatly facilitate the design of high-quality development candidates.

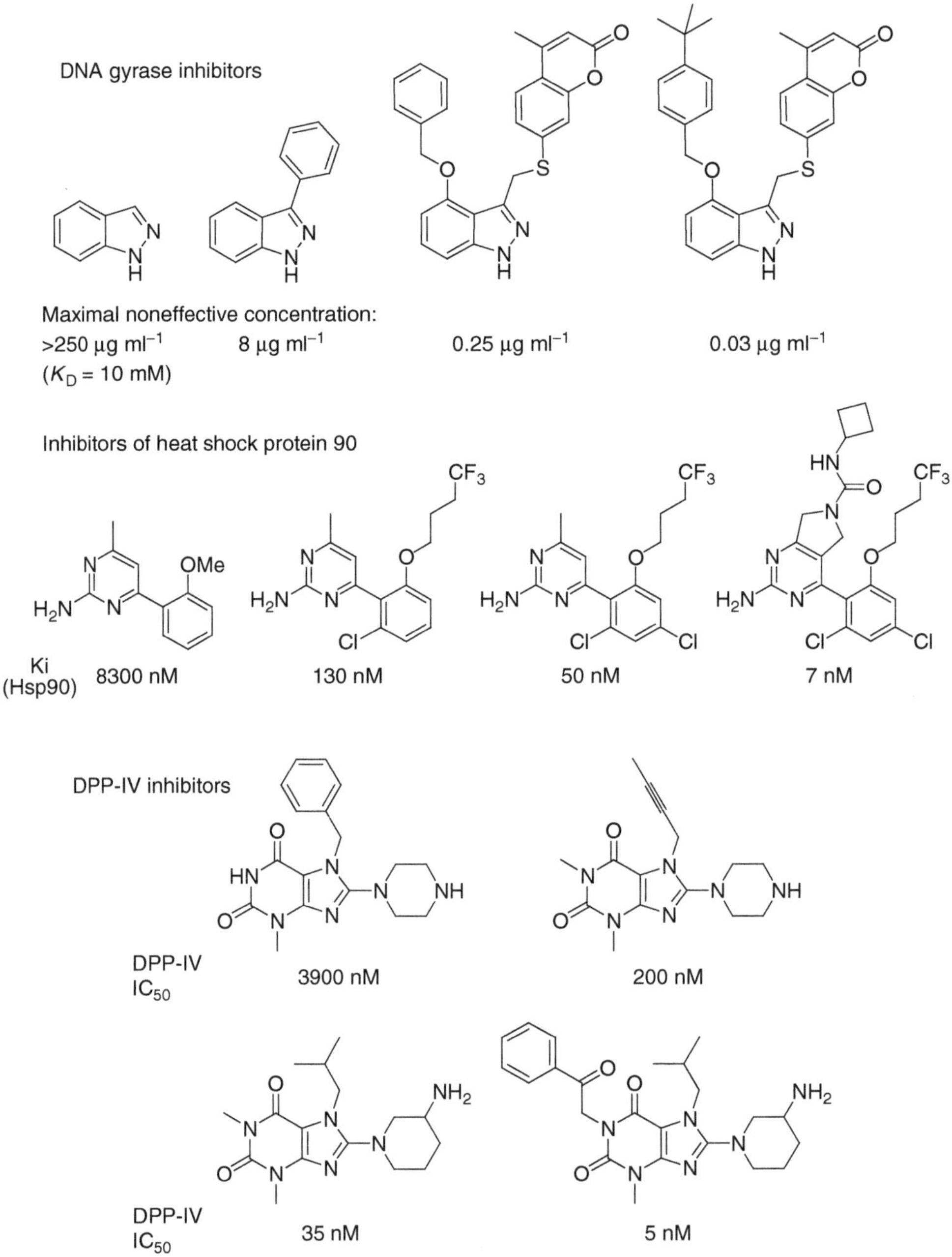

Scheme 1.2 Examples of the enhancement of binding affinity of small molecules to specific proteins.

1.2 The Future of Small-Molecule Drugs

The number of new drugs reaching the market has steadily declined in recent decades, despite steeply increasing R&D expenditures. Thus, while 53 new molecular entities were approved by the US Food and Drug Administration (FDA) in 1996, only 17 attained approval in 2007. Moreover, many "new drugs" are not really new but just resolved racemates, metabolites, prodrugs, or new formulations of older drugs. Thus, the proton pump inhibitor nexium, which ranged among the top 10 best-selling drugs worldwide in 2006, is just resolved omeprazole, launched 1988, and thus no true innovation.

The reasons for this decline in productivity are manifold:

1) Because we have gained a deep understanding of biochemistry and pharmacology, many drug discovery programs start with a target protein and a hypothetical biochemical mechanism. This requires the development of a potent, selective ligand before any proof-of-principle is possible. Because of the high complexity and redundancy of biochemical pathways, however, most hypothetical therapeutic principles will not work in animals.
2) Instead of testing compounds directly in animals, we rely too much on *in vitro* assays. Earlier, the discovery of new drugs was mainly based on *in vivo* assays, as only few *in vitro* assays were available. An *in vivo* assay will only yield a positive result if compounds are sufficiently soluble and lipophilic to reach the target protein and are neither metabolized too rapidly nor removed by active transport mechanisms and if the whole therapeutic concept works. Toxicity and unwanted side effects will also be rapidly recognized in an *in vivo* assay. Thus, by testing compounds directly in animals, efficacy data and a large amount of additional information are obtained. In fact, many older drugs were successfully launched before their biochemical mode of action was understood, and many successful drugs, for example, neuroleptics such as top-selling olanzapine or aripiprazole are highly unselective and would probably have been discarded by modern screening plans.
   To limit oneself to *in vitro* assays or to neglect disappointing *in vivo* results ("a more potent compound should do it") can quickly cost a lot of time, which may have been spent on a more promising project. If a compound cannot reach its target or if the underlying biochemical hypothesis is flawed, an increase in potency or selectivity will not do the job.
3) Advances in molecular modeling and the availability of X-ray structural analyses of proteins cocrystallized with small ligands have ignited further hopes for "*in silico*" drug discovery. Results up to now have mostly been disappointing. Molecular modeling may be a useful source of inspiration for medicinal chemists and has on occasions succeeded to improve potency and selectivity of a ligand, but the large number of failures remains unreported [8]. Although proteins are flexible and readily change their conformation and adapt to new ligands, most current molecular modeling packages keep the protein

structure fixed while docking new ligands [9] and do not automatically include water or ions in the docking process [10]. Even if faster computers and more sophisticated software would overcome these limitations, that would not be sufficient. The prediction of solubility, ADME properties, and chemical stability would also be required [11]. Progress in this field is rapid, but there is still a long way to reliable *in silico* drug development.

4) Development time and costs have increased significantly, because more potential side effects of drugs are continuously being discovered, requiring the identification of much more selective compounds than earlier. While the discovery of paroxetin (1992, Ferrosan) only required the synthesis and testing of ~130 compounds to optimize few parameters, today much larger numbers of compounds are prepared in most discovery projects because a larger number of parameters must be optimized. Between identification of a drug and FDA approval, long development times are often required: taxol, 1971–1992; omeprazole, 1979–1988; fluoxetine, 1972–1985 [12]. The normal time of patent protection (20 years) is, therefore, inadequate for drugs. Unless the time of patent protection for drugs is extended, prices of new drugs will continue skyrocketing.
5) The advent of parallel synthesis in the mid-1990s has enabled medicinal chemists to prepare many compounds quickly, using unsophisticated chemistry. Little time is left, however, to consider test results carefully and to design and prepare the right compound, no matter how difficult or automatable its synthesis. Despite the large numbers of test compounds prepared, the output of new drugs keeps falling.
6) Computers have invaded our offices and laboratories. Instead of training their key skills, chemists and technicians are wasting their time learning how to use databases, virtual screening software, electronic laboratory journals, and trying to keep up with the ever-growing flood of irrelevant emails. In some companies, chemists and technicians are even burdened with accounting tasks and Six Sigma drills. With all these distractions, no sustained focus is possible any longer, and we are losing our expertise in classical medicinal chemistry and organic synthesis. Today, too few hours are spent in front of the hood, and too few experiments are performed [13].

In their assay "The role of the medicinal chemist in drug discovery – then and now" [14], Lombardino and Lowe suggest that the efficiency of drug discovery may be improved by (i) performing more *in vivo* assays earlier in the projects; (ii) by including in each discovery team a "drug champion," that is, an older, experienced scientist to provide historical perspective ("institutional memory") and background information (this book can help); and (iii) by continuous training of young medicinal chemists. I would also propose a return to less molecular modeling, less software but more thinking instead, and less parallel synthesis, in particular during the lead optimization phase, where not chemistry but only structure–property relationships should be the driving force of the project.

For a number of devastating diseases (cancer, multiple sclerosis, bacterial infections with antibiotic-resistant strains, Alzheimer's disease, etc.), no cure is available today. Governmental "regulation" of drug prices is forcing pharmaceutical companies to focus only on high-margin areas but will ultimately preclude any commercial development of new drugs. If politicians believe that medical treatment must be free of charge then the government will also have to provide the new drugs. Only the taxpayer can be forced to finance loss-making ventures forever. However, do politicians really have incentives to improve and extend the life of pensioners even further?

The development of new drugs requires a huge and sustained effort and will only occur in a free society with strong economic incentives and a strong protection of individual property rights. As our democracies continue to degenerate toward socialistic kleptocracies, and governmental size, harassment, and plunder expand, the number of new innovative drugs will keep shrinking [15].

References

1. Newman, D.J. (2008) Natural products as leads to potential drugs: an old process or the new hope for drug discovery? *J. Med. Chem.*, **51**, 2589–2599.
2. Wermuth, C.G. (2004) Selective optimization of side activities: another way for drug discovery. *J. Med. Chem.*, **47**, 1303–1314.
3. Proudfoot, J.R. (2002) Drugs, leads, and drug-likeness: an analysis of some recently launched drugs. *Bioorg. Med. Chem. Lett.*, **12**, 1647–1650.
4. Rolan, P.E. (1994) Plasma protein binding displacement interactions – why are they still regarded as clinically important? *Br. J. Clin. Pharmacol.*, **37**, 125–128.
5. (a) Hollósy, F., Valkó, K., Hersey, A., Nunhuck, S., Kéri, G., and Bevan, C. (2006) Estimation of volume of distribution in humans from high throughput HPLC-based measurements of human serum albumin binding and immobilized artificial membrane partitioning. *J. Med. Chem.*, **49**, 6958–6971; (b) Lombardo, F., Obach, R.S., DiCapua, F.M., Bakken, G.A., Lu, J., Potter, D.M., Gao, F., Miller, M.D., and Zhang, Y. (2006) A hybrid mixture discriminant analysis–random forest computational model for the prediction of volume of distribution of drugs in human. *J. Med. Chem.*, **49**, 2262–2267; (c) Gleeson, M.P., Water, N.J., Paine, S.W., and Davis, A.M. (2006) In silico human and rat V_{ss} quantitative structure – activity relationship models. *J. Med. Chem.*, **49**, 1953–1963; (d) Lombardo, F., Obach, R.S., Shalaeva, M.Y., and Gao, F. (2002) Prediction of volume of distribution values in humans for neutral and basic drugs using physicochemical measurements and plasma protein binding data. *J. Med. Chem.*, **45**, 2867–2876; (2004) *J. Med. Chem.*, **47**, 1242–1250.
6. (a) Congreve, M., Chessari, G., Tisi, D., and Woodhead, A.J. (2008) Recent developments in fragment-based drug discovery. *J. Med. Chem.*, **51**, 3661–3680; (b) Hajduk, P.J. (2006) Fragment-based drug design: how big is too big? *J. Med. Chem.*, **49**, 6972–6976; (c) Boehm, H. *et al.* (2000) Novel inhibitors of DNA gyrase: 3D structure based biased needle screening, hit validation by biophysical methods, and 3D guided optimization. A promising alternative to random screening. *J. Med. Chem.*, **43**, 2664–2674; (d) Maly, D.J., Choong, I.C., and Ellman, J.A. (2000) Combinatorial target-guided ligand assembly: identification of potent subtype-selective c-Src inhibitors. *Proc. Natl. Acad. Sci. U.S.A.*, **97**, 2419–2424; (e) Hajduk, P.J., Meadows, R.P., and Fesik, S.W. (1997) Discovering high-affinity ligands for proteins. *Science*, **278**, 497–499; (f) Hann, M.M., Leach, A.R., and Harper, G. (2001) Molecular complexity and its impact on the probability of finding leads for drug discovery. *J. Chem. Inf. Comput. Sci.*, **41**, 856–864; (g) Oprea, T.I., Davis, A.M., Teague, S.J., and Leeson, P.D. (2001) Is there a difference between leads and drugs? A historical perspective. *J. Chem. Inf. Comput. Sci.*, **41**, 1308–1315.

7. (a) Zehnder, L. *et al.* (2011) Optimization of potent, selective, and orally bioavailable pyrrolodinopyrimidine-containing inhibitors of Heat Shock Protein 90. Identification of development candidate 2-amino-4-{4-chloro-2-[2-(4-fluoro-1*H*-pyrazol-1-yl)ethoxy]-6-methylphenyl}-*N*-(2,2-difluoropropyl)-5,7-dihydro-6*H*-pyrrolo[3,4-*d*]pyrimidine-6-carboxamide. *J. Med. Chem.*, **54**(9), 3368–3385; (b) Eckhardt, M. *et al.* (2007) 8-(3-(*R*)-Aminopiperidin-1-yl)-7-but-3-ynyl-3-methyl-1-(4-methylquinazolin-2-ylmethyl)-3,7-dihydropurine-2,6-dione (BI 1356), a highly potent, selective, long-acting, and orally bioavailable DPP-4 inhibitor for the treatment of type 2 diabetes. *J. Med. Chem.*, **50**, 6450–6453.
8. Warren, G.L. *et al.* (2006) A critical assessment of docking programs and scoring functions. *J. Med. Chem.*, **49**, 5912–5931.
9. Cozzini, P. *et al.* (2008) Target flexibility: an emerging consideration in drug discovery and design. *J. Med. Chem.*, **51**, 6237–6255.
10. Michel, J., Verdonk, M.L., and Essex, J.W. (2006) Protein–ligand binding affinity predictions by implicit solvent simulations: a tool for lead optimization? *J. Med. Chem.*, **49**, 7427–7439.
11. Selick, H.E., Beresford, A.P., and Tarbit, M.H. (2002) The emerging importance of predictive ADME simulation in drug discovery. *Drug Discov. Today*, **7**, 109–116.
12. Wong, D.T., Perry, K.W., and Bymaster, F.P. (2005) The discovery of fluoxetine hydrochloride (prozac). *Nat. Rev. Drug Discov.*, **4**, 764–774.
13. Kubinyi, H. (2003) Drug research: myths, hype and reality. *Nat. Rev. Drug Discov.*, **2**, 665–668.
14. Lombardino, J.G. and Lowe, J.A. III (2004) The role of the medicinal chemist in drug discovery – then and now. *Nat. Rev.*, **3**, 853–862.
15. (a) Ridley, M. (2010) *The Rational Optimist. How Prosperity Evolves*, Fourth Estate, London; (b) Bastiat, F. (1996, first publishing date: 1845) The physiology of plunder, *Economic Sophisms*, Irvington-on-Hudson, The Foundation for Economic Education, Inc., New York http://www.econlib.org/library/Bastiat/basSoph5.html#S.2, Ch.1.

2 Lead Optimization

The development of a new drug is an iterative process in which a hit or lead is systematically modified until a compound with all the desired biological properties is found. After initial identification of a weakly active substance, the first properties to be improved are usually potency and selectivity at the target protein. Such optimization is often quite straightforward because it only requires *in vitro* assays, which can be performed quickly and for a large number of compounds.

While developing a drug, however, most time is often spent with the improvement of pharmacokinetic (PK) parameters of a lead: its oral bioavailability (F), plasma half-life ($t_{1/2}$), and ability to cross the blood–brain barrier (bbb; central nervous system (CNS) drugs only). Moreover, any new drug should ideally neither inhibit nor induce liver enzymes (cytochrome *P*450 (CYP)), be a P-glycoprotein (P-gp) substrate, or interact with human ether-à-go-go related gene (hERG). The optimization of PK parameters is, unfortunately, much slower than the optimization of *in vitro* potency and selectivity because it requires *in vivo* assays and larger amounts of substance.

Because the potency and selectivity of leads is often enhanced with additional lipophilic groups, lead structures tend to become too large and too lipophilic. This may be recognized late because potency and selectivity are usually optimized *in vitro*, where poor solubility is only a minor issue. However, poor solubility is an important reason for the failure of development candidates.

To increase the chances of success, some simple rules of thumb have been proposed to estimate whether a compound is likely to have acceptable absorption, distribution, metabolism, excretion, and toxicity (ADMET) parameters. One is the famous "rule of five," which recommends a molecular weight <500 g mol^{-1}, a log P <5, a number of H-bond donors (acidic X–H groups) <5, and less than 10 hydrogen bond acceptors [1]. Other authors recommend even lower molecular weights (e.g., <400 g mol^{-1}) and log P values <4 [2]. A glance at the tables in later sections of this book confirms that most drugs, in particular orally available drugs, indeed lie within a rather narrow range of molecular weights and log P values [3]. Interestingly, drug candidates with one or more sp^3-hybridized carbon atoms show a higher success rate than "flat" molecules, which may be due to the low solubility of purely aromatic compounds [4].

There are many successful drugs that do not conform to the general recommendations. For instance, antibacterials are often heavy molecules with

Lead Optimization for Medicinal Chemists: Pharmacokinetic Properties of Functional Groups and Organic Compounds, First Edition. Florencio Zaragoza Dörwald.

a large polar surface area (PSA) [5]. Antibacterials must overcome the *bacterial* cell membranes and defense mechanisms, which entails some special physicochemical requirements. When searching for new antibiotics, antifungals, or antiseptics, it may not be a good choice to screen "general-purpose" compound libraries biased toward nonantibiotics.

2.1 What Limits/Reduces Oral Bioavailability?

Oral bioavailability F is defined as the fraction of an orally administered drug reaching systemic circulation. It can be determined by dividing the area under the curve (AUC, plot of concentration *vs* time) of an oral dose by the AUC of the same dose given intravenously. F depends on the extent to which the drug is absorbed from the gastrointestinal tract and on the extent to which it is destroyed during first-pass metabolism and excreted into the intestinal tract in the bile. Because the function of the stomach is mainly digestive and its surface is small, drugs are usually not absorbed from the stomach but from the intestine.

The gastrointestinal (GI) tract may be viewed as a large lipophilic membrane with small, hydrophilic pores with a diameter of 3–8 Å. Critical parameters for absorption of compounds from the gut are the PSA, ionization, molecular weight, solubility, energy required for desolvation from water (i.e., hydrogen bonding), and the number of rotatable bonds [6].

Absorption from the intestine starts to drop at a PSA > 60 Å^2 and becomes negligible at a PSA >140 Å^2. Charged or hydrophilic compounds are only efficiently absorbed if their molecular weight is $<150–200$ g mol^{-1} (the lower the better) or if the molecules are long and thin or by means of an active transport mechanism. Lipophilic, uncharged compounds are more readily absorbed but must be soluble in water to a certain extent, in order to come into close contact with the epithelium of the intestine. Large, lipophilic molecules of low solubility or polymers (no matter if soluble or not) are not significantly absorbed from the gut. It has been suggested that a low number of rotatable bonds may also be conducive to better oral bioavailability [7].

Interestingly, molecules with "intramolecular" charge separation, such as mesoionic compounds (e.g., molsidomine), nitro compounds, pyridine *N*-oxides, azides, and so on, are much more lipophilic than organic cations or anions and are usually well absorbed.

In contrast to the stomach (pH 1–3), the content of the intestine is almost neutral (pH 5–8). Strong acids (sulfonic or phosphonic acids and phosphates) or strong bases (amidines and guanidines) can only be absorbed if their molecular weight is sufficiently low. Carboxylic acids can be absorbed from the stomach but are mainly absorbed from the intestine.

Before reaching systemic circulation, compounds absorbed from the GI tract are directed by the portal vein to the liver, where CYP enzymes catalyze the hydroxylation and oxidation of lipophilic xenobiotics (phase I biotransformations), to speed up their renal excretion. Hydroxylation may be followed by conjugation to glucuronic acid, amino acids, or sulfate (phase II biotransformations), which facilitates further the excretion in urine. High oral bioavailability can only be attained if the drug is resistant to CYP-mediated chemical transformations. This can be achieved either by fine-tuning the log *P* and the PSA or by blocking hydroxylation, for example, by fluorination or introduction of heteroatoms at or near the site of hydroxylation. Alternatively, a drug subject to high first-pass metabolism may be formulated in combination with a compound that blocks liver metabolism (e.g., ritonavir).

Charged compounds formed in the liver are often excreted into the intestine in the bile, reabsorbed from the intestine, and finally, excreted in the urine. Biliary or enterohepatic recirculation refers to a repetitive excretion of a drug into the bile, followed by reabsorption from the small intestine.

Drugs for oral dosing need not be 100% orally bioavailable. High first-pass metabolism may be acceptable for an oral drug if its metabolites are safe and efficiently excreted. Low bioavailability owing to extensive first-pass metabolism may, however, cause a high interindividual variability of F, as well as a high dependency of F on additional factors (food, age, CYP inhibitors/inducers, liver or other diseases). Such drugs are difficult to dose correctly.

2.2 What Limits/Reduces Plasma Half-Life?

Blood is constantly being filtered by the kidneys, at a rate of 7.5 l h^{-1}. Thereby, compounds with molecular weights <50 000 g mol^{-1} are excreted and then partly reabsorbed again into plasma. In addition to passive glomerular filtration, organic cations and organic anions may be renally excreted by an active transport mechanism ("active tubular secretion"). Plasma proteins and the protein-bound fraction of a drug are not eliminated by renal filtration.

The extent to which drugs are reabsorbed varies strongly. Charged organic molecules are generally reabsorbed to a lesser extent than neutral, lipophilic compounds. The only generally applicable strategy to contain renal excretion of a drug is to enhance its binding to plasma proteins or its volume of distribution.

Enzymatic activity capable of destroying a drug is mainly located in the liver, gut, kidneys, lung, and skin. In addition to CYP enzymes, drugs will be exposed to alcohol dehydrogenase, various oxidases and reductases, esterases, phosphatases, and proteases. Moreover, blood contains strong nucleophiles (glutathione, cystein; these are further activated enzymatically, e.g., by glutathione transferases), an electrophilic methylating reagent (*S*-adenosylmethionine), and the electrophilic acetylating reagent acetyl-CoA (in combination with acetyl transferases). All these enzymes and reagents may inactivate a drug or convert a prodrug into the active drug.

The liver receives about 75% of its blood supply from the portal vein and 25% from the systemic circulation. Accordingly, parenteral administration of a drug circumvents to a large extent oxidative enzymatic degradation by the liver. CYP inhibition or induction is also less critical for parenterally dosed compounds.

2.3
How to Improve bbb-Penetration?

The bbb is an additional protection of the CNS from potentially dangerous xenobiotics. Moreover, many drugs are actively excreted from the CNS by P-gp, the product of the multidrug resistance gene, which acts as an active efflux system for a variety of drugs [8].

The bbb is essentially lipophilic and impermeable to large molecules. Only uncharged, lipophilic compounds (log $P > 0$) with a molecular weight $<$400–600 g mol^{-1} and few rotatable bonds can diffuse into the CNS. Hydrogen-bonding ability is highly detrimental for bbb passage, and as a rule of thumb, drugs with a PSA $>$40 Å^2 will not enter the CNS to any significant extent [9]. Even a single carbonyl group (e.g., a ketone, amide, urea, carbamate, etc.) will strongly reduce the ability of a drug to cross the bbb. Strong binding to plasma proteins, which correlates with the log P, may also reduce bbb penetration [10]. Nutrients reach the CNS by means of substrate-specific carriers and active transport mechanisms.

2.4
How to Avoid CYP Inhibition/Induction?

A number of models of the various CYP enzymes have been generated [11] and may be used to predict if a compound is likely to interact strongly with one or more of these enzymes. There is, however, no reliable method to ascertain if the compound will just be a substrate or an inhibitor of the enzyme. This must usually be determined empirically during lead optimization [12].

Molecules capable of strongly coordinating to iron cations (imidazoles and pyridines) may also be CYP inhibitors. In such instances, affinity to CYP can sometimes be lowered by sterically shielding the iron binding site in the inhibitor.

Metabolically resistant molecules can be CYP inducers. For instance, the expression and activity of rat liver enzymes is strongly enhanced if the rat is treated with chlorinated biphenyls. Hence, stabilization of a lead to block metabolic transformations may backfire and should not be exaggerated. Drugs with too long half-lives are, moreover, difficult to dose and may lead to dangerous accumulation, above all in older patients.

Two famous examples of dietary CYP inhibition and induction are grapefruit and St. John's wort. Grapefruit and orange juice contain potent inhibitors of CYP3A4/5, which can cause extended half-lives and dangerously high plasma levels of drugs that are metabolized by CYP3A4/5. St. John's wort and other herbal extracts are CYP inducers and will generally shorten the half-life of many drugs, for example of contraceptives, causing unpleasant surprises here and there. Some fruit juices can decrease the oral bioavailability of drugs as well (e.g., of celiprolol and fexofenadine) [13].

If a valuable lead is a potent CYP inhibitor/inducer or too hepatotoxic, a parenteral formulation may still be viable.

CYP inducers : barbiturates, carbamazepine, dexamethasone, efavirenz, ethanol, felbamate, growth hormone, isoniazid, moricizine, nevirapine, omeprazole, phenobarbital, phenytoin, prednisone, probenecid, rifampin, rifapentine, ritonavir, smoke, troglitazone, and vegetables;

CYP inhibitors : amiodarone, cimetidine, ciprofloxacin, clarithromycin, diltiazem, diphenhydramine, enoxacin, erythromycin, fluconazole, fluoxetine, fluvastatin, fluvoxamine, grapefruit, grepafloxacin, haloperidol, indinavir, itraconazole, ketoconazole, metronidazole, miconazole, nefazodone, paroxetine, quinidine, ritonavir, sertraline, sulfamethoxazole, thioridazine, ticlopidine, and verapamil.

2.5 How to Avoid Interaction with the Human Ether-à-go-go-Related Gene (hERG)?

It is known since long that some drugs can cause cardiac arrhythmias, but these are usually not life threatening. The long QT syndrome, however, may cause torsades de pointes, a ventricular tachyarrhythmia that can degenerate into ventricular fibrillation and sudden death by cardiac arrest.

The QT interval is the time for both the ventricular depolarization and repolarization. QT-prolonging drugs block potassium channels encoded by hERG. A number of drugs had to be withdrawn for causing prolonged QT syndrome while showing high affinity to hERG, for example, clobutinol, astemizole, grepafloxacin, cisapride, and terfenadine. The relationship between hERG affinity and QT prolongation is, however, not yet firmly established. Clemastine, for instance, has a high affinity to hERG (12 nM) but does not induce QT prolongation [14]. Grepafloxacin, on the other hand, which only binds to hERG with micromolar affinity [15], had to be withdrawn one year after launch following reports of severe QT prolongation.

Some SAR (structure–activity relationship) models for the hERG potassium channel have been published and may be used to gain some inspiration if hERG is a problem [16]. hERG counterscreens should be included in every lead optimization program.

2.6 How to Prevent Toxicity?

Strong electrophiles are often toxic or mutagenic because of covalent bond formation with proteins and nucleic acids. Moreover, many nonelectrophilic organic molecules may be converted into electrophilic metabolites in the human body. This reactivity is difficult to predict and often emerges only during clinical trials or after launch.

Drugs may also be phototoxic, that is, sensitize the skin for UV absorption or induce photoallergy (the drug undergoes a photochemical transformation into an allergen). Molecules with suitable chromophores include the quinolone antibiotics, methoxsalen, hypericin, aromatic sulfonamides, tetracyclines, and phenothiazines.

Adverse drug reactions include the typical symptoms of overdosing (skin rashes, agranulocytosis, hepatotoxicity, and aplastic anemia), which are usually observed during preclinical development, and the less frequent, sometimes idiosyncratic reactions, which are first recognized once the drug has been widely marketed.

In the early 1990s, the main reason for stopping drug development was poor PK. By adding ADMET specialists to the discovery teams, this problem became less acute. Today, the major causes of failure during clinical trials and early withdrawal of drugs from the market are lack of efficacy and excessive toxicity [17]. Although toxicity in humans cannot be predicted, the success rate of drug candidates may be enhanced by avoiding substructures known to cause adverse drug reactions. These include electrophilic groups or groups susceptible to the formation of electrophilic metabolites (Table 2.1).

Many drugs contain the functional groups listed in Table 2.1 but do not cause adverse drug reactions because efficient detoxication pathways exist. Moreover, even compounds devoid of any critical functional group can be hepatotoxic (e.g., ximelagatran, Table 55.2). Nevertheless, the risk of excessive toxicity can be lowered by avoiding the groups shown in Table 2.1.

One crucial factor for drug safety is the dose: adverse drug reactions are rarely seen for drugs dosed below 20 mg per day. Thus, drug safety may be enhanced by increasing potency or by improving oral bioavailability.

One example of metabolic activation is the antiinflammatory sudoxicam (Table 63.1), which is metabolized to a hepatotoxic thiourea [18]. This transformation could be blocked by methylation of the thiazole ring. In the resulting meloxicam, the main metabolic path is hydroxylation and oxidation of this newly introduced methyl group.

Table 2.1 Functional groups prone to the formation of electrophilic metabolites [17a].

Haloalkanes	Hydroquinones	Imides
Alkenes	Catechols	Thioureas
Electron-poor alkenes	Aromatic thiols	Sulfonylureas
Alkynes	Cyclopropylamines	Formaldehyde acetals
Phenyl groups	Benzylamines	3-Methylindoles
4-Unsubstituted benzyl ethers	Hydrazines	5-Hydroxyindoles
Ethoxybenzenes	Hydroxylamines	Unsubstituted furans
Bromoarenes	Azides	Unsubstituted thiophenes
Nitroarenes	Hydroxamic acids	Unsubstituted thiazoles
Anilines	Tetrahydropyridines	Unsubstituted oxazoles
Aminophenols	Formamides	Thiazolidinediones

Two further examples of drugs that cause adverse drug reactions because of metabolic activation are clozapine and felbamate. Clozapine contains an electron-rich aniline substructure, which can be oxidized by hydride abstraction to yield an electrophilic, cationic *ortho*-quinone diimine (Scheme 2.1). This metabolite reacts with nucleophiles and can arylate nucleic acids or the thiol groups of vital proteins. A significant number of patients taking this drug develop agranulocytosis [19]. This metabolic activation is also possible in olanzapine but less problematic because of the 30-fold lower dosing of olanzapine. In the clozapine analogs quetiapine and loxapine, the stability toward oxidants was enhanced by replacing the aromatic NH group by S and O.

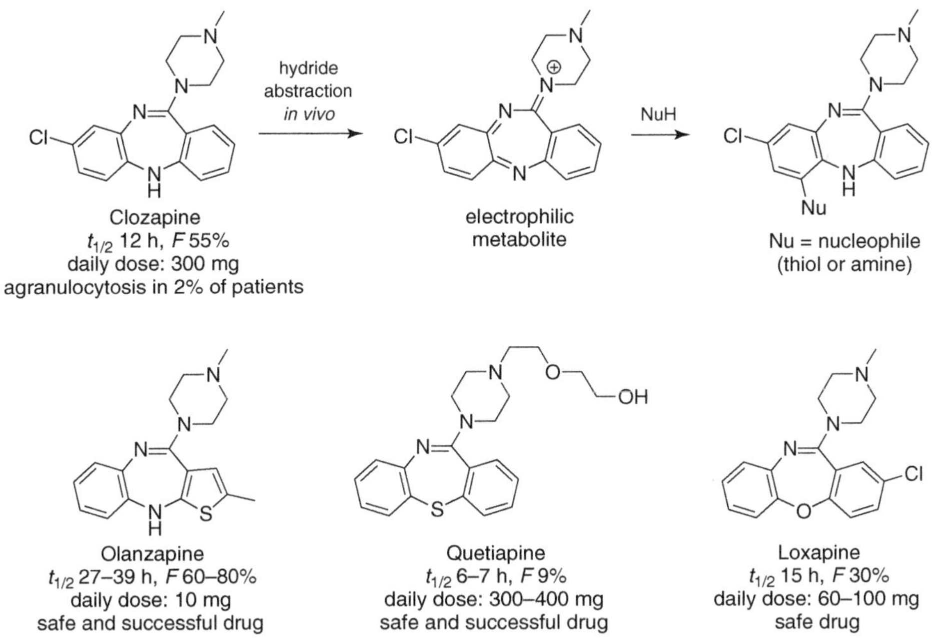

Scheme 2.1 Metabolic activation of clozapine.

The ability of a drug candidate to act as hydride donor should be easy to detect by treatment with a mild oxidant, such as benzoquinone. It would have been more difficult, however, to correctly predict the metabolic fate of felbamate.

Felbamate is an anticonvulsant bis-carbamate. Within a year of its launch in 1993, 34 cases of aplastic anemia and 23 cases of hepatotoxicity occurred, resulting in 18 deaths. This led to a severe restriction of the use of this drug.

The metabolic activation of felbamate proceeds via hydrolysis to a primary alcohol, followed by oxidation to an aldehyde (Scheme 2.2). Elimination to form an α,β-unsaturated aldehyde is particularly easy in this instance because the newly formed alkene is conjugated with the phenyl group. The resulting Michael acceptor, at the high doses required, led to the adverse drug reactions in patients [20].

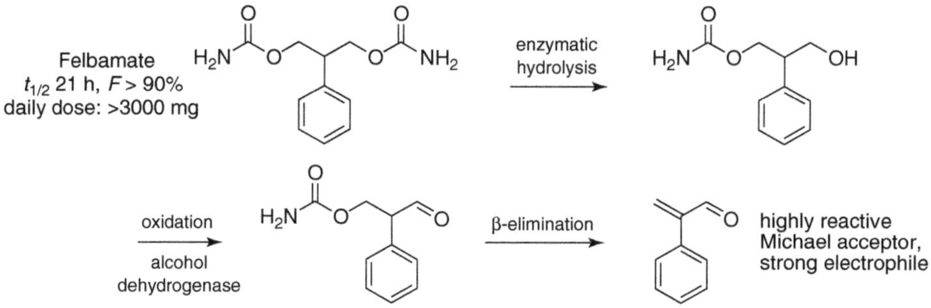

Scheme 2.2 Metabolic activation of felbamate.

With *in vitro* assays, it is possible to predict toxicity with some precision, but false-positives and false-negatives are common [21]. Similarly, animal models do not reveal human poisons accurately. However, our analytical tools are becoming increasingly sensitive, and our knowledge on metabolic pathways is continuously growing. A better understanding of human metabolism and metabolism-dependent toxicity will lead to advanced toxicity models [22], which will improve our ability to exclude questionable compounds early during drug development.

2.7
Examples of PK-Optimization in Animals

2.7.1
Dihydroorotate Dehydrogenase Inhibitors

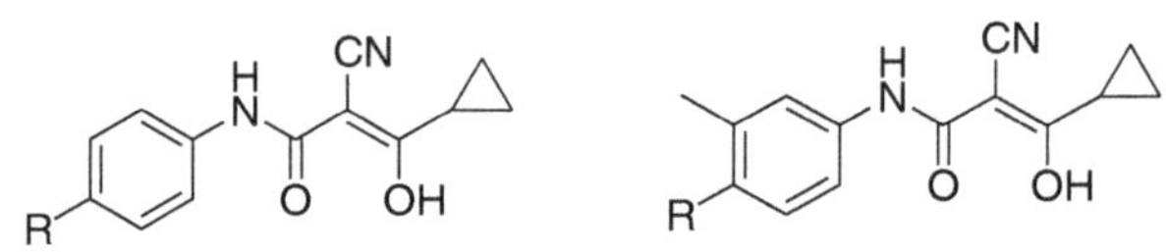

R	$t_{1/2}$ (po, rat)	R	$t_{1/2}$ (po, rat)
MeS-	<1 h	Cl-	2 h
F_3CS-	2 h	F_3C-	4 h
$F_3CS(=O)$-	3 h	Br-	4 h
O_2N-	4 h	F_3CO-	4 h
NC-	5 h	NC-	6 h
MeC(=O)-	6 h	$F_3C–CF_2$-	10 h
F_3C-	8 h	F_3CS-	13 h
$F_3CS(=O)_2$-	14 h		
Br-	20 h		
Cl-	22 h		
F_3CO-	22 h		
4-Cl(C_6H_4)O-	39 h		

Kuo, E.A. *et al.* (1996) Synthesis, structure–activity relationships, and pharmacokinetic properties of dihydroorotate dehydrogenase inhibitors: 2-cyano-3-cyclopropyl-3-hydroxy-*N*-[3′-methyl-4′-(trifluoromethyl)phenyl]propenamide and related compounds. *J. Med. Chem.*, **39**, 4608–4621.

2.7.2
Matrix Metalloproteinase Inhibitors

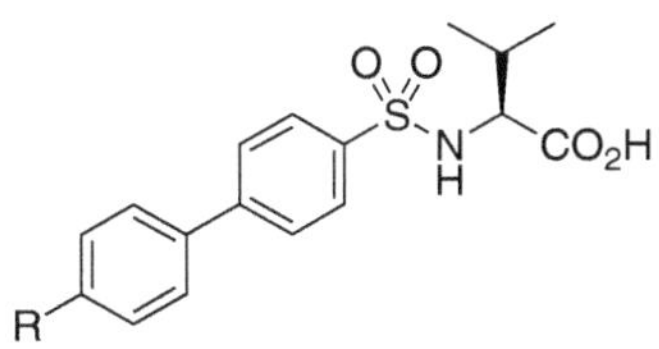

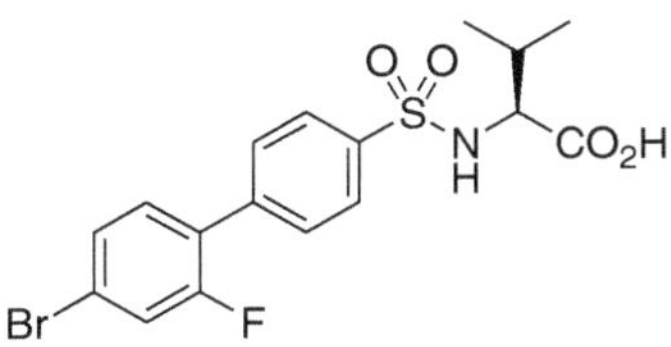

R	$t_{1/2}$ (rat, po)	R	$t_{1/2}$ (rat, po)	$t_{1/2}$ (rat, po)
MeO-	0.36 h	Cl-	34.1 h	117 h
Me-	0.40 h	F_3C-	37.9 h	
H-	3.4 h	Br-	43.6 h	
NC-	12.5 h	F-	45.5 h	
O_2N-	19.5 h			

O'Brien, P.M. *et al.* (2000) Structure-activity relationships and pharmacokinetic analysis for a series of potent, systemically available biphenylsulfonamide matrix metalloproteinase inhibitors. *J. Med. Chem.*, **43**, 156–166.

2.7.3
Antibacterial Aminobenzenesulfonamides

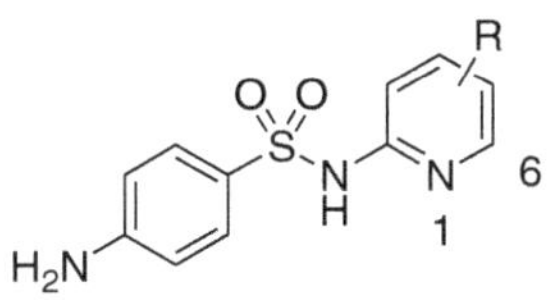

R	$t_{1/2}$ (iv, rat)	R	$t_{1/2}$ (iv, rat)
4-CH_3	1.2 h	5-CH_3	2.4 h
4,6-$(CH_3)_2$	1.2 h	3-CH_3-5-Br	8.2 h
3-CH_3	1.3 h	3-CH_3-5-Cl	8.2 h
3-OCH_3	1.3 h	3-Cl	8.7 h
3-CN	1.4 h	5-NO_2	16.5 h
3-OEt	1.4 h	3,5-Cl_2	21.7 h
H	1.5 h	3,5-Br_2	21.7 h
6-CH_3	1.9 h	5-Cl	21.7 h
5-$N(CH_3)_2$	2.1 h	5-Br	24.8 h
3-CN-5-CH_3	2.2 h		

Seydel, J.K., Trettin, D., and Cordes, H.P. (1980) Quantitative structure–pharmacokinetic relationships derived on antibacterial sulfonamides in rats and its comparison to quantitative structure–activity relationships. *J. Med. Chem.*, **23**, 607–613.

2.7.4
Tyrosine Kinase Inhibitors

Metabolism was the main route of drug elimination of this series of aminopyridines. The major route of metabolism in rat and rhesus monkey (but not in dogs) was N-glucuronidation at the aminopyridine. This biotransformation could be suppressed by introduction of an additional methyl group (last compound).

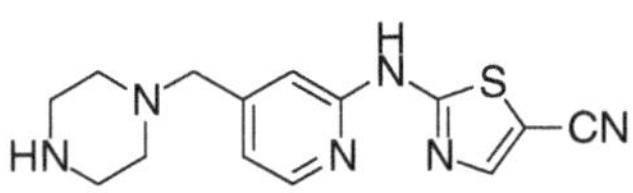

F 100%; $t_{1/2}$ 7.9 h; *V* 22 l kg^{-1}; CL 24 ml min^{-1} kg^{-1} (dog)

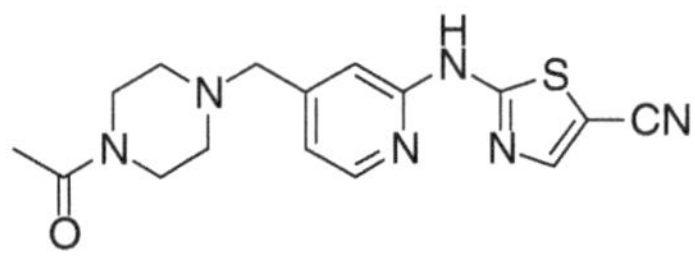

F 87%; $t_{1/2}$ 7.2 h; *V* 0.7 l kg^{-1}; CL 1.2 ml min^{-1} kg^{-1} (dog)

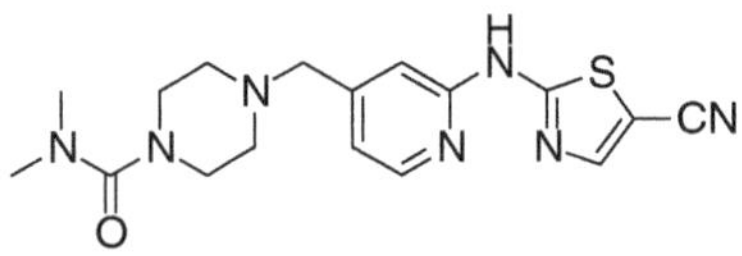

F nd; $t_{1/2}$ 2.0 h; *V* 0.9 l kg^{-1}; CL 7.4 ml min^{-1} kg^{-1} (dog)

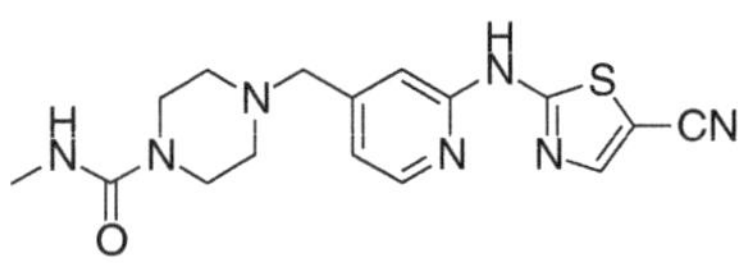

F 99%; $t_{1/2}$ 17 h; *V* 0.7 l kg^{-1}; CL 0.5 ml min^{-1} kg^{-1} (dog)
F 43%; $t_{1/2}$ 1.7 h; *V* 1.3 l kg^{-1}; CL 17.2 ml min^{-1} kg^{-1} (rhesus)

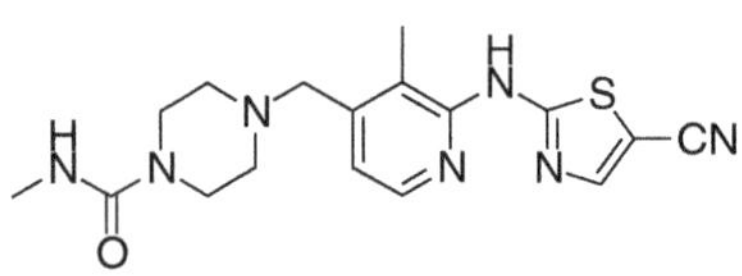

F 92%; $t_{1/2}$ 8.8 h; *V* 1.3 l kg^{-1}; CL 1.7 ml min^{-1} kg^{-1} (rhesus)

Bilodeau, M.T. *et al.* (2004) Potent *N*-(1,3-thiazol-2-yl)pyridin-2-amine vascular endothelial growth factor receptor tyrosine kinase inhibitors with excellent pharmacokinetics and low affinity for the hERG ion channel. *J. Med. Chem.*, **47**, 6363–6372.

2.7.5
5-HT$_{1A}$ Agonists

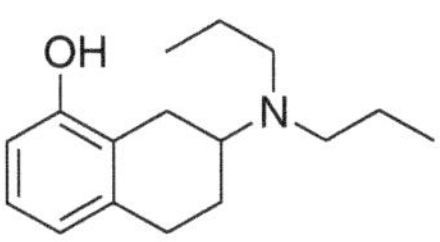

$t_{1/2}$ 60 min (rat)
F 2.4% (rat)

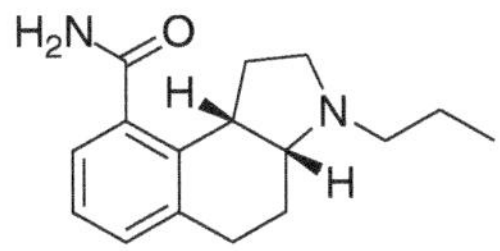

$t_{1/2}$ 120 min (rat)
F 46% (rat)

Lin, C. *et al.* (1993) Synthesis and biological activity of *cis*-(3a*R*)-(−) -2,3,3a,4,5,9b-hexahydro-3-propyl-1*H*-benz[e]indole-9-carboxamide: a potent and selective 5-HT$_{1A}$ receptor agonist with good oral availability. *J. Med. Chem.*, **36**, 2208–2218.

2.7.6
Dipeptidyl Peptidase IV Inhibitors; Structural Variations of Sitagliptin

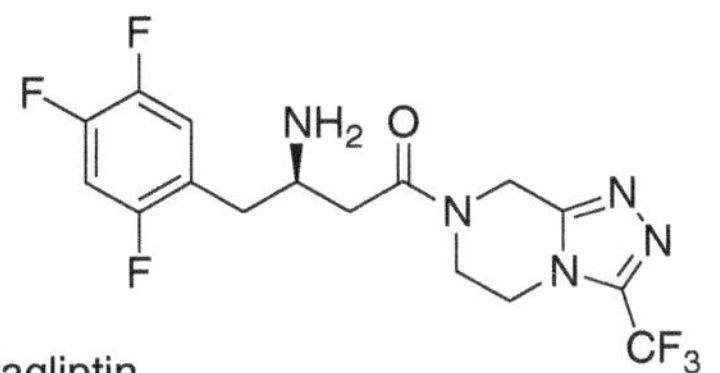

Sitagliptin
$t_{1/2}$ 1.7 h; *F* 76%; CL 60 ml min^{-1} kg^{-1} (rat)

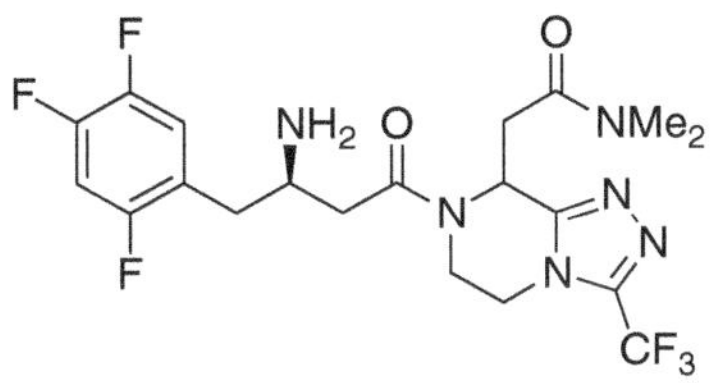

$t_{1/2}$ 2.7 h; *F* 9%; CL 87 ml min^{-1} kg^{-1} (rat)

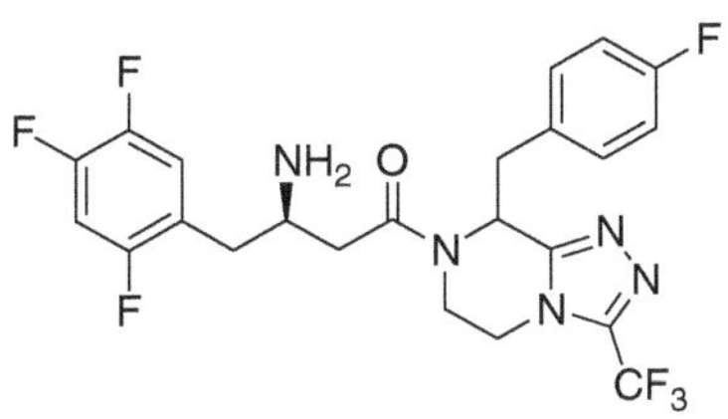

$t_{1/2}$ 1.5 h; *F* 76%; CL 29 ml min^{-1} kg^{-1} (rat)

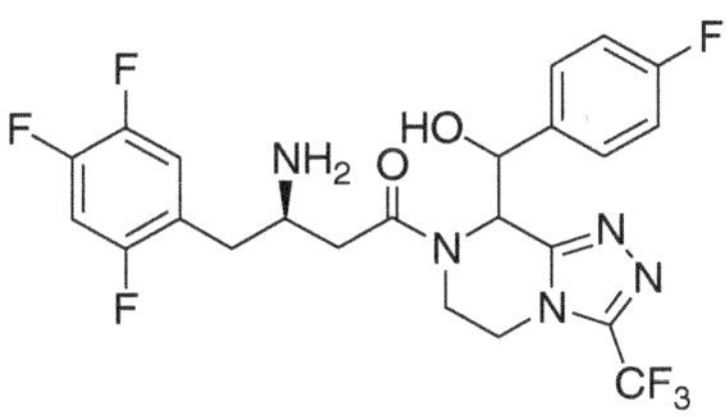

$t_{1/2}$ 2.6 h; *F* 8%; CL 63 ml min^{-1} kg^{-1} (rat)

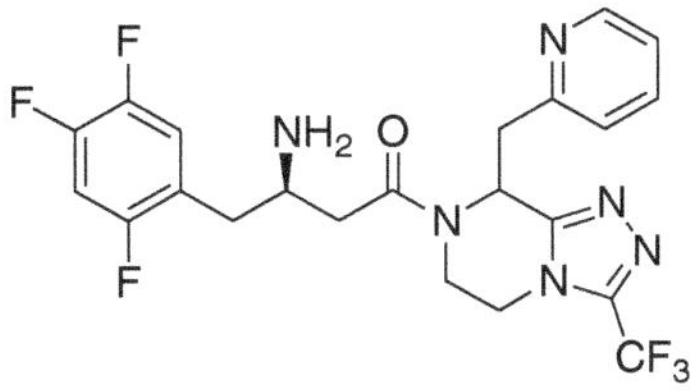

$t_{1/2}$ 1.9 h; *F* 9%; CL 94 ml min^{-1} kg^{-1} (rat)

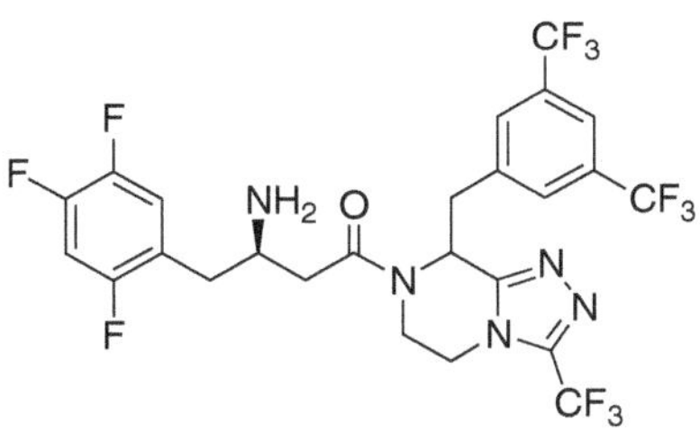

$t_{1/2}$ 4.1 h; *F* 19%; CL 50 ml min^{-1} kg^{-1} (rat)

Kim, D. *et al.* (2008) Discovery of potent and selective dipeptidyl peptidase IV inhibitors derived from β-aminoamides bearing substituted triazolopiperazines. *J. Med. Chem.*, **51**, 589–602.

2.7.7
P-Selectin Inhibitors

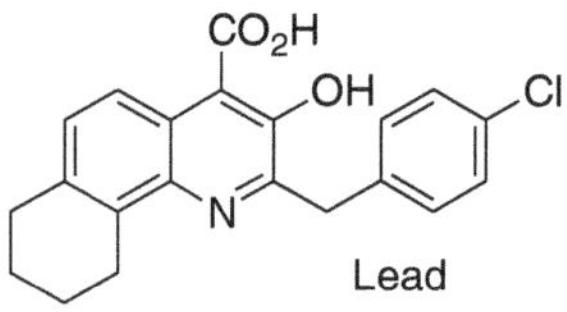

Lead

F 27%; CL 11 ml min^{-1} kg^{-1} (rat)

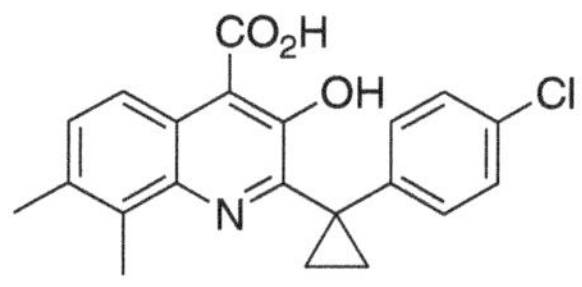

F 75%; CL 7 ml min^{-1} kg^{-1} (rat)

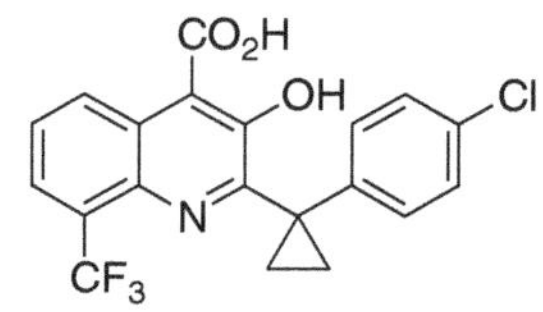

F 67%; CL 0.7 ml min^{-1} kg^{-1} (rat)

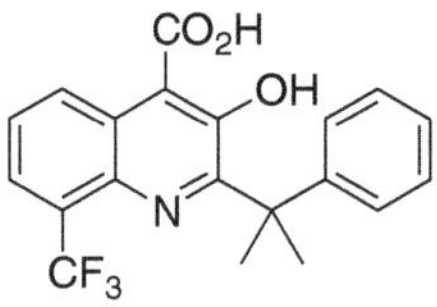

F 93%; CL 3.7 ml min^{-1} kg^{-1} (rat)

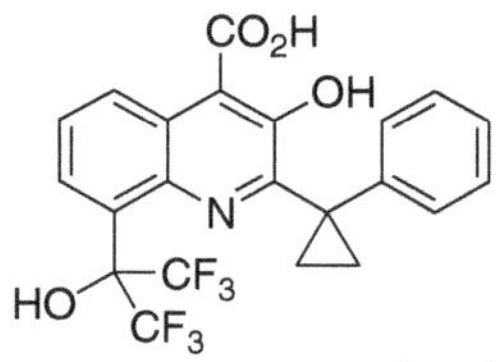

F 44%; CL 1 ml min^{-1} kg^{-1} (rat)

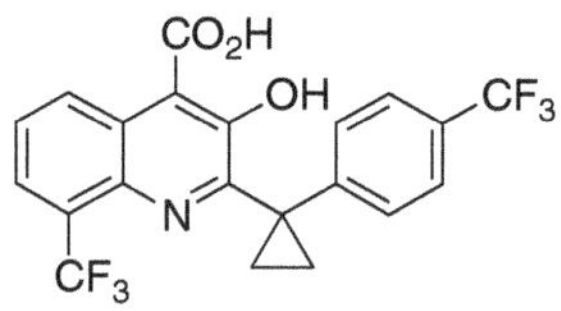

F 31%; CL 0.4 ml min^{-1} kg^{-1} (rat)

Huang, A. *et al.* (2010) Discovery of 2-[1-(4-chlorophenyl)cyclopropyl]-3-hydroxy-8-(trifluoromethyl)quinoline-4-carboxylic acid (PSI-421), a P-selectin inhibitor with improved pharmacokinetic properties and oral efficacy in models of vascular injury. *J. Med. Chem.*, **53**, 6003–6017.

2.7.8
β_3-Adrenergic Agonists

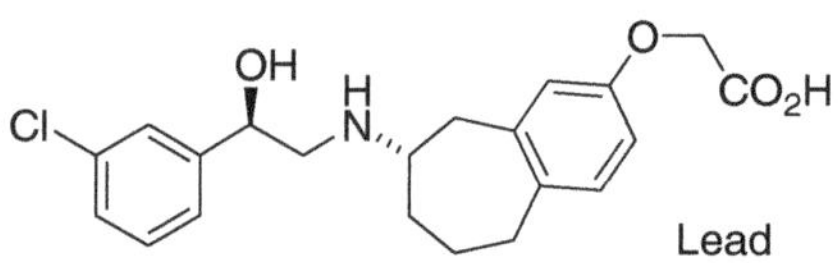

Lead

$t_{1/2}$ 2.3 h (iv); F 35%; CL 12 ml min^{-1} kg^{-1} (iv) (monkey)

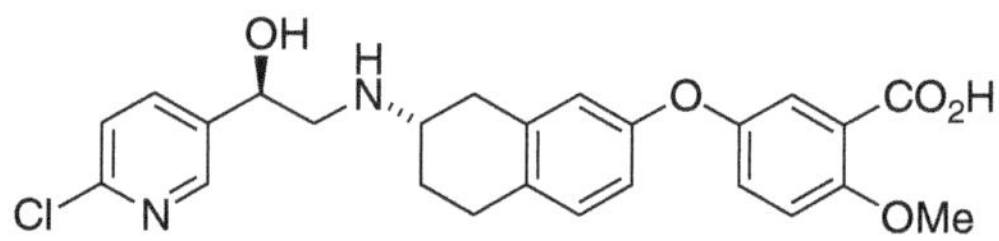

$t_{1/2}$ 8.2 h (iv); F 29%; CL 16 ml min^{-1} kg^{-1} (iv) (monkey)

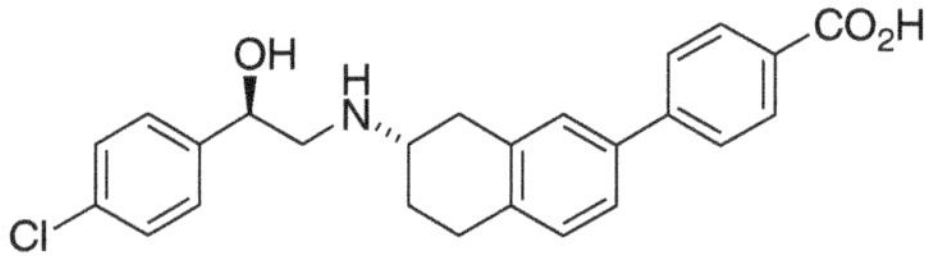

$t_{1/2}$ 7.7 h (iv); F 82%; CL 6.7 ml min^{-1} kg^{-1} (iv) (monkey)

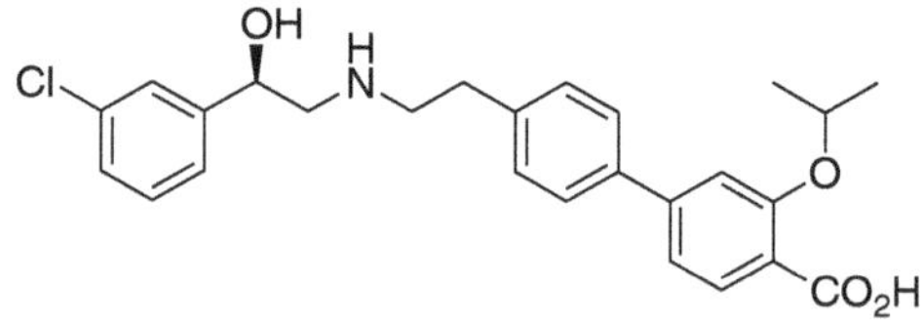

$t_{1/2}$ 19 h (iv); F 83%; CL 0.9 ml min^{-1} kg^{-1} (iv) (monkey)

Imanishi, M. *et al.* (2008) Discovery of a novel series of benzoic acid derivatives as potent and selective human β_3 adrenergic receptor agonists with good oral bioavailability. *J. Med. Chem.*, **51**, 1925–1944 and 4804–4822.

2.7.9
Vanilloid-1 Antagonists

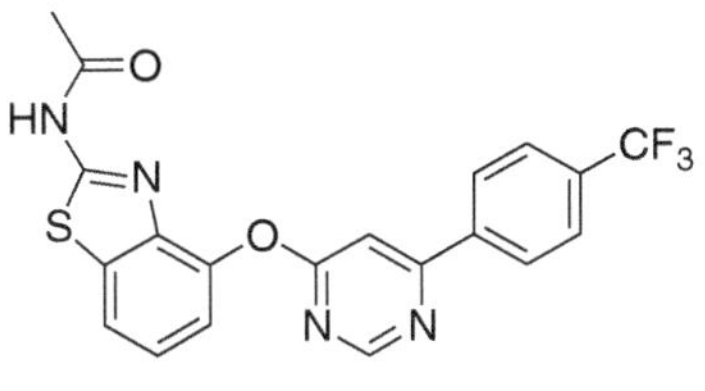

Lead

$t_{1/2}$ 6.3 h (iv); *F* 32%; CL 3.2 ml min^{-1} kg^{-1} (iv); *V* 1.6 l kg^{-1}; (rat) half-life too long!

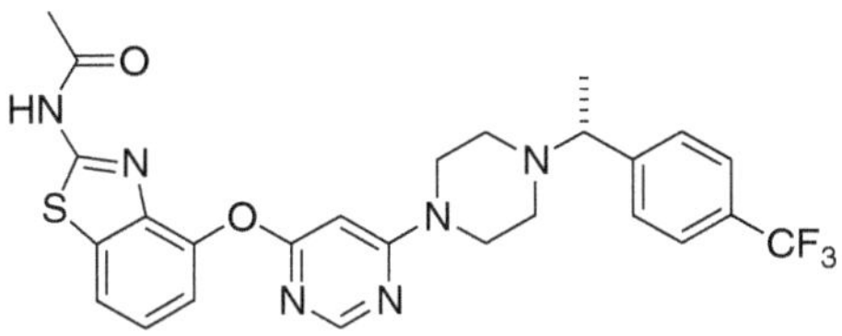

$t_{1/2}$ 2.7 h (iv); *F* 3%; CL 23 ml min^{-1} kg^{-1} (iv); *V* 4.5 l kg^{-1}; (rat)

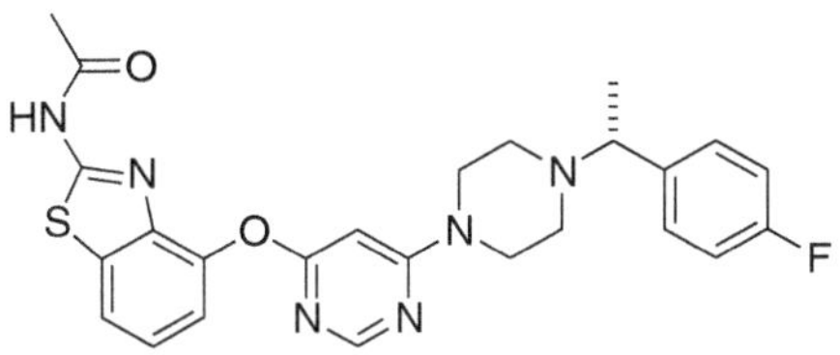

$t_{1/2}$ 2.7 h (iv); *F* 51%; CL 13 ml min^{-1} kg^{-1} (iv); *V* 2.8 l kg^{-1}; (rat)

Wang, H. *et al.* (2007) Novel vanilloid receptor-1 antagonists: 3. The identification of a second-generation clinical candidate with improved physicochemical and pharmacokinetic properties. *J. Med. Chem.*, **50**, 3528–3539.

2.7.10
HIV Protease Inhibitors

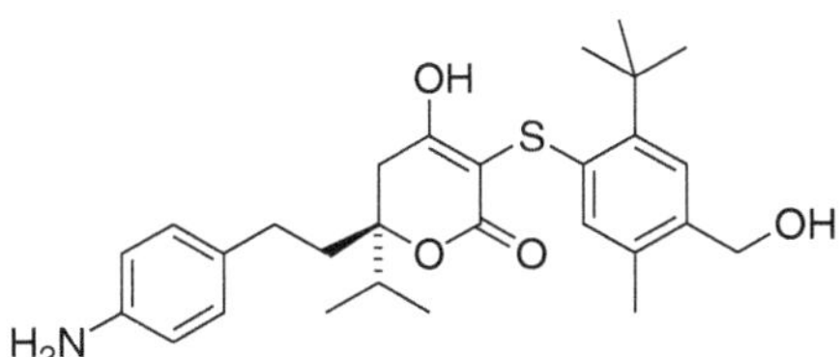

Lead

$t_{1/2}$ 1.2 h; *F* 33% (dog)

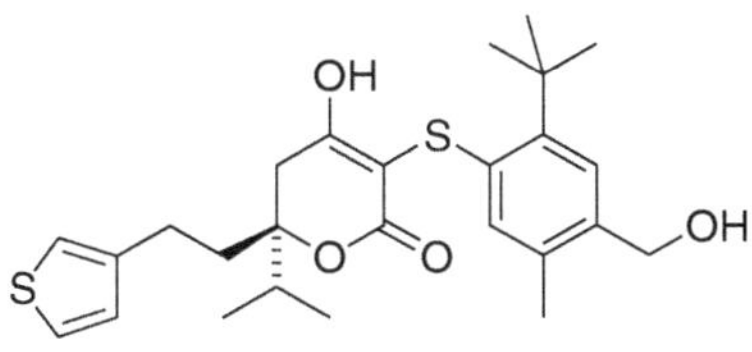

$t_{1/2}$ 5.4 h; *F* 88% (dog)

Hagen, S.E. *et al.* (2001) 4-Hydroxy-5,6-dihydropyrones as inhibitors of HIV protease: the effect of heterocyclic substituents at C-6 on antiviral potency and pharmacokinetic parameters. *J. Med. Chem.*, **44**, 2319–2332.

2.7.11
Matrix Metalloproteinase Inhibitors (Tested in Rats)

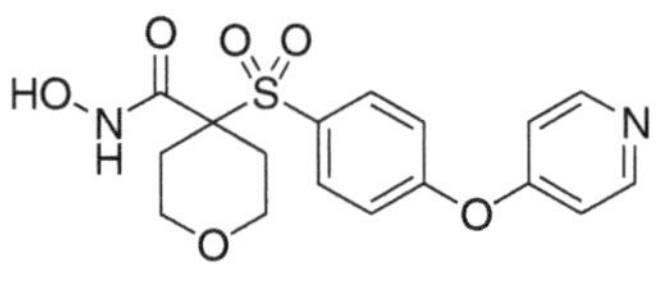

$t_{1/2}$ 0.3 h
F 42%

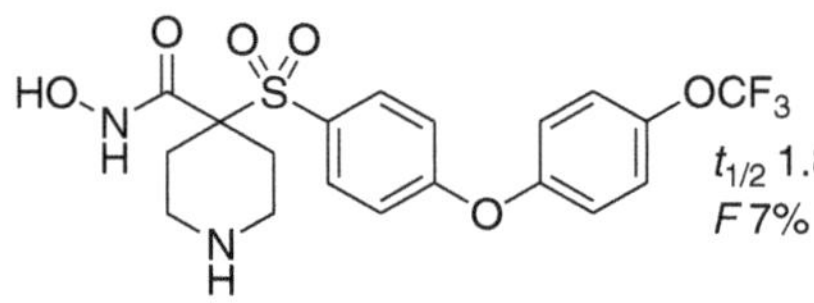

$t_{1/2}$ 1.8 h
F 7%

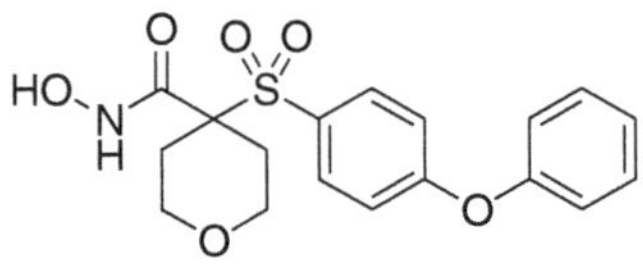

$t_{1/2}$ 1.7 h
F 6%

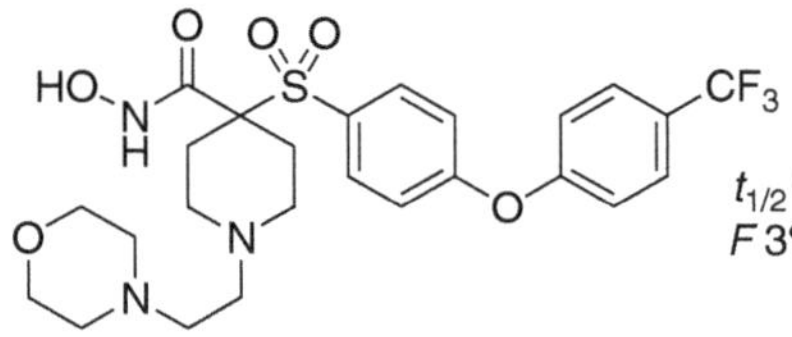

$t_{1/2}$ 0.9 h
F 3%

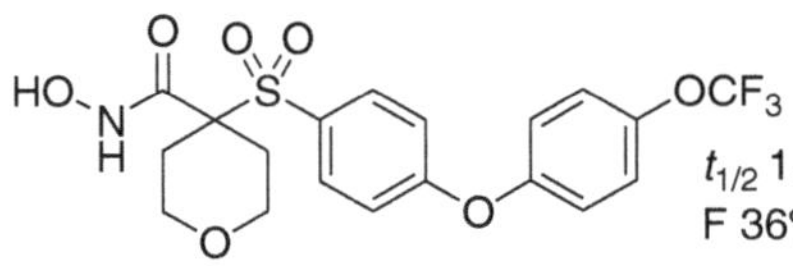

$t_{1/2}$ 1.9 h
F 36%

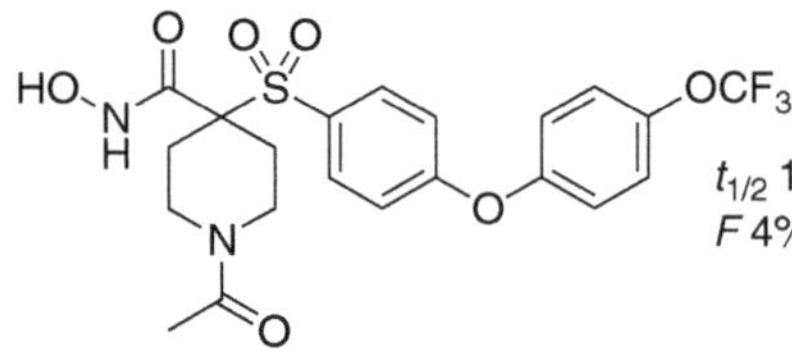

$t_{1/2}$ 1.1 h
F 4%

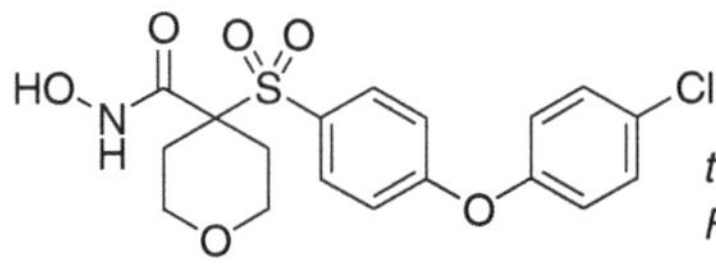

$t_{1/2}$ 1.5 h
F 46%

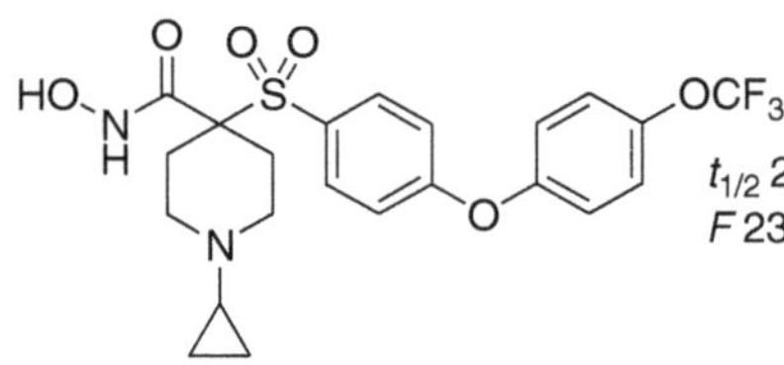

$t_{1/2}$ 2.9 h
F 23%

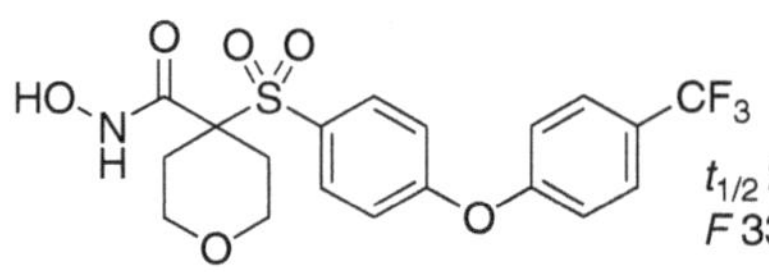

$t_{1/2}$ 2.7 h
F 33%

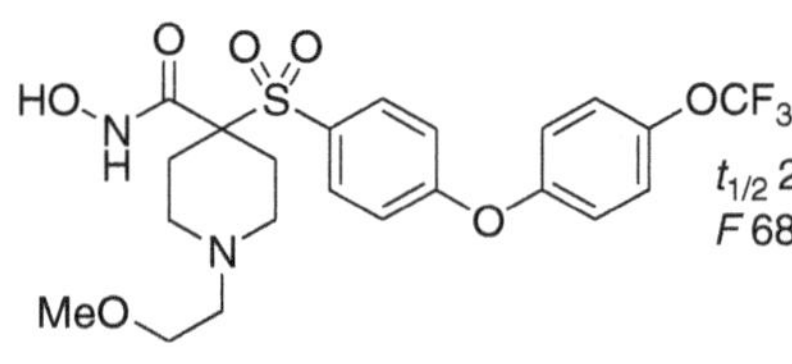

$t_{1/2}$ 2.9 h
F 68%

Becker, D.P. *et al.* (2010) Orally active MMP-1 sparing α-tetrahydropyranyl and α-piperidinyl sulfone matrix metalloproteinase (MMP) inhibitors with efficacy in cancer, arthritis, and cardiovascular disease. *J. Med. Chem.*, **53**, 6653–6680.

2.7.12
HIV Integrase Inhibitors

Both thiophene derivatives were potent in an enzymatic assay but too toxic and weak in a cellular assay. These problems could be solved by replacing the thienyl group by various aminomethyl groups.

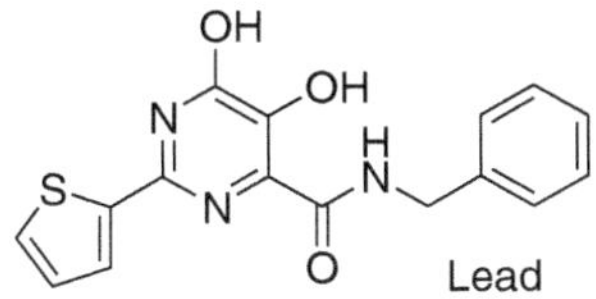

Lead

$t_{1/2}$ 3 h, *F* 15%, CL 5 ml min^{-1} kg^{-1} (rat)

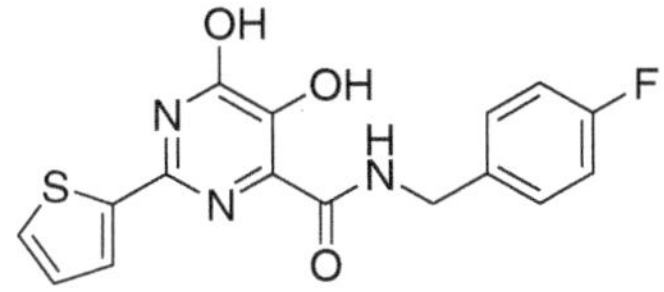

$t_{1/2}$ 1.3 h, *F* 29%, CL 11 ml min^{-1} kg^{-1} (rat)

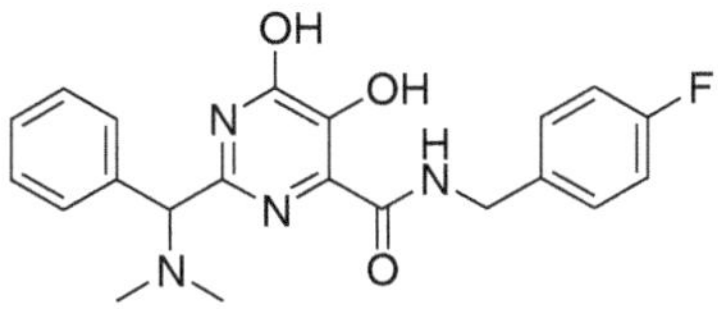

$t_{1/2}$ 1.7 h, *F* 59%, CL 14 ml min^{-1} kg^{-1} (rat)

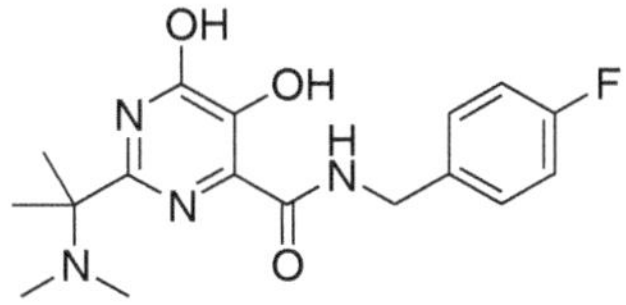

$t_{1/2}$ 2 h, *F* 28%, CL 16 ml min^{-1} kg^{-1} (rat)

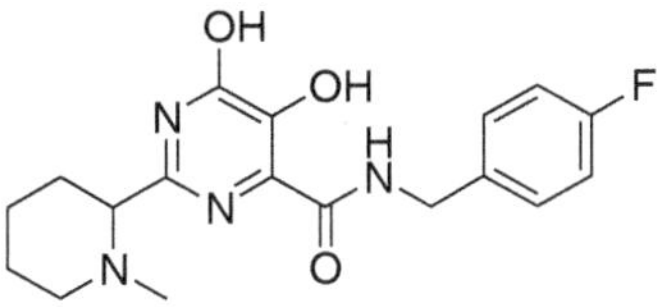

$t_{1/2}$ 0.5 h, *F* 27%, CL 44 ml min^{-1} kg^{-1} (rat)

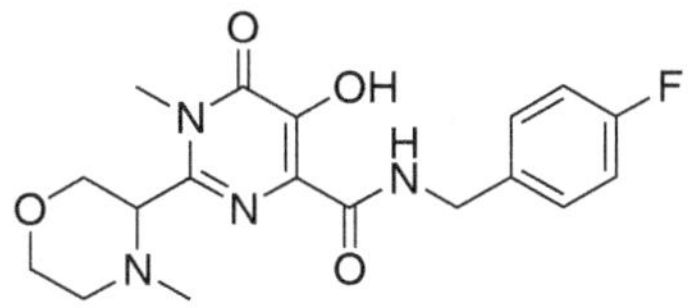

$t_{1/2}$ 1.1 h, *F* 56%, CL 9 ml min^{-1} kg^{-1} (rat)

Summa, V. *et al.* (2006) 4,5-Dihydroxypyrimidine carboxamides and *N*-alkyl-5-hydroxypyrimidinone carboxamides are potent, selective HIV integrase inhibitors with good pharmacokinetic profiles in preclinical species. *J. Med. Chem.*, **49**, 6646–6649.

2.7.13
Dopamine D$_3$ Antagonists

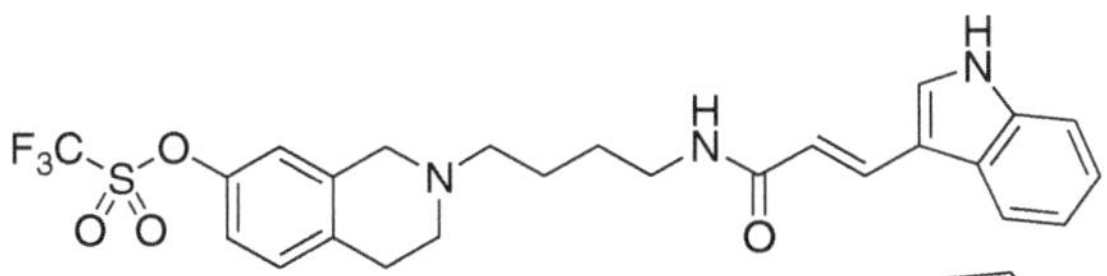

$t_{1/2}$ 2.5 h; *F* 7%; CL 27 ml min^{-1} kg^{-1} (rat)

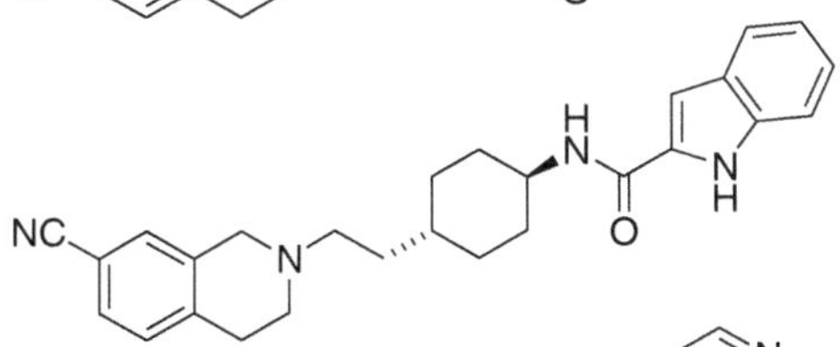

$t_{1/2}$ 6.3 h; *F* 43%; CL 6 ml min^{-1} kg^{-1} (rat)

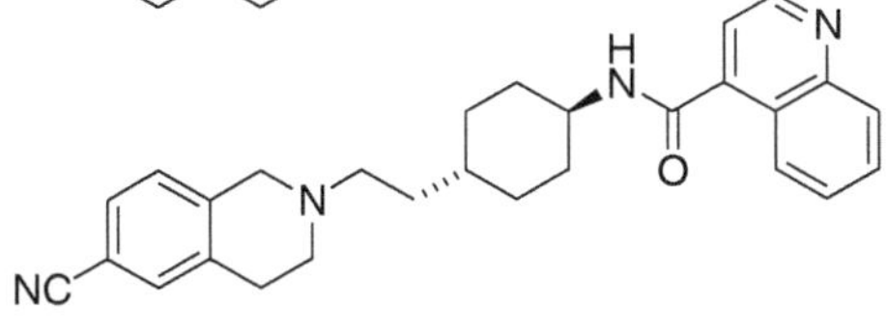

$t_{1/2}$ 2 h; *F* 43%; CL 19 ml min^{-1} kg^{-1} (rat)

Stemp, G. *et al.* (2000) Design and synthesis of *trans-N*-[4-[2-(6-cyano-1,2,3,4-tetrahydroisoquinolin-2-yl)ethyl]cyclohexyl]-4-quinolinecarboxamide (SB-277011): a potent and selective dopamine D3 receptor antagonist with high oral bioavailability and CNS penetration in the rat. *J. Med. Chem.*, **43**, 1878–1885.

2.7.14
Inhibitors of Soluble Epoxide Hydrolase (Tested in Mice, po)

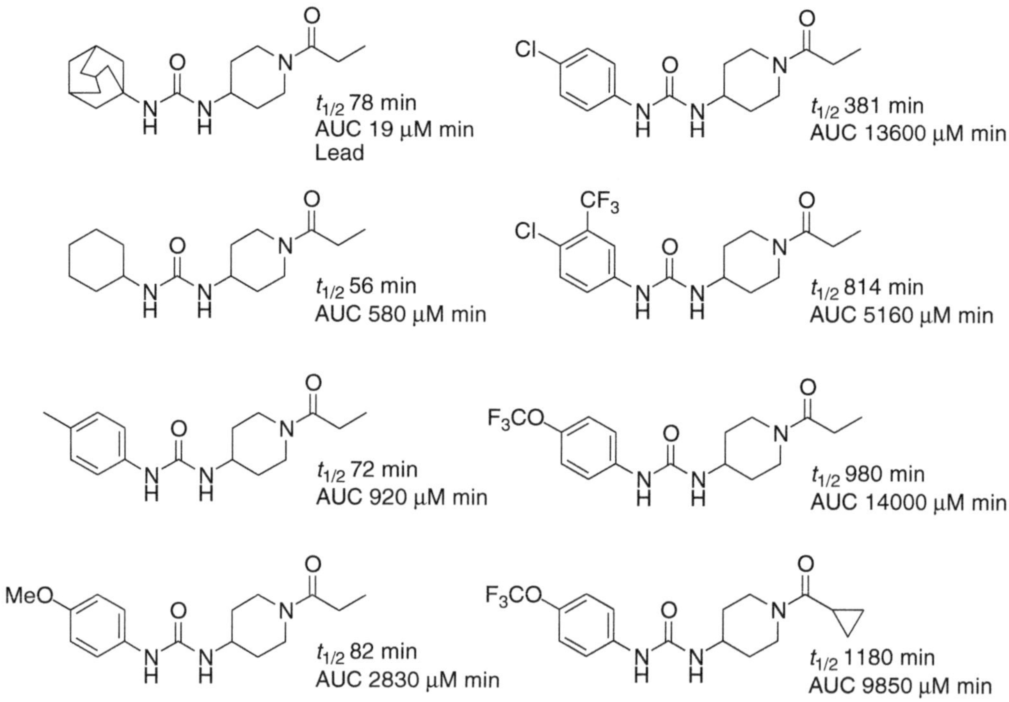

Rose, T.E., Morisseau, C., Liu, J.-Y., Inceoglu, B., Jones, P.D., Sanborn, J.R., and Hammock, B.D. (2010) 1-Aryl-3-(1-acylpiperidin-4-yl)urea inhibitors of human and murine soluble epoxide hydrolase: structure-activity relationships, pharmacokinetics, and reduction of inflammatory pain. *J. Med. Chem.*, **53**, 7067–7075.

2.7.15
Melanin-Concentrating Hormone Receptor-1 Antagonists

Rat AUC (0–6 h) (ng h ml^{-1})

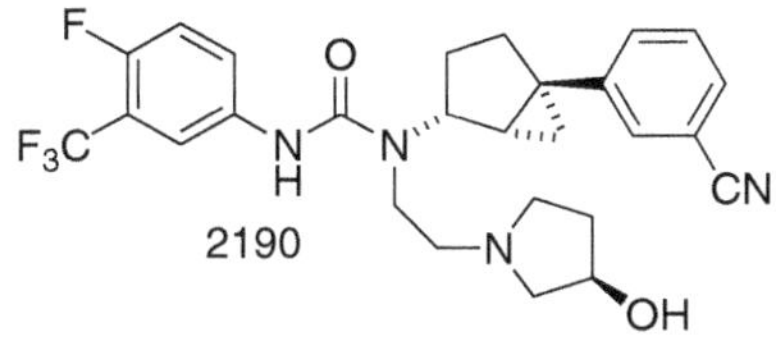

2190

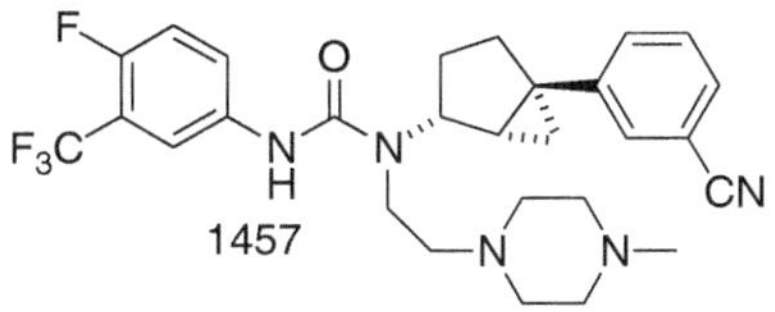

1457

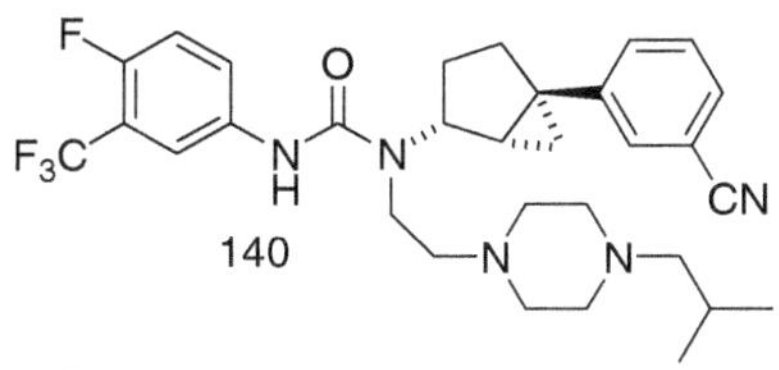

140

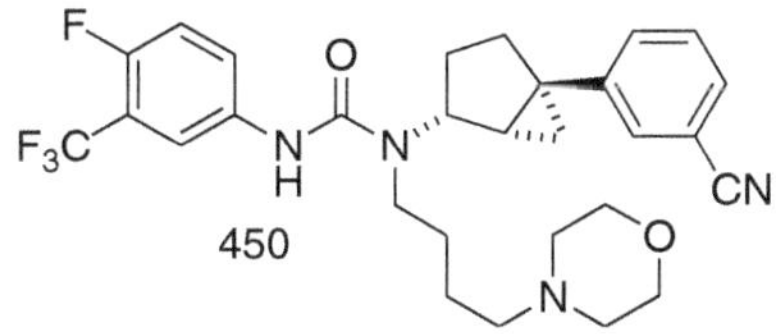

450

Rat AUC (0–6 h) (ng h ml^{-1})

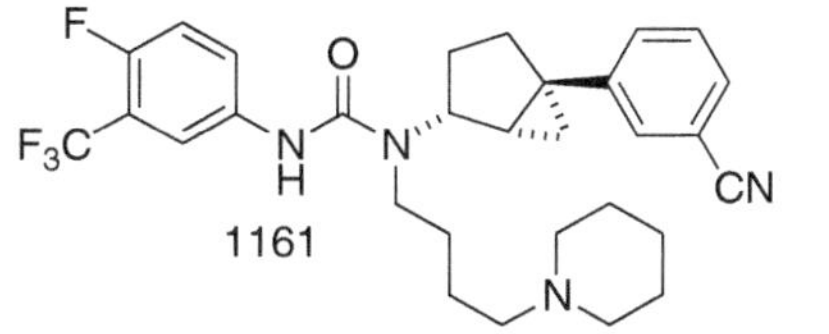

1161

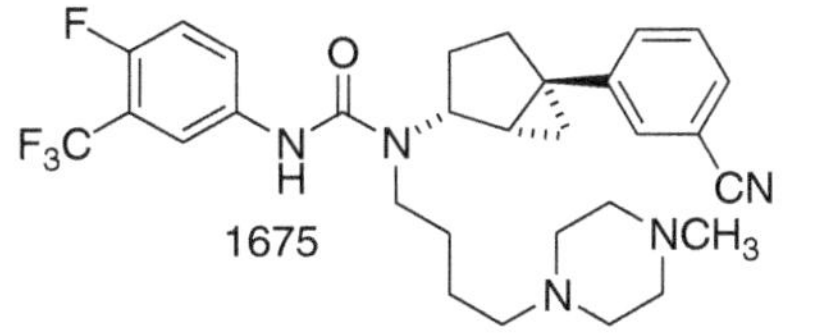

1675

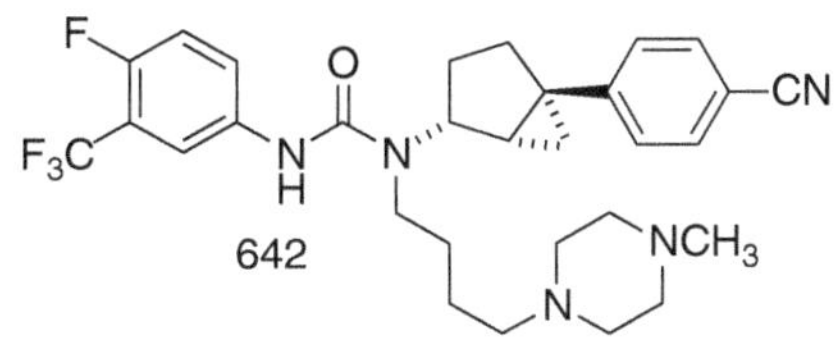

642

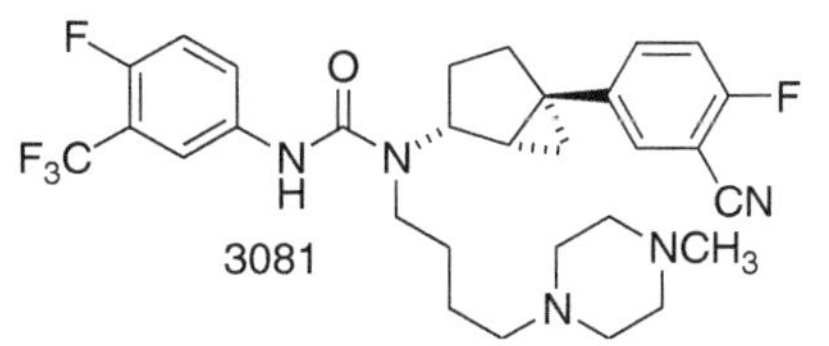

3081

McBriar, M.D. *et al.* (2006) Discovery of orally efficacious melanin-concentrating hormone receptor-1 antagonists as antiobesity agents. Synthesis, SAR, and biological evaluation of bicyclo[3.1.0]hexyl ureas. *J. Med. Chem.*, **49**, 2294–2310.

References

1. Lipinski, C.A., Lombardo, F., Dominy, B.W., and Feeney, P.J. (1997) Experimental and computational approaches to estimate solubility and permeability in drug discovery and development settings. *Adv. Drug Deliv. Rev.*, **23**, 3–25.
2. Gleeson, M.P. (2008) Generation of a set of simple, interpretable ADMET rules of thumb. *J. Med. Chem.*, **51**, 817–834.
3. Leeson, P.D. and Davis, A.M. (2004) Time-related differences in the physical property profiles of oral drugs. *J. Med. Chem.*, **47**, 6338–6348.
4. (a) Lovering, F., Bikker, J., and Humblet, C. (2009) Escape from flatland: increasing saturation as an approach to improving clinical success. *J. Med. Chem.*, **52**, 6752–6756; (b) Wassvik, C.M., Holmén, A.G., Draheim, R., Artursson, P., and Bergström, C.A.S. (2008) Molecular characteristics for solid-state limited solubility. *J. Med. Chem.*, **51**, 3035–3039.

5. O'Shea, R. and Moser, H.E. (2008) Physicochemical properties of antibacterial compounds: implications for drug discovery. *J. Med. Chem.*, **51**, 2871–2878.
6. (a) Varma, M.V.S. *et al.* (2010) Physicochemical space for optimum oral bioavailability: contribution of human intestinal absorbtion and first-pass elimination. *J. Med. Chem.*, **53**, 1098–1108; (b) Linnankoski, J., Mäkelä, J.M., Ranta, V.-P., Urtti, A., and Yliperttula, M. (2006) Computational prediction of oral drug absorbtion based on absorbtion rate constants in humans. *J. Med. Chem.*, **49**, 3674–3681; (c)Yoshida, F. and Topliss, J.G. (2000) QSAR model for drug human oral bioavailability. *J. Med. Chem.*, **43**, 2575–2785; (d) Krämer, S.D. (1999) Absorbtion prediction from physicochemical parameters. *Pharm. Sci. Technol. Today*, **2**, 373–380; (e) Palm, K., Luthman, K., Ungell, A.-L., Strandlund, G., Beigi, F., Lundahl, P., and Artursson, P. (1998) Evaluation of dynamic polar molecular surface area as predictor of drug absorbtion: comparison with other computational and experimental predictors. *J. Med. Chem.*, **41**, 5382–5392; (f) Winiwarter, S., Bonham, N.M., Ax, F., Hallberg, A., Lennernäs, H., and Karlén, A. (1998) Correlation of human jejunal permeability (*in vivo*) of drugs with experimentally and theoretically derived parameters. A multivariate data analysis approach. *J. Med. Chem.*, **41**, 4939–4949; (g) von Geldern, T.W. *et al.* (1996) Azole endothelin antagonists .3. Using $\Delta \log P$ as a tool to improve absorbtion. *J. Med. Chem.*, **39**, 982–991.
7. (a) Veber, D.F., Johnson, S.R., Cheng, H., Smith, B.R., Ward, K.W., and Kopple, K.D. (2002) Molecular properties that influence the oral bioavailability of drug candidates. *J. Med. Chem.*, **45**, 2615–2623; (b) Wenlock, M.C., Austin, R.P., Barton, P., Davis, A.M., and Leeson, P.D. (2003) A comparison of physicochemical property profiles of development and marketed oral drugs. *J. Med. Chem.*, **46**, 1250–1256; See also (c) Lu, J.J. *et al.* (2004) Influence of molecular flexibility and polar surface area metrics on oral bioavailability in the rat. *J. Med. Chem.*, **47**, 6104–6107.
8. (a) Broccatelli, F., Carosati, E., Neri, A., Frosini, M., Goracci, L., Oprea, T.I., and Cruciani, G. (2011) A novel approach for predicting p-glycoprotein (ABCB1) inhibition using molecular interaction fields. *J. Med. Chem.*, **54**, 1740–1751; (b) Pajouhesh, H., and Lenz, G.R. (2005) Medicinal chemical properties of successful central nervous system drugs. *J. Am. Soc. Exp. Neurotherapeutics*, **2**, 541–553; (c) Pardridge, W.M. (1997) Drug delivery to the brain. *J. Cereb. Blood Flow Metabol.*, **17**, 713–731; (d) Begley, D.J. (1996) The blood–brain barrier: principles for targeting peptides and drugs to the central nervous system. *J. Pharm. Pharmacol.*, **48**, 136–146.
9. (a) Hitchcock, S.A. and Pennington, L.D. (2006) Structure–brain exposure relationships. *J. Med. Chem.*, **49**, 7559–7583; (b) Kelder, J. *et al.* (1999) Polar molecular surface as a dominating determinant for oral absorbtion and brain penetration of drugs. *Pharm. Res.*, **16**, 1514–1519.
10. Rowley, M. *et al.* (1997) Effect of plasma protein binding on *in vivo* activity and brain penetration of glycine/NMDA receptor antagonists. *J. Med. Chem.*, **40**, 4053–4068.
11. (a) Shaik, S., Cohen, S., Wang, Y., Chen, H., Kumar, D., and Thiel, W. (2010) P450 enzymes: their structure, reactivity, and selectivity-modeled by QM/MM calculations. *Chem. Rev.*, **110**, 949–1017; (b) Yano, J.K., Denton, T.T., Cerny, M.A., Zhang, X., Johnson, E.F., and Cashman, J.R. (2006) Synthetic inhibitors of cytochrome P-450 2A6: inhibitory activity, difference spectra, mechanism of inhibition, and protein cocrystallization. *J. Med. Chem.*, **49**, 6987–7001; (c) Burton, J., Ijjaali, I., Barberan, O., Petitet, F., Vercauteren, D.P., and Michel, A. (2006) Recursive partitioning for the prediction of cytochromes P450 2D6 and 1A2 inhibition: importance of the quality of the dataset. *J. Med. Chem.*, **49**, 6231–6240; (d) Hudelson, M.G. and Jones, J.P. (2006) Line-walking method for predicting the inhibition of P450 drug metabolism. *J. Med. Chem.*, **49**, 4367–4373; (e) de Groot, M., Ackland, M.J., Horne, V.A., Alex, A.A., and Jones, B.C. (1999) A novel approach to predicting P450 mediated drug metabolism. CYP2D6 catalyzed N-dealkylation reactions and qualitative metabolite predictions using a combined protein and pharmacophore model for CYP2D6. *J. Med. Chem.*, **42**, 4062–4070; (f) Smith, D.A., Ackland, M.J., and Jones, B.C. (1997) Properties of cytochrome P450

isoenzymes and their substrates Part 1: active site characteristics, and Part 2: properties of cytochrome P450 substrates. *Drug Discov. Today*, **2**, 406–414, 479–486.

12. Riley, R.J. (2001) The potential pharmacological and toxicological impact of P450 screening. *Curr. Opin. Drug Discov. Dev.*, **4**, 45–54.
13. (a) Wilkinson, G.R. (2005) Drug metabolism and variability among patients in drug response. *New Engl. J. Med.*, **352**, 2211–2221; (b) Lilja, J.J., Juntti-Patinen, L., and Neuvonen, P.J. (2004) Orange juice substantially reduces the bioavailability of the β-adrenergic blocking agent celiprolol. *Clin. Pharmacol. Ther.*, **75**, 184–190; (c) Dresser, G.K., Bailey, D.G., Leake, B.F., Schwarz, U.I., Dawson, P.A., Freeman, D.J., and Kim, R.B. (2002) Fruit juices inhibit organic anion transporting polypeptide-mediated drug uptake to decrease the oral availability of fexofenadine. *Clin. Pharmacol. Ther.*, **71**, 11–20.
14. Ridley, J.M., Milnes, J.T., Hancox, J.C., and Witchel, H.J. (2006) Clemastine, a conventional antihistamine, is a high potency inhibitor of the hERG K^+ channel. *J. Mol. Cell. Cardiol.*, **40**, 107–118.
15. Bischoff, U., Schmidt, C., Netzer, R., and Pongs, O. (2000) Effects of fluoroquinolones on hERG currents. *Eur. J. Pharmacol.*, **406**, 341–343.
16. (a) Aronov, A.M. (2006) Common pharmacophores for uncharged human ether-a-go-go-related gene (hERG) blockers. *J. Med. Chem*, **49**, 6917–6921; (b) Jamieson, C., Moir, E.M., Rankovic, Z., and Wishart, G. (2006) Medicinal chemistry of hERG optimizations: highlights and hang-ups. *J. Med. Chem.*, **49**, 5029–5046; (c) Cavalli, A., Poluzzi, E., De Ponti, F., and Recanatini, M. (2002) Toward a pharmacophore for drugs inducing the long QT syndrome: insights from a CoMFA study of HERG K^+ channel blockers. *J. Med. Chem.*, **45**, 3844–3853.
17. (a) Kalgutkar, A. (2010) Metabolic activation and idiosyncratic drug toxicity – by avoiding structural alerts, do we mitigate risks? Lecture given at the New England Drug Metabolism Discussion Group Summer Symposium 2010.; (b) Kola, I. and Landis, J. (2004) Can the pharmaceutical industry reduce attrition rates? *Nat. Rev. Drug Discov.*, **3**, 711–716; (c) Roberts, S.A. (2003) Drug metabolism and pharmacokinetics in drug discovery. *Curr. Opin. Drug Discov. Dev.*, **6**, 66–80.
18. Obach, R.S., Kalgutkar, A.S., Ryder, T.F., and Walker, G.S. (2008) *In vitro* metabolism and covalent binding of enol-carboxamide derivatives and anti-inflammatory agents sudoxicam and meloxicam: insights into the hepatotoxicity of sudoxicam. *Chem. Res. Toxicol.*, **21**, 1890–1899.
19. (a) Uetrecht, J., Zahid, N., Tehim, A., Fu, J.M., and Rakhit, S. (1997) Structural features associated with reactive metabolite formation in clozapine analogues. *Chem. Biol. Interact.*, **104**, 117–129; (b) Liu, Z.C. and Uetrecht, J.P. (1995) Clozapine is oxidized by activated human neutrophils to a reactive nitrenium ion that irreversibly binds to the cells. *J. Pharmacol. Exp. Ther.*, **275**, 1476–1483.
20. (a) Leone, A.M., Kao, L.M., McMillan, M.K., Nie, A.Y., Parker, J.B., Kelley, M.F., Usuki, E., Parkinson, A., Lord, P.G., and Johnson, M.D. (2007) Evaluation of felbamate and other antiepileptic drug toxicity potential based on hepatic protein covalent binding and gene expression. *Chem. Res. Toxicol.*, **20**, 600–608; (b) Dieckhaus, C.M., Thompson, C.D., Roller, S.G., and Macdonald, T.L. (2002) Mechanisms of idiosyncratic drug reactions: the case of felbamate. *Chem. Biol. Interact.*, **142**, 99–117; (c) Thompson, C.D. *et al.* (1999) Quantification of patient urine samples of felbamate and three metabolites: acid carbamate and two mercapturic acids. *Epilepsia*, **40**, 769–776.
21. (a) Bauman, J.N., Kelly, J.M., Tripathy, S., Zhao, S.X., Lam, W.W., Kalgutkar, A.S., and Obach, R.S. (2009) Can *in vitro* metabolism-dependent covalent binding data distinguish hepatotoxic from nonhepatotoxic drugs? An analysis using human hepatocytes and liver S-9 fraction. *Chem. Res. Toxicol.*, **22**, 332–340; (b) Obach, R.S., Kalgutkar, A.S., Soglia, J.R., and Zhao, S.X. (2008) Can *in vitro* metabolism-dependent covalent binding data in liver microsomes distinguish hepatotoxic from nonhepatotoxic drugs? An analysis of 18 drugs with consideration of intrinsic clearance and daily dose. *Chem. Res. Toxicol.*, **21**, 1814–1822.

22. (a) Schwöbel, J.A.H., Koleva, Y.K., Enoch, S.J., Bajot, F., Hewitt, M., Madden, J.C., Roberts, D.W., Schultz, T.W., and Cronin, M.T.D. (2011) Measurement and estimation of electrophilic reactivity for predictive toxicology. *Chem. Rev.*, **111**, 2562–2596; (b) Benigni, R. and Bossa, C. (2011) Mechanisms of chemical carcinogenicity and mutagenicity: a review with implications for predictive toxicology. *Chem. Rev.*, **111**, 2507–2536; (c) Benigni, R. (2005) Structure-activity relationship studies of chemical mutagens and carcinogens: mechanistic investigations and prediction approaches. *Chem. Rev.*, **105**, 1767–1800.

Part II
The Pharmacokinetic Properties of Compound Classes

Lead Optimization for Medicinal Chemists: Pharmacokinetic Properties of Functional Groups and Organic Compounds, First Edition. Florencio Zaragoza Dörwald.

3
Alkanes

Alkyl groups are often introduced into leads to improve potency and selectivity, or to increase lipophilicity. If the lead is small and hydrophilic, alkyl groups may not only improve the affinity to the target protein but also enhance the oral availability of the compound by improving its absorption from the small intestine. Moreover, higher lipophilicity will improve tissue distribution and thereby increase the volume of distribution, which, at constant clearance, will also extend the half-life of a drug. PK (pharmacokinetic) improvement by enhancement of lipophilicity is exemplified by the adrenergic agonists oxedrine, bamethan, and clenbuterol (Scheme 3.1).

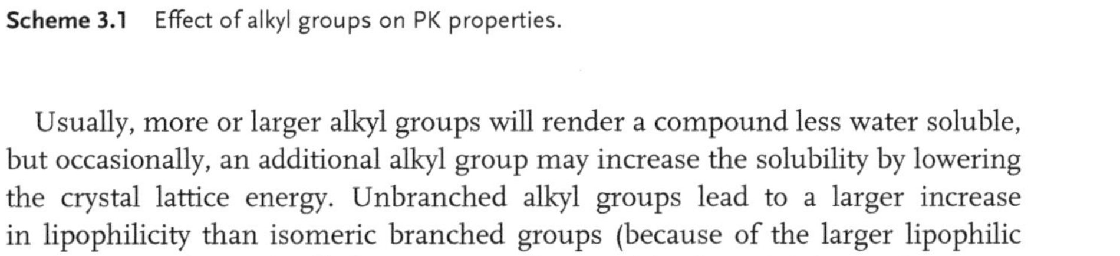

Scheme 3.1 Effect of alkyl groups on PK properties.

Usually, more or larger alkyl groups will render a compound less water soluble, but occasionally, an additional alkyl group may increase the solubility by lowering the crystal lattice energy. Unbranched alkyl groups lead to a larger increase in lipophilicity than isomeric branched groups (because of the larger lipophilic surface area of straight alkyl groups), as illustrated by the solubilities of isomeric pentanols and hexanols shown in Scheme 3.2 [1].

Lead Optimization for Medicinal Chemists: Pharmacokinetic Properties of Functional Groups and Organic Compounds, First Edition. Florencio Zaragoza Dörwald.

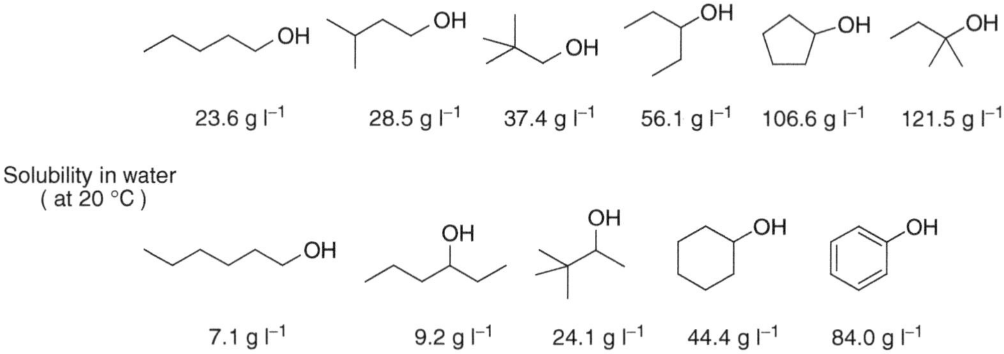

Scheme 3.2 Solubility of pentanols, hexanols, and phenol.

Alkyl groups exert a significant electron-donating effect, and can lower the acidity of phenols or carboxylic acids, and enhance the electron density at arenes or heteroarenes and the basicity of amines. Such effects may have dramatic consequences for the selectivity, potency, and metabolic stability of a drug. One exception is the cyclopropyl group, which has a slight electron-withdrawing effect because of the sp^2-like hybridization of its carbon atoms. Thus, cyclopropylamines or (cyclopropylmethyl)amines are less basic than isopropyl- or isobutylamines (see Chapter 23).

3.1 Metabolism

Alkyl groups can be hydroxylated by liver enzymes. Particularly reactive are benzylic or allylic positions. Methyl groups directly attached to an arene or heteroarene are often hydroxylated or even oxidized to a carboxyl group. As illustrated by the examples in Scheme 3.3, the metabolic fate of alkyl groups is not determined by the structure of the alkyl group alone, but by the structure of the whole molecule. Thus, cyclohexyl groups are often rapidly hydroxylated in positions 4 or 3, but a number of small cyclohexane derivatives, some with long plasma half-lives, do not undergo C-hydroxylation at all.

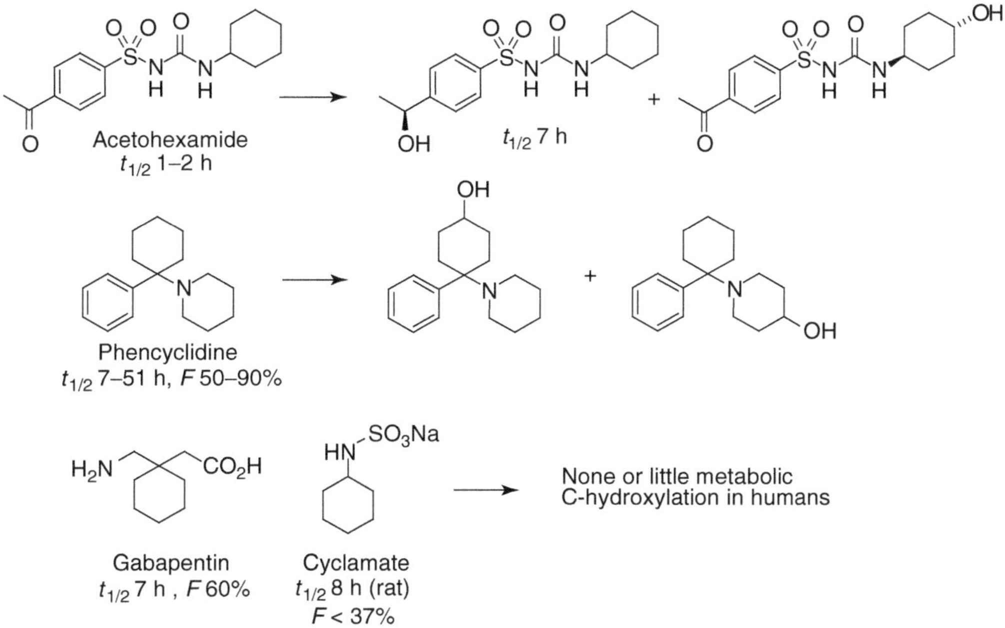

Scheme 3.3 Metabolism of substituted cyclohexanes.

Similarly, amantadine (Scheme 3.4) is hardly metabolized [2], while the closely related rimantadine undergoes extensive hydroxylation at the adamantyl group. More than 20% of orally administered rimantadine is excreted in the urine as hydroxylated metabolites.

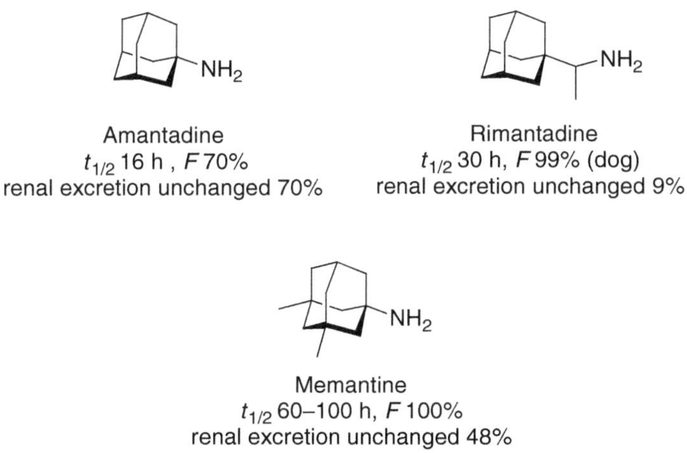

Scheme 3.4 Metabolic stability of adamantylamines.

Long alkyl chains are usually hydroxylated at the penultimate carbon ($\omega - 1$), and may even be oxidized to carboxylic acids. The latter may be α-hydroxylated, decarboxylated, and oxidized further to yield shorter carboxylic acids.

As illustration of typical metabolic reactions of alkyl groups, the human metabolites of nateglinide and irbesartan are shown in Scheme 3.5 [3].

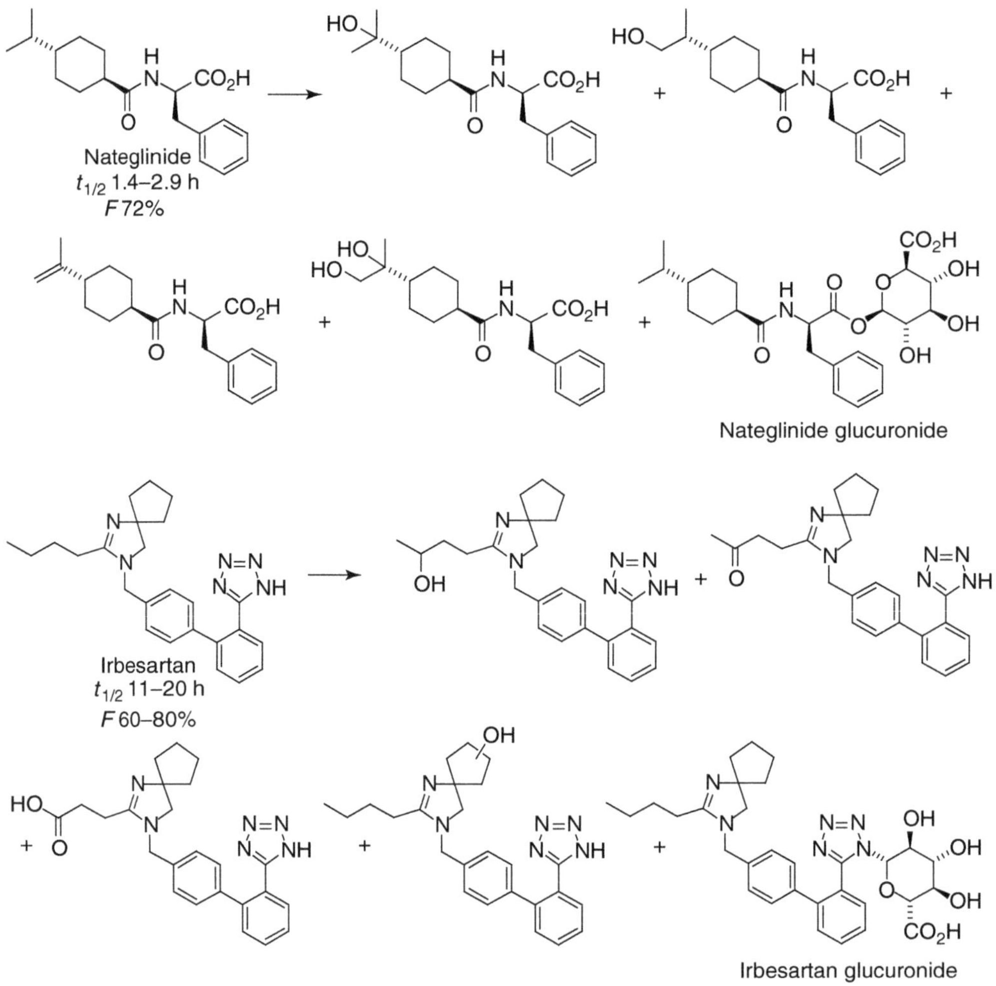

Scheme 3.5 Metabolism of nateglinide and irbesartan.

The metabolic hydroxylation of alkyl groups may be avoided by keeping them small, by substituting hydrogen by fluorine or other heteroatoms, by substituting the alkyl group by chlorine or an alkoxy group, or by rendering the molecule more hydrophilic. Occasionally, structural modification of the alkyl group, for example, the introduction of a methyl, cyclopropyl, or ethinyl group, may also block metabolic hydroxylation, as in the β-adrenergic antagonist betaxolol (Scheme 3.6).

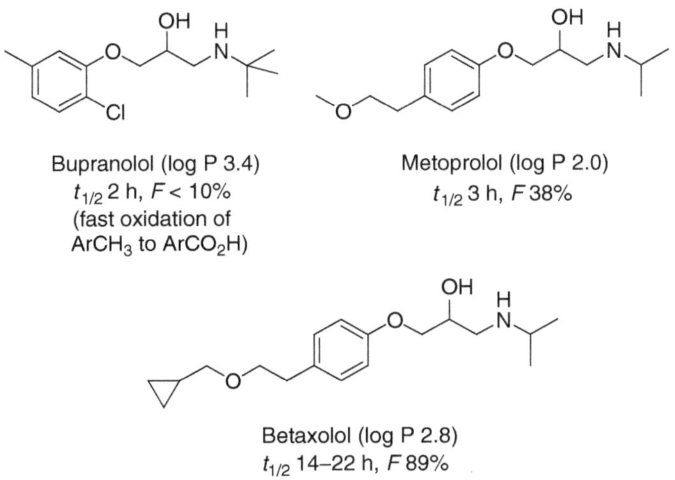

Scheme 3.6 Metabolic stabilization of methyl ethers by conversion to cyclopropylmethyl ethers.

References

1. (a) Ishikawa, M. and Hashimoto, Y. (2011) Improvement in aqueous solubility in small molecule drug discovery programs by disruption of molecular planarity and symmetry. *J. Med. Chem.*, **54**, 1539–1554; (b) Ginnings, P.M. and Baum, R. (1937) Aqueous solubilities of the isomeric pentanols. *J. Am. Chem. Soc.*, **59**, 1111–1113; (c) Stephenson, R. and Stuart, J. (1986) Mutual binary solubilities: water-alcohols and water-esters. *J. Chem. Ing.*, **31**, 56–70.
2. Köppel, C. and Tenczer, J. (1984) A revision of the metabolic disposition of amantadine. *Biol. Mass Spectrom.*, **12**, 499–501.
3. Weaver, M.L. *et al.* (2001) Pharmacokinetics and metabolism of nateglinide in humans. *Drug Metab. Dispos.*, **29**, 415–421.

4
Alkenes and Alkynes

Because alkenes are chemically reactive and may be metabolized to potentially toxic epoxides, medicinal chemists tend to avoid them. Many olefins such as styrene, allyl alcohol, allylamine, 1,3-butadiene, and vinyl chloride are irritant, hepatotoxic agents, and potential carcinogens. However, several drugs with vinyl or ethynyl groups have been a success, and unsaturation should be considered for pharmacokinetic (PK) fine-tuning and patent avoidance. Toxicity will be less of an issue if a compound is highly potent, and few milligrams will suffice for a one-day treatment.

Owing to the higher polarizability of olefins and alkynes, compared to the corresponding saturated compounds, the former tend to be more soluble in water.

Vinyl or ethynyl groups are electron withdrawing and can be used to decrease the electron density at arenes or heteroarenes. Allylamines and propargylamines are significantly less basic than the corresponding propylamines, and allyl alcohols are more acidic than propanols (Scheme 4.1 [1]).

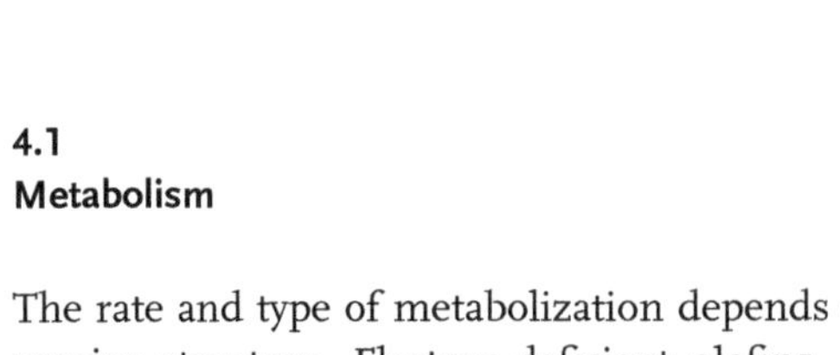

Scheme 4.1 The effect of unsaturation on pK_a values.

4.1
Metabolism

The rate and type of metabolization depends on the olefin's electron density and precise structure. Electron-deficient olefins, such as cinnamic acids, are often hydrogenated *in vivo* to alkanes. Electron-rich olefins and arenes, however, are mainly epoxidized by liver enzymes. Depending on their structure, these epoxides may be more or less reactive toward nucleophiles. Some epoxides are rather unreactive and can be readily detected in plasma or urine, or even be used as drugs, such as the bronchodilator tiotropium ($t_{1/2}$ 5–6 d), the scopolamines, or the

Lead Optimization for Medicinal Chemists: Pharmacokinetic Properties of Functional Groups and Organic Compounds, First Edition. Florencio Zaragoza Dörwald.

epoxide of carbamazepine. Other epoxides, in particular arene oxides, are highly reactive and will either alkylate P450 enzymes and lead to irreversible cytochrome *P*450 (CYP) inhibition, or alkylate other proteins or nucleic acids in the liver, leading to hepatotoxicity or hepatocarcinogenicity. The carcinogenicity of benzpyrene, vinyl chloride, and the aflatoxins is caused by metabolic epoxidation.

Epoxides do not necessarily lead to enzyme inactivation or to the alkylation of DNA, but may also be hydrated enzymatically to 1,2-diols. Furthermore, the ubiquitous sulfur nucleophile glutathione can also be alkylated by sufficiently reactive epoxides and thereby mediate their detoxification.

Unsaturated hydrocarbons may be quickly metabolized to 1,2-diols (via epoxides). This transformation has been used to design drugs with a short half-life. The barbiturate methohexital, for instance, is metabolized faster than other, related barbiturates and has been used as a short-acting, injectable anesthetic (Scheme 4.2). The long half-life and high oral bioavailability of secobarbital, however, shows that metabolic degradation of vinyl groups is not always fast. Particularly long-lived olefins are the piperazine derivatives flunarizine and almitrine.

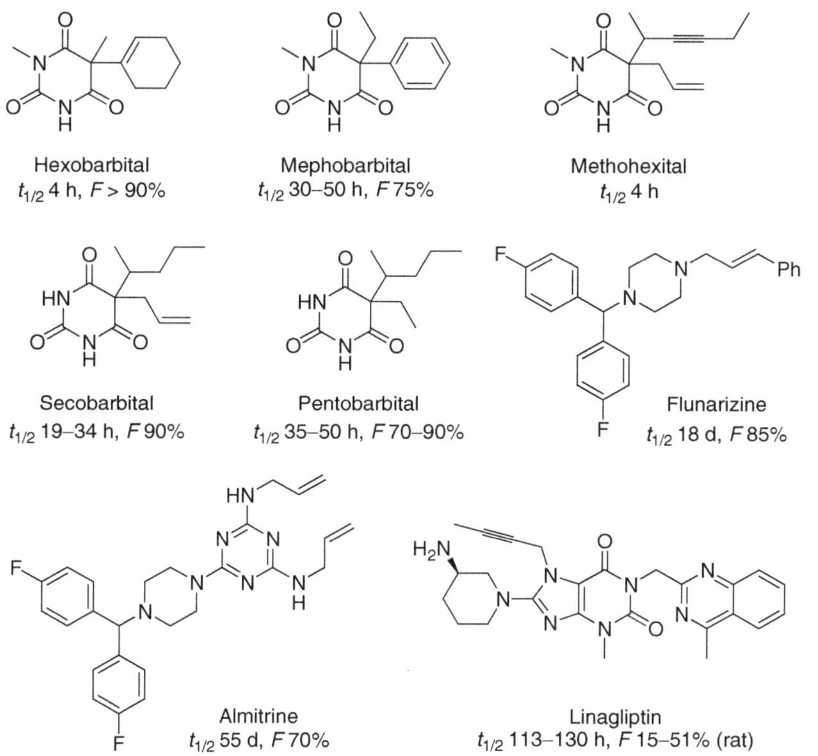

Scheme 4.2 Alkene- and alkyne-containing drugs.

Epoxidation is not necessarily the fastest metabolic reaction of an alkene. In some olefins, allylic hydroxylation or other metabolic pathways may effectively compete with epoxidation, as, for instance, in the antiarrhythmic quinidine (Scheme 4.3).

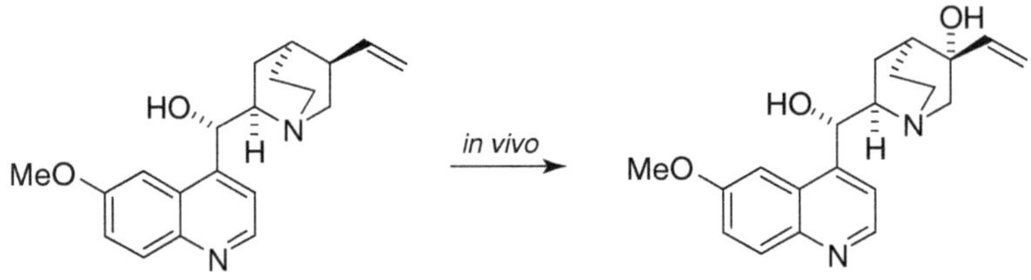

Quinidine
$t_{1/2}$ 6 h, F 80%

3-Hydroxyquinidine
$t_{1/2}$ 12 h

Scheme 4.3 Metabolism of quinidine.

Allylbenzenes such as estragole or safrole are similarly activated by hydroxylation at the allylic/benzylic position (Scheme 4.4). Sulfation of the resulting allylic alcohol leads to strongly alkylating carcinogens.

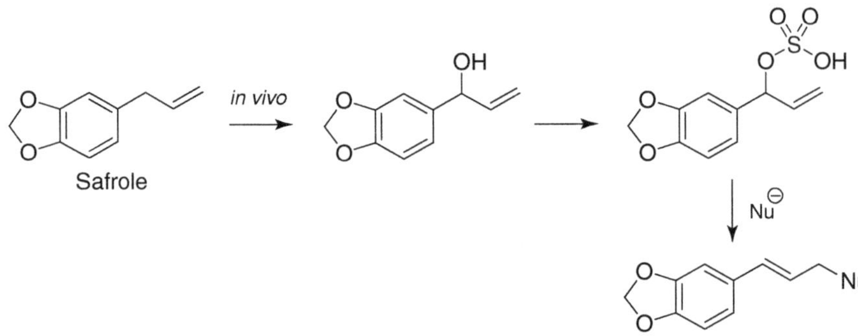

Safrole

Scheme 4.4 Metabolism of safrole.

Not all allylbenzenes are metabolized to allylic sulfates. The short-acting anesthetic acetamidoeugenol, withdrawn for causing local irritation, is converted *in vivo* to a cinnamic acid, presumably via isomerization and oxidation of an intermediate allylic alcohol (Scheme 4.5 [2]).

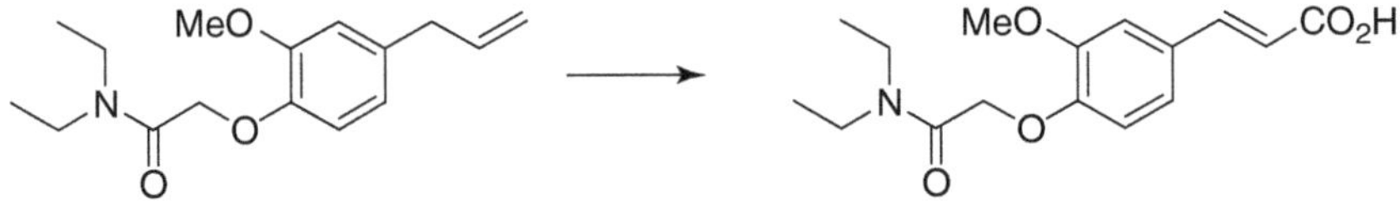

Acetamidoeugenol
$t_{1/2}$ < 10 min; F: low

Main metabolite of acetamidoeugenol
$t_{1/2}$ several hours

Scheme 4.5 Metabolism of acetamidoeugenol.

The different metabolic fates of safrole and acetamidoeugenol suggest that metabolism predictions are far from trivial. Several drugs containing olefins or alkynes have been successfully launched and do not show any excessive toxicity.

Therefore, if a lead with a larger alkyl group is too lipophilic or insoluble, an unsaturated analog may be an option to test. Small, hydrophilic drugs containing an alkene, such as seletracetam, vigabatrin, or tilidine (Scheme 4.6), seem not to be metabolized to toxic epoxides. The main metabolic pathway of the irreversible monoamine oxidase (MAO) inhibitor mofegiline in humans and dogs is carbamate formation by carboxylation and intramolecular vinylic displacement of fluoride [3].

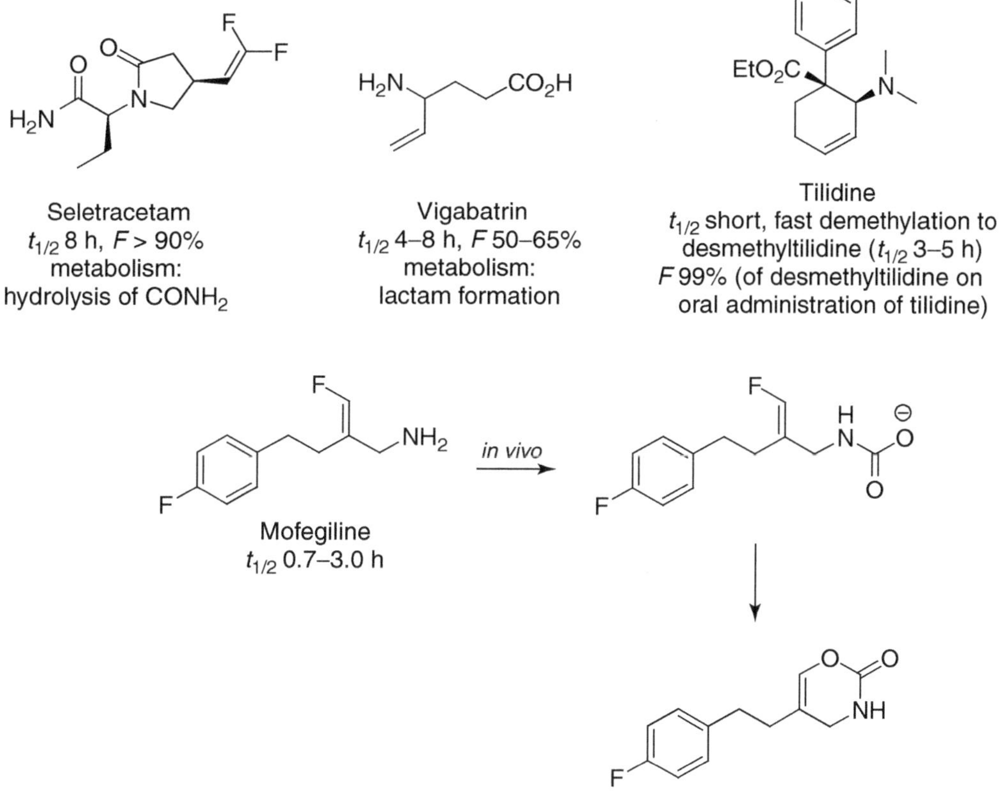

Scheme 4.6 Alkene-containing drugs and their metabolic transformations.

Relatively few alkyne-containing drugs have been marketed. Notable examples include the obsolete oral hypnotics ethinamate and ethchlorvynol, the contraceptive ethinyl estradiol, and the antineoplastic kinase inhibitor erlotinib (Scheme 4.7). In ethinylestradiol, the ethinyl group effectively blocks enzymatic oxidation to estrone and thus first-pass metabolism. Ethinylestradiol has much higher oral availability and longer half-life than the parent estradiol. Metabolic transformations of the ethinyl group of erlotinib are shown in Scheme 4.7.

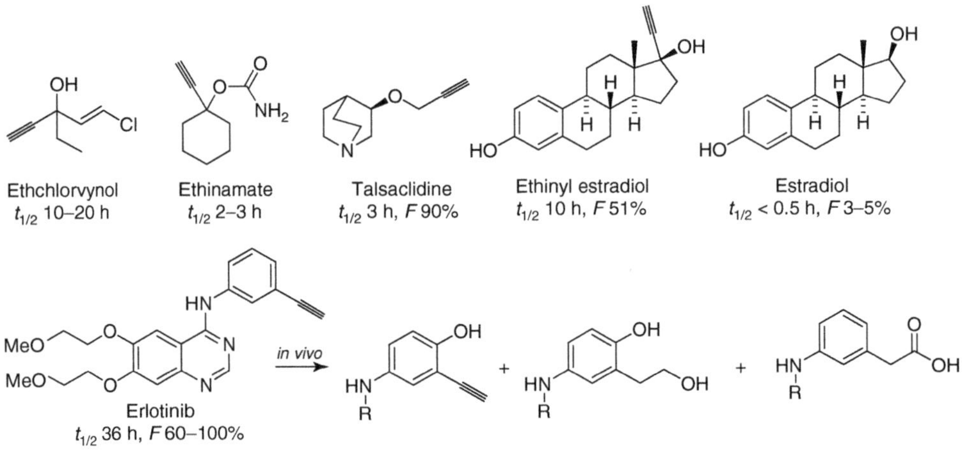

Scheme 4.7 Alkyne-containing drugs.

References

1. (a) Ballinger, P. and Long, F.A. (1960) Acid ionization constants of alcohols. II. Acidities of some substituted methanols and related compounds. *J. Am. Chem. Soc.*, **82**, 795–798; (b) Hine, J., Yeh, C.Y., and Schmalstieg, F.C. (1970) Polar effects on the formation of imines from isobutyraldehyde and primary aliphatic amines. *J. Org. Chem.*, **35**, 340–344.
2. Pulver, R. and Litvan, F. (1960) Über den Stoffwechsel des 2-Methoxy-4-allylphenoxyessigsäure-*N*,*N*-diäthylamid. *Arzneimittel-Forsch.*, **10**, 111–113.
3. Dow, J., Piriou, F., Wolf, E., Dulery, B.D., and Haegele, K.D. (1994) Novel carbamate metabolites of mofegiline, a primary amine monoamine oxidase B inhibitor, in dogs and humans. *Drug Metab. Dispos.*, **22**, 738–749.

5
Arenes

Arenes and heteroarenes are ubiquitous structural elements of drugs. Their popularity is mainly due to their good synthetic accessibility. Substituted arenes and heteroarenes are often easier to prepare than saturated cyclic compounds because no stereoisomers are formed and because the regioselectivity of aromatic substitutions is predictable. Moreover, arenes dissolve better in water than cyclohexanes, mainly because of their higher polarizability.

5.1
Metabolism

The metabolic fate of arenes depends not only on the precise substitution pattern but also on the overall structure of the arene-containing compound. The most common metabolic transformation of arenes is hydroxylation via the intermediate formation of an arene oxide (Scheme 5.1).

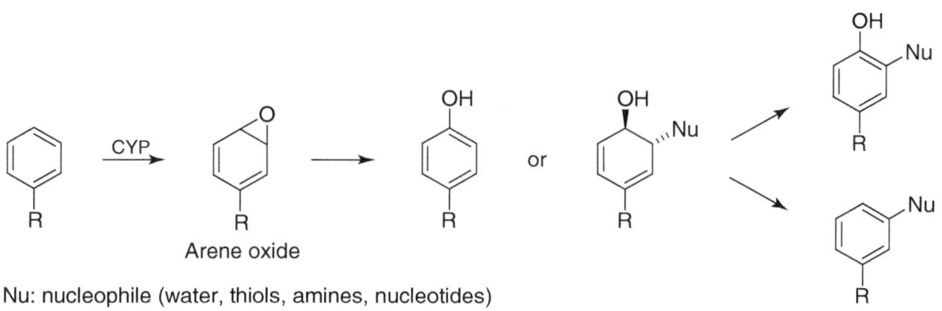

Scheme 5.1 Metabolism of arenes.

The reactivity and thus the toxicity and mutagenicity of arenes and arene-derived metabolic intermediates depends on their substituents. While the arene oxides of benzene or benzpyrene are cancerogenic, the arene oxide of bromobenzene mainly reacts with proteins in the liver leading to necrosis. Alkylbenzenes,

Lead Optimization for Medicinal Chemists: Pharmacokinetic Properties of Functional Groups and Organic Compounds, First Edition. Florencio Zaragoza Dörwald.

such as toluene, are usually not mutagenic but are mainly metabolized at the benzylic positions; phenylalanine is converted to tyrosin. Arenes with electron-withdrawing substituents, such as benzoic acid, or highly substituted arenes are often resistant to metabolic epoxidation. Electron-rich arenes, such as arylethers, alkylarenes, or anilines, are usually metabolized faster than electron-poor arenes (haloarenes, arylketones, and benzoic acid derivatives). The rate of hydroxylation of monosubstituted, electron-rich benzenes can, however, be sufficiently slow for their use in drugs (Scheme 5.2).

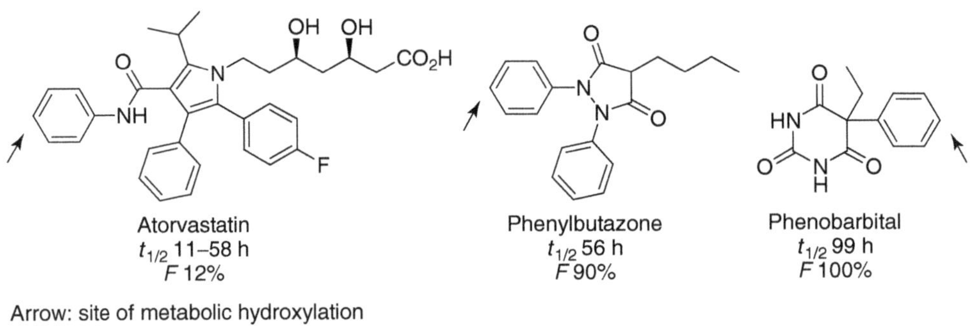

Scheme 5.2 Hydroxylation of drugs with electron-rich phenyl groups.

If one's lead is too short-lived, and aromatic hydroxylation is the fastest metabolic reaction, fluorination of the arene may be an option. Alternatively, if the target permits it, chlorine, alkoxy, acetamido, cyclopropyl, or ethinyl groups may be attached to the site of metabolic hydroxylation. An electron-withdrawing substituent at any position of the arene may also block or retard hydroxylation, as illustrated by the β-adrenergic antagonists in Scheme 5.3. Chlorination of the para position of the dichlorophenoxy group in diclofenac leads to a 30-fold increase of plasma half-life.

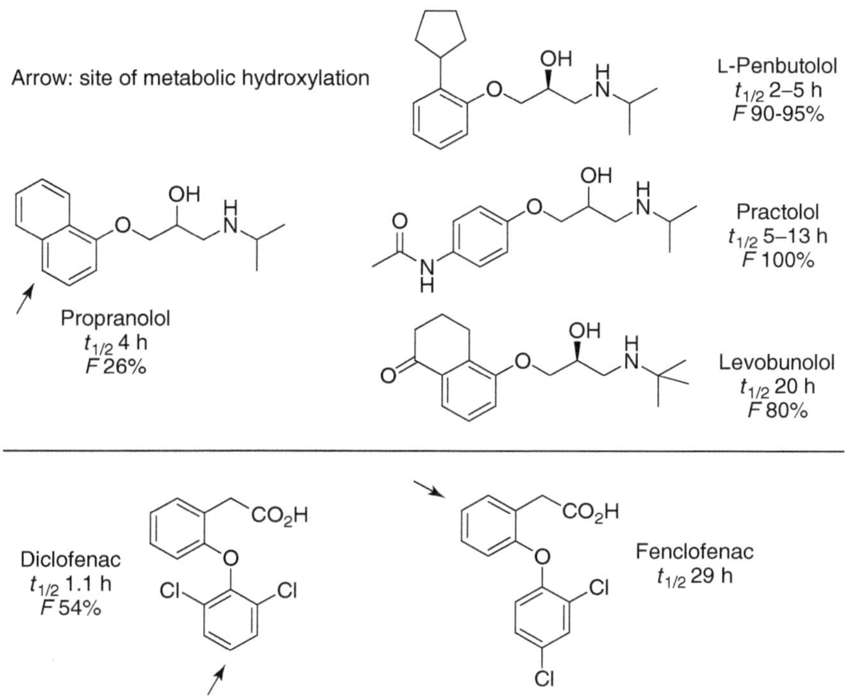

Scheme 5.3 Metabolism and half-life of arylether-containing drugs.

Large, polycyclic arenes can be highly carcinogenic, and lead to tumors when repeatedly applied to the skin. This is, however, not the case for smaller hydrocarbons (Scheme 5.4). Chronic exposure to benzene, however, can cause leukemia.

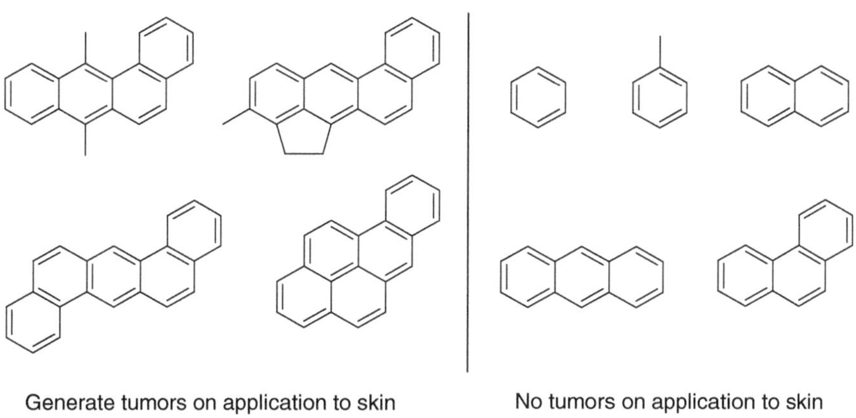

Scheme 5.4 Carcinogenic and non-carcinogenic arenes.

Naphthalene is significantly more reactive than benzene or toluene, and few drugs contain unsubstituted 1- or 2-naphthyl groups (Scheme 5.5). Naphthyloxy groups, however, are frequently used and appear to yield only unproblematic metabolites (e.g., 4-hydroxylated 1-naphthyloxy). One metabolite of cinacalcet is a 1,2-dihydroxylated dihydronaphthalene. 1-Naphthylamine has been used to develop anticholinesterase agents (e.g., 1-naphthyl *N*-methylcarbamate, an insecticide) and rodenticides (1-naphthylthiourea). 2-Naphthylamine and its derivatives are potent carcinogens and are unlikely to be of any use in drug development.

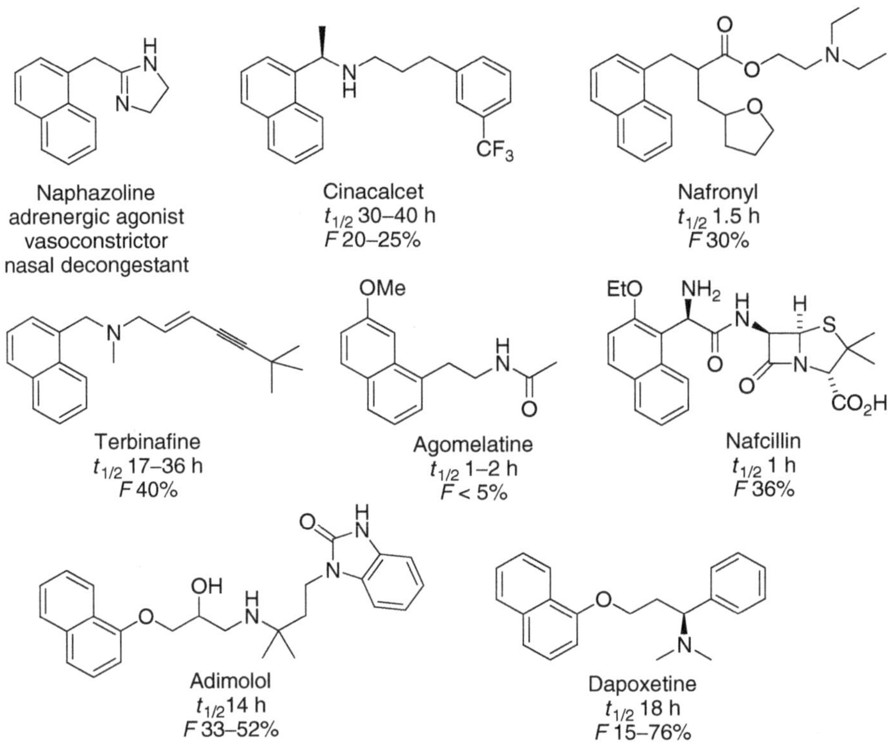

Scheme 5.5 Drugs containing naphthyl groups.

The few monoalkylnaphthalene-containing drugs cannot, in view of the examples in Scheme 5.5, be related to synthetic inaccessibility or biological inactivity alone. A more likely reason may be toxicity or CYP (cytochrome *P*450) inhibition issues. Suitable alternatives to naphthyl are benzannulated heterocycles, such as benzimidazole, benzpyrazole, benzoxazole, benzothiazole, indole, quinoline, quinazoline, and so on.

6
Halides

The properties of the four halogens diverge so much as to merit individual chapters. The only shared property is their ability to act as a leaving group for nucleophilic displacement: organic halides are electrophilic compounds capable of reacting irreversibly with nucleophiles such as thiols, amines, or nucleic acids. However, their ability to do so depends to a large extent on the precise structure of the organic halide.

Highly reactive halides, such as benzylic, allylic, α-alkoxy, α- or β-alkylthio, or α-carbonyl halides, or non-neopentylic alkyl bromides or iodides, are too reactive to be suitable as drugs, unless irreversible alkylation of a protein or nucleic acid is the aim. Vinylic halides may be metabolically activated by epoxidation. Such halides can alkylate DNA and be mutagenic and carcinogenic. A selection of carcinogenic organic halides is shown in Scheme 6.1.

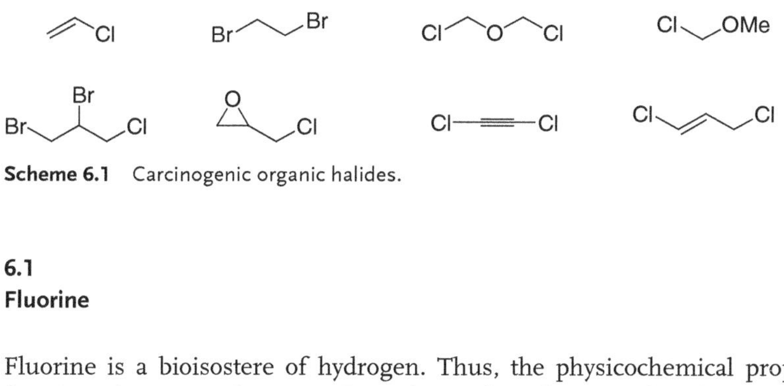

Scheme 6.1 Carcinogenic organic halides.

6.1
Fluorine

Fluorine is a bioisostere of hydrogen. Thus, the physicochemical properties of fluorinated compounds are similar to those of nonfluorinated analogs (Scheme 6.2) [1].

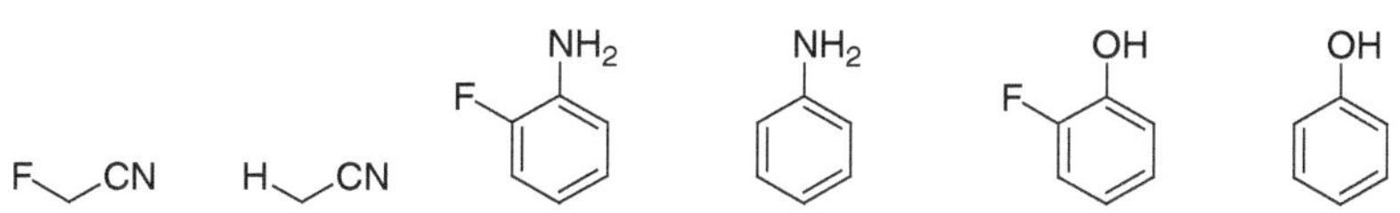

bp	79 °C	82 °C	182 °C	183 °C	172 °C	180 °C
pK_a (H_2O)			3.20	4.58	8.81	10.0
log P	−0.11	−0.33	1.47	1.14	1.82	1.54

Scheme 6.2 Effects of fluorination.

Lead Optimization for Medicinal Chemists: Pharmacokinetic Properties of Functional Groups and Organic Compounds, First Edition. Florencio Zaragoza Dörwald.

However, chemically, hydrogen and fluorine atoms behave quite differently. Because C—F bonds are strong, homolytic cleavage is usually not possible; therefore fluorine atoms may be used to block oxidative degradation by liver enzymes. Aromatic nucleophilic substitution or β-elimination of fluoride can, however, readily occur, also *in vivo*.

Single fluorine atoms at sp^3-carbon have been used to modulate the selectivity and metabolism of antiinflammatory cortisone analogs and of antineoplastic nucleotide analogs (Scheme 6.3).

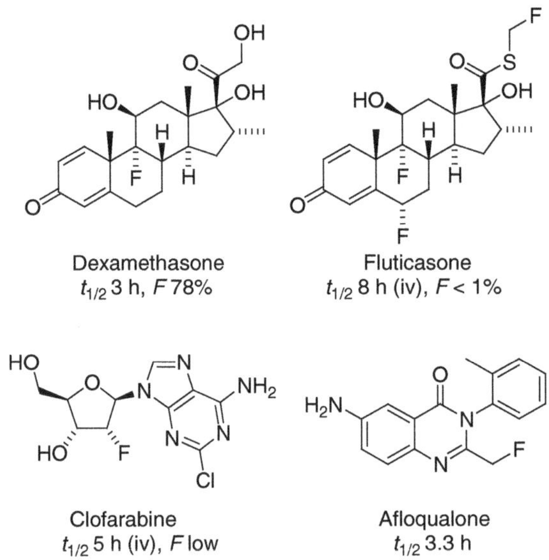

Scheme 6.3 Fluorinated drugs.

Because alkyl fluorides are not easy to prepare, most examples of single fluorine substitution are at vinylic or aromatic positions. One of the most useful properties of single fluorine atoms is their ability to block metabolic hydroxylation. If this is the rate-determining metabolic pathway, selective fluorination may significantly enhance the plasma half-life and oral availability of a lead. A comparison of cinnarizine with flunarizine clearly illustrates this point. Cinnarizine is mainly metabolized by aromatic hydroxylation of the benzhydryl group, which is no longer possible in flunarizine (Scheme 6.4).

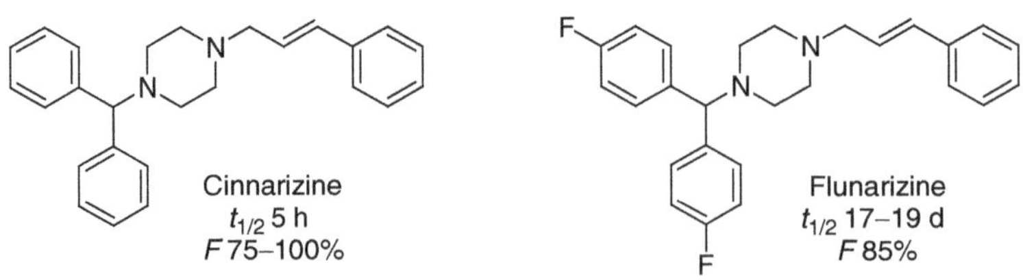

Scheme 6.4 Effect of fluorination.

Trifluoromethyl groups are metabolically stable and have been used as highly hydrophobic substituents, similar to methyl groups or chlorine atoms. Also, difluoromethyl, difluoromethoxy, and trifluoromethoxy groups can be used for this purpose. Such substituents not only increase the stability of a compound but also modify its solubility and affinity to proteins. C—F bonds are less polarizable than C—H bonds, and perfluoroalkanes are therefore not attracted by alkanes through induced-dipole–induced-dipole interactions (dispersion or London interactions). Thus, perfluoroalkyl groups are hydrophobic without being lipophilic.

6.2 Chlorine

The chlorine atom is metabolically inert and lipophilic, and well suited to enhance the affinity of a lead to its target protein. Alkyl chlorides are more reactive toward nucleophiles than alkyl fluorides, but often sufficiently unreactive to resist metabolic degradation. Thus, the sedative clomethiazole, the artificial sweetener sucralose, the antiprotozoal ornidazole, and the plant-growth regulator chlormequat are stable and do not react as electrophiles *in vivo* (Scheme 6.5). However, one of the metabolites of ornidazole is a diol, resulting from the displacement of chloride by hydroxide.

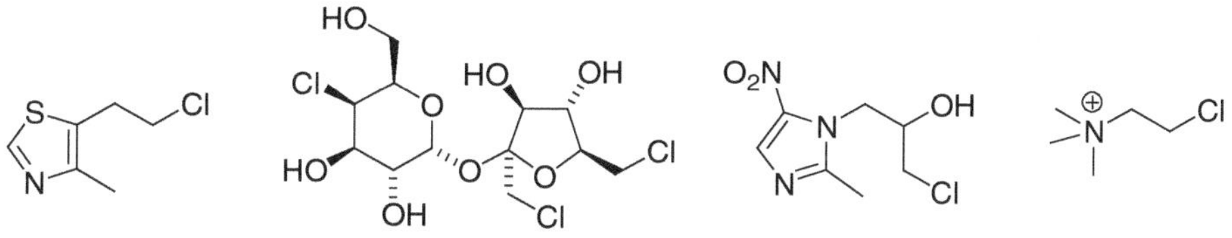

Scheme 6.5 Alkyl chlorides of low toxicity.

Only benzylic, allylic, or similarly activated chlorides will alkylate nucleophiles under physiological conditions. Bis(2-chloroethyl)amine derivatives, for instance, in which the nitrogen atom assists nucleophilic displacement, are useful as antineoplastics by virtue of their ability to alkylate nucleic acids (Scheme 6.6). The most reactive of them must be given intravenously, but the more unreactive ones, such as chlorambucil, are sufficiently stable to permit oral dosing. The irreversible α-adrenergic antagonist phenoxybenzamine is used for treating tumors of the adrenal medulla and sympathetic neurons (pheochromocytomas).

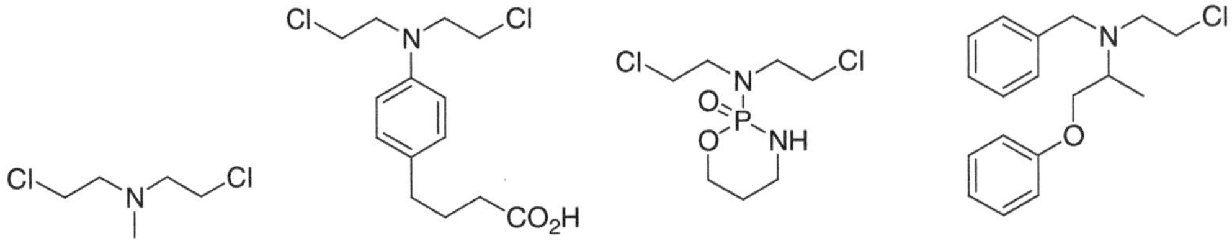

Mechlorethamine $t_{1/2} < 1$ min

Chlorambucil $t_{1/2}$ 1.3 h, *F* 87%

Cyclophosphamide $t_{1/2}$ 7.5 h, *F* 74%

Phenoxybenzamine $t_{1/2} < 24$ h

Scheme 6.6 Alkylating drugs.

Vinyl chlorides are less stable than the corresponding vinyl fluorides and can lead to toxic metabolites. Aryl chlorides, however, are metabolically stable unless strongly electron-withdrawing groups facilitate their substitution by an addition–elimination mechanism. Some chlorinated compounds, such as DDT, chlorinated biphenyls, or tetrachlorodioxin, are extremely inert toward metabolic or chemical degradation, which has forced the authorities of most countries to prohibit their preparation and use.

Arene-bound methyl groups can often be replaced by chlorine to yield metabolically more stable compounds with similar biological activity. This strategy was used to design long-acting analogs of zolpidem and tolbutamide (Scheme 6.7).

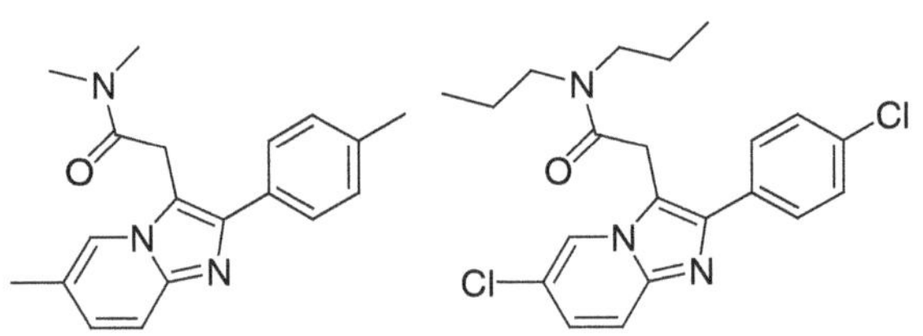

Zolpidem
$t_{1/2}$ 2.2 h, *F* 67%
V 0.54 l kg^{-1}, log P 3.07

Alpidem
$t_{1/2}$ 18–20 h, *F* 13% (rat)
V 5 l kg^{-1} (rat), log P 5.49

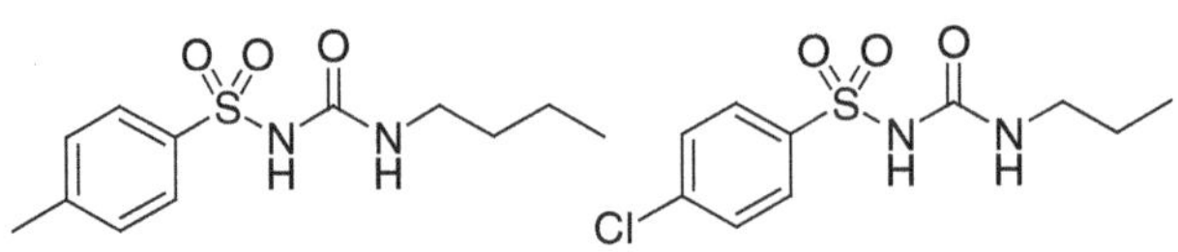

Tolbutamide
$t_{1/2}$ 5.9 h, *F* 93%
V 0.10 l kg^{-1}, log P 2.36

Chlorpropamide
$t_{1/2}$ 33 h, *F* > 90%
V/F 0.1 l kg^{-1}, log P 2.30

Scheme 6.7 Methylated versus chlorinated drugs.

6.3 Bromine

Alkyl bromides are usually too reactive and cannot be used as drugs, unless an alkylating agent is required. One exception is the chemically unreactive inhalation anesthetic halothane ($F_3CCHClBr$), which can, however, cause liver damage on repeated use. Some aryl bromides are readily debrominated *in vivo*, such as the uricosuric benzbromarone (Scheme 6.8). Many aryl bromides are hepatotoxic. Some, however, seem to be safe, such as the expectorant bromhexine and the antiviral brivudine.

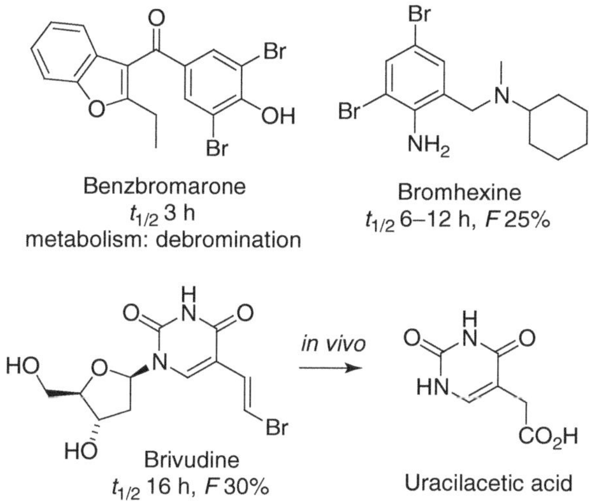

Scheme 6.8

Other drugs containing aryl bromides are brodimoprim, brofaromine, brimonidine, bromocriptine (a 2-bromoindole), etravirine, zimeldine, brompheniramine, and a number of benzodiazepines.

6.4 Iodine

The iodine–carbon bond is rather weak, and most iodine-containing drugs will leak small amounts of iodide. Iodine-containing compounds are only used for special applications, namely, as contrast agents, radioactive tracing agents (^{131}I), topical antiseptics, or for the treatment of thyroid-related diseases. More iodinated drugs include iodoquinol (treatment of parasitic infections) and the antiarrhythmic amiodarone (Scheme 6.9). All these agents can cause acute hypersensitivity reactions because some people become allergic to iodine (iodism).

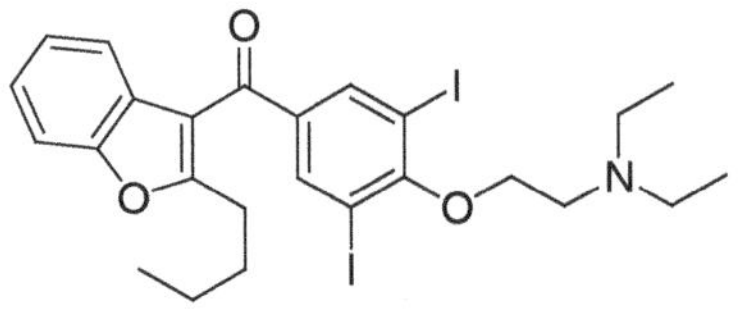

Amiodarone
$t_{1/2}$ 25 h, *F* 46%

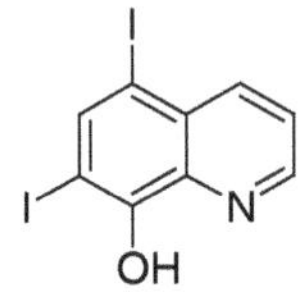

Iodoquinol
$t_{1/2}$ 11–14 h, *F* 8%

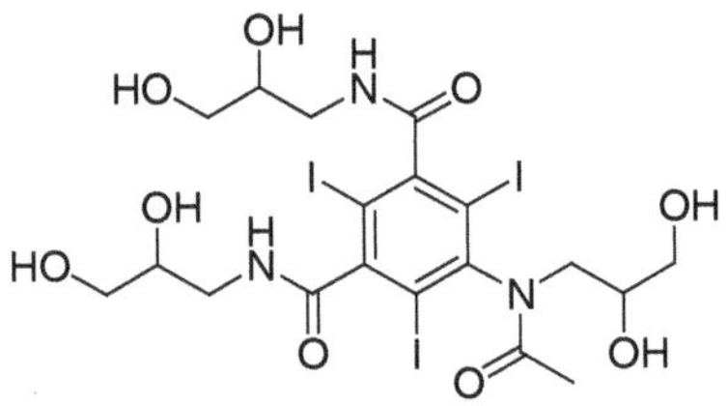

Iohexol
$t_{1/2}$ 3 h

Scheme 6.9 Iodinated drugs.

Contrast agents, such as iohexol, are administered intravenously at high doses in the precise moment of the measurement (e.g., computer tomography) and are excreted mainly unchanged in the urine afterward.

Heavier halogen atoms in a lead may be replaced by azide, cyano, CF_3, OCF_3, $OCHF_2$, SCF_3, SCN, ethinyl, vinyl, or difluorovinyl groups.

6.5 Alkylating Agents

The vesicant properties of sulfur mustard (bis(2-chloroethyl)sulfide), used in World War I for chemical warfare, were already known in 1887. This compound is strongly cytotoxic, in particular to highly proliferative tissue, such as lymphoid tissue, bone marrow, and the intestine. Between the wars, the corresponding nitrogen mustards were shown to be similarly toxic, and in 1942, clinical trials were started to evaluate these alkylating agents in the treatment of various types of cancer. At present, mainly nitrogen mustards, alkyl sulfonates, aziridines, nitrosoureas (Chapter 12), and triazenes (Chapter 10) are used as antineoplastic alkylants (Table 6.1).

One common feature of the alkylating agents for cancer therapy is their ability to mediate two alkylations and thereby crosslink two DNA molecules or DNA with a protein. Such a modification prevents both the replication of DNA and the biosynthesis of proteins.

Reference

1. (a) Hagmann, W.K. (2008) The many roles for fluorine in medicinal chemistry. *J. Med. Chem.*, **51**, 4359–4369; (b) Park, B.K., Kitteringham, N.R., and O'Neill, P.M. (2001) Metabolism of fluorine-containing drugs. *Annu. Rev. Pharmacol. Toxicol.*, **41**, 443–470; (c) Patani, G.A. and LaVoie, E.J. (1996) Bioisosterism: a rational approach in drug design. *Chem. Rev.*, **96**, 3147–3176; (d) Park, B.K. and Kitteringham, N.R. (1994) Effects of fluorine substitution on drug metabolism: pharmacological and toxicological implications. *Drug Metabol. Rev.*, **26**, 605–643.

Table 6.1 Alkylating agents and alkyl chlorides. *V* in l kg^{-1}, CL in ml min^{-1} kg^{-1}, Mwt in g mol^{-1}.

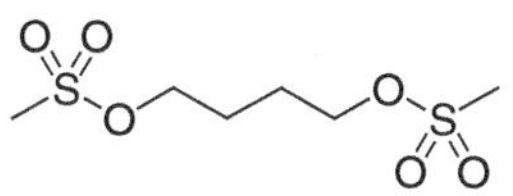	$t^1\!/_2$ *F* pb ur	2.6±0.5 h 70% 3–15% 0.5–3.0%	*V* CL/*F* Mwt PSA log *P*	1.0±0.2 4.5±0.9 246.3 104 Å^2 −0.56	BUSULFAN Antineoplastic; Metabolism: conjugation with glutathione, then conversion to tetrahydrothiophene *S*-oxide, sulfolane, and 3-hydroxysulfolane
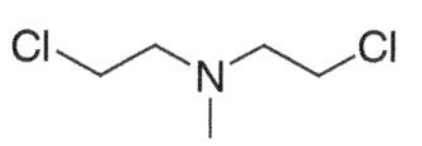	$t^1\!/_2$ *F* pb ur	<1 min – – 0%	*V* CL Mwt PSA log *P*	– – 156.1 3.24 Å^2 0.91	CHLORMETHINE, MECHLORETHAMINE Antineoplastic for iv administration
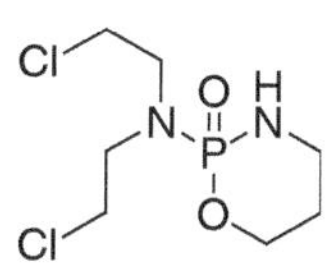	$t^1\!/_2$ *F* pb ur	7.5±4.0 h 74±22% 13% 6.5±4.3%	*V* CL Mwt PSA log *P*	0.9±0.6 1.3±0.5 261.1 51.4 Å^2 0.73	CYCLOPHOSPHAMIDE Antineoplastic, immunosuppressant; activation by hepatic metabolism; Metabolism: N-dealkylation, hydroxylation at CH_2NH, then oxidation to lactam and ring opening to carboxylic acid
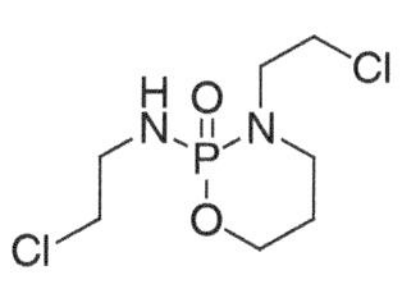	$t^1\!/_2$ *F* pb ur	4–10 h 100% 20% 15%	*V* CL Mwt PSA log *P*	0.4–0.6 0.8–2.3 261.1 51.4 Å^2 0.76	IFOSFAMIDE Antineoplastic, crosses bbb Metabolism: hydroxylation of ring CH_2N to active form of drug, followed by ring-opening and β-elimination to acrolein; hydroxylation of exocyclic CH_2N followed by hydrolysis to chloroacetaldehyde
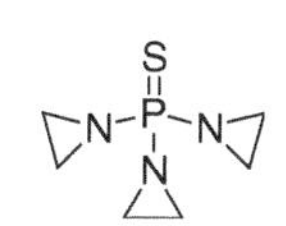	$t^1\!/_2$ *F* pb ur	1–2 h – – <10%	*V* CL Mwt PSA log *P*	– 19 189.2 50.9 Å^2 0.53	THIOTEPA Antineoplastic; Metabolite: phosphoramide (tepa, replacement of S by O; $t^1\!/_2$ 3–24 h)

(*continued overleaf*)

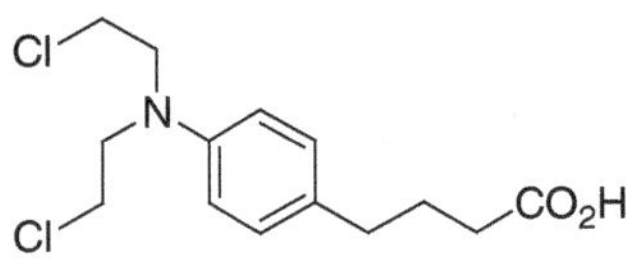

$t_{1/2}$	1.3±0.9 h	*V*	0.29±0.21	CHLORAMBUCIL
F	87±20%	CL	2.6±0.9	Antineoplastic
pb	99%	Mwt	304.2	Metabolism:
ur	<1%	PSA	40.5 Å^2	oxidative degradation to arylacetic
		log *P*	2.61	acid, substitution of Cl by OH

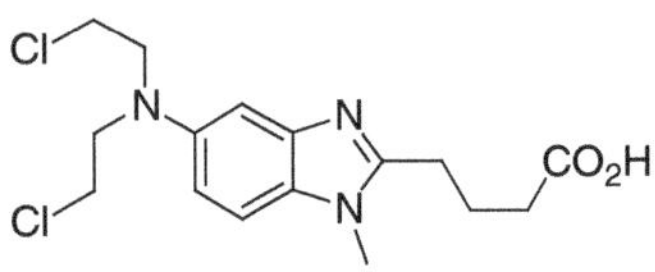

$t_{1/2}$	0.5 h	*V*	0.24–0.32	BENDAMUSTINE
F	–	CL	7.5	Antineoplastic
pb	>95%	Mwt	358.3	Metabolism:
ur	10–45%	PSA	58.4 Å^2	substitution of Cl by OH,
		log *P*	2.38	N-demethylation, conjugation
				with glutathione, hydroxylation of
				propylene

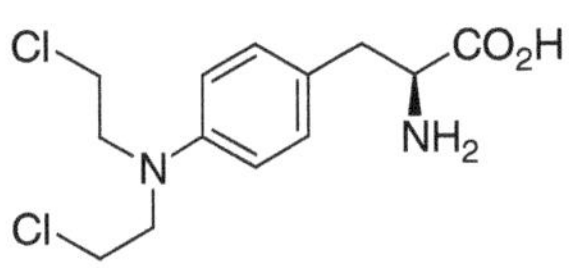

$t_{1/2}$	1.4±0.2 h	*V*	0.5±0.2	MELPHALAN
F	71±23%	CL	5.2±2.9	Antineoplastic
pb	90±5%	Mwt	305.2	Metabolism:
ur	12±7%	PSA	66.6 Å^2	sequential displacement of
		log *P*	0.54	chloride by hydroxide

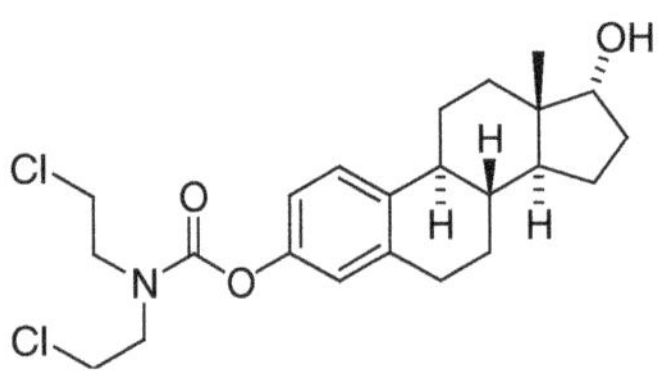

$t_{1/2}$	9–23 h	*V*	0.11	ESTRAMUSTINE
F	44–75%	CL	0.9–1.1	Antineoplastic
pb	99%	Mwt	440.4	Metabolism:
ur	–	PSA	49.8 Å^2	oxidation to ketone, hydrolysis
		log *P*	5.75	of carbamate; oral prodrug:
				phosphoric acid ester

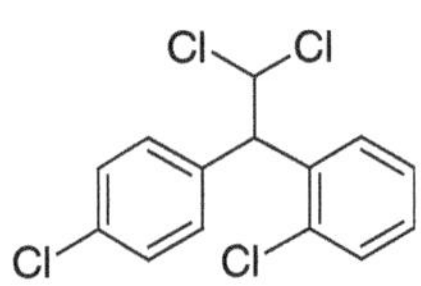

$t_{1/2}$	18–159 d	*V*	–	MITOTANE
F	40%	CL	–	Antiadrenal agent, antineoplastic;
pb	–	Mwt	320.0	blocks 11β-hydroxylation of
ur	0%	PSA	0.0 Å^2	adrenocorticoid precursors
		log *P*	5.41	Metabolite: diarylacetic acid,
				dehydrogenation to
				1,1-diaryl-2,2-dichloroethylene

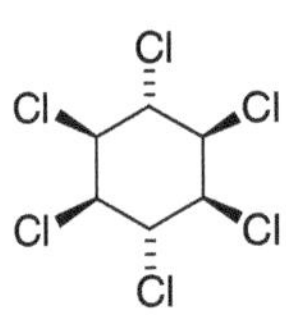

$t_{1/2}$	21 h	*V*	8*	LINDANE
F	–	CL	–	*sheep
pb	91%	Mwt	290.8	Insecticide
ur	–	PSA	0.0 Å^2	Metabolites:
		log *P*	3.82	1,2,4-trichlorobenzene, di- and
				trichlorophenols

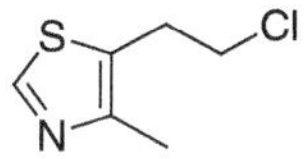	$t\frac{1}{2}$ 5±2 h F 30±10% pb 63±2% ur <1%	V CL Mwt PSA log P	8±5 18±7 161.7 41.1 Å^2 1.71	CLOMETHIAZOLE Sedative, hypnotic, anticonvulsant Metabolism: hydroxylation to 1-hydroxy-2-chloroethyl, then dehydrohalogenation to 5-acetyl-4-methylthiazole, then reduction to alcohol	
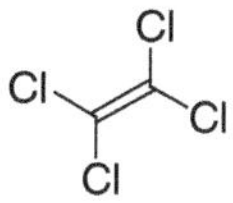	$t\frac{1}{2}$ 144 h F High pb – ur <4%	V CL Mwt PSA log P	8.2 – 165.8 0.0 Å^2 3.07	TETRACHLOROETHYLENE, PERC Dry cleaning solvent, anthelmintic Metabolites: trichloroacetic acid	

$t_{1/2}$, plasma half-life; F, oral bioavailability; pb, plasma protein binding; ur, excretion of unchanged drug in urine; V, volume of distribution; CL, clearance; Mwt, molecular weight; PSA, polar surface area.

7
Azides

At present, only few azido-group-containing drugs are on the market, for example, zidovudine and azidocillin (Scheme 7.1). Zidovudine is a nucleoside mimetic, the azido group replacing a hydroxyl group of 2′-deoxythymidine. Azidocillin is an older β-lactam antibiotic, closely related to ampicillin and benzylpenicillin, which was developed in the late 1960s.

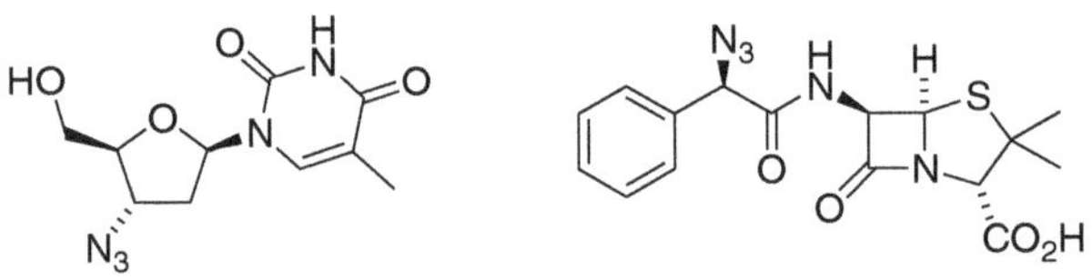

Zidovudine
$t_{1/2}$ 1.1 h, *F* 63%

Azidocillin
$t_{1/2}$ 0.6–1.1 h, *F* 57–64%

Scheme 7.1 Azides as drugs.

Azides are chemically reactive, and may act as electrophilic aminating reagents under acidic conditions, similar to hydroxylamines. High doses of organic azides can therefore cause adverse drug reactions (see Section 2.6). One of the various metabolic transformations of zidovudine and azidocillin is the reduction of the azido group to an amino group.

Lead Optimization for Medicinal Chemists: Pharmacokinetic Properties of Functional Groups and Organic Compounds, First Edition. Florencio Zaragoza Dörwald.

8
Nitro Compounds

Most drugs with nitro groups are nitroarenes, nitroheteroarenes, or nitroethylenes. Nitroalkanes have not been used as drugs and may be metabolically too unstable.

The nitro group has a strong dipole moment, which leads to higher boiling and melting points of nitroarenes when compared to chloro- or bromoarenes (Scheme 8.1). When considering intermolecular attractive forces, nitroarenes are more closely related to acetophenones or benzoic esters than to haloarenes.

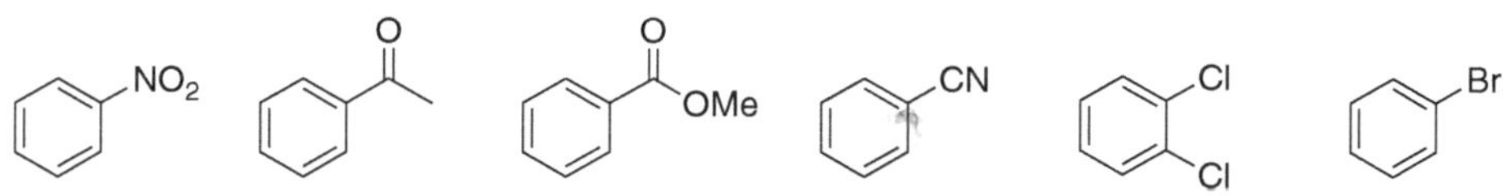

Nitrobenzene	Acetophenone	Methyl benzoate	Benzonitrile	Dichlorobenzene	Bromobenzene
bp 211 °C	bp 203 °C	bp 200 °C	bp 191 °C	bp 181 °C	bp 156 °C
mp 6 °C	mp 21 °C	mp −12 °C	mp −13 °C	mp −17 °C	mp −31 °C
log P 1.92	log P 1.67	log P 2.12	log P 1.58	log P 3.44	log P 2.94

Scheme 8.1 Properties of nitrobenzene and other substituted arenes.

Nitroarenes are rather lipophilic and barely soluble in water. Further, lipophilic compounds with separated charges in their Lewis structures include diazoalkanes, nitrones, azides, and phosphorus- or sulfur-derived ylides. In their noncovalent interactions with proteins, nitroarenes behave as halo- or alkylarenes. For instance, dihydropyridine-based calcium channel antagonists (antihypertensives) often contain nitrophenyl groups, which may be replaced by 1,2-dichlorophenyl without altering the target selectivity (Scheme 8.2). In benzodiazepines, replacement of the 7-chlorine atom by a nitro group is also tolerated without affecting their hypnotic and sedative properties or their ability to cross the blood–brain barrier (bbb). The same is the case for the two anticoagulants acenocoumarol and warfarin.

Lead Optimization for Medicinal Chemists: Pharmacokinetic Properties of Functional Groups and Organic Compounds, First Edition. Florencio Zaragoza Dörwald.

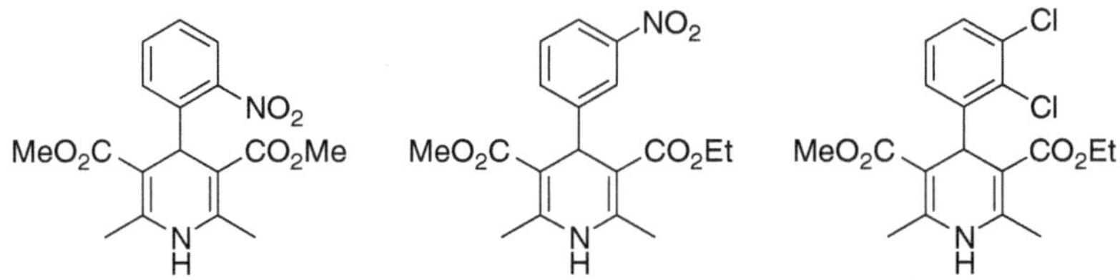

Nifedipine $t_{1/2}$ 2.5 h, F 50% | Nitrendipine $t_{1/2}$ 4 h, F 14–55% | Felodipine $t_{1/2}$ 14 h, F 15%

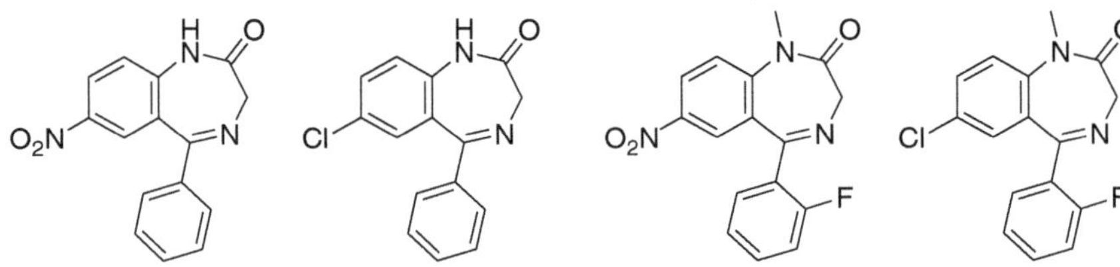

Nitrazepam $t_{1/2}$ 26 h, F 78% | Nordazepam $t_{1/2}$ 73 h, F 99% | Flunitrazepam $t_{1/2}$ 15 h, F 85% | Fludiazepam $t_{1/2}$ 23–48 h

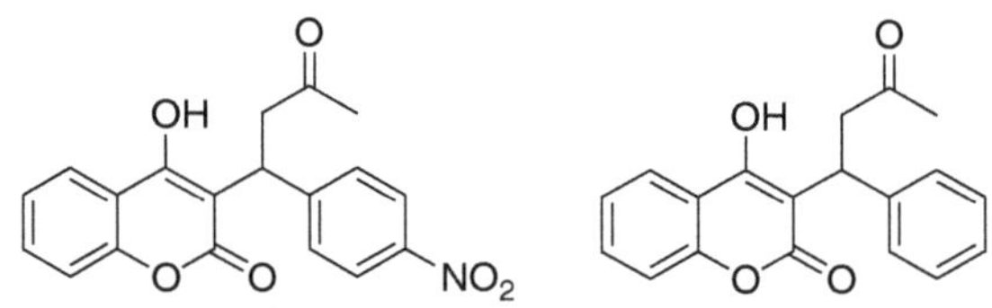

Acenocoumarol $t_{1/2}$ 8–11 h, F 60% | Warfarin $t_{1/2}$ 37 h, F 39%

Scheme 8.2 Impact on pharmacokinetics (PK) of the nitro group.

8.1 Metabolism

The most common metabolic transformation of nitroarenes is reduction to anilines by bacteria in the bowel, followed by N-acetylation. Potentially toxic metabolites include nitrosoarenes and *N*-arylhydroxylamines. The rate of reduction depends on the electron density at the nitroarene and is usually higher at electron-poor nitroarenes. In all the examples in Scheme 8.3, the nitro group undergoes metabolic reduction.

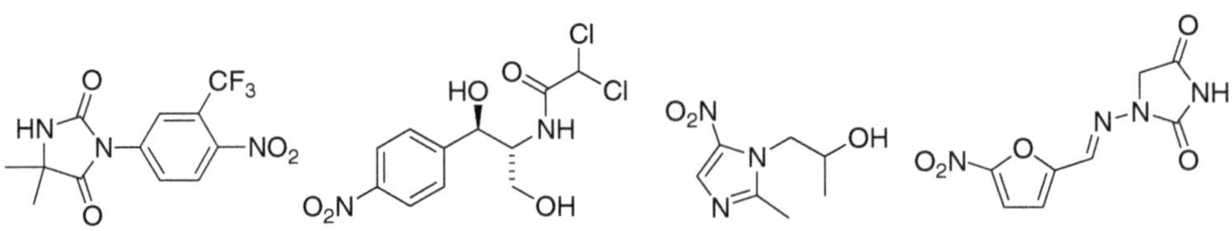

Nilutamide $t_{1/2}$ 38–59 h, F 80–90% | Chloramphenicol $t_{1/2}$ 4 h, F 83% | Secnidazole $t_{1/2}$ 14–20 h, F 99% | Nitrofurantoin $t_{1/2}$ 1 h, F 90%

Scheme 8.3 Nitro group containing drugs.

The antibacterial and antiprotozoal activity of nitroazoles and nitrofurans, such as secnidazole and nitrofurantoin, depends on their reduction to radical anions by the target organism.

Nitro groups that are reduced quickly may serve as prodrugs. Netobimin, for instance, is metabolized by nitro group reduction to albendazole and its active metabolite albendazole sulfoxide (Scheme 8.4).

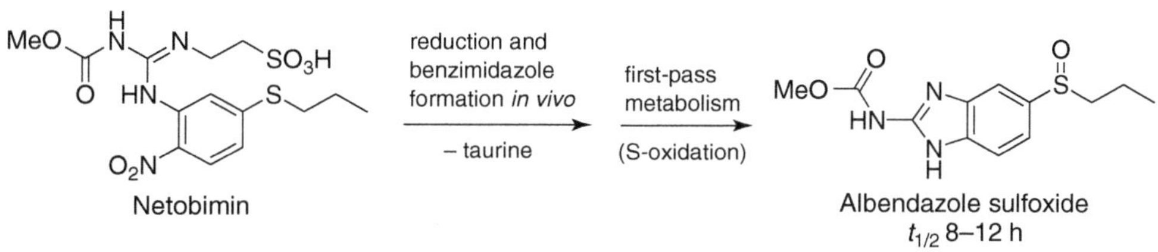

Scheme 8.4 Metabolism of netobimin.

9
Azo Compounds

Paul Ehrlich, working at the Institute for Experimental Therapeutics in Frankfurt am Main, tested the dye methylene blue as potential analgesic because of its high affinity to nerves. In 1891, however, he cured two cases of mild malaria by treatment with methylene blue. Hundreds of dyes were then tested for antibacterial and antiprotozoal activity by Ehrlich and others, which led to the discovery of a number of useful drugs, among others of prontosil by Gerhard Domagk, Fritz Mietzsch, and Josef Klarer in 1932. This substance turned out to be a prodrug of sulfanilamide, a competitive antagonist of 4-aminobenzoic acid, and the active principle of prontosil (Scheme 9.1). The azo group is cleaved in the gut by intestinal bacteria.

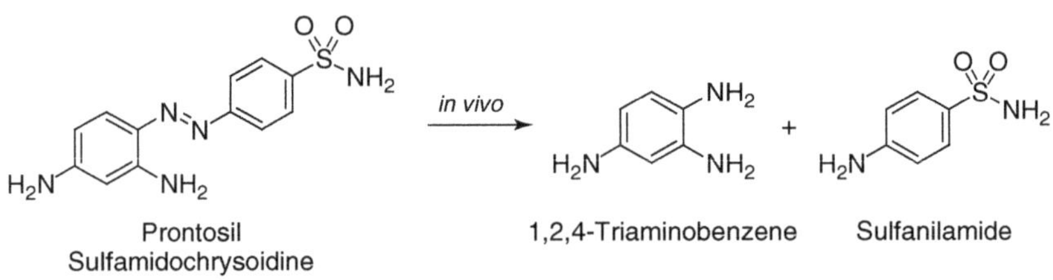

Scheme 9.1 Metabolism of prontosil.

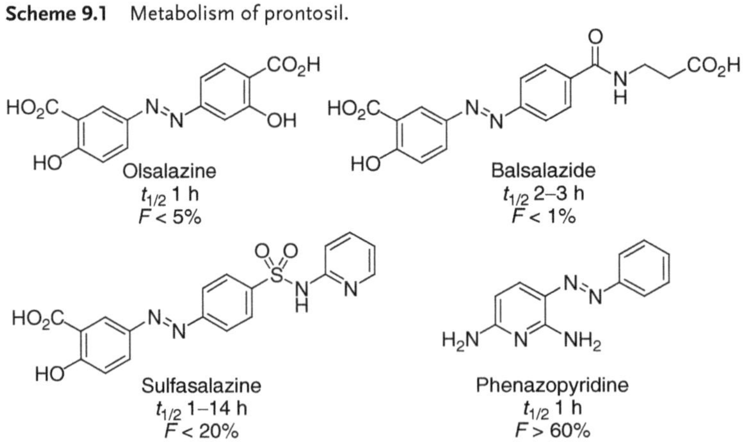

Scheme 9.2 Azo compounds as drugs.

Lead Optimization for Medicinal Chemists: Pharmacokinetic Properties of Functional Groups and Organic Compounds, First Edition. Florencio Zaragoza Dörwald.

Other drugs containing azo groups are shown in Scheme 9.2. These also undergo reductive cleavage in the intestine and are prodrugs for various analgesics and antibacterials.

10
Triazenes

Triazenes result from the reversible addition of secondary amines to diazonium salts. This reaction proceeds readily under basic conditions, but acids quickly regenerate the starting materials. With primary aliphatic amines, diazotization followed by carbocation formation may occur (Scheme 10.1).

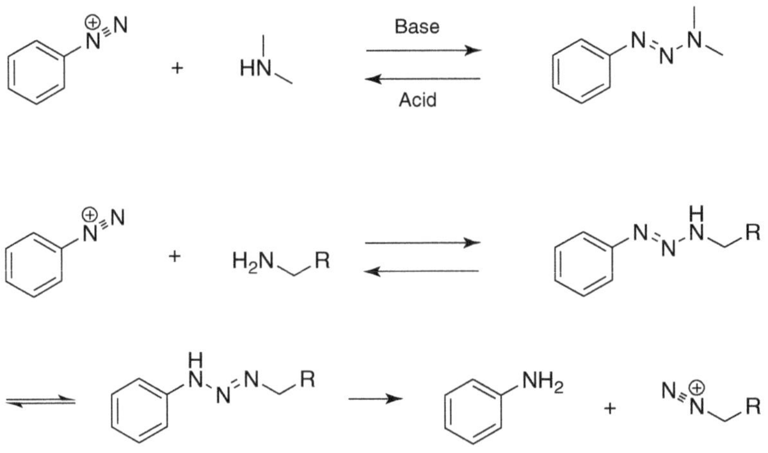

Scheme 10.1 Formation and decomposition of triazenes.

The antineoplastics dacarbazine and CB 10-277 (Scheme 10.2) have to be administered intravenously and are activated by N-demethylation in the liver. The resulting *N*-monomethyltriazenes decompose by the mechanism shown above to yield diazomethane or a methyl diazonium salt, which are exceedingly strong methylating agents. Temozolomide is similarly metabolized to an *N*-monomethyltriazene by urea hydrolysis. All these drugs are used to treat aggressive tumors (melanoma, Hodgkin's disease, and malignant brain tumors). Other drugs that are metabolized into methylating reagents include procarbazine and altretamine.

Lead Optimization for Medicinal Chemists: Pharmacokinetic Properties of Functional Groups and Organic Compounds, First Edition. Florencio Zaragoza Dörwald.

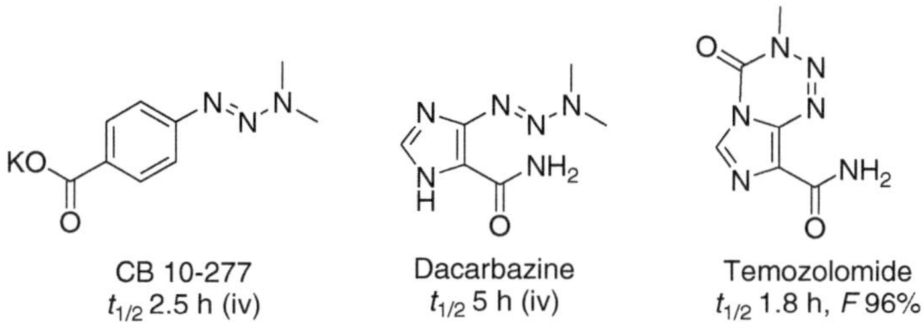

Scheme 10.2 Antineoplastic triazenes.

11
Nitrates and Nitrites

The vasodilating and antianginal properties of alkyl nitrates and nitrites were already discovered in the 1860 s and used since then for the treatment of angina pectoris.

Nitrates and nitrites undergo reductive hydrolysis *in vivo* to yield the parent alcohol and nitric oxide (NO). The latter triggers the biological effect of these substances. The half-life at which nitrates or nitrites generate nitric oxide ranges from a few minutes to several hours (Scheme 11.1). Lipophilic nitrates are hydrolyzed faster and have lower bioavailabilities than hydrophilic nitrates.

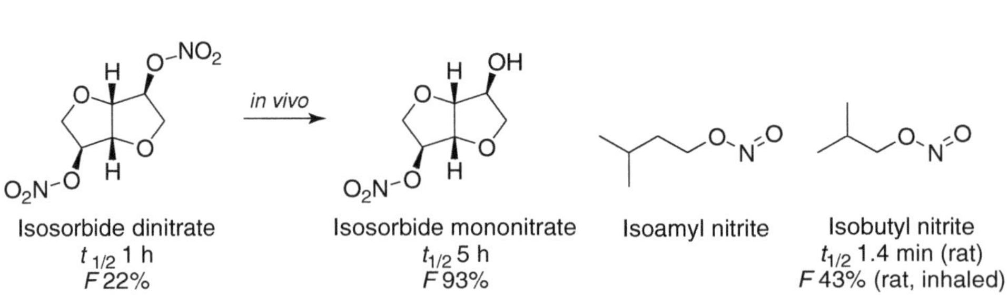

Scheme 11.1 Alkyl nitrates and nitrites as drugs.

Derivatives of *N*-nitrosamines, such as the antineoplastic *N*-nitrosourea carmustine, may also yield nitric oxide *in vivo*. However, the main metabolic pathway of *N*-nitrosoureas is their decomposition to alkyl diazonium salts and eventually carbocations.

Lead Optimization for Medicinal Chemists: Pharmacokinetic Properties of Functional Groups and Organic Compounds, First Edition. Florencio Zaragoza Dörwald.

Further Reading

Sneader, W. (1999) Organic nitrates. *Drug News Perspect.*, **12**, 58–63.

12
N-Nitroso Compounds

Various *N*-(2-chloroethyl)-*N*-nitrosoureas are being used as oral or parenteral drugs for the treatment of aggressive neoplasms, such as brain or pancreatic tumors, or melanoma (Table 12.1). Nitrosoureas are readily cleaved by nucleophiles and are hence rather unstable. Nucleophilic cleavage yields *N*-nitrosamines, which quickly eliminate water and nitrogen to yield strongly alkylating carbocations (Scheme 12.1). The resulting chloroethylated nucleophiles may alkylate additional nucleophilic functional groups and thus cross-link proteins or nucleic acids. As most other antineoplastics, all nitrosoureas are carcinogenic and mutagenic.

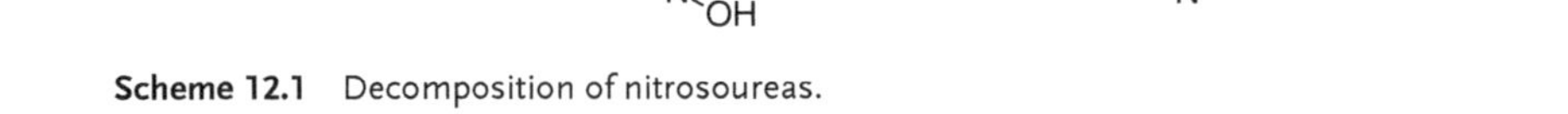

Scheme 12.1 Decomposition of nitrosoureas.

Table 12.1 *N*-Nitrosoureas. *V* in l kg^{-1}; CL in ml min^{-1} kg^{-1}; Mwt in g mol^{-1}.

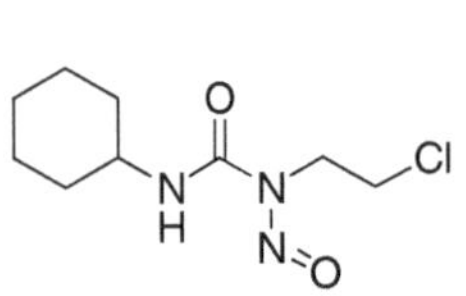	$t\frac{1}{2}$ 1.5 h *F* 100%* pb 50%* ur –		*V* 1.7* CL – Mwt 233.7 PSA 61.8 Å^2 log *P* 2.72		LOMUSTINE Antineoplastic, crosses bbb, carcinogenic, mutagenic Metabolism: hydroxylation of cyclohexyl group (4-*cis* and *trans*-hydroxy; t½ 16–48 h) *metabolites on oral dosing of drug.

Lead Optimization for Medicinal Chemists: Pharmacokinetic Properties of Functional Groups and Organic Compounds, First Edition. Florencio Zaragoza Dörwald.

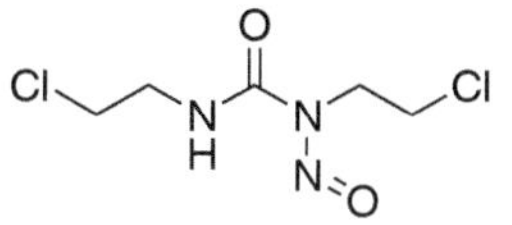	$t^1/_2$	1.5±2.0 h	*V*	3.3±1.7	CARMUSTINE, BCNU
	F	–	CL	56±56	Antineoplastic, carcinogenic,
	pb	77%	Mwt	214.1	mutagenic, crosses bbb
	ur	–	PSA	61.8 Å^2	
			log *P*	1.38	
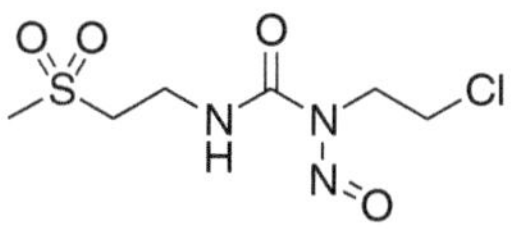	$t^1/_2$	4 min (iv)	*V*	–	CYSTEMUSTINE
	F	Low	CL	–	Antineoplastic
	pb	–	Mwt	257.7	
	ur	–	PSA	104 Å^2	
			log *P*	−0.17	
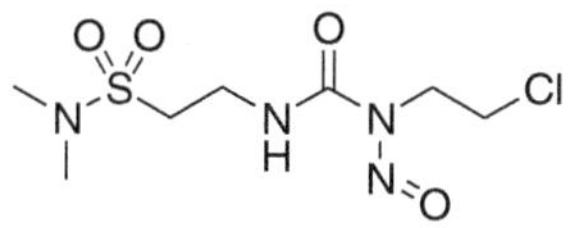	$t^1/_2$	57±22 min	*V*	–	TAUROMUSTINE
	F	26% (dog)	CL	–	Antineoplastic
	pb	–	Mwt	286.7	
	ur	–	PSA	108 Å^2	
			log *P*	0.77	
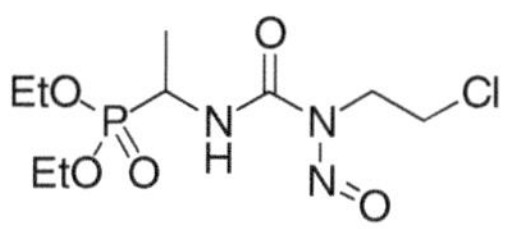	$t^1/_2$	29±12 min*	*V*	–	FOTEMUSTINE
	F	–	CL	28±17	*iv
	pb	–	Mwt	315.7	Antineoplastic
	ur	–	PSA	107 Å^2	
			log *P*	1.37	
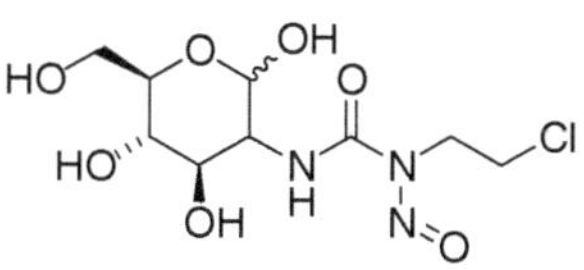	$t^1/_2$	7 min (iv)	*V*	–	CHLOROZOTOCIN
	F	–	CL	–	Antineoplastic
	pb	–	Mwt	313.7	
	ur	–	PSA	160 Å^2	
			log *P*	−0.71	
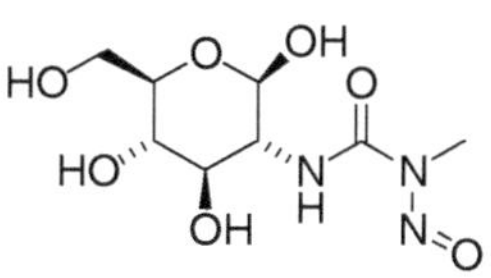	$t^1/_2$	15 min*	*V*	0.7	STREPTOZOTOCIN
	F	Low	CL	–	*iv
	pb	–	Mwt	265.2	Antineoplastic, does not cross
	ur	10–20%*	PSA	160 Å^2	bbb
			log *P*	−1.33	
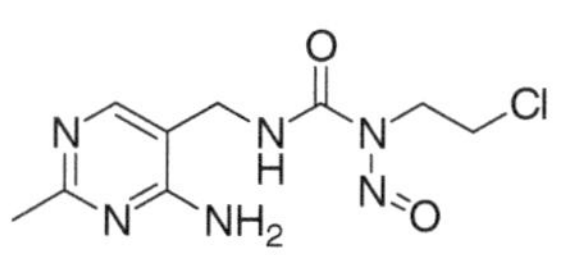	$t^1/_2$	0.2–1.1 h*	*V*	1.6	NIMUSTINE, ACNU
	F	–	CL	35	*iv
	pb	–	Mwt	272.7	Antineoplastic
	ur	–	PSA	114 Å^2	
			log *P*	0.25	

$t_{1/2}$, plasma half-life; *F*, oral bioavailability; pb, plasma protein binding; ur, excretion of unchanged drug in urine; V, volume of distribution; CL, clearance; Mwt, molecular weight; PSA, polar surface area; bbb, blood–brain barrier.

13
N-Oxides

Pyridine or pyrimidine *N*-oxides are common metabolites of pyridines and pyrimidines. As other metabolites, some *N*-oxides also have made their way to the market and are being used as oral drugs. Despite their high dipole moments, these compounds are orally bioavailable (Scheme 13.1). Typical metabolic transformations include O-sulfation, O-glucuronidation, and reductive cleavage of the N–O bond.

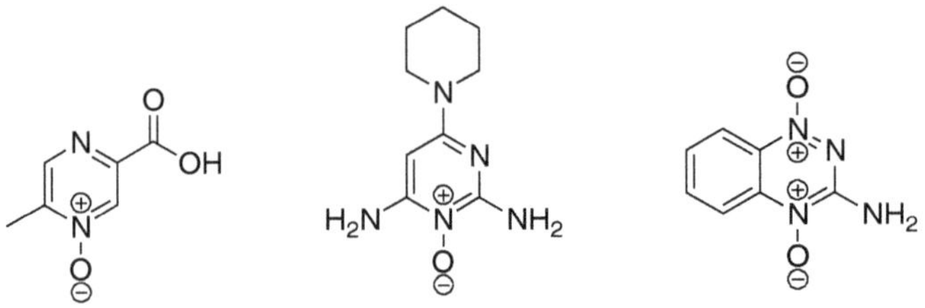

Acipimox $t_{1/2}$ 12–14 h *F* 100%

Minoxidil $t_{1/2}$ 3.1 h *F* 90%

Tirapazamine $t_{1/2}$ 0.8 h (iv) *F* 75%

Scheme 13.1 *N*-Oxides as drugs.

Lead Optimization for Medicinal Chemists: Pharmacokinetic Properties of Functional Groups and Organic Compounds, First Edition. Florencio Zaragoza Dörwald.

14 Alcohols

Hydroxyl groups enhance the solubility and rate of renal excretion of a drug. The lower affinity to liver enzymes may also reduce the first-pass effect and enhance the oral availability of a hydroxylated lead. On the other hand, a high hydrophilicity and large polar surface area can accelerate renal excretion and compromise the ability of a compound to be completely absorbed from the small intestine and to cross membranes, such as the blood–brain barrier (bbb), unless the molecule is very small. Thus, while ethanol ($t_{1/2}$ 0.24 h, F 80%) is well absorbed and even crosses the bbb, polyols such as sorbitol, mannitol, or glycerol have poor oral bioavailabilities. Carbohydrates are usually absorbed by a compound-specific active transport mechanism.

Comparison of fenpiprane, zimeldine, and chlorpheniramine with the tertiary alcohols pridinol, procyclidine, and pirmenol shows that the hydroxyl group can lead to higher oral availability by curbing first-pass metabolism (Scheme 14.1).

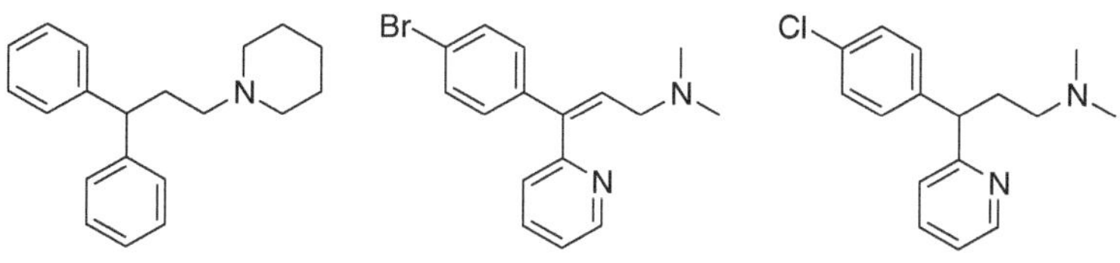

Fenpiprane
active only on
parenteral administration

Zimeldine
$t_{1/2}$ 5 h, F 27%
V 3.2 l kg^{-1}, log P 3.07

Chlorpheniramine
$t_{1/2}$ 20 h, F 41%
V 3.2 l kg^{-1}, log P 2.97

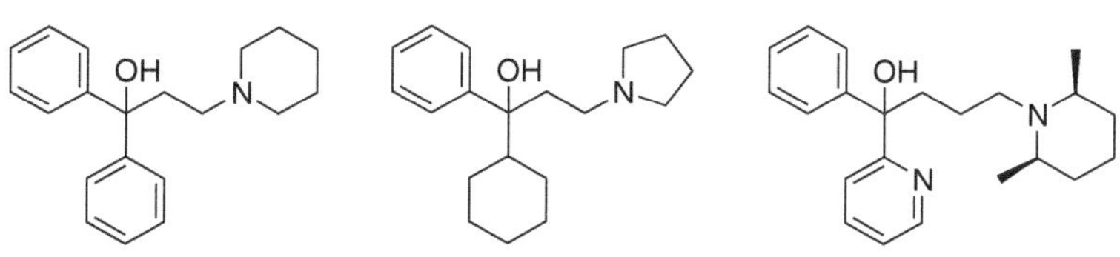

Pridinol
$t_{1/2}$ 14 h, F 77%
log P 3.35

Procyclidine
$t_{1/2}$ 12 h, F 75%
V 0.74 l kg^{-1}, log P 3.86

Pirmenol
$t_{1/2}$ 6–18 h, F 80–90%
V 1.5 l kg^{-1}, log P 3.57

Scheme 14.1 Effect of tertiary hydroxyl groups on pharmacokinetics (PK).

Lead Optimization for Medicinal Chemists: Pharmacokinetic Properties of Functional Groups and Organic Compounds, First Edition. Florencio Zaragoza Dörwald.

14.1 Metabolism

Many hydrophilic alcohols, such as oxiracetam, mannitol, meglumine, miglitol, the bronchodilator salbutamol, or the contrast agent iopamidol, are barely metabolized, but mostly excreted unchanged in the urine (Scheme 14.2).

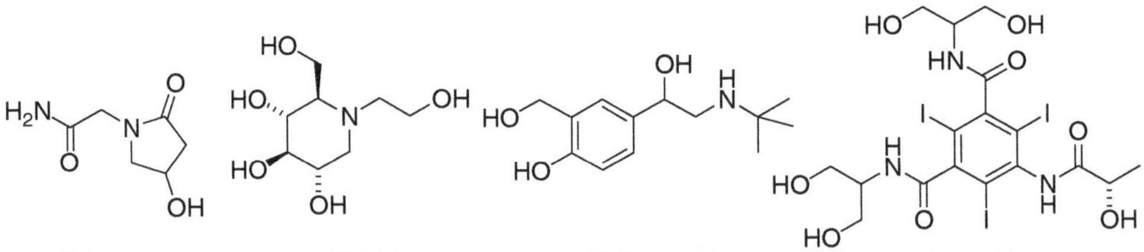

Oxiracetam
$t_{1/2}$ 3–8 h, *F* 75%
renal excretion
unchanged 50–84%

Miglitol
$t_{1/2}$ 2–4 h, *F* 60–90%
renal excretion
unchanged 100%

Salbutamol
$t_{1/2}$ 3–5 h, *F* 43–50%
renal excretion
unchanged 51–64%

Iopamidol
$t_{1/2}$ 23 h, *F* low
renal excretion
unchanged 87% (iv)

Scheme 14.2 Alcohols as drugs.

Larger, more lipophilic alcohols may, however, undergo some sort of metabolism, either to further facilitate their renal excretion or to degrade the molecule into smaller fragments. Primary and secondary alkanols can be oxidized by alcohol dehydrogenase to aldehydes or ketones, which can be further oxidized to carboxylic acids. The oxidation of secondary alcohols to ketones is reversible, and with some substrates, for instance haloperidol, both the ketone and the alcohol can be detected in plasma (Scheme 14.3).

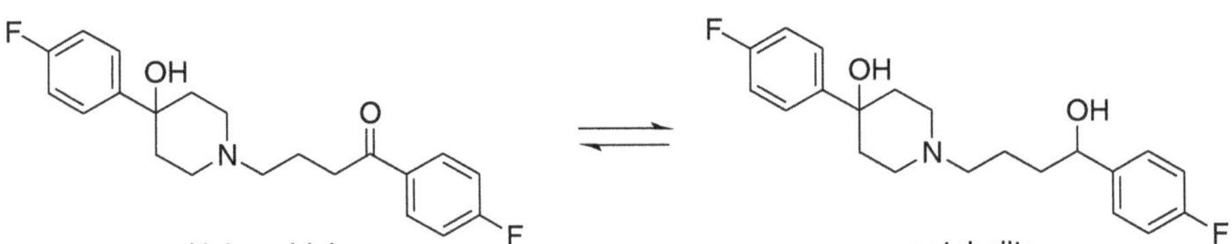

Haloperidol
$t_{1/2}$ 18 h, *F* 60%

metabolite
$t_{1/2}$ 67 h

Scheme 14.3 Reversible reduction of haloperidol.

Because the oxidation of primary, purely aliphatic alcohols is usually fast, few drugs contain this functional group. Alcohols with a heteroatom in β- or γ-position, however, are more resistant to oxidation, as illustrated by the examples shown in Scheme 14.4. Secnidazole undergoes hydroxylation at the vinylic methyl group *in vivo*, chloramphenicol is mainly converted into an inactive glucuronide, and the main metabolite of ciramadol is the phenolic glucuronide (Table 14.1).

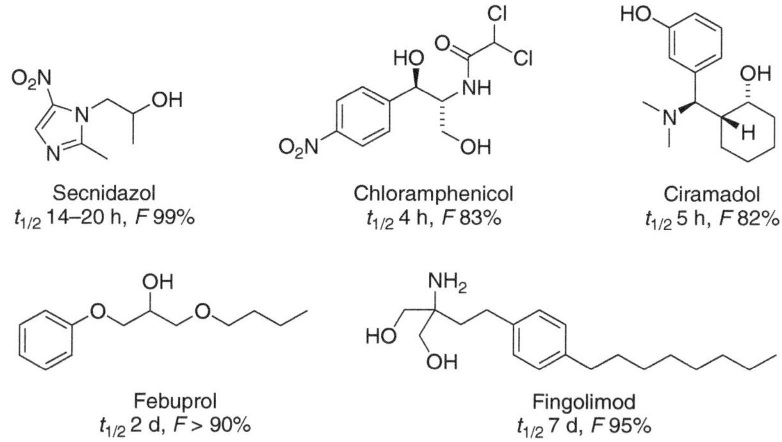

Scheme 14.4 Drugs containing hydroxyl groups.

Drugs with carbohydrate substructures may be metabolized by phosphorylation. This is often the case for antiviral or antineoplastic nucleoside analogs, which are phosphorylated *in vivo* to the corresponding nucleotides and then incorporated into DNA (Scheme 14.5).

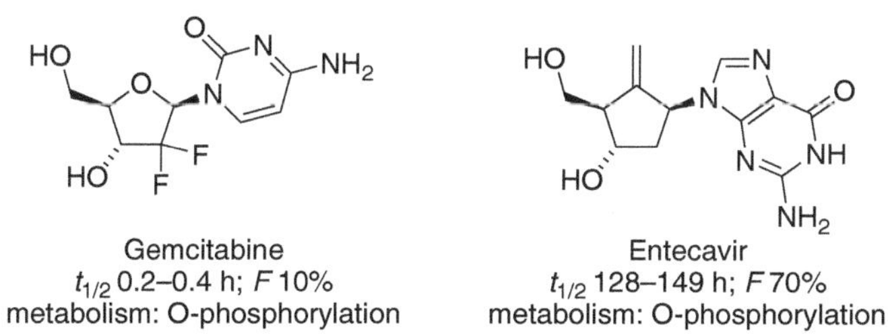

Scheme 14.5 Carbohydrate analogs as drugs.

The oxidation of alcohols may be prevented either by introducing fluorine or other electron-withdrawing atoms or groups at the β-position (to prevent α-hydride abstraction) or by choosing a tertiary alcohol instead. Tertiary alcohols can undergo hydroxylation to 1,2-diols, as in the case of the α_1-adrenergic antagonist trimazosin (Scheme 14.6).

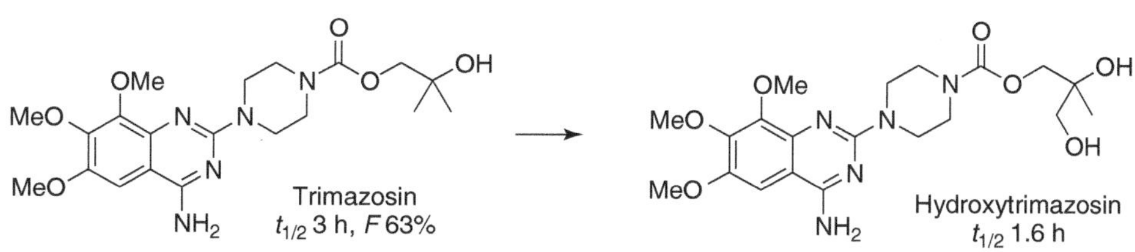

Scheme 14.6 Metabolism of trimazosin.

Table 14.1 Aliphatic alcohols, carbohydrates, and related compounds. *V* in l kg^{-1}; CL in ml min^{-1} kg^{-1}; Mwt in g mol^{-1}.

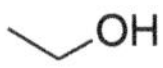	$t_{1/2}$ 0.2±0.1 h *F* 80% pb – ur <3%	*V* 0.54±0.05 CL – Mwt 46.1 PSA 20.2 Å^2 log *P* −0.18	ETHANOL Recreational drug, hypnotic
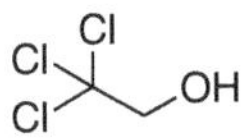	$t_{1/2}$ 4–12 h *F* – pb 70–80% ur –	*V* 0.9 CL – Mwt 149.4 PSA 20.2 Å^2 log *P* 0.97	TRICHLOROETHANOL Metabolite of chloral hydrate
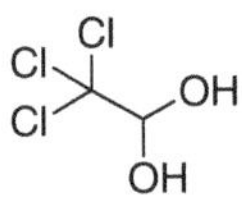	$t_{1/2}$ 8–11 h *F* – pb – ur 0%	*V* 0.6 CL – Mwt 165.4 PSA 40.5 Å^2 log *P* 0.99	CHLORAL HYDRATE Hypnotic, sedative Metabolism: reduction to trichloroethanol, then glucuronidation
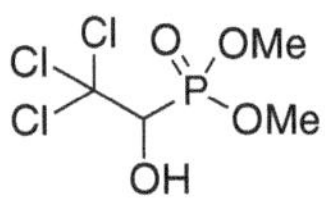	$t_{1/2}$ 2.3–2.5 h *F* – pb – ur 2%	*V* – CL 8.2–9.3 Mwt 257.4 PSA 65.6 Å^2 log *P* 0.49	METRIFONATE Anthelmintic Metabolism: rearrangement to dichlorvos (2,2-dichlorovinyl dimethyl phosphate)
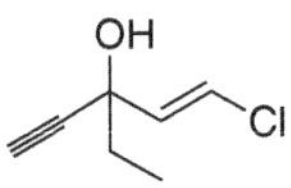	$t_{1/2}$ 10–20 h *F* – pb 35–50% ur Negligible	*V* 3.5 CL 1.7–2.0 Mwt 144.6 PSA 20.2 Å^2 log *P* 1.19	ETHCHLORVYNOL Hypnotic, anticonvulsant, muscle relaxant
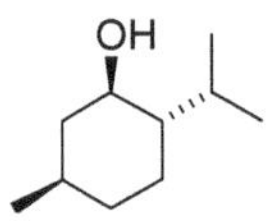	$t_{1/2}$ 1 h* *F* – pb – ur 46–57%*	*V* – CL – Mwt 156.3 PSA 20.2 Å^2 log *P* 3.22	MENTHOL *glucuronide No menthol detectable in plasma after oral dosing
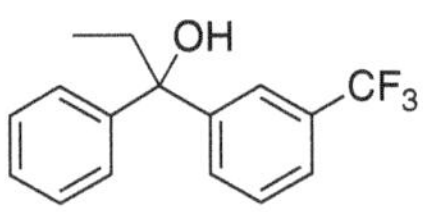	$t_{1/2}$ 21±2 h *F* – pb – ur 0%	*V* 41±18 CL 26±9 Mwt 280.3 PSA 20.2 Å^2 log *P* 4.21	FLUMECINOL Hepatic enzyme inducer Metabolism: phenyl and ethyl hydroxylation, then conjugation
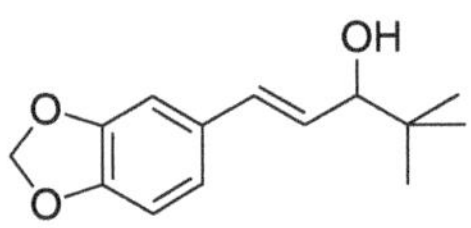	$t_{1/2}$ 4–14 h *F* 21±9% pb 99% ur Negligible	*V* 1 CL 6–29 Mwt 234.3 PSA 38.7 Å^2 log *P* 3.78	STIRIPENTOL Antiepileptic, racemic Metabolism: O-glucuronidation, *tert*-butyl hydroxylation, isomerization of allyl alcohol to propanone, acetal hydrolysis, then catechol O-methylation

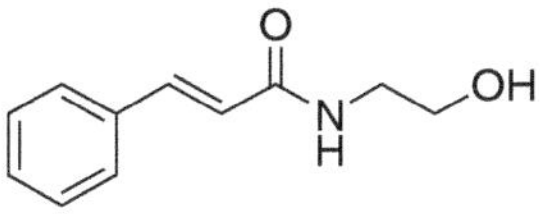	$t_{1/2}$ 1 h *F* 30% pb – ur 1%	*V* – CL – Mwt 191.2 PSA 49.3 Å^2 log *P* 0.44	IDROCILAMIDE Muscle relaxant
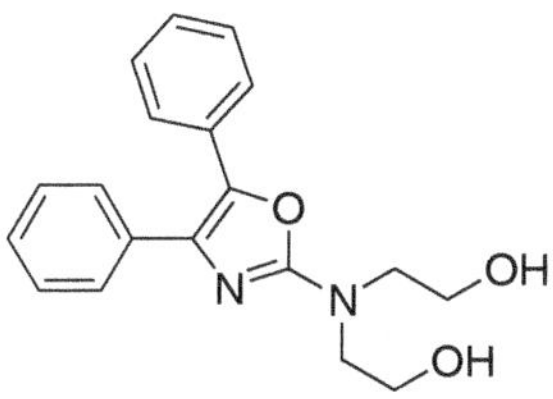	$t_{1/2}$ 41 min (rat) *F* – pb – ur –	*V* 2.1 (rat) CL 34.5 (rat) Mwt 324.4 PSA 69.7 Å^2 log *P* 2.35	DITAZOL Antiinflammatory Metabolites: *N*-desalkyl, *N*, *N*-bisdesalkyl, hydroxy, benzil, hydroxybenzil
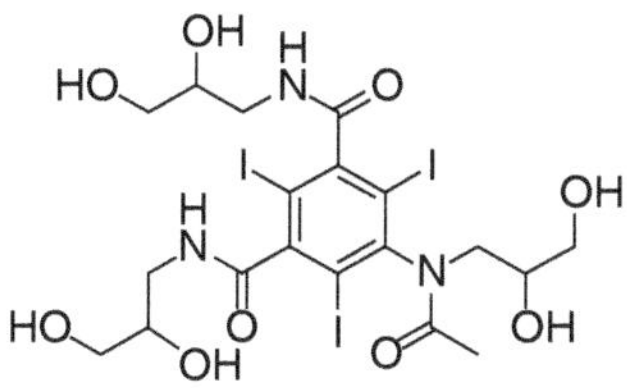	$t_{1/2}$ 2.9 h *F* – pb 0% ur 90%	*V* 0.19 CL 2 Mwt 821.1 PSA 200 Å^2 log *P* −2.92	IOHEXOL X-ray contrast agent; triiodobenzoic acid is too highly plasma bound and too toxic. This toxicity is reduced by conversion to the more hydrophilic iohexol.
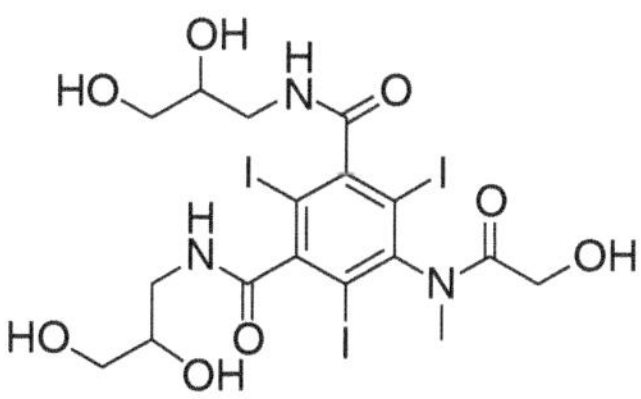	$t_{1/2}$ 1.8 h *F* – pb – ur 87–92%	*V* 0.28 CL 1.6–1.7 Mwt 777.1 PSA 180 Å^2 log *P* −2.93	IOMEPROL X-ray contrast agent
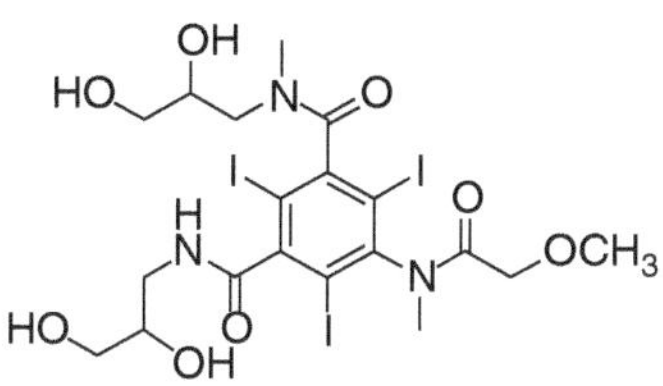	$t_{1/2}$ 4.6 h *F* – pb 1% ur 85%	*V* 0.2 CL 1.4 Mwt 791.1 PSA 169 Å^2 log *P* −2.66	IOPROMIDE X-ray contrast agent
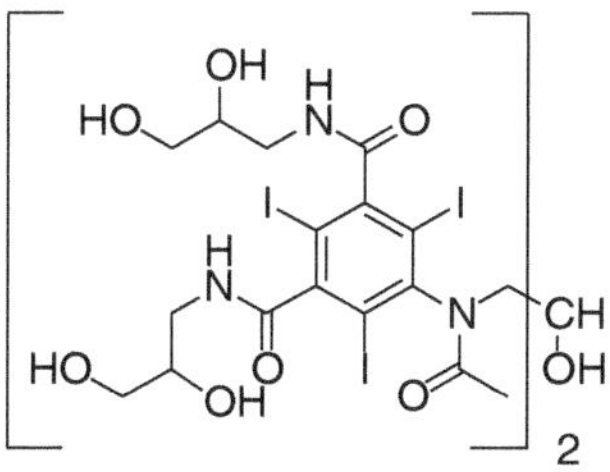	$t_{1/2}$ 2.2 h *F* – pb – ur 97%	*V* 0.28 CL 1.7 Mwt 1550.2 PSA 339 Å^2 log *P* −4.07	IODIXANOL X-ray contrast agent

(*continued overleaf*)

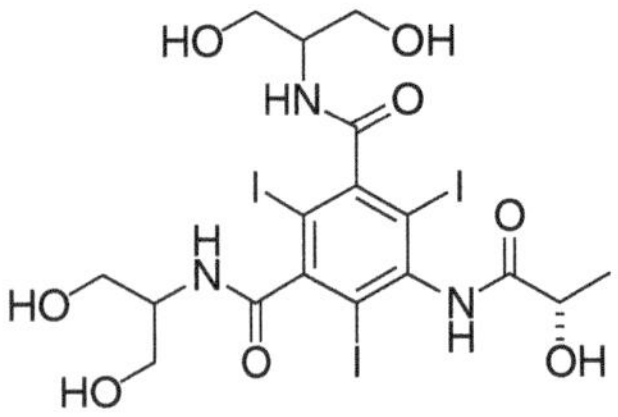	$t^{1}/_{2}$ 2–3 h *F* – pb 0% ur 86%	*V* 0.2–0.3 CL 1.9 Mwt 777.1 PSA 188 Å^2 log *P* −2.55	IOPAMIDOL X-ray contrast agent
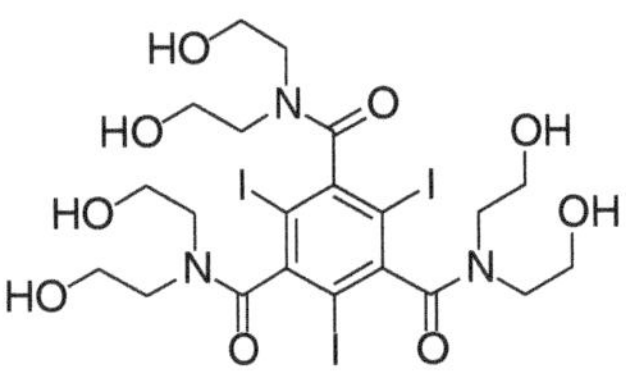	$t^{1}/_{2}$ 2.7 h *F* – pb – ur 87%	*V* 0.2 CL – Mwt 849.2 PSA 182 Å^2 log *P* −3.02	IOSIMIDE X-ray contrast agent
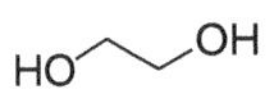	$t^{1}/_{2}$ 3–9 h *F* 92–100% pb – ur 20%	*V* 0.5–0.8 CL – Mwt 62.1 PSA 40.5 Å^2 log *P* −1.36	ETHYLENE GLYCOL Antifreezer Metabolism: oxidation to hydroxyacetaldehyde, glycolic acid, and glyoxylic acid
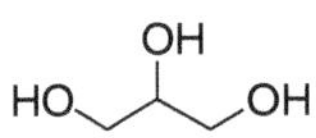	$t^{1}/_{2}$ 0.5–1.2 h *F* >80% pb – ur 7–14%	*V* 0.2–0.3 CL 2 Mwt 92.1 PSA 60.7 Å^2 log *P* −1.85	GLYCEROL Osmotic diuretic
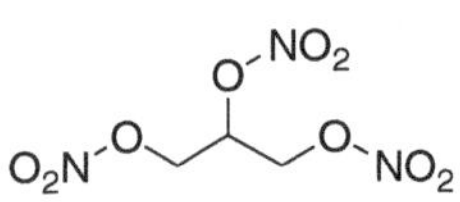	$t^{1}/_{2}$ 2±1 min *F* <1%* pb – ur <1%	*V* 3.3±1.2 CL 230±90 Mwt 227.1 PSA 165 Å^2 log *P* 2.15	NITROGLYCERIN *sublingual: 38±26% Antianginal Metabolism: denitration (to glycerol dinitrate, $t^{1}/_{2}$ 40 min)
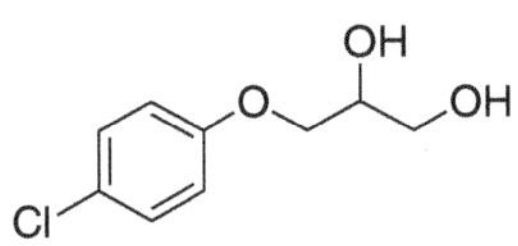	$t^{1}/_{2}$ 3–4 h *F* – pb – ur 0%	*V* 1.3 CL – Mwt 202.6 PSA 49.7 Å^2 log *P* 1.71	CHLORPHENESIN Topical antifungal
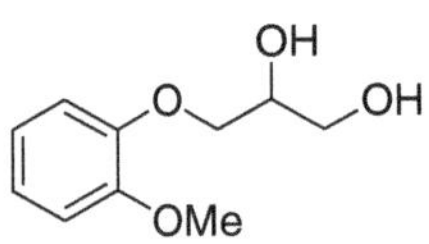	$t^{1}/_{2}$ 1 h *F* High pb 37% ur –	*V* 1 CL – Mwt 198.2 PSA 58.9 Å^2 log *P* 0.46	GUAIFENESIN Expectorant Metabolism: oxidation of primary alcohol to CO_2H

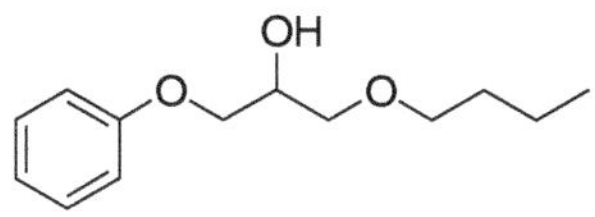	$t^1/_2$ 2 d F >90% pb 60–70% ur Low	V – CL – Mwt 224.3 PSA 38.7 Å^2 log P 2.91	FEBUPROL Choleretic Metabolites: 4-hydroxyphenyl ($t^1/_2$ 3 h), glucuronide
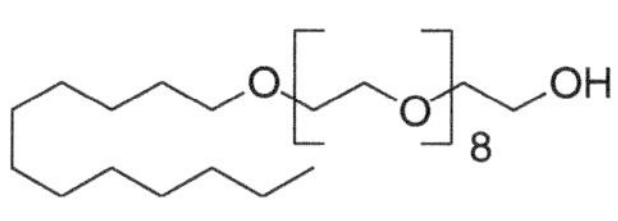	$t^1/_2$ 4 h (iv) F – pb 64% ur –	V 0.4 CL 3 Mwt 582.8 PSA 103.3 Å^2 log P 3.42	POLIDOCANOL Local anesthetic, antipruritic, spermicide Metabolism: oxidative degradation of alkyl chain
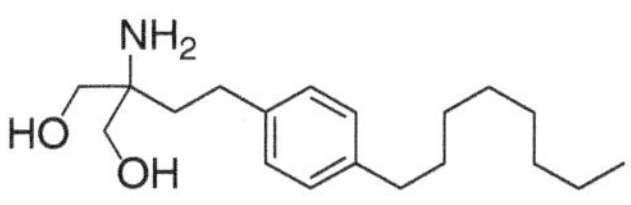	$t^1/_2$ 7±2 d F 95% pb 99.7% ur –	V 18±4 CL 1.6 Mwt 307.5 PSA 66.5 Å^2 log P 4.2	FINGOLIMOD Immunomodulator Active metabolite: phosphate
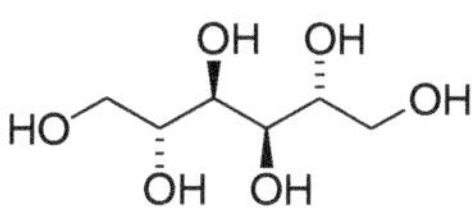	$t^1/_2$ 0.3–1.7 h F 7–20% pb 0% ur 80–90%	V 0.5 CL 7 Mwt 182.2 PSA 121 Å^2 log P −3.26	MANNITOL Osmotic diuretic
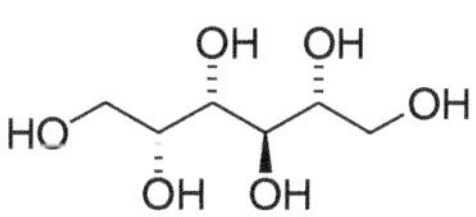	$t^1/_2$ 0.2–0.6 h* F Low pb 0% ur <3%	V 0.2 CL – Mwt 182.2 PSA 121 Å^2 log P −3.26	SORBITOL *iv Sweetener, laxative
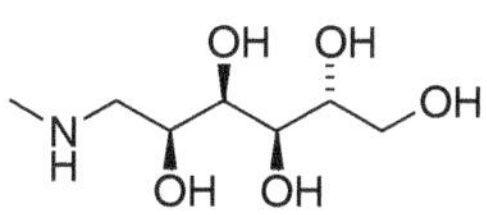	$t^1/_2$ 2.6 h F – pb – ur 91%	V – CL – Mwt 195.2 PSA 113 Å^2 log P −2.51	MEGLUMINE Drug excipient
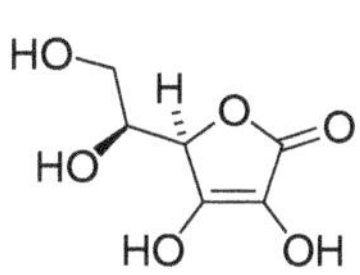	$t^1/_2$ 4–12 h F 100%* pb 24% ur 0%	V – CL 1.1 Mwt 176.1 PSA 107 Å^2 log P −2.41	ASCORBIC ACID, VITAMIN C *for 200 mg dose Larger doses: lower F, higher ur Metabolites: 2-sulfate, dehydroascorbic acid (diketone), oxalic acid
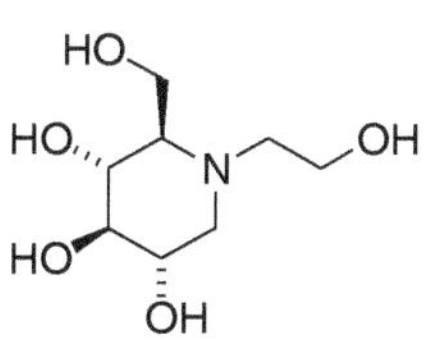	$t^1/_2$ 2–4 h F 60–90% pb <4% ur 100%	V 0.28 CL 1.7 Mwt 207.2 PSA 104 Å^2 log P −2.21	MIGLITOL Antidiabetic; no metabolism

(*continued overleaf*)

Structure					
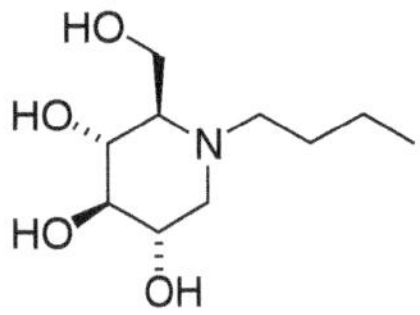	*t*½	6–7 h	*V*	0.7–1.3	MIGLUSTAT
	F	97%	CL	3.5	Antidiabetic
	pb	0%	Mwt	219.3	Metabolism: glucuronidation
	ur	50%	PSA	84.2 Å^2	
			log *P*	0.03	
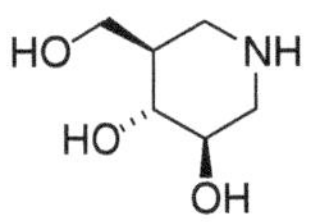	*t*½	14 h	*V*	–	ISOFAGOMINE
	F	–	CL	–	Treatment of Gaucher's disease
	pb	–	Mwt	147.2	
	ur	–	PSA	72.7 Å^2	
			log *P*	−1.80	
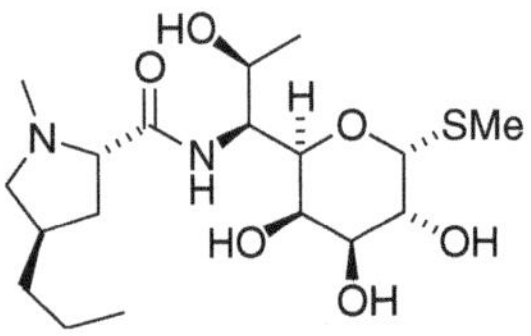	*t*½	5.1±1.5 h	*V*	1.3±0.2	LINCOMYCIN
	F	25±5%	CL	2.1±0.5	Antibacterial
	pb	85±2%	Mwt	406.5	Metabolism: oxidation to
	ur	14±6%	PSA	148 Å^2	sulfoxide, N-demethylation
			log *P*	0.72	
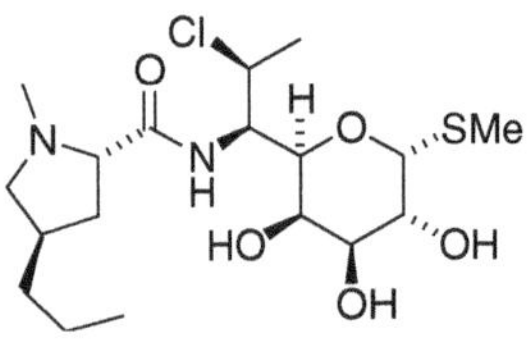	*t*½	2.9±0.7 h	*V*	1.1±0.3	CLINDAMYCIN
	F	87%	CL	4.7±1.3	Bacteriostatic
	pb	93.6±0.2%	Mwt	425.0	Metabolites: sulfoxide,
	ur	13%	PSA	128 Å^2	*N*-desmethyl, glucuronide
			log *P*	1.75	
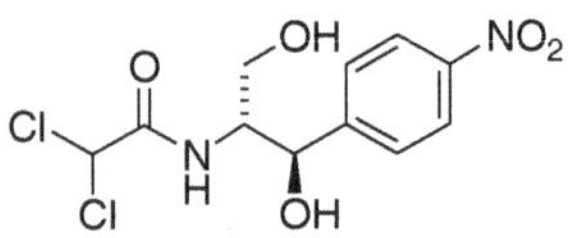	*t*½	4±2 h	*V*	0.94±0.06	CHLORAMPHENICOL
	F	83±8%	CL	2.4±0.2	Antibacterial, antirickettsial,
	pb	53±5%	Mwt	323.1	carcinogenic
	ur	25±15%	PSA	115 Å^2	Metabolism: reduction of
			log *P*	1.10	nitro group by intestinal bacteria
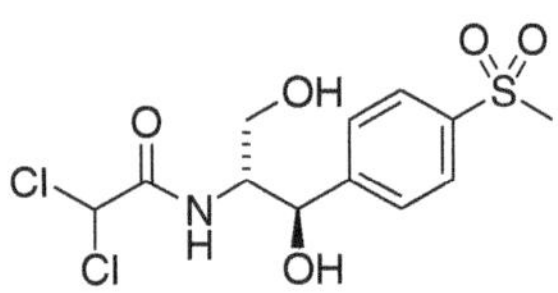	*t*½	1.5–2.4 h	*V*	0.75	THIAMPHENICOL
	F	60%*	CL	–	*calves
	pb	10–20%	Mwt	356.2	Antibacterial
	ur	60%	PSA	112 Å^2	Metabolism: glucuronidation
			log *P*	−0.27	
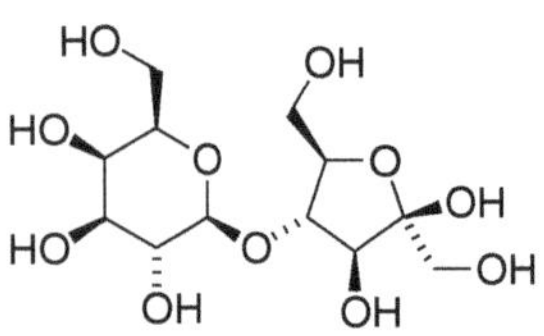	*t*½	2 h	*V*	–	LACTULOSE
	F	0.4–2.0%	CL	–	Laxative
	pb	–	Mwt	342.3	
	ur	<3%	PSA	197 Å^2	
			log *P*	−2.97	

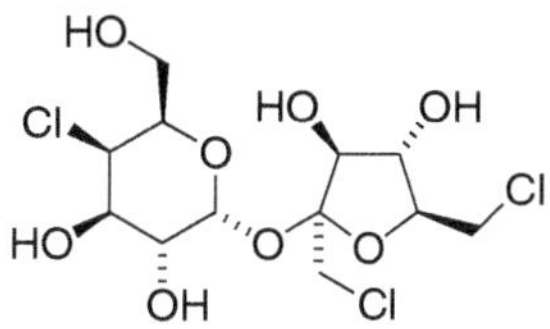	$t_{1/2}$ F pb ur	3–23 h 11–40% – 6–18%	V CL Mwt PSA log P	– – 397.6 129 Å^2 0.23	SUCRALOSE, SPLENDA Artificial sweetener Metabolism: glucuronidation
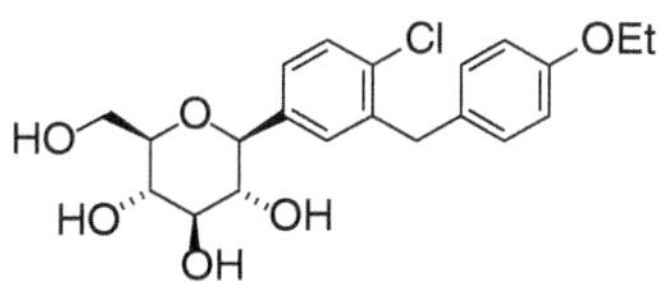	$t_{1/2}$ F pb ur	17 h >75% 96% <2.5%	V CL Mwt PSA log P	0.8 (dog) 1.5 (dog) 408.9 99.4 Å^2 4.42	DAPAGLIFLOZIN Glucose transporter inhibitor, antidiabetic Metabolism: glucuronidation, O-deethylation, hydroxylation
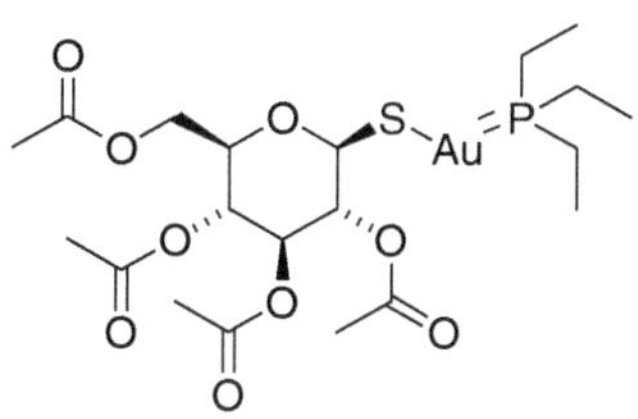	$t_{1/2}$ F pb ur	21±4 h 20±5% 60% 15%	V/F CL/F Mwt PSA log P	0.045 0.025±0.016 678.5 114.5 Å^2 0.81	AURANOFIN Antiarthritic Metabolism: deacetylation
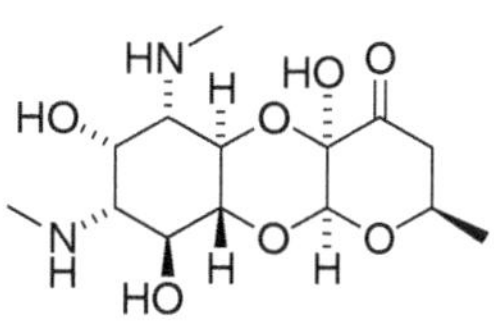	$t_{1/2}$ F pb ur	1–3 h 10% (pig) Low 70–100%*	V CL Mwt PSA log P	0.13 0.99 332.4 130 Å^2 0.75	SPECTINOMYCIN *iv Antibacterial, no metabolism
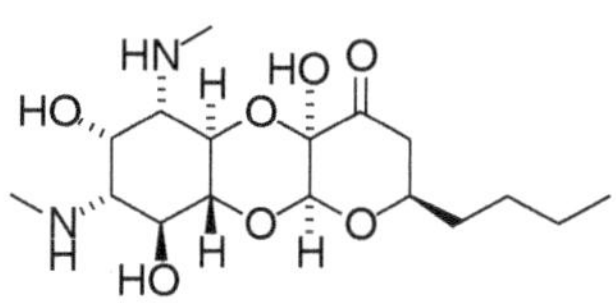	$t_{1/2}$ F pb ur	11 h (iv) 7% – –	V CL Mwt PSA log P	0.7 1.7 374.4 130 Å^2 2.27	TROSPECTOMYCIN Antibacterial
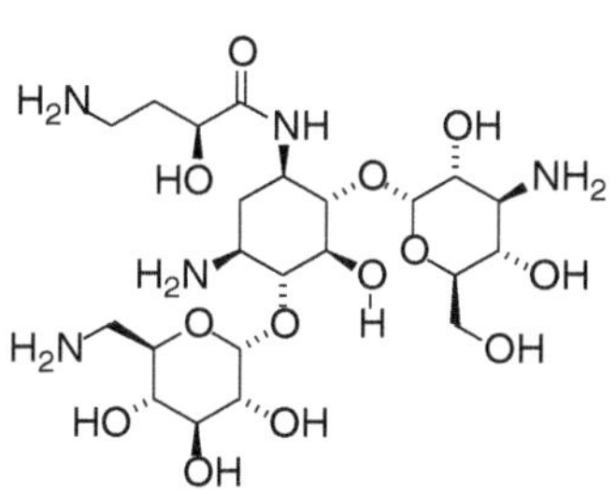	$t_{1/2}$ F pb ur	2–3 h (im) – 0–11% 98% (im)	V CL Mwt PSA log P	0.16 1.1 585.6 332 Å^2 −5.26	AMIKACIN Antibacterial for im or iv administration For review, see: Busscher, G.F., Rutjes, F.P.J.T., van Delft, F.L. 2-Deoxystreptamine: central scaffold of aminoglycoside antibiotics. *Chem. Rev.* 2005, **105**, 775–791.

(*continued overleaf*)

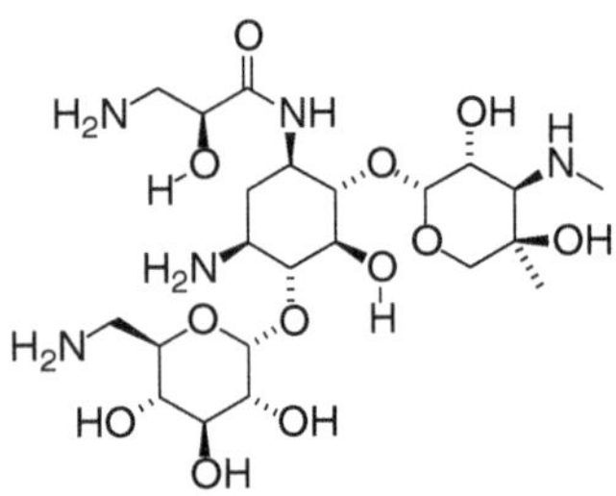

$t_{1/2}$	24 h (iv)	V	0.32	ISEPAMICIN
F	–	CL	1.3	Aminoglycoside antibiotic
pb	5%	Mwt	569.6	
ur	75% (im)	PSA	298 Å^2	
		log P	−4.68	

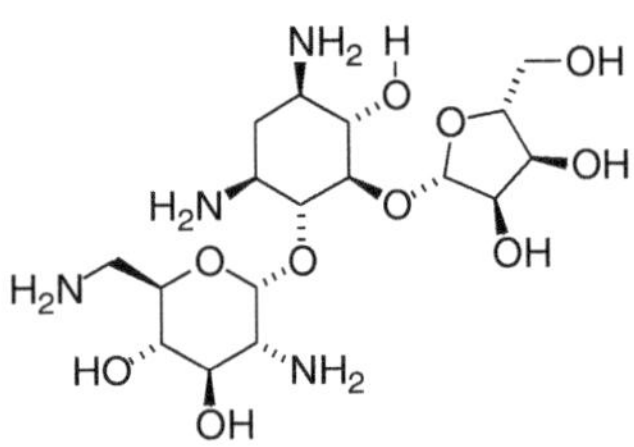

$t_{1/2}$	2.5 h (iv)	V	0.25	RIBOSTAMYCIN
F	–	CL	1.5	Aminoglycoside antibiotic
pb	–	Mwt	454.5	
ur	–	PSA	262 Å^2	
		log P	−3.95	

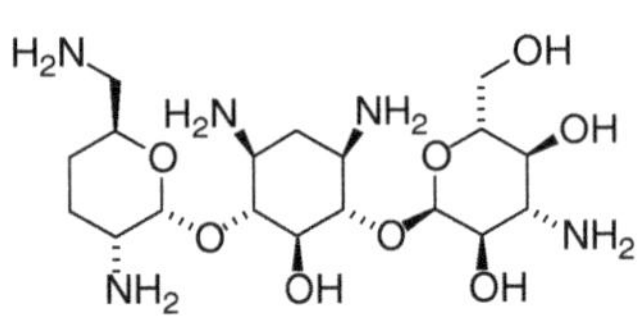

$t_{1/2}$	2.4 h (iv)	V	0.13	DIBEKACIN
F	–	CL	0.80	Antibacterial
pb	–	Mwt	451.5	
ur	High	PSA	248 Å^2	
		log P	−3.58	

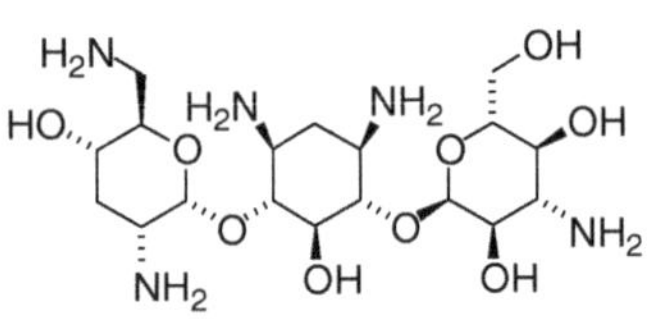

$t_{1/2}$	2 h	V	0.33±0.08	TOBRAMYCIN
F	Low	CL	0.98	Antibacterial
pb	<10%	Mwt	467.5	
ur	60%	PSA	268 Å^2	
		log P	−4.22	

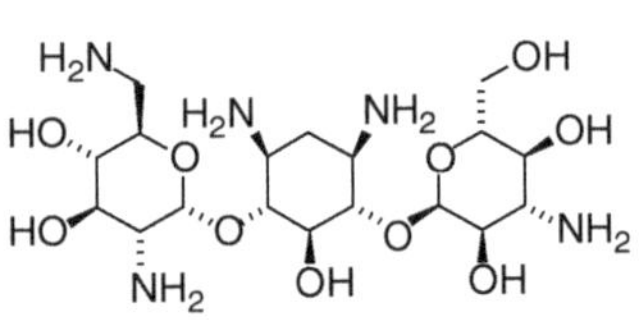

$t_{1/2}$	2.1±0.2 h	V	0.26±0.05	KANAMYCIN B
F	Low	CL	1.4±0.2	*iv
pb	0%	Mwt	483.5	Antibacterial
ur	90%*	PSA	288 Å^2	
		log P	−4.02	

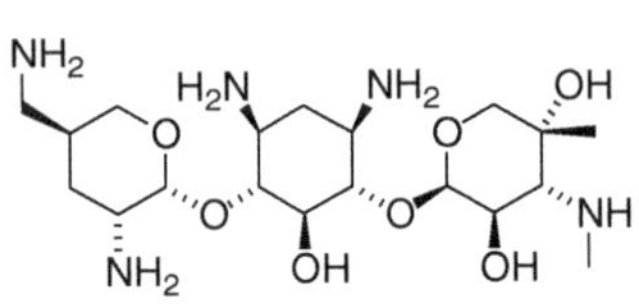

$t_{1/2}$	2–5 h	V	0.3±0.1	GENTAMICIN
F	Low	CL	0.82	*iv
pb	<10%	Mwt	449.6	Antibacterial
ur	>90%*	PSA	213.7 Å^2	
		log P	−4.89	

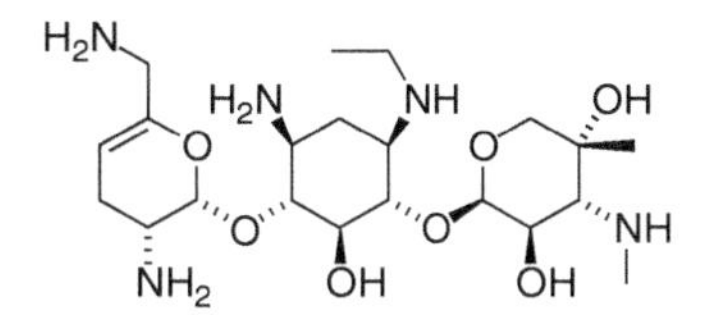	$t^1\!/_2$ 2–4 h *F* 0% pb <10% ur 85±5%*	*V* 0.20±0.02 CL 1.3±0.2 Mwt 475.6 PSA 200 Å^2 log *P* −1.84	NETILMICIN *iv Antibacterial
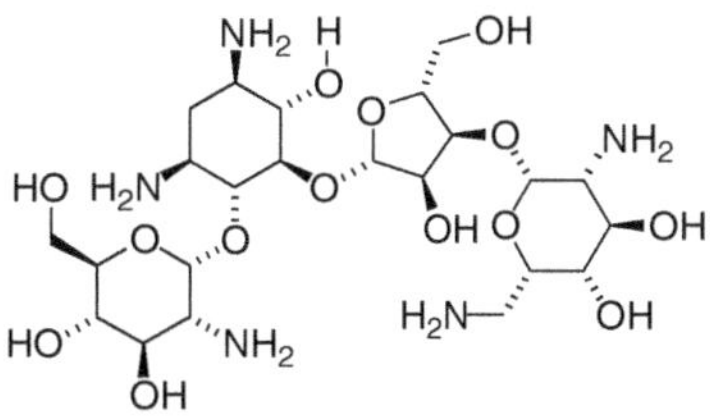	$t^1\!/_2$ 2–3 h *F* Negligible pb – ur High	*V*/*F* 0.4 CL/*F* 1.8 Mwt 615.6 PSA 347 Å^2 log *P* −5.80	PAROMOMYCIN Antibacterial
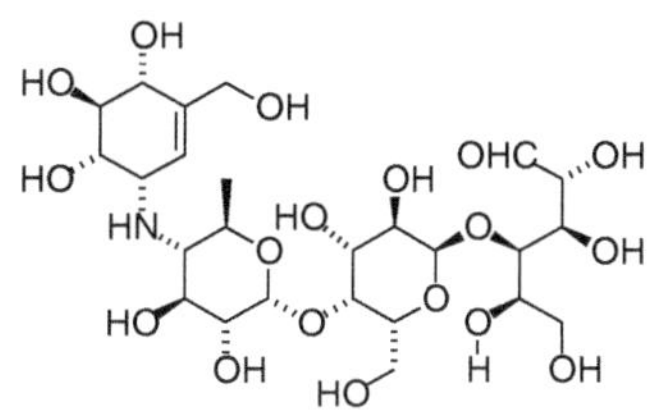	$t^1\!/_2$ 3–9 h *F* 0.5–1.7% pb 15% ur 94% (iv)	*V* 0.32 CL 143 Mwt 645.6 PSA 329 Å^2 log *P* −7.91	ACARBOSE α-Glucosidase, glucoamylase, and sucrase inhibitor, antidiabetic
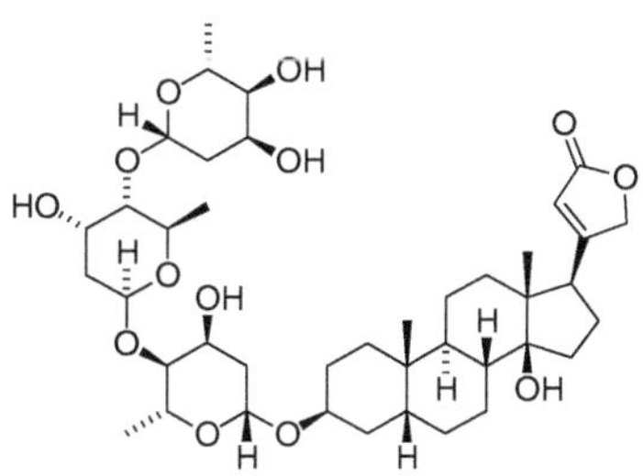	$t^1\!/_2$ 6.7±1.7 d *F* 95% pb 97±0.5% ur 32±15%	*V* 0.54±0.14 CL 0.06±0.02 Mwt 764.9 PSA 183 Å^2 log *P* 2.74	DIGITOXIN Na-K-ATPase membrane pump inhibitor, cardiac digitalis glycoside Active metabolite: digoxin
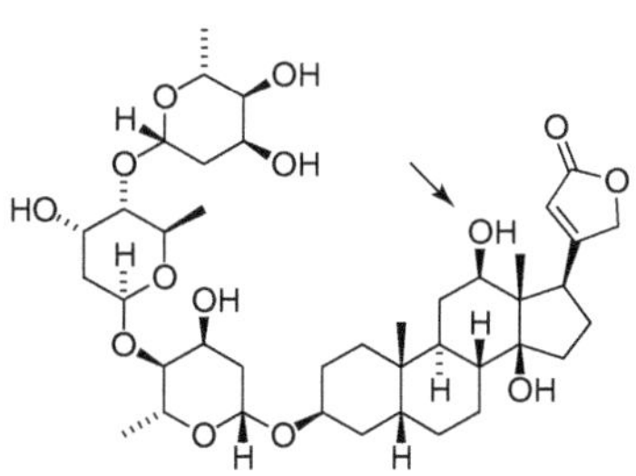	$t^1\!/_2$ 38–42 h *F* 70–80% pb 20–30% ur 70–80%	*V* 4–7 CL 1.8 Mwt 780.9 PSA 203 Å^2 log *P* 1.29	DIGOXIN Na-K-ATPase membrane pump inhibitor, cardiac digitalis glycoside for treatment of atrial fibrillation

$t_{1/2}$, plasma half-life; *F*, oral bioavailability; pb, plasma protein binding; ur, excretion of unchanged drug in urine; V, volume of distribution; CL, clearance; Mwt, molecular weight; PSA, polar surface area.

15
Phenols

Phenol has been used as antiseptic since 1867, when Joseph Lister, professor of surgery at the University of Glasgow, published his first article on the use of phenol in "antiseptic surgery." The antiseptic properties of phenol had, however, already been noticed by the discoverer of phenol ("carbolic acid"), Friedlieb Runge, in 1833. Today, phenol and low-molecular-weight phenols are important antiseptics used in numerous products, such as toothpaste, soap, and other detergents [1]. Other important drugs containing a phenol substructure are antipyretics (paracetamol), steroids, and analogs of catecholamines [2] (see also Table 15.1).

The most common metabolic transformation of phenols is O-glucuronidation and O-sulfation. If the 2- or 4-position of a phenol is unsubstituted, aromatic hydroxylation may occur. Phenols with hydroxyl, alkoxy, or amino group in ortho or para position may be oxidized to quinones. The kinetics of these processes vary widely and are difficult to estimate. Some examples of phenol metabolism are shown in Scheme 15.1.

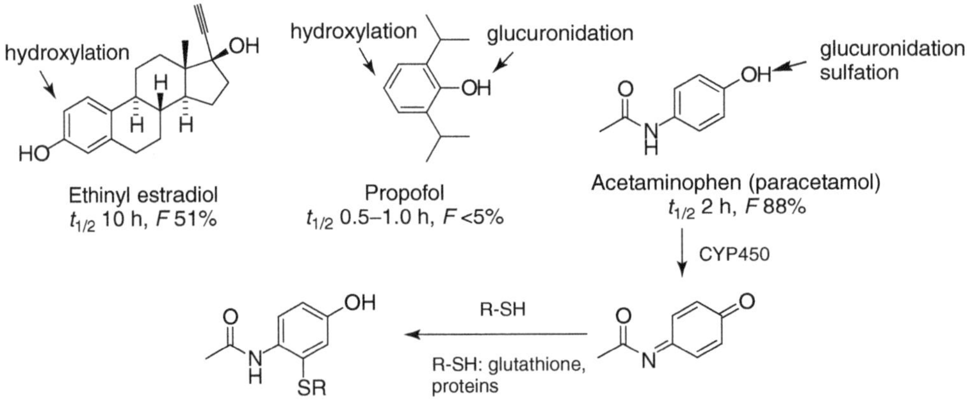

Scheme 15.1 Metabolism of phenols.

Some catechols (1,2-dihydroxybenzenes) can be methylated at high rates by catechol-*O*-methyltransferase (COMT). This is the reason for the short half-lives of dopamine, adrenaline, isoproterenol, and related catecholamines. By replacing the catechol substructure with 1,3-dihydroxybenzene, the plasma half-life of such

Lead Optimization for Medicinal Chemists: Pharmacokinetic Properties of Functional Groups and Organic Compounds, First Edition. Florencio Zaragoza Dörwald.

phenols can be significantly prolonged. Electron-withdrawing substituents may also block methylation of catechols by COMT [3], as does the nitro group in the COMT inhibitor entacapone (Scheme 15.2).

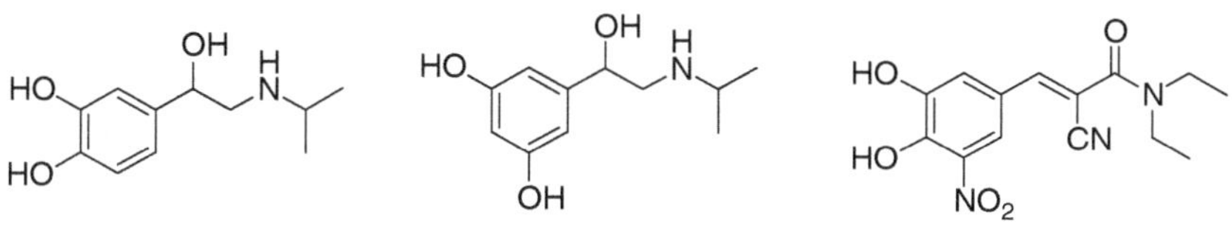

Isoproterenol $t_{1/2}$ 3 min, F 25% | Orciprenaline $t_{1/2}$ 2–6 h, F 10–40% | Entacapone $t_{1/2}$ 2–4 h, F 25–36%

Scheme 15.2 Dihydroxybenzenes as drugs.

Simultaneous dosing of a COMT inhibitor, such as entacapone, is another strategy to increase the half-life of catechols. The effect of levodopa (treatment for Parkinson's disease) can be prolonged by the peripheral decarboxylase inhibitor carbidopa (Scheme 15.3). The latter suppresses the peripheral conversion of levodopa into dopamine, so that larger quantities of levodopa can reach the CNS, where decarboxylation to dopamine occurs. Both the absorption from the small bowel and the crossing of the blood–brain barrier (bbb) of levodopa are mediated by an active transport system for aromatic amino acids. Carbidopa does not cross the bbb.

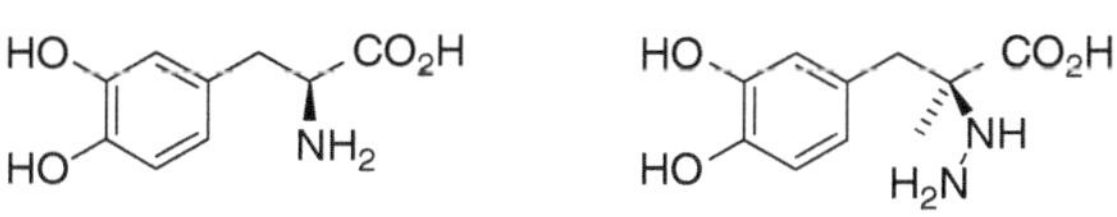

Levodopa $t_{1/2}$ 1.4 h, F 41% | Carbidopa $t_{1/2}$ 2.1 h, F 40–70%

Scheme 15.3

If a metabolically labile phenol is important for biological activity of a lead, the aromatic hydroxyl group may be replaced by a bioisostere. Phenol bioisosteres include benzpyrazoles, benzimidazoles, anilines, sulfonamides of anilines, and sometimes fluorobenzenes. The half-life of the vasodilator bamethan can, for instance, be significantly enhanced by replacing the phenolic hydroxyl group by a methanesulfonylamino group or an amino group (Scheme 15.4).

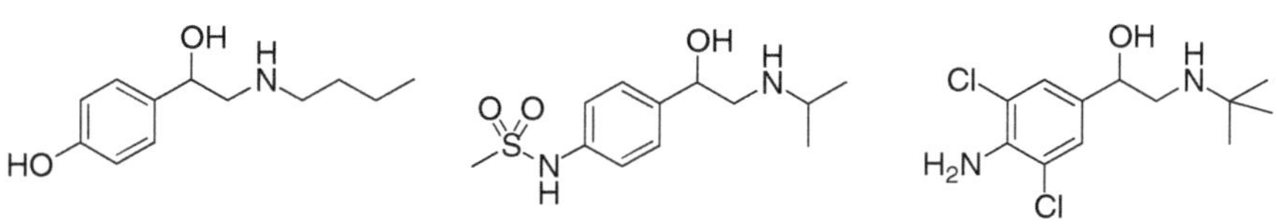

Bamethan $t_{1/2}$ 2.5 h, F 75% | Sotalol $t_{1/2}$ 12 h, F 90–100% | Clenbuterol $t_{1/2}$ 34 h, F 89–98%

Scheme 15.4 Bioisosteres of phenols.

References

1. Sneader, W. (1995) Antiseptics I. *Drug News Perspect.*, **8**, 504–508.
2. Sneader, W. (1996) Antipyretic analgesics. *Drug News Perspect.*, **9**, 61–64.
3. Bird, T.G.C., Arnould, J.C., Bertrandie, A., and Jung, F.H. (1992) Pharmacokinetics of catechol cephalosporins. The effect of incorporating substituents into the catechol moiety on pharmacokinetics in a marmoset model. *J. Med. Chem.*, **35**, 2643–2651.

Table 15.1 Phenols and related compounds. V in l kg^{-1}; CL in ml min^{-1} kg^{-1}; Mwt in g mol^{-1}.

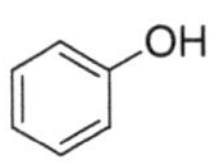	$t^1\!/_2$ F pb ur	1–5 h 90% – 52%	V CL Mwt PSA log P	– – 94.1 20.2 $Å^2$ 1.54	PHENOL Topical anesthetic and antiseptic Metabolism: glucuronidation, sulfation
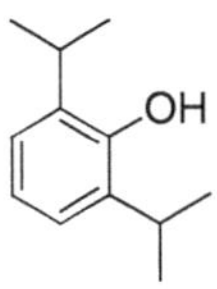	$t^1\!/_2$ F pb ur	0.5–1.0 h <5% 96–99% <0.3%	V CL Mwt PSA log P	3–14 19–33 178.3 20.2 $Å^2$ 3.66	PROPOFOL Anesthetic; Metabolism: aromatic hydroxylation (position 4), sulfation, glucuronidation; Prodrug: fospropofol
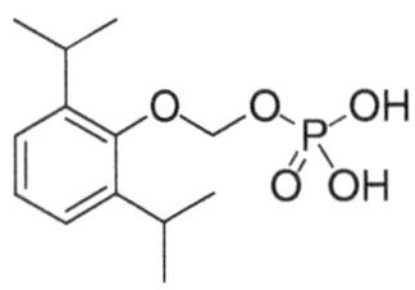	$t^1\!/_2$ F pb ur	0.8–0.9 h* – 98% <0.02%*	V CL Mwt PSA log P	0.33 4.3–6.0 288.3 85.8 $Å^2$ 2.70	FOSPROPOFOL *iv Prodrug of propofol
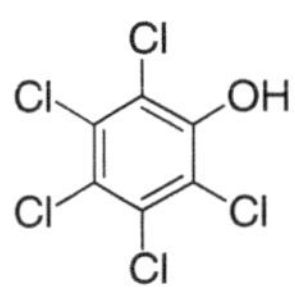	$t^1\!/_2$ F pb ur	27–35 h 91%* >96% 74%	V CL Mwt PSA log P	0.35 0.02 266.3 20.2 $Å^2$ 5.12	PENTACHLOROPHENOL *rat Insecticide, herbicide Metabolism: glucuronidation
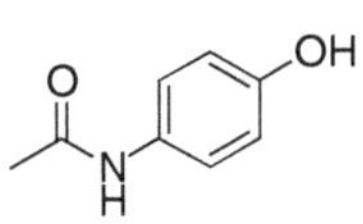	$t^1\!/_2$ F pb ur	2.0±0.4 h 88±15% Negligible 3±1%	V CL Mwt PSA log P	0.95±0.12 5.0±1.4 151.2 49.3 $Å^2$ 0.48	ACETAMINOPHEN, PARACETAMOL Analgesic, hepatotoxic Metabolites: *O*-glucuronide, *O*-sulfate, quinoneimine, conjugate with glutathione

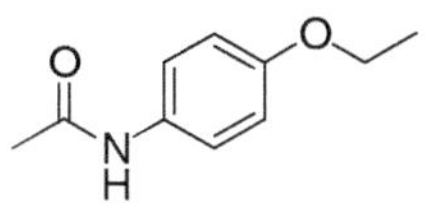	$t_{1/2}$ 0.7–1.5 h *F* Low pb 33% ur 0.1%	*V* 1–2 CL * Mwt 179.2 PSA 38.3 Å^2 log *P* 1.66	PHENACETIN *equal to hepatic blood flow Analgesic; extensive first pass metabolism (O-deethylation, N-deacetylation, N-hydroxylation); withdrawn in 1983 because of hepatotoxicity
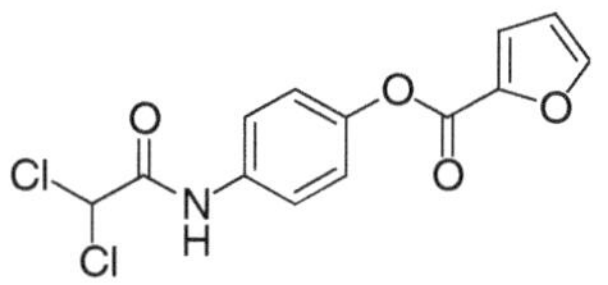	$t_{1/2}$ 3 h* *F* 60–90%* pb – ur –	*V* – CL – Mwt 234.1 PSA 40.5 Å^2 log *P* 1.67	DILOXANIDE FUROATE *phenol on oral dosing of ester Amebicide; Metabolism: ester hydrolysis, then glucuronidation
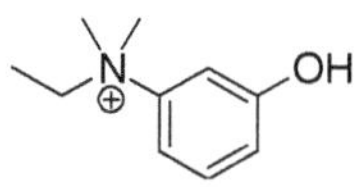	$t_{1/2}$ 1.8±0.6 h *F* Low pb Low ur 65%	*V* 1.1±0.2 CL 9.2±3.2 Mwt 166.2 PSA 20.2 Å^2 log *P* −2.07	EDROPHONIUM Cholinergic, antidote to curare
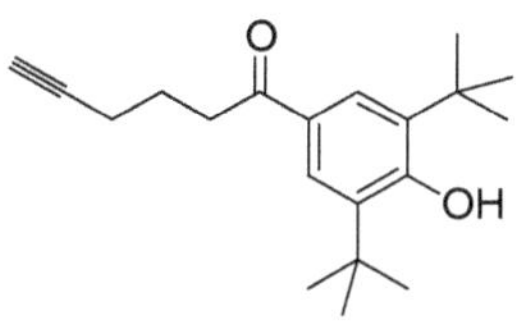	$t_{1/2}$ 36 h (iv) *F* 100% (rat) pb – ur –	*V* 30 CL 9.9 Mwt 300.4 PSA 37.3 Å^2 log *P* 5.8	TEBUFELONE Antiinflammatory, hepatotoxic
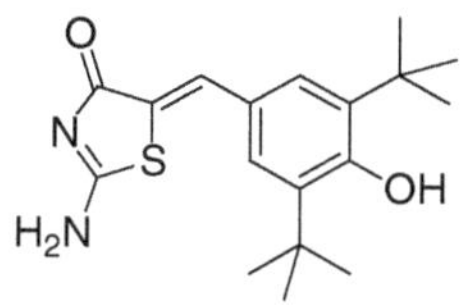	$t_{1/2}$ 100–112 h *F* 99% (rat) pb – ur <0.2%	*V* – CL – Mwt 332.5 PSA 101 Å^2 log *P* 4.74	DARBUFELONE Antiinflammatory
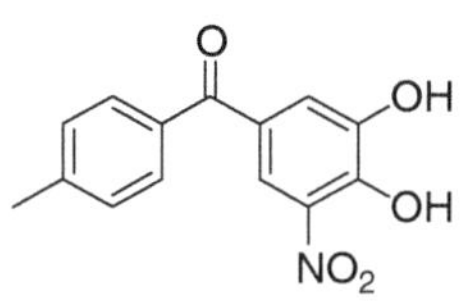	$t_{1/2}$ 1.1–1.3 h *F* 60–72% pb 99.8% ur 0.5%	*V* 0.12 CL 1.7–1.9 Mwt 273.2 PSA 103 Å^2 log *P* 4.15	TOLCAPONE COMT inhibitor, antiparkinsonian Metabolism: glucuronidation, O-methylation of 3-hydroxy, reduction of nitro to amino, hydroxylation of methyl, oxidation of methyl to CO_2H
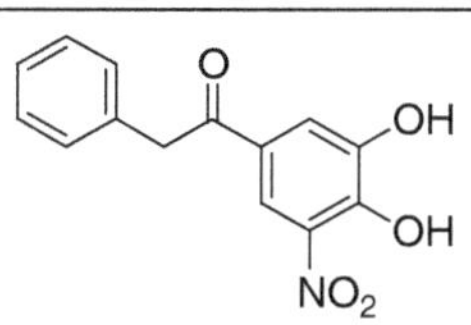	$t_{1/2}$ 4 h *F* >55% (rat) pb >99% ur <1%	*V* – CL – Mwt 273.2 PSA 103 Å^2 log *P* 3.84	NEBICAPONE COMT inhibitor Metabolism: glucuronidation, O-methylation

(*continued overleaf*)

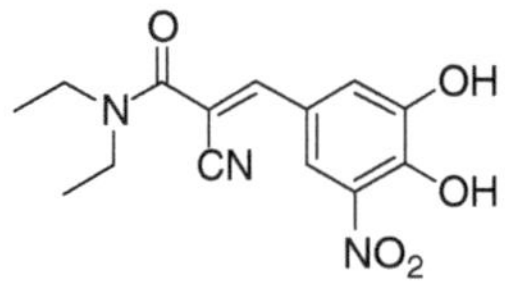

$t^{1/2}$	2.4–3.5 h	*V*	0.28	ENTACAPONE
F	25–36%	CL	12	COMT inhibitor; antiparkinsonian
pb	98%	Mwt	305.3	Metabolism: *E* to *Z* isomerization,
ur	<0.5%	PSA	130 Å^2	glucuronidation, hydrogenation,
		log *P*	2.38	and cleavage of C=C double bond

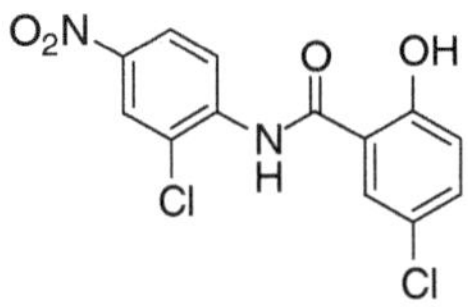

$t^{1/2}$	7 h	*V*	0.9 (rat)	NICLOSAMIDE
F	10% (rat)	CL	20 (rat, iv)	Anthelmintic
pb	–	Mwt	327.1	Metabolism: glucuronidation
ur	–	PSA	95.2 Å^2	
		log *P*	3.77	

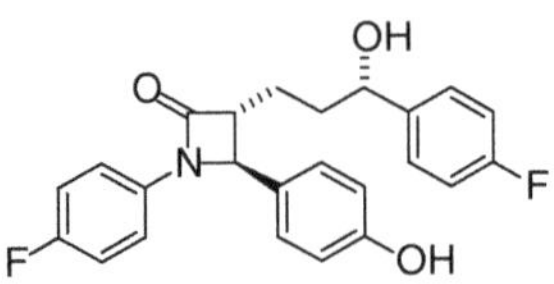

$t^{1/2}$	22 h	*V*	–	EZETIMIBE
F	Unknown	CL	–	Oral antilipemic; active
pb	99.7%	Mwt	409.4	metabolite: phenol glucuronide
ur	2%	PSA	60.8 Å^2	Further metabolism: oxidation of
		log *P*	3.96	secondary alcohol to ketone

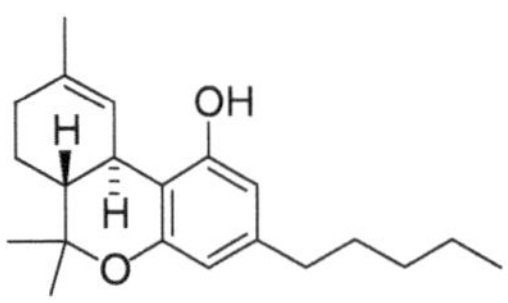

$t^{1/2}$	32±12 h	*V*	8.9±4.2	DRONABINOL, THC
F	8±4%	CL	3.5±0.9	Antiemetic, appetite stimulant
pb	95%	Mwt	314.5	Active metabolite: 11-hydroxy
ur	<1%	PSA	29.5 Å^2	
		log *P*	6.84	

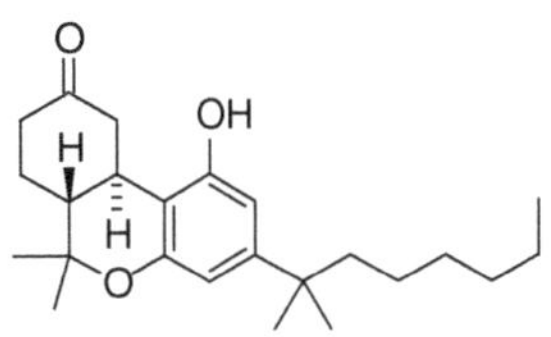

$t^{1/2}$	2 h	*V*	12.5	NABILONE
F	20%	CL	0.7	Antiemetic
pb	–	Mwt	372.5	Metabolism: reduction of ketone to
ur	22% (iv)	PSA	46.5 Å^2	alcohol,
		log *P*	7.25	hydroxylation at CH_2CH_3

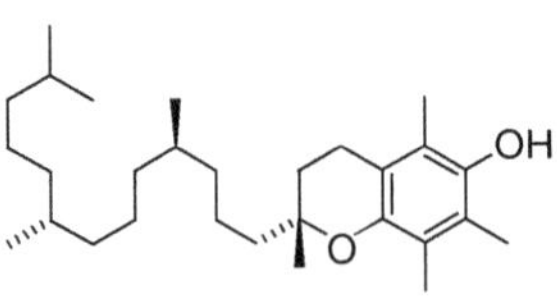

$t^{1/2}$	81±19 h	*V*	–	D-α-TOCOPHEROL
F	35–85%	CL	–	Vitamin
pb	–	Mwt	430.7	Prodrug: acetate
ur	<1%	PSA	29.5 Å^2	
		log *P*	11.0	

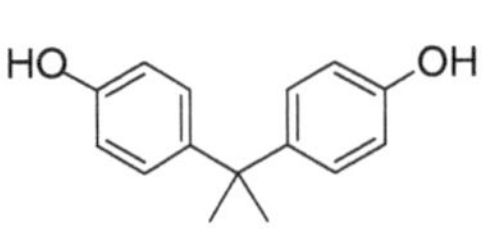

$t^{1/2}$	<6 h	*V*	–	BISPHENOL A
F	Low	CL	–	Monomer for polycarbonates and
pb	–	Mwt	228.3	epoxy resins, fungicide
ur	–	PSA	40.5 Å^2	Metabolism: glucuronidation
		log *P*	3.64	

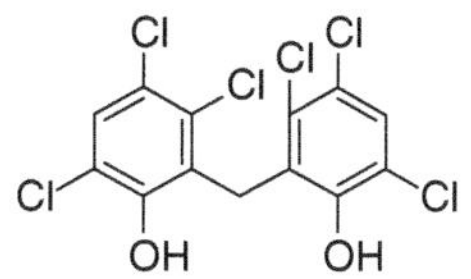	t½ 6–44 h *F* – pb 92% ur –	*V* – CL – Mwt 406.9 PSA 40.5 Å^2 log *P* 7.17	HEXACHLOROPHENE Antiseptic, disinfectant
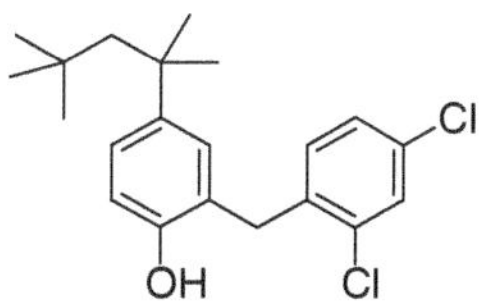	t½ 1.2 h *F* High pb – ur –	*V* – CL – Mwt 365.3 PSA 20.2 Å^2 log *P* 8.25	CLOFOCTOL, OCTOFENE Antibiotic Metabolism: glucuronidation
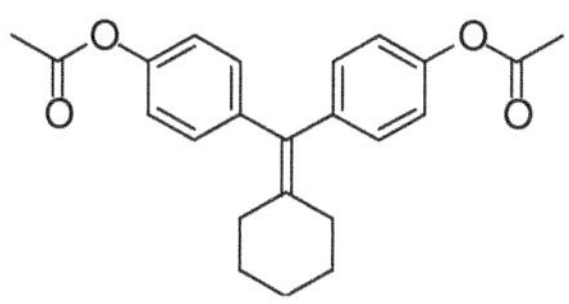	t½ 18–29 h* *F* – pb – ur 0%	*V* – CL – Mwt 364.4 PSA 52.6 Å^2 log *P* 4.66	CYCLOFENIL *diphenol on oral dosing of diacetate; gonad-stimulating principle; withdrawn because of hepatotoxicity
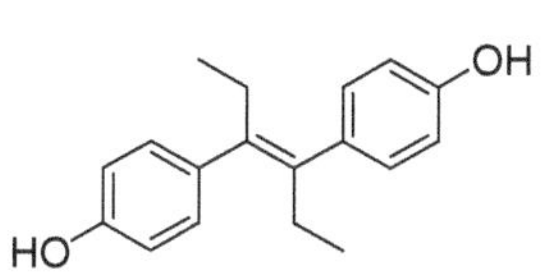	t½ 28 h *F* – pb 50–95% ur –	*V* 5* CL – Mwt 268.4 PSA 40.5 Å^2 log *P* 5.33	DIETHYLSTILBESTROL *monkey; Estrogen; Metabolism: aromatic hydroxylation to catechol, hydroxylation of methyl groups; withdrawn in 1975 as it causes adenocarcinoma of vagina in daughters after use in early pregnancy
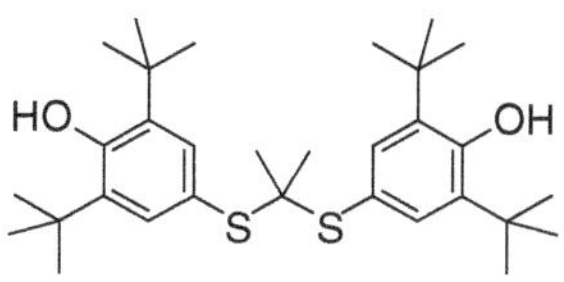	t½ 23–47 d *F* 2–8% pb – ur <2%	*V* – CL – Mwt 516.8 PSA 91.1 Å^2 log *P* 9.00	PROBUCOL Hypolipidemic; poor absorbtion (1–14%) because of poor solubility Metabolism: oxidative cyclization to dispiroquinone (bond formation between the 4-positions of the phenols), then formation of sulfur-free diphenoquinone
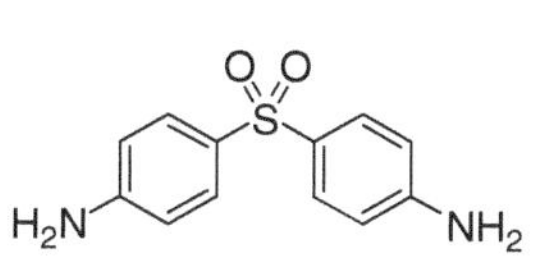	t½ 28±9 h *F* 93±8% pb 73±1% ur 15%	*V* 1.0±0.1 CL 0.60±0.17 Mwt 248.3 PSA 94.6 Å^2 log *P* 0.99	DAPSONE Antibacterial (leprostatic) Metabolism: N-hydroxylation, N-acetylation; causes methemoglobin formation and hemolysis
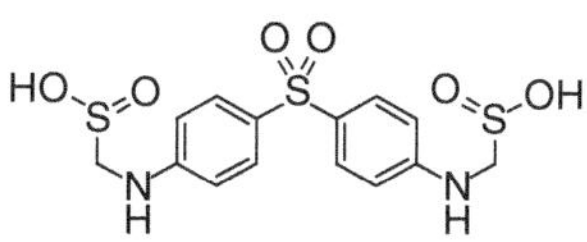	t½ 3–8 h *F* – pb 69% ur –	*V* – CL – Mwt 404.5 PSA 180 Å^2 log *P* −1.91	SULFOXONE Antibacterial (leprostatic) Metabolism: hydrolysis to dapsone

(continued overleaf)

Structure			
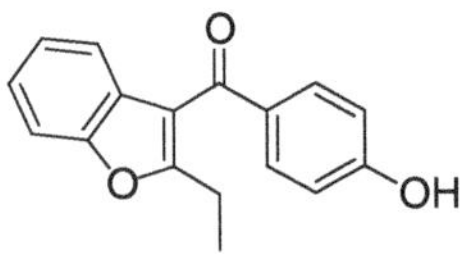	t½ 14±2 h* *F* – pb >99% ur –	*V* – CL – Mwt 266.3 PSA 50.4 Å^2 log *P* 4.73	BENZARONE *after oral administration of benzbromarone Capillary protectant Metabolism: O-sulfation, O-glucuronidation, ethyl hydroxylation; no unchanged benzarone can be detected in plasma after oral administration
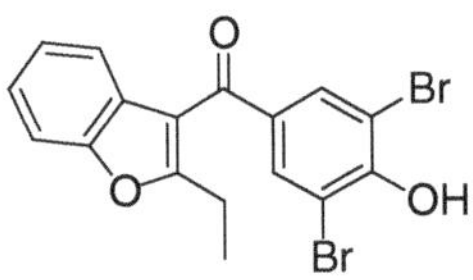	t½ 3±1 h *F* – pb >99% ur 0%	*V* 19 CL – Mwt 424.1 PSA 50.4 Å^2 log *P* 6.65	BENZBROMARONE Uricosuric, hepatotoxic Metabolism: debromination to benzarone
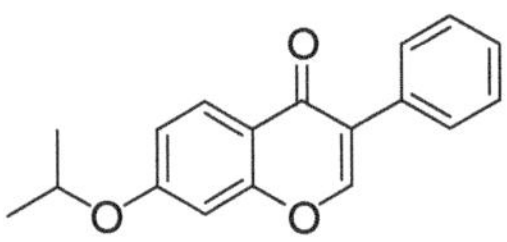	t½ 10 h *F* 24% (rat) pb – ur	*V* – CL – Mwt 280.3 PSA 35.5 Å^2 log *P* 4.25	IPRIFLAVONE Treatment of osteoporosis Metabolism: phenyl 4-hydroxylation, O-deisopropylation
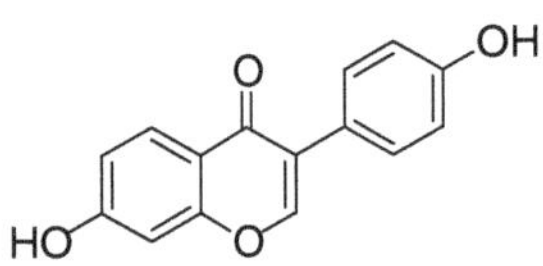	t½ 8 h *F* 2–10%* pb – ur 30–50%	*V* 3–5 CL 7.7 Mwt 254.2 PSA 66.8 Å^2 log *P* 2.63	DAIDZEIN *rat Natural isoflavone, phytoestrogen Metabolism: glucuronidation, sulfation
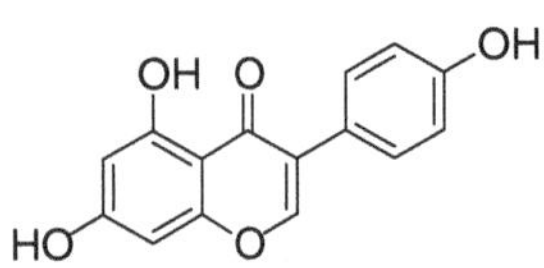	t½ 10 h *F* 20–40%* pb – ur 9%	*V* 2.5 CL – Mwt 270.2 PSA 87.0 Å^2 log *P* 3.11	GENISTEIN *rat Natural isoflavone: phytoestrogen Metabolism: glucuronidation, sulfation
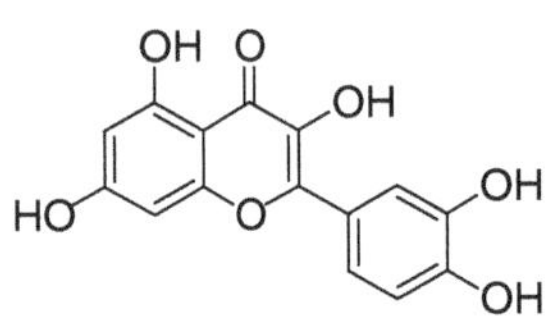	t½ 0.6 h (iv) *F* <1% pb 99% ur 0.5%	*V* 0.1 CL 11 Mwt 302.2 PSA 127 Å^2 log *P* 1.99	QUERCETIN Capillary protectant
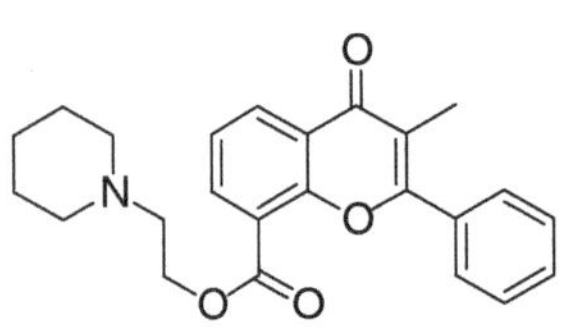	t½ 5 min *F* 100% pb 99.5%* ur 0%	*V* – CL – Mwt 391.5 PSA 55.8 Å^2 log *P* 4.27	FLAVOXATE *acid Antispasmodic: does not cross bbb Metabolism: hydrolysis of ester

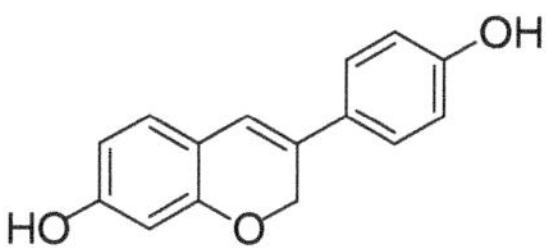	$t_{1/2}$ *F* pb ur	0.7 h (iv) 1% – 0%	*V* CL Mwt PSA log *P*	– – 240.3 49.7 Å^2 3.35	IDRONOXIL Proapoptotic agent: oncolytic
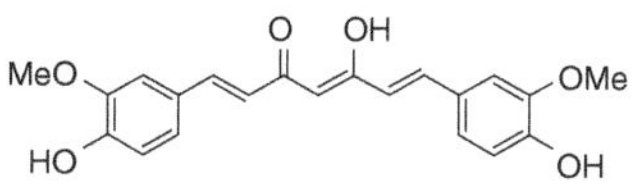	$t_{1/2}$ *F* pb ur	short 0% – 0%	*V* CL Mwt PSA log *P*	– – 368.4 93.1 Å^2 3.07	CURCUMIN Natural dye and antioxidant Metabolism: glucuronidation, sulfation, hydrogenation of C=C double bonds, reduction of ketone to alcohol
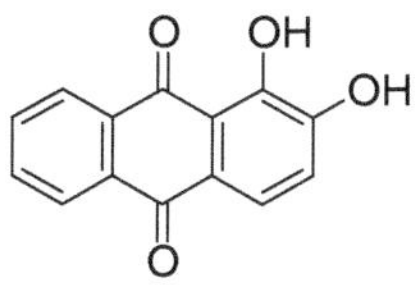	$t_{1/2}$ *F* pb ur	12 h – – 18–36%	*V* CL Mwt PSA log *P*	– 240.2 74.6 Å^2 2.91	ALIZARIN Biological stain Metabolism: O-glucuronidation
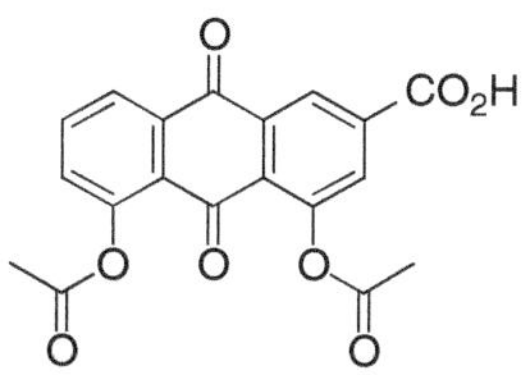	$t_{1/2}$ *F* pb ur	4–8 h* 35%* 99%* 20%*	*V* CL Mwt PSA log *P*	– – 368.3 124 Å^2 3.13	DIACEREIN, DIACETYLRHEIN *rhein on oral dosing of diacerein Antiarthritic Metabolism: deacetylation to rhein, then glucuronidation, sulfation
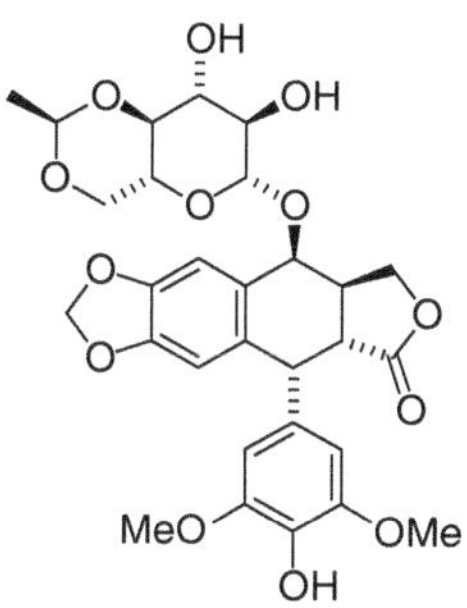	$t_{1/2}$ *F* pb ur	8±4 h 52±17% 96±0.4% 35±5%	*V* CL Mwt PSA log *P*	0.36±0.15 0.68±0.23 588.6 161 Å^2 0.28	ETOPOSIDE Antineoplastic

(continued overleaf)

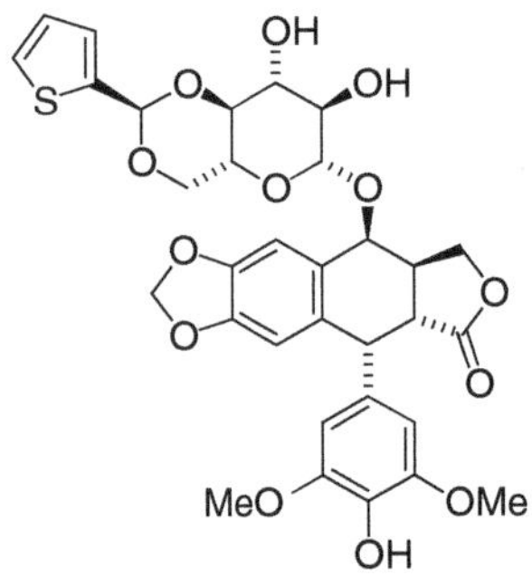

$t_{1/2}$	9±3 h	*V*	0.22±0.05	TENIPOSIDE
F	20–71%	CL	0.37±0.13	Antineoplastic
pb	99%	Mwt	656.7	Metabolism: O-demethylation
ur	8±2%	PSA	189 Å^2	
		log *P*	1.56	

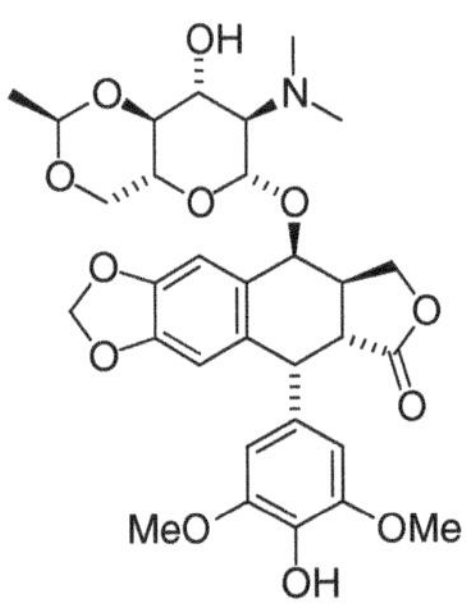

$t_{1/2}$	2–12 h	*V*	0.3	NK-611
F	100%	CL	0.4	Antineoplastic
pb	99%	Mwt	615.6	Metabolites: *N*-desmethyl, *cis*-lactone
ur	3–26% (iv)	PSA	143.9 Å^2	
		log *P*	1.14	

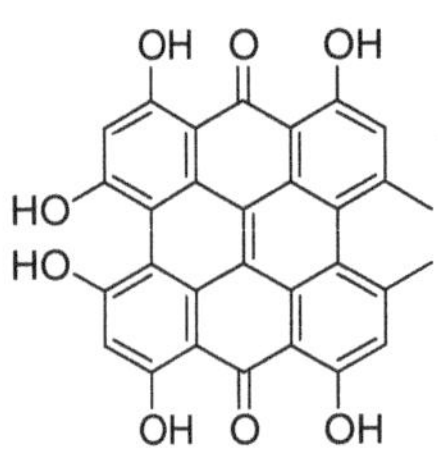

$t_{1/2}$	25–27 h	*V*	0.25	HYPERICIN
F	–	CL	0.13	Antidepressant, phototoxic component of St. John's wort
pb	–	Mwt	504.4	Metabolism: negligible
ur	–	PSA	156 Å^2	
		log *P*	10.8	

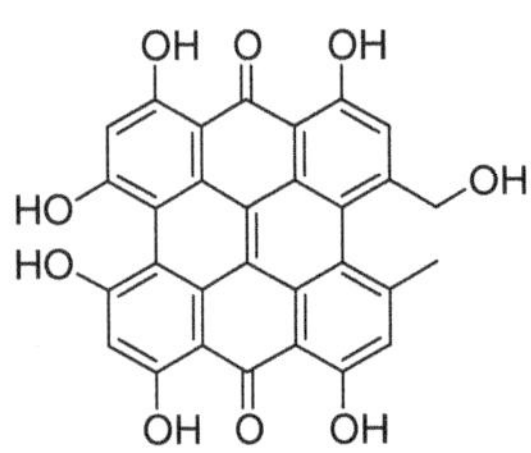

$t_{1/2}$	16–36 h	*V*	0.56	PSEUDOHYPERICIN
F	–	CL	0.62	Dye from *Hypericum perforatum*
pb	–	Mwt	520.4	(St. John's wort)
ur	–	PSA	176 Å^2	
		log *P*	9.29	

$t_{1/2}$, plasma half-life; *F*, oral bioavailability; pb, plasma protein binding; ur, excretion of unchanged drug in urine; V, volume of distribution; CL, clearance; Mwt, molecular weight; PSA, polar surface area; THC, tetrahydrocannabinol.

16
Ethers

Dialkyl ethers are more soluble in water than alkanes and can, therefore, be used to improve the solubility and prevent first-pass metabolism of hydrophobic leads. The replacement of a CH_2 group by oxygen leads to compounds with diverging boiling and melting points, which indicates that oxygen is not a good bioisostere of the methylene group. Such modification may in fact lead to compounds with quite different properties (Scheme 16.1).

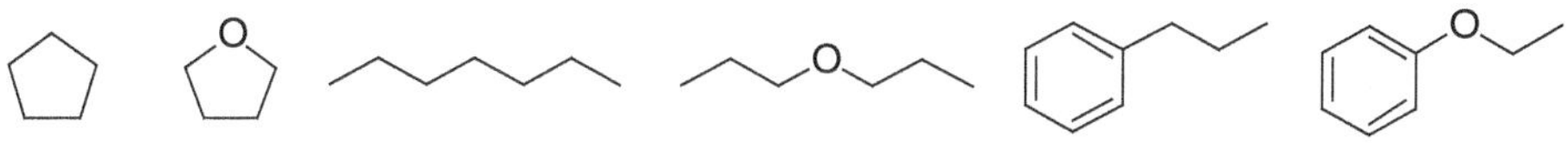

bp	49 °C	67 °C	98 °C	91 °C	159 °C	170 °C
mp	−94 °C	−108 °C	−91 °C	−122 °C	−100 °C	−30 °C
log *P*	2.64	0.47	4.27	2.06	3.74	2.68

Scheme 16.1

If a lead contains a phenol substructure and shows poor absorption or extensive first-pass metabolism, etherification may enhance oral bioavailability. Because alkoxyarenes are often rapidly dealkylated *in vivo*, some ethers may serve as prodrug of alcohols or phenols. Thus, codeine or hydrocodone are more orally bioavailable than their corresponding phenolic metabolites morphine and hydromorphone (Scheme 16.2). For phenacetin, metabolic deethylation is so fast that it can hardly be detected in plasma, and its apparent oral availability is, therefore, low. Because the main metabolite of phenacetin, acetaminophen, has good oral availability, the latter can be used as analgetic as well. In fact, phenacetin is more damaging to the kidneys and liver than acetaminophen and is, therefore, no longer available in many countries. Ethoxybenzenes should be considered as metabolically questionable and, if possible, avoided.

Lead Optimization for Medicinal Chemists: Pharmacokinetic Properties of Functional Groups and Organic Compounds, First Edition. Florencio Zaragoza Dörwald.

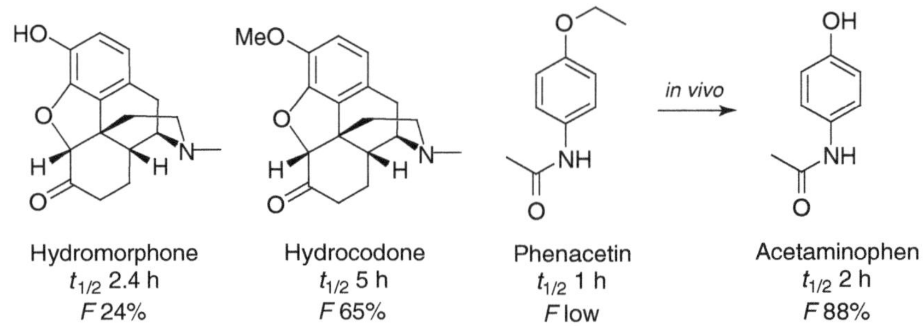

Scheme 16.2 Effect of phenol O-alkylation on PK.

16.1
Metabolism

Alkyl ethers with at least one hydrogen atom in α-position may be α-hydroxylated via hydride abstraction. The resulting hemiacetal can fragment into an alcohol and an aldehyde or a ketone.

With the exception of some methoxy- or ethoxyarenes, most ethers are metabolized rather slowly, and numerous long-acting drugs contain ether substructures (Table 16.1). Ethers of 2-alkoxyethanols (e.g., poly(ethylene glycol) [1]) or 2-aminoethanols are more stable than simple alkyl ethers, as a comparison of some of the metabolites of clemastine and doxylamine shows (Scheme 16.3). The main metabolite of clemastine is the tertiary alcohol resulting from oxidative ether cleavage; no such cleavage occurs in doxylamine.

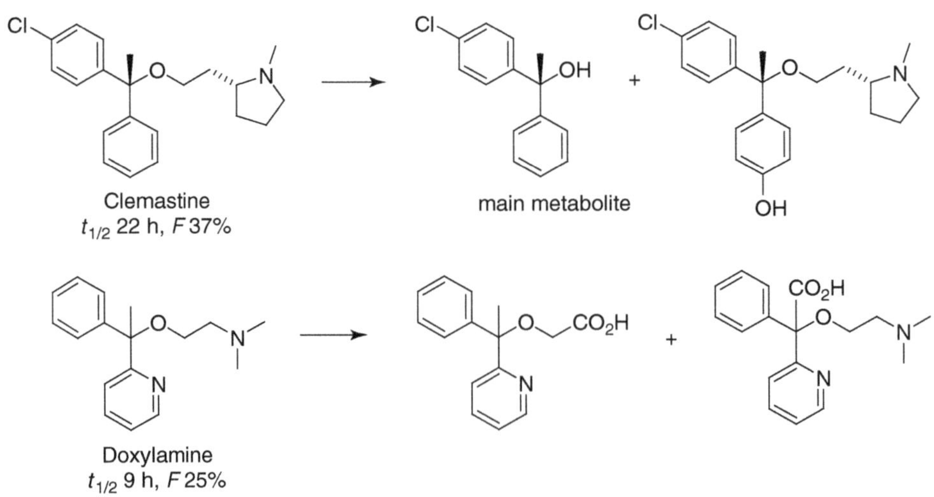

Scheme 16.3 Metabolism of clemastine and doxylamine.

If a lead contains an unstable alkyl ether, its cleavage may be prevented by conversion into a cyclopropylmethylether. One instance of this strategy is the long-acting β-blocker betaxolol (Scheme 16.4).

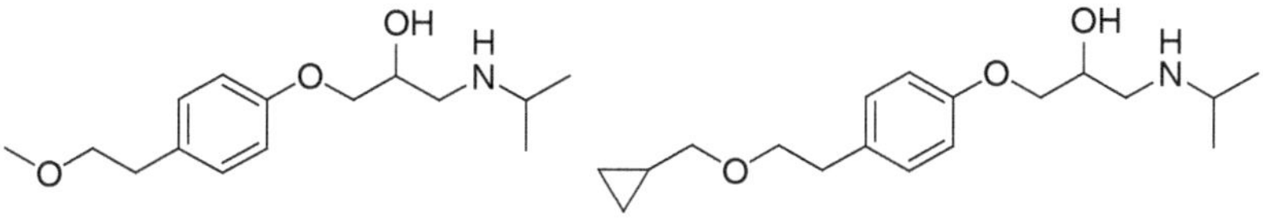

Metoprolol $t_{1/2}$ 3 h, F 38% | Betaxolol $t_{1/2}$ 14–22 h, F 89%

Scheme 16.4 Stabilization of methyl ethers by cyclopropylation.

The conversion of methyl groups into cyclopropylmethyl groups does, however, not always lead to drugs with better pharmacokinetics (PK). The lipophilic benzodiazepines flutoprazepam and prazepam, for instance, are quickly N-dealkylated *in vivo* and have shorter half-lives than their methyl analogs fludiazepam and diazepam (Scheme 16.5).

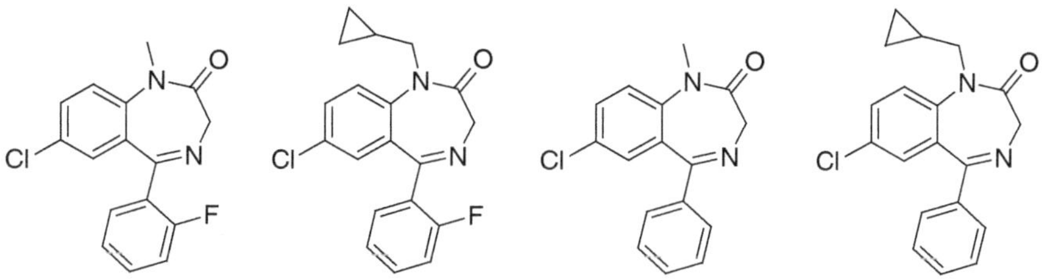

Fludiazepam $t_{1/2}$ 36 h | Flutoprazepam $t_{1/2}$ < 6 h; F 85% | Diazepam $t_{1/2}$ 43 h; F 100% | Prazepam $t_{1/2}$ 1.3 h

Scheme 16.5 Effect of cyclopropylation in benzodiazepines.

Reference

1. Harris, J.M. and Chess, R.B. (2003) Effect of pegylation on pharmaceuticals. *Nat. Rev. Drug Discov.*, **2**, 214–221.

Table 16.1 Ethers, acetals, and related compounds. V in l kg^{-1}; CL in ml min^{-1} kg^{-1}; Mwt in g mol^{-1}.

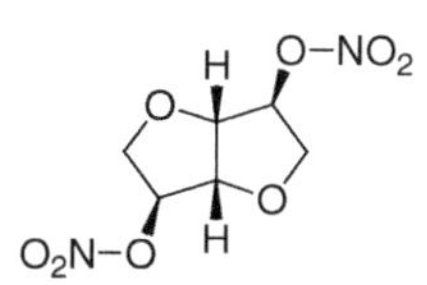	$t^{1/2}$	1.0±0.5 h	V	3.9±1.5	ISOSORBIDE DINITRATE
	F	22±14%	CL	45±20	Vasodilator
	pb	28±12%	Mwt	236.1	
	ur	<1%	PSA	129 Å^2	
			log P	0.95	

(continued overleaf)

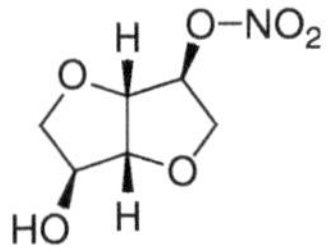	$t^1/_2$ 5±1 h *F* 93±13% pb 0–5% ur <5%	*V* 0.8±0.3 CL 6±2 Mwt 191.1 PSA 93.7 Å^2 log *P* −0.18	ISOSORBIDE MONONITRATE Vasodilator
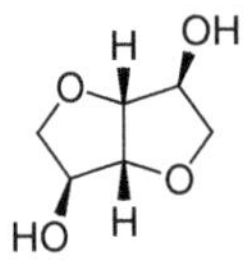	$t^1/_2$ 5–10 h *F* >80% pb – ur High	*V* – CL – Mwt 146.1 PSA 58.9 Å^2 log *P* −1.19	ISOSORBIDE Osmotic agent, treatment of hydrocephalus
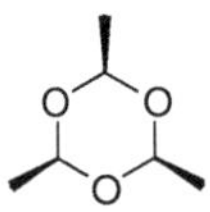	$t^1/_2$ 8±3 h *F* – pb – ur Negligible	*V* 1.7 CL 2.2 Mwt 132.2 PSA 27.7 Å^2 log *P* 1.02	PARALDEHYDE Sedative, hypnotic
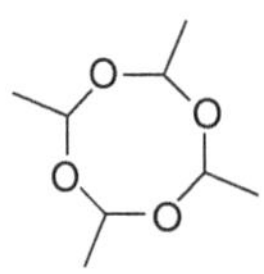	$t^1/_2$ 27 h *F* – pb – ur 1–8%*	*V* – CL – Mwt 176.2 PSA 36.9 Å^2 log *P* 1.36	METALDEHYDE *dog Molluscicide, crosses bbb
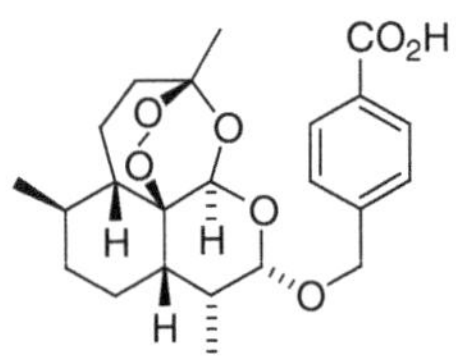	$t^1/_2$ 2.6 h* *F* 80%* pb – ur 0%*	*V* 9.5* CL 15* Mwt 418.5 PSA 83.5 Å^2 log *P* 4.89	ARTELINIC ACID *dog Antimalarial Metabolism: O-debenzylation
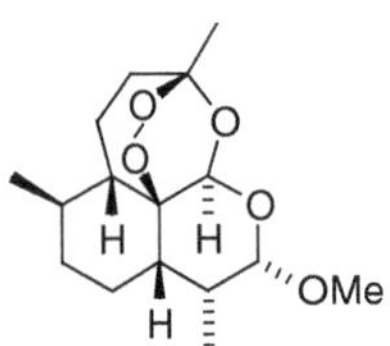	$t^1/_2$ 2–3 h *F* 43% pb 95% ur –	*V* 1–10 CL/*F* 3.7 Mwt 298.4 PSA 46.2 Å^2 log *P* 3.07	ARTEMETHER Antimalarial Metabolism: O-demethylation
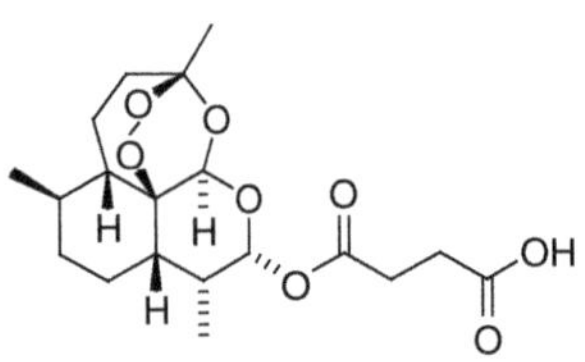	$t^1/_2$ 0.2–0.7 h *F* 30% (rat) pb 75% ur –	*V* 15 CL 1070 Mwt 384.4 PSA 101 Å^2 log *P* 2.94	ARTESUNATE Antimalarial

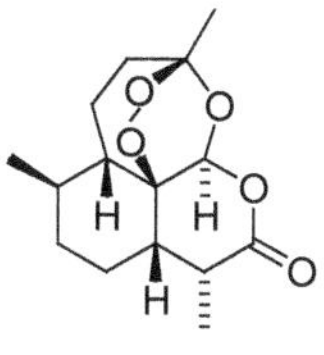	$t_{1/2}$	2.6±0.6 h	V/F	16±9	ARTEMISININ Natural antimalarial
	F	Low	CL/F	66±49	
	pb	77%	Mwt	282.3	
	ur	<1%	PSA	54.0 Å^2	
			log P	2.27	
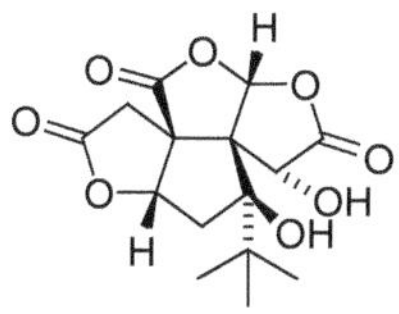	$t_{1/2}$	3.2 h (iv)	V	2.5	BILOBALIDE Neuroprotectant
	F	–	CL	10	
	pb	–	Mwt	326.3	
	ur	30%	PSA	119 Å^2	
			log P	4.46	
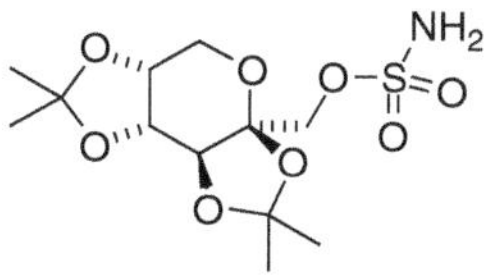	$t_{1/2}$	20–24 h	V	0.55–0.83	TOPIRAMATE Anticonvulsant, antimigraine Metabolism: acetal hydrolysis, hydroxylation of CH_3
	F	80–95%	CL	–	
	pb	10–20%	Mwt	339.4	
	ur	35–80%	PSA	124 Å^2	
			log P	2.16	

$t_{1/2}$, plasma half-life; F, oral bioavailability; pb, plasma protein binding; ur, excretion of unchanged drug in urine; V, volume of distribution; CL, clearance; Mwt, molecular weight; PSA, polar surface area.

17
Epoxides

Epoxides are strong alkylating reagents, but when stabilized by stereochemical or electronic effects, their toxicity may be sufficiently low to be suitable as nonalkylating drugs. Tropane-derived epoxides, for instance, are metabolically stable and only react with nucleophiles under forcing conditions. Stable epoxides can, therefore, be considered as PK-modifying group (e.g., to enhance solubility and metabolic stability) or as a possibility to get around others patents. Examples of nonalkylating epoxides are shown in Scheme 17.1.

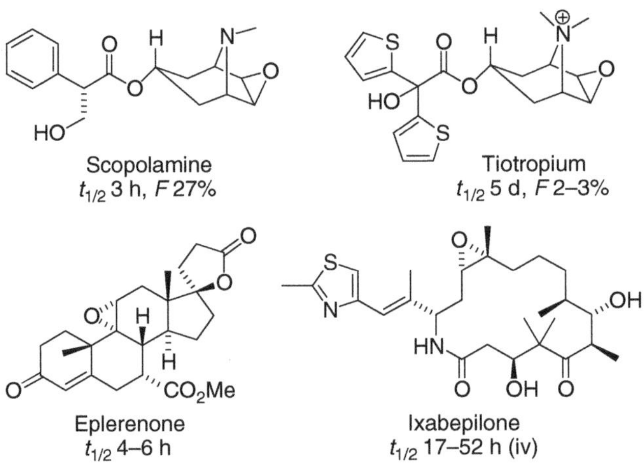

Scheme 17.1 Epoxide-containing drugs.

Lead Optimization for Medicinal Chemists: Pharmacokinetic Properties of Functional Groups and Organic Compounds, First Edition. Florencio Zaragoza Dörwald.

18
Peroxides

Despite being strong oxidants and thermally unstable, organic peroxides are occasionally used as drugs. These include the artemisinin-derived antimalarials (Table 16.1) and the anthelmintic ascaridole (Scheme 18.1). One of the metabolic transformations of these compounds includes reductive O–O bond cleavage.

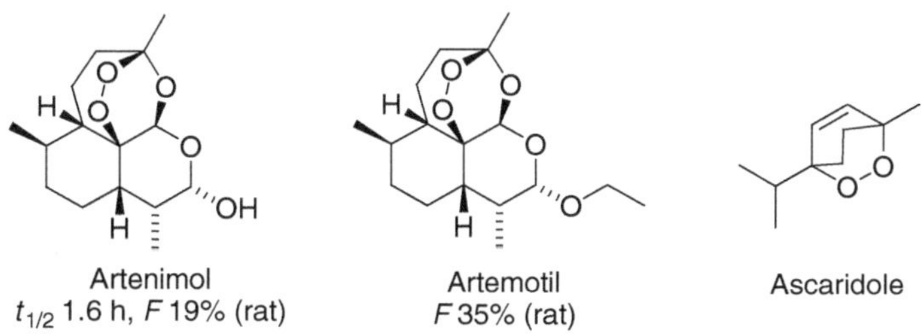

Scheme 18.1 Peroxide-containing drugs.

Lead Optimization for Medicinal Chemists: Pharmacokinetic Properties of Functional Groups and Organic Compounds, First Edition. Florencio Zaragoza Dörwald.

19
Thiols

Because sulfur is less electronegative than oxygen, the S−H bond is weaker and less strongly polarized than the O−H bond. Owing to the weaker hydrogen bonding ability, thiols have lower boiling point and are more lipophilic but more acidic than the corresponding alcohols (Scheme 19.1).

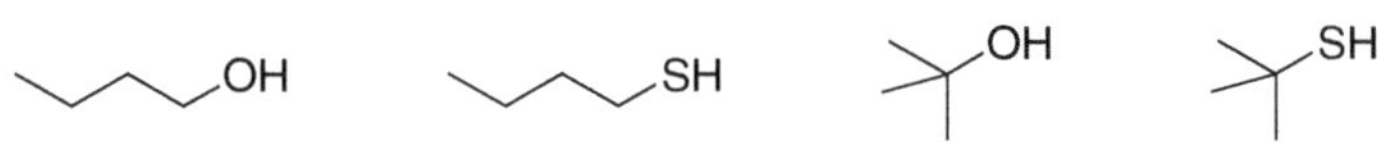

bp	117 °C	98 °C	82 °C	64 °C
mp	−90 °C	−116 °C	26 °C	1 °C
pK_a (in H_2O)	16	10.7	18	11.1
log *P*	0.84	2.46	0.58	2.21

Scheme 19.1 Physicochemical properties of alcohols and thiols.

Free sulfhydryl groups are air sensitive and highly nucleophilic and will only be used as structural element of a drug if indispensable. Thiol-containing drugs include enzyme inhibitors such as captopril or omapatrilate (binding of SH to Zn^{2+} of angiotensin-converting enzyme (ACE)), mucolytics such as *N*-acetylcysteine (as a reducing agent for disulfide bonds in proteins; also as an antidote for glutathione-depleting poisons), antidotes for heavy metal poisoning such as dimercaprol, and organ-protecting adjuvants for chemotherapy such as MESNA (Scheme 19.2). Some gold complexes, used for the treatment of rheumatoid arthritis, contain thiol-based ligands.

Lead Optimization for Medicinal Chemists: Pharmacokinetic Properties of Functional Groups and Organic Compounds, First Edition. Florencio Zaragoza Dörwald.

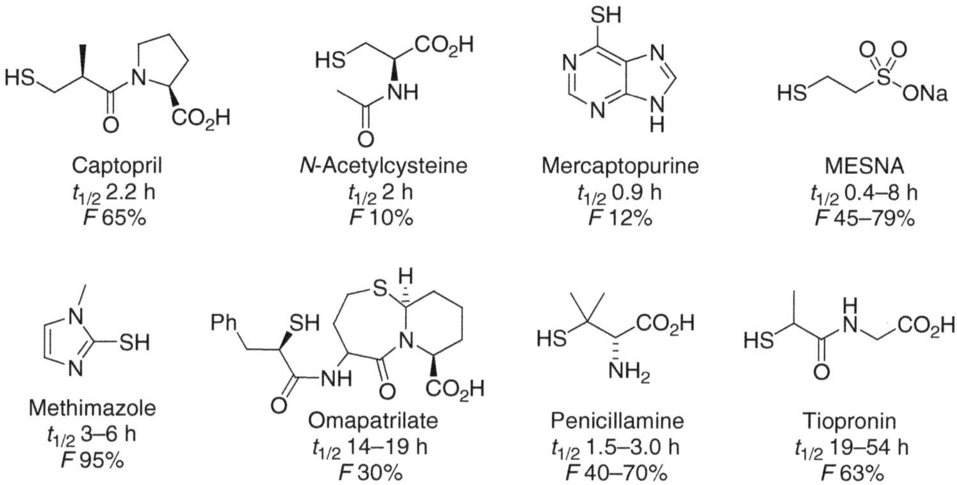

Scheme 19.2 Thiols as drugs.

Typical metabolic transformations of thiols include oxidative disulfide formation, desulfuration, and S-methylation.

20
Thioethers

The thioether group is a structural element of many drugs and does usually not lead to any toxic metabolites. Thioethers have higher boiling points and higher melting points but are less water soluble than the corresponding ethers (Scheme 20.1). Thus, while THF and water are miscible, tetrahydrothiophene and water form two phases.

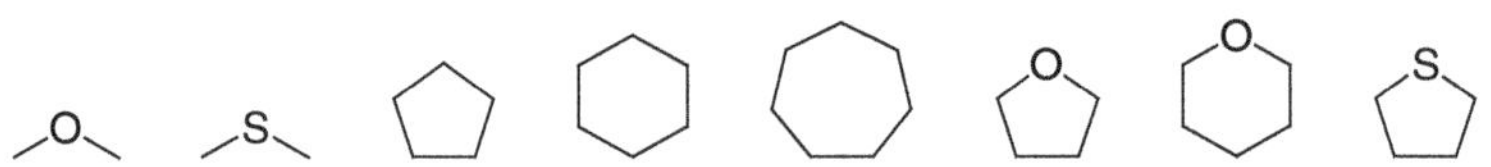

bp	−25 °C	37 °C	49 °C	81 °C	119 °C	67 °C	88 °C	121 °C
mp	−139 °C	−98 °C	−94 °C	7 °C	−12 °C	−108 °C	−45 °C	−96 °C
log *P*	0.02	0.98	2.64	3.16	3.69	0.47	1.00	1.61

Scheme 20.1 Comparison of alkanes, ethers, and thioethers.

If a thioether-containing lead shows poor oral bioavailability or short half-life because of S-oxidation, the thioether may be replaced by the similarly sized, bioisosteric ethylene group (CH_2CH_2) or oxymethylene group (OCH_2). Such modification was realized in phenothiazine-based antipsychotics, which often are poorly orally available. The corresponding dibenzazepines show, however, similar pharmacokinetics (PK) as the phenothiazines, probably because metabolic transformations other than S-oxidation proceed faster (N-dealkylation, N-oxidation, or aromatic hydroxylation). The compounds shown in Scheme 20.2, for instance, have similar plasma half-lives and oral bioavailabilities, no matter if they contain thioethers or not.

Lead Optimization for Medicinal Chemists: Pharmacokinetic Properties of Functional Groups and Organic Compounds, First Edition. Florencio Zaragoza Dörwald.

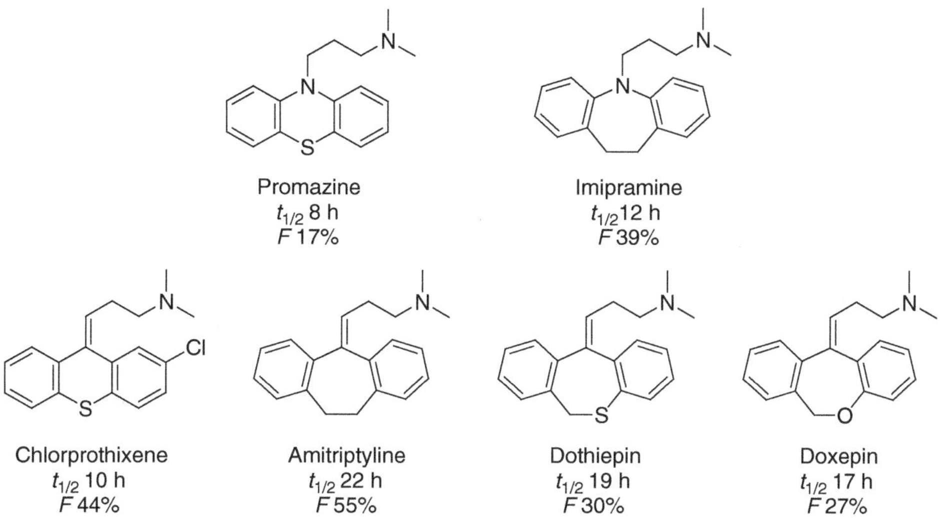

Scheme 20.2 The effect of thioethers on the PK-properties of drugs.

Sulfur is more electronegative than carbon and may be introduced into a lead to lower its basicity. A famous example of this strategy is cimetidine, a histamine H_2 antagonist for gastric acid suppression [1]. The H_2 antagonist burimamide was potent but showed little oral bioavailability (Scheme 20.3). Replacement of a methylene group by sulfur reduced the basicity of the imidazole fragment, and with an additional methyl group, a more lipophilic, orally available drug (metiamide) was obtained. Unfortunately, the thiocarbonyl group caused hepatotoxicity, which was finally overcome by replacement of the thiourea by a cyanoguanidine.

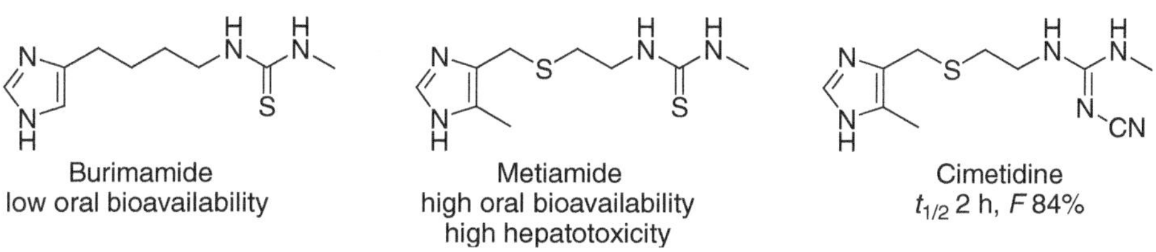

Scheme 20.3 Cimetidine and its leads.

20.1 Metabolism

Typical metabolic transformations of thioethers include dealkylation (same mechanism as dealkylation of ethers or amines) and oxidation to sulfoxides and sulfones. The oxidation to sulfoxides is reversible, and sulfoxides, for example, sulindac or DMSO, can undergo metabolic reduction to thioethers.

The most important metabolic pathway of sulfoxides is, however, their oxidation to sulfones.

The rapid oxidation of thioethers in the liver can lead to low oral bioavailability. Specially sensitive are highly lipophilic thioethers (e.g., phenothiazines) and sterically unshielded thioethers (e.g., arene-bound SCH_3). Occasionally, thioethers have been introduced into drugs to shorten their half-life, for instance, in the barbiturate methitural. If a drug is hydrophilic, however, sulfoxide formation may be acceptably slow. Some examples of thioethers with relatively long half-lives and good oral availabilities are shown in Scheme 20.4.

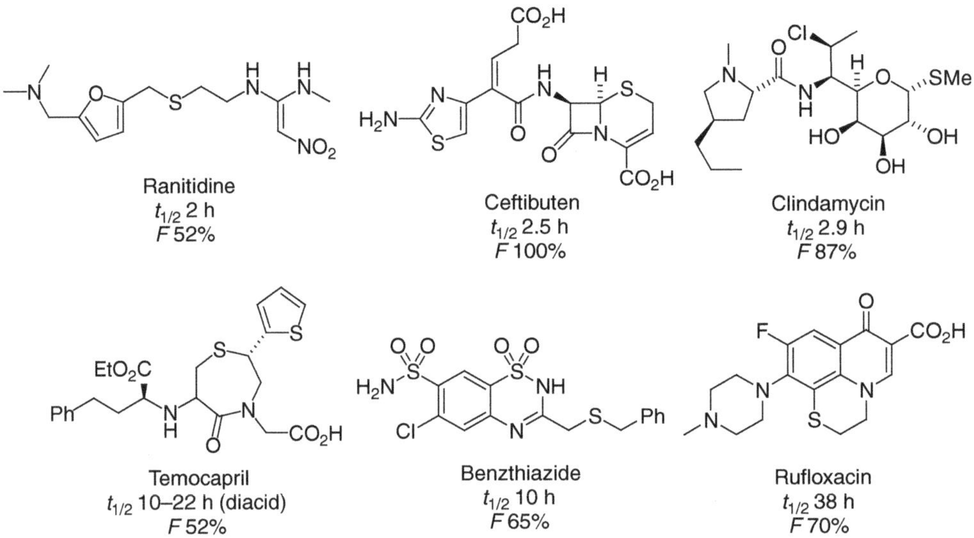

Scheme 20.4 Thioethers as drugs.

Metabolic S-oxidation can on occasions generate electrophilic intermediates that can cause toxicity or an immune response by protein alkylation. The withdrawn diuretic ticrynafen (tienilic acid), for instance, is converted in the liver into a highly reactive thiophene S-oxide, which induces an autoimmune response toward a liver protein (Scheme 20.5). Although any 2-unsubstituted thiophene may undergo these reactions, the required dose of drug is also a critical parameter for adverse drug reactions, and numerous safe, thiophene-containing drugs have been developed.

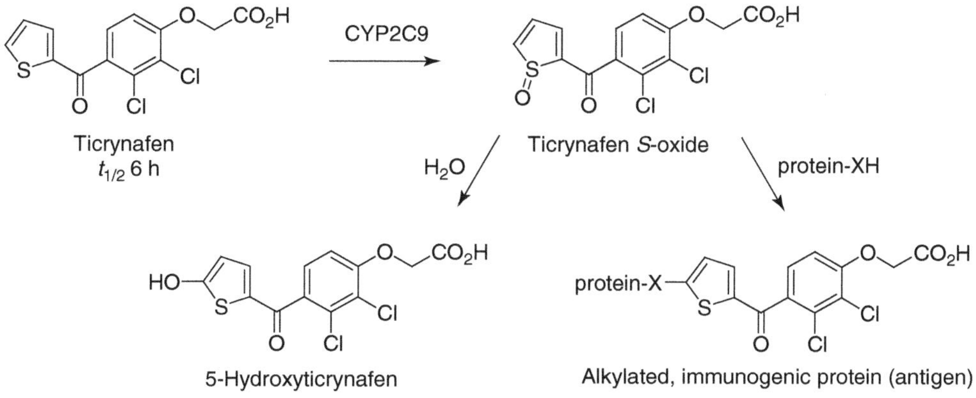

Scheme 20.5 Metabolism of ticrynafen.

Reference

1. Durant, G.J., Emmett, J.C., Ganellin, C.R., Miles, P.D., Parsons, M.E., Prain, H.D., and White, G.R. (1977) Cyanoguanidine–thiourea equivalence in the development of the histamine H_2-receptor antagonist, cimetidine. *J. Med. Chem.*, **20**, 901–906.

21
Sulfoxides

Despite its chemical reactivity, numerous drugs contain a sulfoxide group. In the case of proton pump inhibitors, sulfoxide reactivity is essential to their mechanism of action (Scheme 21.1). Omeprazole, for instance, has a short half-life under acidic conditions (2 min at pH 1–3) and covalently binds to H^+,K^+-ATPase to form a nonfunctional complex with a half-life of about 24 h. The enzyme is reactivated by glutathione-mediated reductive cleavage of the S−S bond.

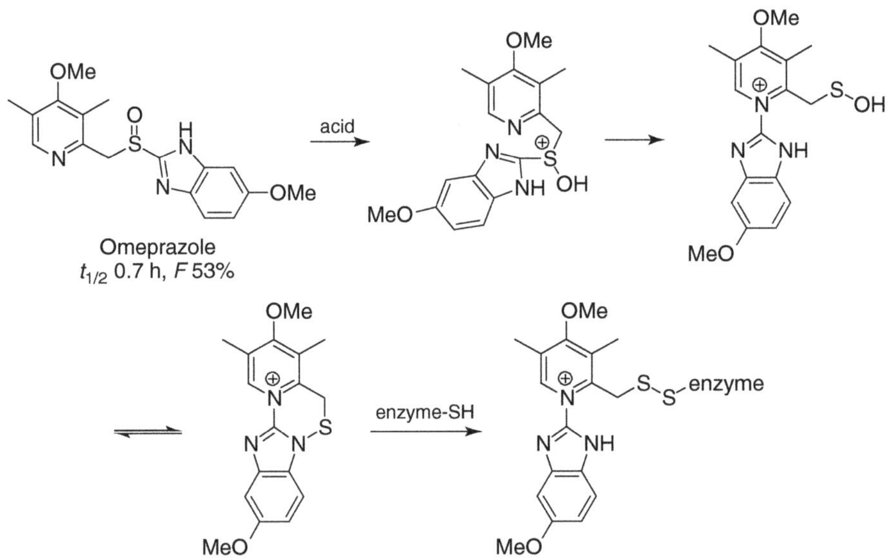

Scheme 21.1 Mechanism of action of omeprazole.

Other sulfoxides with higher metabolic stability than the proton pump inhibitors include the nonsteroidal antiinflammatory drug (NSAID) sulindac and the alcohol metabolite of the immunosuppressant oxisuran (Scheme 21.2).

Lead Optimization for Medicinal Chemists: Pharmacokinetic Properties of Functional Groups and Organic Compounds, First Edition. Florencio Zaragoza Dörwald.

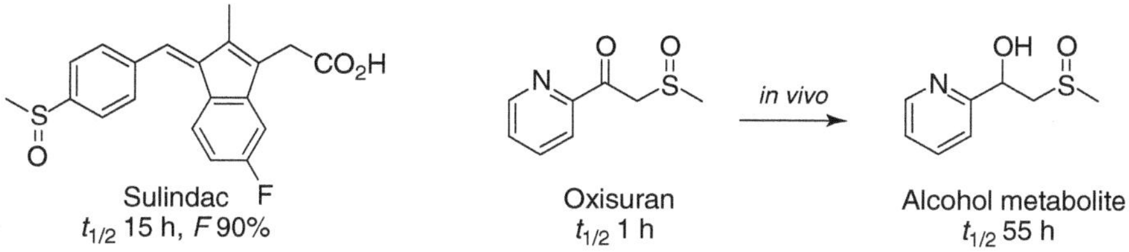

Scheme 21.2 Sulfoxides as drugs.

Besides undergoing Pummerer-type reactions as electrophiles, sulfoxides are mainly oxidized *in vivo* to sulfones or reduced to thioethers.

22
Sulfones

Sulfones are chemically and metabolically stable, highly polar functional groups, and have often been used for lead optimization. This group can be used to enhance solubility and reduce the rate of metabolic transformations or as bioisostere of a sulfonamide or an amide. Examples of sulfone-containing drugs are shown in Scheme 22.1.

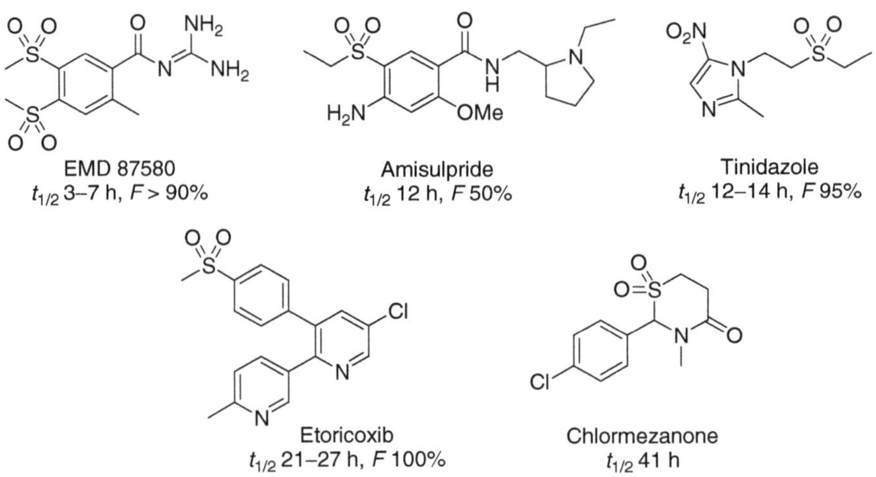

Scheme 22.1 Sulfones as drugs.

Lead Optimization for Medicinal Chemists: Pharmacokinetic Properties of Functional Groups and Organic Compounds, First Edition. Florencio Zaragoza Dörwald.

23
Aliphatic Amines

Aliphatic amines are basic compounds (pK_a 8–10) and will be protonated to a large extent in most organs. As salts, aliphatic amines dissolve more readily in water than free (unprotonated) amines, and drugs containing an amine functionality are, therefore, often formulated as ammonium salts. Further advantages of salts include a higher melting point, lower vapor pressure (no smell), higher stability toward oxidants, light, or carbon dioxide, and an overall better processability.

Natural compounds with an amine functionality, present in our food or body, include amino acids, hormones (serotonin, dopamine), or decarboxylated amino acids such as histamine, tyramine, or tryptamine. A huge number of drugs contain the amine substructure; these include enzyme inhibitors and agonists or antagonists of natural neurotransmitters, in which the amino group is a key pharmacophore (e.g. Table 23.1).

Amino groups may, however, also be introduced into drugs to modify their physicochemical and pharmacokinetic properties. The solubility of highly insoluble, neutral organic compounds may, for instance, be enhanced by linking them to a basic functional group (see, e.g., camptothecin and topotecan, Chapter 60). Moreover, because biological membranes consist of negatively charged phospholipids, basic drugs tend to be distributed more extensively in tissue than neutral or acidic drugs of similar log P and molecular weight. This leads to higher volumes of distribution and, all other factors being constant, to a longer half-life of basic drugs if compared to a similar neutral or acidic drug. Therefore, the half-life of a drug may be enhanced with basic functional groups. Successful examples of this strategy are given in Scheme 23.1.

Lead Optimization for Medicinal Chemists: Pharmacokinetic Properties of Functional Groups and Organic Compounds, First Edition. Florencio Zaragoza Dörwald.

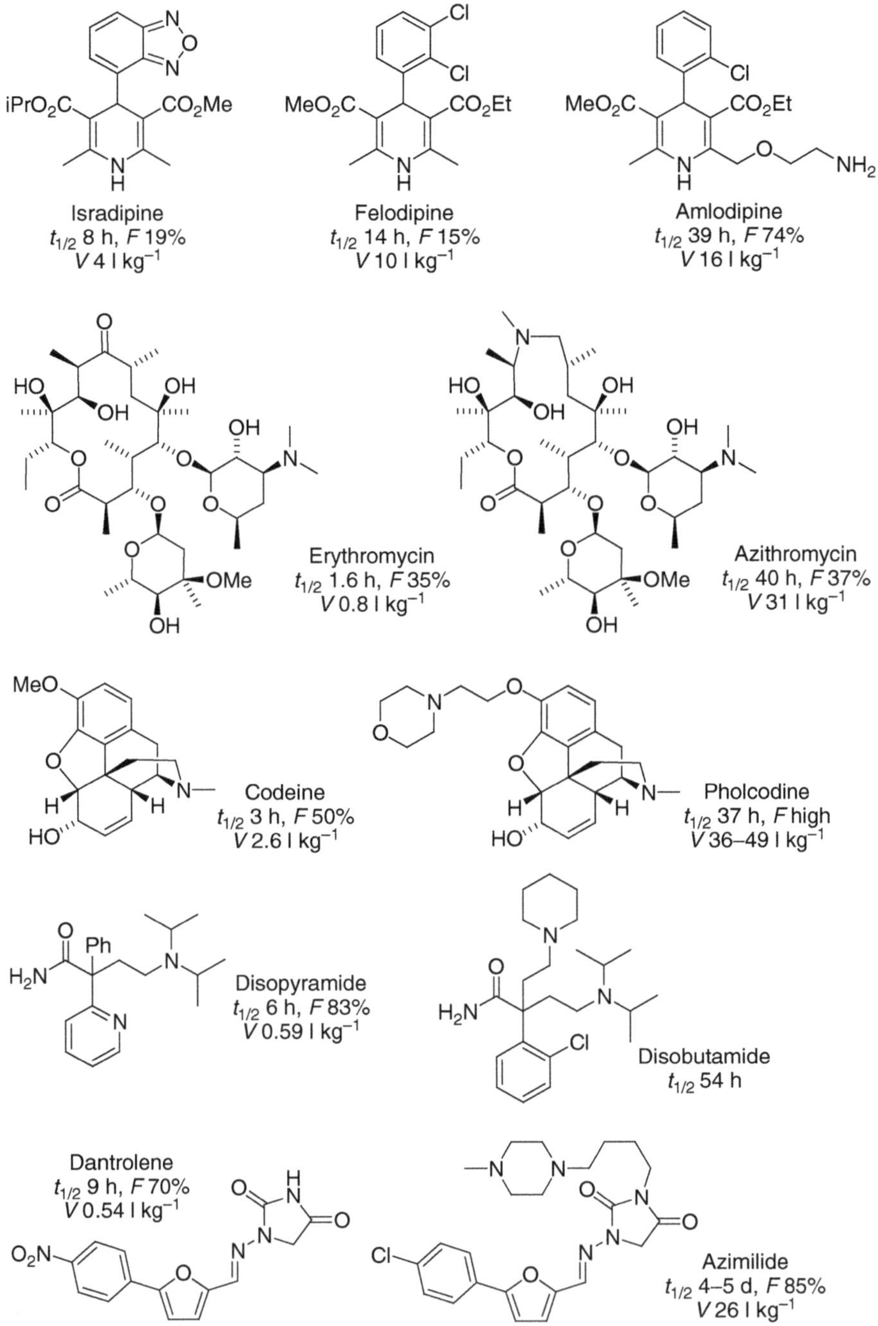

Scheme 23.1 Effect of amines on PK.

Not always does the half-life of a drug increase when introducing an amine functionality. The volume of distribution also depends on the affinity of a drug to plasma proteins. If a modified drug binds stronger to plasma proteins, its volume of distribution and plasma half-life may be lowered if compared to a less protein-bound drug.

Drugs containing two amino groups close to each other (1,2-diaminoethanes, piperazines) are not completely diprotonated in human plasma. The formation of a dication requires a low pH (ethylenediamine: pK_a 10.7, 7.6; piperazine: pK_a 9.8, 5.6), and drugs with these substructures are mainly monoprotonated in plasma and most tissues. Thus, replacement of the dimethylamino group in chlorpromazine by a substituted piperazine does not significantly enhance the volume of distribution (Scheme 23.2).

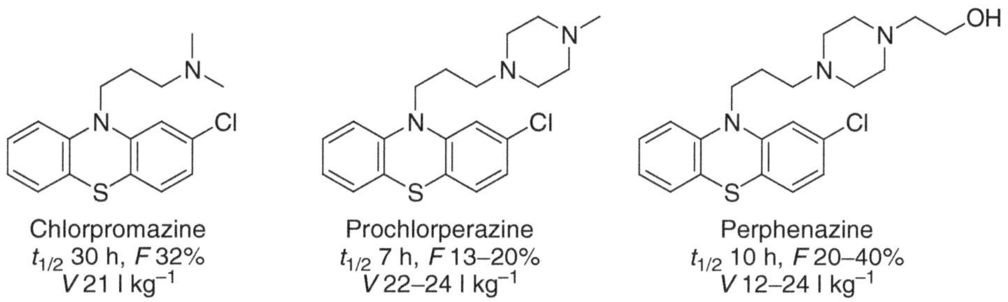

Scheme 23.2 Comparison of monoamines with piperazines.

As illustrated by the examples above, dimethylamino groups can often be replaced by other amine-containing groups, without significant alteration of biological properties. A large number of alternatives for a dimethylamino group have been used with success to fine-tune potency, selectivity, and PK properties. Some of the many possibilities are shown in Scheme 23.3.

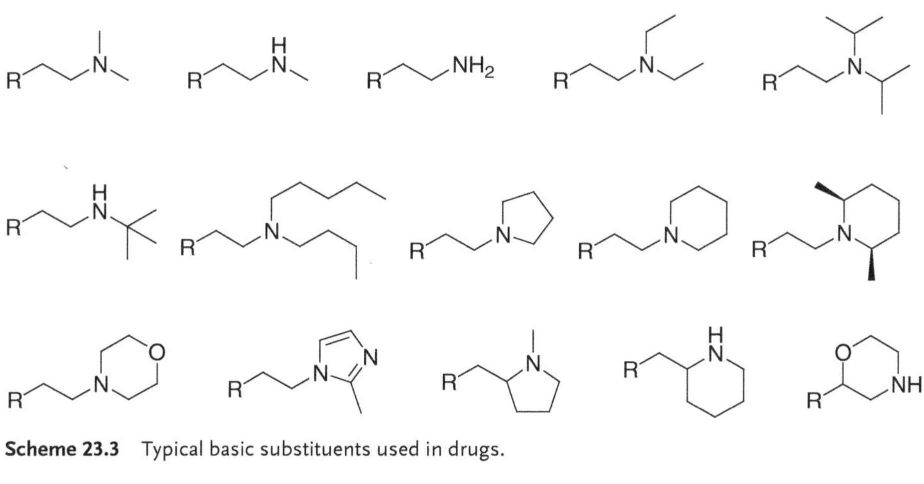

Scheme 23.3 Typical basic substituents used in drugs.

23.1
Basicity

Because only uncharged xenobiotics can cross cell membranes without the aid of a carrier mechanism, the basicity of amines can have an impact on their PK properties. Unfortunately, in many leads, the amino group is a critical pharmacophore and cannot be replaced by an uncharged group without losing affinity to the target protein. It may be possible, however, to slightly lower the basicity of an amine and attain a significant improvement of absorption from the gastrointestinal (GI) tract and blood–brain barrier (bbb) penetration.

The structure of an alkyl substituent may modify the basicity of an amine. Alternatively, a heteroatom or an electron-withdrawing substituent will often decrease the pK_a of an amine. Illustrative examples are given in Scheme 23.4.

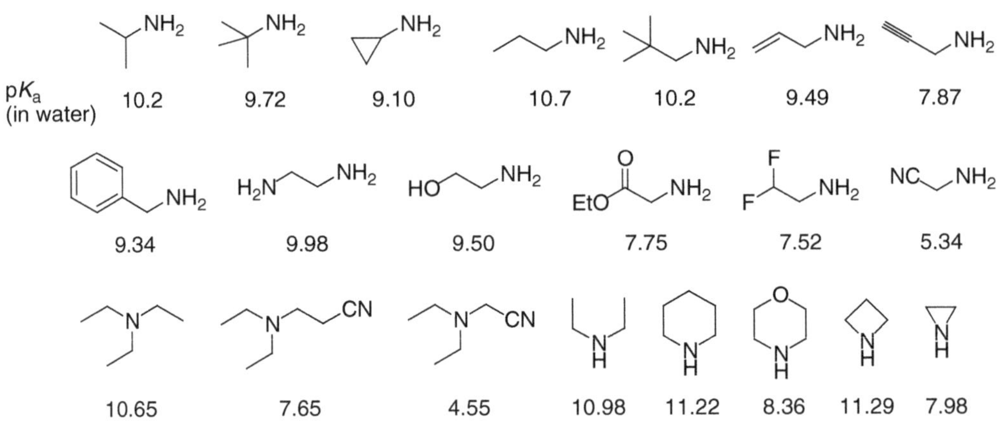

Scheme 23.4 Basicity of amines.

Amines with cyano, sulfonyl, or nitro groups in α- or β-position are not stable and can undergo hydrolysis or β-elimination. The rate of such reactions may, however, be sufficiently low to allow their use in lead optimization.

23.2
Metabolism

Amines are readily oxidized, and typical hepatic transformations include N-dealkylation and N-oxidation. Amines with *tert*-alkyl substituents can be dealkylated by decarboxylation of an intermediate amino acid [1] (Scheme 23.5).

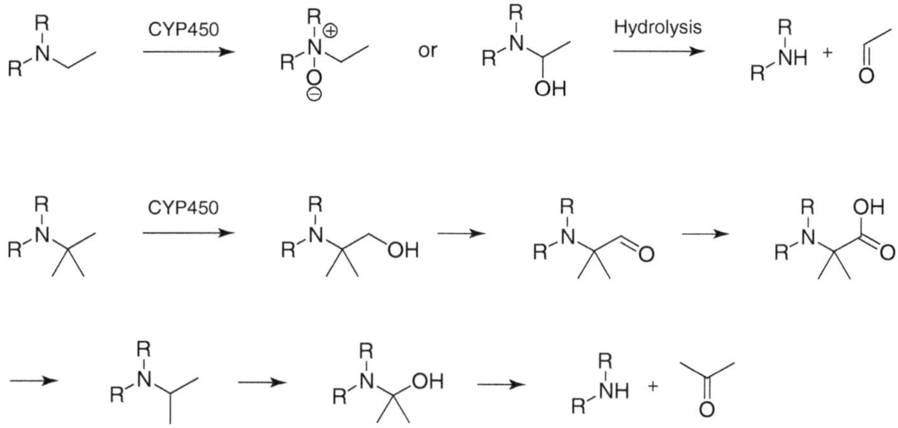

Scheme 23.5 Metabolic N-dealkylation of amines.

The *rate* of metabolic degradation, which ultimately determines the extent of first-pass effect and thus the oral bioavailability and half-life, strongly depends on the overall structure of the drug and cannot be readily predicted. Important factors include the lipophilicity and polar surface area (PSA) of the whole molecule and the basicity of the amino group and its steric accessibility. It is often observed that primary and secondary amines are dealkylated more slowly than similar tertiary amines.

A selection of amines with long plasma half-lives is given in Scheme 23.6.

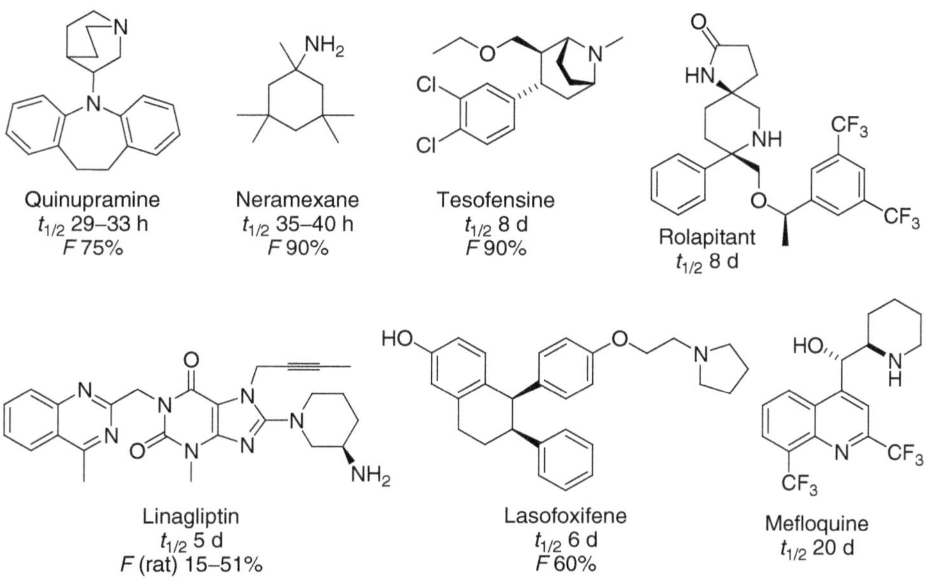

Scheme 23.6 Long-lived amine-containing drugs.

23.3 Rates of N-Dealkylation

The dimethylamino and methylamino groups are present in numerous drugs. Their half-life varies strongly, but in many instances, the half-life increases on demethylation. Thus, secondary methylamines are metabolically more stable than the corresponding tertiary dimethylamines. In addition to the examples shown in Scheme 23.7, this effect has been observed for zimeldine, ruboxistaurine, sertraline, tilidine, and amiodarone (N-deethylation).

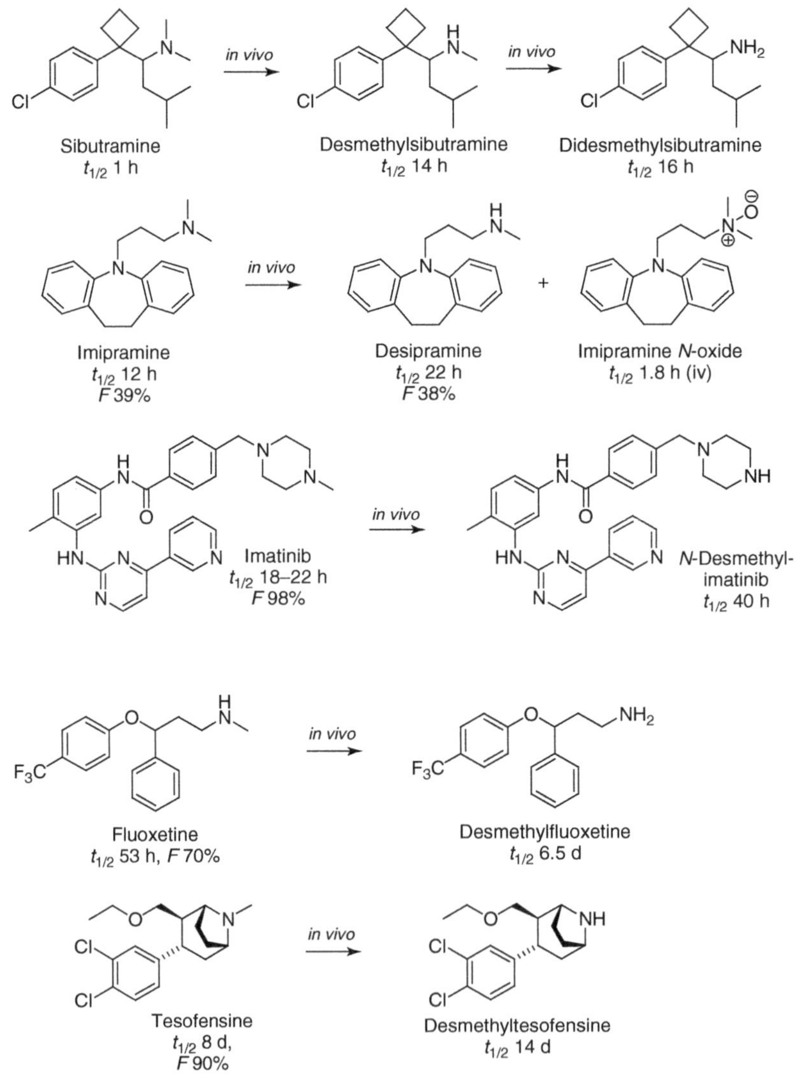

Scheme 23.7 Metabolic N-dealkylation of amines.

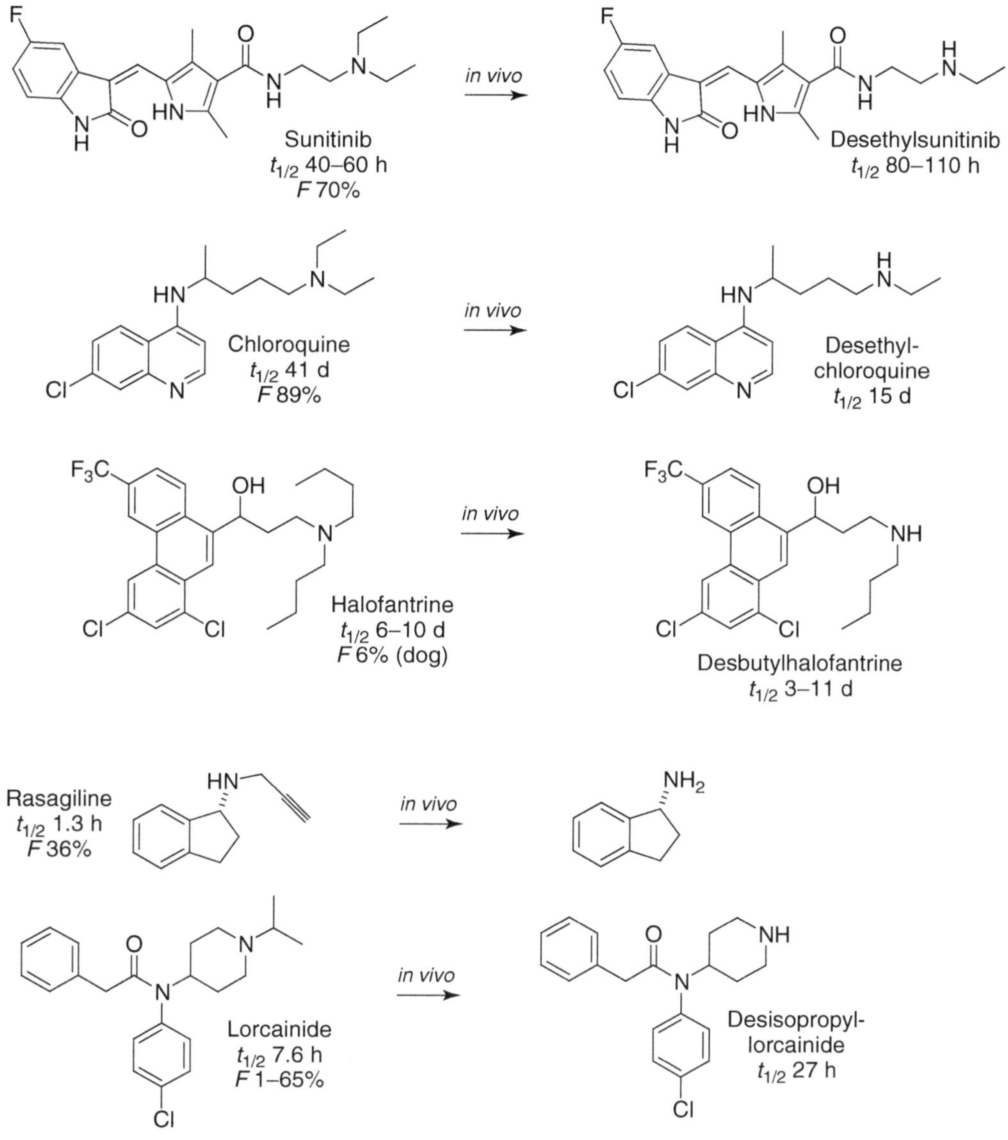

Scheme 23.7 (*Continued*).

Prodrugs are launched occasionally in which the parent drug is formed by metabolic N-dealkylation of an amine. Three such examples are shown in Scheme 23.8. The aim with such prodrugs is to keep the compound inactive during its absorption from the GI tract to avoid unwanted side effects or to enhance its lipophilicity to improve oral bioavailability.

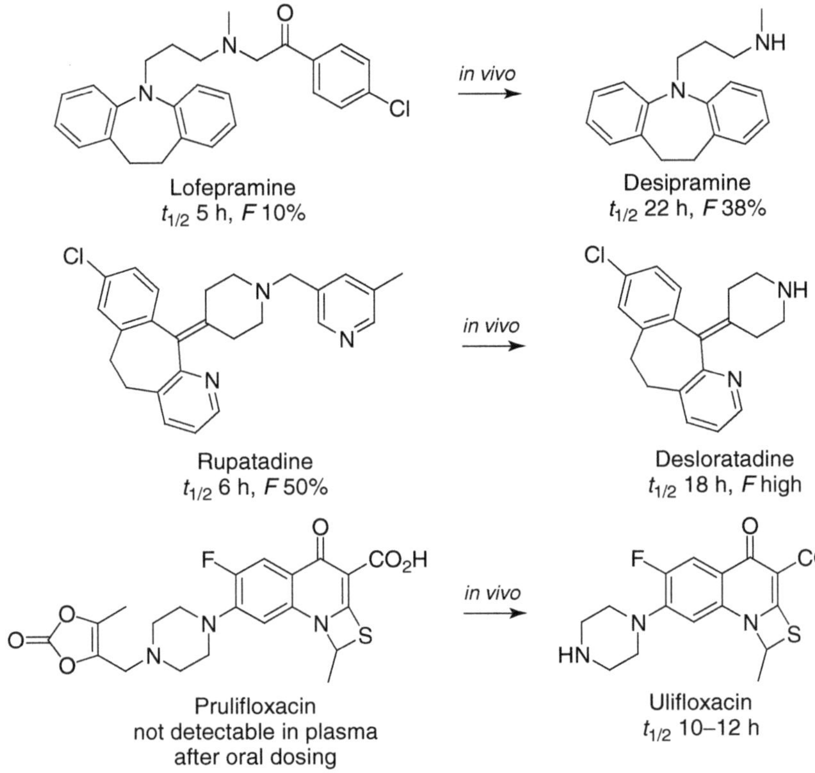

Scheme 23.8 Metabolic N-dealkylation of amine prodrugs.

Highly insoluble primary or secondary amines may be acylated with glycine or short peptides (see also Chapter 33). This can result in a prodrug of higher solubility, which will usually be rapidly deacylated in human plasma.

Reference

1. Kamm, J.J. and Szuna, A. (1973) Studies on an unusual N-dealkylation reaction. II. Characteristics of the enzyme system and a proposed pathway for the reaction. *J. Pharmacol. Exp. Ther.*, **184**, 729–738.

Table 23.1 Aliphatic amines. V in l kg^{-1}; CL in ml min^{-1} kg^{-1}; Mwt in g mol^{-1}.

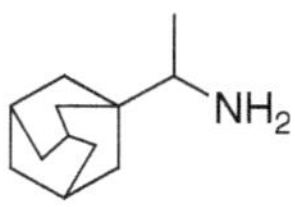	*t*½ 30±14 h *F* 99% (dog) pb 40% ur 9±4%	*V*/*F* 25±14 CL/*F* 10±6 Mwt 179.3 PSA 26.0 Å^2 log *P* 3.31	RIMANTADINE Antiviral Metabolism: hydroxylation of adamantane
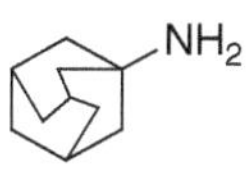	*t*½ 16±3.4 h *F* 70±20% pb 67% ur 70±20%	*V* 6.6±1.5 CL 4.8±0.8 Mwt 151.3 PSA 26 Å^2 log *P* 2.44	AMANTADINE Antiviral, antiparkinsonian No metabolism
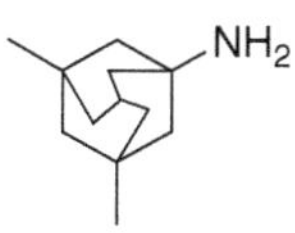	*t*½ 60–100 h *F* 100% pb 45% ur 48%	*V* 10 CL – Mwt 179.3 PSA 26.0 Å^2 log *P* 3.00	MEMANTINE Muscle relaxant, treatment of Alzheimer's disease Metabolites: 4- and 6-hydroxy, dimethylnitrosoadamantane
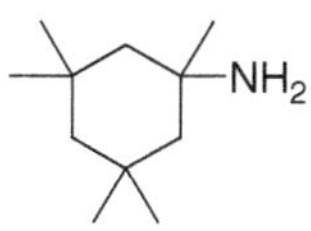	*t*½ 35–40 h *F* 90% pb – ur 30–40%	*V* 11 CL – Mwt 169.3 PSA 26.0 Å^2 log *P* 4.00	NERAMEXANE NMDA antagonist Metabolism: oxidation of CH_3 to CO_2H
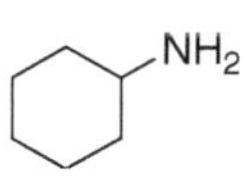	*t*½ 3–5 h *F* – pb – ur –	*V* 2–3 CL – Mwt 99.2 PSA 26.0 Å^2 log *P* 1.50	CYCLOHEXYLAMINE Metabolite of cyclamate (see below) Metabolites: cyclohexanol, *trans*-1,2-dihydroxycyclohexane
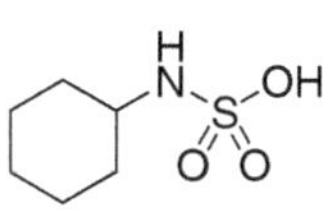	*t*½ 8 h* *F* <37% pb – ur 30–40%	*V* – CL – Mwt 179.2 PSA 74.8 Å^2 log *P* 1.03	CYCLAMATE *rat, dog Artificial sweetener Metabolite: cyclohexylamine
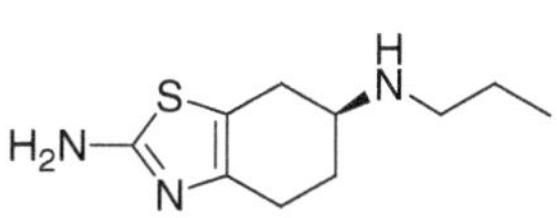	*t*½ 12–14 h *F* 90% pb – ur 90%	*V* 7.0–7.3 CL 0.11 Mwt 211.3 PSA 79.2 Å^2 log *P* 1.42	PRAMIPEXOLE Antiparkinsonian
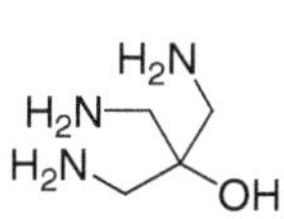	*t*½ 5.6 h *F* – pb – ur 82%	*V* – CL – Mwt 121.1 PSA 86.7 Å^2 log *P* −2.52	THAM For treatment of metabolic acidosis

(*continued overleaf*)

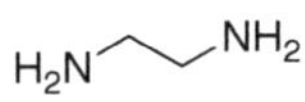	$t^1/_2$ 0.6 h *F* 34% pb – ur 3%	*V* 0.13 CL – Mwt 60.1 PSA 52.0 Å^2 log *P* −2.06	ETHYLENEDIAMINE Hydrochloride as urinary acidifier
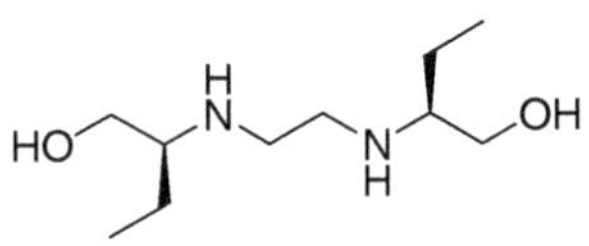	$t^1/_2$ 3.1±0.4 h *F* 77±8% pb <5% ur 79±3%	*V* 1.6±0.2 CL 8.6±0.8 Mwt 204.3 PSA 64.5 Å^2 log *P* −0.29	ETHAMBUTOL Antibacterial Metabolism: oxidation of hydroxymethylene to carboxyl
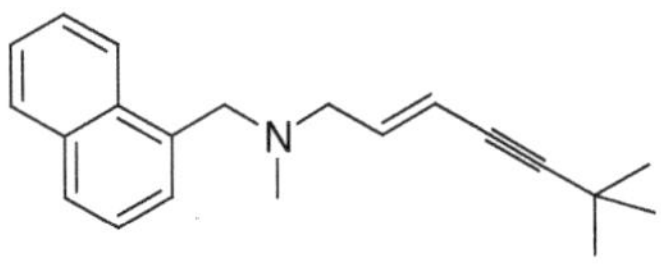	$t^1/_2$ 17–36 h *F* 40% pb >99% ur Negligible	*V* >30 CL – Mwt 291.4 PSA 3.24 Å^2 log *P* 5.58	TERBINAFINE Topical antifungal Metabolism: N-demethylation, hydroxylation of *tert*-butyl, oxidation of *tert*-butyl to carboxyisopropyl, N-de(naphthyl)methylation to naphthaldehyde and naphthoic acid
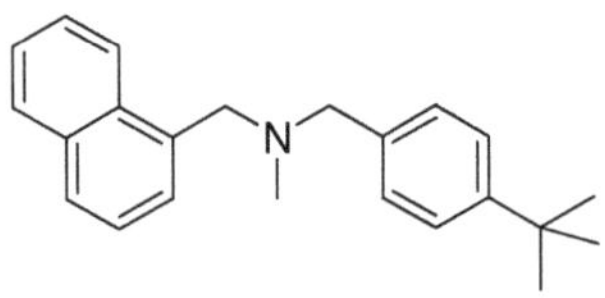	$t^1/_2$ 35–150 h* *F* – pb – ur –	*V* – CL – Mwt 317.5 PSA 3.24 Å^2 log *P* 6.00	BUTENAFINE *after topical administration Topical antifungal Metabolism: hydroxylation of *tert*-butyl
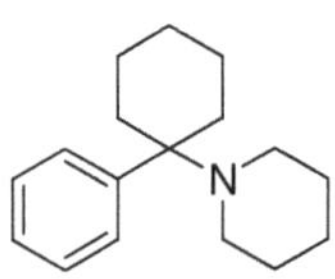	$t^1/_2$ 7–51 h *F* 50–90% pb 60–70% ur 9–10%	*V* 6.2 CL 5.4 Mwt 243.4 PSA 3.24 Å^2 log *P* 4.25	PHENCYCLIDINE Anesthetic, hallucinogenic Metabolism: piperidine 4-hydroxylation, cyclohexane hydroxylation, piperidine ring opening by N-dealkylation
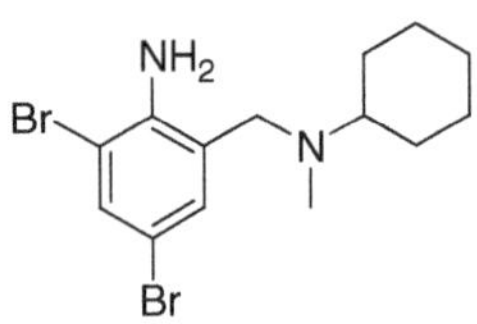	$t^1/_2$ 6–12 h *F* 25% pb 99% ur 0.1%	*V* 5.5–7.0 CL – Mwt 376.1 PSA 29.3 Å^2 log *P* 4.80	BROMHEXINE Expectorant, mucolytic Metabolite: ambroxol (see below)

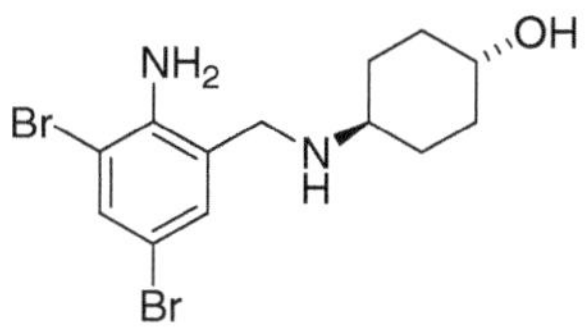	$t_{1/2}$ 7–12 h *F* 70–80% pb 90% ur 5–6%	*V* 1.5 CL 8.7 Mwt 378.1 PSA 58.3 Å^2 log *P* 4.57	AMBROXOL Expectorant, active *N*-desmethyl metabolite of bromhexine Metabolite: dibromoanthranilic acid
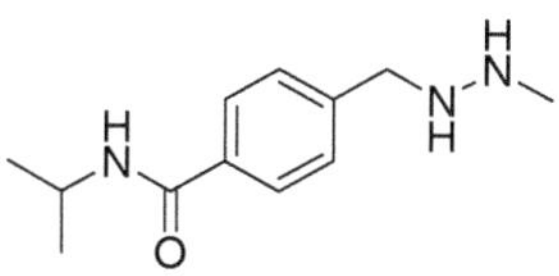	$t_{1/2}$ 7–10 min* *F* 100% pb – ur <5%	*V* – CL – Mwt 221.3 PSA 53.2 Å^2 log *P* 0.06	PROCARBAZINE *iv Antineoplastic; crosses bbb; highly carcinogenic Metabolite: terephthalic *N*-isopropyl monoamide
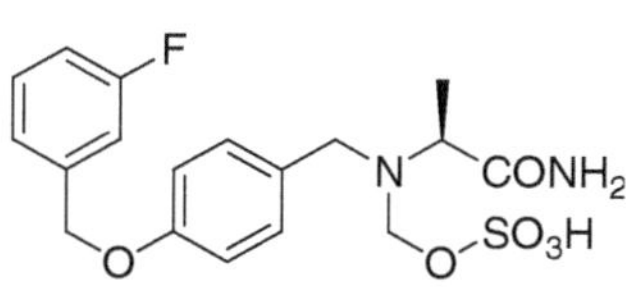	$t_{1/2}$ 21–24 h *F* 80–92% pb 92% ur 7–10%	*V* 2.3 CL – Mwt 302.3 PSA 64.3 Å^2 log *P* 2.20	SAFINAMIDE MAO inhibitor Metabolism: N-dealkylation, hydrolysis of amide, then glucuronidation
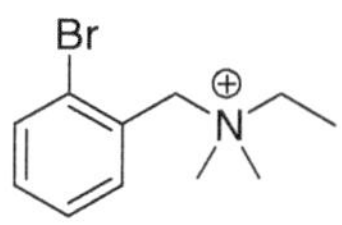	$t_{1/2}$ 8.9±1.8 h *F* 23±9% pb 0–8% ur 77±15%	*V* 5.9±0.8 CL 10±2 Mwt 243.2 PSA 0.0 Å^2 log *P* −0.90	BRETYLIUM Antiadrenergic, antiarrhythmic
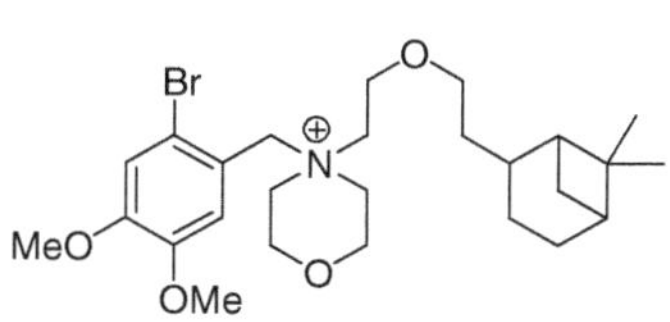	$t_{1/2}$ 1.5 h *F* <0.5% pb 95–98% ur –	*V* – CL – Mwt 511.5 PSA 36.9 Å^2 log *P* 1.25	PINAVERIUM Antispasmodic Metabolism: O-demethylation, hydroxylation at pinanyl, N-debenzylation, then oxidative degradation of morpholine

MAO, monoamine oxidase; NMDA, *N*-methyl-D-aspartic acid; $t_{1/2}$, plasma half-life; *F*, oral bioavailability; pb, plasma protein binding; ur, excretion of unchanged drug in urine; V, volume of distribution; CL, clearance; Mwt, molecular weight; PSA, polar surface area.

24
Quaternary Ammonium Salts

Drugs containing quaternary ammonium groups are mainly designed to interact with the various acetylcholine receptors and with the enzyme acetylcholinesterase. These include muscarinic receptor agonists and antagonists (e.g., quaternized belladonna alkaloids as bronchodilators and antispasmodics), nicotinic receptor agonists and antagonists (neuromuscular blocking agents, ganglionic blocking agents), ganglionic stimulants, and antiacetylcholinesterase agents (cholinergics) [1]. In all these drugs, the quaternary ammonium group is the key pharmacophore. Further, quaternary ammonium salts include the antiarrhythmic bretylium, the opiate antagonist methylnaltrexone, and topical disinfectants, such as benzalkonium (Scheme 24.1).

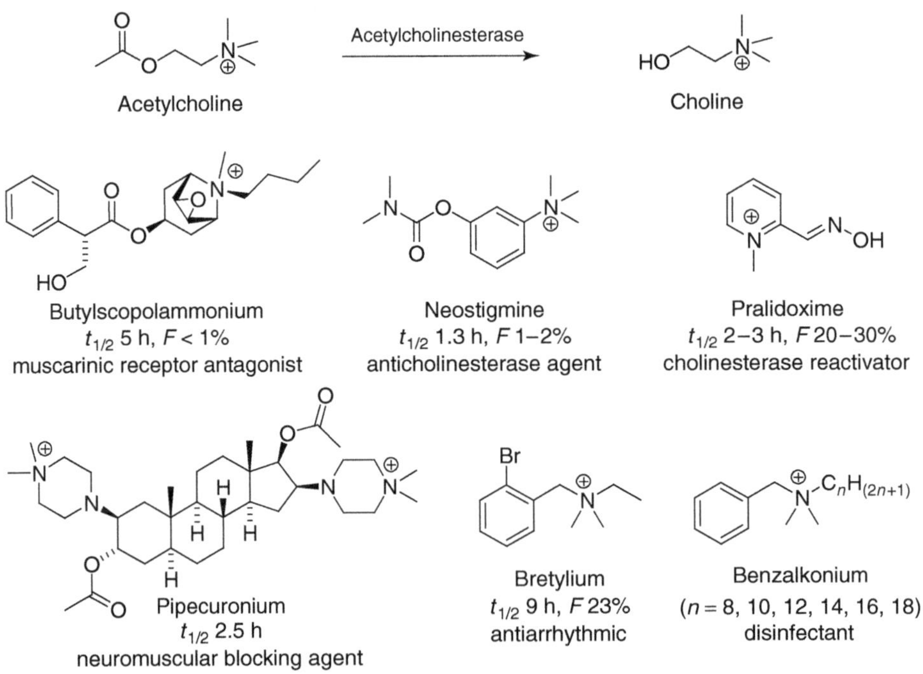

Scheme 24.1 Quaternary ammonium salts as drugs.

Lead Optimization for Medicinal Chemists: Pharmacokinetic Properties of Functional Groups and Organic Compounds, First Edition. Florencio Zaragoza Dörwald.

A common feature of most quaternary ammonium salts is their low oral bioavailability. Because the small intestine behaves essentially as a large, lipophilic membrane, charged molecules cannot be readily absorbed. For this reason, an animal killed by parenteral administration of tubocurarine (a quaternary ammonium salt) can be safely eaten because of the low oral bioavailability of the poison.

Small, charged molecules may nevertheless be absorbed either by means of an active transport mechanism or via small hydrophilic pores. The latter have diameters of 3–8 Å and will enable the passage of either extended, long molecules of up to 400 g mol^{-1} or more spherical molecules of up to 200 g mol^{-1}.

The quaternization of an amino-group-containing drug will usually yield a compound with lower ability to cross the blood–brain barrier (bbb) and lower oral bioavailability. The *in vitro* affinity to the target protein may, however, be similar to that of the tertiary amine. Because of the low oral bioavailability of salts, quaternization is not a common strategy to block central effects of a drug. Examples include the antispasmodic emepronium (Table 46.1) and methylnaltrexone (Table 52.1), a parenteral opiate antagonist designed to prevent some of the side effects of opioids, such as constipation.

The metabolic stability of quaternary ammonium salts depends on the type of substituents, but usually, no metabolic degradation of this functional group is seen *in vivo*. For instance, only 0.06% of a dose of methylnaltrexone is demethylated *in vivo* to naltrexone.

Molecules containing two basic sites (amines or ammonium salts) separated by around 14 Å (~8–12 bonds) may act as neuromuscular blocking agents. Such compounds (e.g., tubocurarine, pipecuronium, atracurium, decamethonium, etc.) interrupt nervous signal transmission and are used as adjuvant muscle relaxants during surgery. High doses of such compounds will, however, cause death by asphyxiation. Before testing diamines in animals, these compounds should, therefore, also be checked for neuromuscular blocking activity.

Reference

1. (a) Sneader, W. (1996) Curare. *Drug News Perspect.*, **9**, 627–630; (b) Sneader, W. (1997) Synthetic muscle relaxants. *Drug News Perspect.*, **10**, 113–117.

25
Amidines

Amidines (pK_a 12) are almost as basic as guanidines and often show poor oral bioavailability and little ability to cross membranes. One exception is low-molecular-weight amidines and imidazolines, which can be highly orally available (Scheme 25.1). Nevertheless, many amidine-containing drugs have to be inhaled or dosed parenterally. These include α-adrenergic agonists and antagonists, antiprotozoals, and antithrombotics (Table 26.1). The benzamidine substructure of the latter is a mimetic of arginine of the natural substrate of the proteases involved in the blood clotting cascade. This amidine is difficult to replace by other functional groups without losing biological activity.

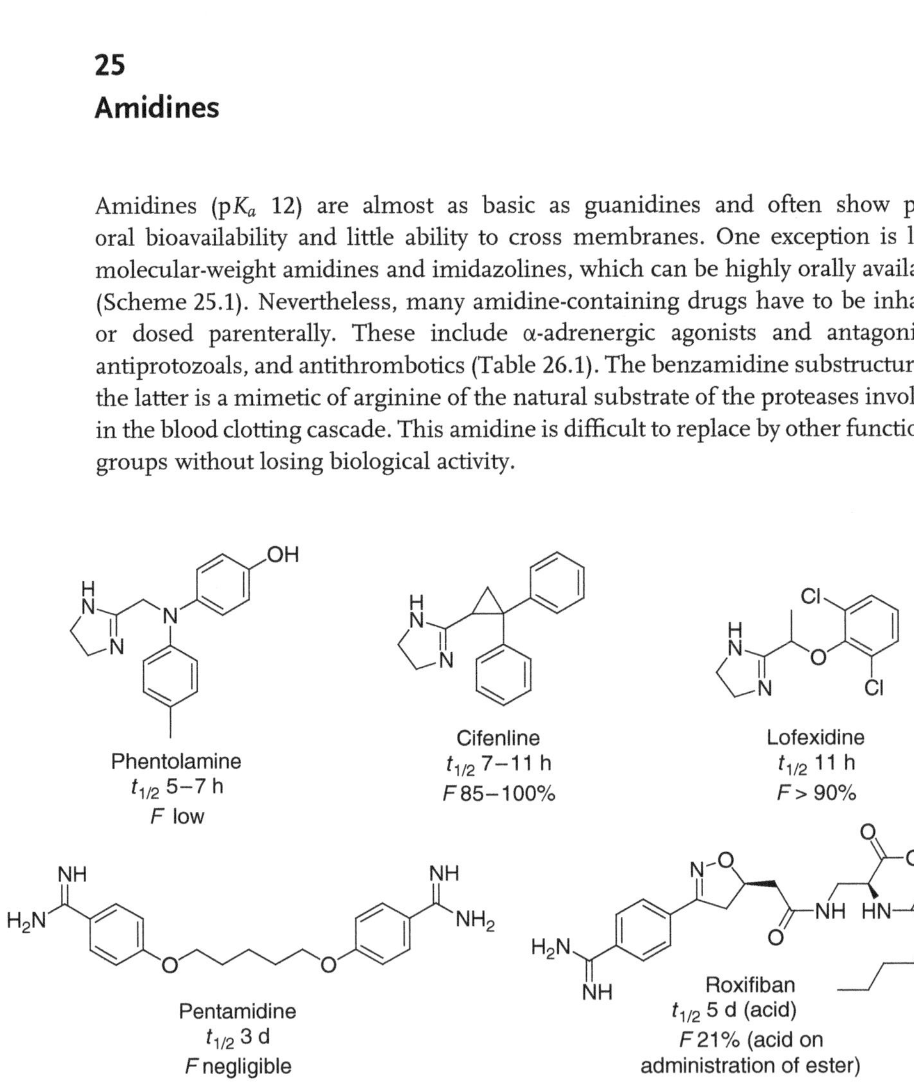

Scheme 25.1 Amidines as drugs.

Because of their poor oral bioavailability, various prodrugs of amidines have been developed (Scheme 25.2). These include *N*-hydroxyamidines, which undergo quick

Lead Optimization for Medicinal Chemists: Pharmacokinetic Properties of Functional Groups and Organic Compounds, First Edition. Florencio Zaragoza Dörwald.

reductive N—O bond cleavage *in vivo*, and carbamates. These amidine precursors are less basic than amidines and show higher oral bioavailability [1]. On oral administration, these prodrugs usually cannot be detected in plasma, and the half-lives given refer to the parent drug.

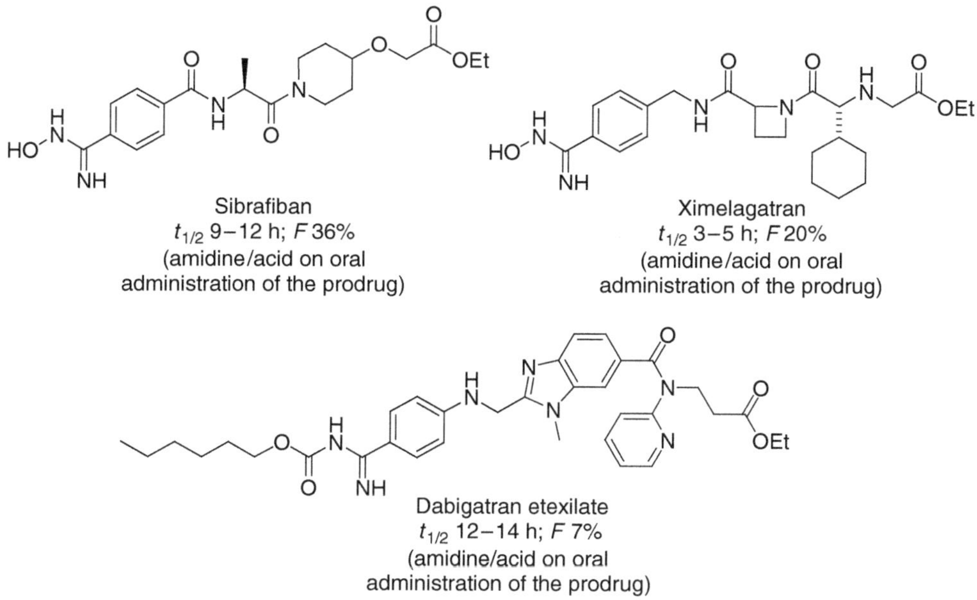

Scheme 25.2 Prodrugs of amidines.

Reference

1. Reeh, C., Wundt, J., and Clement, B. (2007) *N,N'*-Dihydroxyamidines: a new prodrug principle to improve the oral bioavailability of amidines. *J. Med. Chem.*, **50**, 6730–6734.

26
Guanidines, Acylguanidines, and Biguanides

Guanidines are strong bases (pK_a 13) and will be protonated and positively charged in most tissues. Guanidines are highly water soluble, but have poor oral bioavailabilities and cannot efficiently cross membranes or the blood–brain barrier (bbb) (Scheme 26.1). In most guanidine-containing drugs, the guanidine is the key pharmacophore and difficult to replace by other, less basic functional groups. Examples include histamine H_2 antagonists (antiulceratives, e.g., famotidine, ebrotidine), ganglionic blocking agents (antihypertensives, e.g., guanethidine), neuraminidase inhibitors (antivirals, e.g., zanamivir, peramivir), antibacterials (streptomycin), and 5-HT_4 agonist tegaserod (treatment of constipation).

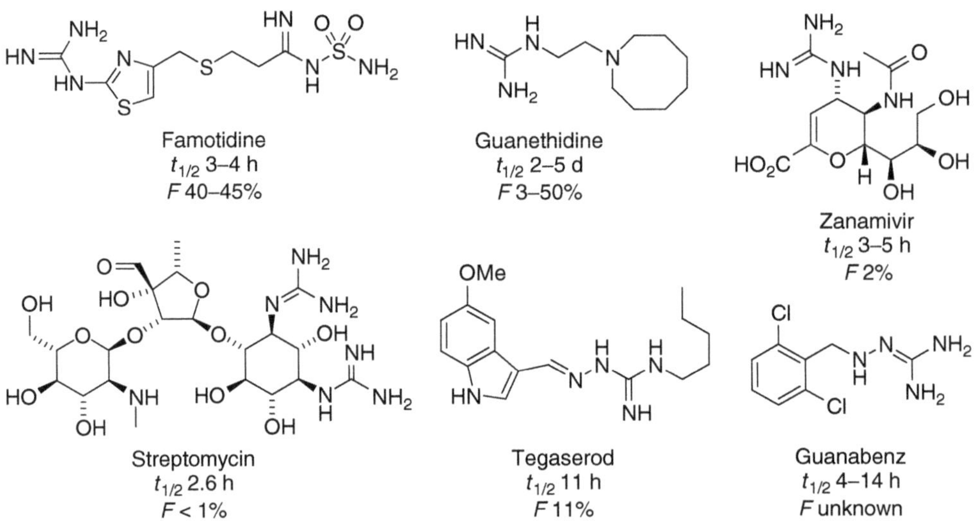

Scheme 26.1 Guanidines as drugs.

Several 2-(arylamino)imidazolines have been developed, mainly as α-adrenergic agonists (Scheme 26.2, Table 26.1). These are slightly less basic and have higher oral bioavailabilities than purely aliphatic guanidines. Some of these compounds can be used as topical, intraocular pressure lowering agents for the treatment of glaucoma [1].

Lead Optimization for Medicinal Chemists: Pharmacokinetic Properties of Functional Groups and Organic Compounds, First Edition. Florencio Zaragoza Dörwald.

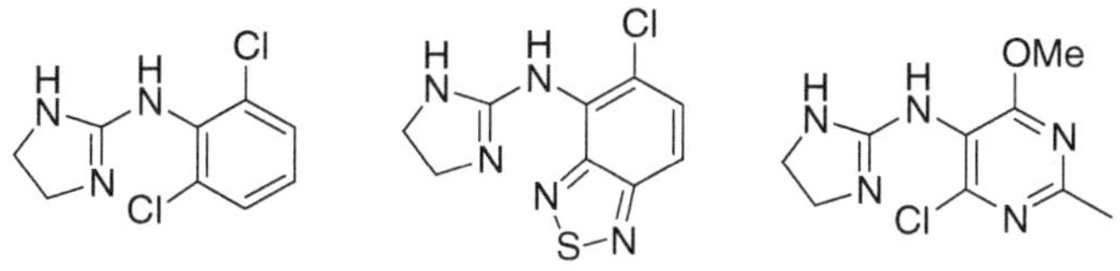

Clonidine $t_{1/2}$ 12 h *F* 75–95%

Tizanidine $t_{1/2}$ 2–4 h *F* 21–40%

Moxonidine $t_{1/2}$ 2–3 h *F* 88%

Scheme 26.2 *N*-arylguanidines as drugs.

26.1 Acylguanidines

Acylated guanidines (pK_a 8–9) are less basic than alkyl- or arylguanidines or aliphatic amines and are readily absorbed from the gastrointestinal (GI) tract. Some, for example, guanfacine (Scheme 26.3), even cross the bbb, despite their large polar surface area (PSA). Most of these compounds are sodium channel blockers and are being used as diuretics, antihypertensives, or cardioprotectants.

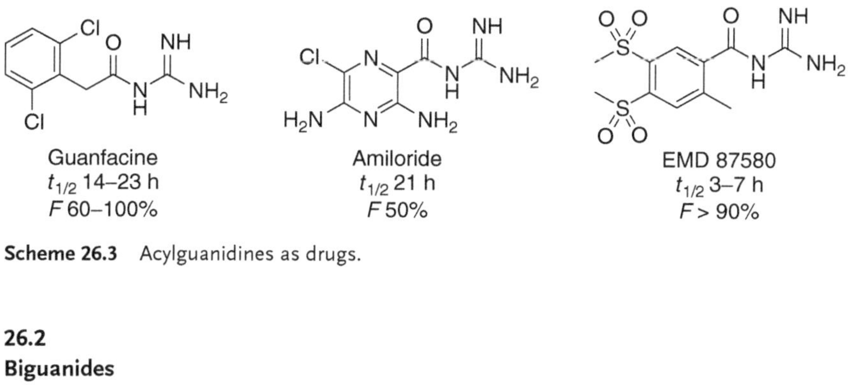

Guanfacine $t_{1/2}$ 14–23 h *F* 60–100%

Amiloride $t_{1/2}$ 21 h *F* 50%

EMD 87580 $t_{1/2}$ 3–7 h *F* > 90%

Scheme 26.3 Acylguanidines as drugs.

26.2 Biguanides

Substituted biguanides are being used as antidiabetics, antiprotozoals, and disinfectants (Scheme 26.4). The antidiabetic effect of biguanides is caused by insulin sensitization of peripheral tissue and by inhibition of hepatic gluconeogenesis. Only biguanides of low molecular weight (<200 g mol^{-1}) are orally bioavailable; larger molecules require either parenteral or topical administration or will, on oral administration, only be active in the GI tract.

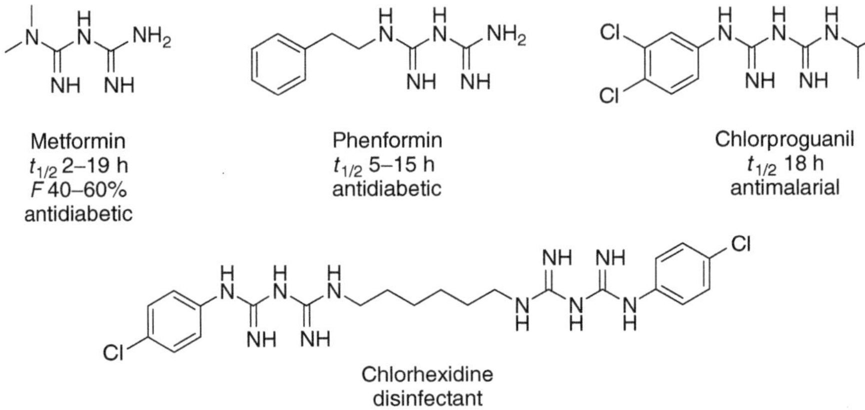

Scheme 26.4 Biguanides as drugs.

Reference

1. Sugrue, M.F. (1997) New approaches to antiglaucoma therapy. *J. Med. Chem.*, **40**, 2793–2809.

Table 26.1 Amidines, guanidines, and related compounds. V in l kg^{-1}; CL in ml min^{-1} kg^{-1}; Mwt in g mol^{-1}.

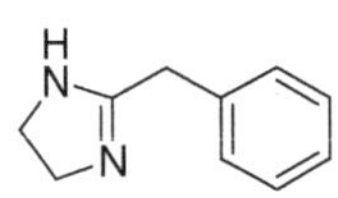	$t½$ F pb ur	3–10 h 90–100% – –	V CL Mwt PSA log P	1.6 – 160.2 24.4 Å^2 1.80	TOLAZOLINE α-Adrenergic antagonist, treatment of pulmonary hypertension in newborns
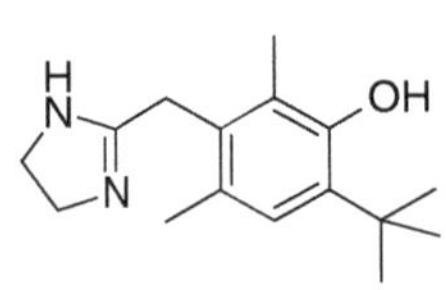	$t½$ F pb ur	35 h* – – –	V CL Mwt PSA log P	– – 260.4 44.6 Å^2 3.76	OXYMETAZOLINE *intranasal α-Adrenergic agonist, nasal decongestant; too toxic for other than topical administration
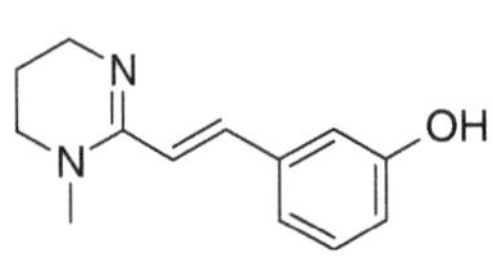	$t½$ F pb ur	– 8–10% – –	V CL Mwt PSA log P	– – 216.3 35.8 Å^2 1.87	OXANTEL Anthelmintic against gut parasites
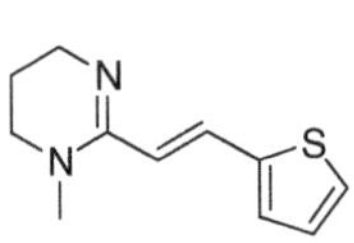	$t½$ F pb ur	– Low – <7%	V CL Mwt PSA log P	– – 206.3 43.8 Å^2 2.08	PYRANTEL Anthelmintic against gut parasites

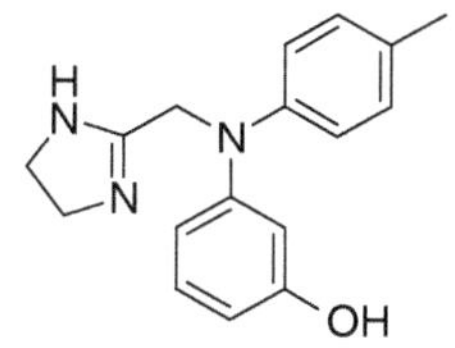

$t_{1/2}$	5–7 h	*V*	–
F	Low	CL	–
pb	54%	Mwt	281.4
ur	3–10%	PSA	47.9 Å^2
		log *P*	4.08

PHENTOLAMINE
α-Adrenergic antagonist, antihypertensive
Metabolism: oxidation of CH_3 to CO_2H

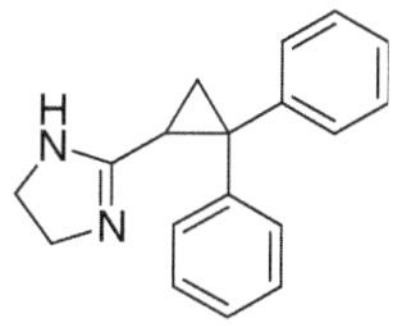

$t_{1/2}$	7–11 h	*V*	4–7
F	85–100%	CL	9–15
pb	50–60%	Mwt	262.4
ur	50–60%	PSA	24.4 Å^2
		log *P*	3.49

CIFENLINE, CIBENZOLINE
Antiarrhythmic
Metabolism: 4-hydroxylation of phenyl, dehydrogenation of imidazolidine

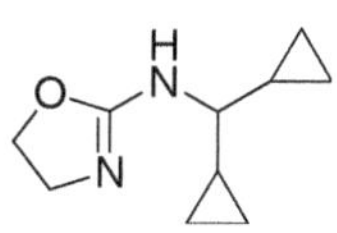

$t_{1/2}$	8–9 h	*V*	4–5
F	100%	CL	6.6
pb	<10%	Mwt	180.3
ur	64%	PSA	33.6 Å^2
		log *P*	1.48

RILMENIDINE
α_2-Adrenergic agonist, antihypertensive, little metabolism

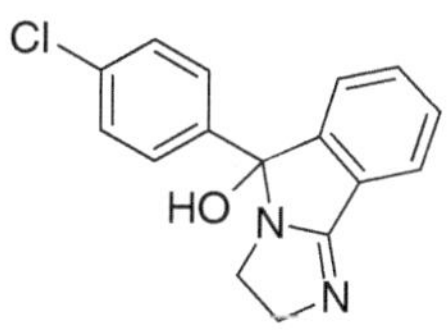

$t_{1/2}$	12–24 h	*V*	–
F	93%	CL	–
pb	–	Mwt	284.7
ur	–	PSA	35.8 Å^2
		log *P*	2.36

MAZINDOL
CNS stimulant, anorexic
Metabolism: oxidation of imidazolidine to imidazolidinone

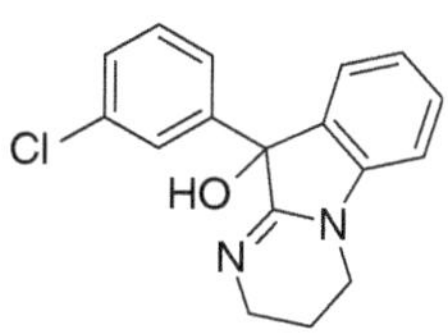

$t_{1/2}$	32 h	*V*	–
F	High	CL	–
pb	–	Mwt	298.8
ur	20%	PSA	35.8 Å^2
		log *P*	1.18

CICLAZINDOL
Antidepressant; Metabolism: hydrolysis of amidine to *N*-(3-aminopropyl)indolone, oxidative N-dealkylation to 3-aminopropionic acids

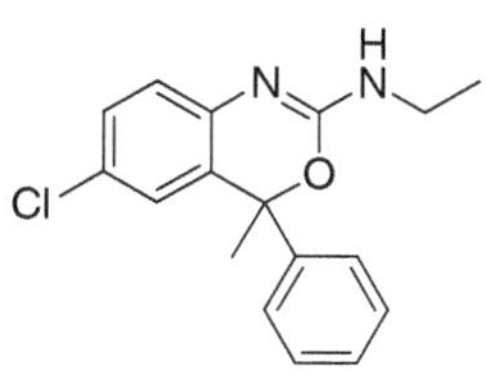

$t_{1/2}$	2–6 h	*V*	–
F	90%	CL	–
pb	–	Mwt	300.8
ur	–	PSA	33.6 Å^2
		log *P*	4.23

ETIFOXINE, ETAFENOXINE
Anxiolytic
Metabolism: N-deethylation, phenyl 4-hydroxylation

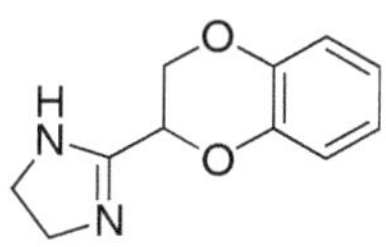

$t_{1/2}$	4.2 h (iv)	*V*	3.3
F	1–23%*	CL	10
pb	–	Mwt	204.2
ur	–	PSA	42.9 Å^2
		log *P*	2.31

IDAZOXAN
*rat
α_2-Adrenergic antagonist
Metabolism (rat): aromatic hydroxylation

(*continued overleaf*)

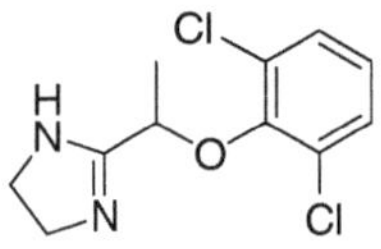

$t_{1/2}$	11 h	V	–	LOFEXIDINE
F	>90%	CL	–	Antihypertensive
pb	80–90%	Mwt	259.1	Main metabolite: 2,6-dichlorophenol
ur	5–20%	PSA	33.6 Å^2	
		log P	3.04	

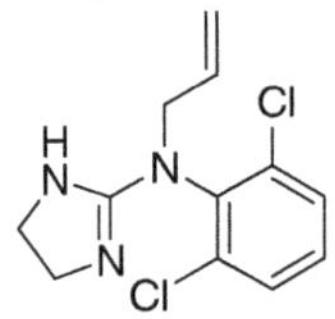

$t_{1/2}$	3–4 h	V	2.9	ALINIDINE
F	100%	CL	10	Antihypertensive
pb	–	Mwt	270.2	
ur	79–90%	PSA	27.6 Å^2	
		log P	3.49	

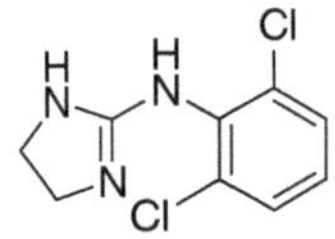

$t_{1/2}$	12±7 h	V	2.1±0.4	CLONIDINE
F	75–95%	CL	3.1±1.2	α_2-Adrenergic agonist,
pb	20–30%	Mwt	230.1	antihypertensive, analgesic, crosses
ur	62±11%	PSA	36.4 Å^2	bbb; Metabolism: aromatic
		log P	2.36	4-hydroxylation

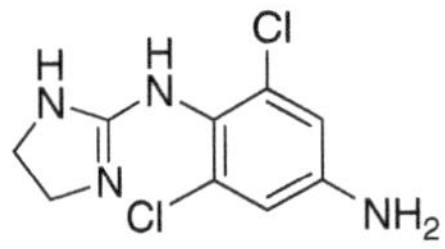

$t_{1/2}$	8 h	V	–	APRACLONIDINE
F	–	CL	–	α_2-Adrenergic agonist to reduce
pb	–	Mwt	245.1	postoperative intraocular pressure
ur	–	PSA	62.4 Å^2	(topical application)
		log P	1.68	

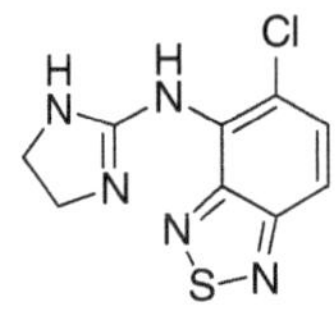

$t_{1/2}$	2.1–4.2 h	V	2.4	TIZANIDINE
F	21–40%	CL	11	α_2-Adrenergic agonist, muscle
pb	30%	Mwt	253.7	relaxant
ur	<3%	PSA	90.4 Å^2	Metabolism: oxidation
		log P	2.39	of imidazoline, N-dealkylation

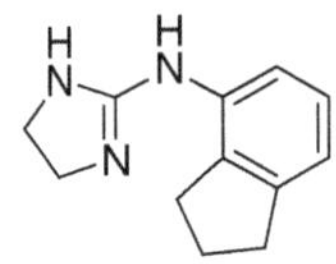

$t_{1/2}$	5–7 h	V	–	INDANAZOLINE
F	–	CL	–	α-Adrenergic agonist,
pb	32%	Mwt	201.3	vasoconstrictor, nasal decongestant
ur	–	PSA	36.4 Å^2	
		log P	2.26	

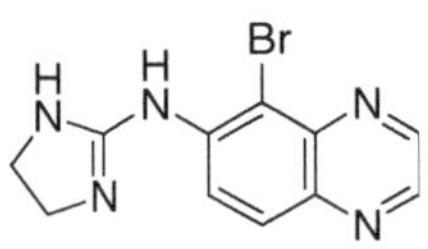

$t_{1/2}$	3 h	V	–	BRIMONIDINE
F	–	CL	–	α_2-Adrenergic agonist,
pb	–	Mwt	292.1	antiglaucoma; Metabolism:
ur	–	PSA	62.2 Å^2	dehydrogenation of imidazolidine,
		log P	1.20	further oxidation to imidazolones

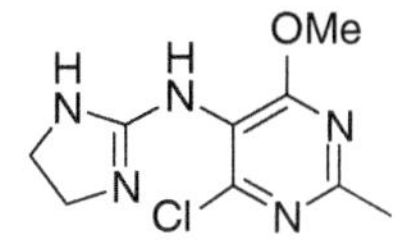	$t_{1/2}$ 1.5–2.5 h	V	1.8–3.0	MOXONIDINE
	F 88%	CL	11–15	Antihypertensive
	pb 6–8%	Mwt	241.7	Metabolism: hydroxylation at CH_3
	ur 46–70%	PSA	71.4 Å^2	and imidazolidine
		log P	0.33	
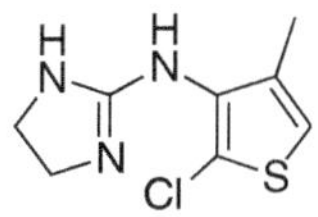	$t_{1/2}$ 3–5 h	V	–	TIAMENIDINE
	F –	CL	–	Antihypertensive
	pb –	Mwt	215.7	
	ur –	PSA	64.7 Å^2	
		log P	1.87	
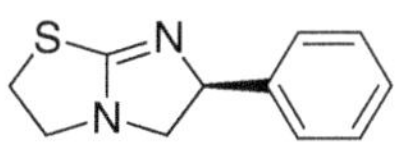	$t_{1/2}$ 4–6 h	V	3–4*	LEVAMISOLE
	F >70%	CL	39–49*	*rabbit
	pb 51%	Mwt	204.3	Anthelmintic
	ur 2–5%	PSA	40.9 Å^2	Metabolite: 4-hydroxyphenyl
		log P	1.84	
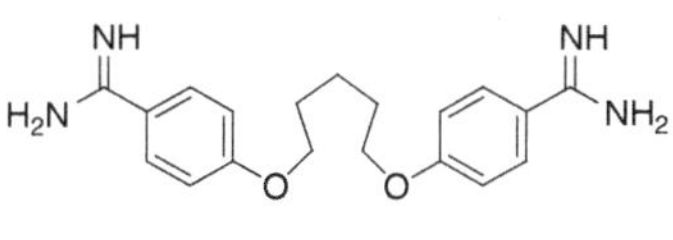	$t_{1/2}$ 3 d	V	190±70	PENTAMIDINE
	F Negligible	CL	18±7	Antiprotozoal
	pb 69%	Mwt	340.4	Metabolism (rat): N-hydroxylation,
	ur 2–18%	PSA	118 Å^2	hydroxylation of alkylene chain,
		log P	2.85	O-dealkylation
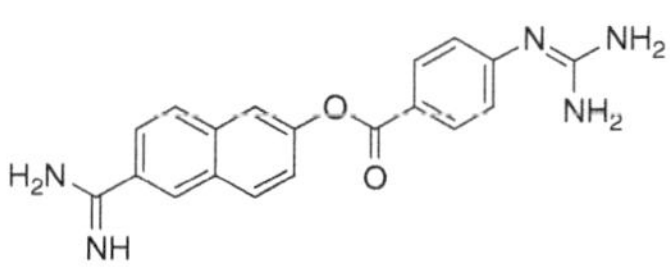	$t_{1/2}$ 23 min	V	–	NAFAMOSTAT
	F Low	CL	–	Serine protease inhibitor,
	pb –	Mwt	347.4	anticoagulant
	ur –	PSA	138 Å^2	Metabolism: ester hydrolysis
		log P	2.06	
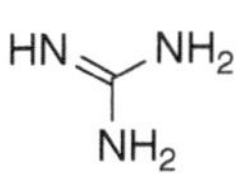	$t_{1/2}$ 7–12 h	V	–	GUANIDINE
	F –	CL	–	Antiviral, antifungal, antipyretic
	pb –	Mwt	59.1	
	ur –	PSA	75.9 Å^2	
		log P	−1.25	
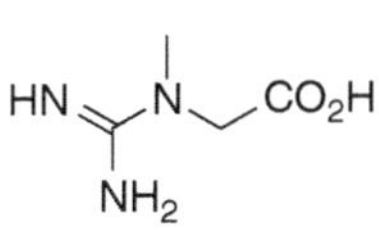	$t_{1/2}$ 6–8 wk	V	–	CREATINE
	F –	CL	3.3	
	pb –	Mwt	131.1	
	ur –	PSA	90.4 Å^2	
		log P	−1.26	
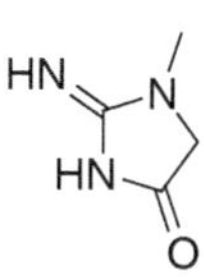	$t_{1/2}$ 4 h	V	–	CREATININE
	F –	CL	3.3	
	pb –	Mwt	113.1	
	ur –	PSA	58.7 Å^2	
		log P	−0.80	

(continued overleaf)

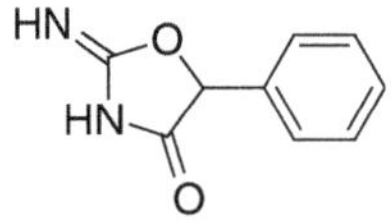

$t^1/_2$	10–18 h	*V*	–	PEMOLINE
F	–	CL	–	CNS stimulant; withdrawn in 2005
pb	30–50%	Mwt	176.2	because of hepatotoxicity
ur	50%	PSA	64.7 Å^2	
		log *P*	0.43	

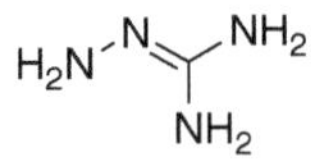

$t^1/_2$	4 h	*V*	–	AMINOGUANIDINE
F	–	CL	–	Nitric oxide synthase inhibitor,
pb	–	Mwt	74.1	antidiabetic
ur	–	PSA	87.9 Å^2	
		log *P*	−1.48	

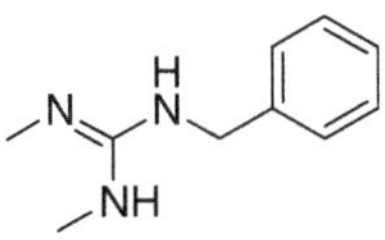

$t^1/_2$	4–9 h	*V*	2.8	BETHANIDINE
F	–	CL	–	Antihypertensive; Metabolism (rat,
pb	–	Mwt	177.3	dog): N-demethylation,
ur	High	PSA	36.4 Å^2	N-debenzylation, N-hydroxylation;
		log *P*	1.73	no significant metabolism in man

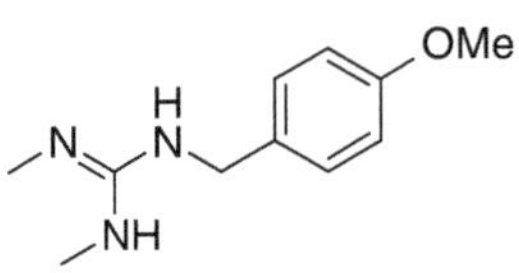

$t^1/_2$	9–20 h	*V*	–	MEOBENTINE
F	–	CL	–	Antiarrhythmic
pb	–	Mwt	207.3	Metabolism: N- and
ur	–	PSA	45.7 Å^2	O-demethylation
		log *P*	1.62	

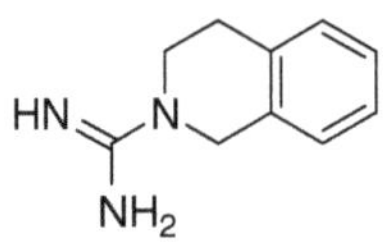

$t^1/_2$	16–30 h	*V*	–	DEBRISOQUIN
F	–	CL	–	Antihypertensive; Metabolism:
pb	–	Mwt	175.2	benzylic 4-hydroxylation, then
ur	–	PSA	53.1 Å^2	elimination of water, aromatic
		log *P*	0.75	hydroxylation, glucuronidation

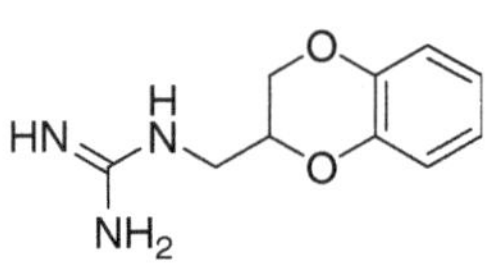

$t^1/_2$	–	*V*	–	GUANOXAN
F	>50%	CL	–	Antihypertensive
pb	–	Mwt	207.2	Metabolism: aromatic 6- and
ur	1–60%	PSA	80.4 Å^2	7-hydroxylation
		log *P*	0.84	

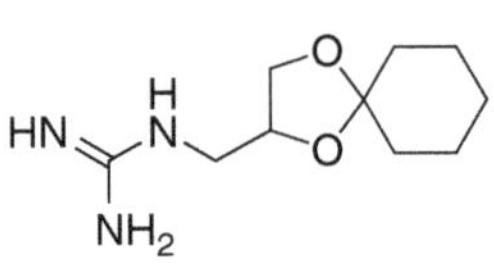

$t^1/_2$	10–12 h	*V*	10	GUANADREL
F	85%	CL	41	Antihypertensive, does not
pb	20%	Mwt	213.3	cross bbb
ur	40–50%	PSA	80.4 Å^2	Metabolism: hydrolysis
		log *P*	0.55	of acetal

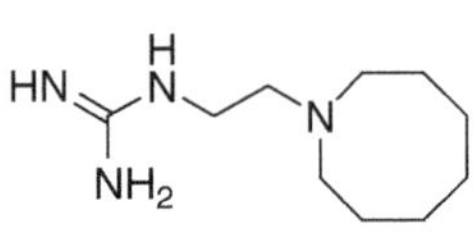

$t^1/_2$	2–5 d	*V*	–	GUANETHIDINE
F	3–50%	CL	–	Antihypertensive
pb	0%	Mwt	198.3	Metabolism: oxidation to *N*-oxide
ur	25–50%	PSA	65.1 Å^2	
		log *P*	1.10	

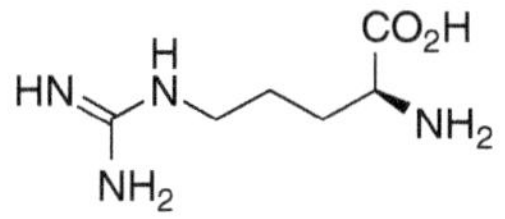	$t^1/_2$	1.3 h	V	0.3	L-ARGININE
	F	68±9%	CL	15±4	Insulin secretagogue
	pb	–	Mwt	174.2	
	ur	–	PSA	125 Å^2	
			log P	−1.65	
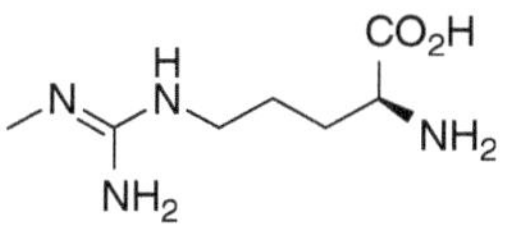	$t^1/_2$	1.2–2.3 h	V	0.6–0.8	L-*N*-METHYLARGININE
	F	–	CL	8.0–9.5	Endogenous nitric oxide synthase inhibitor
	pb	Negligible	Mwt	188.2	
	ur	<1%	PSA	111 Å^2	Metabolism: hydrolysis of guanidine to urea (citrulline)
			log P	−0.08	
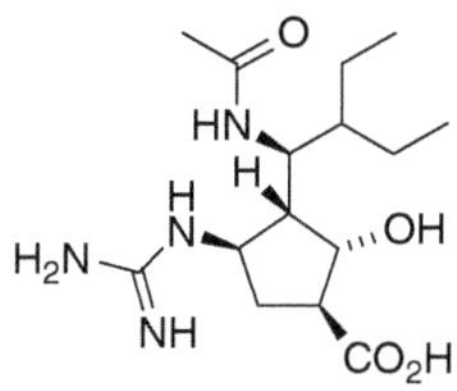	$t^1/_2$	8–21 h (iv)	V	0.5	PERAMIVIR
	F	Low	CL	0.9–1.6	Neuraminidase inhibitor, antiviral
	pb	<30%	Mwt	328.4	
	ur	90%	PSA	149 Å^2	
			log P	−1.37	
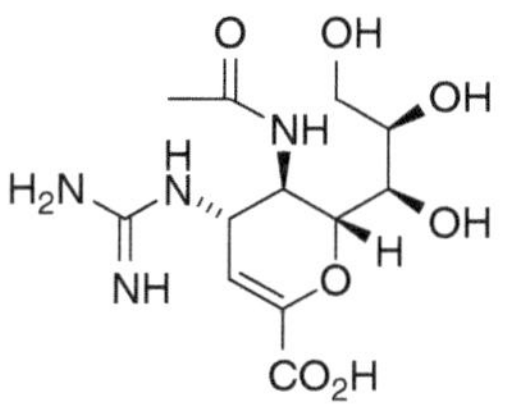	$t^1/_2$	2.6–5.1 h	V	0.3	ZANAMIVIR
	F	2%*	CL	0.6–2.8	*inhalation: 10–20%
	pb	<10%	Mwt	332.3	Neuraminidase inhibitor, antiviral
	ur	80%	PSA	198 Å^2	
			log P	−4.13	
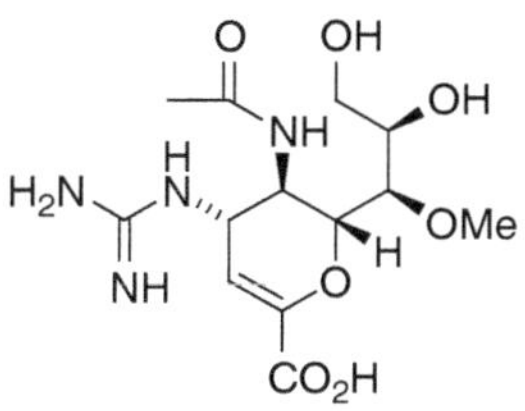	$t^1/_2$	3 d	V	–	LANINAMIVIR
	F	–	CL	–	Methylated zanamivir, antiviral (influenza)
	pb	–	Mwt	346.3	
	ur	–	PSA	187 Å^2	
			log P	-3.06	
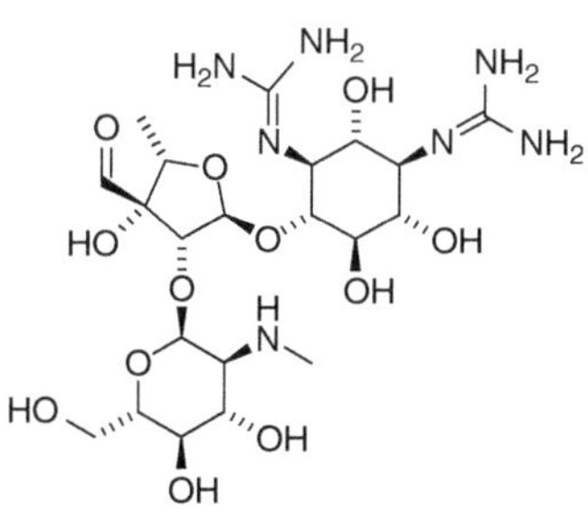	$t^1/_2$	2.6±0.4 h	V	0.25±0.02	STREPTOMYCIN
	F	–	CL	1.2±0.3	Antibacterial
	pb	48±14%	Mwt	581.6	
	ur	55±5%	PSA	331 Å^2	
			log P	−1.22	

(*continued overleaf*)

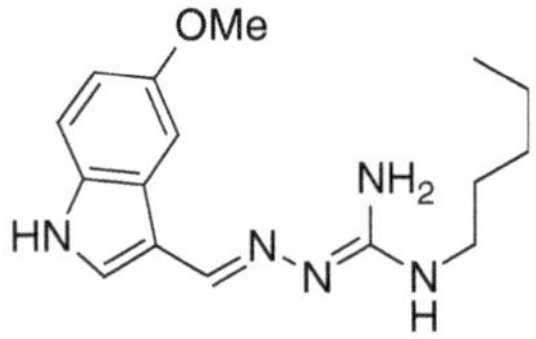	$t_{1/2}$ F pb ur	11±5 h 11% 98% –	V CL Mwt PSA log P	5.2 18.3 301.4 85.3 Å^2 2.91	TEGASEROD 5-HT_4 agonist, treatment of constipation Metabolism: hydrolysis of hydrazone, oxidation to indole-3-carboxylic acid; withdrawn in 2007 because of stroke and heart attack risk
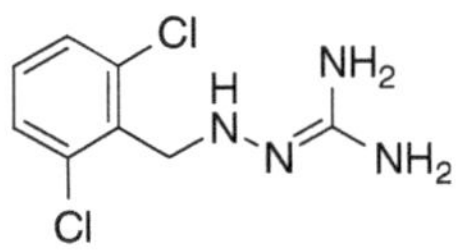	$t_{1/2}$ F pb ur	4–14 h – 90% <1%	V CL Mwt PSA log P	4–17 100–180 231.1 74.3 Å^2 2.66	GUANABENZ α_2-Adrenergic agonist, antihypertensive; Metabolites: 4-hydroxy, 2,6-dichlorobenzylalcohol, 2,6-dichlorobenzaldehyde
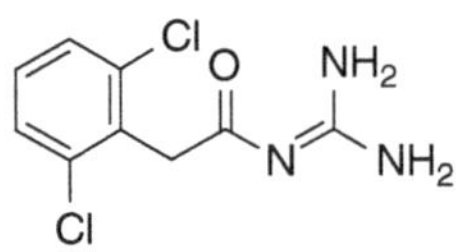	$t_{1/2}$ F pb ur	14–23 h 60–100% 64% 24–37%	V CL Mwt PSA log P	3.9–6.5 2.6–5.2 246.1 79.0 Å^2 1.33	GUANFACINE α_2-Adrenergic agonist, antihypertensive; Metabolism: aromatic 3-hydroxylation, then glucuronidation, sulfation
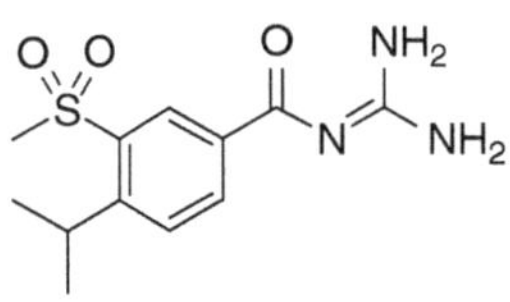	$t_{1/2}$ F pb ur	3.5 h (iv) – – –	V CL Mwt PSA log P	– – 283.4 121 Å^2 −1.01	CARIPORIDE Cardioprotective, sodium/proton exchange inhibitor (NHE-1)
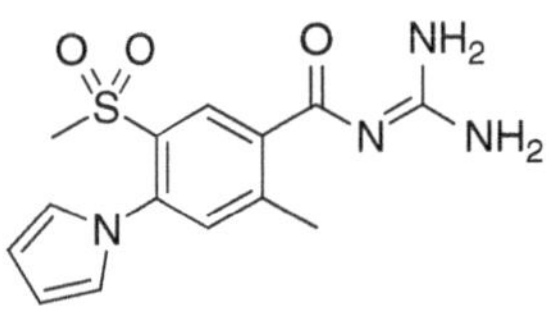	$t_{1/2}$ F pb ur	2.2 h* – 72% 43%*	V CL Mwt PSA log P	1.1–1.5 8.2 320.4 126 Å^2 −1.03	ENIPORIDE *iv Cardioprotective, sodium/proton exchange inhibitor (NHE-1) Metabolism: hydrolysis of pyrrole to 4-aminobenzamide
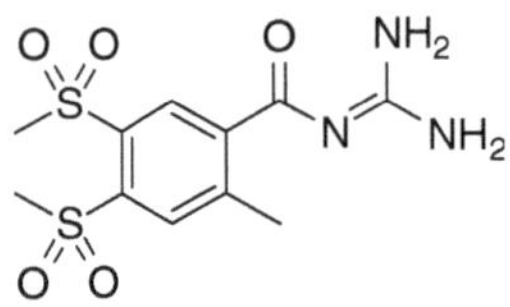	$t_{1/2}$ F pb ur	3–7 h >90% 27% 90%	V CL Mwt PSA log P	0.8–2.1 2.3–5.7 333.4 164 Å^2 −1.66	EMD 87580 Cardioprotective, sodium/proton exchange inhibitor (NHE-1)
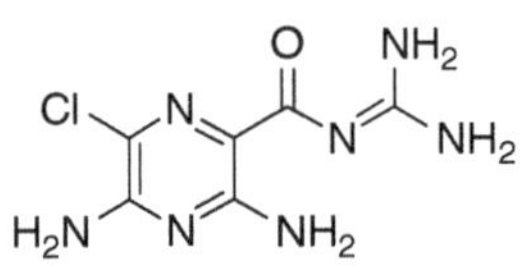	$t_{1/2}$ F pb ur	21±3 h 50% 40% 49±10%	V CL Mwt PSA log P	17±4 10±2 229.6 157 Å^2 0.93	AMILORIDE Sodium channel blocker, diuretic

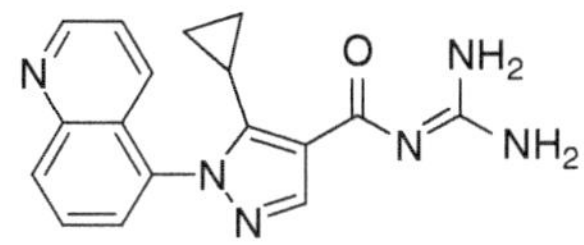	$t^1\!/_2$ 1.5 h* *F* – pb 77% ur –	*V* 3.2* CL 31* Mwt 320.4 PSA 105 Å^2 log *P* 0.20	ZONIPORIDE *iv, monkey Cardioprotective, sodium/proton exchange inhibitor (NHE-1) Metabolism: oxidation to 2-quinolone by aldehyde oxidase
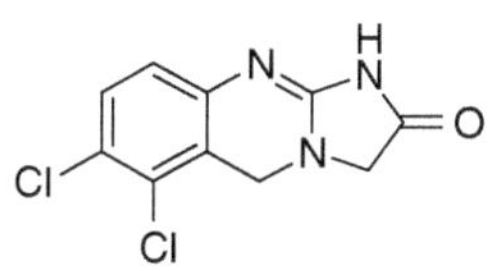	$t^1\!/_2$ 1.3 h *F* 70% pb – ur <1%	*V* 12 CL 2.1 Mwt 256.1 PSA 44.7 Å^2 log *P* 1.13	ANAGRELIDE Antithrombocythemic; Metabolism: hydroxylation of CH_2CO, then hydrolysis to 2-amino-3,4-dihydro-5,6-dichloroquinazoline, then N-glucuronidation
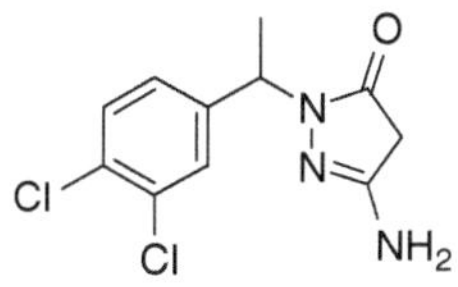	$t^1\!/_2$ 10–17 h *F* – pb 65% ur 5%	*V* 1 CL – Mwt 272.1 PSA 58.7 Å^2 log *P* 2.10	MUZOLIMINE Diuretic Metabolism: hydrolysis of amide, then decarboxylation
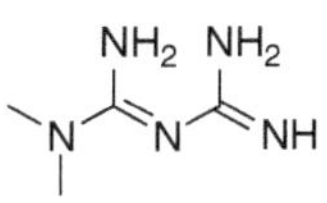	$t^1\!/_2$ 2–19 h *F* 40–60% pb Negligible ur 100%	*V* 0.9–3.9 CL 6.3–10.1 Mwt 129.2 PSA 89 Å^2 log *P* −1.25	METFORMIN Antidiabetic No metabolites known
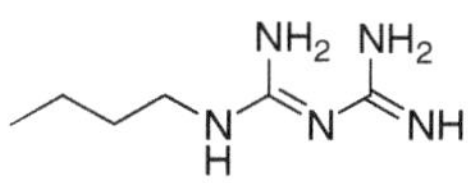	$t^1\!/_2$ 2–6 h *F* – pb 10% ur 35%	*V* 3 CL – Mwt 157.2 PSA 97.8 Å^2 log *P* −1.20	BUFORMIN Antidiabetic, withdrawn in 1978 in most countries for risk of inducing lactic acidosis
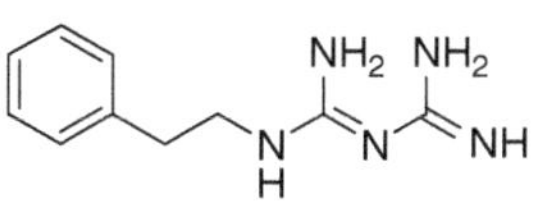	$t^1\!/_2$ 5–15 h *F* 50–60% pb 12–20% ur 54%	*V* 5–10 CL – Mwt 205.3 PSA 102.8 Å^2 log *P* 0.1	PHENFORMIN Antidiabetic; Metabolism: aromatic 4-hydroxylation; withdrawn in 1978 for risk of inducing lactic acidosis
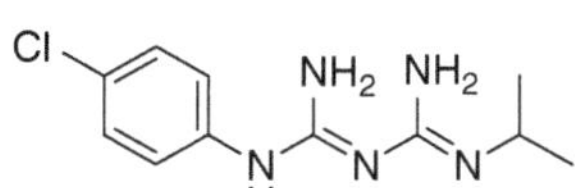	$t^1\!/_2$ 12–15 h *F* >90% pb 75% ur 40–60%	*V* 20–79 CL 14–27 Mwt 253.7 PSA 83.8 Å^2 log *P* 2.49	CHLORGUANIDE, PROGUANIL Antimalarial Metabolism: cyclization to triazine
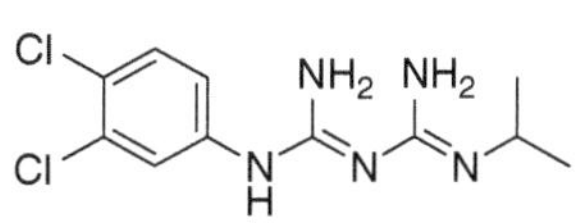	$t^1\!/_2$ 18±7 h *F* – pb 63% ur –	*V* 29 CL 21±2 Mwt 288.2 PSA 83.8 Å^2 log *P* 3.27	CHLORPROGUANIL Antimalarial Metabolism: cyclization to triazine

(*continued overleaf*)

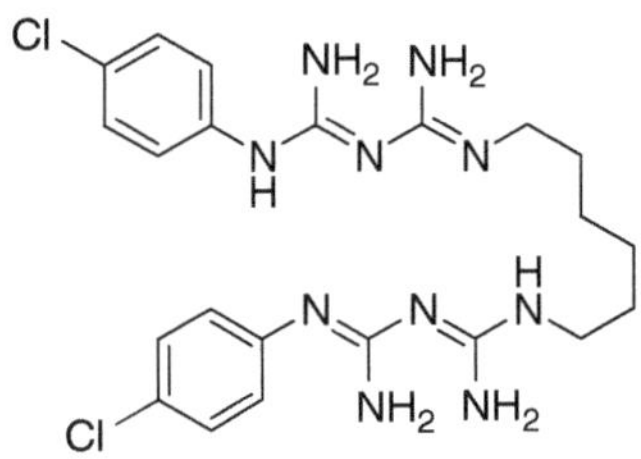	$t^1/_2$ F pb ur	– <1% 87% –	V CL Mwt PSA log P	– – 505.5 168 Å^2 4.58	CHLORHEXIDINE Antiseptic Metabolite: 4-chloroaniline
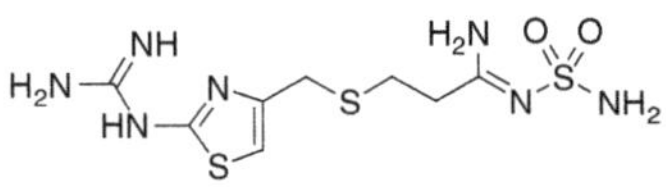	$t^1/_2$ F pb ur	2.6±1.0 h 45±14% 17±7% 67±15%	V CL Mwt PSA log P	1.3±0.2 7.1±1.7 337.5 233 Å^2 −1.66	FAMOTIDINE Histamine H_2 antagonist, antiulcerative
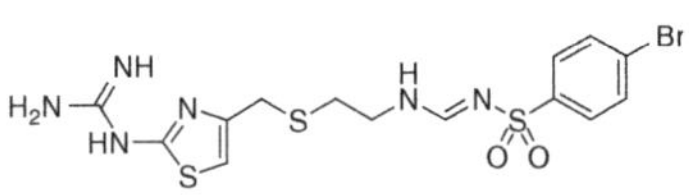	$t^1/_2$ F pb ur	14–20 h – – 1%	V CL Mwt PSA log P	1.7–2.7 – 477.4 195 Å^2 2.83	EBROTIDINE Histamine H_2 antagonist, antiulcerative
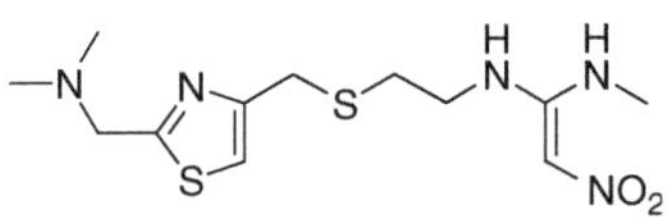	$t^1/_2$ F pb ur	1.3±0.3 h 90% 28±7% 61±3%	V CL Mwt PSA log P	1.2±0.5 10±3 331.5 140 Å^2 −0.49	NIZATIDINE Histamine H_2 antagonist, antiulcerative; active metabolite: *N*-monodesmethyl ($t^1/_2$ 3.8±1.9 h)
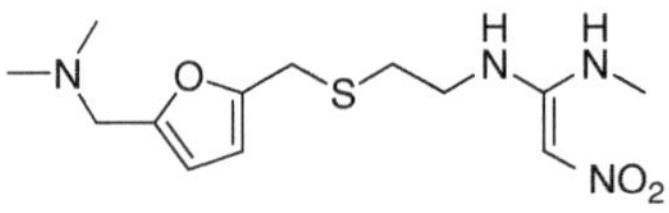	$t^1/_2$ F pb ur	2.1±0.2 h 52±11% 15±3% 69±6%	V CL Mwt PSA log P	1.3±0.4 10.4±1.1 314.4 112 Å^2 −0.07	RANITIDINE Histamine H_2 antagonist, antiulcerative Metabolites: *N*-oxide, *S*-oxide, desmethyl

$t_{1/2}$, plasma half-life; *F*, oral bioavailability; pb, plasma protein binding; ur, excretion of unchanged drug in urine; V, volume of distribution; CL, clearance; Mwt, molecular weight; PSA, polar surface area.

27
Anilines

Although a few unacylated aromatic amines have become successful drugs, medicinal chemists wisely tend to avoid this substructure because of its potential toxicity. Aniline itself is not carcinogenic, but it usually contains small amounts of highly carcinogenic impurities. Aniline is oxidized *in vivo* to *N*-phenylhydroxylamine and nitrosobenzene, which mediate the oxidation of hemoglobin (Fe^{2+}) to methemoglobin (Fe^{3+}) and lead to hemolytic anemia.

Several drugs containing primary aromatic amines, such as nomifensine and phenacetin, had to be withdrawn because of kidney or liver toxicity. Some anilines are potent carcinogens, for instance, those shown in Scheme 27.1.

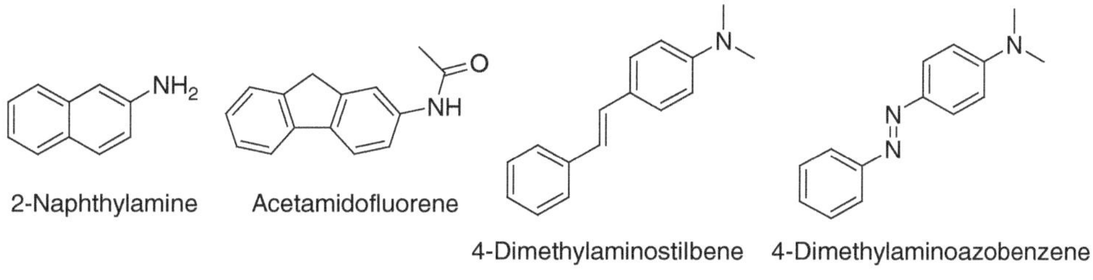

Scheme 27.1 Carcinogenic anilines.

The carcinogenicity of anilines results from metabolic activation. The metabolites of electron-rich anilines include *N*-arylhydroxylamines, their *O*-sulfates, and nitrosoarenes, which are electrophilic arylating reagents capable of N-arylating nucleic acids. Compounds that may be metabolized to aromatic amines, such as *N*-arylamides, -ureas or -carbamates, nitroarenes, azo compounds, or aromatic azides, may also be carcinogenic.

Lead Optimization for Medicinal Chemists: Pharmacokinetic Properties of Functional Groups and Organic Compounds, First Edition. Florencio Zaragoza Dörwald.

Less electron-rich anilines do not readily undergo oxidative metabolism and are therefore less toxic. Derivatives of aminobenzoic acids and aminobenzenesulfonic acids are usually not carcinogenic, and many drugs with these substructures are currently in use. A selection of such "safe" aromatic amines is given in Scheme 27.2. These amines are weak bases (pK_a 1.5–3.5) and will be unprotonated in most organs.

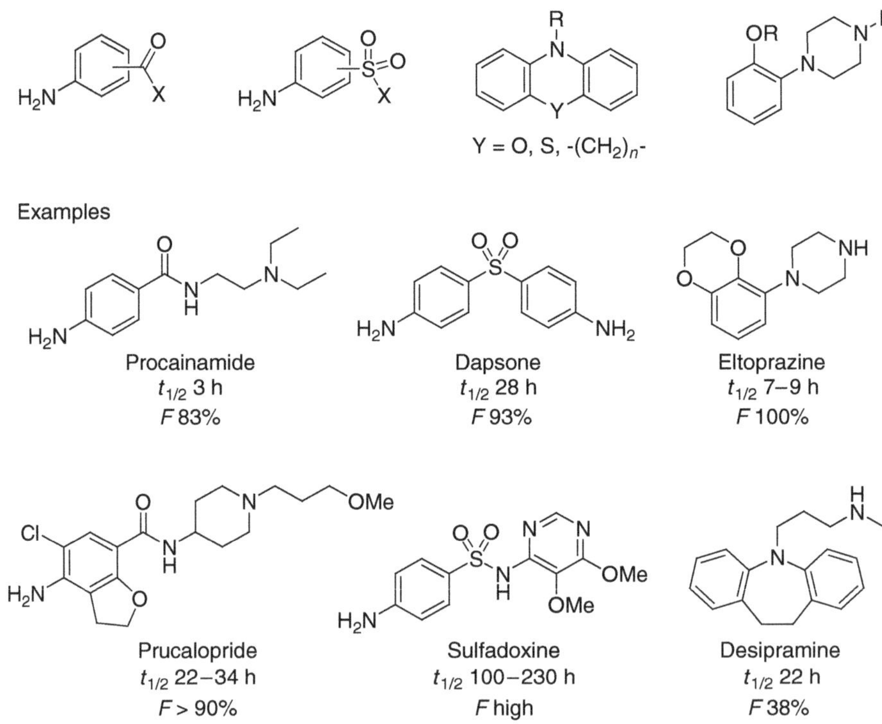

Scheme 27.2 Anilines of low toxicity.

If a lead contains a potentially toxic aromatic amine, it may be replaced by a heteroaromatic amine, such as an aminopyridine, aminopyrimidine, or aminothiazole. These heteroarenes are often metabolically stable and easier to prepare than substituted benzenes. Sterically undemanding aminopyridines may, however, be cytochrome *P*450 (CYP) inhibitors by virtue of their ability to bind to $Fe^{2+/3+}$. Moreover, some polycyclic heteroaromatic amines are proven carcinogens (Chapter 59).

27.1 Metabolism

The typical metabolic transformations of anilines include N-dealkylation, N-oxidation, N-hydroxylation, aromatic hydroxylation, and reversible N-acetylation. The resulting *N*-arylhydroxylamines or *N*-nitrosoarenes are soft electrophiles and electrophilic arylating or aminating agents, in particular if these metabolites are further activated by O-sulfation or O-acetylation (Scheme 27.3). *N*-Acylanilines often undergo hydrolysis *in vivo* to release the unacylated aniline.

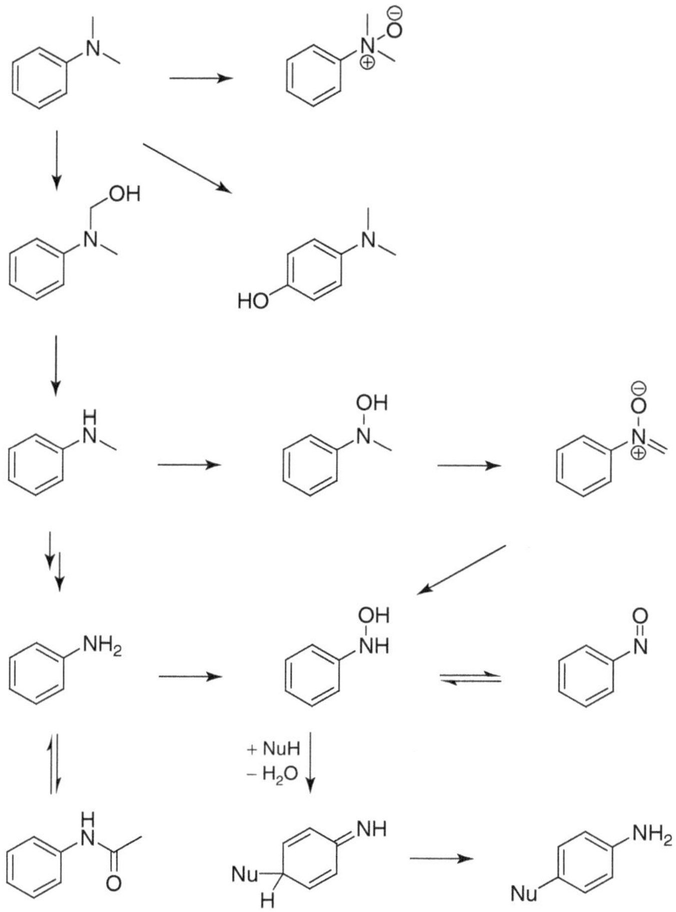

Scheme 27.3 Metabolic transformations of anilines.

The rate and extent to which these chemical transformations occur are highly dependent on the precise structure of the aniline. Electron-poor anilines usually only undergo N-acetylation or no metabolic transformation at all. The examples in Scheme 27.4 illustrate the types of anilines that have acceptably low toxicities for humans (excluding nomifensine) and their plasma half-lives.

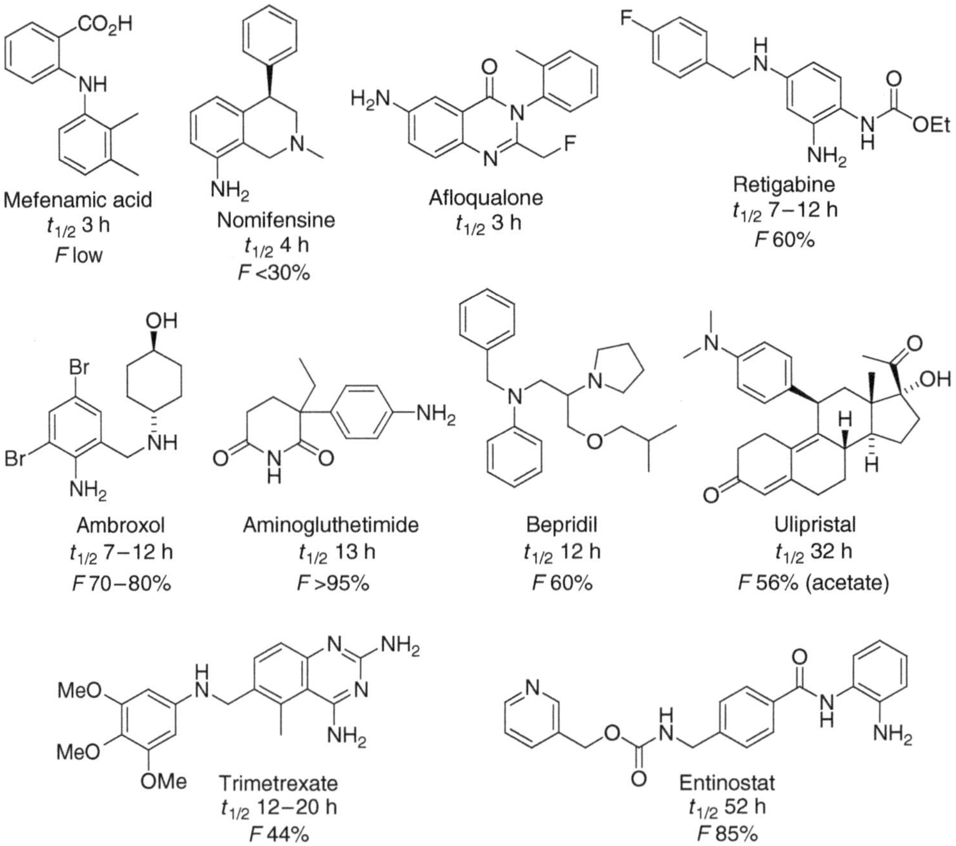

Scheme 27.4 Anilines as drugs.

28
Hydrazines, Acylhydrazines, and Hydrazones

The few alkylhydrazines used as drugs are monoamine oxidase (MAO) inhibitors (phenelzine and iproniazide), cytotoxic agents (procarbazine), antihypertensives (hydralazine), or amino acid decarboxylase inhibitors (carbidopa and benserazide) (Scheme 28.1). Alkylhydrazines (pK_a 6.5–8.0) are less basic but more nucleophilic than amines, and acylhydrazines (pK_a 2–3) or hydrazones (pK_a 3–5) are still less basic than hydrazines.

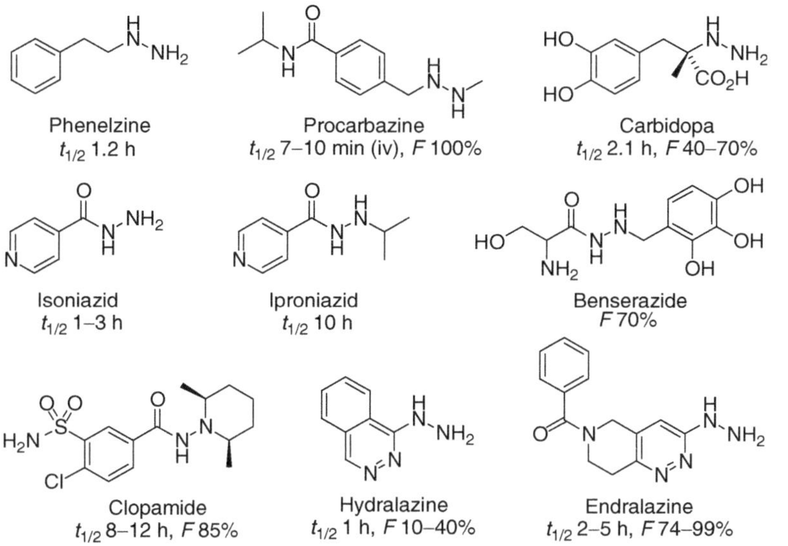

Scheme 28.1 Hydrazine derivatives as drugs.

Typical metabolic transformations of hydrazines include oxidation to azo and azoxy compounds, reductive N–N bond cleavage, and N-acetylation. If acylhydrazines are oxidized to azo compounds, these will usually quickly decompose by either nucleophilic displacement (hydrolysis) or homolytic C–N bond cleavage to yield highly reactive acyl radicals.

Lead Optimization for Medicinal Chemists: Pharmacokinetic Properties of Functional Groups and Organic Compounds, First Edition. Florencio Zaragoza Dörwald.

29
Aldehydes

Because aldehydes are highly electrophilic, few drugs contain a free aldehyde functionality. If they do, the formyl group will usually react quickly *in vivo* with some nucleophile, will be oxidized to an acid, or reduced to an alcohol. Thus, the sedative chloral hydrate is only a prodrug of trichloroethanol. Tucaresol is an electron-rich benzaldehyde of moderate electrophilicity, which acts by forming an imine (a Schiff base) with an amino group of hemoglobin and thereby increases the stability of the oxy-conformation of this protein.

Cyclic acetals or hemiacetals of aldehydes are, however, sufficiently stable and a common substructure of many drugs, for example, in nucleoside or carbohydrate analogs (Scheme 29.1). Oximes (e.g., pralidoxime) and hydrazones (e.g., azimilide) of aldehydes can be found among therapeutic agents as well.

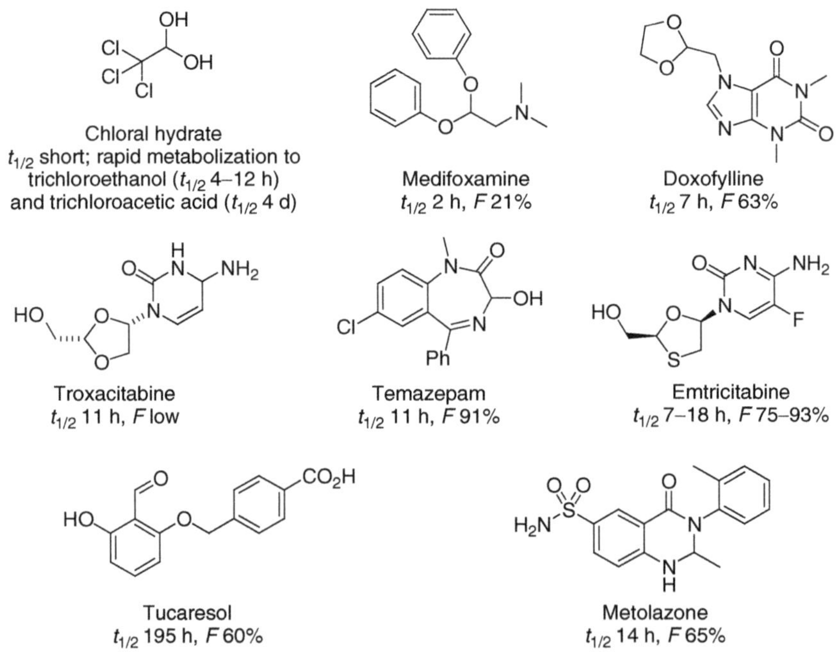

Scheme 29.1 Aldehydes and derivatives thereof as drugs.

Lead Optimization for Medicinal Chemists: Pharmacokinetic Properties of Functional Groups and Organic Compounds, First Edition. Florencio Zaragoza Dörwald.

30
Ketones

Ketones are less electrophilic and chemically reactive than aldehydes, and can be highly orally available and long-lived in human plasma. This functional group is, however, often the weakest point in a molecule. The rate of metabolic transformation depends on the electronic properties of the substituents at the ketone carbonyl group. The more electron donating and sterically demanding the substituents, the higher will be the metabolic stability of a ketone. Benzophenones and alkanoyl benzenes are particularly stable. A selection of therapeutic ketones is sketched in Scheme 30.1 (see also Table 30.1).

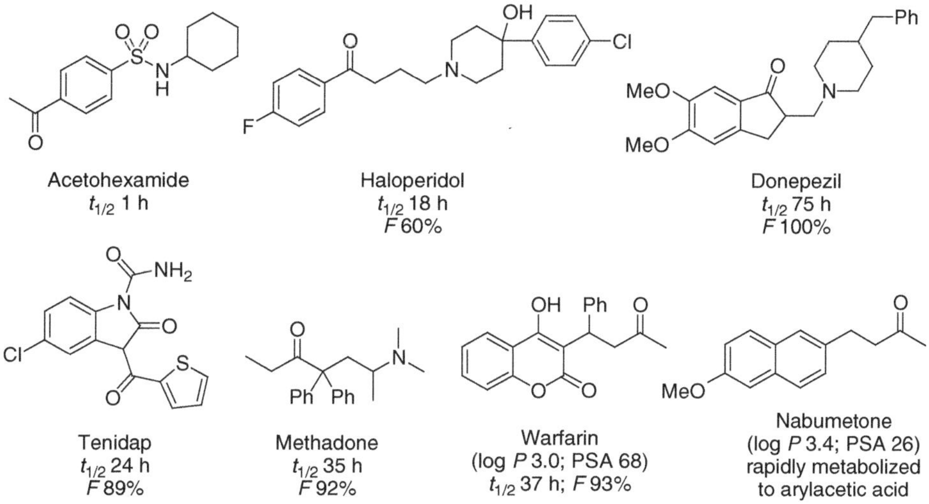

Scheme 30.1 Ketones as drugs.

The different rates of metabolic degradation of warfarin and nabumetone are remarkable. Warfarin is metabolized rather slowly by aromatic hydroxylation and α-hydroxylation and reduction of the ketone. In nabumetone, despite its similarity to warfarin, the ketone functionality is quickly degraded to an arylacetic acid. This

Lead Optimization for Medicinal Chemists: Pharmacokinetic Properties of Functional Groups and Organic Compounds, First Edition. Florencio Zaragoza Dörwald.

could be due to the higher lipophilicity and lower polar surface area (PSA) of nabumetone, which should increase its affinity to liver enzymes and increase its rate of oxidative metabolization.

Ketones are sometimes produced *in vivo* by oxidation of secondary alcohols or reactive methylene groups. The latter may therefore serve as prodrugs of ketones. One such instance is the oral contraceptive desogestrel, which is metabolized to the active ketone etonogestrel (Scheme 30.2).

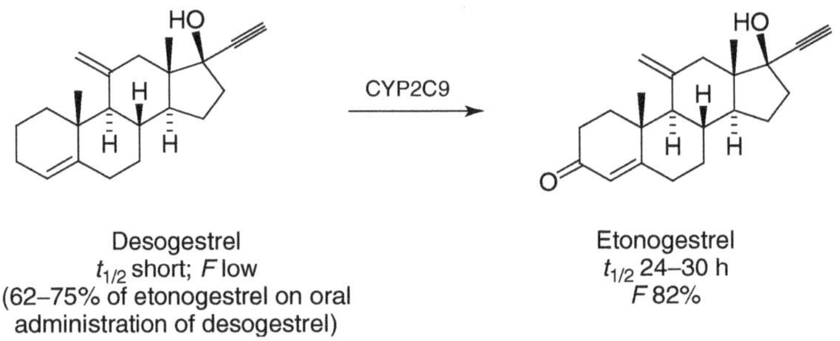

Scheme 30.2 Metabolism of desogestrel.

Some acetals or aminals of ketones can also be sufficiently stable to serve as drugs, as illustrated by the examples in Scheme 30.3. Particularly stable are acetals of ketones with an α-heteroatom (O, S, N, halogen), or with positive charges close to the acetal, which destabilize the intermediate carbocation required for the acid-mediated hydrolysis. Other condensation products of ketones that can be found in drugs include hydrazones, oximes, and enol ethers.

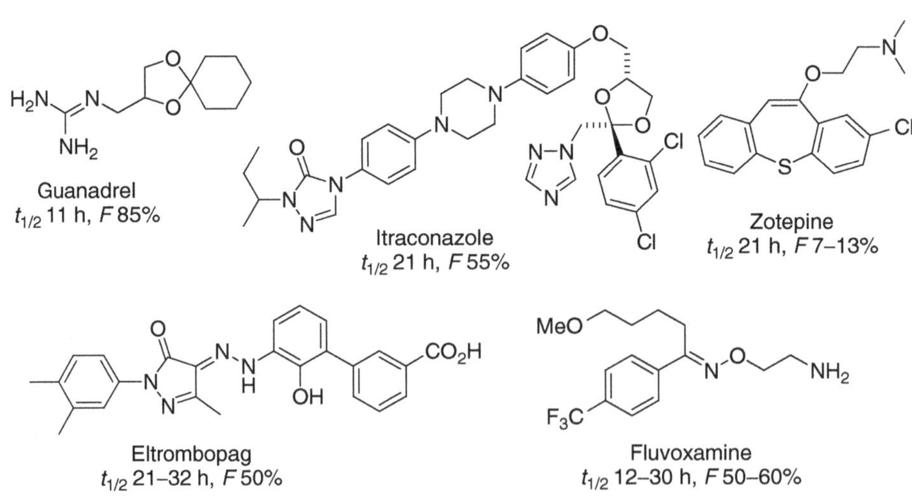

Scheme 30.3 Ketone derivatives as drugs.

Table 30.1 Ketones, lactones, and related compounds. V in l kg^{-1}; CL in ml min^{-1} kg^{-1}; Mwt in g mol^{-1}.

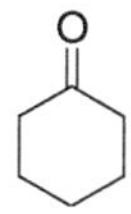	$t^{1}/_{2}$ F pb ur	5 h – – <1%*	V CL Mwt PSA log P	2.6* 27±4* 98.1 17.1 Å^2 0.82	CYCLOHEXANONE *dog Metabolites: cyclohexanol ($t^{1}/_{2}$ 4.8 h), 1,2-cyclohexanediol ($t^{1}/_{2}$ 16 h), 1,4-cyclohexanediol ($t^{1}/_{2}$ 18 h)
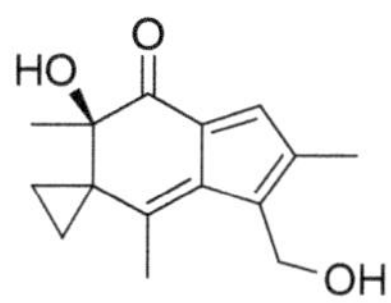	$t^{1}/_{2}$ F pb ur	0.3 h (iv) – 51–52% 0%	V CL Mwt PSA log P	3.1 140 246.3 57.5 Å^2 1.13	IROFULVEN Antineoplastic Metabolism: isomerization to 3-(2-hydroxyethyl)phenol, hydroxylation of both vinylic CH_3 groups
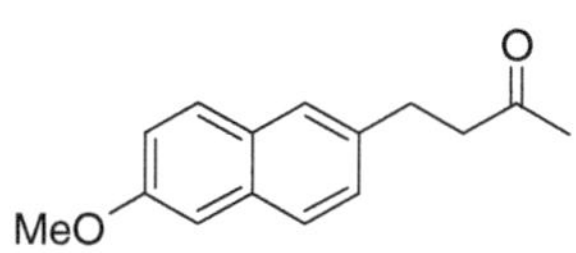	$t^{1}/_{2}$ F pb ur	23±4 h 35% >99% 50%	V/F CL/F Mwt PSA log P	0.79±0.38 0.37±0.25 228.29 26.3 Å^2 3.40	NABUMETONE Prodrug for methoxynaphthylacetic acid, all values given for acid upon oral dosing of prodrug
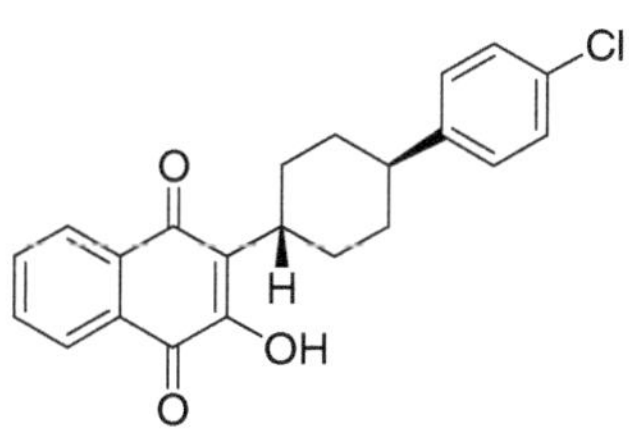	$t^{1}/_{2}$ F pb ur	62–80 h 47% >99% Negligible	V CL Mwt PSA log P	0.6±0.2 0.15 366.8 54.4 Å^2 6.18	ATOVAQUONE Antiprotozoal No metabolites known; excreted unchanged in bile
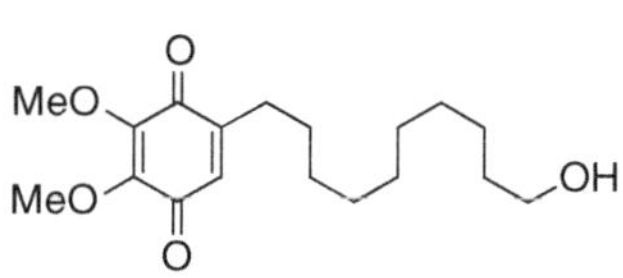	$t^{1}/_{2}$ F pb ur	12–13 h Low 96% Negligible	V CL Mwt PSA log P	– – 338.4 72.8 Å^2 3.91	IDEBENONE Nootropic Metabolism: oxidative degradation of side chain, reduction to hydroquinone
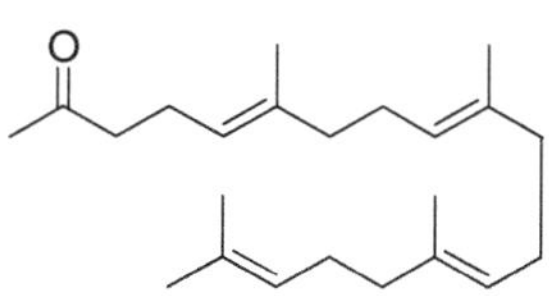	$t^{1}/_{2}$ F pb ur	3.4±1.2 h – – –	V CL Mwt PSA log P	– – 330.6 17.1 Å^2 7.61	TEPRENONE Gastroprotective
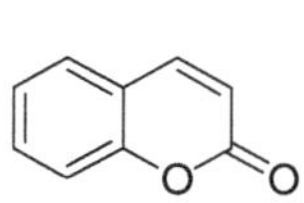	$t^{1}/_{2}$ F pb ur	2 h (iv) 1–3% 86% Low	V CL Mwt PSA log P	– – 146.1 26.3 Å^2 1.39	COUMARIN Flavor; extensive first-pass metabolism to 7-hydroxycoumarin and 2-hydroxyphenylacetic acid, then glucuronidation

(continued overleaf)

Structure					Name
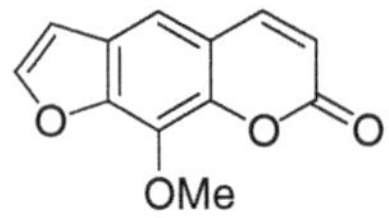	$t_{1/2}$ *F* pb ur	1.4–3.2 h Low 80–90% Negligible	*V* CL Mwt PSA log *P*	0.5–3.2 10–157 216.2 48.7 $Å^2$ 1.52	METHOXSALEN, 8-METHOXYPSORALEN Photosensitizer, pigmentation agent Metabolism: oxidative ring opening of furan to arylacetic acid, demethylation
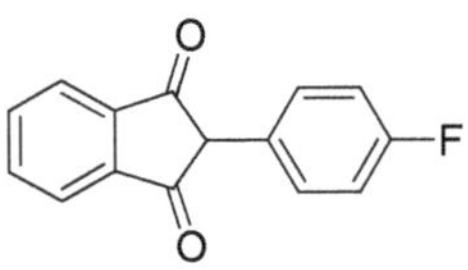	$t_{1/2}$ *F* pb ur	31–69 h High 99% 35%	*V* CL Mwt PSA log *P*	– – 240.2 34.1 $Å^2$ 3.42	FLUINDIONE Anticoagulant Metabolism: hydroxylation
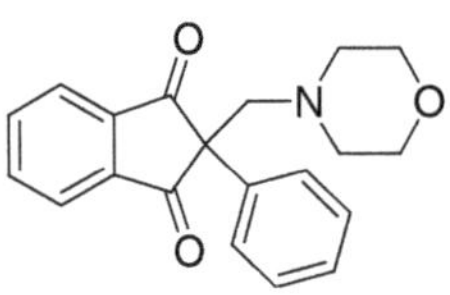	$t_{1/2}$ *F* pb ur	15 h – High –	*V* CL Mwt PSA log *P*	– – 321.4 46.6 $Å^2$ 1.08	OXAZIDIONE Oral anticoagulant
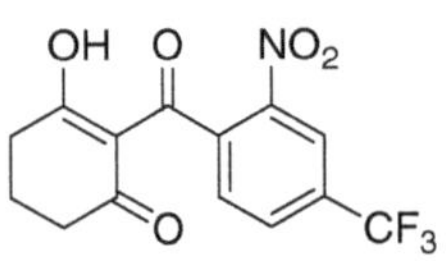	$t_{1/2}$ *F* pb ur	54 h High – –	*V* CL Mwt PSA log *P*	0.1–0.5 0.07 329.2 97.0 $Å^2$ 0.99	NITISINONE, NTBC Treatment of tyrosinemia, herbicide Metabolism: cleavage to 2-nitro-4-trifluoromethylbenzoic acid, hydroxylation of cyclohexane
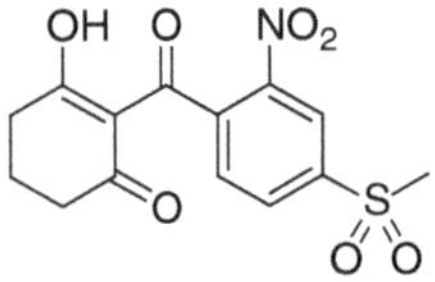	$t_{1/2}$ *F* pb ur	0.9–1.3 h >95% (rat) – 49–72%	*V* CL Mwt PSA log *P*	– – 339.3 140 $Å^2$ −0.99	MESOTRIONE Herbicide
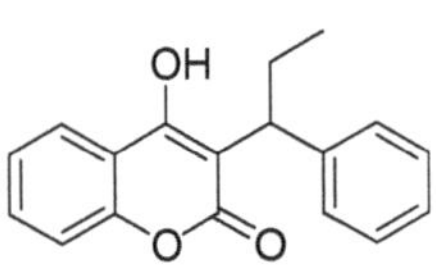	$t_{1/2}$ *F* pb ur	90–140 h 100% 99% 2%	*V* CL Mwt PSA log *P*	0.18 – 280.3 46.5 $Å^2$ 4.64	PHENPROCOUMON Anticoagulant Metabolism: glucuronidation, aromatic 4′-, 6-, and 7-hydroxylation
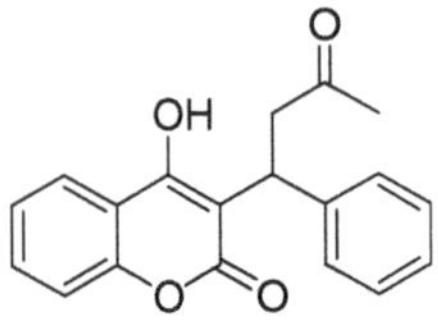	$t_{1/2}$ *F* pb ur	30–40 h 93±8% 99±1% <2%	*V* CL Mwt PSA log *P*	0.14±0.06 0.05±0.02 308.33 67.51 $Å^2$ 3.03	WARFARIN Anticoagulant, CYP2C9 substrate Metabolism: aromatic and benzylic hydroxylation

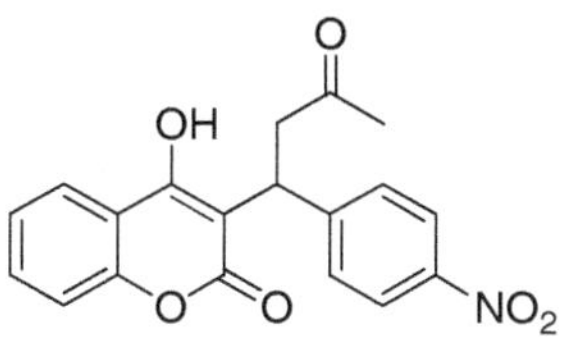	$t_{1/2}$ 3–9 h *F* 83% pb 99% ur <0.2%	*V* 0.1–0.3 CL/*F* 2 Mwt 353.3 PSA 109 Å^2 log *P* 2.97	ACENOCOUMAROL, NICOUMALONE Anticoagulant Metabolism: reduction to aniline, then N-acetylation, reduction of ketone to alcohol, aromatic hydroxylation of chromenone
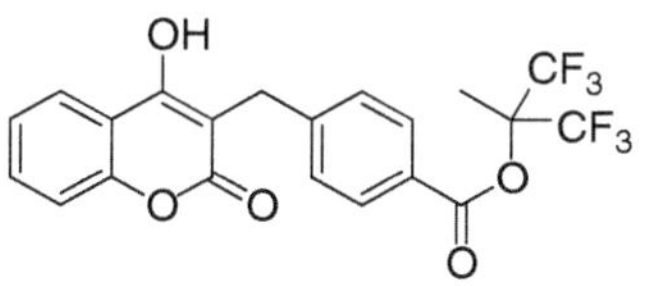	$t_{1/2}$ 107–140 h *F* 80–90%* pb 99% ur –	*V* 0.49 (dog) CL – Mwt 460.3 PSA 72.8 Å^2 log *P* 7.40	TECARFARIN *rat Anticoagulant (same mechanism as warfarin) Metabolism: ester hydrolysis
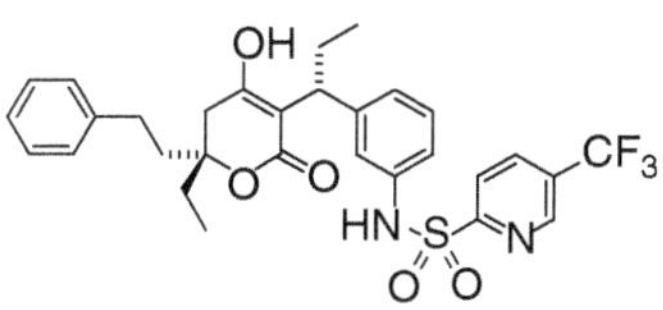	$t_{1/2}$ 5–6 h *F* 42% (rat) pb >99% ur 0.5%	*V* 0.1 CL 0.3 Mwt 602.7 PSA 114 Å^2 log *P* 6.92	TIPRANAVIR Antiviral Metabolite: glucuronide
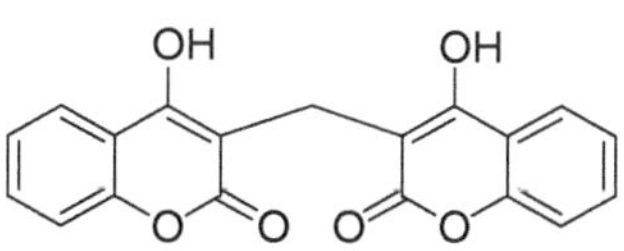	$t_{1/2}$ 7–100 h *F* – pb >99% ur <1%	*V* – CL – Mwt 336.3 PSA 93.1 Å^2 log *P* 2.05	DICOUMAROL Vitamin K antagonist, anticoagulant

$t_{1/2}$, plasma half-life; *F*, oral bioavailability; pb, plasma protein binding; ur, excretion of unchanged drug in urine; V, volume of distribution; CL, clearance; Mwt, molecular weight; PSA, polar surface area.

31
Carboxylic Acids

With a pK_a of 3–5 in water, carboxylic acids are deprotonated and negatively charged in most parts of the body, excluding the stomach. The carboxyl group is an important pharmacophore of many drugs: fibrates, statins, nonsteroidal anti-inflammatory drugs (NSAIDs), angiotensin-converting enzyme (ACE) inhibitors, and β-lactam and quinolone antibacterials contain essential carboxyl groups (Tables 31.1, 31.2). These compounds are usually well absorbed, but have low volumes of distribution (poor distribution in body tissues), and few cross the blood–brain barrer (bbb). Their volume of distribution is generally much smaller than that of amines, because negatively charged carboxylate groups show little affinity to the negatively charged phospholipid bilayers.

The metabolic degradation of carboxylic acids (usually acyl glucuronidation or amidation) is slow, as illustrated by the long-lived drugs in Scheme 31.1.

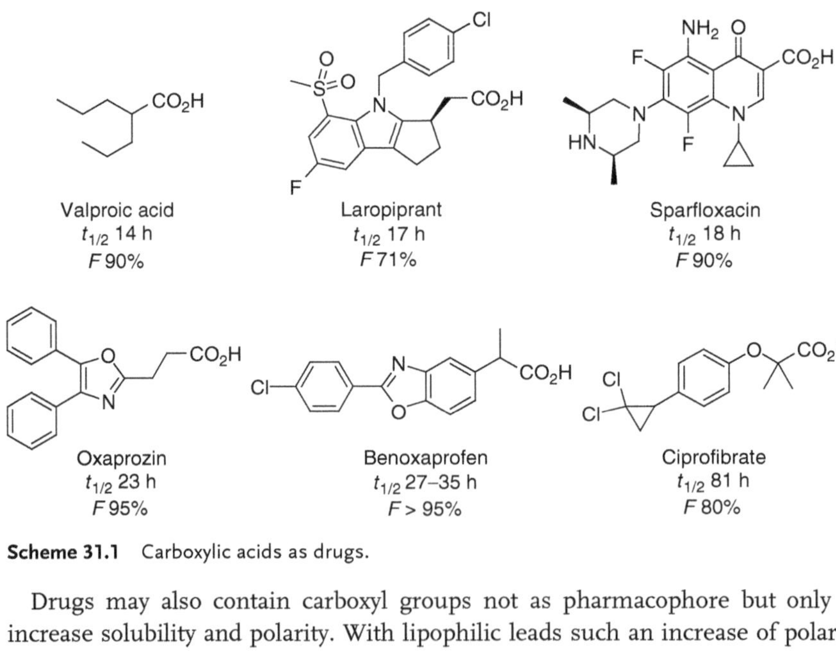

Scheme 31.1 Carboxylic acids as drugs.

Drugs may also contain carboxyl groups not as pharmacophore but only to increase solubility and polarity. With lipophilic leads such an increase of polarity

Lead Optimization for Medicinal Chemists: Pharmacokinetic Properties of Functional Groups and Organic Compounds, First Edition. Florencio Zaragoza Dörwald.

can suppress the first-pass effect and thereby enhance oral bioavailability. Moreover, an increased polar surface area (PSA) may keep compounds from crossing the bbb. For instance, the bioavailability of the histamine H_1 antagonists triprolidine or doxepin can be enhanced, and their CNS (central nervous system)-mediated sedative effect suppressed by introducing a carboxyl group (Scheme 31.2).

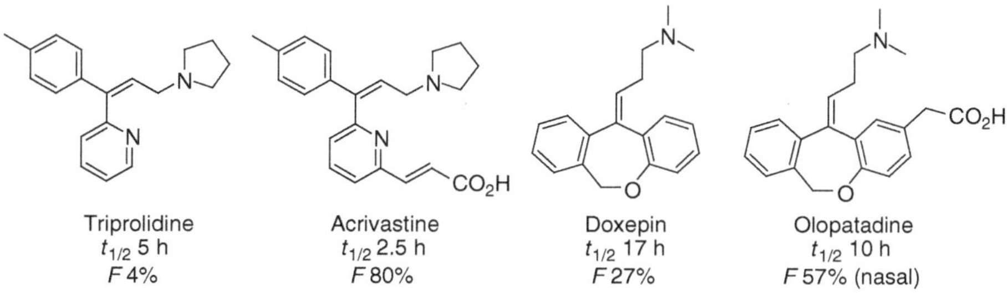

Scheme 31.2 Effect of the carboxyl group on pharmacokinetic (PK) properties.

31.1 Metabolism

Common metabolic transformations of carboxyl groups are activation by thioester formation with coenzyme A, followed by acylation of glycine, glutamine, or other amino acids [1]. Esterification of carboxylic acids with glucuronic acid is mediated by uridine diphosphate glucuronosyltransferases, enzymes which also glucuronylate phenols, alcohols, thiols, and amines. These enzymes are mainly located in the liver and also in the intestine, kidneys, brain, and skin. Fatty acids or related compounds often undergo oxidative degradation to carboxylic acids of lower molecular weight. All these transformations can also occur with carboxylic acids formed by phase I metabolism of other functional groups.

Illustrative examples of the metabolic fate of carboxylic acids are given in Scheme 31.3.

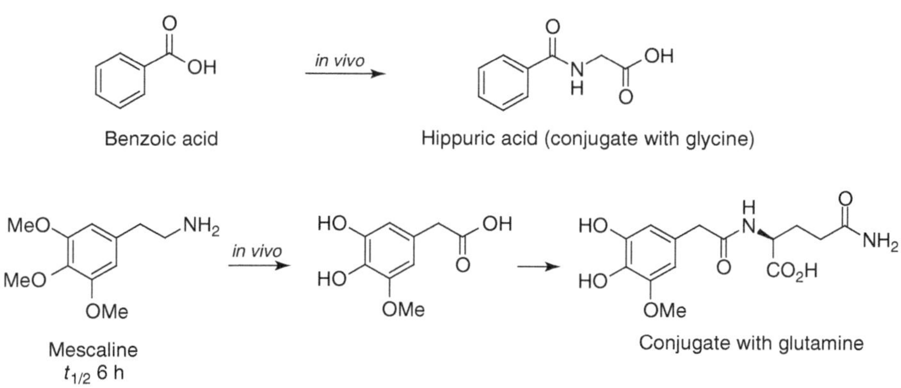

Scheme 31.3 Metabolism of carboxylic acids.

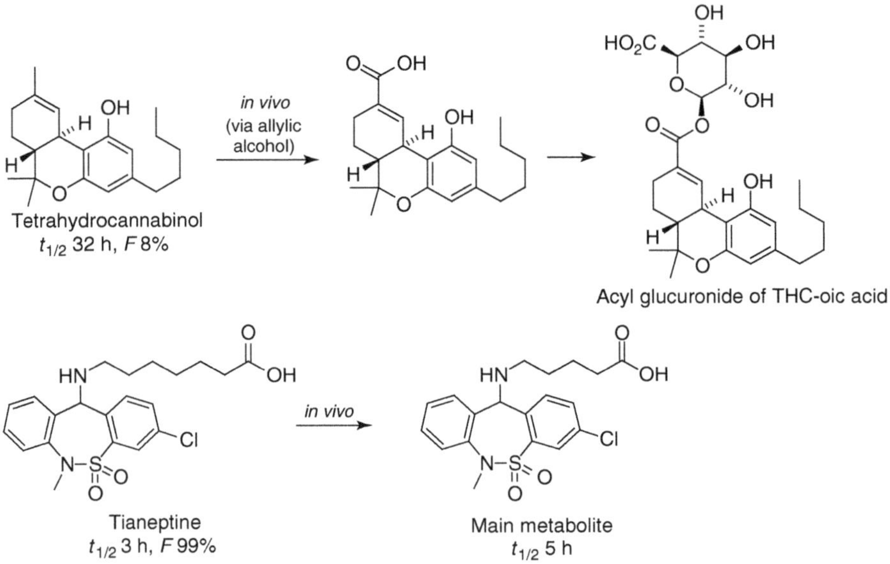

Scheme 31.3 (*Continued*).

31.2
Bioisosteres of Carboxylic Acids

If the carboxyl group of a lead is metabolized too quickly, a metabolically more resistant bioisostere may be tested. Various functional groups are similarly acidic as carboxylic acids but sometimes more resistant to metabolic derivatization [2]. These include sulfonic acids, phosphonic acids, hydroxamic acids, *N*-acylcyanamides, *N*-acylsulfonamides, and tetrazoles. Imides and ureides (*N*-acylureas) are not acidic enough to act as carboxylic acid bioisostere (Scheme 31.4). Hydroxamic

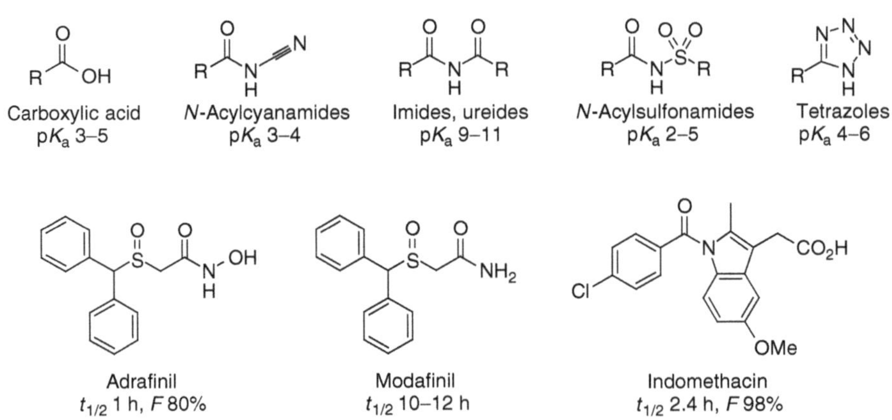

Scheme 31.4 Bioisosteres of carboxylic acids.

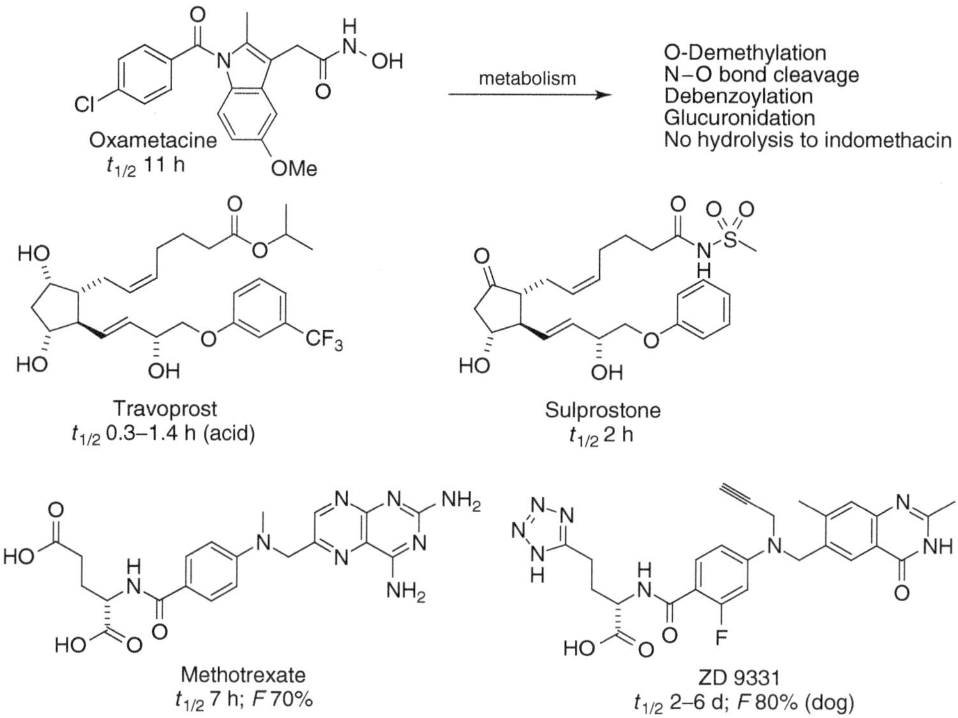

Scheme 31.4 (*Continued*).

acids are metabolically unstable, but may serve as prodrugs of amides (e.g., adrafinil as prodrug of modafinil) or of carboxylic acids (e.g., ibuproxam as prodrug of ibuprofen). Oxametacine, however, is not a prodrug but a real bioisostere of indomethacin.

31.3
Amino Carboxylic Acids, *N*-Acyl Amino Acids, and Related Compounds

Amino acids show minimal solubility in water at their isoelectric point, that is, at the pH where an internal salt is formed and the molecule has no net electric charge. In contrast to peptides, low-molecular-weight amino acids are orally bioavailable, and some even cross the bbb. Many of these hydrophilic molecules undergo little metabolism and are mostly assimilated or excreted unchanged in urine.

Unnatural analogs of amino acids can be antimetabolites or prodrugs, agonists, or antagonists of natural amino acids, including those that act as neurotransmitters (γ-aminobutyric acid (GABA), glycine, glutamic acid, and aspartic acid) (Table 31.2).

References

1. Stachulski, A.V., Harding, J.R., Lindon, J.C., Maggs, J.L., Park, B.K., and Wilson, I.D. (2006) Acyl glucuronides: biological activity, chemical reactivity, and chemical synthesis. *J. Med. Chem.*, **49**, 6931–6945.
2. Meanwell, N.A. (2011) Synopsis of some recent tactical applications of bioisosteres in drug design. *J. Med. Chem.*, **54**, 2529–2591.

Table 31.1 Carboxylic acids, esters, and amides, and hydroxamic acids. *V* in l kg^{-1}; CL in ml min^{-1} kg^{-1}; Mwt in g mol^{-1}.

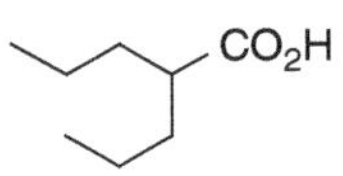	$t^{1}/_{2}$ *F* pb ur	14±3 h 90±10% 93±1% 1.8±2.4%	*V* CL Mwt PSA log *P*	0.22±0.07 0.11±0.02 144.2 37.3 Å^2 2.58	VALPROIC ACID Anticonvulsant, antimigraine, antimanic; Metabolism: dehydrogenation (single, to 2-propyl-2-pentenoic acid ($t^{1}/_{2}$ 1–2 d), and twofold), 3-hydroxylation
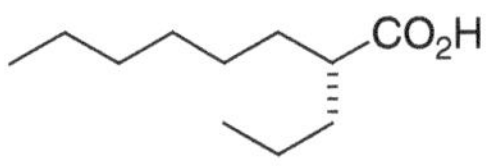	$t^{1}/_{2}$ *F* pb ur	2–3 h (iv) – 99% –	*V* CL Mwt PSA log *P*	0.1–0.2 0.7–1.0 186.3 37.3 Å^2 4.31	ARUNDIC ACID Astrocycte modulating agent for treatment of stroke
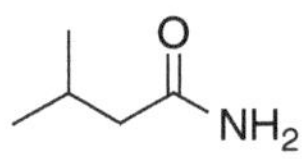	$t^{1}/_{2}$ *F* pb ur	2–4 h High Negligible <4%	*V* CL Mwt PSA log *P*	– – 101.2 43.1 Å^2 0.67	ISOVALERAMIDE Anticonvulsant
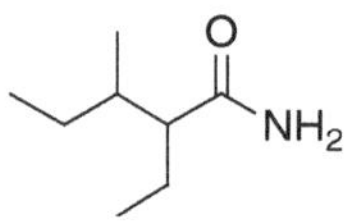	$t^{1}/_{2}$ *F* pb ur	6–10 h – – –	*V* CL Mwt PSA log *P*	– – 143.2 43.1 Å^2 1.89	VALNOCTAMIDE Antiepileptic No metabolism to acid
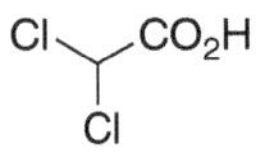	$t^{1}/_{2}$ *F* pb ur	3.4 h – – <2%	*V* CL Mwt PSA log *P*	0.2–0.4 2–5 128.9 37.3 Å^2 0.67	DICHLOROACETIC ACID Caustic, keratolytic, topical astringent Metabolism: hydrolysis to glyoxylate
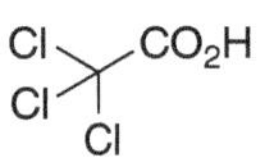	$t^{1}/_{2}$ *F* pb ur	90–100 h – 82–87% 93%	*V* CL Mwt PSA log *P*	– – 163.4 37.3 Å^2 1.73	TRICHLOROACETIC ACID Metabolite of chloral, hemostatic, caustic for chemical skin peeling

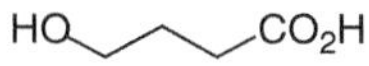	$t_{1/2}$ 0.5–1.0 h *F* 25% pb <1% ur <5%	*V* 0.2–0.4 CL – Mwt 104.1 PSA 57.5 Å^2 log *P* −0.76	4-HYDROXYBUTYRIC ACID (Na-salt: OXYBATE SODIUM) Sedative, narcotic Metabolism: oxidation to succinic acid
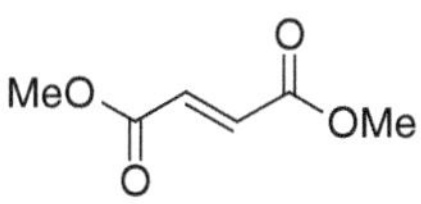	$t_{1/2}$ 12 min *F* – pb – ur –	*V* – CL – Mwt 144.1 PSA 52.6 Å^2 log *P* 0.43	DIMETHYL FUMARATE NF-κB activation inhibitor for the treatment of multiple sclerosis Metabolite: monomethyl fumarate ($t_{1/2}$ 36 h)
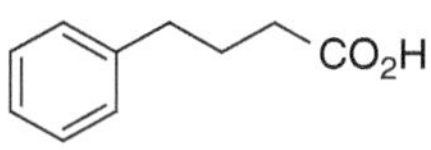	$t_{1/2}$ 0.8 h *F* 78% pb 81% ur –	*V* 0.2 CL 4 Mwt 164.2 PSA 37.3 Å^2 log *P* 2.31	4-PHENYLBUTYRIC ACID Treatment of hyperammonemia (urea cycle disorders) Metabolism: oxidative degradation to phenylacetic acid ($t_{1/2}$ 1.2 h, pb 56%) and *N*-phenylacetylglutamine ($t_{1/2}$ 2.4 h)
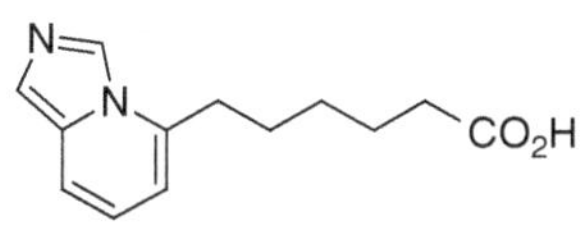	$t_{1/2}$ 1.2 h (iv) *F* – pb – ur –	*V* 0.5 CL 5 Mwt 232.3 PSA 54.6 Å^2 log *P* 2.11	PIRMAGREL Thromboxane synthase inhibitor
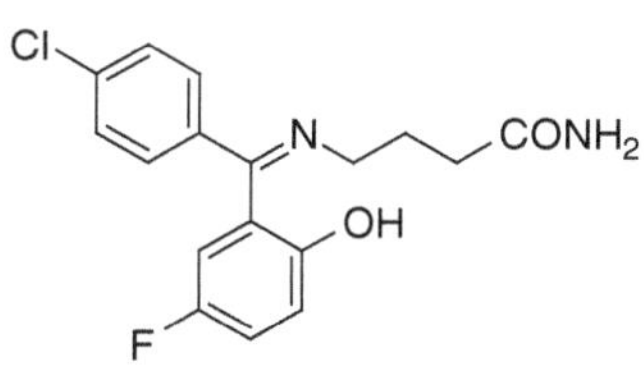	$t_{1/2}$ 4 h *F* 60% pb 98% ur –	*V* – CL – Mwt 334.8 PSA 75.7 Å^2 log *P* 3.14	PROGABIDE GABA analog Metabolism: hydrolysis of amide and imine
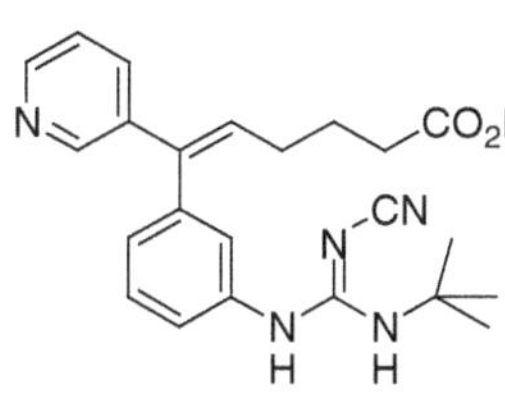	$t_{1/2}$ 10–11 h *F* – pb – ur –	*V* – CL Mwt 405.5 PSA 110 Å^2 log *P* 3.95	TERBOGREL Thromoboxane A_2 receptor and synthase inhibitor
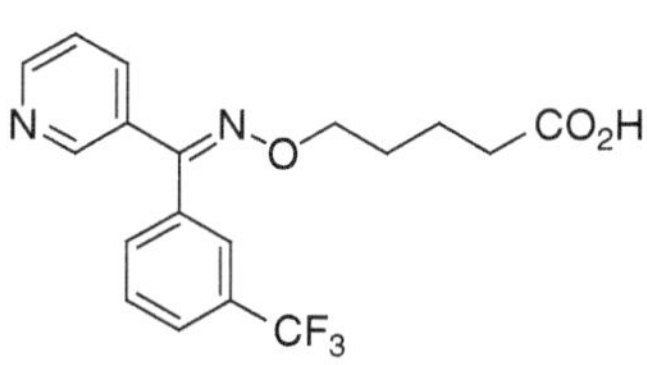	$t_{1/2}$ 6–9 h *F* 90% pb 60% ur –	*V* 0.1–0.2 CL – Mwt 366.3 PSA 71.8 Å^2 log *P* 3.63	RIDOGREL Antithrombotic, thromboxane A_2 synthase inhibitor

(*continued overleaf*)

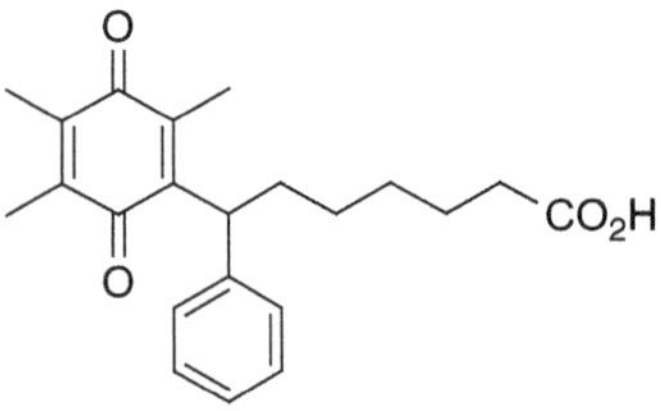

$t_{1/2}$	22 h	V	0.04
F	89%*	CL	0.2
pb	>96%	Mwt	354.4
ur	–	PSA	71.4 Å^2
		log P	4.63

SERATRODAST
*monkey; thromboxane A_2 antagonist, antiasthmatic
Metabolism: glucuronidation, reduction to hydroquinone, then glucuronidation or sulfation, oxidative degradation of alkylene chain, hydroxylation of methyl

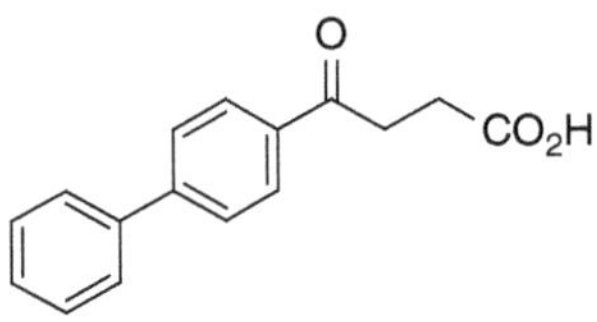

$t_{1/2}$	8–17 h	V	2–4
F	–	CL	–
pb	>99%	Mwt	254.3
ur	<2%	PSA	54.4 Å^2
		log P	3.12

FENBUFEN
NSAID
Metabolism: reduction of ketone to alcohol, oxidative degradation to biphenylylacetic acid

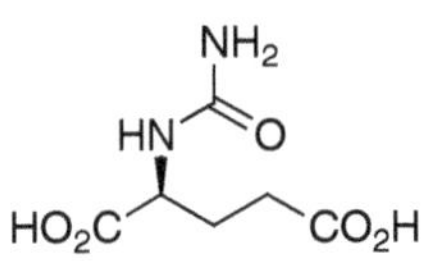

$t_{1/2}$	4–10 h	V	25–89
F	–	CL	88
pb	–	Mwt	190.2
ur	9%	PSA	130 Å^2
		log P	−0.74

CARGLUMIC ACID
Carbamoyl phosphate I synthase activator for treatment of hyperammonemia

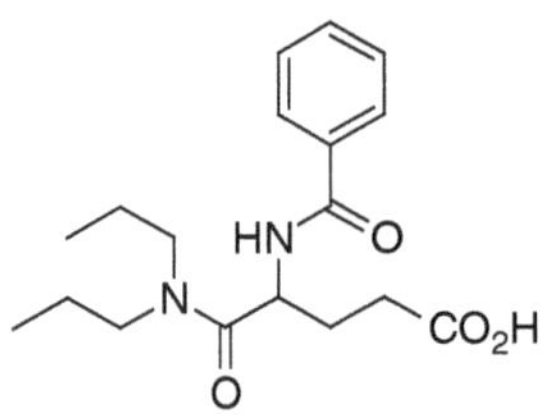

$t_{1/2}$	42 h (rat)	V	–
F	–	CL	–
pb	–	Mwt	334.4
ur	–	PSA	86.7 Å^2
		log P	2.69

PROGLUMIDE
Cholecystokinin antagonist, anticholinergic
Metabolism: N-depropylation

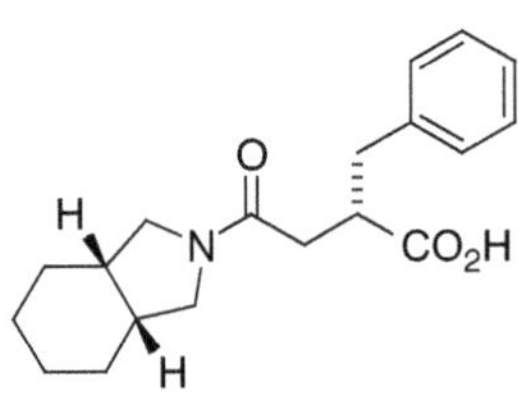

$t_{1/2}$	1–5 h	V	–
F	–	CL	–
pb	–	Mwt	315.4
ur	–	PSA	57.6 Å^2
		log P	2.73

MITIGLINIDE
K_{ATP} channel blocker, antidiabetic
Metabolism: acyl glucuronidation

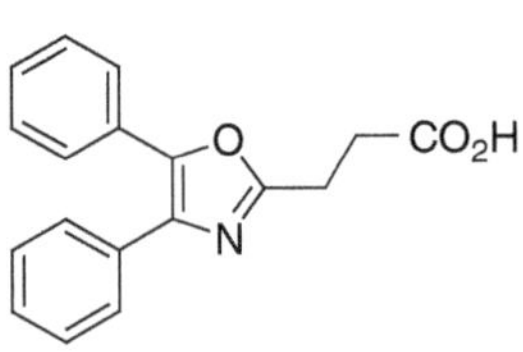

$t_{1/2}$	23±2 h	V	0.19±0.05
F	95–100%	CL	0.035±0.007
pb	99.5%	Mwt	293.3
ur	<1%	PSA	63.3 Å^2
		log P	3.15

OXAPROZIN
Antiinflammatory
Metabolism: glucuronidation, hydroxylation of phenyl groups

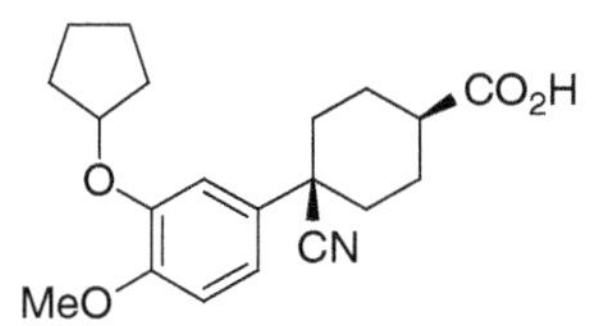	$t_{1/2}$ *F* pb ur	7–8 h 100% 99.5% <1%	*V* CL Mwt PSA log *P*	0.23 0.5 343.4 79.6 Å^2 3.30	CILOMILAST Antiasthmatic Metabolism: glucuronidation, 3-hydroxylation of cyclopentyl, O-decyclopentylation
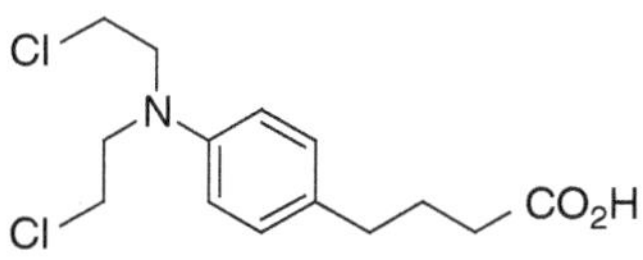	$t_{1/2}$ *F* pb ur	1.3±0.9 h 87±20% 99% <1%	*V* CL Mwt PSA log *P*	0.29±0.21 2.6±0.9 304.2 40.5 Å^2 2.61	CHLORAMBUCIL Antineoplastic Metabolism: oxidative degradation to arylacetic acid, substitution of Cl by OH
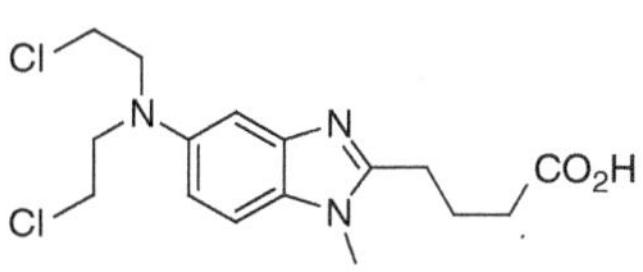	$t_{1/2}$ *F* pb ur	0.5 h – >95% 10–45%	*V* CL Mwt PSA log *P*	0.24–0.32 7.5 358.3 58.4 Å^2 2.38	BENDAMUSTINE Antineoplastic Metabolism: substitution of Cl by OH, N-demethylation, conjugation with glutathione, hydroxylation of propylene
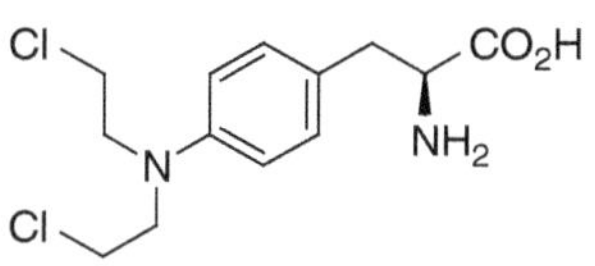	$t_{1/2}$ *F* pb ur	1.4±0.2 h 71±23% 90±5% 12±7%	*V* CL Mwt PSA log *P*	0.5±0.2 5.2±2.9 305.2 66.6 Å^2 0.54	MELPHALAN Antineoplastic Metabolism: sequential displacement of chloride by hydroxide
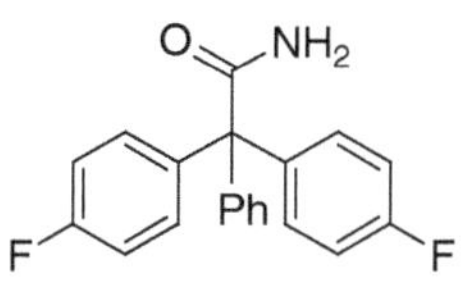	$t_{1/2}$ *F* pb ur	10–13 d – – –	*V* CL Mwt PSA log *P*	– – 323.3 43.1 Å^2 3.97	SENICAPOC, ICA-17043 Gardos channel inhibitor for treatment of sickle cell disease
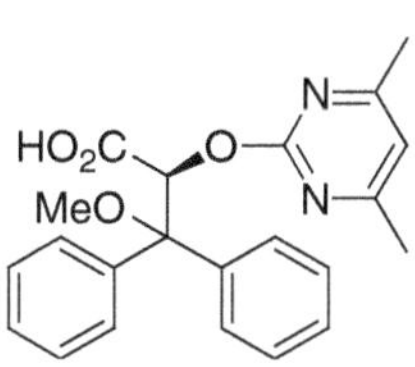	$t_{1/2}$ *F* pb ur	14–17 h 80% 99% 3.3%	*V* CL Mwt PSA log *P*	– – 378.4 81.5 Å^2 6.24	AMBRISENTAN Endothelin antagonist, antihypertensive Metabolism: hydroxylation of methyl groups at pyrimidine; glucuronidation
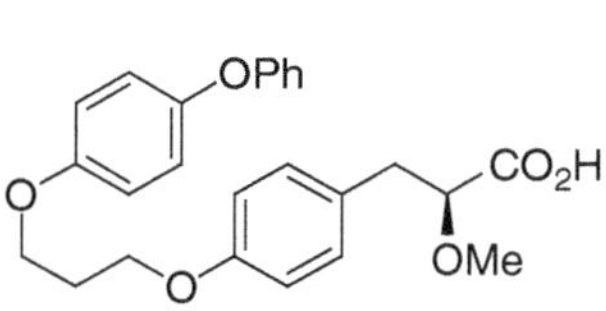	$t_{1/2}$ *F* pb ur	199±75 h* 100%* >99%* 0%*	*V/F* CL/*F* Mwt PSA log *P*	0.16* 0.16* 422.5 74.2 Å^2 4.81	NAVEGLITAZAR *monkey PPAR-α/γ agonist, antidiabetic Metabolism: racemization, dehydrogenation to methoxycinnamic acid, 4-hydroxylation of phenoxy group, then sulfation

(*continued overleaf*)

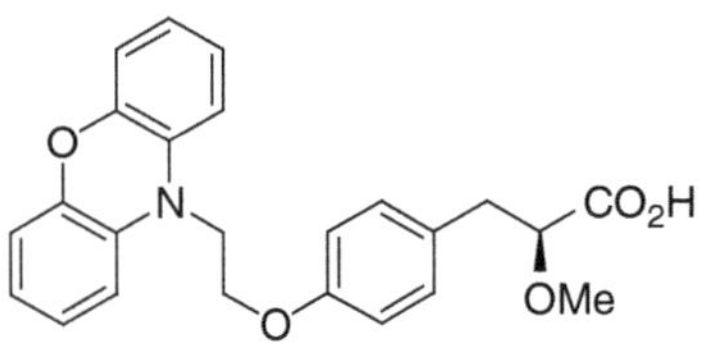

$t_{1/2}$	68–124 h	*V*	0.5*
F	68–93%*	CL	2.3*
pb	–	Mwt	419.5
ur	0%	PSA	68.2 Å^2
		log *P*	3.88

RAGAGLITAZAR
*rat
PPAR-α/γ agonist, antidiabetic; discontinued because of cancerogenicity

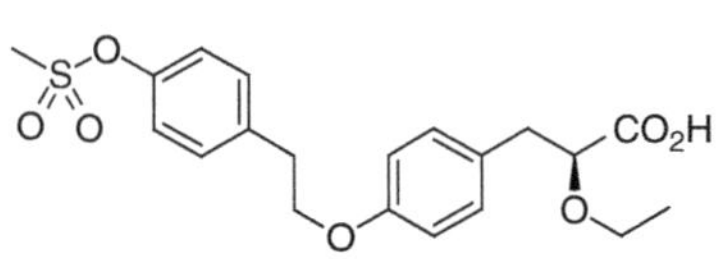

$t_{1/2}$	45 h	*V*	0.14
F	100%	CL	0.04
pb	99.9%	Mwt	408.5
ur	20%	PSA	108 Å^2
		log *P*	2.64

TESAGLITAZAR
PPAR-α/γ agonist, antidiabetic
Metabolism: glucuronidation

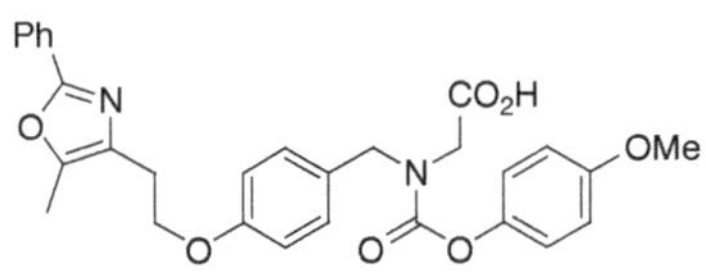

$t_{1/2}$	19–27 h	*V*	0.5–1.7**
F	79%*	CL	12–14***
pb	–	Mwt	516.5
ur	<4%	PSA	111 Å^2
		log *P*	5.59

MURAGLITAZAR
*monkey **rat ***estimate
PPAR-α/γ agonist, antidiabetic
Metabolism: glucuronidation, O-demethylation, O-dealkylation, hydroxylation of oxazolyl-CH_2R, 4-hydroxylation of phenyl group

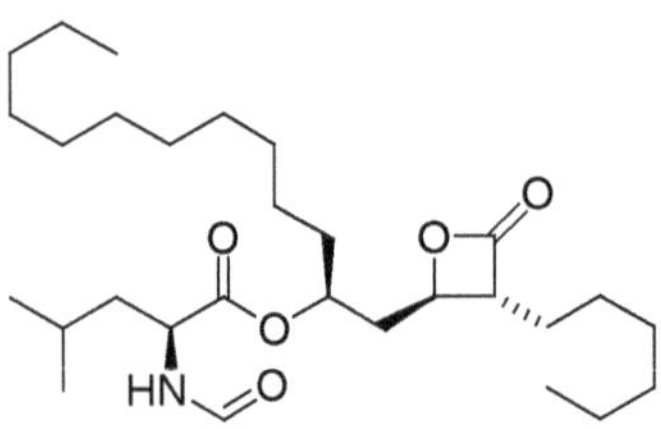

$t_{1/2}$	1–2 h	*V*	–
F	<1%	CL	–
pb	99%	Mwt	495.7
ur	<2%	PSA	81.7 Å^2
		log *P*	7.00

ORLISTAT
Pancreatic lipase inhibitor, antiobesity agent
Metabolism: lactone hydrolysis, ester hydrolysis

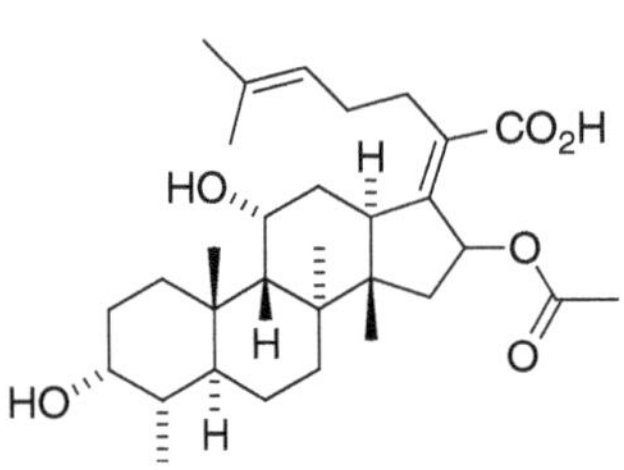

$t_{1/2}$	4–15 h	*V*	0.3
F	91%	CL	
pb	98–99%	Mwt	516.7
ur	<0.1%	PSA	104 Å^2
		log *P*	5.76

FUSIDIC ACID
Antibiotic
Metabolism: oxidation of allylic CH_3 groups, glucuronidation

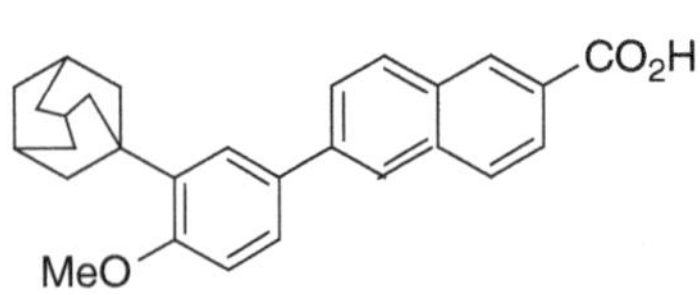

$t_{1/2}$	17±10 h	*V*	–
F	Low	CL	–
pb	>99%	Mwt	412.5
ur	–	PSA	46.5 Å^2
		log *P*	8.21

ADAPALENE
Retinoid for treatment of acne
Metabolism: O-demethylation, hydroxylation

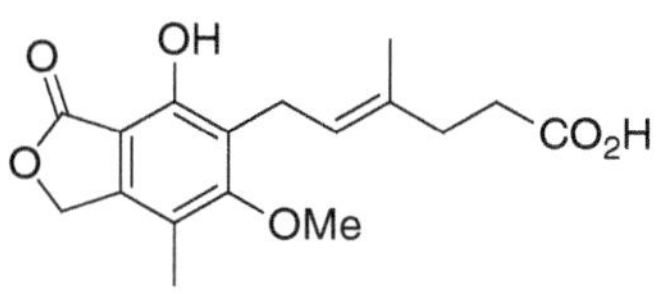

$t_{1/2}$	17–22 h	V	3.6	MYCOPHENOLIC ACID
F	71–100%	CL	2.8	Immunosuppressant
pb	98.7%	Mwt	320.3	Metabolism: glucuronidation
ur	<1%	PSA	93.1 Å^2	
		log P	3.84	

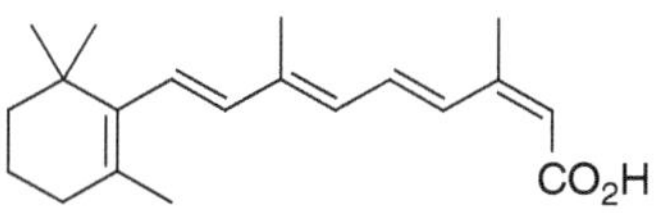

$t_{1/2}$	14±5 h	V/F	7	ISOTRETINOIN
F	25%	CL/F	5.5±2.8	Antiacne agent
pb	99.9%	Mwt	300.4	Metabolite: acyl glucuronide
ur	<1%	PSA	37.3 Å^2	
		log P	6.26	

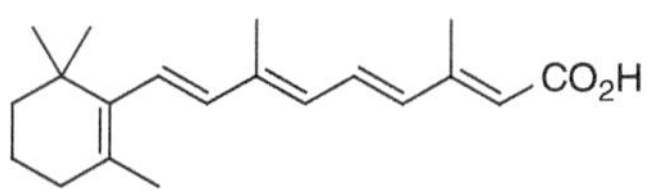

$t_{1/2}$	0.6–1.0 h	V	1.1*	TRETINOIN
F	50%	CL	–	*calculated
pb	–	Mwt	300.4	Acid form of vitamin A
ur	–	PSA	37.3 Å^2	
		log P	6.26	

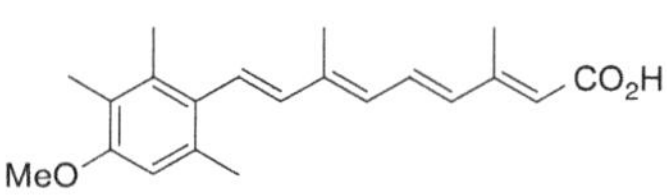

$t_{1/2}$	50–60 h	V	10.2	ACITRETIN
F	60%	CL	–	Metabolite of etretinate,
pb	–	Mwt	326.44	antipsoriatic
ur	>99%	PSA	46.5 Å^2	Metabolism: trans–cis
		log P	6.00	isomerization

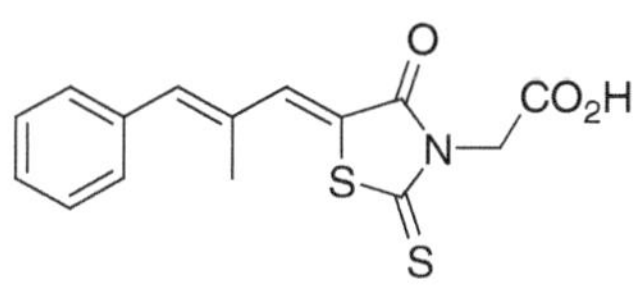

$t_{1/2}$	1.1±0.4 h	V	–	EPALRESTAT
F	–	CL	–	Aldose reductase inhibitor
pb	90%	Mwt	319.4	Metabolism: aromatic
ur	–	PSA	115 Å^2	hydroxylation
		log P	2.02	

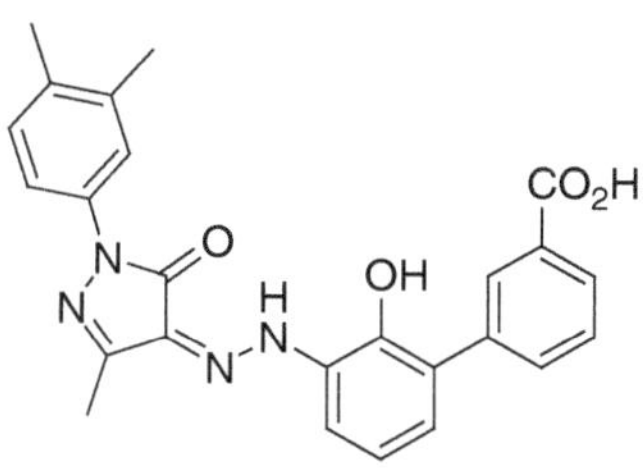

$t_{1/2}$	21–32 h	V	0.13	ELTROMBOPAG
F	50%	CL	0.18	Trombopoetin agonist
pb	99%	Mwt	442.5	Metabolism: oxidation,
ur	0%	PSA	115 Å^2	glucuronidation
		log P	3.71	

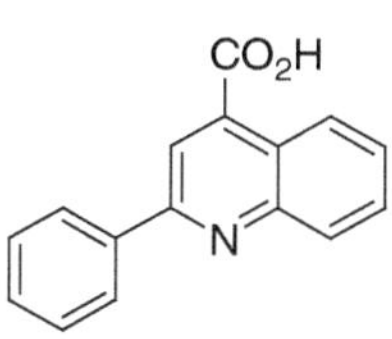

$t_{1/2}$	4 h	V	–	CINCHOPHEN
F	High	CL	–	Analgesic, induces ulcers,
pb	–	Mwt	249.3	hepatotoxic
ur	<5%	PSA	50.2 Å^2	Metabolism: aromatic
		log P	4.18	hydroxylation

(*continued overleaf*)

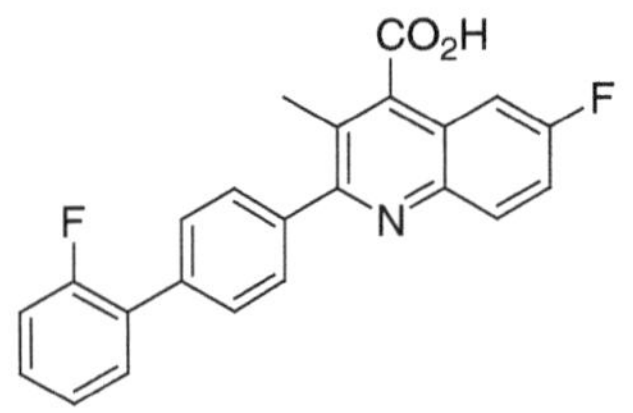	$t^{1}/_{2}$ F pb ur	13–18 h High >98% 0.1–6%*	V CL Mwt PSA log P	0.24±0.08 0.5±0.2 375.4 50.2 Å^2 6.69	BREQUINAR *iv Immunosuppressant
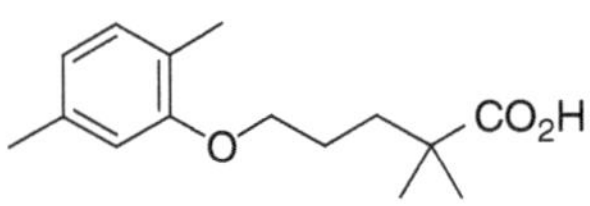	$t^{1}/_{2}$ F pb ur	1.1±0.2 h 98±1% 97–99% <1%	V CL Mwt PSA log P	0.14±0.03 1.7±0.4 250.3 46.5 Å^2 4.30	GEMFIBROZIL Antilipemic Metabolism: hydroxylation and oxidation to CO_2H of aromatic 3-methyl group
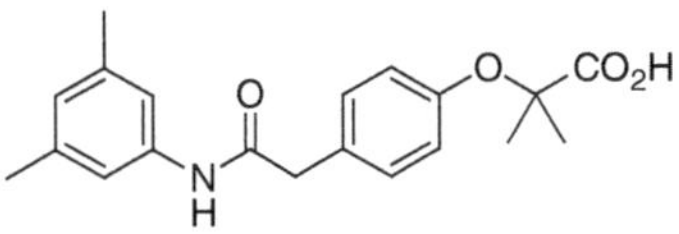	$t^{1}/_{2}$ F pb ur	4–6 h 95% 	V CL Mwt PSA log P	0.2 <1 341.4 75.6 Å^2 3.35	EFAPROXIRAL, RSR-13 Allosteric modifier of hemoglobin Metabolism: glucuronidation
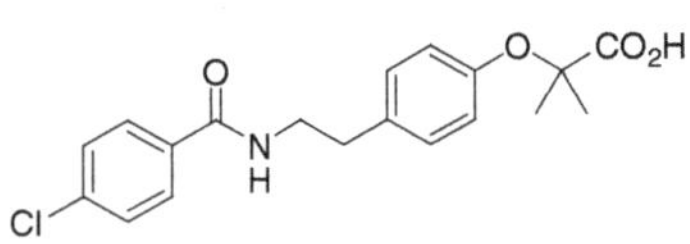	$t^{1}/_{2}$ F pb ur	1.5–3.0 h 100% 95% 35–40%	V CL Mwt PSA log P	0.24–0.35 1.4 361.8 75.6 Å^2 2.50	BEZAFIBRATE Antilipemic Metabolism: glucuronidation
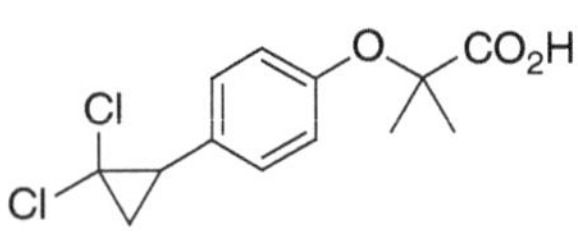	$t^{1}/_{2}$ F pb ur	81 h 80% 99% 7%	V CL Mwt PSA log P	0.15±0.02 0.02 289.2 46.5 Å^2 2.93	CIPROFIBRATE Antilipemic Metabolism: glucuronidation
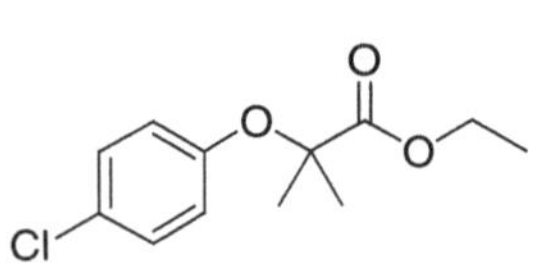	$t^{1}/_{2}$ F pb ur	13±3 h* 95±10%* 97%* 57±2%*	V/F CL/F Mwt PSA log P	0.11±0.02* 0.12±0.01* 242.7 35.5 Å^2 3.88	CLOFIBRATE *acid on oral dosing of ester Antilipemic; prodrug of CPIB
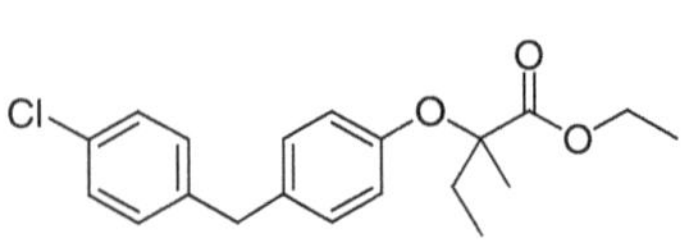	$t^{1}/_{2}$ F pb ur	18±10 h* – >90%* Low	V CL Mwt PSA log P	– 0.5 346.9 35.5 Å^2 6.00	BECLOBRATE *acid on oral dosing of ester Antihyperlipoproteinemic Metabolism: fast hydrolysis of ester, glucuronidation

Structure					
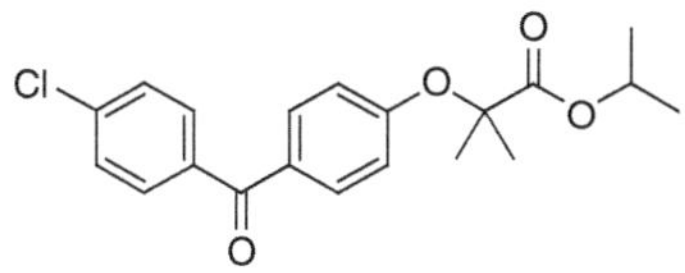	$t^1/_2$ *F* pb ur	20–27 h* 60%* >99%* Negligible	*V* CL Mwt PSA log *P*	0.9* 0.17* 360.8 52.6 Å^2 5.80	FENOFIBRATE *acid on oral dosing of ester Antilipemic, prodrug; ester is not detectable in plasma Further metabolism: glucuronidation
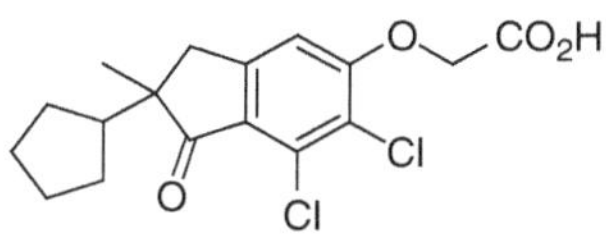	$t^1/_2$ *F* pb ur	2 h – – <10%	*V* CL Mwt PSA log *P*	– – 357.2 63.6 Å^2 3.81	MK-473 Diuretic, uricosuric, antihypertensive Metabolism: cyclopentyl hydroxylation, then oxidation to ketones
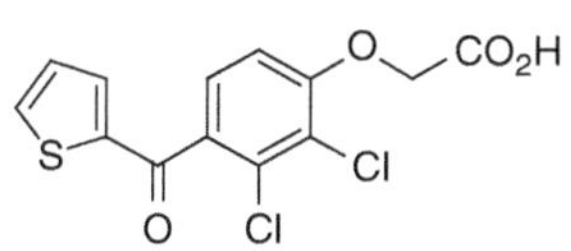	$t^1/_2$ *F* pb ur	6 h – 95–98% –	*V* CL Mwt PSA log *P*	0.1 – 331.2 91.8 Å^2 3.04	TIENILIC ACID, TICRYNAFEN Diuretic, uricosuric, antihypertensive Metabolism: aromatic hydroxylation, S-oxidation to electrophilic thiophene *S*-oxide; withdrawn in 1982 for inducing autoimmune hepatitis
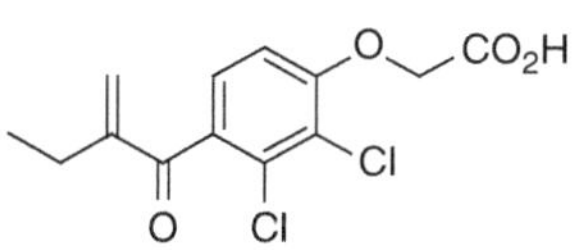	$t^1/_2$ *F* pb ur	2–4 h 7–35% >98% –	*V* CL Mwt PSA log *P*	0.26 8.7 303.1 63.6 Å^2 2.84	ETHACRYNIC ACID Diuretic Active metabolite: addition product to cysteine
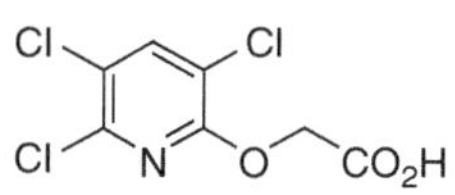	$t^1/_2$ *F* pb ur	5 h – 94–99%* 80%	*V* CL Mwt PSA log *P*	0.1–0.2 0.3 256.5 59.4 Å^2 2.35	TRICLOPYR *dog Herbicide
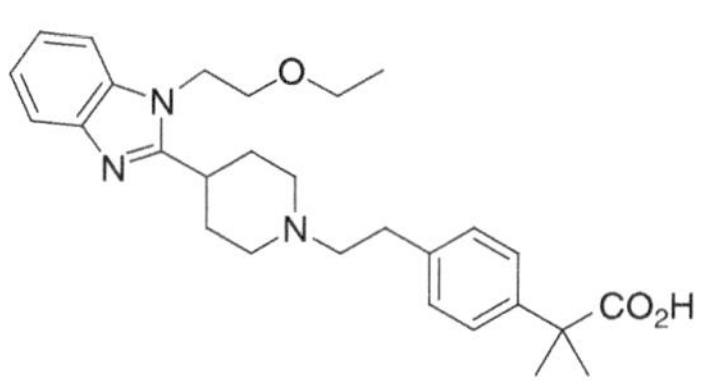	$t^1/_2$ *F* pb ur	5–14 h 25–69%* 85% 33%	*V* CL Mwt PSA log *P*	0.5 4.6 463.6 62.7 Å^2 5.06	BILASTINE *animals Histamine H_1 antagonist No metabolism in humans
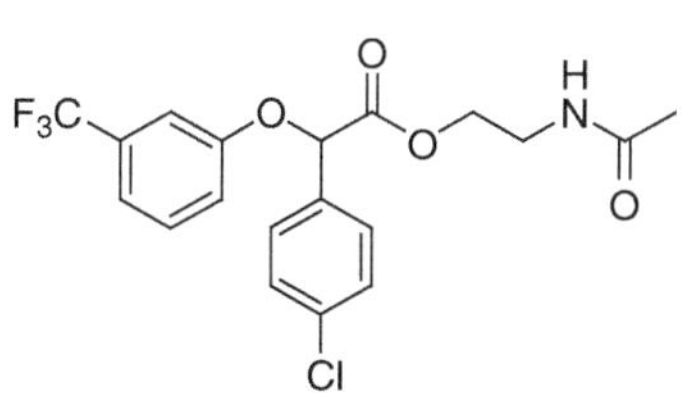	$t^1/_2$ *F* pb ur	24–48 h* High >99%* 39%*	*V* CL Mwt PSA log *P*	– – 415.8 64.6 Å^2 3.77	HALOFENATE *acid on oral dosing of ester Prodrug of halofenic acid

(*continued overleaf*)

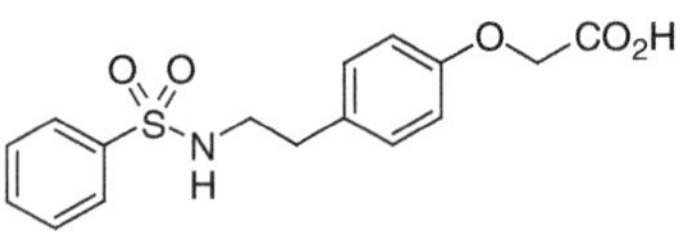

$t^{1}/_{2}$	0.7–2.9 h	V	0.85±0.09
F	–	CL	10
pb	–	Mwt	335.4
ur	52–62%	PSA	101 Å^2
		log P	1.38

SULOTROBAN
Thromboxane A_2 antagonist

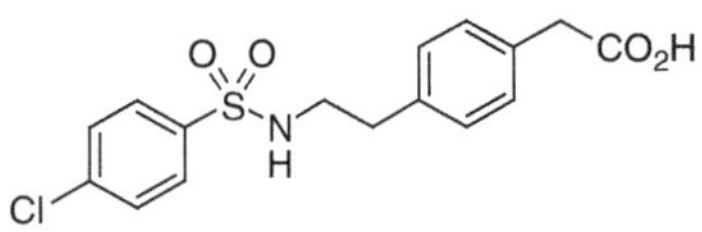

$t^{1}/_{2}$	7 h	V	–
F	High	CL	–
pb	–	Mwt	353.8
ur	40%	PSA	91.9 Å^2
		log P	2.68

DALTROBAN
Thromboxane A_2 antagonist

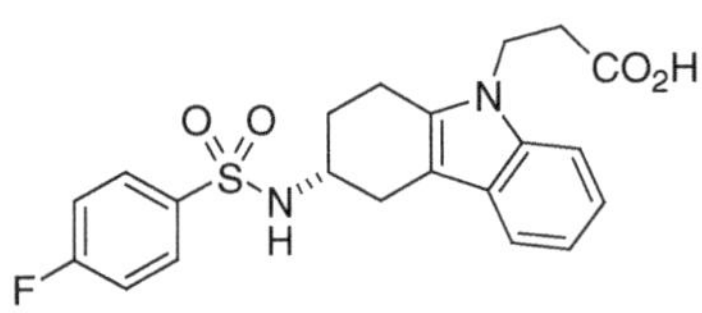

$t^{1}/_{2}$	2–3 h	V	–
F	100%*	CL	12*
pb	>95%	Mwt	416.5
ur	<8%	PSA	96.8 Å^2
		log P	3.02

RAMATROBAN
*dog
Prostaglandin (PGD_2, PGH_2) and thromboxan (TxA_2) antagonist
Metabolism: acyl glucuronidation

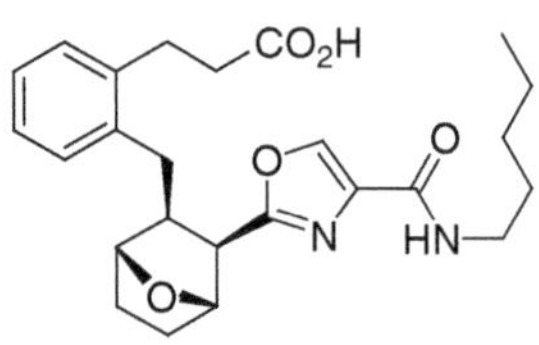

$t^{1}/_{2}$	22 h	V	4.4
F	48%	CL	6.4
pb	–	Mwt	440.5
ur	Low	PSA	102 Å^2
		log P	2.49

IFETROBAN
Thromboxane antagonist
Metabolism: acyl glucuronidation

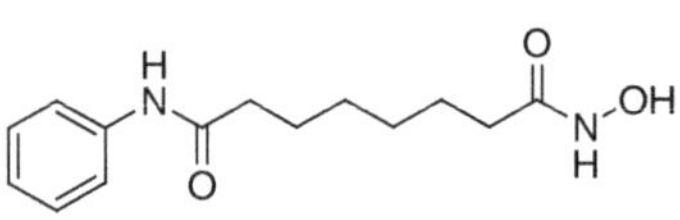

$t^{1}/_{2}$	1–2 h	V	0.5
F	43%	CL	28
pb	71%	Mwt	264.3
ur	Negligible	PSA	78.4 Å^2
		log P	0.86

VORINOSTAT
Histone deacylase inhibitor for treatment of cutaneous T-cell lymphoma
Metabolism: O-glucuronidation, oxidative degradation to succinoyl anilide ($t^{1}/_{2}$ 11 h)

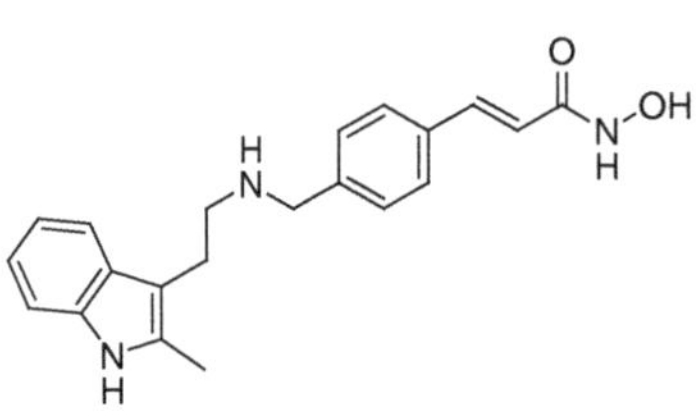

$t^{1}/_{2}$	11–31 h	V	0.4
F	30%	CL	7.3
pb	–	Mwt	349.4
ur	–	PSA	77.2 Å^2
		log P	3.62

PANOBINOSTAT
Histone deacylase inhibitor
Metabolism: reductive N–O bond cleavage, amide hydrolysis, oxidative degradation of cinnamic acid, glucuronidation

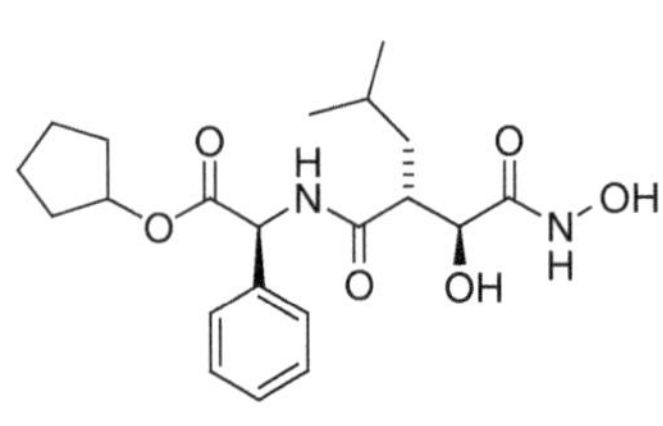

$t^{1}/_{2}$	1–4 h	V	–
F	–	CL	–
pb	–	Mwt	406.5
ur	–	PSA	125 Å^2
		log P	2.24

TOSEDOSTAT
Orally available antineoplastic, aminopeptidase inhibitor
Active metabolite: acid ($t^{1}/_{2}$ 6–11 h)

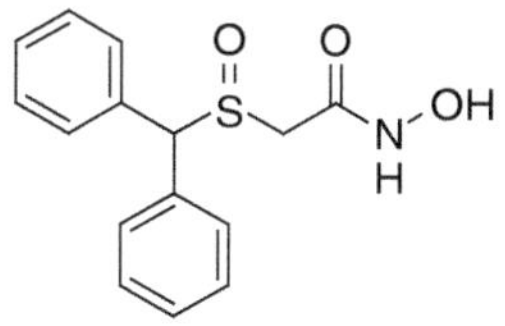	$t^1/_2$ 1 h F 80% pb – ur –	V – CL – Mwt 289.4 PSA 85.6 Å^2 log P 0.77	ADRAFINIL CNS stimulant Metabolism: N-deoxygenation to modafinil
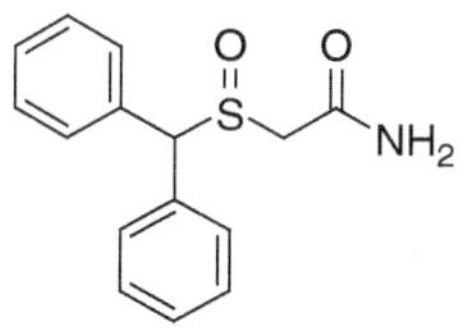	$t^1/_2$ 10–12 h F – pb 62% ur 10%	V 0.9 CL 1.3 Mwt 273.4 PSA 79.4 Å^2 log P 1.18	MODAFINIL CNS stimulant, analeptic Metabolism: hydrolysis to carboxylic acid Armodafinil = *R*-modafinil ($t^1/_2$ 15 h)

$t_{1/2}$, plasma half-life; *F*, oral bioavailability; pb, plasma protein binding; ur, excretion of unchanged drug in urine; V, volume of distribution; CL, clearance; Mwt, molecular weight; PSA, polar surface area; CPIB, *p*-chlorophenoxyisobutyric acid; PPAR, peroxisome proliferator-activated receptor.

Table 31.2 Amino acids, hydroxy acids, and related compounds. *V* in l kg^{-1}; CL in ml min^{-1} kg^{-1}; Mwt in g mol^{-1}.

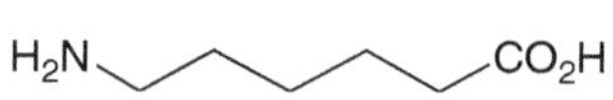	$t^1/_2$ 4.9 h F – pb 8% ur 70–86%	V 0.39 CL 2.7 Mwt 131.2 PSA 63.3 Å^2 log P 0.03	6-AMINOHEXANOIC ACID ε-AMINOCAPROIC ACID Hemostatic Metabolism: oxidative deamination to adipic acid
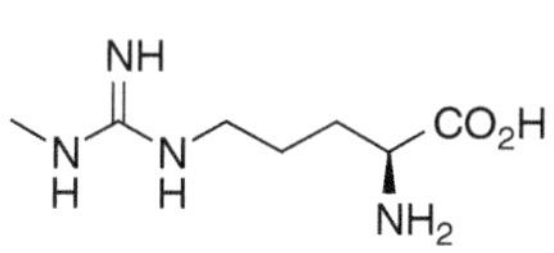	$t^1/_2$ 1.2–2.3 h F – pb Negligible ur <1%	V 0.6–0.8 CL 8.0–9.5 Mwt 188.2 PSA 111 Å^2 log P −0.08	L-*N*-METHYLARGININE Endogenous nitric oxide synthase inhibitor Metabolism: hydrolysis of guanidine to urea (citrulline)
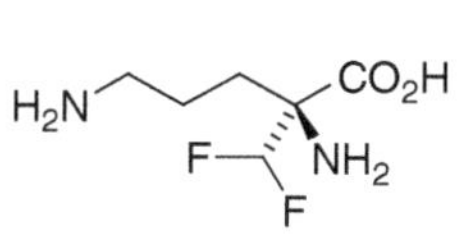	$t^1/_2$ 3–8 h F 54% pb 0% ur 80% (iv)	V 0.34 CL – Mwt 182.2 PSA 89.3 Å^2 log P 0.30	EFLORNITHINE Antineoplastic, antiprotozoal
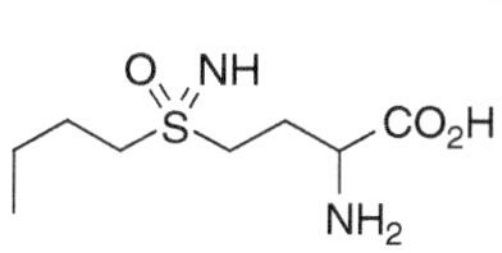	$t^1/_2$ 37 min (iv)* F 2%* pb – ur 60% (iv)	V 0.3–0.5 CL 3–4 Mwt 222.3 PSA 113 Å^2 log P 0.52	BUTHIONINE SULFOXIMINE *mouse Inhibits glutathione biosynthesis

(*continued overleaf*)

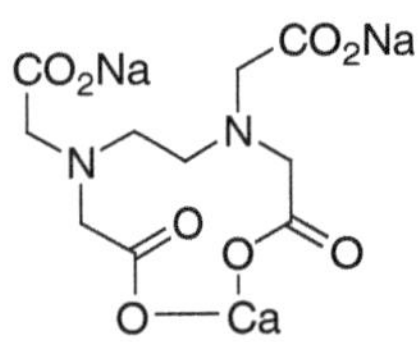	$t^1/_2$	0.5–1.0 h*	V	0.05–0.23	CALCIUM DISODIUM EDETATE,
	F	<5%	CL	0.8	$CaNa_2EDTA$
	pb	0%	Mwt	288.2	*iv
	ur	>95%	PSA	167 Å^2	Chelator for heavy-metal poisoning
			log P	−4.65	
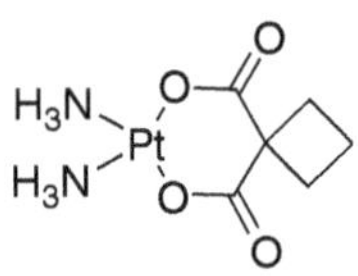	$t^1/_2$	2.0±0.2 h	V	0.24±0.03	CARBOPLATIN
	F	–	CL	1.5±0.3	Antineoplastic
	pb	0%	Mwt	371.3	
	ur	77±5%	PSA	107.9 Å^2	
			log P	−5.20	
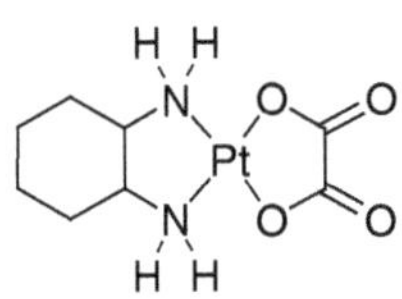	$t^1/_2$	25–27 h	V	8.3–11.6	OXALIPLATIN
	F	–	CL	2.2–2.4	Antineoplastic
	pb	>90%	Mwt	397.3	
	ur	40–50%	PSA	85.8 Å^2	
			log P	−4.96	
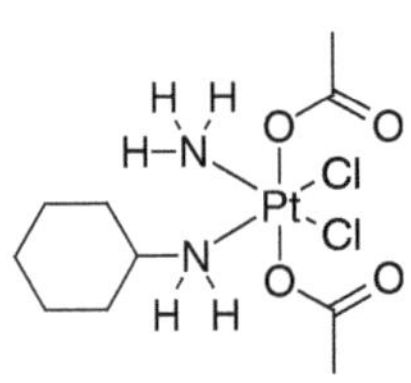	$t^1/_2$	5–12 h	V	–	SATRAPLATIN
	F	–	CL	5.2–6.2*	*monkey
	pb	90–94%	Mwt	500.3	Antineoplastic
	ur	10%	PSA	96.9 Å^2	Metabolism: reductive abstraction
			log P	−3.18	of both acetoxy groups
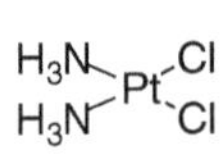	$t^1/_2$	0.5±0.1 h	V	0.3	CISPLATIN
	F	–	CL	6	Antineoplastic
	pb	>90%	Mwt	300.1	
	ur	23±9% (iv)	PSA	55.3 Å^2	
			log P	−4.89	
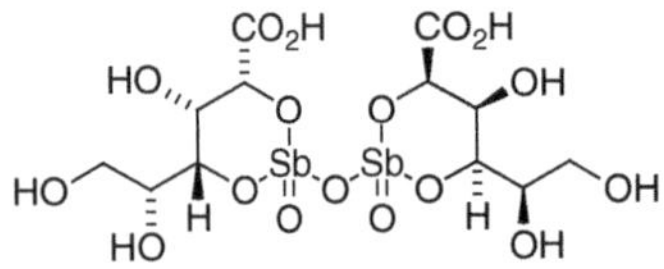	$t^1/_2$	33–76 h	V	0.22	STIBOGLUCONATE
	F	–	CL	–	Anthelmintic
	pb	–	Mwt	679.8	iv or im administration only
	ur	>80%	PSA	276.3 Å^2	
			log P	−5.34	
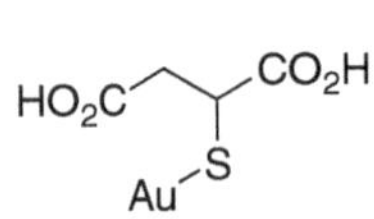	$t^1/_2$	25±5 d	V	0.26±0.05	GOLD SODIUM THIOMALATE
	F	–	CL	7.0±0.6	AUROTHIOMALATE
	pb	>95%	Mwt	346.1	(Acid shown)
	ur	70%	PSA	74.6 Å^2	Antipruritic, treatment of
			log P	−1.18	rheumatoid arthritis

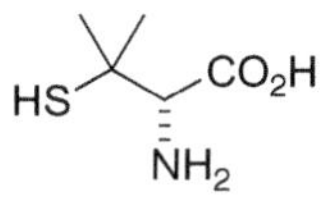	$t^{1/2}$ 1.5–3.0 h *F* 40–70% pb 80% ur 10–40%	*V* 0.9 CL 10.7 Mwt 149.2 PSA 102 Å^2 log *P* 0.85	PENICILLAMINE Antirheumatic, chelating agent (copper) Metabolism: disulfide formation with itself and cysteine, S-methylation
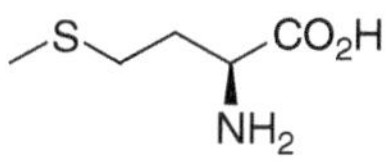	$t^{1/2}$ 1.0–1.5 h *F* 100% pb – ur 5–10%	*V* – CL – Mwt 149.2 PSA 88.6 Å^2 log *P* 0.22	METHIONINE Hepatoprotectant, antidote, urinary acidifier, nutritional supplement
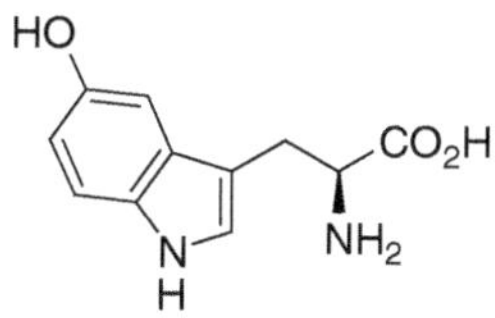	$t^{1/2}$ 2–7 h *F* 50–85% pb 60% ur –	*V* – CL 1.7–3.8 Mwt 220.2 PSA 99.3 Å^2 log *P* −0.29	OXITRIPTAN, 5-HYDROXYTRYPTOPHAN Prodrug of serotonin, antidepressant, antimigraine; crosses bbb Metabolism (in CNS): decarboxylation, then oxidation to hydroxyindolylacetic acid
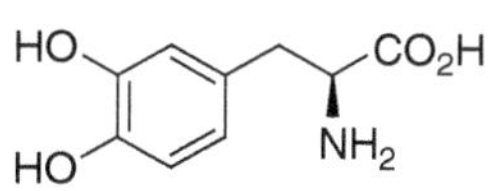	$t^{1/2}$ 1.4±0.4 h *F* 41±16% pb – ur <1%	*V* 1.7±0.4 CL 23±4 Mwt 197.2 PSA 104 Å^2 log *P* −1.15	LEVODOPA Antiparkinsonian
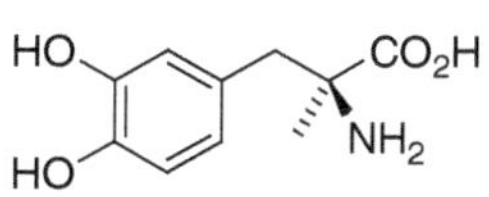	$t^{1/2}$ 1.8±0.6 h *F* 42±16% pb 1–16% ur 40±13%	*V* 0.46±0.15 CL 3.7±1.0 Mwt 211.2 PSA 104 Å^2 log *P* −0.74	METHYLDOPA Antihypertensive
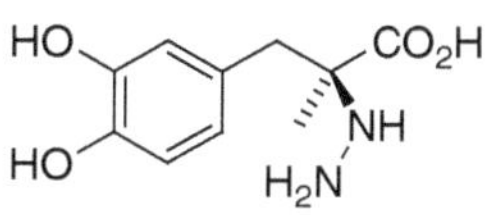	$t^{1/2}$ 2.1 h *F* 40–70% pb 76% ur 30%	*V* – CL – Mwt 226.2 PSA 116 Å^2 log *P* −1.75	CARBIDOPA Antiparkinsonian (in combination with levodopa); does not cross bbb Metabolism: reductive N−N bond cleavage
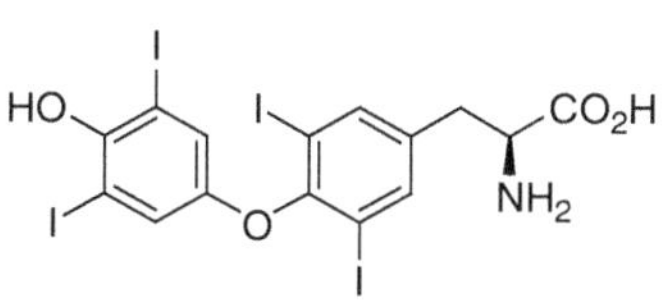	$t^{1/2}$ 5–7 d *F* 91% pb >99.9% ur –	*V* 10 CL – Mwt 776.9 PSA 92.8 Å^2 log *P* 4.72	THYROXINE, T_4 Thyroid hormone

(continued overleaf)

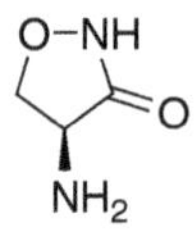	$t_{1/2}$ *F* pb ur	4–30 h 70–90% <20% 60–70%	*V* CL Mwt PSA log *P*	0.2 – 102.1 64.3 Å^2 −2.99	CYCLOSERINE Bacteriostatic, crosses bbb
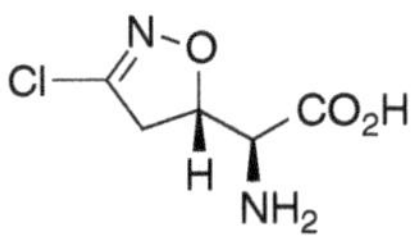	$t_{1/2}$ *F* pb ur	10±4 h (iv) – Negligible 14–19% (iv)	*V* CL Mwt PSA log *P*	0.5 0.69 178.6 84.9 Å^2 −0.78	ACIVICIN Antineoplastic
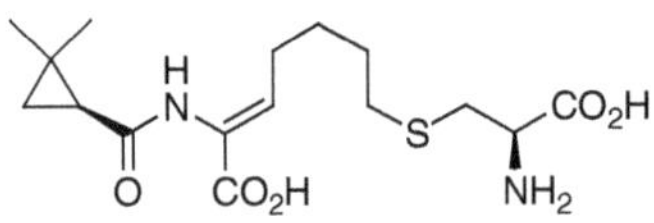	$t_{1/2}$ *F* pb ur	0.8±0.1 h Low 35% 70±3%	*V* CL Mwt PSA log *P*	0.20±0.03 3.0±0.3 358.5 155 Å^2 2.42	CILASTATIN Values for simultaneous administration with imipenem Dipeptidase inhibitor
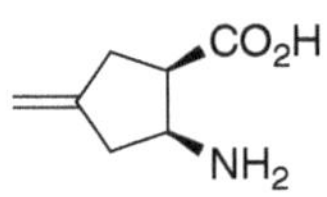	$t_{1/2}$ *F* pb ur	6–7 h 60–100%* <1% 70–90%	*V* CL Mwt PSA log *P*	0.6–0.9** 1.7–5.0** 141.2 63.3 Å^2 −0.16	ICOFUNGIPEN, PLD-118 *animals **rabbit Antifungal
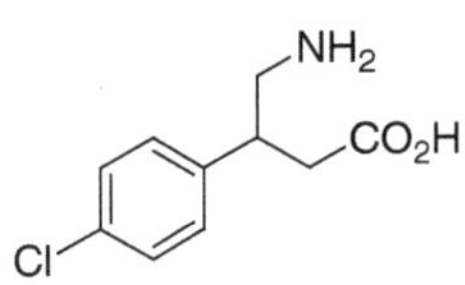	$t_{1/2}$ *F* pb ur	3–7 h 95% 30% 70–80%	*V* CL Mwt PSA log *P*	0.84 2.6 213.7 63.3 Å^2 0.78	BACLOFEN Muscle relaxant, GABA agonist Metabolism: deamination to 4-hydroxy-3-arylbutyric acid
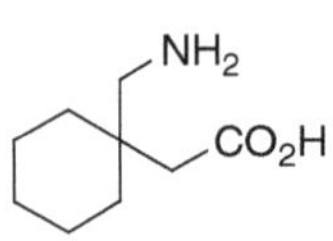	$t_{1/2}$ *F* pb ur	6.5±1.0 h 60% 0% 100%	*V* CL Mwt PSA log *P*	0.80±0.09 1.6±0.3 171.2 63.3 Å^2 0.62	GABAPENTIN Antiepileptic
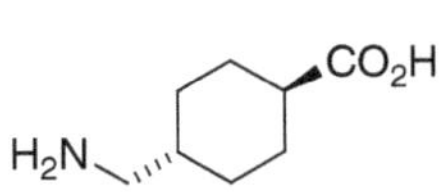	$t_{1/2}$ *F* pb ur	2–3 h 33–35% 3% >95%	*V* CL Mwt PSA log *P*	0.1–0.2 1.6–1.8 157.2 63.3 Å^2 0.50	TRANEZAMIC ACID TRANEXAMIC ACID Hemostatic (antifibrinolytic); crosses bbb
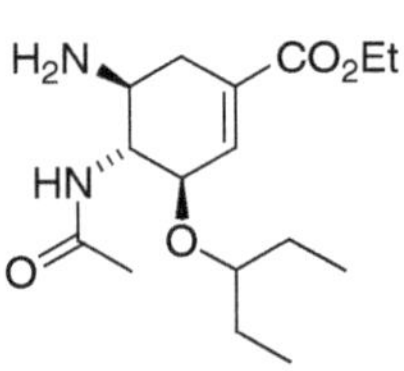	$t_{1/2}$ *F* pb ur	6–10 h* 75%* 3%* 3%	*V* CL Mwt PSA log *P*	0.4* 4.8* 312.4 90.7 Å^2 1.71	OSELTAMIVIR Prodrug of acid, neuraminidase inhibitor, antiviral *acid on oral administration of ester

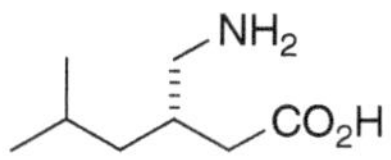	$t^1\!/_2$ F pb ur	5–7 h >90% 0% 99%	V CL Mwt PSA log P	0.5 – 159.2 63.3 Å^2 1.12	PREGABALIN Anticonvulsant
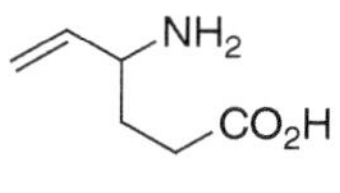	$t^1\!/_2$ F pb ur	4–8 h 50–65% – 50–70%	V CL Mwt PSA log P	0.8 1.7 129.2 63.3 Å^2 −0.10	VIGABATRIN Antiepileptic Racemate (only the S enantiomer is active) Metabolism: lactam formation
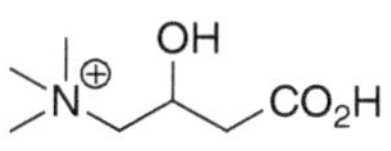	$t^1\!/_2$ F pb ur	2–15 h 9–25% 0% 70–80%	V CL Mwt PSA log P	0.26–0.82 1.5–1.7 147.2 60.4 Å^2 −5.56	CARNITINE Antihyperlipoproteinemic Metabolite: trimethylamine
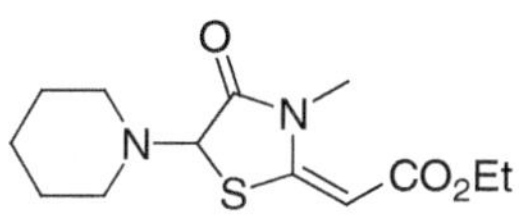	$t^1\!/_2$ F pb ur	2 h >90% – –	V CL Mwt PSA log P	1.8 – 284.4 75.2 Å^2 2.75	ETOZOLIN Diuretic Metabolism: hydrolysis of ester to acid (ozolinon, $t^1\!/_2$ 6–10 h), then glucuronidation
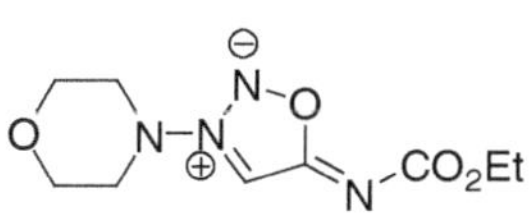	$t^1\!/_2$ F pb ur	1–2 h 44–59% 3–11% <2%	V CL Mwt PSA log P	1.0–1.9 11–17 242.2 84.0 Å^2 −3.32	MOLSIDOMINE Prodrug of nitrous oxide Metabolism: hydrolysis and decarboxylation of carbamate, then rearrangement to *N*-morpholino-*N*-nitrosoaminoacetonitrile
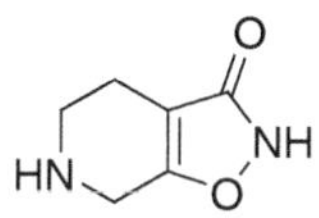	$t^1\!/_2$ F pb ur	1.5–2.0 h 84–93% <2% 50–60%	V CL Mwt PSA log P	– 6.9 140.1 50.4 Å^2 0.71	GABOXADOL Hypnotic Metabolism: O-glucuronidation
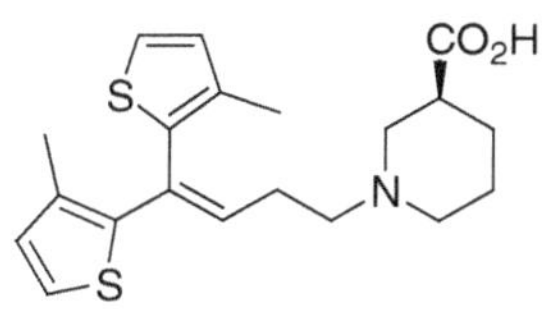	$t^1\!/_2$ F pb ur	5–9 h 90% 96% 2%	V CL Mwt PSA log P	1 3 375.6 97.0 Å^2 4.03	TIAGABINE Antiepileptic Metabolism (CYP3A4): glucuronidation, thiophene oxidation to 2-thiophenone
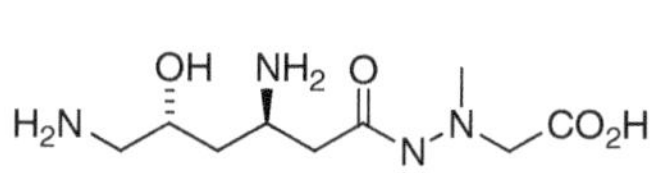	$t^1\!/_2$ F pb ur	2 h – – –	V CL Mwt PSA log P	0.35 – 248.3 142 Å^2 −4.14	NEGAMYCIN Antibacterial

(*continued overleaf*)

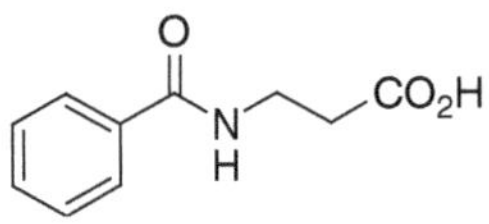	$t^{1}/_{2}$ 0.6 h (iv) F – pb 83% ur 40%	V 0.28 CL 9.2 Mwt 193.2 PSA 66.4 Å^2 log P 0.93	BETAMIPRON Antibacterial adjunct (renal protection)
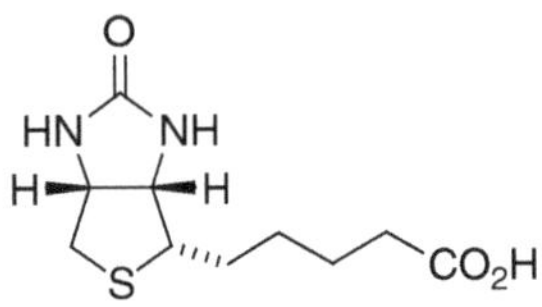	$t^{1}/_{2}$ 26 h F – pb 80% ur 40–76%	V – CL – Mwt 244.3 PSA 104 Å^2 log P 0.86	BIOTIN Vitamin Metabolism: oxidation to sulfoxide, oxidative degradation of 4-carboxybutyl to 2-carboxyethyl (bisnorbiotin)
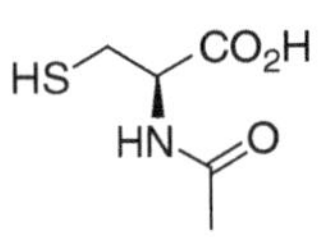	$t^{1}/_{2}$ 2 h* F 10% pb 83% ur 20–30% (iv)	V 0.55 CL 3.1 Mwt 163.2 PSA 105 Å^2 log P −0.70	*N*-ACETYLCYSTEINE *iv: 5.5 h Mucolytic, high first-pass metabolism (N-deacetylation)
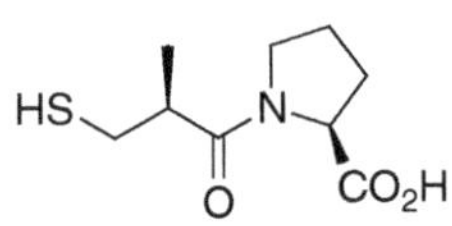	$t^{1}/_{2}$ 2.2±0.5 h F 65% pb 30±6% ur 38±11%	V 0.8±0.2 CL 12.0±1.4 Mwt 217.3 PSA 96.4 Å^2 log P 0.27	CAPTOPRIL ACE inhibitor, antihypertensive Metabolism: oxidative disulfide formation, S-methylation
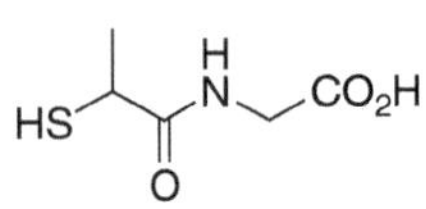	$t^{1}/_{2}$ 19–53 h F 63% pb – ur –	V 5±2 CL 2.6 Mwt 163.2 PSA 105 Å^2 log P −0.67	TIOPRONIN Antidote (heavy metals), mucolytic, hepatoprotectant Metabolite: mixed disulfide with cysteine
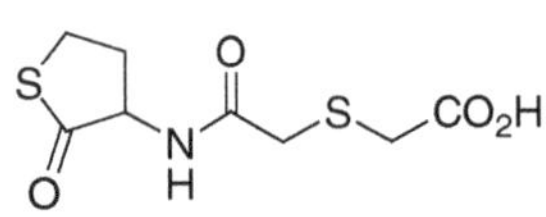	$t^{1}/_{2}$ 1–3 h F – pb 65% ur –	V – CL 41 Mwt 249.3 PSA 134 Å^2 log P −0.32	ERDOSTEINE Active metabolites: *N*-acetylhomocysteine, homocysteine, hydrolyzed thiolactone
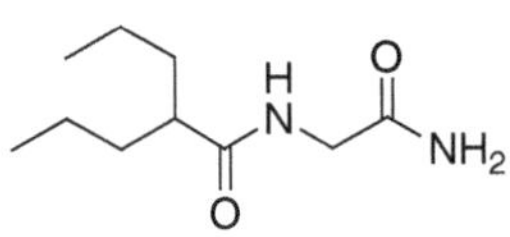	$t^{1}/_{2}$ 6–9 h F – pb – ur 10–20%	V – CL 1.3 Mwt 200.3 PSA 72.2 Å^2 log P 1.02	VALROCEMIDE Anticonvulsant Metabolites: *N*-valproylglycine, valproic acid
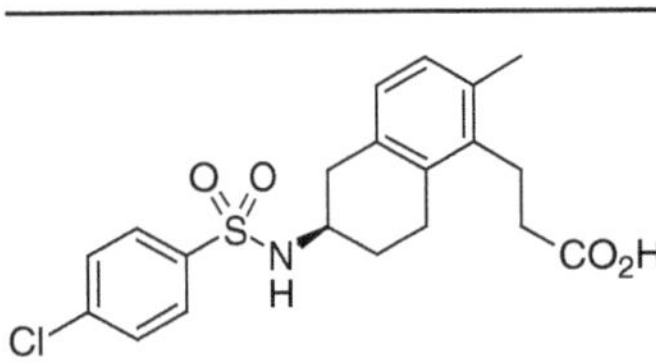	$t^{1}/_{2}$ 6–10 h F – pb – ur –	V – CL – Mwt 407.9 PSA 91.9 Å^2 log P 3.35	TERUTROBAN Antiplatelet agent, antithrombotic

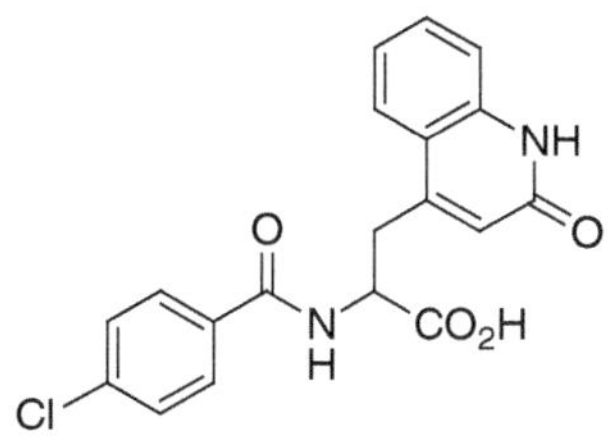	$t_{1/2}$ F pb ur	1–2 h 5% 98% 10%	V CL Mwt PSA log P	0.3 (rat) 21±3 (rat) 370.8 95.5 Å^2 2.56	REBAMIPIDE Antiulcerative Metabolism: 8-hydroxylation
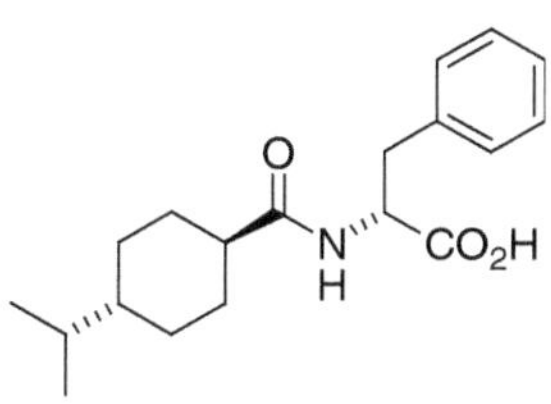	$t_{1/2}$ F Pb ur	1.4–2.9 h 72% 97% 13–16%	V CL Mwt PSA log P	0.13 1.5–2.0 317.4 66.4 Å^2 2.35	NATEGLINIDE Antidiabetic Metabolism: acyl glucuronidation, conversion of isopropyl group to isopropenyl, 1-hydroxy-*i*Pr, 2-hydroxy-*i*Pr, and 1,2-dihydroxy-*i*Pr
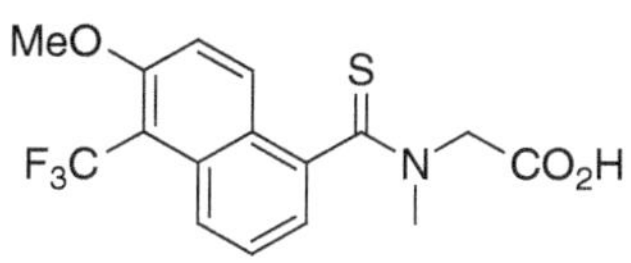	$t_{1/2}$ F pb ur	10–15 h 68% (dog) >99% 30–40%	V CL Mwt PSA log P	0.23 0.5 357.4 81.9 Å^2 3.59	TOLRESTAT Aldose reductase inhibitor for treatment of diabetic neuropathy Metabolism: oxidation to *S*-oxide, then to amide, O-demethylation, then O-sulfation; withdrawn in 1997 because of hepatotoxicity

$t_{1/2}$, plasma half-life; *F*, oral bioavailability; pb, plasma protein binding; ur, excretion of unchanged drug in urine; V, volume of distribution; CL, clearance; Mwt, molecular weight; PSA, polar surface area.

32
Carboxylic Esters

Because of their metabolic lability, carboxylic esters are often used as prodrugs for carboxylic acids, phenols, or alcohols. Esters are more lipophilic than carboxylic acids, and esters of hydrophilic acids will usually be absorbed more readily from the gastrointestinal tract than the parent acid. Ethyl or methyl esters, or acetic acid esters of primary aliphatic alcohols are usually hydrolyzed *in vivo* within minutes and can hardly be detected in plasma (Scheme 32.1).

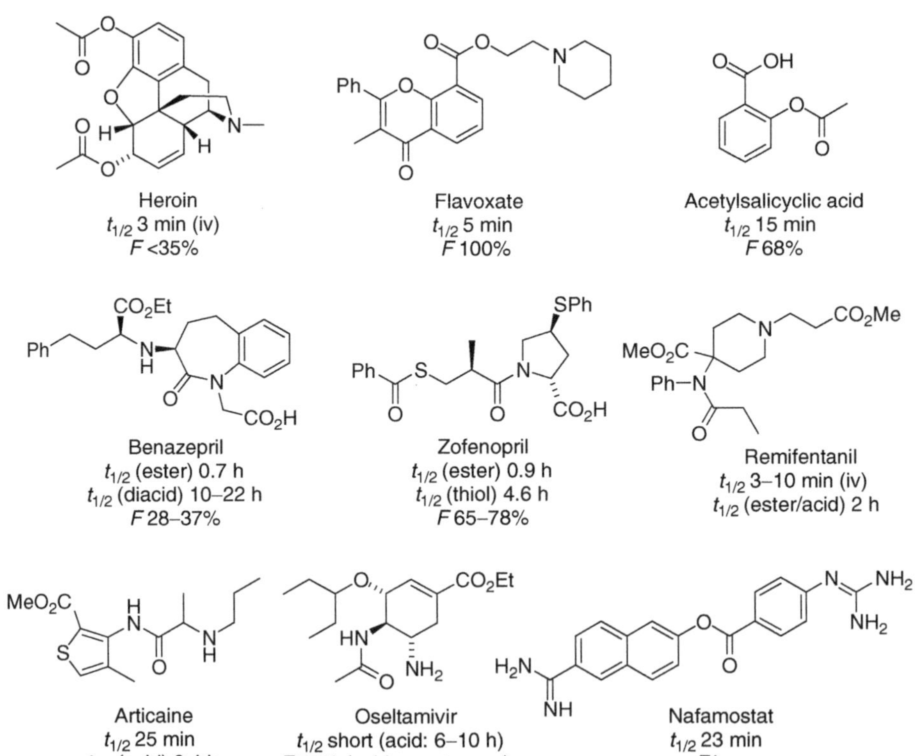

Scheme 32.1 Carboxylic esters as drugs.

Lead Optimization for Medicinal Chemists: Pharmacokinetic Properties of Functional Groups and Organic Compounds, First Edition. Florencio Zaragoza Dörwald.

Some types of ester are, however, sufficiently stable to serve as drugs. Also, for topical or ophthalmic applications, where esterase activity is less intense, carboxylic esters can be sufficiently stable. Esters of tertiary alcohols or alcohols containing tertiary amines (e.g., denaverine) are particularly stable toward enzymatic hydrolysis. Similarly, esters with a carbonyl group in conjugation with electron-donating groups (e.g., vinylogous carbonates or carbamates, electron-rich arylcarboxylic or heteroarylcarboxylic esters), or esters of sterically crowded carboxylic acids, are resistant toward nucleophilic attack and less prone to hydrolysis *in vivo* [1] (Scheme 32.2).

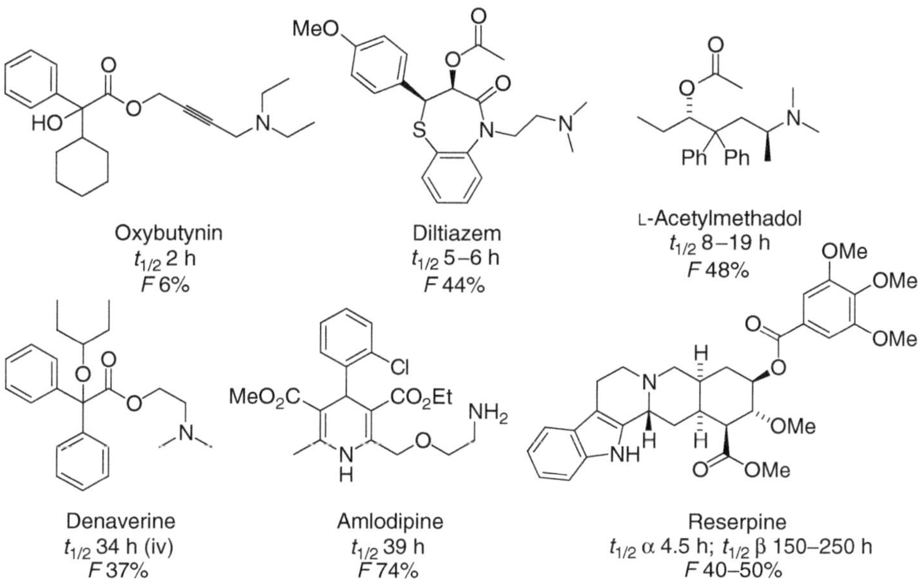

Scheme 32.2 Metabolically resistant esters.

Lactones are chemically and enzymatically more labile than noncyclic esters because of the high-energy *cis* conformation of the alkoxycarbonyl group in lactones. However, five- and six-membered lactones will often be formed spontaneously from the corresponding hydroxy acids at neutral or slightly acidic pH. Accordingly, the metabolic hydrolysis of lactones is often reversible, for example, for the anti-inflammatory rofecoxib, for the statins, or for the tecans. β-Lactones are reactive acylating and alkylating agents that undergo irreversible ring opening.

Reference

1. Buchwald, P. and Bodor, N. (1999) Quantitative structure–metabolism relationships: steric and nonsteric effects in the enzymatic hydrolysis of noncongener carboxylic esters. *J. Med. Chem.*, **42**, 5160–5168.

33
Amides

Amides are a ubiquitous functional group in living organisms (peptides, proteins). Most unnatural amides are sufficiently stable to serve as oral drugs, and only rarely quick enzymatic hydrolysis occurs (Table 33.1). Amides that are chemically reactive toward nucleophiles, such as those derived from highly electrophilic carboxylic acids (e.g., trifluoroacetamides), will also undergo fast metabolic hydrolysis *in vivo*. If an amide resembles the final or first amide bond in a peptide with a positive or negative charge close to the amide bond, quick hydrolysis by an aminopeptidase or carboxypeptidase may occur (Scheme 33.1). Similarly, endopeptidases may cleave uncharged peptidomimetics. Peptidases are, however, highly selective, and most unnatural amides will not be substrates of these enzymes. As in esters and carbamates, steric shielding will also inhibit the enzymatic hydrolysis of amides.

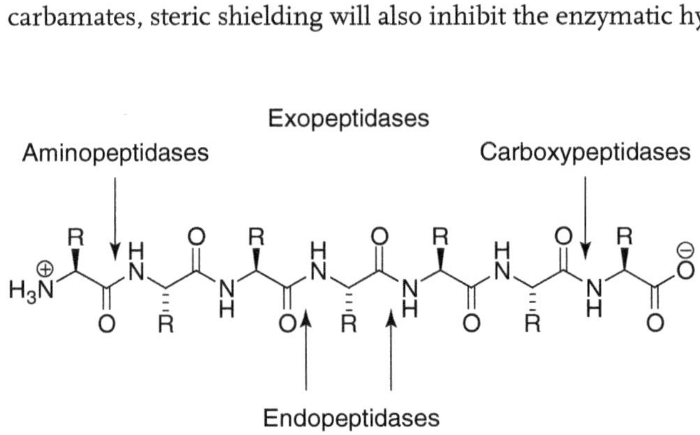

Scheme 33.1 Metabolism of peptides.

The absorption of highly insoluble amines or alcohols from the gastrointestinal tract can be enhanced by acylation with glycine. Aminopeptidases present in the epithelium of the intestine will rapidly deacylate the prodrug and regenerate the active compound, as exemplified by the benzodiazepine prodrug rilmazafone and the antihypotensive midodrine (Scheme 33.2).

Lead Optimization for Medicinal Chemists: Pharmacokinetic Properties of Functional Groups and Organic Compounds, First Edition. Florencio Zaragoza Dörwald.

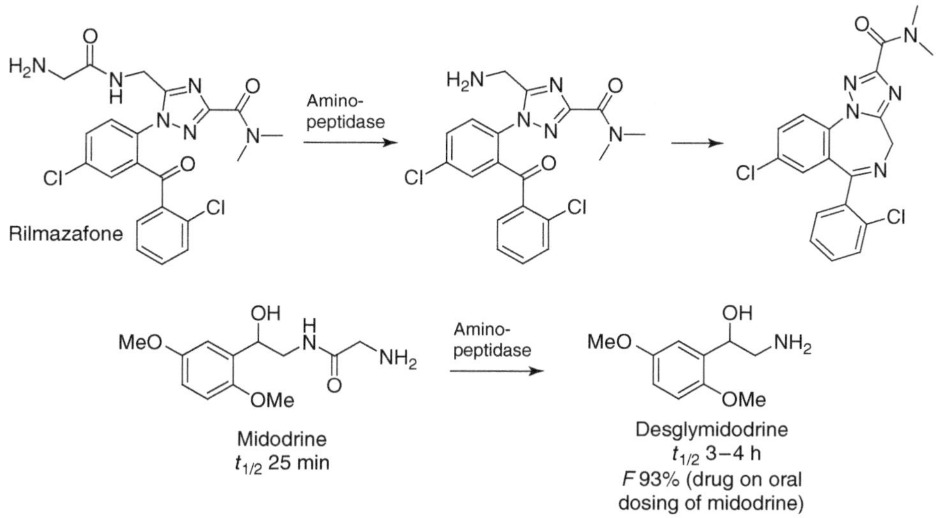

Scheme 33.2 Amides as prodrugs.

Acylation with glycine, alanine, or short peptides can be used to enhance solubility of a compound for a liquid, parenteral formulation (Scheme 33.3). A parenteral formulation may be useful even if the drug is orally bioavailable, for instance, if the drug is hepatotoxic or must be given to unconscious patients. Parenteral dosing also evades first-pass metabolism and CYP (cytochrome *P*450) induction or inhibition. Amide-based prodrugs can also be used as slow-release formulations of short-lived drugs.

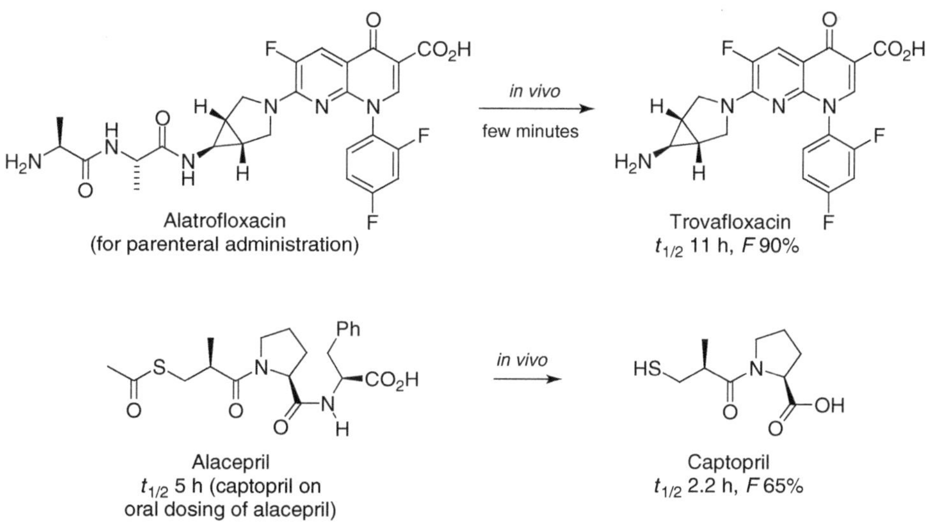

Scheme 33.3 Peptides as prodrugs.

The metabolic stability of N-unsubstituted amides ($RCONH_2$) is difficult to predict. Sterically accessible, lipophilic amides are often hydrolyzed quickly, whereas hydrophilic or sterically shielded amides will be more stable. Some of the amides sketched in Scheme 33.4 have low oral bioavailabilities and half-lives because of amide-bond hydrolysis, but some closely related amides and esters are not extensively hydrolyzed to carboxylic acids.

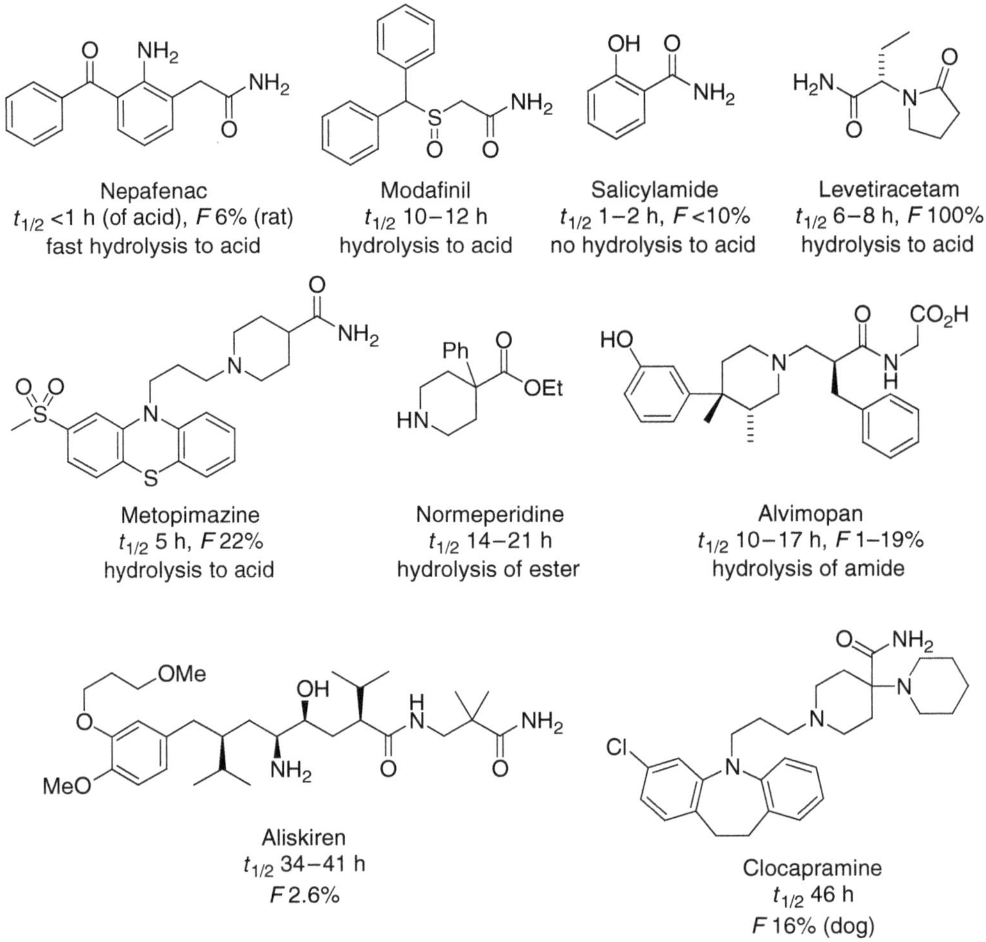

Scheme 33.4 N-Unsubstituted amides as drugs.

N-Alkylamides may be dealkylated via C-hydroxylation followed by hemiaminal hydrolysis, similar to the oxidative dealkylation of amines. This metabolic transformation is often observed in lactams, such as benzodiazepines (Scheme 33.5), and in cyclic imides and ureides (e.g., in barbiturates).

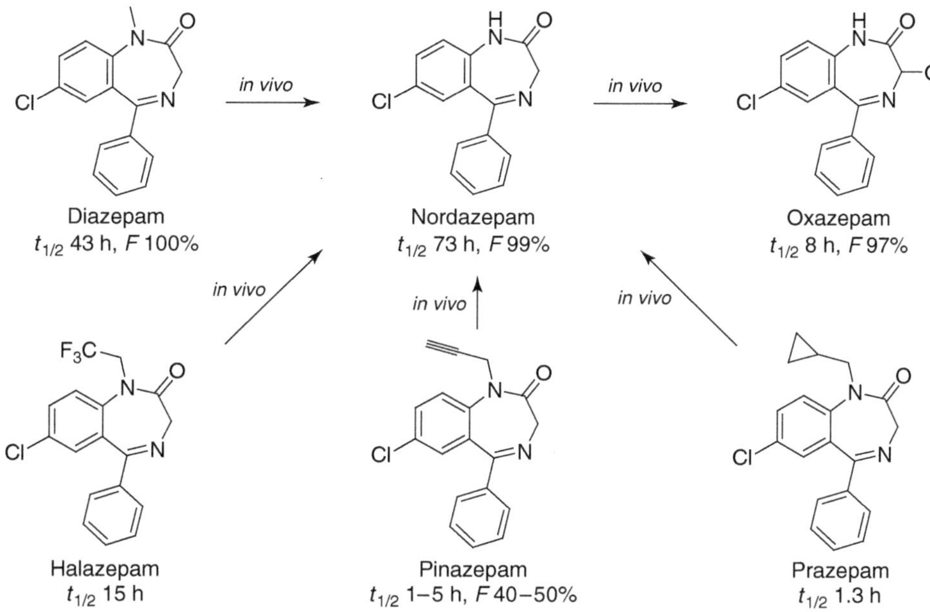

Scheme 33.5 N-Dealkylation of benzodiazepines.

Table 33.1 Amides. V in l kg^{-1}; CL in ml min^{-1} kg^{-1}; Mwt in g mol^{-1}.

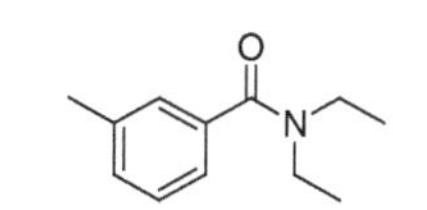	$t^1/_2$ F pb ur	2.5 h 48%* – 10–14%	V CL Mwt PSA log P	2.1 – 191.3 20.3 Å^2 2.42	DEET, AUTAN *from dermal absorption Insect repellant Metabolism: deethylation, oxidation of CH_3 to CO_2H
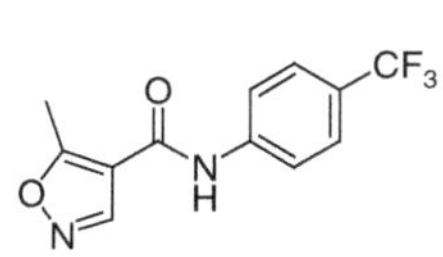	$t^1/_2$ F pb ur	14 d* 82–94%* 99.4%* Negligible	V CL Mwt PSA log P	0.08–0.44* 0.007–0.012* 270.2 55.1 Å^2 2.29	LEFLUNOMIDE *teriflunomide (see below) on oral dosing of leflunomide; antirheumatic, immunomodulator; leflunomide is a prodrug of its active cyanoketone metabolite teriflunomide
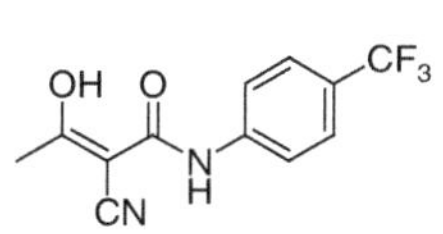	$t^1/_2$ F pb ur	15–18 d <40% >99% –	V CL Mwt PSA log P	– – 270.2 73.1 Å^2 1.52	TERIFLUNOMIDE Dihydroorotate dehydrogenase inhibitor, agent for multiple sclerosis; active metabolite of leflunomide

(continued overleaf)

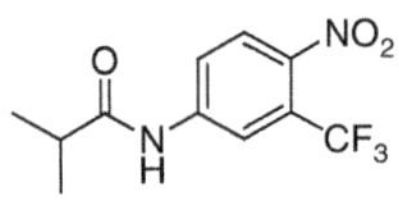

$t_{1/2}$	4–22 h*	V	–
F	–	CL	–
pb	94–96%	Mwt	276.2
ur	Negligible	PSA	74.9 Å^2
		log P	3.52

FLUTAMIDE
*2-hydroxyisopropyl metabolite; Nonsteroidal antiandrogen, prodrug of the active 2-hydroxy-isopropyl metabolite; Metabolism: single, twofold, and threefold hydroxylation of isopropyl group, reduction of nitro to amino, hydrolysis of amide, aromatic hydroxylation

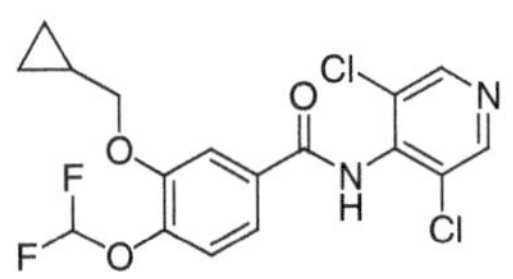

$t_{1/2}$	12–20 h	V	–
F	69–92%	CL	–
pb	99%	Mwt	403.2
ur	–	PSA	60.5 Å^2
		log P	2.31

ROFLUMILAST
PDE4 inhibitor, bronchodilator
Metabolite: pyridine-*N*-oxide

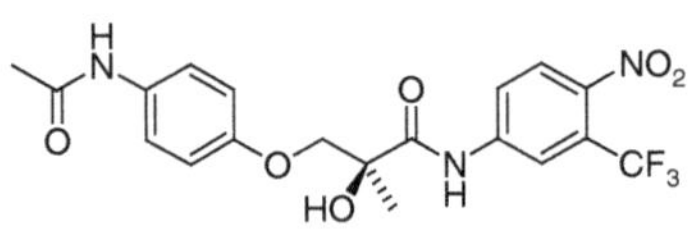

$t_{1/2}$	4 h	V	1.4*
F	38–91%*	CL	3–7*
pb	–	Mwt	441.4
ur	–	PSA	105 Å^2
		log P	4.01

ANDARINE
*dog
Selective androgen receptor modulator
Metabolism: amide hydrolysis (both), reduction of NO_2 to NH_2

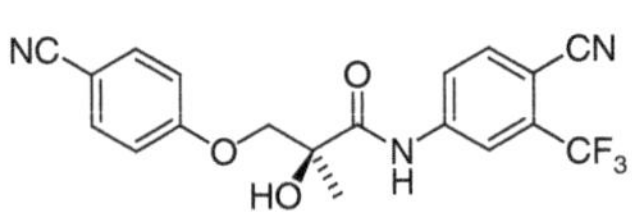

$t_{1/2}$	4 h	V	–
F	–	CL	–
pb	–	Mwt	389.3
ur	–	PSA	106 Å^2
		log P	2.93

OSTARINE
Selective androgen receptor modulator

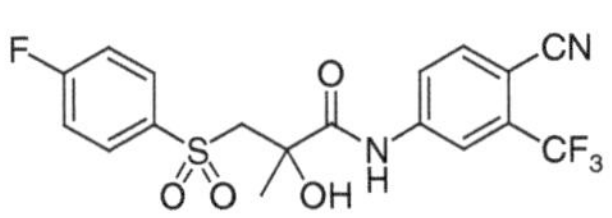

$t_{1/2}$	2–16 d	V	1.2–1.3
F	100%	CL	0.8
pb	96%	Mwt	430.4
ur	0%	PSA	116 Å^2
		log P	4.94

BICALUTAMIDE
Antiandrogen, hormonal antineoplastic
Metabolism: aromatic hydroxylation (ortho to F), glucuronidation

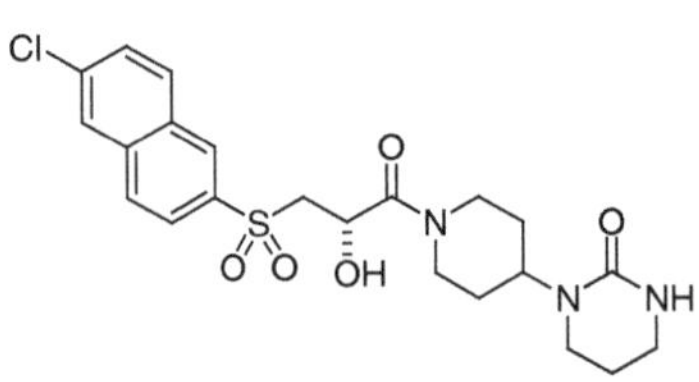

$t_{1/2}$	7 h*	V	0.6
F	53%*	CL	12
pb	–	Mwt	480.0
ur	–	PSA	115 Å^2
		log P	2.77

TAK-442
*monkey
Factor Xa inhibitor, antithrombotic

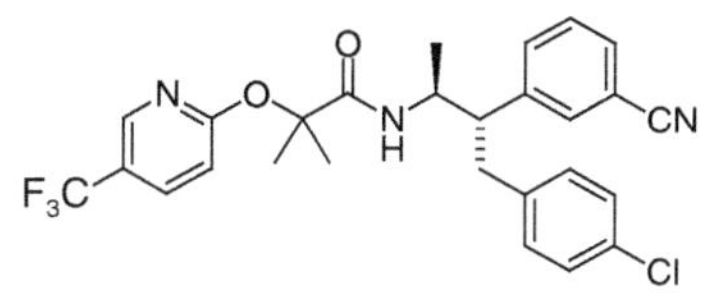	$t_{1/2}$ 3–4 d	*V*	40	TARANABANT
	F 31%*	CL	6.5	*monkey
	pb 100%	Mwt	516.0	Cannabinoid
	ur Negligible	PSA	75.0 Å^2	antagonist, antiobesity
		log *P*	7.13	agent; Metabolism: benzylic
				hydroxylation (at C_3CH)
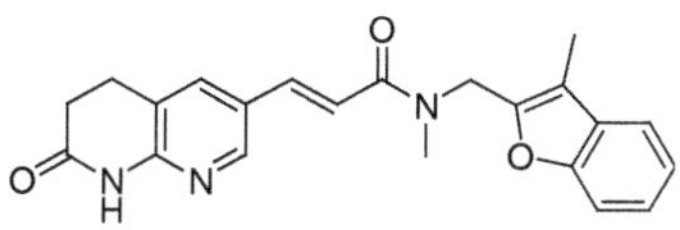	$t_{1/2}$ 8–12 h	*V*	–	AFN-1252
	F 82%	CL	–	Bacterial fatty acid
	pb 99%	Mwt	375.4	biosynthesis inhibitor
	ur –	PSA	74.9 Å^2	
		log *P*	0.85	

$t_{1/2}$, plasma half-life; *F*, oral bioavailability; pb, plasma protein binding; ur, excretion of unchanged drug in urine; V, volume of distribution; CL, clearance; Mwt, molecular weight; PSA, polar surface area; PDE, phosphodiesterase.

34
Lactams and Imides

N-Unsubstituted imides are acidic (pK_a of phthalimide in H_2O: 8.30) but not enough to undergo extensive deprotonation under physiological conditions. Lactams are also more reactive than noncyclic amides because the cis conformation of the amide group in a lactam has higher energy than the trans conformer of an amide. Lactams and imides react with nucleophiles significantly faster than amides. Most imide-containing drugs are five- or six-membered cyclic imides (Table 34.1), which are more stable than open-chain imides (Scheme 34.1).

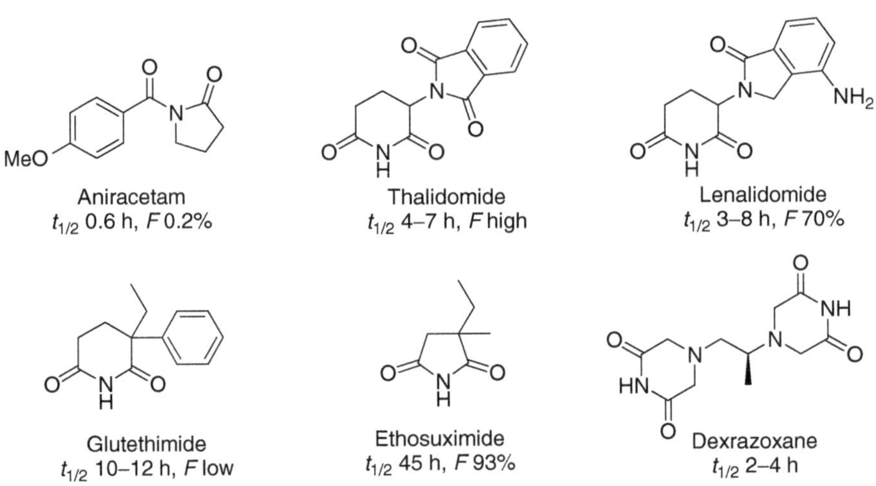

Scheme 34.1 Imides as drugs.

34.1
Pyrazolone Antipyretics

Wilhelm Koenigs and Otto Fischer, working at the University of Munich in the 1870s, believed that the antimalarial/antipyretic quinine had a tetrahydroquinoline substructure [1]. With the aim of discovering new antipyretic drugs, they prepared

Lead Optimization for Medicinal Chemists: Pharmacokinetic Properties of Functional Groups and Organic Compounds, First Edition. Florencio Zaragoza Dörwald.

some tetrahydroquinolines and gave them to Wilhelm Filehne at the University of Erlangen, who found that one of them, kairine, had indeed antipyretic properties (Scheme 34.2). The compound was launched by Hoechst 1881 but turned out to be too toxic. It was replaced by the antipyretics kairoline A and later by thalline, but these compounds still had a low therapeutic index.

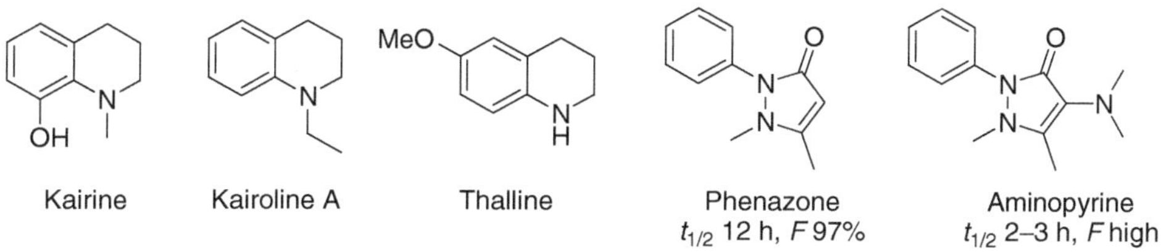

Scheme 34.2 Older antipyretic agents.

In 1883, Ludwig Knorr, also at the University of Erlangen, attempted to prepare a new tetrahydroisochinoline by treating phenylhydrazine with an acetoacetate. The product turned out to be inactive in Filehnes assay, but he recommended to N-alkylate the compound because N-alkylation had reduced toxicity in kairine and kairoline. The resulting product, phenazone, turned out to be a potent and safe antipyretic. The correct pyrazolone structure was established by Knorr in 1887. This compound was marketed by Hoechst under the trade name antipyrine but was later renamed phenazone. It became the best-selling drug worldwide and even outsold sodium salicylate because it tasted better and did not cause gastric irritation. Phenazone can, however, cause agranulocytosis, a drug-induced, often irreversible depletion of neutrophilic granulocytes, which can be fatal. To improve potency and reduce toxicity, further analogs of phenazone were developed, such as aminopyrine, phenylbutazone, oxyphenbutazone, azapropazone, and sulfinpyrazone.

34.2
Five-Membered Lactams as Nootropics

The first drugs for the treatment of epileptic seizures, bromide salts, were available as early as 1857. Later, phenobarbital and phenytoin and a number of other, more potent and selective antiepileptics were developed. In the early 1960s, various γ-butyrolactams were prepared as cyclic γ-aminobutyric acid (GABA) analogs and tested in animal models, with the aim of discovering new sedatives. However, the compounds turned out to improve memory in rodents, and piracetam and other, related lactams, are used today to improve memory and cognitive function. In 1992, it was discovered at UCB that alkylated derivatives of piracetam had an anticonvulsant effect in mice, which led to the development of the γ-lactam antiepileptics levetiracetam and brivaracetam [2].

References

1. Sneader, W. (1996) Antipyretic analgesics. *Drug News Perspect.*, **9**, 61–64.
2. Rogawski, M.A. (2008) Brivaracetam: a rational drug discovery success story. *Br. J. Pharmacol.*, **154**, 1555–1557.

Table 34.1 Cyclic five-membered amides, carbamates, ureas, imides, and related compounds. V in l kg^{-1}; CL in ml min^{-1} kg^{-1}; Mwt in g mol^{-1}.

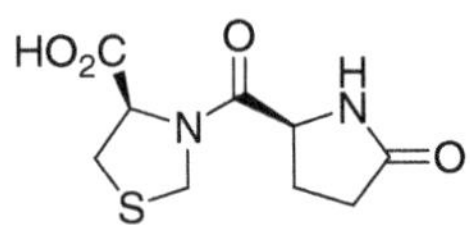	$t^1\!/_2$ F pb ur	4 h 45% 82%	V CL Mwt PSA log P	0.8 2.7 244.3 112 Å^2 −2.38	PIDOTIMOD Immunomodulator
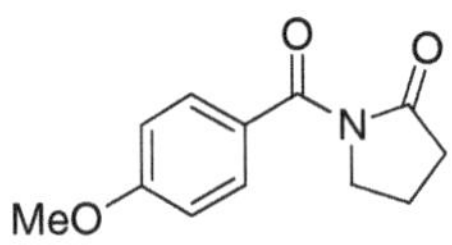	$t^1\!/_2$ F pb ur	0.6 h 0.2% – –	V CL Mwt PSA log P	2–4 141 219.2 46.6 Å^2 1.35	ANIRACETAM Nootropic Metabolites: *N*-(4-methoxybenzoyl)glycine, 4-methoxybenzoic acid
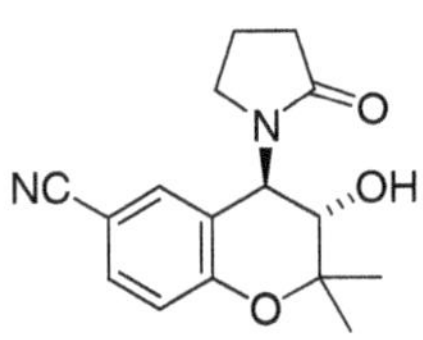	$t^1\!/_2$ F pb ur	18–28 h – – –	V CL Mwt PSA log P	– – 286.3 73.6 Å^2 1.27	CROMAKALIM Potassium channel opener, antihypertensive; Metabolism: glucuronidation, oxidation and hydrolysis of pyrrolidone to succinamide
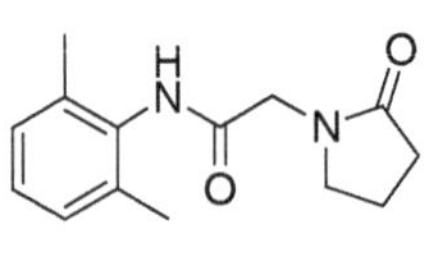	$t^1\!/_2$ F pb ur	4±2 h – – 5%	V CL Mwt PSA log P	0.5–0.7 1.5–2.2 246.3 49.4 Å^2 0.64	NEFIRACETAM Nootropic Metabolism: hydroxylation of pyrrolidone (mainly at position 5), then ring fission
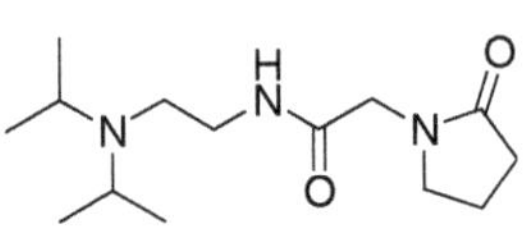	$t^1\!/_2$ F pb ur	5–7 h – 30% –	V CL Mwt PSA log P	2–3 4–5 269.4 52.7 Å^2 −1.40	PRAMIRACETAM Nootropic
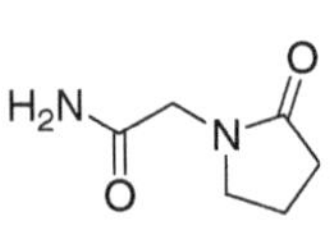	$t^1\!/_2$ F pb ur	5±1 h 100% 0% 80–100%	V CL Mwt PSA log P	0.6±0.1 – 142.2 63.4 Å^2 −1.75	PIRACETAM Nootropic No metabolism

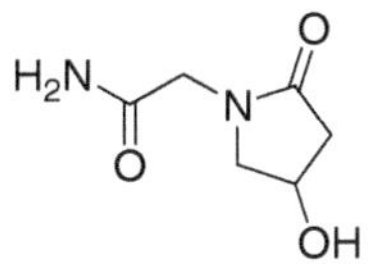	$t\frac{1}{2}$ 3–8 h *F* 75±7% pb – ur 50–84%	*V* 1–2 CL 1.8 Mwt 158.2 PSA 83.6 Å^2 log *P* −2.48	OXIRACETAM Nootropic Almost no metabolites formed in rats
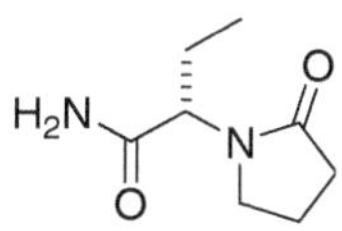	$t\frac{1}{2}$ 6–8 h *F* 100% pb <10% ur 70%	*V* 0.6±0.1 CL 0.96 Mwt 170.2 PSA 63.4 Å^2 log *P* −0.88	LEVETIRACETAM Antiepileptic Metabolism: hydrolysis of $CONH_2$
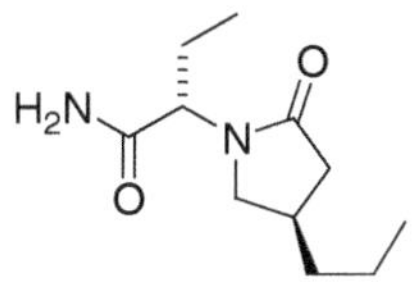	$t\frac{1}{2}$ 8±1 h *F* 100% pb 18% ur 5–9%	*V* 0.5 CL 0.8 Mwt 212.3 PSA 63.4 Å^2 log *P* 0.88	BRIVARACETAM Antiepileptic Metabolism: hydrolysis of $CONH_2$
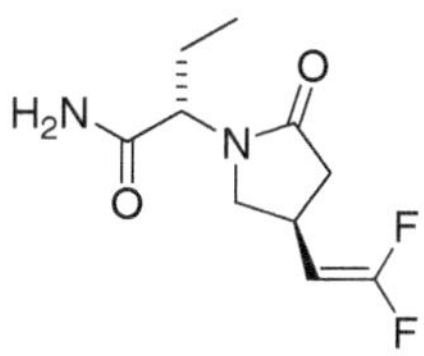	$t\frac{1}{2}$ 8 h *F* >90% pb <10% ur 30%	*V* 0.6 CL – Mwt 232.2 PSA 63.4 Å^2 log *P* 0.71	SELETRACETAM Anticonvulsant Metabolism: hydrolysis of $CONH_2$
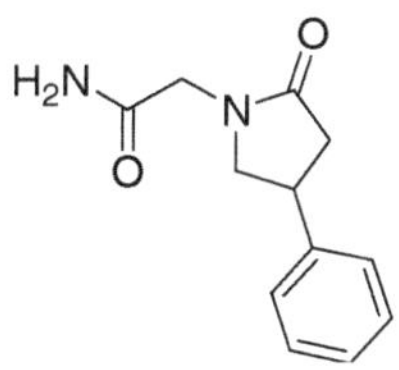	$t\frac{1}{2}$ 3–5 h *F* 100% pb – ur 40%	*V* – CL – Mwt 218.3 PSA 63.4 Å^2 log *P* 0.70	PHENYLPIRACETAM Nootropic
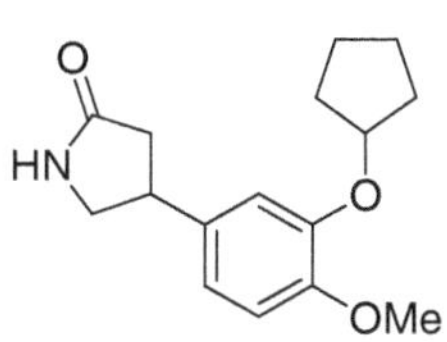	$t\frac{1}{2}$ 6–8 h *F* 74% pb – ur Low	*V* 0.5 CL 2–6 Mwt 275.3 PSA 47.6 Å^2 log *P* 2.00	ROLIPRAM Antidepressant; Metabolism: O-dealkylation, 2- and 3-hydroxylation of cyclopentyl, 5-hydroxylation of pyrrolidone; PK of both isomers is similar, only *R*(−) isomer is active
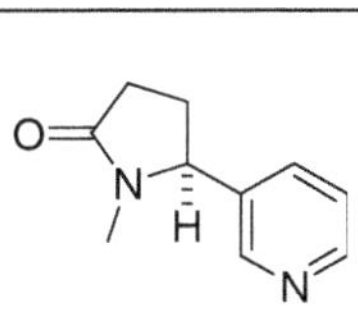	$t\frac{1}{2}$ 12–16 h *F* 84–100% pb – ur 12%	*V* 0.9 CL 0.9 Mwt 176.2 PSA 33.2 Å^2 log *P* 0.08	COTININE Metabolite of nicotine; Metabolism: N-methylation and oxidation of pyridine, hydroxylation of pyrrolidinone, demethylation, glucuronidation

(*continued overleaf*)

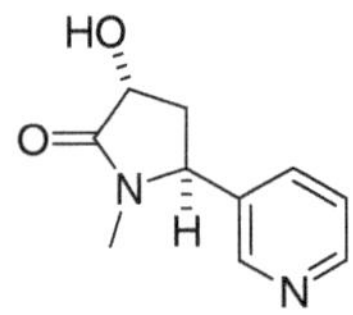	$t^1/_2$ 4–10 h (iv) *F* – pb – ur –	*V* 0.5–1.1 CL 1.1–2.6 Mwt 192.2 PSA 53.4 Å^2 log *P* −1.12	3-HYDROXYCOTININE Metabolite of cotinine
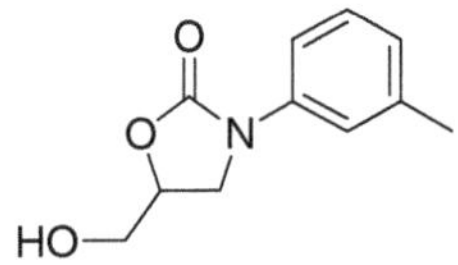	$t^1/_2$ 0.9–2.5 h *F* 50–62% pb 51% ur <1%	*V* 1.1–1.6 CL 8–14 Mwt 207.2 PSA 49.8 Å^2 log *P* 1.53	TOLOXATONE Antidepressant, MAO inhibitor Metabolism: glucuronidation, oxidation of CH_3 to CO_2H
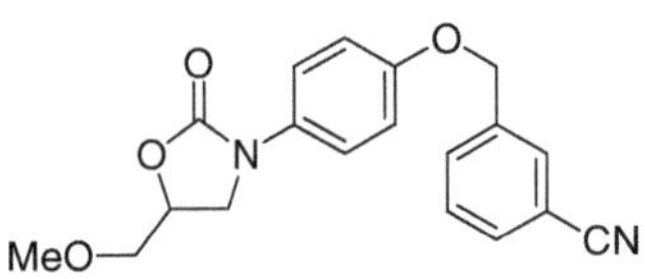	$t^1/_2$ 9–16 h *F* – pb 93–96% ur <1%	*V* 0.5–0.7 CL 0.4–0.8 Mwt 338.4 PSA 71.8 Å^2 log *P* 2.25	CIMOXATONE Antidepressant, MAO inhibitor Active metabolite: *O*-desmethyl ($t^1/_2$ 36 h)
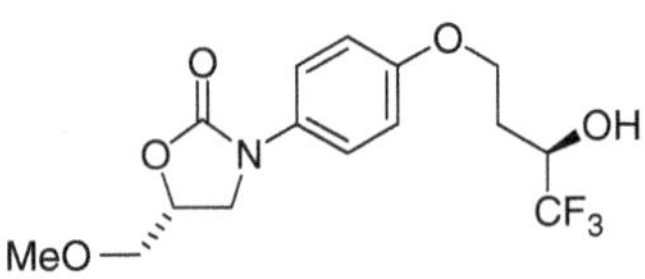	$t^1/_2$ 11–22 h *F* – pb 58% ur <0.2%	*V* – CL – Mwt 349.3 PSA 68.2 Å^2 log *P* 1.91	BEFLOXATONE Antidepressant, MAO inhibitor Active metabolite: *O*-desmethyl ($t^1/_2$ 22 h)
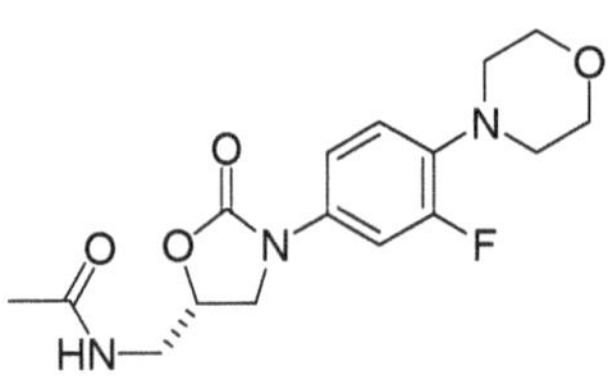	$t^1/_2$ 4–7 h *F* 80–100% pb 31% ur 30–40%	*V* 0.6–0.9 CL 1.6–2.2 Mwt 337.4 PSA 71.1 Å^2 log *P* 0.45	LINEZOLID Antibacterial Metabolism: oxidative ring opening of morpholine
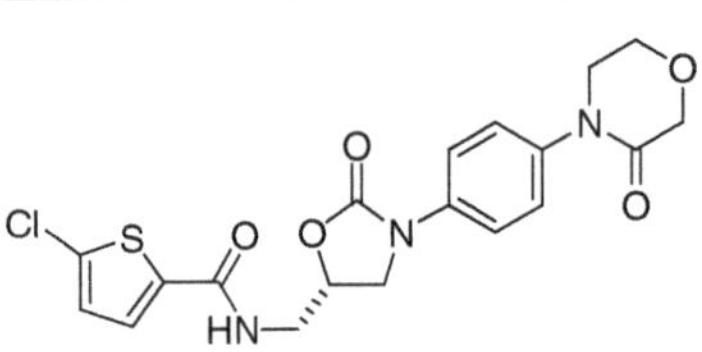	$t^1/_2$ 7–11 h *F* 80–100% pb 92–95% ur 36%	*V* 0.8 CL 2.6 Mwt 435.9 PSA 116 Å^2 log *P* 1.84	RIVAROXABAN Factor Xa inhibitor, antithrombotic Metabolism: oxidative degradation of morpholinone to *N*-(2-hydroxyethyl)oxalamide, amide hydrolysis

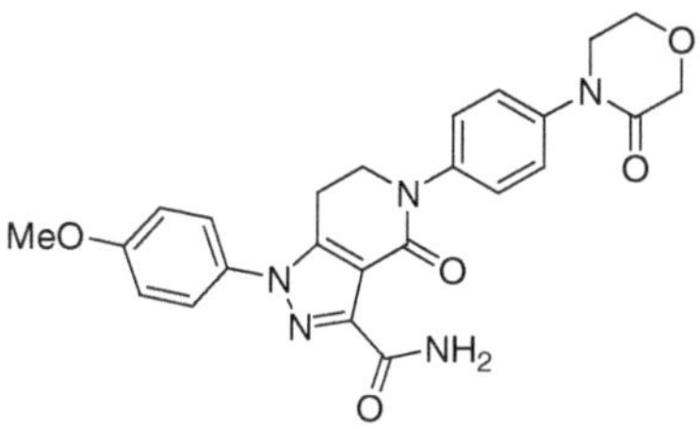

$t^1\!/_2$	4–27 h	*V*	0.17*	APIXABAN
F	51%*	CL	0.8–1.3	*chimpanzee
pb	87%	Mwt	459.5	Anticoagulant, factor
ur	13%	PSA	111 Å^2	Xa inhibitor
		log *P*	0.56	Metabolism: hydroxylation,
				O-demethylation, then
				O-sulfation

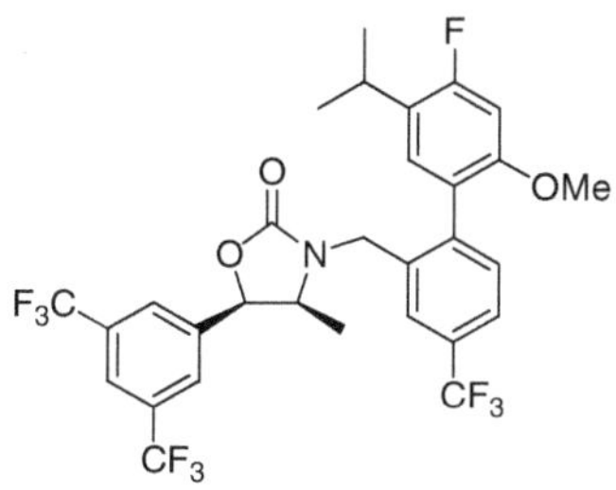

$t^1\!/_2$	9–83 h	*V*	0.31*	ANACETRAPIB
F	13%*	CL	0.11*	*monkey
pb	>99%	Mwt	637.5	Cholesterol ester transfer
ur	<0.1%	PSA	38.8 Å^2	protein inhibitor, antilipemic
		log *P*	8.81	Metabolism: O-demethylation,
				hydroxylation

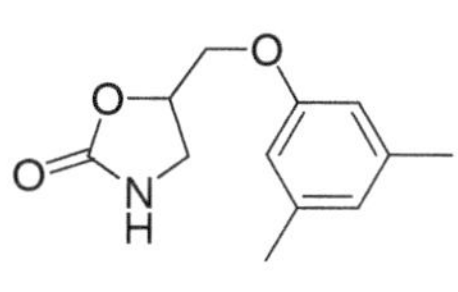

$t^1\!/_2$	9±5 h	*V*/*F*	12.3	METAXALONE
F	–	CL	17	Muscle relaxant
pb	–	Mwt	221.3	Metabolism: oxidation of
ur	27%	PSA	47.6 Å^2	CH_3 to CO_2H,
		log *P*	1.18	O-dealkylation to phenol

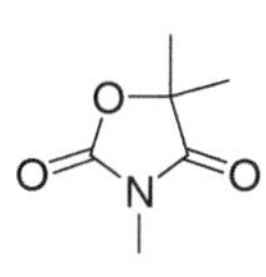

$t^1\!/_2$	14±2 h	*V*	0.7±0.1	TRIMETHADIONE
F	–	CL	–	Anticonvulsant, teratogenic
pb	0%	Mwt	143.1	Metabolism: N-demethylation
ur	<3%	PSA	46.6 Å^2	to dimethadione
		log *P*	0.79	

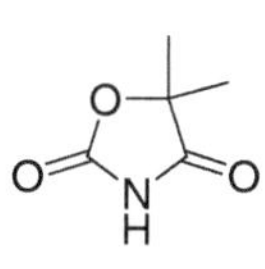

$t^1\!/_2$	6–13 d	*V*	0.4–0.5	DIMETHADIONE
F	–	CL	–	Metabolite of trimethadione
pb	Low	Mwt	129.1	
ur	–	PSA	55.4 Å^2	
		log *P*	0.77	

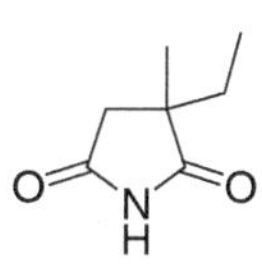

$t^1\!/_2$	45±8 h	*V*	0.7±0.2	ETHOSUXIMIDE
F	93±2%	CL/*F*	0.19±0.04	Anticonvulsant
pb	0%	Mwt	141.2	Metabolism: hydroxylation
ur	25±15%	PSA	46.2 Å^2	at ethyl (both carbons)
		log *P*	0.25	and CH_2CO

(*continued overleaf*)

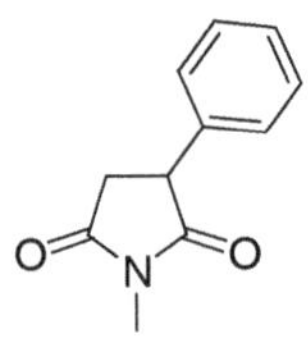	$t^{1}/_{2}$ 7.8 h *F* – pb 21% ur Low	*V* – CL – Mwt 189.2 PSA 37.4 Å^2 log *P* 0.04	PHENSUXIMIDE Anticonvulsant Metabolism: N-demethylation, aromatic 4-hydroxylation
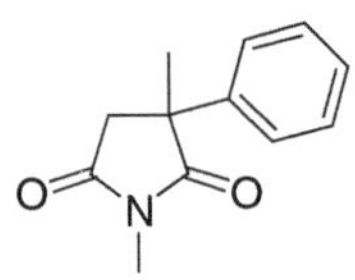	$t^{1}/_{2}$ 2.5±1.5 h *F* – pb Low ur <1%	*V* – CL – Mwt 203.2 PSA 37.4 Å^2 log *P* 0.42	METHSUXIMIDE Anticonvulsant; active metabolite: *N*-desmethyl ($t^{1}/_{2}$ 2–3 d; pb 50%) Further metabolism: aromatic and methyl hydroxylation
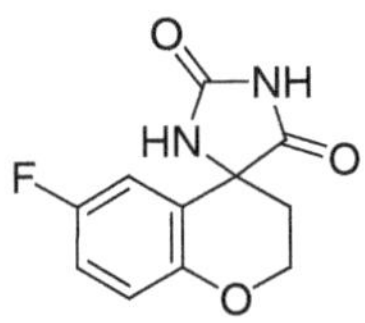	$t^{1}/_{2}$ 40–70 h *F* – pb – ur 26±7%	*V* 0.8 CL – Mwt 236.2 PSA 67.4 Å^2 log *P* 0.49	SORBINIL Aldose reductase inhibitor Metabolites: 2-hydroxy, 5-(2-hydroxyethyl)-5-(2-hydroxy-5-fluorophenyl)-imidazolidinedione
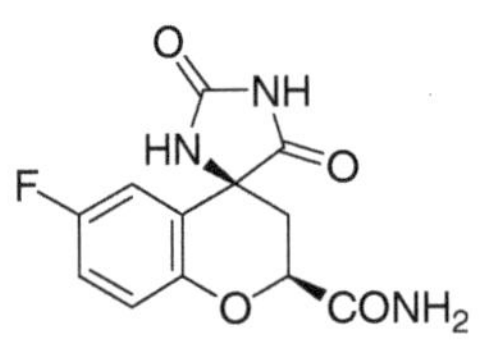	$t^{1}/_{2}$ 1–2 h *F* – pb – ur –	*V* – CL – Mwt 279.2 PSA 111 Å^2 log *P* −0.13	FIDARESTAT Aldose reductase inhibitor, treatment of diabetic neuropathy
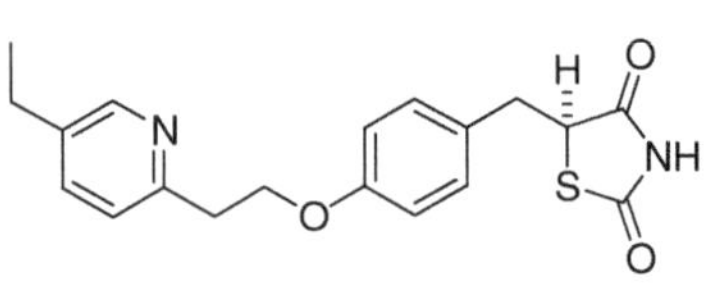	$t^{1}/_{2}$ 5–6 h *F* >80% pb >99% ur Low	*V* 0.25 CL – Mwt 356.4 PSA 93.6 Å^2 log *P* 3.50	PIOGLITAZONE PPAR-γ agonist, insulin sensitizer Metabolism: thiazolidinone hydrolysis, benzylic hydroxylation of ethyl, then oxidation to ketone ($t^{1}/_{2}$ 16–23 h); withdrawn in 2011 because of carcinogenicity
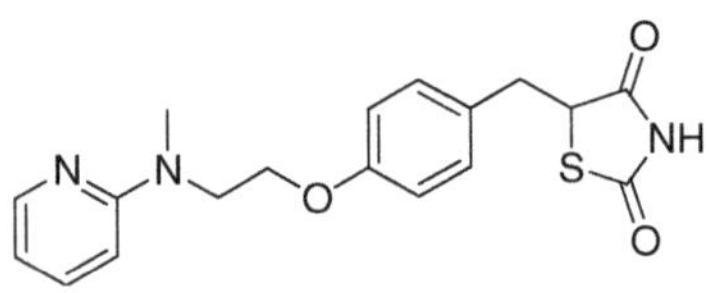	$t^{1}/_{2}$ 3.6–4.2 h *F* 86–100% pb 99.7% ur 0%	*V* 0.19 CL 0.6 Mwt 357.4 PSA 96.8 Å^2 log *P* 2.56	ROSIGLITAZONE PPAR-γ agonist, insulin sensitizer; Metabolism: N-demethylation, 3- and 5-hydroxylation of pyridine, then sulfation and glucuronidation; withdrawn in 2010 in Europe for increasing heart attack risk

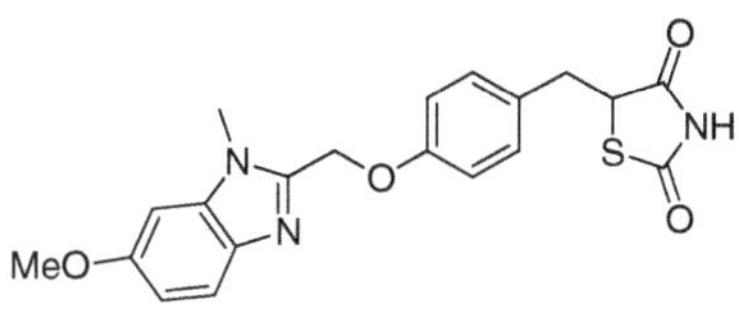	$t^1\!/_2$ 10–15 h F 76%* pb – ur 0.3%*	V 0.14–0.17* CL 0.31–0.37* Mwt 397.5 PSA 103 Å^2 log P 2.95	RIVOGLITAZONE *monkey; PPAR-γ agonist, insulin sensitizer Metabolism (monkey): O- and N-demethylation, thiazoline hydroxylation, thiazoline ring opening
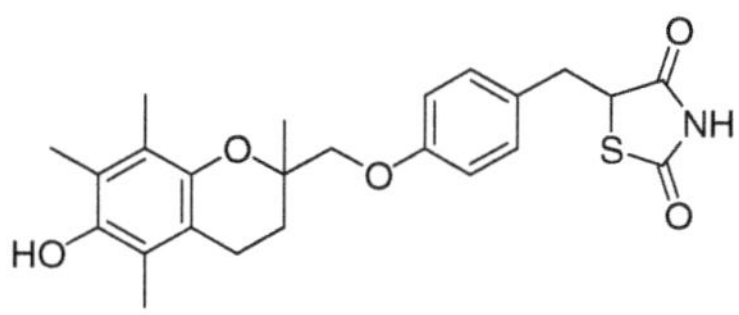	$t^1\!/_2$ 16–34 h F 40–50% pb >99% ur 0%	V 2.5 CL 8.5 Mwt 441.5 PSA 110 Å^2 log P 4.69	TROGLITAZONE PPAR-γ agonist Metabolism: O-sulfation, O-glucuronidation, oxidation to benzoquinone (chromane ring scission); withdrawn in 2000 because of hepatotoxicity
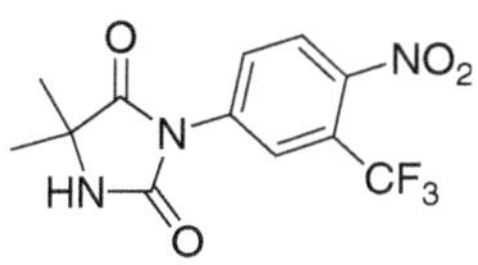	$t^1\!/_2$ 38–59 h F 80–90% pb 84% ur <2%	V – CL 2.5* Mwt 317.2 PSA 95.2 Å^2 log P 3.35	NILUTAMIDE *rat Antiandrogen, antineoplastic Metabolism: CH_3-hydroxylation, reduction of NO_2 to NH_2, then aromatic hydroxylation (ortho to NH_2), reduction of C=O to CHOH
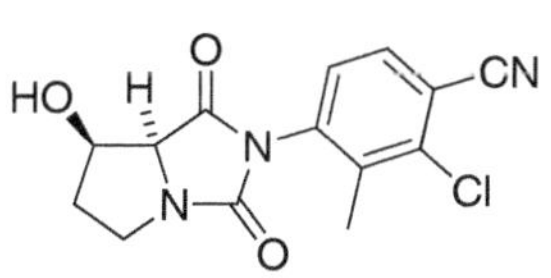	$t^1\!/_2$ 8–14 h F – pb – ur –	V – CL – Mwt 305.7 PSA 84.6 Å^2 log P 0.43	BMS-564929 Selective androgen receptor modulator
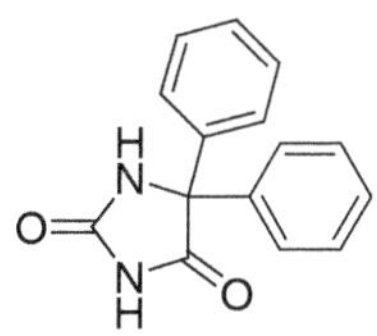	$t^1\!/_2$ 15±9 h F 90±3% pb 89±23% ur 2±8%	V 0.64±0.04 CL – Mwt 252.3 PSA 58.2 Å^2 log P 1.42	PHENYTOIN Anticonvulsant CYP2C9 substrate, carcinogen Metabolism: aromatic 4-hydroxylation, then glucuronidation
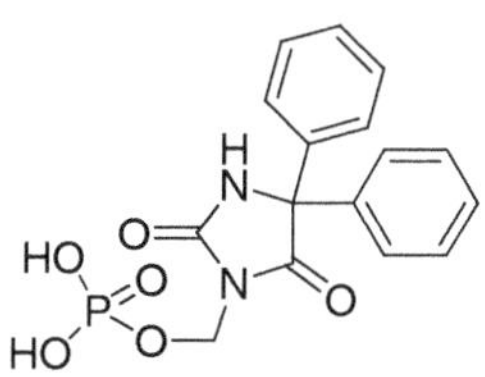	$t^1\!/_2$ 8–15 min F Low pb 95–99% ur –	V 0.07–0.17 CL – Mwt 362.3 PSA 126 Å^2 log P −1.06	FOSPHENYTOIN Prodrug of phenytoin

(*continued overleaf*)

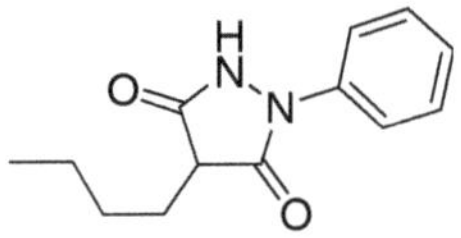	$t^1/_2$ 1.9 h (iv) F >90% pb 99% ur 8%	V – CL – Mwt 232.3 PSA 49.4 Å^2 log P 2.26	MOFEBUTAZONE Antiinflammatory Metabolism: glucuronidation
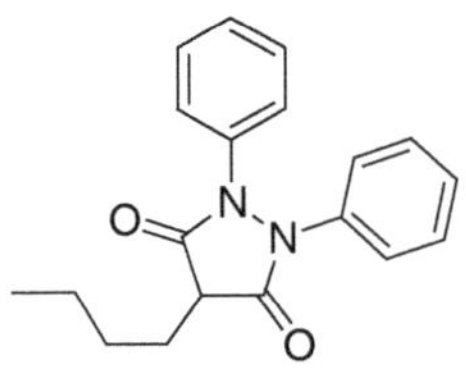	$t^1/_2$ 56±8 h F 90±10% pb 96±1% ur 1%	V 0.1 CL 0.02 Mwt 308.4 PSA 40.6 Å^2 log P 3.38	PHENYLBUTAZONE Antiinflammatory Metabolism: aromatic 4-hydroxylation (to oxyphenbutazone, see below), hydroxylation at CH_2CH_3
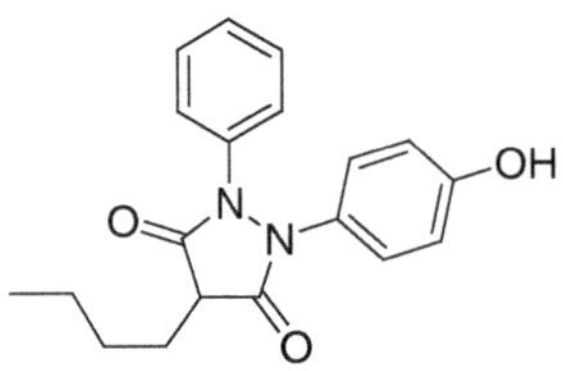	$t^1/_2$ 27–64 h F – pb >98% ur <2%	V 0.15 CL 0.03 Mwt 324.4 PSA 60.9 Å^2 log P 2.74	OXYPHENBUTAZONE Active metabolite of phenylbutazone; withdrawn in 1985 for causing blood dyscrasias
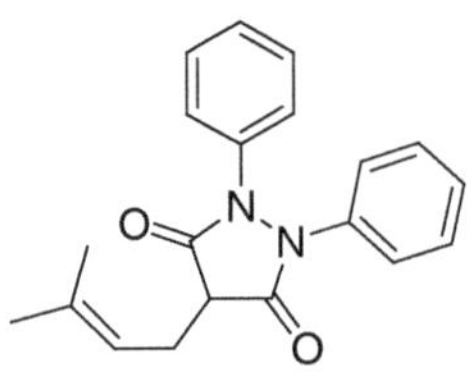	$t^1/_2$ 22–33 h F – pb 90% ur <1%	V 0.3–0.5 CL 0.1–0.2 Mwt 320.4 PSA 40.6 Å^2 log P 3.05	FEPRAZONE Antiinflammatory Metabolism: hydroxylation of methyl and phenyl groups, C-glucuronidation
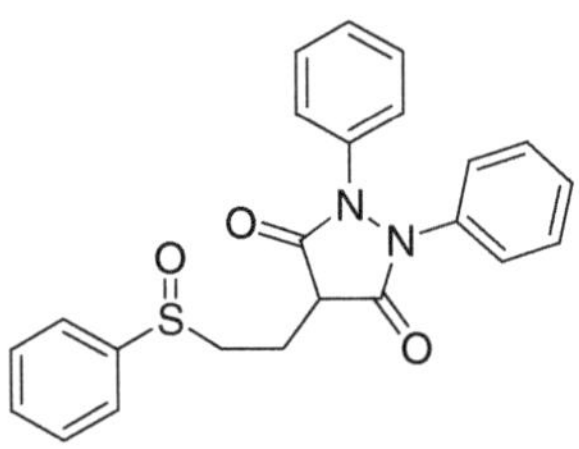	$t^1/_2$ 4±1 h F 100% pb 98.3±0.5% ur 39±9%	V 0.74±0.23 CL 2.4±0.6 Mwt 404.5 PSA 76.9 Å^2 log P 1.89	SULFINPYRAZONE Antithrombotic, uricosuric Metabolism: oxidation to sulfone, reduction to sulfide ($t^1/_2$ 14±5 h)
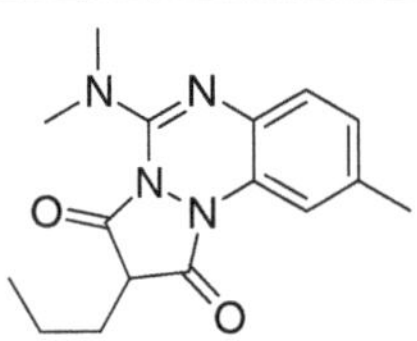	$t^1/_2$ 10–24 h F 83±19% pb 99.5% ur 63%	V 0.15 CL 0.14 Mwt 300.4 PSA 56.2 Å^2 log P 1.78	AZAPROPAZONE, APAZONE Antiinflammatory Metabolism: aromatic hydroxylation

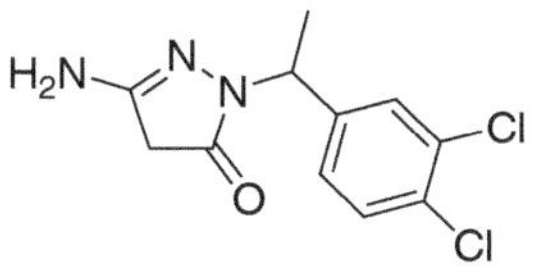	$t^1/_2$ 10–17 h *F* – pb 65% ur 5%	*V* 1 CL – Mwt 272.1 PSA 58.7 $Å^2$ log *P* 2.10	MUZOLIMINE Diuretic Metabolism: hydrolysis of amide, then decarboxylation
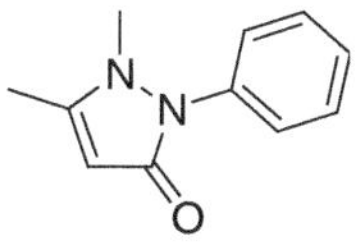	$t^1/_2$ 12 h (iv) *F* 97% pb 7% ur –	*V* 0.77 CL 0.64 Mwt 188.2 PSA 23.6 $Å^2$ log *P* 0.44	ANTIPYRINE, PHENAZONE Analgesic Metabolism: hydroxylation of vinylic methyl group
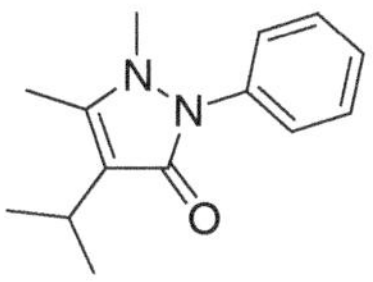	$t^1/_2$ 1.0–1.5 h *F* – pb Low ur 0.6%	*V* 1.3–2.0 CL 11 Mwt 230.3 PSA 23.6 $Å^2$ log *P* 1.72	PROPYPHENAZONE Analgesic, antipyretic Metabolism: N-demethylation, hydroxylation of *i*Pr and phenyl
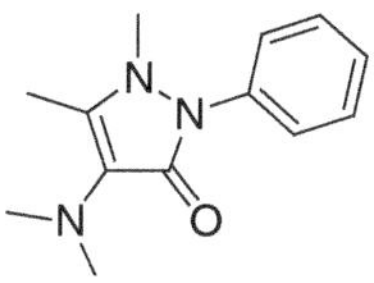	$t^1/_2$ 2.1–3.2 h *F* High pb 15% ur 3–10%	*V* 0.7 CL 2 Mwt 231.3 PSA 26.8 $Å^2$ log *P* 0.85	AMINOPYRINE, AMIDOPYRINE Analgesic, antipyretic Metabolism: N-demethylation of NMe_2, then N-acetylation and N-formylation; withdrawn in 1970 because of bone marrow suppression
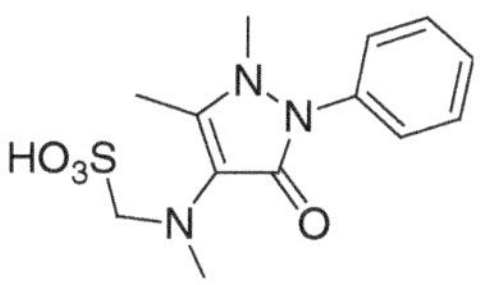	$t^1/_2$ 2.6–3.5 h *F* 85%* pb 58% ur 2–4%	*V* – CL 2.8 Mwt 311.4 PSA 84.6 $Å^2$ log *P* −1.50	METAMIZOL, DIPYRONE *metabolites only Analgesic, antipyretic Metabolites: 4-(methylamino)antipyrine ($t^1/_2$ 3.8 h), 4-(formylamino)antipyrine ($t^1/_2$ 10 h), 4-aminoantipyrine $t^1/_2$ 4–6 h), 4-(acetylamino)antipyrine ($t^1/_2$ 11 h); withdrawn in 1977 for causing agranulocytosis
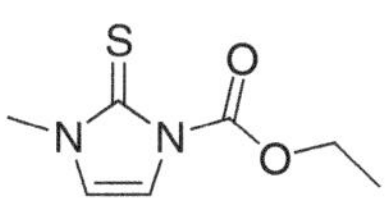	$t^1/_2$ 4–12 h* *F* – pb 40% ur 7%	*V* 0.5–0.9 CL 3.7 Mwt 186.2 PSA 64.9 $Å^2$ log *P* 0.16	CARBIMAZOLE *prodrug of methimazole (see below); all values: methimazole on oral dosing of carbimazole; antihyperthyroid; complete absorption

(*continued overleaf*)

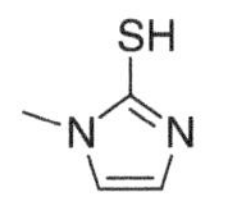	$t^{1}/_{2}$ 3–6 h *F* 95% pb – ur 7%	*V* 0.6–0.9 CL 1.4–3.7 Mwt 114.2 PSA 47.4 Å^2 log *P* −0.34	METHIMAZOLE, THIAMAZOLE Antihyperthyroid Metabolite: 3-methyl-2-thiohydantoin ($t^{1}/_{2}$ 9 h)
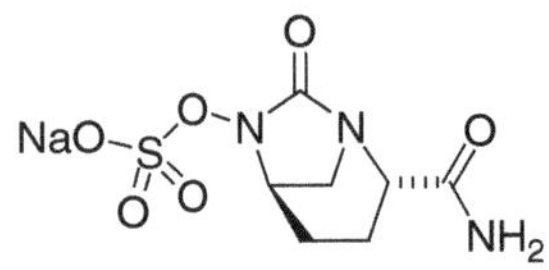	$t^{1}/_{2}$ 2–3 h (iv) *F* – pb – ur –	*V* 0.3–0.4 CL 2–3 Mwt 265.2 PSA 139 Å^2 log *P* −1.87	AVIBACTAM, NXL-104 β-Lactamase inhibitor, antibacterial
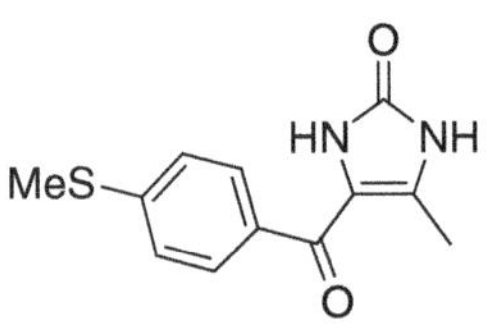	$t^{1}/_{2}$ 1–2 h *F* 45–70% pb 70–85% ur 0.5%	*V* 1.8 CL 6.8–30 Mwt 248.3 PSA 83.5 Å^2 log *P* 3.72	ENOXIMONE Cardiotonic Metabolism: reversible oxidation to sulfoxide
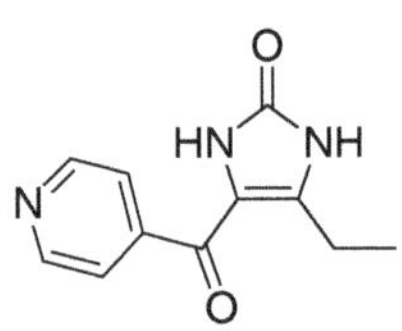	$t^{1}/_{2}$ 1.3 h *F* >80% pb 85% ur 64%	*V* 0.6–1.7 CL 7–9 Mwt 217.2 PSA 71.1 Å^2 log *P* 2.15	PIROXIMONE Inotropic Metabolism (dog): reduction of ketone to alcohol, oxidative cleavage to isonicotinic acid
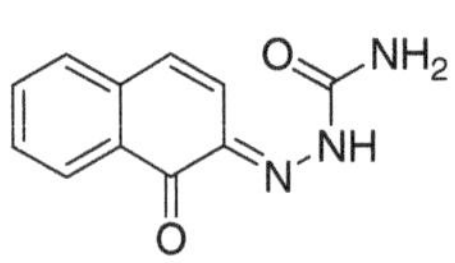	$t^{1}/_{2}$ 1.5 h *F* High pb 0% ur –	*V* – CL – Mwt 215.2 PSA 84.6 Å^2 log *P* 0.95	NAFTAZONE Hemostatic, vasoprotectant Metabolism: reduction of ketone to alcohol, glucuronidation
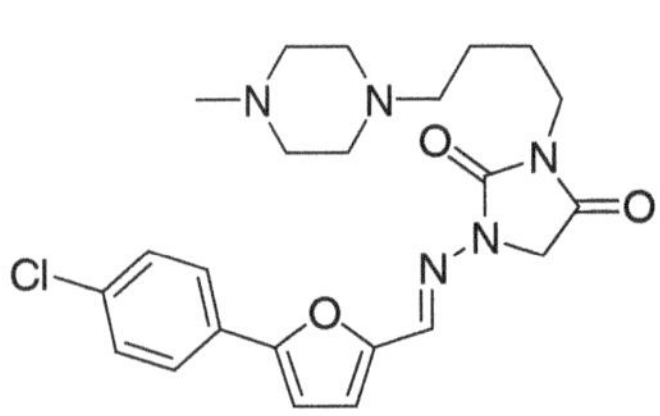	$t^{1}/_{2}$ 4–5 d *F* 85% pb 94% ur <10%	*V* 26 CL 2.5 Mwt 458.0 PSA 72.6 Å^2 log *P* 3.33	AZIMILIDE Antiarrhythmic Metabolism: oxidation to *N*-oxide, N-demethylation, hydrolysis of hydrazone and oxidation of CHO to CO_2H, N-dealkylation and oxidation to butanoic acid

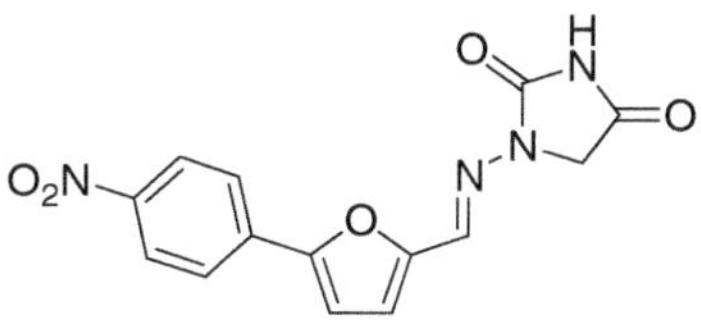	*t*½ *F* pb ur	9 h 70–88% High 1–4%	*V* CL Mwt PSA log *P*	0.54 – 314.3 121 Å^2 1.07	DANTROLENE Muscle relaxant Metabolism: reduction of NO_2 to NH_2, then N-acetylation, hydroxylation of CH_2N
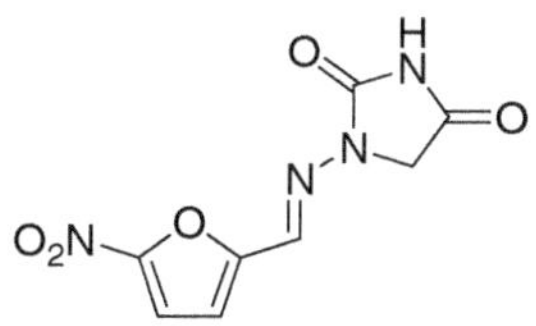	*t*½ *F* pb ur	1.0±0.2 h 90±13% 62±2% 47±13%	*V* CL Mwt PSA log *P*	0.58±0.12 9.9±0.9 238.2 121 Å^2 −0.47	NITROFURANTOIN Antibacterial Metabolism: reduction of NO_2
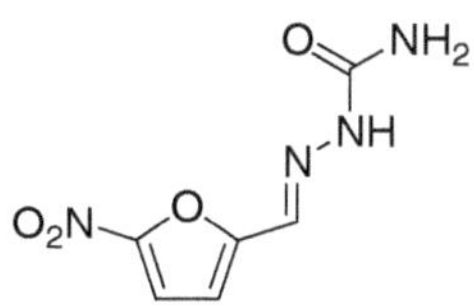	*t*½ *F* pb ur	5 h (cattle) – – –	*V* CL Mwt PSA log *P*	– – 198.1 126 Å^2 0.47	NITROFURAZONE Antibacterial Metabolism: reduction of NO_2; withdrawn in 1974 because of carcinogenicity
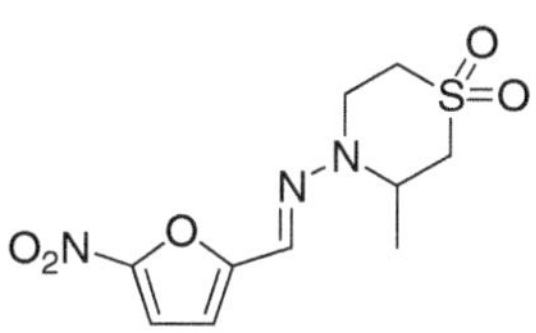	*t*½ *F* pb ur	3±1 h Low – <1%	*V* CL Mwt PSA log *P*	12 – 287.3 117 Å^2 0.73	NIFURTIMOX Antibacterial Metabolite: nitrite
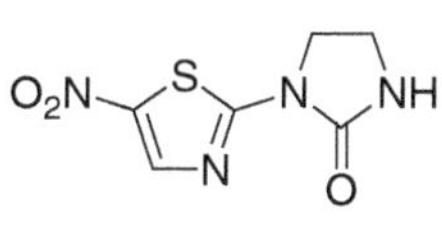	*t*½ *F* pb ur	12–15 h – – <1%	*V* CL Mwt PSA log *P*	– – 214.2 119 Å^2 0.95	NIRIDAZOLE Anthelmintic Metabolism: hydroxylation, oxidation to 4-ketoniridazole, hydrolysis of nitrothiazole to 1-thiocarbamoyl-2-imidazolidinone
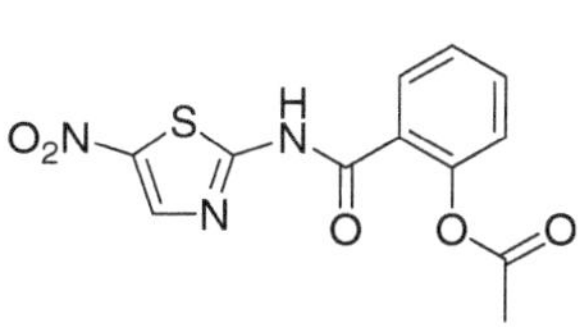	*t*½ *F* pb ur	6 min* High* >99%* –	*V* CLF Mwt PSA log *P*	30* 1.3–2.2* 307.3 142 Å^2 2.74	NITAZOXANIDE *tizoxanide on oral administration of prodrug Prodrug of tizoxanide (*t*½ 1.8 h); antiprotozoal Metabolism: deacetylation to tizoxanide, then glucuronidation

$t_{1/2}$, plasma half-life; *F*, oral bioavailability; pb, plasma protein binding; ur, excretion of unchanged drug in urine; V, volume of distribution; CL, clearance; Mwt, molecular weight; PSA, polar surface area; MAO, monoamine oxidase; PPAR, peroxisome proliferator-activated receptor.

35
Nitriles

The cyano group is strongly electron-withdrawing and may be used to lower the electron density of an arene or heteroarene, to reduce the basicity of amines, or to enhance the polar surface area and solubility of a lead. Nitriles are metabolically quite stable, as shown by the examples in Scheme 35.1 [1].

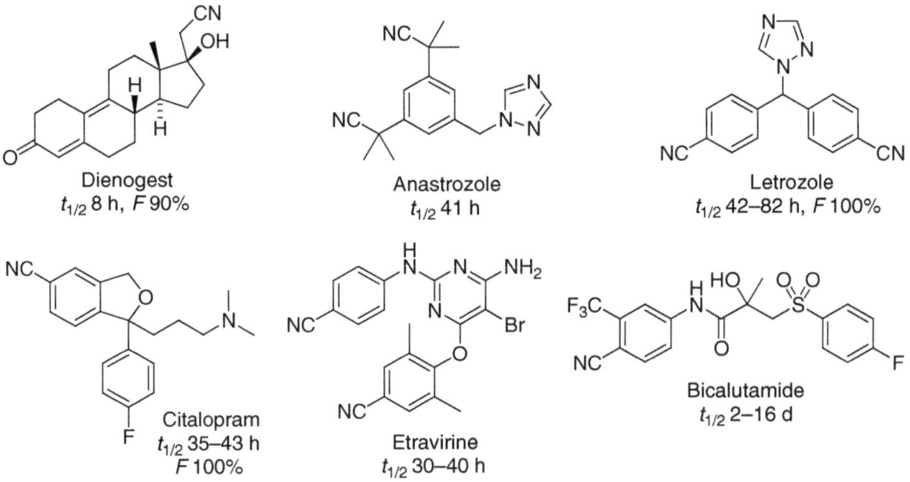

Scheme 35.1 Therapeutic nitriles.

The *N*-cyanoguanidine group has been used as nontoxic thiourea bioisostere in cimetidine. The potassium channel opener pinacidil also contains a cyanoguanidine substructure.

In principle, nitriles may be hydrolyzed *in vivo* to amides or carboxylic acids, but such metabolites are rarely observed. Interestingly, simple nitriles such as RCH_2CN or R_2CHCN (R = alkyl; not *tert*-butyl cyanide) undergo displacement and release of cyanide in rats [2]. If this metabolic transformation also occurs in humans, this would be a problem only in compounds that must be dosed in large quantities. Small amounts of cyanide are goitrogenic (cyanide is metabolized

Lead Optimization for Medicinal Chemists: Pharmacokinetic Properties of Functional Groups and Organic Compounds, First Edition. Florencio Zaragoza Dörwald.

to thiocyanate, which interferes with the iodination step during thyroid hormone biosynthesis), but do not otherwise cause irreversible damage on chronic, sublethal dosing.

References

1. Fleming, F.F., Yao, L., Ravikumar, P.C., Funk, L., and Shook, B.C. (2010) Nitrile-containing pharmaceuticals: efficacious roles of the nitrile pharmacophore. *J. Med. Chem.*, **53**, 7902–7917.
2. (a) Silver, E.H., Kuttab, S.H., Hasan, T., and Hassan, M. (1982) Structural considerations in the metabolism of nitriles to cyanide in vivo. *Drug Metabol. Dispos.*, **10**, 495–498;
(b) Dahl, A.R. and Waruszewski, B.A. (1989) Metabolism of organonitriles to cyanide by rat nasal tissue enzymes. *Xenobiotica*, **19**, 1201–1205.

36
Carbonates

Carbonates have often been used as prodrugs, but rarely as a structural element of drugs. Carbonates are chemically more stable than esters, but this seems not to be the case *in vivo*. One problem may be that effective steric shielding of the carbonyl group in carbonates is more difficult than in esters.

Examples of typical carbonate-based prodrugs for carboxylic acids or phenols are sketched in Scheme 36.1.

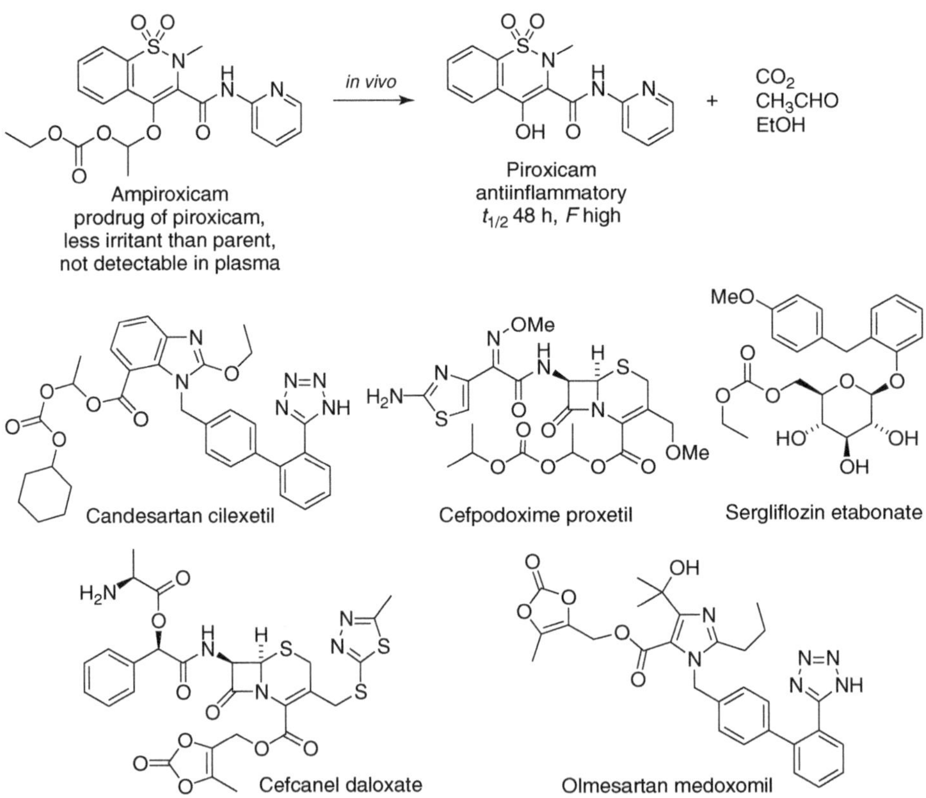

Scheme 36.1 Carbonate prodrugs.

Lead Optimization for Medicinal Chemists: Pharmacokinetic Properties of Functional Groups and Organic Compounds, First Edition. Florencio Zaragoza Dörwald.

37
Carbamates

The resistance of carbamates toward nucleophilic attack lies between that of carboxylic esters and amides. Carbamates of primary amines ($RNHCO_2R'$) can also fragment into isocyanates and alcohols on treatment with bases, a further potential pathway of metabolic degradation.

Short-lived carbamates have been used as prodrugs of heteroaromatic amines (capecitabine), amidines (lefradafiban, dabigatran), and phenols (bambuterol). Examples of such compounds and other short-lived carbamates, in which carbamate hydrolysis is the main metabolic pathway, are shown in Scheme 37.1.

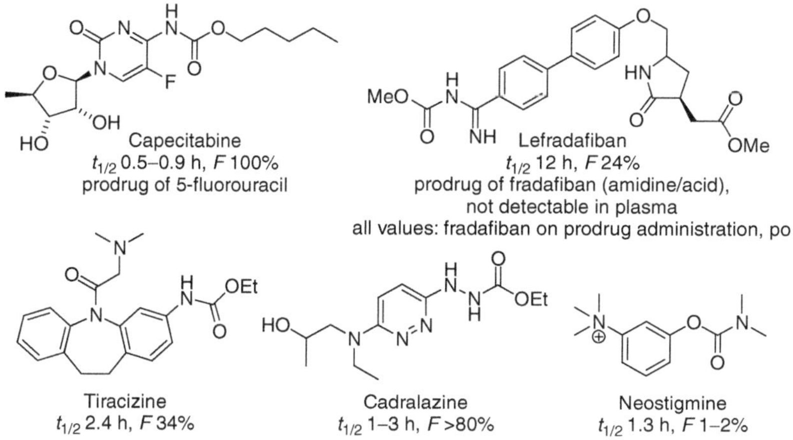

Scheme 37.1 Carbamates as prodrugs or short-acting drugs.

Sterically undemanding carbamates derived from nonbasic amines (anilines, heteroarylamines), hydrazines, and amidines are less stable than those derived from ammonia or aliphatic amines. The latter can be sufficiently long lived to serve as oral drugs, as illustrated by the examples in Scheme 37.2. Cyclic five- or six-membered carbamates are quite stable and do not usually undergo metabolic ring opening. In the examples in Scheme 37.2, carbamate hydrolysis is *not* necessarily the half-life-determining metabolic reaction.

Lead Optimization for Medicinal Chemists: Pharmacokinetic Properties of Functional Groups and Organic Compounds, First Edition. Florencio Zaragoza Dörwald.

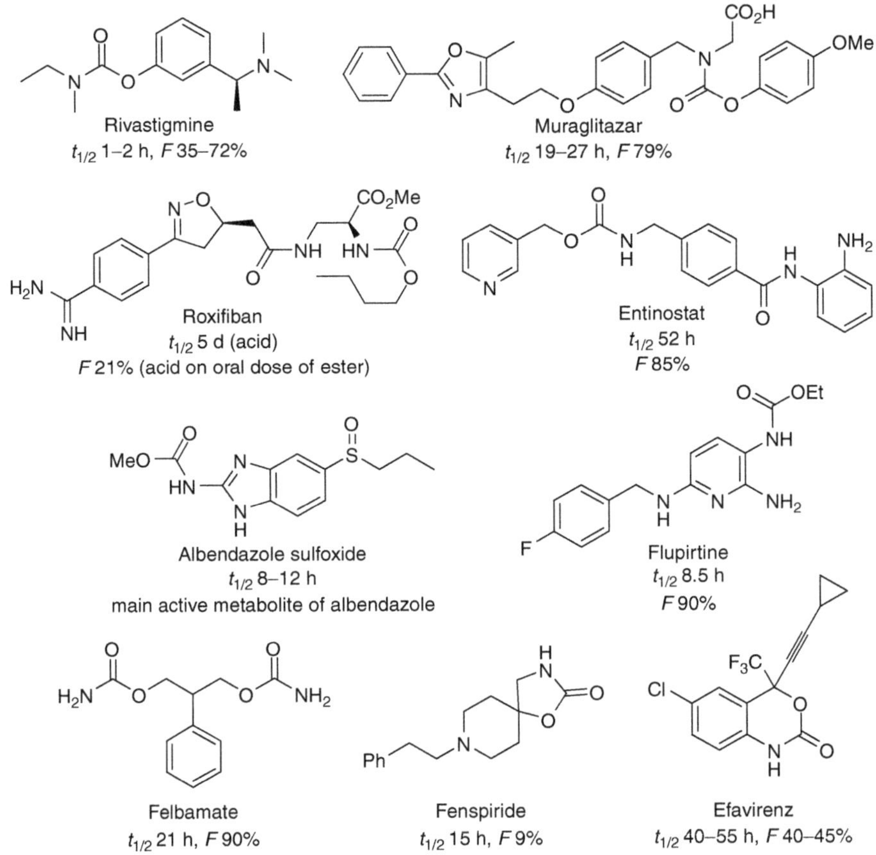

Scheme 37.2 Therapeutic carbamates.

37.1 Carbamates as Hypnotics

The first substances known to induce sleep were bromide and ethanol. In 1869, Oscar Liebreich discovered the hypnotic properties of chloral (trichloroacetaldehyde) that became widely used for this purpose. Because of the foul taste and irritating effect on the stomach, alternatives were sought, and Josef von Mering and Oswald Schmiedeberg at the University of Strasbourg discovered ethyl carbamate (urethane) as a weak, nonirritating hypnotic. The hypothesis behind their discovery was that urethane would be decomposed in the body to ethanol and the respiratory stimulants carbon dioxide and ammonia, what should provide for a safe hypnotic. It turned out, however, that intact ethyl carbamate was the hypnotic agent.

Further structural variations and combinations of low-molecular-weight alcohols, aldehydes, carbamates, ureas, and cyclic imides led to increasingly potent hypnotics and anticonvulsants, such as the acylureas, barbiturates, glutarimides, succinimides, hydantoins, and quinazolones (Tables 34.1, 37.1, and 38.1) [1].

A particularly tragic event was the launch of racemic thalidomide (Contergan) as sedative by the German company Grünenthal in 1957. Hardly any clinical trials had been performed before commercialization. Thalidomide causes peripheral neuritis, and it is also strongly teratogenic. Until its withdrawal in 1961, more than 1000 babies with horrible malformations, typically shortened or absent long bones of arms and legs (phocomelia), were born in Germany and Australia. The thalidomide disaster shattered confidence in the pharmaceutical industry and led to much stricter regulations and to a significant improvement of the quality of new drugs [2].

References

1. (a) Sneader, W. (1997) Anticonvulsants. *Drug News Perspect.*, **10**, 156–160; (b) Sneader, W. (1993) Hypnotics. *Drug News Perspect.*, **6**, 182–186.
2. Botting, J. (2002) The history of thalidomide. *Drug News Perspect.*, **15**, 604–611.

Table 37.1 Noncyclic carbamates. V in l kg^{-1}; CL in ml min^{-1} kg^{-1}; Mwt in g mol^{-1}.

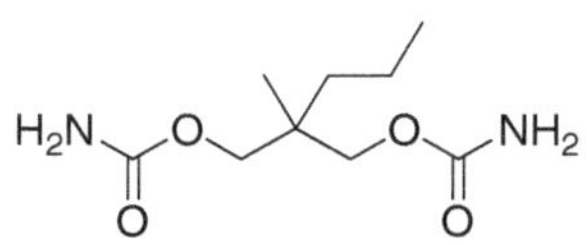	$t½$ F pb ur	10–11 h – 15% 8–20%	V CL Mwt PSA log P	0.75 2–3 218.3 105 Å^2 0.70	MEPROBAMATE Anxiolytic Metabolism: hydroxylation to 2-hydroxypropyl metabolite
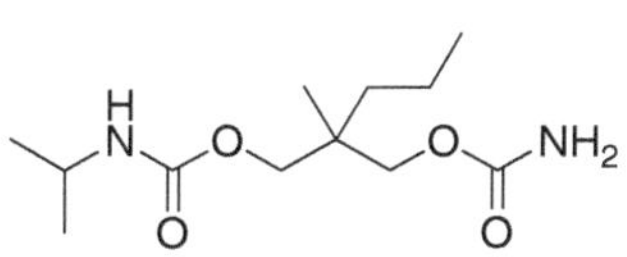	$t½$ F pb ur	8 h – 58% <1%	V CL Mwt PSA log P	– – 260.3 90.7 Å^2 2.10	CARISOPRODOL Muscle relaxant Metabolism: hydroxylation, N-deisopropylation to meprobamate
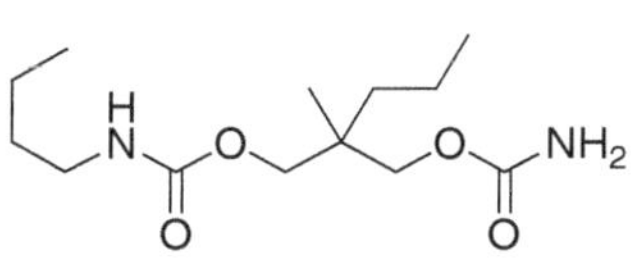	$t½$ F pb ur	3 h – 77% –	V CL Mwt PSA log P	– – 274.4 90.7 Å^2 2.76	TYBAMATE Anxiolytic Metabolism (rat, dog): propyl 2-hydroxylation, N-debutylation
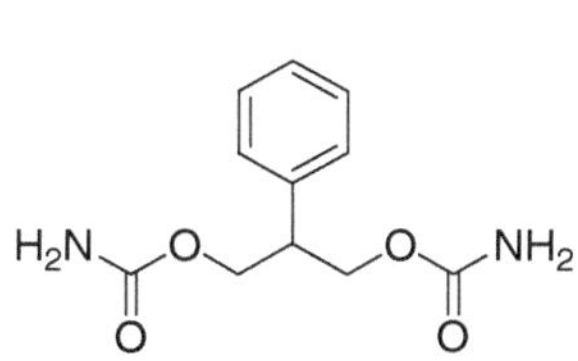	$t½$ F pb ur	21±2 h >90% 22–25% 45±5%	V/F CL/F Mwt PSA log P	0.76±0.08 0.50±0.13 238.2 105 Å^2 0.73	FELBAMATE Anticonvulsant; Metabolism: aromatic 4-hydroxylation, hydrolysis of carbamoyl, then oxidation to aldehyde, then elimination to α,β-unsaturated aldehyde; hepatotoxic

(continued overleaf)

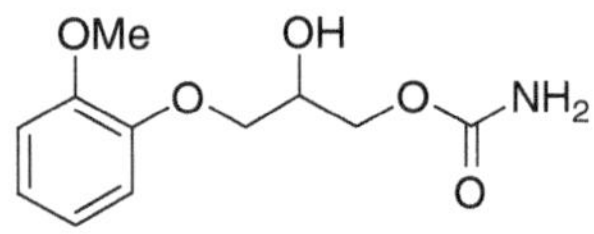	$t^1/_2$ F pb ur	0.9–2.0 h 32% 46–50% 10–15%	V CL Mwt PSA log P	– 3–13 241.2 91.0 Å^2 0.40	METHOCARBAMOL Muscle relaxant Metabolism: O-demethylation, aromatic hydroxylation
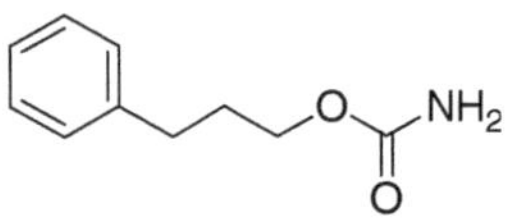	$t^1/_2$ F pb ur	5–8 h – 81% 7%	V CL Mwt PSA log P	– – 179.2 52.3 Å^2 1.93	PHENPROBAMATE Anxiolytic, muscle relaxant Metabolism: aromatic hydroxylation, oxidative degradation of carbamate
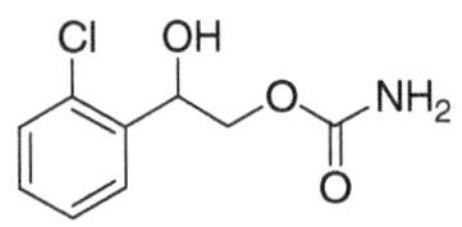	$t^1/_2$ F pb ur	10–14 h 94% <60% <5%	V CL Mwt PSA log P	1 0.7–1.0 215.6 72.6 Å^2 1.38	CARISBAMATE Antiepileptic Metabolism: glucuronidation
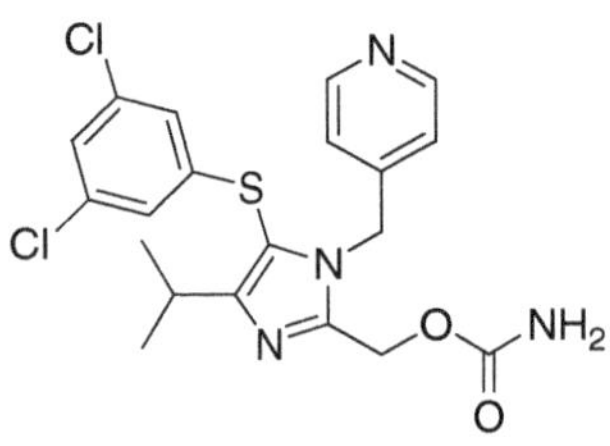	$t^1/_2$ F pb ur	2–3 h Good – Negligible	V CL Mwt PSA log P	4.5 – 451.4 108 Å^2 4.20	CAPRAVIRINE Antiviral (discontinued) Metabolites: pyridine-*N*-oxide, sulfoxide, sulfone
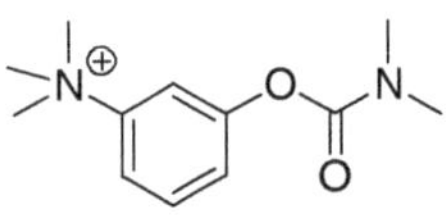	$t^1/_2$ F pb ur	1.3±0.8 h 10–20% Negligible 67%	V CL Mwt PSA log P	0.7±0.3 8.4±2.7 223.3 29.5 Å^2 −2.34	NEOSTIGMINE Cholinergic, miotic, antidote for neuromuscular blocking agents (e.g., curare); Metabolism: hydrolysis of carbamate
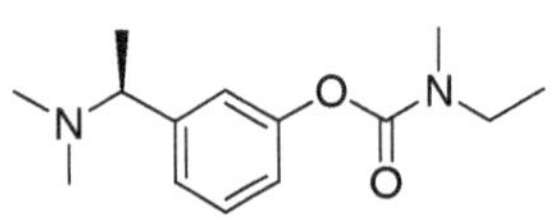	$t^1/_2$ F pb ur	1–2 h 35–72% 40% 0%	V CL Mwt PSA log P	0.7–3.2 5.2–20 250.3 32.8 Å^2 2.14	RIVASTIGMINE Nootropic, crosses bbb Metabolism: hydrolysis of carbamate, N-demethylation
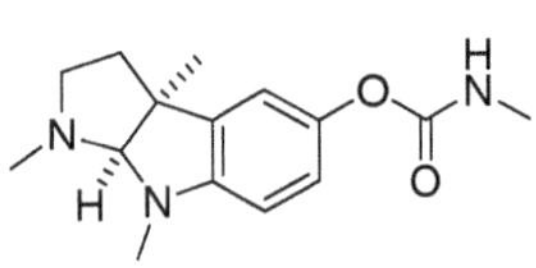	$t^1/_2$ F pb ur	0.5 h 3%* 30–40% <4%	V CL Mwt PSA log P	1–2 22–190 275.4 44.8 Å^2 1.27	PHYSOSTIGMINE *36% (transdermal) Cholinergic, miotic, crosses bbb

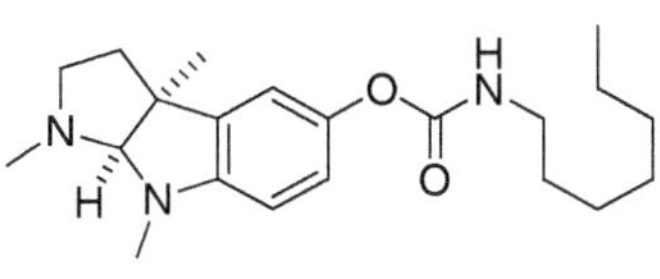	$t^1\!/\!_2$ 12 h *F* – pb – ur –	*V* 1.1 CL 12 Mwt 359.5 PSA 44.8 Å^2 log *P* 4.41	EPTASTIGMINE Cholinergic
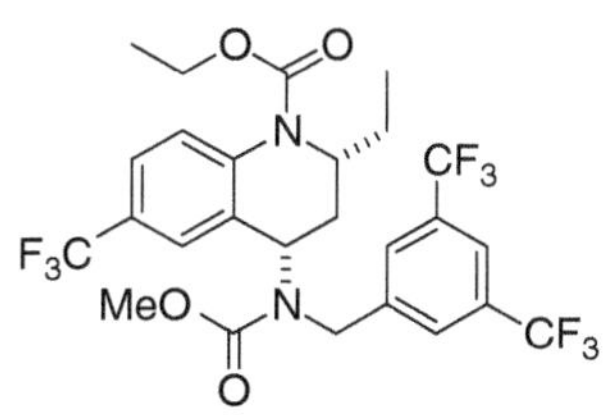	$t^1\!/\!_2$ 211 h *F* 33–45%* pb – ur 0%	*V* 1.1–2.5* CL 6–12* Mwt 600.5 PSA 59.1 Å^2 log *P* 8.16	TORCETRAPIB *rat, monkey CETP inhibitor for treatment of atherosclerosis; development discontinued

$t_{1/2}$, plasma half-life; *F*, oral bioavailability; pb, plasma protein binding; ur, excretion of unchanged drug in urine; V, volume of distribution; CL, clearance; Mwt, molecular weight; PSA, polar surface area; bbb, blood–brain barrier; CETP, cholesteryl ester transfer protein.

38
Ureas

Acyclic ureas are similarly stable as amides toward enzymatic hydrolysis but have been used less frequently in drug design. Illustrative examples are shown in Scheme 38.1. The most important drugs with a urea substructure are *N*-nitroso-*N*-(2-chloroethyl)ureas (Chapter 12), *N*-sulfonylureas (Chapter 64), and barbiturates.

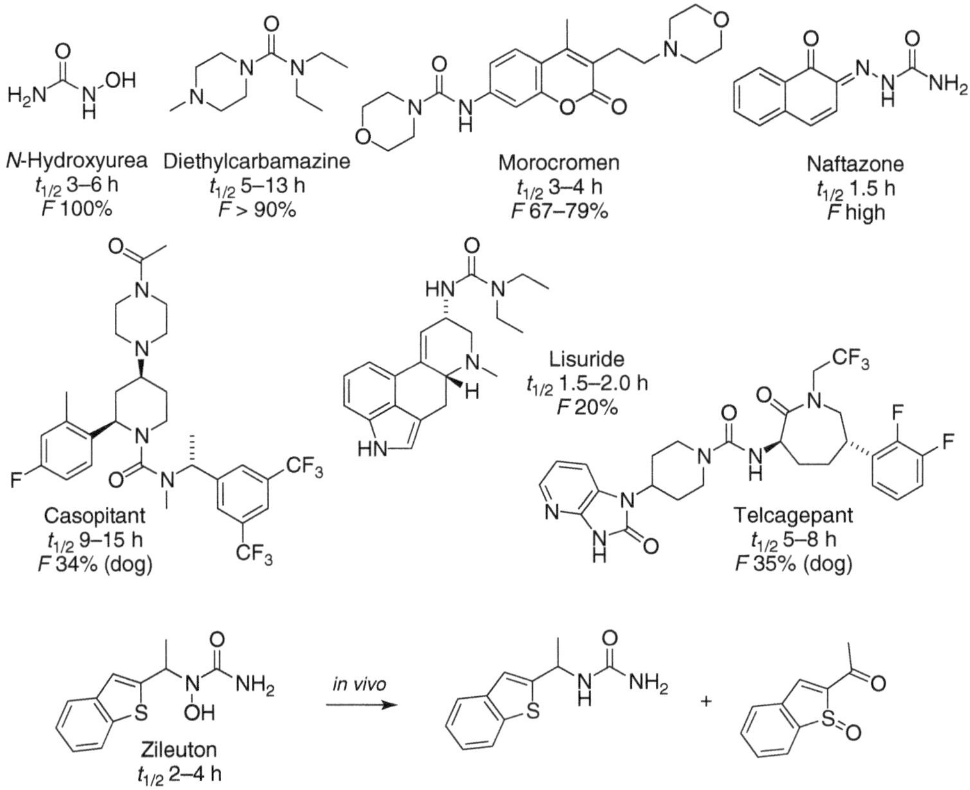

Scheme 38.1 Therapeutic ureas.

Lead Optimization for Medicinal Chemists: Pharmacokinetic Properties of Functional Groups and Organic Compounds, First Edition. Florencio Zaragoza Dörwald.

The parent, unsubstituted urea is used as osmotic diuretic. It must be dosed intravenously because its oral bioavailability is too low. The antineoplastic hydroxyurea, however, is readily absorbed from the gastrointestinal tract and even crosses the blood–brain barrier (bbb). *N*-Hydroxyureas may undergo metabolic N−O bond cleavage, as shown below for the antiasthmatic zileuton.

Among the first hypnotics were *N*-acylureas such as carbromal, bromisovalum, and pheneturide (Scheme 38.2), but no complete pharmacokinetics (PK) profile was determined for these drugs. Noncyclic acylureas are less potent than barbiturates, and are no longer used as hypnotics.

Scheme 38.2 Older hypnotic ureas.

After discovering the weak hypnotic ethylcarbamate, Josef von Mering, working at the University of Strasbourg, continued to prepare carbamates and ureas with the aim of finding better hypnotics. After discovering the potent hypnotic *N*-(2-ethylbutanoyl)urea, he decided to prepare cyclic *N*-acylureas. One of these was 5,5-diethylbarbituric acid (barbital). Because he had doubts about the real structure of his product, he asked his friend Emil Fischer for help. Fischer together with his nephew Alfred Dilthey resynthesized the compound, and it turned out to be the most potent hypnotic ever seen. It was marketed by Bayer as Veronal in 1903 [1].

Hypnotics such as ethanol and barbiturates act by suppressing polysynaptic responses in the central nervous system (CNS), by interacting with various γ-aminobutyric acid (GABA) receptors, and by inhibiting glutamate-activated ion currents, without binding specifically to any receptor. Interestingly, despite their large polar surface area (PSA), barbiturates and related hypnotics are able to cross the bbb and enter the CNS.

Barbiturates are more potent and show fewer side effects than the older hypnotics chloral, paraldehyde, sulfonal, or the simple aliphatic carbamates hedonal or ethyl carbamate. Barbiturates were extensively used for many years as sedatives, hypnotics, anticonvulsants, and anesthetics (Table 38.1). From the 1960s, however, barbiturates have been largely displaced by benzodiazepines, which have a higher therapeutic index and a lower liability for drug–drug interactions, tolerance, and abuse.

Reference

1. Sneader, W. (1993) Hypnotics. *Drug News Perspect.*, **6**, 182–186.

Table 38.1 Thioamides, thioureas, and six-membered lactams, imides, and ureas (barbiturates). V in l kg^{-1}; CL in ml min^{-1} kg^{-1}; Mwt in g mol^{-1}.

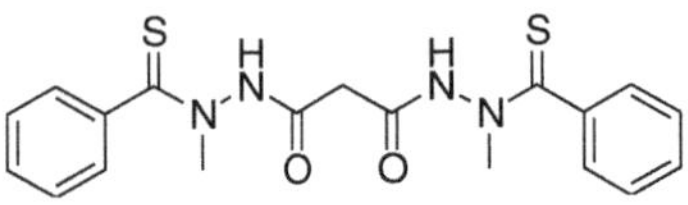	$t^1/_2$ F pb ur	1 h – – –	V CL Mwt PSA log P	0.7 13 400.5 129 Å^2 1.98	ELESCLOMOL Apoptosis inducer, inducer of oxidative stress, HSP70 inducer, oncolytic
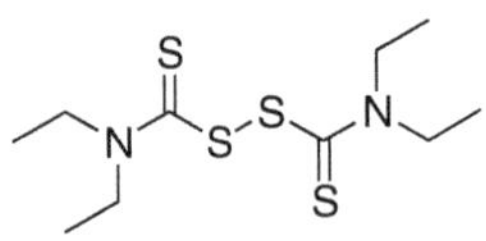	$t^1/_2$ F pb ur	60–120 h >80% 96% –	V CL Mwt PSA log P	– 9–12 296.5 121 Å^2 3.82	DISULFIRAM Aldehyde dehydrogenase inhibitor for treatment of alcoholism; crosses bbb; CYP2E1 inhibitor, hepatotoxic Active metabolite: diethylthiomethylcarbamate ($Et_2NCO(SCH_3)$)
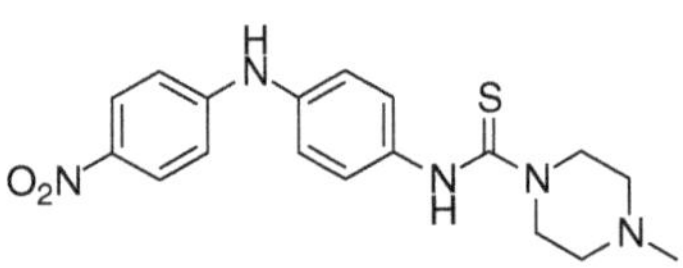	$t^1/_2$ F pb ur	3 h <40% – 1.5%	V CL Mwt PSA log P	– – 371.5 108 Å^2 2.32	AMOCARZINE Anthelmintic Metabolism: N-oxidation
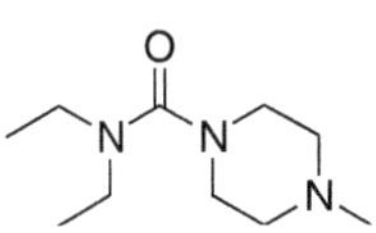	$t^1/_2$ F pb ur	5–13 h >90% Negligible 50%	V CL Mwt PSA log P	1.7–5.7 – 199.3 26.8Å^2 1.66	DIETHYLCARBAMAZINE Anthelmintic Metabolite: *N*-oxide
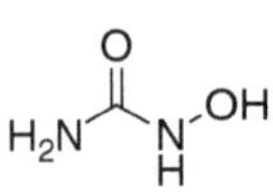	$t^1/_2$ F pb ur	3–6 h 100% 0% 80%	V CL Mwt PSA log P	0.52 1.5 76.1 75.3 Å^2 −1.80	HYDROXYUREA Antineoplastic, crosses bbb
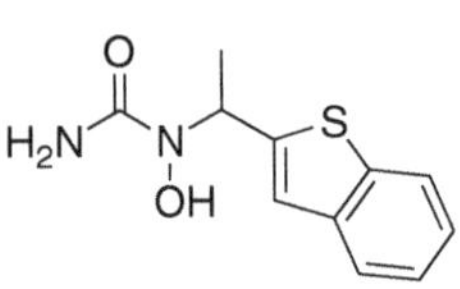	$t^1/_2$ F pb ur	1.8–3.9 h – 94% Negligible	V CL Mwt PSA log P	0.9–2.7 7.8–9.6 236.3 94.8 Å^2 3.74	ZILEUTON Antiasthmatic, antiinflammatory, 5-lipoxygenase inhibitor
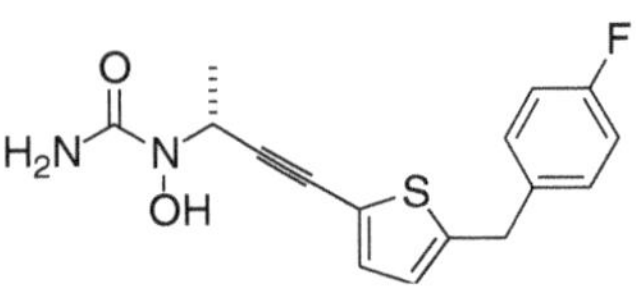	$t^1/_2$ F pb ur	15–18 h High 99% <0.05%	V CL Mwt PSA log P	70–79 – 318.4 94.8 Å^2 4.56	ABT-761 5-Lipoxygenase inhibitor Metabolism: benzylic hydroxylation, reductive cleavage of N–O bond

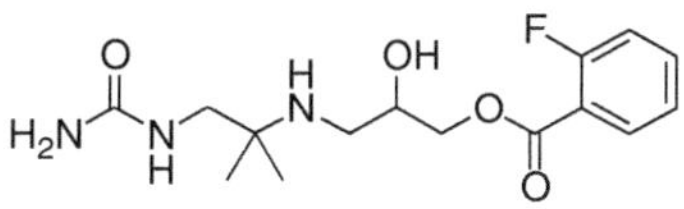

$t^1/_2$	1–7 min*	V	–	FLESTOLOL
F	–	CL	154–260*	*iv
pb	–	Mwt	327.4	β-Adrenoceptor antagonist
ur	–	PSA	114 Å^2	Metabolism: ester hydrolysis
		log P	0.91	

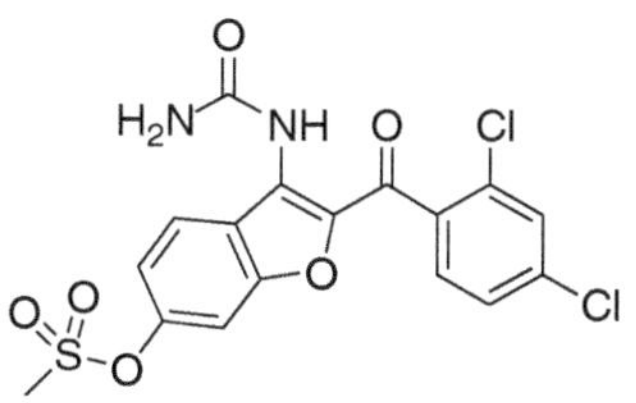

$t^1/_2$	25 h	V	–	LIRIMILAST
F	–	CL	–	PDE$_4$ inhibitor
pb	–	Mwt	443.3	
ur	–	PSA	137 Å^2	
		log P	2.93	

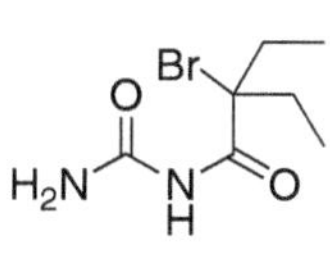

$t^1/_2$	7–15 h	V	–	CARBROMAL
F	–	CL	0.15	Sedative, hypnotic
pb	40%	Mwt	237.1	Metabolism: hydrolysis of urea
ur	–	PSA	72.2 Å^2	to 2-bromo-2-ethylbutyramide,
		log P	1.62	reductive debromination

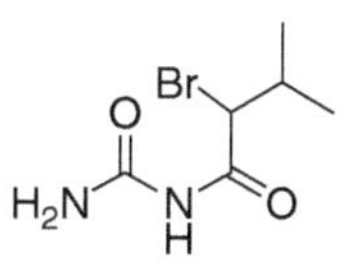

$t^1/_2$	13 h	V	–	BROMISOVALUM
F	–	CL	0.14 (*R*)*	* *S*: 0.012; Sedative, hypnotic
pb	–	Mwt	223.1	Metabolism: reductive
ur	–	PSA	72.2 Å^2	debromination, S-alkylation of
		log P	1.06	cysteine and of *N*-acetylcysteine

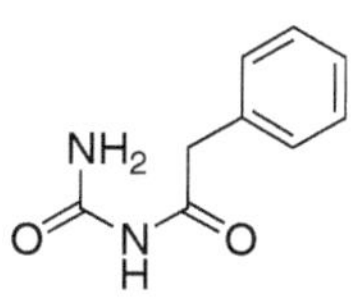

$t^1/_2$	22–25 h	V	–	PHENACEMIDE
F	–	CL	–	Antiepileptic
pb	24%	Mwt	178.2	Metabolism: aromatic
ur	Low	PSA	72.2 Å^2	4-hydroxylation
		log P	0.87	

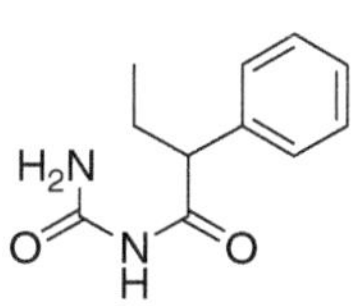

$t^1/_2$	60±30 h	V	3.3	PHENETURIDE
F	–	CL	0.68±0.24	Anticonvulsant
pb	–	Mwt	206.2	Metabolism: aromatic
ur	Low	PSA	72.2 Å^2	4-hydroxylation, hydrolysis of
		log P	1.73	ureide to 2-phenylbutyric acid

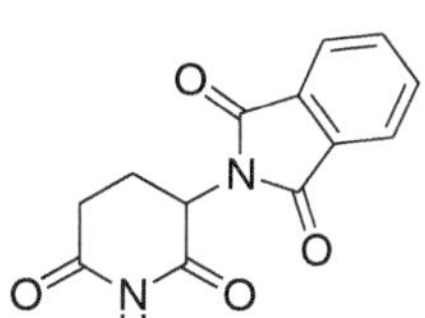

$t^1/_2$	4.1–7.3 h	V	1	THALIDOMIDE
F	High	CL	2.4 (*R*)**	* *S*: 66%; ** *S*: 5.0
pb	55% (*R*)*	Mwt	258.2	Sedative; Metabolism:
ur	<1%	PSA	83.6 Å^2	hydrolysis of both imides,
		log P	0.33	aromatic 4-hydroxylation,
				hydroxylation of CH_2CO

(*continued overleaf*)

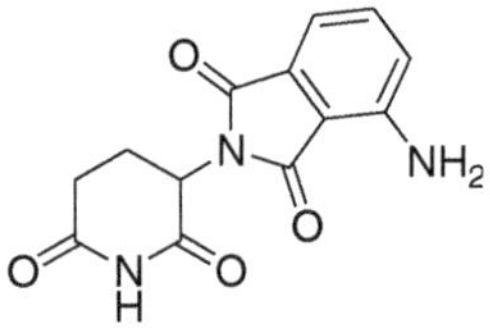

$t^1/_2$	8 h	V	–	POMALIDOMIDE
F	–	CL	–	Immunomodulator
pb	–	Mwt	273.2	
ur	–	PSA	110 Å^2	
		log P	−0.71	

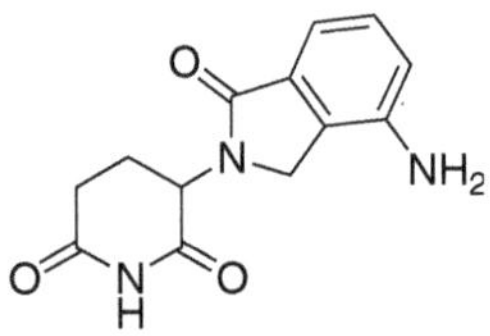

$t^1/_2$	3–8 h	V	1.2	LENALIDOMIDE
F	70%	CL	4.6	Antineoplastic
pb	19–29%	Mwt	259.3	Metabolism: N-acetylation,
ur	65%	PSA	92.5 Å^2	hydrolysis
		log P	−1.09	

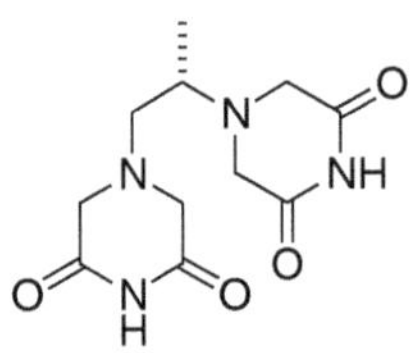

$t^1/_2$	2–4 h	V	0.6–1.1	DEXRAZOXANE
F	–	CL	3.1–4.8	Cardioprotectant (racemate:
pb	<2%	Mwt	268.3	antineoplastic)
ur	46%	PSA	98.8 Å^2	Metabolism: hydrolysis
		log P	−0.91	of imide

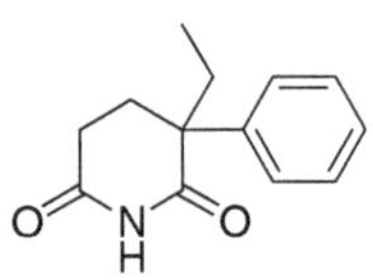

$t^1/_2$	10–12 h	V	2.7	GLUTETHIMIDE
F	–	CL	2.3	Sedative, hypnotic
pb	50%	Mwt	217.3	Metabolites: glucuronides,
ur	<2%	PSA	46.2 Å^2	4-hydroxyphenyl, 4-hydroxy,
		log P	1.75	1-hydroxyethyl

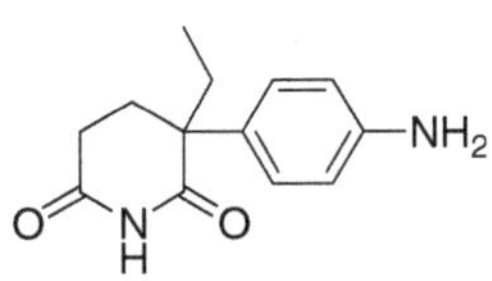

$t^1/_2$	13±2 h	V	0.8–1.1	AMINOGLUTETHIMIDE
F	>95%	CL	0.8–1.0	Anticonvulsant, inhibitor of
pb	20–25%	Mwt	232.3	cortisol biosynthesis
ur	35–50%	PSA	72.2 Å^2	Metabolism: acetylation and
		log P	0.54	hydroxylation of NH_2,
				hydroxylation of CH_2CO,
				1-hydroxylation of ethyl

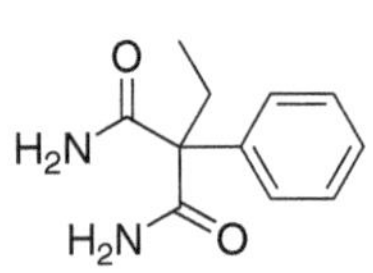

$t^1/_2$	16±3 h	V	0.7±0.1	ETHYLPHENYLMALONAMIDE
F	91±4%	CL	0.5±0.1	Metabolite of primidone
pb	8±1%	Mwt	206.2	(see below)
ur	79±5%	PSA	86.2 Å^2	
		log P	0.35	

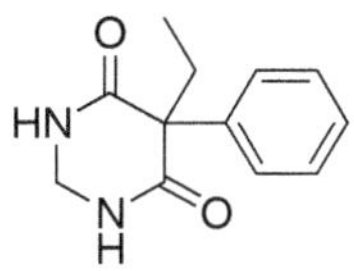	$t^1\!/_2$ F pb ur	15±4 h 92±18% 19% 46±16%	V CL Mwt PSA log P	0.7±0.2 0.55±0.12 218.3 58.2 Å^2 0.83	PRIMIDONE Anticonvulsant Metabolites: phenobarbital, ethylphenylmalonamide
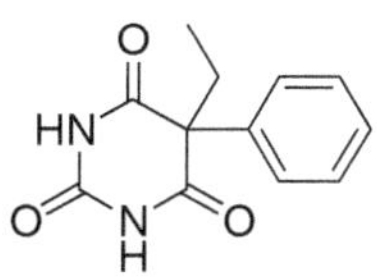	$t^1\!/_2$ F pb ur	99±18 h 100±11% 51±3% 24±5%	V CL Mwt PSA log P	0.54±0.03 0.06±0.01 232.2 75.3 Å^2 0.53	PHENOBARBITAL Anticonvulsant, sedative, hypnotic; Metabolism: N-glucosylation, N-glucuronidation, aromatic 4-hydroxylation
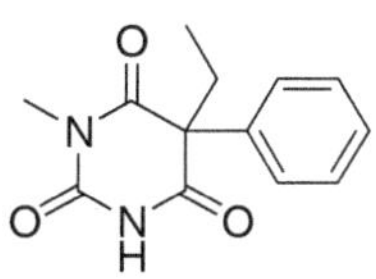	$t^1\!/_2$ F pb ur	30–50 h 75% 65% –	V CL Mwt PSA log P	– – 246.3 66.5 Å^2 1.23	MEPHOBARBITAL Anticonvulsant, sedative, hypnotic, CYP inducer Metabolism: aromatic 4-hydroxylation, N-demethylation
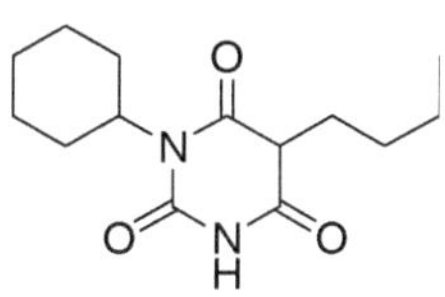	$t^1\!/_2$ F pb ur	25 h 100% 90% –	V CL Mwt PSA log P	– – 266.3 66.5 Å^2 3.61	BUCOLOME Antiinflammatory, CYP2C9 inhibitor Metabolism: N-glucuronidation
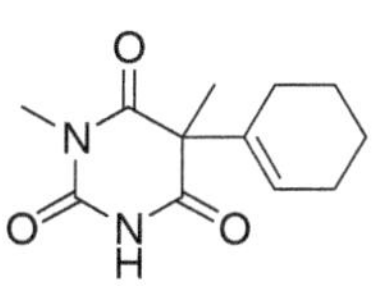	$t^1\!/_2$ F pb ur	4±1 h >90% 47±5% <1%	V CL Mwt PSA log P	1.2±0.3 3.9±0.7 236.3 66.5 Å^2 1.46	HEXOBARBITAL Sedative, hypnotic Metabolism: 3′-(allylic)-hydroxylation, then oxidation to 3′-ketone, little N-demethylation
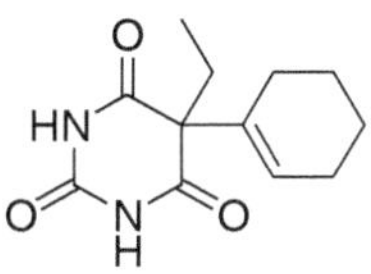	$t^1\!/_2$ F pb ur	13±5 h – 70% <5%	V CL Mwt PSA log P	0.5–0.7 0.5–0.6 236.3 75.3 Å^2 1.50	CYCLOBARBITAL Sedative, hypnotic
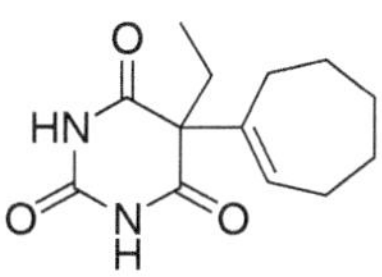	$t^1\!/_2$ F pb ur	8±3 h – – <1%	V CL Mwt PSA log P	– – 250.3 75.3 Å^2 2.03	HEPTABARBITAL Sedative, hypnotic Metabolism: dehydrogenation to cycloheptadiene

(*continued overleaf*)

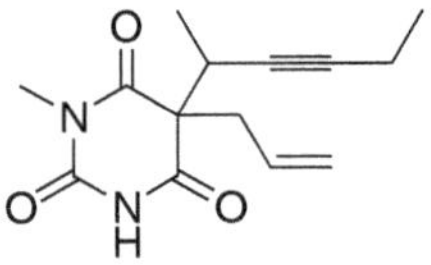	$t_{1/2}$ F pb ur	4±2 h – 73% <1%	V CL Mwt PSA log P	2.2±0.7 11±3 262.3 66.5 Å^2 2.01	METHOHEXITAL Anesthetic Metabolism: 4′-(propargylic)-hydroxylation (no N-demethylation)
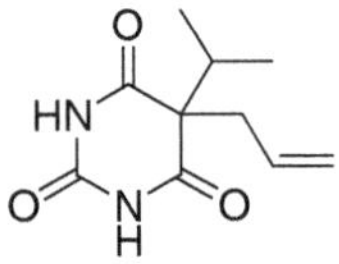	$t_{1/2}$ F pb ur	24±10 h – – –	V CL Mwt PSA log P	– – 210.2 75.3 Å^2 1.28	APROBARBITAL Sedative, hypnotic Metabolism: epoxidation
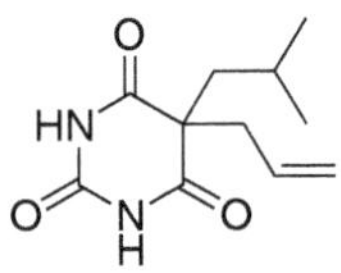	$t_{1/2}$ F pb ur	35–88 h – – –	V CL Mwt PSA log P	– – 224.3 75.3 Å^2 1.79	BUTALBITAL Sedative, hypnotic Metabolism (dog): dihydroxylation of alkene, terminal hydroxylation of isobutyl
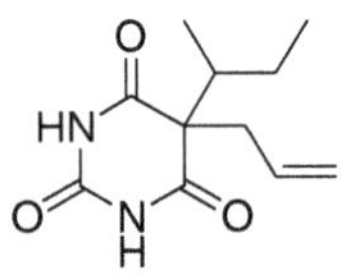	$t_{1/2}$ F pb ur	15 h – – –	V CL Mwt PSA log P	– – 224.3 75.3 Å^2 1.79	TALBUTAL Sedative, hypnotic Metabolism: dihydroxylation of alkene
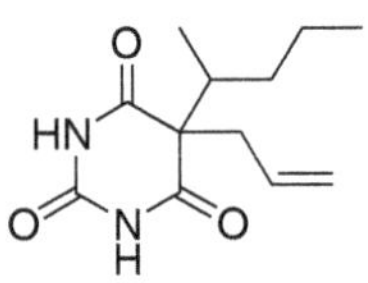	$t_{1/2}$ F pb ur	19–34 h 90% 50–70% 10–20%	V CL Mwt PSA log P	1.5 0.86 238.3 75.3 Å^2 2.30	SECOBARBITAL Sedative, hypnotic Metabolism: dihydroxylation of alkene, hydroxylation at CH_2CH_3
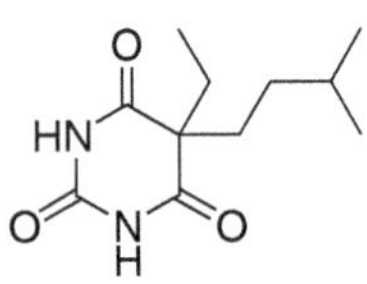	$t_{1/2}$ F pb ur	15–40 h 50±10% 1%	V CL Mwt PSA log P	0.9–1.4 0.5 226.3 75.3 Å^2 2.18	AMOBARBITAL Sedative, hypnotic Metabolism: 3′-hydroxylation (at tertiary CH), N-glucosylation
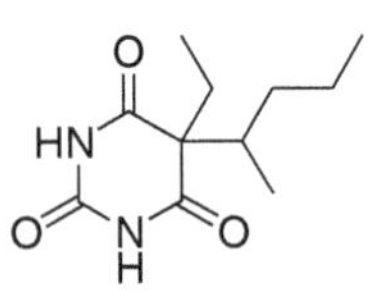	$t_{1/2}$ F pb ur	35–50 h – 50–65% <1%	V CL Mwt PSA log P	0.5–1.0 0.50–0.65 226.3 75.3 Å^2 2.18	PENTOBARBITAL Sedative, hypnotic Metabolites: 1-methyl-3-hydroxybutyl, 1-methyl-3-carboxypropyl

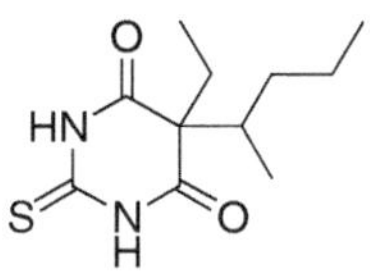	$t_{1/2}$ *F* pb ur	9.0±1.6 h – 85±4% <1%	*V* CL Mwt PSA log *P*	2.3±0.5 3.9±1.2 242.3 90.3 Å^2 3.05	THIOPENTAL Anesthetic Metabolites: 1-methyl-3-hydroxybutyl, 1-methyl-3-carboxypropyl, pentobarbital
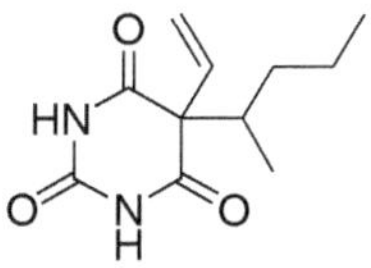	$t_{1/2}$ *F* pb ur	18–34 h – – 1.6%	*V* CL Mwt PSA log *P*	– – 224.3 75.3 Å^2 1.18	VINYLBITAL Sedative, hypnotic Metabolism: devinylation (via diol)
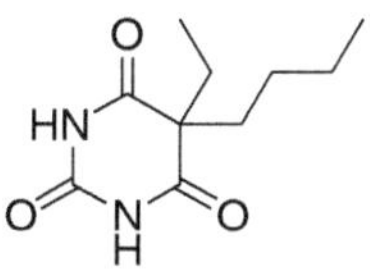	$t_{1/2}$ *F* pb ur	34–42 h – 26% 5–9%	*V* CL Mwt PSA log *P*	0.8 0.23 212.3 75.3 Å^2 1.82	BUTETHAL Sedative, hypnotic Metabolism: hydroxylation at $CH_2\mathit{CH_2}CH_3$
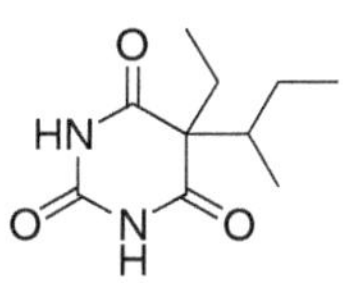	$t_{1/2}$ *F* pb ur	35–50 h – – 3–5%*	*V* CL Mwt PSA log *P*	– – 212.3 75.3 Å^2 1.67	BUTABARBITAL *dog Sedative, hypnotic Metabolism (dog): oxidative degradation of 2-butyl to 1-carboxy-2-propyl
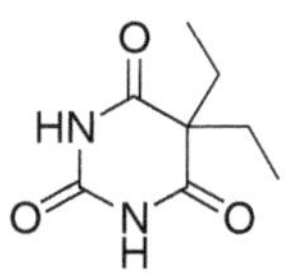	$t_{1/2}$ *F* pb ur	70 h – – 80%	*V* CL Mwt PSA log *P*	– – 184.2 75.3 Å^2 0.80	BARBITAL Sedative, hypnotic Metabolite of *N*-ethyl- and *N*, *N'*-dimethylbarbital
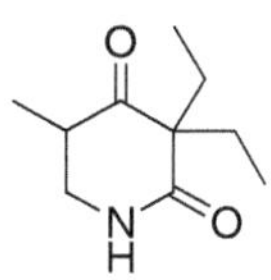	$t_{1/2}$ *F* pb ur	3–6 h – – –	*V* CL Mwt PSA log *P*	– – 183.3 46.2 Å^2 0.65	METHYPRYLON Sedative, hypnotic Metabolism: dehydrogenation to 1,2,3,4-tetrahydropyridine
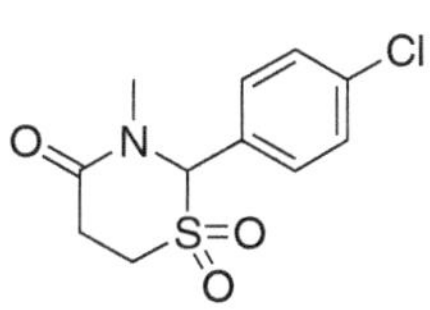	$t_{1/2}$ *F* pb ur	41±4 h – 48% 3%	*V* CL Mwt PSA log *P*	– – 273.7 62.8 Å^2 0.48	CHLORMEZANONE Anxiolytic Metabolism: amide hydrolysis, oxidation and conjugation to 4-chlorobenzoylglycine; withdrawn in 1996 for causing epidermal necrolysis

(*continued overleaf*)

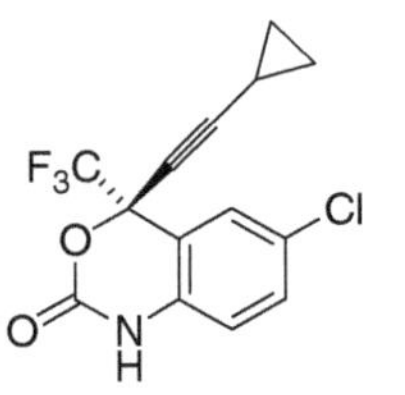	$t^1/_2$ *F* pb ur	40–55 h 40–45% >99% <1%	*V* CL Mwt PSA log *P*	4 2.4 315.7 38.8 Å^2 4.84	EFAVIRENZ Antiviral (HIV) Metabolism: hydroxylation at arene (position 8, ortho of NH) and at tertiary cyclopropyl CH (propargylic position)
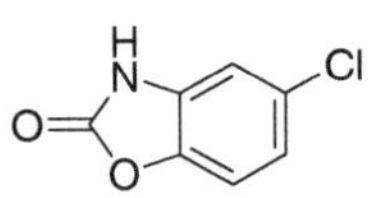	$t^1/_2$ *F* pb ur	1.1 h 100% – –	*V* CL Mwt PSA log *P*	13.7 – 169.6 38.3 Å^2 1.82	CHLORZOXAZONE Muscle relaxant Metabolism: aromatic 6-hydroxylation, then hydrolysis of carbamate and N-acetylation, reductive dechlorination of 6-hydroxychlorzoxazone
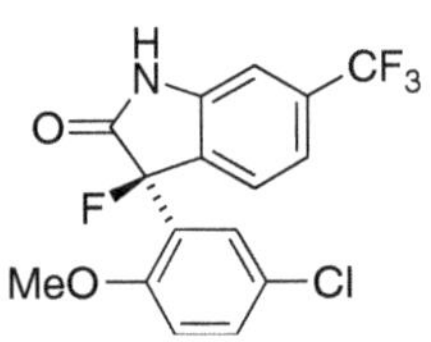	$t^1/_2$ *F* pb ur	37 h (iv) 55% (rat) >99% –	*V* CL Mwt PSA log *P*	10 15 359.7 38.3 Å^2 2.97	MAXIPOST, BMS 204352 Potassium channel opener Metabolism: demethylation, then fluoride displacement by nucleophiles

$t_{1/2}$, plasma half-life; *F*, oral bioavailability; pb, plasma protein binding; ur, excretion of unchanged drug in urine; V, volume of distribution; CL, clearance; Mwt, molecular weight; PSA, polar surface area.

39
Thiocarbonyl Compounds

Despite their high chemical reactivity and frequent toxicity, a surprisingly large number of drugs contain thiocarbonyl groups. These include tuberculostatics (ethionamide, prothionamide), hypnotics (quazepam), anthelmintics (amocarzine), antihyperthyroids (carbimazole, propylthiouracil), anesthetics (thiopental, methitural), and aldose reductase inhibitors for the prevention of eye and nerve damage in diabetics (epalrestat, tolrestat) (Scheme 39.1).

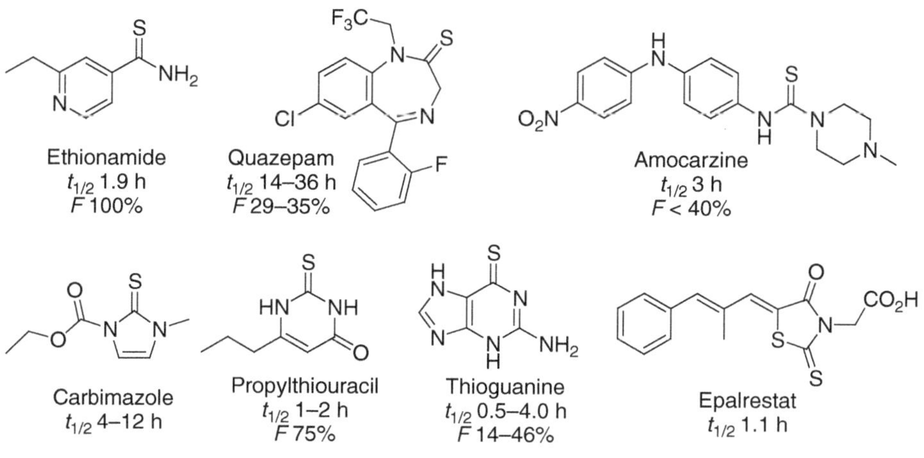

Scheme 39.1 Thiocarbonyl compounds as drugs.

Thiocarbonyl compounds are readily oxidized and then hydrolyzed to the corresponding carbonyl compounds (Scheme 39.2). Thiocarbonyl groups also absorb UV light at longer wavelengths than similar carbonyl compounds (e.g., acetamide: $\lambda_{max} < 210$ nm; thioacetamide: $\lambda_{max} = 358$ nm) and are often light sensitive. Further metabolic transformations of thiocarbonyl compounds include S-methylation and the reduction of $CSNH_2$ to hydroxymethyl.

Lead Optimization for Medicinal Chemists: Pharmacokinetic Properties of Functional Groups and Organic Compounds, First Edition. Florencio Zaragoza Dörwald.

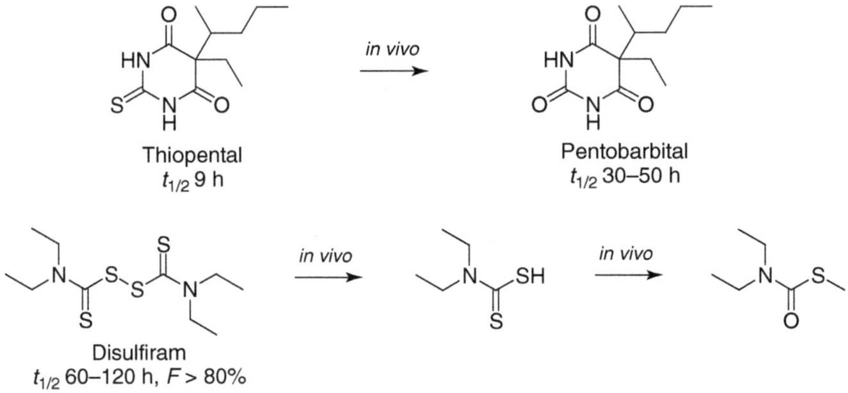

Scheme 39.2 Metabolism of thiocarbonyl compounds.

Oxidized thiocarbonyl derivatives are reactive, potentially toxic electrophiles. One example is metiamide, a potent and orally available histamine H_2-antagonist (for suppression of gastric acid formation, Scheme 20.3). Metiamide was developed from burimamide, which had low oral bioavailability because of its high basicity and hydrophilicity. Metiamide, however, was too toxic, and the final drug resulting from this program was cimetidine in which the thiourea had been replaced by a less reactive cyanoguanidine group.

40
Sulfonic Acids

Sulfonic acids are strong acids (pK_a −2), completely deprotonated at physiological pH. Only sulfonic acids of low molecular weight can be absorbed from the intestine and are orally bioavailable. On the other hand, sulfonic acids are highly water soluble and well suited for liquid, parenteral formulations. A selection of therapeutic sulfonic acids is shown in Scheme 40.1 and Table 40.1.

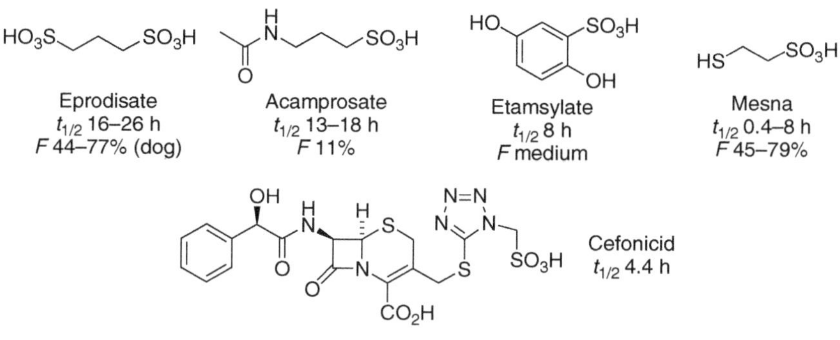

Scheme 40.1 Sulfonic acids as drugs.

α-Aminosulfonic acids are unstable and readily decompose into sulfite, aldehydes, and amines, like the bisulfite adducts of aldehydes (α-hydroxysulfonic acids). Accordingly, the solubility of amino-group-containing drugs may be enhanced by converting them into aminosulfonic acid salts (by treatment with formaldehyde and sulfite). This strategy has, for instance, been realized in dipyrone (Scheme 40.2).

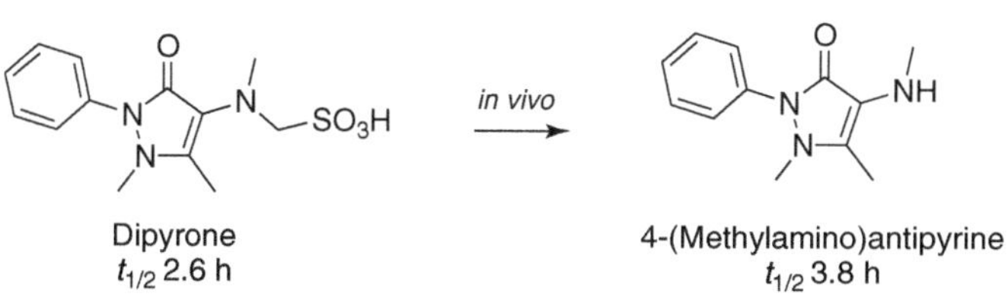

Scheme 40.2 Metabolism of dipyrone.

Lead Optimization for Medicinal Chemists: Pharmacokinetic Properties of Functional Groups and Organic Compounds, First Edition. Florencio Zaragoza Dörwald.

A comparison of PK (pharmacokinetic) data of various analogs of benzylpenicillin is given in Scheme 40.3. The amino group leads to higher oral bioavailability, while the corresponding carboxylic and sulfonic acids show poor oral bioavailability but are better suited for liquid formulations for parenteral administration.

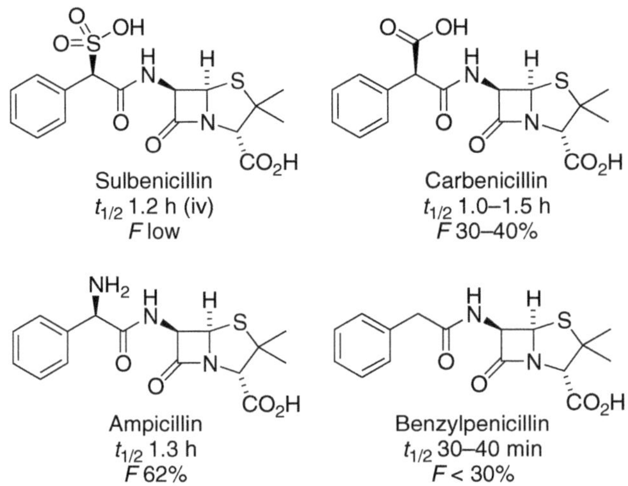

Scheme 40.3 Impact of various functional groups on the PK properties of penicillins.

Table 40.1 Sulfonic acids. V in l kg^{-1}; CL in ml min^{-1} kg^{-1}; Mwt in g mol^{-1}.

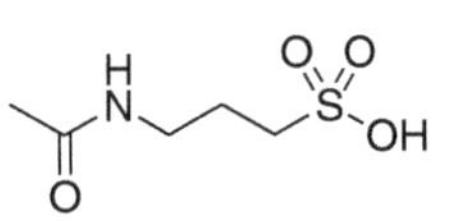	$t^1/_2$ F pb ur	13–18 h 11% Negligible 50%	V CL Mwt PSA log P	1.1 3.4–5.7 181.2 91.9 Å^2 −2.63	ACAMPROSATE GABA agonist for treatment of alcoholism; Metabolism: N-deacetylation to homotaurine (tramiprosate, an antiamyloidogenic agent for treatment of Alzheimer's disease; $t^1/_2$ 2–4 h)
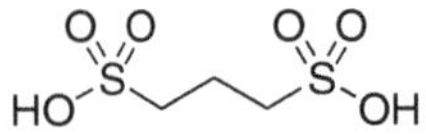	$t^1/_2$ F pb ur	16–26 h 44–77%* 0% 6.5%	V CL Mwt PSA log P	Small – 204.2 126 Å^2 −4.10	EPRODISATE *dog Treatment of amyloidosis

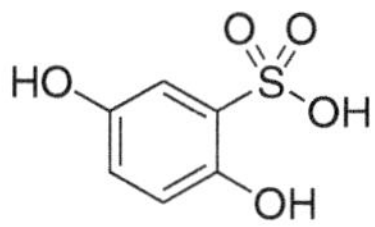	$t^1\!/_2$	8 h	V	0.4*	ETAMSYLATE
	F	Medium	CL	3.5*	*calves, iv
	pb	95%	Mwt	190.2	Hemostatic
	ur	72%	PSA	94.8 Å^2	
			log P	−1.84	
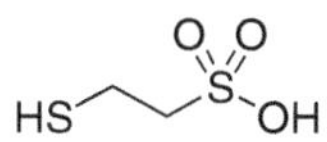	$t^1\!/_2$	0.4–8 h	V	2.1	MESNA
	F	45–79%	CL	12.6	Scavenger of acrolein formed by
	pb	–	Mwt	142.2	metabolism of cyclophosphamide and
	ur	36%	PSA	54.4 Å^2	ifosphamide; Metabolism:
			log P	−2.28	oxidative dimerization to disulfide
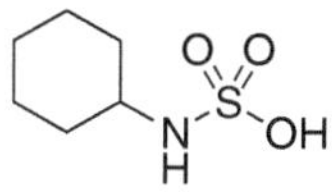	$t^1\!/_2$	8 h*	V	–	CYCLAMATE
	F	<37%	CL	–	*rat, dog
	pb	–	Mwt	179.2	Artificial sweetener
	ur	30–40%	PSA	74.8 Å^2	Metabolite: cyclohexylamine
			log P	1.03	
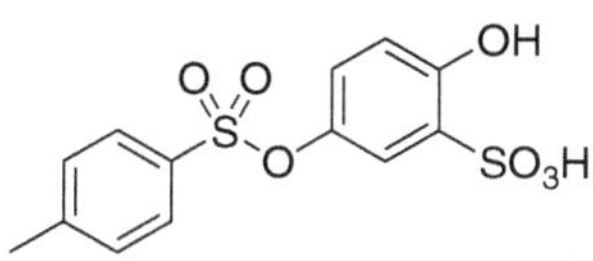	$t^1\!/_2$	18 h	V	–	SULTOSILIC ACID
	F	>60%	CL	0.8	Antihyperlipidemic
	pb	–	Mwt	344.4	Metabolism: hydroxylation and
	ur	28%	PSA	135 Å^2	oxidation of methyl
			log P	0.75	

$t_{1/2}$, plasma half-life; F, oral bioavailability; pb, plasma protein binding; ur, excretion of unchanged drug in urine; V, volume of distribution; CL, clearance; Mwt, molecular weight; PSA, polar surface area; GABA, γ-aminobutyric acid.

41
Sulfonic Esters

Alkyl sulfonates are strong alkylating agents, similar to alkyl chlorides or bromides, and are usually too reactive to act as reversible ligands of proteins. However, alkylating agents, in particular cross-linking agents, may be useful for the treatment of certain types of cancer. Thus, busulfan, the dimesylate of 1,4-butanediol, is a valuable drug for the treatment of chronic granulocytic leukemia (Scheme 41.1).

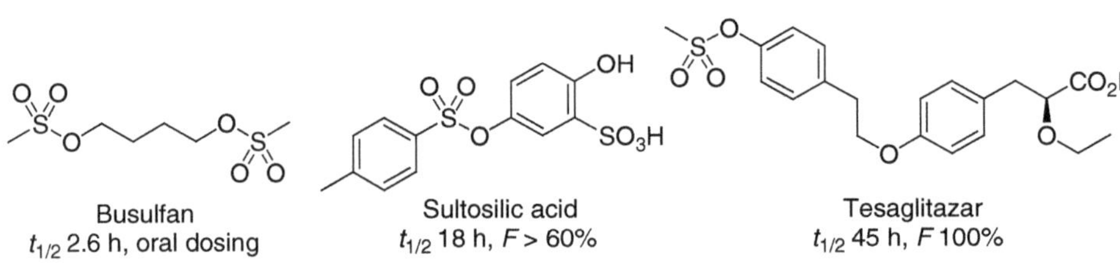

Scheme 41.1 Sulfonic esters as drugs.

Neopentylic or aryl sulfonates, on the other hand, cannot readily undergo nucleophilic substitution and may, therefore, be used as structural element of nonalkylating drugs.

Lead Optimization for Medicinal Chemists: Pharmacokinetic Properties of Functional Groups and Organic Compounds, First Edition. Florencio Zaragoza Dörwald.

42
Sulfates and Sulfamic Acids

The O-sulfation of phenols is a typical metabolic transformation to enhance the solubility of xenobiotics and lower their ability to cross biological membranes. Medicinal chemists can, similarly, enhance the solubility of phenols and alcohols by O-sulfation. However, the oral bioavailability will usually be lowered by O-sulfation. This strategy has only rarely been used in drug design.

Monoaryl- and monoalkylsulfates ($ArOSO_3H$, $ROSO_3H$) are chemically quite stable and usually less toxic than the corresponding phenols and alcohols. A number of drugs are metabolized to biologically active *O*-sulfates (e.g., triamterene, minoxidil), and in the future, an increasing number of such sulfated metabolites will probably be developed as liquid formulations for parenteral administration.

Picosulfate and bisacodyl are laxatives for oral administration (Scheme 42.1). Bisacodyl is deacetylated by intestinal bacteria. These compounds are poorly absorbed from the gastrointestinal tract and act by stimulating peristaltic mobility and by increasing the excretion of water into the intestine. Both compounds show similar biological properties, a further indication that phenols and their sulfates can be bioisosteres. Sulfates will, however, be less orally bioavailable and less membrane penetrating because of their complete ionization at physiological pH.

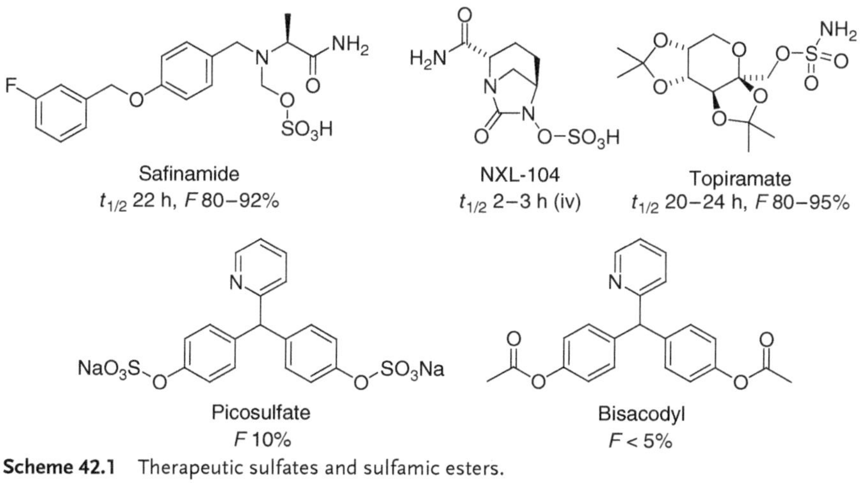

Scheme 42.1 Therapeutic sulfates and sulfamic esters.

Lead Optimization for Medicinal Chemists: Pharmacokinetic Properties of Functional Groups and Organic Compounds, First Edition. Florencio Zaragoza Dörwald.

Closely related to sulfates are sulfamic acids (R_2NSO_3H) and their esters (Scheme 42.2). These compounds can be metabolically rather stable, such as the artificial sweeteners acesulfame and cyclamate. The latter is, though, partially metabolized to cyclohexylamine.

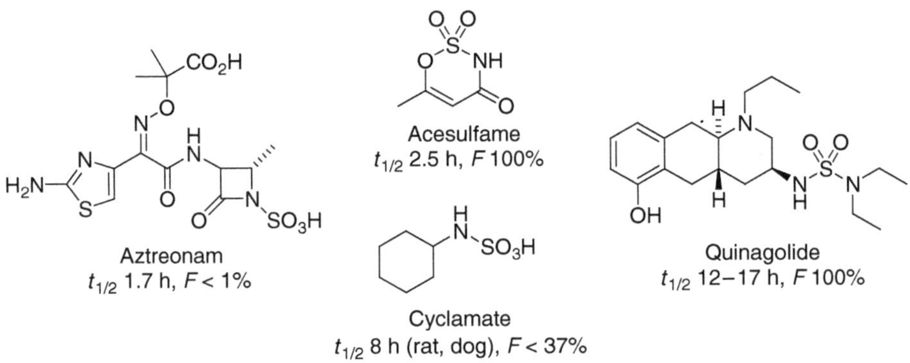

Scheme 42.2 Sulfamic acids and sulfamic esters.

43
Phosphonic Acids

Phosphonic acids (pK_a 2–3 and 8–9) are strong acids and will be completely ionized under physiological conditions. Their oral bioavailability is low because of high fraction of anionic phosphonate. Most of these strong acids can cause esophagitis, gastritis, or peptic ulcers. The phosphonic acid group can be used as mimetic of phosphate monoesters (such as nucleotides or pyrophosphates) in antiviral drugs or as carboxylate analog (Scheme 43.1). Phosphonate monoesters, such as ceronapril, are stable if phosphate-hydrolyzing enzymes (e.g., DNAses or RNAses) do not accept them as substrates.

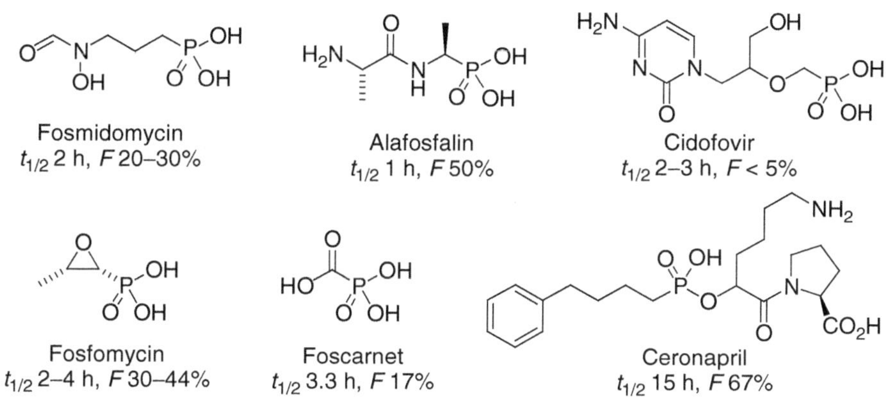

Scheme 43.1 Therapeutic phosphonic acids.

To improve oral bioavailability, prodrugs of phosphonic acids have been developed. Two examples of such prodrugs (antivirals) are shown in Scheme 43.2.

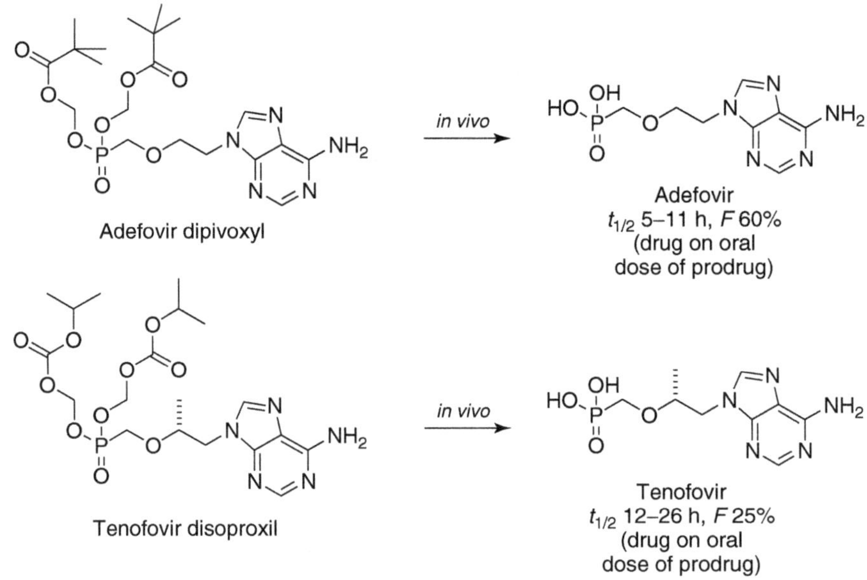

Scheme 43.2 Prodrugs of antiviral phosphonic acids.

A special group of phosphonic acids are the bisphosphonates, for instance, alendronic acid (Table 43.1). These compounds inhibit bone resorption and are used to treat Paget's disease (abnormal bone formation) and osteoporosis. Because of their strong acidity and low lipophilicity, most bisphosphonates are poorly absorbed from the intestine and are usually dosed parenterally. Despite their variable plasma half-lives, once incorporated into bones, bisphosphonates remain there for a long time.

Table 43.1 Phosphonic acids. V in l kg^{-1}; CL in ml min^{-1} kg^{-1}; Mwt in g mol^{-1}.

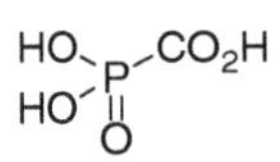	$t^1\!/_2$	3.3±1.5 h	V	0.43±0.13	FOSCARNET
	F	17±4%	CL	2.0±0.6	Antiviral
	pb	15±2%	Mwt	126.0	
	ur	82±13%	PSA	105 Å^2	
			log P	−1.90	
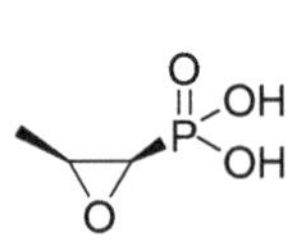	$t^1\!/_2$	12 h (iv)	V	0.32	FOSFOMYCIN
	F	–	CL	2	Antibacterial
	pb	0%	Mwt	138.1	
	ur	–	PSA	79.9 Å^2	
			log P	−1.23	

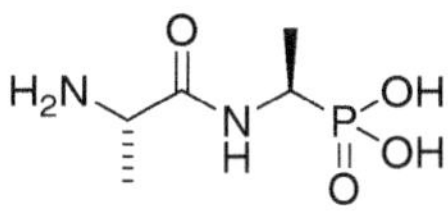	$t^{1/2}$ 1 h *F* 50% pb <10% ur 6–17%	*V* – CL – Mwt 196.1 PSA 122 Å^2 log *P* −2.89	ALAFOSFALIN Antibacterial Metabolism: amide hydrolysis
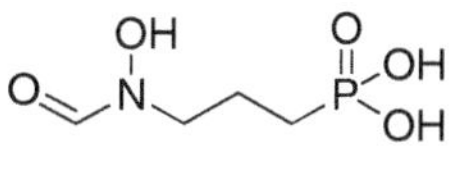	$t^{1/2}$ 1.5–2.0 h *F* 20–30% pb <3% ur 26%	*V* – CL – Mwt 183.1 PSA 98.1 Å^2 log *P* −1.58	FOSMIDOMYCIN Antimalarial
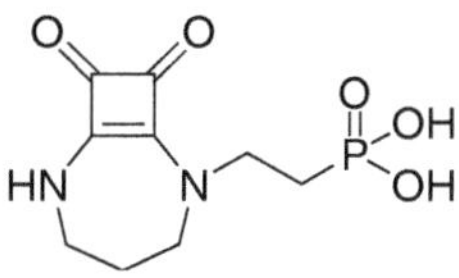	$t^{1/2}$ 5.7 h *F* 3–5% pb – ur –	*V* 0.43 CL 2.5 Mwt 260.2 PSA 117 Å^2 log *P* −1.84	PERZINFOTEL, EAA 090 NMDA antagonist, neuroprotectant, analgesic
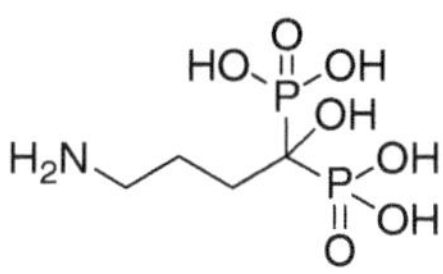	$t^{1/2}$ 10.5 years* *F* 0.8–1.8% pb 78% ur 40–60%	*V* 0.4 CL 2.8 Mwt 249.1 PSA 161 Å^2 log *P* −3.46	ALENDRONIC ACID *in bone Bone resorption inhibitor
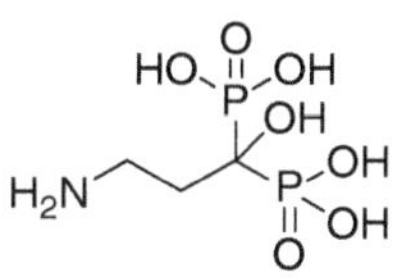	$t^{1/2}$ 28±7 h *F* 0.2% pb 54% ur 100%	*V* – CL 6.7 Mwt 235.1 PSA 181 Å^2 log *P* −0.89	PAMIDRONIC ACID Bone resorption inhibitor
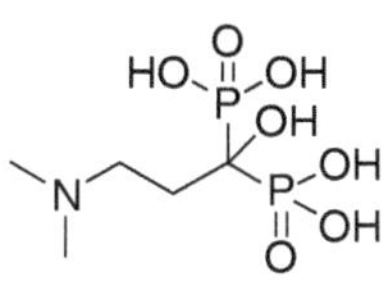	$t^{1/2}$ – *F* 2–4% pb – ur –	*V* – CL – Mwt 263.1 PSA 158 Å^2 log *P* −2.77	OLPADRONIC ACID Bone resorption inhibitor
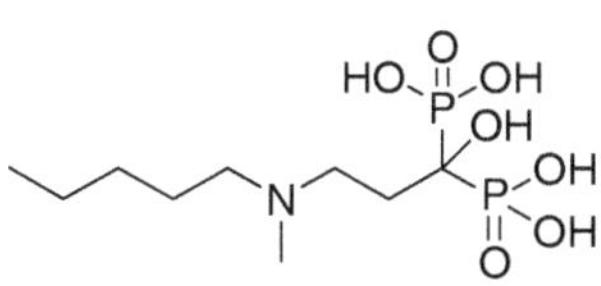	$t^{1/2}$ 10–60 h *F* 0.6% pb 90–99% ur 60%	*V* 1.3–2.0 CL 0.17–0.18 Mwt 319.2 PSA 158 Å^2 log *P* 2.33	IBANDRONIC ACID Bone resorption inhibitor
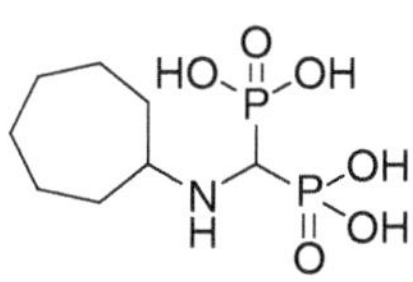	$t^{1/2}$ 1.6–2.0 h* *F* Low pb – ur 55–70%	*V* – CL – Mwt 287.2 PSA 147 Å^2 log *P* −0.97	INCADRONIC ACID *iv Bone resorption inhibitor

(*continued overleaf*)

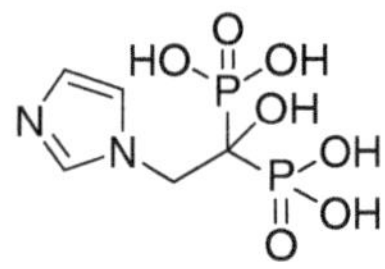	$t½$	146–167 h	V	0.09–0.15	ZOLEDRONIC ACID
	F	<1%	CL	1.0–1.3	Bone resorption inhibitor
	pb	43–56%	Mwt	272.1	
	ur	39%	PSA	168 $Å^2$	
			log P	−2.28	
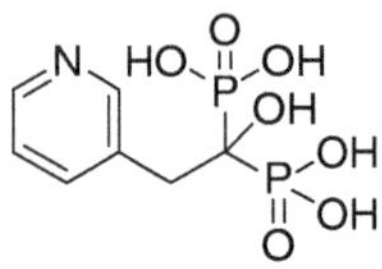	$t½$	66–480 h	V	6.3	RISEDRONIC ACID
	F	<1%	CL	3.8	Bone resorption inhibitor
	pb	24%	Mwt	283.1	
	ur	87%	PSA	168 $Å^2$	
			log P	0.55	
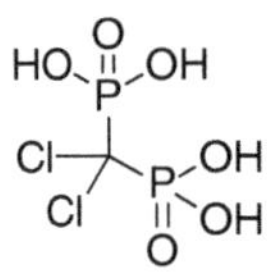	$t½$	13–15 h	V	0.3–0.8	CLODRONIC ACID
	F	1–2%	CL	1.4–1.7	Bone resorption inhibitor,
	pb	2–36%	Mwt	244.9	calcium regulator
	ur	70–90%	PSA	135 $Å^2$	
			log P	1.11	
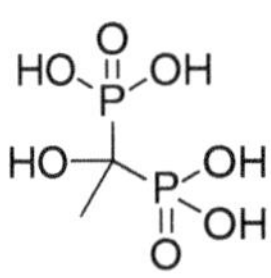	$t½$	1–6 h	V	0.3–1.3	ETIDRONIC ACID
	F	1.5–3.5%	CL	1.5	Calcium regulator for
	pb	–	Mwt	206.0	treatment of hypercalcemia
	ur	100%	PSA	155 $Å^2$	
			log P	−0.75	
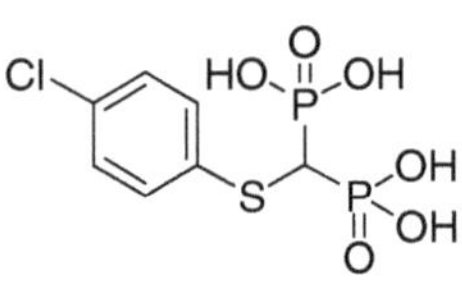	$t½$	50±10 h	V	–	TILUDRONIC ACID
	F	6%	CL	–	Bone resorption inhibitor
	pb	90%	Mwt	318.6	
	ur	60% (iv)	PSA	160 $Å^2$	
			log P	2.00	

$t_{1/2}$, plasma half-life; F, oral bioavailability; pb, plasma protein binding; ur, excretion of unchanged drug in urine; V, volume of distribution; CL, clearance; Mwt, molecular weight; PSA, polar surface area; NMDA, N-methyl D-aspartic acid.

44
Phosphoric Acid Derivatives

Phosphoric acid monoesters or -amides ($XPO(OH)_2$; X = RO, RS, R_2N) usually have short half-lives in plasma because of rapid hydrolysis by ubiquitous phosphatases. Phosphates may, therefore, be used as prodrugs of alcohols, thiols, or NH groups (Scheme 44.1). Phosphates can enhance the solubility of drugs in water (to improve absorption or enable liquid formulations for parenteral administration), and, by conversion into salts, raise their melting points. The latter aspect is particularly useful for drugs that are liquid or low-melting alcohols or phenols.

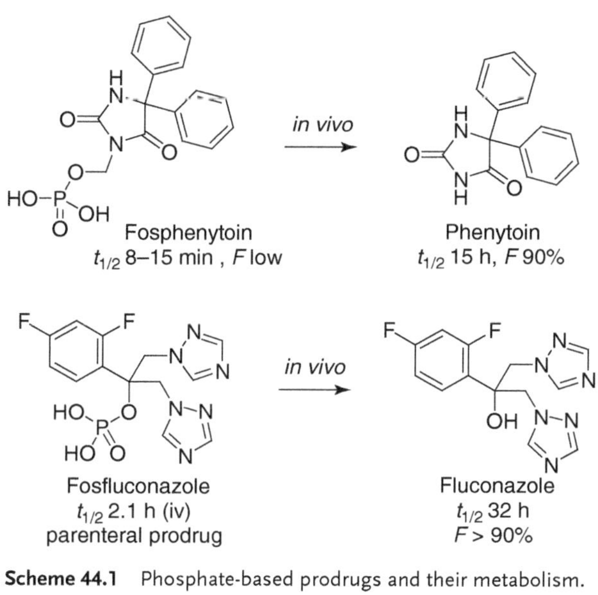

Scheme 44.1 Phosphate-based prodrugs and their metabolism.

Lead Optimization for Medicinal Chemists: Pharmacokinetic Properties of Functional Groups and Organic Compounds, First Edition. Florencio Zaragoza Dörwald.

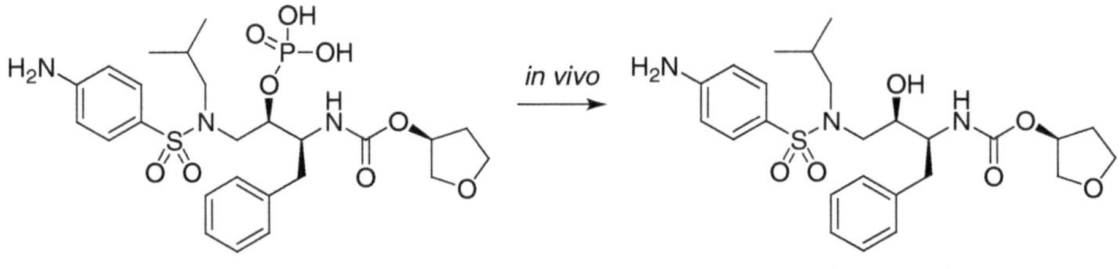

Fosamprenavir
oral prodrug

Amprenavir
$t_{1/2}$ 7–11 h, F 35–90% (dog)

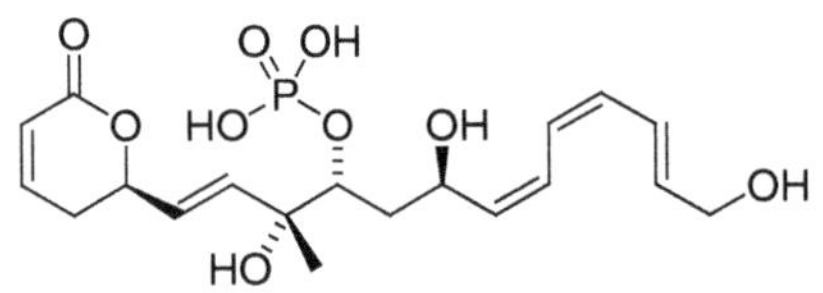

Fostriecin
$t_{1/2}$ 1.5 h (iv)

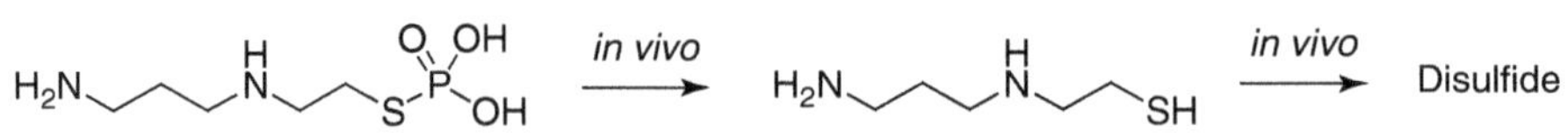

Amifostine
$t_{1/2}$ < 10 min

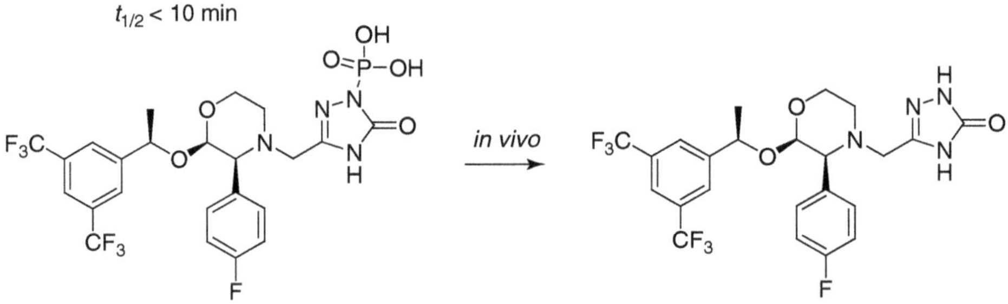

Fosaprepitant
parenteral prodrug

Aprepitant
$t_{1/2}$ 9–13 h, F 60–65%

Scheme 44.1 (*Continued*).

45
N-(Aminoalkyl)benzamides, -Benzoates, and Related Compounds

It was known for long that oral cocaine numbs the mouth. In 1883, Carl Koller, a student of ophthalmology at the University of Vienna, discovered that cocaine was well suited as a topical anesthetic for the eye. If a solution of cocaine is trickled into the eye of a frog or human, the cornea can be touched without causing the normal eye-closing reflex [1].

This finding led to intensive search for more readily accessible, less toxic, and less addicting alternatives to cocaine. Some of the earliest analogs used as local anesthetics are shown in Scheme 45.1. The main initial problems of these drugs were low potency, skin irritation, and low solubility in water. Most of these issues were finally solved with procaine, which became one of the most widely used local anesthetics, until the discovery of lidocaine in 1948 (Chapter 54).

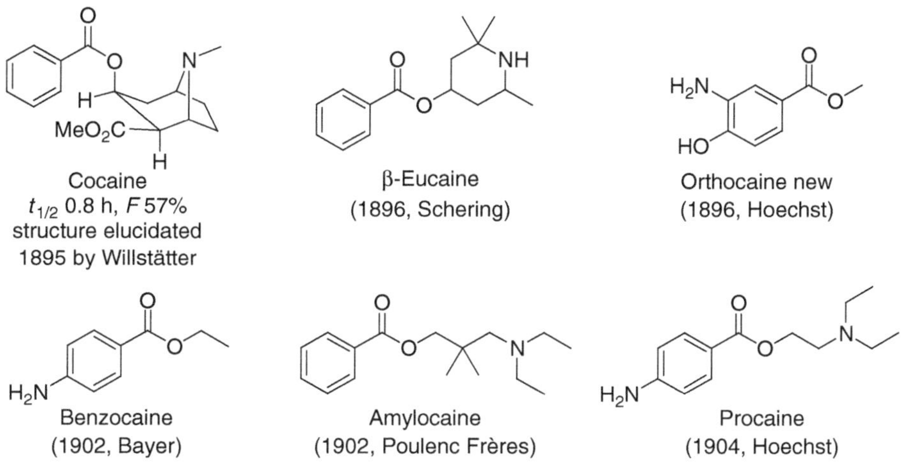

Scheme 45.1 Local anesthetics.

The metabolically more stable amide analog of procaine, procainamide, is not only a local anesthetic but also an antiarrhythmic, by virtue of its Na^+-channel

Lead Optimization for Medicinal Chemists: Pharmacokinetic Properties of Functional Groups and Organic Compounds, First Edition. Florencio Zaragoza Dörwald.

blocking properties. Interestingly, additional small structural variations can lead to a loss of anesthetic and antiarrhythmic activity and give rise to dopamine antagonism, as in metoclopramide, and to serotonin agonism. A large number of drugs for a variety of indications contain the *N*-(aminoalkyl)benzamide substructure (Table 45.1). These compounds are mainly monoamine oxidase (MAO) inhibitors (antidepressants), 5-hydroxytryptamine (5-HT) agonists (antiemetics and gastroprokinetics), and dopamine antagonists (antipsychotics and gastroprokinetics).

Aminoalkylbenzoates such as procaine have short plasma half-lives (<5 min) owing to rapid ester hydrolysis. *N*-(Aminoalkyl)benzamides are more stable and are mainly metabolized by N-dealkylation, N-oxidation, and aromatic hydroxylation. Aminobenzoic acid derivatives are N-acetylated reversibly.

Reference

1. Sneader, W. (1994) Local anesthetics. *Drug News Perspect.*, 7, 60–64.

Table 45.1 Aminoalkylbenzamides, aminoalkylbenzoates, and related compounds. *V* in l kg^{-1}; CL in ml min^{-1} kg^{-1}; Mwt in g mol^{-1}.

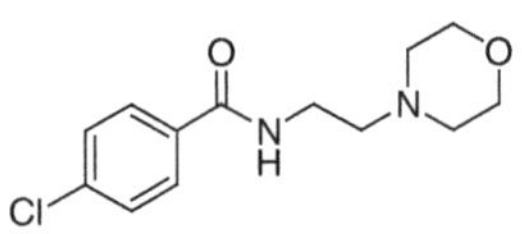	$t^1/_2$	3±1 h	*V*	1–2	MOCLOBEMIDE
	F	60±20%	CL	5–13	MAO inhibitor, antidepressant
	pb	50%	Mwt	268.7	Metabolism: morpholine
	ur	<1%	PSA	41.6 $Å^2$	N-oxidation, aromatic
			log *P*	1.34	hydroxylation, and morpholine
					N-dealkylation
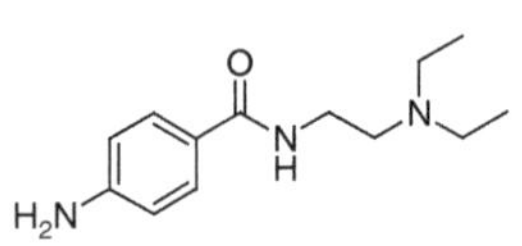	$t^1/_2$	3.0±0.6 h	*V*	1.9±0.3	PROCAINAMIDE
	F	83±16%	CL	2.7	Na channel blocker, antiarrhythmic
	pb	16±5%	Mwt	235.3	Metabolism: N-acetylation (to
	ur	67±8%	PSA	58.4 $Å^2$	acecainide), N-deethylation, and
			log *P*	1.32	N-oxidation, only 1% amide
					hydrolysis
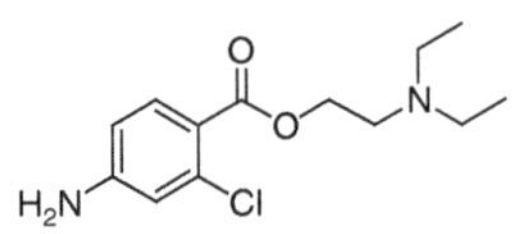	$t^1/_2$	20–25 s	*V*	–	CHLOROPROCAINE
	F	–	CL	–	Local anesthetic
	pb	–	Mwt	270.8	Metabolism: ester hydrolysis
	ur	–	PSA	55.6 $Å^2$	
			log *P*	2.83	
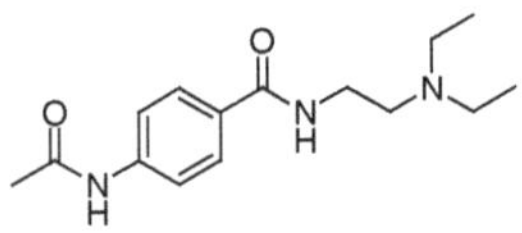	$t^1/_2$	6.0±0.2 h	*V*	1.4±0.2	ACECAINIDE,
	F	83±12%	CL	3.1±0.4	*N*-ACETYLPROCAINAMIDE
	pb	10±9%	Mwt	277.4	Antiarrhythmic
	ur	81±1%	PSA	61.4 $Å^2$	Metabolism: N-deethylation
			log *P*	1.51	and N-deacetylation

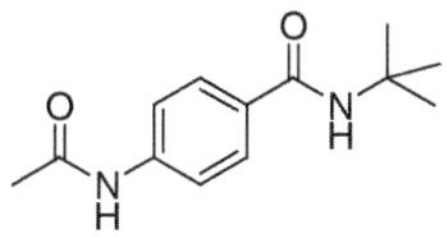	$t^1/_2$ F pb ur	40 h 55% (rat) – –	V CL Mwt PSA log P	– – 234.3 58.2 Å^2 1.54	CPI-1189 Treatment of dementia and irritable bowel disease, crosses bbb
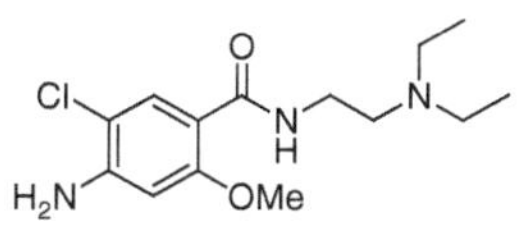	$t^1/_2$ F pb ur	5.0±1.4 h 76±38% 40±4% 20±9%	V CL Mwt PSA log P	3.4±1.3 6.2±1.3 299.8 67.6 Å^2 2.16	METOCLOPRAMIDE Dopamine antagonist, antiemetic, gastroprokinetic; Metabolism: aniline sulfation and glucuronidation, oxidation of CH_2NEt_2 to CO_2H
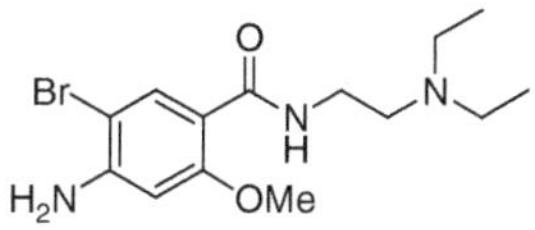	$t^1/_2$ F pb ur	3–5 h 54–75% 40% 10–14%	V CL Mwt PSA log P	3.1 13 344.3 67.6 Å^2 2.54	BROMOPRIDE Dopamine antagonist, antiemetic
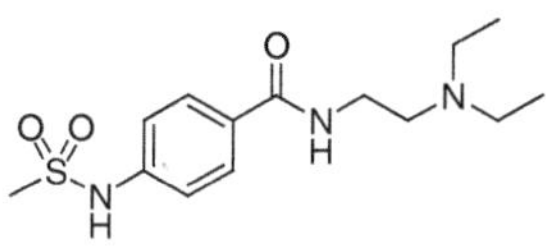	$t^1/_2$ F pb ur	3.6±0.8 h 47±15% 4% 36±12%	V CL Mwt PSA log P	0.8 3.7 313.4 86.9 Å^2 1.15	SEMATILIDE Antiarrhythmic
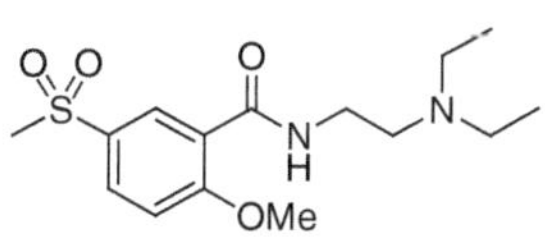	$t^1/_2$ F pb ur	2–5 h 75% Negligible 54–76%	V CL Mwt PSA log P	1.4 4.3 328.4 84.1 Å^2 1.22	TIAPRIDE Antidyskinetic Metabolism: N-deethylation
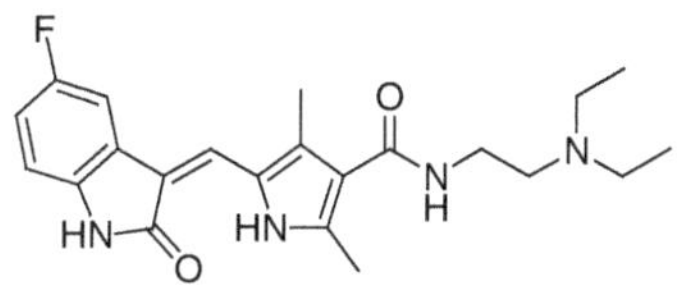	$t^1/_2$ F pb ur	40–60 h 70% 95% <16%	V CL Mwt PSA log P	34 9–16 398.5 77.2 Å^2 3.16	SUNITINIB Kinase inhibitor, antineoplastic; active metabolite: *N*-desethyl ($t^1/_2$ 80–110 h, pb 90%)
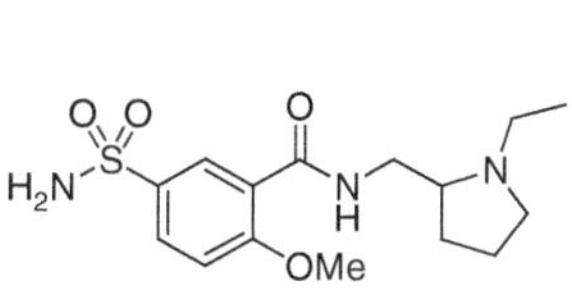	$t^1/_2$ F pb ur	6–8 h 25–36% 11–18% 15%	V CL Mwt PSA log P	0.94 1.9 341.4 110 Å^2 0.78	SULPIRIDE Dopamine D_2 antagonist, antipsychotic, antidepressant, antiemetic; Metabolism: oxidation of pyrrolidine to 2-pyrrolidone, no amide hydrolysis

(*continued overleaf*)

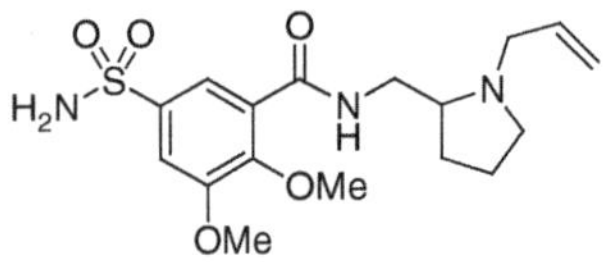	$t^1/_2$ 4 h *F* – pb – ur –		*V* – CL – Mwt 383.5 PSA 119 Å^2 log *P* 0.73		VERALIPRIDE Suppression of menopausal hot flushes; withdrawn
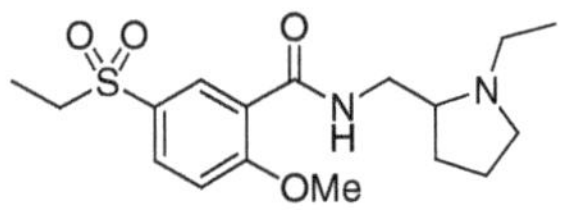	$t^1/_2$ 4–11 h *F* – pb <25% ur 90%		*V* – CL – Mwt 354.5 PSA 84.1 Å^2 log *P* 1.39		SULTOPRIDE Dopamine D_2 antagonist, antipsychotic; Metabolism (rat): O-demethylation, S-deethylation, oxidation of pyrrolidine, amide hydrolysis
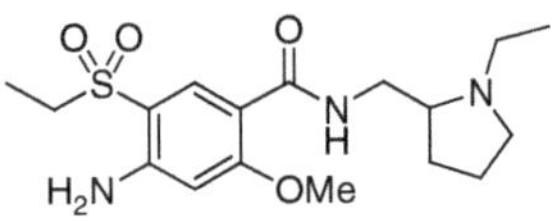	$t^1/_2$ 12 h *F* 50±2% pb 16% ur 26–50%		*V* 5.8 CL 20 Mwt 369.5 PSA 110 Å^2 log *P* 1.50		AMISULPRIDE Dopamine D_2/D_3 antagonist, antipsychotic; Metabolism: oxidation of pyrrolidine to 2-pyrrolidone, N-deethylation
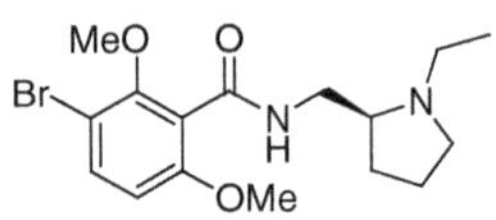	$t^1/_2$ 4–7 h *F* >90% pb 84% ur 10–40%		*V* 0.7 CL 1.5–1.7 Mwt 371.3 PSA 50.8 Å^2 log *P* 2.19		REMOXIPRIDE Dopamine D_2 antagonist, antipsychotic; Metabolism: oxidation to lactam, N-deethylation, hydroxylation of pyrrolidone
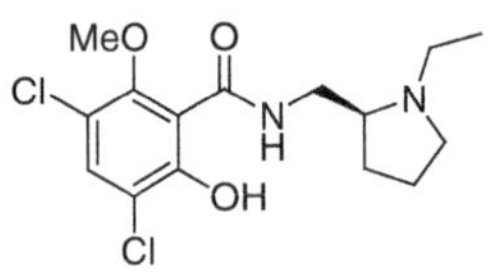	$t^1/_2$ 6–13 h *F* 65–67% pb – ur <1%		*V* 1.5 CL/*F* 2.2±0.5 Mwt 347.2 PSA 61.8 Å^2 log *P* 3.58		RACLOPRIDE Dopamine D_2 antagonist, neuroleptic
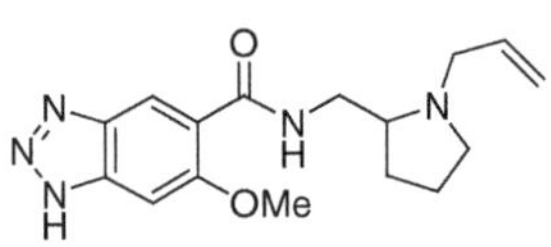	$t^1/_2$ 2–3 h *F* 75–100% pb 75% ur High		*V* 1.6 CL 6.6 Mwt 315.4 PSA 83.1 Å^2 log *P* 0.84		ALIZAPRIDE Antipsychotic, antiemetic
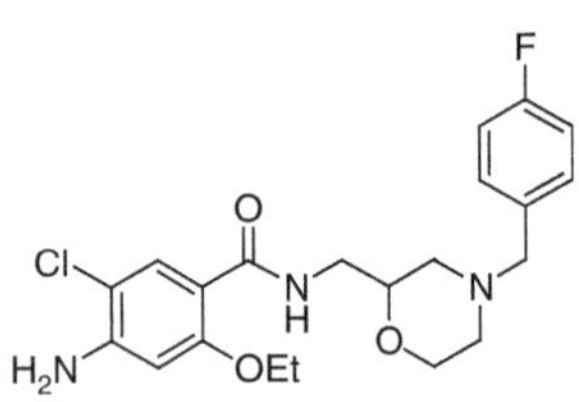	$t^1/_2$ 1.4–2.0 h *F* 14%* pb 99% ur <1%		*V* 1.7–3.5 CL 14–21 Mwt 421.9 PSA 76.8 Å^2 log *P* 2.82		MOSAPRIDE *monkey 5-HT_4 agonist, gastroprokinetic Metabolite: des-4-fluorobenzyl

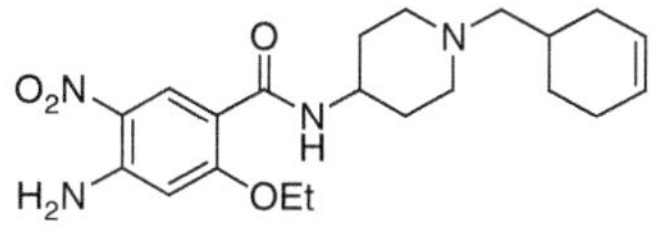

$t_{1/2}$	5–8 h	V	–	CINITAPRIDE
F	–	CL	–	$5\text{-}HT_1$ and $5\text{-}HT_4$ agonist,
pb	–	Mwt	402.5	gastroprokinetic
ur	–	PSA	113 Å^2	Metabolism: N-dealkylation
		log P	3.76	

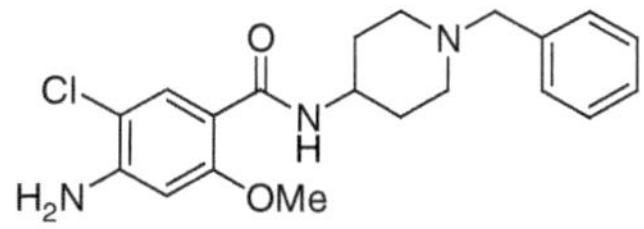

$t_{1/2}$	24 h	V	1.6–3.2*	CLEBOPRIDE
F	–	CL	–	*rat, dog
pb	–	Mwt	373.9	Antiemetic, antispasmodic
ur	–	PSA	67.6 Å^2	Metabolites: glucuronide,
		log P	2.70	desbenzyl; no amide hydrolysis

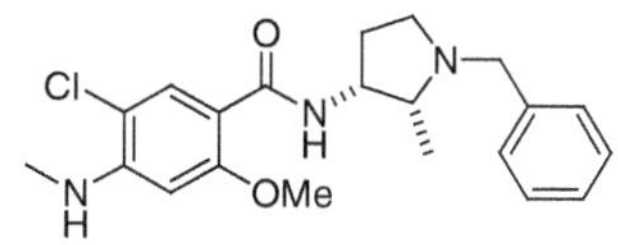

$t_{1/2}$	6–9 h	V	–	NEMONAPRIDE
F	–	CL	–	D_2 antagonist, $5\text{-}HT_{1A}$ agonist
pb	–	Mwt	387.9	Metabolism: N-demethylation
ur	–	PSA	53.6 Å^2	
		log P	3.78	

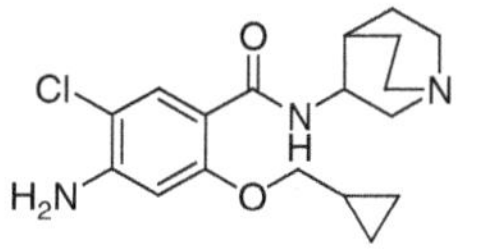

$t_{1/2}$	16±3 h	V	–	PANCOPRIDE
F	82–92%	CL	–	$5\text{-}HT_3$ antagonist, antiemetic
pb	–	Mwt	349.9	
ur	10–17%	PSA	67.6 Å^2	
		log P	2.85	

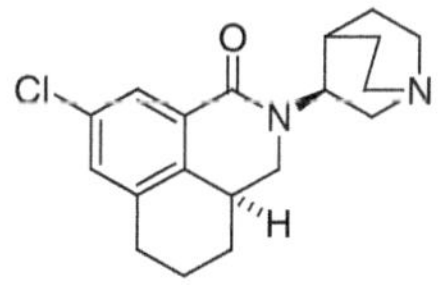

$t_{1/2}$	37±12 h	V	6.9–7.9	PALONOSETRON
F	97%	CL	2.7	$5\text{-}HT_3$ antagonist, antiemetic
pb	62%	Mwt	296.4	Metabolism: N-oxidation (tertiary
ur	40%	PSA	23.6 Å^2	amine), benzylic
		log P	2.61	(6) hydroxylation

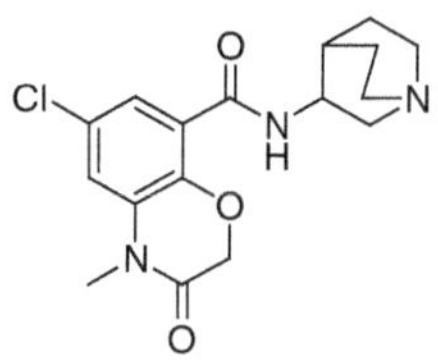

$t_{1/2}$	6–8 h	V	–	AZASETRON
F	90%	CL	–	$5\text{-}HT_3$ antagonist, antiemetic
pb	–	Mwt	349.8	
ur	60–70%	PSA	61.9 Å^2	
		log P	1.83	

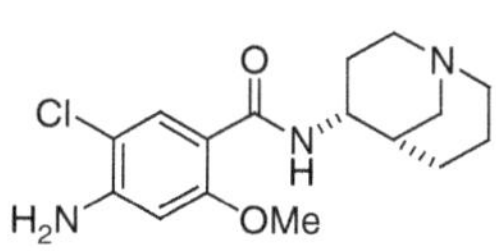

$t_{1/2}$	10 h	V	–	RENZAPRIDE
F	–	CL	–	$5\text{-}HT_4$ agonist, $5\text{-}HT_3$ antagonist,
pb	–	Mwt	323.8	gastroprokinetic
ur	–	PSA	67.6 Å^2	
		log P	1.83	

(*continued overleaf*)

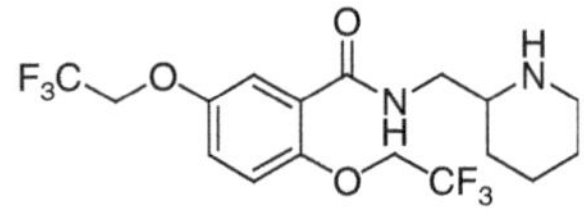	$t_{1/2}$ *F* pb ur	11±3 h 70±11% 61±10% 43±3%	*V* CL Mwt PSA log *P*	4.9±0.4 5.6±1.3 414.3 59.6 Å^2 3.19	FLECAINIDE Antiarrhythmic Metabolism: *m*-O-dealkylation, oxidation to lactam
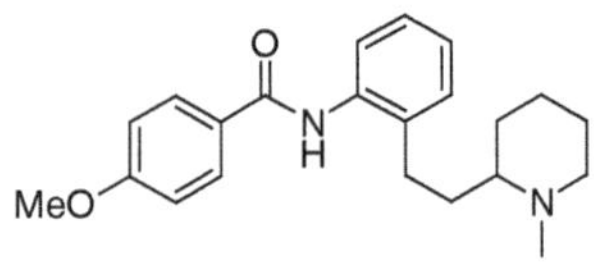	$t_{1/2}$ *F* pb ur	2–12 h 30–55% 70–80% <10%	*V* CL Mwt PSA log *P*	2–4 11–13 352.5 41.6 Å^2 3.75	ENCAINIDE Antiarrhythmic Metabolism: O-demethylation; withdrawn in 1991 because of excessive patient mortality
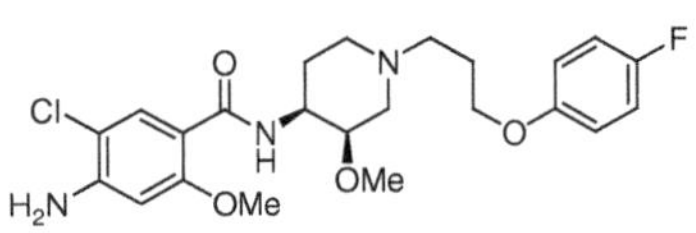	$t_{1/2}$ *F* pb ur	7–10 h 40–50% 98% 1–3%	*V* CL Mwt PSA log *P*	2.4 – 466.0 86.1 Å^2 2.83	CISAPRIDE Peristaltic stimulant; Metabolism: N-dealkylation of piperidine, hydroxylation of phenoxy, cleavage of fluorophenyl ether, no amide hydrolysis; withdrawn in 2000 for causing cardiac arrhythmias
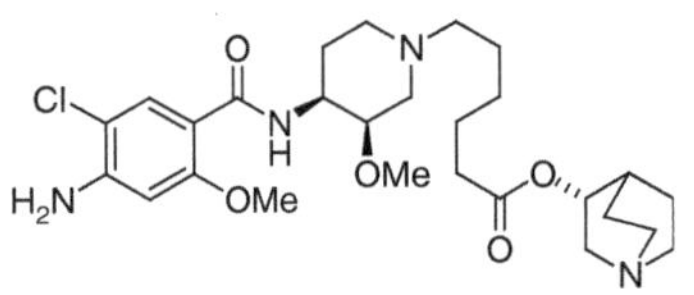	$t_{1/2}$ *F* pb ur	5.4 h – 30–40% –	*V* CL Mwt PSA log *P*	– – 537.1 106 Å^2 2.02	NARONAPRIDE 5-HT_4 agonist, gastroprokinetic Metabolism: ester hydrolysis, aniline glucuronidation, and oxidative degradation of alkanoic acid side chain
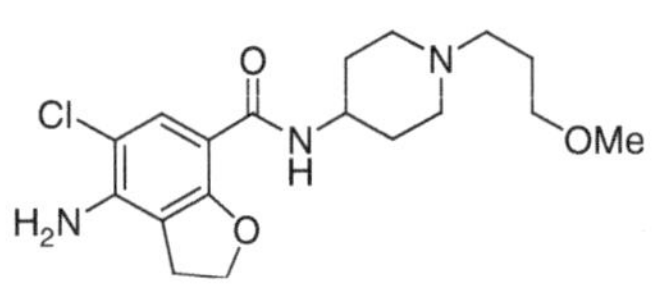	$t_{1/2}$ *F* pb ur	22–34 h >90% 30% 60%	*V* CL Mwt PSA log *P*	8.7 4.9 367.9 76.8 Å^2 1.44	PRUCALOPRIDE 5-HT_4 agonist, gastroprokinetic Metabolism: O-demethylation, then oxidation of alcohol to carboxylic acid
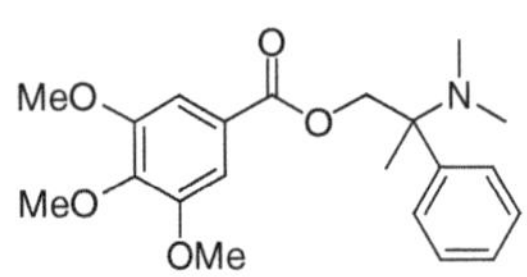	$t_{1/2}$ *F* pb ur	9±3 h 4–6% <5% <2.4%	*V* CL Mwt PSA log *P*	1.4 – 387.5 57.2 Å^2 4.84	TRIMEBUTINE Antispasmodic Metabolism: N-demethylation, ester hydrolysis
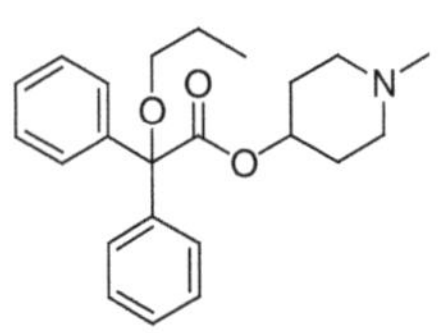	$t_{1/2}$ *F* pb ur	20 h 36–42% 90% 6–8%	*V* CL Mwt PSA log *P*	– – 367.5 38.8 Å^2 4.37	PROPIVERINE Anticholinergic, treatment of urinary incontinence Metabolites: *N*-oxide, *N*-desmethyl ($t_{1/2}$ 20 h), *O*-despropyl

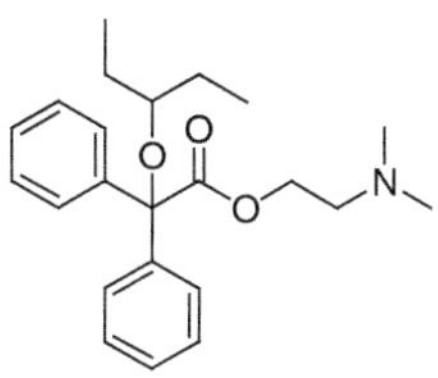	$t_{1/2}$	34 h (iv)	V	7.1	DENAVERINE
	F	37%	CL	5.7	Antispasmodic
	pb	–	Mwt	383.5	Metabolism: N-demethylation
	ur	33% (rat)	PSA	38.8 Å^2	(first-pass), ester hydrolysis,
			log P	4.88	and O-dealkylation
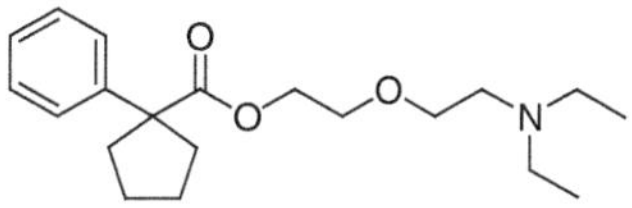	$t_{1/2}$	2.3 h	V	–	CARBETAPENTANE,
	F	<50%	CL	–	PENTOXYVERINE
	pb	–	Mwt	333.5	Antitussive
	ur	<0.5%	PSA	38.8 Å^2	Metabolism: ester hydrolysis
			log P	4.11	
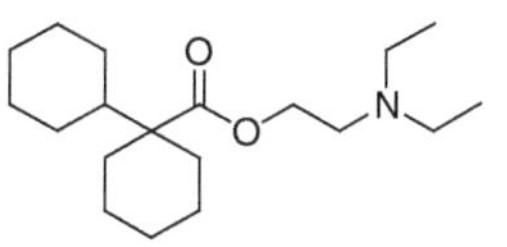	$t_{1/2}$	5–10 h	V	3.7	DICYCLOMINE, DICYCLOVERINE
	F	67%	CL	–	Anticholinergic
	pb	>99%	Mwt	309.5	
	ur	–	PSA	29.5 Å^2	
			log P	4.64	
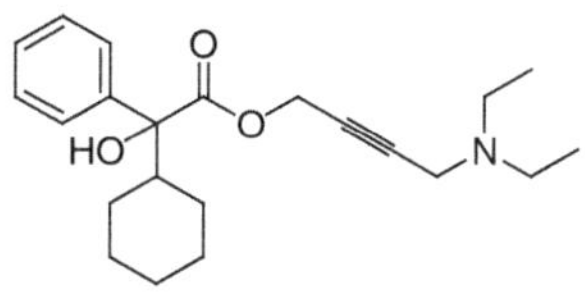	$t_{1/2}$	2 h	V	1–3	OXYBUTYNIN
	F	6%	CL	5–6	Anticholinergic
	pb	>99%	Mwt	357.5	Metabolism: N-deethylation,
	ur	<0.1%	PSA	49.8 Å^2	hydrolysis of ester, and
			log P	5.05	hydroxylation of cyclohexane
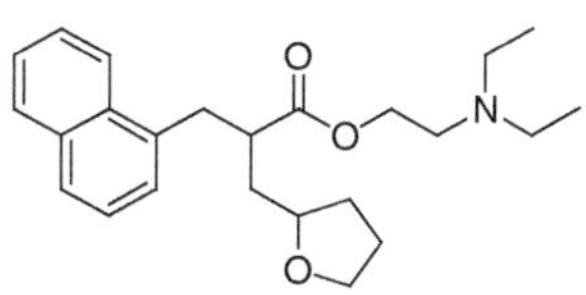	$t_{1/2}$	2–5 h	V	0.9	NAFRONYL, NAFTIDROFURYL
	F	–	CL	–	Vasodilator, crosses bbb
	pb	60–80%	Mwt	383.5	Metabolism: ester hydrolysis,
	ur	<1%	PSA	38.8 Å^2	oxidation to 2-furanone
			log P	3.91	
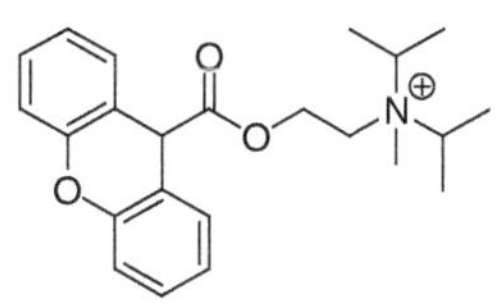	$t_{1/2}$	1.6 h	V	–	PROPANTHELINE
	F	5–15%	CL	–	Anticholinergic
	pb	–	Mwt	368.5	Metabolism: ester hydrolysis,
	ur	3–5%	PSA	35.5 Å^2	aromatic hydroxylation
			log P	0.79	
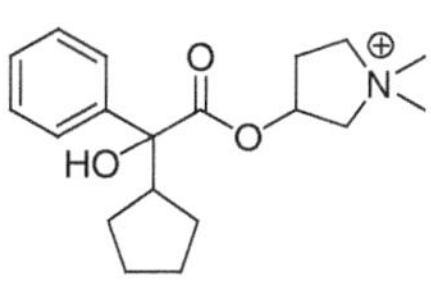	$t_{1/2}$	0.6–1.2 h	V	0.4–1.8	GLYCOPYRROLATE
	F	1–13%	CL	9	Anticholinergic
	pb	–	Mwt	318.4	
	ur	65–80%	PSA	46.5	
			log P	−0.85	

(*continued overleaf*)

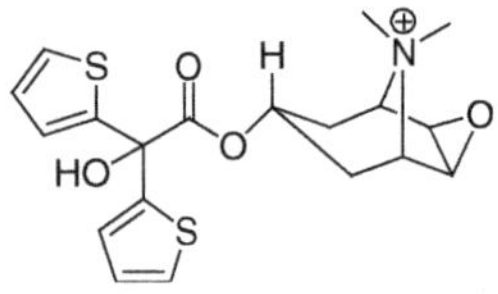	$t_{1/2}$ F pb ur	5 d 2–3%* 71–79% 74% (iv)**	V CL Mwt PSA log P	32 10.9 392.5 59.1 Å^2 −1.94	TIOTROPIUM *20% on inhalation **14% on inhalation Muscarinic antagonist, bronchodilator Metabolism: ester hydrolysis
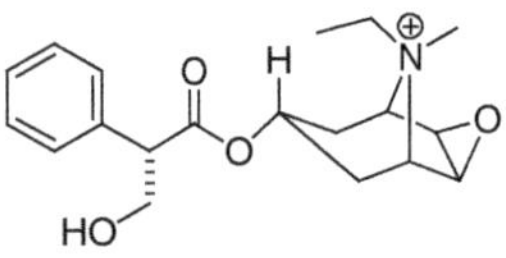	$t_{1/2}$ F pb ur	2.5 h 0.5%* 7% 0.3%	V CL Mwt PSA log P	3.6 31 332.4 59.1 Å^2 −2.56	OXITROPIUM *12% on inhalation Bronchodilator Metabolism: ester hydrolysis
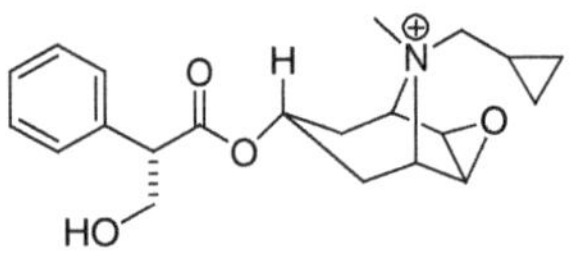	$t_{1/2}$ F pb ur	2 h (iv) 1–4% – 50% (iv)	V CL Mwt PSA log P	– – 358.5 59.1 Å^2 −2.06	CIMETROPIUM Antispasmodic Metabolism: aromatic hydroxylation, ester hydrolysis
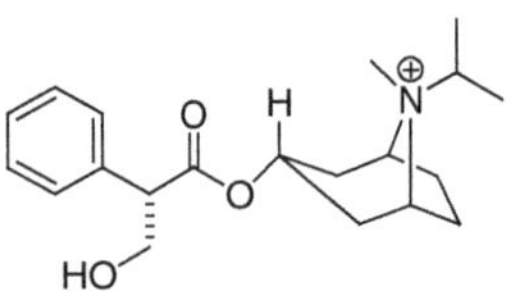	$t_{1/2}$ F pb ur	1.5–4.0 h* 1–6% <20% 24%**	V CL Mwt PSA log P	4.6 31 332.5 46.5 Å^2 −1.47	IPRATROPIUM *iv; **50% (iv) Muscarinic antagonist, bronchodilator, antiarrhythmic Metabolism: ester hydrolysis, elimination of water to α-phenylacrylic acid, dehydroxymethylation to phenylacetic ester
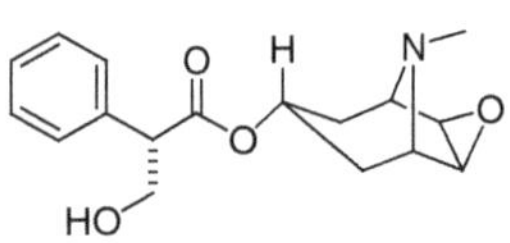	$t_{1/2}$ F pb ur	3±1 h 27±12% – 6±4%	V CL Mwt PSA log P	1.4±0.7 16±13 303.4 62.3 Å^2 1.05	SCOPOLAMINE Anticholinergic, treatment of motion sickness; Metabolism: aromatic 3- and 4-hydroxylation, then phenolic O-methylation, ester hydrolysis, N-demethylation, and dehydration
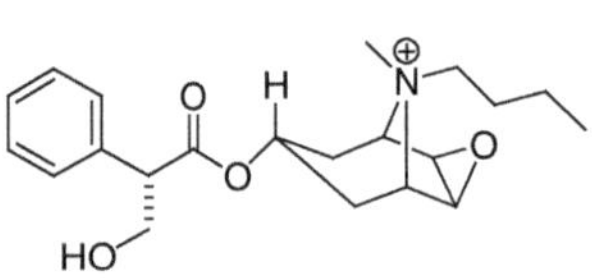	$t_{1/2}$ F pb ur	5 h <1% Low 50%	V CL Mwt PSA log P	2 19 360.5 59.1 Å^2 −1.49	BUTYLSCOPOLAMMONIUM, BUTYLSCOPOLAMINE Antispasmodic Metabolism: ester hydrolysis

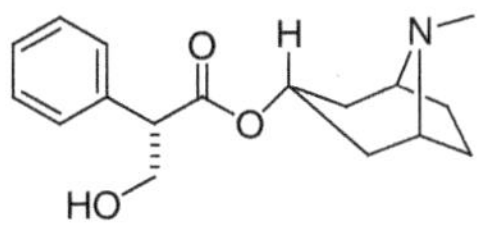	$t^1/_2$ 2–5 h F 50% pb 14–40% ur 50%	V 2–4 CL 7.6 Mwt 289.4 PSA 49.8 Å^2 log P 1.38	ATROPINE (racemate) Anticholinergic; Metabolism: N-demethylation, N-oxidation, and ester hydrolysis; the shown enantiomer (hyoscyamine) has similar PK as atropine
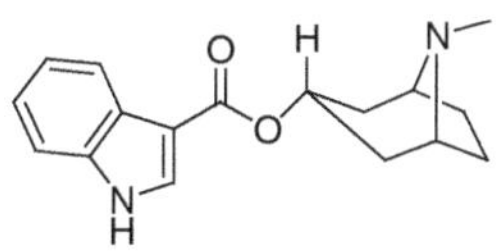	$t^1/_2$ 6–8 h F 27–99% pb 59–71% ur 8–10%	V 6.6–9.7 CL 3–26 Mwt 284.4 PSA 45.3 Å^2 log P 3.03	TROPISETRON 5-HT$_3$ antagonist, antiemetic, CYP2D6 substrate; Metabolism: aromatic 5-, 6-, and 7-hydroxylation, N-demethylation
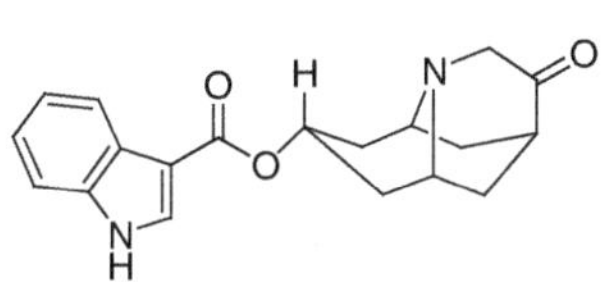	$t^1/_2$ 4–9 h* F 70–89% pb 69–77% ur 19–38%	V 1.6–7.3 CL 6–14 Mwt 324.4 PSA 62.4 Å^2 log P 2.82	DOLASETRON 5-HT$_3$ antagonist, antiemetic; *prodrug ($t^1/_2$ < 10 min): fast *in vivo* reduction of ketone to alcohol; all values for alcohol on oral dosing of ketone; further metabolism: aromatic hydroxylation, N-oxidation of tertiary amine
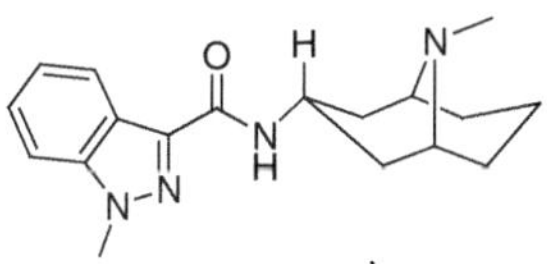	$t^1/_2$ 5.3±3.5 h F 60% pb 65±9% ur 16±14%	V 3.0±1.5 CL 11±9 Mwt 312.4 PSA 45.2 Å^2 log P 1.47	GRANISETRON 5-HT$_3$ antagonist, antiemetic Metabolism: N-demethylation of indazole, aromatic 6- and 7-hydroxylation, then glucuronidation
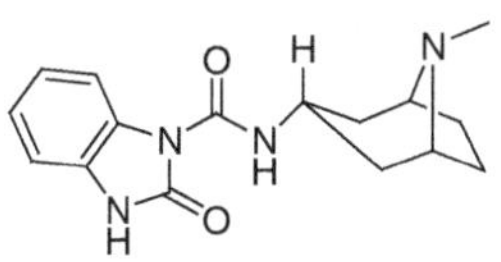	$t^1/_2$ 11–14 h F 64–72% pb – ur 59–73%	V – CL – Mwt 300.4 PSA 64.7 Å^2 log P 1.31	ITASETRON 5-HT$_3$ antagonist, antiemetic
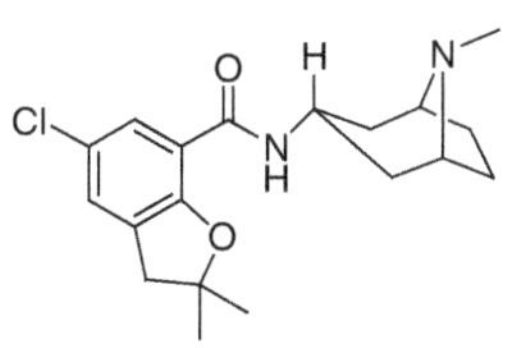	$t^1/_2$ 25–37 h F – pb 75% ur 36%	V – CL – Mwt 348.9 PSA 41.6 Å^2 log P 3.19	ZATOSETRON 5-HT$_3$ antagonist, antiemetic Metabolism: N-oxidation, N-demethylation, benzylic hydroxylation, then oxidation of alcohol to ketone

(*continued overleaf*)

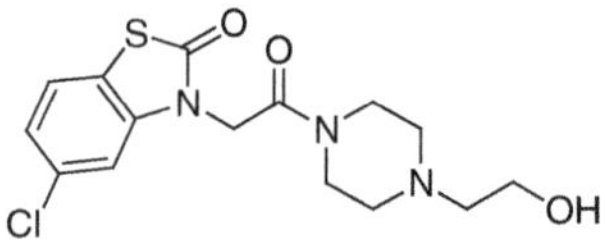	$t_{1/2}$ 1.3 h *F* – pb 24–34% ur <1%		*V* – CL – Mwt 355.8 PSA 89.4 Å^2 log *P* 1.18		TIARAMIDE Antiasthmatic, antiinflammatory Metabolism: O-glucuronidation, N-de(hydroxyethylation), oxidation to *N*-oxide, and *N*-acetic acid
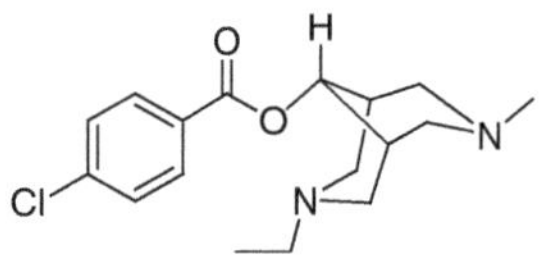	$t_{1/2}$ 9±4 h *F* 56±20% pb – ur –		*V* 9 CL 17 Mwt 322.8 PSA 32.8 Å^2 log *P* 2.05		BISARAMIL Antiarrhythmic Metabolism: ester hydrolysis
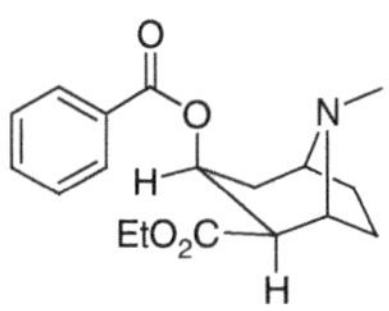	$t_{1/2}$ 1.5 h (iv) *F* – pb 36% ur –		*V* 1.6 CL 13 Mwt 317.4 PSA 55.8 Å^2 log *P* 2.79		COCAETHYLENE Recreational drug (highly addictive)
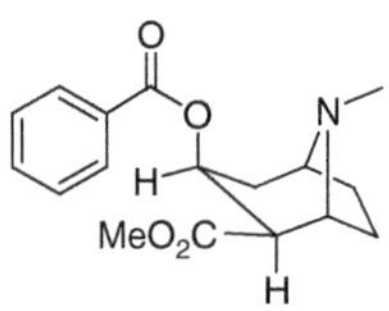	$t_{1/2}$ 0.8±0.2 h *F* 57±19% pb 91% ur <2%		*V* 2.0±0.2 CL 32±6 Mwt 303.4 PSA 55.8 Å^2 log *P* 2.28		COCAINE Local anesthetic, recreational drug (highly addictive)

$t_{1/2}$, plasma half-life; *F*, oral bioavailability; pb, plasma protein binding; ur, excretion of unchanged drug in urine; V, volume of distribution; CL, clearance; Mwt, molecular weight; PSA, polar surface area; bbb, blood–brain barrier; PK, pharmacokinetics.

46
Arylalkylamines

For each amine-containing natural neurotransmitter, several different receptors are known: four histamine receptors, five dopamine receptors, seven serotonin receptors, and five adrenaline receptors (two α and three β receptors). Each receptor is expressed to a different extent in different tissues and elicits a different biological response. Because of the close structural relationship between the hormones, many amine-based unnatural ligands at these receptors have similar structures, and many are nonselective. For most of these receptors, both agonists and antagonists have been developed.

In the following section, most known arylalkylamines are listed, to illustrate pharmacokinetics (PK) changes on minimal and continuous structural modification. Therefore, no arrangement according to the target protein or indication was attempted.

Amines with one, two, or more aryl/heteroaryl groups are useful drugs for a wide variety of indications. The most important drugs of this type include antiallergics, antipruritics, nootropics, antipsychotics, antidepressants, anxiolytics, sedatives, hypnotics, antiepileptics, antiemetics, analgesics, vasodilators, and antiarrhythmics.

46.1
Antihistaminics: History

In the 1930s, several academic laboratories and pharmaceutical companies began seeking compounds that could antagonize the most obvious biological effects of histamine, namely, bronchoconstriction and vasodilation, to gain a better understanding of the physiological role of histamine. Among the first compounds to be marketed as "antihistaminics" (histamine H_1 antagonists) were pyrilamine (1944), tripelenamine (1946), diphenhydramine (1946), and metapyrilene (1947). These drugs, followed by many analogs, became hugely popular for the suppression of the symptoms of common cold and hay fever and are still used for this purpose today. Because of the sedating effect of these "first generation" antihistaminics, caused by their ability to cross the blood–brain barrier (bbb) and block the H_1 receptors in the central nervous system (CNS), these drugs were also used as

Lead Optimization for Medicinal Chemists: Pharmacokinetic Properties of Functional Groups and Organic Compounds, First Edition. Florencio Zaragoza Dörwald.

hypnotics. One patient treated with diphenhydramine reported that she did no longer get motion sick on the streetcar, and this effect could be confirmed in a group of sailors in 1947. Since then, antihistaminics were also used as remedy for motion sickness and as antiemetics (e.g., cyclizine, 1949) [1].

To overcome the sedating effect of these bbb-crossing H_1 antagonists, more hydrophilic drugs were developed, the first ones being terfenadine (1982), astemizole (1983), and loratadine (1986). Their large polar surface areas (>40 Å^2) keep them from entering the CNS.

The most common metabolic transformations of arylalkylamines include N-dealkylation, N-oxidation, and aromatic hydroxylation.

Despite the large number of drugs containing an arylalkylamine substructure, most of them can be assigned to one of the generic structures given in Scheme 46.1. In these, Ar represents optionally substituted aryl or heteroaryl groups, R is alkyl, and X and Y are C, N, O, or CO. The compounds listed in Tables 46.1–46.4 appear in the same order as the generic structures in the sketch.

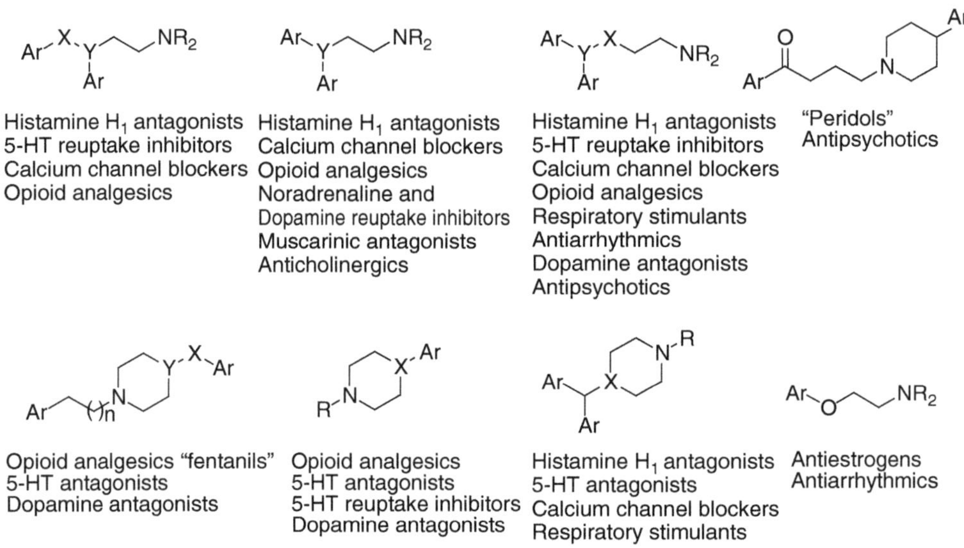

Scheme 46.1 Typical generic structures of arylalkylamine drugs.

Reference

1. Sneader, W. (2001) Histamine and the classic antihistamines. *Drug News Perspect.*, **14**, 618–624.

Table 46.1 (Arylalkyl)amines. *V* in l kg^{-1}; CL in ml min^{-1} kg^{-1}; Mwt in g mol^{-1}.

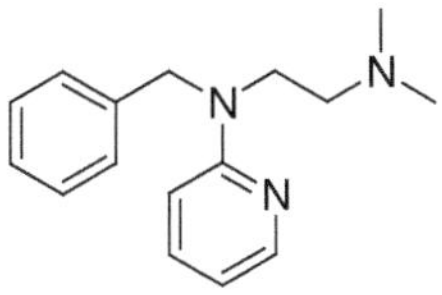	$t^1/_2$ 3.0–4.5 h *F* – pb – ur –	*V* 10 CL 32 Mwt 255.4 PSA 19.4 Å^2 log *P* 2.78	TRIPELENNAMINE Histamine H_1 antagonist Metabolism: aromatic *p*-hydroxylation, benzylic C-hydroxylation, N-depyridylation, N-debenzylation
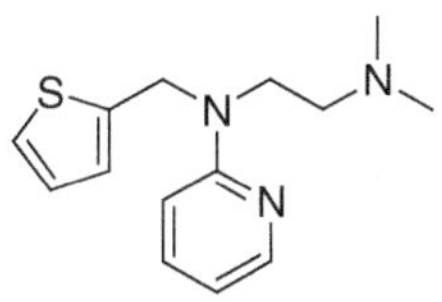	$t^1/_2$ 1.6 h (iv) *F* 4–46% pb – ur –	*V* 3.3 CL 28 Mwt 261.4 PSA 47.6 Å^2 log *P* 2.78	METAPYRILENE, THIONYLAN Histamine H_1 antagonist Metabolites: hydroxymethylthiophene, thiophene-2-carboxylic acid, 2-(2-thienylmethyl)aminopyridine), *N*-desmethyl; withdrawn in 1979 for being hepatocarcinogenic
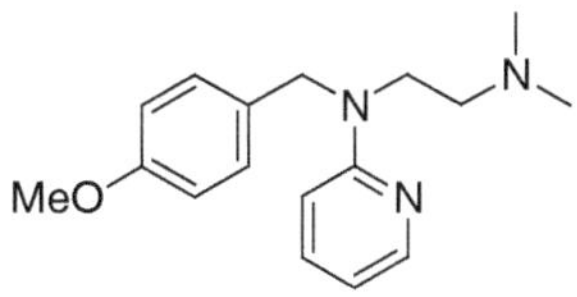	$t^1/_2$ 14–25 h *F* – pb – ur –	*V* – CL – Mwt 285.4 PSA 28.6 Å^2 log *P* 2.67	PYRILAMINE, MEPYRAMINE Histamine H_1 antagonist Duration of action: 4–6 h Metabolism: O-demethylation
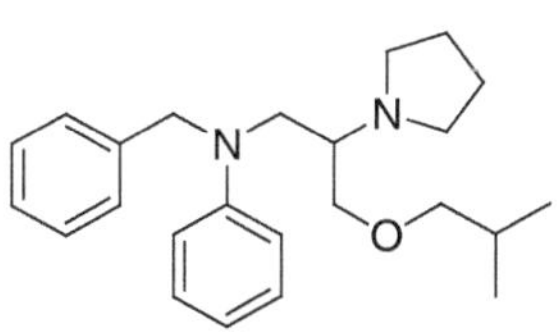	$t^1/_2$ 12 h *F* 60% pb >99% ur <1%	*V* 8±5 CL 5.3±2.5 Mwt 366.5 PSA 15.7 Å^2 log *P* 5.80	BEPRIDIL Calcium channel blocker, antianginal Metabolism: aromatic hydroxylation to 4-aminophenols, N-debenzylation, then N-acetylation, isobutyl hydroxylation, pyrrolidine hydroxylation
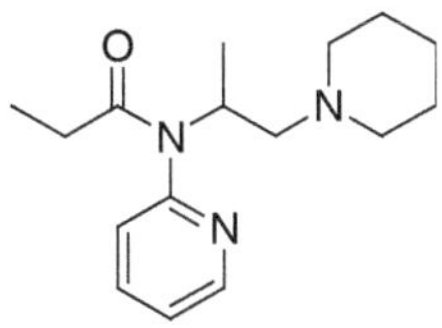	$t^1/_2$ 5–7 h *F* >97% pb – ur 24%	*V* 2.3 CL 6.3 Mwt 275.4 PSA 36.4 Å^2 log *P* 3.01	PROPIRAM Analgesic, narcotic Metabolism: piperidine hydroxylation, then oxidation to lactam, then hydrolysis of lactam
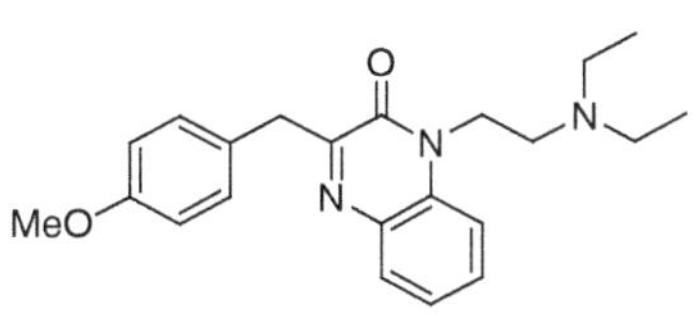	$t^1/_2$ 25 min *F* Low pb – ur –	*V* – CL – Mwt 365.5 PSA 45.1 Å^2 log *P* 3.61	CAROVERINE Calcium channel blocker, antispasmodic

(*continued overleaf*)

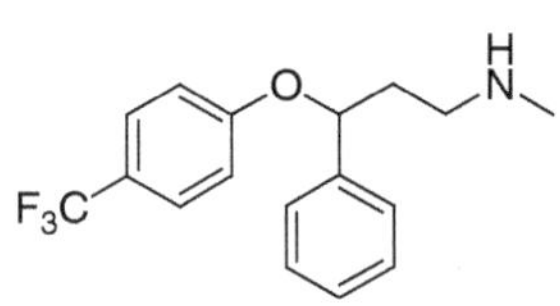

t½	53±41 h	*V/F*	35±21	FLUOXETINE
F	70%	CL/*F*	9.6±6.9	5-HT reuptake inhibitor, antidepressant
pb	94%	Mwt	309.3	Substrate and inhibitor of CYP2D6
ur	2–5%	PSA	21.3 Å^2	Active metabolite: N-desmethyl (t½ 6.5±2.5 d, CL 2 ml min^{-1} kg^{-1})
		log *P*	3.93	

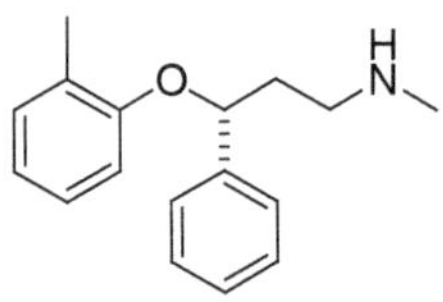

t½	5.2 h	*V*	0.85	ATOMOXETINE
F	63–94%	CL	6–9	Noradrenaline reuptake inhibitor for treatment of attention deficite hyperactivity disorder
pb	98%	Mwt	255.4	Metabolism: 4-hydroxylation of aryloxy, N-demethylation
ur	<3%	PSA	21.3 Å^2	
		log *P*	3.28	

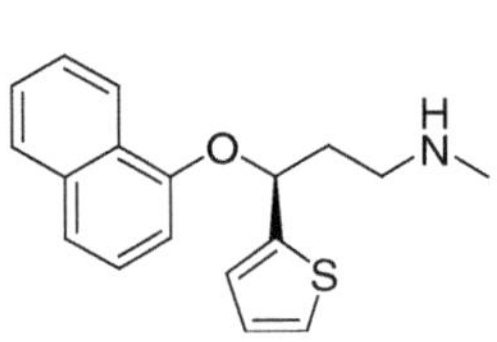

t½	8–17 h	*V*	–	DULOXETINE
F	32–80%	CL	16	Serotonin/noradrenaline reuptake inhibitor, antidepressant, analgesic
pb	>90%	Mwt	297.4	Metabolism (CYP2D6, CYP1A2): 4-, 5-, and 6-naphthyl hydroxylation, then glucuronidation, sulfation, methylation
ur	–	PSA	49.5 Å^2	
		log *P*	4.81	

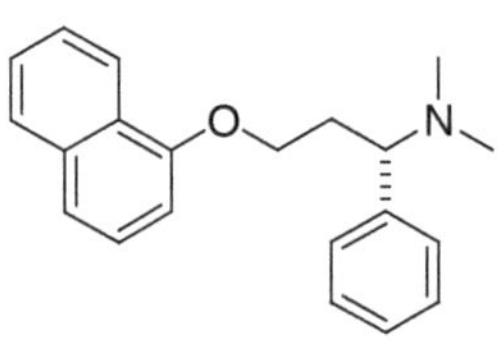

t½	18 h	*V*	2.5±0.9	DAPOXETINE
F	15–76%	CL	–	Serotonin transporter inhibitor
pb	>99%	Mwt	305.4	Metabolism: N-demethylation, N-oxidation, naphthyl 4-hydroxylation
ur	Negligible	PSA	12.5 Å^2	
		log *P*	5.13	

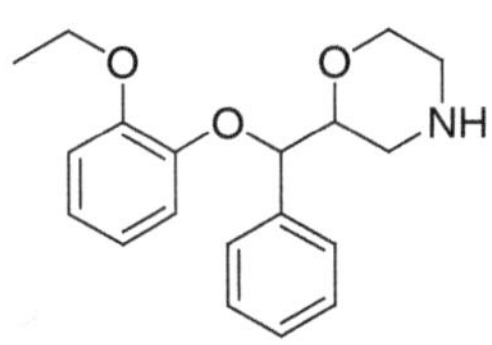

t½	13 h	*V*	0.7–0.9	REBOXETINE
F	95%	CL	0.4–0.8	Noradrenaline reuptake inhibitor, antidepressant
pb	97%	Mwt	313.4	Metabolism: O-deethylation, hydroxylation of aryloxy
ur	8–9%	PSA	39.7 Å^2	
		log *P*	2.45	

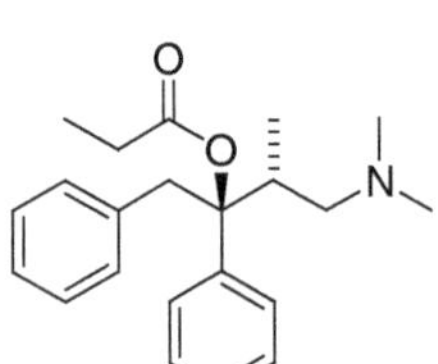

t½	8–24 h	*V*	12–16	PROPOXYPHENE
F	29–70%	CL	14–15	Opioid analgesic
pb	76–80%	Mwt	339.5	Metabolism: N-demethylation, ester hydrolysis
ur	Negligible	PSA	29.5 Å^2	Active metabolite: norpropoxyphene (see below), which is cardiotoxic; withdrawn in 2010 for inducing cardiac arrhythmias
		log *P*	4.10	

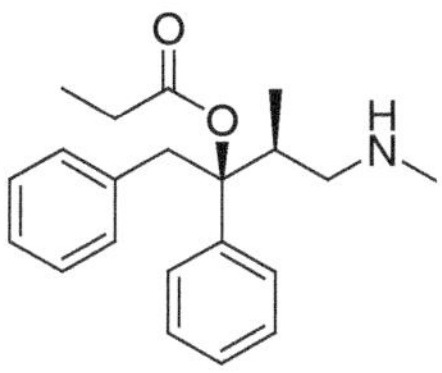	$t^{1}/_{2}$ 18–37 h *F* – pb – ur 20–25%	*V* 10–25 CL 0.3 Mwt 325.4 PSA 38.3 Å^2 log *P* 3.74	NORPROPOXYPHENE Metabolite of propoxyphene
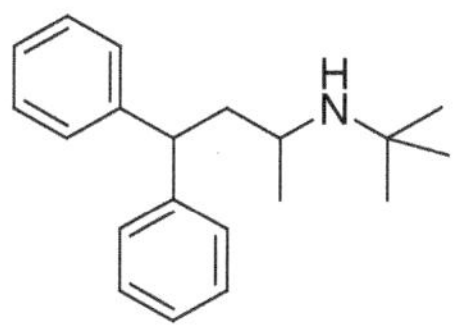	$t^{1}/_{2}$ 60±5 h *F* 92±29% pb 85% ur 15%	*V* 3.7–12.6 CL 1.1 Mwt 281.4 PSA 12.0 Å^2 log *P* 4.68	TERODILINE Calcium antagonist, antianginal Metabolism: aromatic 4-hydroxylation, benzylic hydroxylation; withdrawn in 1991 for QT interval prolongation
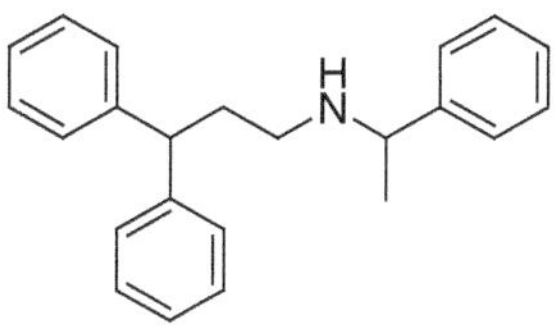	$t^{1}/_{2}$ 20–35 h *F* 2–16% pb – ur Traces	*V* – CL – Mwt 315.5 PSA 12.0 Å^2 log *P* 5.20	FENDILINE Calcium blocker, vasodilator Metabolism: aromatic hydroxylation, debenzylation
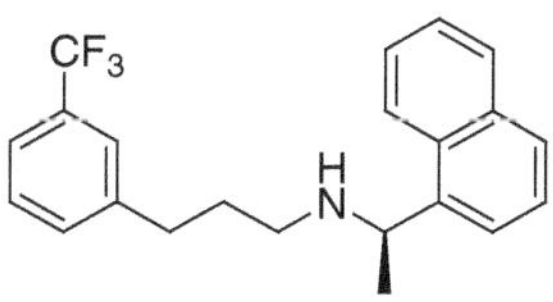	$t^{1}/_{2}$ 30–40 h *F* 20–25% pb 97% ur –	*V* 15.4 CL – Mwt 357.4 PSA 12 Å^2 log *P* 5.74	CINACALCET Calcium mimetic Metabolism: dihydroxylation of naphthalene, oxidation to 3-(trifluoromethyl)cinnamide, oxidative degradation to 3-trifluoromethylbenzoic acid and glycine conjugate thereof
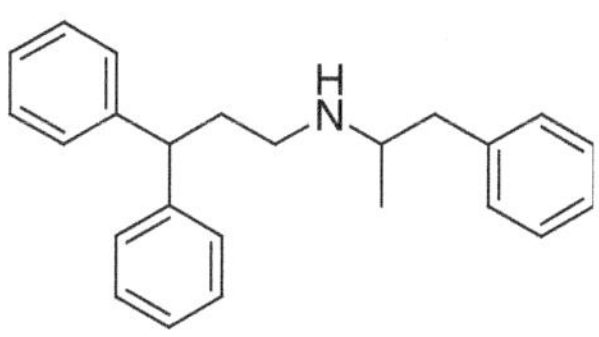	$t^{1}/_{2}$ 14±7 h *F* 15% pb 97% ur <0.1%	*V* – CL – Mwt 329.5 PSA 12.0 Å^2 log *P* 5.7	PRENYLAMINE Calcium channel blocker, coronary vasodilator Metabolism: aromatic hydroxylation, N-dealkylation
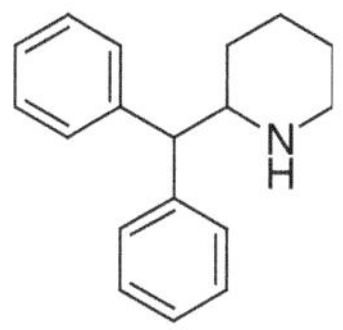	$t^{1}/_{2}$ 16–20 h *F* >90% pb – ur –	*V* – CL – Mwt 251.4 PSA 12.0 Å^2 log *P* 4.07	DESOXYPIPRADROL Noradrenaline/dopamine reuptake inhibitor for treatment of attention deficite hyperactivity disorder

(*continued overleaf*)

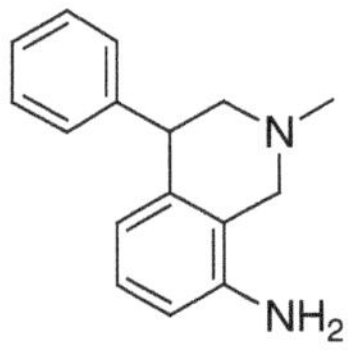	$t^{1}/_{2}$ 3–5 h F <30% pb 60% ur <5%	V 6 CL 22 Mwt 238.3 PSA 29.3 Å^2 log P 1.01	NOMIFENSINE Dopamine transport blocker, antidepressant Metabolism: N-glucuronidation, aromatic hydroxylation, aniline oxidation followed by glutathione conjugation; withdrawn in 1986 because of kidney and liver toxicity and induction of hemolytic anemia
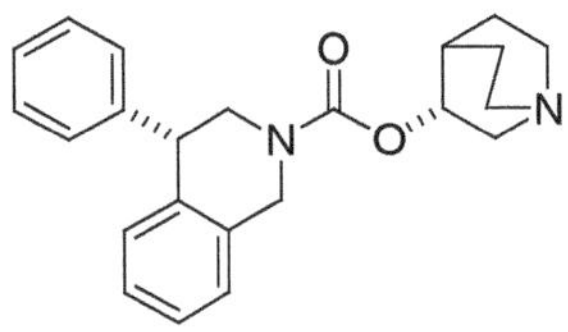	$t^{1}/_{2}$ 45–68 h F 90% pb 98% ur 11%	V 9.2 CL 2.4 Mwt 362.5 PSA 32.8 Å^2 log P 3.70	SOLIFENACIN Anticholinergic for treatment of overactive bladder Metabolism: benzylic hydroxylation (4*R*-hydroxy), N-oxidation, N-glucuronidation
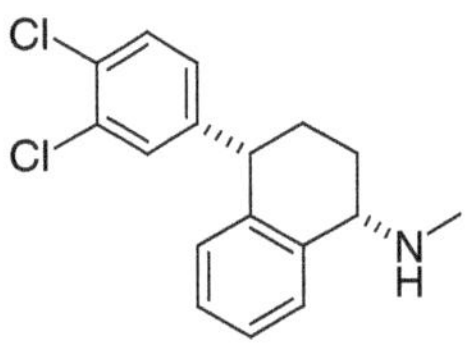	$t^{1}/_{2}$ 23 h F >44% pb >99% ur <0.2%	V/F 76±26 CL/F 38±14 Mwt 306.2 PSA 12.0 Å^2 log P 5.08	SERTRALINE Serotonin reuptake inhibitor, antidepressant Metabolite (20–60 times less potent than parent): *N*-desmethyl ($t^{1}/_{2}$ 77±2 h)
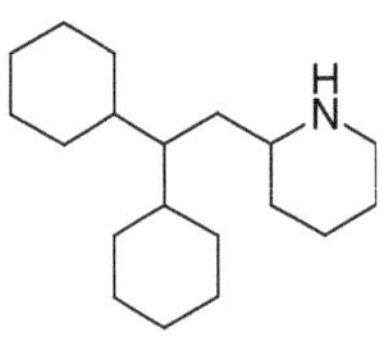	$t^{1}/_{2}$ 2–12 d F 88% pb – ur 0.1%	V/F 4–23 CL/F 0.5–5.6 Mwt 277.5 PSA 12.0 Å^2 log P 6.47	PERHEXILINE Carnitine palmitoyltransferase inhibitor, vasodilator, diuretic, antianginal Metabolism (CYP2D6); 4-hydroxylation of cyclohexyl; high interpatient variability of metabolism
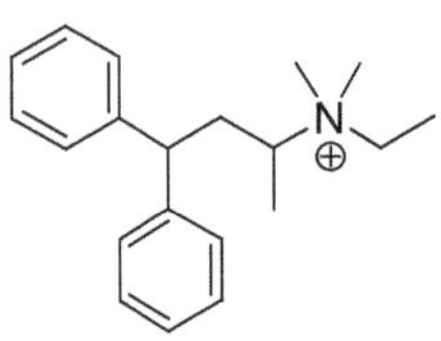	$t^{1}/_{2}$ 1.9±0.4 h F Low pb – ur –	V – CL 6.2–9.7 Mwt 282.5 PSA 0.0 Å^2 log P 0.56	EMEPRONIUM Anticholinergic, antispasmodic Metabolism: aromatic 4-hydroxylation
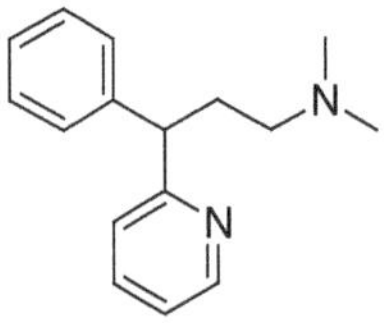	$t^{1}/_{2}$ 17±1 h F – pb – ur –	V – CL – Mwt 240.3 PSA 16.1 Å^2 log P 2.20	PHENIRAMINE Histamine H_1 antagonist Metabolites: desmethyl, didesmethyl

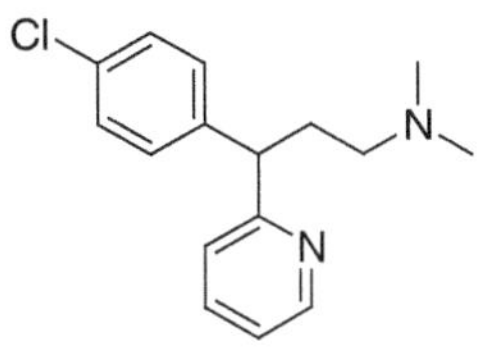

$t_{1/2}$	20±5 h	V	3.2±0.3
F	41±16%	CL	1.7±0.1
pb	70±3%	Mwt	274.8
ur	0.3–26%	PSA	16.1 $Å^2$
		log P	2.97

CHLORPHENIRAMINE
Histamine H_1 antagonist
Duration of action: 5±1 h
Metabolism: N-demethylation

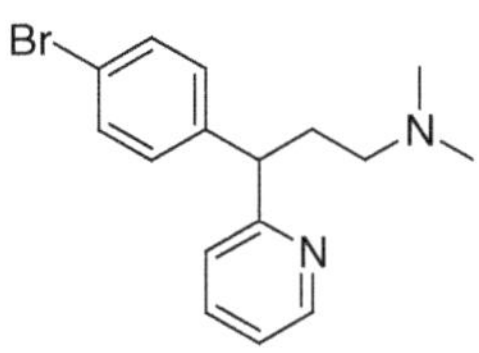

$t_{1/2}$	25±9 h	V	11.7±3.1
F	–	CL	6.0±2.3
pb	72%	Mwt	319.2
ur	3%	PSA	16.1 $Å^2$
		log P	3.24

BROMPHENIRAMINE
Histamine H_1 antagonist
Duration of action: 5±1 h
Metabolism: N-demethylation, oxidative deamination to 3-(2-pyridyl)-3-(4-bromophenyl)-propionic acid, then conjugation with glycine

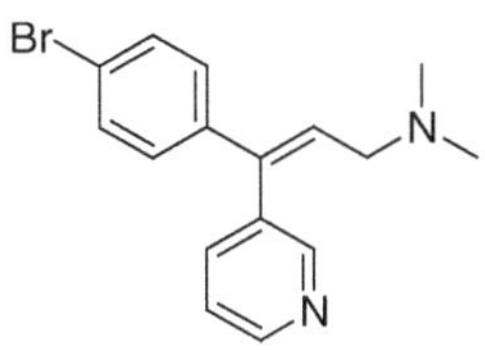

$t_{1/2}$	5±2 h	V	3.2±2.0
F	27±16%	CL	–
pb	90%	Mwt	317.2
ur	–	PSA	16.1 $Å^2$
		log P	3.07

ZIMELDINE
Antidepressant
Metabolism: N-demethylation, oxidation to pyridine-*N*-oxide and acrylic acid; withdrawn in 1983 for causing Guillain-Barré syndrome

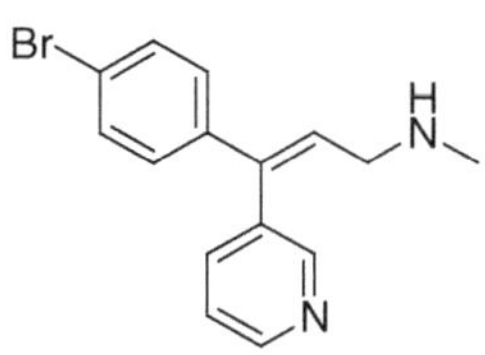

$t_{1/2}$	16±6 h	V	9.6±1.8
F	64±27%	CL	–
pb	80%	Mwt	303.2
ur	–	PSA	24.9 $Å^2$
		log P	2.45

NORZIMELDINE
Metabolite of zimeldine

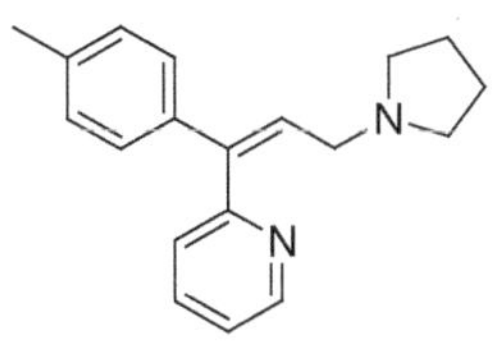

$t_{1/2}$	5±1 h	V	5.8
F	4%	CL	15.4
pb	90%	Mwt	278.4
ur	–	PSA	16.1 $Å^2$
		log P	3.25

TRIPROLIDINE
Histamine H_1 antagonist
Metabolism: CH_3 hydroxylation and oxidation to carboxylic acid

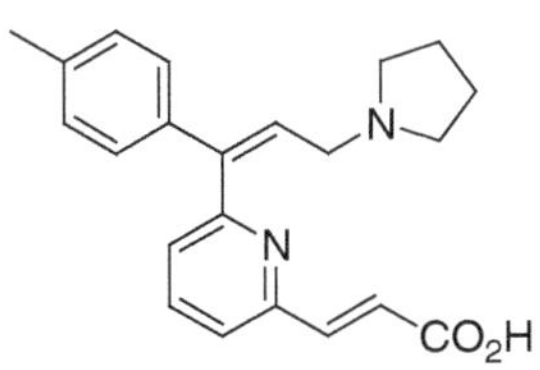

$t_{1/2}$	2.5±1 h	V	0.64
F	80%	CL	4.3±0.6
pb	50%	Mwt	348.4
ur	59%	PSA	53.4 $Å^2$
		log P	2.41

ACRIVASTINE
Histamine H_1 antagonist
Duration of action: 7±1 h
Active metabolite: propionic acid formed by hydrogenation of alkene

(*continued overleaf*)

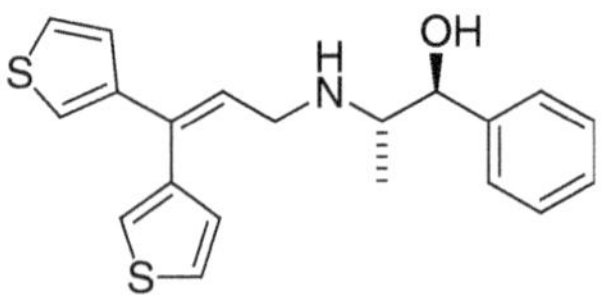	$t^{1/2}$ 3.8 h* *F* 100%* pb – ur –	*V* – CL – Mwt 355.5 PSA 88.7 Å^2 log *P* 3.42	TINOFEDRINE *dog
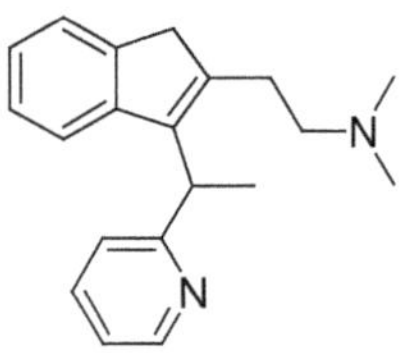	$t^{1/2}$ 6.3 h (iv) *F* – pb 90% ur 1% (po)*	*V* – CL – Mwt 292.4 PSA 16.1 Å^2 log *P* 3.84	DIMETHINDENE *5–10% (iv) Histamine H_1 antagonist Metabolism: N-demethylation
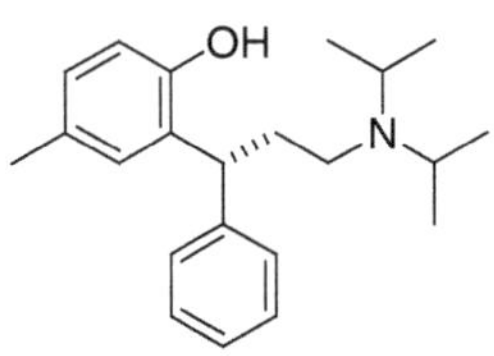	$t^{1/2}$ 2–3 h *F* 10–74% pb 96% ur 1%	*V* 0.9–1.6 CL 8–11 Mwt 325.5 PSA 23.5 Å^2 log *P* 5.77	TOLTERODINE Antimuscarinic for treatment of urinary incontinence Metabolism: N-dealkylation by CYP3A, hydroxylation by CYP2D6 (see below)
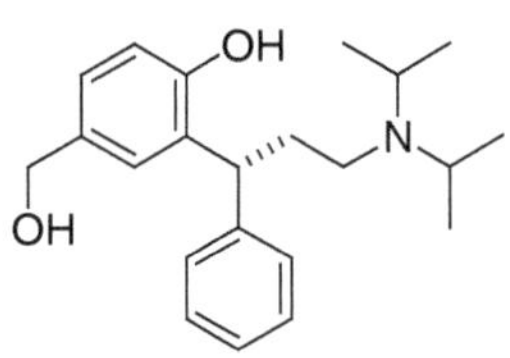	$t^{1/2}$ 7–9 h *F* 52%* pb 64% ur 4%	*V* 2.6 CL 2.1 Mwt 341.5 PSA 43.7 Å^2 log *P* 4.12	5-HYDROXYMETHYL TOLTERODINE Active metabolite of tolterodine *on oral dosing of fesoterodine (prodrug: arylester of isobutyric acid)
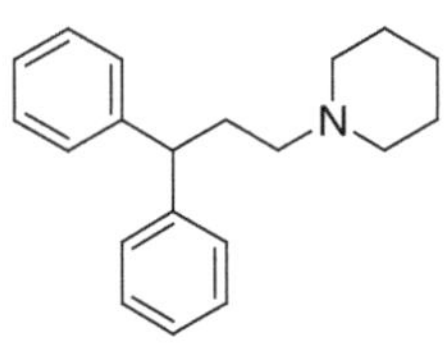	$t^{1/2}$ 13 h* *F* – pb – ur –	*V* – CL – Mwt 279.4 PSA 3.24 Å^2 log *P* 4.62	FENPIPRANE *horse, iv Antispasmodic sc administration to humans (probably because of low *F*)
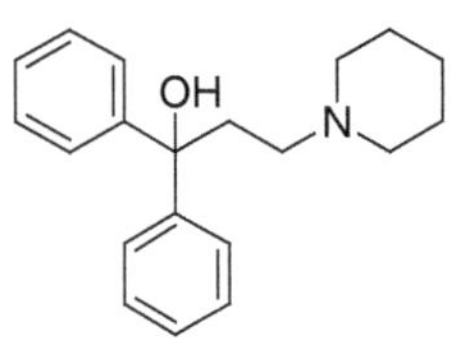	$t^{1/2}$ 14 h *F* 77% pb – ur 9%	*V* – CL – Mwt 295.4 PSA 23.5 Å^2 log *P* 3.35	PRIDINOL, MYOSON Anticholinergic, antiparkinsonian, analgesic Metabolism: piperidine and aromatic hydroxylation

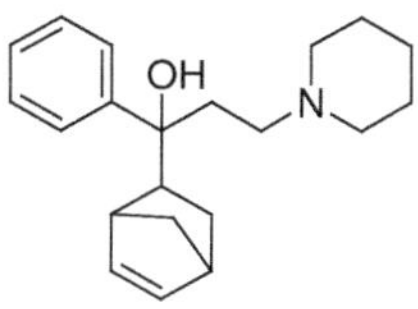	$t_{1/2}$ 11–21 h *F* 33±5% pb 91% ur 0%	*V* 24±4 CL 12±1 Mwt 311.5 PSA 23.5 Å^2 log *P* 4.27	BIPERIDEN Anticholinergic, antiparkinsonian Metabolism: hydroxylation of bicycloheptene (60%) and piperidine (40%)
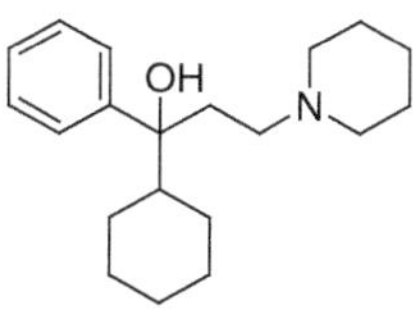	$t_{1/2}$ 5–10 h *F* – pb – ur –	*V* – CL 6.3 Mwt 301.5 PSA 23.5 Å^2 log *P* 4.39	TRIHEXYPHENIDYL Anticholinergic, antiparkinsonian Metabolism: hydroxylation, N-oxidation
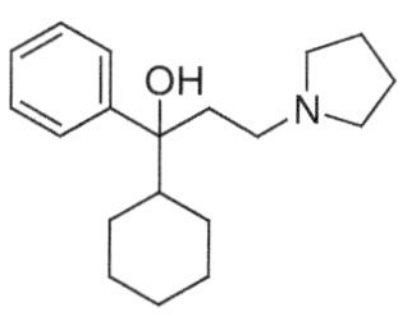	$t_{1/2}$ 12 h *F* 75% pb 100% ur –	*V* 0.74 CL 0.86 Mwt 287.4 PSA 23.5 Å^2 log *P* 3.86	PROCYCLIDINE Anticholinergic, antiparkinsonian Metabolism (rat): cyclohexyl 3- and 4-hydroxylation
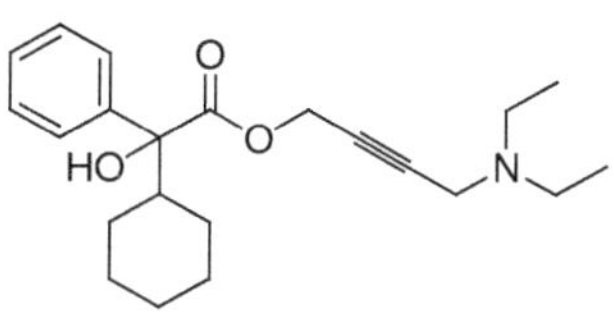	$t_{1/2}$ 2–3 h *F* 2–11% pb >99% ur <1%	*V* 2.8 CL 6 Mwt 357.5 PSA 49.8 Å^2 log *P* 5.05	OXYBUTYNIN Anticholinergic Metabolism: N-deethylation, ester hydrolysis
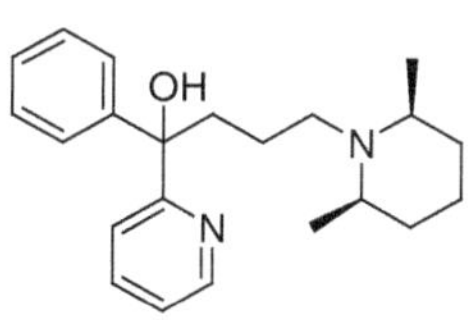	$t_{1/2}$ 6–18 h *F* 80–90% pb 80–90% ur 20–30%	*V* 1.5 CL 1.8–2.9 Mwt 338.5 PSA 36.4 Å^2 log *P* 3.57	PIRMENOL Antiarrhythmic Metabolism: oxidation of piperidine to iminium, then reversible tautomerization to enamine or hydrolysis to aminoketone
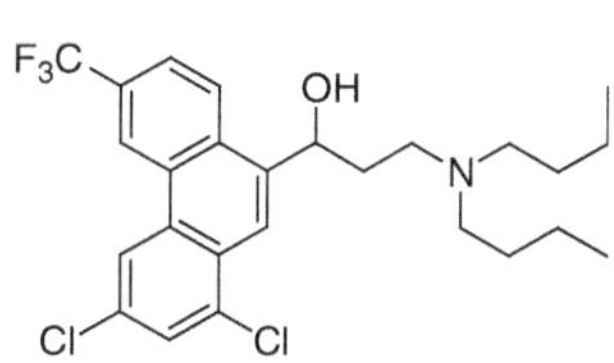	$t_{1/2}$ 6–10 d *F* 6±5%* pb >99% ur 0%	*V* 73 CL – Mwt 500.4 PSA 23.5 Å^2 log *P* 8.90	HALOFANTRINE *dog Antimalarial Metabolite: desbutyl ($t_{1/2}$ 3–11 d)
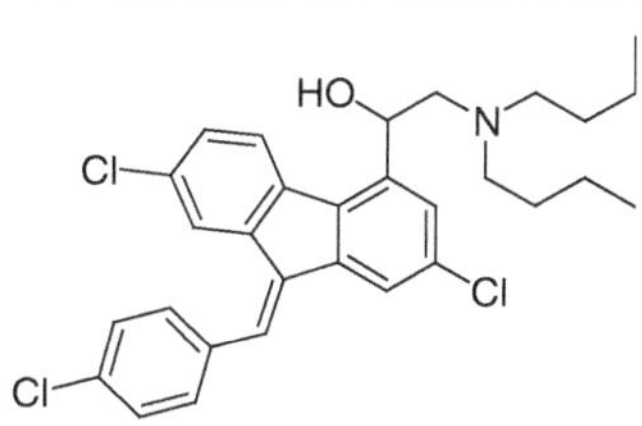	$t_{1/2}$ 3–5 d *F* Variable pb 99.7% ur –	*V* 2–4 CL – Mwt 528.9 PSA 23.5 Å^2 log *P* 8.67	LUMEFANTRINE Antimalarial Metabolism: N-debutylation

(continued overleaf)

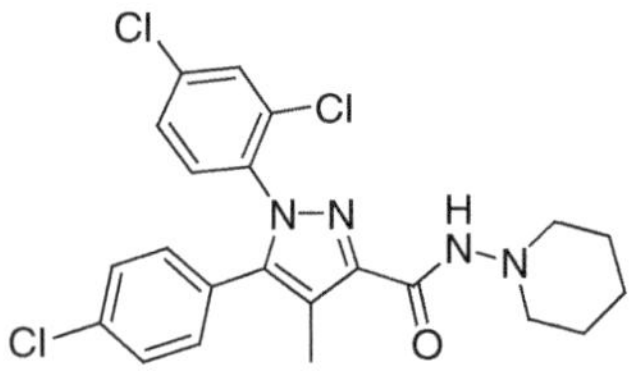

$t^{1}/_{2}$	6–9 d	V	–
F	–	CL/F	1.3–2.3
pb	>99.9%	Mwt	463.8
ur	3%	PSA	50.2 Å^{2}
		log P	4.85

RIMONABANT
Cannabinoid CB_1 antagonist, appetite suppressant; withdrawn in 2008 because of depression and suicide risk

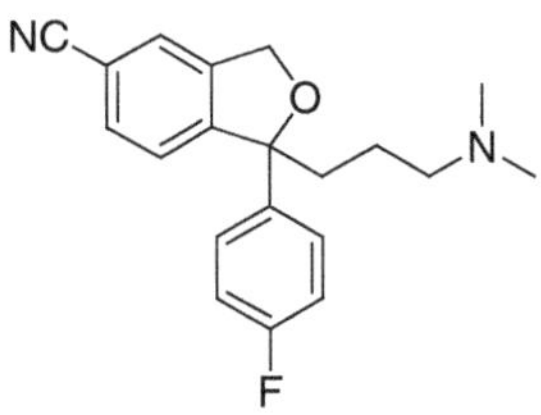

$t^{1}/_{2}$	19–45 h	V	12–17
F	80–100%	CL	5.7–6.3
pb	50–80%	Mwt	324.4
ur	6–31%	PSA	36.3 Å^{2}
		log P	3.48

CITALOPRAM
5-HT reuptake inhibitor, antidepressant
Metabolites: desmethyl (CYP2C19), didesmethyl (CYP2D6), *N*-oxide, deaminated propionic acid

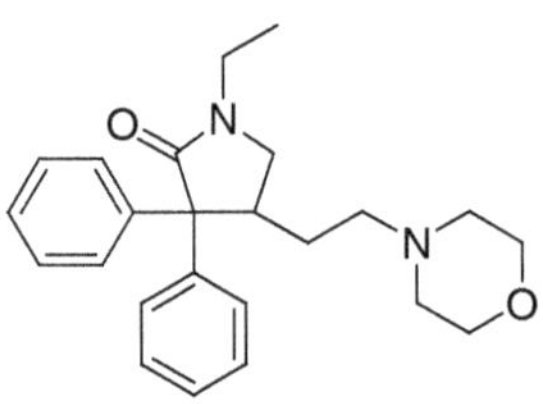

$t^{1}/_{2}$	2–4 h	V	1.5
F	60%	CL	5.7
pb	76–85%*	Mwt	378.5
ur	<5%	PSA	32.8 Å^{2}
		log P	5.64

DOXAPRAM
*horse
Respiratory stimulant
Metabolism: oxidative ring opening of morpholine, oxidation of morpholine to lactam ("2-ketodoxapram")

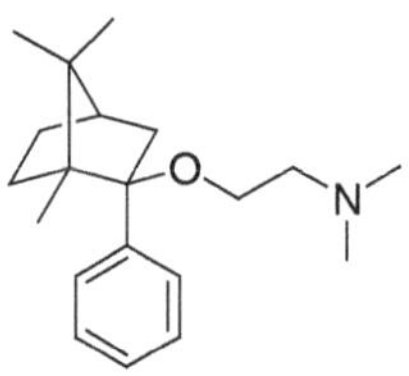

$t^{1}/_{2}$	21–31 h	V	32–37 (dog)
F	23–50%	CL	–
pb	–	Mwt	301.5
ur	Traces (iv)	PSA	12.5 Å^{2}
		log P	3.72

DERAMCICLANE
$5\text{-}HT_{2A/2C}$ antagonist, anxiolytic
Metabolite: *N*-desmethyl ($t^{1}/_{2}$ 38±7 h)

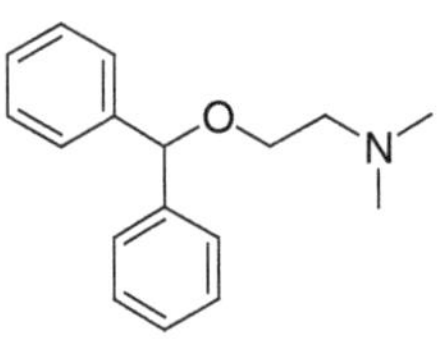

$t^{1}/_{2}$	9±3 h	V	4.5±2.8
F	61±25%	CL	6.2±1.7
pb	78±3%	Mwt	255.4
ur	2±1%	PSA	12.5 Å^{2}
		log P	3.00

DIPHENHYDRAMINE
Histamine H_1 antagonist
Duration of action: 5±1 h
Metabolism: N-demethylation, oxidation of CH_2NMe_2 to CO_2H

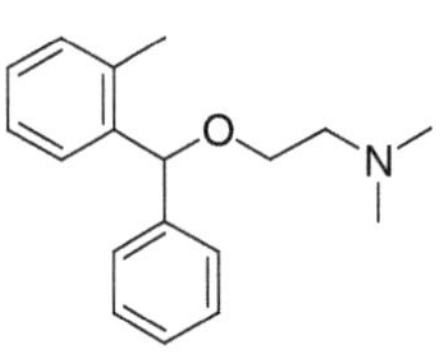

$t^{1}/_{2}$	16±3 h	V	4.5±2.8
F	90%	CL	6.2±1.7
pb	95%	Mwt	269.4
ur	8%	PSA	12.5 Å^{2}
		log P	3.33

ORPHENADRINE
Histamine H_1 antagonist, muscle relaxant, CYP2B inducer
Metabolism: N-demethylation, N-oxidation to *N*-oxide, oxidation of CH_2NMe_2 to CO_2H, O-dealkylation to 2-methylbenzhydrol

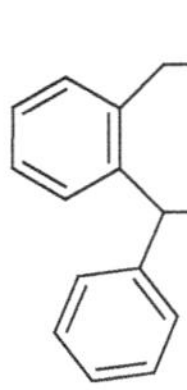	$t^{1/2}$ 3–8 h *F* 40% pb 75% ur 5%	*V* 5.6 CL 12 Mwt 253.3 PSA 12.5 $Å^2$ log *P* 1.91	NEFOPAM Analgesic, antidepressant Metabolites: *N*-desmethyl, *N*-oxide
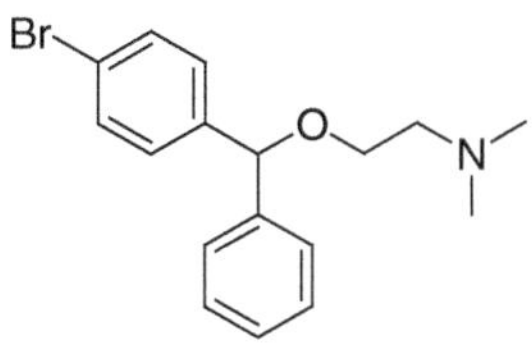	$t^{1/2}$ 3±1 h *F* 65% pb 96% ur –	*V* 0.9 CL – Mwt 334.3 PSA 12.5 $Å^2$ log *P* 4.03	BROMODIPHENHYDRAMINE, BROMAZINE Histamine H_1 antagonist Metabolism: N-oxidation, N-demethylation
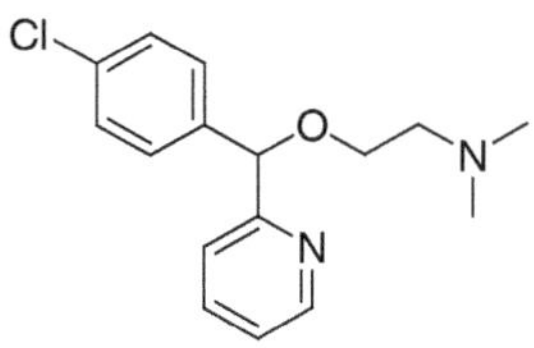	$t^{1/2}$ 15±5 h *F* – pb – ur 5–8%	*V* – CL – Mwt 290.8 PSA 25.4 $Å^2$ log *P* 2.69	CARBINOXAMINE Histamine H_1 antagonist; duration of action: 3–6 h Metabolism: N-demethylation
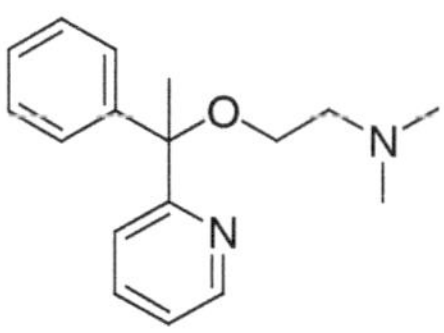	$t^{1/2}$ 9±3 h *F* 25% pb – ur 60%	*V* 2.5 CL 3.3 Mwt 270.4 PSA 25.4 $Å^2$ log *P* 2.34	DOXYLAMINE Histamine H_1 antagonist, hypnotic Metabolism: N-oxidation, N-demethylation, oxidation of CCH_3 to CCO_2H and of CH_2NMe_2 to CO_2H
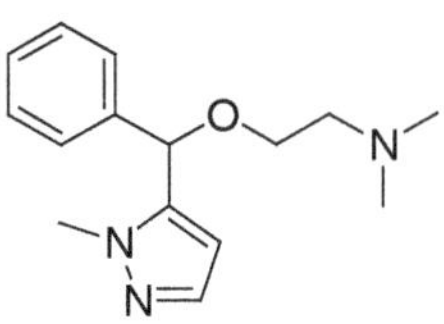	$t^{1/2}$ 6.7–7.1 h *F* 60% (dog) pb – ur –	*V* – CL – Mwt 259.4 PSA 25.4 $Å^2$ log *P* 1.61	CIZOLIRTINE Analgesic Metabolism: N-oxidation, N-demethylation, O-dealkylation
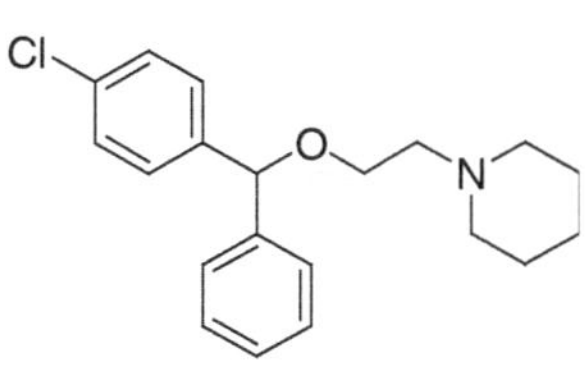	$t^{1/2}$ 2–7 h *F* 40% pb >97% ur –	*V* – CL 14 Mwt 329.9 PSA 12.5 $Å^2$ log *P* 4.58	CLOPERASTINE Histamine H_1 antagonist

(*continued overleaf*)

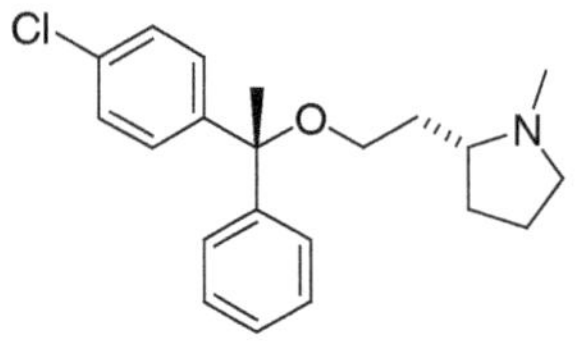

$t_{1/2}$	22±13 h	V	13±5	CLEMASTINE
F	37±12%	CL	8.3±1.6	Histamine H_1 antagonist
pb	95%	Mwt	343.9	Duration of action: 18±6 h
ur	<2%	PSA	12.5 Å^2	Metabolism: O-dealkylation to tertiary alcohol, aromatic hydroxylation
		log P	5.30	

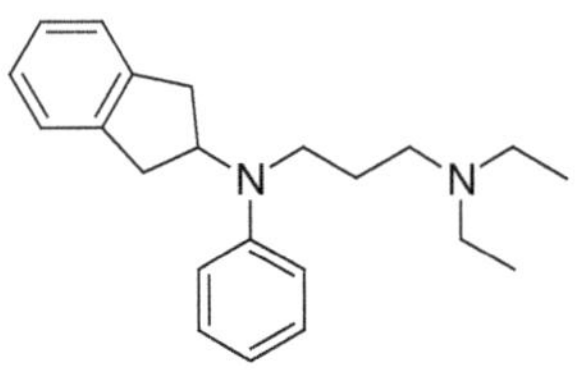

$t_{1/2}$	12–66 h	V	4	APRINDINE
F	75%	CL	1	Antiarrhythmic, local anesthetic
pb	85–95%	Mwt	322.5	Metabolism: N-deethylation, de-indanylation, hydroxylation
ur	2%	PSA	6.48 Å^2	
		log P	5.68	

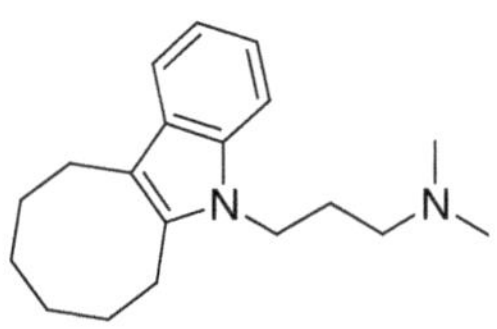

$t_{1/2}$	53 h	V	–	IPRINDOLE
F	–	CL	11	Antidepressant
pb	–	Mwt	284.4	Metabolism: N-demethylation, benzylic oxidation
ur	–	PSA	8.17 Å^2	
		log P	5.02	

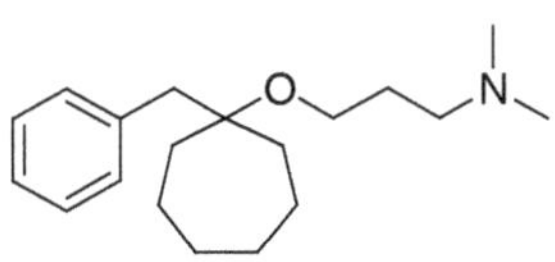

$t_{1/2}$	12–18 h	V	7–9	BENCYCLANE
F	20–80%	CL	11	Vasodilator
pb	–	Mwt	289.5	Metabolism: hydroxylation, oxidation of cycloheptane to 4-hydroxy and 4-oxo
ur	3%	PSA	12.5 Å^2	
		log P	4.19	

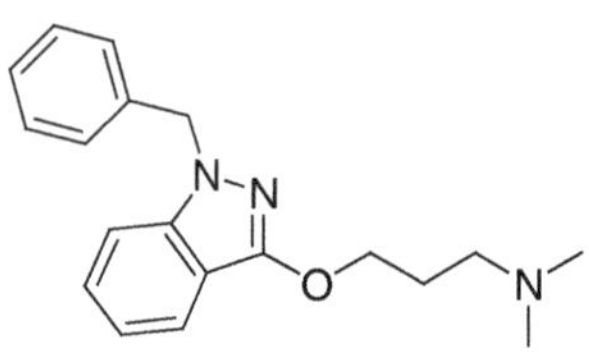

$t_{1/2}$	10–13 h	V	1.7–3.3	BENZYDAMINE
F	87%	CL	2.5–3.0	Analgesic, antiinflammatory
pb	10–15%	Mwt	309.4	Metabolism: N-oxidation, aromatic hydroxylation
ur	5%	PSA	30.3 Å^2	
		log P	3.56	

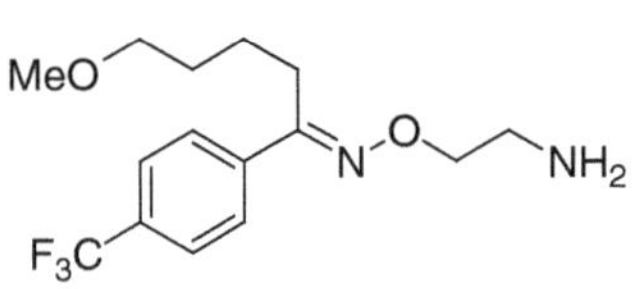

$t_{1/2}$	12–30 h	V	20–25	FLUVOXAMINE
F	50–60%	CL	15.2	5-HT reuptake inhibitor, antidepressant
pb	77%	Mwt	318.3	Metabolism: oxidation of CH_2OMe to CO_2H, N-acetylation, replacement of NH_2 by OH
ur	3%	PSA	56.8 Å^2	
		log P	3.71	

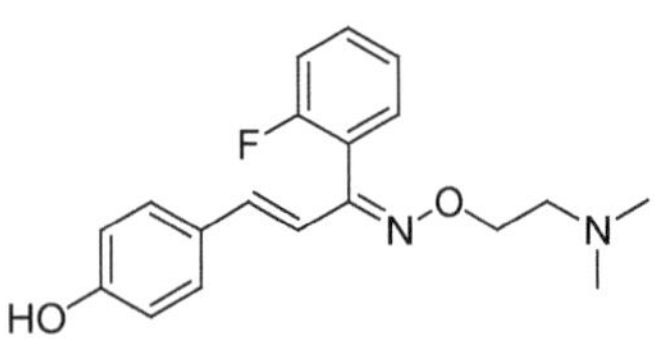

$t_{1/2}$	50 h	V	6*	EPLIVANSERIN
F	>70%	CL	33*	*baboon
pb	80%	Mwt	328.4	5-HT_{2A} antagonist, hypnotic
ur	–	PSA	45.1 Å^2	Metabolite: *N*-desmethyl ($t_{1/2}$ 70 h)
		log P	4.22	

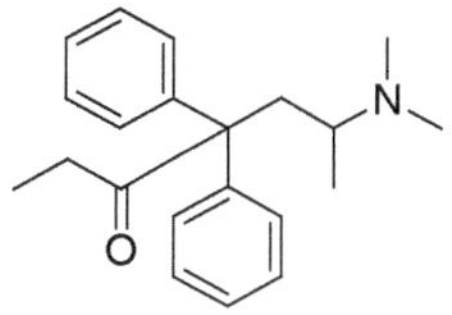	$t_{1/2}$ 35±12 h *F* 92±21% pb 89±1.4% ur 24±10%	*V* 3.8±0.6 CL 1.4±0.5 Mwt 309.5 PSA 20.3 Å^2 log *P* 3.93	METHADONE Analgesic, racemate Metabolism: N-demethylation, then cyclization, aromatic hydroxylation
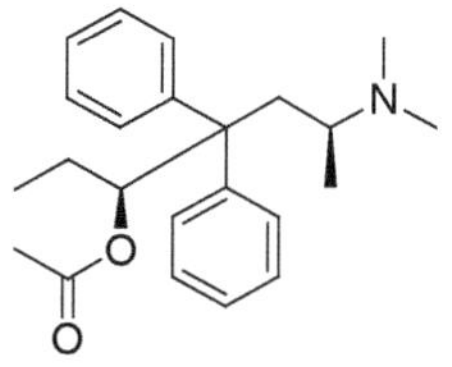	$t_{1/2}$ 8–19 h *F* 48% pb 80% ur 20–25%	*V* 8.3 CL 5.4 Mwt 353.5 PSA 29.5 Å^2 log *P* 4.24	L-ACETYLMETHADOL Treatment of narcotic addiction Metabolism: N-demethylation, ester hydrolysis
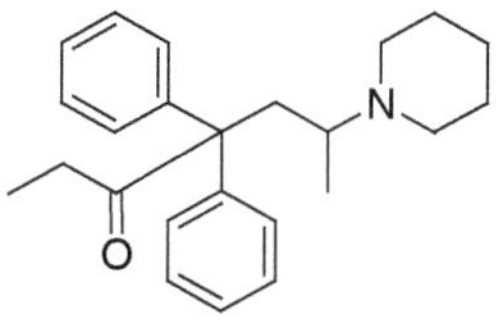	$t_{1/2}$ 3.2–3.8 h *F* – pb – ur 1%	*V* – CL – Mwt 349.5 PSA 20.3 Å^2 log *P* 5.10	DIPIPANONE Analgesic
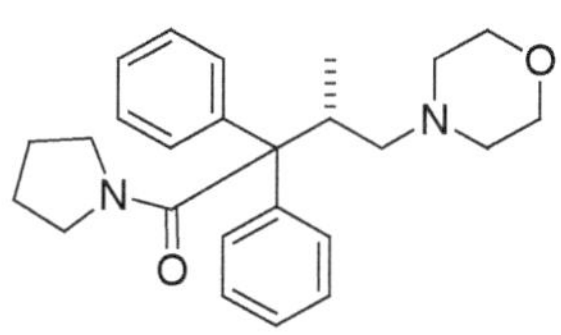	$t_{1/2}$ 4±1 h *F* – pb – ur <0.1%	*V* – CL 3.8 Mwt 392.5 PSA 32.8 Å^2 log *P* 2.53	DEXTROMORAMIDE Analgesic Metabolism (horse): aromatic 4-hydroxylation, N-dealkylation of amide
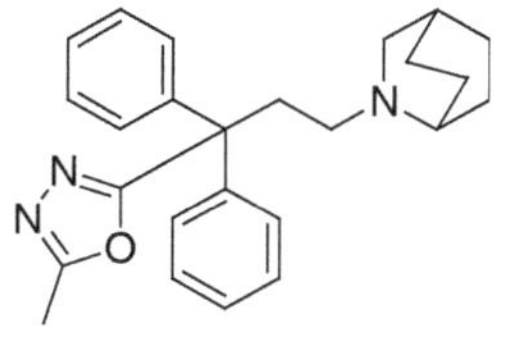	$t_{1/2}$ 27–36 h *F* 100% pb 90% ur –	*V* – CL 0.44 Mwt 387.5 PSA 42.2 Å^2 log *P* 3.73	NUFENOXOLE Antidiarrheal Metabolism: isoquinuclidine hydroxylation
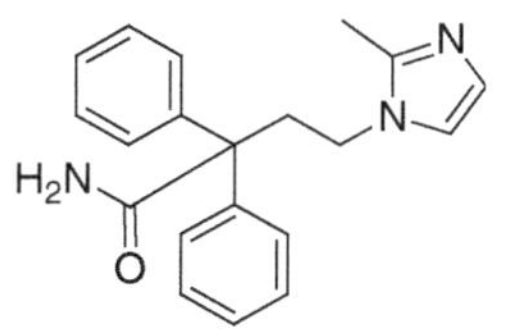	$t_{1/2}$ 3.0±0.3 h *F* 58% pb 88% ur 8%	*V* 1.9±0.4 CL 7.6±1.6 Mwt 319.4 PSA 56.0 Å^2 log *P* 2.42	IMIDAFENACIN Muscarinic antagonist Metabolism: N-glucuronidation, dihydroxylation and oxidative degradation of imidazole
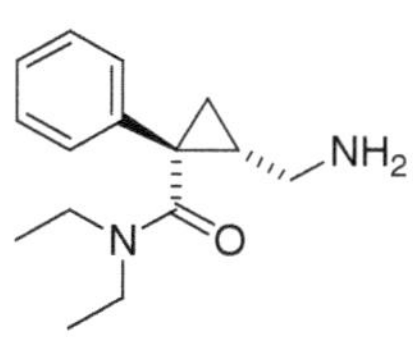	$t_{1/2}$ 6–8 h *F* 85–90% pb 13% ur 50–60%	*V* 5.3 CL 23 Mwt 246.4 PSA 46.3 Å^2 log *P* 0.83	MILNACIPRAN 5-HT reuptake inhibitor, Antidepressant Metabolism: N-deethylation

(*continued overleaf*)

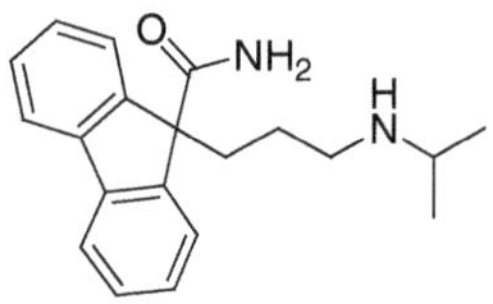

t½	3–5 h*	V	–
F	–	CL	–
pb	–	Mwt	308.4
ur	–	PSA	55.1 Å^2
		log P	3.87

INDECAINIDE
*monkey
Antiarrhythmic
Metabolite: *N*-desisopropyl

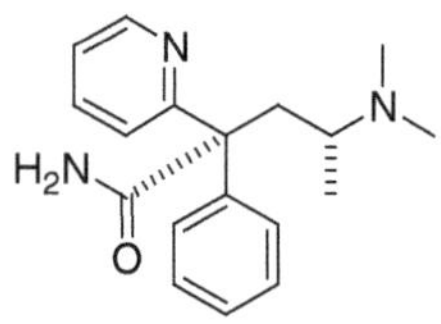

t½	5.2 h	V	–
F	–	CL	–
pb	–	Mwt	297.4
ur	>80%	PSA	59.2 Å^2
		log P	1.45

VAMICAMIDE
Anticholinergic

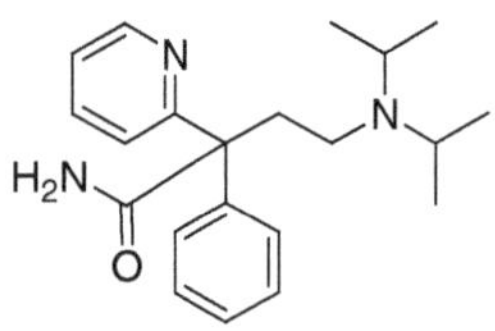

t½	6±1 h	V	0.6±0.2
F	83±11%	CL	1.2±0.4
pb	68–89%	Mwt	339.5
ur	55±6%	PSA	59.2 Å^2
		log P	2.33

DISOPYRAMIDE
Antiarrhythmic
Metabolite: *N*-desisopropyl

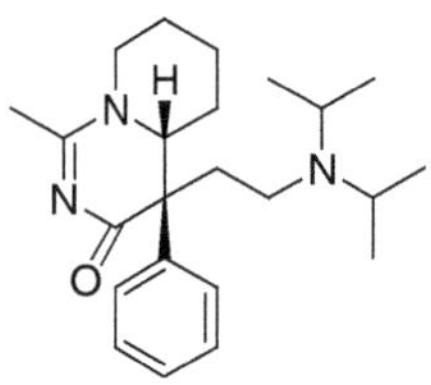

t½	1.1–1.9 h	V	1
F	0–43%	CL	6.8±1.1
pb	–	Mwt	369.5
ur	26%	PSA	35.9 Å^2
		log P	2.93

ACTISOMIDE
Antiarrhythmic
Metabolism: hydroxylation of piperidine, N-deisopropylation and pyrrolidone formation

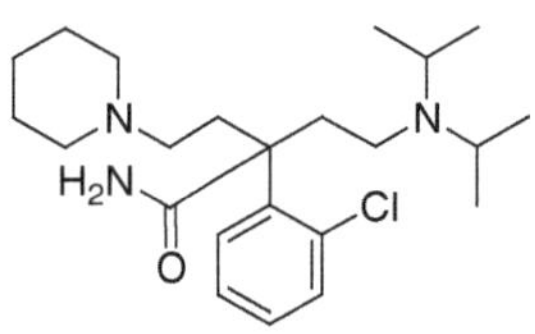

t½	54±18 h	V	–
F	–	CL	–
pb	–	Mwt	408.0
ur	–	PSA	49.6 Å^2
		log P	4.31

DISOBUTAMIDE
Antiarrhythmic
Metabolism: N-deisopropylation

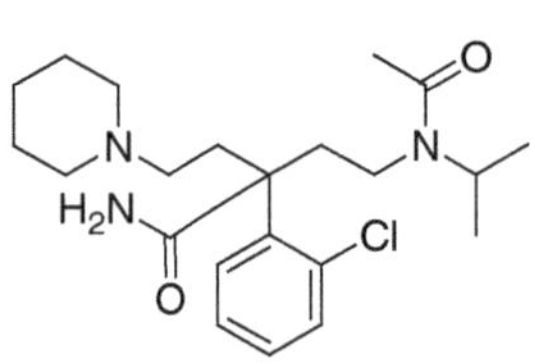

t½	12 h	V	4.3
F	43–62%	CL	4
pb	–	Mwt	408.0
ur	9%	PSA	66.6 Å^2
		log P	2.84

BIDISOMIDE
Antiarrhythmic
Metabolism: piperidine hydroxylation, N-deisopropylation, deacetylation and cyclization to 2-pyrrolidone

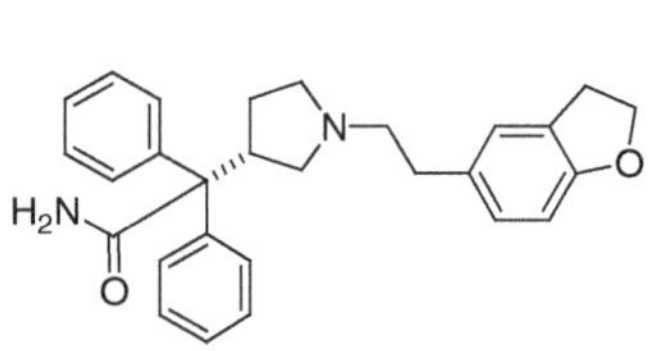

t½	13–19 h	V	2.6
F	15–19%	CL	12
pb	98%	Mwt	426.6
ur	3%	PSA	55.6 Å^2
		log P	4.50

DARIFENACIN
Muscarinic acetylcholine M_3 antagonist for treatment of urinary incontinence
Metabolism: hydroxylation and ring opening of dihydrobenzofuran, N-dealkylation of pyrrolidine

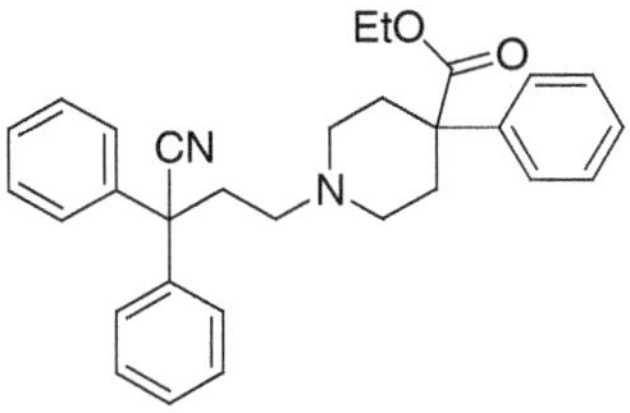

$t_{1/2}$	2–3 h	V	3.8
F	–	CL	–
pb	74–95%	Mwt	452.6
ur	<1%	PSA	53.3 Å^2
		log P	4.51

DIPHENOXYLATE
Opiate receptor agonist, antidiarrheal
Metabolism: ester hydrolysis, aromatic hydroxylation (PhC-CN)
Active metabolite: acid (difenoxin, $t_{1/2}$ 4–12 h)

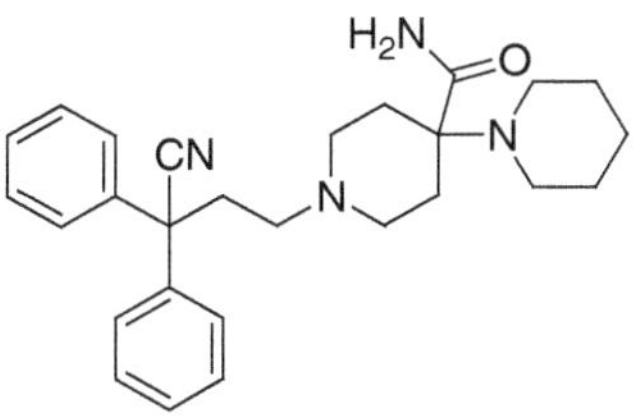

$t_{1/2}$	8–10 h (iv)	V	5–6
F	–	CL	6–9
pb	94%	Mwt	430.6
ur	Negligible	PSA	73.4 Å^2
		log P	2.78

PIRITRAMIDE
Analgesic
Mostly administered iv

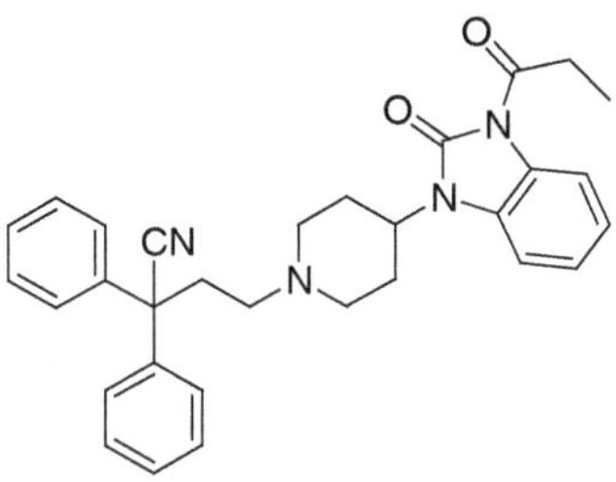

$t_{1/2}$	11–24 h	V	–
F	–	CL	–
pb	–	Mwt	492.6
ur	<0.3%	PSA	67.7 Å^2
		log P	4.16

BEZITRAMIDE
Analgesic
Metabolism: hydrolytic removal of propionyl group

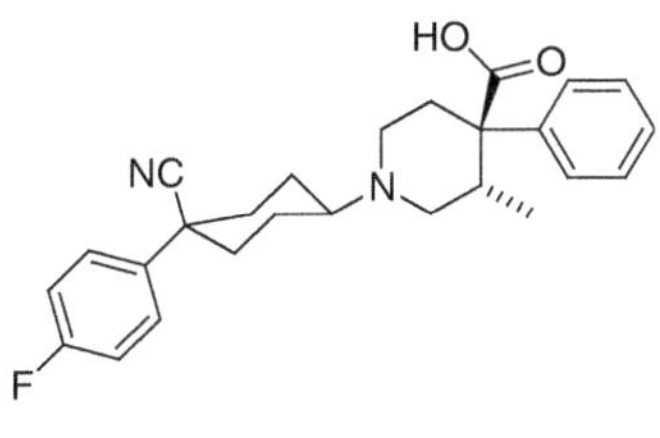

$t_{1/2}$	35–40 h	V	1.1
F	100%	CL	0.43
pb	55%	Mwt	420.5
ur	65–70%	PSA	64.3 Å^2
		log P	4.05

LEVOCABASTINE
Histamine H_1 antagonist; duration of action: 20±4 h
Metabolism: glucuronidation

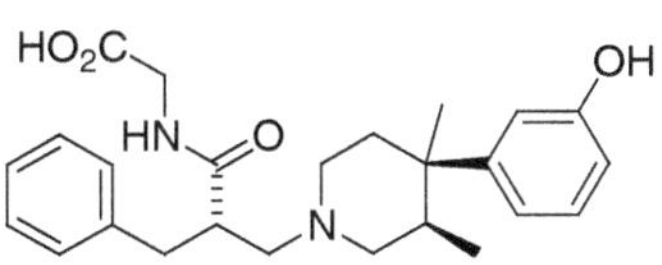

$t_{1/2}$	10–17 h	V	0.3–0.6
F	1–19%	CL	6
pb	80–90%	Mwt	424.5
ur	2%	PSA	89.9 Å^2
		log P	3.38

ALVIMOPAN
Opiate antagonist; does not cross bbb
Metabolism: amide hydrolysis

(*continued overleaf*)

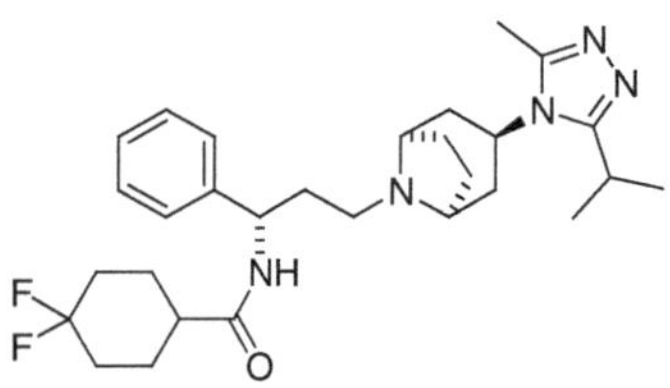

$t_{1/2}$	16 h	V	3
F	23–33%	CL	11.3
pb	76%	Mwt	513.7
ur	8%	PSA	63.1 Å^2
		log P	5.30

MARAVIROC
Antiviral (HIV)
Metabolism: N-dealkylation of tropane

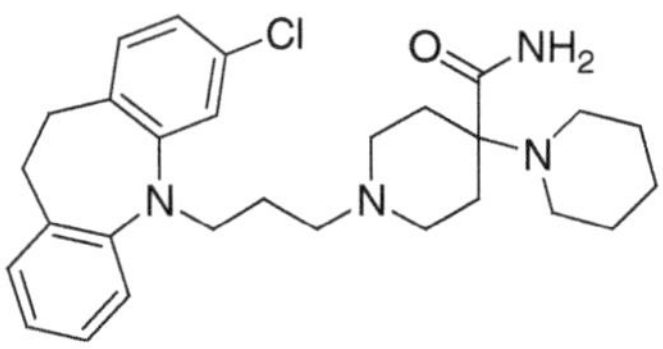

$t_{1/2}$	3 h (dog)	V	–
F	16%	CL	–
pb	–	Mwt	481.1
ur	–	PSA	52.8 Å^2
		log P	5.28

CLOCAPRAMINE
Antipsychotic
Metabolite: mosapramine

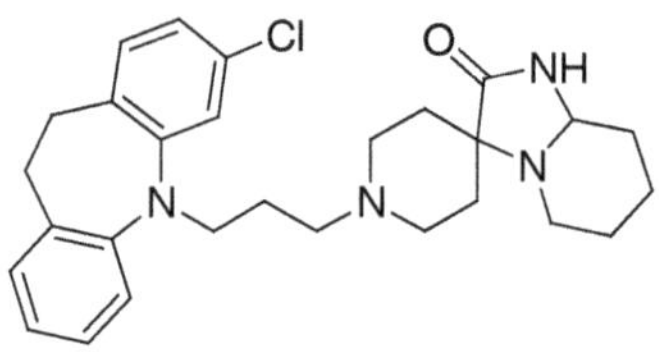

$t_{1/2}$	15±2 h	V	–
F	29±7%	CL	–
pb	–	Mwt	479.1
ur	–	PSA	38.8 Å^2
		log P	7.36

MOSAPRAMINE
Antipsychotic, metabolite of clocapramine
Metabolism: aromatic and benzylic hydroxylation, oxidation to imidazolone

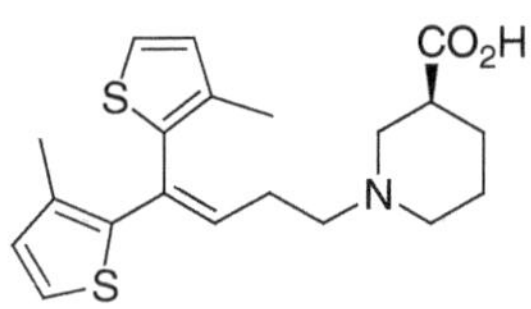

$t_{1/2}$	5–9 h	V	1
F	90%	CL	3
pb	96%	Mwt	375.6
ur	2%	PSA	97.0 Å^2
		log P	4.03

TIAGABINE
GABA reuptake inhibitor, antiepileptic
Metabolism (CYP3A4): glucuronidation, thiophene oxidation to 2-thiophenone

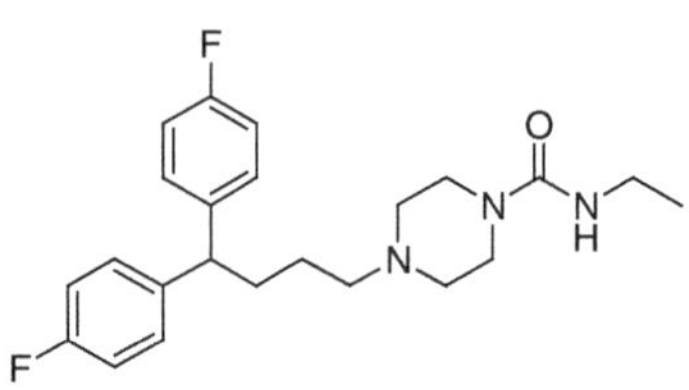

$t_{1/2}$	4 h (pig)	V	–
F	–	CL	–
pb	–	Mwt	401.5
ur	–	PSA	35.6 Å^2
		log P	4.12

AMPEROZIDE
Mixed $D_{2/5}$–HT_2 antagonist, atypical antipsychotic
Metabolism: N-deethylation

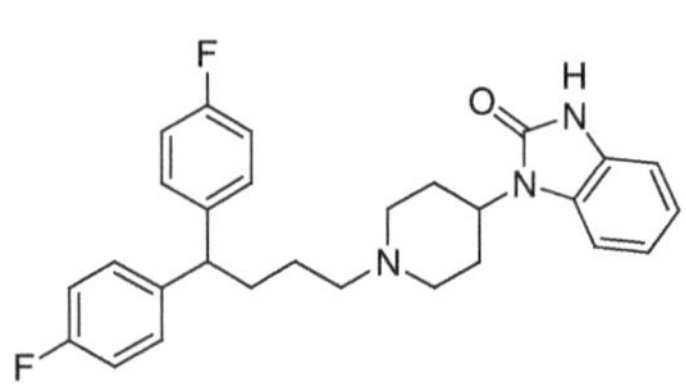

$t_{1/2}$	111±57 h	V/F	28±18
F	40–50%	CL	4.1±3.8
pb	99%	Mwt	461.6
ur	<1%	PSA	35.6 Å^2
		log P	5.76

PIMOZIDE
Dopamine antagonist, antipsychotic
Metabolism: N-dealkylation of piperidine

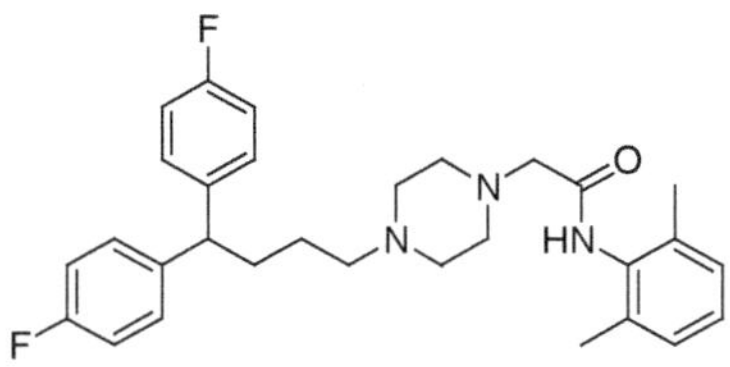

$t^1/_2$	24 h	V/F	–	LIDOFLAZINE
F	–	CL	–	Calcium channel blocker, vasodilator
pb	–	Mwt	491.6	Metabolism: N-dealkylation
ur	–	PSA	35.6 Å^2	
		log P	5.42	

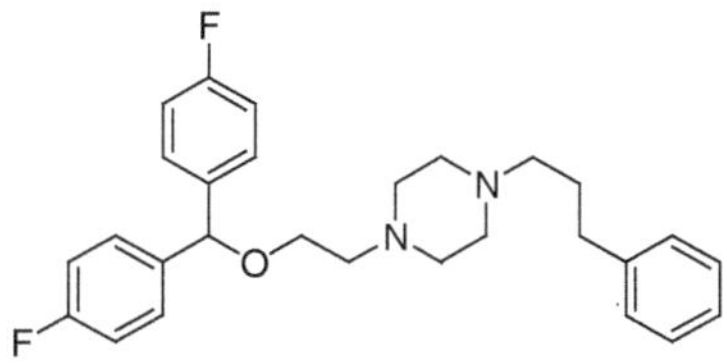

$t^1/_2$	54–66 h	V/F	–	VANOXERINE
F	–	CL	–	Dopamine reuptake inhibitor, antiarrhythmic, treatment of cocaine addiction
pb	>99%	Mwt	450.6	
ur	–	PSA	15.7 Å^2	
		log P	5.35	

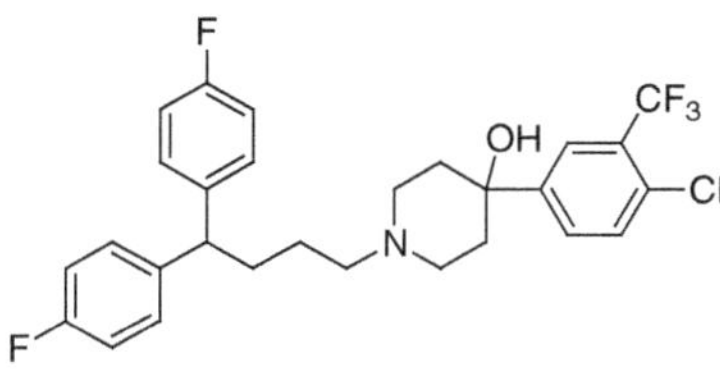

$t^1/_2$	120–199 h	V	–	PENFLURIDOL
F	–	CL	–	Antipsychotic
pb	98%	Mwt	524.0	Metabolism: N-dealkylation of piperidine to diarylbutyric acid
ur	5%	PSA	23.5 Å^2	
		log P	6.95	

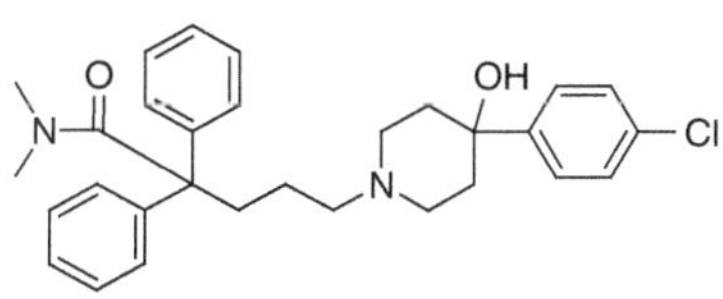

$t^1/_2$	11±2 h	V	–	LOPERAMIDE
F	40%	CL	–	Opioid agonist, antidiarrheal; does not cross bbb
pb	95–97%	Mwt	477.0	Metabolism: N-demethylation, N-oxidation
ur	1–2%	PSA	43.8 Å^2	
		log P	4.15	

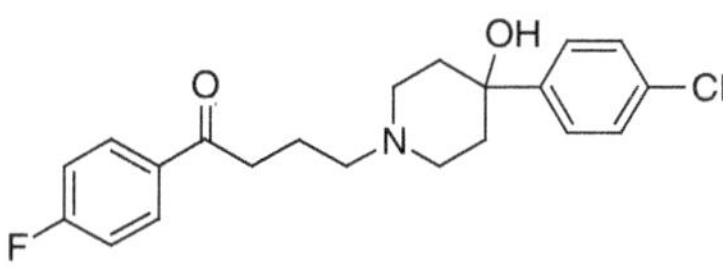

$t^1/_2$	18±5 h	V	18±7	HALOPERIDOL
F	60±18%	CL	12±3	Dopamine antagonist, antipsychotic; CYP2D6 inhibitor and inducer
pb	92±2%	Mwt	375.9	Metabolism: glucuronidation, N-dealkylation, dehydratation, and oxidation to pyridine, reversible reduction of ketone to alcohol ($t^1/_2$ 67±51 h, CL 10±5)
ur	1%	PSA	40.5 Å^2	
		log P	3.76	

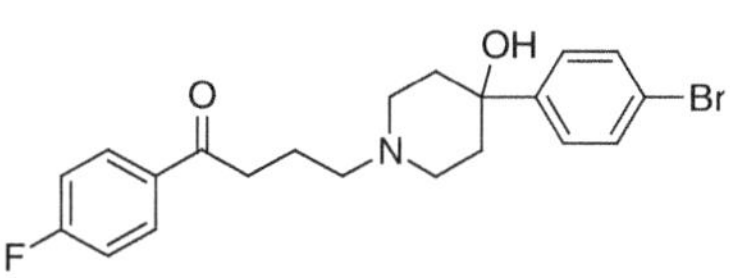

$t^1/_2$	25±10 h	V	–	BROMPERIDOL
F	50%	CL	–	Antipsychotic
pb	90%	Mwt	420.3	Metabolism: similar to haloperidol
ur	<1%	PSA	40.5 Å^2	
		log P	4.03	

(*continued overleaf*)

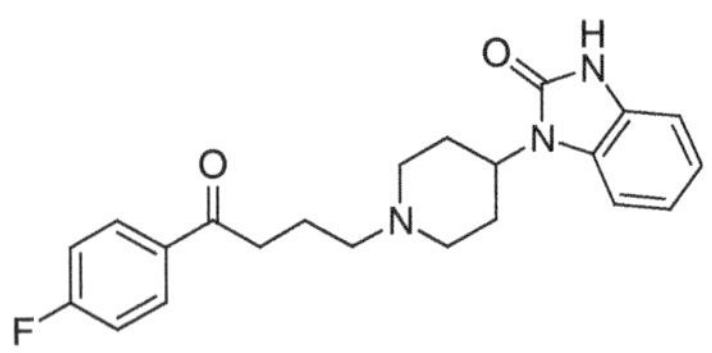	$t_{1/2}$ 5±3 h F 45±20% pb >90% ur 0.1%	V 5±2 CL 8±5 Mwt 381.4 PSA 52.7 Å^2 log P 3.50	BENPERIDOL Antipsychotic Metabolism: reversible reduction of ketone to alcohol
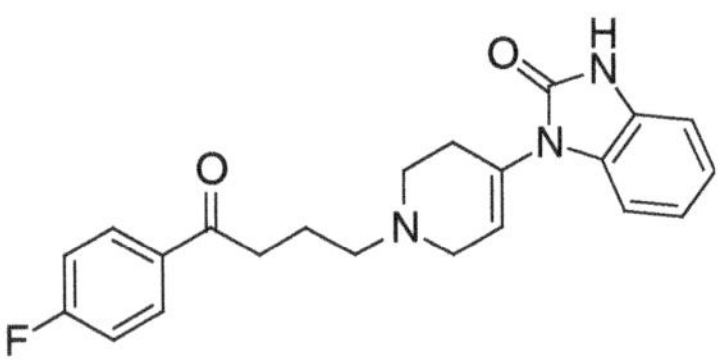	$t_{1/2}$ 1.7–2.2 h* F Low pb 85–90% ur 10%*	V 1.4±0.3 CL 13.9 Mwt 379.4 PSA 52.7 Å^2 log P 3.10	DROPERIDOL *iv Dopamine antagonist, antipsychotic Metabolism: N-dealkylation of tetrahydropyridine; withdrawn in 2001
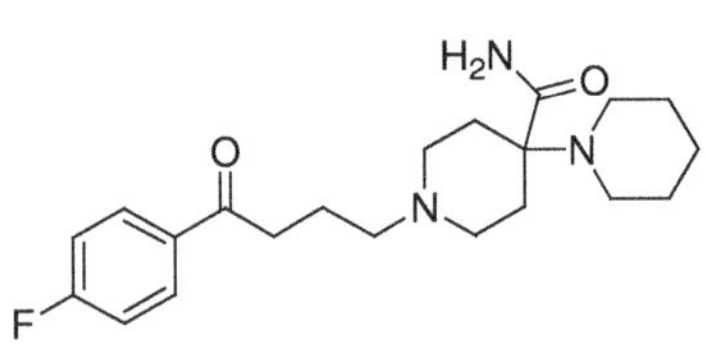	$t_{1/2}$ 13±5 h F – pb – ur –	V – CL – Mwt 375.5 PSA 66.6 Å^2 log P 2.47	PIPAMPERONE Antipsychotic Metabolism: N-dealkylation of piperidine, N-oxidation, reduction of ketone
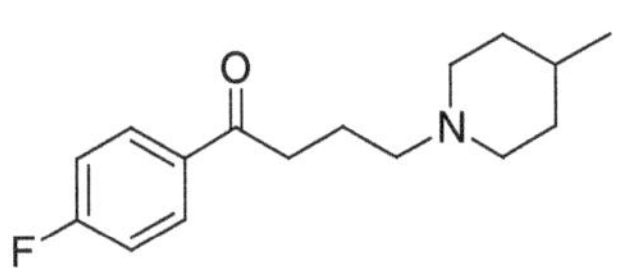	$t_{1/2}$ 4±2 h F 60% pb – ur –	V 14 CL 46 Mwt 263.4 PSA 20.3 Å^2 log P 3.76	MELPERONE Antipsychotic CYP2D6 inhibitor
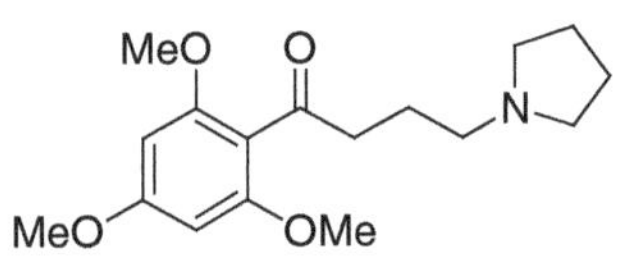	$t_{1/2}$ 1.5–4.3 h F 30–80% pb 60–80% ur 12–28%	V 1.2–1.6 CL 3.8–9.0 Mwt 307.4 PSA 48.0 Å^2 log P 3.13	BUFLOMEDIL Vasodilator Metabolism: 4-O-demethylation
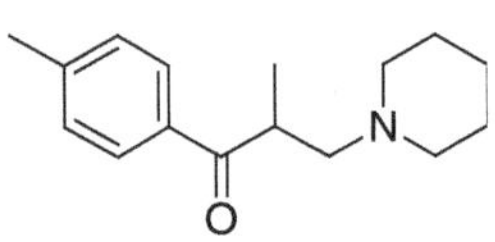	$t_{1/2}$ 1.6±0.7 h F 17–22% pb – ur –	V – CL 36±9 Mwt 245.4 PSA 20.3 Å^2 log P 3.01	TOLPERISONE Muscle relaxant Metabolism: benzylic hydroxylation, reduction of ketone to alcohol
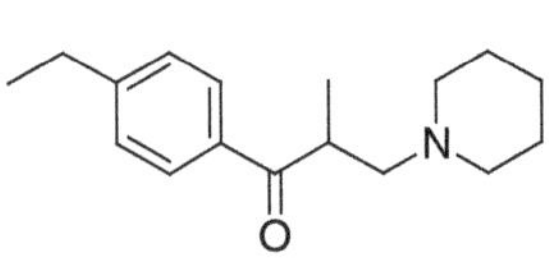	$t_{1/2}$ 1.6–1.8 h F Low pb – ur –	V – CL – Mwt 259.4 PSA 20.3 Å^2 log P 4.34	EPERISONE Muscle relaxant Metabolism: benzylic hydroxylation

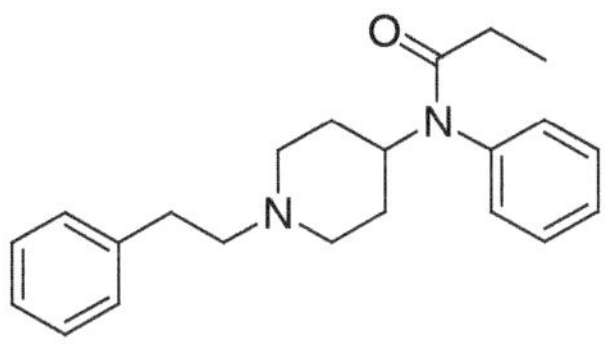	$t^1/_2$ 3.7±0.4 h *F* 50% pb 84±2% ur 6–8%	*V* 4.0±0.4 CL 13±2 Mwt 336.5 PSA 23.6 Å^2 log *P* 3.68	FENTANYL Analgesic, narcotic Metabolism: N-dealkylation of piperidine, amide hydrolysis
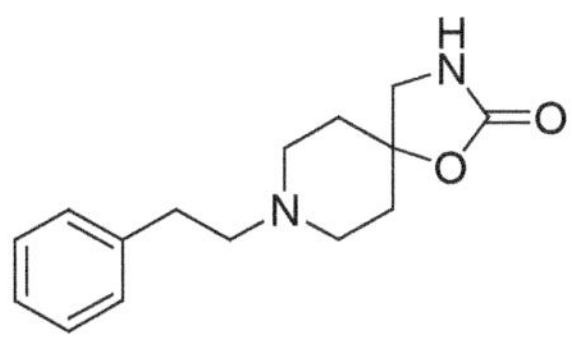	$t^1/_2$ 14–16 h *F* 90% pb – ur –	*V* 3.0 CL 2.6 Mwt 260.3 PSA 41.6 Å^2 log *P* 0.89	FENSPIRIDE Bronchodilator Metabolism: aromatic hydroxylation
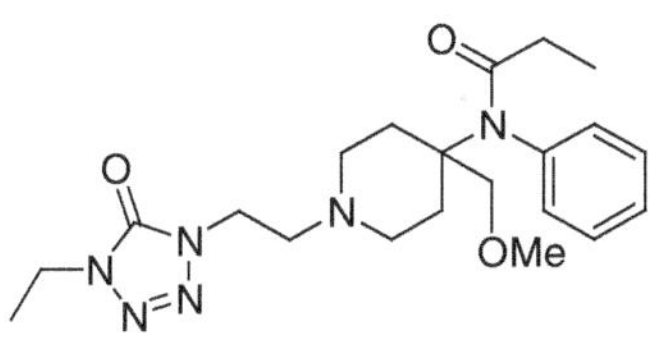	$t^1/_2$ 1.6±0.2 h *F* 42±15% pb 92±2% ur <1%	*V* 0.8±0.3 CL 6.7±2.4 Mwt 416.5 PSA 81.1 Å^2 log *P* 2.16	ALFENTANIL Analgesic Metabolism: N-dealkylation of piperidine and aniline
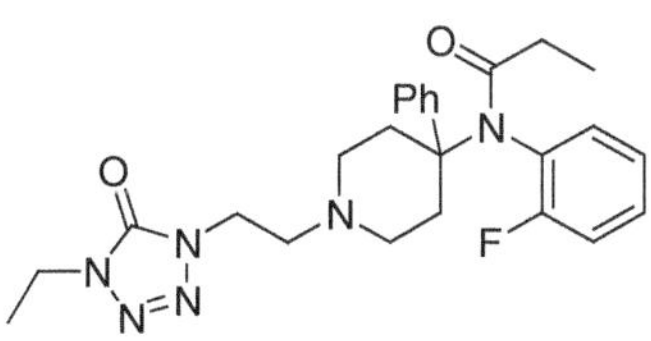	$t^1/_2$ 2 h *F* – pb – ur –	*V* 0.46 CL 5.6 Mwt 466.6 PSA 71.8 Å^2 log *P* 3.30	TREFENTANIL Analgesic
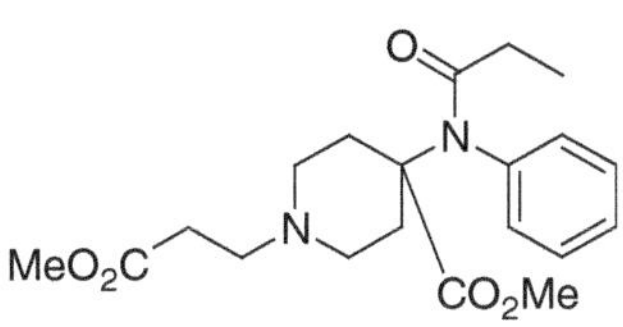	$t^1/_2$ 3–10 min *F* – pb 70% ur Negligible	*V* 0.3–0.6 CL 36–54 Mwt 376.5 PSA 76.2 Å^2 log *P* 2.05	REMIFENTANIL Opioid agonist, anesthetic for iv dosing Metabolite: carboxylic acid ($t^1/_2$ 2 h)
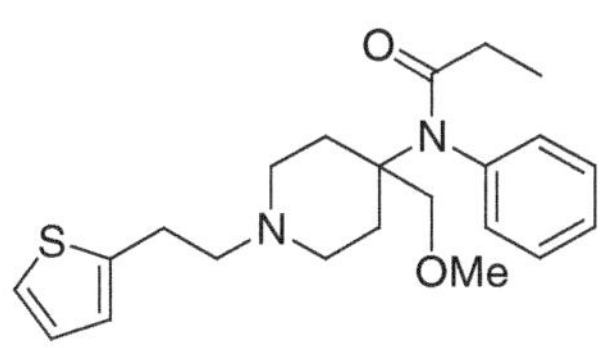	$t^1/_2$ 2.7±1.2 h *F* 90% pb 93±1% ur 6%	*V* 1.7±0.6 CL 12.7±2.5 Mwt 386.6 PSA 61.0 Å^2 log *P* 3.95	SUFENTANYL Analgesic Metabolism: N-dealkylation of piperidine and aniline, O-demethylation, aromatic hydroxylation

(*continued overleaf*)

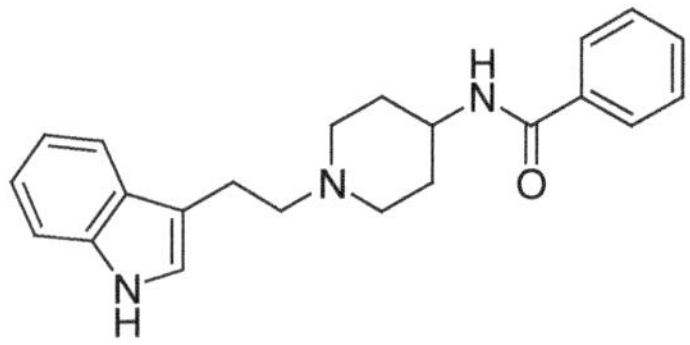

$t^{1}/_{2}$ 4–12 h	*V* 5–8		INDORAMIN
F 24–31%	CL 18–26		Antihypertensive
pb 72–92%	Mwt 347.5		Active metabolite: 6-hydroxylated indole
ur <11%	PSA 48.1 $Å^2$		
	log *P* 2.87		

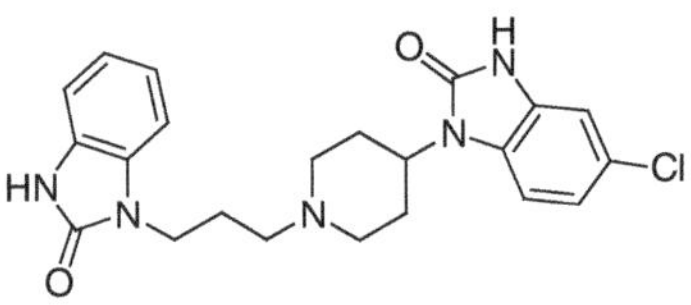

$t^{1}/_{2}$ 7–16 h	*V* 3–6	DOMPERIDONE
F 12–24%	CL 7–10	Dopamine antagonist, antiemetic, does not cross bbb readily
pb 91–93%	Mwt 425.9	Metabolism: aromatic 5-hydroxylation at nonchlorinated arene, oxidative cleavage to 1*H*-piperidine and 3-substituted propionic acid
ur 1.4%	PSA 67.9 $Å^2$	
	log *P* 4.05	

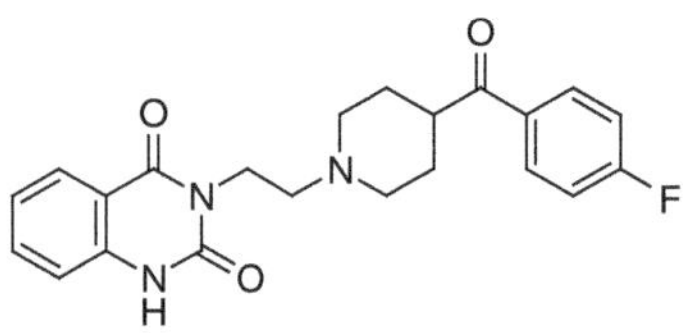

$t^{1}/_{2}$ 6–14 h	*V* 5–10	KETANSERIN
F 53%	CL 6–10	Serotonin antagonist, antihypertensive
pb 94%	Mwt 395.4	Metabolism: reversible reduction of ketone, N-dealkylation of piperidine, aromatic hydroxylation of quinazolinedione
ur <4%	PSA 69.7 $Å^2$	
	log *P* 3.56	

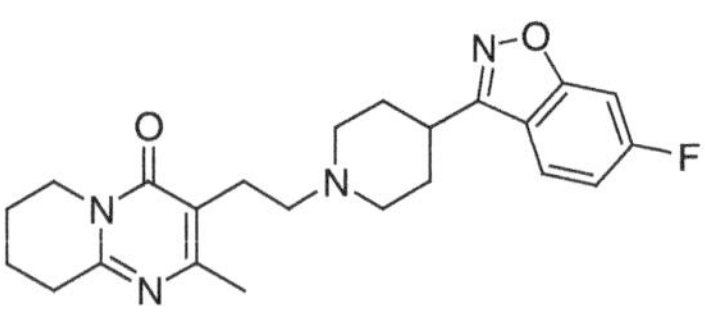

$t^{1}/_{2}$ 3.2±0.8 h	*V* 1.1±0.2	RISPERIDONE
F 66±28%	CL 5.4±1.4	Antipsychotic, CYP2D6 substrate
pb 89%	Mwt 410.5	Metabolism: hydroxylation to paliperidone (see below)
ur 3±2%	PSA 61.9 $Å^2$	
	log *P* 2.68	

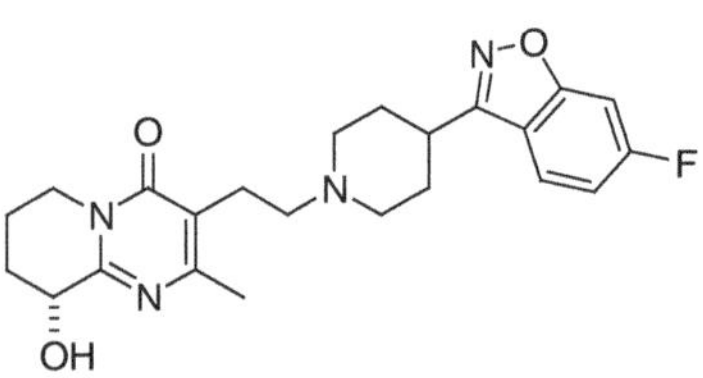

$t^{1}/_{2}$ 20±3 h	*V* 7.5	PALIPERIDONE
F 28%	CL –	Active metabolite of risperidone
pb 74–77%	Mwt 426.5	Metabolism: N-dealkylation, N–O bond cleavage, oxidation to ketone, aliphatic hydroxylation
ur 51–67%	PSA 82.2 $Å^2$	
	log *P* 1.52	

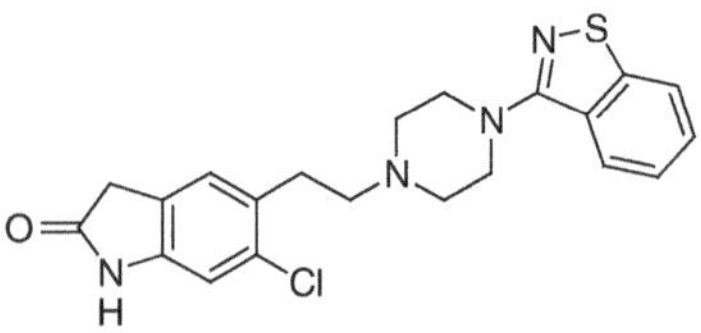

$t^{1}/_{2}$ 6±3 h	*V* 1.5–2.3	ZIPRASIDONE
F 45±15%	CL 11.7	Antipsychotic
pb >99%	Mwt 412.9	Metabolism: reductive N–S bond cleavage, then S-methylation, oxidation to sulfoxide, N-dealkylation of piperazine
ur <1%	PSA 76.7 $Å^2$	
	log *P* 4.00	

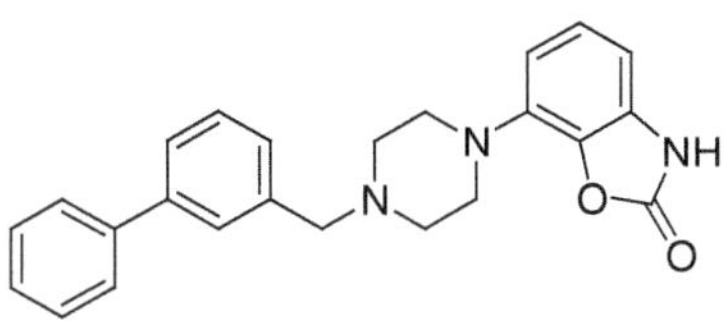	$t^{1}/_{2}$ 14.4 h *F* 54% pb >99% ur 2%	*V/F* 20 CL – Mwt 385.5 PSA 44.8 Å^2 log *P* 3.39	BIFEPRUNOX 5-HT$_{1A}$ agonist, antipsychotic
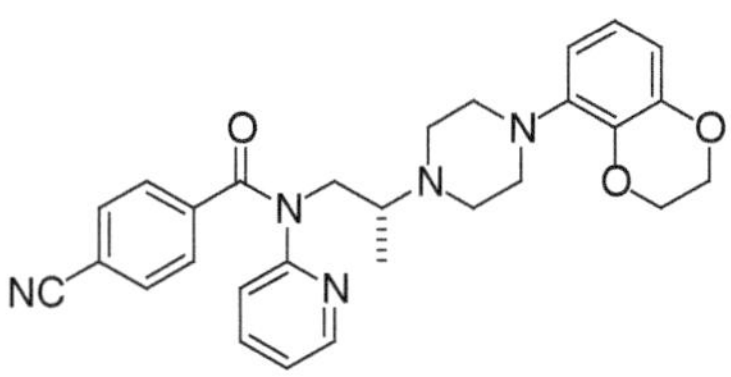	$t^{1}/_{2}$ 6–9 h *F* – pb >99.5% ur <1%	*V* 0.4–0.6 CL 0.7–0.8 Mwt 483.6 PSA 81.9 Å^2 log *P* 1.94	LECOZOTAN 5-HT$_{1A}$ antagonist for treatment of Alzheimer's disease Metabolism: N-dealkylation to eltoprazine
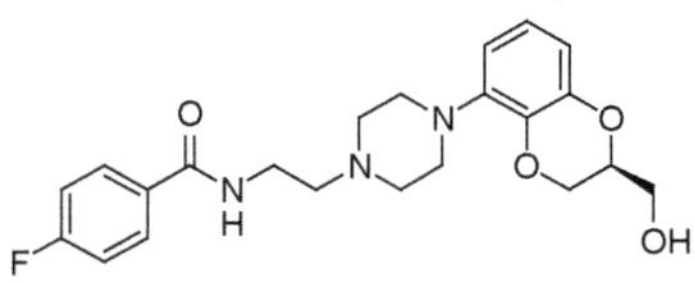	$t^{1}/_{2}$ 6–9 h *F* – pb – ur –	*V* – CL – Mwt 415.5 PSA 74.3 Å^2 log *P* 2.36	FLESINOXAN Antidepressant
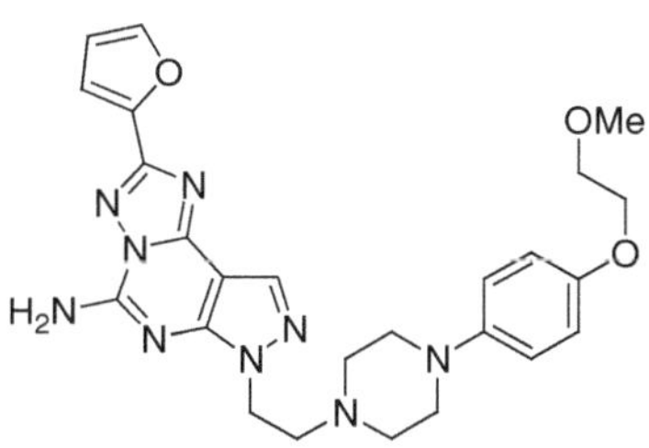	$t^{1}/_{2}$ 1.1 h* *F* 4%* pb – ur –	*V* 2.6** CL 37** Mwt 503.6 PSA 125 Å^2 log *P* 1.45	PRELADENANT *monkey **rat Adenosin A$_{2A}$ antagonist for treatment of Alzheimer's disease Metabolism: O-demethylation
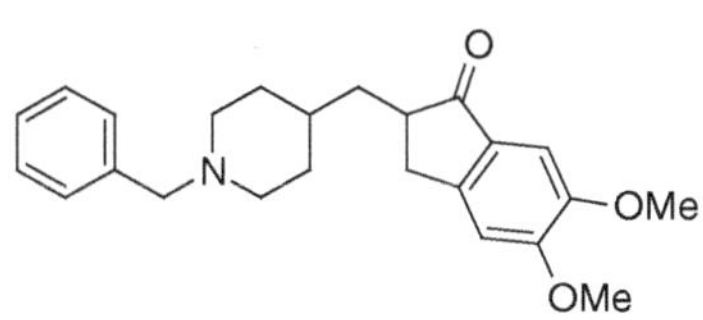	$t^{1}/_{2}$ 67–83 h *F* 100% pb 93–96% ur 10%	*V* 9–13 CL 1.8–2.2 Mwt 379.5 PSA 38.8 Å^2 log *P* 4.71	DONEPEZIL, E-2020 Acetylcholinesterase inhibitor for treatment of Alzheimer's disease Metabolism: O-dealkylation, 4-hydroxylation of phenyl, N-debenzylation
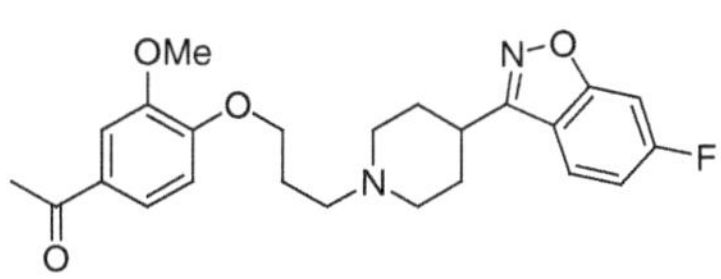	$t^{1}/_{2}$ 18–33 h *F* 96% pb 95% ur <1%	*V* 20–43 CL 12–26 Mwt 426.5 PSA 64.8 Å^2 log *P* 3.81	ILOPERIDONE Dopamine and serotonin antagonist, antipsychotic Metabolism: reduction of ketone to alcohol, O-demethylation, hydroxylation

(*continued overleaf*)

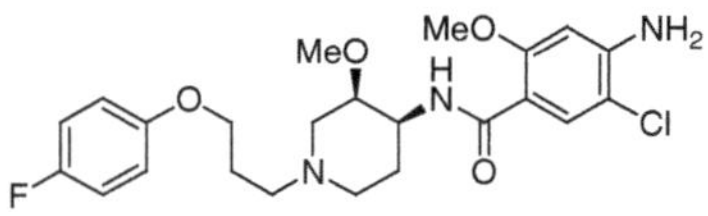	$t_{1/2}$ 7–10 h *F* 40–50% pb 98% ur 1–3%	*V* 2.4 CL – Mwt 466.0 PSA 86.1 Å^2 log *P* 2.83	CISAPRIDE Peristaltic stimulant Metabolism: N-dealkylation of piperidine, hydroxylation of phenoxy, cleavage of fluorophenyl ether, no amide hydrolysis; withdrawn in 2000 for causing cardiac arrhythmias
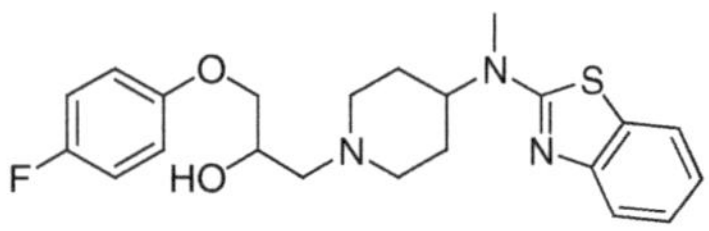	$t_{1/2}$ 18±10 h *F* – pb 99% ur <0.5%	*V* – CL 5.1 Mwt 415.5 PSA 77.1 Å^2 log *P* 4.26	SABELUZOLE Nootropic Metabolism: 6-hydroxylation of benzothiazole, glucuronidation
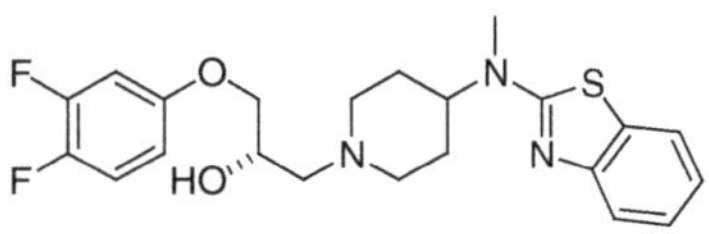	$t_{1/2}$ 15–24 h *F* – pb >99% ur –	*V* 2.5–3.4 CL 2.2–3.9 Mwt 433.5 PSA 77.1 Å^2 log *P* 4.69	LUBELUZOLE Neuroprotectant Metabolism: 6-hydroxylation of benzothiazole, glucuronidation
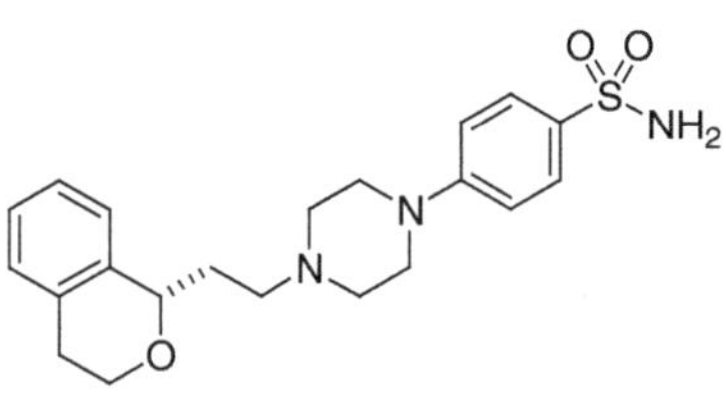	$t_{1/2}$ 14 h* *F* 76%* pb – ur Negligible	*V* – CL – Mwt 401.5 PSA 84.3 Å^2 log *P* 2.42	U-101387 *monkey Dopamine D_4 antagonist
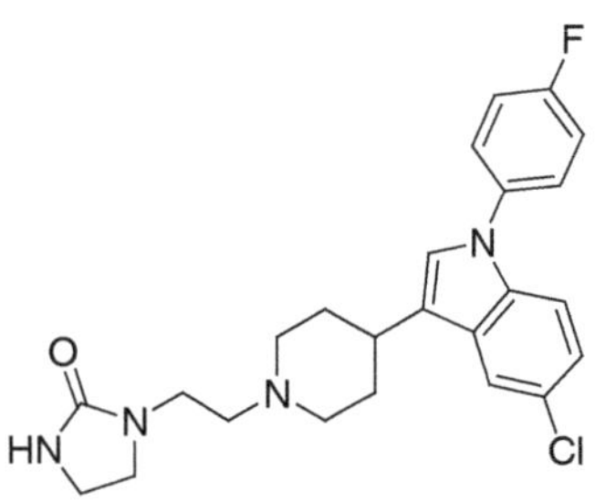	$t_{1/2}$ 85–102 h *F* 74% pb >99% ur <1%	*V* 20 CL 3.6 Mwt 440.9 PSA 40.5 Å^2 log *P* 3.67	SERTINDOLE Antipsychotic Metabolism: N-dealkylation of piperidine, dehydrogenation of imidazolinone
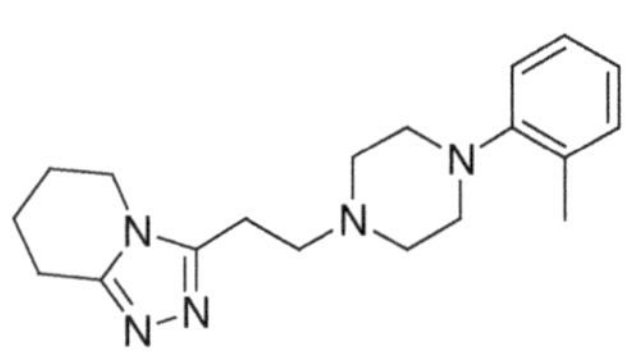	$t_{1/2}$ 2 h *F* – pb – ur <10%	*V* – CL – Mwt 325.5 PSA 37.2 Å^2 log *P* 2.89	DAPIPRAZOLE α-Adrenergic blocker, antiglaucoma

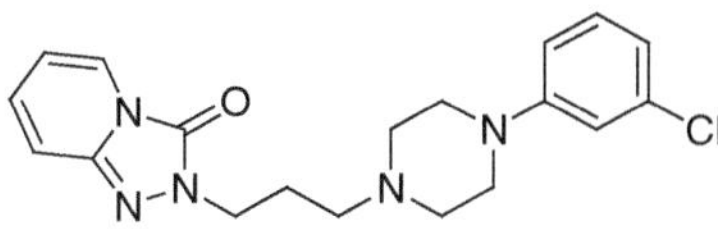	$t_{1/2}$ 6.5±1.8 h *F* 75±30% pb 93% ur <1%	*V* 1.0±0.3 CL 1.8±0.6 Mwt 371.9 PSA 42.4 Å^2 log *P* 2.76	TRAZODONE Antidepressant Metabolism: N-oxidation, hydroxylation, N-dealkylation to 1-(3-chlorophenyl)piperazine ($t_{1/2}$ 4–8 h)
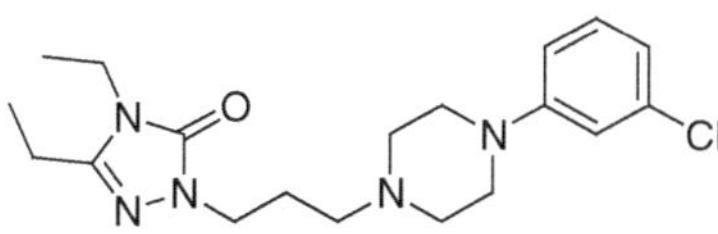	$t_{1/2}$ <22 h *F* – pb – ur Negligible	*V* – CL – Mwt 377.9 PSA 42.4 Å^2 log *P* 2.74	ETOPERIDONE Antidepressant Metabolism: similar to metabolism of nefazodone (see below)
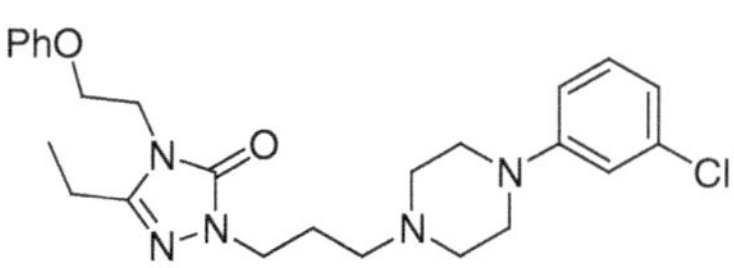	$t_{1/2}$ 3–10 h *F* 15–23% pb 99% ur –	*V* 0.22–0.87 CL 7.5 Mwt 470.0 PSA 51.6 Å^2 log *P* 4.09	NEFAZODONE Antidepressant Metabolism: 1-hydroxylation of ethyl group (to secondary alcohol: $t_{1/2}$ 1.4–6.5 h), deethylation to triazoledione, N-dealkylation of piperazine
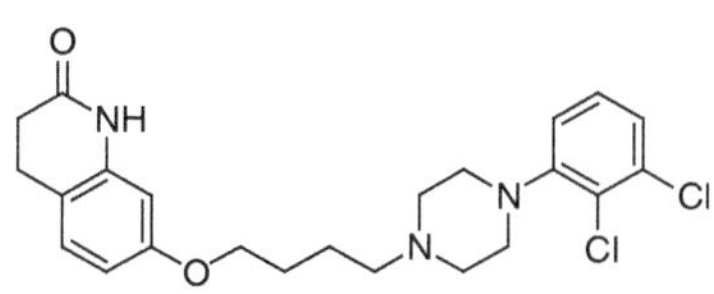	$t_{1/2}$ 75–146 h *F* 87% pb 99% ur <1%	*V* 4.9 CL 0.7 Mwt 448.4 PSA 44.8 Å^2 log *P* 5.60	ARIPIPRAZOLE Antidepressant, antipsychotic, crosses bbb Metabolism: dehydrogenation of piperidinone, N-dealkylation, N-oxidation
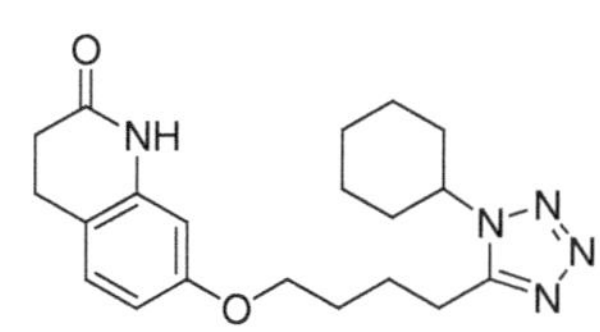	$t_{1/2}$ 9–21 h *F* 87–100% pb 95–98% ur <1%	*V/F* 2.8 CL/*F* 3 Mwt 369.5 PSA 81.9 Å^2 log *P* 3.40	CILOSTAZOL PDE III inhibitor, antithrombotic Metabolites: 4′-*trans*-hydroxy, 3,4-dehydro
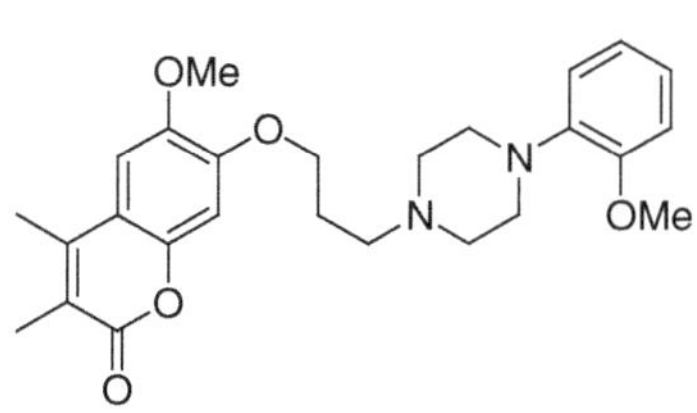	$t_{1/2}$ 11–18 h *F* – pb High ur –	*V* – CL – Mwt 452.6 PSA 64.4 Å^2 log *P* 4.29	ENSACULIN, ANSECULIN, KA-672 Nootropic Metabolism: aromatic hydroxylation
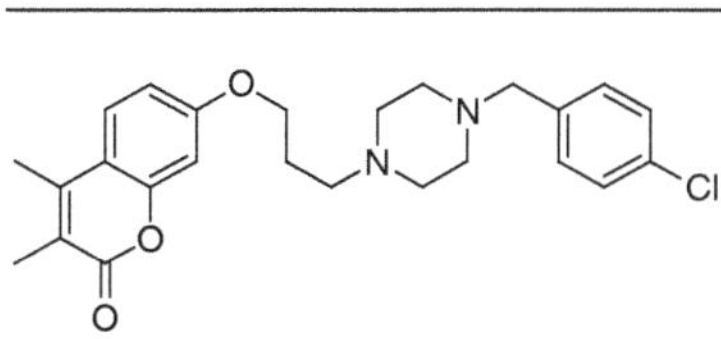	$t_{1/2}$ 10–26 h *F* 57% pb – ur Negligible	*V* 2 CL 1.5 Mwt 441.0 PSA 42.0 Å^2 log *P* 4.13	PICUMAST Antiallergic Metabolism: N-dealkylation, hydroxylation of 3-CH_3, then oxidation to carboxylic acid

(*continued overleaf*)

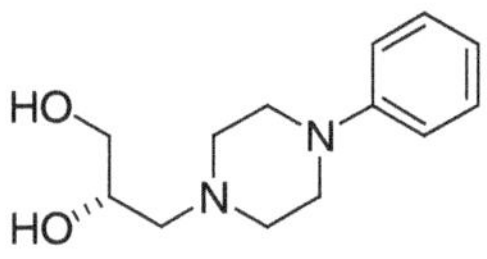	$t^1/_2$ 1–2 h F 75% pb 11–14% ur –	V – CL – Mwt 236.3 PSA 46.9 Å^2 log P −0.03	LEVODROPROPIZINE Antiussive Metabolism: aromatic 4-hydroxylation
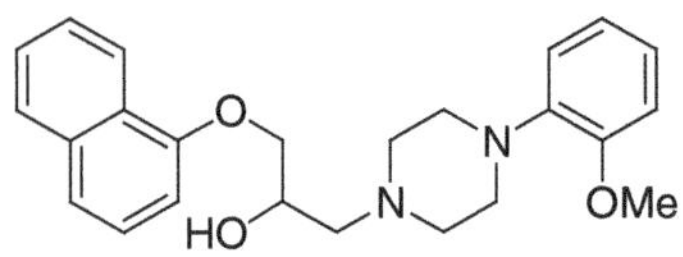	$t^1/_2$ 5.4±3.2 h F 7–30% pb 95–97% ur –	V – CL 11±1.6 Mwt 392.5 PSA 45.2 Å^2 log P 4.1	NAFTOPIDIL Antihypertensive Metabolism: aromatic hydroxylation, piperazine dealkylation, O-demethylation
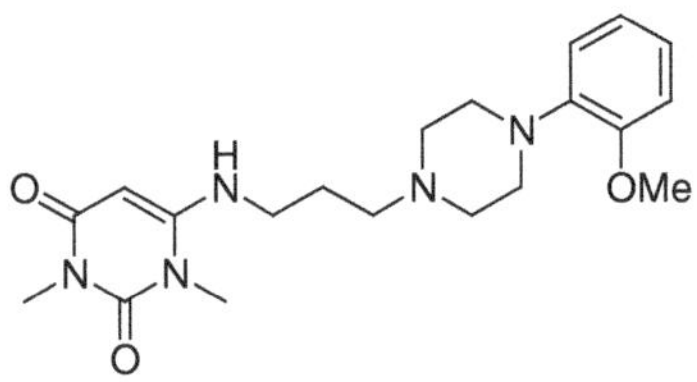	$t^1/_2$ 2–5 h F 63–80% pb 75–80% ur 10–15%	V 0.4–0.8 CL 1.8–3.8 Mwt 387.5 PSA 68.4 Å^2 log P 1.64	URAPIDIL Antihypertensive Metabolites: 4-hydroxy-2-methoxyphenyl (inactive, $t^1/_2$ 6–10 h), *N*-desmethyl (active, $t^1/_2$ 10–16 h)
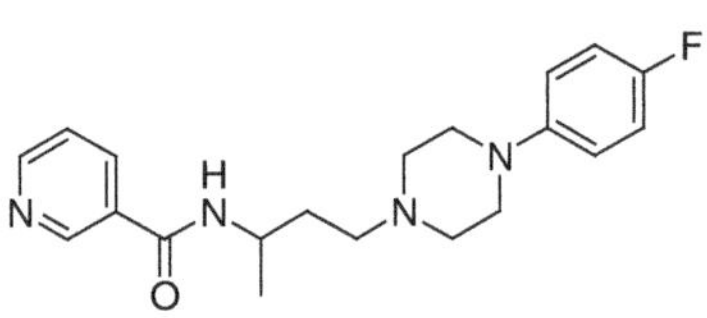	$t^1/_2$ 4.5±1.0 h F – pb – ur –	V – CL – Mwt 356.4 PSA 48.5 Å^2 log P 2.16	NIAPRAZINE Histamine H_1 antagonist Metabolism: N-dealkylation, N-oxidation, N-dearylation, aromatic hydroxylation
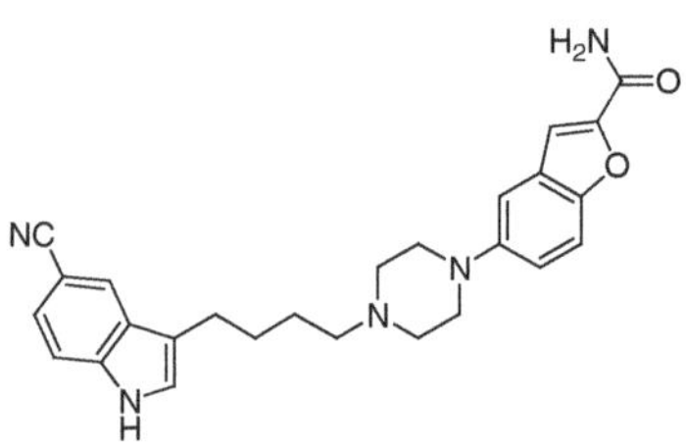	$t^1/_2$ 20–26 h F – pb 96–99% ur 0.2–1.0%	V/F 11–17 CL/F 5.6–6.9 Mwt 441.5 PSA 102 Å^2 log P 2.59	VILAZODONE 5-HT reuptake inhibitor, 5-HT_{1A} agonist, antidepressant Metabolism: hydroxylation to 6-hydroxyindole, then glucuronidation
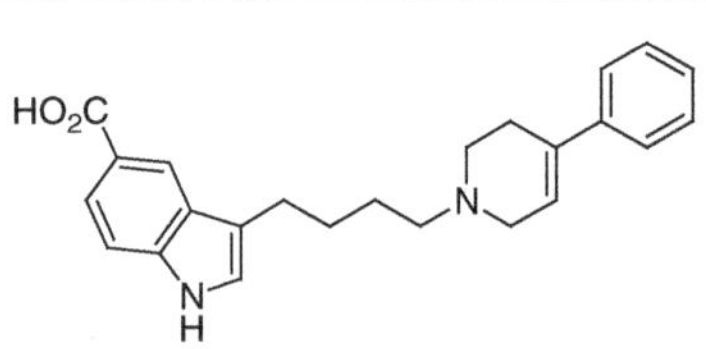	$t^1/_2$ 5.5 h F – pb – ur –	V – CL – Mwt 374.5 PSA 56.3 Å^2 log P 4.60	CARMOXIROLE Antihypertensive, dopamine D_{A2} agonist, does not cross bbb readily Metabolite: glucuronide
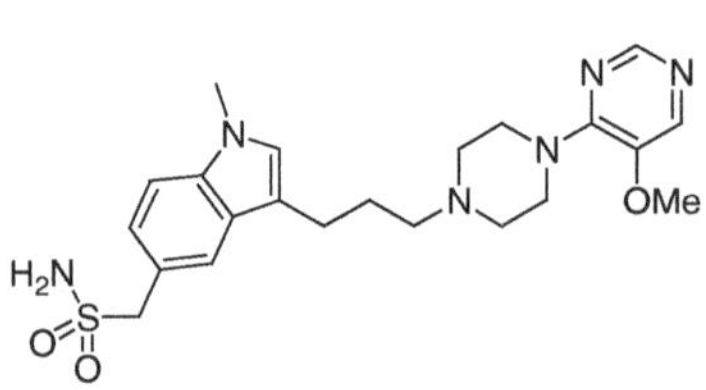	$t^1/_2$ 5–10 h F 17% pb – ur 2–4%	V 0.9 CL 6.4 Mwt 458.6 PSA 112 Å^2 log P 1.58	AVITRIPTAN Antimigraine, 5-HT_1-receptor agonist Active metabolites: *N*-desmethyl ($t^1/_2$ 11 h), *O*-desmethyl ($t^1/_2$ 5.2 h)

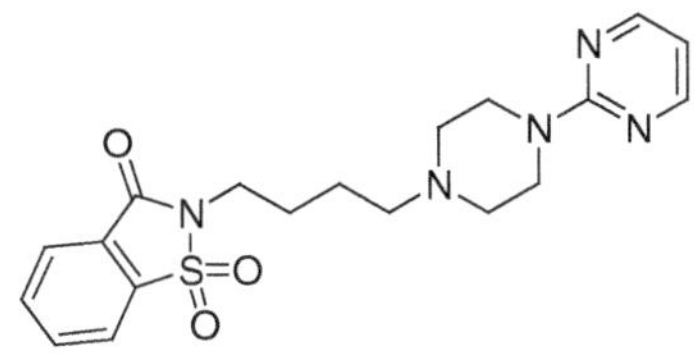	$t_{1/2}$ 1–3 h *F* 10% (dog) pb – ur –	*V* – CL – Mwt 401.5 PSA 95.1 Å^2 log *P* 0.97	IPSAPIRONE Anxiolytic Metabolism: N-dealkylation of piperazine
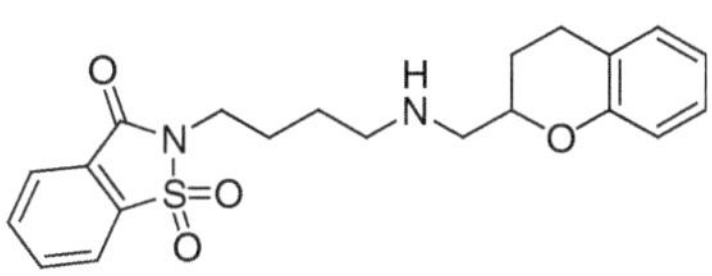	$t_{1/2}$ 0.8–1.8 h* *F* – pb – ur –	*V* 0.2 CL 2.2 Mwt 400.5 PSA 84.1 Å^2 log *P* 2.85	REPINOTAN *iv 5-HT_{1A} agonist, neuroprotectant, crosses bbb Metabolism: chroman 6-hydroxylation
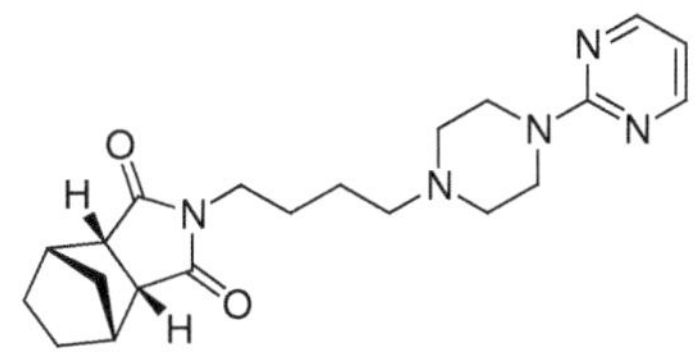	$t_{1/2}$ 1.2–1.4 h *F* Low pb – ur –	*V* – CL – Mwt 383.5 PSA 69.6 Å^2 log *P* 1.09	TANDOSPIRONE Anxiolytic, antidepressant Metabolism: hydroxylation of pyrimidine and norbornane, N-dealkylation of piperazine
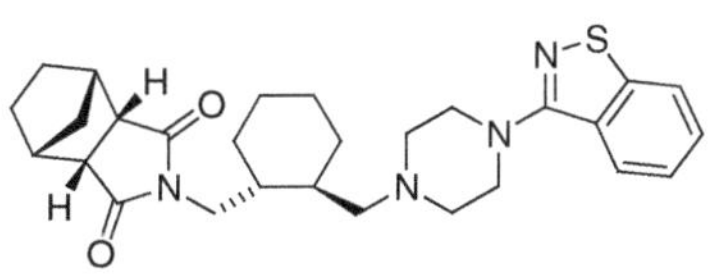	$t_{1/2}$ 12–18 h *F* 9–19% pb >99% ur 0.1%	*V* 95 CL 60 Mwt 492.7 PSA 85.0 Å^2 log *P* 4.52	LURASIDONE Antipsychotic Metabolism: N-dealkylation, hydroxylation of norbornane, S-oxidation
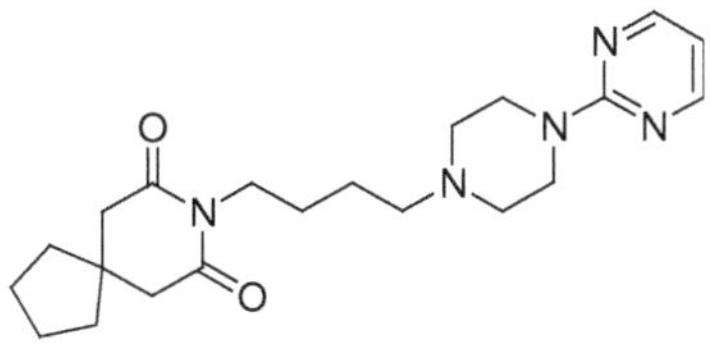	$t_{1/2}$ 2–11 h *F* 4% pb 95% ur <1%	*V* 5.3±2.6 CL 28–45 Mwt 385.5 PSA 69.6 Å^2 log *P* 1.59	BUSPIRONE Anxiolytic Metabolism: hydroxylation of pyrimidine at position 5 and α to the carbonyl groups, N-dealkylation of piperazine
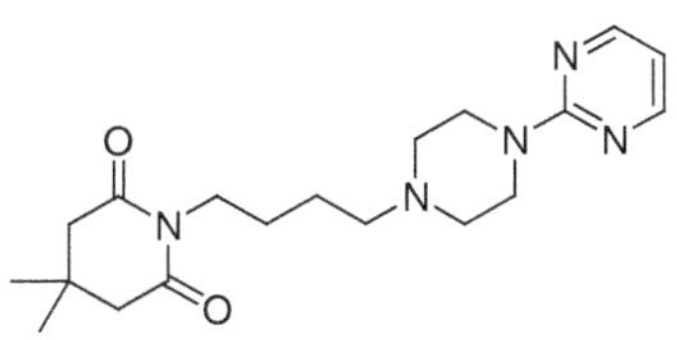	$t_{1/2}$ 3±2 h *F* 15% pb – ur –	*V* 4.8 CL – Mwt 359.5 PSA 69.6 Å^2 log *P* 1.06	GEPIRONE Anxiolytic Metabolism: as buspirone
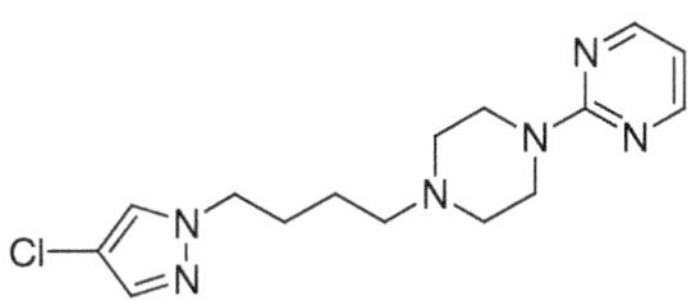	$t_{1/2}$ 1–6 h *F* 10% (dog) pb – ur –	*V* – CL – Mwt 320.8 PSA 45.2 Å^2 log *P* 2.68	LESOPITRON Anxiolytic Metabolism: pyrimidine 5-hydroxylation

(*continued overleaf*)

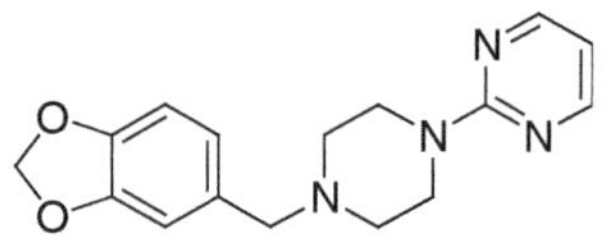	$t^1/_2$ 2–7 h *F* 10% pb – ur 0%	*V* – CL – Mwt 298.3 PSA 50.7 Å^2 log *P* 2.43	PIRIBEDIL Dopamine agonist, vasodilator Metabolism: acetal hydrolysis to catechol, pyrimidine hydroxylation to 4-hydroxypyrimidine, N-oxidation of pyrimidine
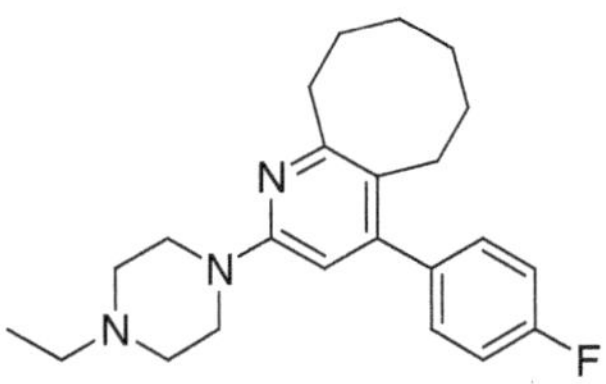	$t^1/_2$ 11±4 h *F* – pb >99.7% ur 0%	*V* – CL – Mwt 367.5 PSA 19.4 Å^2 log *P* 6.03	BLONANSERIN 5-HT_{2A}/D_2 antagonist, atypical antipsychotic Metabolism: hydroxylation at $(CH_2)_2\mathit{CH_2}(CH_2)_2$
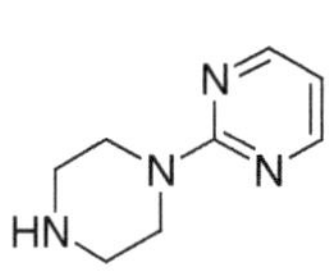	$t^1/_2$ 6±2 h *F* – pb – ur –	*V* – CL – Mwt 164.2 PSA 41.1 Å^2 log *P* 0.49	1-(2-PYRIMIDINYL)PIPERAZINE Active metabolite of ipsapirone, gepirone, tandospirone, buspirone
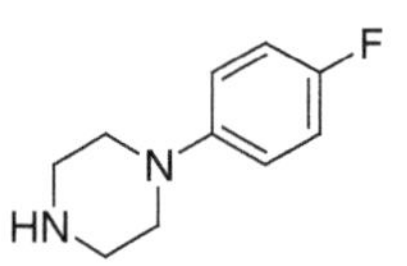	$t^1/_2$ 6–8 h *F* – pb – ur –	*V* – CL – Mwt 180.2 PSA 15.3 Å^2 log *P* 1.44	1-(4-FLUOROPHENYL) PIPERAZINE Active metabolite of niaprazine
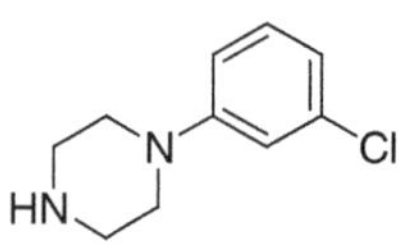	$t^1/_2$ 3–6 h *F* 12–84% pb – ur –	*V* 2.5 CL 6.1 Mwt 196.7 PSA 15.3 Å^2 log *P* 1.74	1-(3-CHLOROPHENYL) PIPERAZINE Recreational drug, causes headaches Metabolism: aromatic hydroxylation, oxidative degradation of piperazine
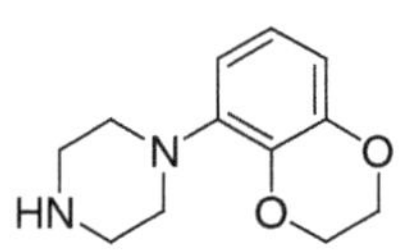	$t^1/_2$ 7–9 h *F* 100% pb 15% ur 40%	*V* 3.5 CL 8 Mwt 220.3 PSA 33.7 Å^2 log *P* 0.86	ELTOPRAZINE 5-$HT_{1/2C}$ agonist, serenic
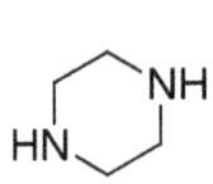	$t^1/_2$ variable *F* 75% pb 60–70% ur 20%	*V* – CL – Mwt 86.1 PSA 24.1 Å^2 log *P* −1.50	PIPERAZINE Anthelmintic Metabolites: *N*-nitroso, *N*-nitroso-3-hydroxy

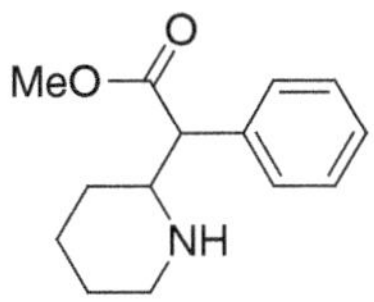	$t_{1/2}$ 1.9–3.5 h *F* 11–52% pb 10–33% ur <1%	*V* 6 CL 0.78–4.8 Mwt 233.3 PSA 38.3 Å^2 log *P* 2.31	METHYLPHENIDATE CNS stimulant, crosses bbb Metabolism: ester hydrolysis to ritalinic acid, oxidation to lactam
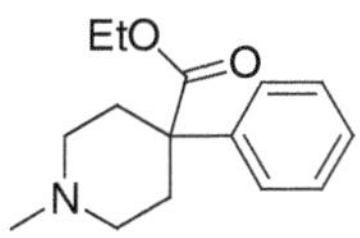	$t_{1/2}$ 3–5 h *F* 50–60% pb 65–75% ur 1–25%	*V* 4.4±0.9 CL 17±5 Mwt 247.33 PSA 29.5 Å^2 log *P* 2.19	MEPERIDINE, PETHIDINE Analgesic, narcotic; Metabolism: N-demethylation (to normeperidine, $t_{1/2}$ 14–21 h), hydrolyis of ester, aromatic 4-hydroxylation
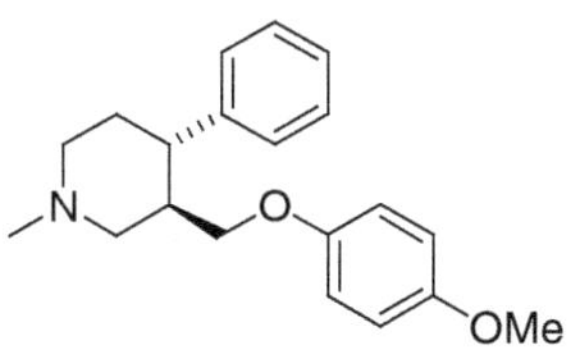	$t_{1/2}$ 7–27 h *F* 5–10% pb – ur <2%	*V* – CL – Mwt 311.4 PSA 21.7 Å^2 log *P* 3.89	FEMOXETINE Antidepressant (SRI); high first-pass effect Metabolism: N- and O-demethylation, aromatic hydroxylation
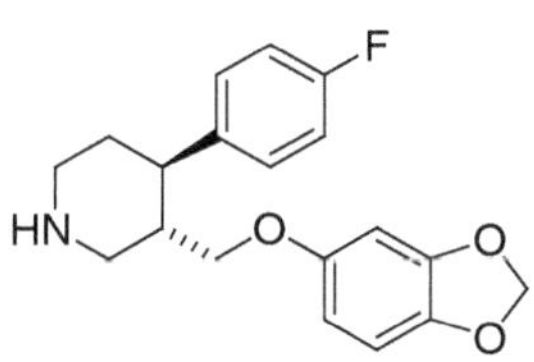	$t_{1/2}$ 17±3 h *F* 50% pb 95% ur <2%	*V* 17±10 CL/*F* 8.6±3.2 Mwt 329.4 PSA 39.7 Å^2 log *P* 3.70	PAROXETINE Antidepressant; CYP2D6 substrate and inhibitor Metabolism: O-demethylenation, then O-methylation (COMT), O-dearylation to hydroxymethylpiperidine
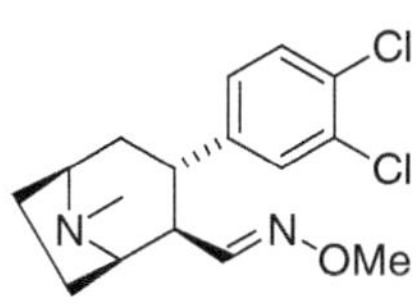	$t_{1/2}$ 24 h *F* 0.8%* pb – ur 0%	*V* 46* CL 32* Mwt 327.3 PSA 24.8 Å^2 log *P* 3.15	BRASOFENSINE *monkey Dopamine reuptake inhibitor for treatment of Parkinson's disease Metabolism: *E* to *Z* isomerization of oxime, N- and O-demethylation
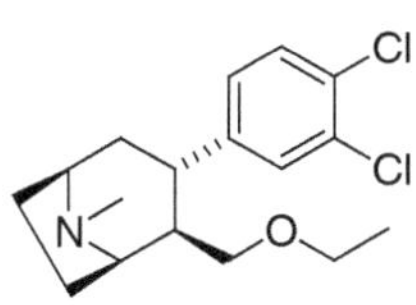	$t_{1/2}$ 8 d *F* >90% pb – ur –	*V* 8.6 CL 0.5–0.6 Mwt 328.3 PSA 12.5 Å^2 log *P* 3.94	TESOFENSINE Monoamine reuptake inhibitor; appetite suppressant Active metabolite: *N*-desmethyl ($t_{1/2}$ 14 d)

(*continued overleaf*)

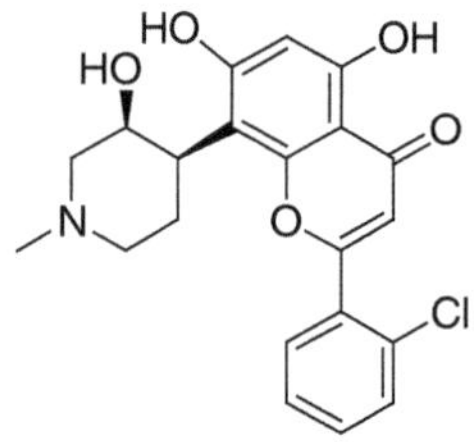	$t^1\!/_2$ 5.2 h (iv) F 20% (rat) pb 94% ur –	V 1.5 CL 6.1 Mwt 401.8 PSA 90.2 Å^2 log P 1.93	FLAVOPIRIDOL Antineoplastic Metabolism: glucuronidation

$t_{1/2}$, plasma half-life; F, oral bioavailability; pb, plasma protein binding; ur, excretion of unchanged drug in urine; V, volume of distribution; CL, clearance; Mwt, molecular weight; PSA, polar surface area; COMT, catechol-*O*-methyltransferase; GABA, γ-aminobutyric acid; 5-HT, 5-hydroxytryptamine; PDE, phosphodiesterase; sc, subcutaneous.

Table 46.2 (Arylalkyl)piperazines and -piperidines. V in l kg^{-1}; CL in ml min^{-1} kg^{-1}; Mwt in g mol^{-1}.

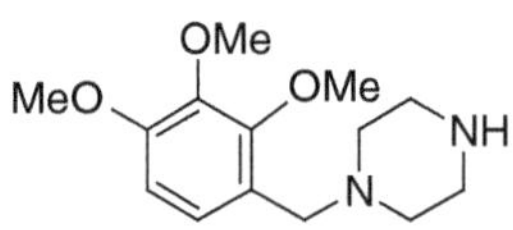	$t^1\!/_2$ 6 h F 89% pb 16% ur 62%	V 4.6 CL 0.9 Mwt 266.3 PSA 43.0 Å^2 log P −0.19	TRIMETAZIDINE Antianginal Metabolism: O-demethylation
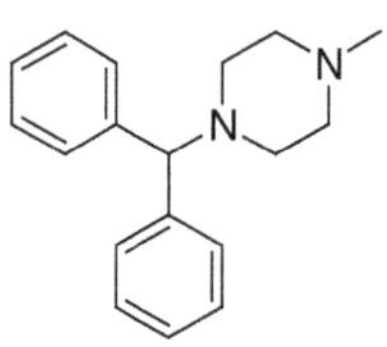	$t^1\!/_2$ 7–24 h F 75% pb 60% ur <1%	V 16±3 CL 0.2 Mwt 266.4 PSA 6.48 Å^2 log P 2.47	CYCLIZINE Antiemetic Duration of action: 5±1 h Metabolism: N-demethylation to inactive norcyclizine ($t^1\!/_2$ 20 h, pb 59%)
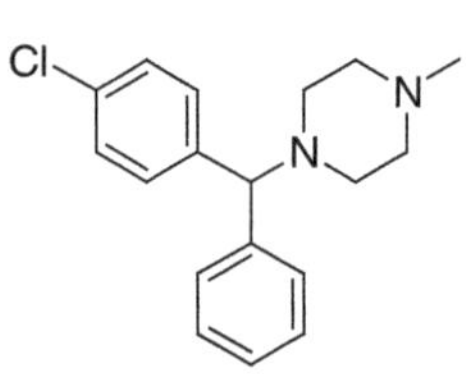	$t^1\!/_2$ 12 h F – pb – ur –	V – CL – Mwt 300.8 PSA 6.48 Å^2 log P 3.24	CHLORCYCLIZINE Histamine H_1 antagonist Metabolism: N-demethylation to norchlorcyclizine ($t^1\!/_2$ 6–9 d, pb 88%)
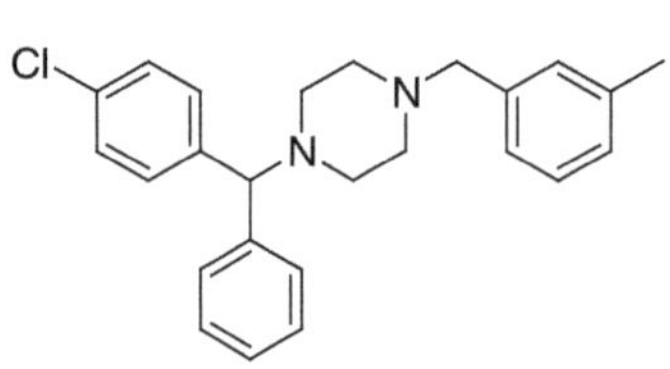	$t^1\!/_2$ 6 h F 22% (dog) pb – ur –	V – CL – Mwt 391.0 PSA 6.48 Å^2 log P 5.28	MECLIZINE, MECLOZINE Histamine H_1 antagonist, antiemetic Metabolism: N-debenzylation

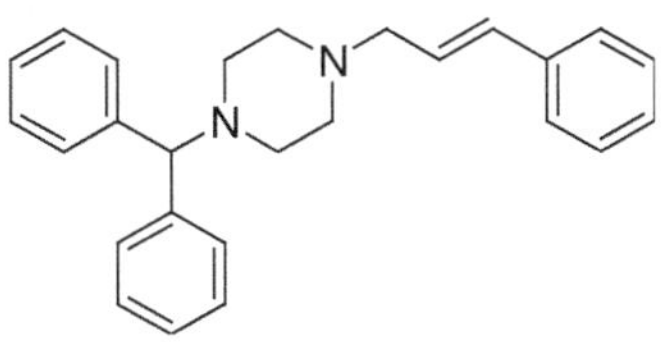

$t_{1/2}$	5±2 h	V	–	CINNARIZINE Histamine H_1 antagonist, antiemetic Metabolism: aromatic 4-hydroxylation, N-oxidation, piperazine dealkylation
F	75–100%	CL	–	
pb	91%	Mwt	368.5	
ur	<1%	PSA	6.48 Å^2	
		log P	5.03	

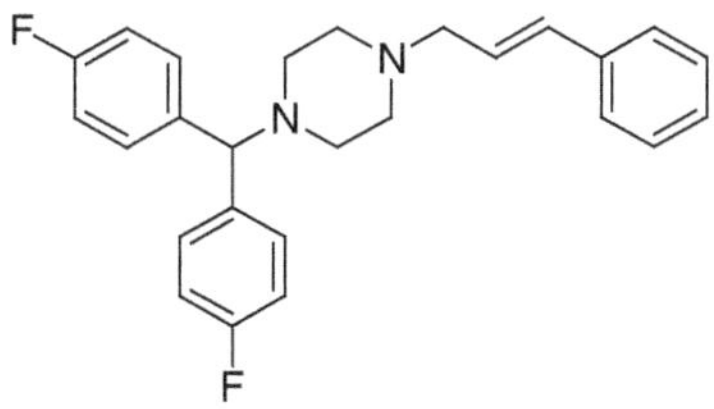

$t_{1/2}$	17–19 d	V	43–78	FLUNARIZINE Ca channel blocker, antimigraine Metabolism: aromatic 4-hydroxylation (cinnamyl group), N-oxidation, piperazine dealkylation
F	85%	CL	6.8	
pb	90%	Mwt	404.5	
ur	<0.2%	PSA	6.48 Å^2	
		log P	5.59	

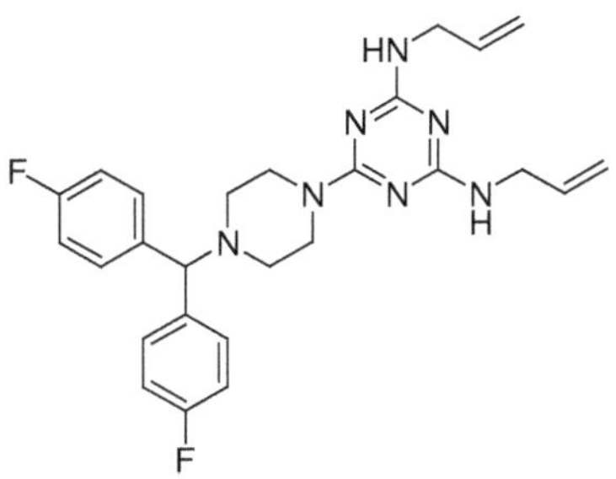

$t_{1/2}$	55±16 d	V	15	ALMITRINE Respiratory stimulant Metabolism: dihydroxylation of vinyl groups, deallylation
F	70%	CL	4±2	
pb	99.5%	Mwt	477.6	
ur	–	PSA	69.2 Å^2	
		log P	6.11	

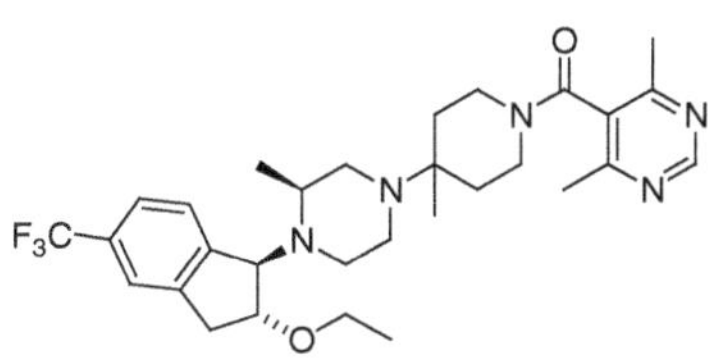

$t_{1/2}$	42–56 h	V	–	INCB 9471 CC chemokine receptor 5 (CCR5) antagonist, antiviral (HIV) Metabolism: O-deethylation
F	>95%	CL	–	
pb	84%	Mwt	559.7	
ur	–	PSA	61.8 Å^2	
		log P	3.45	

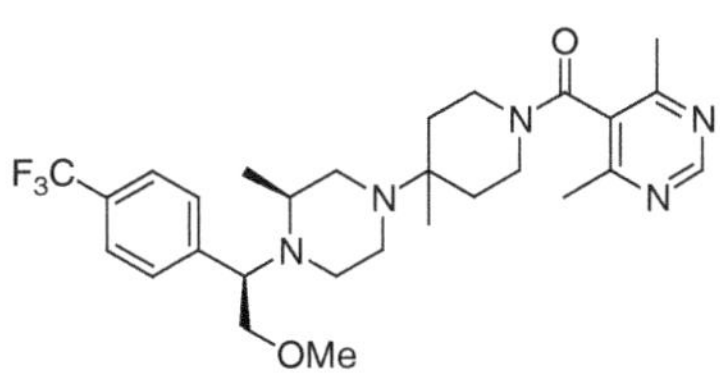

$t_{1/2}$	28–33 h	V	12	VICRIVIROC *monkey CCR5 antagonist, antiviral (HIV) Metabolism: N-oxidation of pyrimidine, conversion of pyrimidine to pyrazole-*N*-oxide, O-demethylation, ring opening of piperazine by N-dealkylation, benzylic C–N bond cleavage, oxidation of CH_2OMe to CO_2H
F	89–100%*	CL	5.4	
pb	84%	Mwt	533.6	
ur	4%	PSA	61.8 Å^2	
		log P	2.00	

(*continued overleaf*)

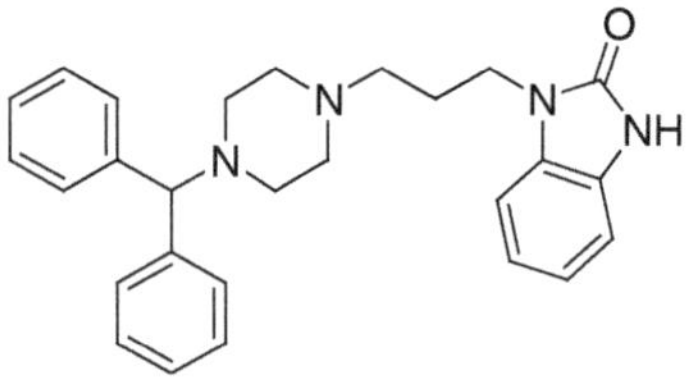

$t^1/_2$	17±3 h	V	37±4
F	–	CL	–
pb	91–98%	Mwt	426.6
ur	<0.1%	PSA	38.8 Å^2
		log P	4.34

OXATOMIDE
Histamine H_1 antagonist, antiemetic
Metabolism: N-dealkylation, aromatic hydroxylation of benzimidazolone

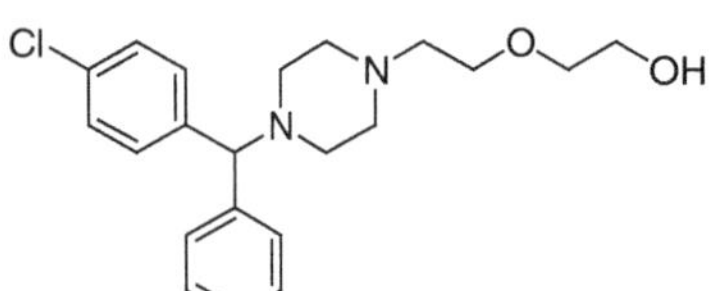

$t^1/_2$	7–20 h	V	16±3
F	–	CL	13±4
pb	93%	Mwt	374.9
ur	0.8%	PSA	35.9 Å^2
		log P	2.32

HYDROXYZINE
Histamine H_1 antagonist, antipruritic
Main metabolite: cetirizine (see below)

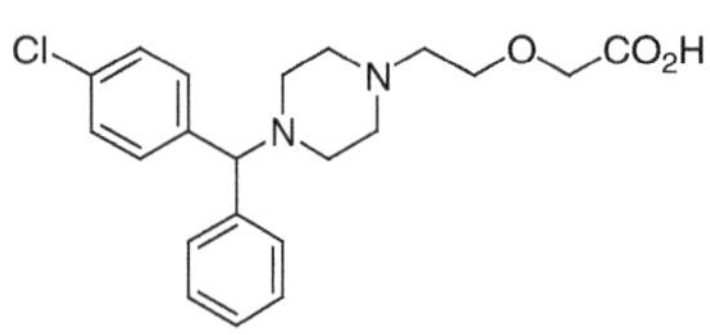

$t^1/_2$	9±2 h	V	0.5–0.8
F	60–70%	CL	0.6–1.2
pb	93%	Mwt	388.9
ur	65±5%	PSA	53.0 Å^2
		log P	1.62

CETIRIZINE
Histamine H_1 antagonist, antipruritic
Duration of action: 18±6 h
Metabolism: O-dealkylation

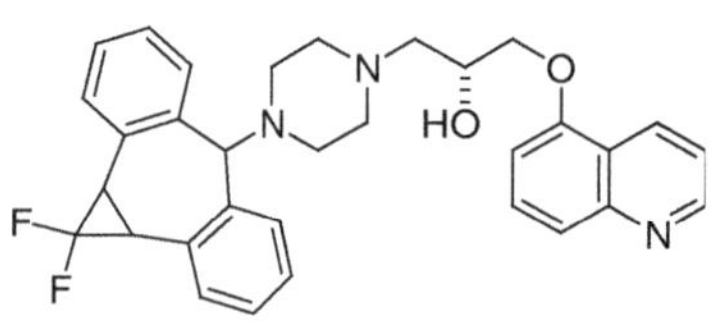

$t^1/_2$	25 h (iv)	V	268–350
F	–	CL	108–158
pb	–	Mwt	527.6
ur	–	PSA	48.8 Å^2
		log P	4.30

ZOSUQUIDAR
P-Glycoprotein inhibitor, adjunct antineoplastic
Metabolism: N-oxidation and hydroxylation of quinoline

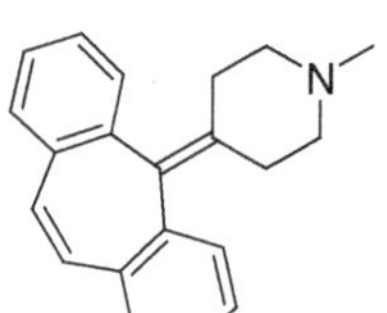

$t^1/_2$	1–4 h	V	–
F	–	CL	–
pb	96–99%	Mwt	287.4
ur	0%	PSA	3.2 Å^2
		log P	5.1

CYPROHEPTADINE
Histamine H_1 antagonist, antipruritic
Metabolites: quaternary glucuronide, *N*-desmethyl, epoxide

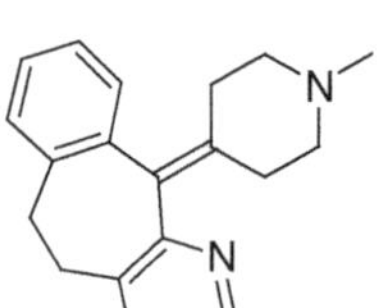

$t^1/_2$	9–12 h	V	–
F	–	CL	–
pb	Low	Mwt	290.4
ur	20%	PSA	16.1 Å^2
		log P	3.5

AZATADINE
Histamine H_1 antagonist, antipruritic

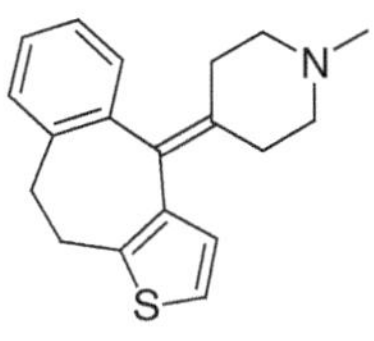	$t_{1/2}$ 23 h F 78% pb 91% ur <1%	V – CL – Mwt 295.4 PSA 31.5 Å^2 log P 2.71	PIZOTYLINE, PIZOTIFEN 5-HT antagonist for migraine prophylaxis Main metabolite: *N*-glucuronide ($t_{1/2}$ 23 h, V 1 l kg^{-1})
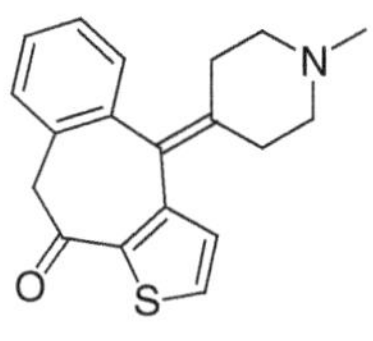	$t_{1/2}$ 18±7 h F 50% pb 75% ur 1%*	V 56 CL – Mwt 309.4 PSA 48.6 Å^2 log P 2.19	KETOTIFEN *10% of *N*-desmethyl metabolite; histamine H_1 antagonist, antipruritic Metabolism: N-demethylation, N-oxidation, N-glucuronidation, reduction of keto group Active metabolite: desmethyl
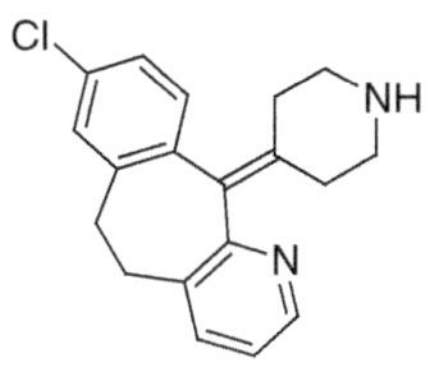	$t_{1/2}$ 18±6 h F High pb 73–87% ur 0%	V 49 CL – Mwt 310.8 PSA 24.9 Å^2 log P 3.5	DESLORATADINE Metabolite of loratadine (see below) Metabolism: 3-hydroxylation of pyridine, then glucuronidation
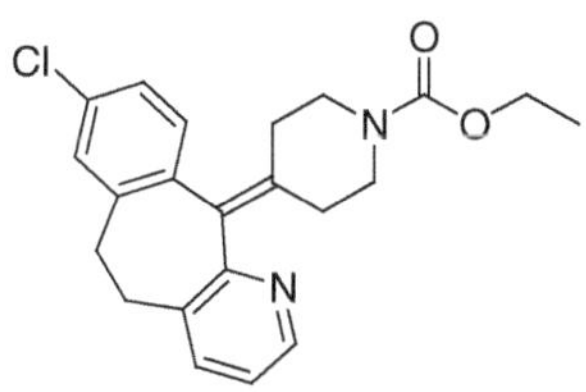	$t_{1/2}$ 8±6 h F – pb 97% ur –	V 120±80 CL 142±57 Mwt 382.9 PSA 42.4 Å^2 log P 4.6	LORATADINE Histamine H_1 antagonist Duration of action: 24 h Metabolism: hydrolysis of carbamate to desloratadine
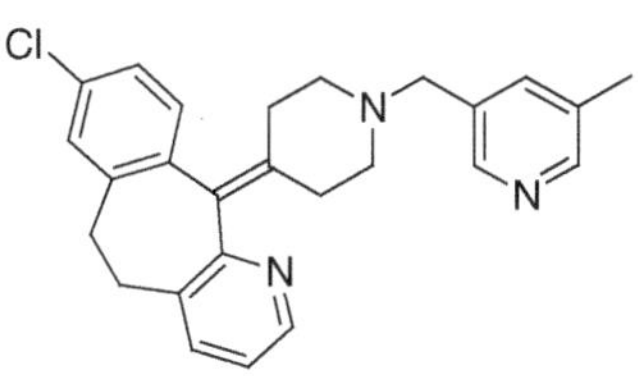	$t_{1/2}$ 6 h F 50% pb >95% ur 0%	V 143 CL – Mwt 416.0 PSA 29.0 Å^2 log P 6.11	RUPATADINE Histamine H_1 antagonist, antipruritic Metabolism: piperidine N-dealkylation to desloratadine, then 3-hydroxylation of pyridine to 3-hydroxydesloratadine
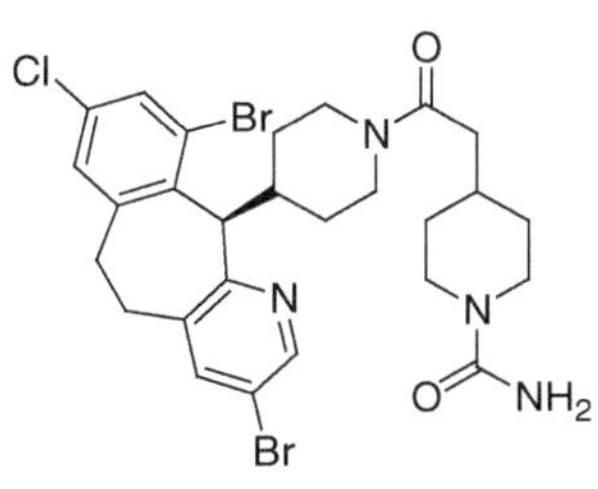	$t_{1/2}$ 4–11 h F 30–70% pb – ur <0.02%	V – CL/F 3–6 Mwt 638.8 PSA 79.5 Å^2 log P 5.03	LONAFARNIB Farnesyltransferase inhibitor, antineoplastic Metabolism: hydroxylation at CH_2NCONH_2, then elimination of H_2O

(*continued overleaf*)

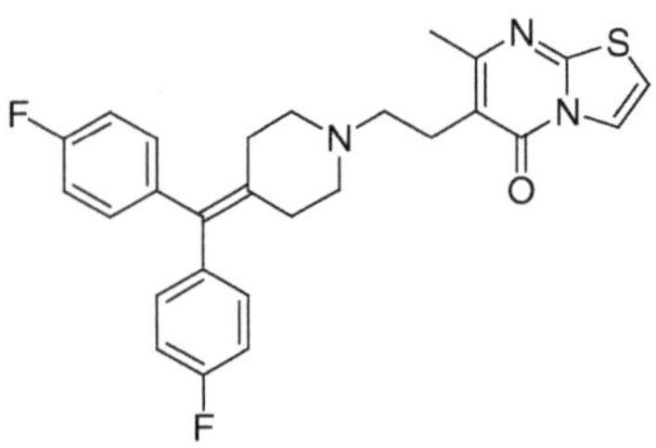	$t_{1/2}$ 43±13 h *F* 75–83% pb >98% ur –	*V* – CL – Mwt 477.6 PSA 61.2 Å^2 log *P* 5.20	RITANSERIN 5-HT$_2$ antagonist, anxiolytic, antidepressant
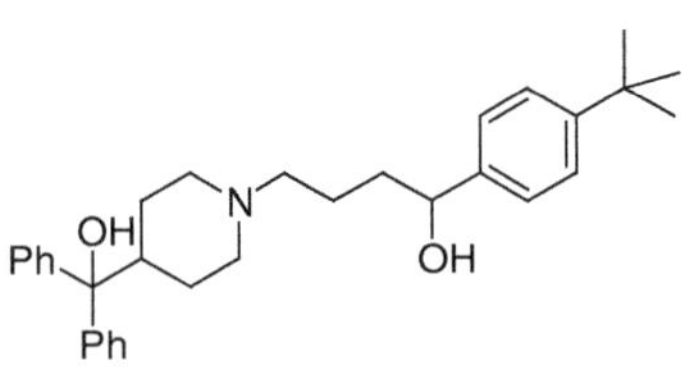	$t_{1/2}$ – *F* <1%* pb 97% ur Negligible	*V* – CL 1133 Mwt 471.7 PSA 43.7 Å^2 log *P* 5.62	TERFENADINE Nonsedating histamine H$_1$ antagonist, prodrug of fexofenadine (see below) *fexofenadine on oral dosing of terfenadine: 70% Other metabolism: N-dealkylation of piperidine; withdrawn in 1998 for causing cardiac arrhythmias
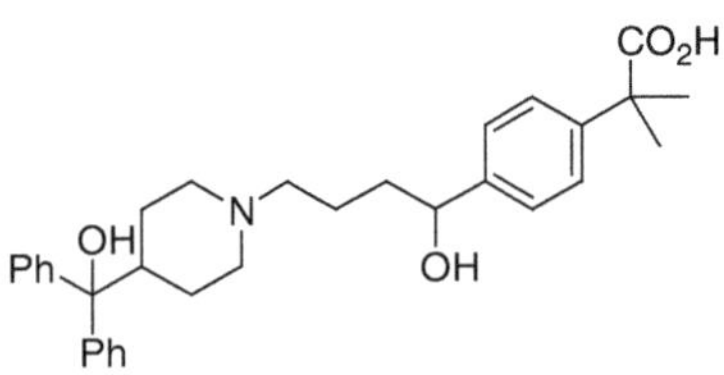	$t_{1/2}$ 12±5 h *F* 33% pb 60–70% ur 19±7%	*V* 5.4–5.8 CL/*F* 9±2 Mwt 501.7 PSA 81.0 Å^2 log *P* 3.73	FEXOFENADINE Nonsedating histamine H$_1$ antagonist, *P*-gp substrate, metabolite of terfenadine (see above)
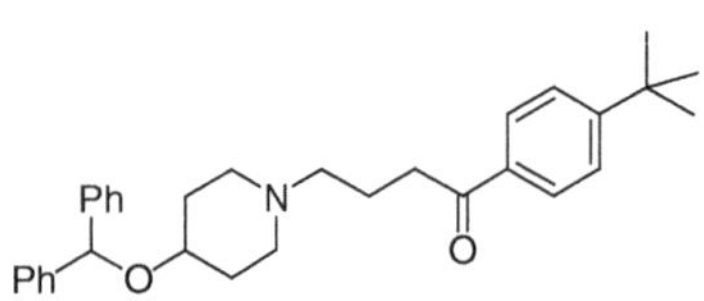	$t_{1/2}$ 6–20 h *F* – pb 98% ur 1.3–1.8%	*V* – CL – Mwt 469.7 PSA 29.5 Å^2 log *P* 6.81	EBASTINE Histamine H$_1$ antagonist Metabolism: hydroxylation of *tert*-butyl, N-dealkylation of piperidine Main active metabolite: carebastine (see below)
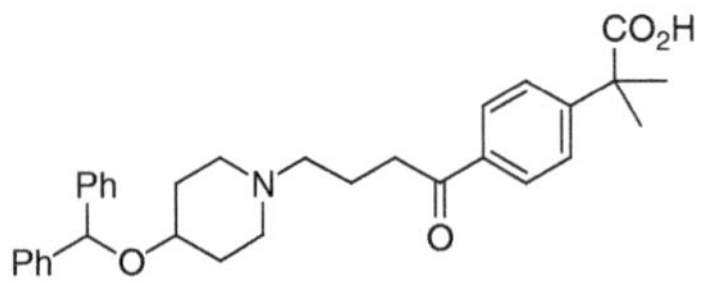	$t_{1/2}$ 17±10 h *F* – pb 98% ur 85±10%	*V* 1.4–2.2 CL 14–15 Mwt 499.6 PSA 66.8 Å^2 log *P* 4.93	CAREBASTINE Metabolite of ebastine
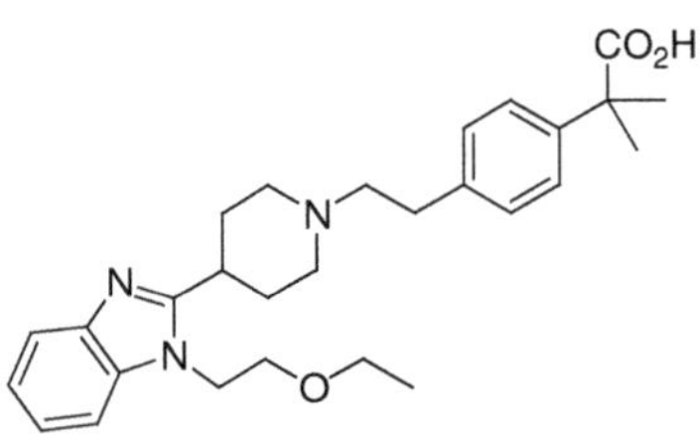	$t_{1/2}$ 5–14 h *F* 25–69%* pb 85% ur 33%	*V* 0.5 CL 4.6 Mwt 463.6 PSA 62.7 Å^2 log *P* 5.06	BILASTINE *animals Histamine H$_1$ antagonist; no metabolism in humans

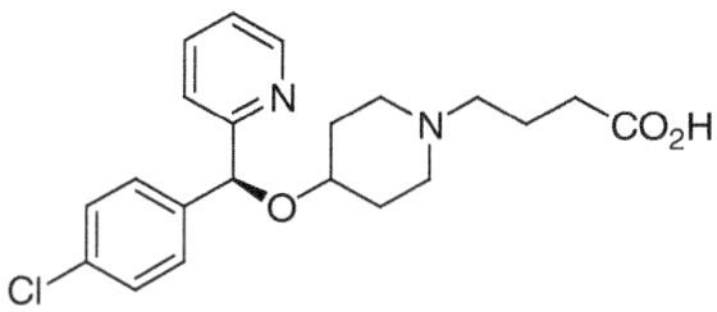

$t_{1/2}$	2.3–3.3 h	V	–
F	>80%	CL	–
pb	55%	Mwt	388.9
ur	>70%	PSA	62.7 Å^2
		log P	3.67

BEPOTASTINE
Nonsedating histamine H_1 antagonist
Metabolism: piperidine N-dealkylation, oxidative degradation of side chain, ether cleavage to benzylic alcohol

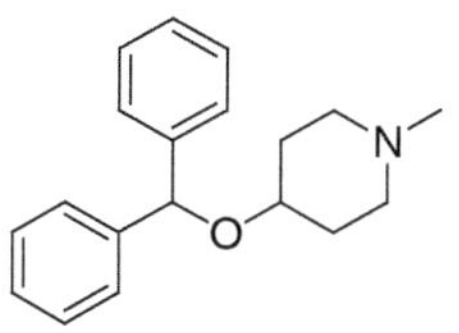

$t_{1/2}$	32±8 h	V	–
F	–	CL	–
pb	>99%	Mwt	281.4
ur	2–7%	PSA	12.5 Å^2
		log P	3.43

DIPHENYLPYRALINE
Histamine H_1 antagonist
Metabolites: *N*-oxide, *N*-desmethyl

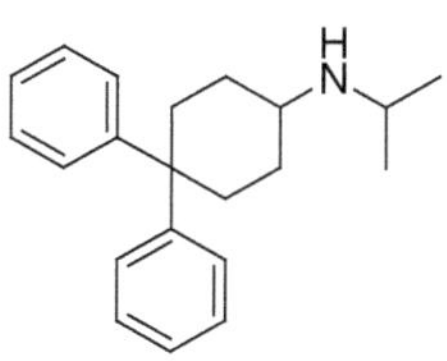

$t_{1/2}$	3 h*	V	–
F	–	CL	–
pb	90%	Mwt	293.5
ur	Low	PSA	12.0 Å^2
		log P	5.53

PRAMIVERINE
*monkey
Antispasmodic
Metabolism: N-dealkylation, hydroxylation

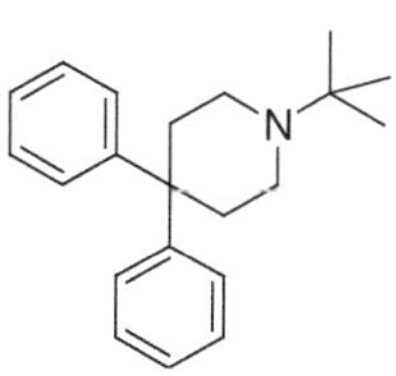

$t_{1/2}$	31 h	V	2.7
F	47%	CL	–
pb	96%	Mwt	293.5
ur	20%	PSA	3.24 Å^2
		log P	4.98

BUDIPINE
Antiparkinsonian
Metabolites: 4-hydroxyphenyl ($t_{1/2}$ 59 h), *N*-oxide, alcohol (hydroxylation of *tert*-butyl)

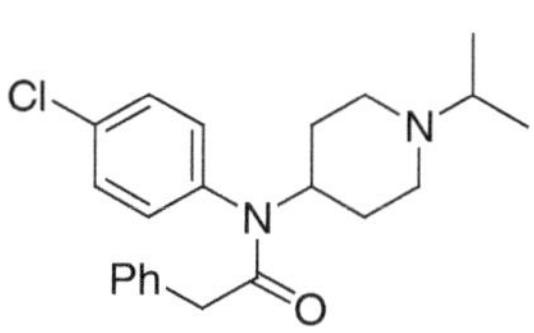

$t_{1/2}$	8±2 h	V	6.4±2.4
F	1–65%*	CL	17.5±2.8
pb	80±5%	Mwt	370.9
ur	<2%	PSA	23.6 Å^2
		log P	3.62

LORCAINIDE
*dose dependent
Antiarrhythmic
Active metabolite: *N*-desisopropyl ($t_{1/2}$ 27±8 h)

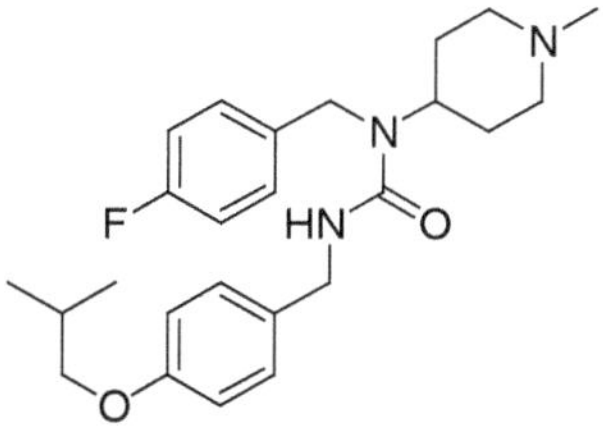

$t_{1/2}$	55 h	V	–
F	>43% (rat)	CL	5.9–8.5
pb	95%	Mwt	427.6
ur	–	PSA	44.8 Å^2
		log P	4.67

PIMAVANSERIN, ACP-103
5-HT_{2a} inverse agonist

(*continued overleaf*)

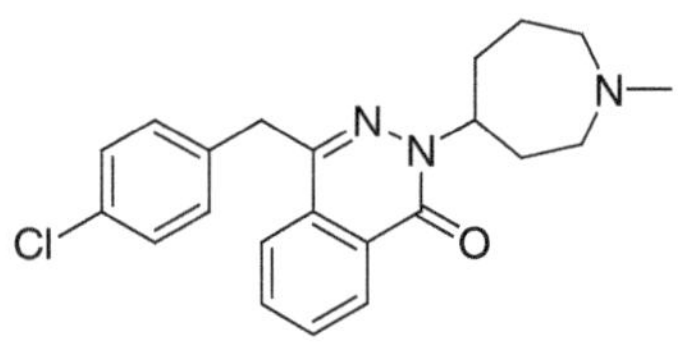	$t^{1}/_{2}$ 22±4 h *F* 82% pb 83±5% ur 2–7%	*V* 14.5 CL 9±3 Mwt 381.9 PSA 35.9 Å^2 log *P* 3.47	AZELASTINE Histamine H_1 antagonist Metabolites: 6-hydroxy (para to carbonyl group), *N*-desmethyl ($t^{1}/_{2}$ 54±15 h)
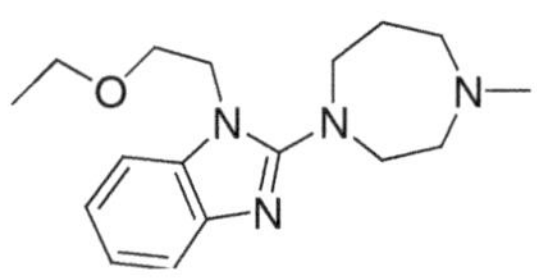	$t^{1}/_{2}$ 3–7 h *F* 46–90% pb – ur 3.6%	*V* – CL 21 Mwt 302.4 PSA 33.5 Å^2 log *P* 2.78	EMEDASTINE Histamine H_1 antagonist; eye drops Metabolism: aromatic 5- and 6-hydroxylation, N-oxidation of diazepane
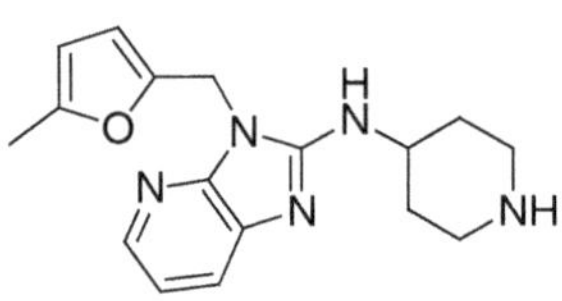	$t^{1}/_{2}$ 9–28 h *F* – pb – ur –	*V* – CL – Mwt 311.4 PSA 67.9 Å^2 log *P* 1.18	NOBERASTINE Histamine H_1 antagonist
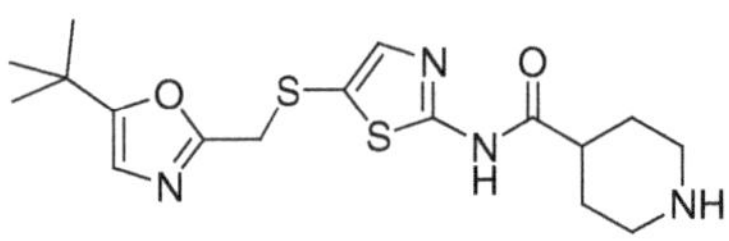	$t^{1}/_{2}$ 5–18 h (iv) *F* 4–33% pb – ur –	*V* 1.5–8.4 CL – Mwt 380.5 PSA 134 Å^2 log *P* 2.79	SNS-032 Cyclin-dependent kinase inhibitor, antineoplastic
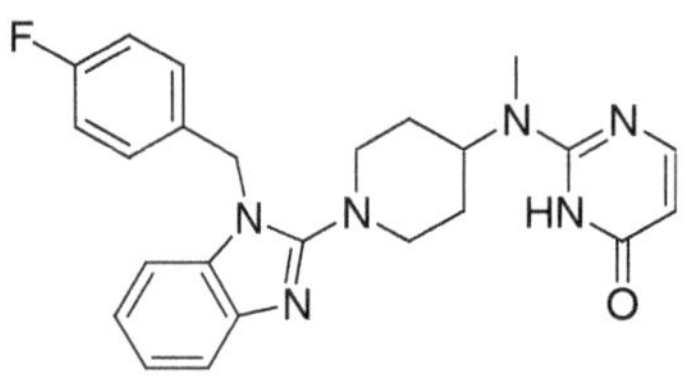	$t^{1}/_{2}$ 8–13 h *F* 58±8% pb 98% ur <0.5%	*V* 1.0–1.4 CL 1.2 Mwt 432.5 PSA 65.8 Å^2 log *P* 3.63	MIZOLASTINE Nonsedating histamine H_1 antagonist Metabolism: glucuronidation, sulfation
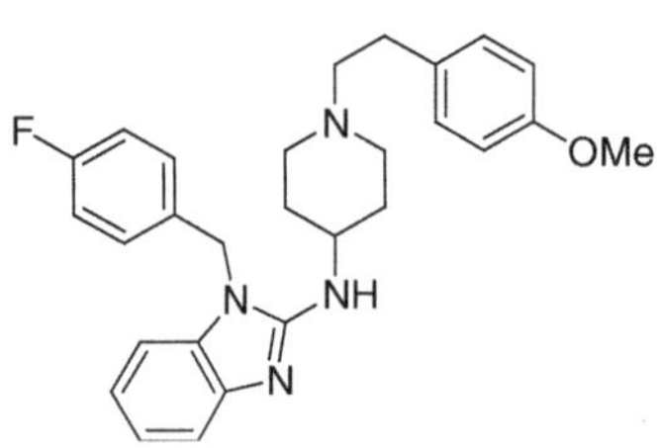	$t^{1}/_{2}$ 1–2 d *F* 95% pb 97% ur 0%	*V/F* 250 CL 11 Mwt 458.6 PSA 42.3 Å^2 log *P* 5.52	ASTEMIZOLE Nonsedating histamine H_1 antagonist, does not cross bbb Active metabolite: *O*-desmethyl ($t^{1}/_{2}$ 7–20 d); withdrawn in 1999 because of affinity to hERG
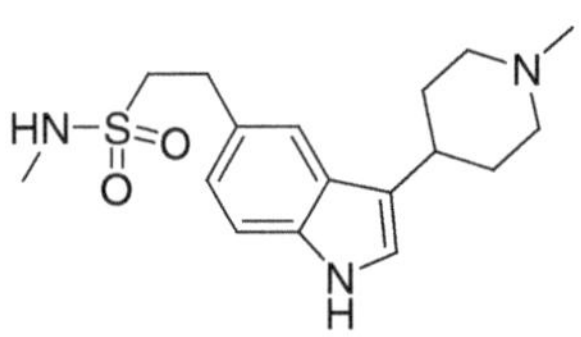	$t^{1}/_{2}$ 5–7 h *F* 63–74% pb 20–31% ur 50%	*V* 2.4 CL 6.6 Mwt 335.5 PSA 73.6 Å^2 log *P* 1.35	NARATRIPTAN Antimigraine, 5-HT$_{1A,B,D,F}$ agonist Metabolite: *N*-oxide

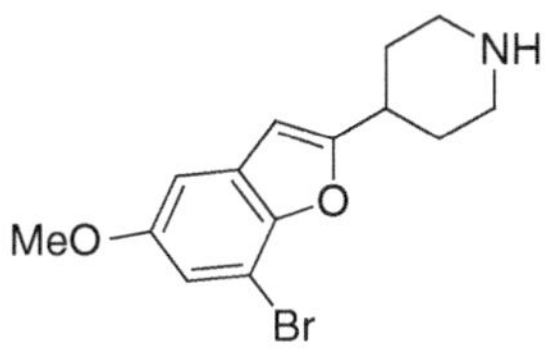	$t_{1/2}$ F pb ur	12–15 h 90% 98% <1%	V CL Mwt PSA log P	4.3–6.0 4.3–4.8 310.2 34.4 Å^2 3.65	BROFAROMINE Antidepressant, MAO inhibitor Metabolism: O-demethylation
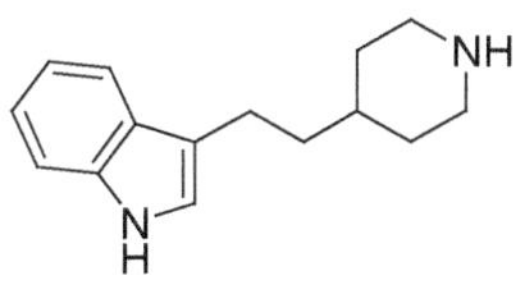	$t_{1/2}$ F pb ur	10 h – – 3%	V CL Mwt PSA log P	– – 228.3 27.8 Å^2 3.27	INDALPINE Antidepressant (SSRI) Metabolism: oxidation to 2-piperidinone ($t_{1/2}$ 12 h)

$t_{1/2}$, plasma half-life; F, oral bioavailability; pb, plasma protein binding; ur, excretion of unchanged drug in urine; V, volume of distribution; CL, clearance; Mwt, molecular weight; PSA, polar surface area; hERG, human ether-à-go-go-related gene; MAO, monoamine oxidase; SSRI, selective serotonin reuptake inhibitor.

Table 46.3 Aryloxyalkylamines. V in l kg^{-1}; CL in ml min^{-1} kg^{-1}; Mwt in g mol^{-1}.

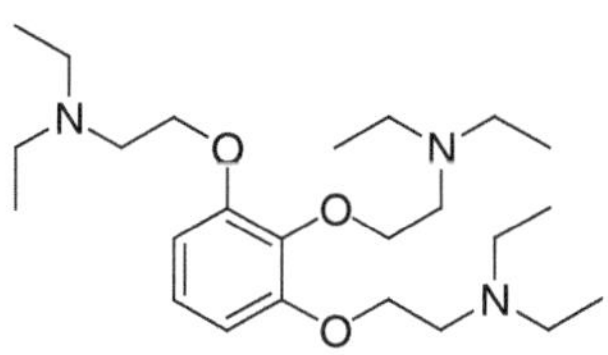	$t_{1/2}$ F pb ur	2.3–2.7 h – 30–70% 85–100%	V CL Mwt PSA log P	0.21–0.24 1.2–1.6 423.6 37.4 Å^2 2.58	GALLAMINE Muscle relaxant
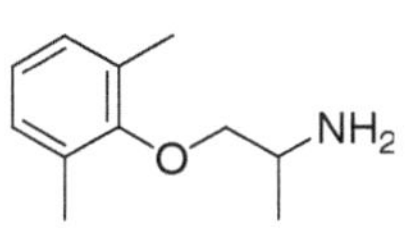	$t_{1/2}$ F pb ur	9±2 h 87±13% 63±3% 4–15%	V CL Mwt PSA log P	4.9±0.5 6±3 179.3 12.7 Å^2 2.6	MEXILETINE Antiarrhythmic Metabolism: aromatic 4-hydroxylation, benzylic hydroxylation, glucuronidation
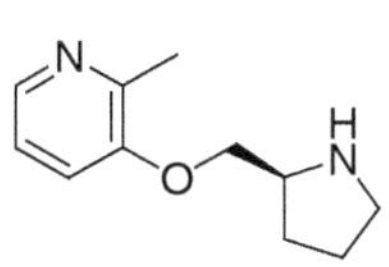	$t_{1/2}$ F pb ur	– – – –	V CL Mwt PSA log P	– – 192.3 34.2 Å^2 1.41	POZANICLINE Nicotinic acetylcholine agonist, nootropic, neuroprotective
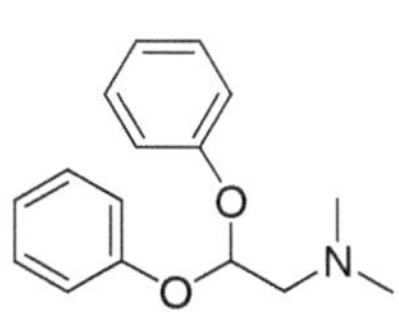	$t_{1/2}$ F pb ur	2±1 h 21% – 0.3%	V CL Mwt PSA log P	14±6 18±3 257.3 21.7 Å^2 2.52	MEDIFOXAMINE Antidepressant Metabolism: N-demethylation, N-oxidation, oxidation to diphenoxyacetic acid

(*continued overleaf*)

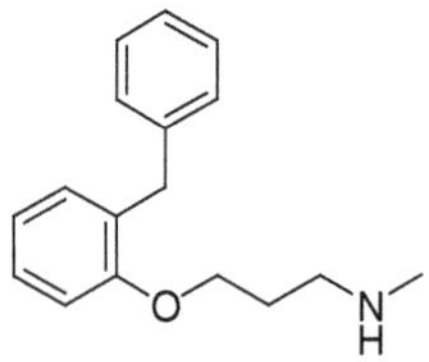

$t_{1/2}$	3.3 h	V	8
F	48%	CL	–
pb	95%	Mwt	269.4
ur	<1%	PSA	21.3 Å^2
		log P	4.19

BIFEMELANE
Nootropic

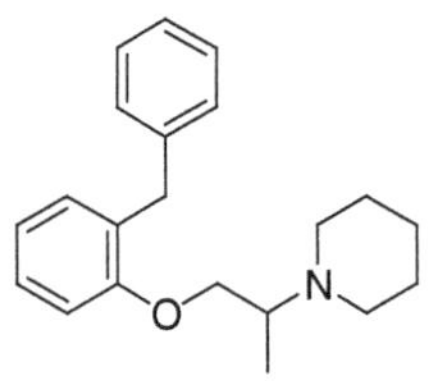

$t_{1/2}$	4–5 h	V	–
F	92%	CL	–
pb	–	Mwt	309.5
ur	–	PSA	12.5 Å^2
		log P	5.15

BENPROPERINE
Antiussive
Metabolites: 4-piperidinol, 3-piperidinol

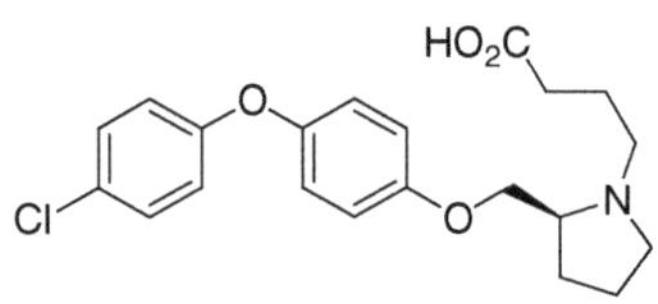

$t_{1/2}$	9 h	V	–
F	>80%*	CL	–
pb	–	Mwt	389.9
ur	–	PSA	59.0 Å^2
		log P	4.63

DG-051
*monkey
Leukotriene A_4 hydrolase inhibitor
Metabolism: N-oxidation, aromatic hydroxylation (ortho to Cl)

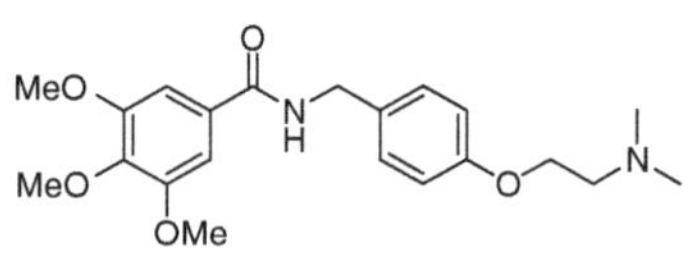

$t_{1/2}$	7–9 h	V	–
F	56–62%	CL	–
pb	–	Mwt	388.5
ur	30–50%	PSA	69.3 Å^2
		log P	1.26

TRIMETHOBENZAMIDE
Dopamine antagonist, antiemetic
Metabolite: *N*-oxide

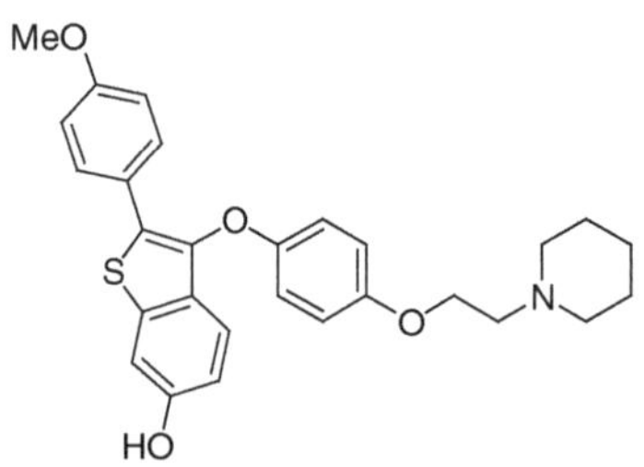

$t_{1/2}$	30–50 h	V	–
F	5%*	CL	–
pb	–	Mwt	475.6
ur	–	PSA	79.4 Å^2
		log P	7.78

ARZOXIFENE
*monkey
Estrogen receptor modulator for treatment of osteoporosis
Metabolism: glucuronidation, O-demethylation; development discontinued

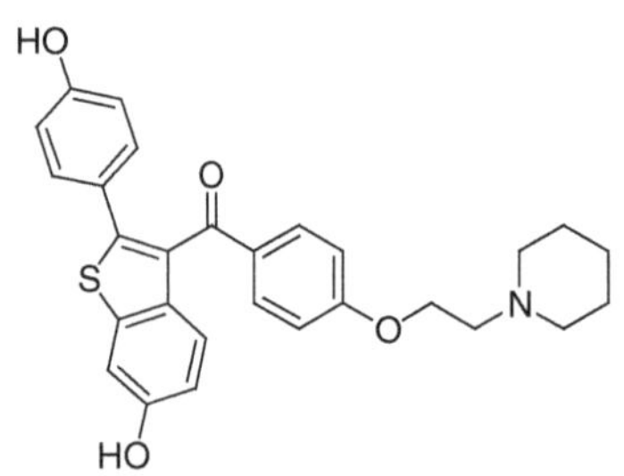

$t_{1/2}$	28 h	V	–
F	2%	CL	–
pb	99%	Mwt	473.6
ur	–	PSA	98.2 Å^2
		log P	4.57

RALOXIFENE
Antiosteoporotic
Metabolism: glucuronidation

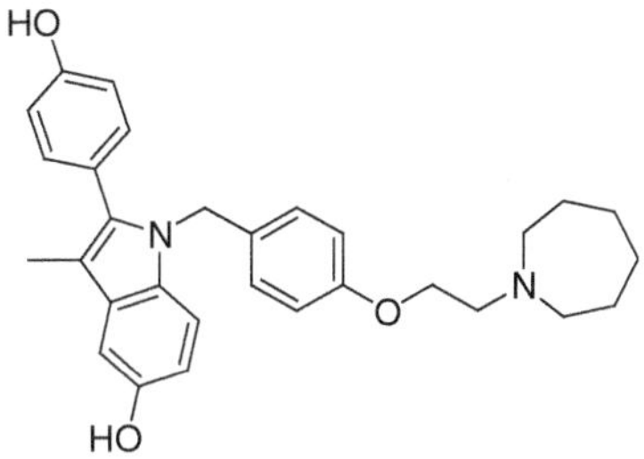

$t^1/_2$	28±11 h	V	15±4	**BAZEDOXIFENE**
F	6%	CL	6	Estrogen receptor modulator for treatment of osteoporosis
pb	96–99%	Mwt	470.6	Metabolism: glucuronidation
ur	<1%	PSA	52.9 Å^2	
		log P	6.59	

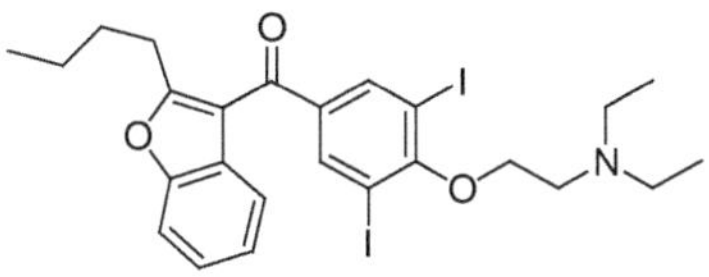

$t^1/_2$	25±12 d	V	66±44	**AMIODARONE**
F	46±22%	CL	1.9±0.4	Antiarrhythmic, antithyroid
pb	100%	Mwt	645.3	Metabolites: *N*-desethyl ($t^1/_2$ 61 d), iodide
ur	0%	PSA	42.7 Å^2	
		log P	7.82	

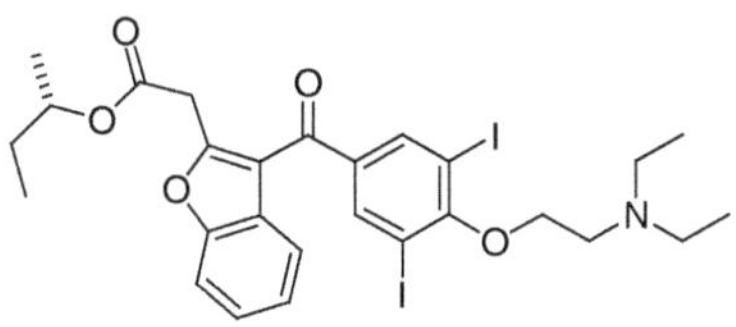

$t^1/_2$	7 h	V	13	**BUDIODARONE**
F	–	CL	–	Antiarrhythmic
pb	–	Mwt	703.4	Metabolism: ester hydrolysis
ur	–	PSA	69.0 Å^2	
		log P	8.00	

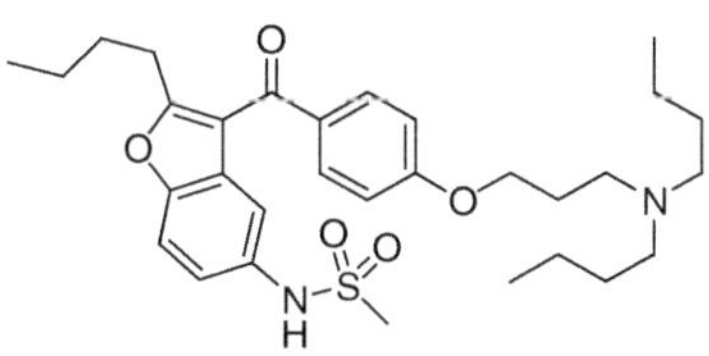

$t^1/_2$	24–30 h	V	19–22	**DRONEDARONE**
F	15–20%	CL	–	Antiarrhythmic
pb	>98%	Mwt	556.8	Main metabolite: *N*-desbutyl ($t^1/_2$ 20–25 h)
ur	0%	PSA	97.2 Å^2	
		log P	7.58	

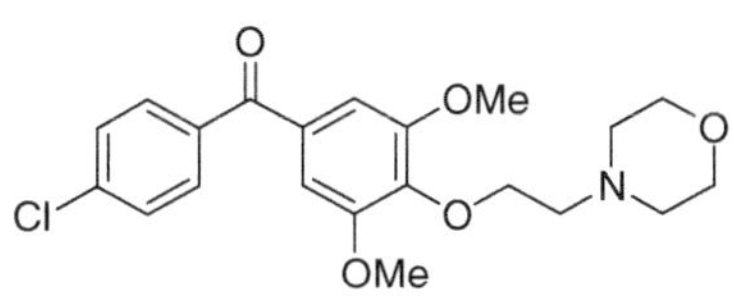

$t^1/_2$	–	V	–	**MORCLOFONE**
F	80%	CL	–	Cough suppressant
pb	99%	Mwt	405.9	
ur	–	PSA	57.2 Å^2	
		log P	2.55	

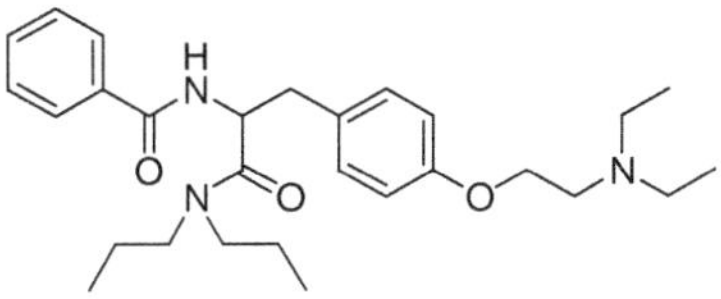

$t^1/_2$	2–3 h	V	3.4	**TIROPRAMIDE**
F	23%	CL	3.3–4.8	Antispasmodic
pb	65%	Mwt	467.6	Metabolism: N-deethylation, N-depropylation
ur	–	PSA	61.9 Å^2	
		log P	4.14	

(*continued overleaf*)

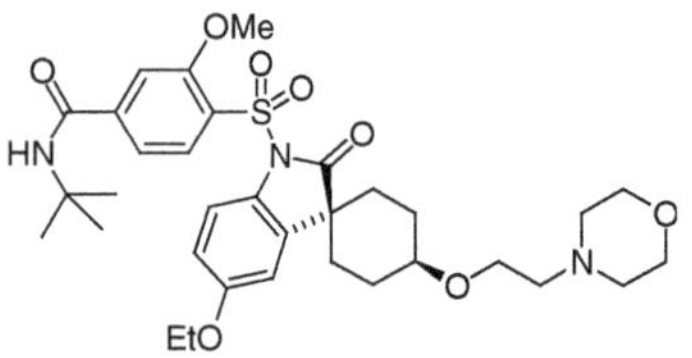	$t^1/_2$ 14–17 h *F* 6–11% pb 95–96% ur <3%	*V* 7–10 CL 13–17 Mwt 643.8 PSA 132 Å^2 log *P* 4.77	SATAVAPTAN, SR 121463 Vasopressin antagonist; discontinued because of low *F* (high first-pass metabolism).
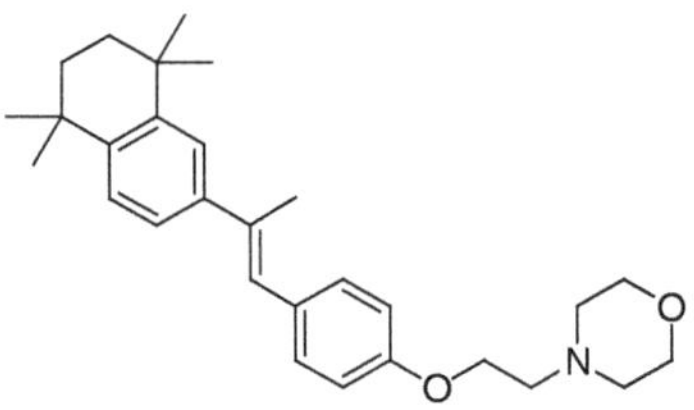	$t^1/_2$ 91±56 h *F* 60%* pb – ur –	*V* 6.5* CL/*F* 8±4 Mwt 433.6 PSA 21.7 Å^2 log *P* 6.76	MOFAROTENE *dog Arotinoid for cancer prevention
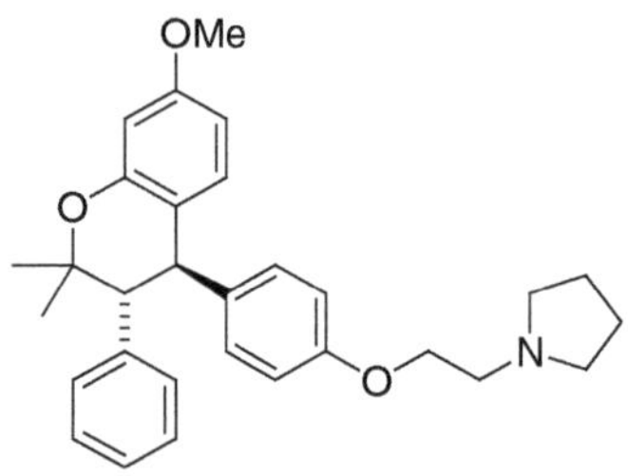	$t^1/_2$ 7 d *F* 60%* pb 90–100% ur –	*V* 22 CL 1.6 Mwt 457.6 PSA 30.9 Å^2 log *P* 6.21	CENTCHROMAN, ORMELOXIFEN *rat Estrogen receptor modulator, contraceptive Metabolism: O-demethylation
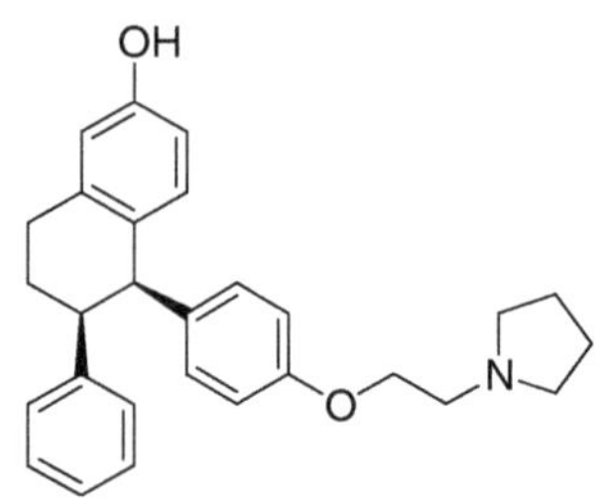	$t^1/_2$ 6 d *F* 60% pb >99% ur <2%	*V* 13 CL 1 Mwt 413.6 PSA 32.7 Å^2 log *P* 6.51	LASOFOXIFENE Estrogen receptor modulator for treatment of osteoporosis Metabolism: sulfation, glucuronidation, oxidation of pyrrolidine, catechol formation followed by O-methylation
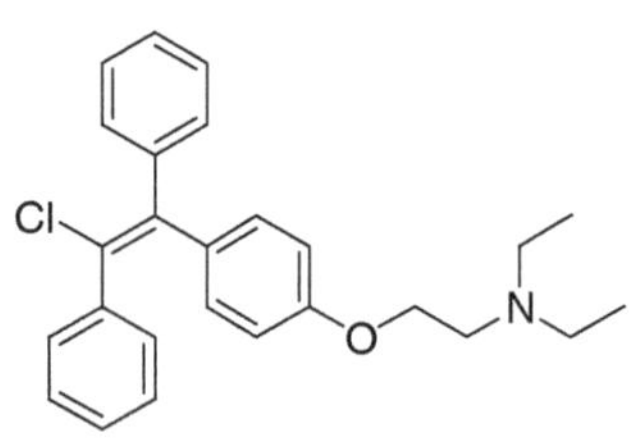	$t^1/_2$ 5–7 d *F* >90% pb – ur –	*V* >60 CL – Mwt 406.0 PSA 12.5 Å^2 log *P* 5.85	CLOMIPHENE Mixture of isomers; enclomiphene shown; gonad-stimulating principle Metabolism: N-deethylation, aromatic 4-hydroxylation

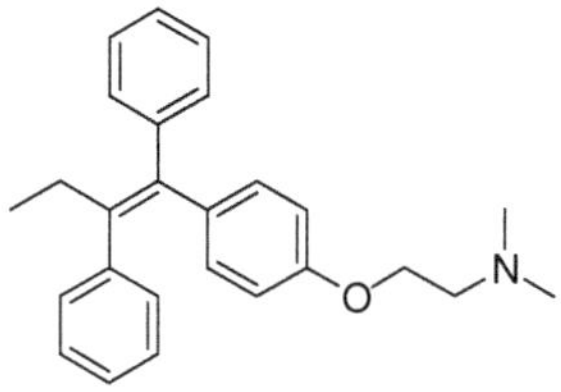	$t_{1/2}$ 8±4 d F 25–100% pb >98% ur <1%	V/F 55±5 CL/F 1.4 Mwt 371.5 PSA 12.5 Å^2 log P 5.13	TAMOXIFEN Antiestrogen, antineoplastic Metabolites: 4-hydroxyphenyl (vicinal to ethyl group), *N*-desmethyl
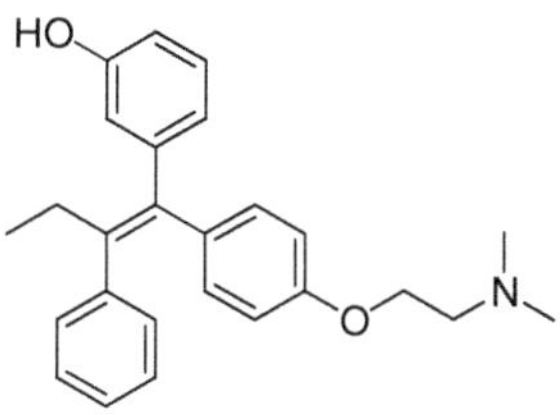	$t_{1/2}$ 1–2 d F 80% pb – ur <1%	V 7 CL 32±4 Mwt 387.5 PSA 32.7 Å^2 log P 7.30	DROLOXIFENE Antineoplastic Metabolites: glucuronide, *N*-desmethyl, *N*-oxide
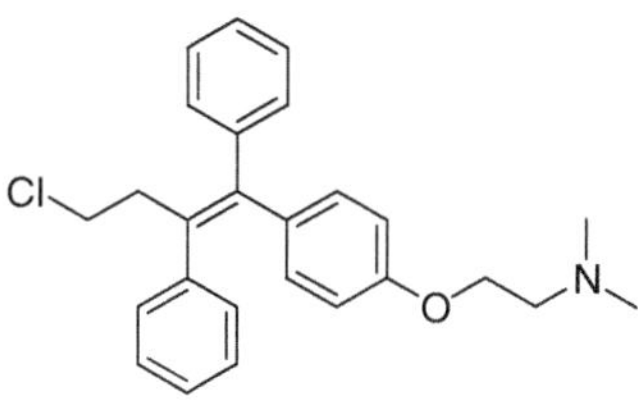	$t_{1/2}$ 4–7 d F 100% pb 99.7% ur –	V 7–10 CL 1 Mwt 406.0 PSA 12.5 Å^2 log P 4.77	TOREMIFENE Antineoplastic Metabolites: 4-hydroxy, *N*-desmethyl
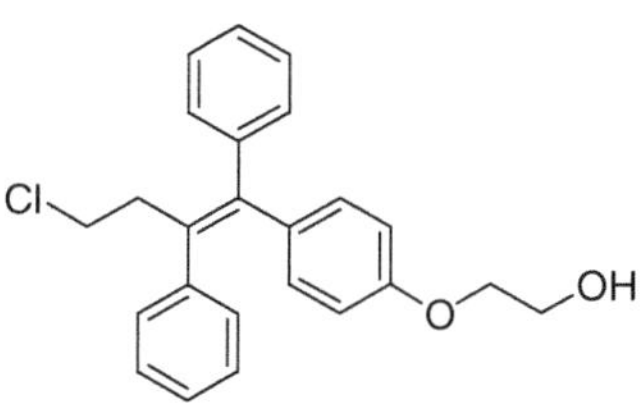	$t_{1/2}$ 30 h F – pb 93–98% ur –	V – CL – Mwt 378.9 PSA 29.5 Å^2 log P 4.43	OSPEMIFENE Metabolite of toremifene; treatment of postmenopausal osteoporosis and urogenital atrophy

Table 46.4 Piperidines and related compounds. V in l kg^{-1}; CL in ml min^{-1} kg^{-1}; Mwt in g mol^{-1}.

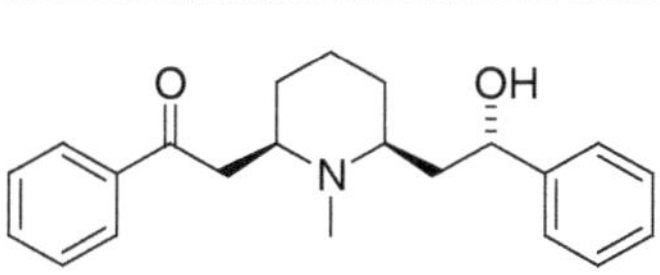	$t_{1/2}$ 1–5 h F Low pb – ur <0.5%	V – CL – Mwt 337.5 PSA 40.5 Å^2 log P 3.61	LOBELINE Respiratory stimulant
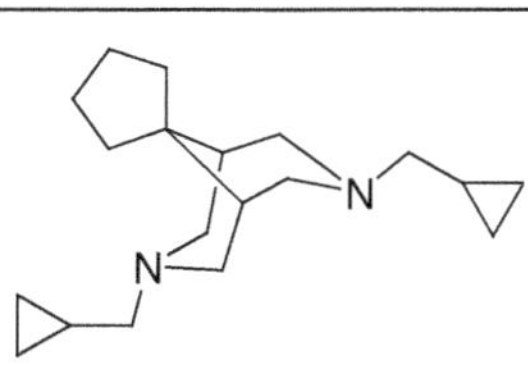	$t_{1/2}$ 8–13 h F 50% pb 96% ur 84% (iv)	V 1 CL 3–4 Mwt 288.5 PSA 6.48 Å^2 log P 3.67	TEDISAMIL Potassium channel blocker, bradycardic, antiarrhythmic Metabolism: hydroxylation

(*continued overleaf*)

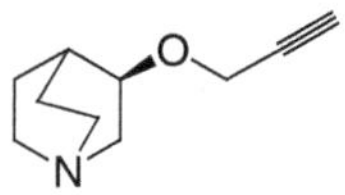	$t_{1/2}$ 3 h *F* 90% pb – ur >86%	*V* 2–5 CL – Mwt 165.2 PSA 12.5 Å^2 log *P* 0.92	TALSACLIDINE Muscarinic agonist
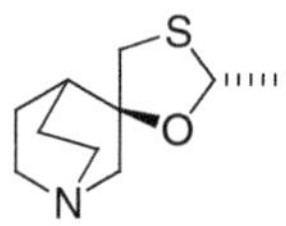	$t_{1/2}$ 3–6 h *F* – pb <20% ur 16%	*V* 6 CL – Mwt 199.3 PSA 37.8 Å^2 log *P* 2.44	CEVIMELINE Muscarinic agonist, parasympathomimetic Metabolism: glucuronidation, oxidation to sulfoxide, *N*-oxide
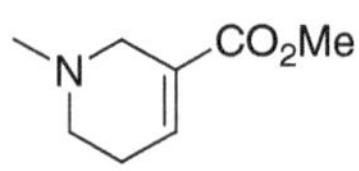	$t_{1/2}$ 0.2 h *F* Low pb – ur 0.3–0.4%	*V* 2.6 CL 170 Mwt 155.2 PSA 29.5 Å^2 log *P* −0.20	ARECOLINE Anthelmintic, cathartic Metabolism: N-oxidation, hydrogenation of alkene, ester hydrolysis, then conjugation with glycine
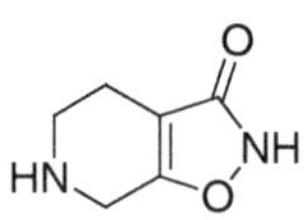	$t_{1/2}$ 1.5–2.0 h *F* 84–93% pb <2% ur 50–60%	*V* – CL 6.9 Mwt 140.1 PSA 50.4 Å^2 log *P* 0.71	GABOXADOL Hypnotic Metabolism: O-glucuronidation
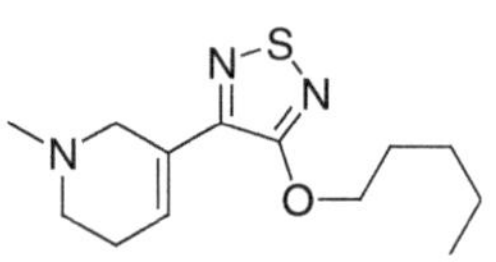	$t_{1/2}$ 3.6 h *F* <1% pb – ur –	*V* – CL – Mwt 281.4 PSA 66.5 Å^2 log *P* 4.28	XANOMELINE Muscarinic agonist, crosses bbb; cholinergic, nootropic Metabolites: *N*-oxide, *N*-desmethyl
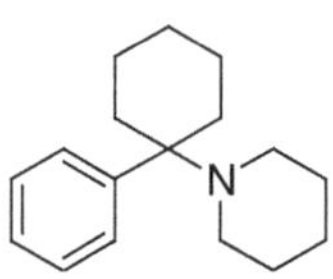	$t_{1/2}$ 7–51 h *F* 50–90% pb 60–70% ur 9–10%	*V* 6.2 CL 5.4 Mwt 243.4 PSA 3.24 Å^2 log *P* 4.25	PHENCYCLIDINE Anesthetic, hallucinogenic Metabolism: piperidine 4-hydroxylation, cyclohexane hydroxylation, piperidine ring opening by N-dealkylation
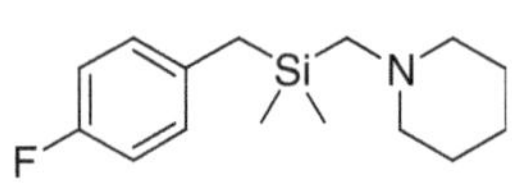	$t_{1/2}$ 12–16 h *F* – pb 86–95% ur –	*V* – CL – Mwt 265.4 PSA 3.24 Å^2 log *P* 4.51	SILPERISONE Muscle relaxant, crosses bbb
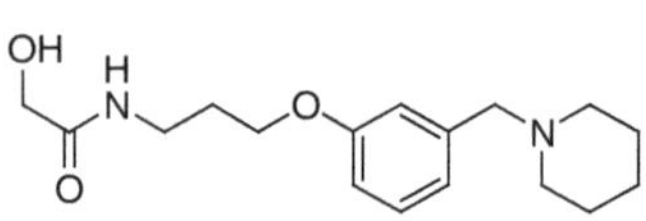	$t_{1/2}$ 7–8 h *F* >95% pb 5–7% ur 55–60%	*V* 3.2 CL 5–7 Mwt 306.4 PSA 61.8 Å^2 log *P* 1.77	ROXATIDINE Histamine H_2 antagonist; roxatidine is dosed as the acetate ester prodrug

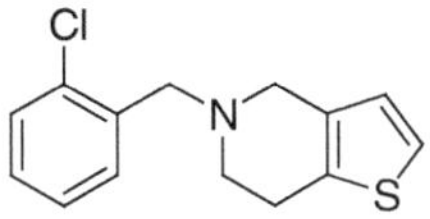	$t_{1/2}$ 13–33 h *F* 80–90% pb 98% ur 2%	*V* – CL – Mwt 263.8 PSA 31.5 Å^2 log *P* 3.23	TICLOPIDINE Antithrombotic Metabolism: oxidation to 2-hydroxythiophene, S-oxidation, then dimerization; may lead to neutro- and thrombocytopenia
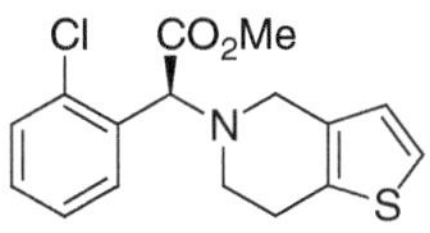	$t_{1/2}$ 2.5 h* *F* – pb 98% ur 0%	*V* – CL – Mwt 321.8 PSA 57.8 Å^2 log *P* 4.23	CLOPIDOGREL *acid: 8 h Platelet inhibitor, antithrombotic Metabolism: hydrolysis of ester, oxidation of thiophene to 2-thiophenone, then hydrolysis of thioester
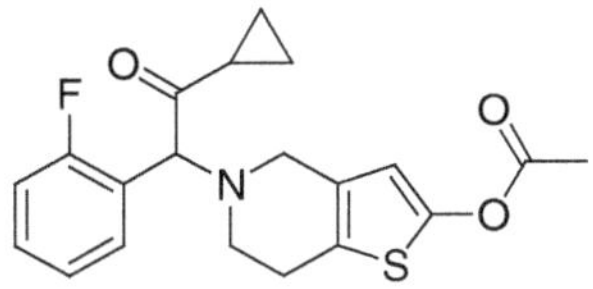	$t_{1/2}$ 2–15 h* *F* 70–80%* pb 98%* ur 0%	*V* 0.7–1.1* CL 29–43* Mwt 373.4 PSA 74.8 Å^2 log *P* 3.17	PRASUGREL *deacetylated metabolite on oral dosing of prasugrel Platelet inhibitor Further metabolism: thiolactone hydrolysis, then S-methylation and conjugation with cystein
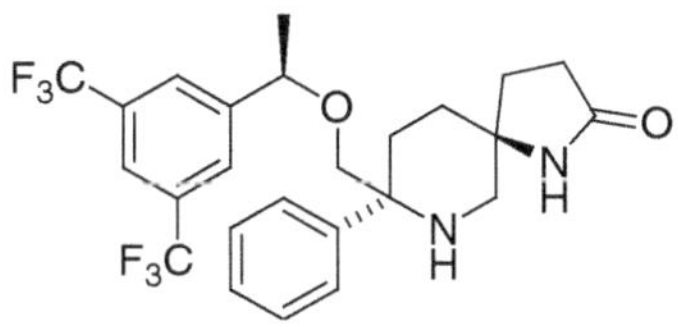	$t_{1/2}$ 180 h *F* – pb – ur –	*V* – CL – Mwt 500.5 PSA 50.4 Å^2 log *P* 4.01	ROLAPITANT NK_1 antagonist
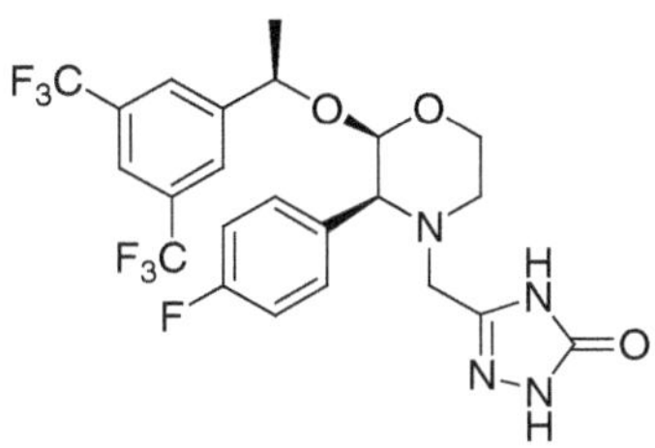	$t_{1/2}$ 9–13 h *F* 60–65% pb >95% ur <5%	*V* 1 CL 0.9–1.3 Mwt 534.4 PSA 75.2 Å^2 log *P* 4.23	APREPITANT NK_1 antagonist, antiemetic Metabolism: oxidation of morpholine
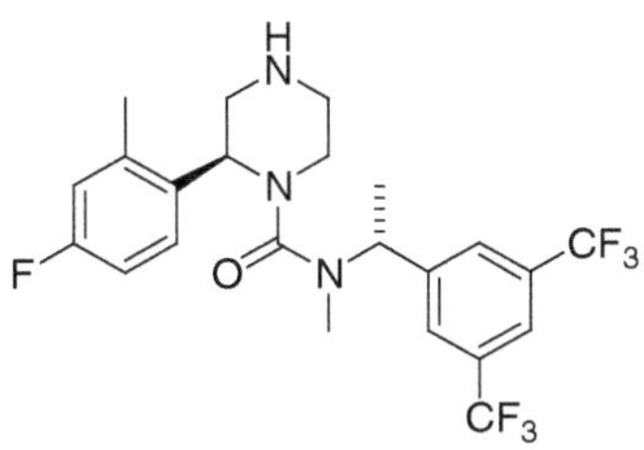	$t_{1/2}$ 11 h* *F* 71%* pb – ur –	*V* – CL 15* Mwt 491.5 PSA 35.6 Å^2 log *P* 5.58	GW597599 *dog NK_1 antagonist

(*continued overleaf*)

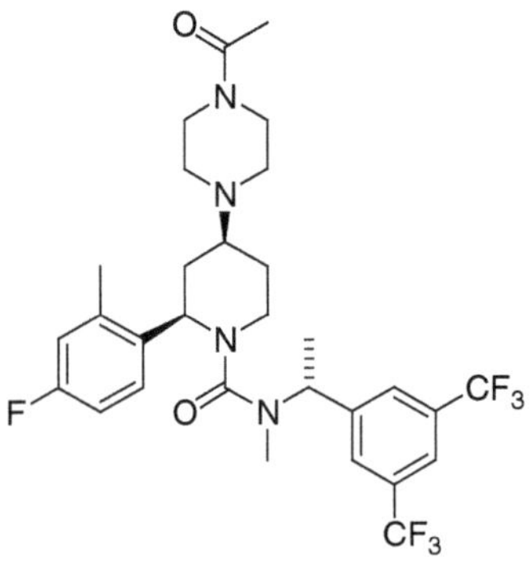

$t_{1/2}$	9–15 h	V	–	CASOPITANT
F	34% (dog)	CL	–	NK_1 antagonist, antiemetic, anxiolytic, antidepressant
pb	High	Mwt	616.6	Metabolism: hydroxylation, N-deacetylation, N-dealkylations
ur	Negligible	PSA	47.1 Å^2	
		log P	5.59	

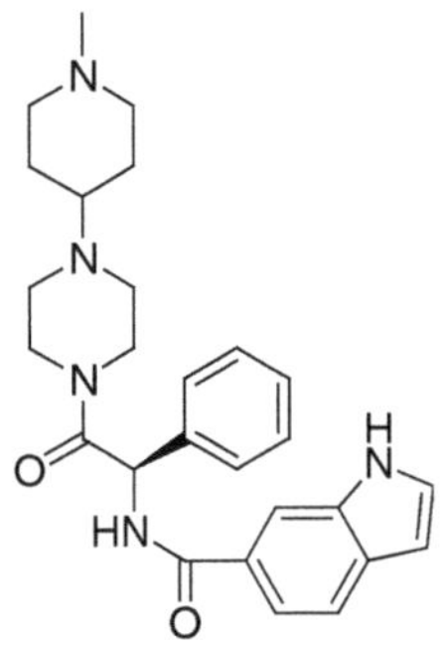

$t_{1/2}$	25 h	V	–	LY-517717
F	25–82%*	CL	–	*animals
pb	–	Mwt	459.6	Factor Xa inhibitor, antithrombotic
ur	–	PSA	71.7 Å^2	
		log P	3.14	

$t_{1/2}$, plasma half-life; F, oral bioavailability; pb, plasma protein binding; ur, excretion of unchanged drug in urine; V, volume of distribution; CL, clearance; Mwt, molecular weight; PSA, polar surface area; NK_1, neurokinin 1.

47
Phenethylamines (2-Phenylethylamines)

The phenethylamine substructure can be found in numerous natural products and drugs. Depending on the precise substitution pattern, phenethylamines show different biological effects and have been used in various therapeutic areas, for example, as central nervous system (CNS) agents (stimulants, antidepressants, and appetite suppressants), cardiovascular agents, or antianginals (Tables 47.1 and 47.2).

47.1
Biological Activity of Phenethylamines

Phenethylamines with hydroxyl or bioequivalent groups, being structurally related to dopamine and adrenaline (epinephrine), are agonists or antagonists at dopamine, adrenaline, or serotonin receptors or have an effect on the metabolism or transport of these natural neurotransmitters (Scheme 47.1).

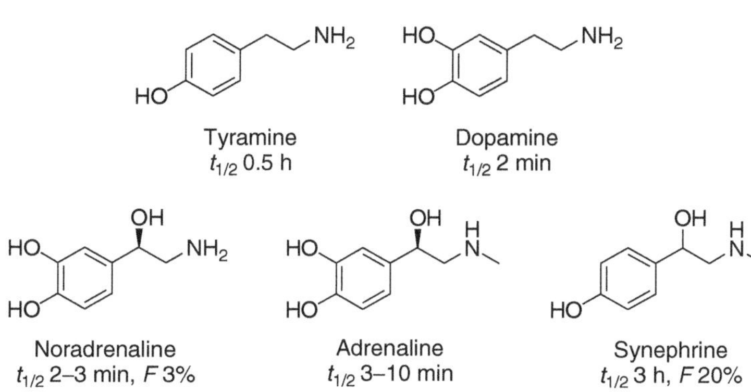

Scheme 47.1 Natural phenethylamines.

Phenethylamines devoid of hydroxyl groups usually have little affinity for adrenaline or dopamine receptors, but they may interact with monoamine oxidase (MAO) and the neuronal transport system for catecholamines. This interaction can enhance the levels of dopamine and noradrenaline in the CNS. Such nonphenolic phenethylamines, for example, amphetamine, are called

Lead Optimization for Medicinal Chemists: Pharmacokinetic Properties of Functional Groups and Organic Compounds, First Edition. Florencio Zaragoza Dörwald.

"indirectly acting sympathomimetic drugs," and their biological effects include CNS stimulation, attention enhancement, appetite suppression, and suppression of fatigue. Adrenaline agonists of low oral bioavailability have been developed for the treatment of glaucoma (as eye drops) or asthma (for inhalation).

47.2 Metabolism

Most natural small molecules present in the human body are metabolized rapidly by substrate-specific enzymes. Therefore, natural metabolites, such as natural amino acids, peptides, carbohydrates, steroids, or eicosanoids, tend to show poor pharmacokinetics and are not well suited as building blocks for drugs, unless short half-lives are the aim (as in prodrugs).

This is also true for most natural phenethylamines. In particular, those with a catechol (*o*-dihydroxyphenol) substructure have short plasma half-lives, because enzymes quickly degrade these compounds. Catecholamines (2-(3,4-dihydroxyphenyl)ethylamines) are rapidly methylated at the phenolic 3-hydroxyl group by catechol-*O*-methyltransferase (COMT) or converted into *O*-sulfates by the sulfotransferases present in the gut wall (Scheme 47.2). Phenethylamines devoid of a benzylic hydroxyl group are oxidatively deaminated by MAO (α-hydroxylation of *N*-alkylamines followed by hydrolysis of the resulting hemiaminals). MAO can also be found in the epithelial mucosa of the intestine and can, therefore, contribute significantly to the first-pass metabolization, and strongly diminish oral bioavailability. Many *N,N*-dimethyl or *N,N*-diethyl phenethylamines are quickly N-dealkylated *in vivo* by MAO. Phenethylamines with an unsubstituted phenyl group may also be hydroxylated at the latter by liver enzymes (CYP 450 (cytochrome *P*450)).

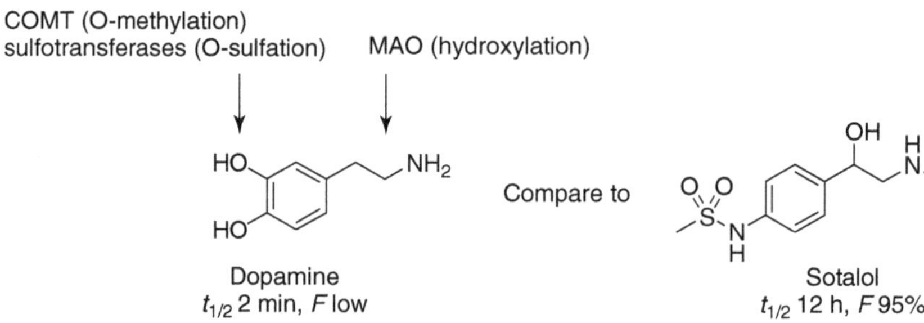

Scheme 47.2 Metabolism of dopamine.

Metabolization by COMT will *not* occur in phenethylamines devoid of a catechol substructure, and oxidation by MAO may be slowed by bulky *N*-alkyl groups (isopropyl, *tert*-butyl), by alkyl groups at the ethylene chain (e.g., α-methylation), or by heteroatoms or hydroxyl groups at the benzylic position or close to the nitrogen atom.

Most phenethylamines devoid of phenolic OH groups can cross the blood–brain barrier. CNS penetration and absorption from the gastrointestinal (GI) tract may be enhanced by increasing lipophilicity or by lowering the polar surface area (PSA). Oral bioavailability may also be enhanced by reducing the number of rotatable bonds.

As for most basic drugs, the rate of renal excretion can be strongly enhanced if the urine is rendered acidic, for example, by administration of ammonium chloride to the patient. When metabolism by MAO and COMT has been blocked, renal excretion of the unchanged drug will often become half-life determining.

47.3 Tetrahydroisochinolines and Related Compounds

Opium, the dried juice of the unripe seed capsules of the poppy plant, *Papaver somniferum*, contains mainly two groups of alkaloids: the phenanthrene derivatives (10% morphine, 0.5% codeine, and 0.2% thebaine) and the 2-benzylisoquinoline derivatives (6% noscapine and 1% papaverine). However, other plants also produce alkaloids with an isoquinoline substructure, and more than 500 benzylisoquinoline alkaloids have been isolated to date. Many of these are biologically active and have been used in medicine directly or as leads for the development of synthetic analogs with improved selectivity or therapeutic index. Interestingly, although plants produce these compounds mainly as a defense against herbivores, isoquinoline alkaloids show a broad spectrum of pharmacological properties (Table 47.3).

Isoquinolines and tetrahydroisoquinolines can be readily synthesized from phenethylamines with the Pictet–Spengler and Bischler–Napieralski reactions (Scheme 47.3).

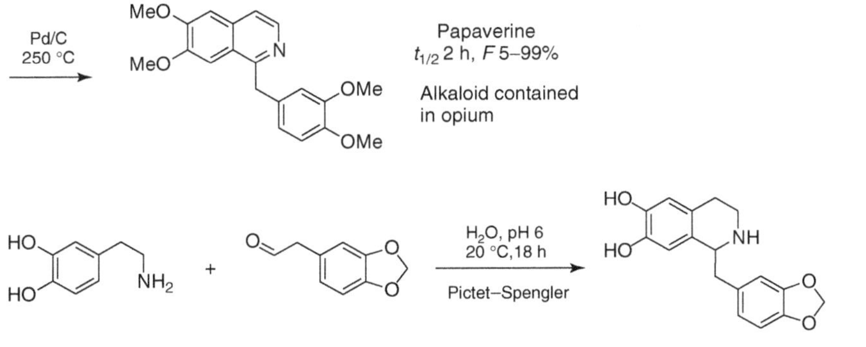

Scheme 47.3 Synthesis of isoquinolines and tetrahydroisoquinolines.

Further Reading

(a) Sneader, W. (2001) The discovery and synthesis of epinephrine. *Drug News Perspect.*, **14**, 491–494; (b) Sneader, W. (2001) Epinephrine analogues. *Drug News Perspect.*, **14**, 539–543.
Zhang, A., Zhang, Y., Branfman, A.R., Baldessarini, R.J., and Neumeyer, J.L. (2007) Advances in development of dopaminergic aporphinoids. *J. Med. Chem.*, **50**, 171–181.

Table 47.1 Phenethylamines and related compounds. *V* in l/kg; CL in ml min^{-1} kg^{-1}; Mwt in g/mol.

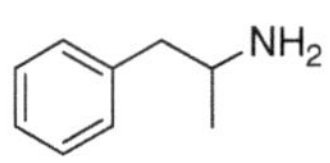	$t^{1}/_{2}$ 18 h *F* 20–25% pb 15–40% ur 20–30%	*V* 3–6 CL 9.7 Mwt 135.2 PSA 26.0 Å^2 log *P* 1.5	AMPHETAMINE CNS-stimulant, anorexic Metabolism: benzylic and aromatic 4-hydroxylation *N*-Lys-amphetamine: $t^{1}/_{2}$ 0.5 h
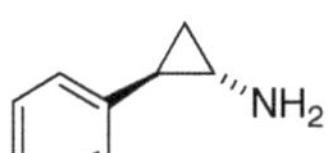	$t^{1}/_{2}$ 1–2 h *F* – pb – ur –	*V* 2.7 CL 14 Mwt 133.2 PSA 26.0 Å^2 log *P* 1.4	TRANYLCYPROMINE (racemic) Irreversible MAO inhibitor, antidepressant, CYP2A6 inhibitor; Metabolism: N-acetylation, hydroxylation
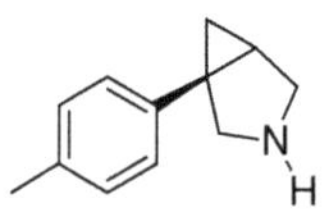	$t^{1}/_{2}$ 1.6 h *F* >89% pb 95–97%* ur 0%	*V/F* 1.58±0.32 CL/*F* 15 Mwt 173.3 PSA 12.0 Å^2 log *P* 2.15	BICIFADINE *monkey; NMDA antagonist, 5-HT/norepinephrine reuptake inhibitor, analgesic Metabolism: oxidation to lactam ($CHCH_2NH$ to CHCONH), oxidation of CH_3 to CO_2H, then acyl glucuronidation
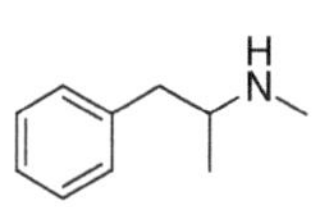	$t^{1}/_{2}$ 6–21 h *F* 60–70% pb – ur 30%	*V* 3–5 CL – Mwt 149.2 PSA 12.0 Å^2 log *P* 2.1	METHAMPHETAMINE CNS-stimulant, anorexic Metabolism: N-demethylation, benzylic and aromatic 4-hydroxylation; withdrawn in 1973 for risk of abuse
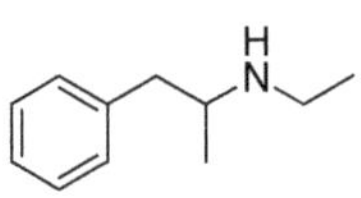	$t^{1}/_{2}$ 5 h *F* – pb – ur 45%	*V* – CL – Mwt 163.3 PSA 12.0 Å^2 log *P* 2.71	ETHYLAMPHETAMINE CNS-stimulant, anorexic Metabolism: aromatic 4-hydroxylation
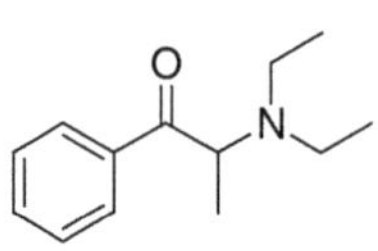	$t^{1}/_{2}$ 2 h *F* – pb – ur <5%	*V* – CL – Mwt 205.3 PSA 20.3 Å^2 log *P* 2.6	DIETHYLPROPION, DIETHYLCATHINONE, AMFEPRAMONE Anorexic; Metabolism: N-deethylation, reduction of ketone, oxidative degradation, and conjugation to hippuric acid

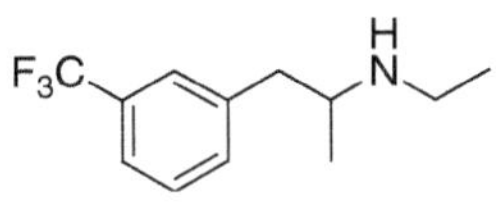	$t_{1/2}$ 14–30 h *F* 60–70% pb 30% ur 13±6%	*V* 8–10 CL 11 Mwt 231.3 PSA 3.4 Å^2 log *P* 3.5	FENFLURAMINE Anorexic; Metabolism: N-deethylation, N-oxidation and nitrone-formation; withdrawn in 1997 because of risk of heart valve damage
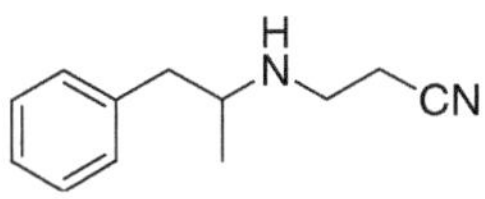	$t_{1/2}$ 2 h *F* – pb – ur 5–9%	*V* – CL – Mwt 188.3 PSA 35.8 Å^2 log *P* 2.04	FENPROPOREX CNS-stimulant, anorexic Metabolism: N-decyanoethylation, aromatic 4-hydroxylation
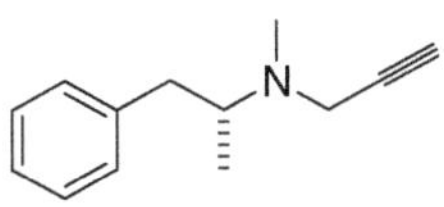	$t_{1/2}$ 2±1 h *F* 4.4% pb 94% ur Negligible	*V* 1.9 CL/*F* 1500 Mwt 187.3 PSA 3.2 Å^2 log *P* 2.7	SELEGILINE, DEPRENYL Irreversible MAO inhibitor, antidepressant; Metabolism: N-demethylation, depropargylation, benzylic and aromatic 4-hydroxylation
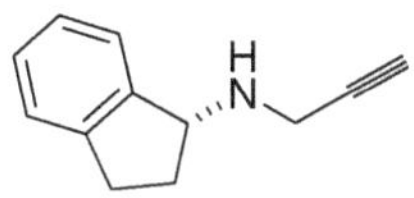	$t_{1/2}$ 1.3±0.7 h *F* 36% pb 60–70% ur <1%	*V* 3.7 CL 24 Mwt 171.2 PSA 12.0 Å^2 log *P* 2.27	RASAGILINE MAO inhibitor, antiparkinsonian Metabolism: N-depropargylation
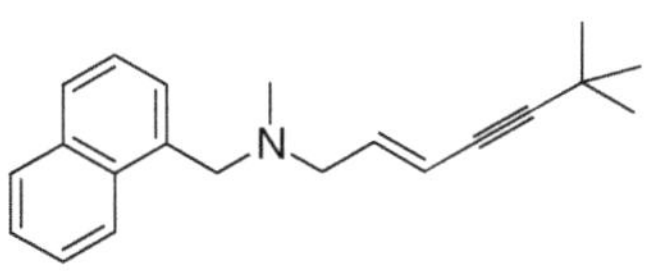	$t_{1/2}$ 19–36 h *F* 40% pb >99% ur Negligible	*V* >30 CL – Mwt 291.4 PSA 3.24 Å^2 log *P* 5.58	TERBINAFINE Antifungal; Metabolism: N-demethylation, hydroxylation of *tert*-butyl, oxidation of *tert*-butyl to carboxyisopropyl, N-de(naphthyl)methylation (to naphthaldehyde and naphthoic acid)
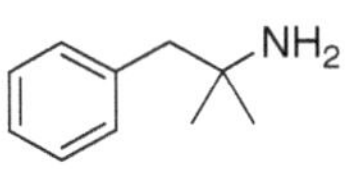	$t_{1/2}$ 24±8 h *F* – pb 96% ur 70–80%	*V* 3–4 CL – Mwt 149.2 PSA 26 Å^2 log *P* 2.3	PHENTERMINE Anorexic; Metabolism (rat): N-hydroxylation, aromatic 4-hydroxylation
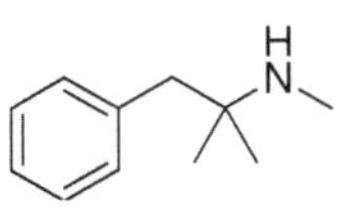	$t_{1/2}$ 17–18 h *F* – pb – ur –	*V* – CL – Mwt 163.3 PSA 12 Å^2 log *P* 2.6	MEPHENTERMINE Anorexic; Metabolism: N-demethylation

(*continued overleaf*)

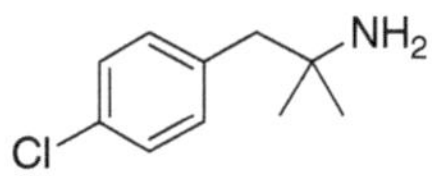	$t^1/_2$ 41 h *F* – pb – ur 17%	*V* – CL – Mwt 183.7 PSA 26 Å^2 log *P* 2.8	CHLORPHENTERMINE Anorexic; Metabolism: N-oxidation to N-hydroxy, nitroso, nitro
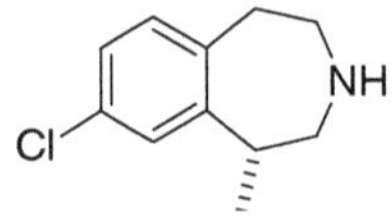	$t^1/_2$ 11±2 h *F* >90% pb 70% ur <1%	*V/F* 2.7 CL – Mwt 195.7 PSA 12.0 Å^2 log *P* 2.89	LORCASERIN 5-HT_{2C} agonist for treatment of obesity; Metabolites: *N*-sulfate ($t^1/_2$ 41±10 h), *N*-carbamoyl glucuronide
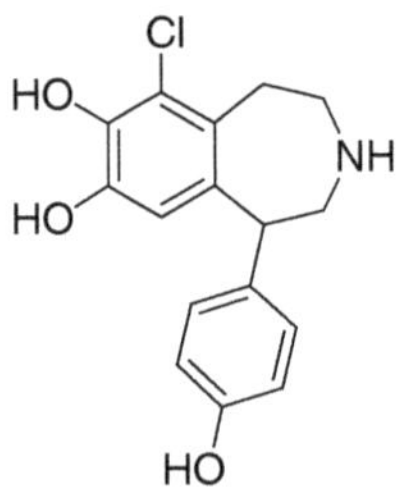	$t^1/_2$ 10 min *F* 5–7% pb 88% ur <1%	*V* 0.23–0.66 CL 25–38 Mwt 305.8 PSA 72.7 Å^2 log *P* 2.06	FENOLDOPAM Dopamine agonist, antihypertensive Metabolism: O-sulfation, O-methylation
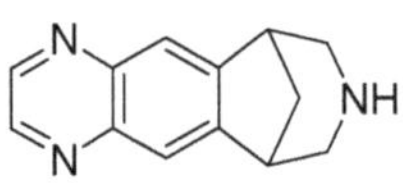	$t^1/_2$ 10–58 h *F* High pb <20% ur 92%	*V* 6.4 CL – Mwt 211.3 PSA 37.8 Å^2 log *P* 0.74	VARENICLINE Nicotinic agonist for treatment of nicotine addiction Metabolism: glucuronidation, N-formylation
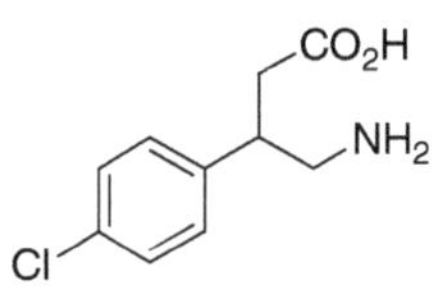	$t^1/_2$ 3–7 h *F* 95% pb 30% ur 70–80%	*V* 0.84 CL 2.6 Mwt 213.7 PSA 63.3 Å^2 log *P* 0.78	BACLOFEN Muscle relaxant, GABA agonist; Metabolism: deamination to 4-hydroxy-3-arylbutyric acid
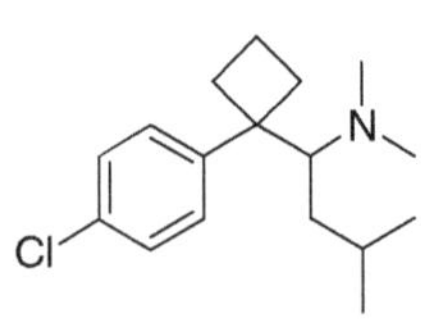	$t^1/_2$ 8–20 h *F* – pb 97% ur 0%	*V* 10.6* CL 35* Mwt 279.8 PSA 3.2 Å^2 log *P* 5.1	SIBUTRAMINE *rat Anorexic; Metabolism: N-demethylation; monodesmethyl: $t^1/_2$ 14 h; didesmethyl: $t^1/_2$ 16 h; withdrawn 2010 for cardiovascular risk
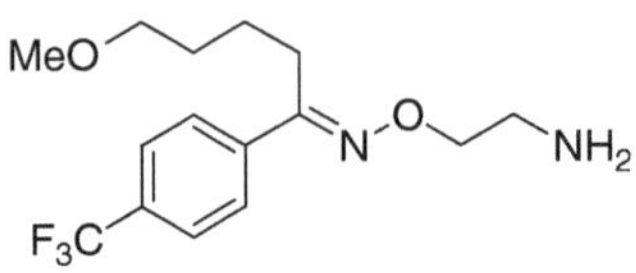	$t^1/_2$ 12–30 h *F* 50–60% pb 77% ur 3%	*V* 20–25 CL 15.2 Mwt 318.3 PSA 56.8 Å^2 log *P* 3.71	FLUVOXAMINE 5-HT reuptake inhibitor, antidepressant; Metabolism: oxidation of CH_2OMe to CO_2H, N-acetylation, replacement of NH_2 by OH

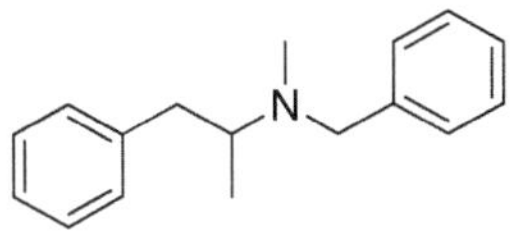	$t_{1/2}$ 24±8 h *F* – pb 75–99% ur –	*V* – CL – Mwt 239.4 PSA 3.2 Å^2 log *P* 3.9	BENZPHETAMINE Anorexic; Metabolism: aromatic 4-hydroxylation, N-dealkylation
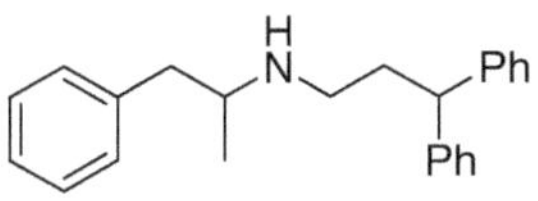	$t_{1/2}$ 14 h *F* 15% pb 97% ur <0.1%	*V* – CL – Mwt 329.5 PSA 12 Å^2 log *P* 5.7	PRENYLAMINE Coronary vasodilator Metabolism: aromatic 4-hydroxylation, N-dealkylation
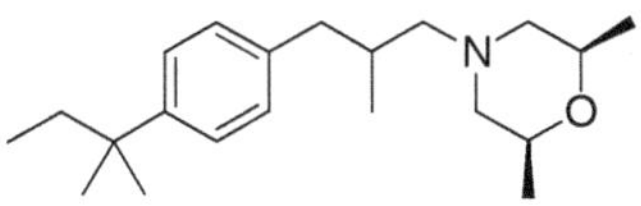	$t_{1/2}$ – *F* 70%* pb High ur –	*V* – CL – Mwt 317.5 PSA 12.5 Å^2 log *P* 5.73	AMOROLFINE *rat, dog Topical antifungal
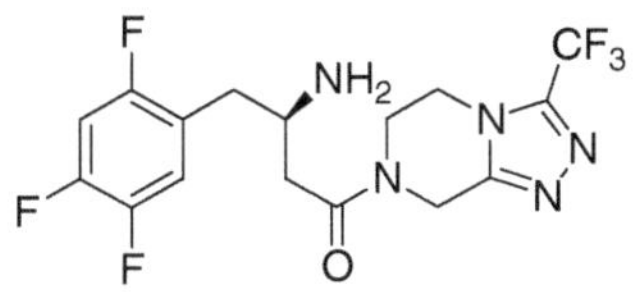	$t_{1/2}$ 8–14 h *F* 87% pb 38% ur 79%	*V* 3 CL – Mwt 407.3 PSA 77.0 Å^2 log *P* 1.30	SITAGLIPTIN DPPIV inhibitor, antidiabetic
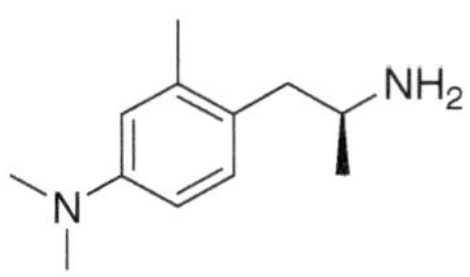	$t_{1/2}$ 5–12 h *F* – pb – ur 0.2%	*V* 4.2–8.4 CL 7–14 Mwt 192.3 PSA 29.3 Å^2 log *P* 1.93	AMIFLAMINE Antidepressant, MAO-inhibitor Metabolism: N-demethylation
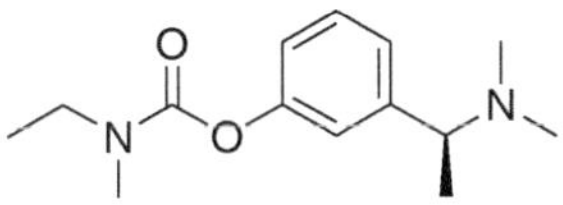	$t_{1/2}$ 1–2 h *F* 35–72% pb 40% ur 0%	*V* 0.7–3.2 CL 5.2–20 Mwt 250.3 PSA 32.8 Å^2 log *P* 2.14	RIVASTIGMINE Nootropic, crosses bbb Metabolism: hydrolysis of carbamate, N-demethylation
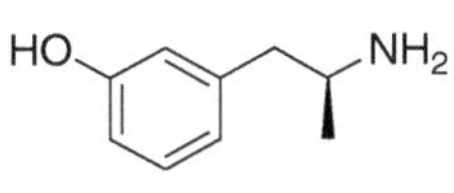	$t_{1/2}$ 3.5–4.0 h *F* 90–100% pb – ur –	*V* – CL 1.8±0.3 Mwt 151.2 PSA 46.3 Å^2 log *P* 1.11	GEPEFRINE Sympathomimetic, antihypotensive
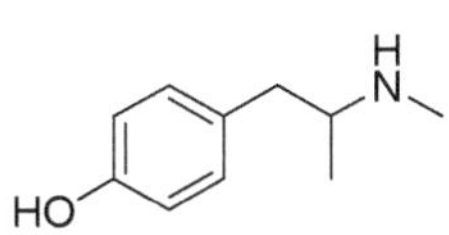	$t_{1/2}$ 4.8 h *F* 20% pb – ur 3%	*V* – CL – Mwt 165.2 PSA 32.3 Å^2 log *P* 1.8	PHOLEDRINE α-Adrenergic agonist, antihypotensive, metabolite of methamphetamine

(*continued overleaf*)

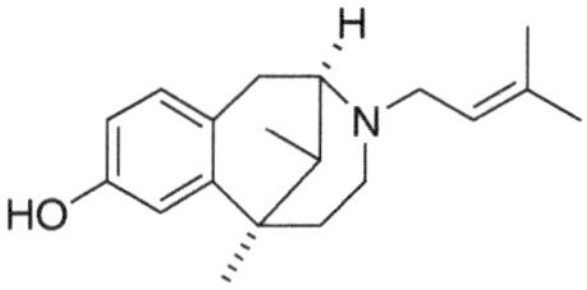	$t_{1/2}$ 4.6±1.0 h *F* 47±15% pb 65% ur 15±7%	*V/F* 7.1±1.4 CL/*F* 17±5 Mwt 285.4 PSA 23.5 Å^2 log *P* 4.15	PENTAZOCINE Analgesic; Metabolism: allylic oxidation of CH_3 to CH_2OH and CO_2H
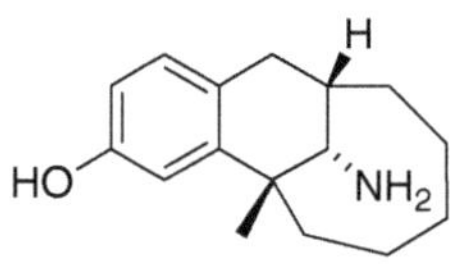	$t_{1/2}$ 2.5±0.3 h *F* – pb – ur <1%	*V* 11–12 CL 50 Mwt 245.4 PSA 46.3 Å^2 log *P* 3.68	DEZOCINE Analgesic; Metabolism: O-glucuronidation, N-oxidation
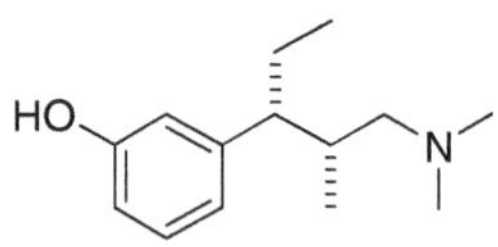	$t_{1/2}$ 2–4 h *F* 32±7% pb 20% ur 3%	*V* 8.3±1.5 CL 24±3 Mwt 221.3 PSA 23.5 Å^2 log *P* 3.02	TAPENTADOL Opiate agonist, analgesic Metabolism: glucuronidation, sulfation, N-demethylation
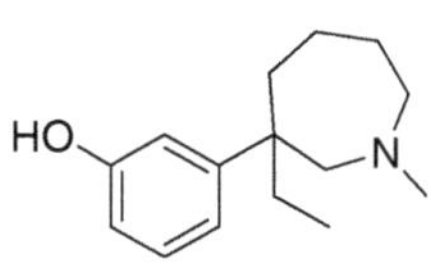	$t_{1/2}$ 2±1 h *F* 9% pb 27% ur 0%	*V* 5 CL 30 Mwt 233.4 g/mol PSA 23.5 Å^2 log *P* 3.40	MEPTAZINOL Opioid analgesic Metabolism: glucuronidation, oxidation to lactam
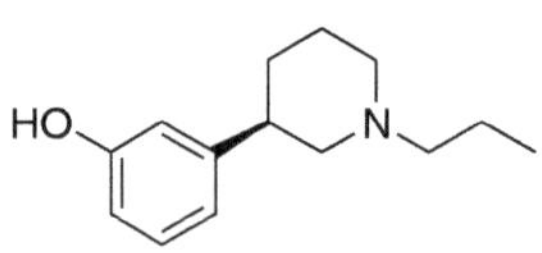	$t_{1/2}$ 2.0–2.5 h* *F* 4% pb – ur –	*V* – CL – Mwt 219.3 PSA 23.5 Å^2 log *P* 2.74	PRECLAMOL *im Dopamine D_2 agonist, antipsychotic; Metabolism: O-sulfation
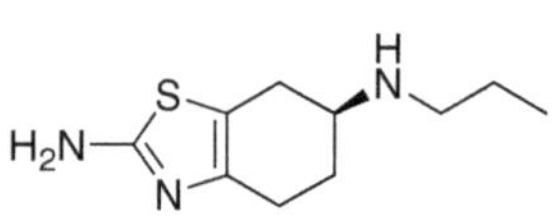	$t_{1/2}$ 12–14 h *F* 90% pb – ur 90%	*V* 7.0–7.3 CL 0.11 Mwt 211.3 PSA 79.2 Å^2 log *P* 1.42	PRAMIPEXOLE Antiparkinsonian
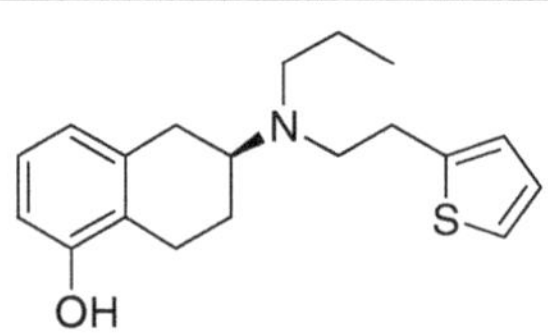	$t_{1/2}$ 5–7 h *F* 0.5%* pb 92% ur –	*V* 2.3 CL – Mwt 211.3 PSA 53.7 Å^2 log *P* 0.6	ROTIGOTINE *37% transdermal Dopamine agonist, antiparkinsonian; high first-pass effect; Metabolism: N-dealkylation, glucuronidation
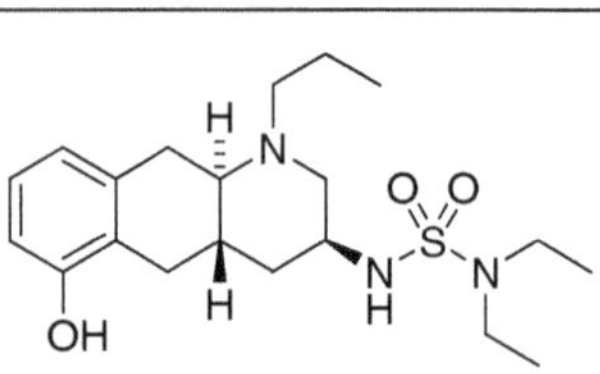	$t_{1/2}$ 12–17 h *F* 100% pb 90% ur –	*V* 1.5 CL – Mwt 395.6 PSA 81.3 Å^2 log *P* 4.12	QUINAGOLIDE Dopamine D_2 agonist for treatment of elevated levels of prolactin; Metabolites: glucuronides, sulfates, *N*-desethyl, *N*-bis-desethyl

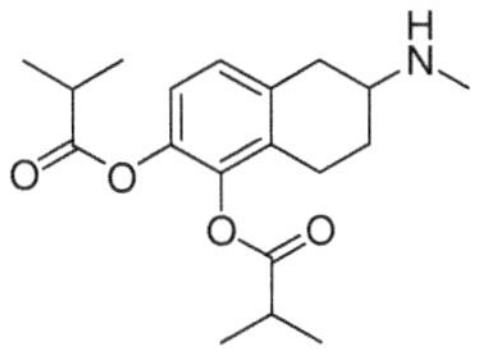	$t_{1/2}$ 3 h *F* – pb – ur –	*V* – CL – Mwt 333.4 PSA 64.6 Å^2 log *P* 1.97	NOLOMIROLE Dopamine D_2 and α_2 adrenergic agonist for treatment of congestive heart failure
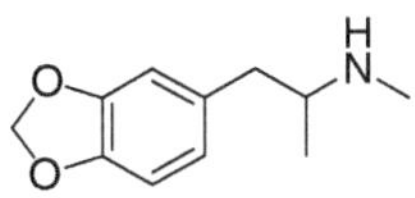	$t_{1/2}$ 6–9 h *F* – pb 20% ur 20%	*V* 6–7 CL – Mwt 193.2 PSA 30.5 Å^2 log *P* 2.1	3,4-METHYLENEDIOXY-METHAMPHETAMINE, MDMA, ECSTASY Recreational drug Metabolism: hydrolysis of acetal
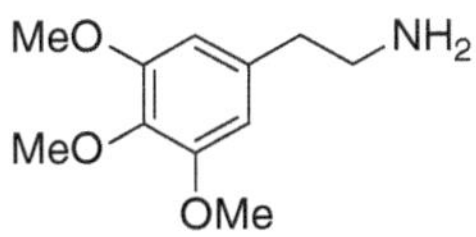	$t_{1/2}$ 6 h *F* – pb – ur 58%	*V* – CL – Mwt 211.3 PSA 53.7 Å^2 log *P* 0.6	MESCALINE 5-HT_{2A} agonist, natural hallucinogenic; Metabolism: oxidation to arylacetic and benzoic acids, O-demethylation, N-acetylation
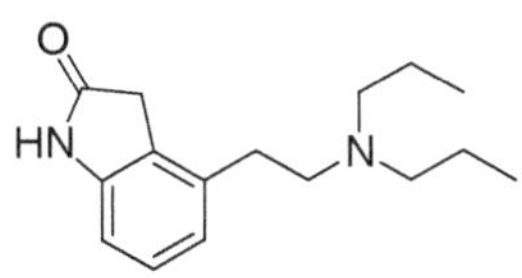	$t_{1/2}$ 2–10 h *F* 55% pb 10–39% ur <5%	*V* 7.5 CL 6–12 Mwt 260.4 PSA 32.3 Å^2 log *P* 2.49	ROPINIROLE Dopamine agonist, antiparkinsonian, crosses bbb; Metabolism: N-depropylation, aromatic 7-hydroxylation
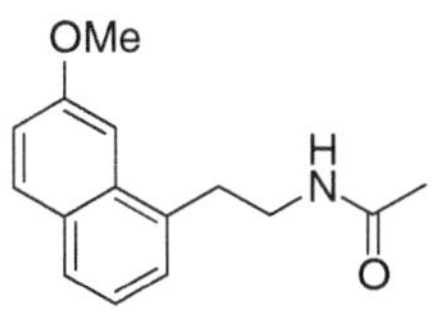	$t_{1/2}$ 1–2 h *F* <5% pb 95% ur 0.01% (iv)	*V* 0.54 CL 17 Mwt 243.3 PSA 38.3 Å^2 log *P* 2.27	AGOMELATINE Antidepressant; Metabolism: O-demethylation, aromatic 3-hydroxylation (*meta* to amidoethyl)
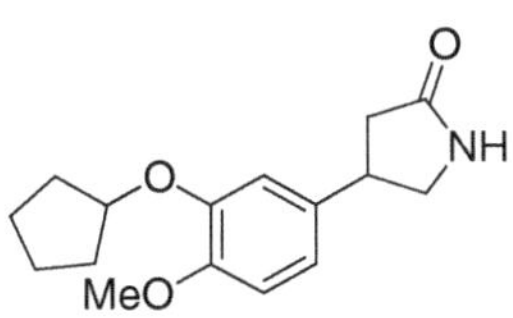	$t_{1/2}$ 8 h *F* 74% pb – ur –	*V* 0.5 CL 6.1 Mwt 275.3 PSA 47.6 Å^2 log *P* 2.00	ROLIPRAM Antidepressant; only the *R*(−) isomer is active Metabolism: O-dealkylation, cyclopentyl 2- and 3-hydroxylation, pyrrolidone 5-hydroxylation
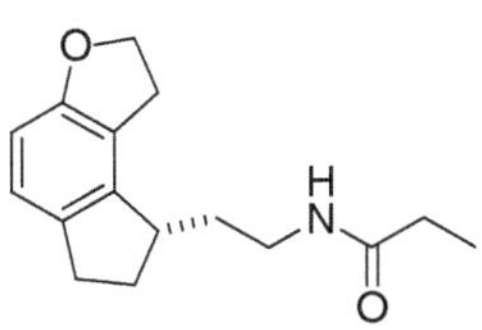	$t_{1/2}$ 0.8–2.6 h *F* 0.5–12% pb 82% ur <0.1%	*V* 2 CL 13 Mwt 259.3 PSA 38.3 Å^2 log *P* 2.29	RAMELTEON, TAK-375 Melatonin MT_1/MT_2 agonist Metabolism: hydroxylation of ethyl group, benzylic hydroxylation

(*continued overleaf*)

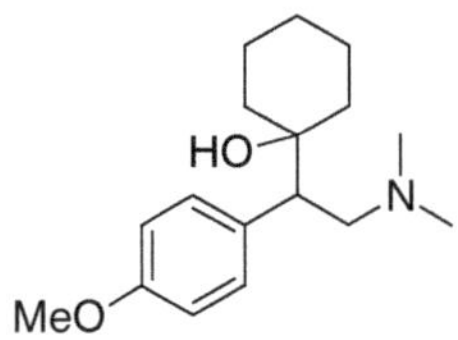	$t_{1/2}$ 4.9±2.4 h F 10–60% pb 27±2% ur 4.6±3%	V/F 8±4 CL/F 22±10 Mwt 277.4 PSA 32.7 Å^2 log P 3.3	VENLAFAXINE Antidepressant, serotonin-norepinephrine reuptake inhibitor, racemate Metabolism: N- and O-demethylation
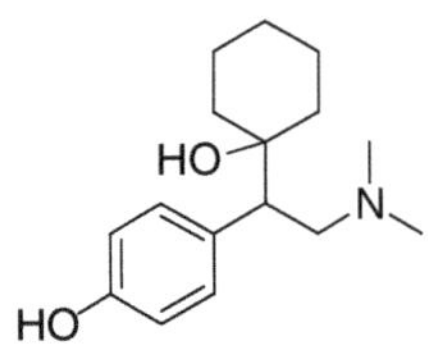	$t_{1/2}$ 10±4 h F – pb 30±12% ur 29±7%	V – CL – Mwt 263.4 PSA 43.7 Å^2 log P 3.3	*O*-DESMETHYL-VENLAFAXINE Main active metabolite of venlafaxine
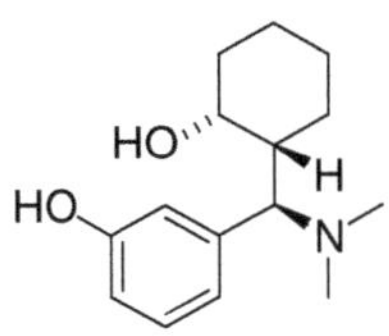	$t_{1/2}$ 5 h F 82% pb – ur 40–50%	V 1.5 CL 4.9 Mwt 249.4 PSA 43.7 Å^2 log P 2.00	CIRAMADOL Analgesic; Metabolite: phenolic glucuronide
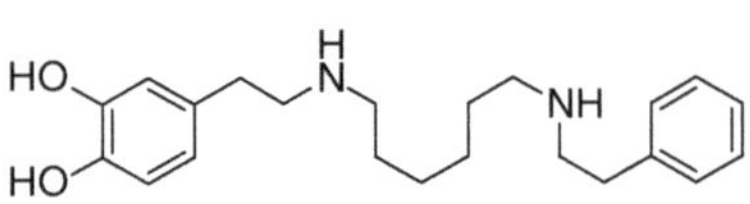	$t_{1/2}$ 7 min F – pb – ur –	V – CL 36 Mwt 356.5 PSA 64.5 Å^2 log P 3.23	DOPEXAMINE Cardiotonic
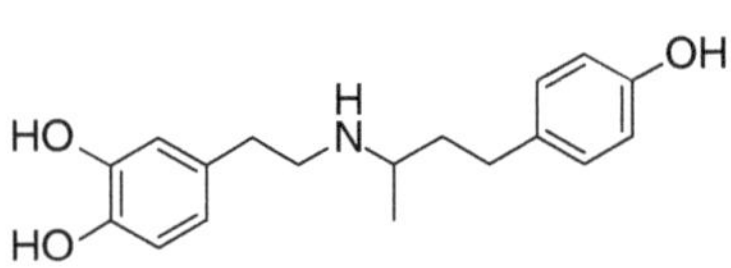	$t_{1/2}$ 2–3 min F – pb – ur –	V 0.2 CL 60 Mwt 301.4 PSA 72.7 Å^2 log P 3.1	DOBUTAMINE Cardiotonic, usually given as continuous infusion (iv) Metabolism: glucuronidation, O-methylation of catechol
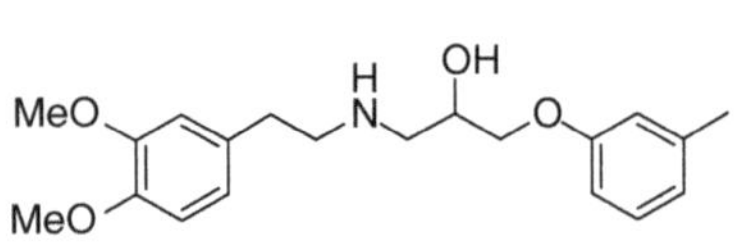	$t_{1/2}$ 1.5–2.0 h F 60% pb >95% ur <1%	V 1.5 CL 4.4–6.9 Mwt 345.4 PSA 60.0 Å^2 log P 3.00	BEVANTOLOL Antianginal, antihypertensive, antiarrhythmic Metabolism: aromatic 4-hydroxylation, oxidation of CCH_3 to CCH_2OH and CCO_2H, then glucuronidation
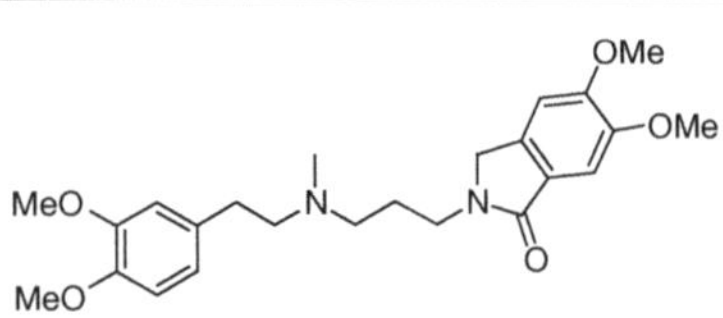	$t_{1/2}$ 1–3 h F – pb 88% ur 14% (iv)	V – CL 17±2 Mwt 428.5 PSA 60.5 Å^2 log P 1.48	FALIPAMIL Calcium channel blocker Metabolism: N-demethylation

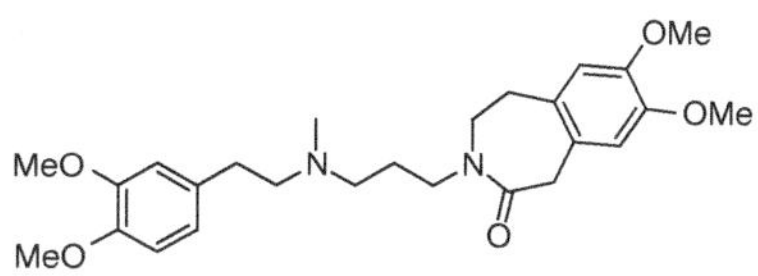	$t_{1/2}$ 2–3 h *F* 50% pb – ur 12%	*V* – CL – Mwt 456.6 PSA 60.5 Å^2 log *P* 4.48	ZATEBRADINE Antianginal Metabolism: N- and O-demethylation
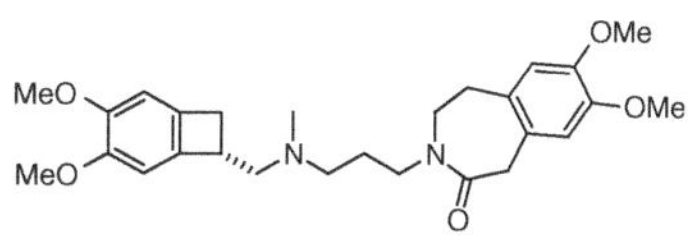	$t_{1/2}$ 2–11 h *F* 40±10% pb 70% ur 4%	*V* 1.4 CL 6.2 Mwt 468.6 PSA 60.5 Å^2 log *P* 3.69	IVABRADINE Antianginal Metabolism: N-demethylation
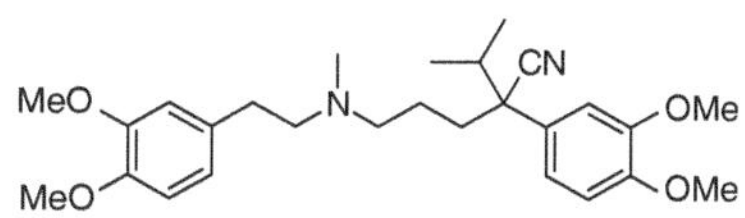	$t_{1/2}$ 4.0±1.5 h *F* 22±8% pb 90±2% ur <3%	*V* 5.0±2.1 CL 15±6 Mwt 454.6 PSA 64.0 Å^2 log *P* 4.5	VERAPAMIL Antihypertensive, antianginal, antiarrhythmic, CYP3A4 substrate; active metabolite: *N*-desmethyl ($t_{1/2}$ 9±3 h)
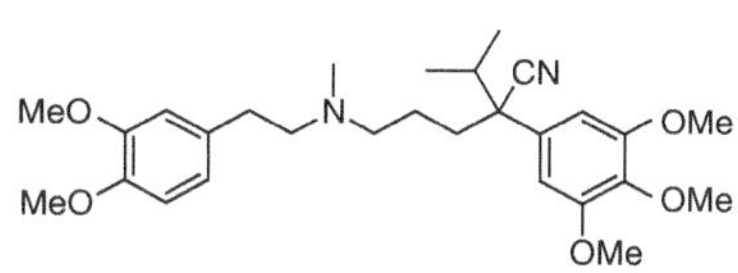	$t_{1/2}$ 2.5–8.0 h *F* 15–23% pb 93% ur <2%	*V* 2.1 CL 6–17 Mwt 484.6 PSA 73.2 Å^2 log *P* 3.83	GALLOPAMIL Antianginal; Metabolism: O-demethylation, N-dealkylation
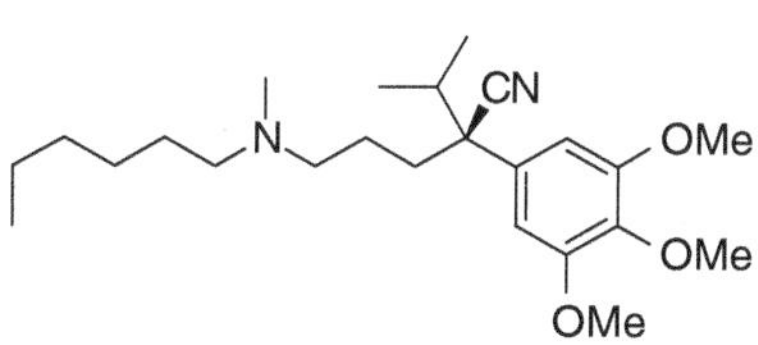	$t_{1/2}$ 13 h *F* 23% pb 95–97% ur –	*V* – CL 26–67 (dog) Mwt 404.6 PSA 54.7 Å^2 log *P* 4.45	NEXOPAMIL Calcium channel blocker, 5-HT$_2$ antagonist, antihypertensive
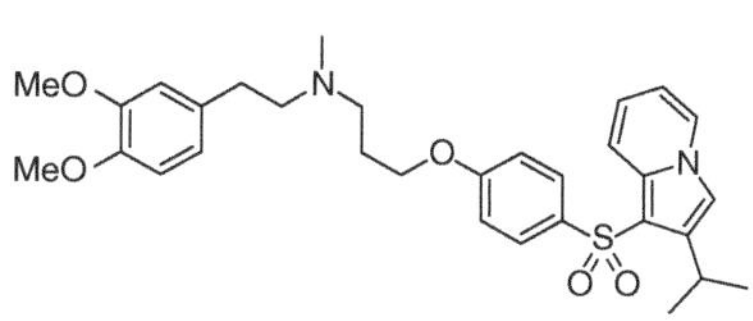	$t_{1/2}$ 5±3 h *F* – pb – ur –	*V* – CL – Mwt 550.7 PSA 77.9 Å^2 log *P* 5.06	FANTOFARONE Calcium channel blocker; antihypertensive Metabolite: *N*-desmethyl ($t_{1/2}$ 4±1 h)

(*continued overleaf*)

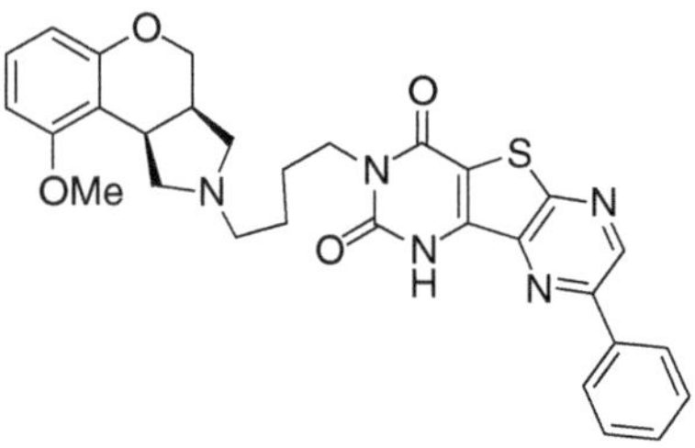	$t^1/_2$ 9–16 h *F* – pb – ur <1%	*V/F* 3.6–14.2 CL/*F* 4.4–10.9 Mwt 555.7 PSA 125 Å^2 log *P* 4.39	FIDUXOSIN α_1-Adrenoceptor antagonist for treatment of benign prostatic hyperplasia; orally available
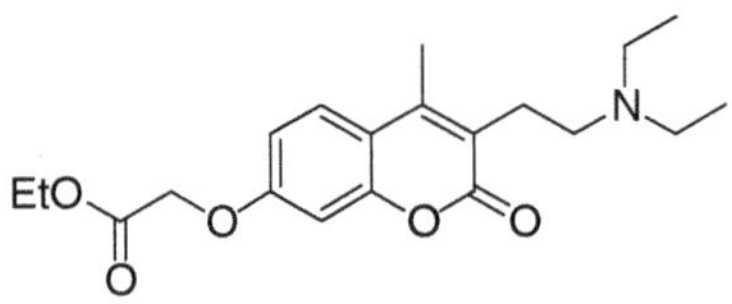	$t^1/_2$ 0.8 h* *F* 25–30% pb – ur 20%*	*V* 0.4* CL – Mwt 361.4 PSA 65.1 Å^2 log *P* 3.70	CARBOCROMEN *acid Vasodilator Metabolite: carboxylic acid
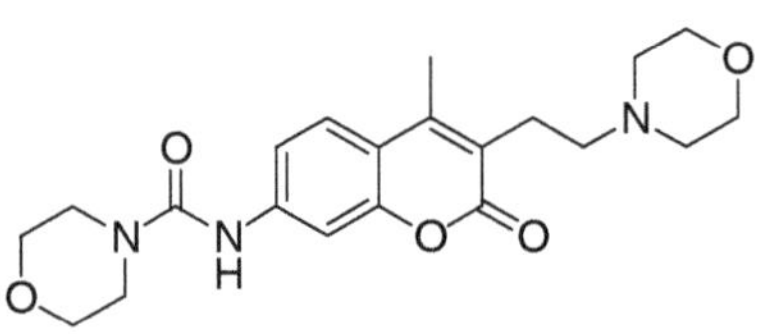	$t^1/_2$ 3–4 h *F* 67–79% pb – ur 54–64%	*V* – CL – Mwt 401.5 PSA 80.3 Å^2 log *P* 0.42	MOROCROMEN Cardiovascular agent
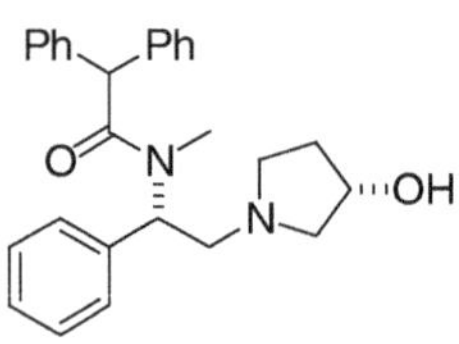	$t^1/_2$ 6–20 h *F* 40–50% pb 95–97% ur –	*V* – CL – Mwt 414.5 PSA 43.8 Å^2 log *P* 3.71	ASIMADOLINE Analgesic; Metabolism: glucuronidation, aromatic hydroxylation, oxidative ring fission of pyrrolidine
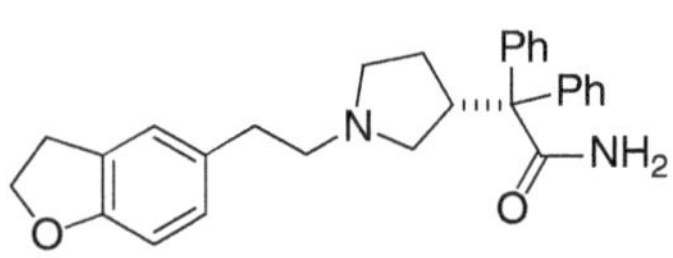	$t^1/_2$ 13–19 h *F* 15–19% pb 98% ur 3%	*V* 2.6 CL 12 Mwt 426.6 PSA 55.6 Å^2 log *P* 4.50	DARIFENACIN Muscarinic acetylcholine M_3 antagonist for treatment of urinary incontinence Metabolism: hydroxylation and ring-opening of dihydrobenzofuran, N-dealkylation of pyrrolidine
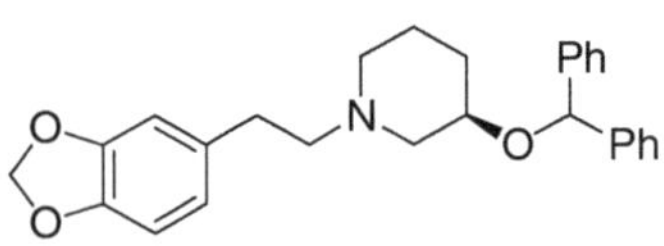	$t^1/_2$ – *F* – pb 99.99% ur –	*V* – CL 0.65 Mwt 415.5 PSA 30.9 Å^2 log *P* 6.34	ZAMIFENACIN Muscarinic acetylcholine M_3 antagonist for treatment of irritable bowel syndrom

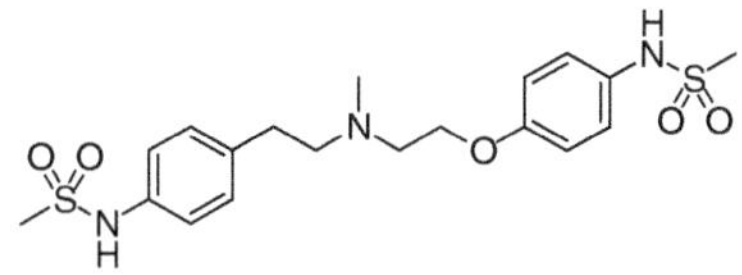	$t_{1/2}$ 7–8 h *F* 92–96% pb 60–70% ur 70%	*V* 3.2–3.6 CL 5.0–6.7 Mwt 441.6 PSA 105 Å^2 log *P* 2.0	DOFETILIDE Antiarrhythmic Metabolism: N-dealkylation, N-oxidation
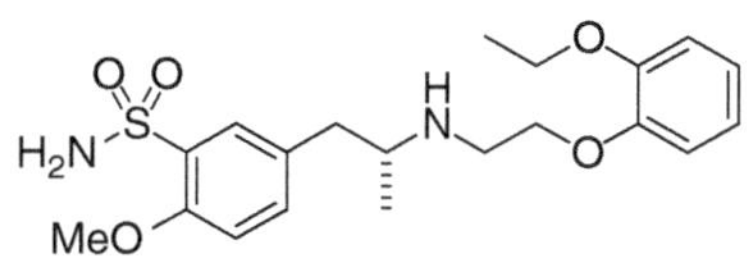	$t_{1/2}$ 9±4 h *F* 100% pb 99% ur 11±3%	*V* 0.23 CL 0.5–0.7 Mwt 408.5 PSA 99.9 Å^2 log *P* 2.6	TAMSULOSIN For treatment of benign prostatic hypertrophy; CYP3A4 and 2D6 substrate Metabolism: aromatic hydroxylation (*para* to OR), O-demethylation, O-deethylation, N-dealkylation (to 2-ethyloxyphenoxyacetic acid)
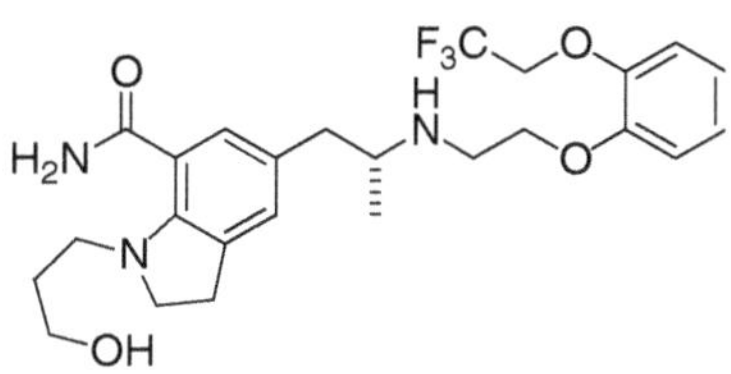	$t_{1/2}$ 13±8 h *F* 32% pb 97% ur <4%	*V* 0.8 CL 2.6 Mwt 495.5 PSA 97.1 Å^2 log *P* 2.33	SILODOSIN α_1-Adrenoceptor antagonist Metabolism: glucuronidation, oxidation of alcohol
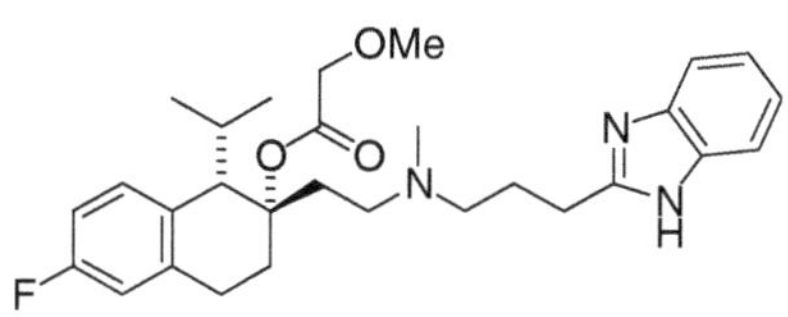	$t_{1/2}$ 17–25 h *F* 70–90% pb >99% ur <3%	*V* 2–3 CL 4–7 Mwt 495.6 PSA 67.5 Å^2 log *P* 6.21	MIBEFRADIL Calcium channel blocker, antihypertensive Metabolism: ester hydrolysis, benzylic hydroxylation to secondary alcohol, then oxidation to ketone, N-dealkylation; withdrawn in 1998 because of drug–drug interactions
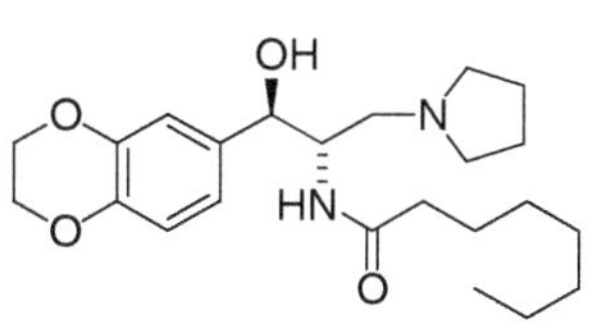	$t_{1/2}$ 7 h *F* High pb – ur <1.5%	*V/F* 133±69 CL – Mwt 404.5 PSA 71.0 Å^2 log *P* 3.61	ELIGLUSTAT Inhibitor of glucosylceramide synthase for treatment of Gaucher and Fabry disease

$t_{1/2}$, plasma half-life; *F*, oral bioavailability; pb, plasma protein binding; ur, excretion of unchanged drug in urine; V, volume of distribution; CL, clearance; Mwt, molecular weight; PSA, polar surface area; bbb, blood–brain barrier; DPPIV, dipeptidylpeptidase IV; GABA, γ-aminobutyric acid; HT, hydroxytryptamine; NMDA, *N*-methyl D-aspartic.

Table 47.2 2-Aryl-2-hydroxyethylamines and related compounds. *V* in l/kg; CL in ml min^{-1} kg^{-1}; Mwt in g/mol.

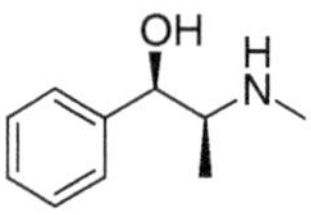	$t^1\!/_2$ 3–6 h *F* 85% pb 0% ur 22–99%	*V*/*F* 2.2–4.5 CL/*F* 4.3–9.8 Mwt 165.2 PSA 32.3 Å^2 log *P* 1.2	EPHEDRINE Bronchodilator Metabolism: N-dealkylation, aromatic 4-hydroxylation, then conjugation
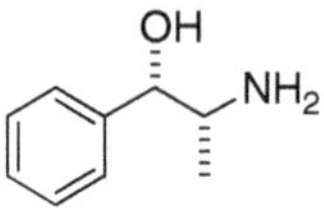	$t^1\!/_2$ 4.7±0.4 h *F* >70% pb – ur 65±10%	*V*/*F* 4.1±0.9 CL/*F* 10±3 Mwt 151.2 PSA 46.3 Å^2 log *P* 0.3	PHENYLPROPANOLAMINE Decongestant, anorexic, bronchodilator; withdrawn in 2000 because of stroke risk
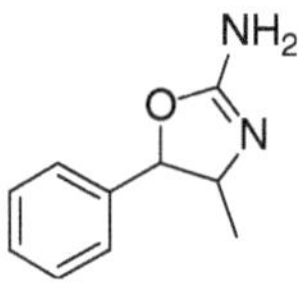	$t^1\!/_2$ 10–19 h *F* 62% pb – ur –	*V* – CL – Mwt 176.2 PSA 47.6 Å^2 log *P* 1.32	METHYLAMINOREX Stimulant, anorexic, and recreational drug (racemic cis isomer)
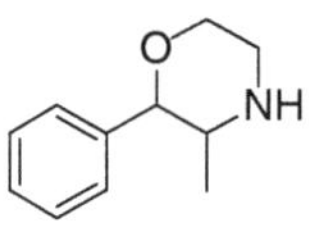	$t^1\!/_2$ 16–31 h *F* – pb – ur 19%	*V* – CL – Mwt 177.3 PSA 21.3 Å^2 log *P* 1.86	PHENMETRAZINE Anorexic Metabolism: N- and O-dealkylation
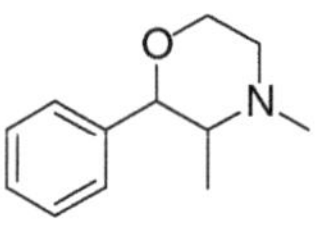	$t^1\!/_2$ 19–24 h *F* – pb – ur –	*V* – CL – Mwt 191.3 PSA 12.5 Å^2 log *P* 2.1	PHENDIMETRAZINE Anorexic Metabolism: N-demethylation, N-oxidation
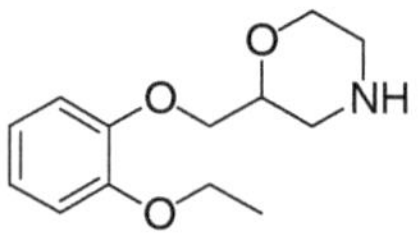	$t^1\!/_2$ 3.3±1.0 h *F* 85±14% pb – ur –	*V* 0.73±0.28 CL 124±11 Mwt 237.3 PSA 39.7 Å^2 log *P* 1.3	VILOXAZINE Antidepressant Metabolism: aromatic 4- and 5-hydroxylation
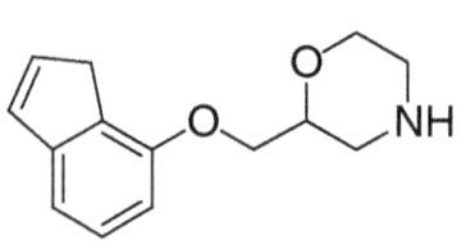	$t^1\!/_2$ 2.2 h (rat) *F* – pb – ur –	*V* – CL – Mwt 231.3 PSA 30.5 Å^2 log *P* 1.79	INDELOXAZINE Antidepressant, nootropic Metabolism (rat): dihydroxylation of alkene, N-acetylation, oxidative degradation of morpholine ring

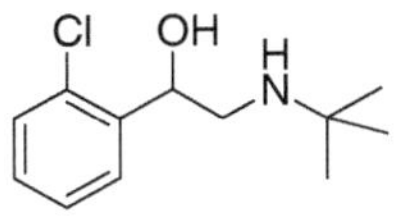	*t*½ 4±2 h *F* 40% pb – ur 14%	*V* – CL – Mwt 227.7 PSA 32.3 Å^2 log *P* 2.7	TULOBUTEROL Bronchodilator Metabolism: aromatic hydroxylation (para to alkyl)
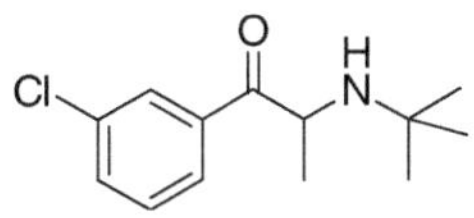	*t*½ 12±4 h *F* 5–20%* pb 84±2% ur <1%	*V*/*F* 7.2±1.6 CL/*F* 35±9 Mwt 239.8 PSA 29.1 Å^2 log *P* 3.4	BUPROPION *animals; antidepressant, dopamine–norepinephrine reuptake inhibitor Metabolism: reduction of ketone to alcohol, hydroxylation of *tert*-butyl, then hemiacetal formation
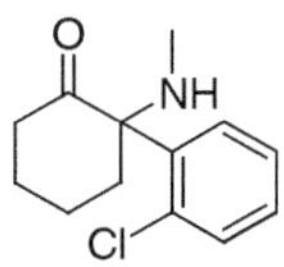	*t*½ 2.3 h *F* 20±7% pb 12–47% ur 4%	*V* 1.8±0.7 CL 15±5 Mwt 237.7 PSA 29.1 Å^2 log *P* 2.27	KETAMINE Anesthetic (iv); NMDA (*N*-methyl aspartate) antagonist Metabolism: N-demethylation, dehydrogenation to conjugated cyclohexenone
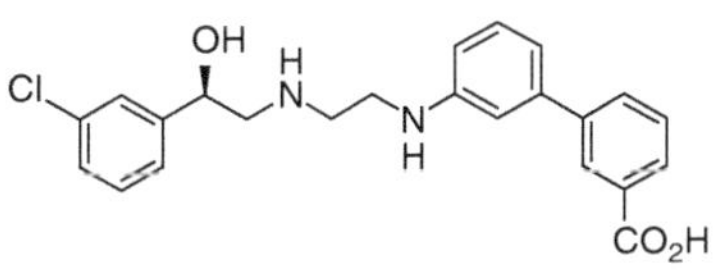	*t*½ 5 h* *F* 46%* pb – ur 14%	*V* – CL – Mwt 227.7 PSA 32.3 Å^2 log *P* 2.7	SOLABEGRON *monkey β3 adrenergic agonist, treatment of overactive bladder and irritable bowel syndrom Metabolism: acyl glucuronidation
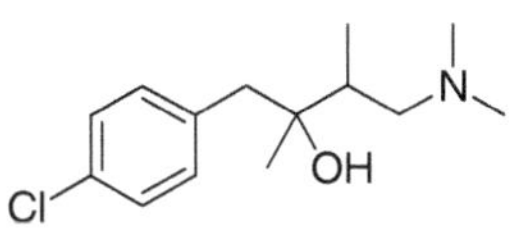	*t*½ 23–32 h *F* – pb Low ur –	*V* – CL – Mwt 255.8 PSA 23.5 Å^2 log *P* 3.26	CLOBUTINOL Antitussive Metabolite: *N*-desmethyl; withdrawn 2007 for causing cardiac arrhythmias
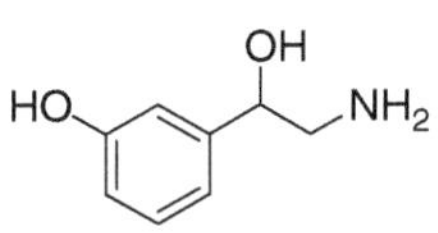	*t*½ 3–7 h *F* 20% pb – ur –	*V* – CL – Mwt 410.9 PSA 81.6 Å^2 log *P* 4.4	NORFENEFRINE Adrenergic Metabolism: glucuronidation, sulfation, oxidative deamination to hydroxymandelic acid and (hydroxyphenyl)ethyleneglycol

(*continued overleaf*)

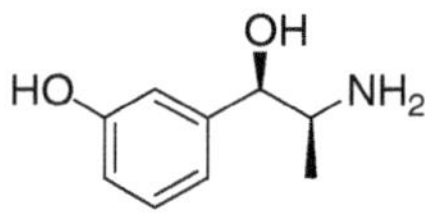	*t*½ 0.3–1.0 h* *F* – pb 45% ur –	*V* – CL – Mwt 167.2 PSA 66.5 Å^2 log *P* −0.2	METARAMINOL *iv Adrenergic
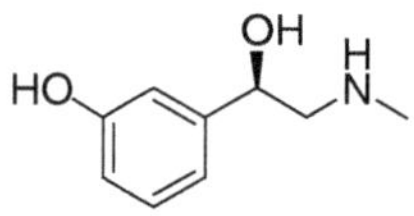	*t*½ 2.1–3.4 h *F* 38% pb – ur 2.6%	*V* 2.6–5.0 CL – Mwt 167.2 PSA 52.5 Å^2 log *P* 0.4	PHENYLEPHRINE Mydriatic, decongestant Metabolism: O-glucuronidation, O-sulfation, oxidation (by MAO) to 3-hydroxymandelic acid, deamination to diol
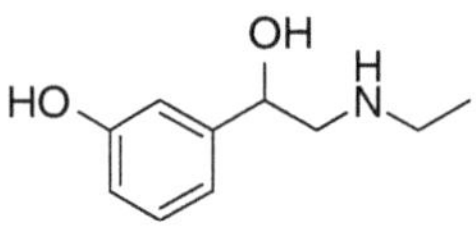	*t*½ 2±1 h *F* 35% pb 23% ur –	*V* 2.1 CL 14 Mwt 181.2 PSA 52.5 Å^2 log *P* 0.4	ETILEFRIN Antihypotensive Metabolism: O-sulfation
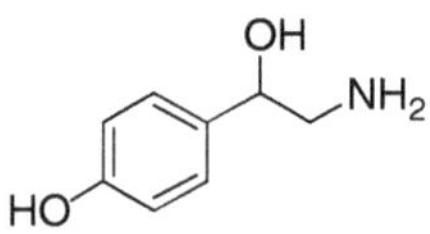	*t*½ 2.5 h *F* 25% pb – ur –	*V* – CL – Mwt 153.2 PSA 66.5 Å^2 log *P* −0.6	OCTOPAMINE Adrenergic Metabolism: deamination, conjugation
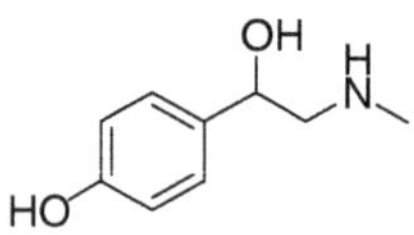	*t*½ 2–3 h *F* 10–20% pb – ur 2.5%	*V* – CL – Mwt 167.2 PSA 52.5 Å^2 log *P* 0.4	OXEDRINE, SYNEPHRINE Adrenergic, vasopressor Metabolism: O-sulfation, O-glucuronidation, deamination to hydroxymandelic acid and 2-(hydroxyphenyl)glycol
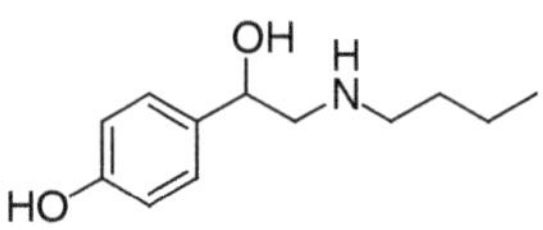	*t*½ 2.5 h *F* 75% pb – ur 30%	*V* 2.5–5.0 CL – Mwt 209.3 PSA 52.5 Å^2 log *P* 1.9	BAMETHAN Vasodilator
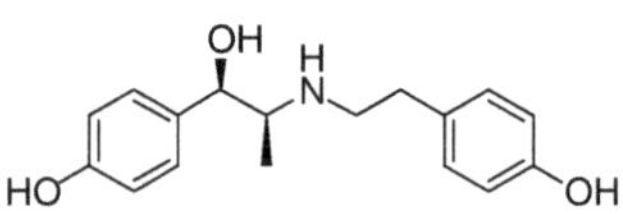	*t*½ 1.7–2.6 h *F* 30% pb 25–35% ur 35%	*V* 7 CL 23 Mwt 287.4 PSA 72.7 Å^2 log *P* 2.1	RITODRINE Tocolytic Metabolism: glucuronidation, sulfation

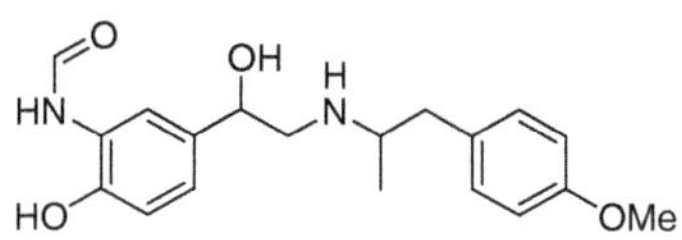	$t^1/_2$ 10 h *F* – pb 61–64% ur 10%	*V* – CL – Mwt 344.4 PSA 90.8 Å^2 log *P* 2.2	FORMOTEROL Antiasthmatic Metabolism: O-demethylation, glucuronidation, sulfation, N-deformylation
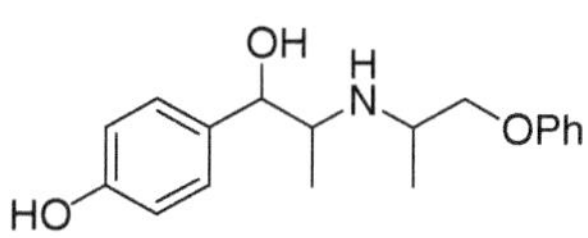	$t^1/_2$ 1.3 h *F* 2.2% pb – ur –	*V* – CL – Mwt 301.4 PSA 61.7 Å^2 log *P* 2.8	ISOXSUPRINE β-Adrenergic agonist, vasodilator
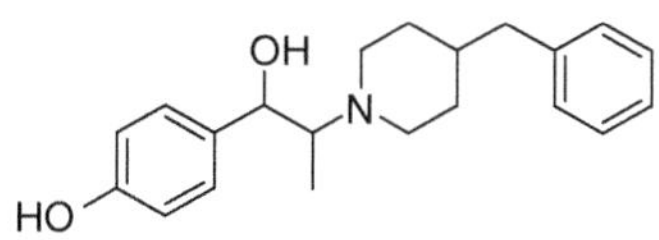	$t^1/_2$ 1 h *F* 5% pb – ur –	*V* – CL – Mwt 325.5 PSA 43.7 Å^2 log *P* 3.6	IFENPRODIL Vasodilator
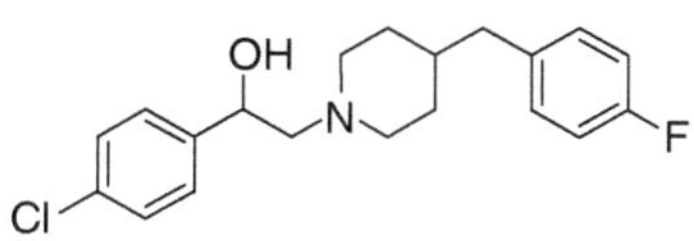	$t^1/_2$ 16–20 h *F* 47% pb – ur –	*V* – CL – Mwt 347.9 PSA 23.5 Å^2 log *P* 4.71	ELIPRODIL NMDA antagonist Metabolism: glucuronidation
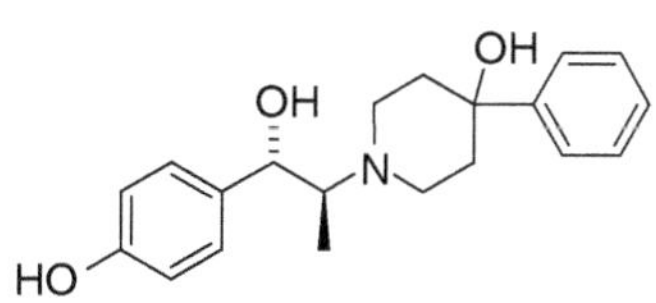	$t^1/_2$ 2–20 h *F* 23–62% pb 37% ur 6%	*V* 4.4–6.5 CL 4–27 Mwt 327.4 PSA 63.9 Å^2 log *P* 1.75	TRAXOPRODIL Glutamate (NMDA) antagonist, neuroprotectant Metabolism: hydroxylation to catechol, then O-methylation, glucuronidation, sulfation
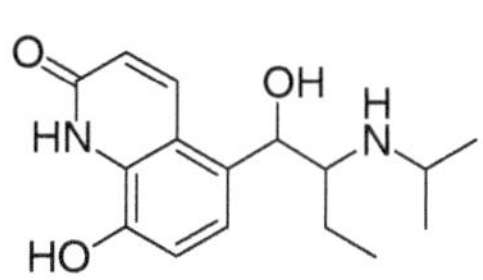	$t^1/_2$ 4 h *F* 40% pb – ur 20%	*V* 1.4 CL – Mwt 290.4 PSA 85.3 Å^2 log *P* 1.7	PROCATEROL Bronchodilator Metabolism: glucuronidation
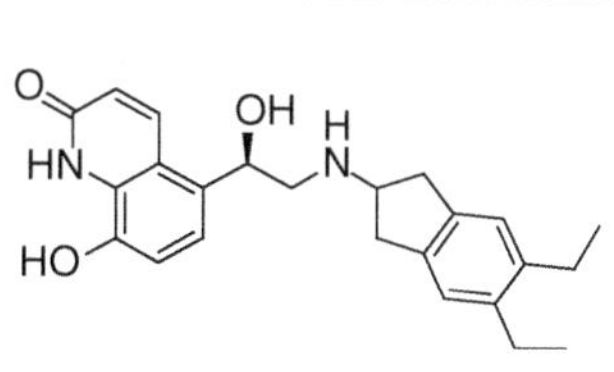	$t^1/_2$ 116–182 h *F* 43%* pb 95% ur <2%	*V* 39 CL 6 Mwt 392.5 PSA 81.6 Å^2 log *P* 3.88	INDACATEROL *inhaled, 33% (dog, po) Adrenergic agonist, bronchodilator Metabolism: hydroxylation, phenol-glucuronidation, N-dealkylation

(*continued overleaf*)

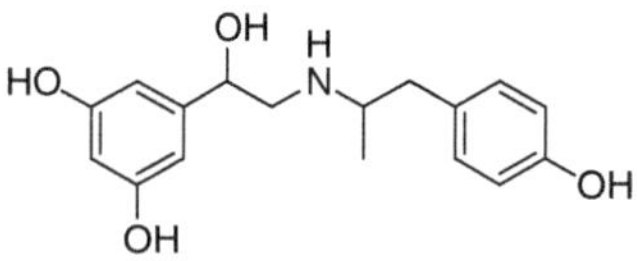	$t_{1/2}$ 3 h *F* 2–5% pb 50–55% ur 60%	*V* 1.9–2.7 CL – Mwt 303.4 PSA 92.9 Å^2 log *P* 1.5	FENOTEROL Bronchodilator, tocolytic Metabolism: O-sulfation, O-glucuronidation
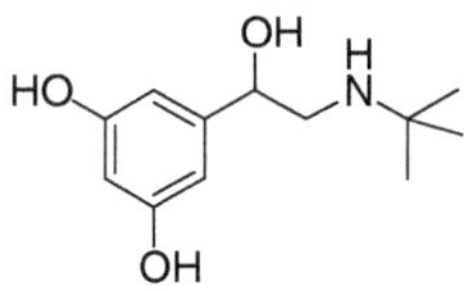	$t_{1/2}$ 14±2 h *F* 14±2% pb 20% ur 56±4%	*V* 1.8±0.2 CL 3.4±0.6 Mwt 225.3 PSA 72.7 Å^2 log *P* 1.4	TERBUTALINE Bronchodilator, tocolytic Metabolism: O-sulfation
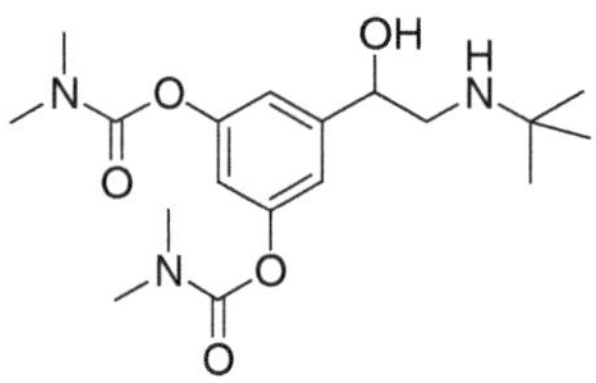	$t_{1/2}$ 9–15 h *F* 12–20% pb 45% ur –	*V* 1.6 CL – Mwt 367.4 PSA 71.6 Å^2 log *P* 1.4	BAMBUTEROL Bronchodilator Metabolism: hydrolysis of carbamates
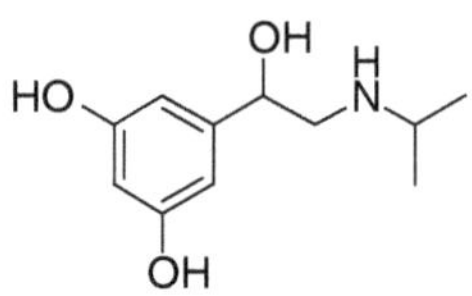	$t_{1/2}$ 2–6 h *F* 10–40% pb 10% ur –	*V* 7.6 CL 14 Mwt 211.3 PSA 72.7 Å^2 log *P* 1.0	ORCIPRENALINE, METAPROTERENOL Bronchodilator Metabolism: O-sulfation
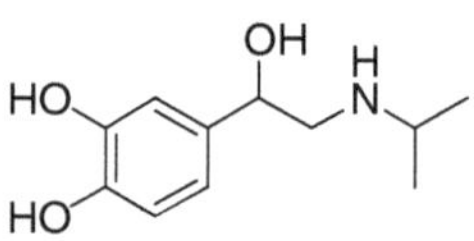	$t_{1/2}$ 3 min *F* 25% pb – ur 40%	*V* – CL – Mwt 211.3 PSA 72.7 Å^2 log *P* 0.32	ISOPROTERENOL Bronchodilator Metabolism: aromatic O-methylation by COMT
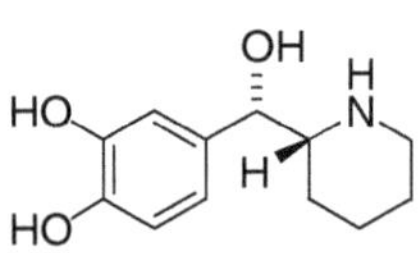	$t_{1/2}$ 3 min (iv) *F* – pb – ur 2%	*V* – CL – Mwt 223.3 PSA 72.7 Å^2 log *P* 0.74	RIMITEROL Bronchodilator Metabolism: O-sulfation, O-methylation
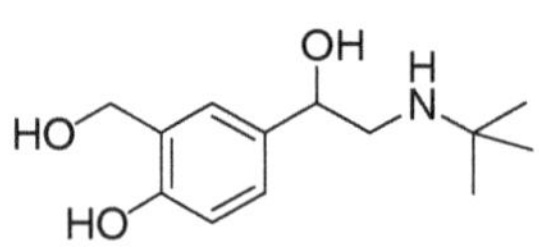	$t_{1/2}$ 2.7–5.0 h *F* 43–50% pb 8% ur 51–64%	*V* 1.6–2.5 CL 6.8–7.7 Mwt 239.3 PSA 72.7 Å^2 log *P* 1.4	SALBUTAMOL, ALBUTEROL Bronchodilator, tocolytic Metabolism: phenol sulfation

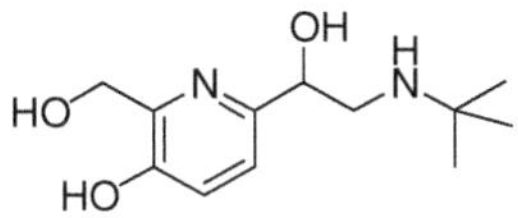	$t^{1}/_{2}$ 2–3 h *F* 60% pb – ur 10%	*V* – CL – Mwt 240.3 PSA 85.6 Å^2 log *P* 0.4	PIRBUTEROL Bronchodilator Metabolism: sulfation
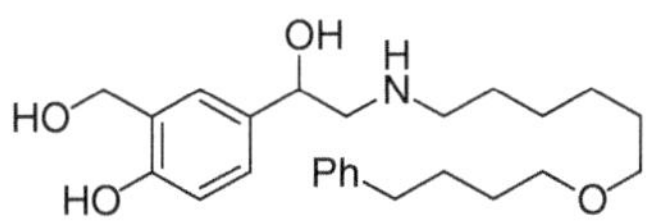	$t^{1}/_{2}$ 9±3 h *F* – pb 96% ur <5%	*V* – CL – Mwt 415.6 PSA 82 Å^2 log *P* 4.2	SALMETEROL Racemate, bronchodilator, CYP3A4 substrate Metabolism: benzylic hydroxylation
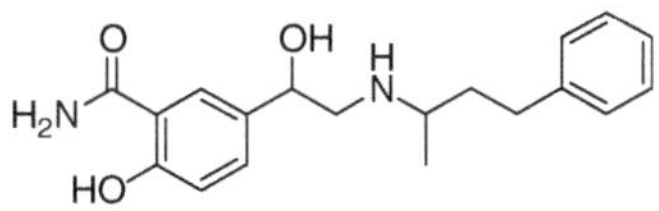	$t^{1}/_{2}$ 4.9±2.0 h *F* 18±5% pb 50% ur <5%	*V* 9.4±3.4 CL 25±10 Mwt 328.4 PSA 95.6 Å^2 log *P* 2.7	LABETALOL Antihypertensive, four stereoisomers Metabolism: phenol glucuronidation, N-dealkylation to 3-amino-1-phenylbutane
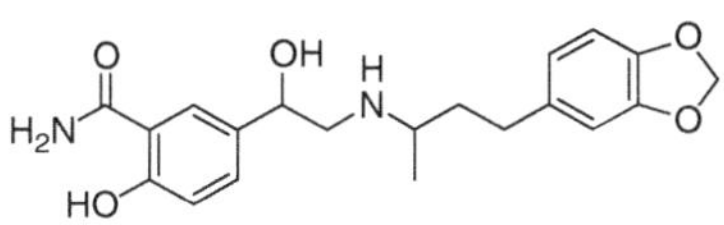	$t^{1}/_{2}$ 11–16 h *F* 30–64% pb – ur 8%	*V* 16 CL 16 Mwt 372.4 PSA 114 Å^2 log *P* 2.57	MEDROXALOL Vasodilator
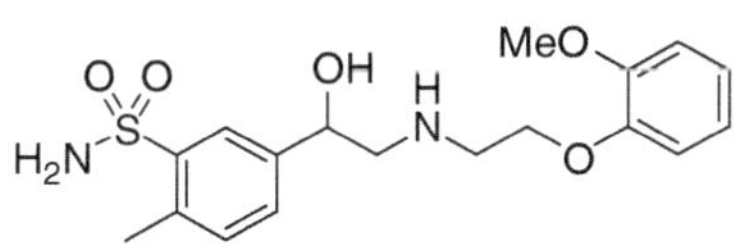	$t^{1}/_{2}$ 3–5 h *F* >93% pb 97–98% ur 35%	*V* 0.75 CL 1.9 Mwt 380.5 PSA 119 Å^2 log *P* 1.01	AMOSULALOL Antihypertensive Metabolism: aromatic hydroxylation (para to alkoxy), benzylic hydroxylation, O-demethylation
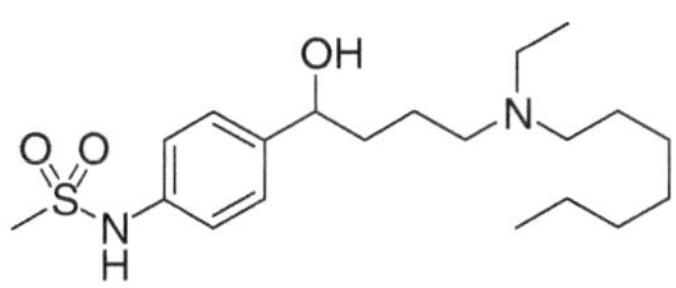	$t^{1}/_{2}$ 6–9 h (iv) *F* 3% pb 41% ur <5%	*V* 11–15 CL 24–31 Mwt 384.6 PSA 78.0 Å^2 log *P* 4.17	IBUTILIDE Antiarrhythmic Metabolism: hydroxylation and oxidative degradation of heptyl group
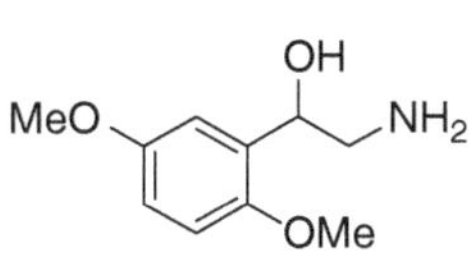	$t^{1}/_{2}$ 3–4 h *F* 93%* pb Low ur 80%	*V* 4–8 CL 20–30 Mwt 197.2 PSA 64.7 Å^2 log *P* −0.03	DESGLYMIDODRINE Antihypotensive, vasoconstrictor, does not cross bbb *drug on oral administration of its prodrug midodrine Metabolism: O-demethylation

(*continued overleaf*)

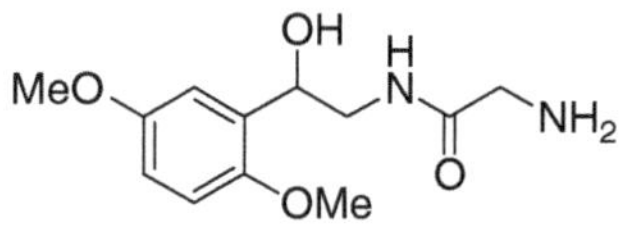	$t^1/_2$ 25 min *F* – pb low ur 2–3%	*V* – CL – Mwt 254.3 PSA 93.8 Å^2 log *P* −0.4	MIDODRINE Prodrug of desglymidodrine
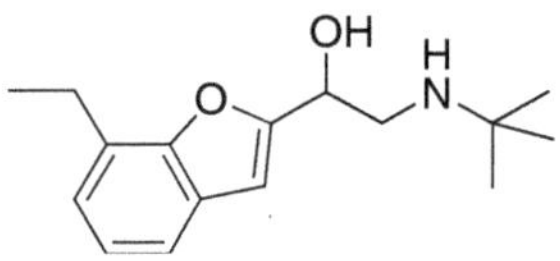	$t^1/_2$ 3.6 h *F* 46±15% pb – ur <1%	*V* 3.3 CL 10.7 Mwt 261.4 PSA 45.4 Å^2 log *P* 3.5	BUFURALOL Antianginal, antihypertensive Metabolism: aromatic and benzylic hydroxylation, then oxidation of benzylic alcohol to ketone
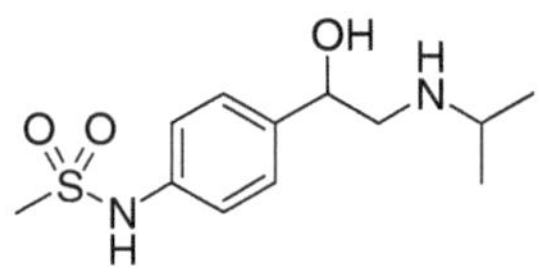	$t^1/_2$ 12±3 h *F* 90–100% pb 0% ur >75%	*V/F* 2.0±0.4 CL/*F* 2.6±0.5 Mwt 272.4 PSA 78.4 Å^2 log *P* 0.8	SOTALOL Antianginal, antiarrhythmic, antihypertensive
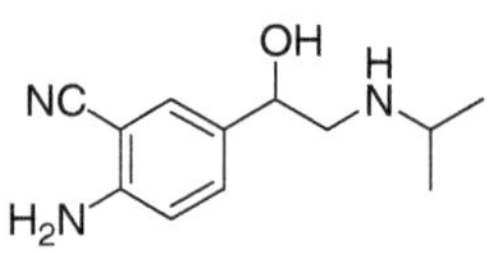	$t^1/_2$ 1 h* *F* – pb – ur –	*V* 4.1* CL – Mwt 219.3 PSA 82.1 Å^2 log *P* 0.65	CIMATEROL *steer, im Repartitioning agent
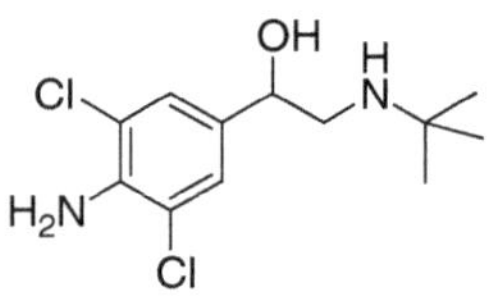	$t^1/_2$ 34 h *F* 89–98% pb >90% ur 67%	*V* – CL – Mwt 277.2 PSA 58.3 Å^2 log *P* 2.8	CLENBUTEROL Bronchodilator, tocolytic Metabolism: aromatic N-hydroxylation
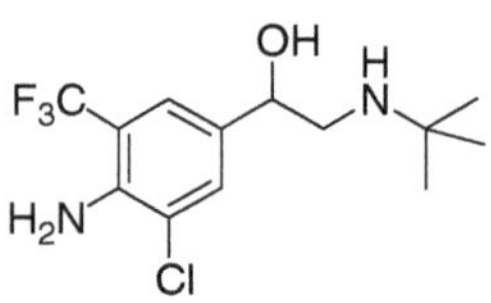	$t^1/_2$ 20–30 h *F* >91% pb – ur 24–64%	*V* 5.8 CL 3.0 Mwt 310.8 PSA 58.3 Å^2 log *P* 3.0	MABUTEROL Bronchodilator, antiasthmatic Metabolism: CH_3 hydroxylation

Table 47.3 Tetrahydrochinolines, tetrahydroisochinolines, and related compounds. *V* in l/kg; CL in ml min^{-1} kg^{-1}; Mwt in g/mol.

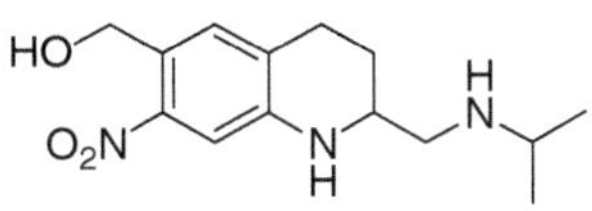	$t^1/_2$ *F* pb ur	1.0–2.5 h Good – –	*V* CL Mwt PSA log *P*	– – 279.3 90.1 Å^2 2.02	OXAMNIQUINE Anthelmintic Metabolism: oxidation of alcohol to carboxyl group
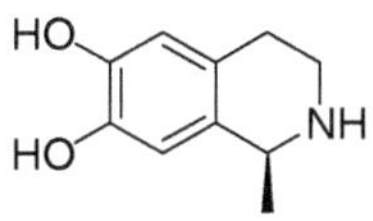	$t^1/_2$ *F* pb ur	12 min – – –	*V* CL Mwt PSA log *P*	– – 179.2 52.5 Å^2 0.85	SALSOLINOL Endogenous neurotoxin, contained in chocolate
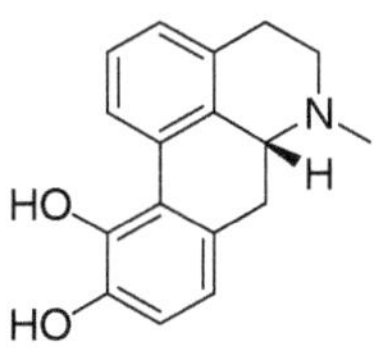	$t^1/_2$ *F* pb ur	41±15 min* 2% 94% <1%	*V* CL Mwt PSA log *P*	1.6±0.5 40±15 267.3 43.7 Å^2 1.70	APOMORPHINE *iv Emetic, crosses bbb Metabolism: N-demethylation, O-sulfation, and O-glucuronidation
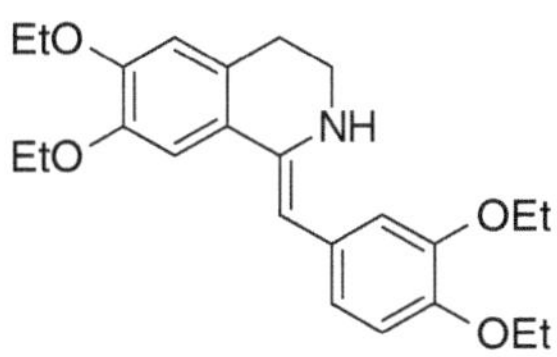	$t^1/_2$ *F* pb ur	7–12 h 25–91% 80–95% –	*V* CL Mwt PSA log *P*	3 3.5 397.5 49.0 Å^2 3.91	DROTAVERINE Phosphodiesterase IV inhibitor, antispasmodic
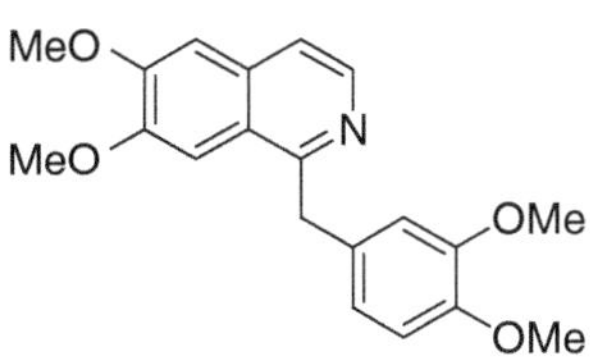	$t^1/_2$ *F* pb ur	2 h (iv) 5–99% 93% <1%	*V* CL Mwt PSA log *P*	1 11 339.4 49.8 Å^2 2.93	PAPAVERINE Vasodilator, opium alkaloid Metabolism: demethylation
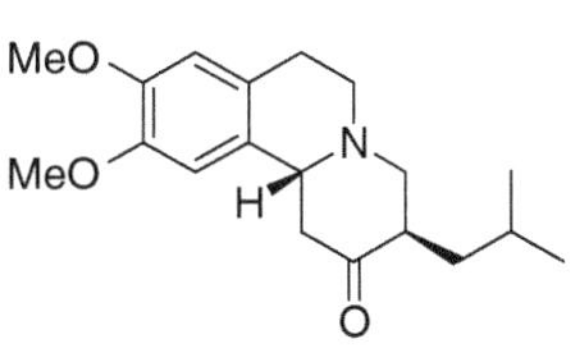	$t^1/_2$ *F* pb ur	2–8 h 5–7% 82–85% 0%	*V* CL Mwt PSA log *P*	– – 317.4 38.8 Å^2 2.93	TETRABENAZINE Antidyskinetic, antipsychotic Metabolism: reduction of ketone to alcohol, O-demethylation

(continued overleaf)

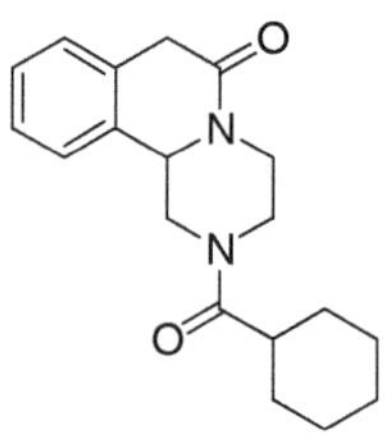

$t^1\!/_2$	1–2 h	*V*	*V*
F	Low	CL	–
pb	–	Mwt	312.4
ur	–	PSA	40.6 Å^2
		log *P*	2.66

PRAZIQUANTEL
Anthelmintic, crosses bbb

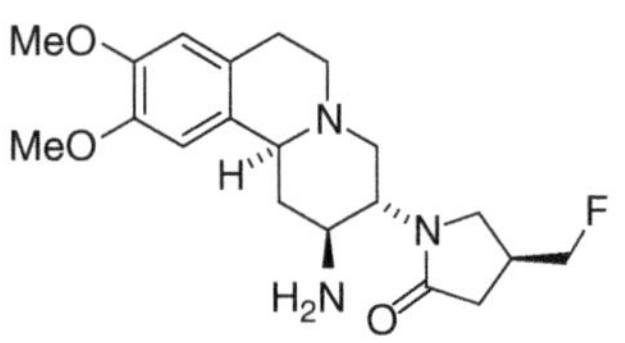

$t^1\!/_2$	10–13 h*	*V*	–
F	28–174%*	CL	–
pb	–	Mwt	377.5
ur	36%*	PSA	68.0 Å^2
		log *P*	1.21

CARMEGLIPTIN
*animals
DPPIV inhibitor, antidiabetic; little metabolism

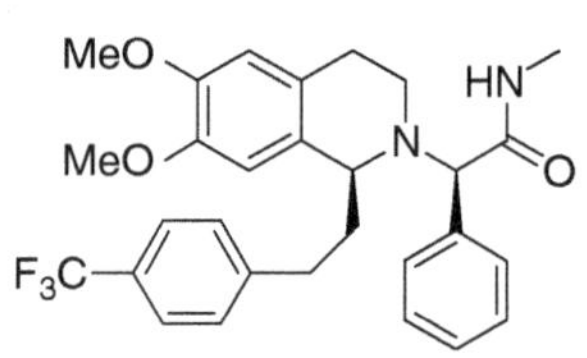

$t^1\!/_2$	13–19 h	*V*	–
F	–	CL	–
pb	–	Mwt	512.6
ur	–	PSA	50.8 Å^2
		log *P*	5.89

ALMOREXANT
Orexin antagonist, hypnotic

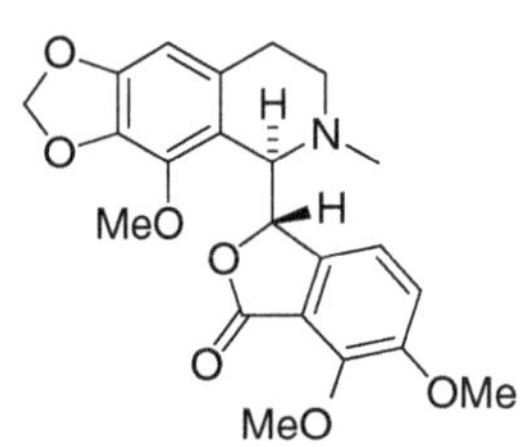

$t^1\!/_2$	2–4 h	*V*	4.7
F	30%	CL	22
pb	–	Mwt	413.4
ur	<1%	PSA	75.7 Å^2
		log *P*	2.39

NOSCAPINE
Antitussive, opium alkaloid
Metabolism: O-demethylation, cleavage of C–C bond between lactone and piperidine

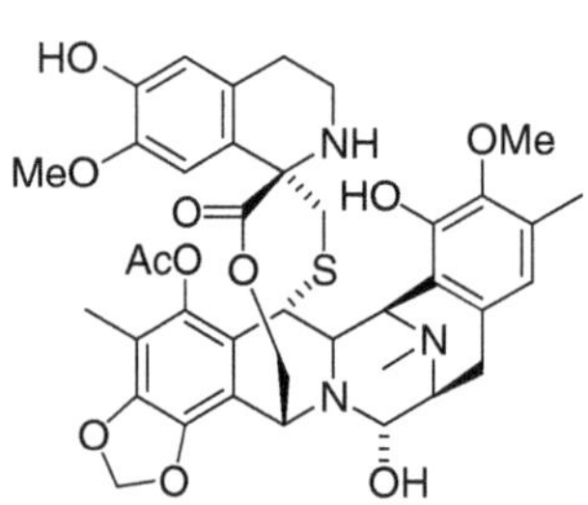

$t^1\!/_2$	17–100 h*	*V*	17–47
F	–	CL	13–35
pb	>95%	Mwt	761.8
ur	<1%*	PSA	194 Å^2
		log *P*	1.87

TRABECTEDIN, ET-743
*iv
Antineoplastic

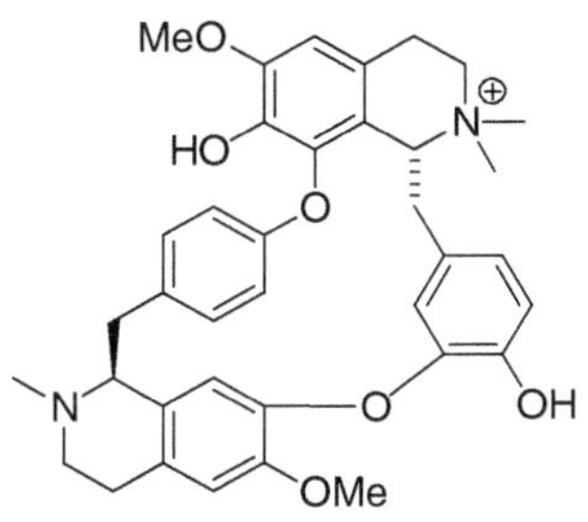

$t^1/_2$	2.0±1.1 h	V	0.39±0.14	D-TUBOCURARINE
F	Low	CL	1.9±0.6	Neuromuscular blocking agent
pb	50±8%	Mwt	609.7	
ur	63±35%	PSA	80.6 Å^2	
		log P	1.99	

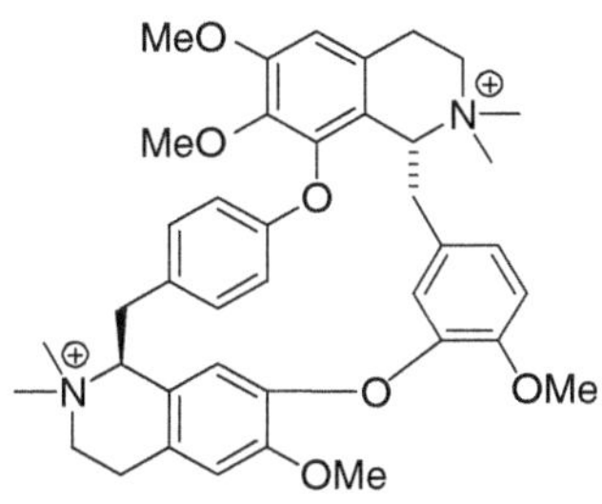

$t^1/_2$	4.7±0.9 h	V	0.35±0.04	METOCURINE
F	–	CL	1.3±0.3	Neuromuscular blocking agent
pb	35±6%	Mwt	652.8	
ur	50%	PSA	55.4 Å^2	
		log P	−1.41	

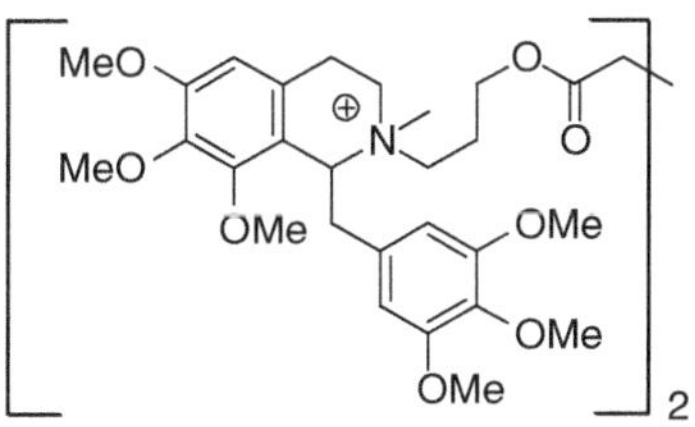

$t^1/_2$	1.2–1.6 h	V	0.12–0.22	DOXACURIUM
F	–	CL	1.5–2.7	Neuromuscular blocking agent
pb	28–42%	Mwt	1035.2	
ur	24–38%	PSA	163.4 Å^2	
		log P	−1.76	

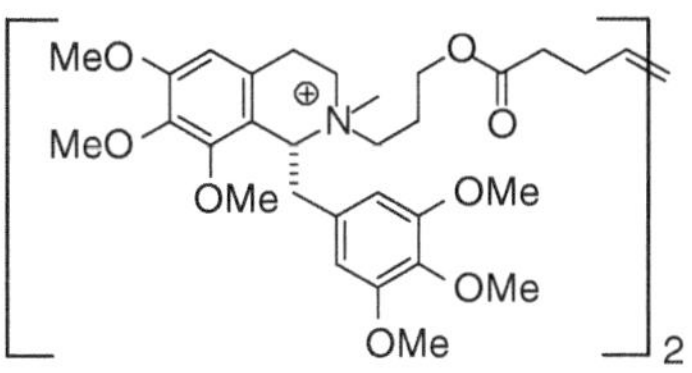

$t^1/_2$	0.8 h (iv)	V	0.2	MIVACURIUM
F	–	CL	55–70	Mixture of stereoisomers
pb	30%	Mwt	1089.3	
ur	0%	PSA	163.4 Å^2	
		log P	−0.23	

(continued overleaf)

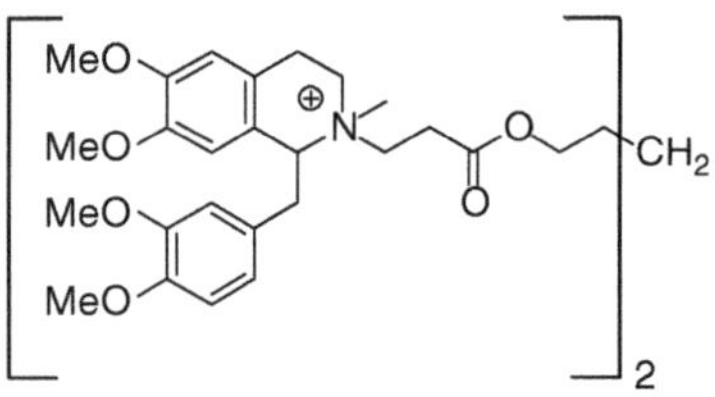

$t\frac{1}{2}$	0.3–0.4 h	*V*	0.16±0.06	ATRACURIUM
F	–	CL	6.2±2.0	Neuromuscular blocking agent; mixture of stereoisomers
pb	38%	Mwt	929.2	
ur	6–11%	PSA	126.5 Å^2	
		log *P*	−1.59	

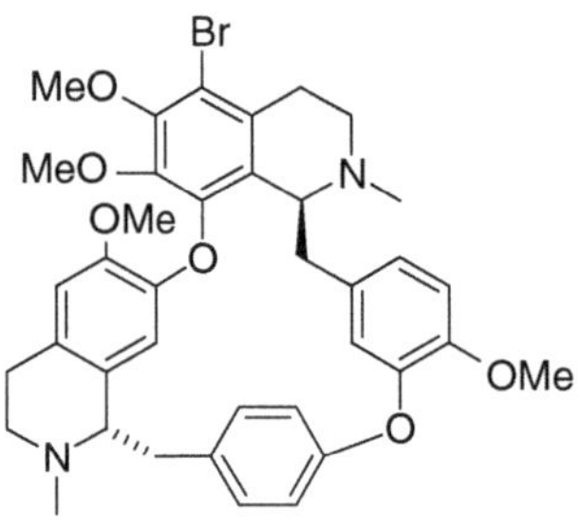

$t\frac{1}{2}$	52–63 h	*V*	3.8–4.2	BROMOTETRANDRINE
F	–	CL	11	Induces reversal of P-gp mediated multidrug resistance
pb	78%	Mwt	701.7	
ur	0.6%	PSA	61.9 Å^2	
		log *P*	4.28	

48
Aminoalkylindoles and Indole Alkaloids

In the 1970s, it was proposed that migraine may be caused by a disturbance of the extracranial circulation. Selective serotonin agonists were then sought, which were able to induce vasoconstriction of the extracranial blood vessels [1]. One of the first antimigraine drugs based on this mechanism of action was sumatriptan (1984), a selective 5-hydroxytryptamine $(5\text{-HT})_{1B/1D}$ agonist. Sumatriptan and other "triptans" (Table 48.1) are more selective than the earlier used ergot alkaloid derivatives (e.g., ergotamine). These derivatives show more side effects than triptans, such as nausea, vomiting, and leg weakness.

Another group of therapeutically relevant indole derivatives are the vinca alkaloids (e.g., vincristine). These are specific tubulin ligands, which block the mitosis of cells and induce apoptosis. Vinca alkaloids can be used to treat a variety of cancers.

Indoles are electron-rich, chemically reactive heteroarenes. Position 3 is particularly nucleophilic, but this position is blocked in most indole-containing drugs. Stabilization toward oxidation or electrophiles may also be achieved with electron-withdrawing substituents. Bioisosteres of indoles include benzimidazoles or benzothiazoles, which are more stable and easier to prepare than substituted indoles.

Reference

1. (a) Nichols, D.E. and Nichols, C.D. (2008) Serotonin receptors. *Chem. Rev.* **108**, 1614–1641; (b) Sneader, W. (1996) Ergot and its alkaloids, *Drug News Perspect.*, **9**, 253–256.

Lead Optimization for Medicinal Chemists: Pharmacokinetic Properties of Functional Groups and Organic Compounds, First Edition. Florencio Zaragoza Dörwald.

Table 48.1 Aminoalkylindoles, indole alkaloids, and related compounds. V in l kg^{-1}; CL in ml min^{-1} kg^{-1}; Mwt in g mol^{-1}.

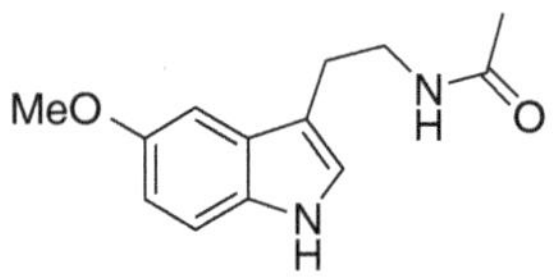	$t^1/_2$	0.5–1.0 h	V	0.6	MELATONIN
	F	100%*	CL	–	*monkey
	pb	61–85%	Mwt	232.3	Natural hormone; induces pigment lightening in skin cells
	ur	–	PSA	54.1 Å^2	Metabolism: aromatic 6-hydroxylation, then O-sulfation
			log P	1.04	
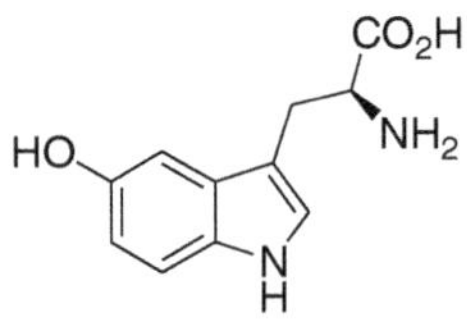	$t^1/_2$	6 h	V	–	OXITRIPTAN, 5-HYDROXYTRYPTOPHAN
	F	50–85%	CL	–	Prodrug of serotonin, antidepressant, antimigraine; crosses bbb
	pb	60%	Mwt	220.2	Metabolism (in CNS): decarboxylation, then oxidation to hydroxyindolylacetic acid
	ur	–	PSA	99.3 Å^2	
			log P	−0.29	
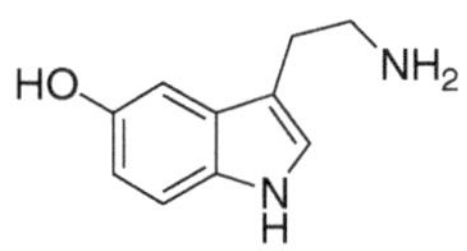	$t^1/_2$	Short	V	–	SEROTONIN
	F	0%	CL	–	Natural hormone
	pb	–	Mwt	176.2	Metabolism: O-methylation and N-acetylation to melatonin, oxidative deamination
	ur	–	PSA	62.0 Å^2	
			log P	0.55	
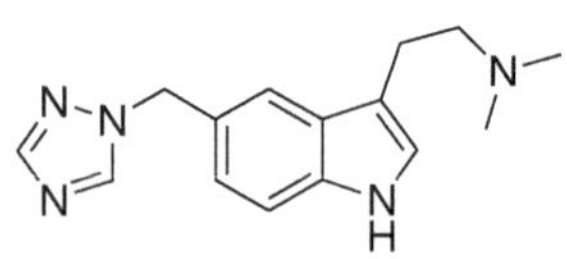	$t^1/_2$	1.5–2.9 h	V	1.3–2.2	RIZATRIPTAN
	F	35–47%	CL	12–19	5-$HT_{1B,1D}$ agonist, antimigraine, crosses bbb
	pb	14%	Mwt	269.3	Metabolism: oxidation to *N*-oxide and triazolomethylindole-3-acetic acid, aromatic 6-hydroxylation, N-demethylation
	ur	8–35%	PSA	49.7 Å^2	
			log P	0.96	
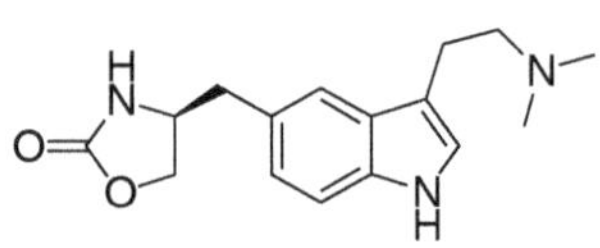	$t^1/_2$	2.3–3.0 h	V	7	ZOLMITRIPTAN
	F	36–60%	CL	8.7	5-$HT_{1B,1D}$ agonist, antimigraine
	pb	14–25%	Mwt	287.4	Metabolites: *N*-oxide, *N*-desmethyl ($t^1/_2$ 2–4 h), oxidation to indole-3-acetic acid
	ur	<10%	PSA	57.4 Å^2	
			log P	1.64	
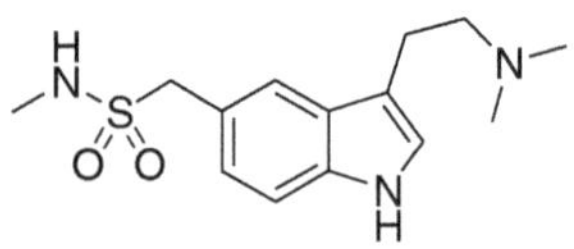	$t^1/_2$	1.9±0.3 h	V	0.65±0.10	SUMATRIPTAN
	F	14±5%	CL	16±2	Antimigraine
	pb	14–21%	Mwt	295.4	Metabolism: oxidation to indole-3-acetic acid (by MAO)
	ur	22±4%	PSA	73.6 Å^2	
			log P	0.44	

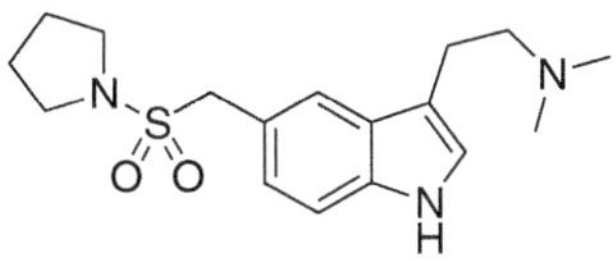	$t_{1/2}$ 3–4 h F 69–80% pb 23–40% ur 40–65%	V 2.6–2.8 CL 8.9–9.5 Mwt 335.5 PSA 64.8 Å^2 log P 1.89	ALMOTRIPTAN Antimigraine Metabolism: oxidation to indole-3-acetic acid (by MAO), N-demethylation, oxidative ring opening of pyrrolidine to 4-aminobutyric acid
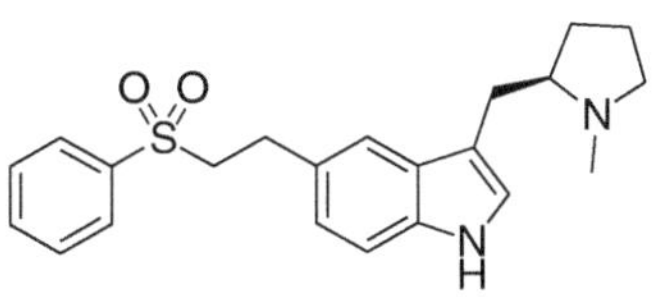	$t_{1/2}$ 3.6–7.0 h F 50% pb 85% ur 85%	V 2–3 CL 9 Mwt 382.5 PSA 61.6 Å^2 log P 3.08	ELETRIPTAN Antimigraine Metabolism: N-demethylation, oxidation to *N*-oxide
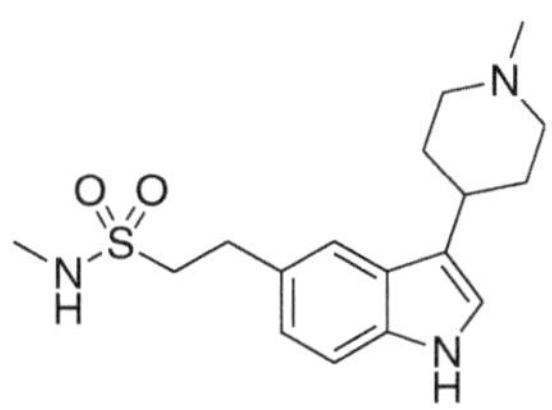	$t_{1/2}$ 5–7 h F 63–74% pb 20–31% ur 50%	V 2.4 CL 6.6 Mwt 335.5 PSA 73.6 Å^2 log P 1.35	NARATRIPTAN 5-HT$_{1A,1B,1D,1F}$ agonist, antimigraine Metabolite: *N*-oxide
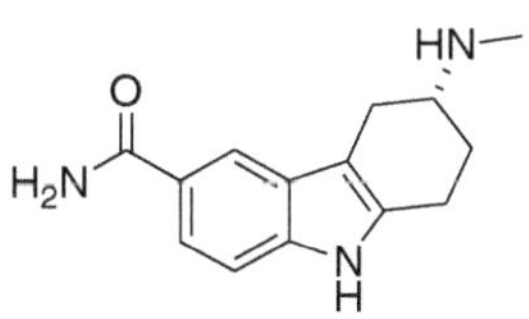	$t_{1/2}$ 24–30 h F 18–30% pb 15% ur 2–10%	V 3.6 CL 2.5 Mwt 243.3 PSA 70.9 Å^2 log P 0.85	FROVATRIPTAN 5-HT$_{1B,1D}$ agonist, antimigraine Metabolism: N-demethylation, then N-acetylation of aliphatic amine, aromatic hydroxylation
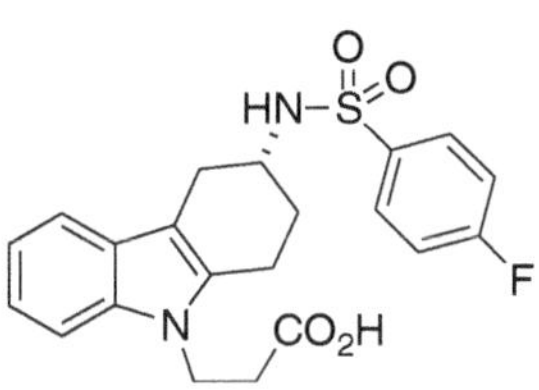	$t_{1/2}$ 2–3 h F 100%* pb >95% ur <8%	V – CL 12* Mwt 416.5 PSA 96.8 Å^2 log P 3.02	RAMATROBAN *dog PGD_2, PGH_2, and TxA_2 antagonist Metabolism: acyl glucuronidation
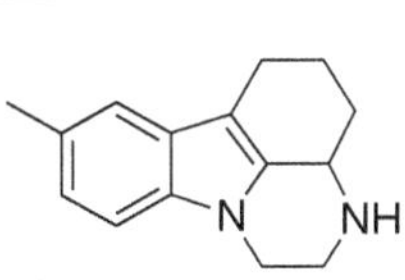	$t_{1/2}$ 1–7 h F 20–30% pb – ur –	V – CL – Mwt 226.3 PSA 17.0 Å^2 log P 3.62	PIRLINDOLE MAO inhibitor, antidepressant; active metabolite: dehydropirlindole (nonchiral imine)

(*continued overleaf*)

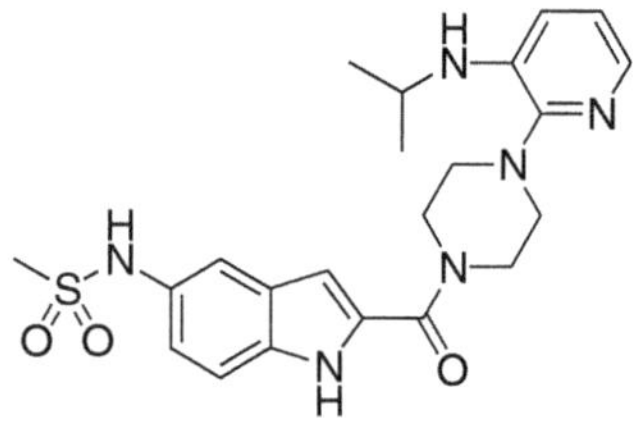

$t^{1}/_{2}$	2–11 h	*V*	–
F	>51%	CL	–
pb	98%	Mwt	456.6
ur	<5%	PSA	119 Å^2
		log *P*	−1.23

DELAVIRDINE
Reverse transcriptase inhibitor, antiviral
Metabolism: N-dealkylation, pyridine hydroxylation

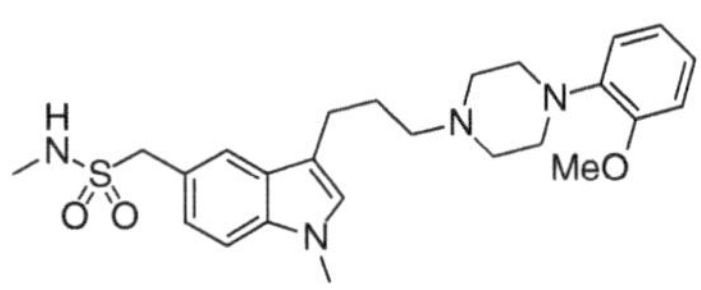

$t^{1}/_{2}$	8–10 h	*V*	0.9
F	17%	CL	6.4
pb	–	Mwt	458.6
ur	–	PSA	112 Å^2
		log *P*	1.58

AVITRIPTAN
Antimigraine
Metabolites: *N*-desmethyl ($t^{1}/_{2}$ 11 h), *O*-desmethyl ($t^{1}/_{2}$ 5 h)
Further metabolism: N-dealkylation

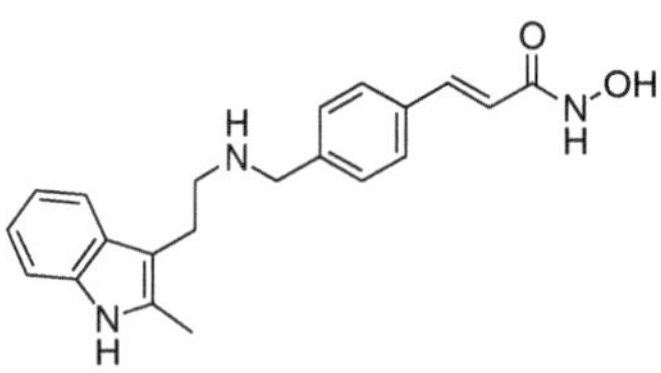

$t^{1}/_{2}$	11–31 h	*V*	0.4
F	30%	CL	7.3
pb	–	Mwt	349.4
ur	–	PSA	77.2 Å^2
		log *P*	3.62

PANOBINOSTAT
Histone deacylase inhibitor
Metabolism: reductive N−O bond cleavage, amide hydrolysis, oxidative degradation of cinnamic acid, glucuronidation

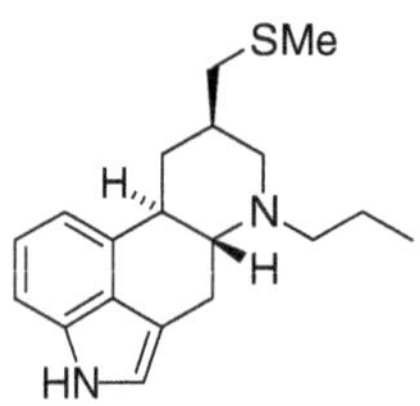

$t^{1}/_{2}$	15–42 h	*V*	17–32
F	20%	CL	22–28
pb	90%	Mwt	314.5
ur	–	PSA	44.3 Å^2
		log *P*	3.90

PERGOLIDE
Antiparkinsonian
Metabolites: despropyl, sulfoxide, and sulfone; withdrawn in 2007 because of hepatotoxicity

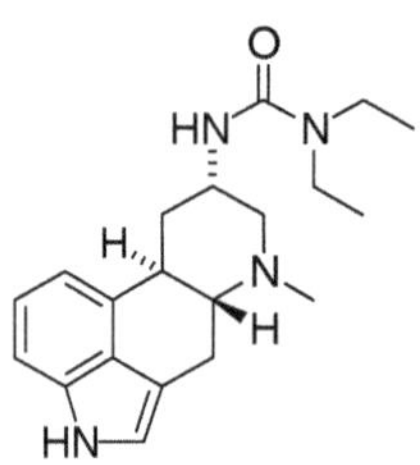

$t^{1}/_{2}$	1.6 h	*V*	1.3–1.7
F	20–32%	CL	17
pb	–	Mwt	340.5
ur	–	PSA	51.4 Å^2
		log *P*	2.20

TERGURIDE
Antiparkinsonian, antihyperprolactinemic

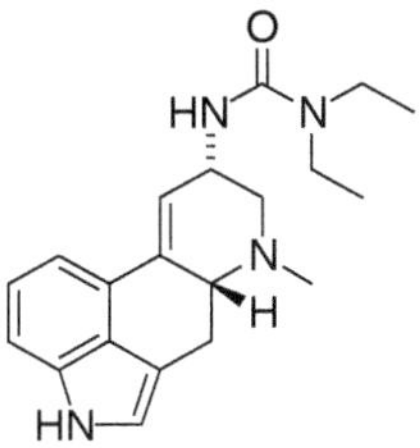

$t_{1/2}$	1.5–2.0 h	*V*	2.3	LISURIDE
F	20%	CL	10–20	Antimigraine, prolactin inhibitor, antiparkinsonian
pb	15%	Mwt	338.5	Metabolism: N-deethylation, aromatic hydroxylation, oxidation of both alkenes
ur	0%	PSA	51.4 Å^2	
		log *P*	3.21	

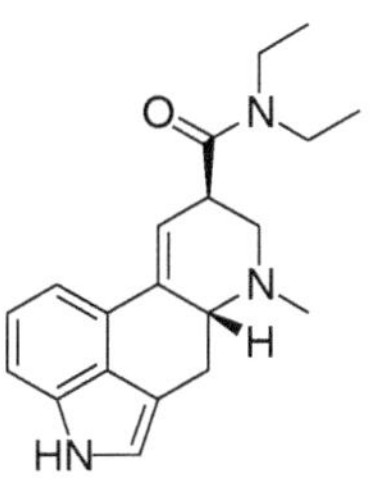

$t_{1/2}$	3–5 h	*V*	0.28	LSD
F	High	CL	–	5-HT$_{2A,2C}$ agonist, synthetic hallucinogenic
pb	90%	Mwt	323.4	Metabolism: oxidation of indole to 2-indolone, N-deethylation, N-demethylation
ur	Low	PSA	39.3 Å^2	
		log *P*	2.71	

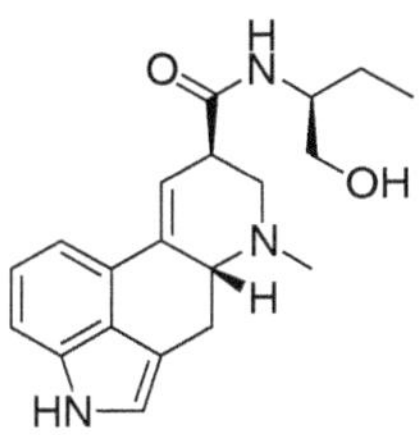

$t_{1/2}$	4±1 h	*V*	0.5–1.0	METHYLERGOMETRINE, METHYLERGONOVINE
F	85%	CL	7.7	Oxytocic, metabolite of methysergide
pb	35%	Mwt	339.4	
ur	3%	PSA	68.4 Å^2	
		log *P*	2.99	

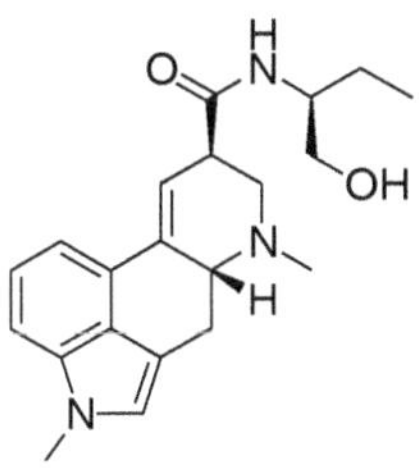

$t_{1/2}$	1 h	*V*	–	METHYSERGIDE
F	13%	CL	72–104 (rat)	Antimigraine
pb	84% (rat)	Mwt	353.5	Metabolism: N-demethylation of indole to methylergometrine
ur	–	PSA	57.5 Å^2	
		log *P*	1.81	

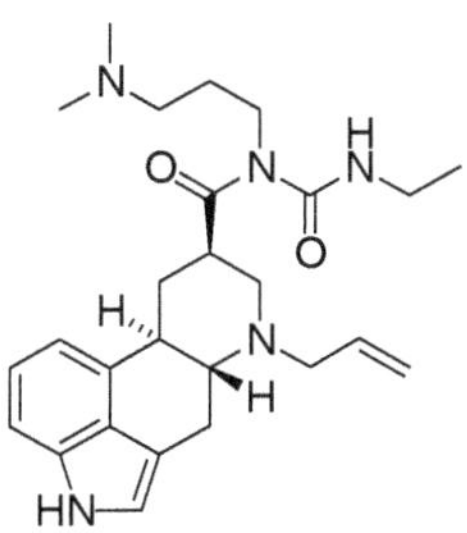

$t_{1/2}$	63–69 h	*V*	–	CABERGOLINE
F	63% (rat)	CL	47	Prolactin inhibitor, crosses bbb
pb	40–42%	Mwt	451.6	Metabolism: hydrolysis of acylurea, N-dealkylation of imide, and piperidine
ur	10%	PSA	71.7 Å^2	
		log *P*	2.43	

(*continued overleaf*)

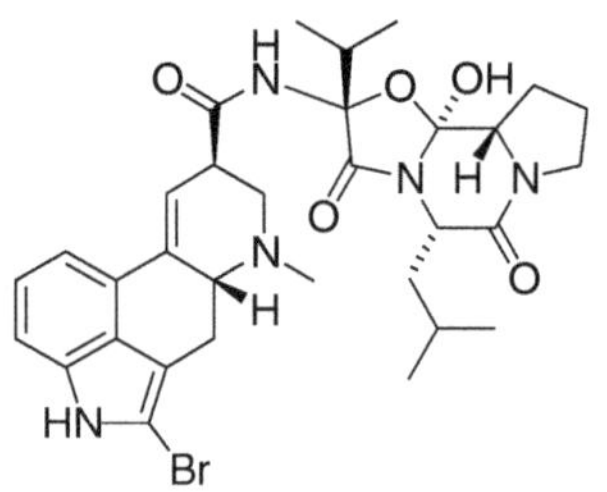

*t*½	7±5 h	*V*	2±1	BROMOCRIPTINE
F	3–6%	CL	5±6	Dopamine agonist for treatment of hyperprolactinemia
pb	93%	Mwt	654.6	Metabolism: hydroxylation of proline
ur	2%	PSA	118 Å^2	
		log *P*	8.6	

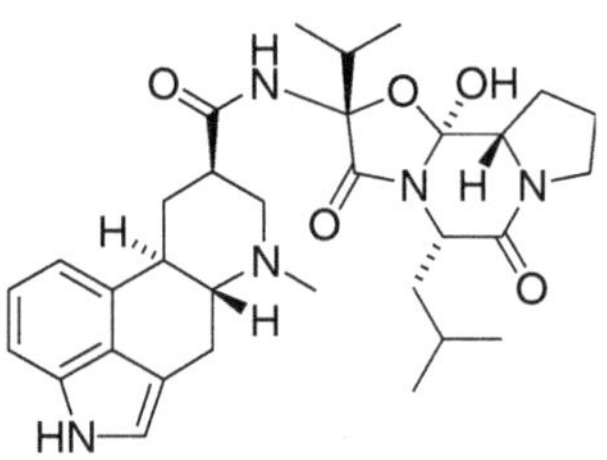

*t*½	15–20 h	*V*	22±4	DIHYDROERGOCRYPTINE
F	2–3%	CL	–	Dopamine agonist for treatment of Parkinson disease; crosses bbb
pb	45–64%	Mwt	577.7	
ur	–	PSA	118 Å^2	
		log *P*	6.37	

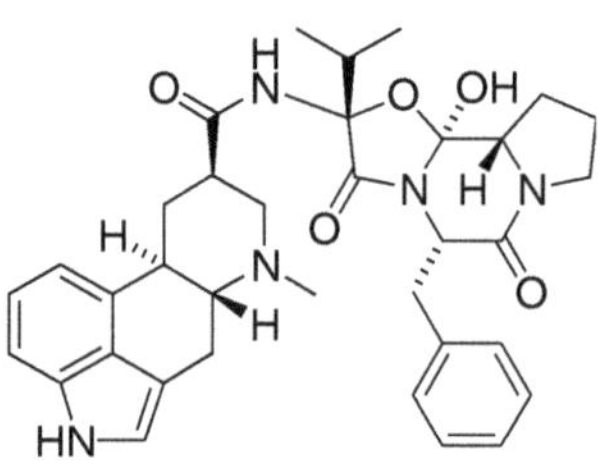

*t*½	14 h	*V*	16	DIHYDROERGOCRISTINE
F	–	CL	28	5-HT antagonist, vasoconstrictor
pb	68%	Mwt	611.7	GI absorption: 25%
ur	<1%	PSA	118 Å^2	
		log *P*	6.55	

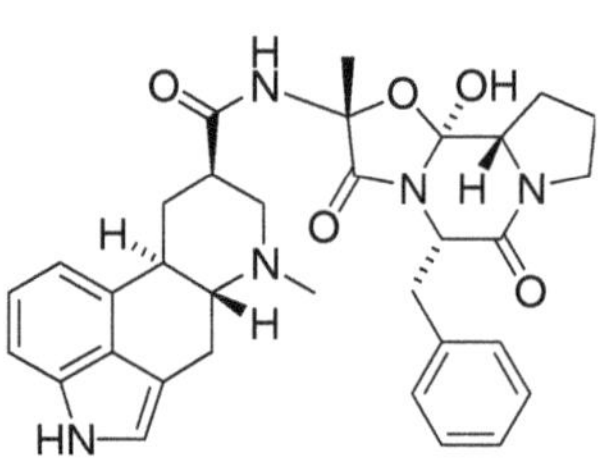

*t*½	9–13 h	*V*	15–30	DIHYDROERGOTAMINE
F	1%*	CL	14.3	*nasal: 38%
pb	93%	Mwt	583.7	Vasoconstrictor; high first-pass effect
ur	–	PSA	118 Å^2	Absorption from GI tract: 30%
		log *P*	5.69	Metabolite: 8-hydroxy

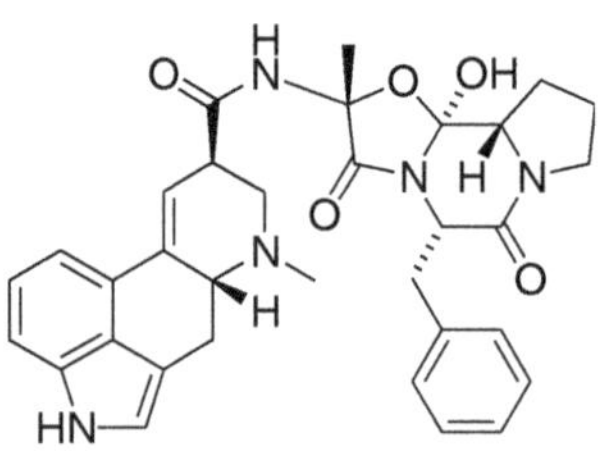

*t*½	2.1 h (iv)	*V*	3.6	ERGOTAMINE
F	2%	CL	15	Antimigraine
pb	98%	Mwt	581.7	
ur	–	PSA	118 Å^2	
		log *P*	7.37	

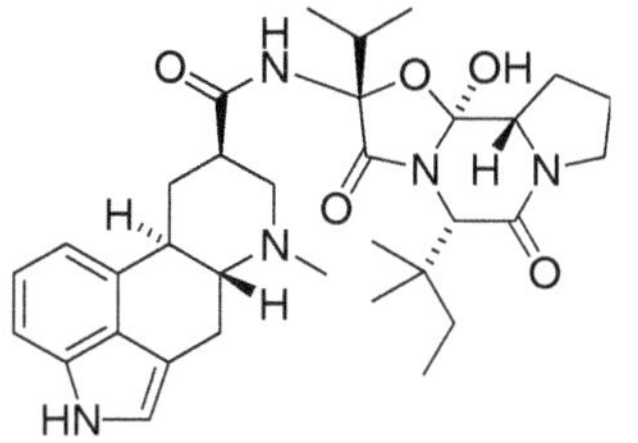

$t^1\!/_2$	3.5 h	*V*	15	ERGOLOID, DIHYDROERGOTOXINE
F	9–25%	CL	25	Nootropic for the treatment of Alzheimer disease
pb	98–99%	Mwt	591.8	
ur	11%	PSA	118.2 Å^2	
		log *P*	3.07	

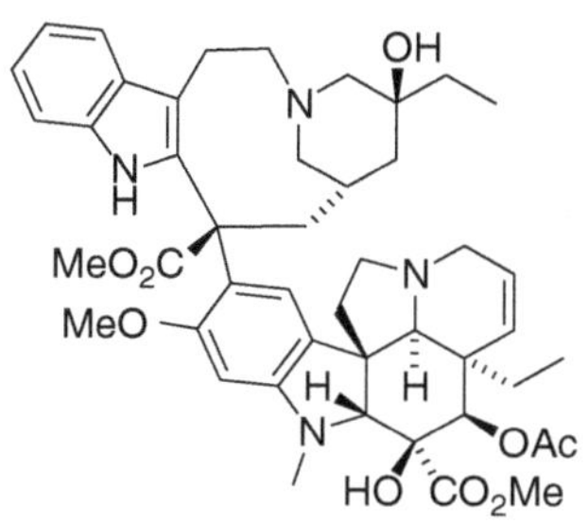

$t^1\!/_2$	1.0–1.5 h	*V*	24	VINBLASTINE
F	–	CL	10.6	Antineoplastic
pb	75%	Mwt	810.97	Metabolism: O-deacetylation
ur	–	PSA	154 Å^2	
		log *P*	5.92	

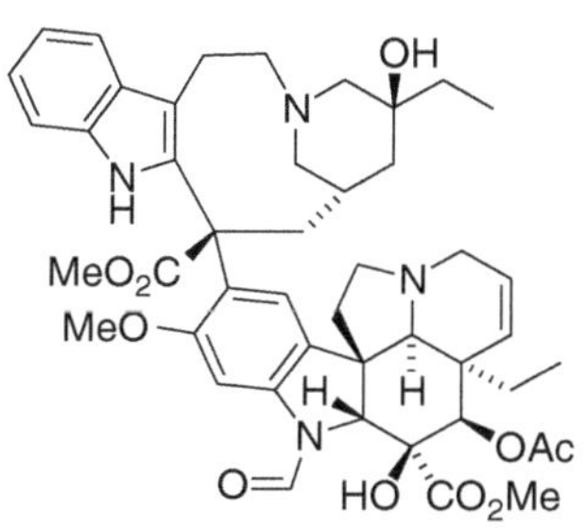

$t^1\!/_2$	1.0–2.5 h	*V*	8.6	VINCRISTINE
F	–	CL	1.8	Antineoplastic
pb	75%	Mwt	825.0	
ur	12%	PSA	171 Å^2	
		log *P*	5.75	

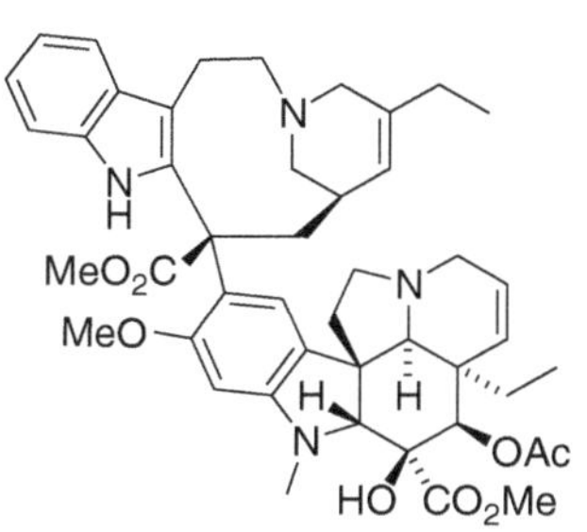

$t^1\!/_2$	20–40 h	*V*	76	VINORELBINE
F	12–59%	CL	17–21	Antineoplastic
pb	98%	Mwt	778.9	
ur	11%	PSA	134 Å^2	
		log *P*	7.08	

(*continued overleaf*)

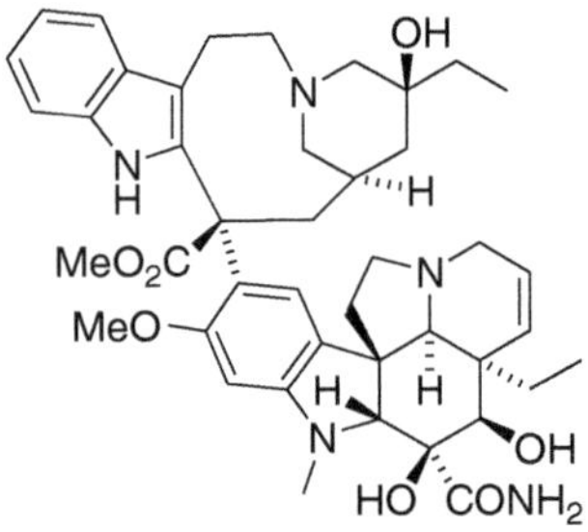

$t_{1/2}$	35 h (iv)	*V*	5	VINDESINE
F	–	CL	2.2	Antineoplastic
pb	–	Mwt	753.9	
ur	–	PSA	165 Å^2	
		log *P*	4.94	

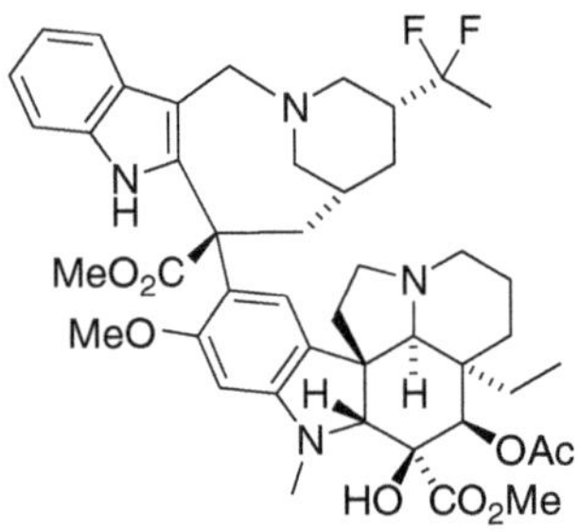

$t_{1/2}$	40 h (iv)	*V*	35	VINFLUNINE
F	57%	CL	10	Antineoplastic
pb	67%	Mwt	816.9	Metabolite: *O*-desacetyl
ur	–	PSA	134 Å^2	($t_{1/2}$ 120 h)
		log *P*	4.12	

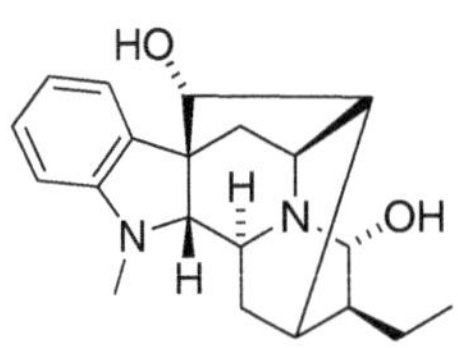

$t_{1/2}$	1.6 h	*V*	6.2	AJMALINE
F	20%	CL	9.8	Antihypertensive, antiarrhythmic
pb	29–61%	Mwt	326.4	Metabolism: aromatic mono- and dihydroxylation, then O-methylation, reduction of aminal to piperidine, oxidation of both hydroxyl groups to carbonyl groups, N-oxidation
ur	4%	PSA	46.9 Å^2	
		log *P*	1.81	

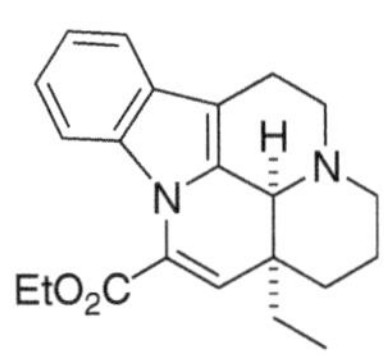

$t_{1/2}$	2.5±0.5 h	*V*	3.8 (rat)	VINPOCETINE
F	57%	CL	13±2	Vasodilator, nootropic
pb	66%	Mwt	350.5	Metabolism: hydrolysis of ester
ur	0%	PSA	34.5 Å^2	
		log *P*	3.98	

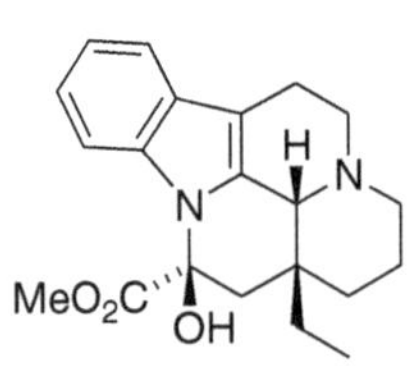

$t_{1/2}$	1–6 h	*V*	2 (rat)	VINCAMINE
F	58% (rat)	CL	–	Vasodilator, crosses bbb
pb	30%	Mwt	354.4	Metabolism (rat): hydrolysis of ester, then decarboxylation and oxidation to lactam, 6-(allylic)-hydroxylation, then oxidation to 6-ketone
ur	5–10%	PSA	54.7 Å^2	
		log *P*	2.94	

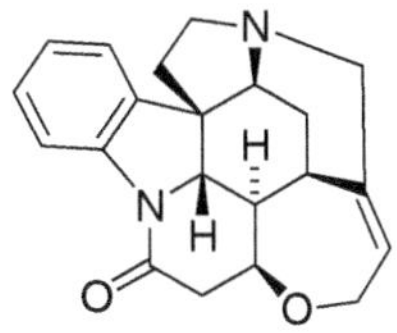	$t^{1}/_{2}$ 10–16 h F – pb 55–59% ur 10–20%	V 13 CL – Mwt 334.4 PSA 32.8 Å^2 log P 1.59	STRYCHNINE Rodenticide, central stimulant
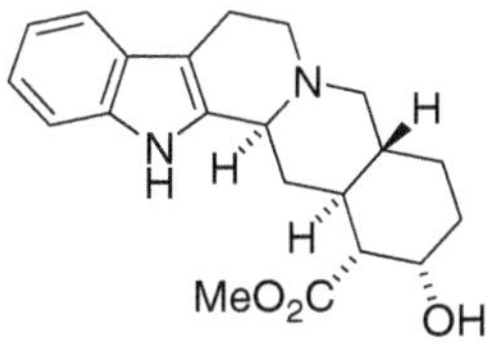	$t^{1}/_{2}$ 0.3–2.5 h F 5–40% pb – ur 0.5–1.0%	V 1 CL 26 Mwt 354.4 PSA 65.6 Å^2 log P 2.87	YOHIMBINE Adrenergic α_2 antagonist, mydriatic Metabolism: aromatic hydroxylation (para to amino group and para to vinyl group)
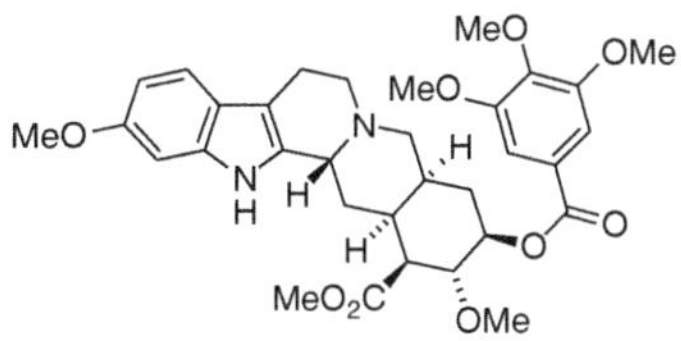	$t^{1}/_{2}$ 168–271 h F 40–50% pb 95% ur 1%	V – CL – Mwt 608.7 PSA 118 Å^2 log P 4.45	RESERPINE Antihypertensive Metabolism: ester hydrolysis, 4-O-demethylation of trimethoxyphenyl
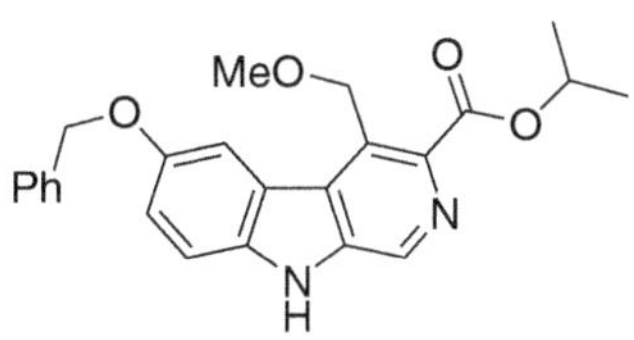	$t^{1}/_{2}$ 4–12 h F 39–65% pb 94.8% ur <2%	V 14 CL 11–13 Mwt 404.5 PSA 73.4 Å^2 log P 4.76	ABECARNIL Benzodiazepine agonist, anxiolytic Metabolism: O-debenzylation, then glucuronidation and sulfation, ester hydrolysis
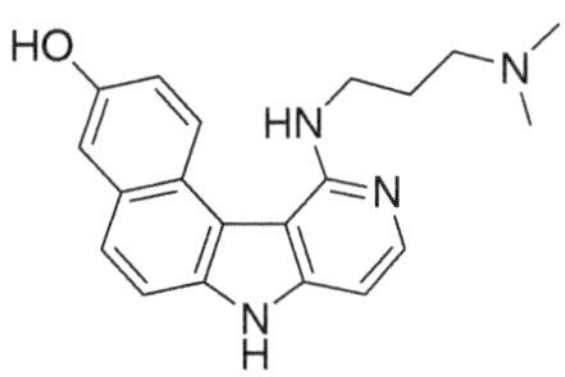	$t^{1}/_{2}$ 19 h (iv) F – pb – ur –	V 11 CL 18 Mwt 348.4 PSA 64.2 Å^2 log P 4.01	INTOPLICINE Topoisomerase I inhibitor, antineoplastic
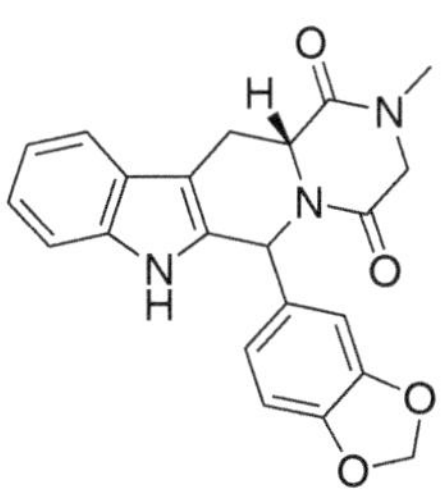	$t^{1}/_{2}$ 17.5 h F 81% pb 94% ur Negligible	V 0.9 CL 0.6 Mwt 389.4 PSA 74.9 Å^2 log P 2.71	TADALAFIL Phosphodiesterase V inhibitor for the treatment of erectile dysfunction Metabolite: methylcatechol glucuronide

(*continued overleaf*)

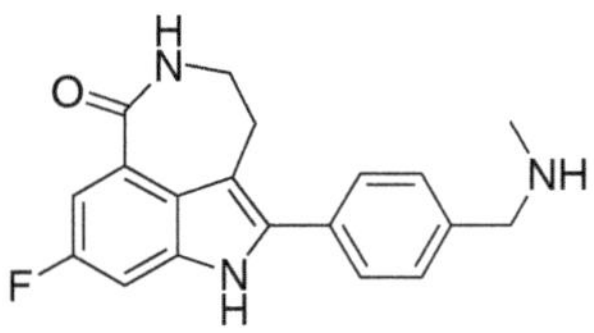	$t^1/_2$ 9.5 h (iv) *F* – pb – ur –	*V* 3.3 CL – Mwt 323.4 PSA 56.9 Å^2 log *P* 2.62	AG-14699 PARP inhibitor, oncolytic
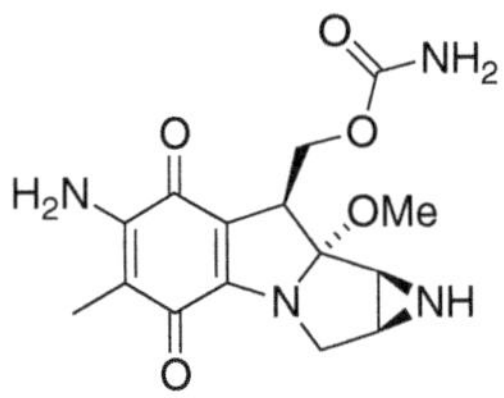	$t^1/_2$ 0.7 h *F* – pb – ur –	*V* 0.5 CL 13.4 Mwt 334.3 PSA 147 Å^2 log *P* −0.30	MITOMYCIN C Antineoplastic Metabolism: elimination of MeOH to form pyrrole, then ring opening of aziridine to aminoalcohol or aminophosphate
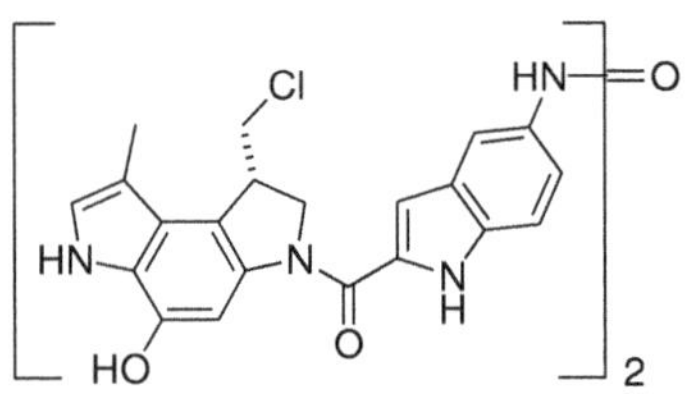	$t^1/_2$ 2 h (iv) *F* – pb – ur –	*V* 0.06 CL 0.7 Mwt 815.7 PSA 185 Å^2 log *P* 9.15	BIZELESIN Alkylant, antineoplastic
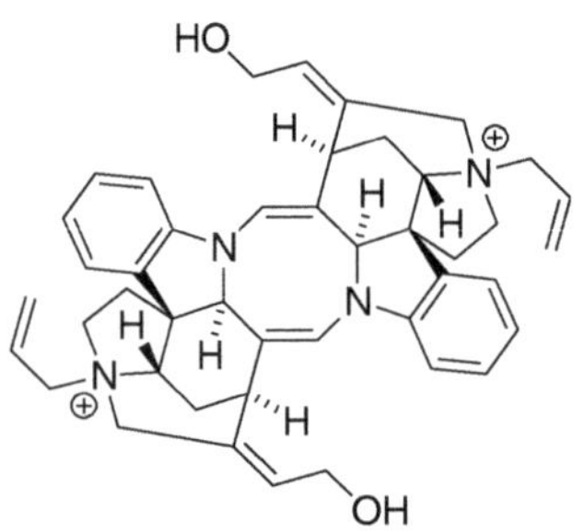	$t^1/_2$ 3–4 h *F* – pb 40% ur 80–85%	*V* 0.28–0.36 CL 1.3–1.4 Mwt 666.9 PSA 46.9 Å^2 log *P* −2.45	ALCURONIUM Muscle relaxant
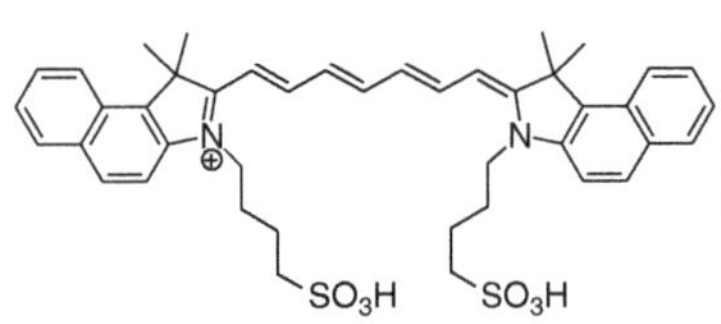	$t^1/_2$ 4 min (iv) *F* – pb 95% ur –	*V* 0.04 CL 6.8 Mwt 754.0 PSA 115.0 Å^2 log *P* 1.38	INDOCYANINE GREEN Diagnostic dye

$t_{1/2}$, plasma half-life; *F*, oral bioavailability; pb, plasma protein binding; ur, excretion of unchanged drug in urine; V, volume of distribution; CL, clearance; Mwt, molecular weight; PSA, polar surface area; bbb, blood–brain barrier; GI, gastrointestinal; LSD, lysergic acid diethylamide; MAO, monoamine oxidase; PARP, poly(ADP-ribose)polymerase; PGD, prostaglandin; TxA, thromboxane.

49
Phenothiazines

The first tricyclic antipsychotics were the phenothiazines [1]. Fenethazine (Scheme 49.1) had been tested as an antimalarial during World War II (WWII), and in the late 1940s, researchers at Rhone-Poulenc discovered its antihistaminic properties (H_1 antagonism). A small structural modification led to the more potent promethazine. Besides showing antihistaminic properties, promethazine was found to prolong the barbiturate-induced sleeping time in rodents.

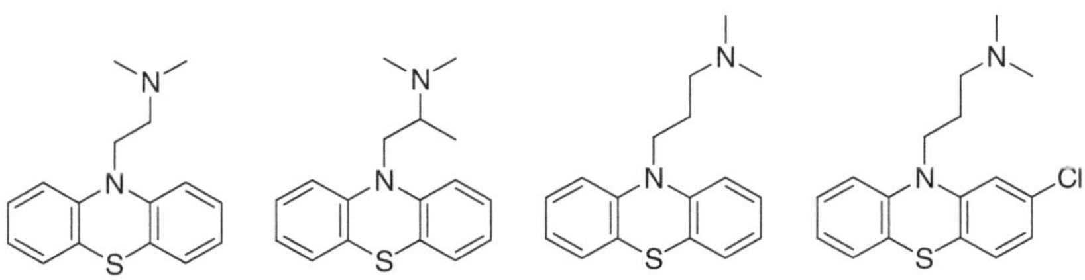

Fenethazine

Promethazine $t_{1/2}$ 18 h, F 25%

Promazine $t_{1/2}$ 8 h, F 17%

Chlorpromazine $t_{1/2}$ 30 h, F 32%

Scheme 49.1 Leads of the antipsychotic chlorpromazine.

Henri Laborit, a military surgeon working in a hospital in Bizerte, Tunisia, often treated his patients with low doses of several drugs simultaneously, instead of administering a large dose of a single drug, in the belief that this would minimize the risk of overdosing. In 1949, while using promethazine to pretreat patients to be anesthetized, he noticed the strong sedating effect of this drug. In 1950, research was initiated at Rhone-Poulenc to discover better phenothiazine-based anxiolytics. An older compound, promazine, prepared as a potential antihistaminic but devoid of H_1 antagonism, was chosen as the lead structure. Shortly thereafter, chlorpromazine was prepared, which became the first antipsychotic drug (marketed 1952) and a breakthrough in the treatment of schizophrenia [2].

Since then, a huge number of different phenothiazines were developed as neuroleptics for the treatment of psychoses and anxiety. Their mechanism of action is assumed to be central dopamine D_2 receptor antagonism. Because

Lead Optimization for Medicinal Chemists: Pharmacokinetic Properties of Functional Groups and Organic Compounds, First Edition. Florencio Zaragoza Dörwald.

phenothiazines are well tolerated by humans, this substructure has also been used in other types of drugs, for example, in antipyretic arylacetic or arylpropionic acids (Table 49.1).

49.1 Metabolism

Owing to their extensive first-pass metabolism, the oral bioavailability of most phenothiazines is low. Moreover, one inherent problem of most phenothiazines is their high affinity for plasma proteins. The free fraction of such drugs can strongly vary with disease or age, and high interindividual variability of pharmacokinetics (PK) is often observed. This complicates the dosing of the drug, and careful monitoring of patients is required.

The typical metabolic pathways of phenothiazines are shown in Scheme 49.2.

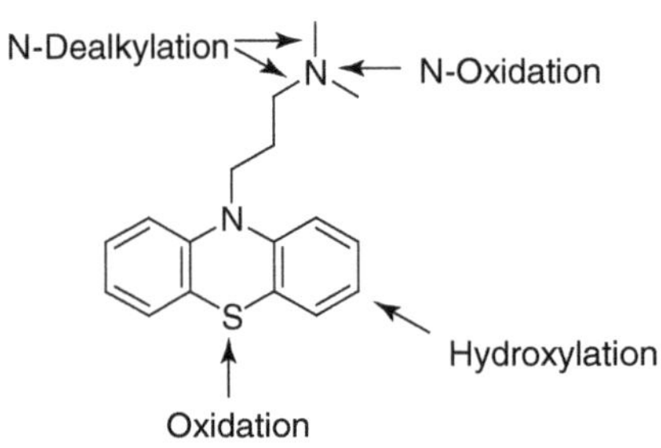

Scheme 49.2 Metabolism of phenothiazines.

References

1. (a) Marino, M.J. and Knutsen, L.J.S., and Williams, M. (2008) Emerging opportunities for antipsychotic drug discovery in the postgenomic era. *J. Med. Chem.*, **51**, 1077–1107;
(b) Sneader, W. (2001) Histamine and the classic antihistamines. *Drug News Perspect.* **14**, 618–624.
2. Sneader, W. (2002) The 50th anniversary of chlorpromazine. *Drug News Perspect.* **5**, 466–471.

Table 49.1 Phenothiazines and related compounds. V in l kg^{-1}; CL in ml min^{-1} kg^{-1}; Mwt in g mol^{-1}.

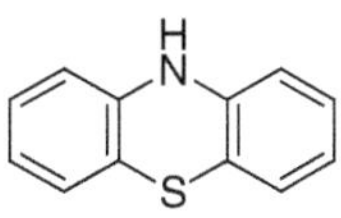	$t\frac{1}{2}$ F pb ur	11±5 h – – –	V CL Mwt PSA log P	– – 199.3 15.8 Å^2 4	PHENOTHIAZINE Anthelmintic Metabolism: N-glucuronidation, oxidation to sulfoxide, aromatic hydroxylation, then oxidation to quinoneimine
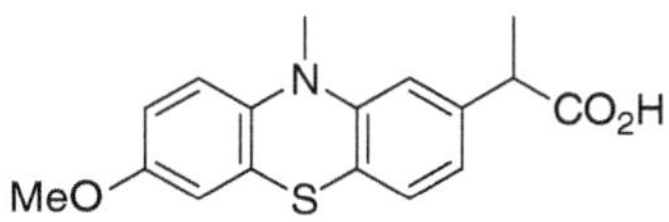	$t\frac{1}{2}$ F pb ur	11 h – – –	V CL Mwt PSA log P	– – 315.4 51.5 Å^2 4.3	PROTIZINIC ACID Antiinflammatory Metabolism: O-demethylation, oxidation to sulfoxide
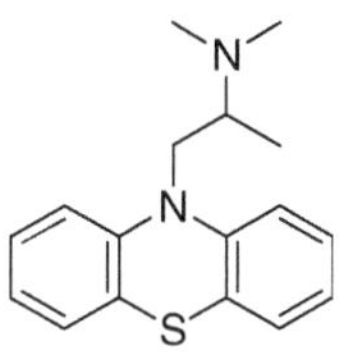	$t\frac{1}{2}$ F pb ur	18±2 h 25% 93% 0.6%	V CL Mwt PSA log P	13.5 16 284.4 8.2 Å^2 4.4	PROMETHAZINE Antihistaminic, antiemetic Duration of action: 5±1 h Metabolism: N-demethylation, oxidation to sulfoxide
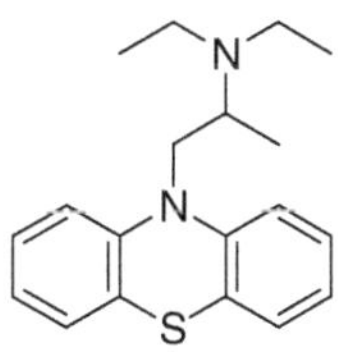	$t\frac{1}{2}$ F pb ur	1–2 h <5% (rat) 93% –	V CL Mwt PSA log P	– – 312.5 8.2 Å^2 5.2	ETHOPROPAZINE Antiparkinsonian, anticholinergic
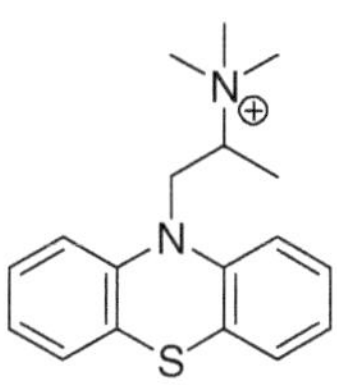	$t\frac{1}{2}$ F pb ur	6 h (im) 10% 40% (iv)	V CL Mwt PSA log P	3–6 12 299.5 4.93 Å^2 0.49	THIAZINAMIUM Antihistaminic for im dosing Metabolism: oxidation to sulfoxide
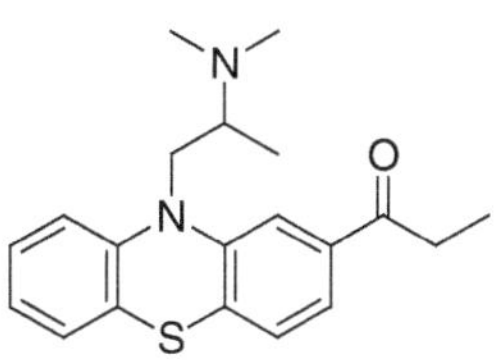	$t\frac{1}{2}$ F pb ur	9±6 h 33% 81% –	V CL Mwt PSA log P	2–3 3–10 340.5 25.2 Å^2 4.8	PROPIOMAZINE Sedative, hypnotic Metabolism: N-demethylation

(*continued overleaf*)

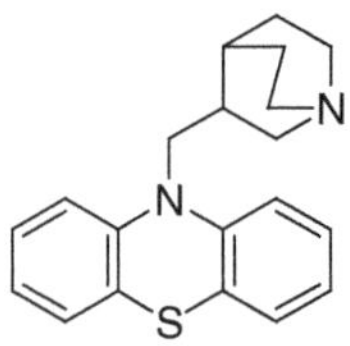

$t^1/_2$	45±26 h	*V*	–
F	–	CL	–
pb	–	Mwt	322.5
ur	–	PSA	8.2 Å^2
		log *P*	5.1

MEQUITAZINE
Antihistaminic, does not cross bbb readily
Metabolism: S-oxidation, aromatic hydroxylation (para to N)

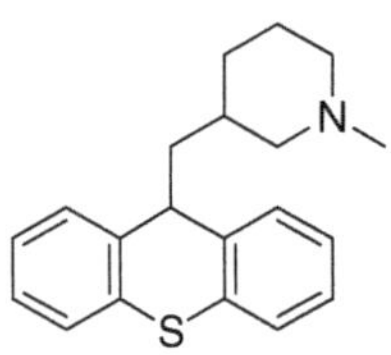

$t^1/_2$	4–14 h	*V*	–
F	–	CL	–
pb	–	Mwt	309.5
ur	–	PSA	3.2 Å^2
		log *P*	4.7

METHIXENE
Anticholinergic
Metabolites: desmethyl, *N*-oxide, sulfoxide, phenols

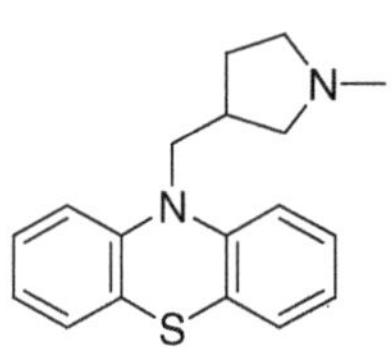

$t^1/_2$	12 h	*V*	–
F	–	CL	–
pb	–	Mwt	296.4
ur	–	PSA	8.2 Å^2
		log *P*	4.6

METHDILAZINE
Antipruritic, analgesic, antimicrobial
Metabolism: oxidation to sulfoxide

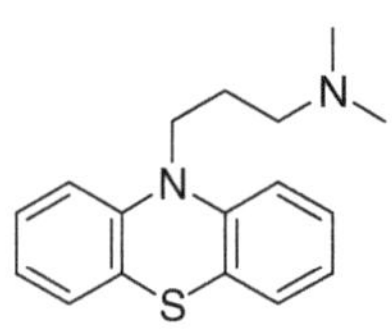

$t^1/_2$	8±7 h (iv)	*V*	8±4
F	17±9%	CL	14±6
pb	94%	Mwt	284.4
ur	–	PSA	8.2 Å^2
		log *P*	4.4

PROMAZINE
Antipsychotic
Metabolism: N-demethylation, S-oxidation by CYP1A2

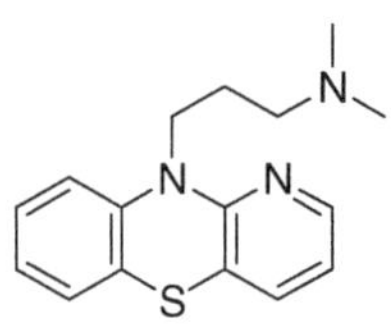

$t^1/_2$	2–3 h	*V*	3
F	–	CL	14
pb	–	Mwt	285.4
ur	–	PSA	21.1 Å^2
		log *P*	3.5

PROTHIPENDYL
Antipsychotic

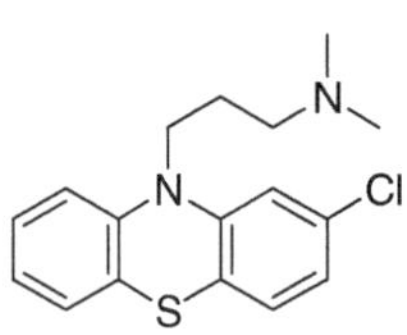

$t^1/_2$	30±7 h	*V*	21±9
F	32±19%	CL	9±3
pb	95–98%	Mwt	318.9
ur	<1%	PSA	8.2 Å^2
		log *P*	5.0

CHLORPROMAZINE
Antiemetic, antipsychotic, induces hepatic and intestinal mucosa enzymes
Metabolism: N-demethylation, 7-hydroxylation, S-oxidation; active metabolites: 7-hydroxy ($t^1/_2$ 25±15 h), *N*-oxide ($t^1/_2$ 6.7±1.4 h)

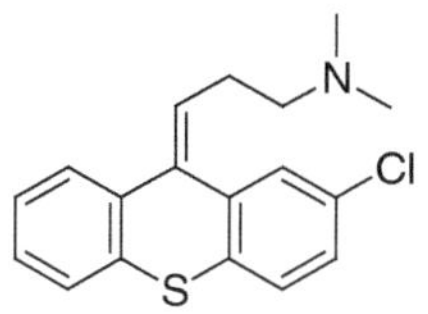

$t_{1/2}$	10±2 h	*V*	11–23	CHLORPROTHIXENE
F	23–64%	CL	14–20	Antipsychotic; single isomer (*Z*)
pb	>95%	Mwt	315.9	Metabolism: S-oxidation to
ur	–	PSA	3.2 Å^2	sulfoxide, N-demethylation,
		log *P*	5.4	addition of glutathione to alkene

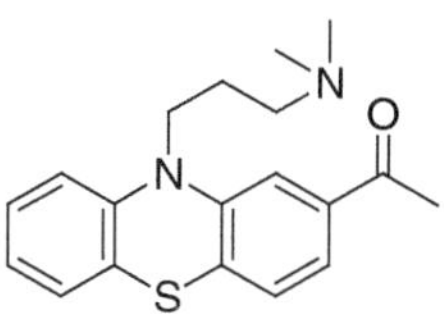

$t_{1/2}$	16 h*	*V*	6.6	ACEPROMAZINE
F	20%*	CL	–	*dog
pb	>99%	Mwt	326.5	Sedative
ur	–	PSA	25.2 Å^2	Metabolism: reduction of ketone to
		log *P*	4.3	alcohol

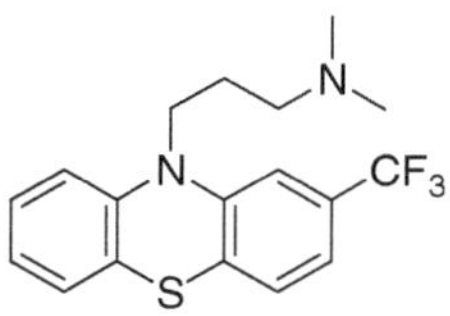

$t_{1/2}$	22±19 h	*V*	–	TRIFLUPROMAZINE
F	–	CL	–	Antipsychotic
pb	85–95%	Mwt	352.4	
ur	<1%	PSA	8.2 Å^2	
		log *P*	5.3	

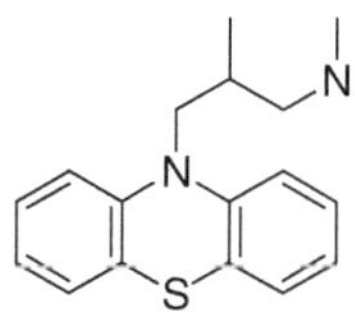

$t_{1/2}$	4.8±0.6 h	*V*	–	TRIMEPRAZINE
F	–	CL	–	Antipruritic
pb	>90%	Mwt	298.5	Metabolism: oxidation to sulfoxide
ur	20%	PSA	8.2 Å^2	and sulfone, aromatic
		log *P*	4.9	hydroxylation, N-dealkylation

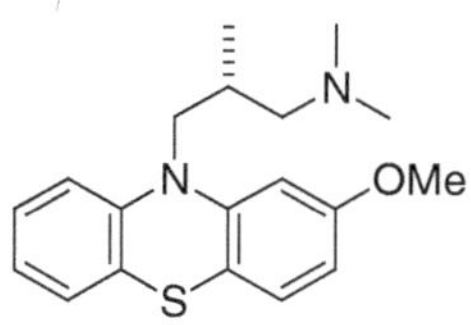

$t_{1/2}$	23±8 h	*V*	23–42	METHOTRIMEPRAZINE,
F	50–60%	CL	–	LEVOMEPROMAZINE
pb		Mwt	328.5	Analgesic
ur	1%	PSA	17.4 Å^2	Metabolites: *N*- and *O*-desmethyl,
		log *P*	4.9	sulfoxide, glucuronide

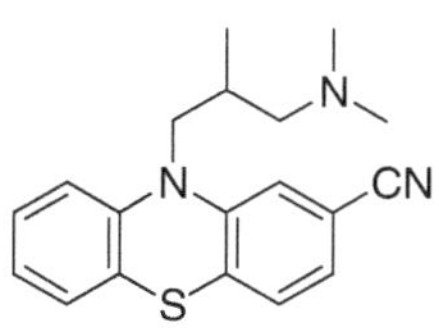

$t_{1/2}$	13±3 h	*V*	–	CYAMEMAZINE
F	10–70%	CL	–	Antipsychotic
pb	–	Mwt	323.5	Metabolism: N-demethylation,
ur	–	PSA	32.0 Å^2	sulfoxide formation
		log *P*	4.6	

(*continued overleaf*)

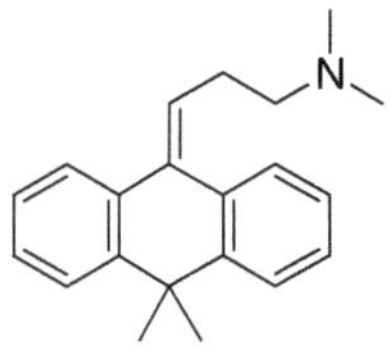	$t_{1/2}$ 18±6 h *F* – pb 89% ur –	*V* – CL – Mwt 291.4 PSA 3.2 Å^2 log *P* 4.6	MELITRACENE Antidepressant; active metabolite: *N*-desmethyl (litracene)
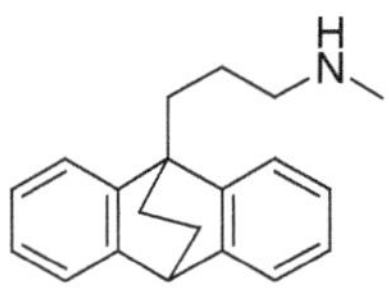	$t_{1/2}$ 36–108 h *F* 37–67% pb 88% ur 3–4%	*V* 52±18 CL 14 Mwt 277.4 PSA 12.0 Å^2 log *P* 4.0	MAPROTILINE Antidepressant Metabolism: N-demethylation, hydroxylation
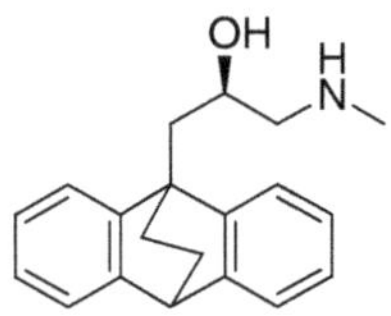	$t_{1/2}$ 19 h *F* 40% pb – ur <1%	*V* 24 CL 14.6 Mwt 293.4 PSA 32.3 Å^2 log *P* 3.13	LEVOPROTILINE Antidepressant Metabolism: glucuronidation
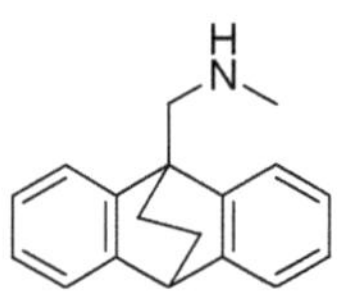	$t_{1/2}$ 2–3 h *F* – pb – ur –	*V* – CL – Mwt 249.4 PSA 12.0 Å^2 log *P* 3.5	BENZOCTAMINE Anxiolytic
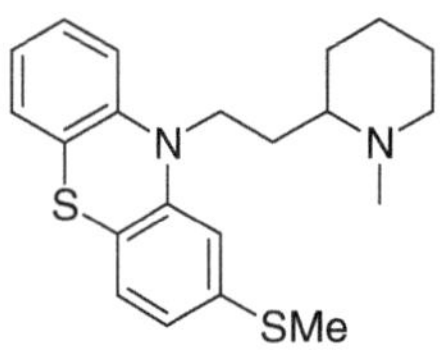	$t_{1/2}$ 10–36 h *F* 25–32% pb 95% ur <1%	*V* 21±9 CL 10–14 Mwt 370.6 PSA 8.2 Å^2 log *P* 5.6	THIORIDAZINE Antipsychotic Metabolites: sulfoxide ($t_{1/2}$ 9–12 h), sulfone ($t_{1/2}$ 9–11 h) Further metabolism: aromatic hydroxylation, N-oxidation, N-demethylation; withdrawn in 2005 because of cardiotoxicity
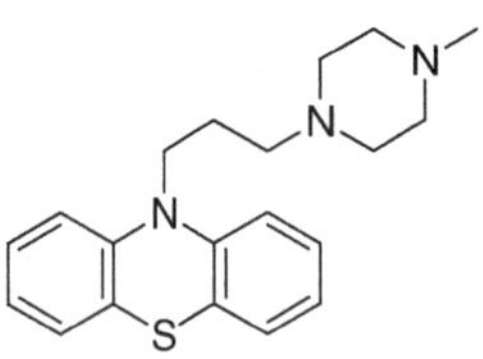	$t_{1/2}$ 6 h *F* 3% pb 94–98% ur –	*V* – CL – Mwt 339.5 PSA 11.4 Å^2 log *P* 4.3	PERAZINE Antipsychotic Metabolism: N-demethylation

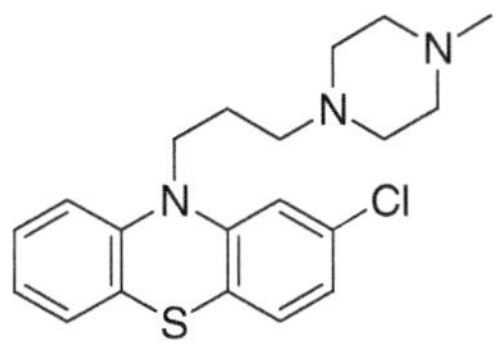

$t_{1/2}$	7±1 h	V	22–24
F	13–20%	CL	–
pb	91–99%	Mwt	373.9
ur	<1%	PSA	11.4 Å^2
		log P	4.9

PROCHLORPERAZINE
Antipsychotic
Metabolism: N-demethylation

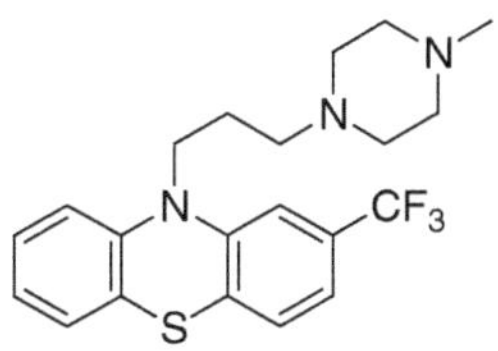

$t_{1/2}$	15±5 h	V	–
F	–	CL	–
pb	91–99%	Mwt	407.5
ur	–	PSA	11.4 Å^2
		log P	5.1

TRIFLUOPERAZINE
Antipsychotic
Metabolite: sulfoxide ($t_{1/2}$ 5.8±1.3 h)
Further metabolism: aromatic hydroxylation, N-oxidation

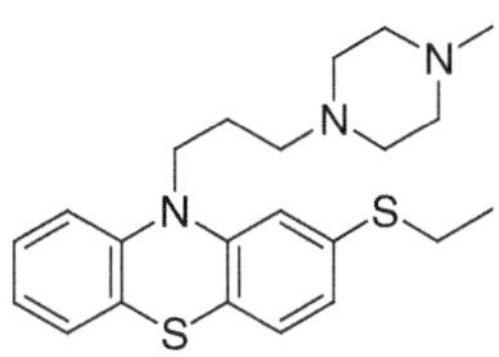

$t_{1/2}$	12 h	V	2.7
F	–	CL	–
pb	60%	Mwt	399.6
ur	3%	PSA	11.4 Å^2
		log P	5.1

THIETHYLPERAZINE
Antiemetic
Metabolism: aromatic hydroxylation, S-oxidation, dealkylation

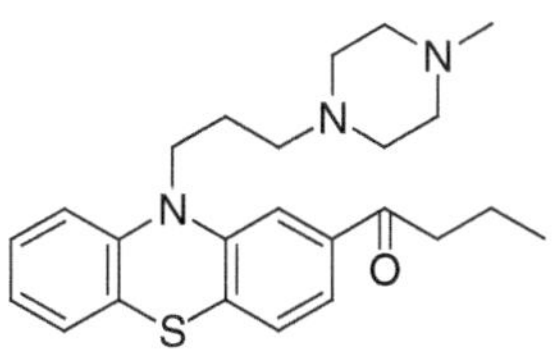

$t_{1/2}$	5–30 h	V	–
F	–	CL	–
pb	–	Mwt	409.6
ur	–	PSA	52.1 Å^2
		log P	4.57

BUTAPERAZINE
Antipsychotic
Metabolites: sulfoxide, sulfone

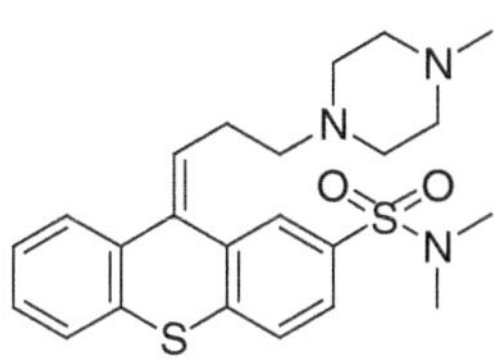

$t_{1/2}$	34 h	V	–
F	–	CL	–
pb	>99%	Mwt	443.6
ur	–	PSA	43.9 Å^2
		log P	4.0

THIOTHIXENE
Antipsychotic
Metabolism: addition of glutathione to alkene

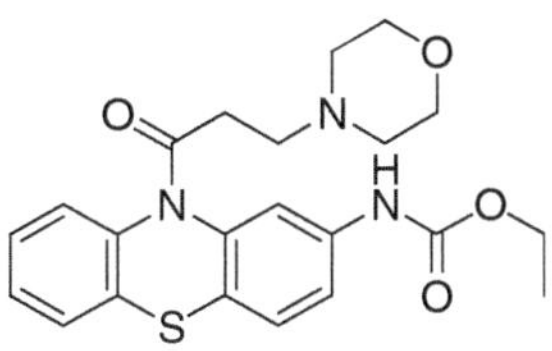

$t_{1/2}$	4±2 h	V	4–11
F	38%	CL	20–42
pb	80–95%	Mwt	427.5
ur	<1%	PSA	72.8 Å^2
		log P	4.4

MORICIZINE
Antiarrhythmic
Metabolism: hydrolysis of carbamate and amide, O-deethylation to carbamic acid, morpholine cleavage to 3-aminopropionyl and hydroxyethylaminopropionyl, oxidation to sulfoxide

(*continued overleaf*)

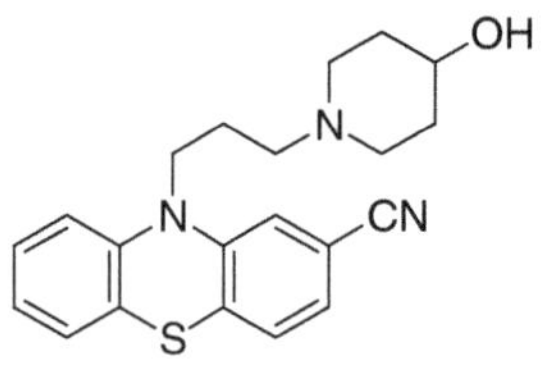

$t_{1/2}$	12 h	*V*	–
F	–	CL	–
pb	–	Mwt	365.5
ur	–	PSA	52.2 Å^2
		log *P*	3.1

PERICYAZINE, PROPERICIAZINE
Antipsychotic

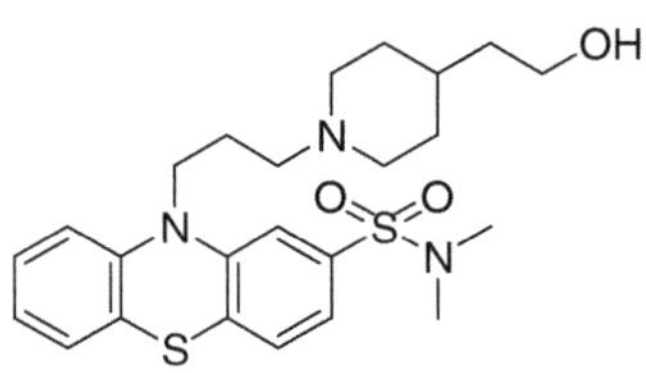

$t_{1/2}$	11 h	*V*	8.4
F	–	CL	–
pb	–	Mwt	475.7
ur	–	PSA	65.8 Å^2
		log *P*	4.3

PIPOTIAZINE
Antipsychotic
Prodrug: palmitate ($t_{1/2}$ 12–24 h, im)
Metabolism: hydroxylation, sulfoxide formation

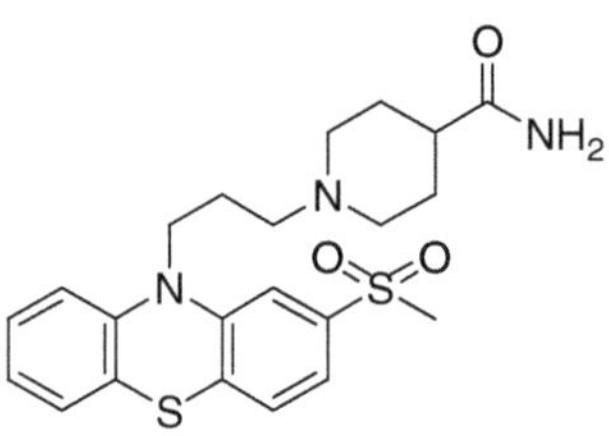

$t_{1/2}$	5 h	*V*	–
F	22%	CL	13
pb	–	Mwt	445.6
ur	–	PSA	85.4 Å^2
		log *P*	2.7

METOPIMAZINE
Antiemetic
Metabolism: amide hydrolysis to carboxylic acid

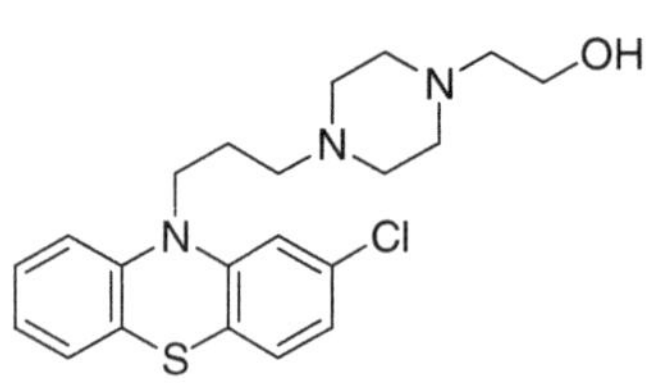

$t_{1/2}$	10±2 h	*V*	10–35
F	20–40%	CL	12–38
pb	92%	Mwt	404.0
ur	Low	PSA	31.6 Å^2
		log *P*	4.3

PERPHENAZINE
Antipsychotic, crosses bbb
Prodrug: enanthate ($t_{1/2}$ 2–3 d, im)
Metabolites: sulfoxide, 7-hydroxy ($t_{1/2}$ 10–19 h), *N*-oxide, glucuronide

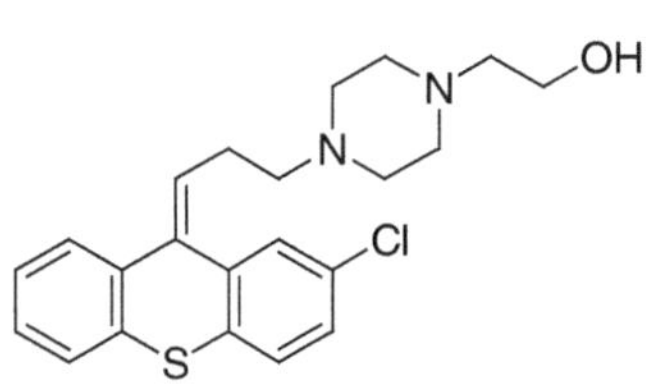

$t_{1/2}$	12–29 h	*V*	12–24
F	50%	CL	12.8
pb	98%	Mwt	401.0
ur	–	PSA	26.7 Å^2
		log *P*	4.7

CLOPENTHIXOL (*E*/*Z* mixture)
Zuclopenthixol (*Z* isomer, shown)
Antipsychotic
Prodrug: decanoate ($t_{1/2}$ 19 d)
Metabolism: N-dealkylation, S-oxidation

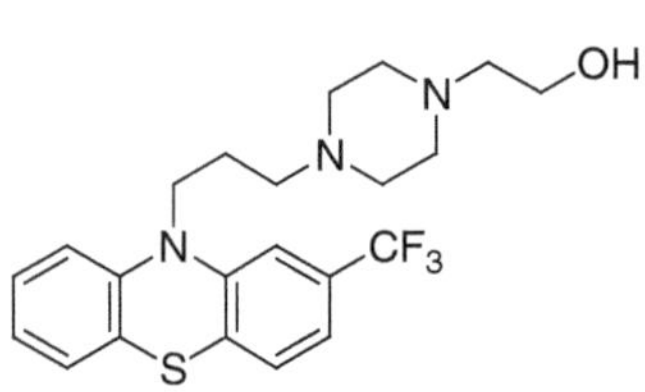

$t_{1/2}$	25±9 h	*V*	10
F	–	CL	12.8
pb	>99%	Mwt	437.5
ur	–	PSA	31.6 Å^2
		log *P*	4.5

FLUPHENAZINE
Antipsychotic
Prodrug: decanoate (Dapotum®, $t_{1/2}$ 132–206 h, im)
Metabolism: aromatic 7-hydroxylation, then glucuronidation

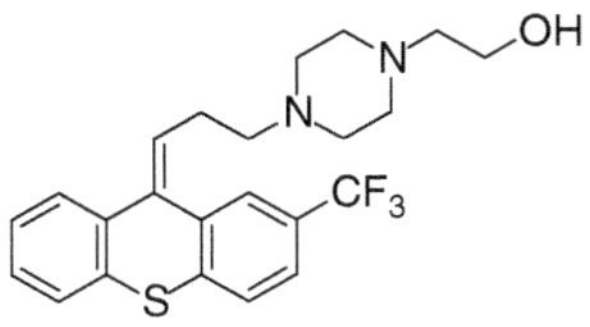	$t_{1/2}$ 29±10 h *F* 48–60% pb >95% ur –	*V* 15±2 CL 6±1 Mwt 434.5 PSA 26.7 Å^2 log *P* 4.9	FLUPENTIXOL Antipsychotic; mixture of isomers Prodrug: decanoate ($t_{1/2}$ 3–7 d, im) Metabolism: oxidation to sulfoxide, addition of glutathione to alkene, glucuronidation, sulfation		

$t_{1/2}$, plasma half-life; *F*, oral bioavailability; pb, plasma protein binding; ur, excretion of unchanged drug in urine; V, volume of distribution; CL, clearance; Mwt, molecular weight; PSA, polar surface area; bbb, blood–brain barrier.

50
Dibenzazepines and Related Tricyclic Compounds

The antidepressant properties of imipramine, a close analog of promazine, were discovered in 1957 by Roland Kuhn, a swiss psychiatrist. Compared to phenothiazines, dibenzazepines (Table 50.1) are less sedating but more mood enhancing and act as norepinephrine and serotonin reuptake inhibitors. Typical metabolic transformations are shown in Scheme 50.1.

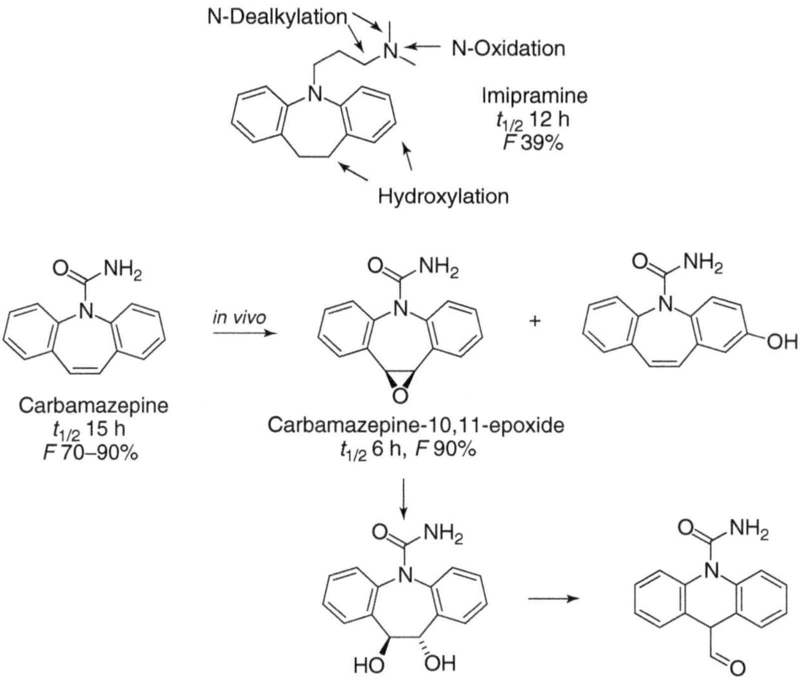

Scheme 50.1 Metabolism of dibenzazepines.

Lead Optimization for Medicinal Chemists: Pharmacokinetic Properties of Functional Groups and Organic Compounds, First Edition. Florencio Zaragoza Dörwald.

Table 50.1 Dibenzazepines and related compounds. *V* in l kg^{-1}; CL in ml min^{-1} kg^{-1}; Mwt in g mol^{-1}.

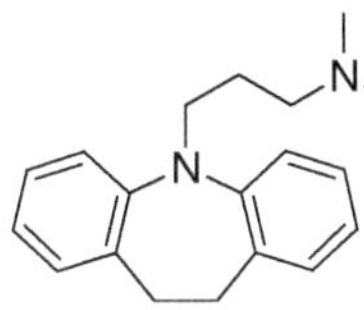	$t^{1}/_{2}$ *F* pb ur	12±5 h 39±7% 89–91% <2%	*V* CL Mwt PSA log *P*	18±3 15±4 280.4 6.5 Å^2 4.2	IMIPRAMINE Antidepressant Metabolism: aromatic and benzylic hydroxylation, N-demethylation, N-oxidation
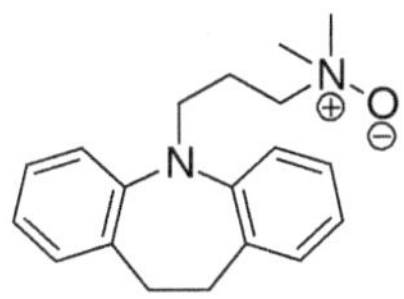	$t^{1}/_{2}$ *F* pb ur	1.8 h (iv) – – –	*V* CL Mwt PSA log *P*	1.9 12 296.4 20.3 Å^2 1.71	IMIPRAMINE *N*-OXIDE Antidepressant
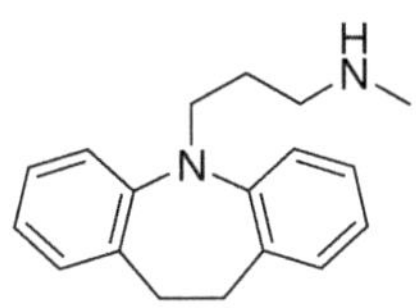	$t^{1}/_{2}$ *F* pb ur	22±5 h 38±13% 82±2% 2%	*V* CL Mwt PSA log *P*	20±3 10±2 266.4 15.3 Å^2 3.9	DESIPRAMINE Antidepressant Metabolism: aromatic and benzylic hydroxylation
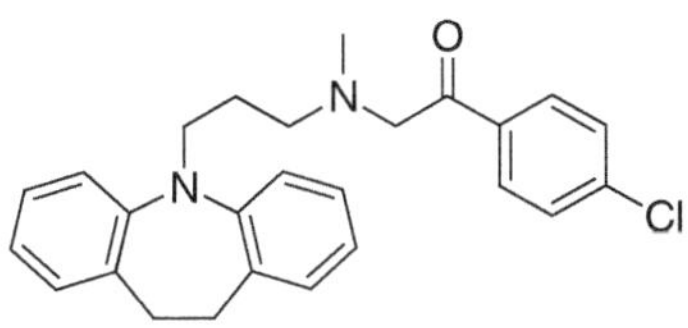	$t^{1}/_{2}$ *F* pb ur	5±1 h 10% 99% –	*V* CL Mwt PSA log *P*	– 163 419.0 23.5 Å^2 6.0	LOFEPRAMINE Antidepressant Metabolites: desipramine, 2-hydroxydesipramine
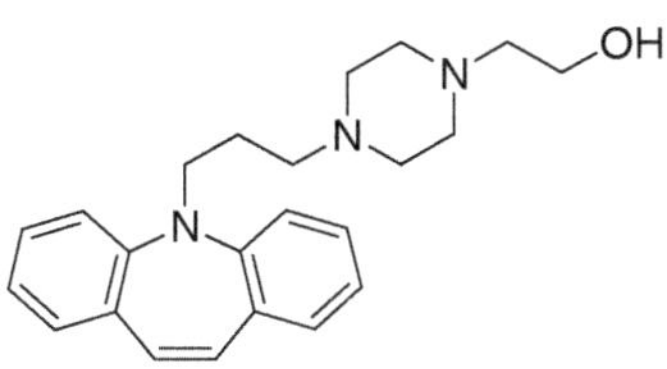	$t^{1}/_{2}$ *F* pb ur	9±2 h 95% 91% 7%	*V* CL Mwt PSA log *P*	10 – 363.5 31.6 Å^2 3.1	OPIPRAMOL Antidepressant, antipsychotic Metabolites: des(hydroxyethyl), *N*-carboxymethyl, *N*-oxide, dibenzazepine
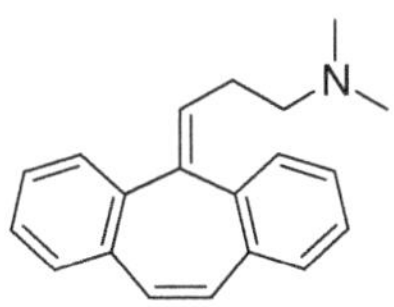	$t^{1}/_{2}$ *F* pb ur	22±14 h 55% 93% –	*V* CL Mwt PSA log *P*	– 9.8 275.4 3.2 Å^2 4.9	CYCLOBENZAPRINE Muscle relaxant Metabolism: N-demethylation, aromatic hydroxylation, *trans*-dihydroxylation of olefin, addition of glutathione to alkene

(*continued overleaf*)

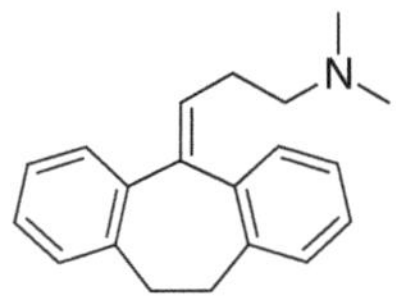	$t_{1/2}$ *F* pb ur	21±5 h 48±11% 95% <2%	*V* CL Mwt PSA log *P*	15±3 12±3 277.4 3.2 Å^2 4.2	AMITRIPTYLINE Antidepressant Metabolism: N-demethylation to nortriptyline and benzylic hydroxylation by CYP2D6
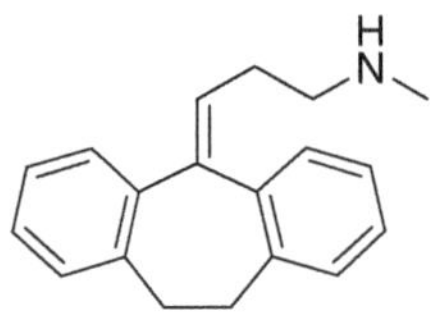	$t_{1/2}$ *F* pb ur	31±13 h 51±5% 92±2% 2±1%	*V* CL Mwt PSA log *P*	18±4 7.2±1.8 263.4 12.0 Å^2 3.9	NORTRIPTYLINE Antidepressant Metabolite: 10-hydroxynortriptyline
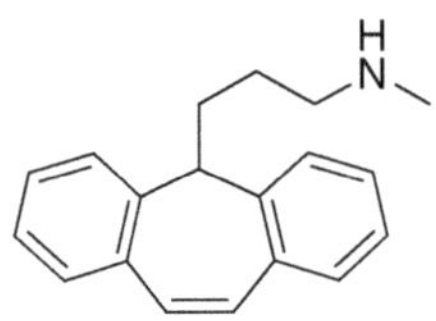	$t_{1/2}$ *F* pb ur	78±11 h 77–93% 91–93% –	*V* CL Mwt PSA log *P*	21–23 3–4 263.4 12.0 Å^2 4.2	PROTRIPTYLINE Antidepressant Metabolism: N-oxidation, N-demethylation, dihydroxylation of alkene, oxidative rearrangement to formylanthracene
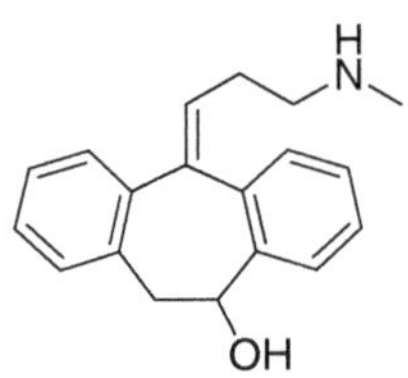	$t_{1/2}$ *F* pb ur	8 h – – –	*V* CL Mwt PSA log *P*	7.7 11 279.4 32.3 Å^2 3.0	HYDROXYNORTRIPTYLINE Metabolite of nortriptyline
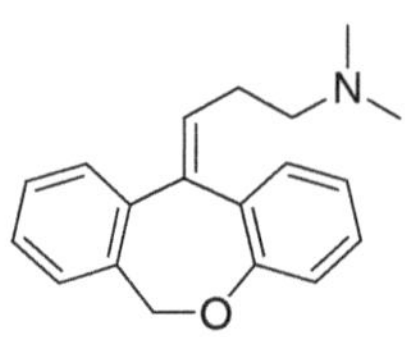	$t_{1/2}$ *F* pb ur	17±6 h 27±10% 78% 0%	*V* CL Mwt PSA log *P*	20±8 14±3 279.4 12.5 Å^2 3.9	DOXEPIN Antidepressant Active metabolite: *N*-desmethyldoxepin ($t_{1/2}$ 37±15 h)
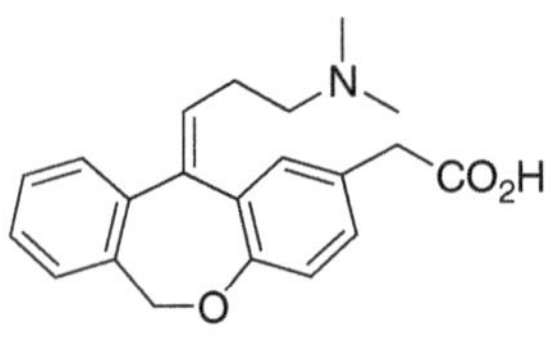	$t_{1/2}$ *F* pb ur	10±2 h* 57%* 55% 60–70%*	*V* CL Mwt PSA log *P*	– – 337.4 49.8 Å^2 2.96	OLOPATADINE *on intranasal dosing Histamine H_1 antagonist Metabolites: *N*-desmethyl, *N*-oxide

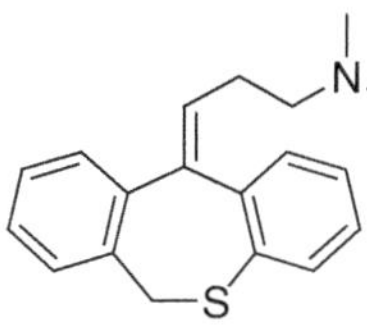

$t^1/_2$	19±5 h	V	11–78
F	30%	CL	23–63
pb	–	Mwt	295.5
ur	0.5%	PSA	3.2 Å^2
		log P	4.2

DOTHIEPIN, DOSULEPIN
Antidepressant; mixture of isomers
Metabolism: N-demethylation, S-oxidation

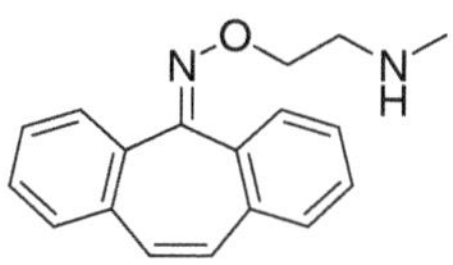

$t^1/_2$	35 h	V	–
F	–	CL	–
pb	–	Mwt	278.4
ur	–	PSA	33.6 Å^2
		log P	4.3

DEMEXIPTILINE
Antidepressant
Metabolism: N-demethylation

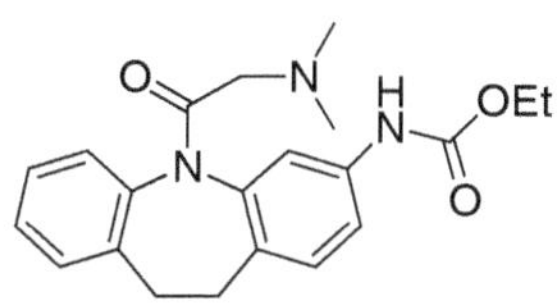

$t^1/_2$	2.4 h	V	–
F	34%	CL	–
pb	–	Mwt	367.4
ur	3%	PSA	61.9 Å^2
		log P	2.23

TIRACIZINE
Antiarrhythmic
Metabolism: carbamate hydrolysis (to des(ethoxycarbonyl), $t^1/_2$ 14 h), N-demethylation

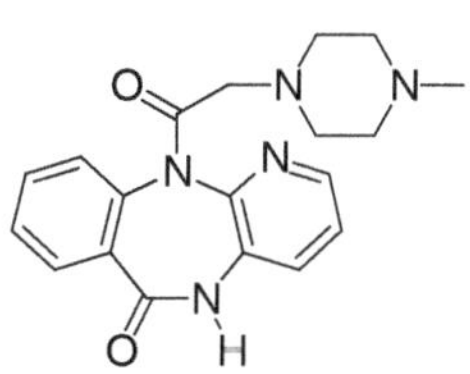

$t^1/_2$	11–14 h	V	1.3–3.4
F	21–33%	CL	2.8–3.6
pb	12%	Mwt	351.4
ur	50%	PSA	74.2 Å^2
		log P	1.3

PIRENZEPINE
Antiulcerative, muscarinic acetylcholine M_1 antagonist; does not cross bbb
Metabolism: N-dealkylation

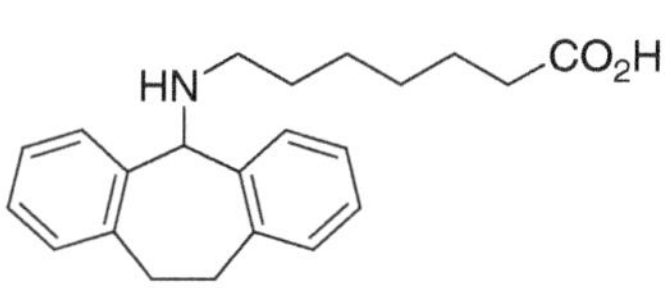

$t^1/_2$	0.8 h	V	2.4
F	–	CL	32
pb	–	Mwt	337.5
ur	–	PSA	49.3 Å^2
		log P	3.98

AMINEPTINE
CNS stimulant
Metabolism: oxidative degradation of side chain

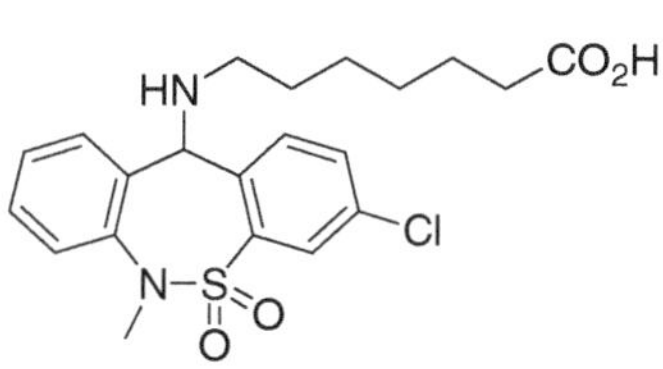

$t^1/_2$	2.5–3.3 h	V	0.5–1.2
F	99%	CL	3.4–5.5
pb	95%	Mwt	437.0
ur	<3%	PSA	95.1 Å^2
		log P	4.13

TIANEPTINE
Antidepressant
Metabolism: oxidative degradation to C_5-aminoacid, little first-pass metabolism

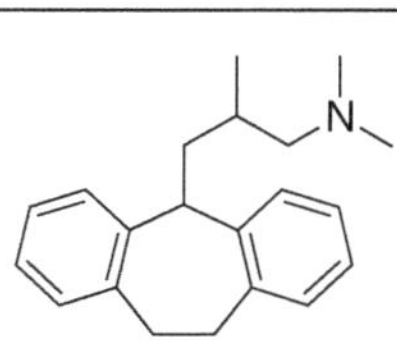

$t^1/_2$	20 h	V	–
F	–	CL	–
pb	>90%	Mwt	293.5
ur	<2%	PSA	3.24 Å^2
		log P	5.30

BUTRIPTYLINE
Antidepressant
Metabolism: N-demethylation to norbutriptyline

(continued overleaf)

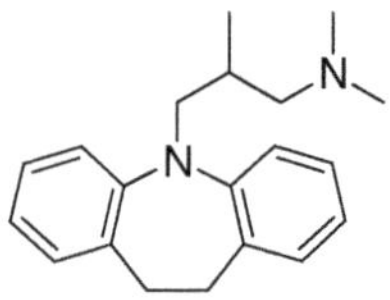

*t*½	17±6 h	*V*	31
F	41%	CL	16
pb	95%	Mwt	294.4
ur	–	PSA	6.5 Å^2
		log *P*	4.6

TRIMIPRAMINE
Antidepressant
Metabolism: N-demethylation, aromatic hydroxylation

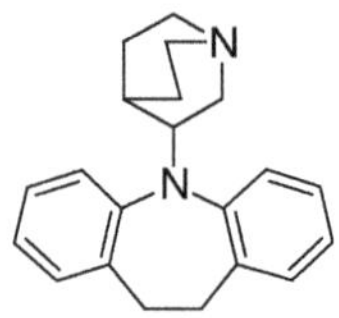

*t*½	29–33 h	*V*	34±5
F	75%	CL	–
pb	–	Mwt	304.4
ur	–	PSA	6.5 Å^2
		log *P*	4.98

QUINUPRAMINE
Antidepressant

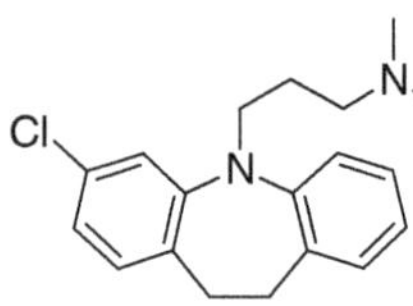

*t*½	27±9 h	*V*	7–20
F	55±15%	CL	6–11
pb	98%	Mwt	314.9
ur	1–3%	PSA	6.5 Å^2
		log *P*	4.8

CLOMIPRAMINE
Antidepressant
Metabolites: *N*-desmethyl (*t*½ 66±11 h), *N*-oxide, phenols, benzylic alcohols

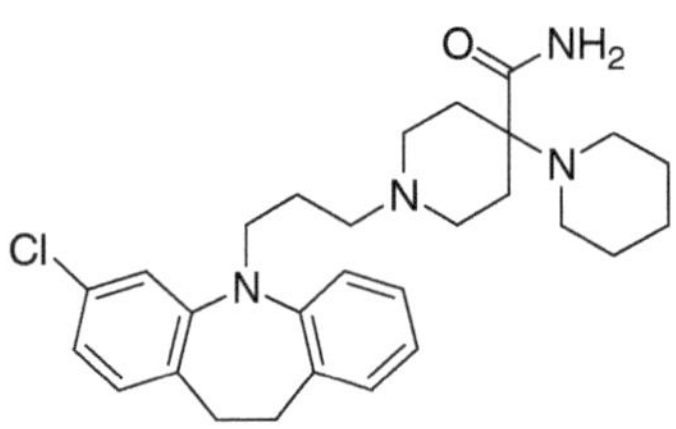

*t*½	3 h (dog)	*V*	–
F	16%	CL	–
pb	–	Mwt	481.1
ur	–	PSA	52.8 Å^2
		log *P*	5.28

CLOCAPRAMINE
Antipsychotic
Metabolite: mosapramine

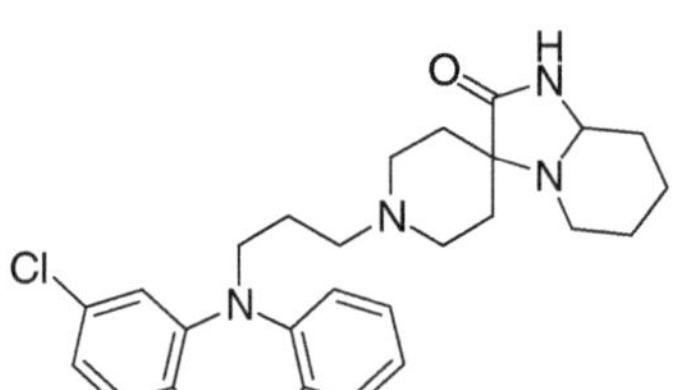

*t*½	15±2 h	*V*	–
F	29±7%	CL	–
pb	–	Mwt	479.1
ur	–	PSA	38.8 Å^2
		log *P*	7.36

MOSAPRAMINE
Antipsychotic, metabolite of clocapramine
Metabolism: aromatic and benzylic hydroxylation, oxidation to imidazolone

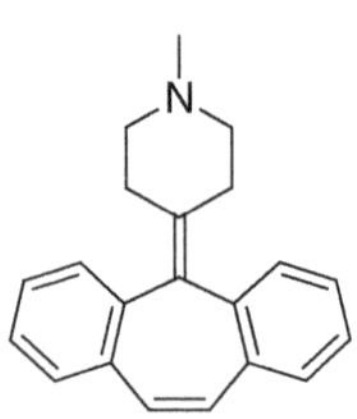

*t*½	1–4 h	*V*	–
F	–	CL	–
pb	96–99%	Mwt	287.4
ur	0%	PSA	3.2 Å^2
		log *P*	5.1

CYPROHEPTADINE
Histamine H_1 antagonist, antipruritic
Metabolites: quaternary glucuronide, *N*-desmethyl, epoxide

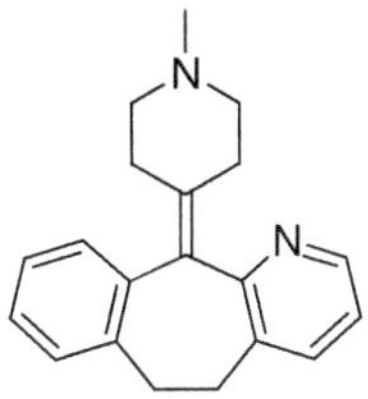	$t^{1}/_{2}$ *F* pb ur	9–12 h – Low 20%	*V* CL Mwt PSA log *P*	– – 290.4 16.1 Å^2 3.5	AZATADINE Histamine H_1 antagonist
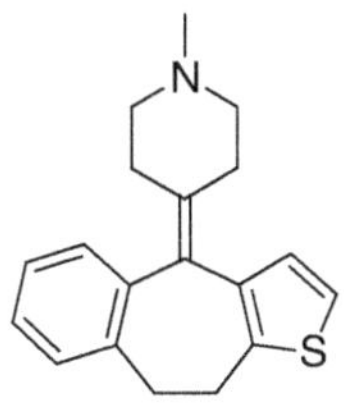	$t^{1}/_{2}$ *F* pb ur	23 h 78% 91% <1%	*V* CL Mwt PSA log *P*	– – 295.4 31.5 Å^2 2.71	PIZOTYLINE, PIZOTIFEN 5-HT antagonist for migraine prophylaxis; main metabolite: *N*-glucuronide ($t^{1}/_{2}$ 23 h, *V* 1 l kg^{-1})
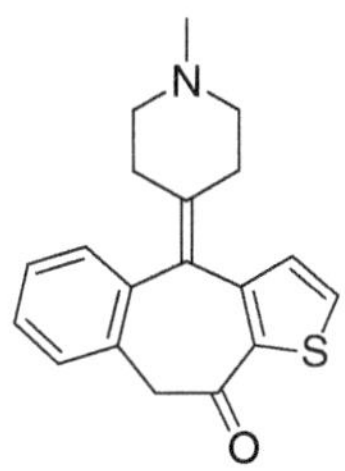	$t^{1}/_{2}$ *F* pb ur	18±7 h 50% 75% 1%*	*V* CL Mwt PSA log *P*	56 – 309.4 48.6 Å^2 2.19	KETOTIFEN *10% of *N*-desmethyl metabolite Antiasthmatic Metabolism: N-demethylation, N-oxidation, N-glucuronidation, reduction of keto group; active metabolite: desmethyl
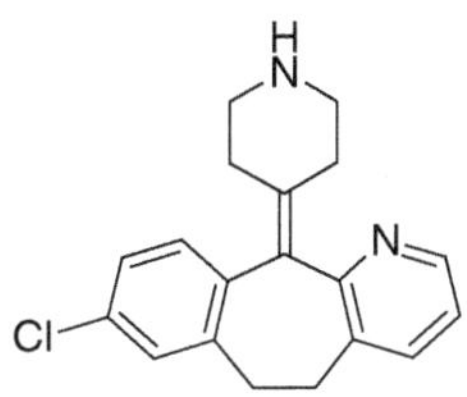	$t^{1}/_{2}$ *F* pb ur	18±6 h High 73–87% 0%	*V* CL Mwt PSA log *P*	49 – 310.8 24.9 Å^2 3.5	DESLORATADINE Metabolite of loratadine Metabolism: 3-hydroxylation of pyridine, then glucuronidation
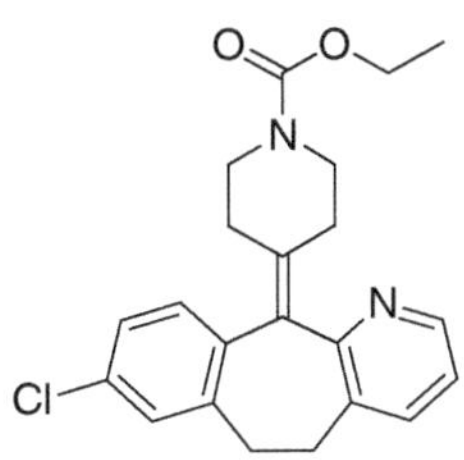	$t^{1}/_{2}$ *F* pb ur	8±6 h – 97% –	*V* CL Mwt PSA log *P*	120±80 142±57 382.9 42.4 Å^2 4.6	LORATADINE Histamine H_1 antagonist Duration of action: 24 h Metabolism: hydrolysis of carbamate

(*continued overleaf*)

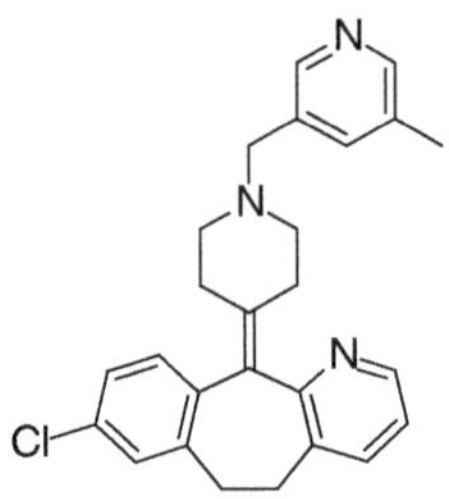	$t^1/_2$ 6 h	V	143	RUPATADINE	
	F 50%	CL	–	Histamine H_1 antagonist	
	pb >95%	Mwt	416.0	Metabolism:	
	ur 0%	PSA	29.0 Å^2	piperidine-N-dealkylation (to desloratadine), then 3-hydroxylation of pyridine to 3-hydroxydesloratadine	
		log P	6.11		
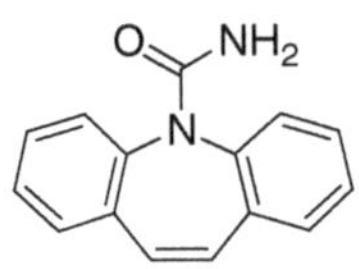	$t^1/_2$ 15±5 h	V/F	1.4±0.4	CARBAMAZEPINE	
	F 70–90%	CL/F	1.3±0.5	Analgesic, anticonvulsant; half-life decreases on chronic administration	
	pb 74±3%	Mwt	236.3	Metabolite: 10,11-epoxide	
	ur <1%	PSA	48.0 Å^2		
		log P	2.8		
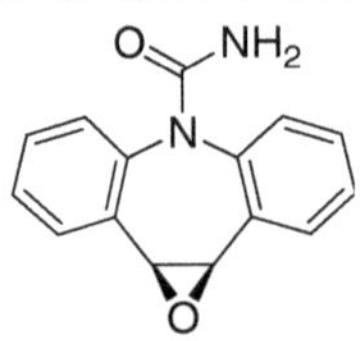	$t^1/_2$ 6 h	V/F	1.1±0.2	CARBAMAZEPINE-10,11-EPOXIDE	
	F 90±11%	CL/F	1.7±0.3	Metabolite of carbamazepine	
	pb 50%	Mwt	252.3		
	ur <1%	PSA	58.9 Å^2		
		log P	2.5		
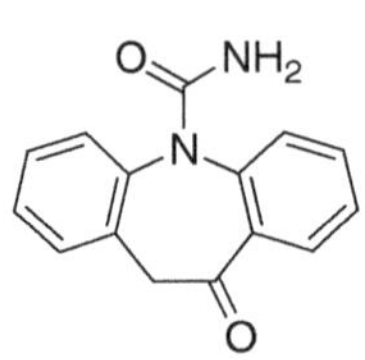	$t^1/_2$ 2 h	V	0.7–0.8	OXCARBAZEPINE	
	F –	CL	–	Anticonvulsant	
	pb 40%	Mwt	252.3	Metabolites:	
	ur <1%	PSA	63.4 Å^2	10-hydroxycarbamazepine ($t^1/_2$ 9 h), glucuronide thereof, 10,11-dihydroxy	
		log P	2.5		
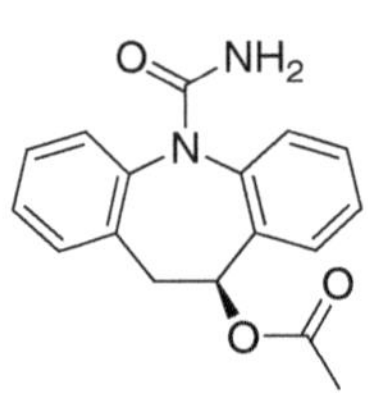	$t^1/_2$ 9–13 h*	V	0.8*	ESLICARBAZEPINE Acetate	
	F >90%*	CL	0.9*	*eslicarbazepine on oral dosing of ester; prodrug of eslicarbazepine, sodium channel blocker	
	pb 40%*	Mwt	296.3	Metabolism: glucuronidation	
	ur 60%*	PSA	72.6 Å^2		
		log P	1.70		
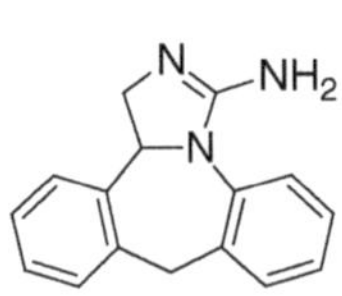	$t^1/_2$ 8–12 h	V	5.2	EPINASTINE	
	F 40%	CL	14	Antihistaminic (racemic; eye-drops); does not cross bbb, little metabolism	
	pb 64%	Mwt	249.3		
	ur 55%	PSA	41.6 Å^2		
		log P	2.9		

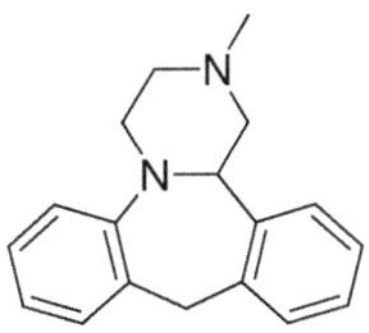

$t^{1}/_{2}$	14±3 h	*V*	16	MIANSERIN
F	20%	CL	9.9±1.2	Antidepressant
pb	90–95%	Mwt	264.4	Metabolism: N-demethylation, hydroxylation
ur	–	PSA	6.5 Å^2	
		log *P*	3.6	

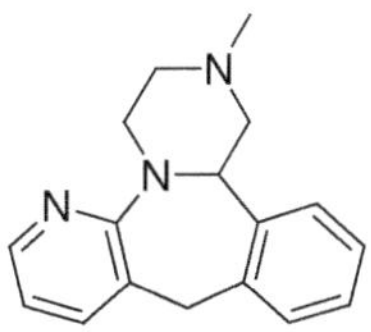

$t^{1}/_{2}$	13–40 h	*V*	4.8	MIRTAZAPINE
F	48–50%	CL	7.4	Antidepressant
pb	85%	Mwt	265.4	Metabolism: pyridine hydroxylation (para to piperazine), N-demethylation, N-oxidation
ur	<4%	PSA	19.4 Å^2	
		log *P*	2.7	

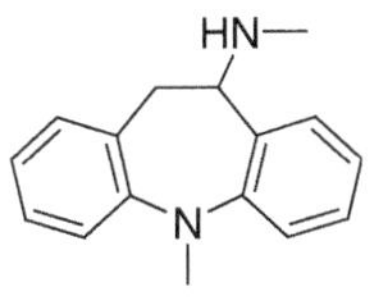

$t^{1}/_{2}$	8.3 h	*V*	0.6–1.4	METAPRAMINE
F	–	CL	17–27	Antidepressant
pb	–	Mwt	238.3	Metabolism: N-demethylation
ur	–	PSA	15.3 Å^2	
		log *P*	3.4	

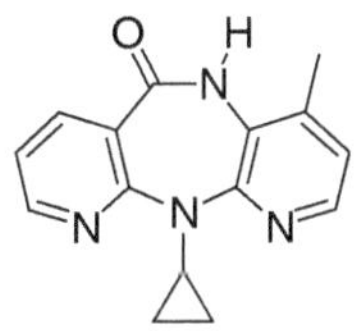

$t^{1}/_{2}$	25–45 h	*V*	1.2	NEVIRAPINE
F	93±9%	CL	0.3	Antiviral (HIV)
pb	60%	Mwt	266.3	Metabolism: hydroxylation and oxidation of CH_3 to CO_2H, hydroxylation of pyridines, then glucuronidation
ur	<3%	PSA	58.1 Å^2	
		log *P*	1.84	

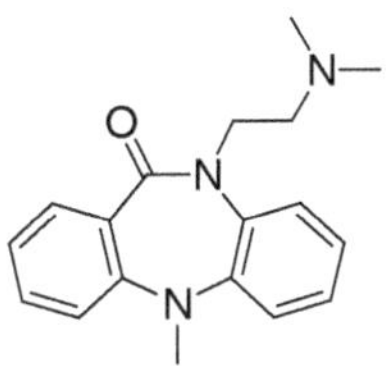

$t^{1}/_{2}$	4–5 h	*V*	4.6	DIBENZEPIN
F	25%	CL	–	Antidepressant
pb	80%	Mwt	295.4	Metabolism: demethylation at side chain
ur	1%	PSA	30.2 Å^2	
		log *P*	3.0	

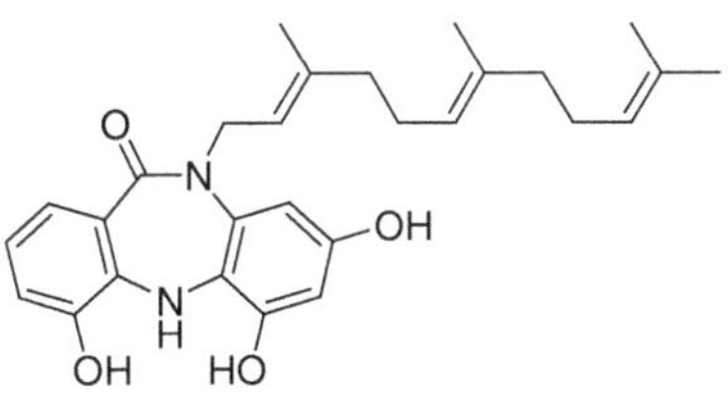

$t^{1}/_{2}$	9.4 h (iv)	*V*	–	DIAZEPINOMICIN
F	–	CL	–	Antibacterial
pb	67%	Mwt	462.6	Metabolite: glucuronide
ur	–	PSA	93.0 Å^2	
		log *P*	8.11	

(*continued overleaf*)

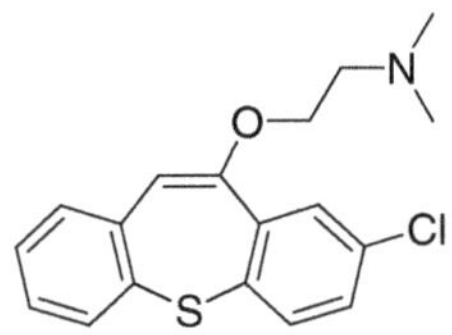

$t^{1}/_{2}$	21±9 h	V	109±59	ZOTEPINE
F	7–13%	CL	77±70	Atypical antipsychotic
pb	97%	Mwt	331.9	Metabolites: *N*-desmethyl,
ur	–	PSA	12.5 Å^2	*N*-oxide, sulfoxide, phenols
		log P	5.5	

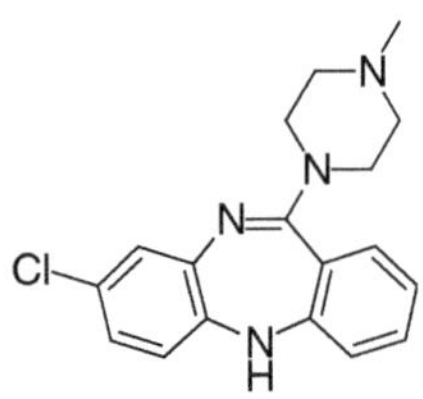

$t^{1}/_{2}$	12±4 h	V	5.4±3.5	CLOZAPINE
F	55±12%	CL	6.1±1.6	Antipsychotic
pb	>95%	Mwt	326.8	Metabolism: N-demethylation,
ur	<1%	PSA	30.9 Å^2	aromatic hydroxylation,
		log P	3.94	oxidative degradation of piperazine ring

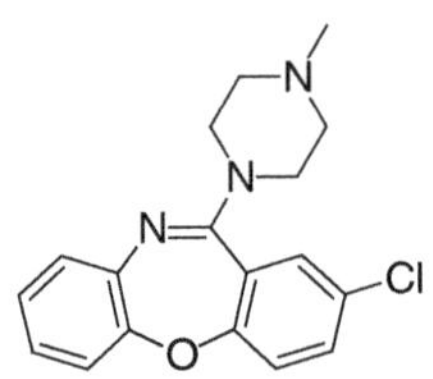

$t^{1}/_{2}$	15±3 h	V	–	LOXAPINE
F	30%	CL	–	Anxiolytic
pb	–	Mwt	327.8	Metabolism: aromatic
ur	–	PSA	32.5 Å^2	hydroxylation, N-demethylation
		log P	4.2	to amoxapine

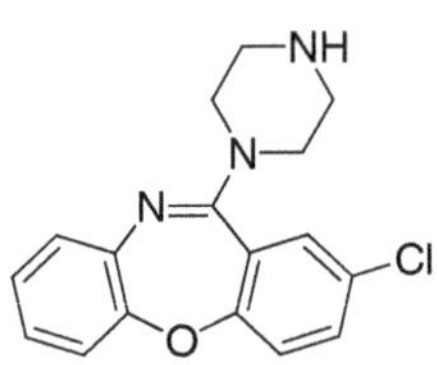

$t^{1}/_{2}$	8–11 h	V	–	AMOXAPINE
F	–	CL	15	Antidepressant
pb	–	Mwt	313.8	Metabolite of loxapine
ur	Negligible	PSA	41.3 Å^2	Metabolites: 7- and
		log P	3.7	8-hydroxyamoxapine ($t^{1}/_{2}$ 6 and 30 h)

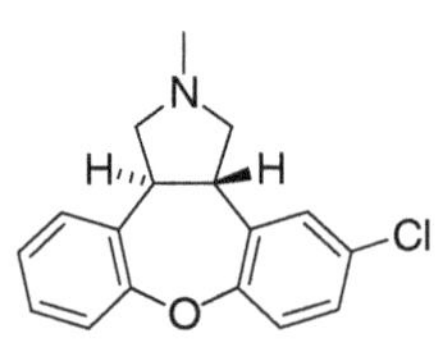

$t^{1}/_{2}$	24 h	V	20–25	ASENAPINE
F	<2%*	CL	13	*35% buccal
pb	95%	Mwt	285.8	Antipsychotic
ur	Low	PSA	12.5 Å^2	Metabolites: desmethyl,
		log P	4.31	*N*-glucuronide, *N*-carbamoylglucuronide

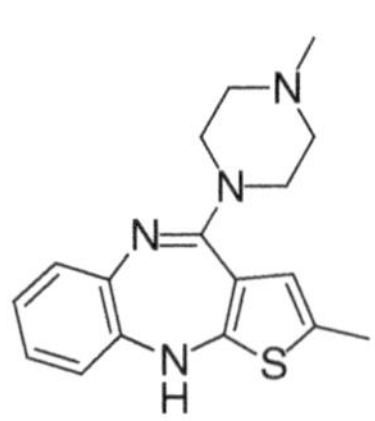

$t^{1}/_{2}$	27–39 h	V	10–22	OLANZAPINE
F	60–80%	CL	4.5–5.1	Antipsychotic
pb	93%	Mwt	312.4	Metabolism: N-demethylation,
ur	7%	PSA	35.2 Å^2	hydroxylation of C*Me*
		log P	3.5	

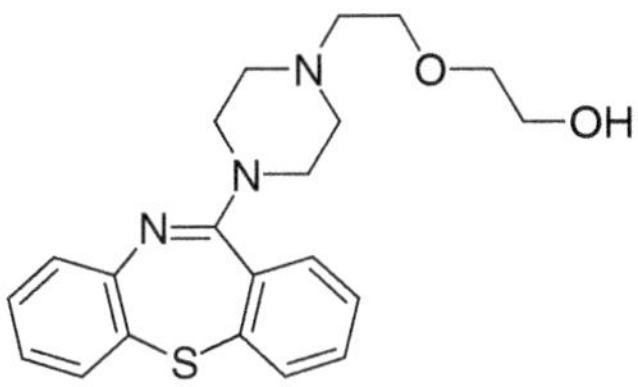

$t½$	6–7 h	V	10±4	QUETIAPINE
F	9±4%	CL	–	Atypical antipsychotic
pb	83%	Mwt	383.5	Metabolism: piperazine
ur	1–5%	PSA	48.8 Å^2	N-dealkylation (to
		log P	3.5	desalkylquetiapine, $t½$ 9–12 h)

$t_{1/2}$, plasma half-life; F, oral bioavailability; pb, plasma protein binding; ur, excretion of unchanged drug in urine; V, volume of distribution; CL, clearance; Mwt, molecular weight; PSA, polar surface area; bbb, blood–brain barrier; CNS, central nervous system; 5-HT, 5-hydroxytryptamine.

51
3-Aryloxy-2-Hydroxypropylamines (β-Adrenergic Antagonists; "β-Blockers")

Five adrenaline receptors have been identified: two α- and three β-receptors. α-Adrenoceptor antagonists were developed as early as 1933 (piperoxan), and the more potent tolazoline (early 1940s) and phentolamine (1950s) are still used today as nasal decongestants by virtue of their vasoconstricting properties.

In 1958, James Black joined the pharmaceutical division of ICI and attempted to develop a β-adrenaline antagonist for the treatment of angina pectoris. The result of this research was the β-blocker propranolol, launched in 1965. Since then, a large number of 3-aryloxy-2-hydroxypropylamines were developed and marketed for use in humans, mainly as antianginals, antihypertensives, antiarrhythmics, and, topically, for the treatment of glaucoma (lowering the intraocular pressure).

A comparison of the structures and pharmacokinetic (PK) properties of these compounds gives useful insight into how PK properties may be influenced by structural modifications and which chemical substructures are acceptable for a drug. The 3-aryloxy-2-hydroxypropylamine substructure is essential for biological activity and conserved in most of these compounds. Half-lives and oral bioavailabilities of β-blockers range from 1 to 22 h and from 5 to 100%, respectively, and β-blockers include highly lipophilic compounds, which readily cross the blood–brain barrier (such as propranolol), and more hydrophilic molecules, which are unable to attain high concentrations in the central nervous system (CNS) (e.g., atenolol).

51.1
Metabolism

The hydrophilic 2-hydroxypropylamine substructure neither is metabolized by specific hormone-degrading enzymes nor is a good substrate of the less substrate-specific liver enzymes (CYP). Therefore, metabolism of β-blockers occurs mostly at the lipophilic aryloxy group. For instance, propranolol, alprenolol, penbutolol, and tertatolol are mainly hydroxylated at the 4-position of the aryloxy group (by liver enzymes) to yield monoalkylated hydroquinones (Scheme 51.1). Other metabolic pathways include dealkylation of the aryloxy and amino groups. Thus, first-pass

Lead Optimization for Medicinal Chemists: Pharmacokinetic Properties of Functional Groups and Organic Compounds, First Edition. Florencio Zaragoza Dörwald.

metabolism may be blocked (and oral bioavailability enhanced) by introducing polar or hydrophilic groups into the aryl substructure of β-blockers (Table 51.1).

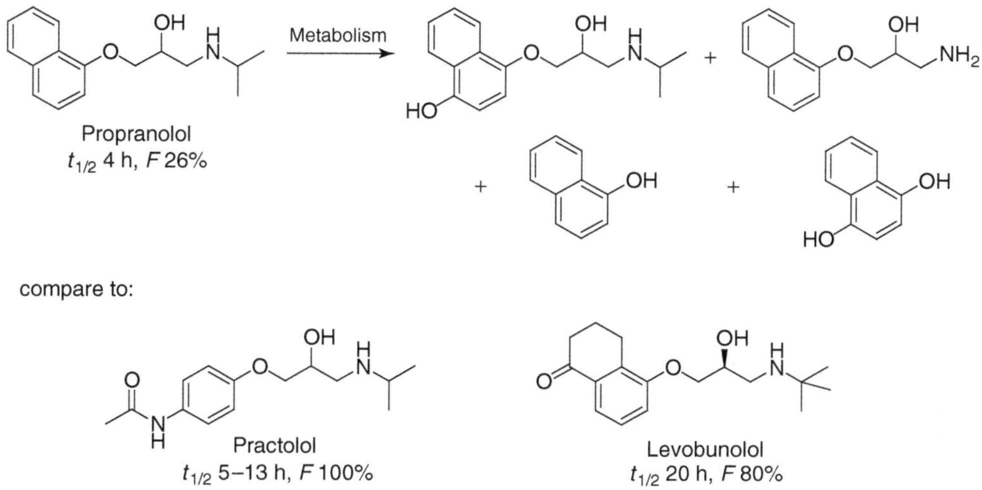

Scheme 51.1 Metabolism of propranolol.

Table 51.1 3-Aryloxy-2-hydroxypropylamines (β-adrenergic antagonists; antianginals, antihypertensives, antiarrhythmics, and antiglaucoma agents). *V* in l kg^{-1}; CL in ml min^{-1} kg^{-1}; Mwt in g mol^{-1}.

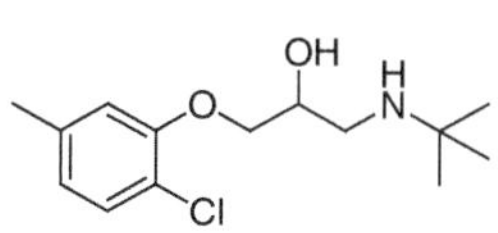	$t½$ *F* pb ur	2.0±0.3 h <10% 80–90% <10%	*V* CL Mwt PSA log *P*	– – 271.8 41.5 Å^2 3.4	BUPRANOLOL Metabolism: oxidation of ArMe to $ArCO_2H$
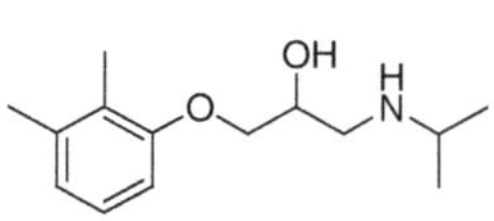	$t½$ *F* pb ur	3 h – – 1%	*V* CL Mwt PSA log *P*	– – 237.3 41.5 Å^2 2.6	XIBENOLOL Metabolism: aromatic 4-hydroxylation ($t½$ 3 h), 3-CH_3 hydroxylation, oxidation of 3-CH_3 to CO_2H
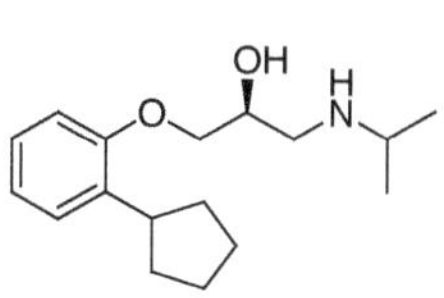	$t½$ *F* pb ur	2–5 h 90–95% 80–98% 4–6%	*V* CL Mwt PSA log *P*	2.3–7.2 16.6 277.4 41.5 Å^2 3.1	L-PENBUTOLOL Metabolites: glucuronide, 4-hydroxyphenoxy

(*continued overleaf*)

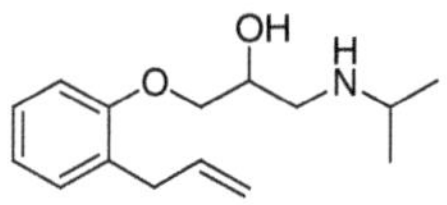	t½	2–3 h	V	3	ALPRENOLOL
	F	10–15%	CL	1.2	Metabolism: glucuronidation,
	pb	80–90%	Mwt	249.4	N-dealkylation, then
	ur	<1%	PSA	41.5 Å^2	N-methylation, 4-hydroxylation,
			log P	2.6	oxidative degradation
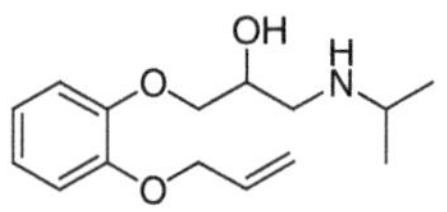	t½	1–2 h	V	1.6	OXPRENOLOL
	F	24–60%	CL	–	Metabolism: glucuronidation,
	pb	70–80%	Mwt	265.4	aromatic 4- and 5-hydroxylation
	ur	2–5%	PSA	50.7 Å^2	
			log P	2.1	
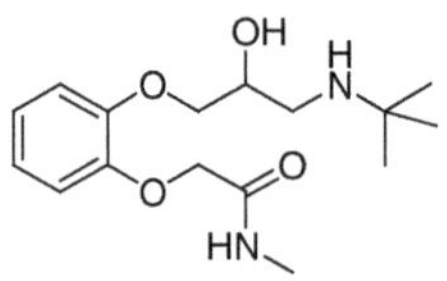	t½	6–8 h	V	2–3	CETAMOLOL
	F	–	CL	4–6	Metabolism: aromatic
	pb	–	Mwt	310.4	hydroxylation, N-demethylation,
	ur	30–50%	PSA	79.8 Å^2	O-dealkylation, oxidative
			log P	1.36	deamination to aryloxyacetic acid
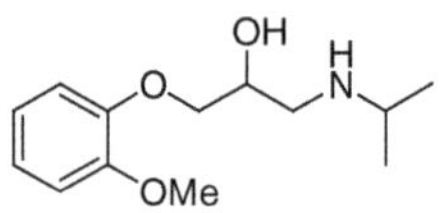	t½	2.1±0.3 h	V	–	MOPROLOL
	F	–	CL	–	
	pb	–	Mwt	239.3	
	ur	–	PSA	50.7 Å^2	
			log P	1.42	
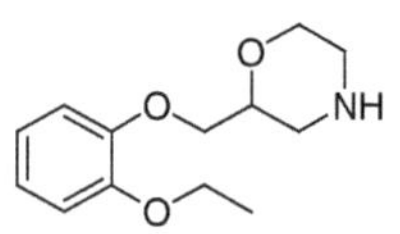	t½	3.3±1.0 h	V	0.73±0.28	VILOXAZINE
	F	85±14%	CL	2.0±0.2	Antidepressant
	pb	88±3%	Mwt	237.3	Metabolism: aromatic 4- and
	ur	12–15%	PSA	39.7 Å^2	5-hydroxylation
			log P	1.3	
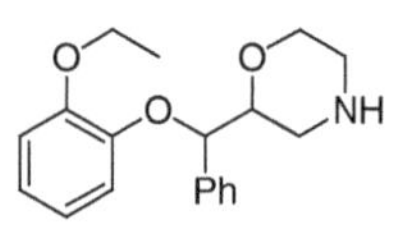	t½	13 h	V	0.92	REBOXETINE
	F	95%	CL	0.41–0.53	Antidepressant
	pb	92–97%	Mwt	313.4	Metabolism: O-deethylation,
	ur	8–9%	PSA	39.7 Å^2	aromatic hydroxylation (para to
			log P	2.45	both alkoxy groups)
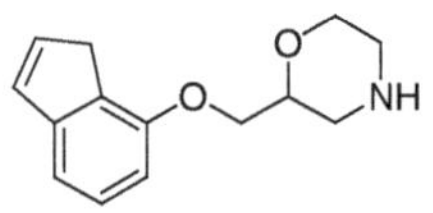	t½	2.2 h (rat)	V	–	INDELOXAZINE
	F	–	CL	–	Antidepressant, nootropic
	pb	–	Mwt	231.3	Metabolism (rat): dihydroxylation
	ur	–	PSA	30.5 Å^2	of alkene, N-acetylation, oxidative
			log P	1.79	degradation of morpholine ring
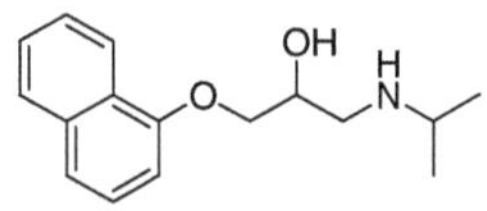	t½	3.9±0.4 h	V	4.3±0.6	PROPRANOLOL
	F	26±10%	CL	16±5	Metabolism: 4-, 2-, 5-, and
	pb	87±6%	Mwt	259.4	7-hydroxylation of naphthyl,
	ur	<1%	PSA	41.5 Å^2	N-deisopropylation
			log P	3.0	

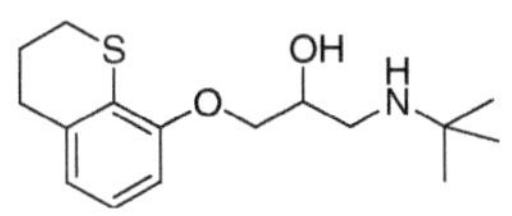	*t*½ *F* pb ur	9±7 h 60% 94% <1%	*V* CL Mwt PSA log *P*	0.4–0.7 – 295.5 41.5 Å^2 3.2	TERTATOLOL Metabolism: aromatic 4-hydroxylation, benzylic hydroxylation, sulfoxide formation
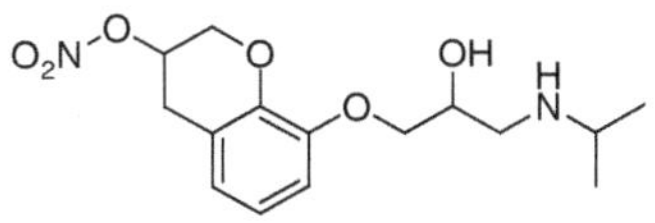	*t*½ *F* pb ur	3.7 h 35% 34% 6.3%	*V* CL Mwt PSA log *P*	5.6 – 326.5 105.8 Å^2 2.2	NIPRADILOL Metabolism: hydrolysis of nitrate to alcohol, N-deisopropylation
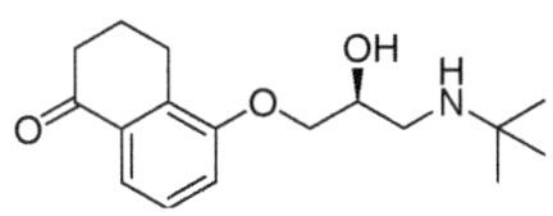	*t*½ *F* pb ur	20 h 80% – 14%	*V* CL Mwt PSA log *P*	5.5 0.16 291.4 58.6 Å^2 2.6	LEVOBUNOLOL Metabolism: reduction of keto group to alcohol
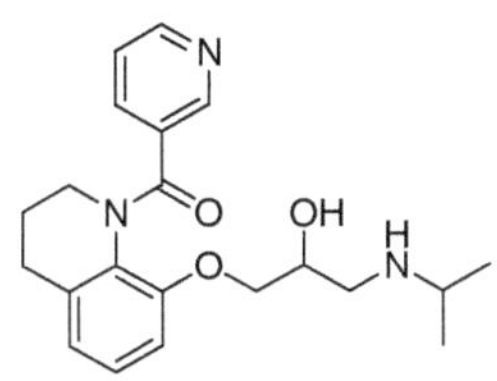	*t*½ *F* pb ur	1–3 h 70% 70% 9%	*V* CL Mwt PSA log *P*	0.67 3.6 369.5 74.7 Å^2 1.58	NICAINOPROL Antiarrhythmic
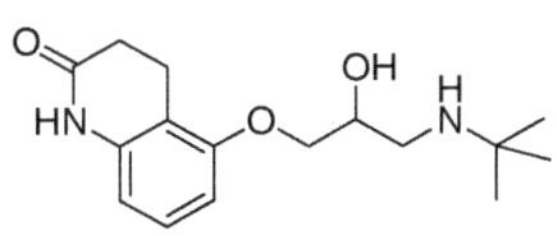	*t*½ *F* pb ur	6–8 h 85% 15% 50–70%	*V* CL Mwt PSA log *P*	4.1 8–10 292.4 70.6 Å^2 2.0	CARTEOLOL Metabolism: benzylic hydroxylation, glucuronidation
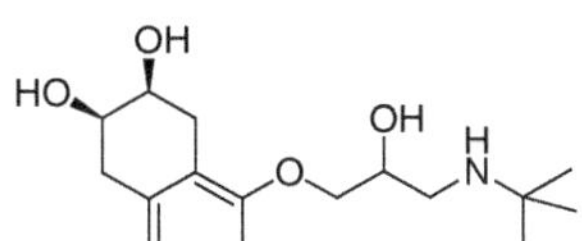	*t*½ *F* pb ur	16±2 h 34±5% 20±4% 73±4%	*V* CL Mwt PSA log *P*	1.9±0.2 2.9±0.6 309.4 82.0 Å^2 1.2	NADOLOL
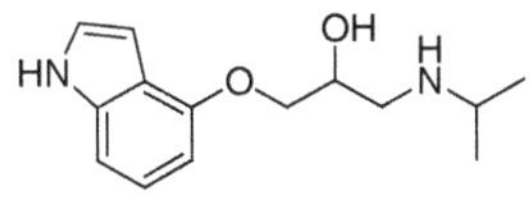	*t*½ *F* pb ur	3.6±0.6 h 75±9% 51±3% 54±9%	*V* CL Mwt PSA log *P*	2.3±0.9 8.3±1.8 248.3 57.3 Å^2 2.0	PINDOLOL Metabolism: glucuronidation, aromatic hydroxylation (to indolin-2-one and 3-hydroxy), then sulfation, oxidative ring scission to anthranilic acid, oxidative deamination

(*continued overleaf*)

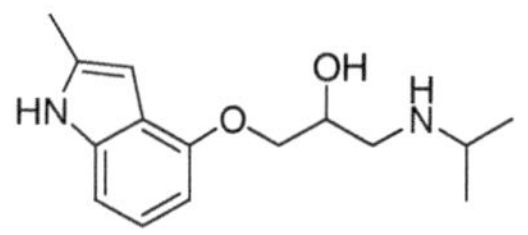	$t_{1/2}$ 4.0±1.5 h *F* >95% pb 50–60% ur 2%	*V* 5.7 CL – Mwt 262.4 PSA 57.3 Å^2 log *P* 2.2	MEPINDOLOL Prodrug of mepindolol: *O*-benzoylmepindolol (bopindolol; $t_{1/2}$ mepindolol on oral dosing of bopindolol: 6–10 h)
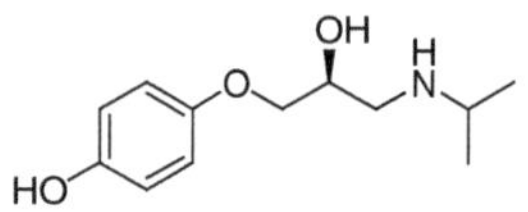	$t_{1/2}$ 2.4 h *F* 45–55% pb <5% ur 30–60%	*V* 2.6–3.4 CL 21 Mwt 225.3 PSA 61.7 Å^2 log *P* 1.3	PRENALTEROL Cardiotonic Metabolism: O-sulfation, O-glucuronidation, oxidative deamination to 4-(hydroxyphenoxy)lactic acid
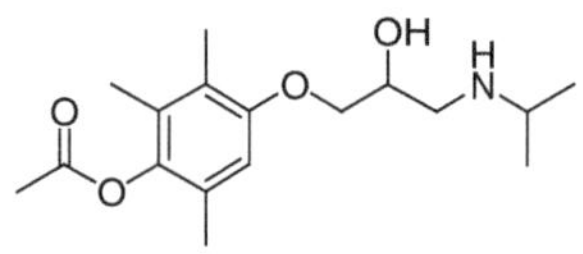	$t_{1/2}$ 2.5–4.0 h* *F* 50%* pb 70%* ur 12%*	*V* 3.5* CL 15.7* Mwt 309.4 PSA 67.8 Å^2 log *P* 2.28	METIPRANOLOL *rapid deacetylation *in vivo*. values for desacetyl metabolite on oral dosing of metipranolol
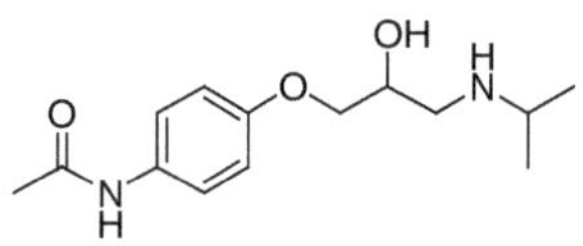	$t_{1/2}$ 5–13 h *F* 100% pb 32% ur >90%	*V* 1.6 CL 0.14 Mwt 266.3 PSA 70.6 Å^2 log *P* 1.0	PRACTOLOL Metabolism: deacetylation, then N-hydroxylation of aniline
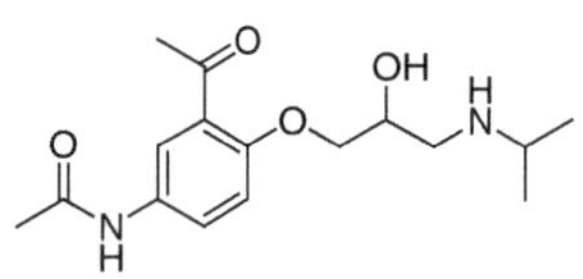	$t_{1/2}$ 10±2 h *F* 30–43% pb 6–9% ur 65%	*V* – CL 16±1 Mwt 308.4 PSA 87.7 Å^2 log *P* 0.85	DIACETOLOL Metabolite of acebutolol (see below)
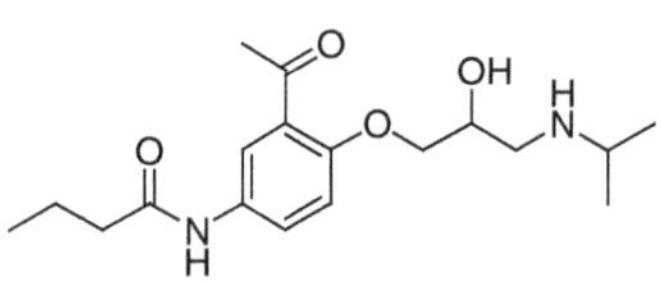	$t_{1/2}$ 2.7±0.4 h *F* 37±12% pb 26±3% ur 30%	*V* 1.2±0.3 CL 6.8±0.8 Mwt 336.4 PSA 87.7 Å^2 log *P* 2.3	ACEBUTOLOL Metabolism: hydrolysis of amide, then acetylation to diacetolol (see above)
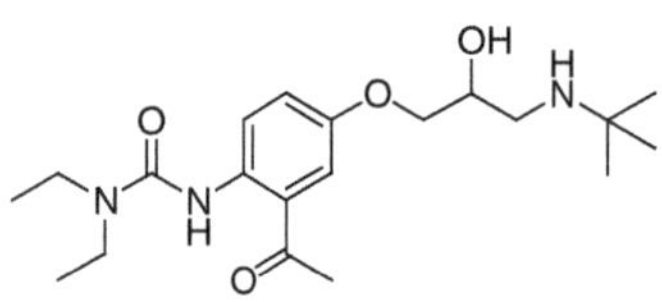	$t_{1/2}$ 4–6 h *F* 55–74% pb 25–30% ur 12–18%	*V* 4.5 CL – Mwt 379.5 PSA 90.9 Å^2 log *P* 2.7	CELIPROLOL Metabolism: N-deethylation

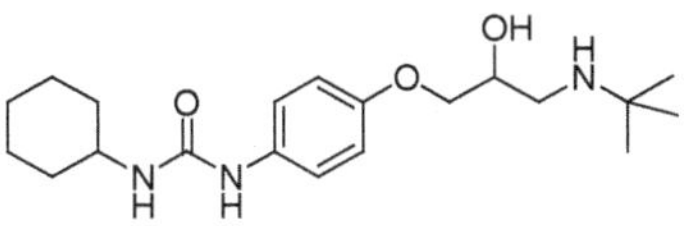	$t_{1/2}$ *F* pb ur	9–11 h 52% 60% 28%	*V* CL Mwt PSA log *P*	3.3 4.9 363.5 82.6 Å^2 3.7	TALINOLOL Metabolism: hydroxylation of cyclohexyl group
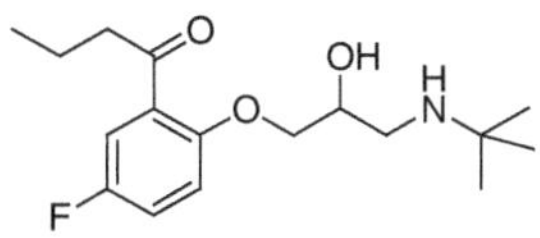	$t_{1/2}$ *F* pb ur	7–10 h – – 4%	*V* CL Mwt PSA log *P*	3–8 10–18 311.4 58.6 Å^2 3.31	BUTOFILOLOL Antihypertensive
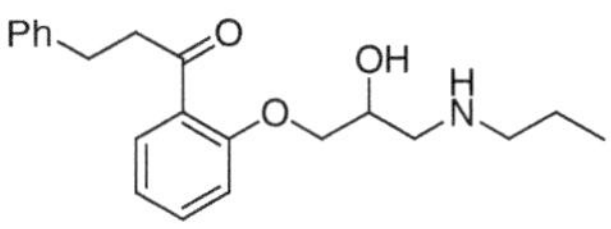	$t_{1/2}$ *F* pb ur	5.5±2.1 h 5–50% 85–95% <1%	*V* CL Mwt PSA log *P*	3.6±2.1 17±8 341.5 58.6 Å^2 3.5	PROPAFENONE CYP2D6 substrate Metabolism: N-depropylation, aromatic hydroxylation (para to OR)
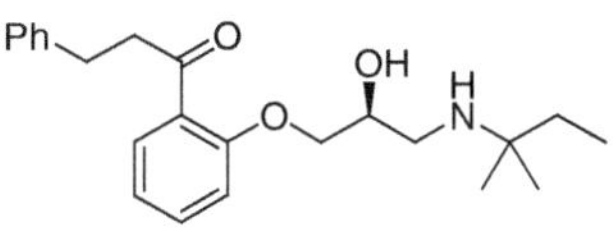	$t_{1/2}$ *F* pb ur	1.5 h (iv) 11–33% 98% –	*V* CL Mwt PSA log *P*	1.2 11 369.5 58.6 Å^2 4.12	(*S*)-DIPRAFENONE Metabolism: aromatic hydroxylation (para to OR)
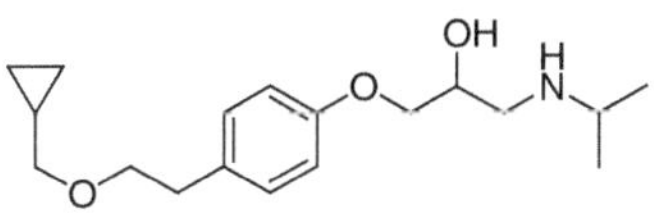	$t_{1/2}$ *F* pb ur	14–22 h 89±5% 55% 15%	*V* CL Mwt PSA log *P*	4.9–9.8 4.7 307.4 50.7 Å^2 2.8	BETAXOLOL Metabolism: oxidative deamination to aryloxyacetic acid, benzylic hydroxylation
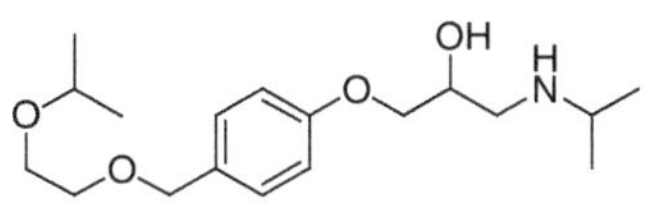	$t_{1/2}$ *F* pb ur	11±3 h 91±10% 35±5% 63±8%	*V* CL Mwt PSA log *P*	3.2±0.5 3.7±0.7 325.5 60.0 Å^2 2.3	BISOPROLOL Metabolism: O-deisopropylation
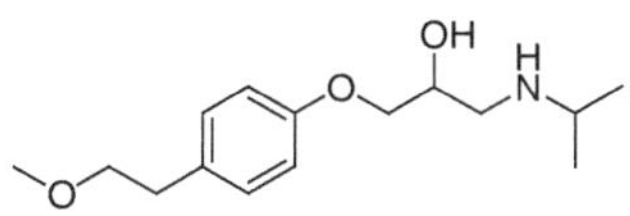	$t_{1/2}$ *F* pb ur	3.2±0.2 h 38±14% 11±1% 10±3%	*V* CL Mwt PSA log *P*	4.2±0.7 15±3 267.4 50.7 Å^2 2.0	METOPROLOL Metabolism: benzylic hydroxylation, O-demethylation
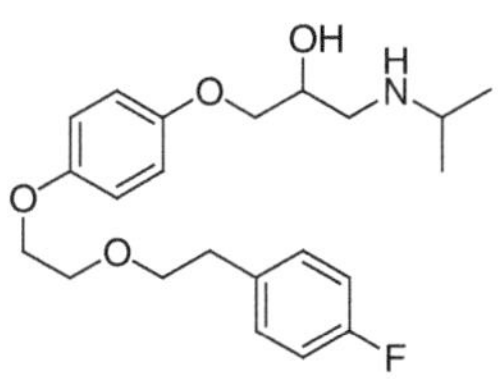	$t_{1/2}$ *F* pb ur	20 h 57% – –	*V* CL Mwt PSA log *P*	9 – 391.5 60.0 Å^2 3.7	RO 31-1118 *S*-enantiomer: flusoxolol

(*continued overleaf*)

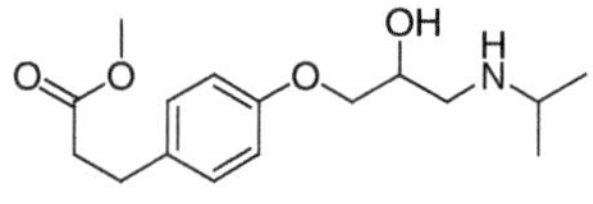	$t^1/_2$ 5–10 min *F* – pb 41–55% ur <1%	*V* 1.9±1.3 CL 170±70 Mwt 295.4 PSA 67.8 Å^2 log *P* 1.92	ESMOLOL Metabolism: ester hydrolysis
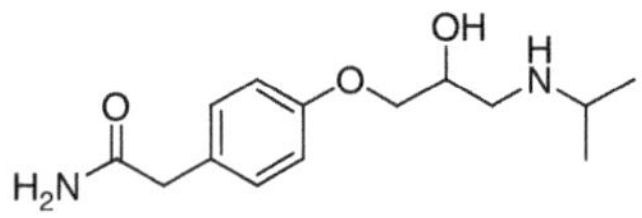	$t^1/_2$ 6±2 h *F* 56±30% pb 5–16% ur >90%	*V* 0.95±0.15 CL 1.2–2.1 Mwt 266.3 PSA 84.6 Å^2 log *P* 0.7	ATENOLOL Metabolism: hydroxylation of benzylic CH_2
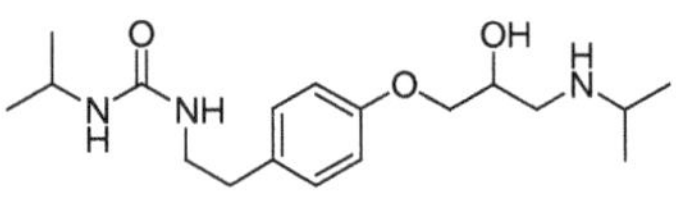	$t^1/_2$ 6 h *F* 27–46% pb – ur 56% (iv)	*V* 1.1 CL 4.5 Mwt 337.5 PSA 82.6 Å^2 log *P* 1.66	PAFENOLOL Metabolism: benzylic hydroxylation
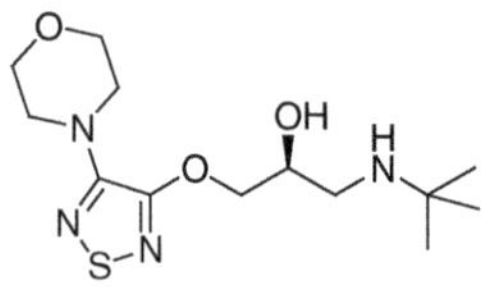	$t^1/_2$ 4±1 h *F* 30–50% pb 60±3% ur 15%	*V* 2.1±0.8 CL 7.3±3.3 Mwt 316.4 PSA 79.8 Å^2 log *P* 1.1	TIMOLOL Metabolism: oxidation of morpholine to lactam, *N*-(2-hydroxyethyl)glycolamido, and to 2-hydroxyethylamino group, hydroxylation of *tert*-butyl
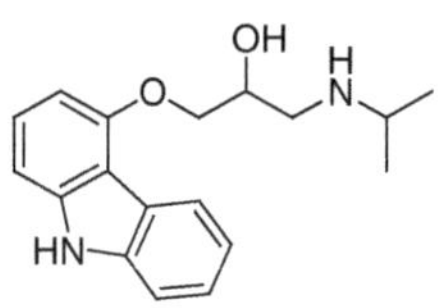	$t^1/_2$ 8 h *F* <10% pb 81% ur <0.2%	*V* 10.9 CL – Mwt 298.4 PSA 57.3 Å^2 log *P* 3.3	CARAZOLOL Metabolism (dog): O-glucuronidation, oxidative deamination, aromatic hydroxylation
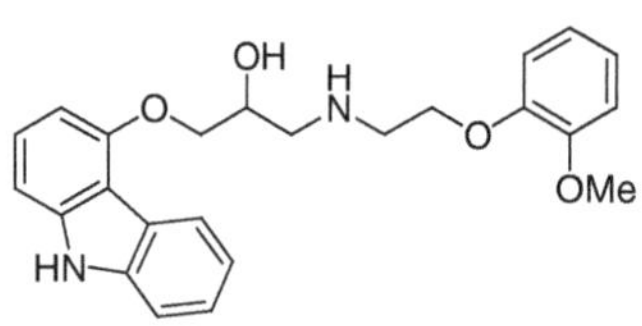	$t^1/_2$ 4±2 h *F* 25% pb 95% ur <2%	*V* 1.5±0.3 CL 8.7±1.7 Mwt 406.5 PSA 75.7 Å^2 log *P* 3.6	CARVEDILOL Antihypertensive Metabolism: glucuronidation, O-demethylation, aryloxy hydroxylation, carbazole 2-hydroxylation
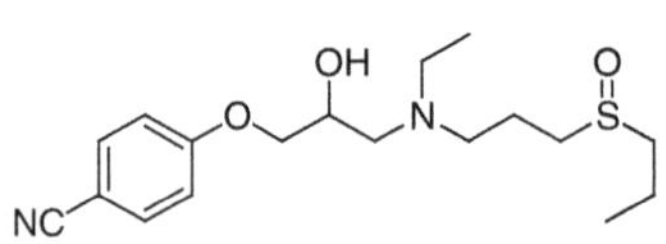	$t^1/_2$ 2.3–3.6 h *F* 55% pb 20% ur 10%	*V* 2.6 CL 11±1 Mwt 352.5 PSA 92.8 Å^2 log *P* 1.79	ALMOKALANT Antiarrhythmic Metabolism: N-dealkylation, O-glucuronidation, oxidation to sulfone, CH_3 hydroxylation of propyl chain

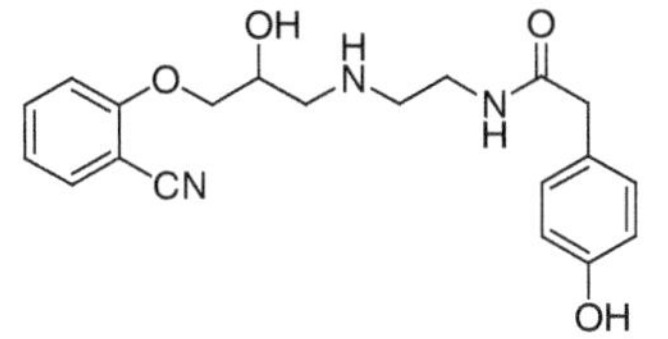	$t^{1}/_{2}$ *F* pb ur	17±9 h 10±8% 50% 1–26%	*V* CL Mwt PSA log *P*	6±3 28±12 369.4 114.6 Å^2 0.92	EPANOLOL Antihypertensive, antianginal
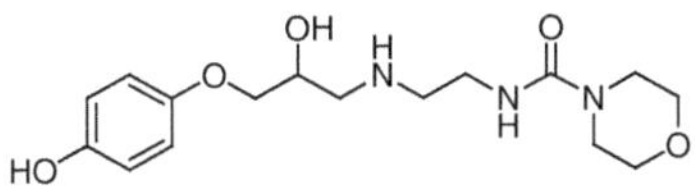	$t^{1}/_{2}$ *F* pb ur	16±3 h 5% 3% 73%	*V* CL Mwt PSA log *P*	1.1–2.0 3.7 339.4 103.3 Å^2 0.38	XAMOTEROL Metabolism: O-sulfation
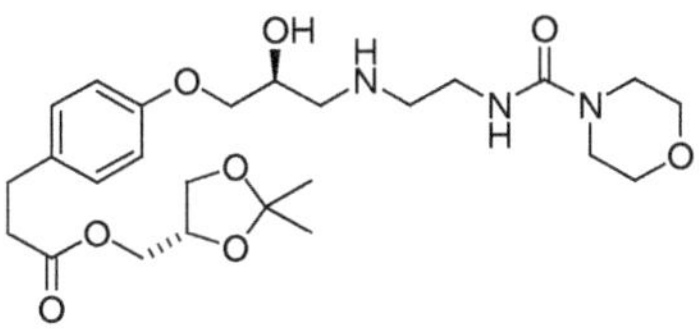	$t^{1}/_{2}$ *F* pb ur	2–4 min – – –	*V* CL Mwt PSA log *P*	– – 509.6 128 Å^2 0.80	LANDIOLOL Metabolism: ester hydrolysis
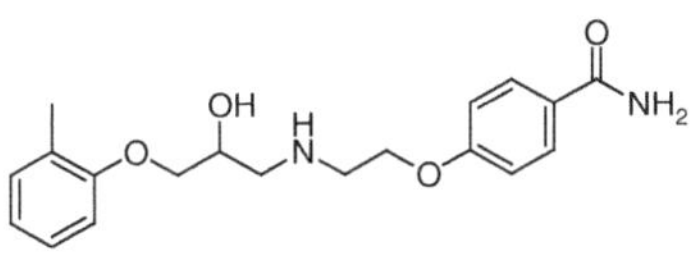	$t^{1}/_{2}$ *F* pb ur	2.6 h – 91% –	*V* CL Mwt PSA log *P*	4 14 344.4 93.8 Å^2 1.7	TOLAMOLOL Metabolism: aromatic 4-hydroxylation
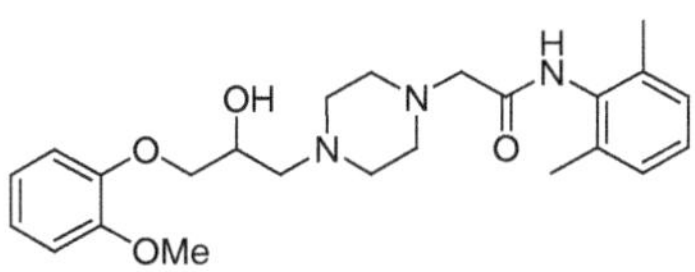	$t^{1}/_{2}$ *F* pb ur	1–2 h 35–50% 63% <5%	*V* CL Mwt PSA log *P*	1.3–2.8 10 427.5 74.3 Å^2 1.31	RANOLAZINE Antianginal Metabolism: hydroxylation, glucuronidation, N-dealkylation, O-demethylation, O-dearylation, amide hydrolysis
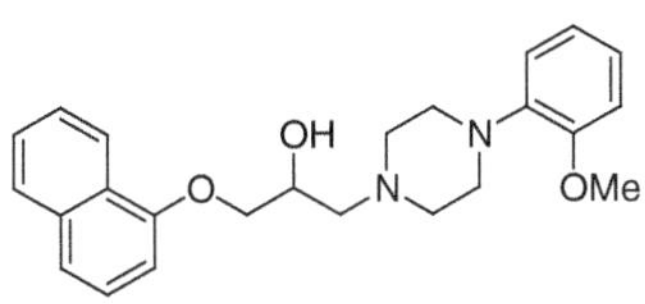	$t^{1}/_{2}$ *F* pb ur	5.4±3.2 h 7–30% 95–97% –	*V* CL Mwt PSA log *P*	– 11±1.6 392.5 45.2 Å^2 4.1	NAFTOPIDIL Antihypertensive Metabolism: aromatic hydroxylation, piperazine dealkylation, O-demethylation
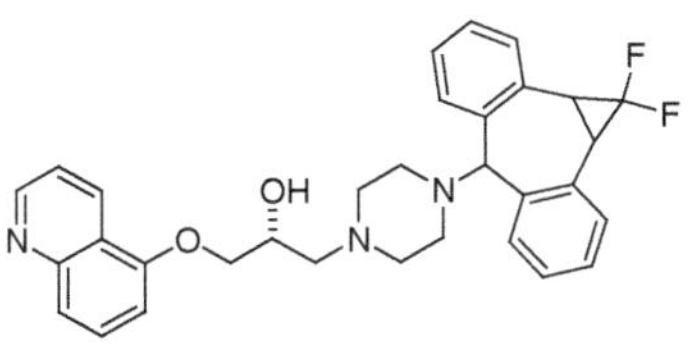	$t^{1}/_{2}$ *F* pb ur	25 h (iv) – – –	*V* CL Mwt PSA log *P*	268–350 108–158 527.6 48.8 Å^2 4.30	ZOSUQUIDAR P-glycoprotein inhibitor, adjunct antineoplastic Metabolism: N-oxidation and hydroxylation of quinoline

(continued overleaf)

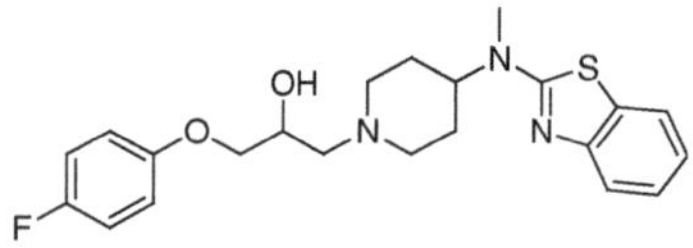	$t^1/_2$ *F* pb ur	18±10 h – 99% <0.5%	*V* CL Mwt PSA log *P*	– 5.1 415.5 77.1 Å^2 4.26	SABELUZOLE Nootropic Metabolism: 6-hydroxylation of benzothiazole, glucuronidation
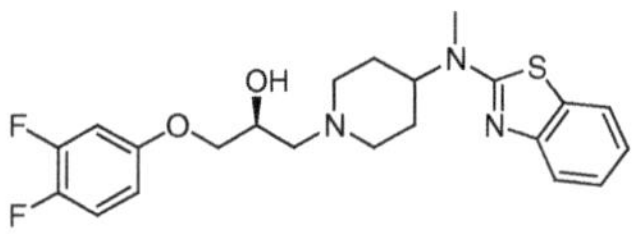	$t^1/_2$ *F* pb ur	15–24 h – >99% –	*V* CL Mwt PSA log *P*	2.5–3.4 2.2–3.9 433.5 77.1 Å^2 4.69	LUBELUZOLE Neuroprotectant Metabolism: 6-hydroxylation of benzothiazole, glucuronidation
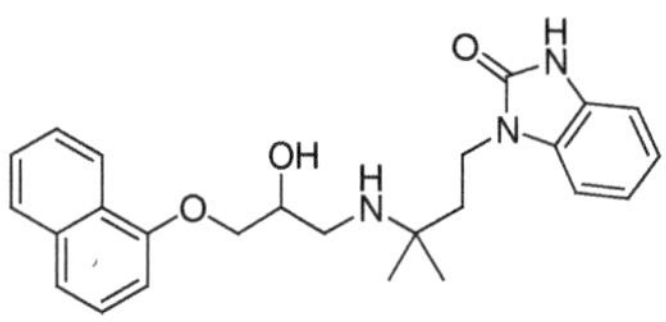	$t^1/_2$ *F* pb ur	14±5 h 33–52% – –	*V* CL Mwt PSA log *P*	– 12±6 474.0 79.3 Å^2 4.20	ADIMOLOL Antihypertensive
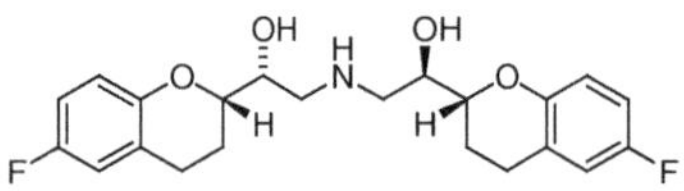	$t^1/_2$ *F* pb ur	10–50 h 12–96% 98% <1%	*V* CL Mwt PSA log *P*	10–39 12.3 405.4 71 Å^2 3.1	NEBIVOLOL Metabolism: aromatic and benzylic hydroxylation, N-dealkylation, glucuronidation
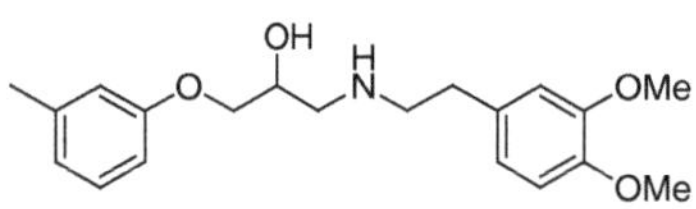	$t^1/_2$ *F* pb ur	1.5–2.0 h 60% >95% <1%	*V* CL Mwt PSA log *P*	1.5 4.4–6.9 345.4 60.0 Å^2 3.00	BEVANTOLOL Antianginal, antihypertensive, antiarrhythmic Metabolism: aromatic 4-hydroxylation, oxidation of CCH_3 to CCH_2OH and CCO_2H, then glucuronidation

$t_{1/2}$, plasma half-life; *F*, oral bioavailability; pb, plasma protein binding; ur, excretion of unchanged drug in urine; V, volume of distribution; CL, clearance; Mwt, molecular weight; PSA, polar surface area.

52
Opiates

Enkephalins, endorphins, and dynorphins are the three classes of known endogenous opioid peptides [1]. The N-terminus of all these peptides is H-Tyr-Gly-Gly-, and morphine and related strong analgesics are mimetics of this peptide fragment (Scheme 52.1).

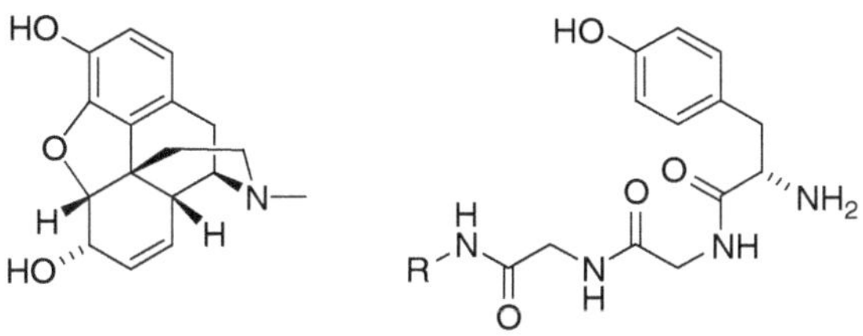

Morphine
$t_{1/2}$ 1.9 h, F 24%

Common N-terminus
of opioid peptides

Scheme 52.1

The opiate and opioid peptide binding receptors have been classified into three subtypes: μ-, κ-, and δ-receptors. These are mainly expressed in the central and peripheral nervous system (e.g., in the spinal cord) and in the ileum. Morphine, the main alkaloid of opium, is a μ-receptor agonist and its main effects in humans are analgesia, drowsiness, changes in mood, and mental clouding. Numerous analogs of morphine have been prepared with the aim of identifying less addicting and less constipating drugs, but these efforts have largely been unsuccessful. Opiates remain the treatment of choice for strong pain and, at low dose, for the treatment of cough. Table 52.1 includes only those compounds that are structurally related to morphine.

Reference

1. (a) Trescot, A.M., Datta, S., Lee, M., and Hansen, H. (2008) Opioid pharmacology. *Pain Physician*, **11**, S133–S153; (b) Sneader, W. (1998) The discovery of heroin. *Lancet*, **352**, 1697–1699.

Lead Optimization for Medicinal Chemists: Pharmacokinetic Properties of Functional Groups and Organic Compounds, First Edition. Florencio Zaragoza Dörwald.

Table 52.1 Opiates and related compounds. *V* in l kg^{-1}, CL in ml min^{-1} kg^{-1}, Mwt in g mol^{-1}.

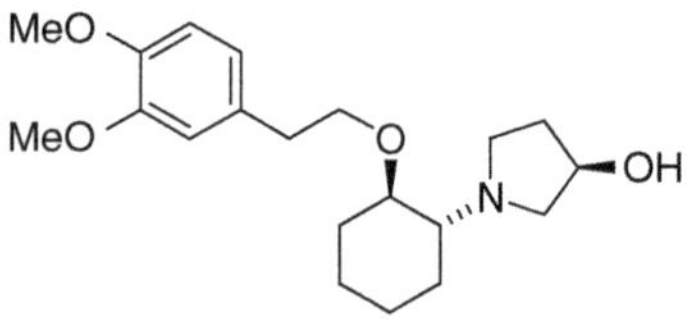	$t^1/_2$ *F* pb ur	4–8 h 20% 53–63% –	*V* CL Mwt PSA log *P*	0.4–0.8 – 349.5 51.2 Å^2 2.40	VERNAKALANT Treatment of atrial fibrillation Metabolism: O-demethylation
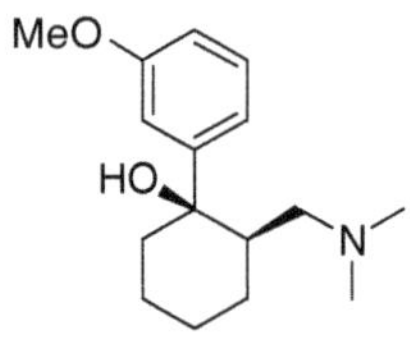	$t^1/_2$ *F* pb ur	5–6 h 70–90% 20% 10–30%	*V* CL Mwt PSA log *P*	2.9–3.7 6.7–8.3 263.4 32.7 Å^2 2.32	TRAMADOL Analgesic, crosses bbb; active metabolite: *O*-desmethyl ($t^1/_2$ 5–10 h); other metabolites: *N*-desmethyl
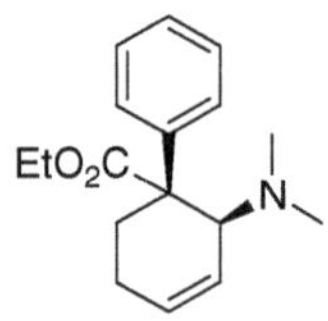	$t^1/_2$ *F* pb ur	– 8±5% 79% 0.2%	*V* CL Mwt PSA log *P*	3.7 18±4 273.4 29.5 Å^2 4.62	TILIDINE Analgesic, racemic Metabolism: N-demethylation to nortilidine
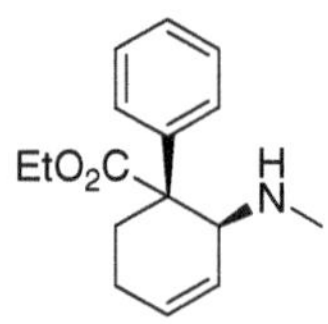	$t^1/_2$ *F* pb ur	3–5 h 99%* 25% 2–3%	*V* CL Mwt PSA log *P*	3±1 12±2 259.3 38.3 Å^2 4.75	NORTILIDINE *on oral administration of tilidine; metabolite of tilidine Metabolism: N-demethylation to bisnortilidine ($t^1/_2$ 5 h)
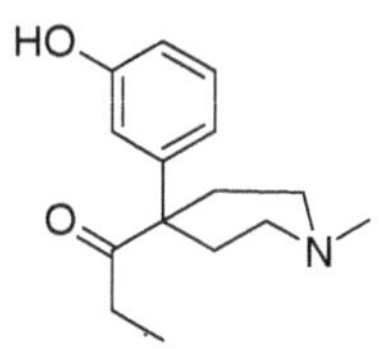	$t^1/_2$ *F* pb ur	3±1 h 34±16% – 3–10%	*V/F* CL/*F* Mwt PSA log *P*	11.2 26 247.3 40.5 Å^2 0.80	KETOBEMIDONE Analgesic; Metabolism: glucuronidation, N-demethylation, N-oxidation, aromatic 4-hydroxylation
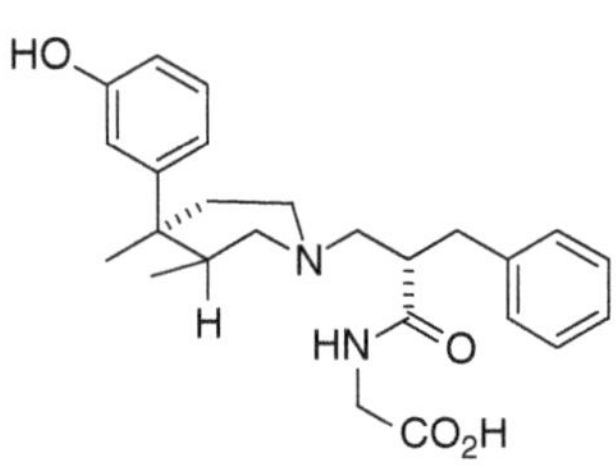	$t^1/_2$ *F* pb ur	10–17 h 1–19% 80–90% 2%	*V* CL Mwt PSA log *P*	0.3–0.6 6 424.5 89.9 Å^2 3.38	ALVIMOPAN Opiate antagonist; does not cross bbb Metabolism: amide hydrolysis

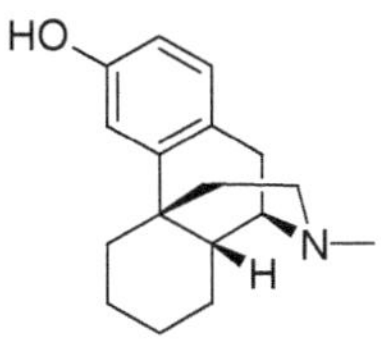	$t_{1/2}$ *F* pb ur	11–15 h* – 40% 5%	*V* CL Mwt PSA log *P*	10–13 10.5 257.4 23.5 Å^2 3.26	LEVORPHANOL *iv Analgesic Metabolism: glucuronidation, N-demethylation
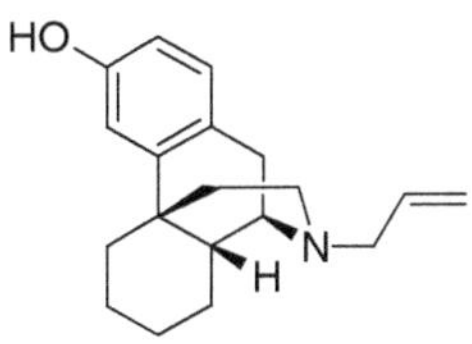	$t_{1/2}$ *F* pb ur	1 h Low – –	*V* CL Mwt PSA log *P*	– – 283.4 23.5 Å^2 3.85	LEVALLORPHAN Opioid antagonist Metabolism: glucuronidation, N-deallylation
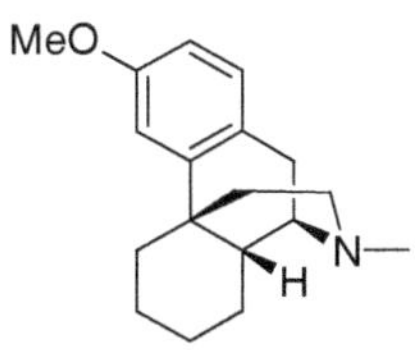	$t_{1/2}$ *F* pb ur	3–6 h – – –	*V* CL Mwt PSA log *P*	– – 271.4 12.5 Å^2 3.89	LEVOMETHORPHAN Analgesic Metabolism: O-demethylation to levorphanol
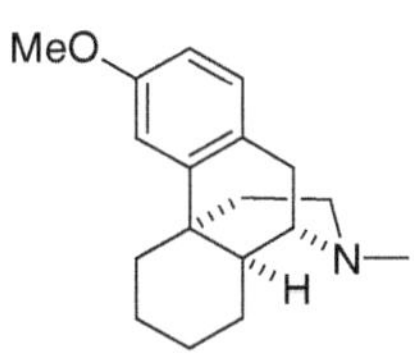	$t_{1/2}$ *F* pb ur	1–4 h 11% – –	*V* CL Mwt PSA log *P*	5.0–6.4* – 271.4 12.5 Å^2 3.89	DEXTROMETHORPHAN *dog Antitussive Metabolism: O-demethylation to dextrorphanol ($t_{1/2}$ 3–4 h), N-demethylation
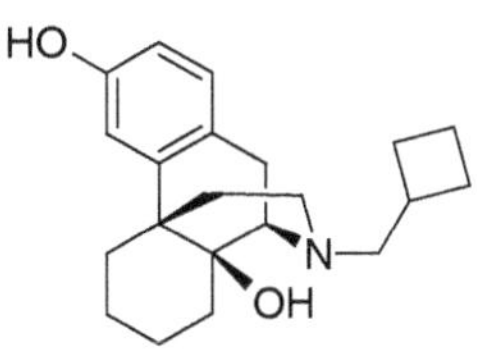	$t_{1/2}$ *F* pb ur	2.5–5.8 h 5–17%* 83% 2–10%	*V* CL Mwt PSA log *P*	4–13 41–43 327.5 43.7 Å^2 3.54	BUTORPHANOL *intranasal: 48–70% Analgesic, antiussive; crosses bbb Metabolism: hydroxylation of cyclobutyl, N-dealkylation
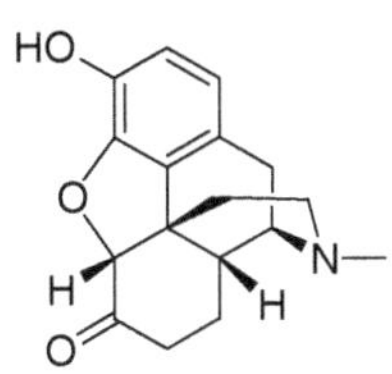	$t_{1/2}$ *F* pb ur	2–3 h 22–26% 8–19% 6%	*V* CL Mwt PSA log *P*	3–4 15–28 285.3 49.8 Å^2 2.13	HYDROMORPHONE Analgesic, narcotic Metabolism: glucuronidation, reduction of ketone to alcohol

(continued overleaf)

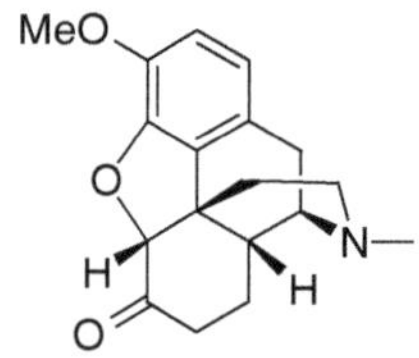	$t^1/_2$ F pb ur	4–6 h 60–70% – 12%	V CL Mwt PSA log P	3.3–4.7 9.6 299.4 38.8 Å^2 2.57	HYDROCODONE Antiussive Metabolism: N- and O-demethylation, reduction of ketone to alcohol, glucuronidation
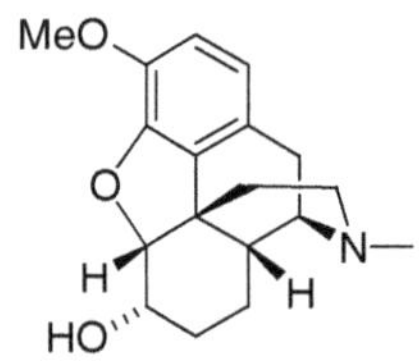	$t^1/_2$ F pb ur	3.3–4.5 h 12–34% – 13–22%	V CL Mwt PSA log P	1.0–1.3 4 301.4 41.9 Å^2 0.61	DIHYDROCODEINE Analgesic, antiussive Metabolism: O- and N-demethylation, glucuronidation
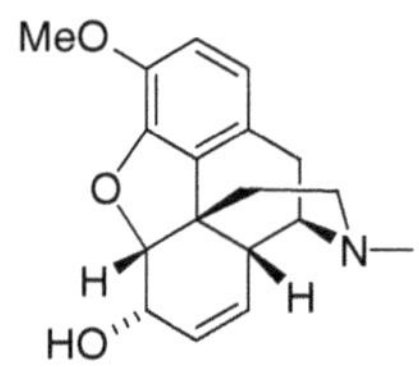	$t^1/_2$ F pb ur	2.9±0.7 h 50±7% 7% <10%	V/F CL/F Mwt PSA log P	2.6±0.3 11±2 299.4 41.9 Å^2 1.39	CODEINE Analgesic, antiussive Metabolism: O- and N-demethylation, glucuronidation
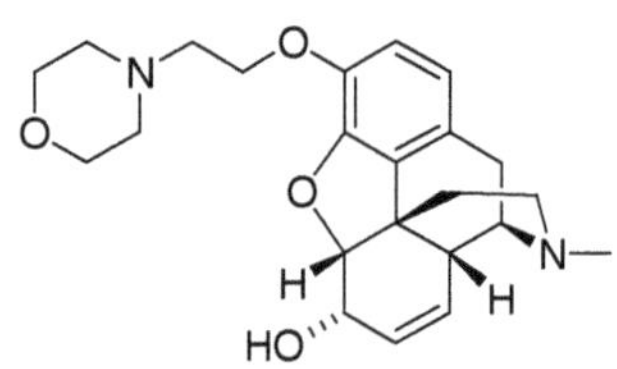	$t^1/_2$ F pb ur	37±4 h High 26% 30–50%	V CL Mwt PSA log P	36–49 10 398.5 54.4 Å^2 0.56	PHOLCODINE Antiussive Metabolism: N-demethylation, N-dealkylation of morpholine (to hydroxyethylmorphine)
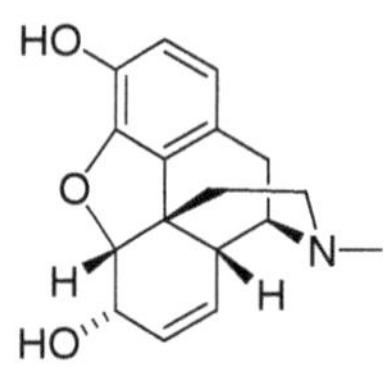	$t^1/_2$ F pb ur	1.9±0.5 h 24±12% 35±2% 4±5%	V CL Mwt PSA log P	3.3±0.9 24±10 285.3 52.9 Å^2 0.87	MORPHINE Analgesic, narcotic; active metabolite: 6-glucuronide ($t^1/_2$ 4.5±1.5 h, ur 14±7%)

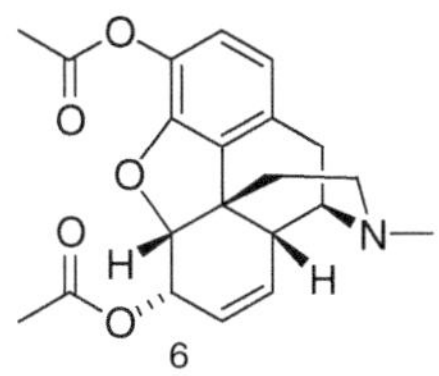	$t_{1/2}$ F pb ur	3 min (iv) <35% – Negligible	V CL Mwt PSA log P	1.0±0.5 175±74 369.4 65.1 Å^2 1.58	HEROIN, DIACETYLMORPHINE Prodrug of morphine Metabolites: 6-monoacetylmorphine ($t_{1/2}$ 20 min), morphine, 6-glucuronide
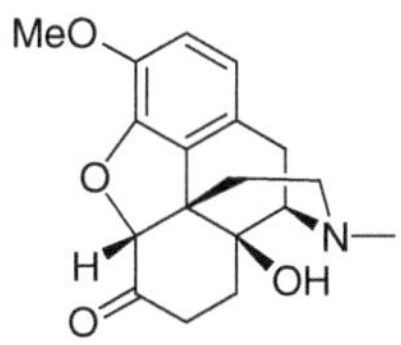	$t_{1/2}$ F pb ur	2–6 h 60–87% 38–45% 13–19%	V CL Mwt PSA log P	2–4 11–19 315.4 59.0 Å^2 1.59	OXYCODONE Analgesic Metabolism: N- and O-demethylation, glucuronidation
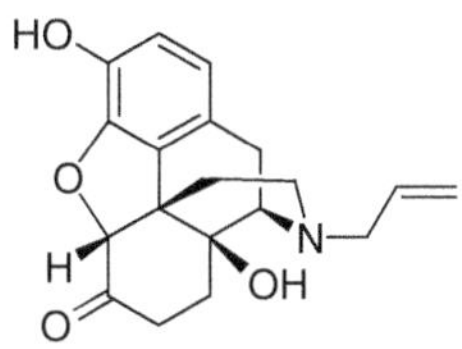	$t_{1/2}$ F pb ur	1.1±0.6 h 2% 32–46% Negligible	V CL Mwt PSA log P	2.1 22 327.4 70.0 Å^2 1.78	NALOXONE Narcotic antagonist Metabolism: glucuronidation, N-dealkylation, reduction of ketone to alcohol
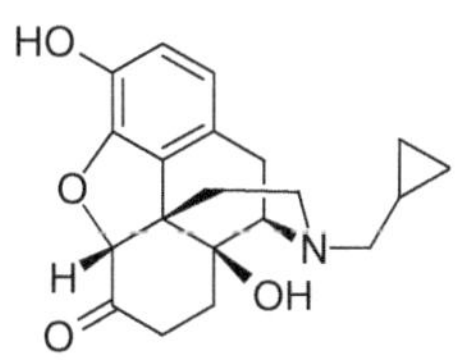	$t_{1/2}$ F pb ur	2.7±1.0 h 5–40% 21% <1%	V CL Mwt PSA log P	19±5 48±6 341.4 70.0 Å^2 2.05	NALTREXONE Narcotic antagonist Metabolism: glucuronidation, N-dealkylation, reduction of ketone to alcohol ($t_{1/2}$ 9 h)
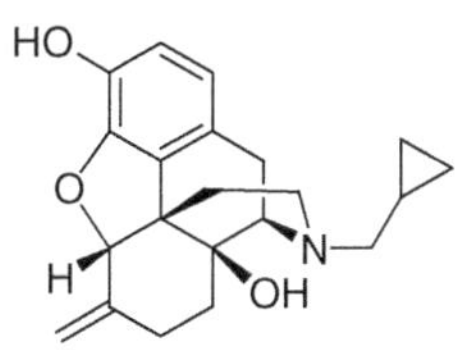	$t_{1/2}$ F pb ur	8–11 h 40–50% 34–45% 3–8%	V CL Mwt PSA log P	8.2 15 339.4 52.9 Å^2 1.63	NALMEFENE Narcotic antagonist Metabolism: glucuronidation, N-dealkylation
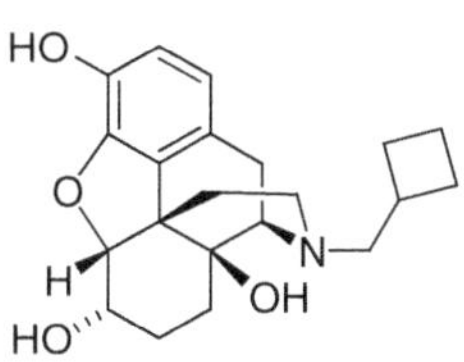	$t_{1/2}$ F pb ur	2.3±1.2 h 16±8% 50% 4±2%	V CL Mwt PSA log P	3.8±1.1 22±5 357.4 73.2 Å^2 1.22	NALBUPHINE Analgesic, narcotic Metabolism: N-dealkylation, oxidation of secondary alcohol to ketone

(*continued overleaf*)

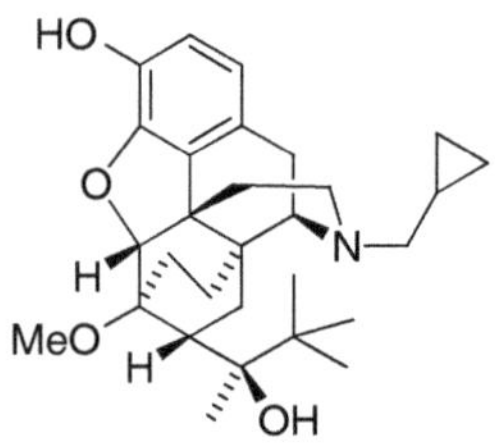	$t_{1/2}$ *F* pb ur	2–7 h 30%* 96% Negligible	*V* CL Mwt PSA log *P*	1–5 13–19 467.6 62.2 Å^2 2.83	BUPRENORPHINE *sublingual Analgesic Metabolism: N-dealkylation, glucuronidation
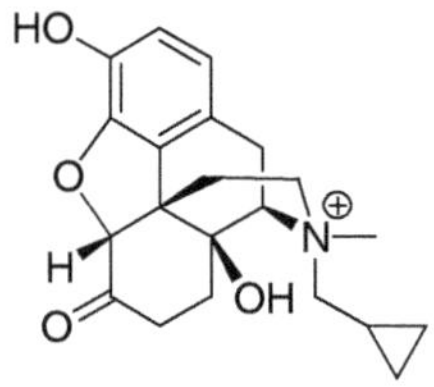	$t_{1/2}$ *F* pb ur	2.0–2.3 h Low 11–15% 50% (sc)	*V* CL Mwt PSA log *P*	1.1–2.6 6.5–8.6 356.4 66.8 Å^2 −2.63	METHYLNALTREXONE Narcotic antagonist, does not cross bbb; Metabolism: reduction of ketone to alcohol, N-demethylation (<0.06% of dose)
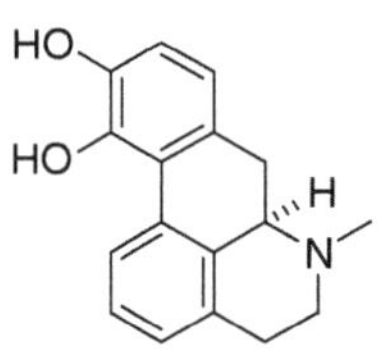	$t_{1/2}$ *F* pb ur	41±15 min* 2% 94% <1%	*V* CL Mwt PSA log *P*	1.6±0.5 40±15 267.3 43.7 Å^2 1.70	APOMORPHINE *iv Emetic Metabolism: N-demethylation, O-sulfation, O-glucuronidation
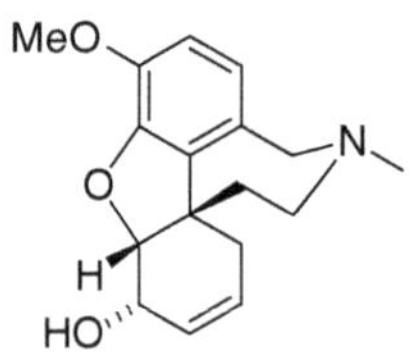	$t_{1/2}$ *F* pb ur	5–8 h 85–100% 17–18% 18–25%	*V* CL Mwt PSA log *P*	2.6 5.7 287.4 41.9 Å^2 −0.05	GALANTAMINE Cholinesterase inhibitor Metabolism (by CYP2D6 and CYP3A4): N- and O-demethylation, N-oxidation

$t_{1/2}$, plasma half-life; *F*, oral bioavailability; pb, plasma protein binding; ur, excretion of unchanged drug in urine; V, volume of distribution; CL, clearance; Mwt, molecular weight; PSA, polar surface area; bbb, blood–brain barrier.

53
N-(Carboxyalkyl)-α-Amino Acid Amides (Prils)

The endogenous peptide angiotensin II strongly raises arterial blood pressure by inhibiting the excretion of water and Na^+ by the kidneys and by increasing the total peripheral vascular resistance. In the 1960s, some peptides present in viper venom were found to inhibit angiotensin converting enzyme (ACE). One of these peptides (teprotide, pyrGlu-Trp-Pro-Arg-Pro-Gln-Ile-Pro-Pro-OH) turned out to lower blood pressure more consistently than peptidic angiotensin II antagonists such as saralasin (1-sarcosine,8-isoleucine-angiotensin II). This discovery led to the search for orally available ACE inhibitors as antihypertensives.

Angiotensin II is the product of the ACE-mediated cleavage of angiotensin I, which has only about 1% of the blood-pressure-increasing potency of angiotensin II (Scheme 53.1). Angiotensin III, which results from the degradation of angiotensin II by an aminopeptidase, is only 25% as potent as angiotensin II. The angiotensin receptors (AT_1 and AT_2) are G-protein-coupled receptors (GPCRs) located at the surface of cells in various tissues.

Angiotensinogen H_2N-Asp-Arg-Val-Tyr-Ile-His-Pro-Phe-His-Leu-Val-Ile-His-Ser-R

↓ renin

Angiotensin I H_2N-Asp-Arg-Val-Tyr-Ile-His-Pro-Phe-His-Leu-OH

↓ ACE

Angiotensin II H_2N-Asp-Arg-Val-Tyr-Ile-His-Pro-Phe-OH

↓ aminopeptidase

Angiotensin III H_2N-Arg-Val-Tyr-Ile-His-Pro-Phe-OH

Scheme 53.1 Biosynthesis of angiotensin peptides.

The first ACE inhibitors were peptidomimetics such as captopril (1977) and enalapril, which are analogs of the carboxy terminus of angiotensin I (Scheme 53.2). The mercapto and carboxyl groups in captopril/enalaprilate bind to a Zn^{2+} cation

Lead Optimization for Medicinal Chemists: Pharmacokinetic Properties of Functional Groups and Organic Compounds, First Edition. Florencio Zaragoza Dörwald.

in ACE, which coordinates to the carbonyl group of the Phe−His amide bond in angiotensin I to render it more electrophilic.

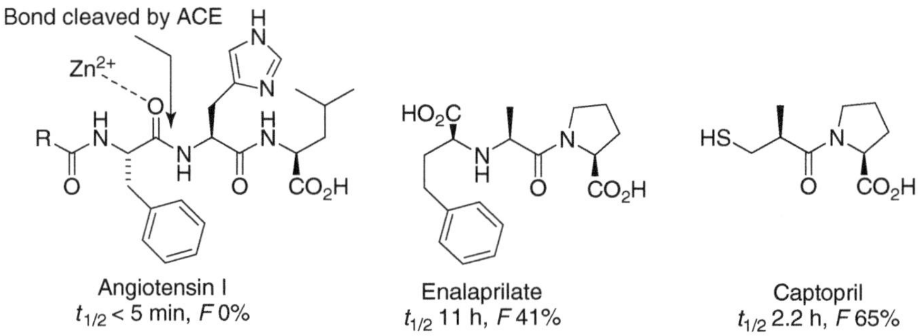

Scheme 53.2

Today, in addition to the ACE inhibitors given in Table 53.1, a large number of nonpeptidic angiotensin II antagonists have been developed (the "sartans," Chapter 72) and several renin inhibitors ("kirens," e.g., enalkiren) are being developed as antihypertensives (Chapter 55) [1].

Reference

1. Webb, R.L., Schiering, N., Sedrani, R., and Maibaum, J. (2010) Direct renin inhibitors as a new therapy for hypertension. *J. Med. Chem.*, **53**, 7490–7520.

Table 53.1 ACE inhibitors (prils) and related compounds. *V* in l kg^{-1}, CL in ml min^{-1} kg^{-1}, Mwt in g mol^{-1}.

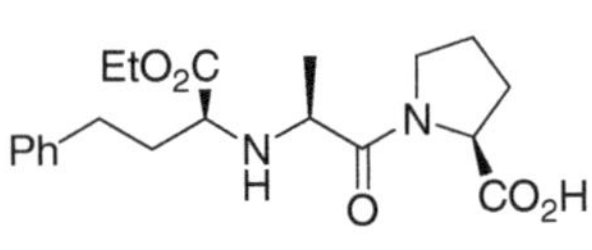	$t_{1/2}$ 11 h *F* 41±15% pb <50% ur 88±7%*	*V/F* 1.7±0.7 CL/*F* 4.9±1.5 Mwt 376.5 PSA 95.9 Å^2 log *P* 3.25	ENALAPRIL *diacid iv Metabolism: hydrolysis to enalaprilate (diacid)
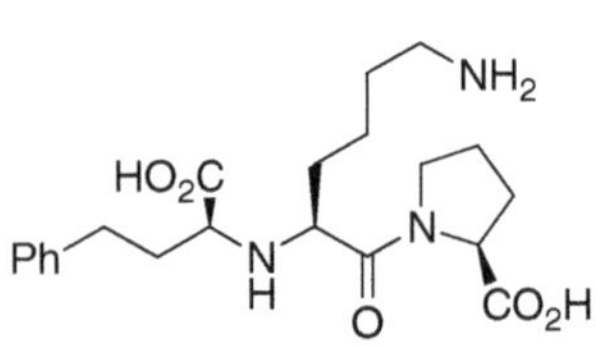	$t_{1/2}$ 21±9 h *F* 25±20% pb < 5% ur 94±6%	*V/F* 2.4±1.4 CL/*F* 4.2±2.2 Mwt 405.5 PSA 133 Å^2 log *P* 3.47	LISINOPRIL

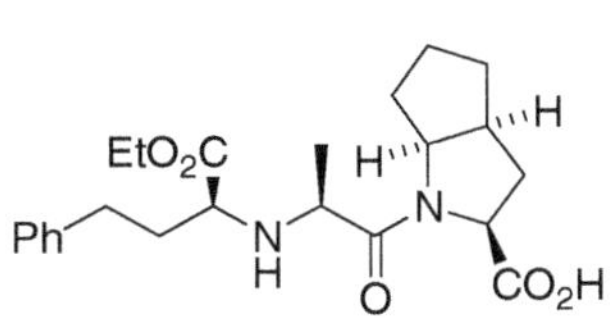	$t_{1/2}$ 14±7 h *F* 28–44% pb 73%* ur 39±17%	*V* 1.4** CL 1.1±0.4 Mwt 416.5 PSA 95.9 Å^2 log *P* 4.37	RAMIPRIL *ester; diacid: 56%; **ester; diacid: 7.7 l kg^{-1}
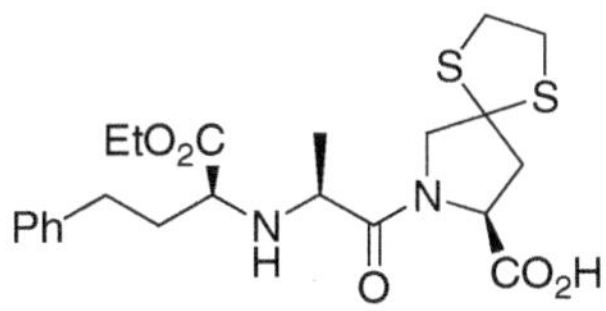	$t_{1/2}$ 1.5–2.2 h* *F* 50% pb – ur 12%	*V* 0.6 CL 0.15 Mwt 466.6 PSA 147 Å^2 log *P* 5.88	SPIRAPRIL *terminal $t_{1/2}$ 30–40 h
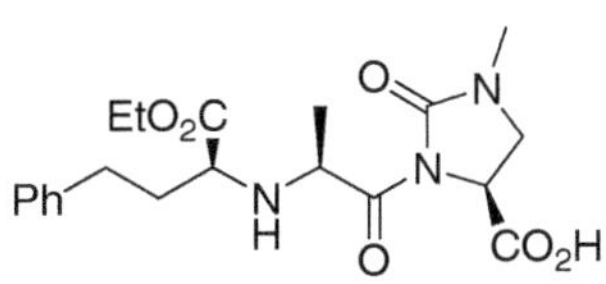	$t_{1/2}$ 7.6–14 h *F* 40% pb 85%* ur 9%	*V* – CL – Mwt 405.4 PSA 116 Å^2 log *P* 2.28	IMIDAPRIL *ester
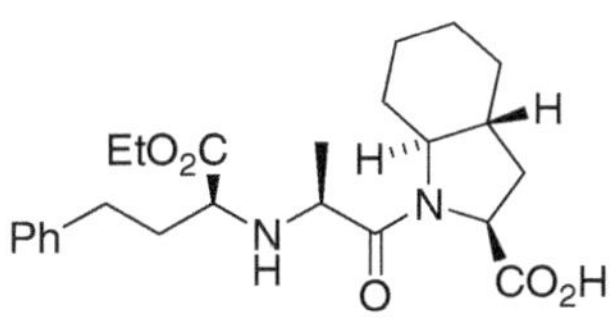	$t_{1/2}$ 6 h (ester)* *F* 40–70% pb 80%** ur 33%	*V* 0.3 (ester) CL Mwt 430.5 PSA 95.9 Å^2 log *P* 4.90	TRANDOLAPRIL *diacid: 10–24 h **ester; diacid: 65–95%
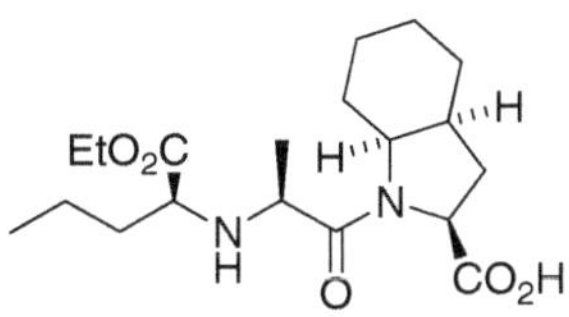	$t_{1/2}$ 1.5–2.9 h* *F* 75%* pb 60%* ur 4–12%*	*V* 0.22* CL 1.2–2.3* Mwt 368.5 PSA 95.9 Å^2 log *P* 4.10	PERINDOPRIL *ester
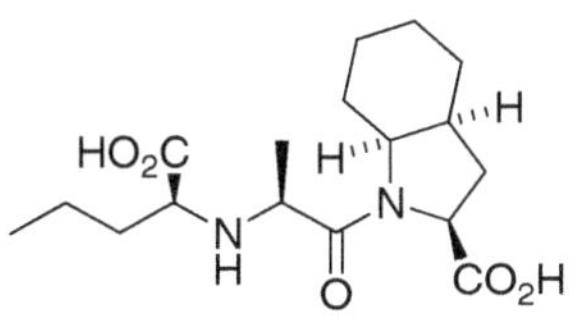	$t_{1/2}$ 25–30 h *F* 17–20% pb 10–20% ur 60–70%	*V* 0.16 CL – Mwt 340.4 PSA 107 Å^2 log *P* 2.37	PERINDOPRILATE

(continued overleaf)

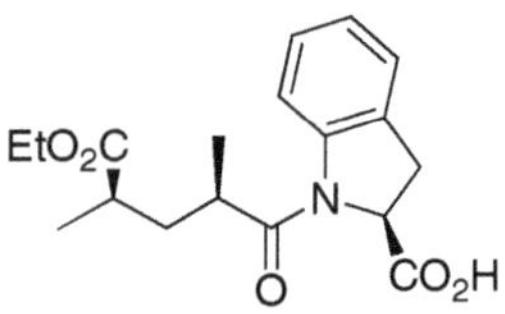

$t^1/_2$	0.8 h*	*V*	0.56	PENTOPRIL
F	–	CL	3.1	*ester; diacid: 3.9 h
pb	–	Mwt	333.4	**ester; diacid: 40%
ur	21%**	PSA	83.9 Å²	
		log *P*	4.09	

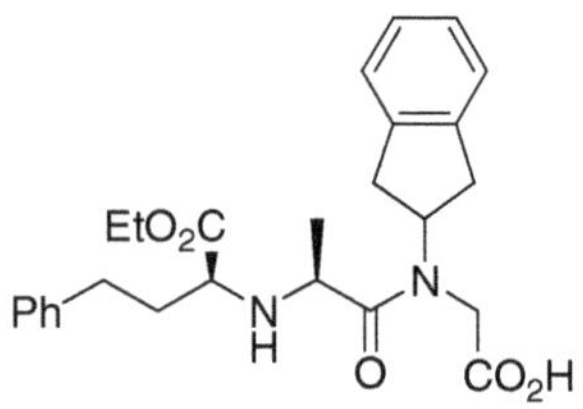

$t^1/_2$	0.4 h*	*V*	–	DELAPRIL
F	>55%	CL	13.3	*ester; diacid: 1.2 h
pb	–	Mwt	452.5	Metabolism to diacid and
ur	–	PSA	95.9 Å²	5-hydroxyindane diacid ($t^1/_2$ 1.4 h).
		log *P*	4.48	*F*: diacid after oral dosing of ester

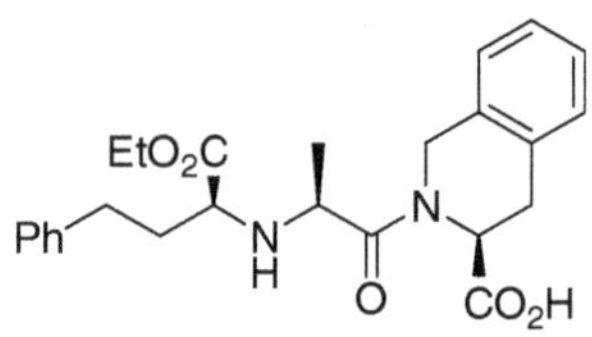

$t^1/_2$	2.2±0.2 h	*V*/*F*	0.9**	QUINAPRIL
F	49±11%	CL/*F*	26**	*ester; diacid: 28±9%
pb	97%	Mwt	438.5	**ester; diacid: 2.0±0.6
ur	3%*	PSA	95.9 Å²	
		log *P*	4.79	

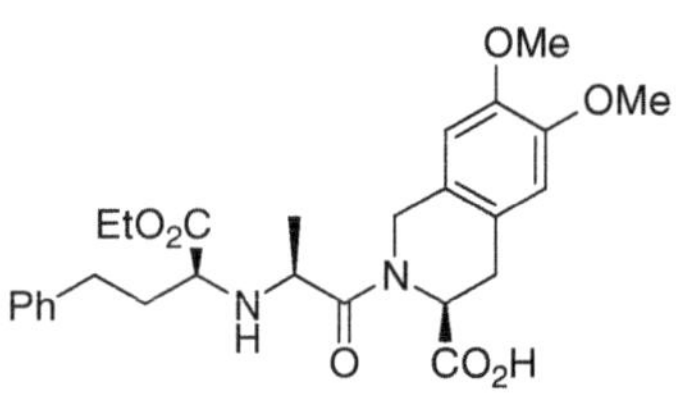

$t^1/_2$	1 h	*V*	–	MOEXIPRIL
F	22%	CL	–	Fast hydrolysis to
pb	90%	Mwt	498.6	moexiprilate in the liver and
ur	1%	PSA	114 Å²	in tissue by carboxyesterases
		log *P*	4.77	(not in the GI tract); *F*: diacid after
				oral administration of ester

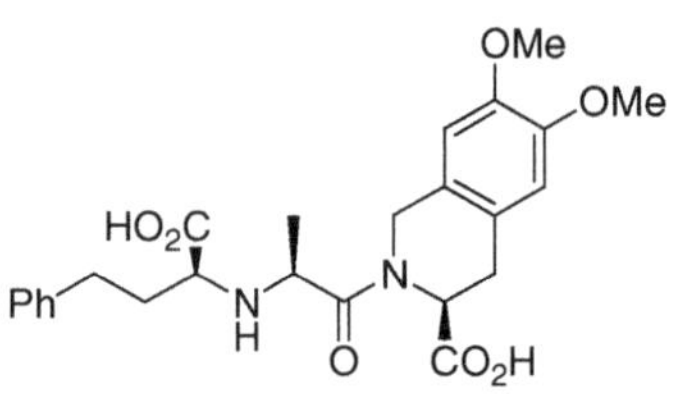

$t^1/_2$	6±4 h	*V*	2.8	MOEXIPRILATE
F	13%	CL	3.3	
pb	50–70%	Mwt	470.5	
ur	40%	PSA	125 Å²	
		log *P*	5.16	

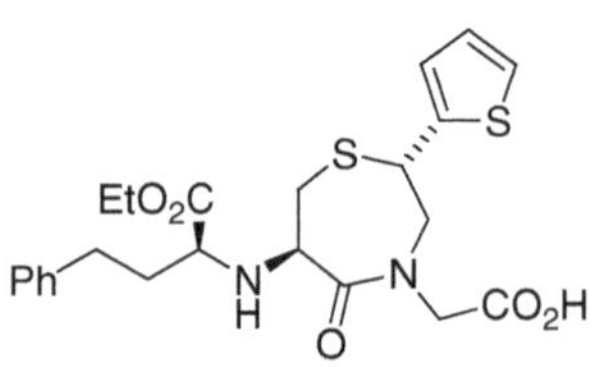

$t^1/_2$	1.5 h*	*V*	–	TEMOCAPRIL
F	65%	CL/*F*	34*	*ester; diacid: 7–15 h
pb	–	Mwt	476.6	**diacid
ur	57%**	PSA	149 Å²	
		log *P*	6.57	

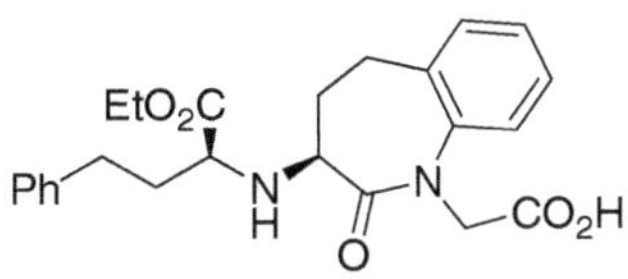	$t^1\!/_2$ 0.7 h* *F* 28–37% pb 95–97% ur <1%**	*V*/*F* 0.12*** CL/*F* 0.3–0.4*** Mwt 424.5 PSA 95.9 Å^2 log *P* 4.54	BENAZEPRIL *ester; diacid: 10–22 h **ester; diacid: 18% ***diacid
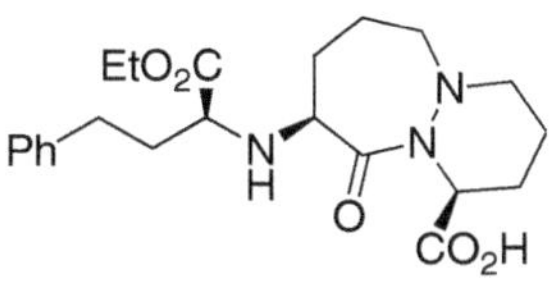	$t^1\!/_2$ 43±7 h *F* 47–75% pb 24%* ur 90–100%*	*V* 0.3–0.6 CL 1.8–4.1 Mwt 417.5 PSA 99.2 Å^2 log *P* 2.21	CILAZAPRIL *diacid
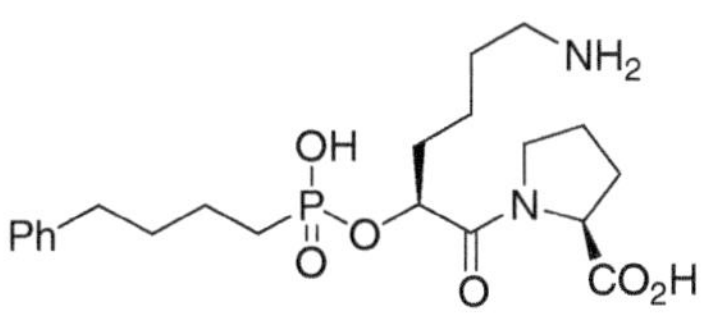	$t^1\!/_2$ 15 h *F* 67% pb – ur –	*V* – CL – Mwt 440.5 PSA 140 Å^2 log *P* 1.79	CERONAPRIL Antihypertensive
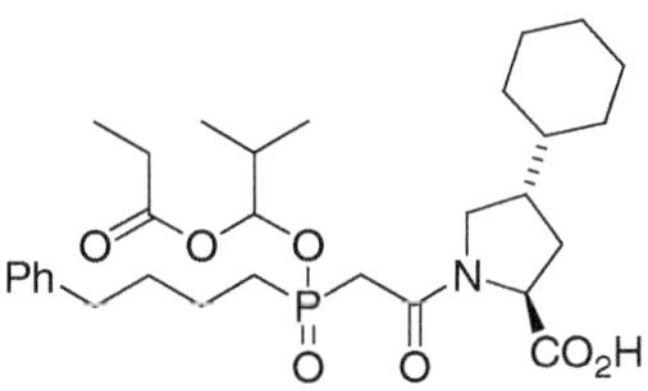	$t^1\!/_2$ 11±1 h* *F* 36±7% pb >95%* ur 43±9%* –	*V* 0.13±0.03 CL 0.5±0.1 Mwt 563.7 PSA 120 Å^2 log *P* 6.47	FOSINOPRIL *diacid
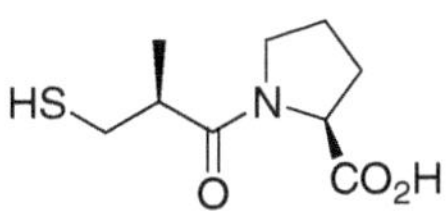	$t^1\!/_2$ 2.2±0.5 h *F* 38–65% pb 30±6% ur 38±11%	*V* 0.8±0.2 CL 12.0±1.4 Mwt 217.3 PSA 96.4 Å^2 log *P* 0.27	CAPTOPRIL ACE inhibitor, antihypertensive; Metabolism: oxidative disulfide formation, S-methylation
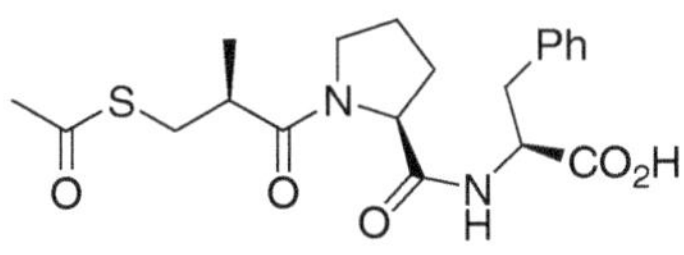	$t^1\!/_2$ 5 h* *F* – pb – ur –	*V* – CL – Mwt 406.5 PSA 129 Å^2 log *P* 0.19	ALACEPRIL Prodrug of captopril *captopril on oral dosing of alacepril
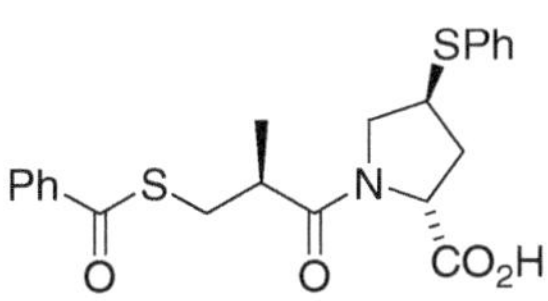	$t^1\!/_2$ 0.9 h* *F* 65–78% pb 88%** ur <10%	*V* 1.5** CL 20** Mwt 429.6 PSA 125 Å^2 log *P* 5.62	ZOFENOPRIL *ester; thiol: 4.6±1.0 h **thiol Metabolism: hydrolysis to thiol *F*: thiol after oral dosing of ester

(*continued overleaf*)

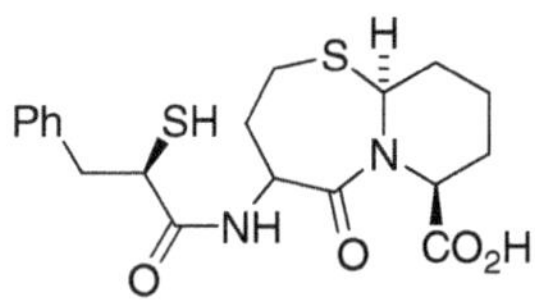	$t_{1/2}$ 14–19 h *F* 30% pb – ur –		V – CL – Mwt 408.5 PSA 151 Å^2 log *P* 1.65		OMAPATRILATE Antihypertensive Metabolism: disulfide formation with cysteine, S-methylation, acyl glucuronidation, hydrolysis of exocyclic amide, oxidation to sulfoxides
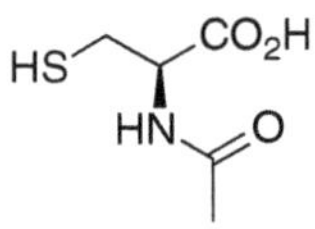	$t_{1/2}$ 2.3–6.0 h *F* 6–10% pb 50% ur 30%		*V* 0.33–0.47 CL 3.5–14 Mwt 163.2 PSA 105 Å^2 log *P* −0.70		*N*-ACETYLCYSTEINE
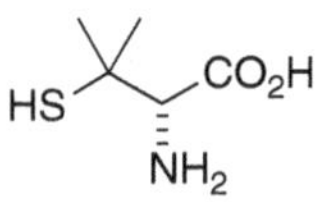	$t_{1/2}$ 1.5–3.0 h *F* 40–70% pb 80% ur 40%		*V* – CL 10.7 Mwt 149.2 PSA 102 Å^2 log *P* 0.85		PENICILLAMINE
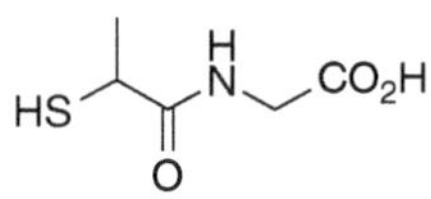	$t_{1/2}$ 19–53 h *F* 63% pb – ur –		*V* – CL – Mwt 163.2 PSA 105 Å^2 log *P* −0.67		TIOPRONIN Antidote for heavy-metal poisoning, mucolytic, hepatoprotectant Metabolite: mixed disulfide with cysteine

$t_{1/2}$, plasma half-life; *F*, oral bioavailability; pb, plasma protein binding; ur, excretion of unchanged drug in urine; V, volume of distribution; CL, clearance; Mwt, molecular weight; PSA, polar surface area; GI, gastrointestinal.
All data except calculated values (unless otherwise noted): acid on oral dosing of ester.

54
Anilides and Amides of Glycine

N-(2,6-Dimethylphenyl)glycine amide derivatives are important local anesthetics that are less irritating and safer than procaine (Table 54.1). The first member of this group, lidocaine, was discovered in the 1940s at the University of Stockholm. The chemist Holger Erdtman, trying to elucidate the structure of the alkaloid gramine, came to prepare isogramine instead. A common practice of organic chemists at that time was to taste newly synthesized products, and while doing so, Erdtman's tongue was numbed. A project was started to identify an analog of isogramine, suitable as local anesthetic. This research led to the discovery and marketing by Astra of lidocaine in 1948 (Scheme 54.1).

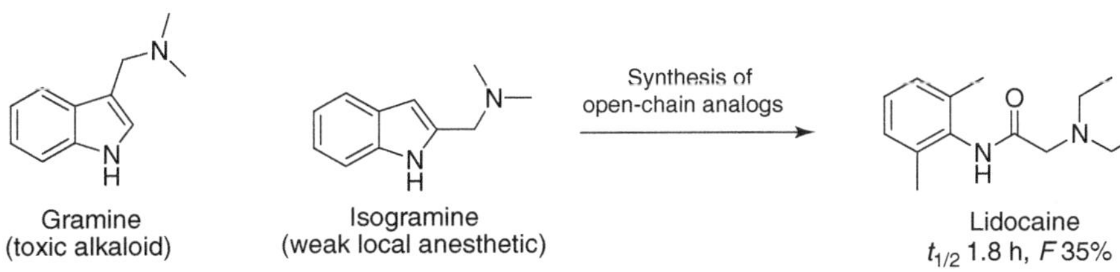

Scheme 54.1 The origin of lidocaine.

Although methylanilines, including 2-methyl- and 2,6-dimethylaniline, are suspected carcinogens, and the typical metabolic reactions of glycylanilides include amide hydrolysis, lidocaine and related anilides are relatively safe drugs, even after parenteral administration. Typical metabolic transformations, exemplified by those of lidocaine, are shown in Scheme 54.2.

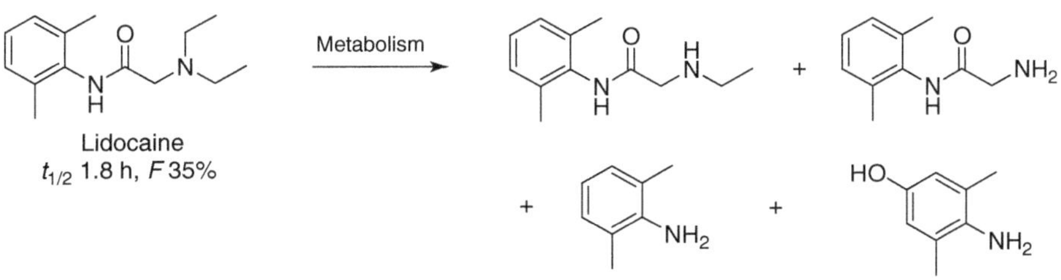

Scheme 54.2 Metabolism of lidocaine.

Lead Optimization for Medicinal Chemists: Pharmacokinetic Properties of Functional Groups and Organic Compounds, First Edition. Florencio Zaragoza Dörwald.

Structurally related to lidocaine are the antiarrhythmic mexiletine and the antihypertensives lofexidine and guanfacine (Scheme 54.3). Their improved pharmacokinetics (PK) is mainly due to the absence of the metabolically labile tertiary amine and the replacement of the xylyl group by a dichlorophenyl group.

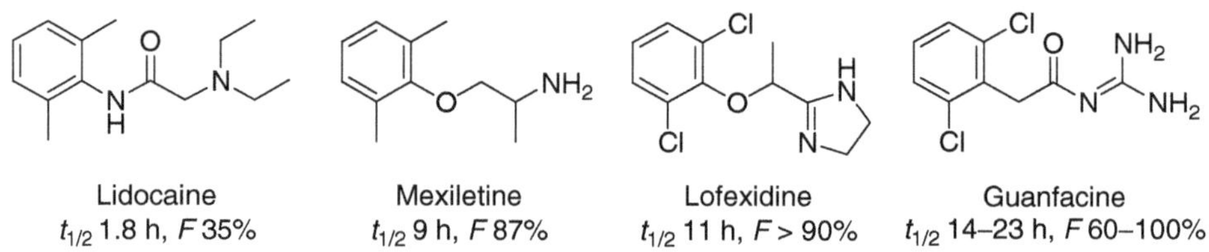

Scheme 54.3 Drugs structurally related to lidocaine.

A further class of therapeutically relevant amides of glycine are the DPPIV (dipeptidylpeptidase IV) inhibitors. DPPIV is the enzyme that deactivates GLP-1 (glucagonlike peptide 1) by cleaving off the N-terminal dipeptide of GLP-1 (H-His-Ala-OH) (Scheme 54.4). GLP-1 is normally released when nutrients reach the intestine and strongly potentiates insulin secretion by acting on receptors at pancreatic β-cells. DPPIV inhibitors enhance the levels of GLP-1 and can, therefore, be used to treat type II diabetes.

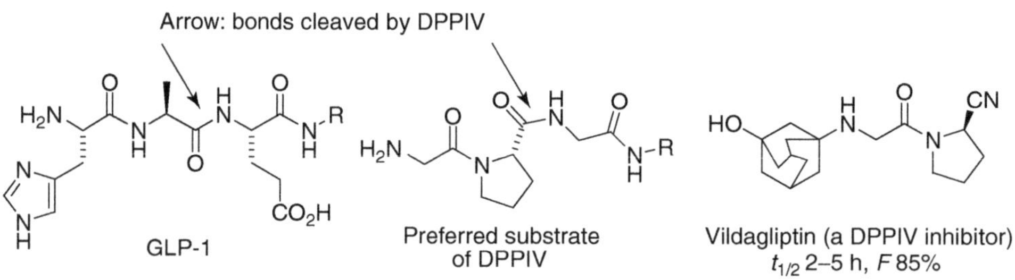

Scheme 54.4

The first DPPIV inhibitors, such as vildagliptin, were mimetics of the N-terminus of preferred substrates of this enzyme (H-Gly-Pro-R, H-His-Ala-R), in which the target amide bond was replaced by a nitrile. DPPIV is a prolyl oligopeptidase, which preferentially cleaves N-terminal dipeptides from peptides with alanine or proline at the penultimate position. Later, high-throughput screening provided peptide-unrelated leads, which led to DPPIV inhibitors such as alogliptin, carmegliptin, and linagliptin.

Table 54.1 Glycylanilides and related compounds. *V* in l kg^{-1}, CL in ml min^{-1} kg^{-1}, Mwt in g mol^{-1}.

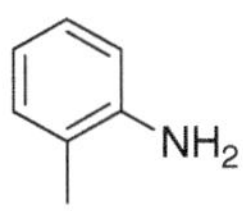	*t*½ *F* pb ur	4±1 h – – –	*V* CL Mwt PSA log *P*	– – 107.2 26.0 Å² 1.38	*o*-TOLUIDINE Metabolite of prilocaine, suspected human carcinogen Metabolism: N-oxidation to hydroxylamine and nitrosotoluene
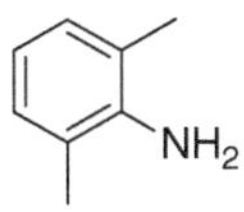	*t*½ *F* pb ur	8±4 h – – –	*V* CL Mwt PSA log *P*	– – 121.2 26.0 Å² 1.63	2,6-DIMETHYLANILINE Metabolite of some local anesthetics, suspected human carcinogen
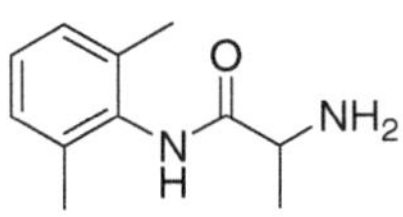	*t*½ *F* pb ur	14±2 h 89±5% 10±15% 38±7%	*V* CL Mwt PSA log *P*	3.0±0.2 2.6±0.5 192.3 55.1 Å² 0.81	TOCAINIDE Antiarrhythmic, racemic; (*R* : *t*½ 17±6 h; *S* : *t*½ 10±4 h) Metabolites: carbamoyl glucuronide (*t*½ 13 h), lactoxylidide (*t*½ 29 h)
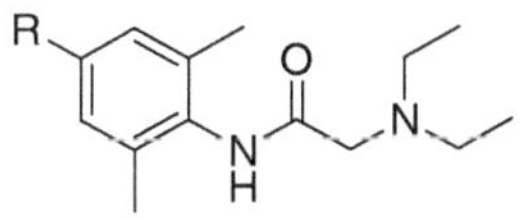	*t*½ *F* pb ur	1.8±0.4 h 35±11% 70±5% 2±1%	*V* CL Mwt PSA log *P*	1.1±0.4 9.2±2.4 234.3 32.3 Å² 2.20	LIDOCAINE, LIGNOCAINE (R = H); TRIMECAINE (R = CH_3; PK identical to lidocaine); antiarrhythmic, local anesthetic; Metabolism: N-deethylation, cyclization to imidazolidone, hydrolysis of amide, aromatic 3- and 4-hydroxylation
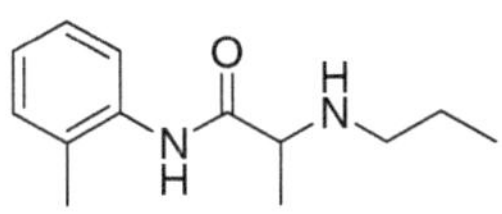	*t*½ *F* pb ur	1–2 h 20–40% 28–40% <5%	*V* CL Mwt PSA log *P*	3–4 29–34 220.3 41.1 Å² 2.03	PRILOCAINE Local anesthetic Metabolism: amide hydrolysis, aromatic hydroxylation
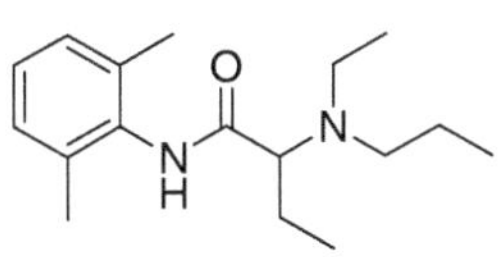	*t*½ *F* pb ur	2–3 h – 94% <1%	*V* CL Mwt PSA log *P*	2 17 276.4 32.3 Å² 3.57	ETIDOCAINE Local anesthetic Metabolism: N-dealkylation
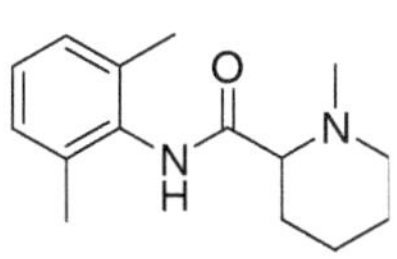	*t*½ *F* pb ur	1–3 h – 78% 5–10%	*V* CL Mwt PSA log *P*	1.0–1.3 12.3 246.4 32.3 Å² 1.78	MEPIVACAINE Local anesthetic Metabolism: aromatic hydroxylation, N-demethylation

(*continued overleaf*)

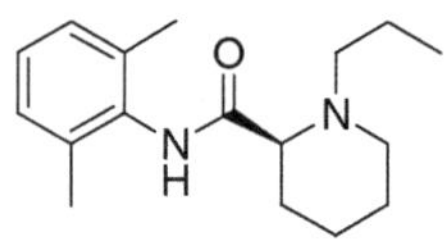	$t^{1}/_{2}$ *F* pb ur	1.8 h (iv)* 56±18%** 90–94% 1%	*V* CL Mwt PSA log *P*	0.6–0.8 5–8 274.4 32.3 Å² 2.80	ROPIVACAINE *4.2 h (epidural), 2.6 h (rectal) **rectal Local anesthetic; does not racemize *in vivo*; Metabolism: aromatic hydroxylation (CYP1A2), N-dealkylation (CYP3A4)
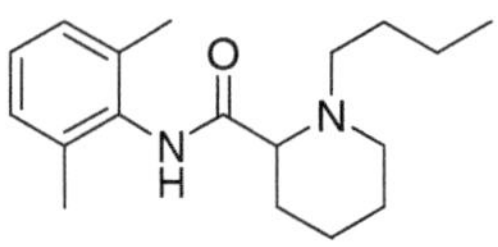	$t^{1}/_{2}$ *F* pb ur	2.4±1.2 h – 95±1% <4%	*V* CL Mwt PSA log *P*	0.9±0.4 7.1±2.8 288.4 32.3 Å² 3.31	BUPIVACAINE Local anesthetic Metabolism: N-dealkylation (CYP3A)
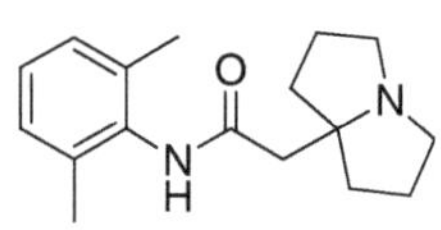	$t^{1}/_{2}$ *F* pb ur	4–9 h – 30% 75–86%	*V* CL Mwt PSA log *P*	1.48 3.8 272.4 32.3 Å² 1.88	PILSICAINIDE Antiarrhythmic
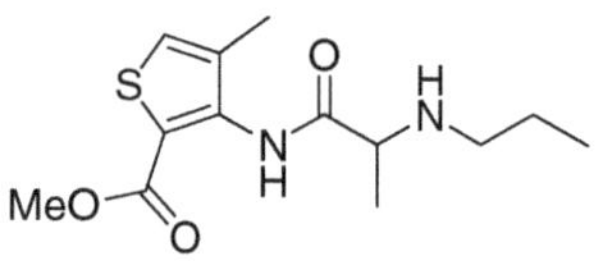	$t^{1}/_{2}$ *F* pb ur	25 min – 57–73% 1.4%	*V* CL Mwt PSA log *P*	5±3 127±54 284.4 95.7 Å² 2.22	ARTICAINE, CARTICAINE Local anesthetic Inactive metabolite: articainic acid
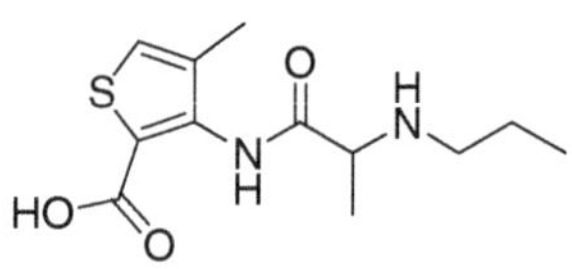	$t^{1}/_{2}$ *F* pb ur	2.4±0.3 h – 60–90% –	*V* CL Mwt PSA log *P*	0.5–0.7 2.7±0.5 270.4 107 Å² 2.38	ARTICAINIC ACID Main metabolite of articaine
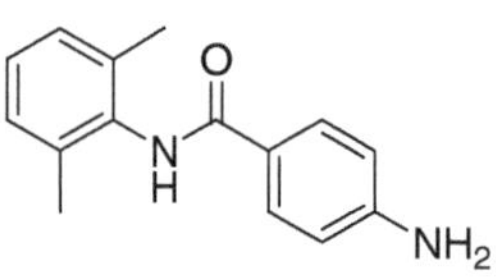	$t^{1}/_{2}$ *F* pb ur	5.5 h (dog) – – –	*V* CL Mwt PSA log *P*	1.2 (dog) – 240.3 55.1 Å² 1.90	AMELTOLIDE Anticonvulsant Metabolism: NH_2 acetylation
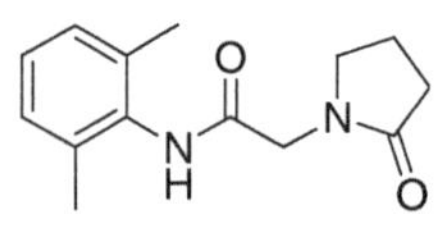	$t^{1}/_{2}$ *F* pb ur	2–6 h – – 5%	*V* CL Mwt PSA log *P*	0.5–0.7 1.5–2.2 246.3 49.4 Å² 0.64	NEFIRACETAM Nootropic Metabolism: hydroxylation of pyrrolidone (mainly at position 5)

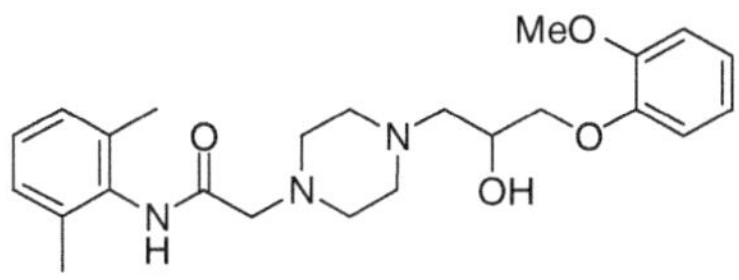	$t\frac{1}{2}$ 1–2 h *F* 35–50% pb 63% ur <5%	*V* 1.3–2.8 CL 10 Mwt 427.5 PSA 74.3 Å^2 log *P* 1.31	RANOLAZINE Antianginal; Metabolism: hydroxylation, glucuronidation, N-dealkylation, O-demethylation, O-dearylation, amide hydrolysis
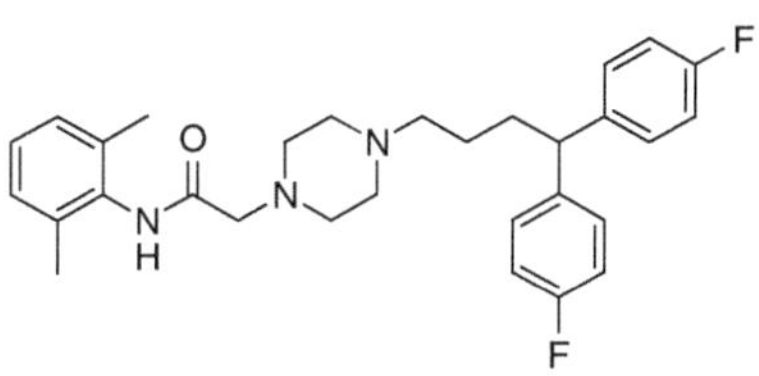	$t\frac{1}{2}$ 22 h *F* 35–40% pb – ur <5%	*V* 10 CL – Mwt 491.6 PSA 35.6 Å^2 log *P* 5.42	LIDOFLAZINE Calcium channel antagonist Metabolism: N-dealkylation
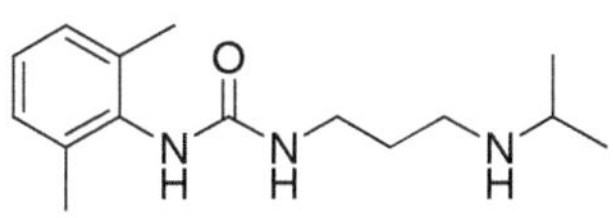	$t\frac{1}{2}$ 3–11 h *F* 73–84% pb 14–20% ur 84%	*V* 1.2–3.4 CL 4.5–11.1 Mwt 263.4 PSA 53.2 Å^2 log *P* 2.78	RECAINAM Antiarrhythmic Metabolites: desisopropyl, *p*-hydroxy
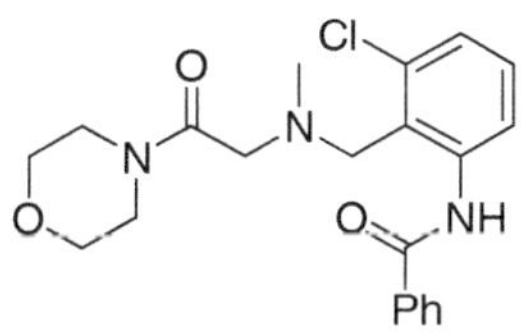	$t\frac{1}{2}$ 7–9 h *F* 65% pb – ur –	*V* – CL – Mwt 401.9 PSA 61.9 Å^2 log *P* 2.73	FOMINOBEN Antiussive Metabolism: N-debenzoylation
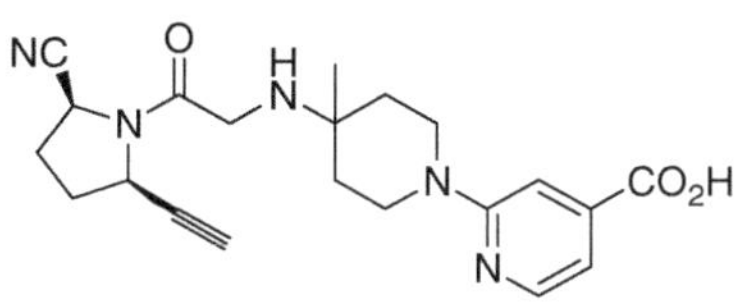	$t\frac{1}{2}$ 2 h* *F* 11%* pb – ur –	*V* – CL – Mwt 395.5 PSA 110 Å^2 log *P* 0.61	ABT-279 *monkey DPPIV inhibitor
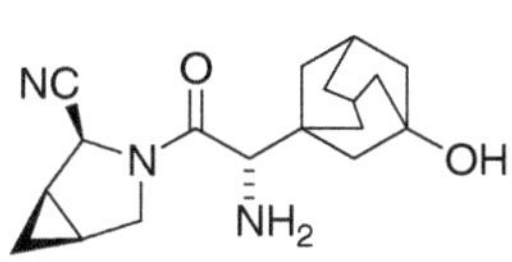	$t\frac{1}{2}$ 2–3 h *F* 51%* pb <30% ur 12–29%	*V* 2.7 CL 14.5* Mwt 315.4 PSA 90.4 Å^2 log *P* −0.14	SAXAGLIPTIN *monkey DPPIV inhibitor, antidiabetic; active metabolite: 5-hydroxy ($t\frac{1}{2}$ 3–7 h) (hydroxylation at C_3CH)

(*continued overleaf*)

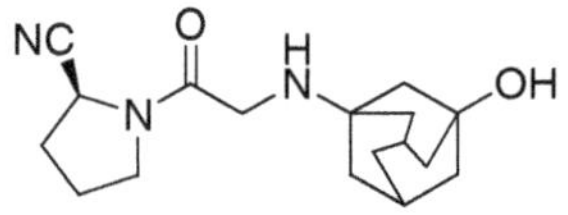	$t^1/_2$ 2–5 h *F* 85% pb 9% ur 22%	*V* 1 CL 12 Mwt 303.4 PSA 76.4 Å^2 log *P* −0.14	VILDAGLIPTIN DPPIV inhibitor, antidiabetic Metabolism: hydrolysis of nitrile to acid, amide hydrolysis, glucuronidation, oxidation of pyrrolidine
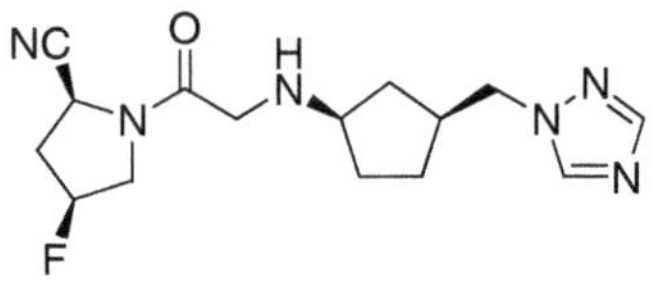	$t^1/_2$ 2.2 h* *F* 94%* pb – ur –	*V* – CL 16 Mwt 320.4 PSA 86.8 Å^2 log *P* −1.34	MELOGLIPTIN *monkey DPPIV inhibitor, antidiabetic
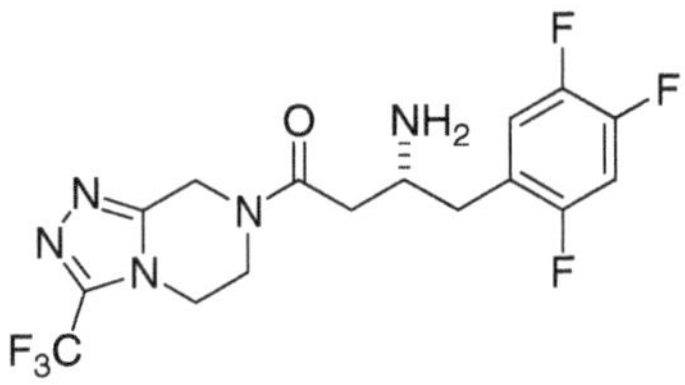	$t^1/_2$ 8–14 h *F* 87% pb 38% ur 79%	*V* 3 CL – Mwt 407.3 PSA 77.0 Å^2 log *P* 1.30	SITAGLIPTIN DPPIV inhibitor, antidiabetic

$t_{1/2}$, plasma half-life; *F*, oral bioavailability; pb, plasma protein binding; ur, excretion of unchanged drug in urine; V, volume of distribution; CL, clearance; Mwt, molecular weight; PSA, polar surface area; DPPIV, dipeptidylpeptidase IV.

55
Peptides, Peptidomimetics, and Related Oligoamides

The use of peptides as drugs seems attractive because of their predictable and safe metabolism and their high potency and selectivity. However, few clinically viable peptides have been identified, and most have to be administered parenterally. A notable exception is the gonadotropin agonists (e.g., buserelin), which are modified decapeptides suitable for nasal dosing [1]. Chemically modified peptides are mainly used as hormone analogs or as protease inhibitors.

Reasons for the scant use of peptide-based drugs are mainly their poor pharmacokinetic (PK) properties, in particular, their low oral bioavailability, poor membrane penetration, and quick enzymatic degradation. Moreover, unnatural peptides with a molecular weight above 700–1000 g mol^{-1} can elicit an immune response. Proteins, though, can have plasma half-lives of several days (e.g., albumin or antibodies).

Peptide bonds can be stabilized toward enzymatic cleavage with the aid of unnatural amino acids, such as *N*-alkylamino acids or aminoisobutyric acid, or by conformational constraints, for example, by cyclization [2]. Alternatively, amino- or carboxypeptidase activity may be blocked by modifying the terminal functional groups of a peptide, if the target allows this. Many natural, peptide-based poisons (e.g., amanitin, a mushroom poison, Scheme 55.1) are mono- or polycyclic peptides with high oral bioavailability.

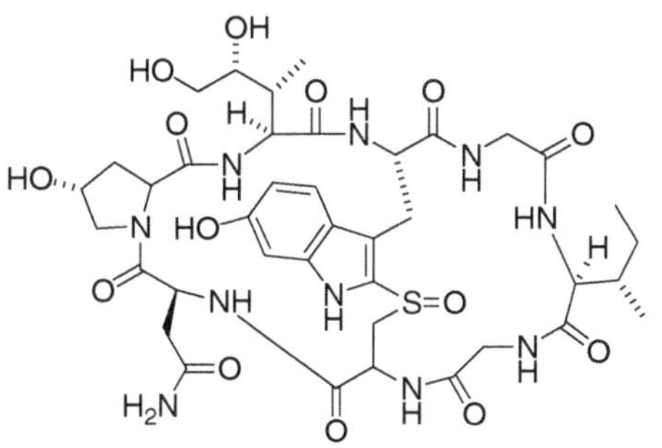

α-Amanitin
$t_{1/2}$ (dog) 0.5–1.0 h

Scheme 55.1

Lead Optimization for Medicinal Chemists: Pharmacokinetic Properties of Functional Groups and Organic Compounds, First Edition. Florencio Zaragoza Dörwald.

Renal excretion may be delayed and membrane permeability enhanced by lipophilization. Acylation with fatty acids, for instance, increases the affinity of peptides to albumin and keeps them from being excreted renally. This lipophilization can be permanent or transient, that is, the acylated peptide may be a prodrug of the peptide. The absorption of peptides from the small intestine may be enhanced by alkylating the NH groups or replacing them with non-hydrogen-bond-donating groups [3].

The membrane permeability and enzymatic stability of peptides may also be enhanced by glycosylation. Although glycosylated peptides are less lipophilic than the corresponding nonglycosylated peptides, transport through membranes can be brought about by active transport mechanisms or endocytosis.

Finally, the PK and other properties of peptides (and proteins) may be improved by conjugation to a polymer, for example, poly(ethylene glycol) ("pegylation") [4]. Pegylation delays renal clearance and enzymatic degradation, enhances solubility, and reduces immunogenicity of peptides and proteins.

55.1
Peptidomimetics

Peptides based on proteinogenic amino acids alone are usually not well suited as drugs, because of their poor absorption from the gastrointestinal tract and low metabolic stability. Moreover, the high polar surface area (PSA) of peptides also keeps them from crossing other membranes or the blood–brain barrier (bbb). Unnatural peptides with a molecular weight > 700 g mol^{-1} may, furthermore, induce an immune response.

When the only lead available in a project is a peptide, medicinal chemists often end up spending huge resources on the discovery of a peptidomimetic with acceptable PK properties, often without success.

One strategy to improve the PK properties of peptides is to replace peptide bonds by an unnatural substructure, not prone to enzymatic degradation. The simplest way to achieve this is by using *N*-methylamino acids; all other strategies usually involve multistep, expensive syntheses of structurally complex intermediates. Particularly difficult is the rational design of inhibitors for endopeptidases, which often recognize several amino acid side chains and are highly selective. Four examples of hydroxyethylene peptidomimetics are shown in Scheme 55.2. The development of such drugs is complicated not only by the difficulty of attaining acceptable PK properties but also by the high cost of producing the final drug.

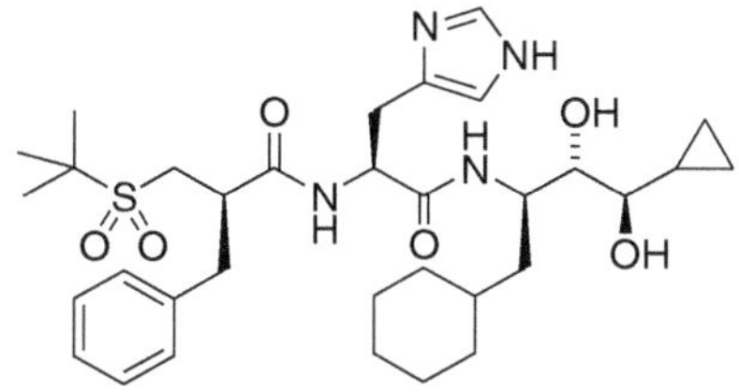

Remikiren (renin inhibitor)
$t_{1/2}$ 9 h, F 1–2%

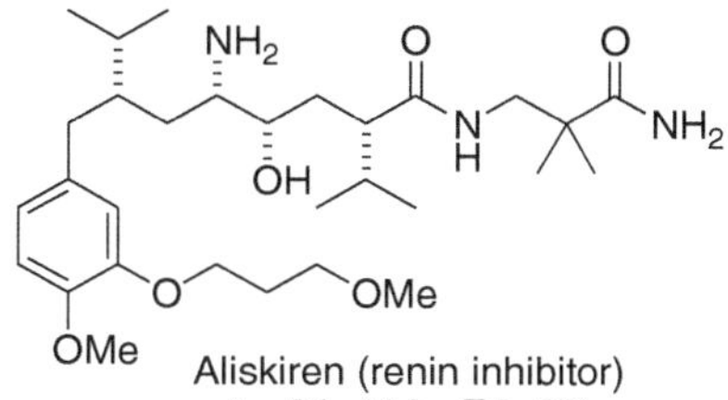

Aliskiren (renin inhibitor)
$t_{1/2}$ 34–41 h, F 2–3%

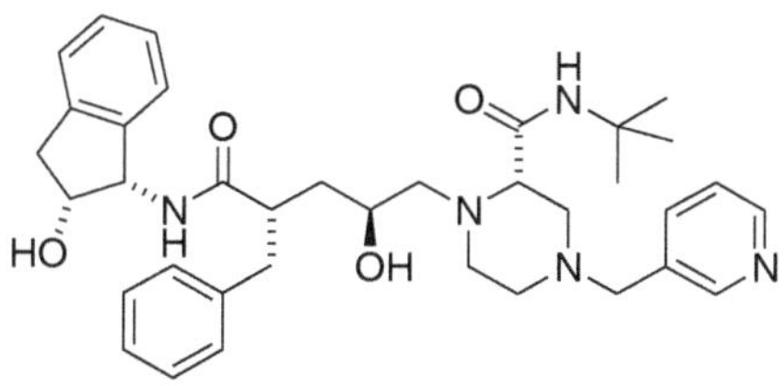

Indinavir (HIV protease inhibitor)
$t_{1/2}$ 1.8 h, F 65%

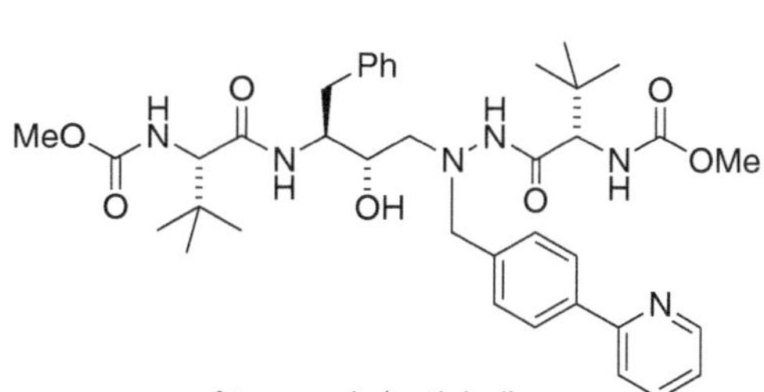

Atazanavir (antiviral)
$t_{1/2}$ 6.5 h, F 60–68%

Scheme 55.2 Peptidomimetics.

Because it is so difficult to turn a peptide into an orally available small molecule with good PK (see Table 55.1), even weak but structurally less complex hits resulting from a high-throughput screening should always be evaluated carefully. It is usually much easier to improve potency, selectivity, and the pharmacokinetics of a weak, nonpeptidic hit than to convert a potent and selective peptide into a compound with acceptable pharmacokinetics (see, e.g., the development of losartan, Chapter 72). In general, weak hits devoid of problematic functional groups are more promising than potent hits containing such groups (e.g., highly polar, basic, acidic, chemically reactive, or metabolically questionable functional groups).

55.2
Thrombin Inhibitors and Related Compounds

The oldest, but still popular, antithrombotics are warfarin, heparin, and aspirin. With the goal of identifying more selective antithrombotics with fewer side effects, potent thrombin inhibitors with suitable PK are being sought [5]. Unfortunately, the drugs resulting from these endeavors are mainly peptidomimetics, most of them amidines or guanidines of low oral bioavailability (Table 55.2).

Blood coagulation is a complex process, involving various different enzymes. In the last step of the coagulation cascade, the protease thrombin cleaves two peptides from (soluble) fibrinogen to generate fibrin, which undergoes oligomerization to form the clot. As most coagulation factors, thrombin cleaves an Arg-X bond in fibrinogen, but not Lys-X or other closely related peptide bonds. Therefore, it has been difficult to find thrombin inhibitors devoid of an amidine or guanidine group. Most of the drugs known today must be dosed parenterally.

One way to improve the oral bioavailability of these highly basic compounds is to lower the basicity of the amidine/guanidine group. This can, for instance, be achieved by alkoxycarbonylation or by using *N*-hydroxyamidines, which are converted *in vivo* into the corresponding amidines.

References

1. Conn, P.M. and Crowley, W.F. (1991) Gonadotropin-releasing hormone and its analogs. *New Engl. J. Med.*, **324**, 93–103.
2. (a) Malakoutikhah, M., Prades, R., Teixido, M., and Giralt, E. (2010) N-Methyl phenylalanine-rich peptides as highly versatile blood–brain barrier shuttles. *J. Med. Chem.*, **53**, 2354–2363; (b) Biron, E. *et al.*. (2008) Improving oral bioavailability of peptides by multiple N-methylation: somatostatin analogues. *Angew. Chem. Int. Ed.*, **47**, 2595–2599; (c) Witt, K.A., Gillespie, T.J., Huber, J.D., Egleton, R.D., and Davis, T.P. (2001) Peptide drug modifications to enhance bioavailability and blood–brain barrier permeability. *Peptides*, **22**, 2329–2343; (d) Fauch ere, J. and Thurieau, C. (1992) Evaluation of the stability of peptides and pseudopeptides as a tool in peptide drug design. *Adv. Drug Res.*, **23**, 127–159.
3. von Geldern, T. W. *et al.*. (1996) Azole endothelin antagonists. 3. Using ΔlogP as a tool to improve absorbtion. *J. Med. Chem.*, **39**, 982–991.
4. Harris, J.M. and Chess, R.B. (2003) Effect of pegylation on pharmaceuticals. *Nat. Rev. Drug Discov.*, **2**, 214–221.
5. (a) Pinto, D.J.P., Smallheer, J.M., Cheney, D.L., Knabb, R.M., and Wexler, R.R. (2010) Factor Xa inhibitors: next-generation antithrombotic agents. *J. Med. Chem.*, **53**, 6243–6274; (b) Chackalamannil, S. (2006) Thrombin receptor (protease activated receptor-1) antagonists as potent antithrombotic agents with strong antiplatelet effects. *J. Med. Chem.*, **49**, 5389–5403.

Table 55.1 Peptides and peptidelike compounds. *V* in l kg^{-1}, CL in ml min^{-1} kg^{-1}, Mwt in g mol^{-1}.

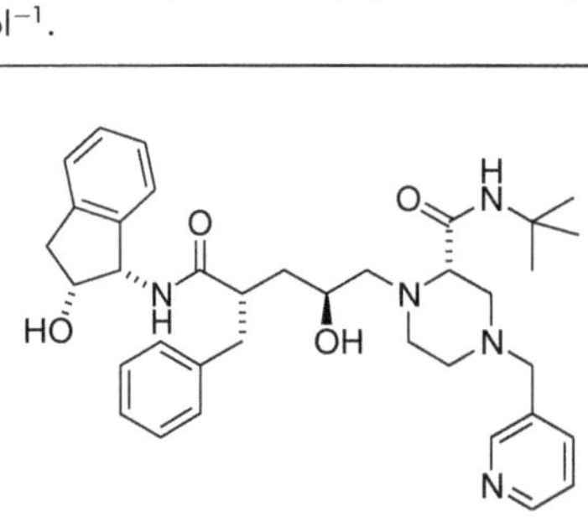	$t_{1/2}$ *F* pb ur	1.8±0.4 h 65% 61% 10–12%	*V* CL Mwt PSA log *P*	0.82 18 613.8 118 Å^2 2.88	INDINAVIR Antiviral Metabolism: pyridine N-glucuronidation, pyridine N-oxidation, 3′-hydroxylation at indane, 4-hydroxylation of phenylmethyl, N-depyridomethylation of piperazine
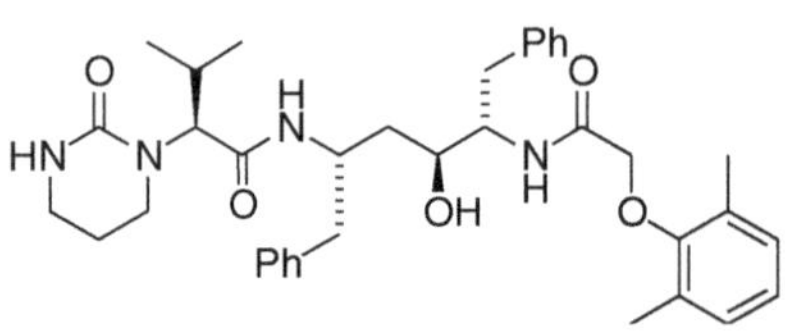	$t_{1/2}$ *F* pb ur	5–6 h Low* 98–99% 2.2%	*V* CL Mwt PSA log *P*	1.3 1.5–1.8 628.8 120 Å^2 6.26	LOPINAVIR *rat: 25% Antiviral Active metabolites: 4-oxo, 4-hydroxy

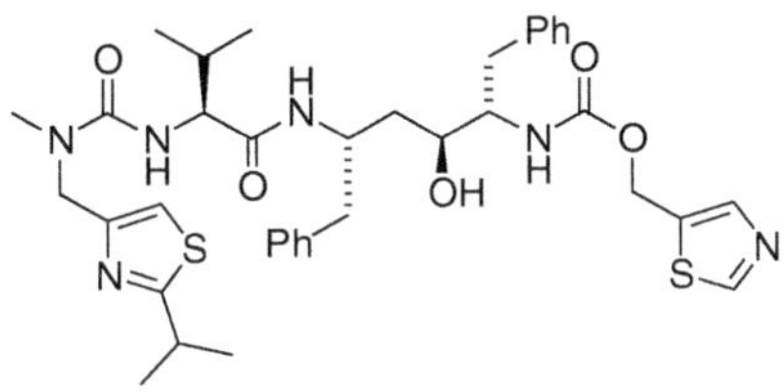	$t^{1}/_{2}$ 3–5 h *F* 75% (rat) pb 98–99% ur <10%	*V* 0.3–0.6 CL – Mwt 720.9 PSA 202 Å^2 log *P* 5.28	RITONAVIR Antiviral Metabolism: N-demethylation of urea, carbamate hydrolysis, hydroxylation at $(CH_3)_2CHC(=N)S$
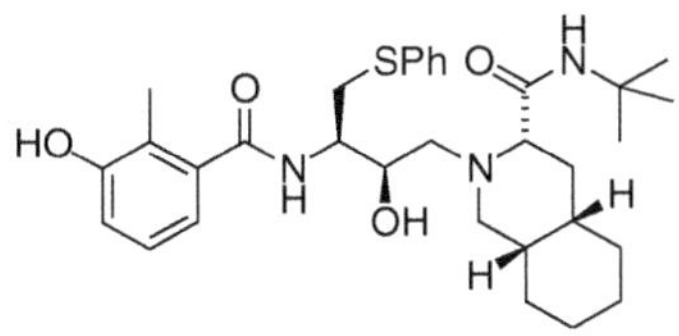	$t^{1}/_{2}$ 4–5 h *F* 47–88% pb >98% ur 1–2%	*V* 2–7 CL 6–16 Mwt 567.8 PSA 127 Å^2 log *P* 7.28	NELFINAVIR, AG-1343 Antiviral (HIV) Metabolism: hydroxylation of *tert*-butyl, 4′-hydroxylation to catechol, then 3′-O-methylation, oxidation to sulfoxide (minor pathway)
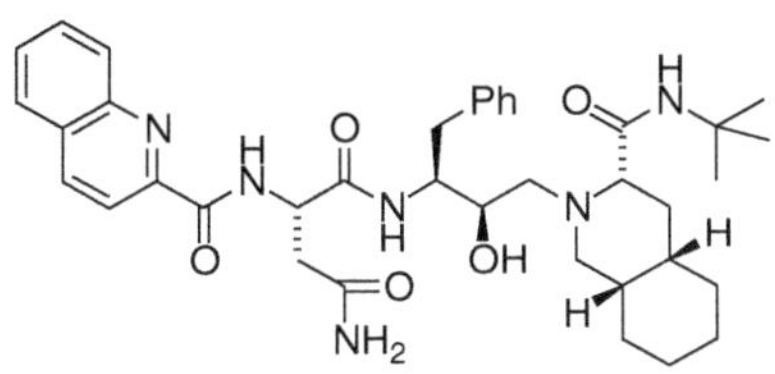	$t^{1}/_{2}$ 7–15 h (iv) *F* 1–9% pb 97% ur 1–3%	*V* 3.6 CL 13 Mwt 670.8 PSA 167 Å^2 log *P* 4.44	SAQUINAVIR Antiviral (HIV), low absorption from GI, high first-pass effect Metabolism: hydroxylation of cyclohexane and *tert*-butyl groups
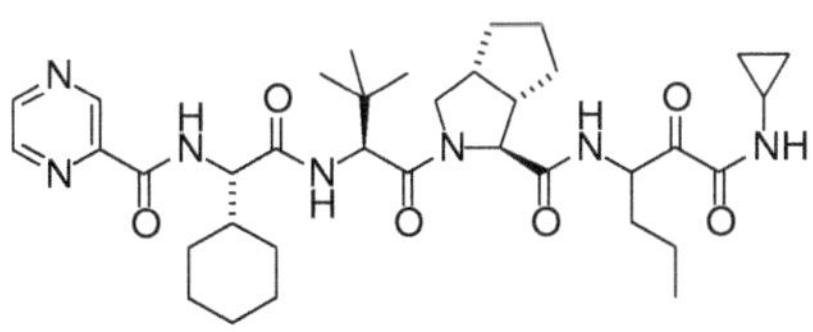	$t^{1}/_{2}$ 9–11 h *F* – pb 59–76% ur 1%	*V* 3.9 CL – Mwt 679.9 PSA 180 Å^2 log *P* 3.93	TELAPREVIR Antiviral (hepatitis C) Metabolite: pyrazinecarboxylic acid
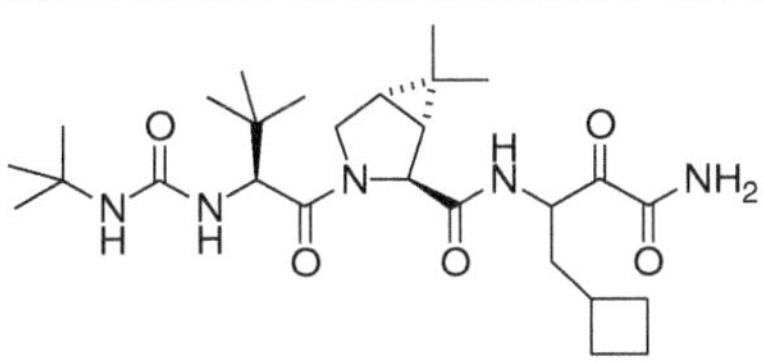	$t^{1}/_{2}$ 7–15 h *F* 30% (dog) pb 75% ur 3%	*V/F* 12 CL 41 Mwt 519.7 PSA 151 Å^2 log *P* 2.05	BOCEPREVIR Hepatitis C virus protease inhibitor
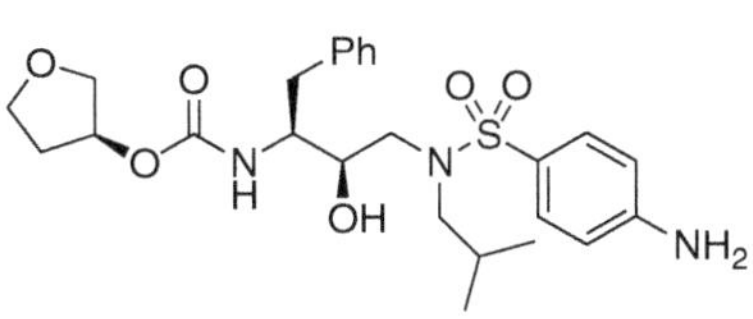	$t^{1}/_{2}$ 7–11 h *F* 35–90%* pb 90% ur <3%	*V* 5–7 CL – Mwt 505.6 PSA 140 Å^2 log *P* 4.20	AMPRENAVIR *dog Antiviral (HIV) Metabolism: oxidation of tetrahydrofuran and aniline, N-dealkylation, carbamate hydrolysis; prodrug: *O*-phosphate (fosamprenavir)

(*continued overleaf*)

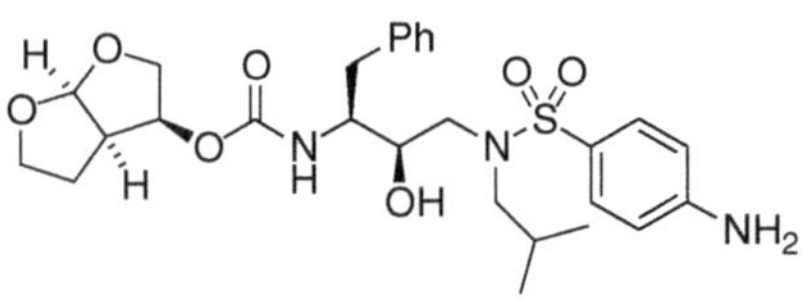	$t^{1}/_{2}$ 15 h *F* 37% pb 95% ur 8% *V* 1.4±0.9 CL 8 Mwt 547.7 PSA 149 Å^2 log *P* 3.94	DARUNAVIR Antiviral (HIV) Metabolism: carbamate hydrolysis, isobutyl hydroxylation, aniline hydroxylation, benzylic hydroxylation, glucuronidation
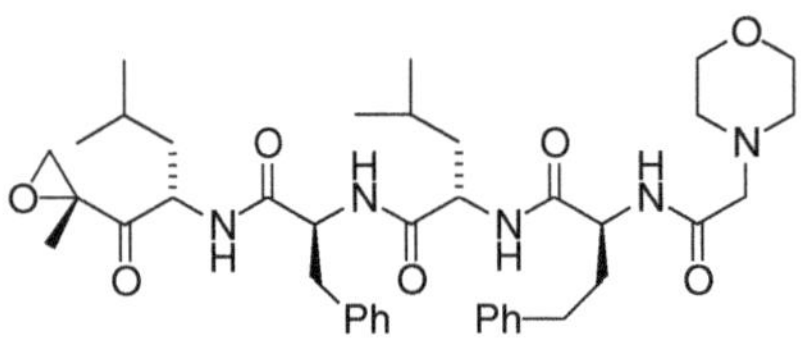	$t^{1}/_{2}$ 10–40 min *F* Low pb 95–99% ur – *V* 0.7–15 CL 13–285 Mwt 719.9 PSA 158 Å^2 log *P* 6.71	CARFILZOMIB Proteasome inhibitor, antineoplastic
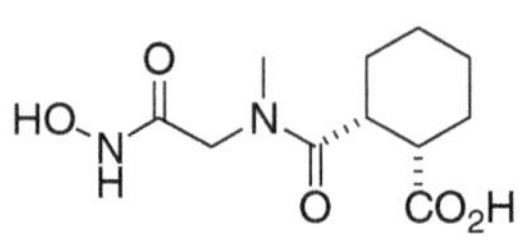	$t^{1}/_{2}$ 1.5–3.0 h *F* 30% pb 0% ur 8% *V* 0.8* CL – Mwt 258.3 PSA 107 Å^2 log *P* −1.49	IDRAPRIL *dog iv ACE inhibitor, antihypertensive Metabolism: reductive cleavage of N-O bond, hydrolysis of amide
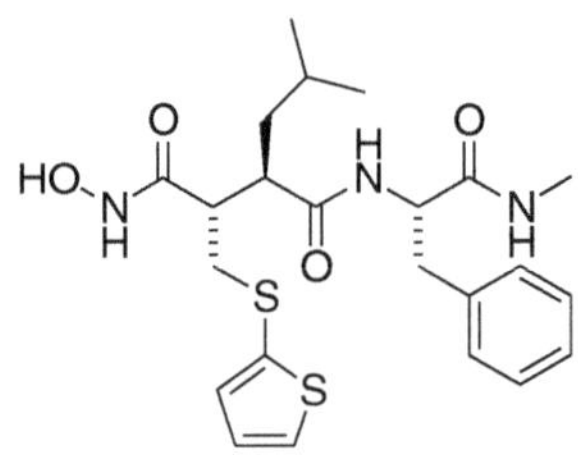	$t^{1}/_{2}$ 349–743 h* *F* Low pb 96% ur – *V* – CL – Mwt 477.6 PSA 161 Å^2 log *P* 3.30	BATIMASTAT *ip Matrix metalloproteinase inhibitor, antineoplastic
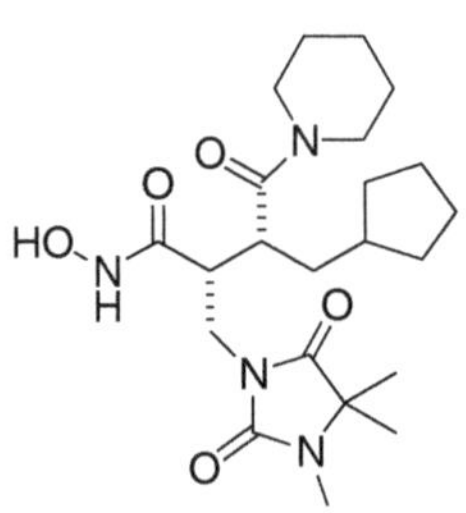	$t^{1}/_{2}$ 8–10 h *F* 43% pb 90% ur 9% *V* – CL – Mwt 436.6 PSA 110 Å^2 log *P* 2.06	CIPEMASTAT, TROCADE Matrix metalloproteinase inhibitor for treatment of rheumatoid arthritis Metabolism: glucuronidation

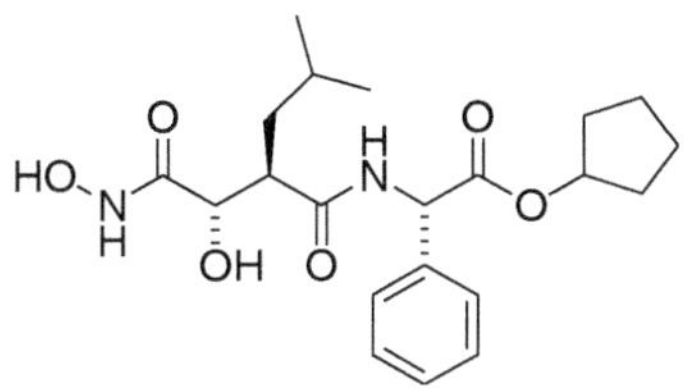	$t^1/_2$ 1–4 h *F* – pb – ur –	*V* – CL – Mwt 406.5 PSA 125 Å^2 log *P* 2.24	TOSEDOSTAT Orally available antineoplastic, aminopeptidase inhibitor; active metabolite: acid ($t^1/_2$ 6–11 h)
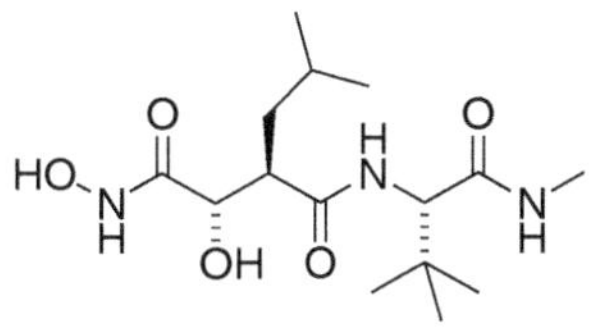	$t^1/_2$ 5–10 h *F* Good pb – ur 2–3%	*V* 3.4 CL – Mwt 331.4 PSA 128 Å^2 log *P* 1.31	MARIMASTAT Matrix metalloproteinase inhibitor, antineoplastic
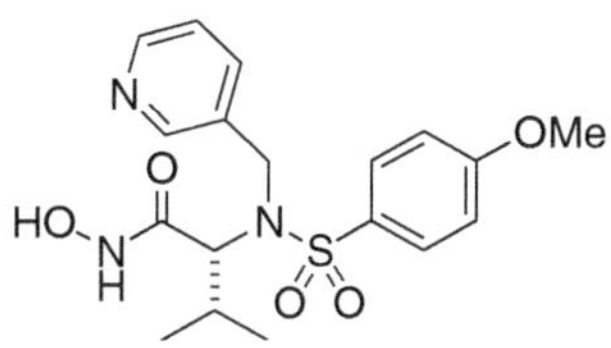	$t^1/_2$ 1–8 h *F* 44% (rat) pb 57–80% ur –	*V* – CL – Mwt 393.5 PSA 108.8 Å^2 log *P* 1.37	MMI-270 Matrix metalloproteinase inhibitor, antineoplastic
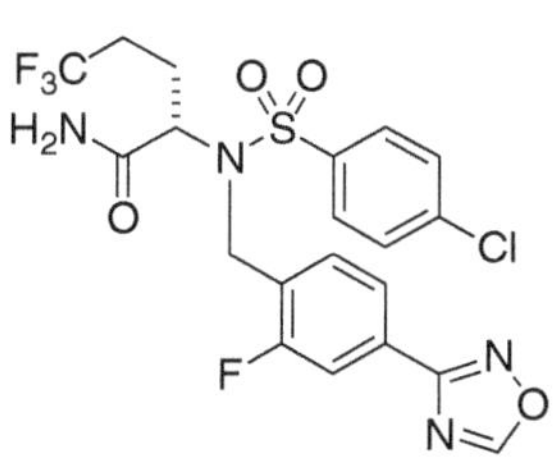	$t^1/_2$ 40 h *F* 42%* pb – ur –	*V* 2.2** CL 3.4** Mwt 520.9 PSA 128 Å^2 log *P* 3.97	BMS-708163 *dog **estimate based on allometric scaling; γ-secretase inhibitor (treatment of Alzheimer disease)
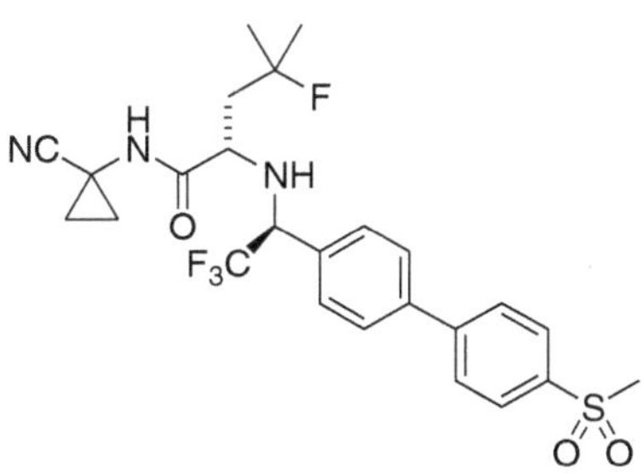	$t^1/_2$ 66–93 h *F* 18%* pb – ur –	*V* 0.7* CL – Mwt 525.6 PSA 107 Å^2 log *P* 2.92	ODANACATIB *monkey Cathepsin K inhibitor for treatment of osteoporosis Metabolism: CH_3-hydroxylation, demethylation of sulfone, addition of glutathione to nitrile

(*continued overleaf*)

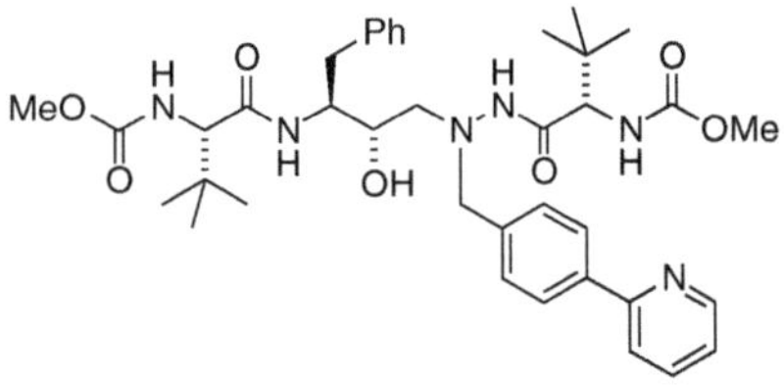	$t^1/_2$ 6.5 h *F* 60–68% pb 86% ur 6%	*V* 1.4 CL 3.3 Mwt 704.9 PSA 171 Å^2 log *P* 5.20	ATAZANAVIR Antiviral Metabolism: N-debenzylation, carbamate hydrolysis, hydroxylation, alcohol oxidation to ketone
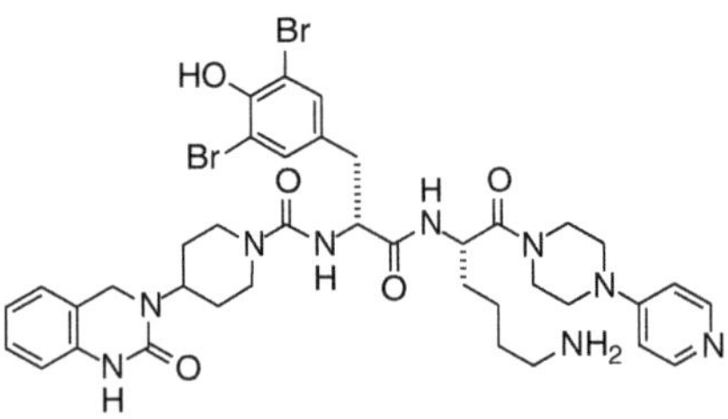	$t^1/_2$ 2.5 h (iv) *F* Low pb – ur –	*V* 0.3 CL 2.6 Mwt 869.7 PSA 176 Å^2 log *P* 3.55	OLCEGEPANT CGRP antagonist, antimigraine
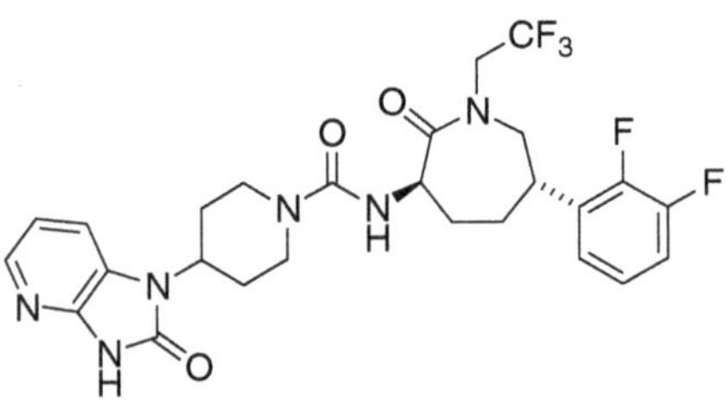	$t^1/_2$ 5–8 h *F* 35% (dog) pb 94% ur <5%	*V* – CL – Mwt 566.5 PSA 97.4 Å^2 log *P* 2.26	TELCAGEPANT, MK-0974 CGRP antagonist, antimigraine Metabolism: oxidation to pyridine-N-oxide
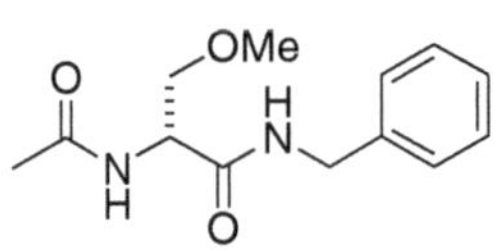	$t^1/_2$ 13 h *F* 100% pb <15% ur 40%	*V* 0.6 CL – Mwt 250.3 PSA 67.4 Å^2 log *P* 0.90	LACOSAMIDE Antiepileptic Metabolism: O-demethylation
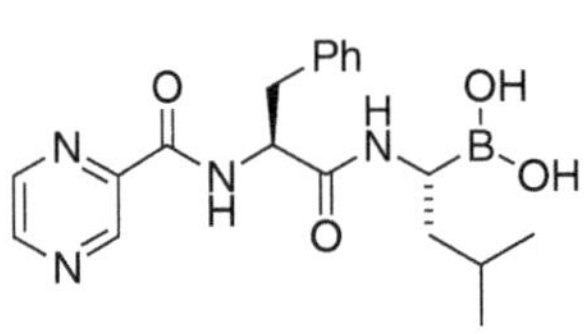	$t^1/_2$ 6–19 h (iv) *F* – pb 70–85% ur –	*V* 5–12 CL 19 Mwt 384.2 PSA 124 Å^2 log *P* 2.45	BORTEZOMIB Antineoplastic Metabolism: substitution of $B(OH)_2$ by OH, hydroxylation of *CH*(Me)$_2$, benzylic hydroxylation, phenylalaninamide hydrolysis

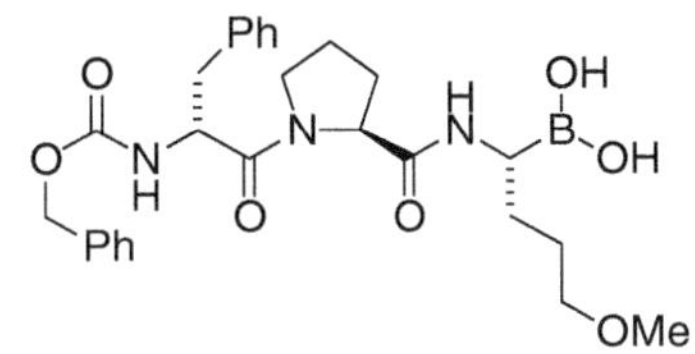	$t^1/_2$ 1.7–2.3 h* *F* – pb 90% ur 14%*	*V* 0.6 CL – Mwt 525.4 PSA 137 Å^2 log *P* 3.54	FLOVAGATRAN *iv Thrombin inhibitor
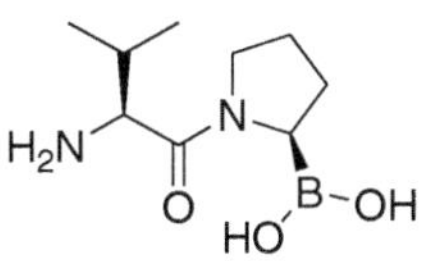	$t^1/_2$ 2.8 h *F* – pb – ur –	*V* – CL – Mwt 214.1 PSA 86.8 Å^2 log *P* 0.01	TALABOSTAT Orally available DPPIV inhibitor, antidiabetic
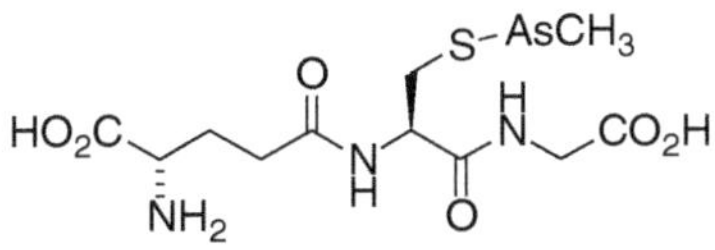	$t^1/_2$ 14–18 h *F* 58% pb – ur –	*V* – CL – Mwt 411.3 PSA 184 Å^2 log *P* −5.14	DARINAPARSIN Antineoplastic, apoptosis inducer
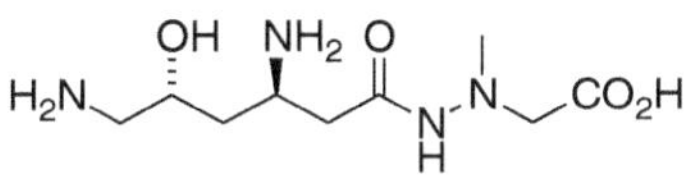	$t^1/_2$ 2 h *F* – pb – ur –	*V* 0.35 CL – Mwt 248.3 PSA 142 Å^2 log *P* −4.14	NEGAMYCIN Antibacterial
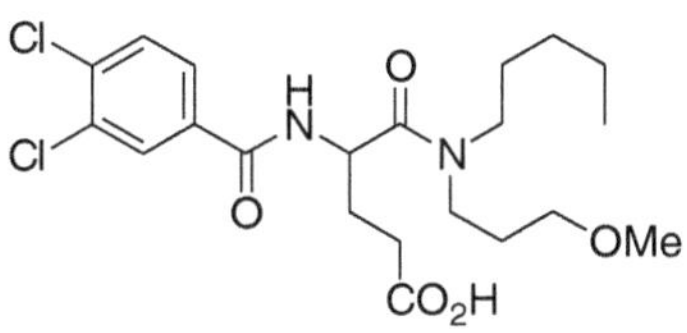	$t^1/_2$ 5–8 h *F* 97% pb 94–98% ur 1%	*V* 0.24 CL 1.0 Mwt 461.4 PSA 95.9 Å^2 log *P* 4.37	LOXIGLUMIDE CCK antagonist, gastroprokinetic, racemic Metabolism: hydroxylation and oxidative degradation of pentyl, O-demethylation, then oxidation of primary alcohol to CO_2H, hydrolysis of both amides
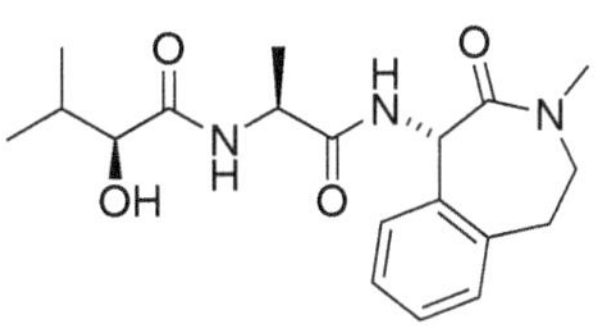	$t^1/_2$ 2–3 h *F* – pb – ur 44%	*V* – CL/*F* 4.3 Mwt 361.4 PSA 98.7 Å^2 log *P* 2.23	SEMAGACESTAT Oral γ-secretase inhibitor for treatment of Alzheimer disease Metabolism: hydrolysis of the amide bond proximal to the benzazepine ring, benzylic hydroxylation of the benzazepine ring

(*continued overleaf*)

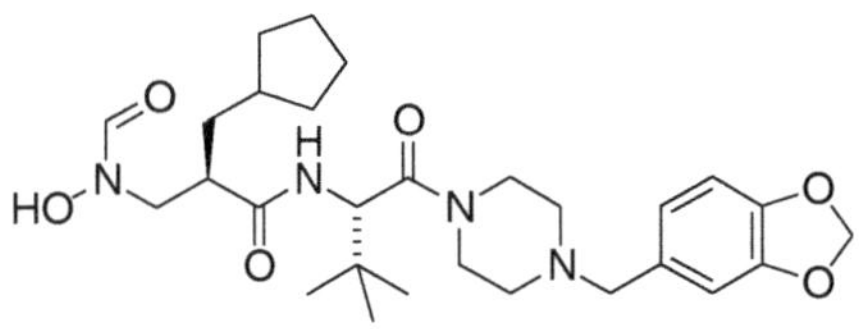

$t^{1}/_{2}$	5–17 h	V	0.9–1.6	BB 83698
F	–	CL	3.7±2.0	Peptide deformylase inhibitor, antibacterial
pb	78–82%	Mwt	530.7	
ur	–	PSA	112 Å^2	
		log P	3.56	

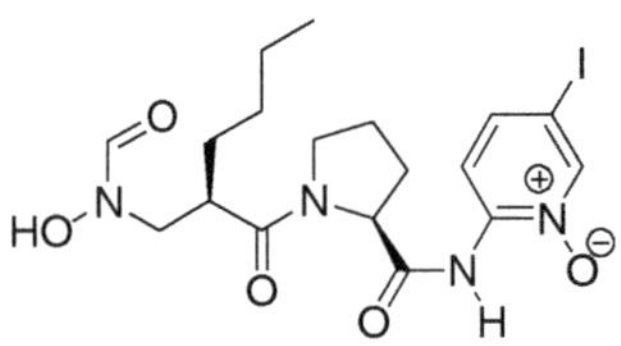

$t^{1}/_{2}$	2.0–4.2 h	V	–	LBM 415
F	22–100%*	CL	31*	*rodents
pb	–	Mwt	396.4	Peptide deformylase inhibitor, antibacterial
ur	–	PSA	94.9 Å^2	
		log P	0.04	

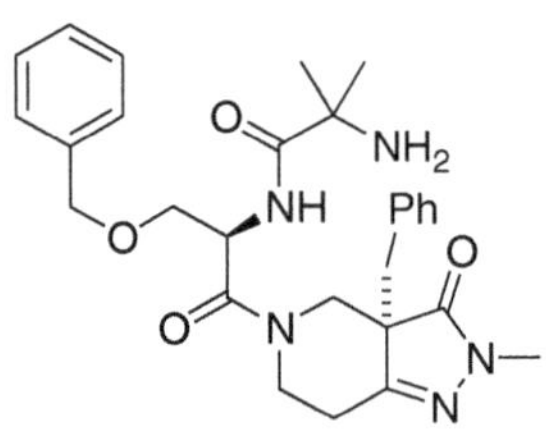

$t^{1}/_{2}$	2.4 h*	V	–	CAPROMORELIN
F	65%*	CL	–	*rat
pb	–	Mwt	505.6	Growth hormone secretagogue
ur	2–5%*	PSA	117 Å^2	Metabolism (rat): O-debenzylation, aromatic hydroxylation of benzyl ether, N-demethylation
		log P	4.07	

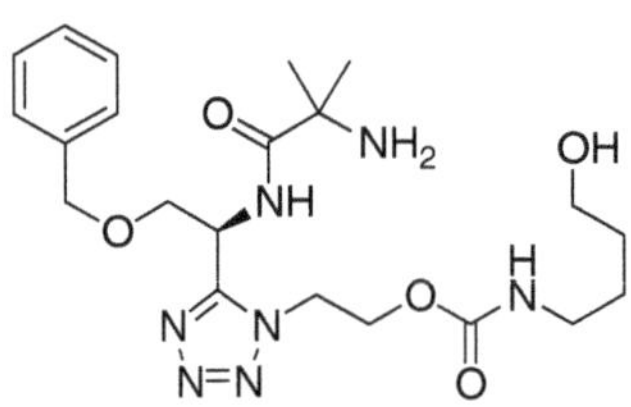

$t^{1}/_{2}$	1.1 h*	V	–	BMS-317180
F	12%*	CL	–	*monkey
pb	–	Mwt	463.5	Growth hormone secretagogue
ur	–	PSA	162 Å^2	
		log P	0.93	

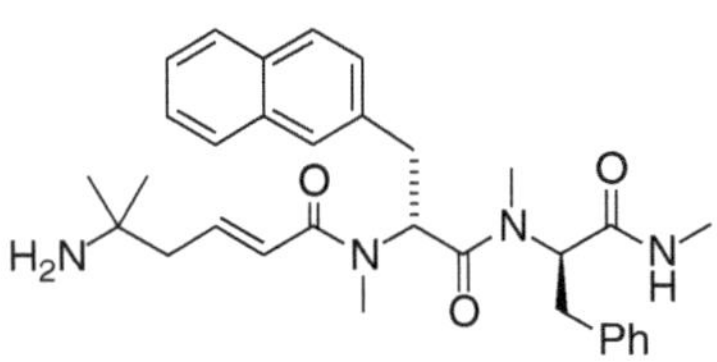

$t^{1}/_{2}$	4 h*	V	–	TABIMORELIN
F	30–35%*	CL	–	*dog
pb	–	Mwt	528.7	Growth hormone secretagogue; CYP3A4 inhibitor
ur	–	PSA	95.7 Å^2	
		log P	3.80	

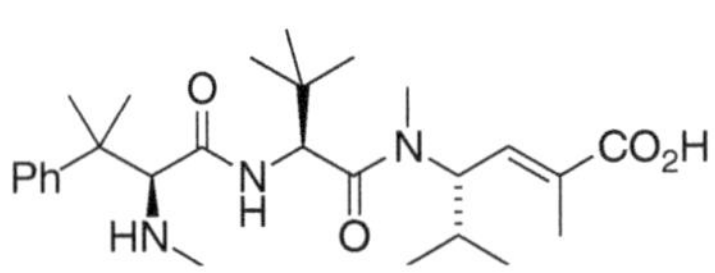

$t^{1}/_{2}$	9 h	V	2.7	TALTOBULIN
F	–	CL	4	Tubulin polymerization inhibitor, antineoplastic
pb	–	Mwt	473.7	
ur	–	PSA	98.7 Å^2	
		log P	4.80	

Structure							Compound
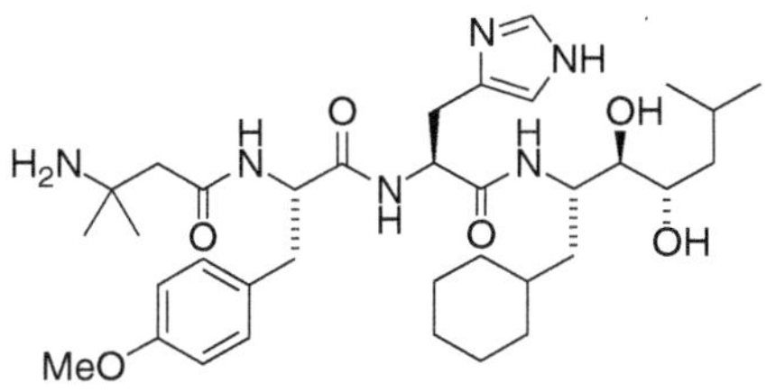	$t^{1}/_{2}$ *F* pb ur	1.6 h 0% 94% –		*V* CL Mwt PSA log *P*	0.2 – 656.9 192 Å^2 3.52		ENALKIREN Renin inhibitor, antihypertensive
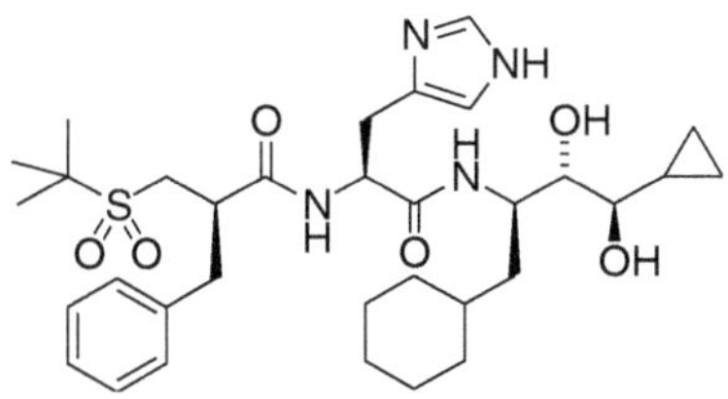	$t^{1}/_{2}$ *F* pb ur	9 h 1–2% 83% –		*V* CL Mwt PSA log *P*	1 12.8 630.8 170 Å^2 3.38		REMIKIREN Renin inhibitor, antihypertensive Metabolism: hydroxylation
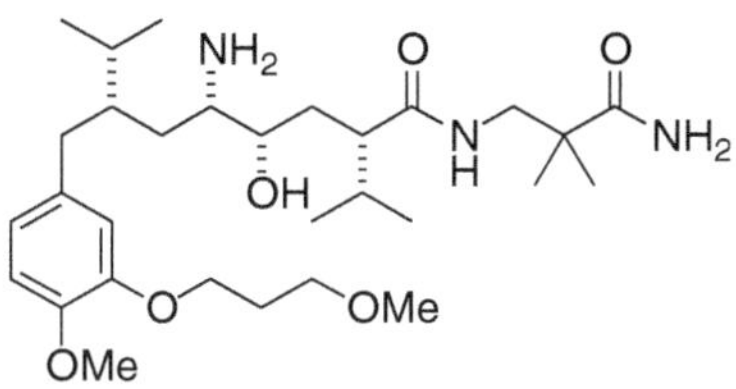	$t^{1}/_{2}$ *F* pb ur	34–41 h 2.6% 47–51% <1%		*V* CL Mwt PSA log *P*	2.1 2.3 (iv) 551.8 146 Å^2 2.74		ALISKIREN Renin inhibitor, antihypertensive; 91% of dose is excreted unchanged in feces
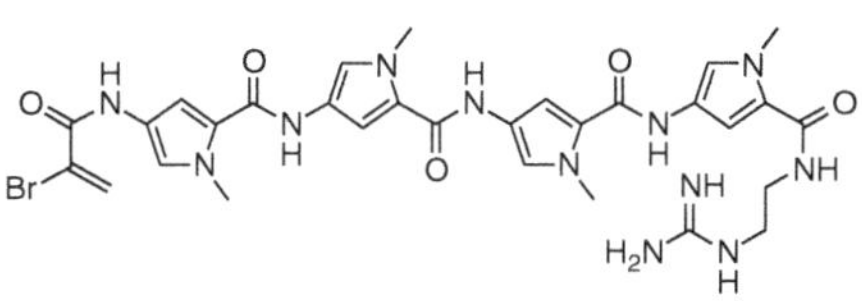	$t^{1}/_{2}$ *F* pb ur	5±2 h (iv) – – –		*V* CL Mwt PSA log *P*	0.2 4 723.6 207 Å^2 −2.61		BROSTALLICIN DNA ligand, antineoplastic
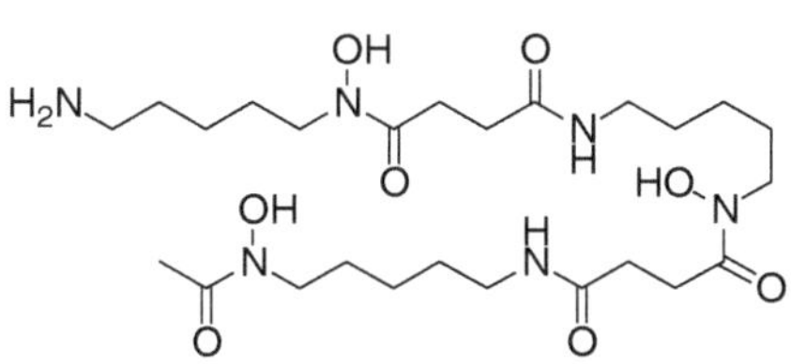	$t^{1}/_{2}$ *F* pb ur	6 h (iv) <2% <10% 30–35%		*V* CL Mwt PSA log *P*	2.0–2.5 4–5 560.7 206 Å^2 −0.61		DESFERRIOXAMIN, DEFEROXAMINE Chelating agent for iron Metabolism: transamination, oxidation to carboxylic acid, N-hydroxylation
Pyroglutamic acid-His-Trp-Ser- -Tyr-(O-tBu)ser-Leu- -Arg-Pro-NHEt	$t^{1}/_{2}$ *F* pb ur	1.2 h 2.5–3.3%* 15% 17%		*V* CL Mwt PSA log *P*	– – 1239.4 438 Å^2 0.76		BUSERELIN *intranasal; po (pig): 1–5% GnRH agonist, treatment of hormone-sensitive cancers

(*continued overleaf*)

Pyroglutamic acid-His-Trp-Ser- -(O-tBu)Ser-Tyr-(O-tBu)ser-Leu- -Arg-Pro-NH-NHMe	$t^1/_2$ F pb ur	4–5 h (sc) – 20–28% 20% (sc)	V CL Mwt PSA log P	0.2–0.6 1.9 1269.4 493 Å^2 −0.26	GOSERELIN GnRH agonist, treatment of hormone-sensitive cancers
Pyroglutamic acid-His-Trp-Ser- -Tyr-Nal-Leu-Arg-Pro-Gly-NH_2	$t^1/_2$ F pb ur	2–3 h 4–21%* 80% 3%*	V CL Mwt PSA log P	0.4–0.5 – 1322.5 472 Å^2 0.53	NAFARELIN *intranasal dosing GnRH agonist
Pyroglutamic acid-His-Trp-Ser- -Tyr-leu-Leu-Arg-Pro-NHEt	$t^1/_2$ F pb ur	2.9–3.6 h <1% 50% –	V CL Mwt PSA log P	0.37–0.53 2.0–2.2 1209.4 429 Å^2 0.55	LEUPRORELIN GnRH analog, treatment of hormone-sensitive cancers
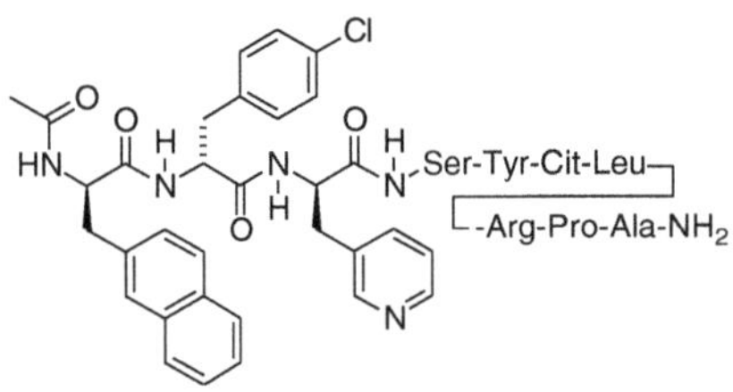	$t^1/_2$ F pb ur	8–12 h (iv) – 86% 2–4% (sc)	V CL Mwt PSA log P	0.4–1.2 1.3 1431.0 496 Å^2 1.38	CETRORELIX (Cit: citrulline) GnRH antagonist
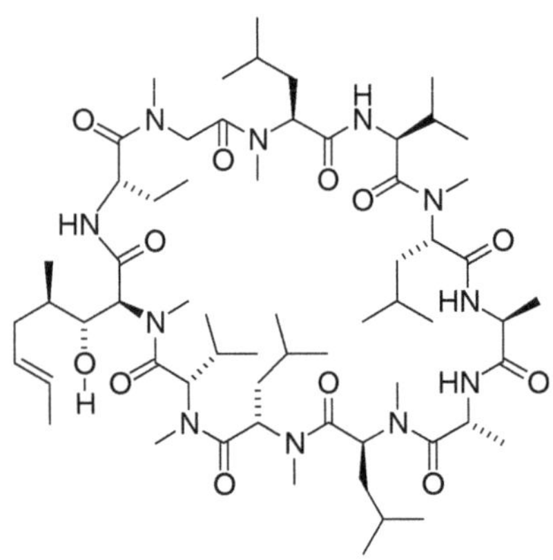	$t^1/_2$ F pb ur	5.6±2.0 h 27±9% 93±2% <1%	V CL Mwt PSA log P	1.3±0.3 5.3±1.5 1202.6 279 Å^2 2.79	CYCLOSPORINE Natural immunosuppressant

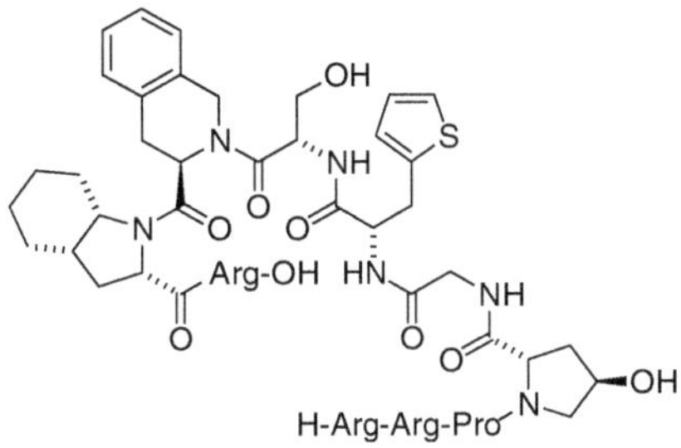

$t^1/_2$	2 h	V	0.3–0.4	ICATIBANT
F	Low	CL	4–5	Bradykinin B_2 antagonist for treatment of hereditary angioedema
pb	44%	Mwt	1304.5	Metabolism: peptide bond hydrolysis (Thi-Ser and Gly-Thi)
ur	<10%	PSA	544 Å^2	
		log P	−3.27	

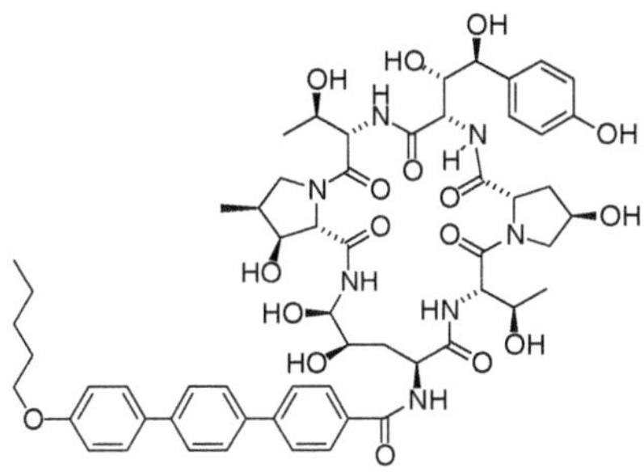

$t^1/_2$	40–50 h	V	0.77	ANIDULAFUNGIN
F	2–7%	CL	0.64	Antifungal
pb	84%	Mwt	1140.2	
ur	–	PSA	377 Å^2	
		log P	4.34	

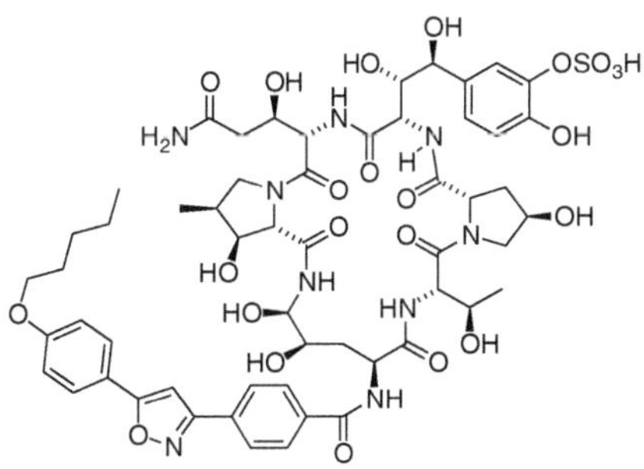

$t^1/_2$	11–17 h	V	0.21	MICAFUNGIN
F	Low	CL	0.17	Antifungal
pb	>99%	Mwt	1270.3	Metabolized by aryl sulfatase and COMT
ur	<1%	PSA	519 Å^2	
		log P	−7.49	

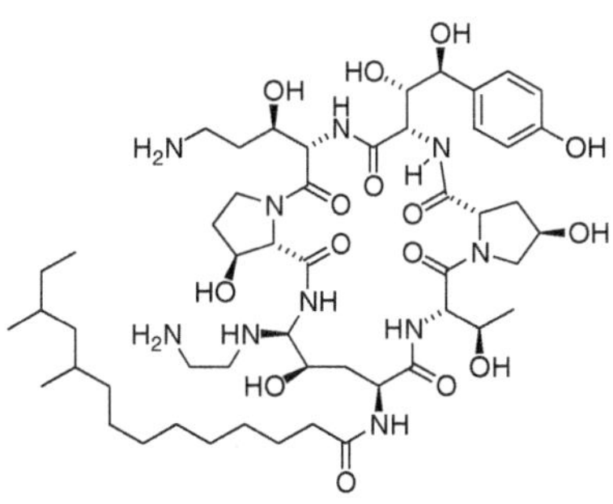

$t^1/_2$	27 h (iv)	V	0.13	CASPOFUNGIN
F	–	CL	0.14	Antifungal
pb	96%	Mwt	1093.3	
ur	–	PSA	412 Å^2	
		log P	3.53	

(*continued overleaf*)

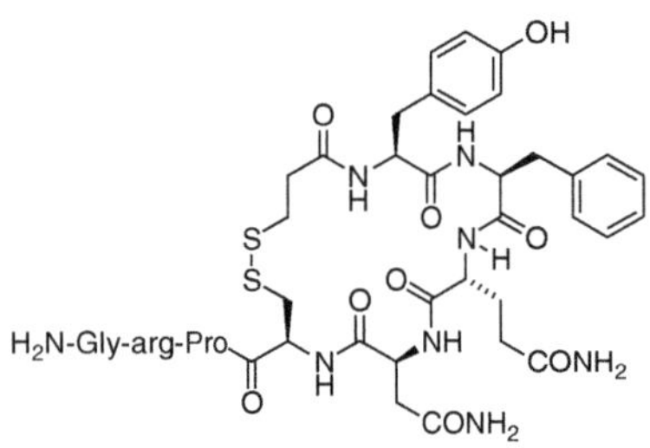

$t_{1/2}$	2.7–4.6 h	*V*	0.2–0.4	DESMOPRESSIN
F	0.1%*	CL	0.3–0.7	*intranasal : 2–3%
pb	–	Mwt	1069.2	Vasopressin agonist,
ur	47%	PSA	486 Å^2	antidiuretic
		log *P*	−2.95	

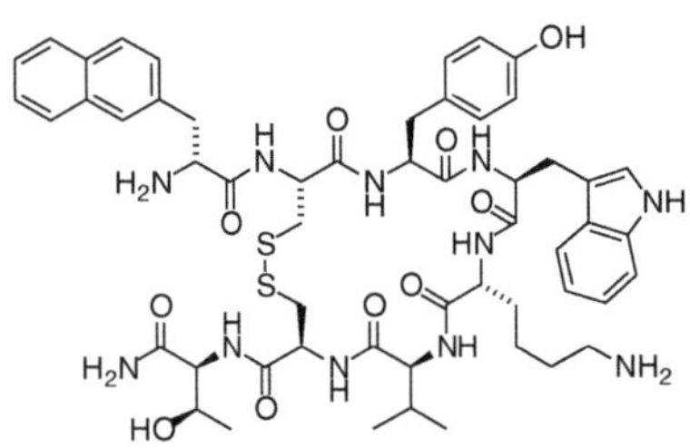

$t_{1/2}$	1.3 h	*V*	0.46	LANREOTIDE
F	–	CL	4.1	Somatostatin analog for
pb	80%	Mwt	1096.3	treatment of acromegaly
ur	<5% (sc)	PSA	406 Å^2	and carcinoid syndrome
		log *P*	4.99	

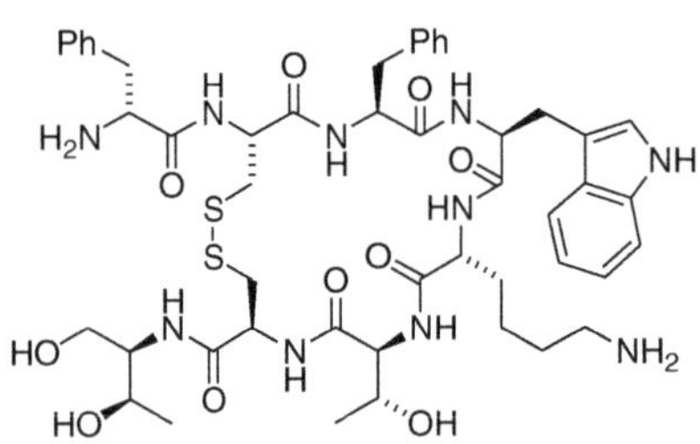

$t_{1/2}$	1.5±0.4 h	*V*	0.35±0.22	OCTREOTIDE
F	1–3%	CL	2.7±1.4	Somatostatin analog for sc
pb	65%	Mwt	1019.2	dosing
ur	11–20%	PSA	383 Å^2	Treatment of carcinoid
		log *P*	3.40	syndrome

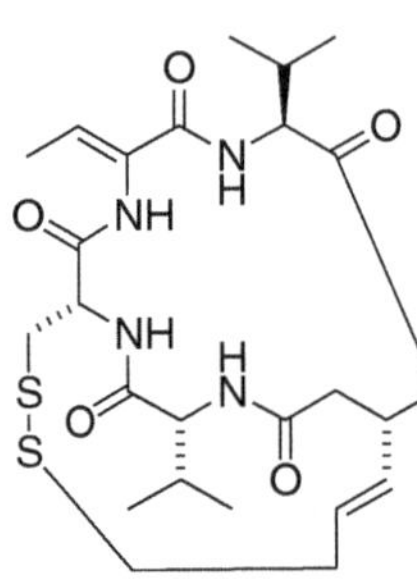

$t_{1/2}$	3 h (iv)	*V*	1.2	ROMIDEPSIN
F	–	CL	7.4	Histone deacetylase
pb	92–94%	Mwt	540.7	inhibitor, antineoplastic
ur	–	PSA	193 Å^2	
		log *P*	3.60	

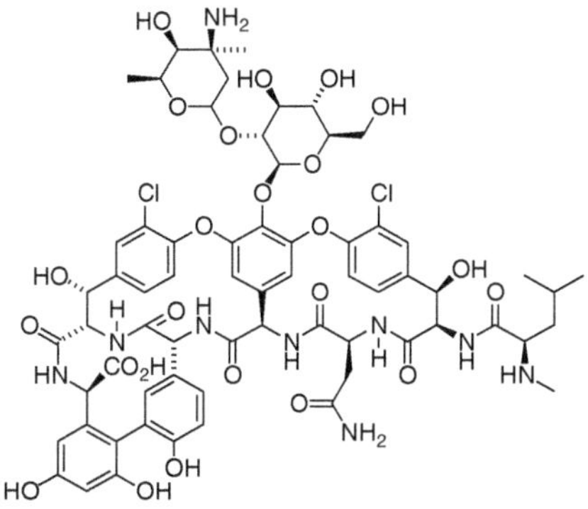	$t_{1/2}$ F pb ur	5.6±1.8 h Low 30±10% 79±11%	V CL Mwt PSA log P	0.39±0.06 1.4±0.1 1449.3 530 Å^2 −1.41	VANCOMYCIN Antibacterial

$t_{1/2}$, plasma half-life; F, oral bioavailability; pb, plasma protein binding; ur, excretion of unchanged drug in urine; V, volume of distribution; CL, clearance; Mwt, molecular weight; PSA, polar surface area; ACE, angiotensin converting enzyme; CGRP, calcitonin gene related peptide; CCK, cholecystokinin; COMT, catechol-O-methyltransferase; DPPIV, dipeptidylpeptidase IV; GI, gastrointestinal; GnRH, gonadotropin releasing hormone.

Table 55.2 Antithrombotic amino-, amidino-, and guanidino carboxylic acids. V in l kg^{-1}, CL in ml min^{-1} kg^{-1}, Mwt in g mol^{-1}.

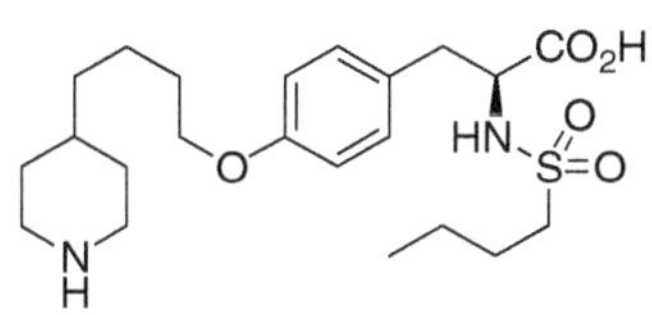	$t_{1/2}$ F pb ur	1.5–3.0 h Low 64% 38–80%	V CL Mwt PSA log P	0.3–0.7 3.3–4.8 440.6 113 Å^2 4.14	TIROFIBAN Fibrinogen antagonist, antithrombotic for iv administration Metabolism: oxidation to 2-piperidinone, O-dealkylation
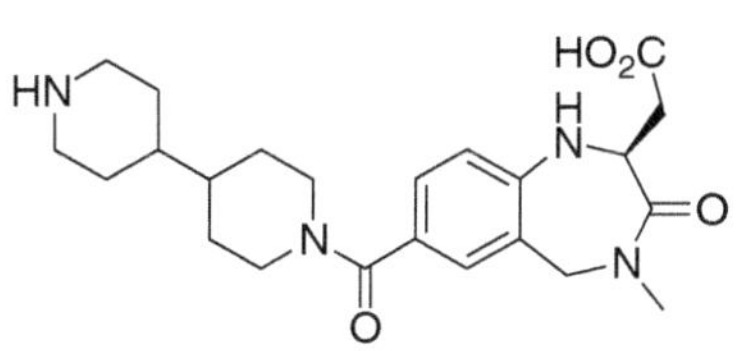	$t_{1/2}$ F pb ur	12–20 h 2% <13% High	V CL/F Mwt PSA log P	– 97 428.5 102 Å^2 −0.30	LOTRAFIBAN Platelet glycoprotein IIb/IIIa blocker, antithrombotic
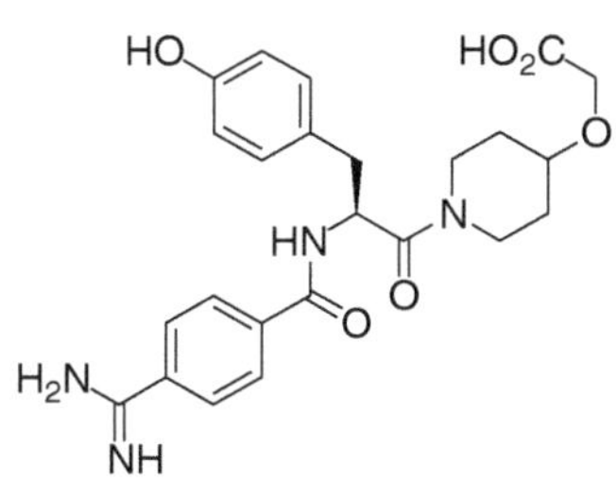	$t_{1/2}$ F pb ur	2–4 h 1%* 6% 90% (iv)	V CL Mwt PSA log P	0.3 1.9 468.5 166 Å^2 0.08	LAMIFIBAN *mouse Antithrombotic for iv administration

(continued overleaf)

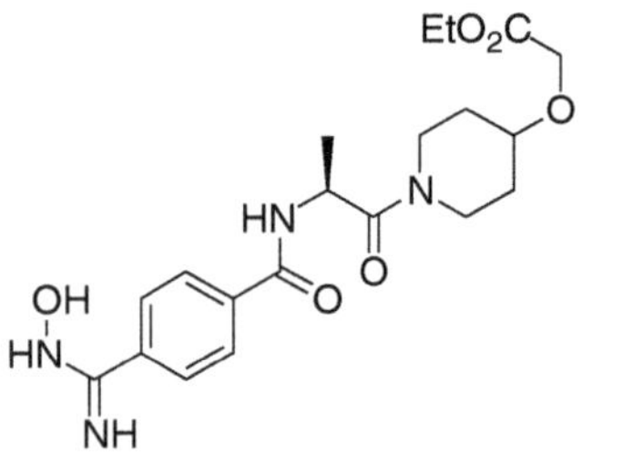	$t\frac{1}{2}$ 9–12 h* *F* 36%* pb – ur 57%*	*V* – CL 5.2* Mwt 420.5 PSA 144 Å^2 log *P* 1.04	SIBRAFIBAN *acid/amidine on oral dosing of ester/hydroxyamidine; oral antithrombotic
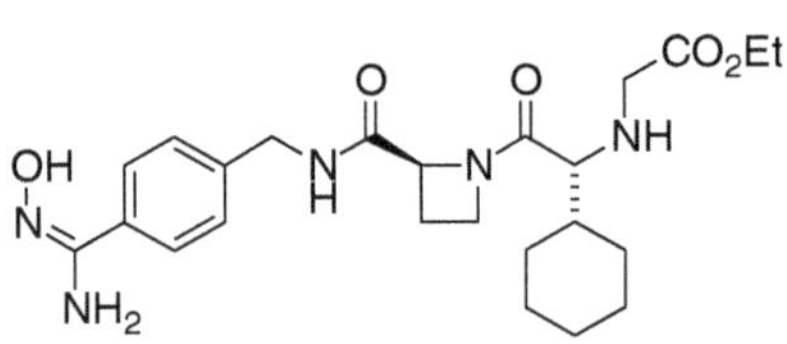	$t\frac{1}{2}$ 3–5 h* *F* 20%* pb <15%* ur 83%**	*V* 0.22** CL 145±15** Mwt 473.6 PSA 144 Å^2 log *P* 2.08	XIMELAGATRAN *amidine/acid (melagatran) on oral dosing of ximelagatran **melagatran on iv dosing Withdrawn in 2006 because of hepatotoxicity
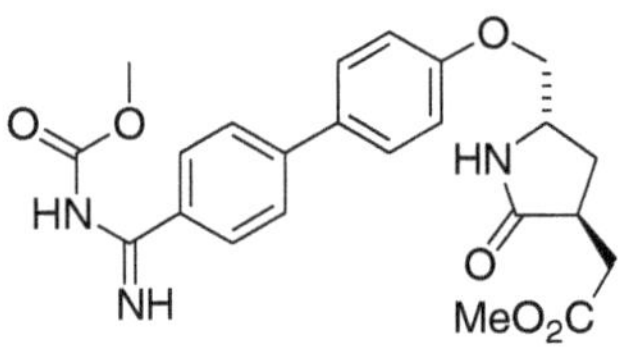	$t\frac{1}{2}$ 11–13 h* *F* 24%* pb Low ur 17%*	*V* – CL 6.3* Mwt 439.5 PSA 127 Å^2 log *P* 0.40	LEFRADAFIBAN *amidine/acid (fradafiban) on oral dosing of lefradafiban
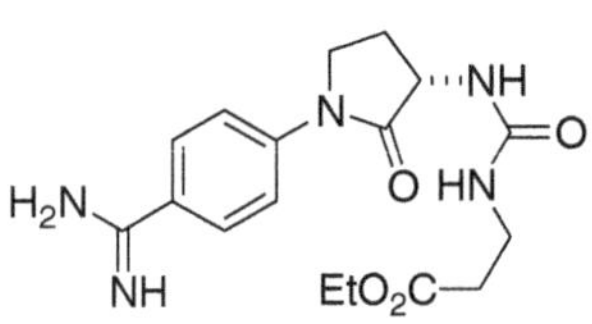	$t\frac{1}{2}$ 10–18 h* *F* 9%* pb – ur –	*V* – CL – Mwt 361.4 PSA 138 Å^2 log *P* −0.43	ORBOFIBAN *acid on oral dosing of ester Antithrombotic
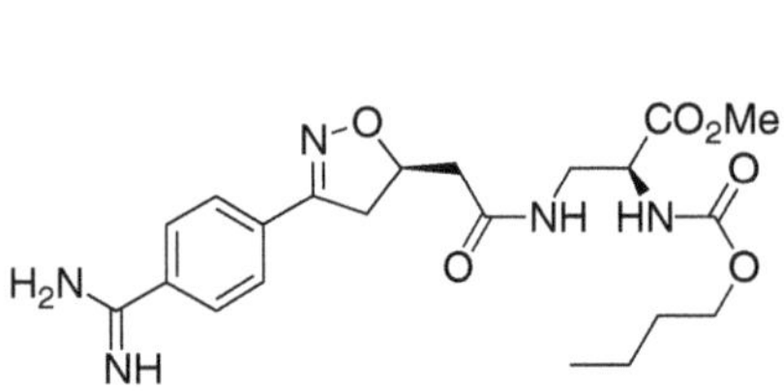	$t\frac{1}{2}$ 5 d* *F* 21%* pb – ur –	*V* – CL – Mwt 447.5 PSA 165 Å^2 log *P* 1.60	ROXIFIBAN *acid on oral dosing of ester Platelet GPIIb/IIIa antagonist, antithrombotic Metabolism: ester hydrolysis, 2- and 3-hydroxylation of butyl, 4- and 5-hydroxylation of isoxazoline

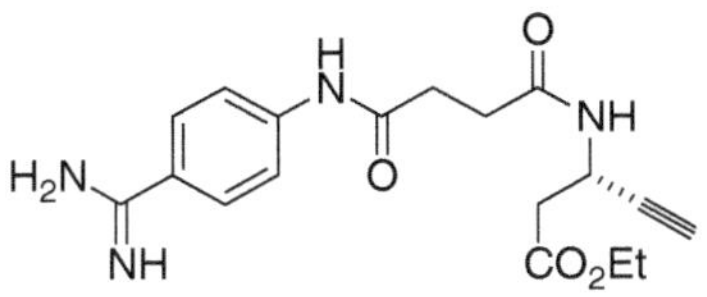	$t^{1}/_{2}$ 4–6 h* *F* 13%* pb – ur 29%*	*V* – CL 0.08 (iv) Mwt 358.4 PSA 134 Å^2 log *P* 0.46	XEMILOFIBAN *acid on oral dosing of ester Antithrombotic
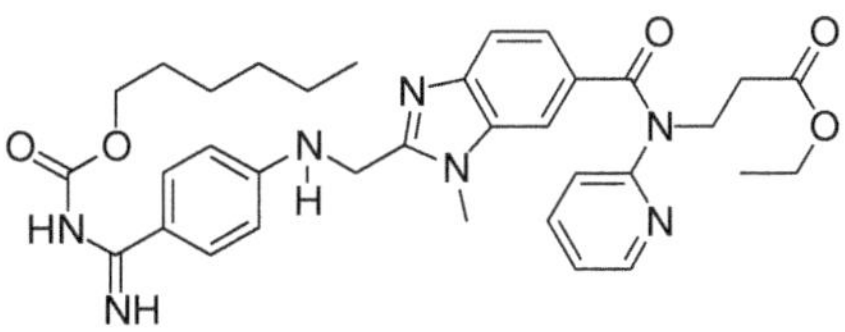	$t^{1}/_{2}$ 12–14 h* *F* 6.5%* pb 35%* ur 77%*	*V* 0.9–1.1* CL – Mwt 471.5* PSA 145 Å^{2}* log *P* −0.19*	DABIGATRAN ETEXILATE *acid/amidine (dabigatran) on oral dosing of prodrug
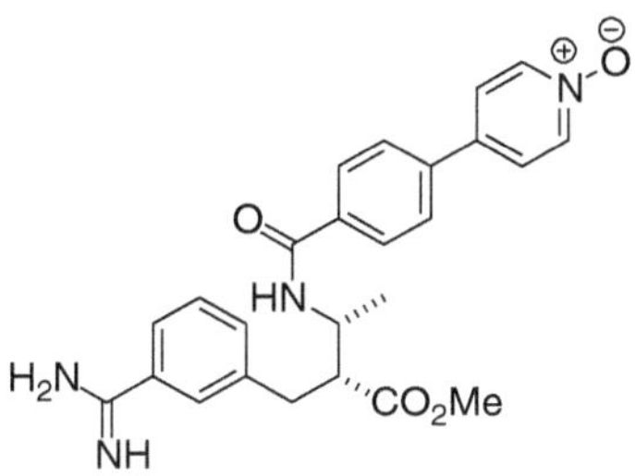	$t^{1}/_{2}$ 2–3 h* *F* Low pb – ur 13–32%*	*V* 0.65 CL – Mwt 446.5 PSA 127 Å^2 log *P* 0.58	OTAMIXABAN *iv Factor Xa inhibitor
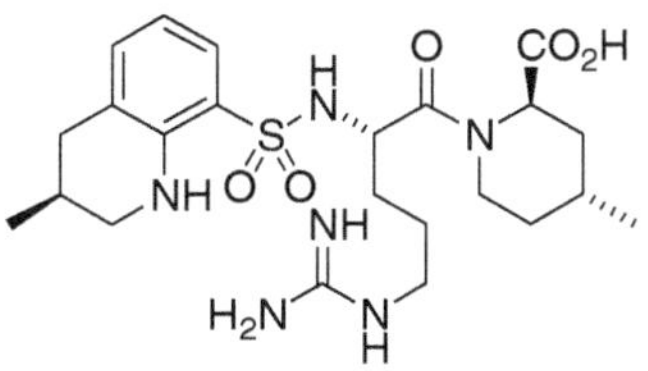	$t^{1}/_{2}$ 0.4–0.8 h* *F* Low pb 54% ur 16%*	*V* 0.2 CL 5 Mwt 508.6 PSA 186 Å^2 log *P* 2.06	ARGATROBAN *iv Thrombin inhibitor, antithrombotic; CYP3A4 substrate Metabolism: oxidation to quinoline and hydroxylation; reason for low *F*: too basic, poor absorption due to charges
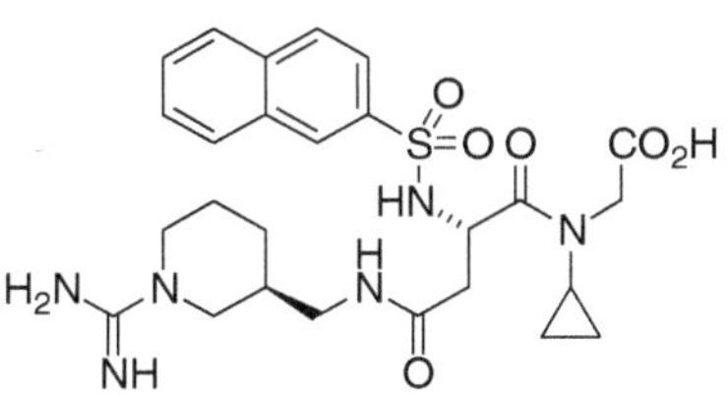	$t^{1}/_{2}$ 1.9 h (iv) *F* – pb – ur –	*V* 0.36 CL 6.4 Mwt 558.7 PSA 194 Å^2 log *P* 0.27	NAPSAGATRAN Thrombin inhibitor

(*continued overleaf*)

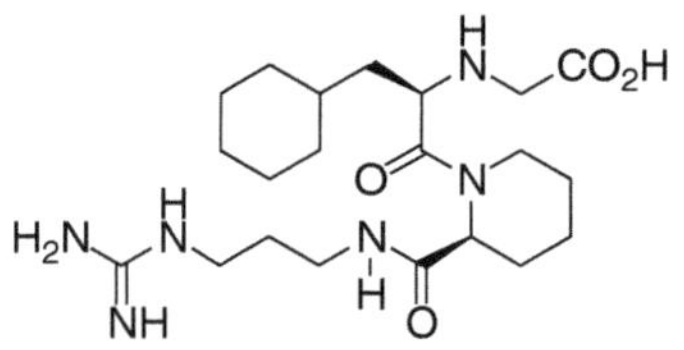	$t^1/_2$ 1 h *F* 2%* pb 20–28% ur –	*V* 0.26 CL 6.1 Mwt 438.6 PSA 161 Å^2 log *P* 0.98	INOGATRAN *monkey Little metabolism in humans
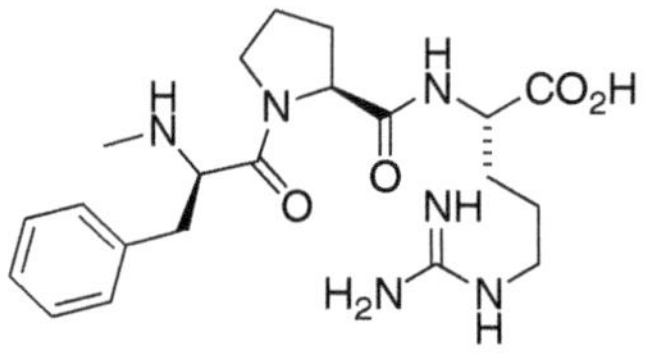	$t^1/_2$ 0.5 h *F* <6% pb – ur –	*V* – CL 6 Mwt 416.5 PSA 140 Å^2 log *P* −0.44	EFEGATRAN Thrombin inhibitor

$t_{1/2}$, plasma half-life; *F*, oral bioavailability; pb, plasma protein binding; ur, excretion of unchanged drug in urine; V, volume of distribution; CL, clearance; Mwt, molecular weight; PSA, polar surface area.

56
Oligoarylamines, Oligoarylamides, Oligoarylcarbamates, and Oligoarylureas

In recent years, several enzyme inhibitors have been developed with the aid of a thorough understanding of the synthetic accessibility of molecules and molecular modeling. Typical target enzymes included HIV protease and the various tyrosine kinases involved in the growth of tumors [1]. Arenes and heteroarenes are easy to synthesize, nonchiral, and easy to functionalize with various substituents. Therefore, these drugs often end up being short oligoarenes or -heteroarenes linked by one- to three-atom spacers (diarylureas, diarylethers, diarylamines, diarylethylenes, benzanilides, etc.) (Table 56.1). Because such compounds do not readily dissolve in water [2], solubility-enhancing, basic substituents are often included, such as 1-piperazinyl or other aminoalkyl substituents. Amines may also prolong the plasma half-life of these oligoarenes by increasing their volume of distribution (Chapter 23).

References

1. Liao, J.J.-L. (2007) Molecular recognition of protein kinase binding pockets for design of potent and selective kinase inhibitors. *J. Med. Chem.*, **50**, 409–424.
2. Lovering, F., Bikker, J., and Humblet, C. (2009) Escape from flatland: increasing saturation as an approach to improving clinical success. *J. Med. Chem.*, **52**, 6752–6756.

Table 56.1 Oligoarylamines and related compounds. V in l kg^{-1}, CL in ml min^{-1} kg^{-1}, and Mwt in g mol^{-1}.

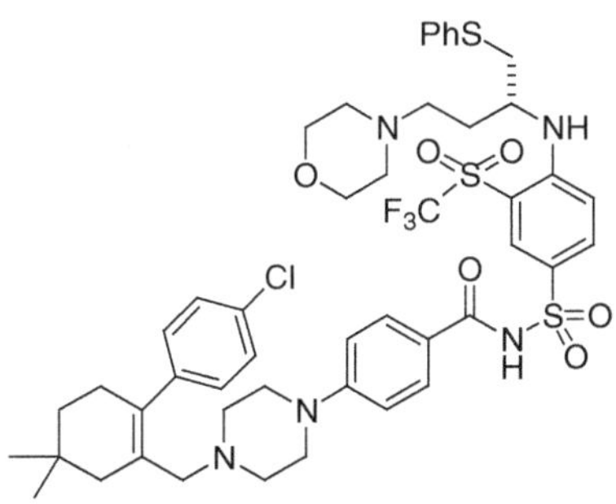	$t^1\!/\!_2$ F pb ur	14–20 h 20%* – –	V CL/F Mwt PSA log P	– 0.8–1.9 974.6 170 Å^2 9.70	NAVITOCLAX, ABT-263 *monkey Bcl-2 inhibitor, apoptosis inducer, oncolytic

(*continued overleaf*)

Lead Optimization for Medicinal Chemists: Pharmacokinetic Properties of Functional Groups and Organic Compounds, First Edition. Florencio Zaragoza Dörwald.

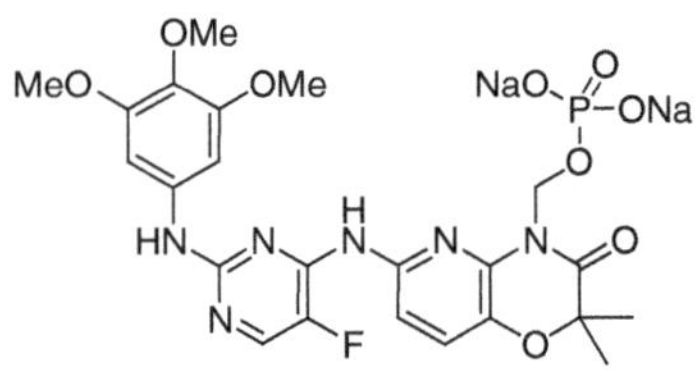

$t_{1/2}$	11–16 h*	V	–	**FOSTAMATINIB**
F	–	CL	–	Orally available prodrug of R406; spleen tyrosine kinase inhibitor;
pb	98%	Mwt	580.5	*R406 on oral dosing of prodrug; further metabolism: N-glucuronidation, 4-O-demethylation, then O-sulfation, O-glucuronidation
ur	–	PSA	197 Å^2	
		log P	1.48	

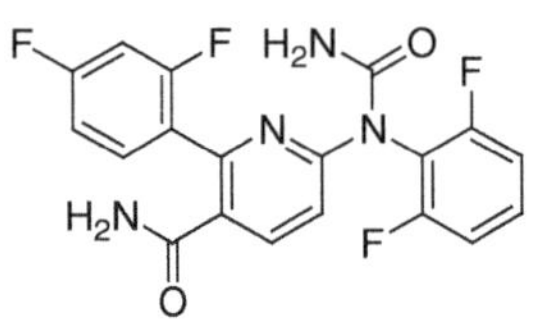

$t_{1/2}$	16–20 h	V	73	**VX-702**
F	–	CL	1	MAP kinase inhibitor, treatment of rheumatoid arthritis and acute coronary syndrome
pb	–	Mwt	404.3	
ur	66%	PSA	102 Å^2	
		log P	0.76	

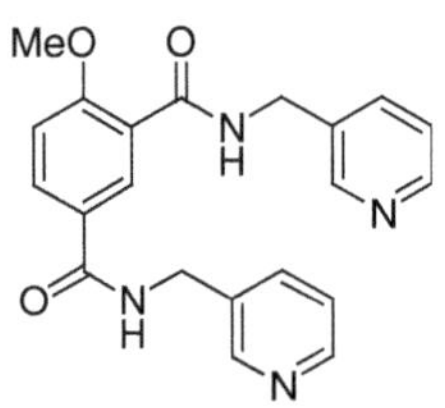

$t_{1/2}$	1–3 h	V	–	**PICOTAMIDE**
F	>46%	CL	–	Antithrombotic, platelet aggregation inhibitor
pb	–	Mwt	376.4	
ur	30%	PSA	93.2 Å^2	
		log P	1.88	

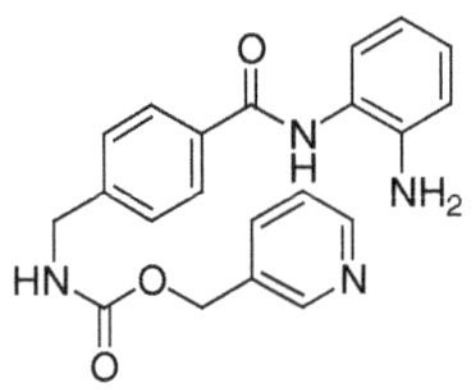

$t_{1/2}$	52±22 h	V	–	**ENTINOSTAT, MS-275**
F	85%	CL	–	Histone deacetylase inhibitor, antineoplastic
pb	–	Mwt	376.4	
ur	–	PSA	106 Å^2	
		log P	1.46	

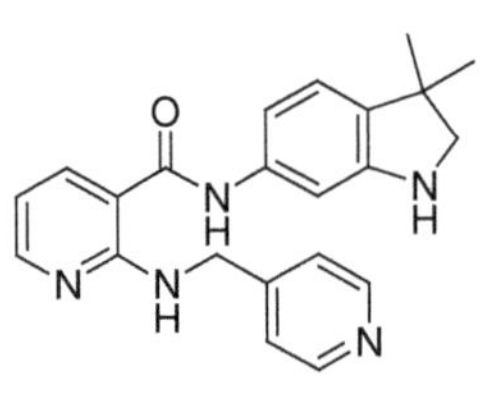

$t_{1/2}$	5–8 h	V	–	**MOTESANIB, AMG 706**
F	43% (dog)	CL	10–19	Antiangiogenic, antineoplastic
pb	–	Mwt	373.5	Metabolism: N-carbamoyl-glucuronidation, N-glucuronidation, indoline CH_2 hydroxylation, then oxidation to lactam, N-oxidation of 4-pyridyl, N-depicolinylation
ur	–	PSA	78.9 Å^2	
		log P	3.56	

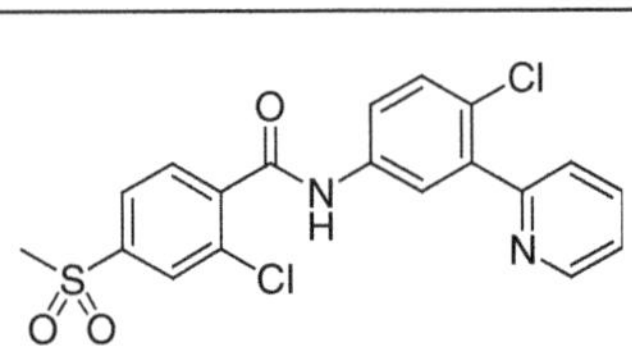

$t_{1/2}$	10–14 d	V	0.77**	**VISMODEGIB**
F	13–53%*	CL	–	*animals
pb	>94%	Mwt	421.3	**calculated
ur	–	PSA	84.5 Å^2	SMO antagonist, hedgehog signaling inhibitor, oncolytic
		log P	2.98	

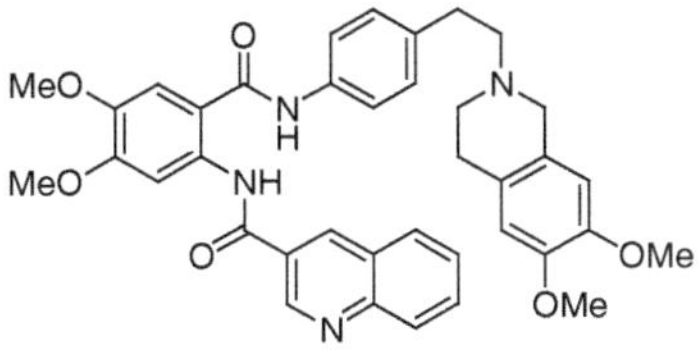

$t_{1/2}$	34±14 h*	V	4	TARIQUIDAR
F	–	CL	4.4±3.0	*iv
pb	–	Mwt	646.7	*P*-glycoprotein inhibitor,
ur	–	PSA	111 Å^2	adjunct antineoplastic
		log P	6.38	

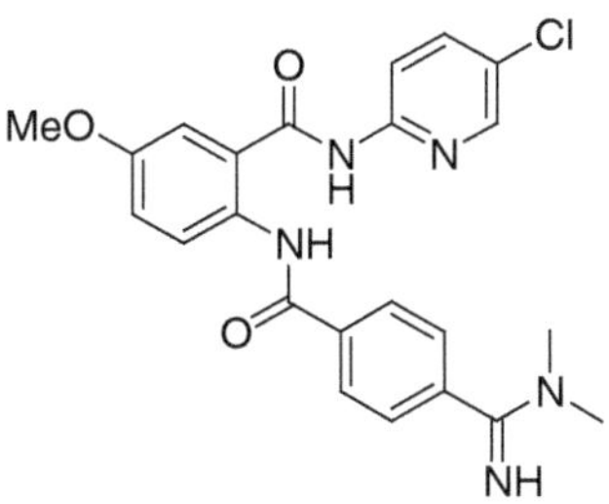

$t_{1/2}$	19 h	V	–	BETRIXABAN, PRT 054021
F	47%	CL	–	Factor Xa inhibitor,
pb	–	Mwt	451.9	antithrombotic
ur	–	PSA	107 Å^2	
		log P	2.93	

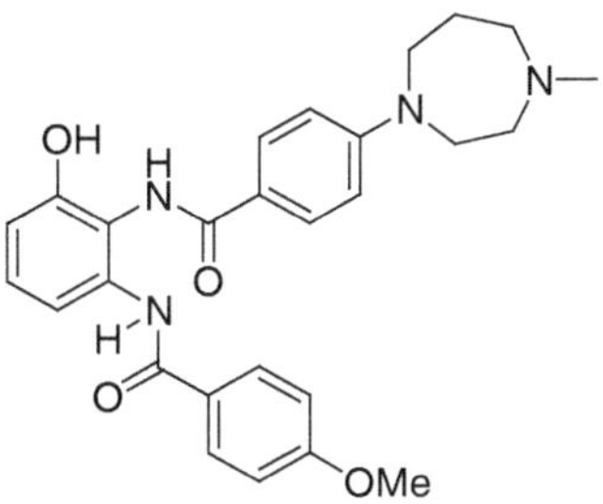

$t_{1/2}$	14–18 h	V	–	DAREXABAN, YM-150
F	–	CL	–	Factor Xa inhibitor,
pb	–	Mwt	474.6	antithrombotic
ur	–	PSA	94.1 Å^2	
		log P	4.08	

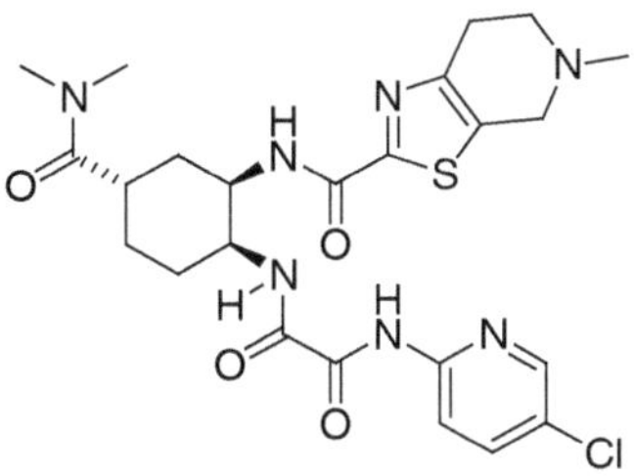

$t_{1/2}$	8–11 h	V	4.5	EDOXABAN
F	45%	CL	–	Factor Xa inhibitor,
pb	40–59%	Mwt	548.1	antithrombotic
ur	36–45%	PSA	165 Å^2	Metabolism: hydrolysis of oxal
		log P	1.24	pyridylamide,
				N-demethylation of piperidine,
				hydrolysis of dimethylamide

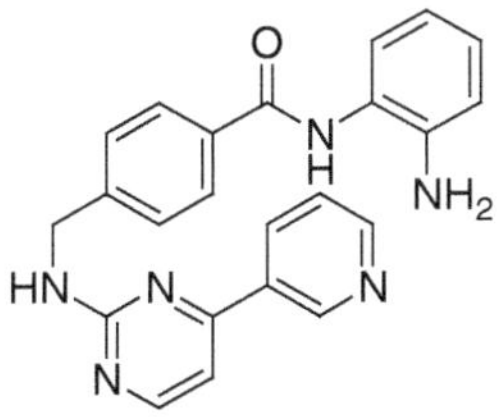

$t_{1/2}$	7–11 h	V	17	MGCD-0 103
F	42% (rat)	CL	24	Histone deacetylase inhibitor,
pb	–	Mwt	396.4	apoptosis inducer, oncolytic
ur	–	PSA	106 Å^2	
		log P	1.88	

(*continued overleaf*)

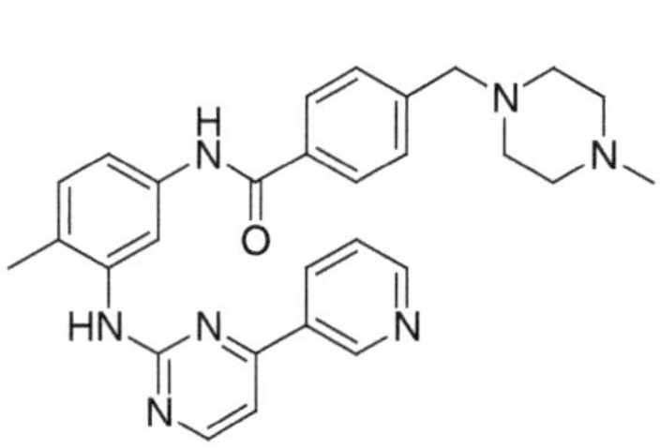

$t^1\!/_2$	18–22 h	V	5–6
F	98%	CL	2.3–3.2
pb	95%	Mwt	493.6
ur	5%	PSA	86.3 Å^2
		log P	2.48

IMATINIB
Antineoplastic, substrate and inhibitor of CYP3A4
Metabolism: N-demethylation (*N*-desmethyl: $t^1\!/_2$ 40 h), N-oxidation of pyridine, hydroxylation of CH_3C, oxidation of piperazine to lactam, conversion of CCH_2N to CO_2H

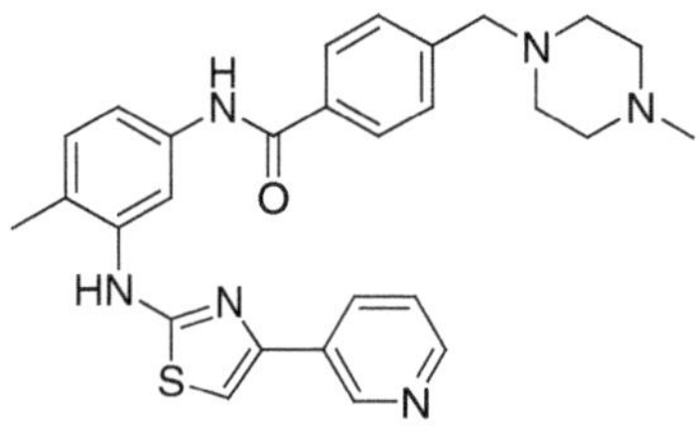

$t^1\!/_2$	13 h	V	17**
F	60%*	CL	14**
pb	93%	Mwt	498.6
ur	–	PSA	102 Å^2
		log P	2.89

MASITINIB
*cat
**dog
Antineoplastic
Metabolism: N-dealkylation, hydroxylation, amide hydrolysis

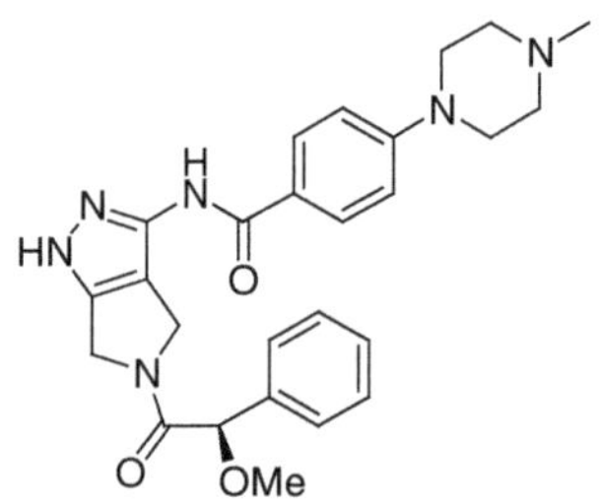

$t^1\!/_2$	18–26 h*	V	10–16
F	–	CL	5.5–7.7
pb	–	Mwt	474.6
ur	–	PSA	93.8 Å^2
		log P	2.34

DANUSERTIB
*iv
Antineoplastic
Metabolite: *N*-oxide of piperazine

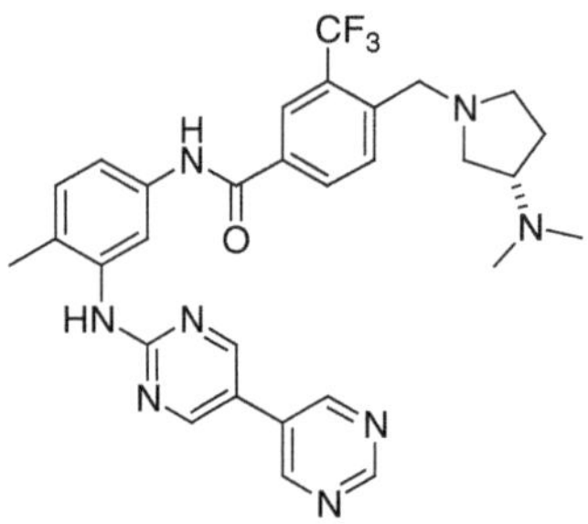

$t^1\!/_2$	1 h	V	–
F	32%	CL	–
pb	–	Mwt	576.6
ur	–	PSA	99.2 Å^2
		log P	3.03

BAFETINIB
Kinase inhibitor, antineoplastic

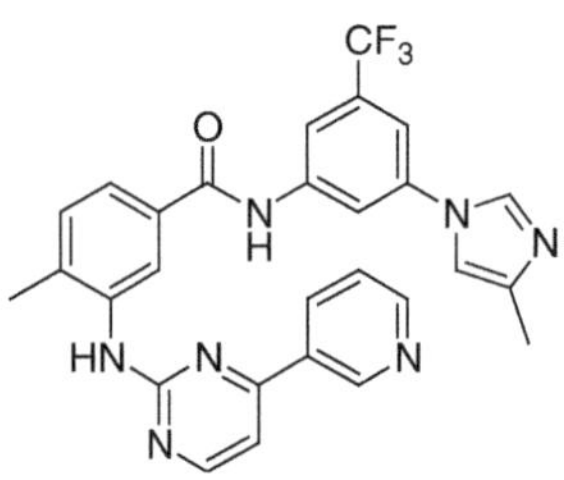

$t^1\!/_2$	17 h	V	2.7
F	30%	CL	7.5
pb	98%	Mwt	529.5
ur	<4%	PSA	97.6 Å^2
		log P	5.15

NILOTINIB
Kinase inhibitor, antineoplastic
Metabolism: hydroxylation of methyl in imidazole, then oxidation to carboxylic acid

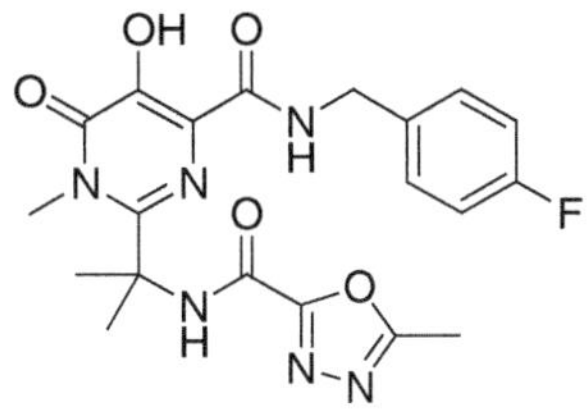

$t_{1/2}$	9 h	*V*	6–7	RALTEGRAVIR
F	32%	CL	8.3	Antiviral (HIV), does not cross bbb
pb	83%	Mwt	444.4	Metabolism: glucuronidation
ur	9%	PSA	150 Å^2	
		log *P*	−0.68	

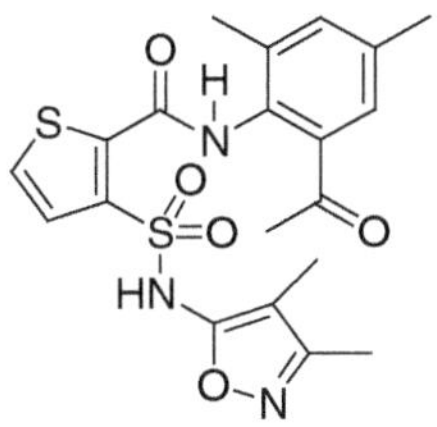

$t_{1/2}$	6–7 h	*V*	–	TBC-3711
F	>80%	CL	–	Endothelin ETA antagonist for treatment of pulmonary hypertension
pb	–	Mwt	447.5	
ur	–	PSA	118.4 Å^2	
		log *P*	2.34	

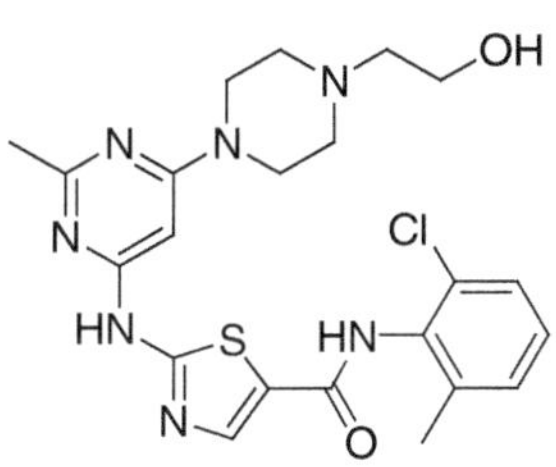

$t_{1/2}$	5–6 h	*V*	39	DASATINIB
F	34%*	CL	34**	*dog
pb	96%	Mwt	488.0	**monkey
ur	0.1%	PSA	135 Å^2	Kinase inhibitor, antineoplastic
		log *P*	0.14	Metabolism: hydroxylation by CYP3A4, N-oxidation, N-dealkylation, oxidation to carboxylic acid

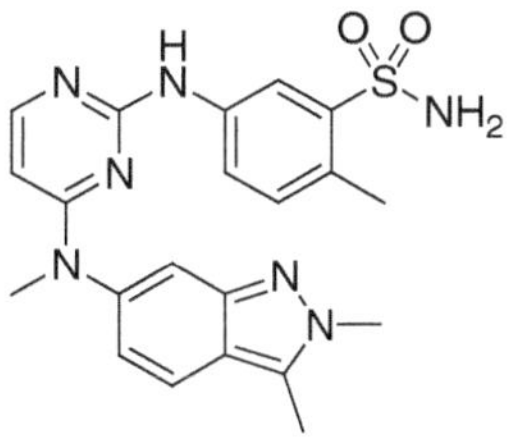

$t_{1/2}$	31 h	*V*	0.3	PAZOPANIB
F	14–39%	CL	1.4	Tyrosine kinase inhibitor, antiangiogenic
pb	>99%	Mwt	437.5	
ur	<4%	PSA	127 Å^2	
		log *P*	1.98	

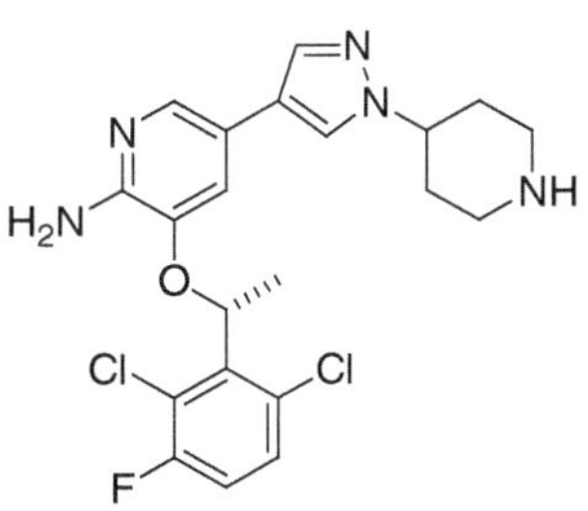

$t_{1/2}$	42 h	*V*	26	CRIZOTINIB
F	32–66%	CL/*F*	15–26	Kinase inhibitor, antineoplastic
pb	91%	Mwt	450.3	Metabolism: piperidine oxidation to lactam, O-dealkylation
ur	2.3%	PSA	78.0 Å^2	
		log *P*	3.55	

(*continued overleaf*)

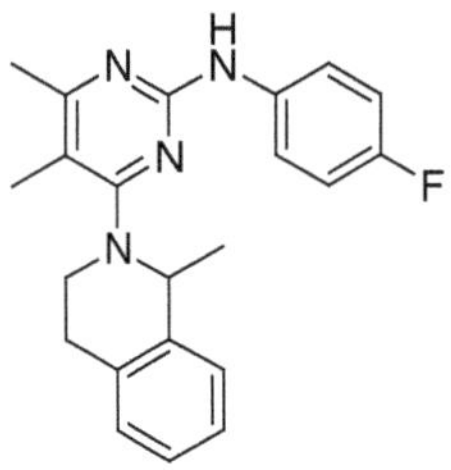

$t_{1/2}$	2.2–2.4 h	V	–	REVAPRAZAN, YH1885
F	40% (rat)	CL	–	Proton pump inhibitor,
pb	–	Mwt	362.4	antiulcerative
ur	0%	PSA	41.1 Å^2	
		log P	5.29	

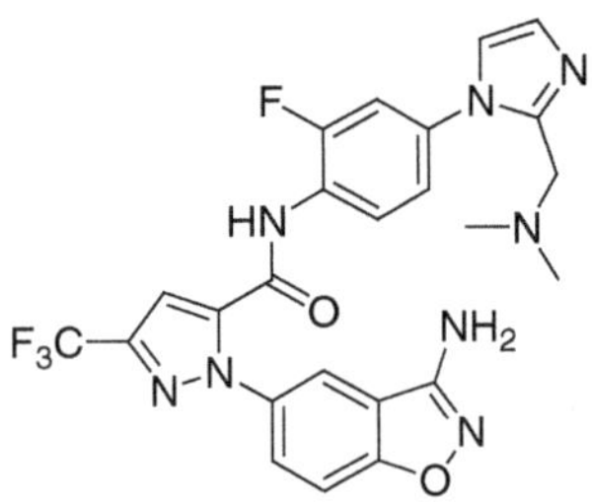

$t_{1/2}$	5.3 h*	V	3.4*	RAZAXABAN
F	84%*	CL	18*	*dog
pb	–	Mwt	528.5	Factor Xa inhibitor,
ur	–	PSA	120 Å^2	anticoagulant
		log P	1.87	Metabolism (rat, dog): reductive N—O bond cleavage, then glucuronidation of phenol, N-dealkylation

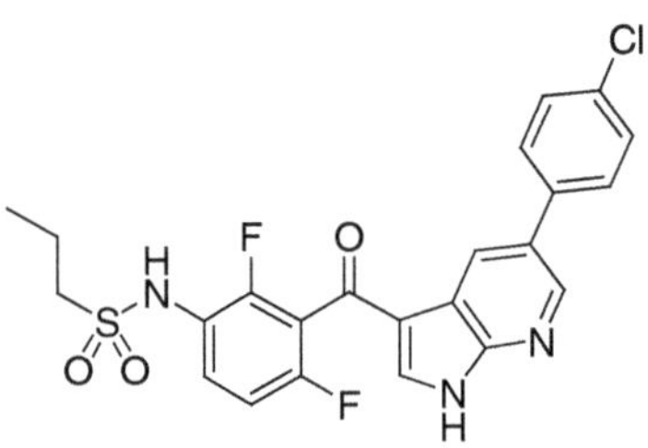

$t_{1/2}$	30–120 h	V	1.6	VEMURAFENIB
F	–	CL	0.3	B-raf enzyme inhibitor,
pb	>99%	Mwt	489.9	antineoplastic
ur	<1%	PSA	96.0 Å^2	
		log P	3.94	

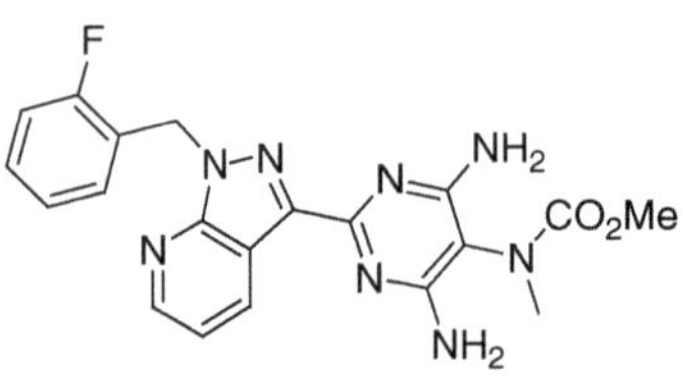

$t_{1/2}$	5–12 h	V/F	0.4	RIOCIGUAT
F	–	CL/F	0.4	Guanylate cyclase stimulator
pb	–	Mwt	422.4	for treatment of pulmonary
ur	–	PSA	138 Å^2	hypertension
		log P	–0.31	

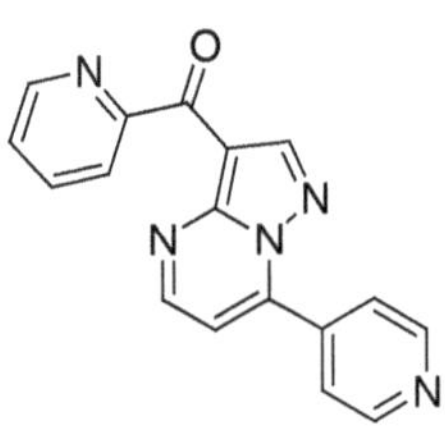

$t_{1/2}$	1 h (rat)	V	–	OCINAPLON
F	–	CL	–	Anxiolytic
pb	–	Mwt	301.3	Metabolite: 4-pyridyl-N-oxide
ur	–	PSA	73.0 Å^2	
		log P	2.16	

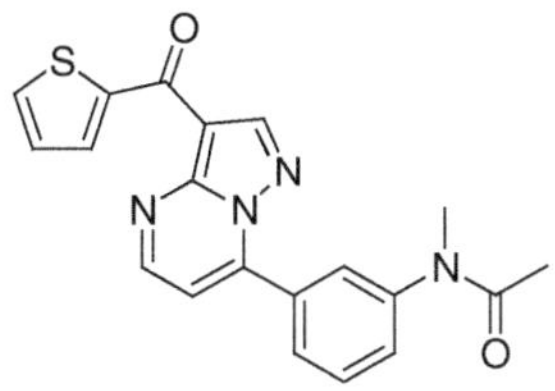	$t^1/_2$	1.5–2.0 h	V	–	INDIPLON
	F	–	CL	–	Hypnotic
	pb	–	Mwt	376.4	Metabolism: N-deacetylation,
	ur	<1%	PSA	95.8 Å^2	N-demethylation
			log P	0.83	
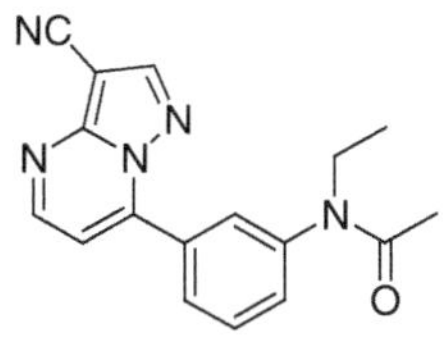	$t^1/_2$	1.1 h	V	1.3	ZALEPLON
	F	31±10%	CL	16	Hypnotic
	pb	60%	Mwt	305.3	Metabolites: 5-oxo
	ur	<1%	PSA	74.3 Å^2	(4-pyrimidone), *N*-desethyl
			log P	1.40	
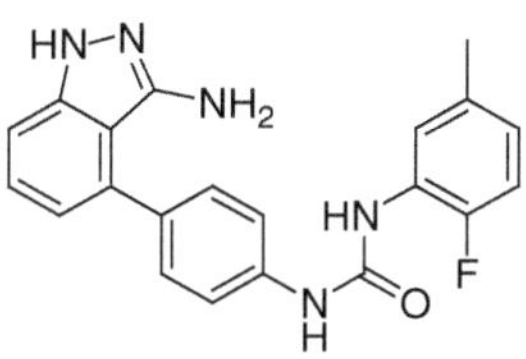	$t^1/_2$	17±5 h	V	1.2*	LINIFANIB, ABT-869
	F	10%*	CL	0.7±0.3	*monkey
	pb	99%	Mwt	375.4	VEGF inhibitor,
	ur	4%	PSA	95.8 Å^2	antiangiogenic, antineoplastic
			log P	4.34	Metabolism: oxidation of CH_3
					to CO_2H
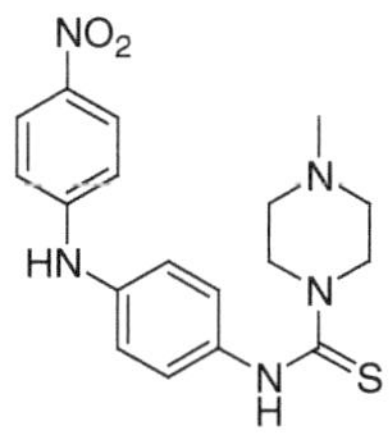	$t^1/_2$	3 h	V	–	AMOCARZINE
	F	<40%	CL	–	Anthelmintic
	pb	–	Mwt	371.5	Metabolism: N-oxidation
	ur	1.5%	PSA	108 Å^2	
			log P	2.32	
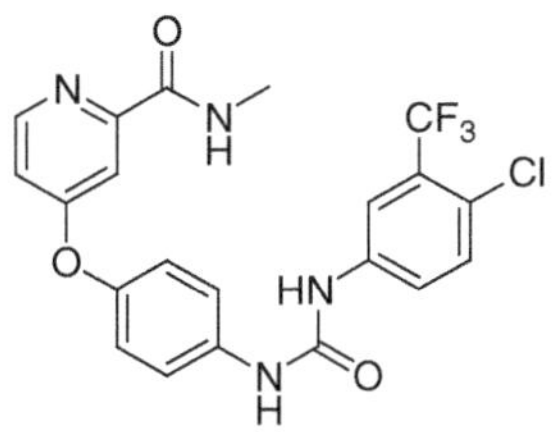	$t^1/_2$	25–48 h	V	0.7	SORAFENIB
	F	40–50%	CL	0.7 (rat)	Tyrosine kinase inhibitor,
	pb	99.5%	Mwt	464.8	antineoplastic
	ur	0%	PSA	92.4 Å^2	Active metabolite:
			log P	5.16	pyridine-*N*-oxide; further
					metabolism: glucuronidation
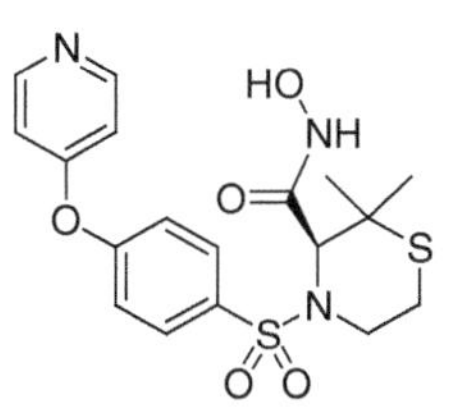	$t^1/_2$	1–5 h	V	–	PRINOMASTAT
	F	High	CL	–	Matrix metalloproteinase
	pb	69–90%	Mwt	423.5	inhibitor, antineoplastic
	ur	<2%	PSA	143 Å^2	Metabolite: pyridine-*N*-oxide
			log P	4.29	($t^1/_2$ 6–10 h)

(*continued overleaf*)

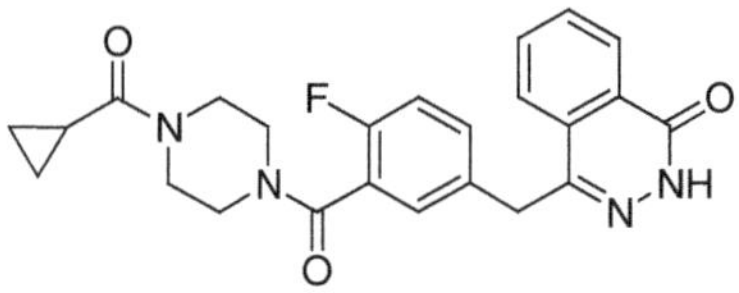	$t_{1/2}$ *F* pb ur	5–10 h 100%* – –	*V* CL Mwt PSA log *P*	16–87 0.3–2.4 434.5 82.1 Å^2 0.00	OLAPARIB *rat, dog PARP inhibitor, antineoplastic
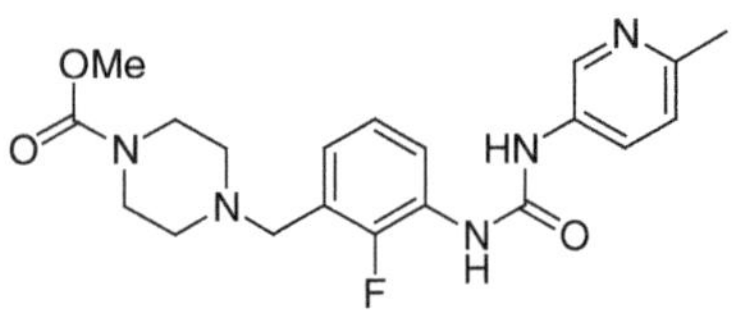	$t_{1/2}$ *F* pb ur	18–27 h 100% – –	*V* CL Mwt PSA log *P*	– – 401.4 86.8 Å^2 1.36	OMECAMTIV MECARBIL Cardiac myosin activator for treatment of heart failure
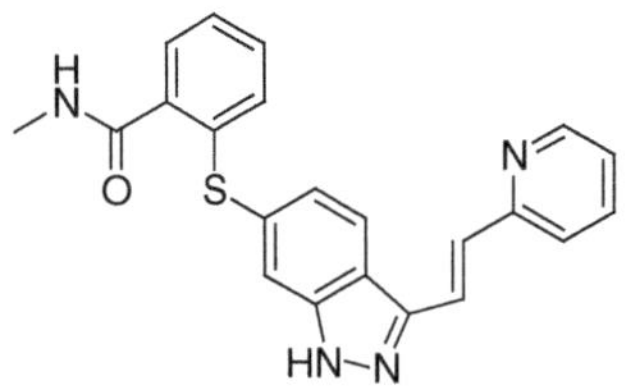	$t_{1/2}$ *F* pb ur	2–5 h 58% 99.6% <1%	*V* CL/*F* Mwt PSA log *P*	1.9 6 386.5 96.0 Å^2 4.15	AXITINIB Antineoplastic Metabolism: glucuronidation
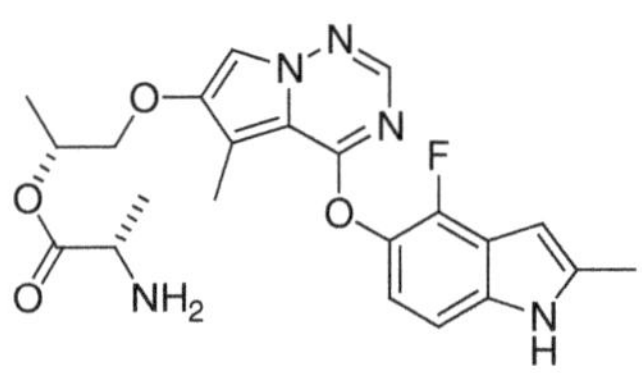	$t_{1/2}$ *F* pb ur	14 h* 55–97%* – 0%	*V* CL Mwt PSA log *P*	– – 370.4 84.7 Å^2 2.48	BRIVANIB ALANINATE *brivanib on oral dosing of prodrug Prodrug of brivanib with improved solubility Kinase inhibitor, antineoplastic Metabolism: O-dealkylation, sulfation,
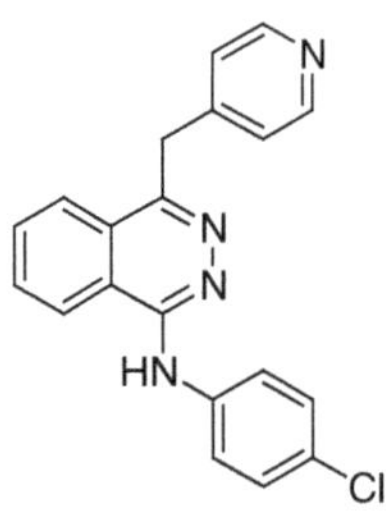	$t_{1/2}$ *F* pb ur	4.6±1.1 h 18–57% – 0.4%	*V* CL Mwt PSA log *P*	3.3 – 346.8 50.7 Å^2 3.80	VATALANIB Antiangiogenic tyrosine kinase inhibitor Metabolism: pyridine N-oxidation, aromatic hydroxylation, hydroxylation of CH_2
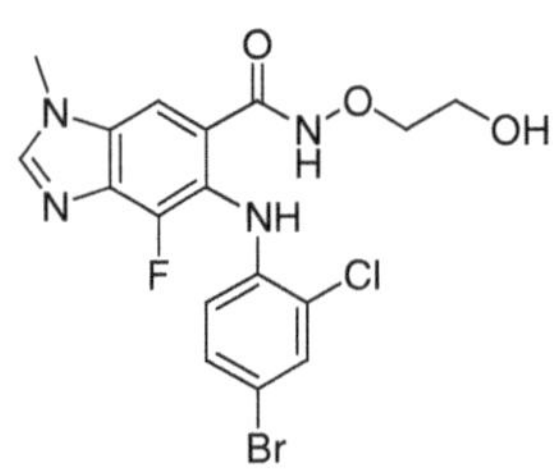	$t_{1/2}$ *F* pb ur	4–14 h – 97% –	*V*/*F* CL/*F* Mwt PSA log *P*	0.9–4.6 8–31 457.7 88.4 Å^2 5.55	SELUMETINIB, AZD 6244 MEK inhibitor Metabolism: N-demethylation

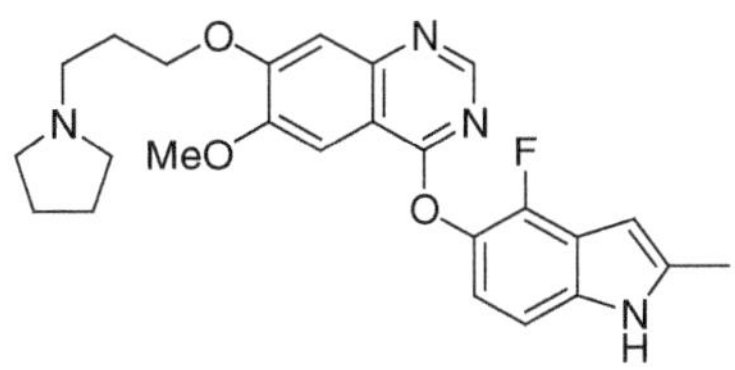

$t^{1}/_{2}$	12–35 h	*V*	–
F	>40% (rat)	CL	–
pb	–	Mwt	450.5
ur	–	PSA	72.5 Å^2
		log *P*	4.80

CEDIRANIB
VEGF tyrosine kinase inhibitor, antineoplastic, crosses bbb
Metabolism: N-glucuronidation (at pyrrolidine)

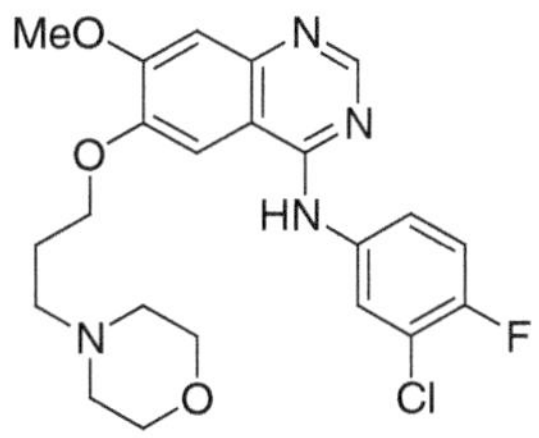

$t^{1}/_{2}$	34–65 h	*V*	23
F	59%	CL	7–12
pb	91%	Mwt	446.9
ur	<0.5%	PSA	68.7 Å^2
		log *P*	4.11

GEFITINIB
Kinase inhibitor, antineoplastic
Metabolism (CYP3A4): demethylation, oxidative defluorination, N-dealkylation of tertiary amine

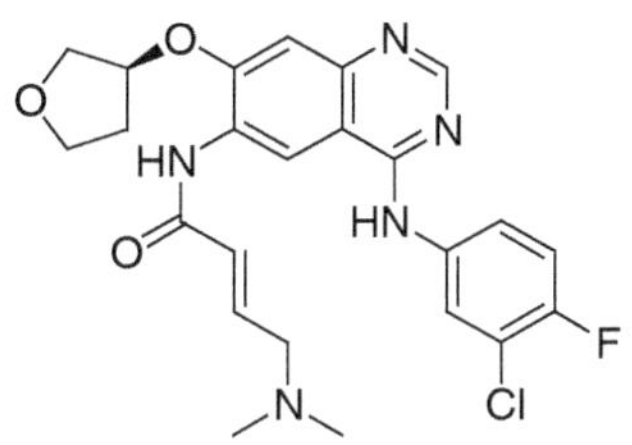

$t^{1}/_{2}$	34 h	*V*/*F*	19–73
F	–	CL/*F*	11
pb	–	Mwt	485.9
ur	–	PSA	88.6 Å^2
		log *P*	3.59

AFATINIB, BIBW 2992
Kinase inhibitor, antineoplastic

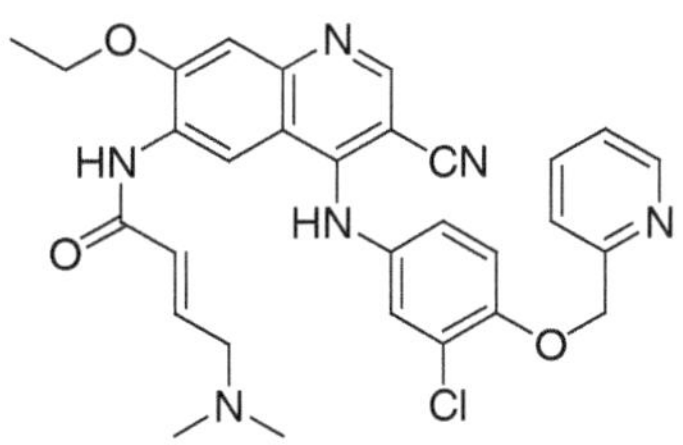

$t^{1}/_{2}$	10–17 h	*V*	63–95
F	–	CL/*F*	47–105
pb	–	Mwt	557.0
ur	–	PSA	112 Å^2
		log *P*	5.46

NERATINIB
Tyrosine kinase inhibitor, oncolytic

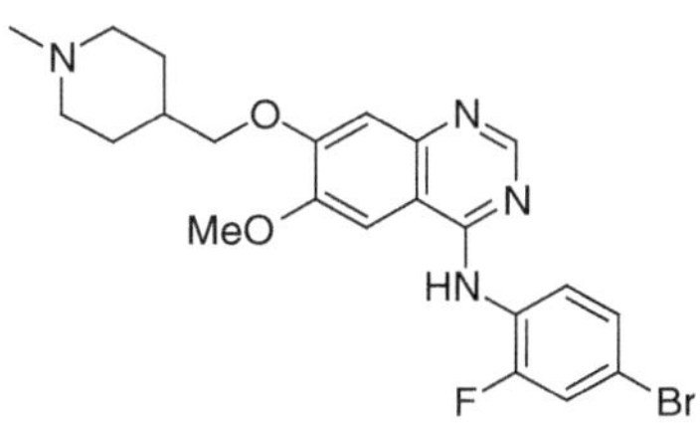

$t^{1}/_{2}$	120 h	*V*	115
F	Good	CL	3.4
pb	92–96%	Mwt	475.4
ur	<5%	PSA	59.5 Å^2
		log *P*	5.51

VANDETANIB
Kinase inhibitor, antineoplastic
Metabolism: N-oxidation, N-demethylation, glucuronidation

(continued overleaf)

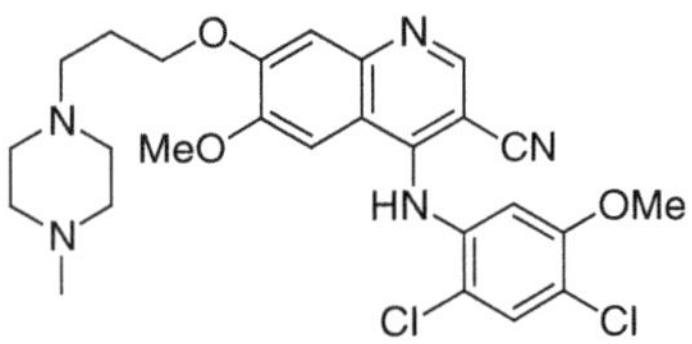

$t_{1/2}$	17–39 h	V	18.6	BOSUTINIB
F	18%	CL/F	38–64	Tyrosine kinase inhibitor, antineoplastic
pb	–	Mwt	530.5	
ur	1%	PSA	82.9 Å^2	
		log P	5.48	

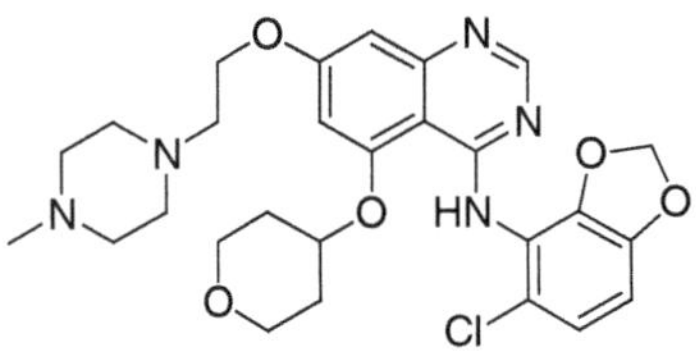

$t_{1/2}$	40 h	V/F	>30	SARACATINIB
F	–	CL/F	22	Antineoplastic
pb	–	Mwt	542.0	
ur	–	PSA	90.4 Å^2	
		log P	2.74	

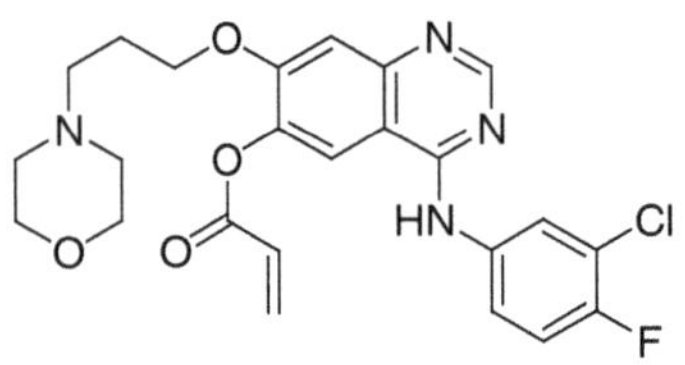

$t_{1/2}$	4 h	V	3–10	CANERTINIB
F	33%	CL	68	Irreversible tyrosine kinase inhibitor, antineoplastic
pb	–	Mwt	485.9	
ur	–	PSA	88.6 Å^2	
		log P	3.65	

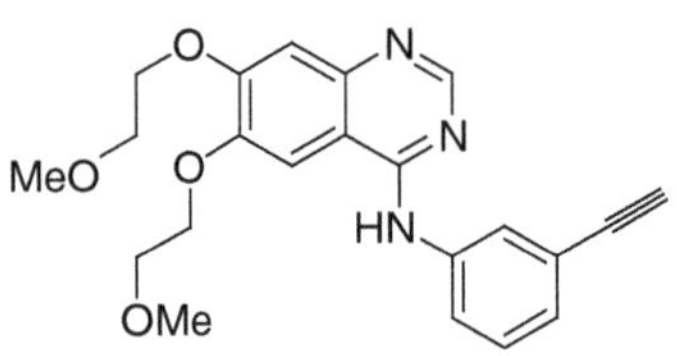

$t_{1/2}$	36 h	V	3.6	ERLOTINIB
F	60–100%	CL	1.1	Kinase inhibitor, antineoplastic
pb	95%	Mwt	393.4	Metabolism: O-demethylation, aromatic hydroxylation, oxidation of ethynylbenzene to arylacetic acid
ur	<2%	PSA	74.7 Å^2	
		log P	2.39	

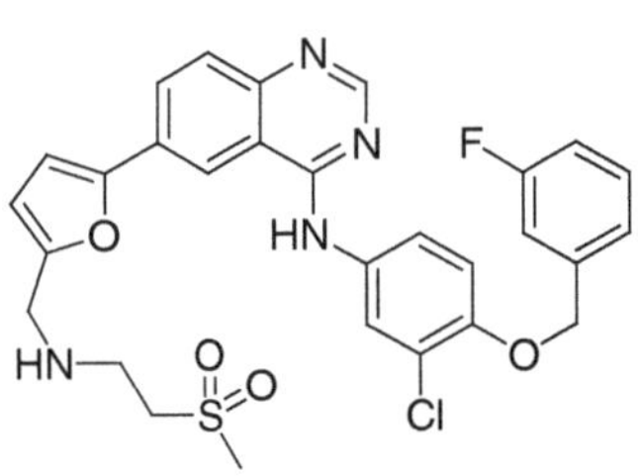

$t_{1/2}$	24 h	V	34	LAPATINIB
F	50–100%	CL	–	Kinase inhibitor, antineoplastic
pb	99%	Mwt	581.1	
ur	<2%	PSA	115 Å^2	
		log P	5.14	

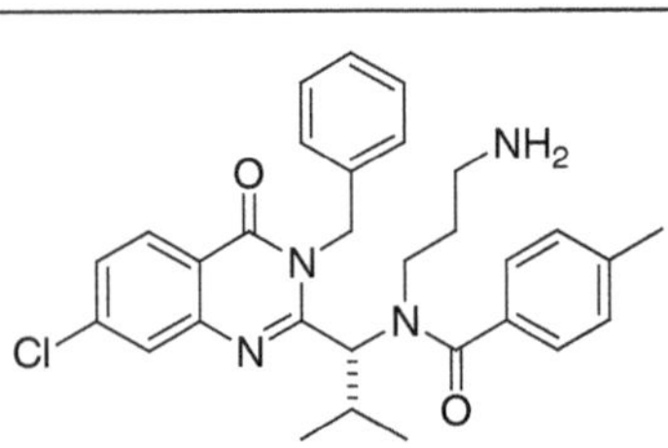

$t_{1/2}$	29–54 h	V	–	ISPINESIB
F	–	CL	0.5–6.3	Kinesin spindle protein inhibitor, antineoplastic
pb	81–96%	Mwt	517.1	
ur	–	PSA	79.0 Å^2	
		log P	5.20	

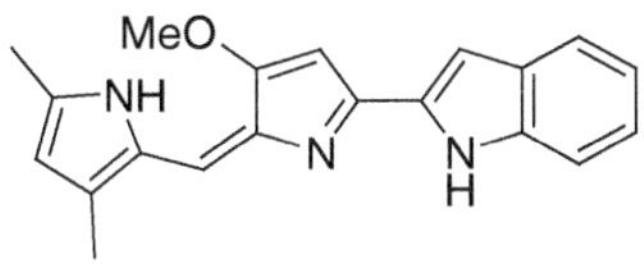	$t_{1/2}$ F pb ur	39–60 h (iv) – – –	V CL Mwt PSA log P	23–52 32–72 317.4 53.2 Å^2 3.52	OBATOCLAX Bcl-2 inhibitor, antineoplastic
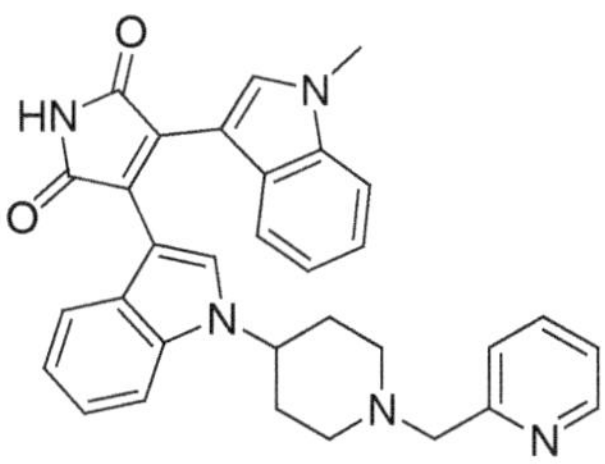	$t_{1/2}$ F pb ur	40–54 h* – 95% –	V CL Mwt PSA log P	– 11.2 515.6 62.3 Å^2 4.43	ENZASTAURIN *parent + metabolites Kinase inhibitor Metabolism: N-depicolylation, N-demethylation
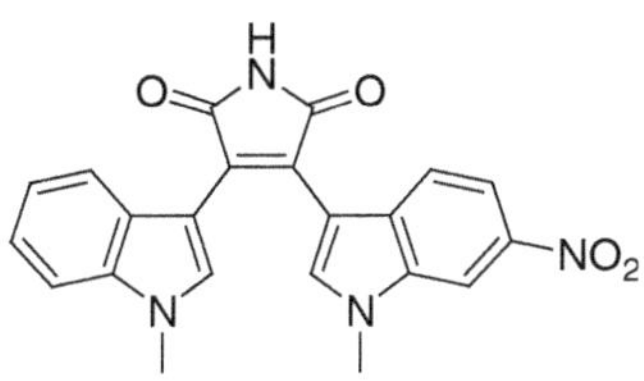	$t_{1/2}$ F pb ur	9 h 12%* – <1%	V CL Mwt PSA log P	– – 400.4 92.0 Å^2 3.73	MKC-1, RO 31–7453 *monkey Antimitotic, antineoplastic Metabolism: N-demethylation
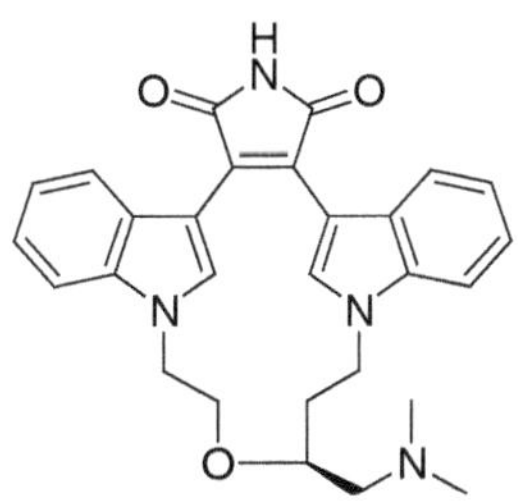	$t_{1/2}$ F pb ur	5 h – – <1%	V/F CL/F Mwt PSA log P	15* 33 468.6 58.6 Å^2 3.92	RUBOXISTAURIN Protein kinase C inhibitor for treatment of diabetic retinopathy Metabolites: *N*-desmethyl ($t_{1/2}$ 24 h), 6-hydroxyindole
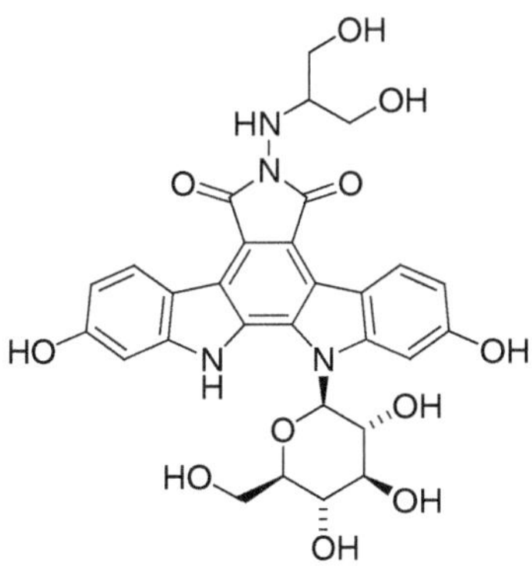	$t_{1/2}$ F pb ur	20–25 h – – 1–4%	V CL Mwt PSA log P	17–23 20–24 608.6 241 Å^2 0.75	EDOTECARIN Antineoplastic

(*continued overleaf*)

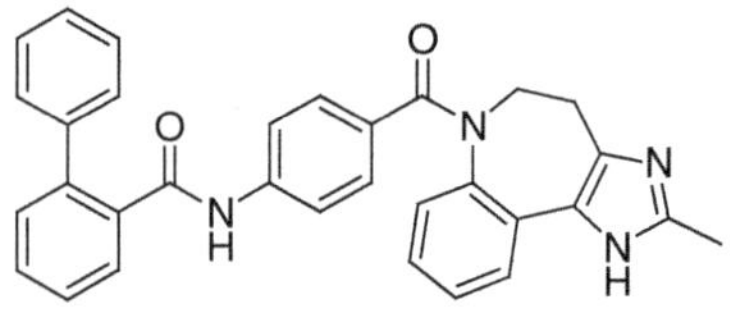	$t_{1/2}$ 6 h *F* 34–55% pb 99% ur 1% (iv)	*V* 0.8 CL 3–7 Mwt 498.6 PSA 78.1 Å^2 log *P* 4.40	CONIVAPTAN Vasopressin antagonist, due to CYP-inhibition the development of an oral formulation was discontinued and an iv formulation was chosen
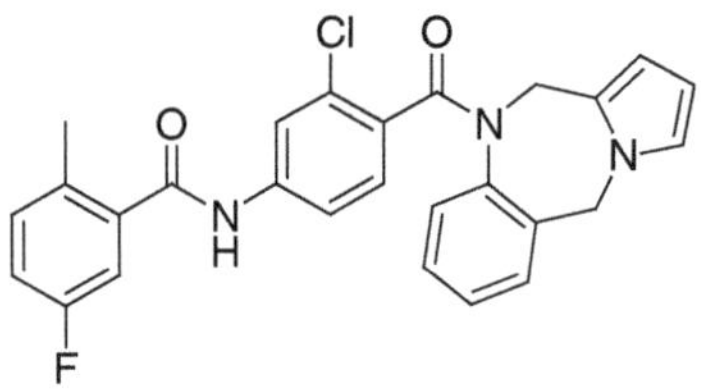	$t_{1/2}$ 7–10 h *F* – pb – ur –	*V* – CL – Mwt 473.9 PSA 49.4 Å^2 log *P* 7.23	LIXIVAPTAN Vasopressin V_2 antagonist Metabolism: oxidation of pyrrole to lactam
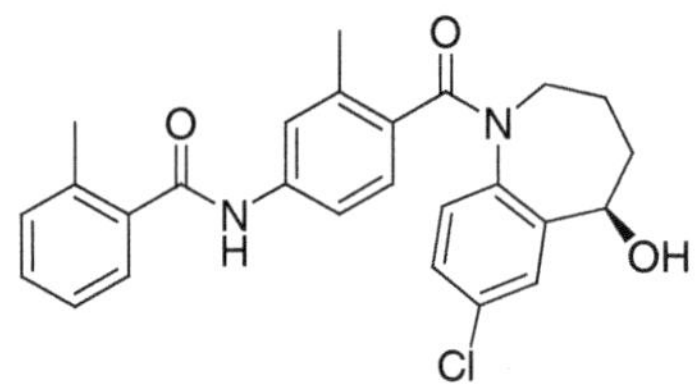	$t_{1/2}$ 8–12 h *F* 56% pb 98% ur <1%	*V* 3 CL – Mwt 448.9 PSA 69.6 Å^2 log *P* 4.33	TOLVAPTAN Vasopressin antagonist for treatment of hyponatremia Metabolism: hydroxylation

$t_{1/2}$, plasma half-life; *F*, oral bioavailability; pb, plasma protein binding; ur, excretion of unchanged drug in urine; V, volume of distribution; CL, clearance; Mwt, molecular weight; PSA, polar surface area; CYP, cytochrome P; MAP, mitogen-activated protein; MEK, mitogen-activated protein kinase kinase; PARP, poly(ADP-ribose)polymerase; VEGF, vascular endothelial growth factor.

57
Imidazoles

Imidazole is a weak organic base (pK_a 7.0) and an important structural element of various drugs. Histamine, biosynthesized by decarboxylation of the amino acid histidine, is the natural agonist of four receptors, and selective agonists and antagonists have been developed for each of them (Scheme 57.1). Histamine H_1 antagonists (the classic antihistaminics, Chapter 46), used as antipruritics, do not usually contain an imidazole substructure, but an aliphatic amine instead. Many H_2 antagonists, however, contain an imidazole as the basic pharmacophore. These drugs suppress the production of gastric acid and can be used for the treatment of esophageal reflux or gastric ulcers. H_3 antagonists (imidazole derivatives or amines) are being considered as appetite suppressants and for the treatment of attention deficit disorders. Potential indications for H_4 antagonists (amines) include asthma and allergic rhinitis, that is, the same as for H_1 antagonists [1].

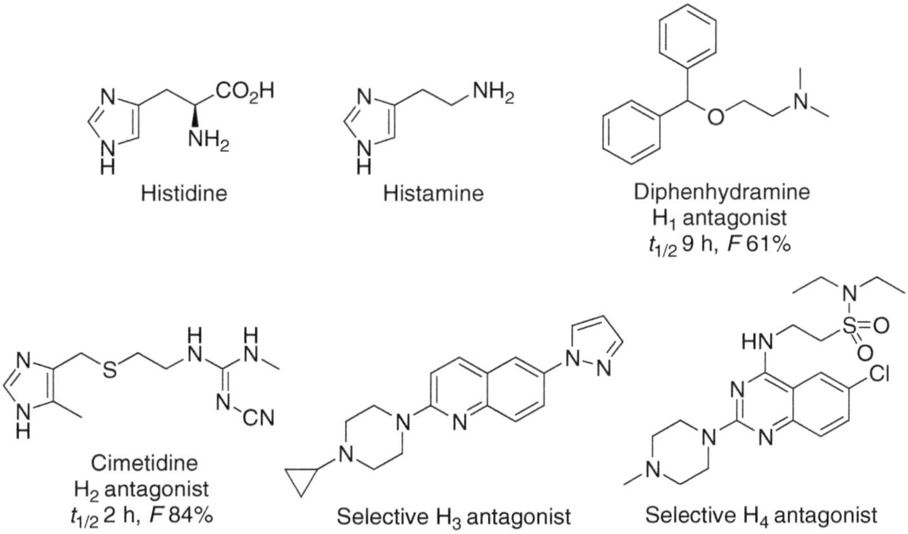

Scheme 57.1 Histamine receptor agonists and antagonists [1,2].

Lead Optimization for Medicinal Chemists: Pharmacokinetic Properties of Functional Groups and Organic Compounds, First Edition. Florencio Zaragoza Dörwald.

Further imidazole-containing drugs are the α_2-adrenergic agonists (e.g., dexmedetomidine) and the muscarinic acetylcholine agonist pilocarpine. One group of serotonin 5-HT_3 antagonists, the "setrons," also contain imidazole as the pharmacophore. These drugs are mainly used as antiemetics for patients undergoing chemotherapy (Table 57.1).

The trichomonacidal properties of 2-nitroimidazole (azomycin) were discovered in 1955 and led to the development of numerous nitroimidazoles as antiprotozoals. Metronidazole, for instance, is active against a wide variety of anaerobic bacteria and anaerobic protozoal parasites. Nitroimidazoles are less toxic to aerobic than to anaerobic organisms. *In vivo*, these compounds are reduced by ferredoxin to hydroxylamines and aminoimidazoles, which can damage DNA and interfere with the multienzyme complex pyruvate dehydrogenase, which is critical for the production of acetyl-CoA under anaerobic conditions [3].

Lipophilic imidazoles and 1,2,4-triazoles can inhibit the fungal sterol 14-α-demethylase and, thereby, impair the biosynthesis of ergosterol. These compounds are used to suppress the growth of fungi for the treatment of fungal infections.

Benzimidazoles such as thiabendazole, discovered in 1961, are highly active against gastrointestinal nematodes and are used as anthelmintics for humans and in veterinary medicine. Benzimidazoles bind strongly and selectively to parasite β-tubulin and, thereby, inhibit microtubule polymerization in nematodes. Their low oral bioavailability is an advantage because the target organ of these drugs is the gastrointestinal tract.

Imidazoles form strong hydrogen bonds, have a large polar surface area, and do not readily cross the blood–brain barrier (bbb). Typical metabolic transformations include N-methylation and N-glucuronidation. Sterically unencumbered, lipophilic imidazoles (and pyridines) may coordinate strongly to metals (e.g., iron in cytochrome P (CYP) enzymes) and act as CYP inhibitors. Cimetidine, for instance, inhibits both CYP2D6 and CYP2E1.

References

1. Smits, R.A. *et al.* (2010) Synthesis and QSAR of quinazoline sulfonamides as highly potent human histamine H_4 receptor inverse agonists. *J. Med. Chem.*, **53**, 2390–2400.
2. Zaragoza, F., Stephensen, H., Peschke, B., and Rimvall, K. (2005) 2-(4-Alkylpiperazin-1-yl)quinolines as a new class of imidazole-free histamine H_3 receptor antagonists. *J. Med. Chem.*, **48**, 306–311.
3. (a) Kim, P., Zhang, L., Manjunatha, U.H., Singh, R., Patel, S., Jiricek, J., Keller, T.H., Boshoff, H.I., Barry, C.E., and Dowd, C.S. (2009) Structure–activity relationships of antitubercular nitroimidazoles. 1. Structural features associated with aerobic and anaerobic activities of 4- and 5-nitroimidazoles. *J. Med. Chem.*, **52**, 1317–1328; (b) Kim, P. *et al.* (2009) Structure–activity relationships of antitubercular nitroimidazoles. 2. Determinants of aerobic activity and quantitative structure–activity relationships. *J. Med. Chem.*, **52**, 1329–1344; (c) Singh, S., Bharti, N., and Mohapatra, P.P. (2009) Chemistry and biology of synthetic and naturally occurring antiamoebic agents. *Chem. Rev.*, **109**, 1900–1947.

Table 57.1 Imidazoles and related compounds (see also Table 58.1 (triazoles)). *V* in l kg^{-1}, CL in ml min^{-1} kg^{-1}, Mwt in g mol^{-1}.

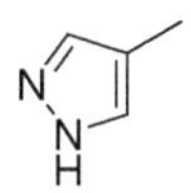	*t*½ *F* pb ur	15–40 h 100% Low 1–4%	*V* CL Mwt PSA log *P*	0.6–1.0 – 82.1 28.7 Å^2 1.10	FOMEPIZOLE Alcohol dehydrogenase inhibitor, antidote to methanol, and ethylene glycol poisoning Metabolites: 4-(hydroxymethyl)pyrazole, 4-carboxypyrazole, *N*-glucuronides
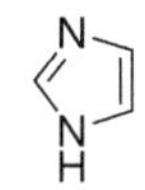	*t*½ *F* pb ur	2–3 h 100% 5–15% –	*V* CL Mwt PSA log *P*	– – 68.1 28.7 Å^2 −0.10	IMIDAZOLE
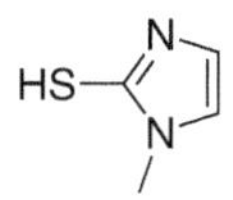	*t*½ *F* pb ur	3–6 h 95% – 7%	*V* CL Mwt PSA log *P*	0.6–0.9 1.4–3.7 114.2 47.4 Å^2 −0.34	METHIMAZOLE, THIAMAZOLE Antihyperthyroid Metabolite: 3-methyl-2-thiohydantoin (*t*½ 9 h)
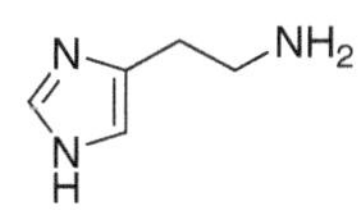	*t*½ *F* pb ur	3 min – – 2–3%	*V* CL Mwt PSA log *P*	– – 111.2 54.7 Å^2 −0.70	HISTAMINE Natural neurotransmitter Metabolism: N-methylation of imidazole, oxidation to imidazolylacetic acid
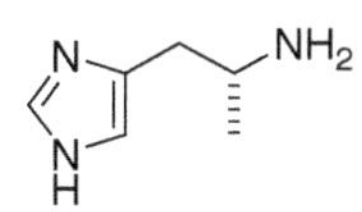	*t*½ *F* pb ur	3 min – – –	*V* CL Mwt PSA log *P*	– – 125.2 54.7 Å^2 −0.35	(*R*)-α-METHYLHISTAMINE Synthetic histamine H_3 agonist
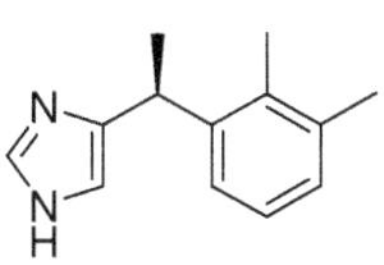	*t*½ *F* pb ur	2–5 h 12–20% 94% 0%	*V* CL Mwt PSA log *P*	1.3–1.6 8–11 200.3 28.7 Å^2 2.79	DEXMEDETOMIDINE α_2-receptor agonist, sedative, analgesic, CYP2D6 inhibitor Metabolism: N-glucuronidation, hydroxylation, oxidation of aromatic CH_3 to CO_2H
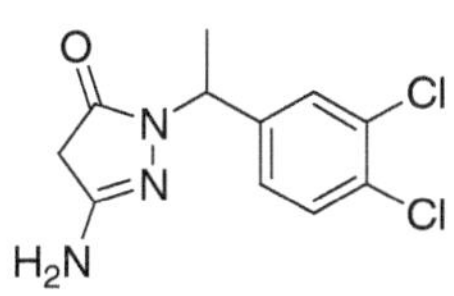	*t*½ *F* pb ur	10–17 h – 65% 5%	*V* CL Mwt PSA log *P*	1 – 272.1 58.7 Å^2 2.10	MUZOLIMINE Diuretic Metabolism: hydrolysis of amide, then decarboxylation

(*continued overleaf*)

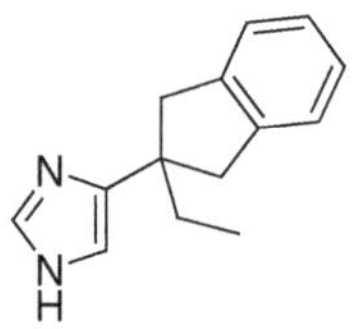

$t^{1}/_{2}$	1.7–2.0 h	V	3.0–3.5	ATIPAMEZOLE
F	2%	CL	18–25	α_2 Adrenoceptor antagonist,
pb	–	Mwt	212.3	antidote to α_2
ur	–	PSA	28.7 Å^2	adrenoceptor-induced sedation
		log P	2.40	

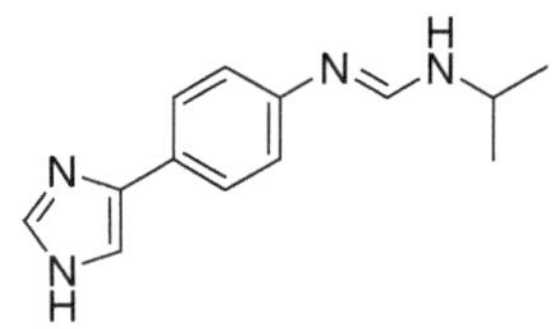

$t^{1}/_{2}$	9–19 h	V	–	MIFENTIDINE
F	–	CL	10–12	Histamine H_2 antagonist,
pb	–	Mwt	228.3	antiulcerative
ur	13%	PSA	53.1 Å^2	Metabolism: hydrolysis to
		log P	1.98	formamide and aniline,
				oxidation to urea

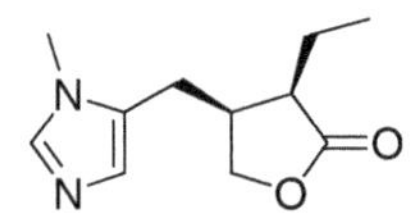

$t^{1}/_{2}$	0.8–1.4 h	V	2.9	PILOCARPINE
F	100%	CL	–	Antiglaucoma agent, miotic,
pb	0%	Mwt	208.3	diaphoretic (stimulates
ur	16%	PSA	44.1 Å^2	sweating and salivation)
		log P	−0.24	Metabolism: lactone hydrolysis

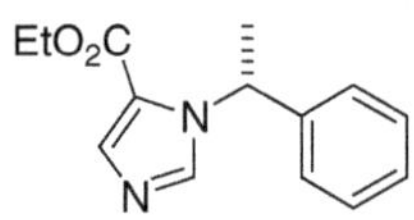

$t^{1}/_{2}$	4–6 h	V	2–5	ETOMIDATE
F	Low	CL	12–25	Hypnotic
pb	77%	Mwt	244.3	Metabolism: hydrolysis of ester,
ur	2%	PSA	44.1 Å^2	N-dealkylation
		log P	3.05	

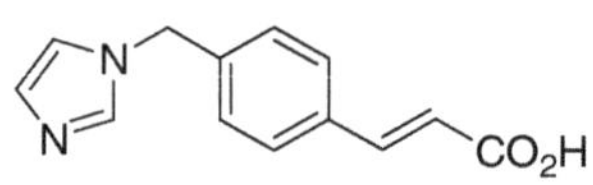

$t^{1}/_{2}$	1–2 h	V	0.7±0.4	OZAGREL
F	–	CL	10±6	Inhibitor of thromboxane
pb	–	Mwt	228.3	synthase, antithrombotic
ur	42±16%	PSA	55.1 Å^2	
		log P	0.53	

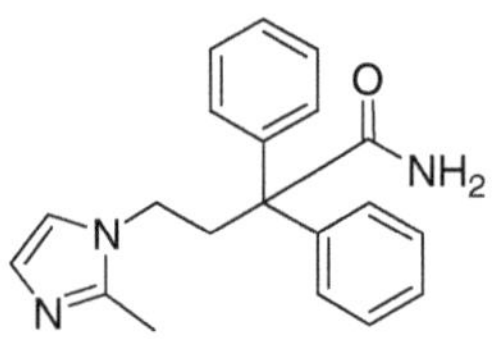

$t^{1}/_{2}$	3.0±0.3 h	V	1.9±0.4	IMIDAFENACIN
F	58%	CL	7.6±1.6	Muscarinic antagonist
pb	88%	Mwt	319.4	Metabolism:
ur	8%	PSA	56.0 Å^2	N-glucuronidation,
		log P	2.42	dihydroxylation and oxidative
				degradation of imidazole

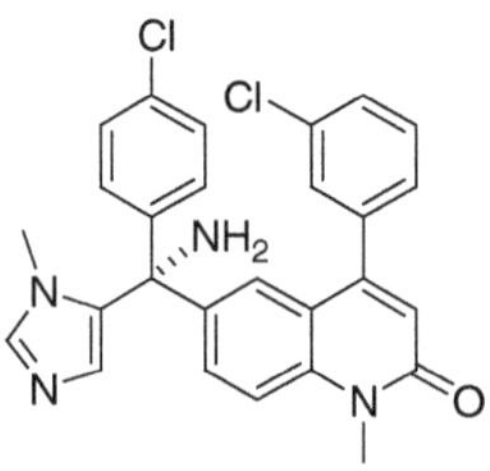

$t^{1}/_{2}$	2–3 h	V/F	7–11	TIPIFARNIB, R115777
F	Yes	CL	–	Farnesyltransferase inhibitor,
pb	99%	Mwt	489.4	antineoplastic
ur	–	PSA	64.2 Å^2	
		log P	4.01	

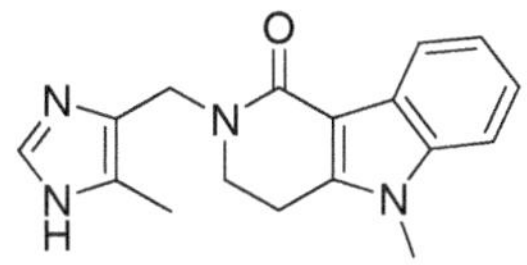	$t_{1/2}$ 1.5–1.7 h *F* 47–60% pb 82% ur 6%	*V* 1.1 CL 8.7 Mwt 294.4 PSA 53.9 Å^2 log *P* 0.88	ALOSETRON Treatment of irritable bowel syndrome Metabolism: aromatic 6-hydroxylation (ortho to indole N) and 7-hydroxylation, then O-sulfation and O-glucuronidation, N-demethylation, hydroxylation of CMe, hydroxylation of imidazole; withdrawn in 2000 for causing ischemic colitis and constipation
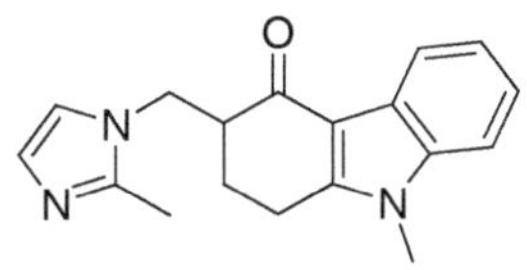	$t_{1/2}$ 3.5±1.2 h *F* 62±15% pb 73±2% ur 5%	*V* 1.9±0.5 CL 5.9±2.6 Mwt 293.4 PSA 39.8 Å^2 log *P* 1.56	ONDANSETRON Antiemetic, 5-HT$_3$ antagonist Metabolism: aromatic 7- and 8-hydroxylation
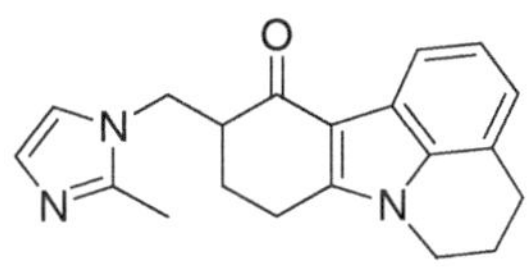	$t_{1/2}$ 1.6–1.9 h *F* 87% (rat) pb – ur –	*V* – CL – Mwt 319.4 PSA 39.8 Å^2 log *P* 2.65	CILANSETRON 5-HT$_3$ antagonist for treatment of irritable bowel syndrome Metabolism: benzylic hydroxylation
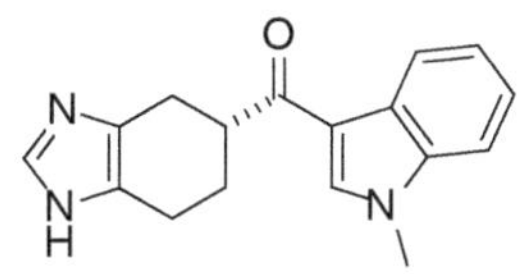	$t_{1/2}$ 5.8±1.2 h *F* – pb 89–91% ur 16–22%*	*V* 2 CL 4.5–5.0 Mwt 279.3 PSA 50.7 Å^2 log *P* 1.99	RAMOSETRON *iv 5-HT$_3$ antagonist, antiemetic Metabolism: N-demethylation, hydroxylation
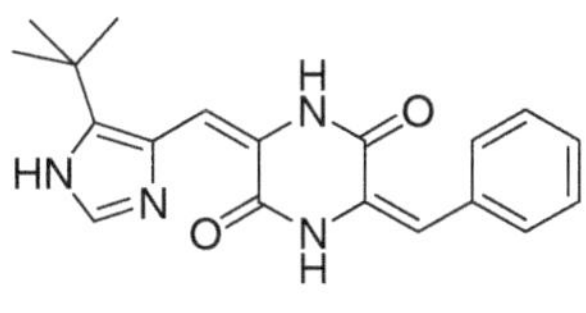	$t_{1/2}$ 6.4 h (iv) *F* – pb – ur –	*V* 3.2 CL 8 Mwt 336.4 PSA 86.9 Å^2 log *P* 2.66	PLINABULIN Tubulin polymerization inhibitor, vascular disrupting agent, oncolytic
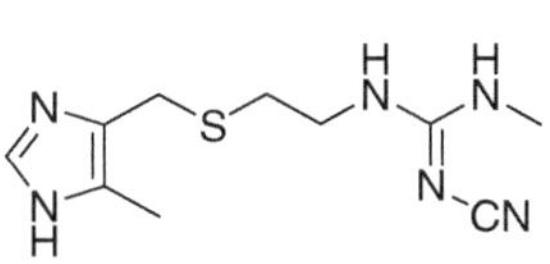	$t_{1/2}$ 2.0±0.3 h *F* 84±13% pb 19% ur 62±20%	*V* 1.0±0.2 CL 8.3±2.0 Mwt 252.3 PSA 114 Å^2 log *P* −0.07	CIMETIDINE Histamine H$_2$ antagonist, antiulcerative Metabolism: S-oxidation to sulfoxide, hydroxylation of CCH_3, hydrolysis to guanylurea

(continued overleaf)

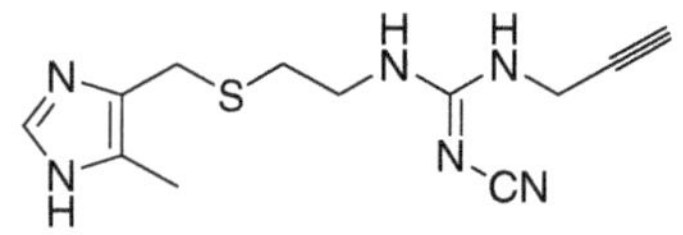

$t^{1}/_{2}$	1.2–1.6 h	V	2
F	–	CL	15
pb	–	Mwt	276.4
ur	35–40%	PSA	114 Å^2
		log P	0.28

ETINTIDINE
Histamine H_2 antagonist, antiulcerative
Metabolism: N-glucuronidation

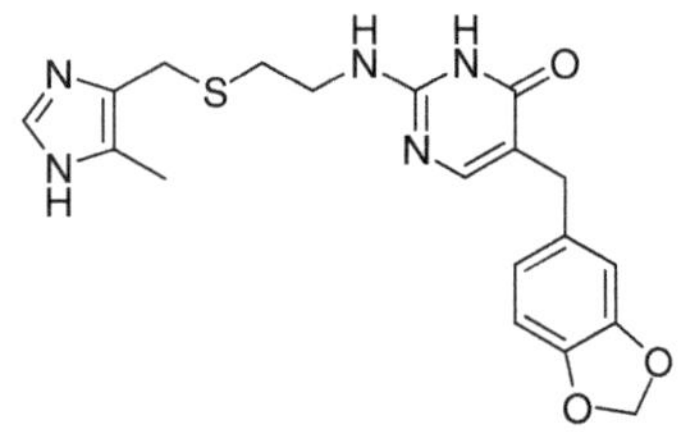

$t^{1}/_{2}$	2–3 h	V	0.7
F	53–91%	CL	3.2
pb	–	Mwt	399.5
ur	3–5%	PSA	126 Å^2
		log P	2.21

OXMETIDINE
Histamine H_2 antagonist, antiulcerative
Metabolism: glucuronidation

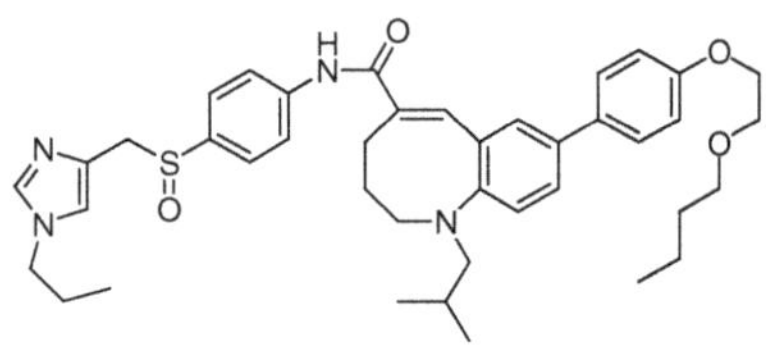

$t^{1}/_{2}$	35–40 h	V/F	30
F	Good	CL	15
pb	–	Mwt	697.0
ur	–	PSA	85.7 Å^2
		log P	7.57

TBR-652
Chemokine 5 (CCR 5) antagonist, antiviral (HIV)

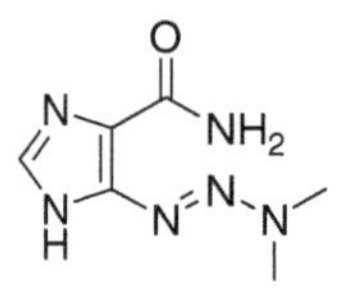

$t^{1}/_{2}$	5 h (iv)	V	1.5
F	–	CL	2.6
pb	5%	Mwt	182.2
ur	9–30%	PSA	99.7 Å^2
		log P	0.03

DACARBAZINE
Methylating reagent, antineoplastic
Metabolism: N-demethylation, reductive cleavage of N=N double bond

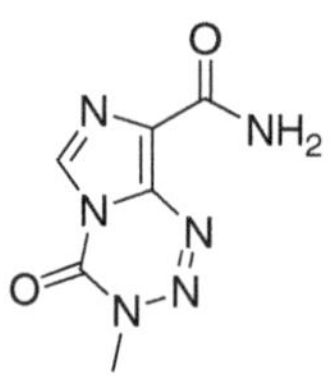

$t^{1}/_{2}$	1.8 h	V/F	0.6
F	96%	CL	4.5±1.3
pb	15%	Mwt	194.2
ur	6±3%	PSA	106 Å^2
		log P	−1.27

TEMOZOLOMIDE
Antineoplastic, crosses bbb
Metabolism: hydrolysis of amide and urea, then formation of diazomethane and 4-aminoimidazole-5-carboxylate

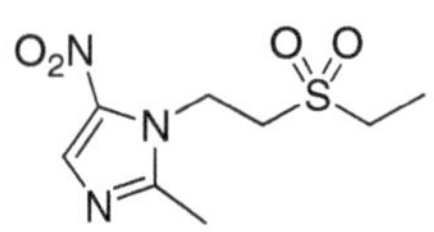

$t^{1}/_{2}$	12–14 h	V	0.6
F	95%	CL	0.6
pb	12–20%	Mwt	247.3
ur	20–25%	PSA	106 Å^2
		log P	−0.29

TINIDAZOLE
Antiprotozoal
Metabolism: hydroxylation to 2-(hydroxymethyl), N-oxidation of imidazole, hydroxylation and nitro group migration (to 2-methyl-4-nitro-5-hydroxy)

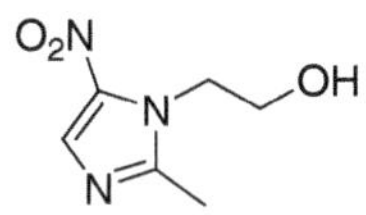	$t^{1/2}$ 8.5±2.9 h *F* 99±8% pb 11±3% ur 10±2%	*V* 0.74±0.10 CL 1.3±0.3 Mwt 171.2 PSA 83.9 Å^2 log *P* −0.14	METRONIDAZOLE Antiprotozoal, antibacterial, crosses bbb Metabolism: hydroxylation of CH_3, oxidation of CH_2OH to CO_2H, imidazole N-oxidation
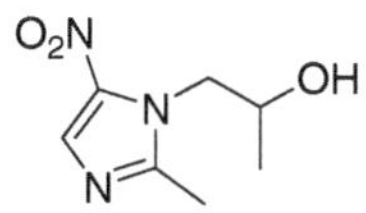	$t^{1/2}$ 14–20 h *F* 99±11% pb 15% ur 22%	*V* 0.8 CL 0.4 Mwt 185.2 PSA 83.9 Å^2 log *P* 0.22	SECNIDAZOLE Antiprotozoal Metabolism: hydroxylation, glucuronidation
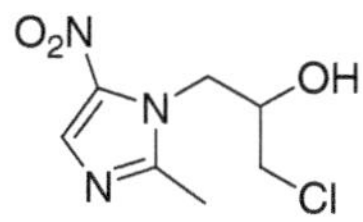	$t^{1/2}$ 11–14 h *F* >97% pb 11–13% ur 4%	*V* 0.7–1.0 CL 0.8 Mwt 219.6 PSA 83.9 Å^2 log *P* 0.60	ORNIDAZOLE Antiprotozoal Metabolism: hydroxylation, hydrolysis of CH_2Cl
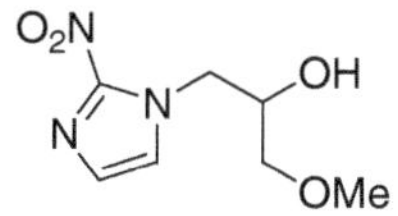	$t^{1/2}$ 11±1 h *F* 91% pb Low ur 10–20%	*V* 0.6 CL 0.6 Mwt 201.2 PSA 93.1 Å^2 log *P* −0.41	MISONIDAZOLE Radiosensitizer Metabolism: O-demethylation, glucuronidation, reduction of NO_2 to NH_2
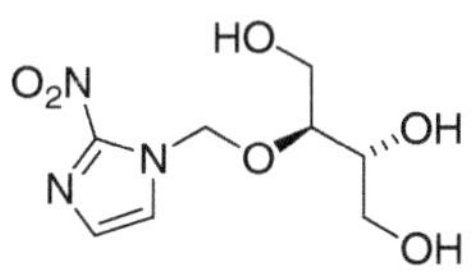	$t^{1/2}$ 4.2–4.6 h* *F* – pb – ur 79%*	*V* – CL – Mwt 247.2 PSA 129 Å^2 log *P* −1.41	DORANIDAZOLE *iv Antineoplastic (radiosensitzer) Metabolism: glucuronidation, carbonation
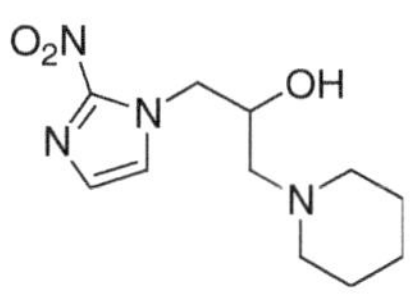	$t^{1/2}$ 5.6 h (iv) *F* – pb – ur –	*V* 4–6 CL 8–15 Mwt 254.3 PSA 87.1 Å^2 log *P* 0.97	PIMONIDAZOLE, Ro 03-8799 Metabolism: oxidation to piperidine *N*-oxide
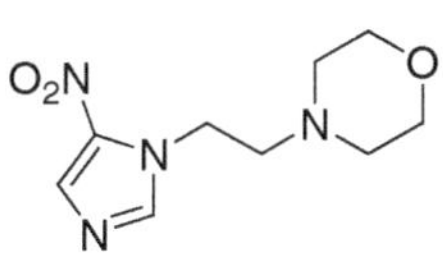	$t^{1/2}$ 2–5 h *F* – pb – ur –	*V* 3 CL 0.8 Mwt 226.2 PSA 76.1 Å^2 log *P* −0.15	NIMORAZOLE Antiprotozoal, hypoxic radiosensitizer
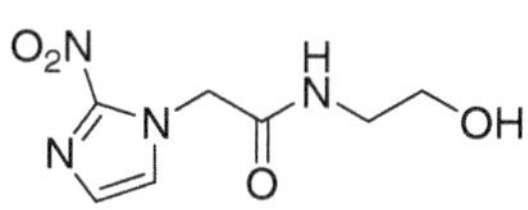	$t^{1/2}$ 1–6 h *F* – pb – ur –	*V* 0.5–0.9 CL 1.2–2.5 Mwt 214.2 PSA 113 Å^2 log *P* −1.37	ETANIDAZOLE Antineoplastic (radiosensitzer); Metabolism: reduction

(*continued overleaf*)

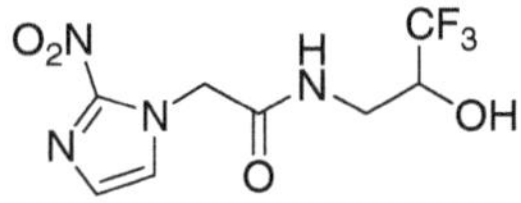	$t_{1/2}$	3.5±0.6 h*	*V*	0.75±0.12	SR-4554
	F	–	CL	3.2±0.9	*iv
	pb	–	Mwt	282.2	Hypoxia probe
	ur	87±9%*	PSA	108 Å^2	
			log *P*	0.29	
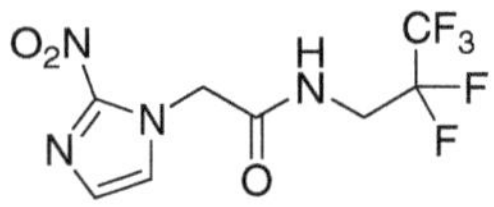	$t_{1/2}$	12±3 h*	*V*	0.56	EF5
	F	–	CL	–	*iv
	pb	–	Mwt	302.2	Hypoxia probe
	ur	70%*	PSA	87.8 Å^2	
			log *P*	1.35	
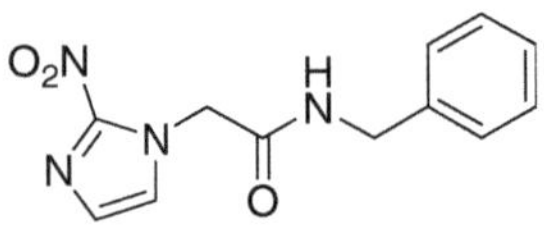	$t_{1/2}$	11–14 h	*V*	0.6	BENZNIDAZOLE
	F	100%	CL	–	Antiprotozoal
	pb	44%	Mwt	260.3	Reduction of NO_2 leads to
	ur	5%	PSA	92.7 Å^2	reactive electrophile
			log *P*	0.91	
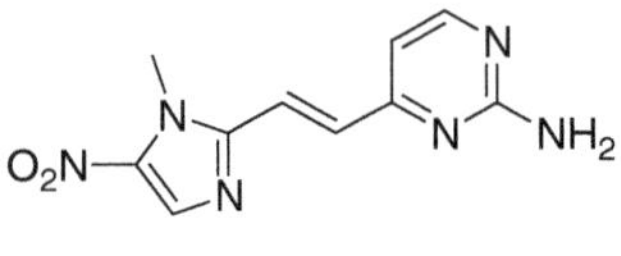	$t_{1/2}$	0.8 h	*V*	–	AZANIDAZOLE
	F	–	CL	–	Antiprotozoal
	pb	–	Mwt	246.2	
	ur	<0.5%	PSA	111 Å^2	
			log *P*	0.67	
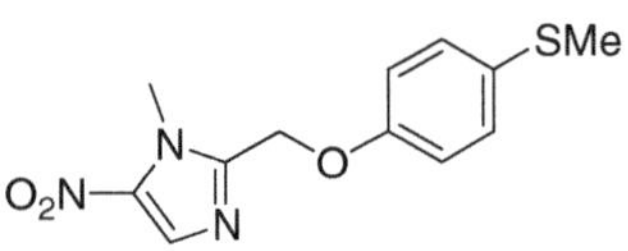	$t_{1/2}$	13 min	*V*	3.3*	FEXINIDAZOLE
	F	10–14%*	CL	273±121*	*dog
	pb	95%	Mwt	279.3	Antiprotozoal
	ur	–	PSA	98.2 Å^2	Metabolism: oxidation to
			log *P*	2.13	sulfoxide and sulfone; crosses
					bbb
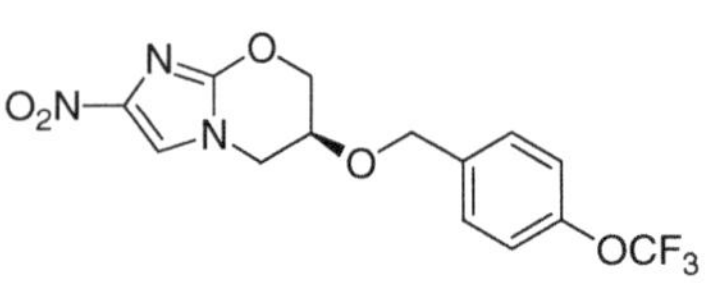	$t_{1/2}$	16–20 h	*V*	–	PA-824
	F	40%	CL	–	Antibacterial
	pb	95%	Mwt	359.3	
	ur	–	PSA	86.4 Å^2	
			log *P*	2.70	

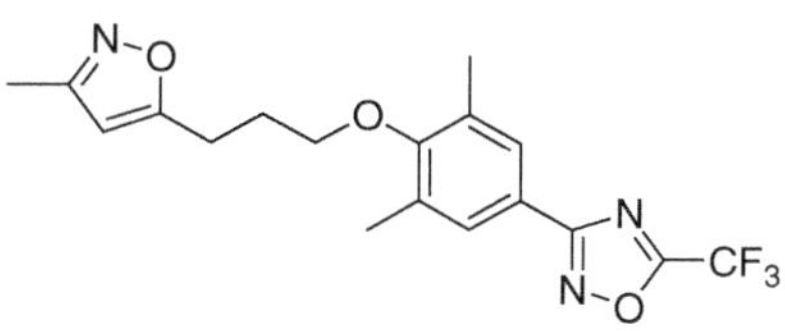	$t\frac{1}{2}$ 3–9 h *F* 70% pb >99% ur <1%	*V* 5±2 CL 5.7 Mwt 381.4 PSA 74.2 Å^2 log *P* 5.77	PLECONARIL Antiviral
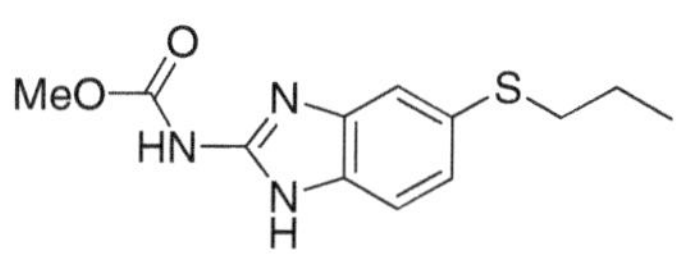	$t\frac{1}{2}$ Short *F* 5% pb 70% ur 0%	*V* – CL – Mwt 265.3 PSA 92.3 Å^2 log *P* 3.00	ALBENDAZOLE Anthelmintic (veterinary); active metabolite: sulfoxide (see below); poor oral bioavailability due to low solubility in water and high first-pass metabolism
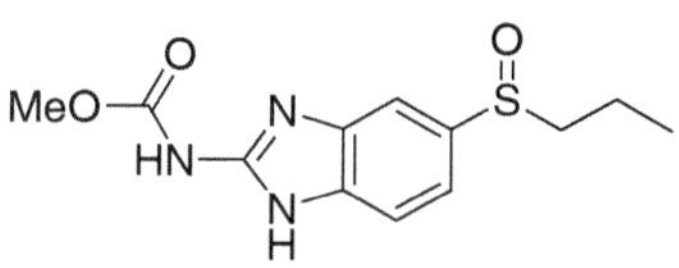	$t\frac{1}{2}$ 8–12 h *F* – pb 70% ur –	*V* – CL – Mwt 281.3 PSA 103 Å^2 log *P* 0.68	ALBENDAZOLE SULFOXIDE Main active metabolite of albendazole and netobimin Metabolism: oxidation to sulfone
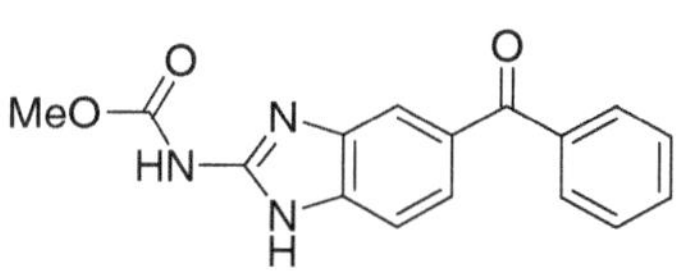	$t\frac{1}{2}$ 3–6 h *F* 22% pb 90–95% ur <2%	*V* 1–2 CL 3.7 Mwt 295.3 PSA 84.1 Å^2 log *P* 2.83	MEBENDAZOLE Anthelmintic; poor solubility Metabolism: carbamate hydrolysis, reduction of ketone to alcohol
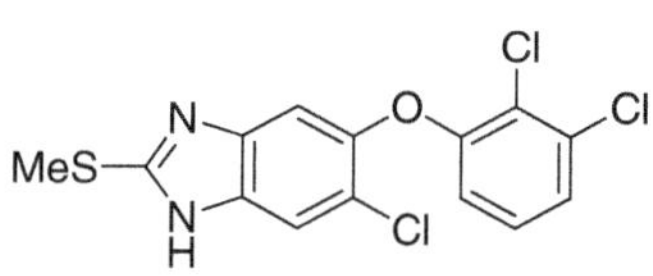	$t\frac{1}{2}$ 6–8 h *F* – pb – ur –	*V* – CL – Mwt 359.7 PSA 63.2 Å^2 log *P* 5.39	TRICLABENDAZOLE Anthelmintic Prodrug of sulfoxide; all values for sulfoxide on oral administration of triclabendazole
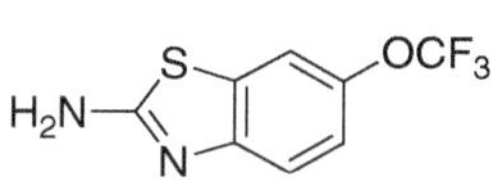	$t\frac{1}{2}$ 9–14 h *F* 30–100% pb 97% ur 0.4–2.0%	*V* 5.6 CL 11.3 Mwt 234.2 PSA 76.4 Å^2 log *P* 2.92	RILUZOLE Neuroprotective Metabolism: N-hydroxylation, then O- and N-glucuronidation
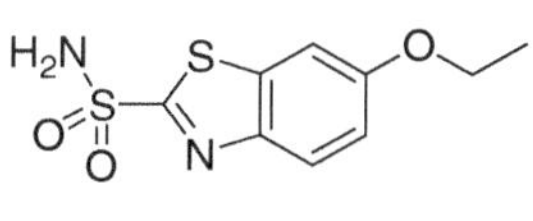	$t\frac{1}{2}$ 4.0±1.5 h *F* 65% pb >95% ur 40%	*V* – CL – Mwt 258.3 PSA 119 Å^2 log *P* 2.14	ETHOXZOLAMIDE Carbonic anhydrase inhibitor, diuretic, antiglaucoma agent

(*continued overleaf*)

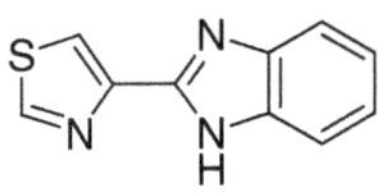	$t^{1/2}$ F pb ur	1–2 h Low – <1%	V CL Mwt PSA log P	2.8* 27* 201.3 69.8 Å^2 2.47	THIABENDAZOLE *anephric patient Anthelmintic Metabolism: aromatic 5-hydroxylation
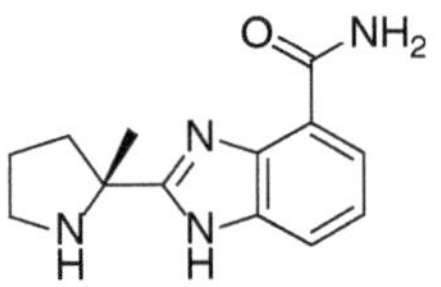	$t^{1/2}$ F pb ur	4–5 h 56–92%* – 70%	V CL/F Mwt PSA log P	– 7 244.3 83.8 Å^2 0.30	VELIPARIB *animals Poly(ADP-ribose)polymerase inhibitor, antineoplastic Metabolism: oxidation to lactam, hydrolysis of amide

$t_{1/2}$, plasma half-life; F, oral bioavailability; pb, plasma protein binding; ur, excretion of unchanged drug in urine; V, volume of distribution; CL, clearance; Mwt, molecular weight; PSA, polar surface area.

58
Triazoles

Lipophilic imidazoles and 1,2,4-triazoles are important antifungal drugs (Table 58.1). Azole antifungals inhibit the fungal sterol 14-α-demethylase, thereby blocking the biosynthesis of ergosterol from lanosterol. This leads to the accumulation of methylsterols in cell membranes and thus to the impairment of membrane-bound enzyme systems, which keeps fungi from growing. Triazoles usually have longer half-lives and a lower effect on human sterol biosynthesis than imidazoles, and most antifungals under development are, therefore, triazoles.

One of the first antifungal azoles launched was clotrimazole (1973). This highly insoluble compound is mainly used topically, but for related lipophilic azoles (e.g., miconazole), a parenteral formulation has been developed. Azoles with higher oral bioavailability became available later. Improved oral bioavailability was attained by lowering their lipophilicity and enhancing their metabolic stability, for example, by conversion to tertiary alcohols. Quick renal excretion was avoided by adjusting the lipophilicity to a value that still provided for a high degree of tubular reabsorption. In other words, high oral bioavailability and long half-lives were attained by careful adjustment of the log *P* to reduce the affinity to CYP enzymes and improve the solubility in water to a degree that does not cause quick renal excretion.

Lipophilic triazoles can also be rendered more soluble by transient N-alkylation to cationic triazolium prodrugs. One example of this strategy is isavuconazonium (Scheme 58.1), an oral prodrug of isavuconazole.

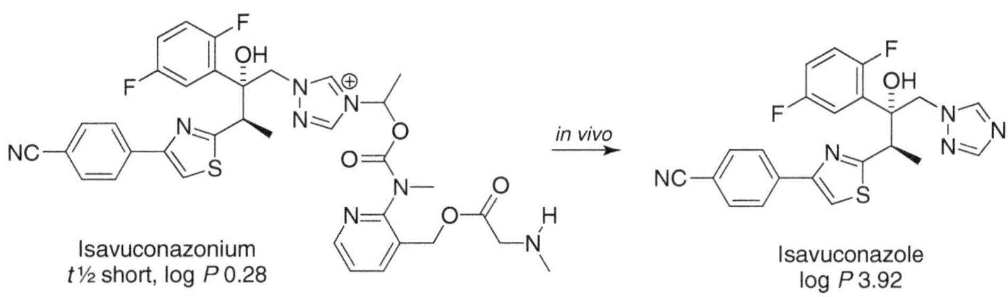

Scheme 58.1 Metabolism of isavuconazonium.

Lead Optimization for Medicinal Chemists: Pharmacokinetic Properties of Functional Groups and Organic Compounds, First Edition. Florencio Zaragoza Dörwald.

Table 58.1 Triazoles and related compounds. V in l kg^{-1}; CL in ml min^{-1} kg^{-1}; Mwt in g mol^{-1}.

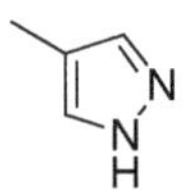	$t½$ 15–40 h F 100% pb Low ur 1–4%	V 0.6–1.0 CL – Mwt 82.1 PSA 28.7 Å^2 log P 1.10	FOMEPIZOLE Alcohol dehydrogenase inhibitor, antidote to methanol and ethylene glycol poisoning Metabolites: 4-(hydroxymethyl)pyrazole, 4-carboxypyrazole, N-glucuronides
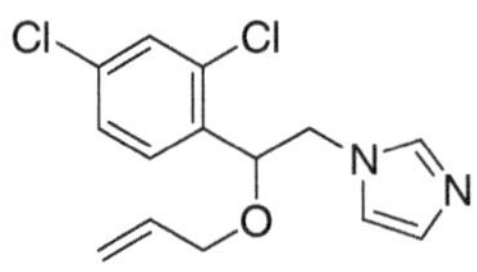	$t½$ 11 h* F 1.4%* pb – ur 0–10%**	V – CL – Mwt 297.2 PSA 27.1 Å^2 log P 3.56	ENILCONAZOLE *cow **rat Antifungal
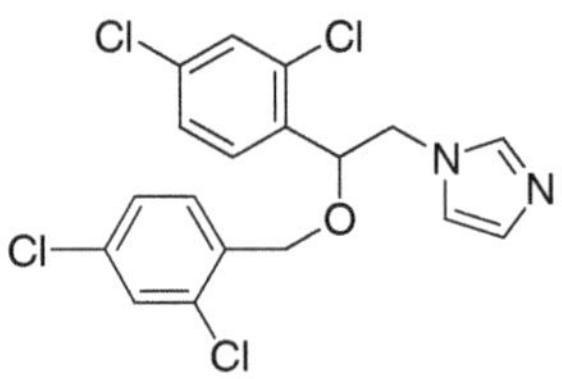	$t½$ 20–25 h F 25–30% pb 88% ur <1%	V 20 CL – Mwt 416.1 PSA 27.1 Å^2 log P 4.97	MICONAZOLE Topical antifungal Metabolism: N- and O-dealkylation
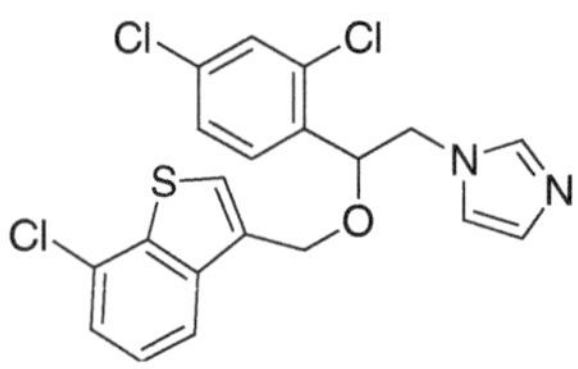	$t½$ 5 h (rat) F 48% (rat) pb >99% ur 0%	V – CL – Mwt 437.8 PSA 55.3 Å^2 log P 5.77	SERTACONAZOLE Topical antifungal
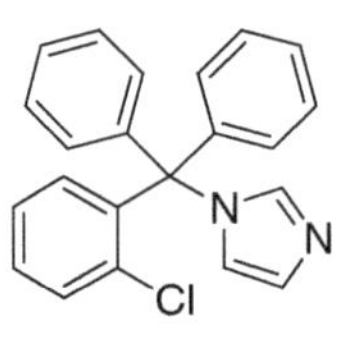	$t½$ 3–5 h F Poor pb 90% ur <1%*	V 5±2 CL – Mwt 344.8 PSA 17.8 Å^2 log P 4.92	CLOTRIMAZOLE *dog iv; antifungal Metabolism: N-dealkylation (to yield imidazole, chlorotritylalcohol, and (2-chlorophenyl)(4-hydroxyphenyl)-phenylmethane
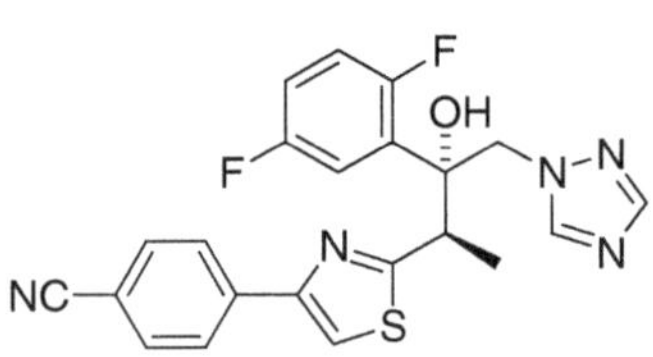	$t½$ 60–120 h* F 100%* pb 98%* ur <1%	V 7 CL 0.5–0.7 Mwt 437.5 PSA 111 Å^2 log P 3.92	ISAVUCONAZOLE Antifungal *drug on oral administration of the prodrug isavuconazolium (see text)

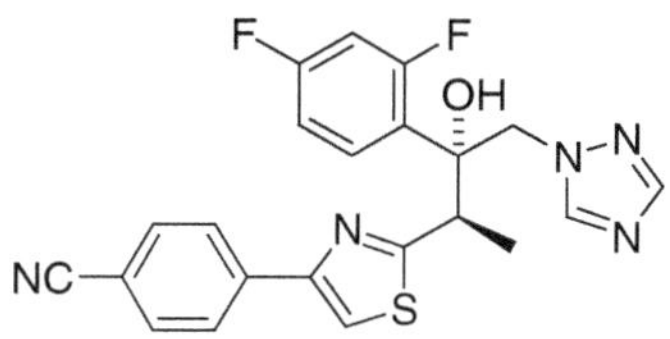

$t^1\!/\!_2$	103–240 h	V	11	RAVUCONAZOLE
F	48–74%*	CL	–	*animals
pb	98%	Mwt	437.5	Antifungal
ur	–	PSA	111 Å^2	
		log P	3.89	

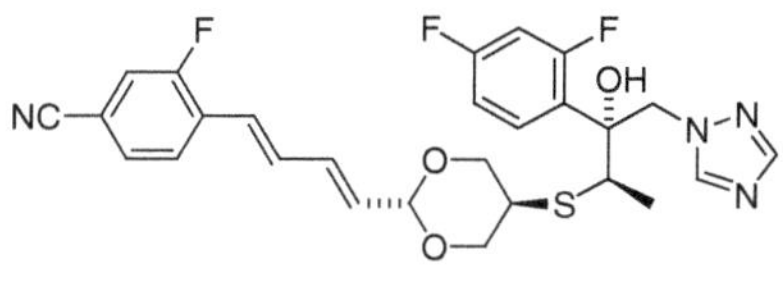

$t^1\!/\!_2$	4.6 h (iv)*	V	2.3*	CS-758
F	100%*	CL	7.7*	*monkey
pb	–	Mwt	542.6	Antifungal
ur	–	PSA	118 Å^2	
		log P	3.70	

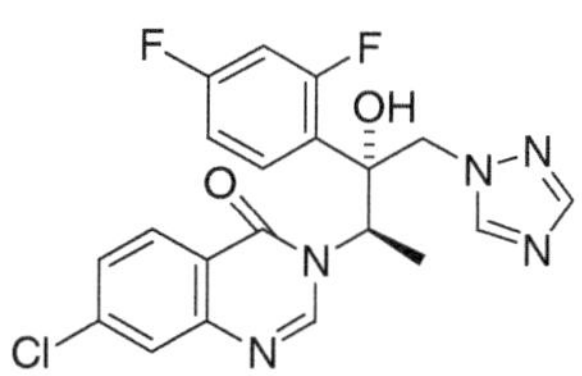

$t^1\!/\!_2$	30–56 h	V	2.4**	ALBACONAZOLE
F	100%*	CL	0.16**	*dog, **rat
pb	–	Mwt	431.8	Antifungal
ur	0%	PSA	78.7 Å^2	Metabolism: hydroxylation, then glucuronidation and sulfation
		log P	3.00	

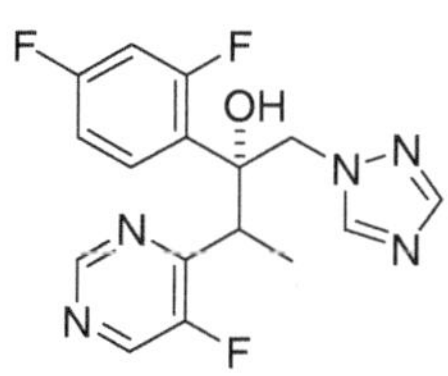

$t^1\!/\!_2$	6 h	V	2.0–4.6	VORICONAZOLE
F	96%	CL	8.3	Antifungal
pb	58%	Mwt	349.3	Metabolism: N-oxidation of pyrimidine, hydroxylation of methyl
ur	<1%	PSA	71.8 Å^2	
		log P	0.93	

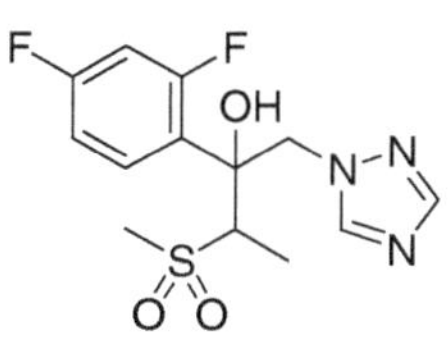

$t^1\!/\!_2$	49 h	V	0.62	GENACONAZOLE
F	100%	CL	0.19	Antifungal
pb	–	Mwt	331.3	Metabolism: negligible
ur	76%	PSA	93.5 Å^2	
		log P	0.97	

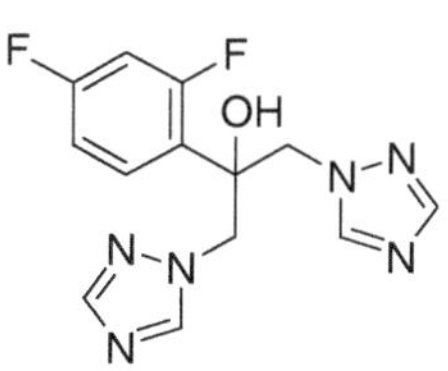

$t^1\!/\!_2$	32±5 h	V	0.60	FLUCONAZOLE
F	>90%	CL	0.27±0.07	Antifungal
pb	11±1%	Mwt	306.3	Metabolism: glucuronidation, N-oxidation
ur	75±9%	PSA	81.7 Å^2	
		log P	0.45	

(continued overleaf)

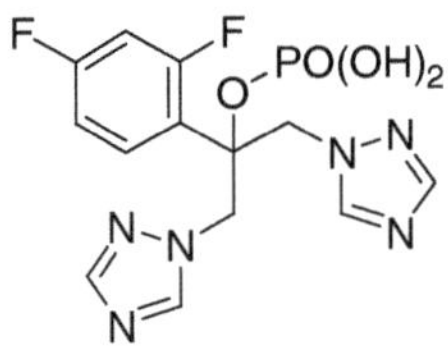	$t_{1/2}$ 2.1 h (iv) F – pb 96% ur <1% (iv)	V 0.15 CL 1.3 Mwt 386.3 PSA 128 Å^2 log P −1.23	FOSFLUCONAZOLE Prodrug of fluconazole for parenteral administration
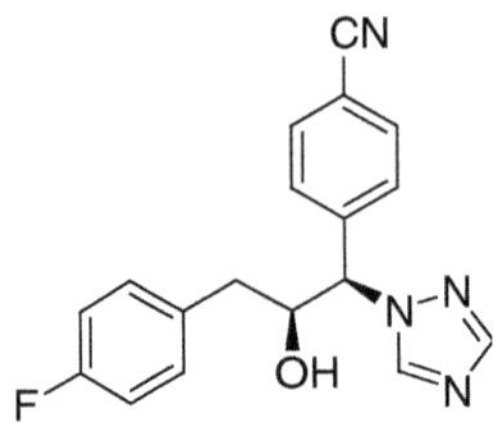	$t_{1/2}$ 2.9–3.4 h F – pb – ur –	V – CL – Mwt 322.3 PSA 69.8 Å^2 log P 2.41	FINROZOLE Aromatase inhibitor, antineoplastic Metabolism: glucuronidation
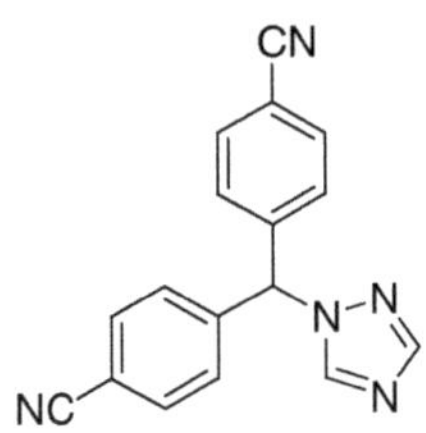	$t_{1/2}$ 42–82 h F 100% pb 59% ur 4–6%	V 1.9–2.2 CL 0.3–1.0 Mwt 285.3 PSA 78.3 Å^2 log P 0.43	LETROZOLE Antineoplastic Metabolism: N-dealkylation to benzhydrol and triazole, then glucuronidation
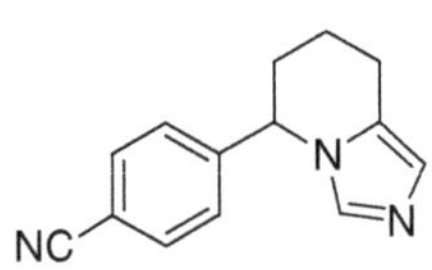	$t_{1/2}$ 9 h F – pb – ur <5%	V – CL 10 Mwt 223.3 PSA 41.6 Å^2 log P 1.48	FADROZOLE Aromatase inhibitor, oral antineoplastic Metabolism: benzylic hydroxylation
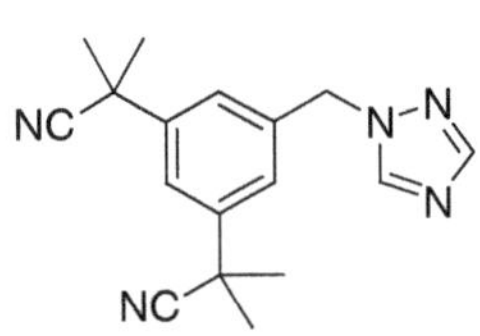	$t_{1/2}$ 41 h F 84–100% pb 40% ur 10%	V 1.4 CL 0.4 Mwt 293.4 PSA 78.3 Å^2 log P 0.97	ANASTROZOLE Aromatase inhibitor, antineoplastic Metabolism: N-dealkylation, hydroxylation, glucuronidation
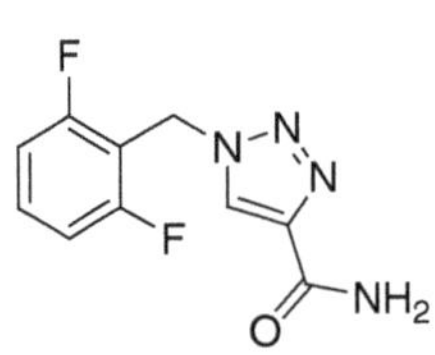	$t_{1/2}$ 8–12 h F 60–85% pb 34% ur <2%	V 0.8–1.2 CL – Mwt 238.2 PSA 73.8 Å^2 log P 1.00	RUFINAMIDE Antiepileptic Metabolite: carboxylic acid

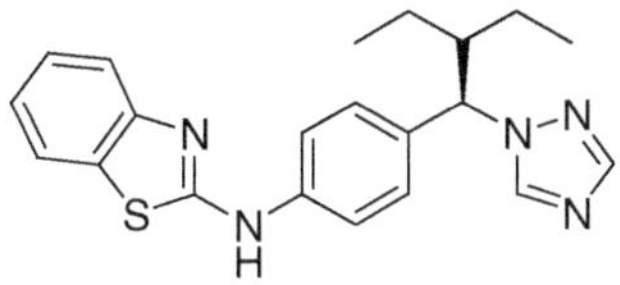	$t^1/_2$ 1.1–2.1 h *F* – pb – ur –	*V* – CL – Mwt 377.5 PSA 78.9 Å^2 log *P* 5.16	RAMBAZOLE, R-115866 Retinoic acid metabolism blocking agent Metabolism: hydroxylation of 3-pentyl, aromatic benzothiazole hydroxylation (para to N)
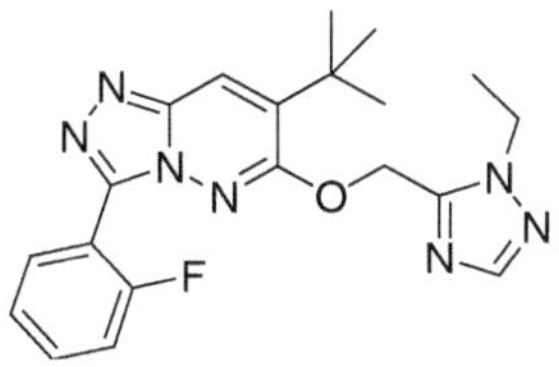	$t^1/_2$ 7 h *F* 36% (rat) pb 90% ur <1%	*V* – CL/*F* 3.8 Mwt 395.4 PSA 78.1 Å^2 log *P* 3.01	TPA-023 GABA agonist, anxiolytic Metabolism: hydroxylation of *tert*-butyl, N-deethylation, N-glucuronidation
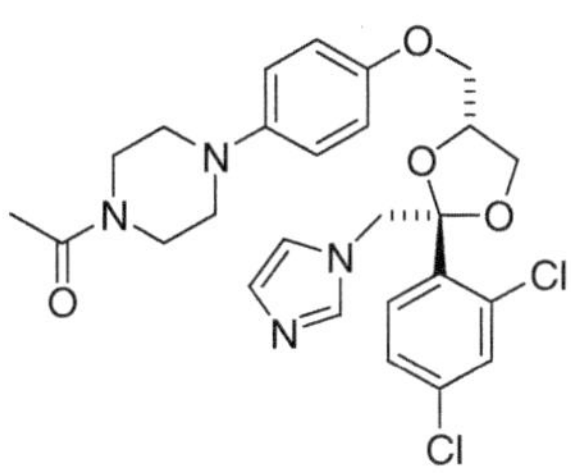	$t^1/_2$ 3.3±1.0 h *F* – pb 99% ur <1%	*V*/*F* 2.4±1.6 CL/*F* 8.4±4.1 Mwt 531.4 PSA 69.1 Å^2 log *P* 4.04	KETOCONAZOLE Antifungal Metabolism: N-deacetylation, oxidative degradation of piperazine, O-dealkylation, aromatic hydroxylation
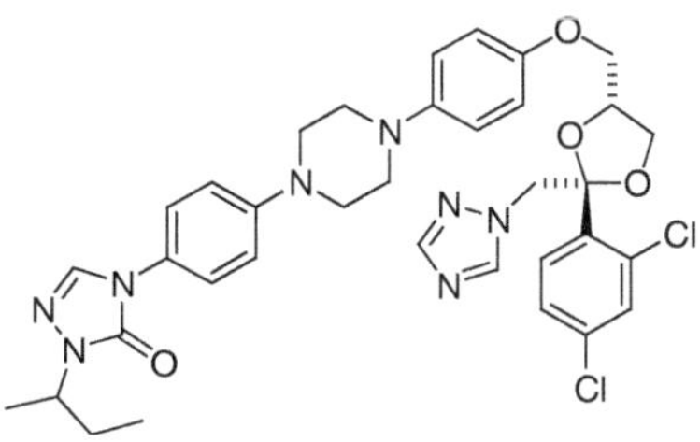	$t^1/_2$ 21±6 h *F* 55% pb 99.8% ur <0.1%	*V*/*F* 14±5 CL/*F* 23±10 Mwt 705.6 PSA 101 Å^2 log *P* 5.00	ITRACONAZOLE Antifungal, CYP3A4 inhibitor Metabolism: hydroxylation of CH_2Me
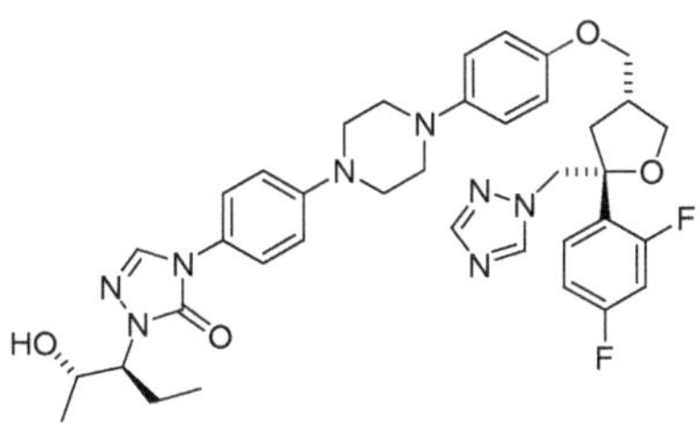	$t^1/_2$ 20–66 h *F* 11–27%* pb >98% ur <0.2%	*V* 27 CL – Mwt 700.8 PSA 107 Å^2 log *P* 2.25	POSACONAZOLE *dog Antifungal Metabolism: glucuronidation

(*continued overleaf*)

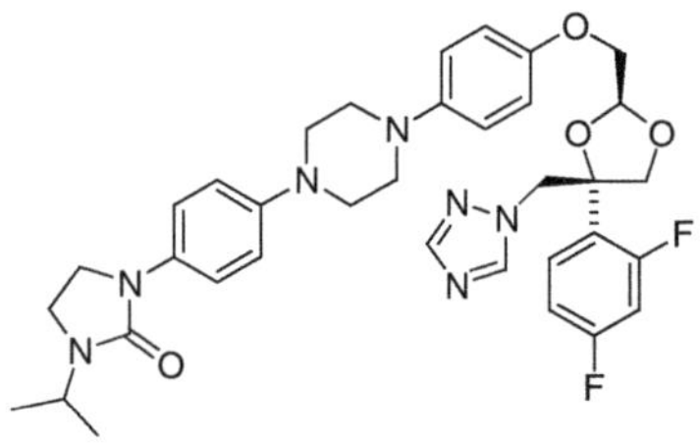	$t_{1/2}$ 56–81 h F – pb – ur –	V – CL – Mwt 659.7 PSA 88.4 Å^2 log P 4.19	PRAMICONAZOLE, R-126638 Antifungal

$t_{1/2}$, plasma half-life; *F*, oral bioavailability; pb, plasma protein binding; ur, excretion of unchanged drug in urine; V, volume of distribution; CL, clearance; Mwt, molecular weight; PSA, polar surface area; GABA, γ-aminobutyric acid.

59
Pyridines, Pyrimidines, and Related Compounds

Pyridines (pK_a 5) are common substructures of drugs (Table 59.1) and natural products (vitamins and alkaloids) and can serve as bioisostere of carbonyl groups or of other basic heterocycles, such as imidazoles. Thus, nicotine is an acetylcholine agonist, the pyridine ring acting as an ester analog, and the vasodilator betahistine is a histamine agonist (Scheme 59.1).

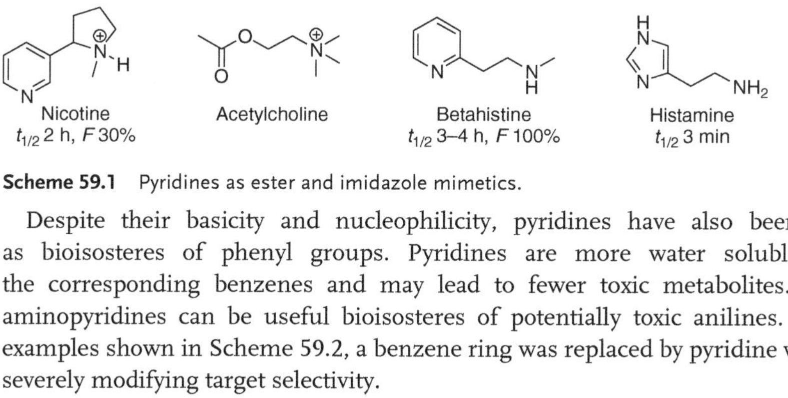

Scheme 59.1 Pyridines as ester and imidazole mimetics.

Despite their basicity and nucleophilicity, pyridines have also been used as bioisosteres of phenyl groups. Pyridines are more water soluble than the corresponding benzenes and may lead to fewer toxic metabolites. Thus, aminopyridines can be useful bioisosteres of potentially toxic anilines. In the examples shown in Scheme 59.2, a benzene ring was replaced by pyridine without severely modifying target selectivity.

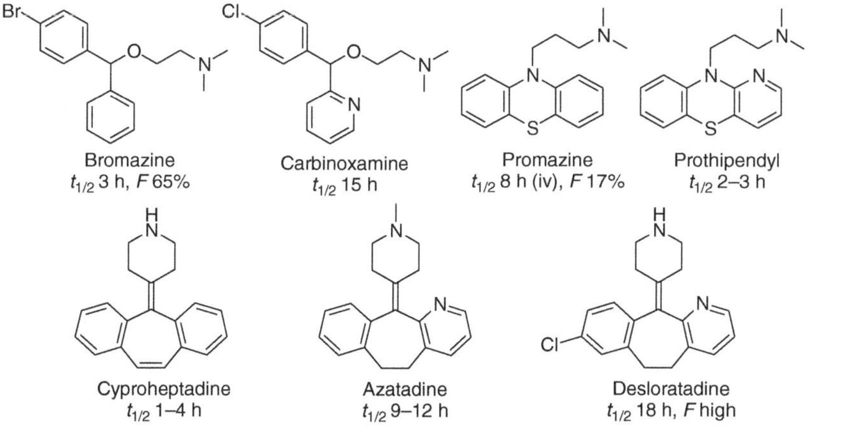

Scheme 59.2 PK-comparison (pharmacokinetics) pyridyl *versus* phenyl.

Lead Optimization for Medicinal Chemists: Pharmacokinetic Properties of Functional Groups and Organic Compounds, First Edition. Florencio Zaragoza Dörwald.

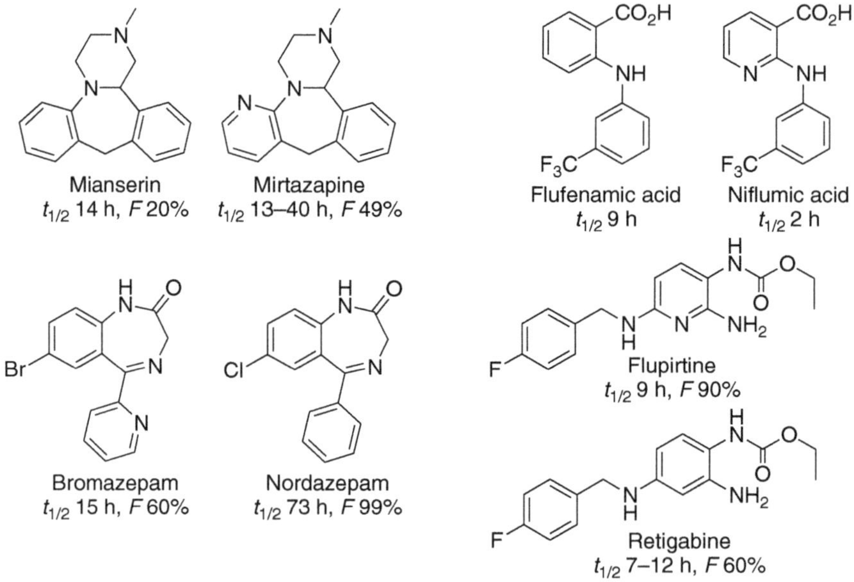

Scheme 59.2 (*Continued*).

However, not all heterocyclic amines are toxicologically safe. The compounds shown in Scheme 59.3 are formed when meat is heated (e.g., on frying) by the reaction of creatine or creatinine with amino acids and carbohydrates (Maillard reaction). All these compounds are carcinogenic for rodents, and the probable cause for the increased risk of colorectal cancer in people who frequently eat meat.

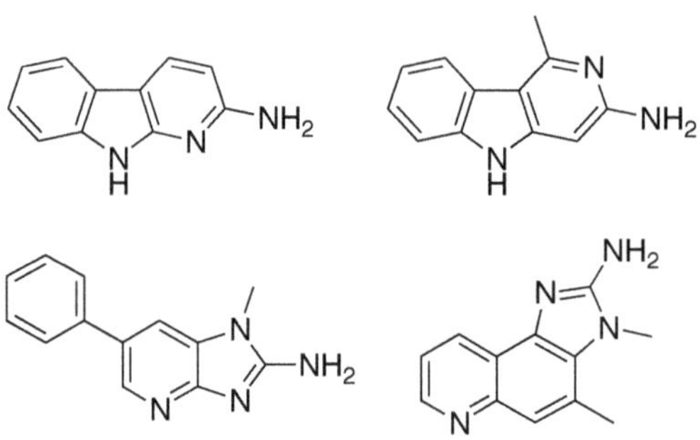

Scheme 59.3 Carcinogenic polycyclic heteroarenes.

One further potential problem of pyridines is their ability to coordinate to iron, which can lead to CYP (cytochrome P) inhibition. This will be important for highly basic and 2,6-unsubstituted pyridines but can be avoided with electron-withdrawing substitutents and by steric shielding of the nitrogen atom. In the example shown in Scheme 59.4 [1], the initial leads were strong inhibitors of CYP1A2. By increasing

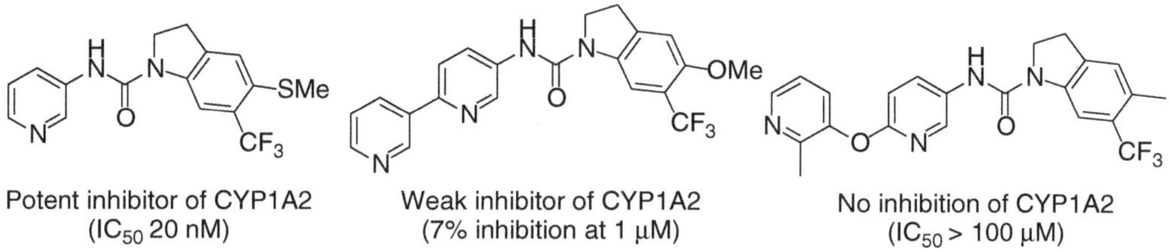

Scheme 59.4 Pyridines as CYP inhibitors.

the steric shielding of the pyridyl nitrogen atoms, the affinity for CYP enzymes could be effectively suppressed.

Typical metabolic transformations of pyridines include N-oxidation, N-methylation, and C-hydroxylation (Scheme 59.5). Interestingly, *N*-methylpyridinium salts, such as trigonelline or the metabolites of bioactive pyridines, do not react as electrophilic methylating agents and show little toxicity.

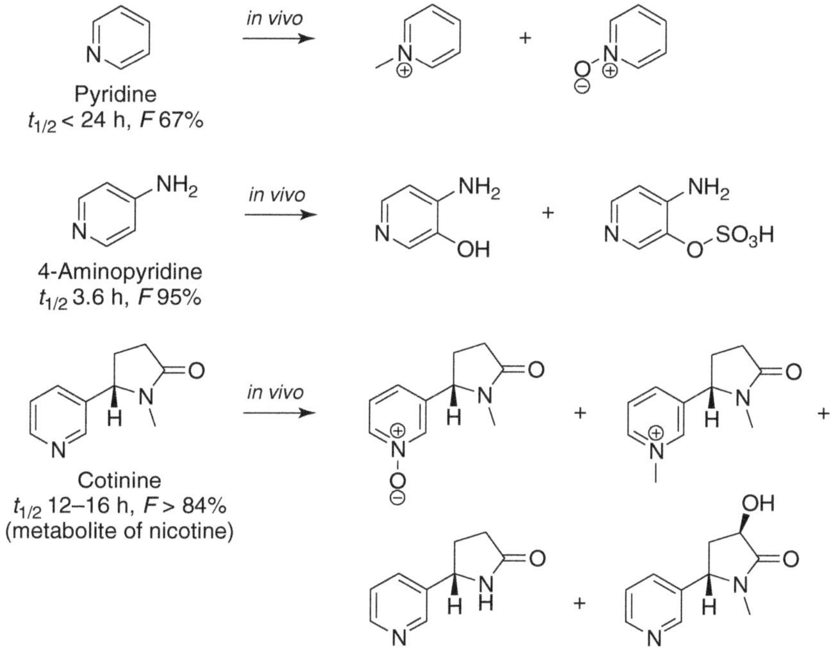

Scheme 59.5 Metabolism of pyridines.

The rate of pyridine metabolism varies strongly and depends on the precise substitution pattern and the overall physicochemical properties of the drug. Examples of long-lived pyridines are shown in Scheme 59.6.

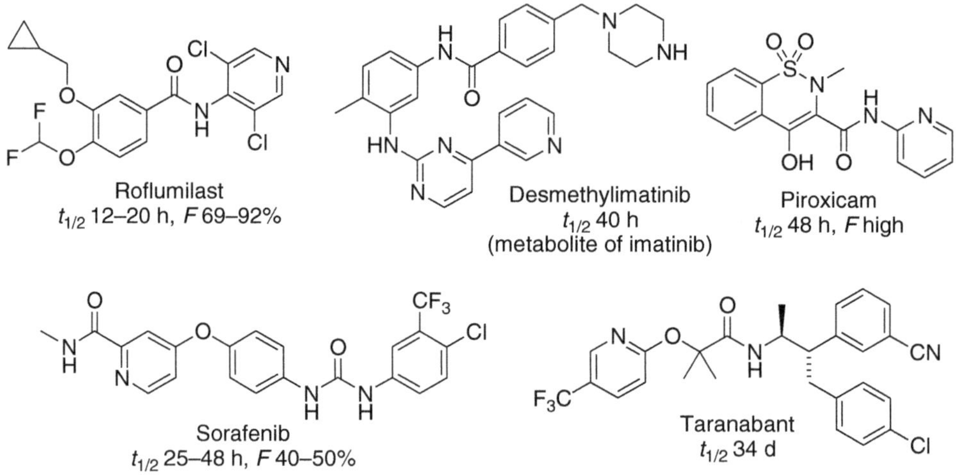

Scheme 59.6 Pyridines as drugs.

59.1 Proton Pump Inhibitors

In 1967, the "gastrin" project was initiated at Astra in Hässle (Sweden) with the aim of developing a drug for suppression of gastric acid secretion. As starting hypothesis, it was assumed that the local anesthetic lidocaine was capable of inhibiting the acid-stimulating hormone gastrin. After five years of development work, however, it turned out that such a gastrin inhibitor had no effect *in vivo* on acid secretion (thus the importance of early proof of concept in animals).

Looking for a new lead, researchers at Astra chose CMN 131 (Scheme 59.7), a weak inhibitor of acid secretion developed at Searle. The compound was too toxic because of the thiocarbonyl group, but served well as lead.

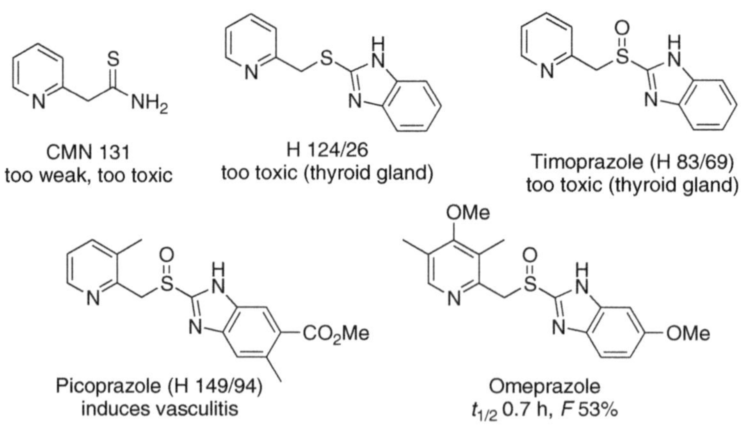

Scheme 59.7 Development of omeprazole.

This time, animal models were used to test new compounds, and within a few years, the benzimidazoles H 124/26 and timoprazole were found. Both were more potent than CMN 131, but showed thyroid-gland-related toxicity. The less toxic picoprazole was discovered in 1976, but disappointingly, this compound induced vasculitis in some dogs. Finally, in 1979, omeprazole was prepared and shown to be a potent and safe inhibitor of gastric acid production. Its new mechanism of action was the inhibition of gastric H^+, K^+-ATPase, and omeprazole turned out to be superior to the previously used histamine H_2 antagonists. Racemic omeprazole was launched 1988 in Sweden, while the enantiomerically pure *S*-isomer, which is about four times as potent as the *R*-isomer, was approved in 2000. The *S*-isomer is prepared by enantioselective oxidation of the corresponding thioether with the aid of a chiral titanium complex [2].

Typical metabolic transformation of omeprazole-type proton pump inhibitors includes oxidation to a sulfone, reduction to a thioether, and O-demethylation (Table 59.2).

References

1. Bromidge, S.M. *et al.* (2000) Biarylcarbamoylindolines are novel and selective 5-HT_{2C} receptor inverse agonists: identification of 5-methyl-1-[[2-[(2-methyl-3-pyridyl)oxy]-5-pyridyl]carbamoyl]-6-trifluoromethylindoline (SB-243213) as a potential antidepressant/anxiolytic agent. *J. Med. Chem.*, **43**, 1123–1134.
2. Carlsson, E., Lindberg, P., and von Unge, S. (May 2002) Two of a kind. *Chem. Br.*, 42–45.

Table 59.1 Pyridines, pyrimidines, and related azines. *V* in $l\ kg^{-1}$, CL in $ml\ min^{-1}\ kg^{-1}$, Mwt in $g\ mol^{-1}$.

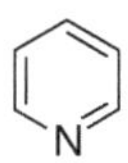	$t_{1/2}$ <24 h *F* 67% pb – ur –	*V* – CL – Mwt 79.1 PSA 12.9 Å^2 log *P* 0.84	PYRIDINE Metabolism: N-methylation, N-oxidation
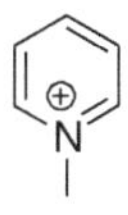	$t_{1/2}$ 2–3 h *F* – pb – ur 60–70%	*V* – CL – Mwt 94.1 PSA 3.89 Å^2 log *P* −4.39	*N*-METHYLPYRIDINIUM Contained in coffee, metabolite of pyridine
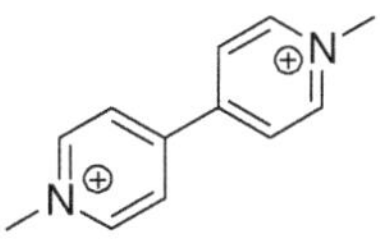	$t_{1/2}$ 12 h *F* <30% pb – ur 90%	*V* 1–3 CL – Mwt 186.3 PSA 7.77 Å^2 log *P* −5.82	PARAQUAT Herbicide

(*continued overleaf*)

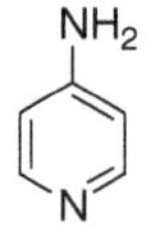	$t_{1/2}$ F pb ur	3.6±0.9 h 95% Negligible >90%	V CL Mwt PSA log P	2.6±0.9 10.2±2.3 94.1 38.9 Å^2 0.34	4-AMINOPYRIDINE, FAMPRIDINE, DALFAMPRIDINE Potassium channel blocker, treatment of multiple sclerosis Metabolites: 3-hydroxy and sulfate thereof
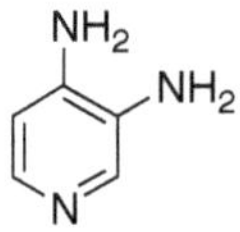	$t_{1/2}$ F pb ur	0.5–2 h 30% – –	V CL Mwt PSA log P	– 30 109.1 64.9 Å^2 −0.34	3,4-DIAMINOPYRIDINE, AMIFAMPRIDINE Potassium channel blocker, treatment of multiple sclerosis
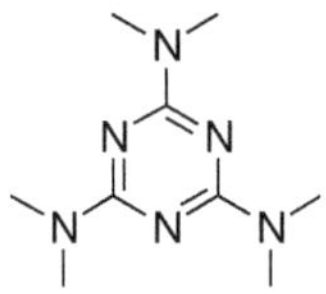	$t_{1/2}$ F pb ur	5–14 h Low* 94% <1%	V CL Mwt PSA log P	– – 210.3 48.4 Å^2 2.53	ALTRETAMINE *variable Antineoplastic (ovarian cancer) Metabolism: hydroxylation at methyl groups followed by demethylation
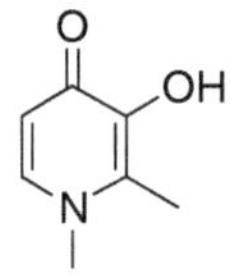	$t_{1/2}$ F pb ur	1–3 h – <10% >80%	V CL Mwt PSA log P	1.6–1.7 8.3–9.4 139.2 40.5 Å^2 −0.65	DEFERIPRONE Chelating agent Metabolism: glucuronidation
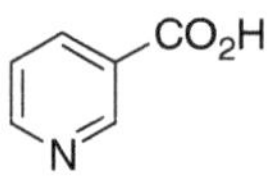	$t_{1/2}$ F pb ur	20–45 min 60–76% <20% 12%	V CL Mwt PSA log P	0.3 – 123.1 50.2 Å^2 0.22	NICOTINIC ACID, NIACIN, VITAMIN B_3 Metabolites: *N*-nicotinoyl glycine, nicotinamide, NAD
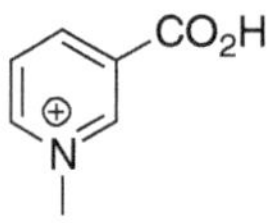	$t_{1/2}$ F pb ur	4–6 h – – 40–60%	V CL Mwt PSA log P	– – 138.1 41.2 Å^2 −3.98	TRIGONELLINE Contained in coffee
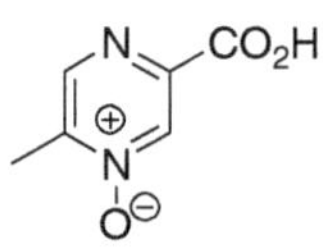	$t_{1/2}$ F pb ur	12–14 h 100% 0% >90%	V CL Mwt PSA log P	0.3–0.4 – 154.12 75.7 Å^2 −1.00	ACIPIMOX Antihyperlipoproteinemic Metabolism: N-deoxygenation
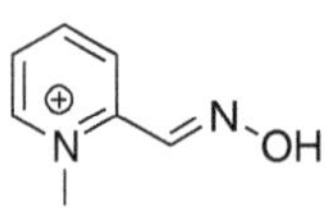	$t_{1/2}$ F pb ur	2–3 h 20–30% Low 80±13%	V CL Mwt PSA log P	3±2 9±4 137.2 36.5 Å^2 −2.85	PRALIDOXIME Cholinesterase reactivator (antidote against cholinesterase inhibitor poisoning), crosses bbb Metabolite: 1-methyl-2-cyanopyridinium

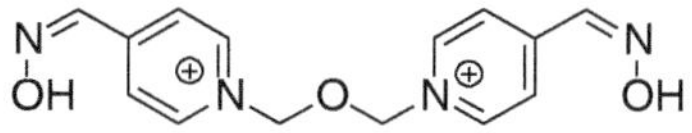	$t^1/_2$ 1.3 h (im) *F* Low pb Low ur 90%	*V* 0.2 CL – Mwt 288.3 PSA 82.2 Å^2 log *P* −5.72	OBIDOXIME Cholinesterase reactivator (antidote against cholinesterase inhibitor poisoning)
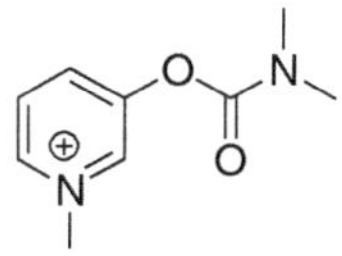	$t^1/_2$ 1.9±0.2 h *F* 14±3% pb Low ur 85±5%	*V* 1.1±0.3 CL 8.5±1.7 Mwt 181.2 PSA 33.4 Å^2 log *P* −3.93	PYRIDOSTIGMINE Cholinergic, does not cross bbb Metabolite: 3-hydroxy-*N*-methylpyridinium
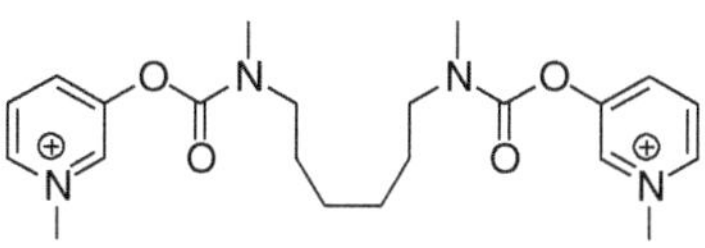	$t^1/_2$ 69 h *F* 5% pb – ur 7%	*V* – CL – Mwt 416.5 PSA 66.9 Å^2 log *P* −5.29	DISTIGMINE Cholinesterase inhibitor
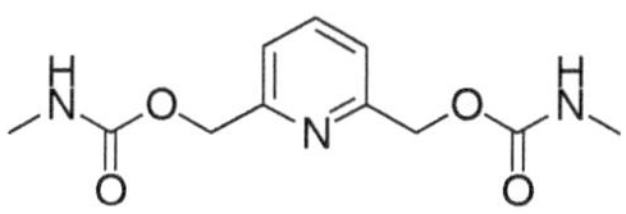	$t^1/_2$ 3–8 h *F* – pb – ur –	*V/F* 0.5–1.3 CL – Mwt 253.3 PSA 89.6 Å^2 log *P* 0.24	PYRIDINOL CARBAMATE Antiarteriosclerotic Metabolism: demethylation
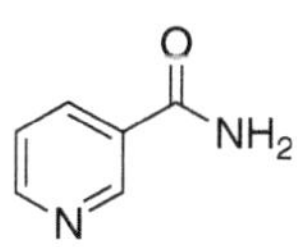	$t^1/_2$ 4–6 h *F* – pb – ur –	*V* – CL – Mwt 122.12 PSA 56 Å^2 log *P* −0.37	NICOTINAMIDE VITAMIN B_3 Metabolism: N-methylation of pyridine, then addition of water at position 4 and para to carbamoyl group
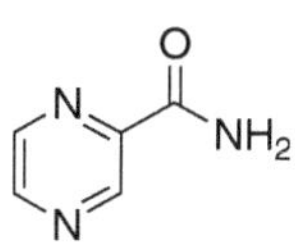	$t^1/_2$ 9±2 h *F* 100% pb 10–50% ur 0–3%	*V/F* 0.70±0.09 CL/*F* 0.90±0.15 Mwt 123.11 PSA 68.9 Å^2 log *P* −0.71	PYRAZINAMIDE Tuberculostatic, crosses bbb Metabolite: pyrazinecarboxylic acid
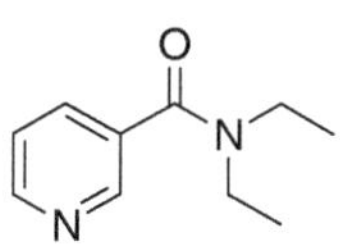	$t^1/_2$ 0.5 h *F* – pb – ur –	*V* – CL – Mwt 178.2 PSA 33.2 Å^2 log *P* 0.92	NIKETHAMIDE Respiratory stimulant Metabolism: N-dealkylation
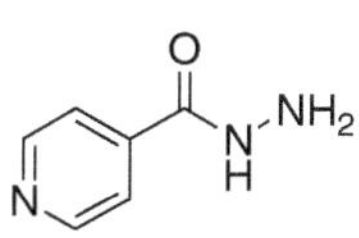	$t^1/_2$ 1–3 h *F* – pb 0–15% ur 10–30%	*V* 0.67±0.15 CL 7.4±2.0* Mwt 137.1 PSA 68.0 Å^2 log *P* −0.77	ISONIAZIDE *Fast acetylators; slow acetylators: 3.7±1.1; antibacterial, hepatotoxic Metabolism: N-acetylation

(*continued overleaf*)

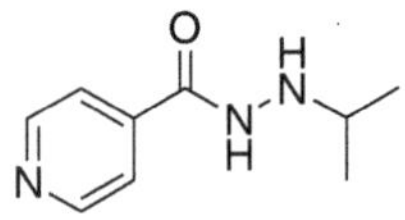

$t_{1/2}$	10 h	V	–
F	–	CL	–
pb	–	Mwt	179.2
ur	15%	PSA	54.0 Å^2
		log P	0.41

IPRONIAZID
MAO inhibitor, antidepressant
Metabolite: 4-pyridinecarboxylic acid

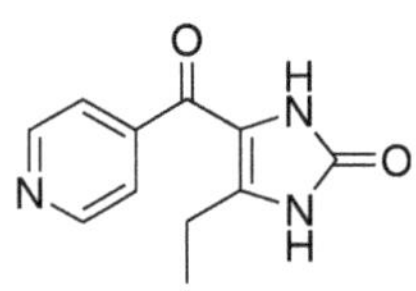

$t_{1/2}$	1.3 h	V	0.6–1.7
F	>80%	CL	7–9
pb	85%	Mwt	217.2
ur	64%	PSA	71.1 Å^2
		log P	2.15

PIROXIMONE
Inotropic
Metabolism (dog): reduction of ketone to alcohol, oxidative cleavage to isonicotinic acid

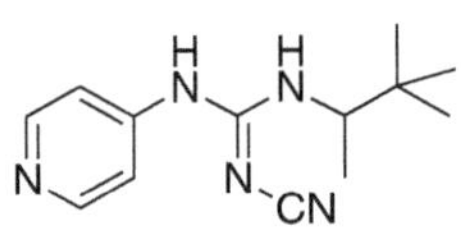

$t_{1/2}$	1.5–3.0 h	V	3.4
F	60%	CL	9.5
pb	60–65%	Mwt	245.3
ur	4–10%	PSA	73.1 Å^2
		log P	2.31

PINACIDIL
Potassium channel opener, antihypertensive
Metabolism: N-oxidation of pyridine ($t_{1/2}$ *N*-oxide: 2.5 h), hydroxylation of *t*-Bu, N-dealkylation

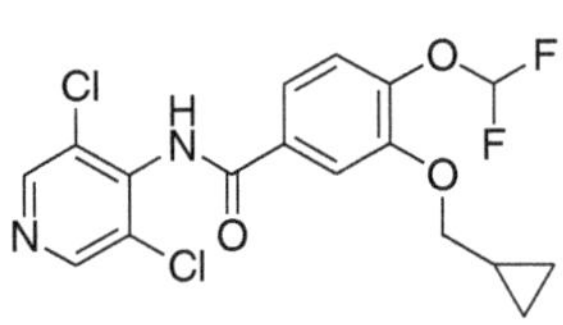

$t_{1/2}$	12–20 h	V	–
F	69–92%	CL	–
pb	99%	Mwt	403.2
ur	–	PSA	60.5 Å^2
		log P	2.31

ROFLUMILAST
PDE4 inhibitor, bronchodilator
Metabolite: pyridine *N*-oxide

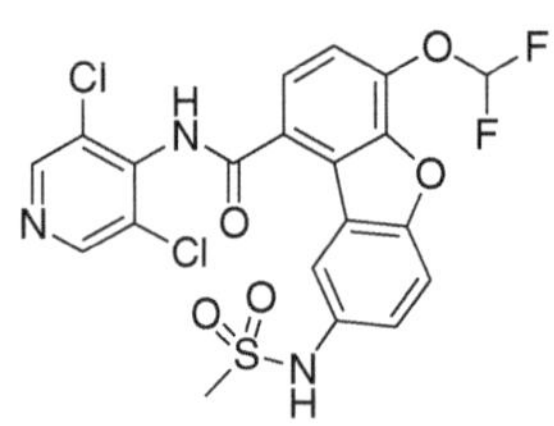

$t_{1/2}$	14–30 h	V	–
F	>50%*	CL	–
pb	–	Mwt	516.3
ur	–	PSA	119 Å^2
		log P	4.56

OGLEMILAST
*mice
PDE4 inhibitor
Metabolite: pyridine *N*-oxide

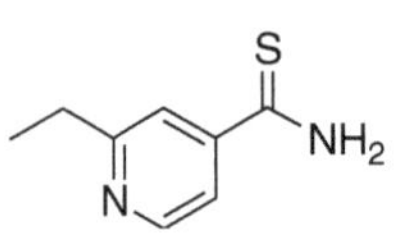

$t_{1/2}$	1.9±0.3 h	V	1.4±0.3
F	100%	CL	–
pb	10–30%	Mwt	166.2
ur	<1%	PSA	71.0 Å^2
		log P	1.17

ETHIONAMIDE
Tuberculostatic
Metabolism: S-oxidation, conversion to carboxamide, reduction of thioamide to hydroxymethyl

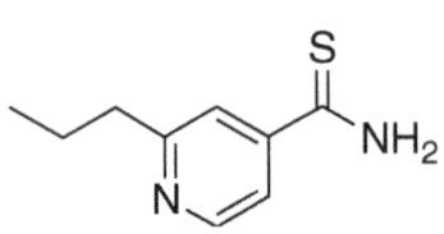

$t_{1/2}$	1.4 h	V	–
F	–	CL	–
pb	–	Mwt	180.3
ur	<1%	PSA	71.0 Å^2
		log P	1.68

PROTIONAMIDE
Tuberculostatic
Metabolism: S-oxidation, conversion to carboxamide and carboxylic acid, N-methylation

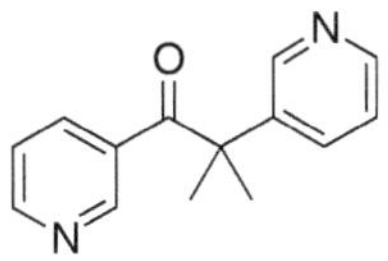	$t^{1}/_{2}$ 0.5–2.5 h *F* – pb – ur <1%	*V* – CL – Mwt 226.3 PSA 42.9 Å^2 log *P* 1.14	METYRAPONE, METOPYRONE Oral diagnostic aid Metabolism: reduction of ketone to alcohol
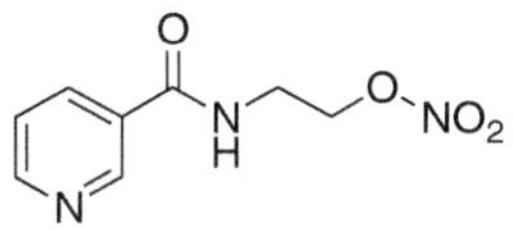	$t^{1}/_{2}$ 0.8–1.5 h *F* 75% pb 24% ur <1%	*V* 0.5–1.4 CL 7.3–16.4 Mwt 211.2 PSA 97.0 Å^2 log *P* 0.74	NICORANDIL Antianginal Metabolism: hydrolysis to nicotinic acid
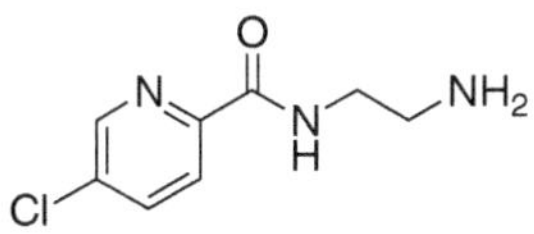	$t^{1}/_{2}$ 8–9 h *F* 78% pb – ur –	*V* 3.6±0.6 CL – Mwt 199.6 PSA 68.0 Å^2 log *P* 0.20	LAZABEMIDE MAO inhibitor Metabolism: oxidation of CH_2NH_2 to CO_2H
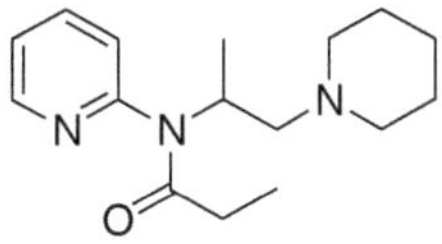	$t^{1}/_{2}$ 5–7 h *F* >97% pb – ur 24%	*V* 2.3 CL 6.3 Mwt 275.4 PSA 36.4 Å^2 log *P* 3.01	PROPIRAM Analgesic, narcotic; Metabolism: piperidine hydroxylation, then oxidation to lactam, then hydrolysis of lactam
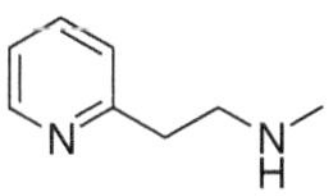	$t^{1}/_{2}$ 3–4 h *F* 100% pb 0–5% ur 0%	*V* – CL – Mwt 136.2 PSA 24.9 Å^2 log *P* −0.02	BETAHISTINE Vasodilator, antivertigo, histamine agonist Metabolites: desmethyl and 2-pyridylacetic acid
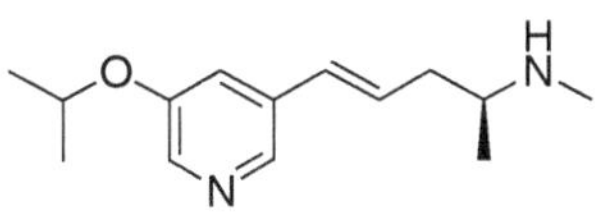	$t^{1}/_{2}$ 3–9 h *F* – pb – ur –	*V* – CL – Mwt 234.3 PSA 34.2 Å^2 log *P* 2.47	ISPRONICLINE Nicotinic acetylcholine partial agonist, nootropic Metabolism: O-dealkylation, N-demethylation, pyridine N-oxidation, oxidative degradation of side chain
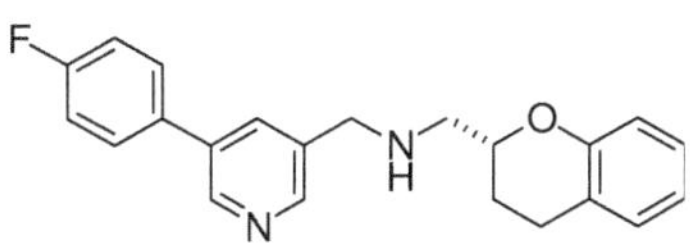	$t^{1}/_{2}$ 5–7 h *F* 89% pb >99.9% ur <5%	*V* 0.3–0.6 CL 0.7 Mwt 348.4 PSA 34.2 Å^2 log *P* 4.38	SARIZOTAN 5-HT_{1A} agonist for treatment of dyskinesia Metabolism: aromatic hydroxylation (para to O), hydroxylation at CH_2CH_2, N-de(pyridylmethylation), then oxidation to pyridine-3-carboxylic acid and chroman-2-carboxylic acid

(continued overleaf)

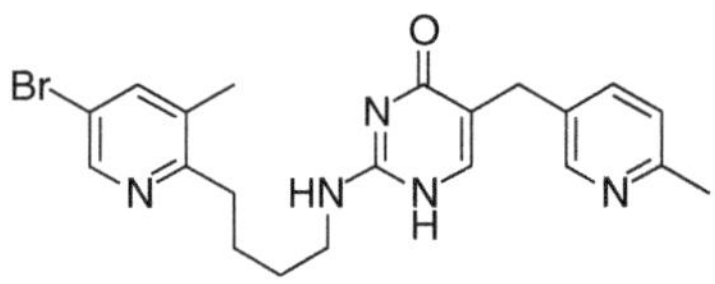	$t^1/_2$ 12 h *F* High pb >99% ur <0.2%	*V* – CL – Mwt 442.4 PSA 79.3 Å^2 log *P* 3.76	TEMELASTINE Histamine H_1 antagonist; does not cross bbb Metabolism: N-glucuronidation
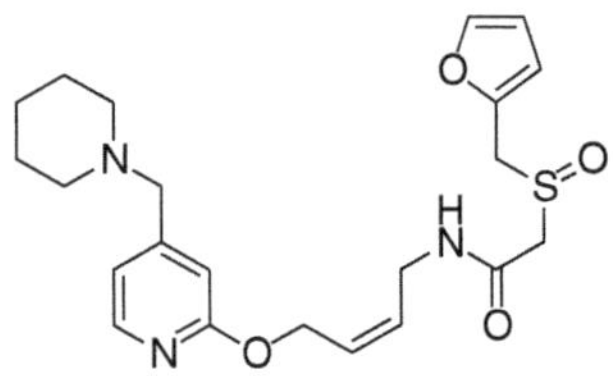	$t^1/_2$ 2 h *F* – pb – ur –	*V* 0.8 CL 4 Mwt 431.6 PSA 104 Å^2 log *P* 2.24	LAFUTIDINE Histamine H_2 antagonist Metabolism: hydroxylation, oxidation to sulfone
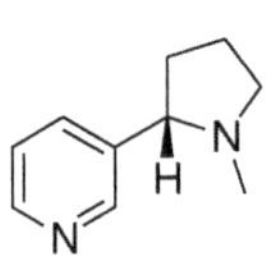	$t^1/_2$ 2.0±0.7 h *F* 30%* pb 5±3% ur 17±9%	*V* 2.6±0.9 CL 18.5±5.4 Mwt 162.2 PSA 16.1 Å^2 log *P* 0.57	NICOTINE *pulmonal: 90% Tobacco alkaloid Metabolism: N-demethylation (to nornicotine, $t^1/_2$ 7–9 h), N-oxidation and hydroxylation of pyrrolidine, N-methylation of pyridine, conversion to cotinine
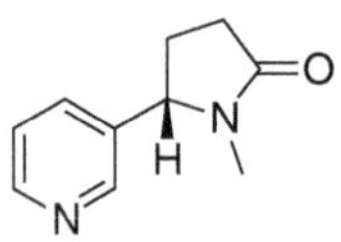	$t^1/_2$ 12–16 h *F* 84–100% pb – ur 12%	*V* 0.9 CL 0.9 Mwt 176.2 PSA 33.2 Å^2 log *P* 0.08	COTININE Metabolite of nicotine Metabolism: N-methylation and oxidation of pyridine, hydroxylation of pyrrolidinone, demethylation, glucuronidation
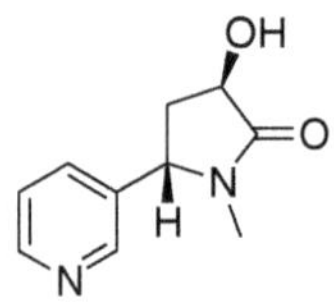	$t^1/_2$ 4–10 h* *F* – pb – ur –	*V* 0.5–1.1 CL 1.1–2.6 Mwt 192.2 PSA 53.4 Å^2 log *P* −1.12	3-HYDROXYCOTININE *iv Metabolite of cotinine
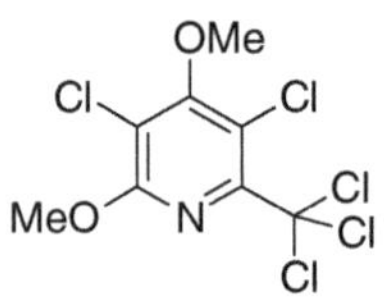	$t^1/_2$ 18 h *F* 49±18% pb – ur –	*V* – CL – Mwt 325.4 PSA 31.4 Å^2 log *P* 5.26	PENCLOMEDINE Antineoplastic Metabolism: demethylation
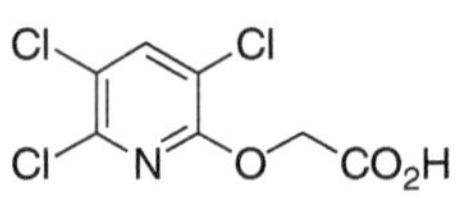	$t^1/_2$ 5 h *F* – pb – ur 80%	*V* – CL – Mwt 256.5 PSA 59.4 Å^2 log *P* 2.35	TRICLOPYR Herbicide

Structure			Drug
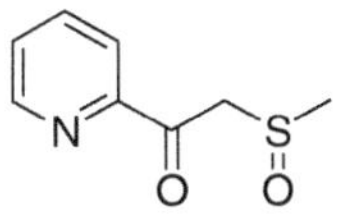	$t_{1/2}$ 1.2 h F – pb – ur <1%	V – CL – Mwt 183.2 PSA 66.2 Å^2 log P 0.02	OXISURAN Immunosuppressor Metabolism: reduction of ketone to alcohol ($t_{1/2}$ 55 h)
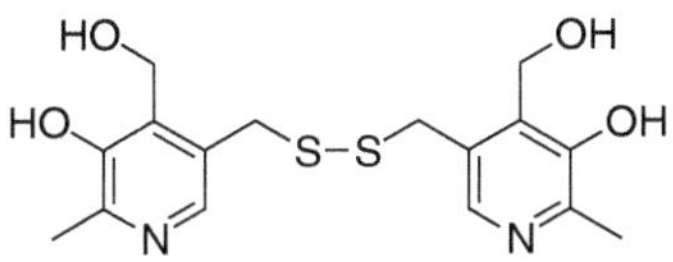	$t_{1/2}$ 2–3 h F – pb – ur –	V – CL – Mwt 368.5 PSA 157 Å^2 log P 0.64	PYRITINOL Nootropic
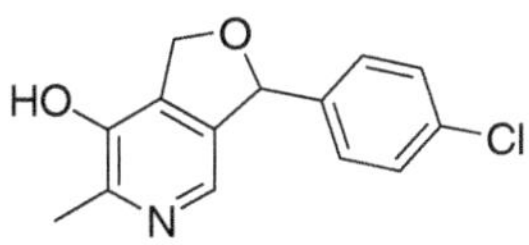	$t_{1/2}$ 5–18 h F 75% (rat) pb 97% ur <1%	V 0.6 CL – Mwt 261.7 PSA 42.4 Å^2 log P 1.78	CICLETANINE Antihypertensive Metabolism: glucuronidation, sulfation
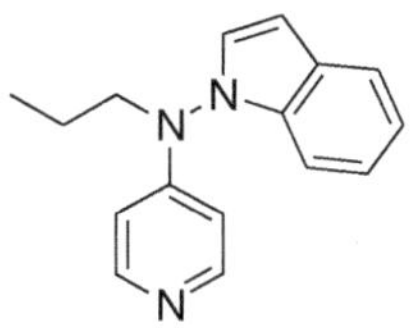	$t_{1/2}$ 7±2 h* F – pb – ur 1%*	V – CL 21±7* Mwt 251.3 PSA 21.1 Å^2 log P 4.18	BESIPIRDINE *monkey Nootropic Metabolism: N-depropylation
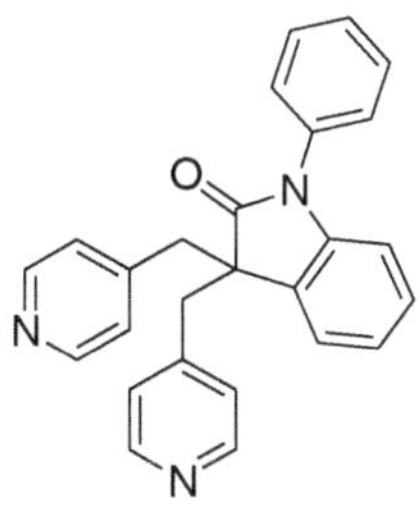	$t_{1/2}$ 1–3 h F <10% (rat) pb – ur Negligible	V 2 (rat) CL 0.6 Mwt 391.5 PSA 46.1 Å^2 log P 1.95	LINOPIRDINE CNS stimulant, nootropic Metabolism: N-oxidation
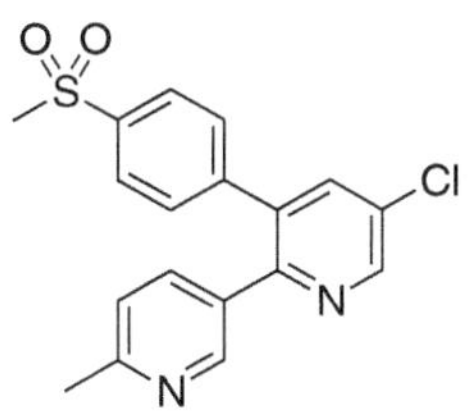	$t_{1/2}$ 21–27 h F 100% pb 92% ur <1%	V 1.6–1.7 CL 0.7–0.8 Mwt 358.8 PSA 68.3 Å^2 log P 2.21	ETORICOXIB Antiinflammatory, analgesic Metabolism (CYP3A4): oxidation of methylpyridine to hydroxymethyl- and carboxypyridine

(continued overleaf)

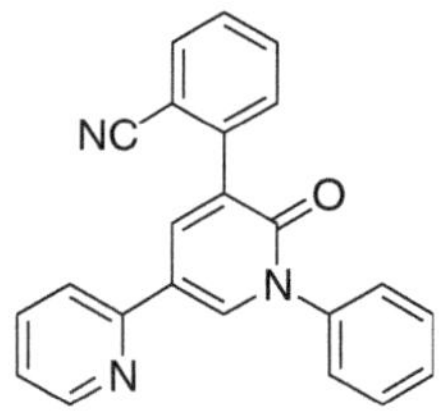

$t^1/_2$	8 h*	V	–	PERAMPANEL
F	75%*	CL	–	*monkey
pb	–	Mwt	349.4	AMPA antagonist, antiepileptic
ur	–	PSA	57.0 Å^2	
		log P	3.70	

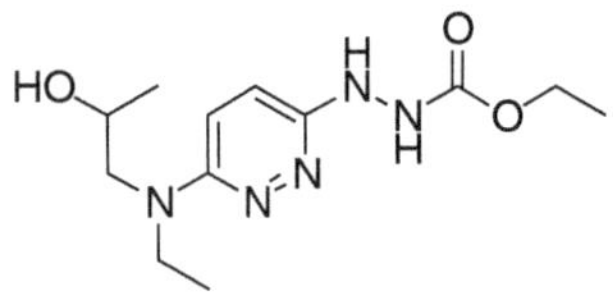

$t^1/_2$	0.8–3.2 h	V	0.67	CADRALAZINE
F	>80%	CL	4.4	Antihypertensive
pb	26%	Mwt	283.3	Metabolism: hydrolysis of carbamate
ur	70–80%	PSA	99.6 Å^2	
		log P	0.54	

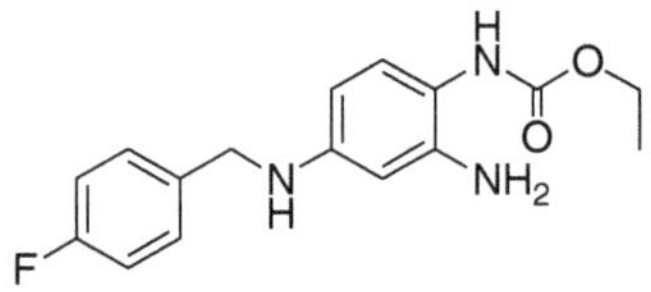

$t^1/_2$	7–12 h	V	6.2	RETIGABINE, EZOGABINE
F	60%	CL	11	Anticonvulsant
pb	80%	Mwt	303.3	Metabolism: N-acetylation,
ur	36%	PSA	76.4 Å^2	N-glucuronidation
		log P	0.96	

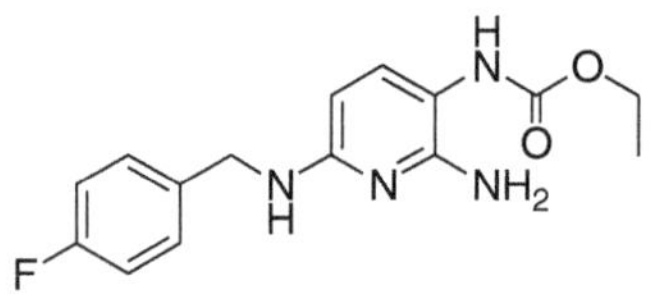

$t^1/_2$	8.5 h	V	1.2	FLUPIRTINE
F	90%*	CL	1.6	*rectal: 72%
pb	80%	Mwt	304.3	Analgesic, crosses bbb
ur	15%	PSA	89.3 Å^2	Metabolism: carbamate hydrolysis,
		log P	1.60	then N-acetylation

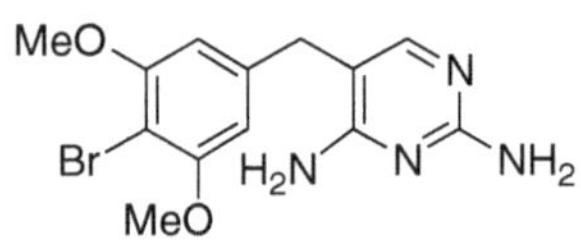

$t^1/_2$	32–35 h	V	1.5	BRODIMOPRIM
F	80–90%	CL	0.6	Antibacterial
pb	93%	Mwt	339.2	Metabolism: N-oxidation,
ur	4–7%	PSA	96.3 Å^2	hydroxylation of methylene group,
		log P	2.29	O-demethylation, then
				O-glucuronidation and sulfation

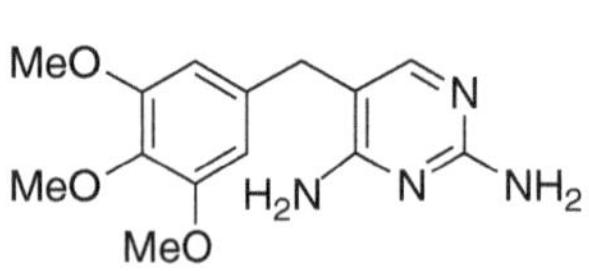

$t^1/_2$	10±2 h	V	1.6±0.2	TRIMETHOPRIM
F	100%	CL	1.9±0.3	Inhibitor of dihydrofolate
pb	37±5%	Mwt	290.3	biosynthesis, antibacterial
ur	63±10%	PSA	106 Å^2	Metabolism: O-demethylation, then
		log P	0.59	oxidation to *o*-quinones, *p*-quinone
				methides, and iminoquinone
				methides

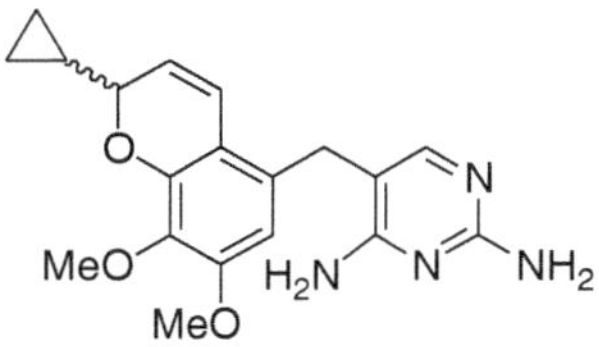	$t_{1/2}$ 2–4 h (iv) *F* 41% pb 92–94% ur 2%	*V* 1.3–1.7 CL 7.8–9.2 Mwt 354.4 PSA 106 Å^2 log *P* 2.20	ICLAPRIM Antibacterial Metabolism: O-demethylation, oxidative ring scission of pyran, dihydroxylation of alkene, N-oxidation, glucuronidation
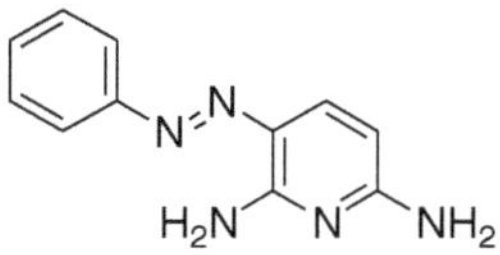	$t_{1/2}$ 43–75 min *F* >60% pb – ur 65%	*V* – CL – Mwt 213.2 PSA 89.7 Å^2 log *P* 2.21	PHENAZOPYRIDINE Analgesic Metabolism: phenyl 2- and 4-hydroxylation, reductive cleavage of N=N bond
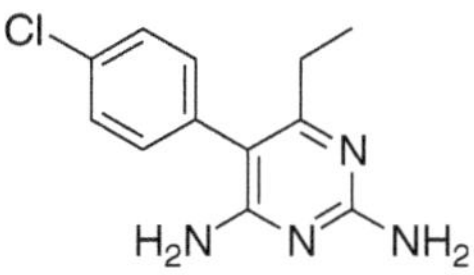	$t_{1/2}$ 81±32 h *F* 100% pb 87±1% ur 65%	*V/F* 2.3±0.6 CL/*F* 0.41±0.06 Mwt 248.7 PSA 77.8 Å^2 log *P* 2.75	PYRIMETHAMINE Antiprotozoal Metabolite: *N*-oxide
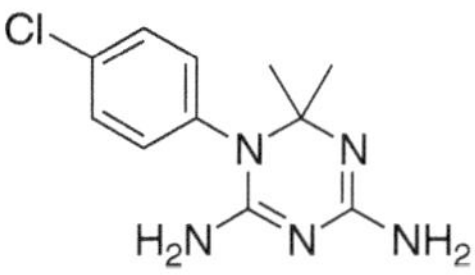	$t_{1/2}$ 12–15 h *F* – pb – ur –	*V* – CL – Mwt 251.7 PSA 80.0 Å^2 log *P* 0.95	CYCLOGUANIL Active metabolite of proguanil
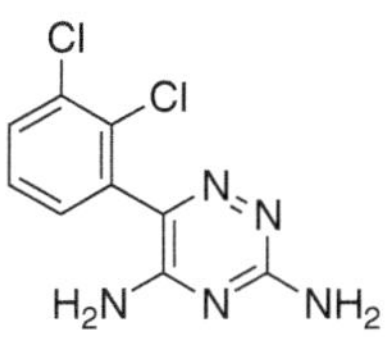	$t_{1/2}$ 18–37 h *F* 100% pb 40–60% ur 5–10%	*V* 0.9–1.3 CL 0.35–0.60 Mwt 256.1 PSA 90.7 Å^2 log *P* 1.24	LAMOTRIGINE Anticonvulsant Metabolites: *N*-glucuronide, *N*-oxide, *N*-methyl
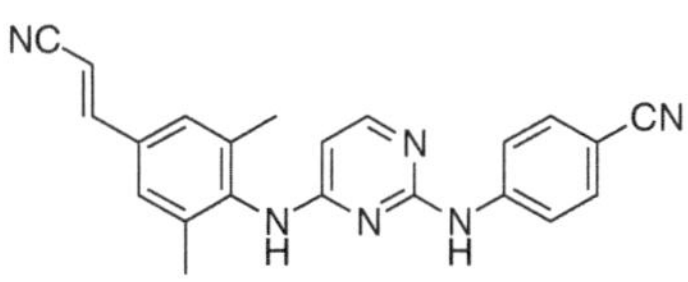	$t_{1/2}$ 36–48 h *F* 100%* pb 99.7% ur <1%	*V* – CL – Mwt 366.4 PSA 97.4 Å^2 log *P* 3.63	RILPIVIRINE *dog Antiviral (HIV)
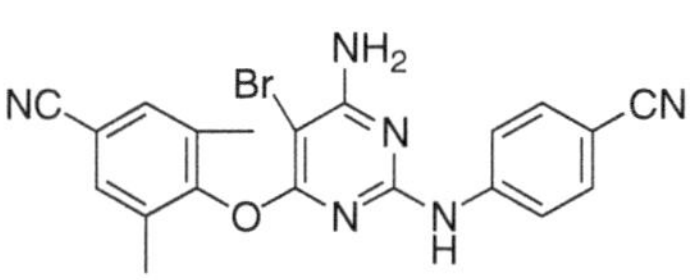	$t_{1/2}$ 30–40 h *F* – pb 99.9% ur 0%	*V* 6.5 CL – Mwt 435.3 PSA 121 Å^2 log *P* 4.19	ETRAVIRINE Antiviral (HIV) Metabolism: hydroxylation of methyl followed by glucuronidation

(*continued overleaf*)

Structure	Pharmacokinetics	Properties	Compound
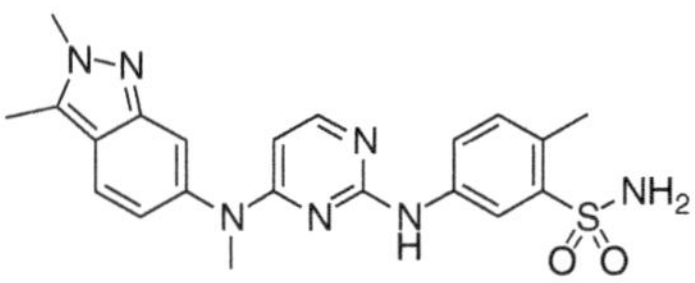	$t_{1/2}$ 31 h *F* 14–39% pb >99% ur <4%	*V* 0.3 CL 1.4 Mwt 437.5 PSA 127 Å^2 log *P* 1.98	PAZOPANIB Tyrosine kinase inhibitor, antiangiogenic
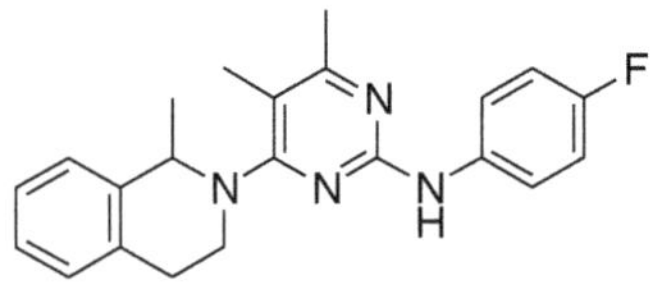	$t_{1/2}$ 2.2–2.4 h *F* 40% (rat) pb – ur 0%	*V* – CL – Mwt 362.4 PSA 41.1 Å^2 log *P* 5.29	REVAPRAZAN, YH1885 Proton pump inhibitor, antiulcerative
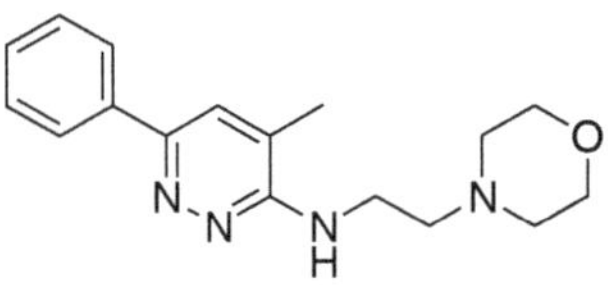	$t_{1/2}$ 2–5 h *F* 5% (rat) pb – ur Trace (rat)	*V* – CL – Mwt 298.4 PSA 50.3 Å^2 log *P* 2.04	MINAPRINE Antidepressant Metabolism: aromatic 4-hydroxylation, N-oxidation, N-dealkylation
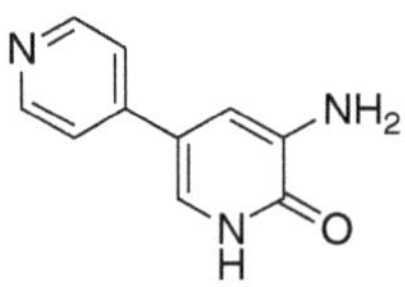	$t_{1/2}$ 1.4–5.8 h *F* 93±12% pb 35–49% ur 10–40%	*V* 1.3±0.3 CL 2.4–11.6 Mwt 187.2 PSA 68.0 Å^2 log *P* −0.21	AMRINONE Phosphodiesterase III inhibitor, cardiotonic Metabolism: N-acetylation, N-acylation with glycolic acid
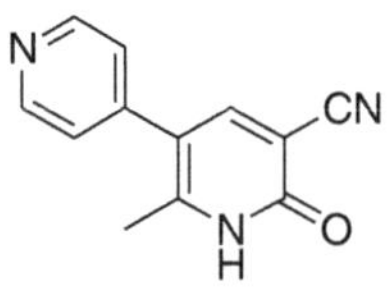	$t_{1/2}$ 0.8±0.2 h *F* 80–92% pb 85±10% ur 85±10%	*V* 0.32±0.08 CL 6.1±1.3 Mwt 211.2 PSA 65.8 Å^2 log *P* 1.27	MILRINONE Phosphodiesterase III inhibitor, cardiotonic Metabolism: O-glucuronidation
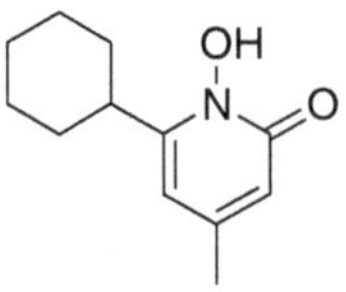	$t_{1/2}$ 1.7 h *F* – pb 96% ur 6%	*V* – CL – Mwt 207.3 PSA 40.5 Å^2 log *P* 2.57	CICLOPIROX Antifungal Metabolism: glucuronidation
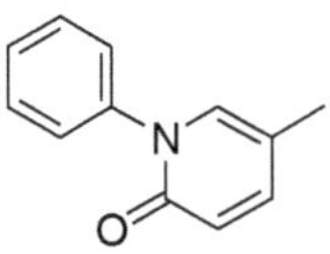	$t_{1/2}$ 2–3 h *F* 56% (dog) pb 58% ur <1%	*V* – CL – Mwt 185.2 PSA 20.3 Å^2 log *P* 1.82	PIRFENIDONE Antifibrotic Metabolism: oxidation of methyl to carboxyl

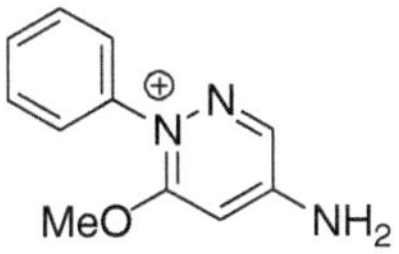	$t_{1/2}$ 9–13 h *F* 40–74% pb – ur 32%	*V* 2–3 CL – Mwt 202.2 PSA 52.0 Å^2 log *P* −1.76	AMEZINIUM Antihypotensive Metabolism: O-demethylation (to uncharged pyridazin-2-one), then hydroxylation and conjugation
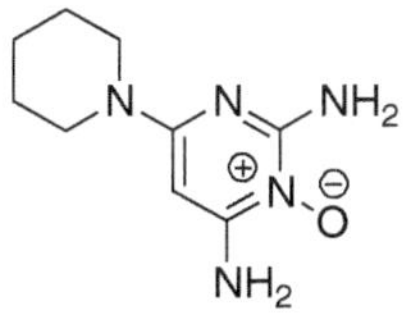	$t_{1/2}$ 3.1±0.6 h *F* 90% pb 0% ur 20±6%	*V/F* 2.7±0.7 CL/*F* 24±6 Mwt 209.3 PSA 93.6 Å^2 log *P* 1.62	MINOXIDIL Potassium channel opener, antihypertensive, antialopecia agent Metabolism: O-sulfation, glucuronidation, 3- and 4-hydroxylation of piperidine, oxidative ring-opening of piperidine (to 4-carboxybutylamino), reductive N−O bond cleavage
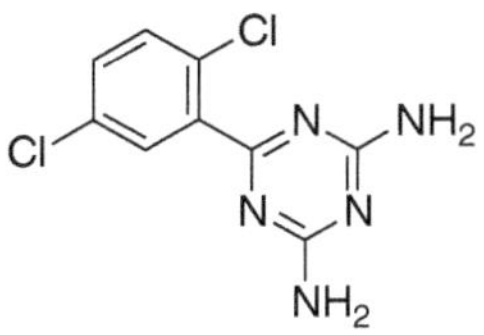	$t_{1/2}$ 6 d *F* – pb – ur 1.4%	*V* – CL – Mwt 256.1 PSA 90.7 Å^2 log *P* 2.08	IRSOGLADINE Antiulcerative Metabolism: N-oxidation
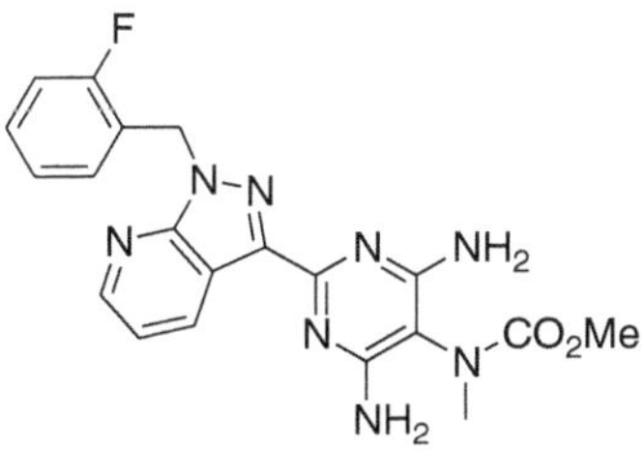	$t_{1/2}$ 5–12 h *F* – pb – ur –	*V/F* 0.4 CL/*F* 0.4 Mwt 422.4 PSA 138 Å^2 log *P* −0.31	RIOCIGUAT Guanylate cyclase stimulator for treatment of pulmonary hypertension
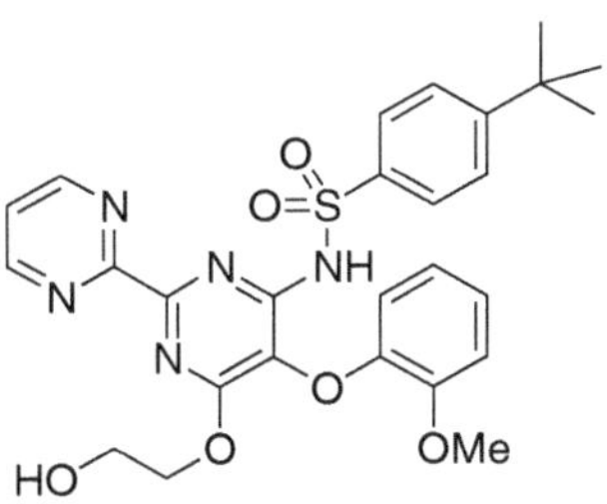	$t_{1/2}$ 4.3–7.3 h *F* 43–50% pb 96–98% ur <1%	*V* 0.2 CL 1.4–2.8 Mwt 551.6 PSA 154 Å^2 log *P* 1.15	BOSENTAN Antihypertensive CYP2C9 and CYP3A4 inducer Metabolism: hydroxylation of *t*-Bu, O-demethylation

(*continued overleaf*)

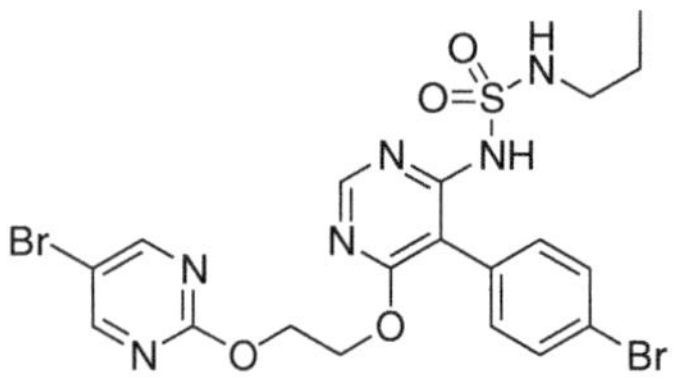	$t^1/_2$ 13–18 h *F* 80%* pb 99%* ur –	*V* – CL – Mwt 588.3 PSA 137 Å^2 log *P* 5.41	MACITENTAN *dog Endothelin antagonist Metabolite: despropyl ($t^1/_2$ 45–66 h)
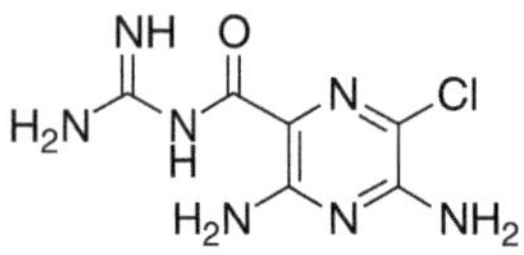	$t^1/_2$ 6–10 h *F* 50% pb 40% ur 49±10%	*V* 17±4 CL 9.7±1.9 Mwt 229.6 PSA 157 Å^2 log *P* 0.93	AMILORIDE Diuretic
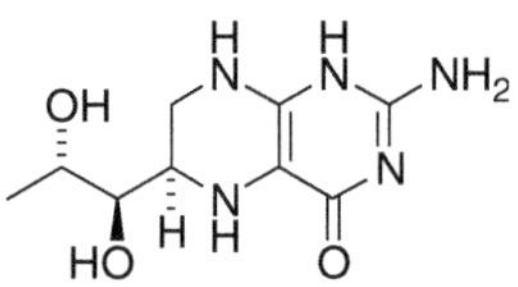	$t^1/_2$ 7 h *F* 10%* pb Low ur –	*V* – CL – Mwt 241.3 PSA 132 Å^2 log *P* −4.22	SAPROPTERIN *animals Essential enzymatic cofactor; treatment of phenylketonuria Metabolites: dihydrobiopterin, biopterin
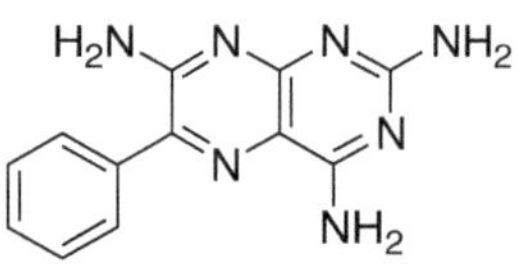	$t^1/_2$ 4.2±0.7 h *F* 54±12% pb 61±2% ur 52±10%	*V* 13.4±4.9 CL 63±20 Mwt 253.3 PSA 130 Å^2 log *P* 1.16	TRIAMTERENE Diuretic Metabolism: hydroxylation, glucuronidation, sulfation
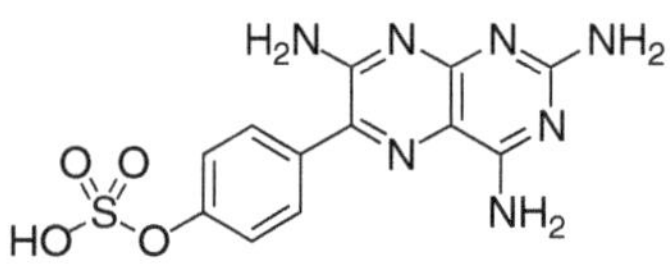	$t^1/_2$ 3.1±1.2 h *F* – pb 90±1% ur –	*V* – CL 2.3±0.6 Mwt 349.3 PSA 202 Å^2 log *P* −0.13	HYDROXYTRIAMTERENE SULFATE Active metabolite of triamterene
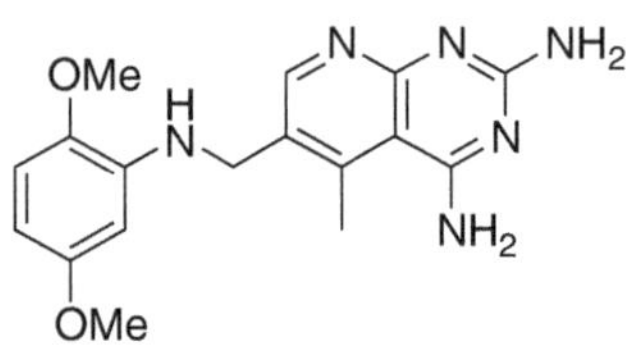	$t^1/_2$ 2–6 h *F* 75–95% pb – ur <3%*	*V* 1.8* CL 10* Mwt 325.4 PSA 109 Å^2 log *P* 1.97	PIRITREXIM *dog Antineoplastic Metabolism: O-demethylation

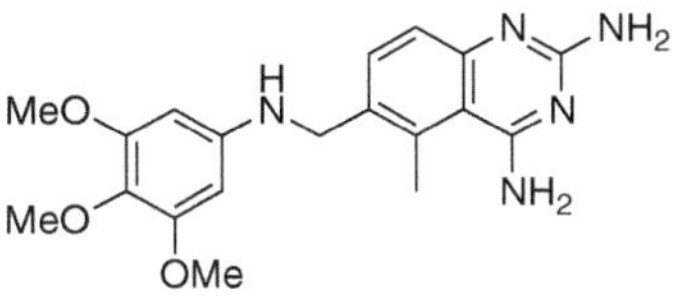	$t_{1/2}$ 12–20 h F 44% pb 98% ur <5%	V 0.6–0.8 CL 0.4–0.7 Mwt 369.4 PSA 118 Å^2 log P 1.76	TRIMETREXATE Antineoplastic Metabolism: O-demethylation, then glucuronidation, sulfation

$t_{1/2}$, plasma half-life; F, oral bioavailability; pb, plasma protein binding; ur, excretion of unchanged drug in urine; V, volume of distribution; CL, clearance; Mwt, molecular weight; PSA, polar surface area; AMPA, α-amino-3-hydroxy-5-methyl-4-isoxazolepropionic acid receptor; bbb, blood–brain barrier; CNS, central nervous system; MAO, monoamine oxidase; NAD, nicotine–adenine-dinucleotide; PDE, phosphodiesterase.

Table 59.2 Proton pump inhibitors and related compounds. V in l kg^{-1}; CL in ml min^{-1} kg^{-1}; Mwt in g mol^{-1}.

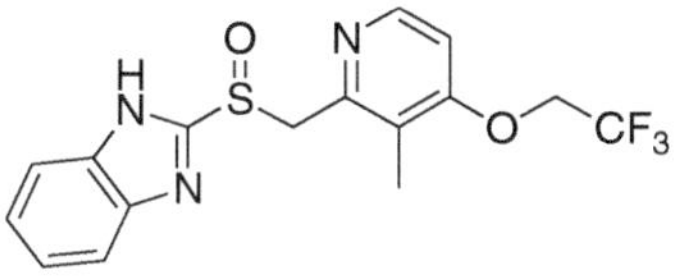	$t_{1/2}$ 0.9–2.1 h F 81–91% pb 97–99% ur <2%	V 0.28–0.45 CL 2.6–7.4 Mwt 369.4 PSA 87.1 Å^2 log P 2.58	LANSOPRAZOLE Racemic, proton pump inhibitor, antiulcerative Metabolism by CYP3A4 and CYP2C19 to sulfone and 5-hydroxybenzimidazole
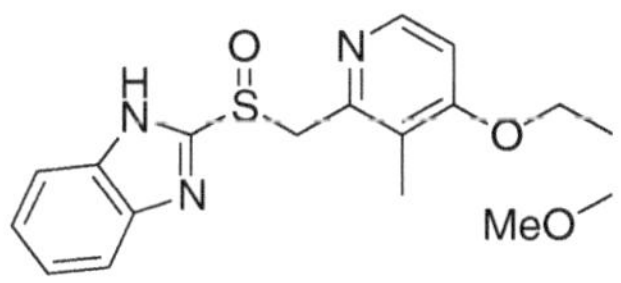	$t_{1/2}$ 1–2 h F 52% pb 95–98% ur 0%	V 0.22 CL 4.3–8.4 Mwt 359.4 PSA 96.3 Å^2 log P 2.10	RABEPRAZOLE Racemic, proton pump inhibitor, antiulcerative Metabolism: demethylation, oxidation to sulfone, reduction to thioether
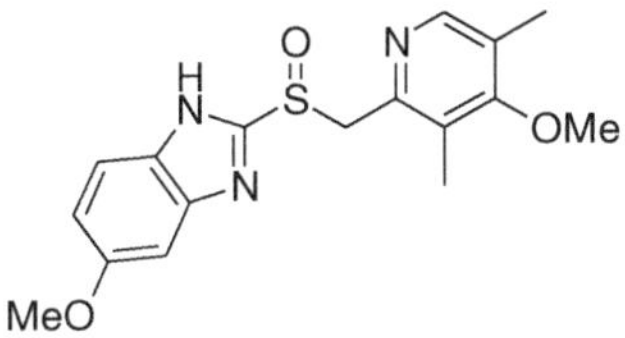	$t_{1/2}$ 0.7±0.5 h F 53±29% pb 95% ur <0.1%	V 0.34±0.09 CL 7.5±2.7 Mwt 345.4 PSA 96.3 Å^2 log P 2.36	OMEPRAZOLE Racemic, proton pump inhibitor, antiulcerative Metabolism by CYP2C19 and CYP3A to 5-(hydroxymethyl), sulfone, *O*-desmethyl (5-hydroxybenzimidazole)
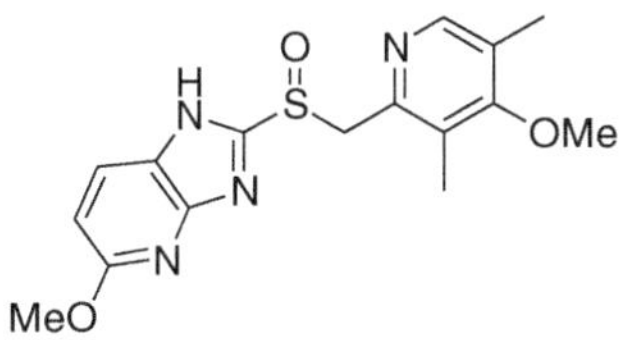	$t_{1/2}$ 5–8 h F 77–88% pb – ur –	V – CL – Mwt 346.4 PSA 104 Å^2 log P 346.4	TENATOPRAZOLE Racemic, proton pump inhibitor, antiulcerative Metabolites: sulfide, sulfone

(*continued overleaf*)

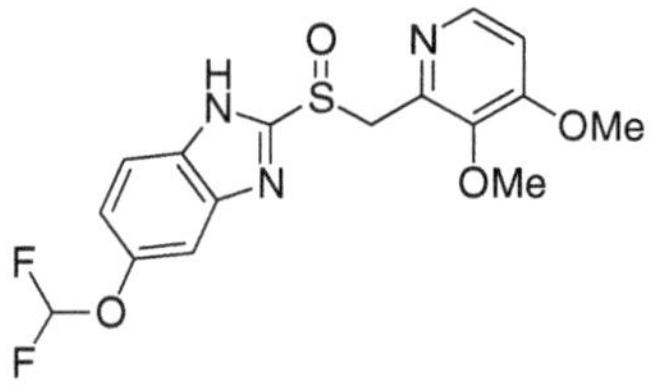	$t_{1/2}$ 0.9–1.9 h F 67–89% pb 98% ur Negligible	V 0.17 CL 2.2 Mwt 383.4 PSA 106 Å^2 log P 1.57	PANTOPRAZOLE Racemic, proton pump inhibitor, antiulcerative Metabolism: oxidation to sulfone, reduction to thioether, demethylation followed by sulfation
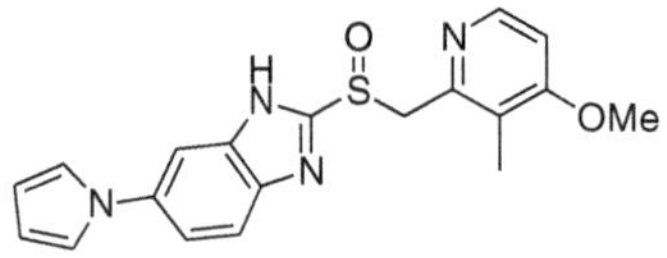	$t_{1/2}$ 4–9 h F – pb – ur –	V/F 1.7–2.5 CL/F 2.4–3.0 Mwt 366.4 PSA 92.0 Å^2 log P 2.66	ILAPRAZOLE Racemic, proton pump inhibitor, antiulcerative Metabolism: oxidation to sulfone, reduction to sulfide, hydroxylation of pyrrole
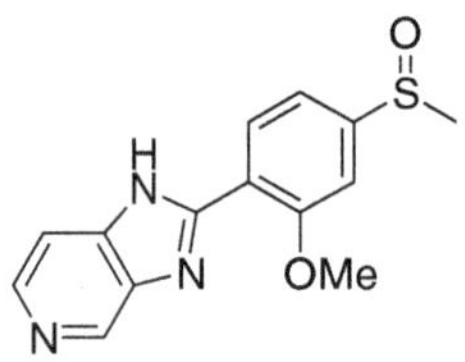	$t_{1/2}$ 1.4 h F – pb – ur –	V 2–3 CL 19–24 Mwt 287.3 PSA 87.1 Å^2 log P 1.14	ISOMAZOLE Cardiotonic Metabolism: oxidation to sulfone, hydroxylation
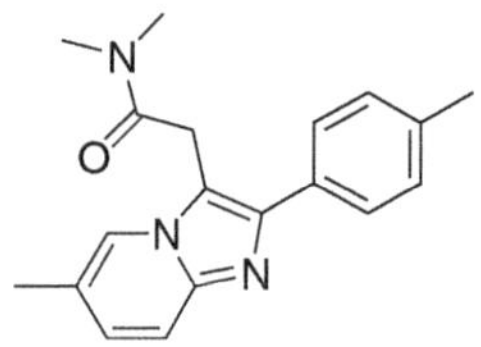	$t_{1/2}$ 2.2±0.4 h F 67±20% pb 92% ur <1%	V 0.54±0.09 CL 4.3±2.2 Mwt 307.4 PSA 37.6 Å^2 log P 3.07	ZOLPIDEM Sedative, hypnotic; CYP3A4 substrate Metabolism: benzylic hydroxylation, then oxidation of CH_2OH to CO_2H
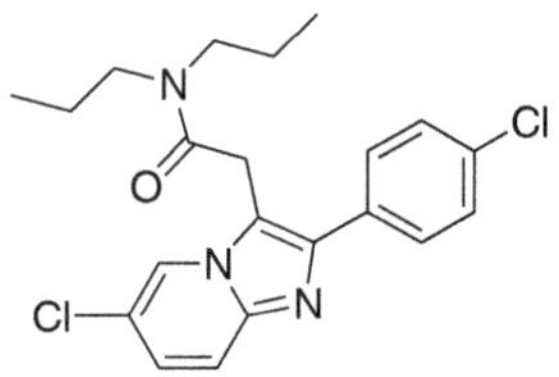	$t_{1/2}$ 18–20 h F 13% (rat) pb 99% ur 0%	V 5 (rat) CL – Mwt 404.3 PSA 37.6 Å^2 log P 5.49	ALPIDEM Anxiolytic Metabolism: N-dealkylation, hydroxylation of CH_2Me; withdrawn in 1996 because of hepatotoxicity
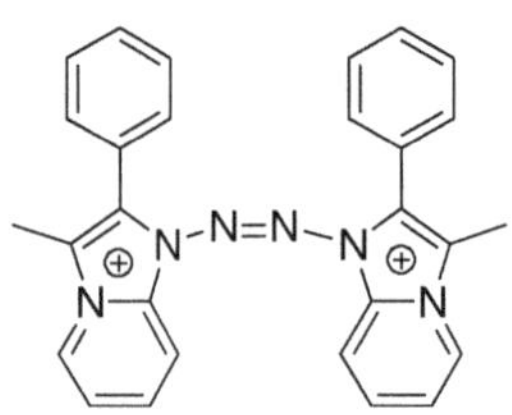	$t_{1/2}$ 1 h F – pb 17% ur 0%	V 0.18–0.23 CL 1.9–2.3 Mwt 444.5 PSA 42.8 Å^2 log P −1.44	FAZADINIUM Muscle relaxant Metabolism: reductive N=N bond cleavage, phenyl 4-hydroxylation

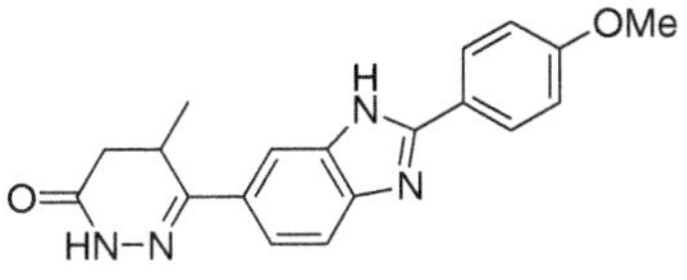	$t_{1/2}$ 0.4–1.5 h *F* 60–65% pb 98% ur –	*V* 2 CL 14 Mwt 334.4 PSA 79.4 Å^2 log *P* 2.08	PIMOBENDAN Cardiotonic Metabolism: N-glucuronidation, O-demethylation, then O-glucuronidation Active metabolite: *O*-desmethyl ($t_{1/2}$ 2 h)
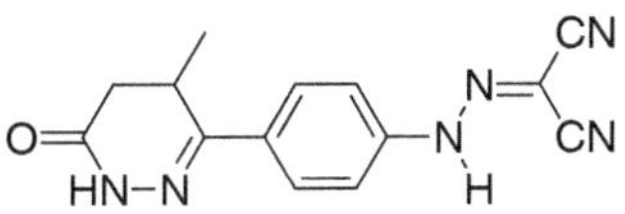	$t_{1/2}$ 0.8–1.8 h *F* 85±6% pb 98% ur 0%	*V* 0.28 CL 4 Mwt 280.3 PSA 113 Å^2 log *P* 1.02	SIMENDAN Calcium sensitizer for treatment of congestive heart failure; levosimendan has similar PK Metabolism: reduction to 4-aminophenyl, then N-acetylation
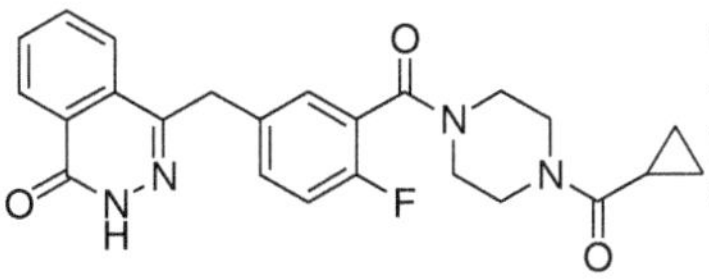	$t_{1/2}$ 5–10 h *F* 100%* pb – ur –	*V* 16–87 CL 0.3–2.4 Mwt 434.5 PSA 82.1 Å^2 log *P* 0.00	OLAPARIB *rat, dog PARP inhibitor, antineoplastic

$t_{1/2}$, plasma half-life; *F*, oral bioavailability; pb, plasma protein binding; ur, excretion of unchanged drug in urine; V, volume of distribution; CL, clearance; Mwt, molecular weight; PSA, polar surface area; PARP = poly(ADP-ribose)polymerase.

60
Quinolines

The quinoline substructure is not essential in all quinoline-containing drugs (Table 60.1). Substituted quinolines and quinazolines are, however, much easier to synthesize and metabolically less questionable than, for instance, substituted naphthalenes.

The most important quinoline-derived drugs are antimalarials, quinolone antibacterials, and the "tecan" antineoplastics. In these drugs, the quinoline substructure is an essential pharmacophore and cannot be replaced by other heteroarenes. In other drugs, however, the quinoline substructure just acts as lipophilic substituent or as scaffold and may be replaced by isoquinolines, quinazolines, or related heteroarenes. Thus, although most antineoplastic tyrosine kinase inhibitors are quinazolines, such as gefitinib, a quinoline scaffold can yield compounds with similar biological activity (e.g., bosutinib, Scheme 60.1).

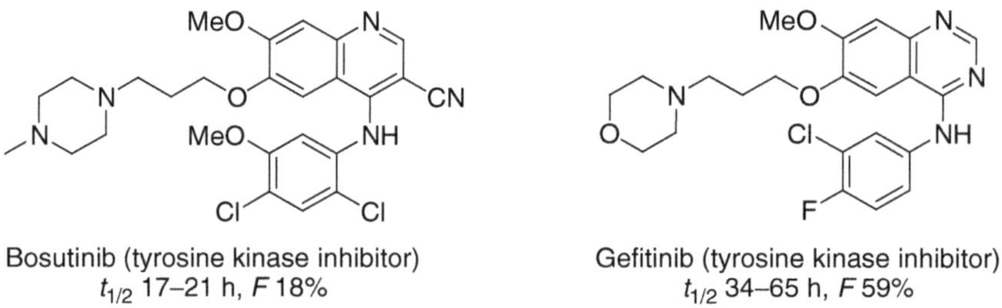

Scheme 60.1 Quinolines versus quinazolines.

Few drugs (mainly topical antifungals) contain unsubstituted naphthyl groups, which may hint at toxic metabolites. Quinolines, being less electron rich, are not readily metabolized to arene oxides and may, therefore, be safe bioisosteres of naphthyl groups.

Lead Optimization for Medicinal Chemists: Pharmacokinetic Properties of Functional Groups and Organic Compounds, First Edition. Florencio Zaragoza Dörwald.

The antimalarial properties of extracts from the chinchona bark were already known in the seventeenth century. The active principle, quinine, was isolated in 1820 by Pelletier and Caventou and used as antimalarial since then.

In the 1930s, cheaper, more readily accessible antimalarials were sought, and thousands of acridine and quinine analogs were prepared and tested. Interestingly, chloroquine (Scheme 60.2) was independently discovered by German (1934) and American chemists (1943) and became one of the most successful antimalarial drugs. Quinoline-derived antimalarials have high oral bioavailabilities and long half-lives and can be used for years without serious side effects.

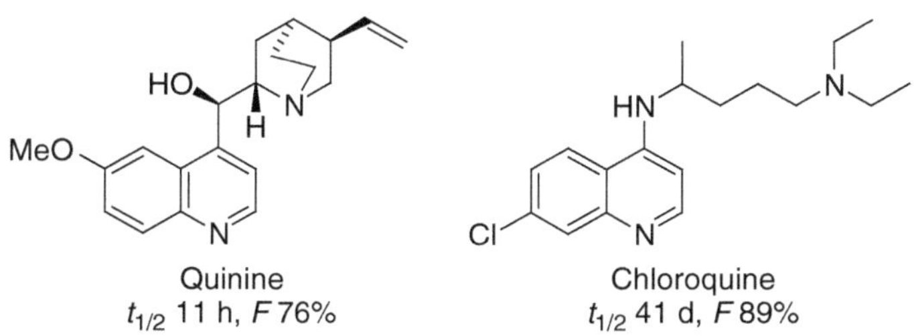

Scheme 60.2 Antimalarial quinolines.

Chloroquine is highly toxic to malaria schizonts, the pre-erythrocytic stage of the malaria parasite *Plasmodium falciparum*. The drug is assumed to act by inhibiting the parasites digestion of hemoglobin and by fragmenting and intercalating into its nucleic acids. Some malaria strains have become chloroquine resistant, and new drugs and treatment regimes are constantly being developed [1].

Quinolines are metabolically stable, and degradation *in vivo* takes place only at the substituents. Only quinolines substituted with electron-donating groups may undergo aromatic hydroxylation *in vivo*. The long half-life of chloroquine is mainly due to its extensive distribution into all kinds of tissue (volume of distribution: 100–1000 l kg^{-1}).

Illustrative examples of quinoline metabolism are sketched in Scheme 60.3.

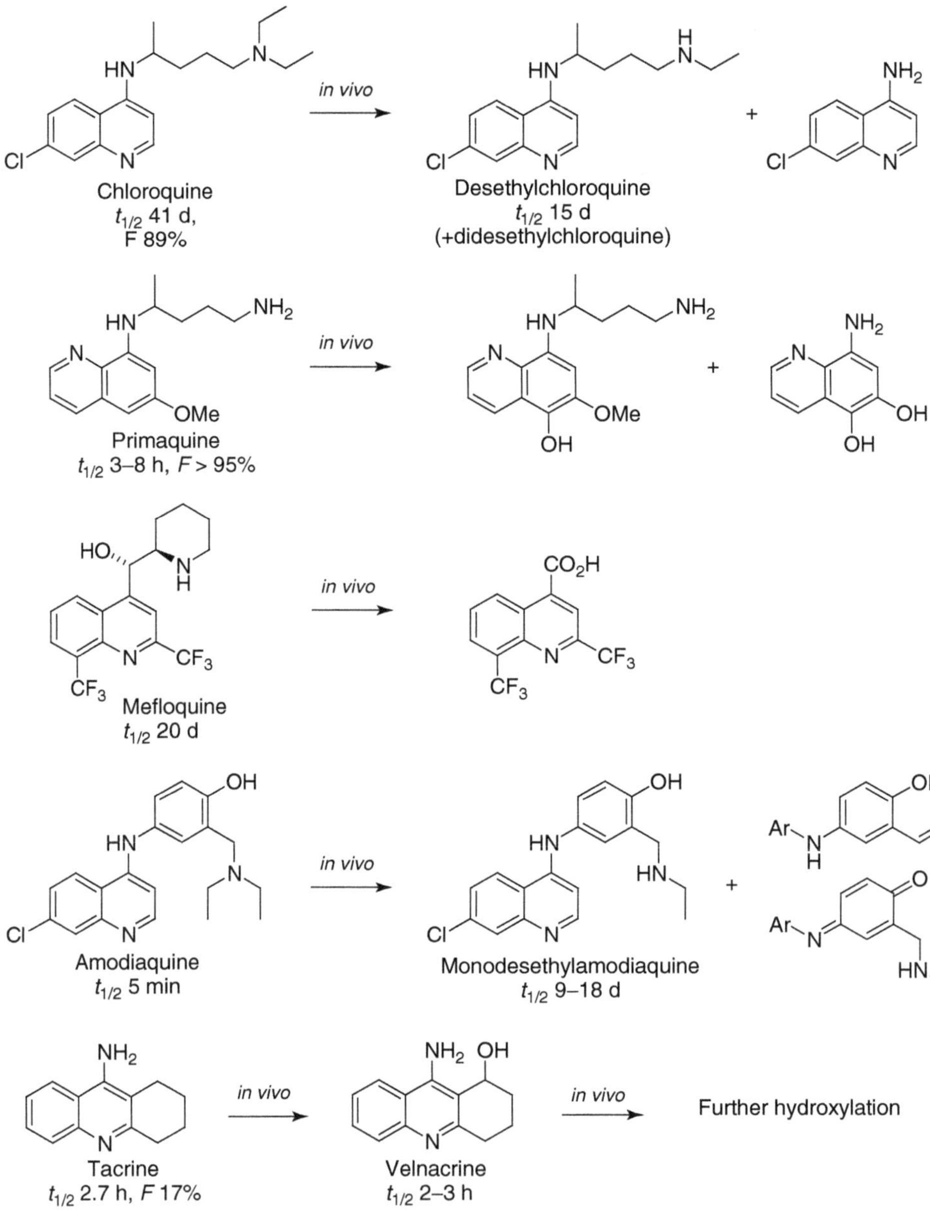

Scheme 60.3 Metabolism of quinolines.

60.1 Tecans

The antineoplastic properties of the tree *Camptotheca acuminata* were discovered 1958, and the active alkaloid camptothecin was isolated 1966 at the Research Triangle Institute (NC). This compound does not readily dissolve in water (0.0025 mg ml^{-1}), not even as sodium salt of the saponified lactone. Despite its new and promising biological activity (topoisomerase I inhibition [2]), the compound could not be advanced to phase II clinical trials because of poor solubility and excessive toxicity.

In the 1980s, the National Cancer Institute (USA) and SmithKline Beecham provided camptothecin to the University of Florida at Gainsville, where a team headed by Warren Ross developed a soluble, semisynthetic analog of camptothecin: topotecan (Scheme 60.4). The remaining development was carried out by SmithKline Beecham, and topotecan was finally launched in 1996 for the treatment of ovarian and small cell lung cancer. Since then, a number of additional camptothecin analogs were successfully developed [3] (Table 60.2).

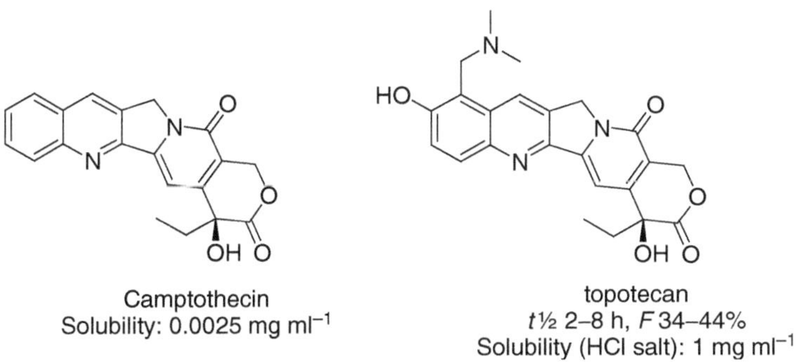

Scheme 60.4 Improvement of the solubility of camptothecin.

60.2 Quinazolines

Structurally closely related to quinolines are the quinazolines, the substructure of various important classes of drugs. These include α_1-adrenergic antagonists (e.g., prazosin), tyrosine kinase inhibitors (e.g., erlotinib), and some antifolates. The quinazoline and quinoline substructures can often be interchanged, with rough retention of biological activity. Other heterocycles that may be used as an alternative to quinolines or quinazolines include pteridines and naphthyridines (Scheme 60.5). Quinazolines are, however, often easiest to prepare (from anthranilic acid derivatives).

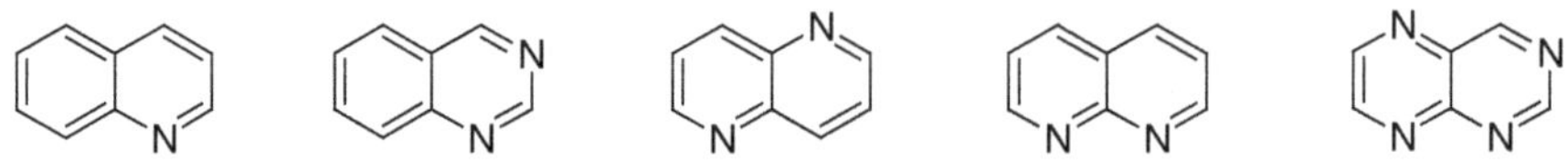

	Quinoline	Quinazoline	1,5-Naphthyridine	1,8-Naphthyridine	Pteridine
mp	−15 °C	48 °C	73 °C	98 °C	138 °C
bp	237 °C	241 °C			265 °C

Scheme 60.5

α_1-Adrenergic antagonists lead to a general vasodilation and are mainly used as antihypertensives and for the treatment of benign prostatic hyperplasia (Table 60.3). Main metabolic pathways include O-demethylation of methoxy groups and N-dealkylation of amines.

References

1. O'Neill, P.M., Bray, P.G., Hawley, S.R., Ward, S.A., and Park, B.K. (1998) 4-Aminoquinolines – past, present, and future: a chemical perspective. *Pharmacol. Ther.*, **77**, 29–58.
2. Pommier, Y. (2009) DNA topoisomerase I inhibitors: chemistry, biology, and interfacial inhibition. *Chem. Rev.*, **109**, 2894–2902.
3. Verma, R.P. and Hansch, C. (2009) Camptothecins: a SAR/QSAR study. *Chem. Rev.*, **109**, 213–235.

Table 60.1 Quinolines and related compounds. *V* in l kg^{-1}; CL in ml min^{-1} kg^{-1}; Mwt in g mol^{-1}.

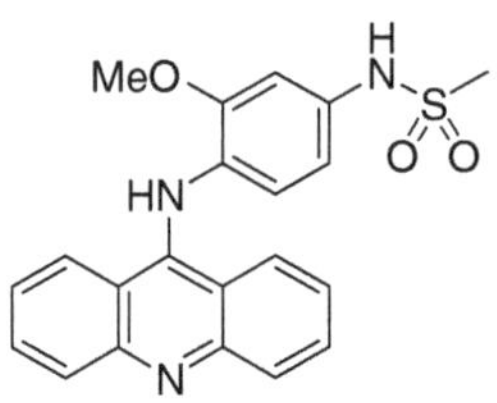	$t^1/_2$ 5–9 h *F* Low pb 97% ur 20%(iv)	*V* 1.7 CL 4.3 Mwt 393.5 PSA 88.7 Å^2 log *P* 3.32	AMSACRINE Antineoplastic Metabolism: glucuronidation, oxidation to quinoneimine
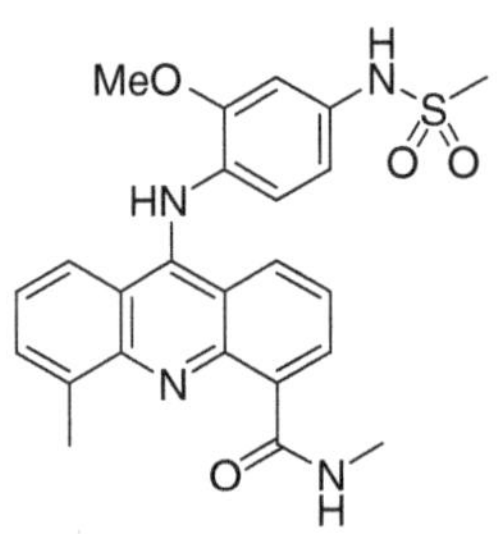	$t^1/_2$ 2.4 h *F* – pb >99% ur –	*V* 0.3 CL 2.6 Mwt 464.5 PSA 118 Å^2 log *P* 0.46	ASULACRINE Topoisomerase II inhibitor, antineoplastic Metabolism: oxidative thiolation with glutathione (meta to OCH_3)

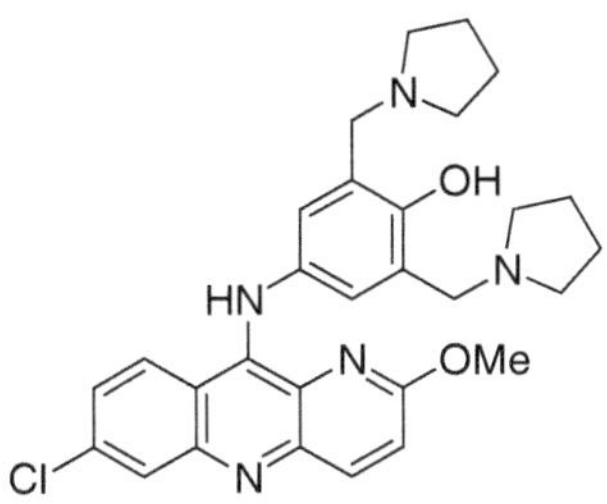	$t^1/_2$	16–17 d	*V*	4.6	PYRONARIDINE
	F	20%	CL	5–7	Antimalarial
	pb	–	Mwt	518.1	Metabolism: dehydrogenation of
	ur	1%	PSA	73.8 Å^2	aminophenol to quinoneimine,
			log *P*	4.86	O-demethylation, hydroxylation
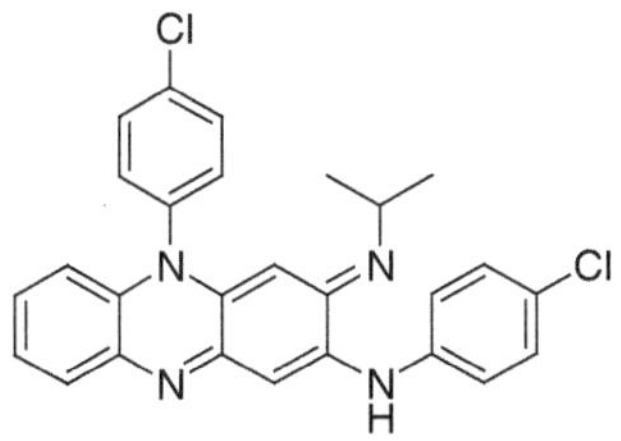	$t^1/_2$	70 d	*V*	–	CLOFAZIMINE
	F	45–62%	CL	–	Tuberculostatic
	pb	–	Mwt	473.4	Metabolism: hydratation,
	ur	<1%	PSA	40.0 Å^2	hydroxydechlorination, hydrolytic
			log *P*	7.46	deamination, then glucuronidation
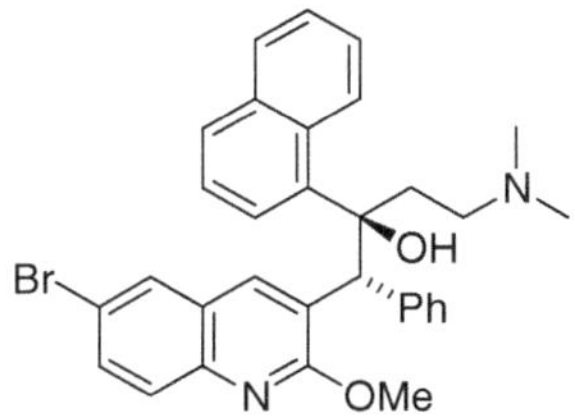	$t^1/_2$	24 h	*V*	–	TMC-207
	F	Reasonable	CL	–	Tuberculostatic
	pb	–	Mwt	555.5	Metabolite: *N*-monodesmethyl
	ur	–	PSA	45.6 Å^2	
			log *P*	7.71	
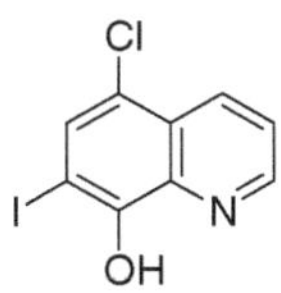	$t^1/_2$	11–14 h	*V*	–	CLIOQUINOL
	F	11–32%	CL	–	Treatment of Alzheimer's disease,
	pb	–	Mwt	305.5	chelator (Zn, Cu), crosses bbb
	ur	Traces	PSA	33.1 Å^2	Metabolite: glucuronide; withdrawn
			log *P*	4.17	in 1973 for causing neurotoxicity
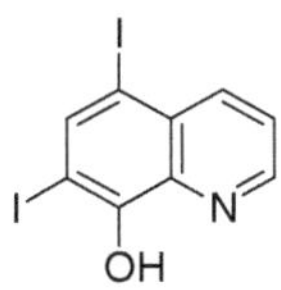	$t^1/_2$	11–14 h	*V*	–	IODOQUINOL
	F	8%	CL	–	Antiamebic
	pb	–	Mwt	397.0	
	ur	–	PSA	33.1 Å^2	
			log *P*	4.10	
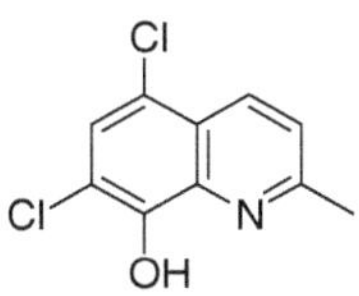	$t^1/_2$	7 h	*V*	–	CHLORQUINALDOL
	F	–	CL	–	Antibacterial
	pb	–	Mwt	228.1	Metabolite: glucuronide
	ur	<1%	PSA	33.1 Å^2	
			log *P*	3.51	

(*continued overleaf*)

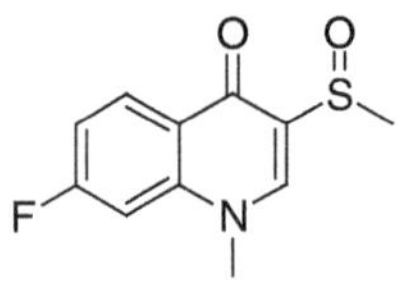	$t^{1}/_{2}$ 1.5–2.0 h *F* 72% pb – ur 1%	*V* 0.8 CL 8.2 Mwt 239.3 PSA 56.6 Å^2 log *P* −1.43	FLOSEQUINAN Vasodilator, antihypertensive Active metabolite: sulfone ($t^{1}/_{2}$ 30–40 h)
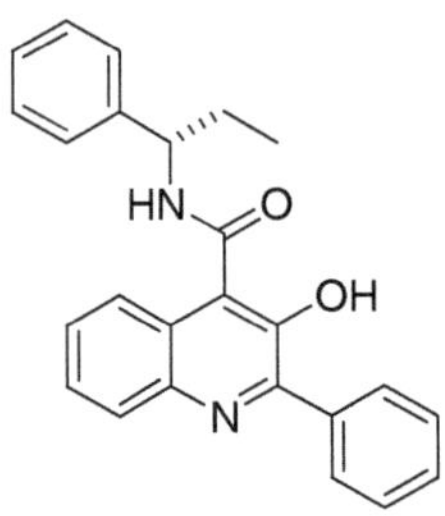	$t^{1}/_{2}$ 35–40 h *F* High pb – ur –	*V* – CL – Mwt 382.5 PSA 62.2 Å^2 log *P* 5.80	TALNETANT NK_3 antagonist for treatment of schizophrenia
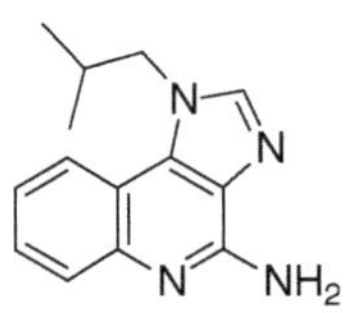	$t^{1}/_{2}$ 2.5 h *F* 47% pb – ur <1%	*V* 5 CL 16 Mwt 240.3 PSA 56.7 Å^2 log *P* 3.01	IMIQUIMOD Antiviral, immunomodulator Metabolism: glucuronidation, aromatic and aliphatic hydroxylation, N-oxidation
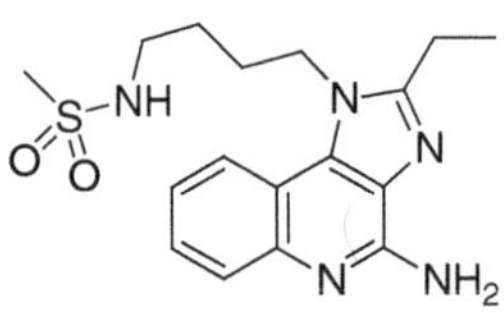	$t^{1}/_{2}$ 9–14 h *F* 27% pb – ur –	*V* 4 CL 16 Mwt 361.5 PSA 111 Å^2 log *P* 1.89	852 A Immunomodulator
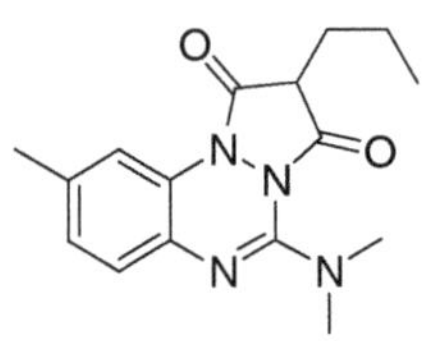	$t^{1}/_{2}$ 10–17 h *F* 83±19% pb 99.5% ur 63%	*V* 0.15 CL 0.14 Mwt 300.4 PSA 56.2 Å^2 log *P* 1.78	AZAPROPAZONE Antiinflammatory, metabolized by CYP2C9 Metabolite: 8-hydroxy
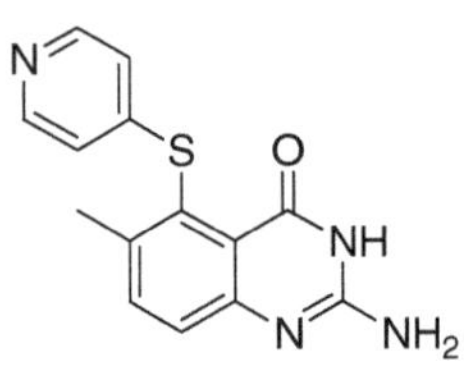	$t^{1}/_{2}$ 2.5±0.9 h* *F* >95% pb – ur 9–35%*	*V* – CL 2–6 Mwt 284.3 PSA 106 Å^2 log *P* 2.10	NOLATREXED, AG-337 *iv Thymidylate synthase inhibitor, antineoplastic

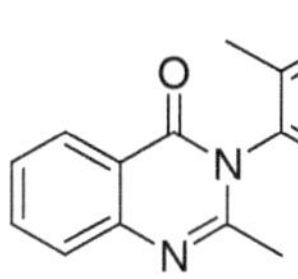	$t_{1/2}$ 20–60 h *F* – pb 70–90% ur Negligible	*V* 6 CL 1.7 Mwt 250.3 PSA 32.7 Å^2 log *P* 2.50	METHAQUALONE Metabolites: 4′-hydroxy, *N*-oxide, 2′-hydroxymethyl, 3′-hydroxy, 6-hydroxy, 2-hydroxymethyl; withdrawn in 1984 for causing addiction
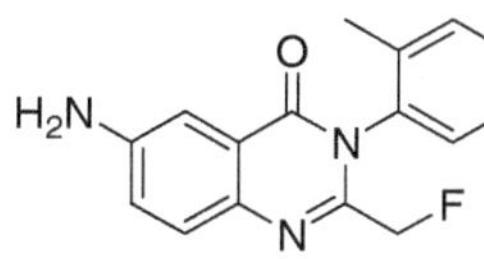	$t_{1/2}$ 3.3 h *F* – pb – ur 4%	*V* – CL – Mwt 283.3 PSA 58.7 Å^2 log *P* 0.85	AFLOQUALONE Muscle relaxant, crosses bbb Metabolites: *N*-acetyl, *N*-acetyl-2′-hydroxymethyl, *N*-glucuronide
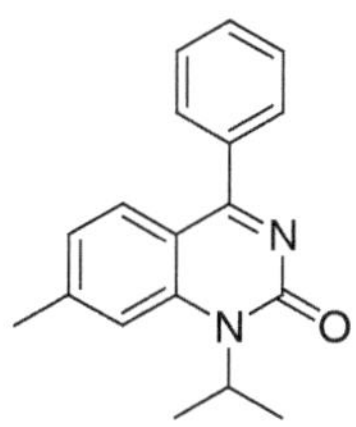	$t_{1/2}$ 0.6–1.3 h *F* 6–8% pb >98% ur Negligible	*V* 0.15 CL 9.5 Mwt 278.4 PSA 32.7 Å^2 log *P* 3.16	PROQUAZONE Antiinflammatory Metabolism: aromatic 3-hydroxylation, then conjugation
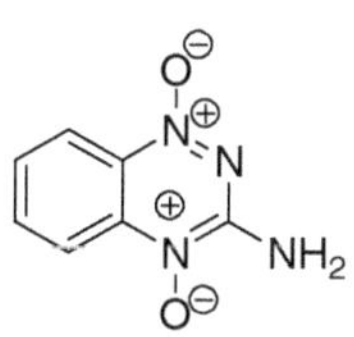	$t_{1/2}$ 0.8 h (iv) *F* 75% pb 19% ur 14–53%	*V* 0.6 CL 8.9 Mwt 178.2 PSA 89.8 Å^2 log *P* −0.29	TIRAPAZAMINE Hypoxia-selective antineoplastic Metabolism: reductive cleavage of N−O bonds
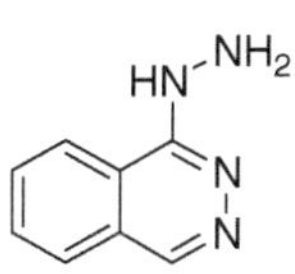	$t_{1/2}$ 1.0±0.3 h *F* 10–40% pb 87% ur 1–15%	*V* 1.5±1.0 CL 56±13 Mwt 160.2 PSA 62.8 Å^2 log *P* 0.95	HYDRALAZINE Antihypertensive Metabolism: N-acetylation and cyclization, N-formylation and cyclization, aromatic hydroxylation, substitution of hydrazine by hydroxyl
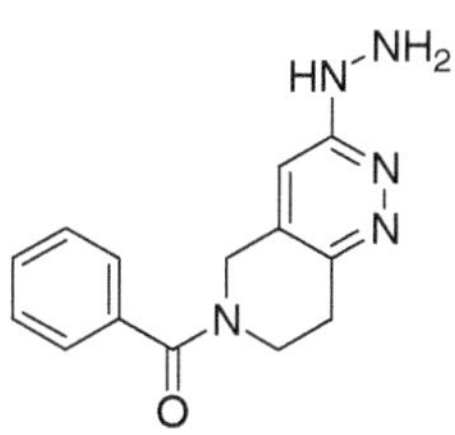	$t_{1/2}$ 2–5 h *F* 74–99% pb – ur –	*V* 2.2 CL 12–13 Mwt 269.3 PSA 83.1 Å^2 log *P* −0.27	ENDRALAZINE Antihypertensive Metabolism: hydrazone formation with pyruvate and α-ketoglutarate

(*continued overleaf*)

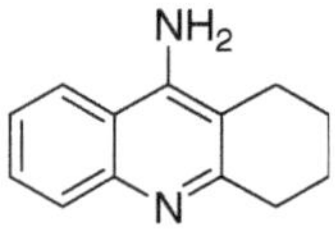	$t^1\!/_2$ 2.7±0.9 h *F* 17±3% pb 55% ur <1%	*V* 5.9±3.0 CL 36±18 Mwt 198.3 PSA 38.9 Å^2 log *P* 2.56	TACRINE Acetylcholinesterase inhibitor, nootropic, crosses bbb Metabolism: 1-hydroxylation (to velnacrine, see below), 2-, 3-, 4-hydroxylation
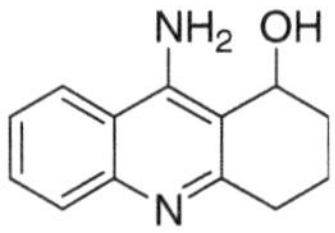	$t^1\!/_2$ 1.7–2.4 h *F* – pb – ur 11–30%	*V* – CL – Mwt 214.3 PSA 59.1 Å^2 log *P* 1.47	VELNACRINE Metabolite of tacrine
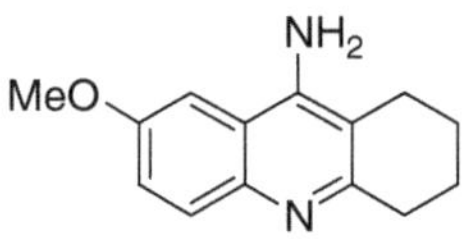	$t^1\!/_2$ 9±4 h *F* – pb – ur –	*V* – CL – Mwt 228.3 PSA 48.1 Å^2 log *P* 2.45	7-METHOXYTACRINE Cholinergic, less hepatotoxic than tacrine
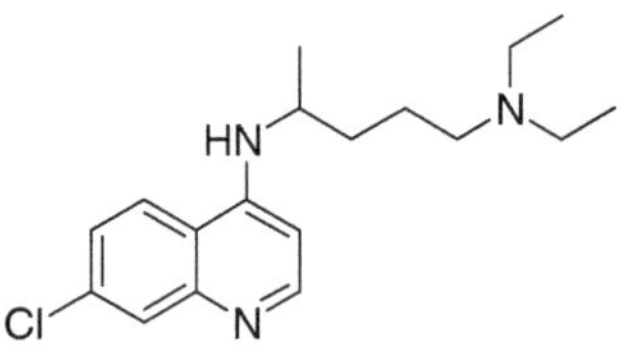	$t^1\!/_2$ 41±14 d *F* 89±16% pb 61±9% ur 61±4%	*V* 115±61 CL 1.8±0.4 Mwt 319.9 PSA 28.2 Å^2 log *P* 4.41	CHLOROQUINE Antimalarial Metabolites: desethyl ($t^1\!/_2$ 15±6 d), bisdesethyl, 4-amino-7-chloroquinoline
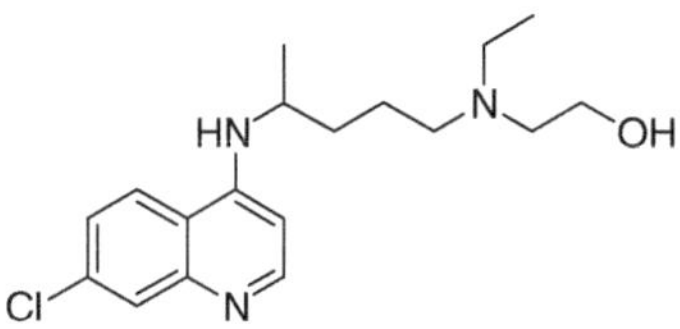	$t^1\!/_2$ 72 h *F* 44–85% pb 50–65% ur 16–52%	*V* 76–630 CL – Mwt 335.9 PSA 48.4 Å^2 log *P* 3.53	HYDROXYCHLOROQUINE Antimalarial Metabolites: desethyl, desethylchloroquine
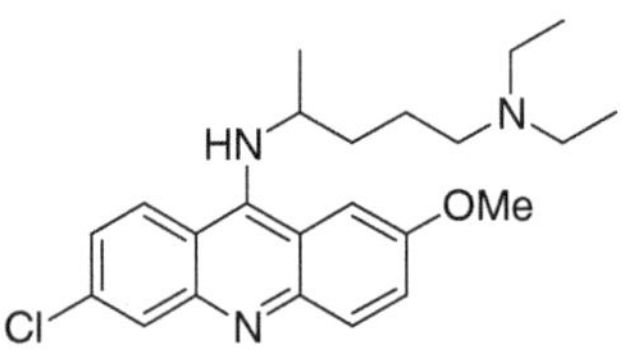	$t^1\!/_2$ 5–14 d *F* Low pb 80–90% ur –	*V* 45 CL 5.1 Mwt 400.0 PSA 37.4 Å^2 log *P* 5.59	QUINACRINE Antimalarial, crosses bbb Metabolism: N-deethylation, then oxidation of CH_2NH_2 to CO_2H, O-demethylation, N-dealkylation (to 4-aminoacridine)
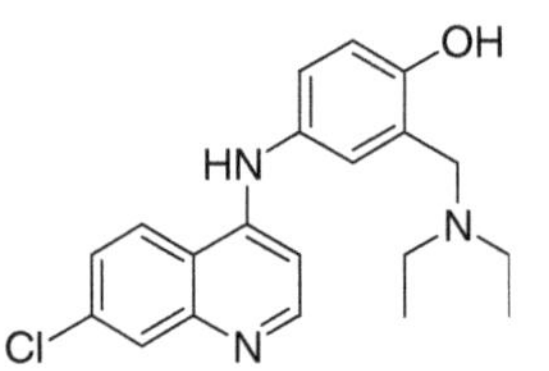	$t^1\!/_2$ 5 min *F* – pb 90% ur 2%	*V* 17–34 CL – Mwt 355.9 PSA 48.4 Å^2 log *P* 3.13	AMODIAQUINE Antimalarial Metabolism: N-dealkylation, oxidation to benzaldehyde and quinoneimine Active metabolite: monodesethyl ($t^1\!/_2$ 9–18 d); hepatotoxic, induces agranulocytosis

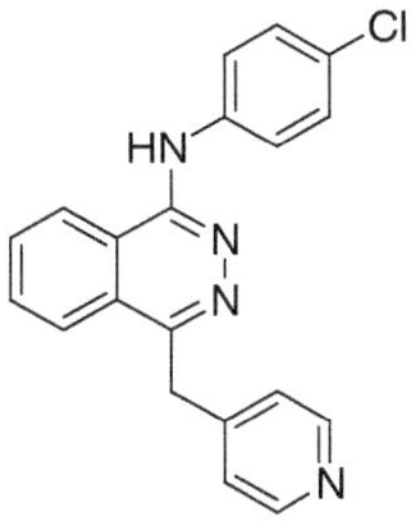	$t^1/_2$ 4.6±1.1 h F 18–57% pb – ur 0.4%	V 3.3 CL – Mwt 346.8 PSA 50.7 Å^2 log P 3.80	VATALANIB Antiangiogenic tyrosine kinase inhibitor Metabolism: pyridine N-oxidation, aromatic hydroxylation, hydroxylation of CH_2
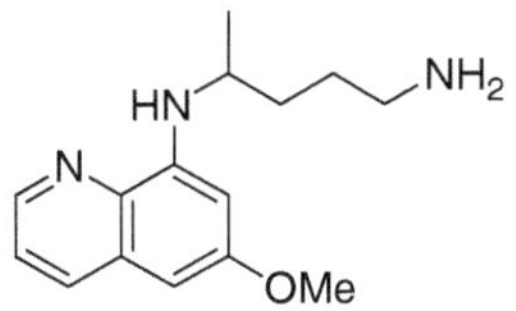	$t^1/_2$ 3–8 h F >95% pb – ur <1%	V 3–8 CL 20 Mwt 259.4 PSA 60.2 Å^2 log P 2.74	PRIMAQUINE Antimalarial Metabolism: aromatic hydroxylation (para to alkylamino), O-demethylation, N-dealkylation (to 5,6-dihydroxy-8-aminoquinoline)
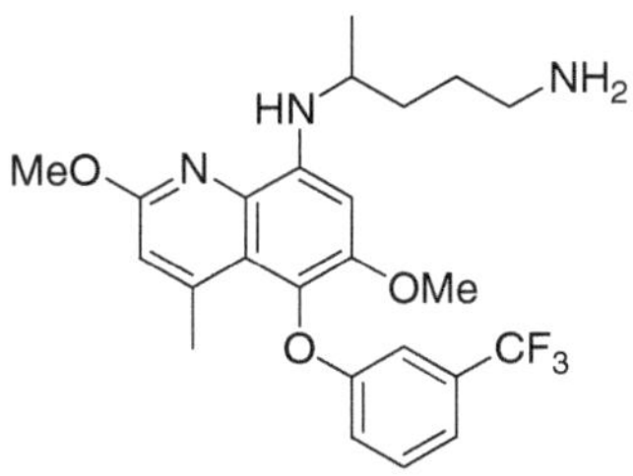	$t^1/_2$ 2–3 weeks F – pb – ur –	V/F 24±5 CL 0.9±0.2 Mwt 463.5 PSA 78.6 Å^2 log P 5.81	TAFENOQUINE Antimalarial Metabolism: hydroxylation of alkyl side chain
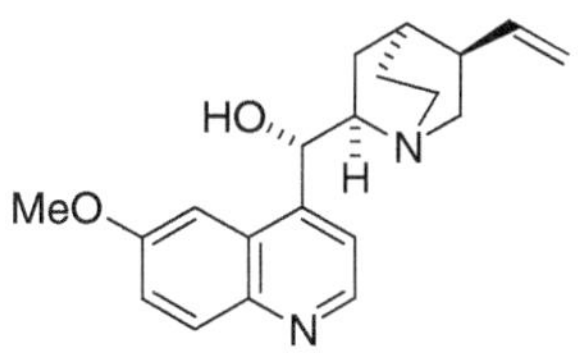	$t^1/_2$ 6.2±1.8 h F 80±15%* pb 87±3% ur 18±5%	V 2.7±1.2 CL 4.7±1.8 Mwt 324.4 PSA 45.6 Å^2 log P 2.82	QUINIDINE *sulfate Antimalarial, antiarrhythmic, CYP3A4 substrate Metabolism: alkene dihydroxylation, oxidation of alcohol to ketone, N-oxidation of quinuclidine; active metabolite: 3-hydroxyquinidine ($t^1/_2$ 12±3 h, pb 60±10%)
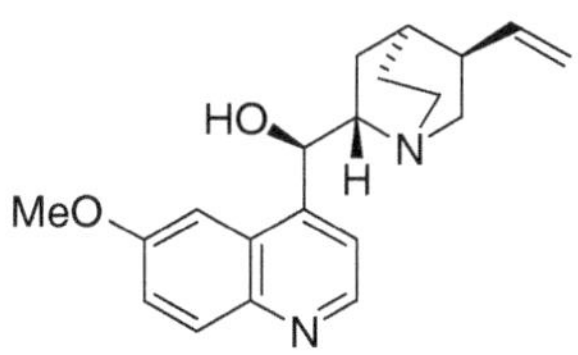	$t^1/_2$ 11±2 h F 76±11% pb 93±3% ur 12%	V 1.8±0.4 CL 1.9±0.5 Mwt 324.4 PSA 45.6 Å^2 log P 2.82	QUININE Antimalarial, muscle relaxant

(*continued overleaf*)

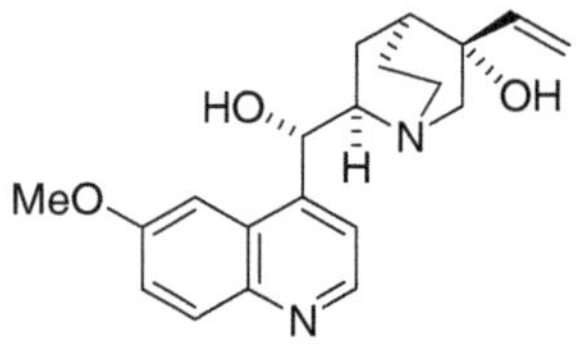	$t^{1/2}$ 11±4 h *F* 50–90% pb 53% ur –	*V* 7±2 CL 12 Mwt 340.4 PSA 65.8 Å^2 log *P* 2.01	3-HYDROXYDIHYDRO-QUINIDINE Antiarrhythmic
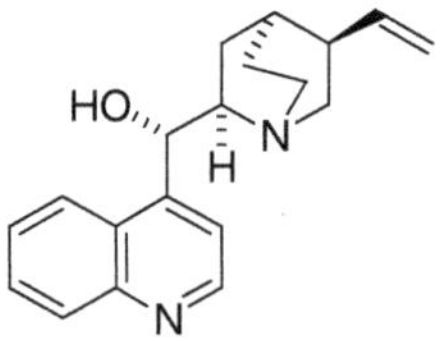	$t^{1/2}$ 2±1 h (iv) *F* 44% (rat) pb – ur <10% (iv)	*V* – CL 39 Mwt 294.4 PSA 36.4 Å^2 log *P* 2.79	CINCHONINE Antimalarial Metabolism: similar to quinine
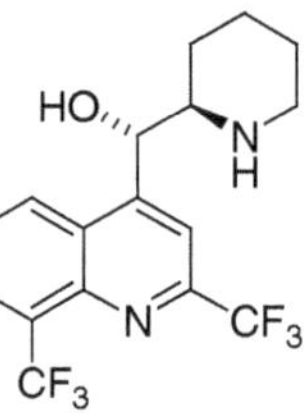	$t^{1/2}$ 20±4 d *F* – pb 98.2% ur <1%	*V/F* 19±6 CL/*F* 0.43±0.14 Mwt 378.3 PSA 45.2 Å^2 log *P* 2.20	MEFLOQUINE Racemic, antimalarial, does not cross bbb
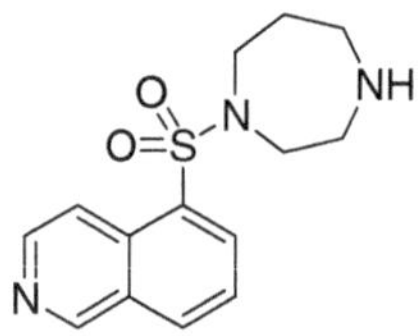	$t^{1/2}$ 0.8 h *F* High pb – ur –	*V* – CL – Mwt 291.4 PSA 70.7 Å^2 log *P* 1.72	FASUDIL Kinase inhibitor, vasodilator Metabolism: hydroxylation to isoquinolin-1-one ($t^{1/2}$ 5 h)
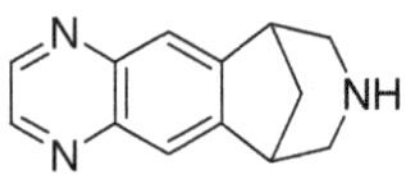	$t^{1/2}$ 10–58 h *F* High pb <20% ur 92%	*V* 6.4 CL – Mwt 211.3 PSA 37.8 Å^2 log *P* 0.74	VARENICLINE Nicotinic agonist for treatment of nicotine addiction Metabolism: glucuronidation, N-formylation
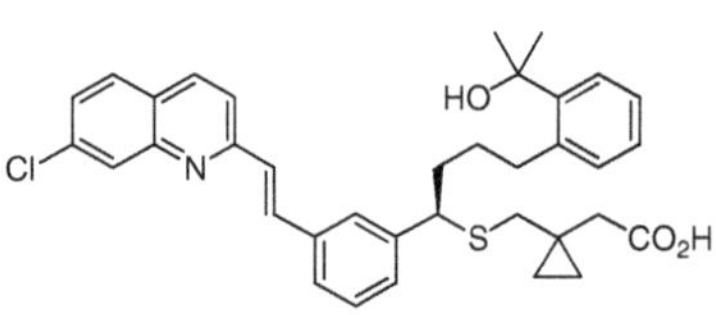	$t^{1/2}$ 2.7–5.5 h *F* 58–66% pb >99% ur <0.2%	*V* 0.13–0.16 CL 0.59–0.72 Mwt 586.2 PSA 95.7 Å^2 log *P* 7.80	MONTELUKAST Antiasthmatic Metabolism: acyl glucuronidation, oxidation to sulfoxide, benyzlic hydroxylation (at CH_2), CH_3 hydroxylation

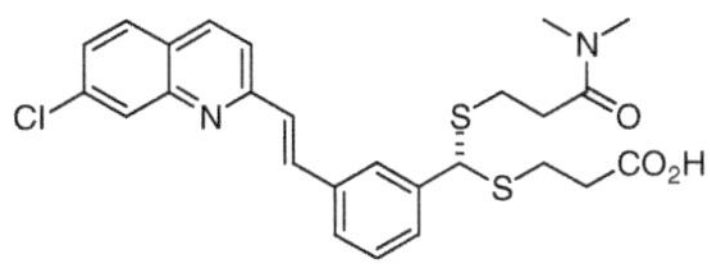	$t^1\!/_2$ 3–5 h F 100% pb >99% ur –	V 0.11 CL 0.68 Mwt 515.1 PSA 121 Å² log P 5.93	VERLUKAST Leukotriene D_4 antagonist Metabolism: oxidation to sulfoxide, hydroxylation of NCH_3, then N-demethylation

$t_{1/2}$, plasma half-life; *F*, oral bioavailability; pb, plasma protein binding; ur, excretion of unchanged drug in urine; V, volume of distribution; CL, clearance; Mwt, molecular weight; PSA, polar surface area; bbb, blood–brain barrier.

Table 60.2 Camptothecin analogs. *V* in l kg^{-1}; CL in ml min^{-1} kg^{-1}; Mwt in g mol^{-1}.

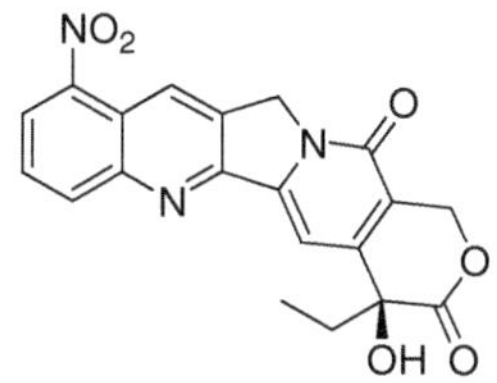	$t^1\!/_2$ 15–18 h F 1–5% (dog) pb 97% ur 10%*	V 0.8 CL 0.4 Mwt 393.4 PSA 126 Å² log P 0.33	RUBITECAN *drug + metabolite Topoisomerase I inhibitor, antineoplastic, enantiomerically pure Active metabolite: 9-aminocamptothecin
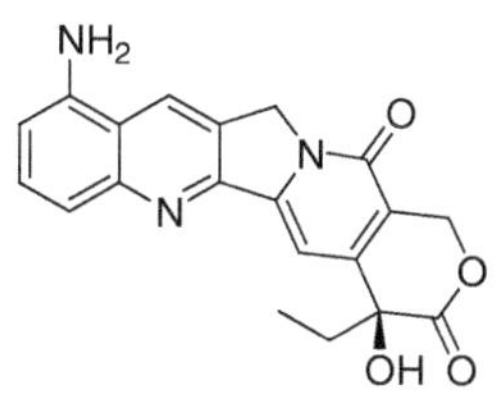	$t^1\!/_2$ 4–13 h F 49% pb >99% ur –	V 2.2 CL 6.5 Mwt 363.4 PSA 106 Å² log P 0.75	9-AMINOCAMPTOTHECIN Antineoplastic, metabolite of rubitecan
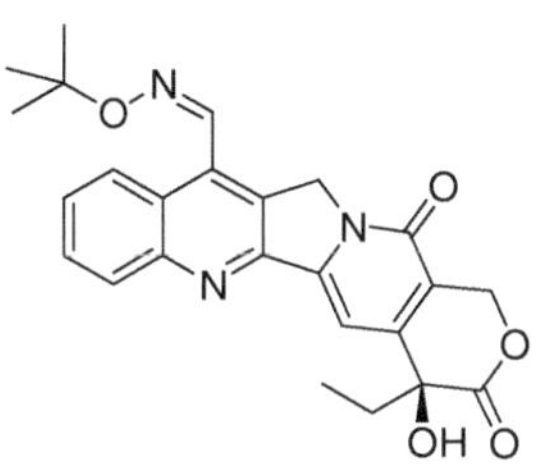	$t^1\!/_2$ 77±37 h F 40–80%* pb – ur <10%	V – CL 0.15±0.10 Mwt 447.5 PSA 101 Å² log P 3.74	GIMATECAN *rat, dog Topoisomerase inhibitor, antineoplastic Metabolism: hydroxylation of *t*Bu, glucuronidation, sulfation
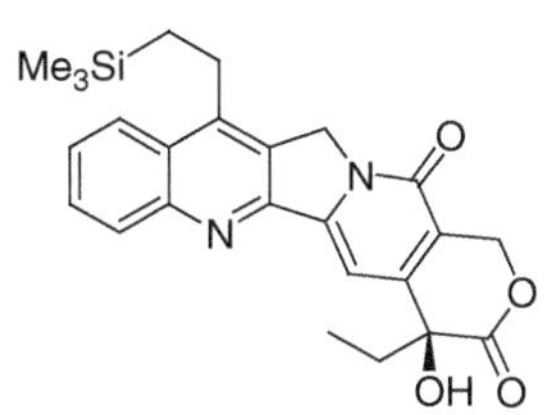	$t^1\!/_2$ 13±4 h F >40% pb 99% ur –	V 3.6±1.1 CL 7.1±4.3 Mwt 448.6 PSA 79.7 Å² log P 4.46	KARENITECIN Antineoplastic

(*continued overleaf*)

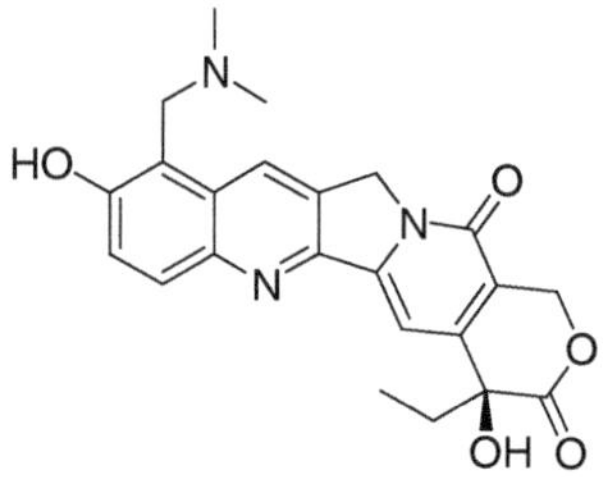

$t^1/_2$	1.7–8.4 h	V	0.4–2.1
F	34–44%	CL	2.7–30
pb	6.6–31%	Mwt	421.5
ur	20–42%	PSA	103 Å^2
		log P	1.39

TOPOTECAN
Antineoplastic
Metabolism: lactone hydrolysis, N-demethylation, phenol glucuronidation

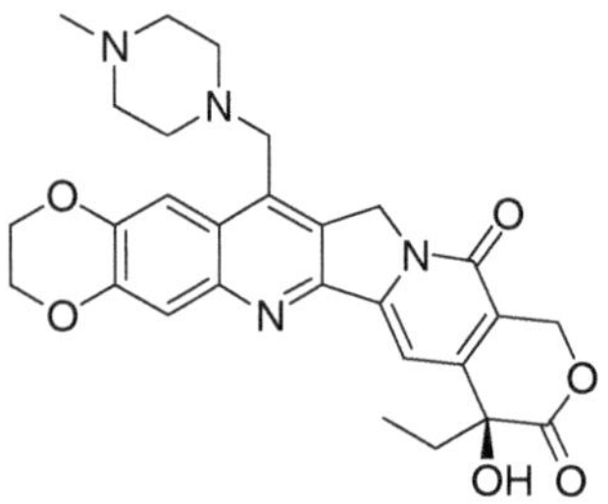

$t^1/_2$	2–17 h*	V	0.1–0.4
F	<6%	CL	0.1–1.3
pb	–	Mwt	518.6
ur	15%	PSA	105 Å^2
		log P	0.97

LURTOTECAN
*iv
Antineoplastic

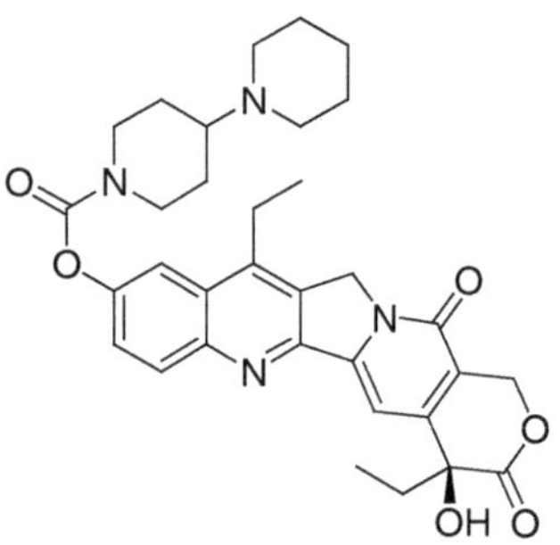

$t^1/_2$	6–21 h	V	2.6–6.3
F	12–21%*	CL	5–18
pb	65%	Mwt	586.7
ur	10–26%	PSA	113 Å^2
		log P	3.73

IRINOTECAN
*mice
Antineoplastic
Metabolism: reversible hydrolysis of lactone, oxidative ring-opening of terminal piperidine to aminopentanoic acid and 4-aminopiperidine, hydrolysis of carbamate to phenol (SN-38; $t^1/_2$ 8–22 h, pb 95%), then phenol glucuronidation

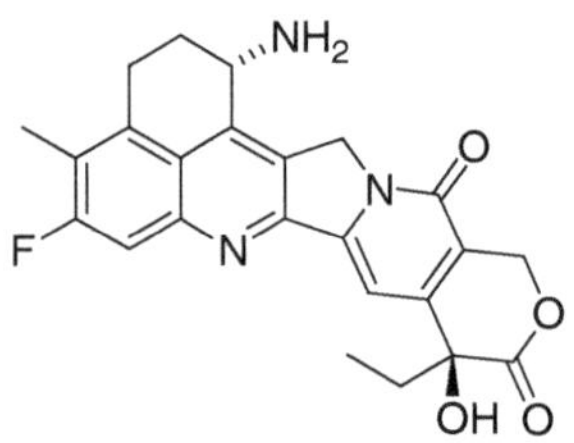

$t^1/_2$	7–8 h*	V	0.44
F	–	CL	0.96
pb	–	Mwt	435.5
ur	1–16%*	PSA	106 Å^2
		log P	2.13

EXATECAN
*iv
Antineoplastic
Metabolism: lactone hydrolysis, benzylic hydroxylation (at CH_3 and at $\mathit{CH_2}CH_2$)

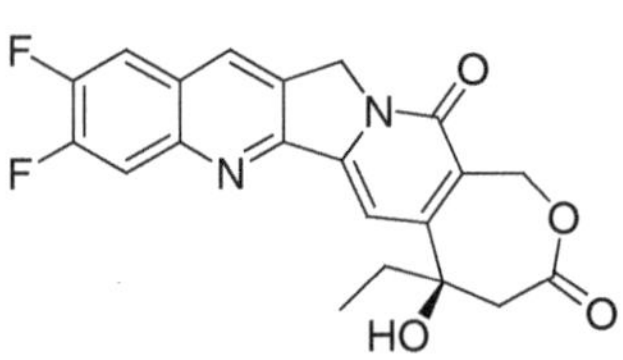

$t^1/_2$	2–4 h	V	1.3–1.8
F	72–95%	CL	6–11
pb	–	Mwt	398.4
ur	9%	PSA	79.7 Å^2
		log P	2.01

DIFLOMOTECAN
Antineoplastic
Metabolism: reversible lactone hydrolysis

Table 60.3 Quinazolines and related compounds. V in l kg^{-1}; CL in ml min^{-1} kg^{-1}; Mwt in g mol^{-1}.

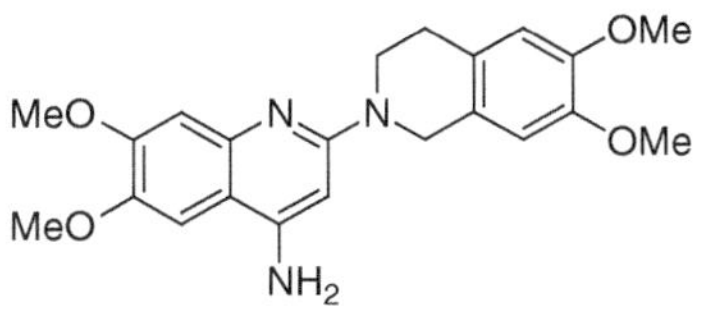	$t^1/_2$ F pb ur	3–4 h – – –	V CL Mwt PSA log P	6.3 14 395.5 79.1 Å^2 3.32	ABANOQUIL α_1-Adrenergic antagonist
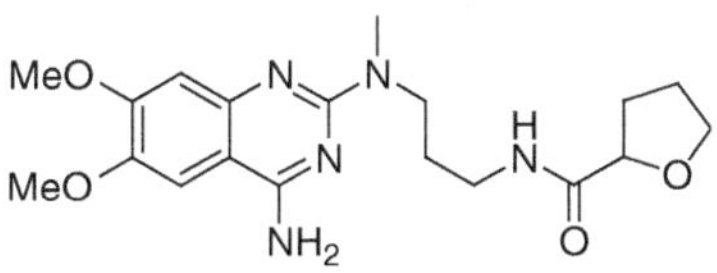	$t^1/_2$ F pb ur	5–9 h 49–100% 82–90% 11%	V CL Mwt PSA log P	1.5–3.2 5.9 389.5 112 Å^2 −1.00	ALFUZOSIN α_1-Adrenergic antagonist, antihypertensive Metabolism: O-demethylation, N-dealkylation, oxidation
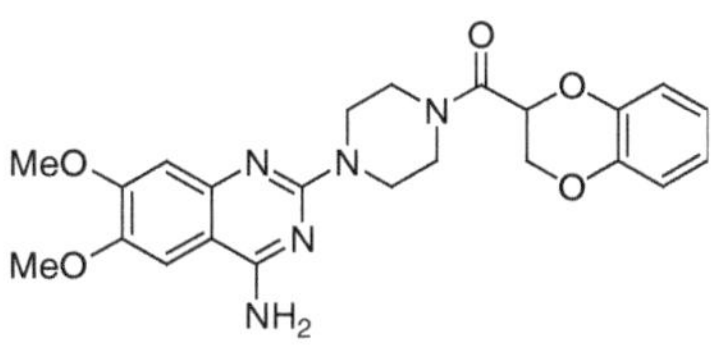	$t^1/_2$ F pb ur	11±2 h 63±14% 98–99% 0.6–1.2%	V CL Mwt PSA log P	1.5±0.3 1.7±0.4 451.5 112 Å^2 2.85	DOXAZOSIN α_1-Adrenergic antagonist, antihypertensive Metabolism: demethylation, aromatic hydroxylation of benzodioxan
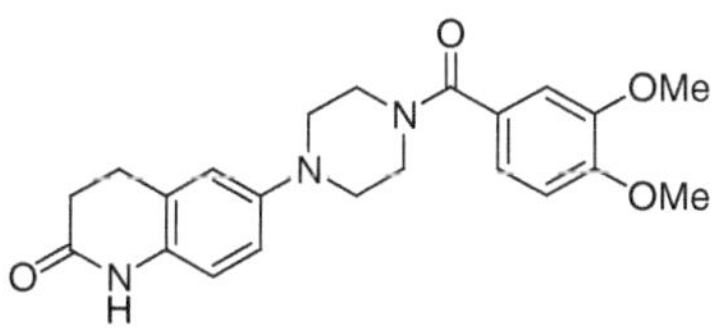	$t^1/_2$ F pb ur	37–46 h – 80% 19–26%	V CL Mwt PSA log P	– 0.08 395.5 71.1 Å^2 1.71	VESNARINONE Cardiotonic Metabolism: oxidation to reactive quinone imine, dehydrogenation of piperazine; causes agranulocytosis
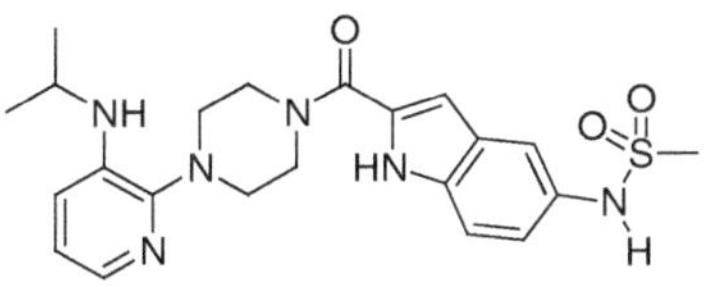	$t^1/_2$ F pb ur	2–11 h >51% 98% <5%	V CL Mwt PSA log P	0.8–1.0 – 456.6 119 Å^2 −1.23	DELAVIRDINE Antiviral (reverse transcriptase inhibitor) Metabolism: N-dealkylation, pyridine hydroxylation
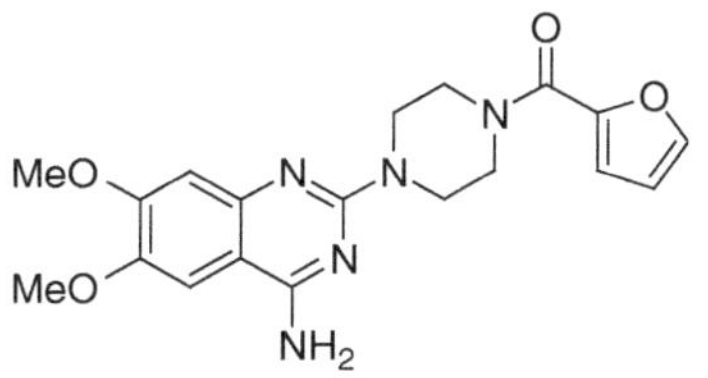	$t^1/_2$ F pb ur	2.9±0.8 h 68±17% 95±1% <1%	V CL Mwt PSA log P	0.60±0.13 3.0±0.3 383.4 107 Å^2 2.14	PRAZOSIN α_1-Adrenergic antagonist, antihypertensive Metabolism: demethylation, N-dealkylation, glucuronidation, amide hydrolysis, hydroxylation of piperazine, then elimination of water, oxidative degradation of piperazine, reductive furan opening, then oxidation to 2-oxoglutarate

(*continued overleaf*)

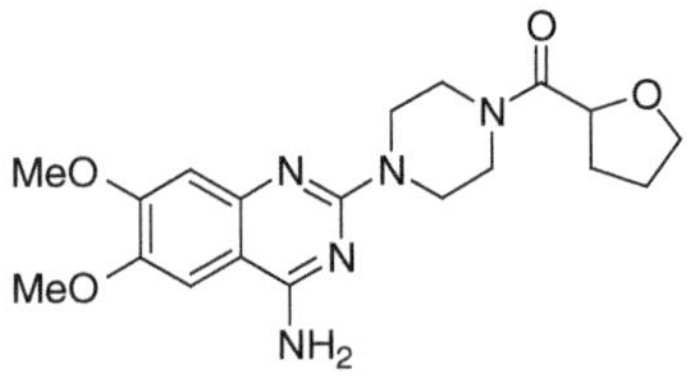

$t_{1/2}$	12±3 h	*V*	0.80±0.18	TERAZOSIN
F	90%	CL	1.1±0.2	α_1-Adrenergic antagonist, antihypertensive
pb	92±2%	Mwt	387.4	Metabolism: demethylation, amide hydrolysis
ur	12±3%	PSA	103 Å^2	
		log *P*	−0.96	

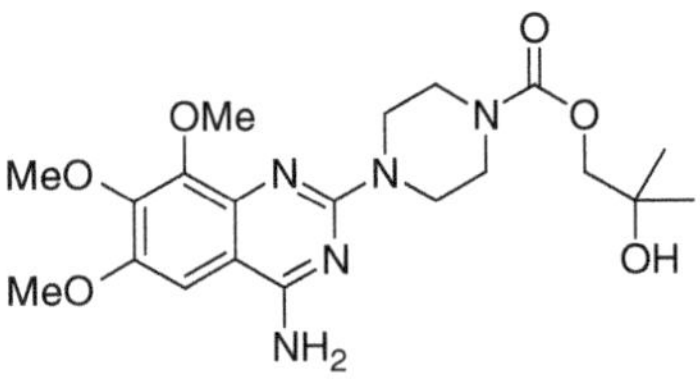

$t_{1/2}$	3 h	*V*	0.18	TRIMAZOSIN
F	63%	CL	1.0	Antihypertensive
pb	–	Mwt	435.5	Metabolism: hydroxylation of *gem*-dimethyl to 1-hydroxytrimazosin ($t_{1/2}$ 1.6±0.5 h)
ur	–	PSA	133 Å^2	
		log *P*	1.13	

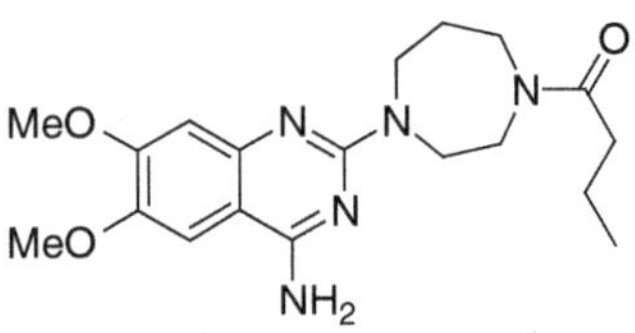

$t_{1/2}$	2–12 h	*V*	0.72	BUNAZOSIN
F	45%	CL	4.8	α_1-Adrenergic antagonist
pb	94–97%	Mwt	373.5	Metabolism: N-glucuronidation, O-demethylation, then O-glucuronidation
ur	1%	PSA	93.8 Å^2	
		log *P*	2.27	

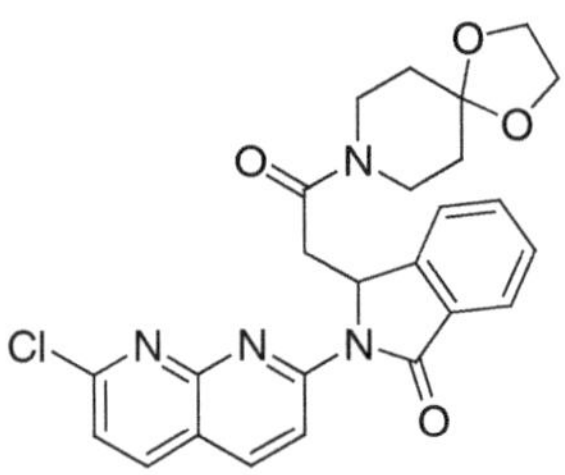

$t_{1/2}$	10.5 h	*V*	–	PAZINACLONE
F	–	CL	–	Sedative, anxiolytic
pb	95%	Mwt	478.9	Metabolism: acetal hydrolysis, then reduction of ketone to alcohol ($t_{1/2}$ 7.6 h)
ur	0%	PSA	84.9 Å^2	
		log *P*	0.52	

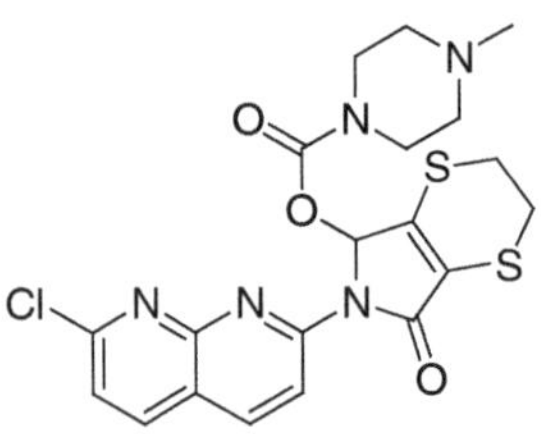

$t_{1/2}$	2 h	*V*	–	SURICLONE
F	–	CL	–	Anxiolytic
pb	–	Mwt	478.0	Metabolism: oxidation to *N*-oxide, *S*-oxide, N-demethylation
ur	–	PSA	129 Å^2	
		log *P*	3.81	

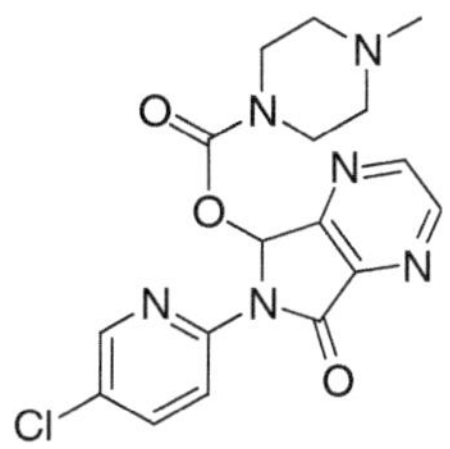	$t^{1}/_{2}$ 4±2 h *F* 80% pb 45% ur 5%	*V* 1.4–1.6 CL 3.6 Mwt 388.8 PSA 91.8 Å^2 log *P* 2.71	ZOPICLONE Sedative, hypnotic Metabolism: N-demethylation, N-oxidation of tertiary amine, hydrolysis of carbamate
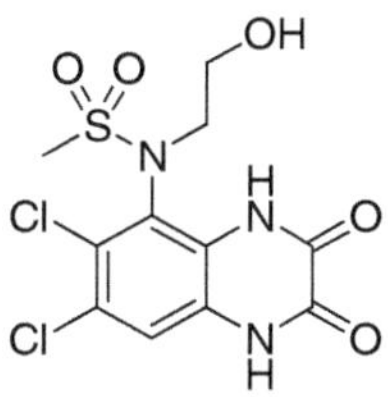	$t^{1}/_{2}$ 1.4 h* *F* 17% pb 87% ur 58–68%*	*V* 0.8 CL 6.0 Mwt 368.2 PSA 124 Å^2 log *P* −1.11	UK-240455 *iv NMDA glycine site antagonist, crosses bbb Metabolism: glucuronidation

$t_{1/2}$, plasma half-life; *F*, oral bioavailability; pb, plasma protein binding; ur, excretion of unchanged drug in urine; V, volume of distribution; CL, clearance; Mwt, molecular weight; PSA, polar surface area; NMDA, *N*-methyl d-aspartic acid.

61
Nucleoside Analogs

Chemically modified nucleosides can disrupt the biosynthesis of RNA and DNA by inhibiting essential enzymes, by blocking the elongation of nucleic acids, or by rendering the resulting RNAs/DNAs nonfunctional. Nucleoside analogs ("antimetabolites") can thus be used to interfere with the replication of cells (e.g., as antineoplastics) and viruses [1] (Tables 61.1 and 61.2). Many of these compounds are mutagenic.

Because of their low lipophilicity, the oral bioavailability of unnatural carbohydrates depends on the availability of suitable active transport mechanisms and can vary strongly.

Reference

1. (a) De Clercq, E. (2010) Highlights in the discovery of antiviral drugs: a personal retrospective. *J. Med. Chem.*, **53**, 1438–1450; (b) Parker, W.B. (2009) Enzymology of purine and pyrimidine antimetabolites used in the treatment of cancer. *Chem. Rev.*, **109**, 2880–2893.

Table 61.1 Nucleoside analogs with monocyclic heteroarenes. V in l kg^{-1}; CL in ml min^{-1} kg^{-1}; Mwt in g mol^{-1}.

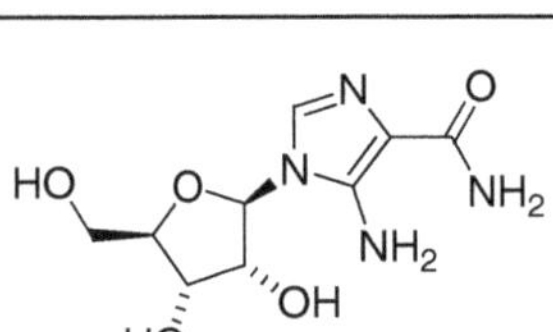	$t^1\!/_2$	Short	V	–	ACADESINE
	F	<5%	CL	37	Cardioprotectant
	pb	1%	Mwt	258.2	Metabolites: uric acid, hypoxanthine
	ur	–	PSA	157 Å^2	
			log P	−2.29	
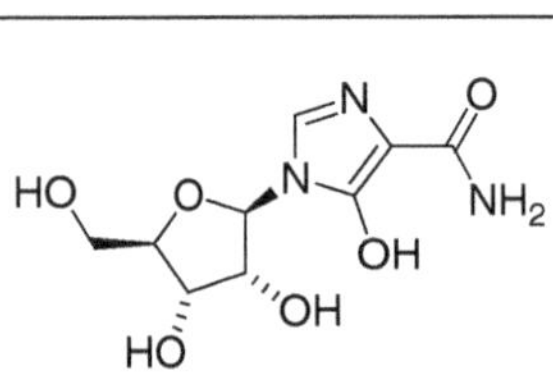	$t^1\!/_2$	3 h	V	0.8	MIZORIBINE
	F	12–81%	CL	2.2–4.8	Immunosuppressant
	pb	–	Mwt	259.2	Metabolism: phosphorylation
	ur	50–100%	PSA	151 Å^2	
			log P	−0.79	

Lead Optimization for Medicinal Chemists: Pharmacokinetic Properties of Functional Groups and Organic Compounds, First Edition. Florencio Zaragoza Dörwald.

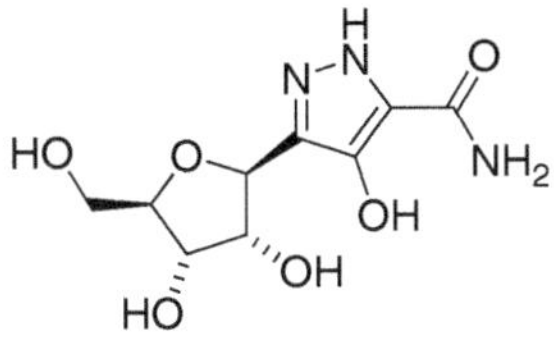	$t^{1}/_{2}$	0.1–0.5 h*	V	–	PYRAZOFURIN
	F	–	CL	–	*iv
	pb	–	Mwt	259.2	Antiviral
	ur	<12%	PSA	162 Å^2	
			log P	−1.75	
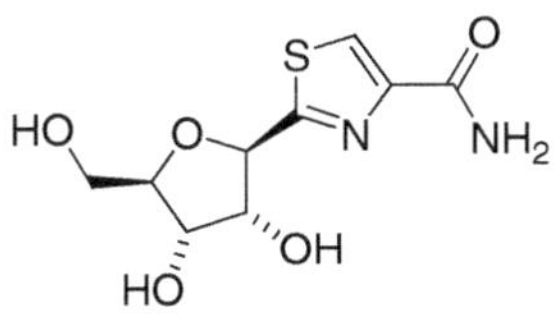	$t^{1}/_{2}$	6–8 h (iv)	V	1.1	TIAZOFURIN
	F	–	CL	2.4	Antineoplastic, crosses bbb
	pb	–	Mwt	260.3	Metabolism: phosphorylation
	ur	50–85% (iv)	PSA	154 Å^2	
			log P	−0.78	
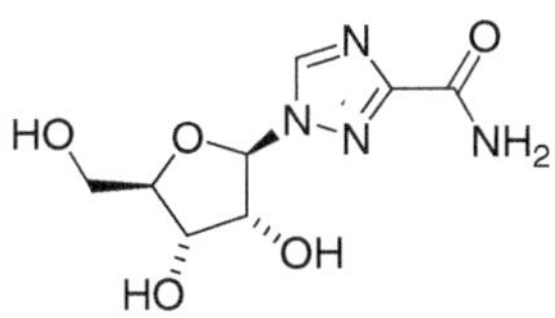	$t^{1}/_{2}$	2 h*	V	14	RIBAVIRIN
	F	45–65%	CL	5.2	*terminal phase: 18–36 h
	pb	0%	Mwt	244.2	Antiviral
	ur	35±8%	PSA	144 Å^2	Metabolism: phosphorylation,
			log P	−1.85	aminal and amide hydrolysis
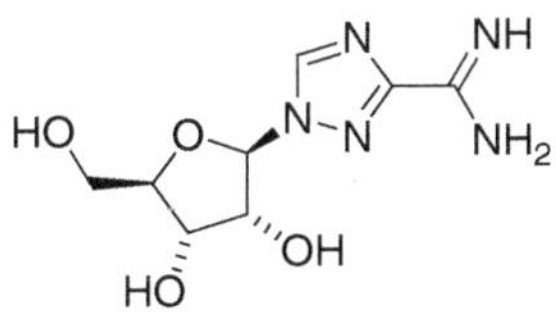	$t^{1}/_{2}$	28–76 h	V	19*	VIRAMIDINE
	F	66%	CL	2.7	*monkey
	pb	7%	Mwt	243.2	Prodrug of ribavirin (see above)
	ur	1–4%	PSA	151 Å^2	Metabolism: hydrolysis of aminal
			log P	−1.72	to free triazole and amidine to
					amide and CO_2H
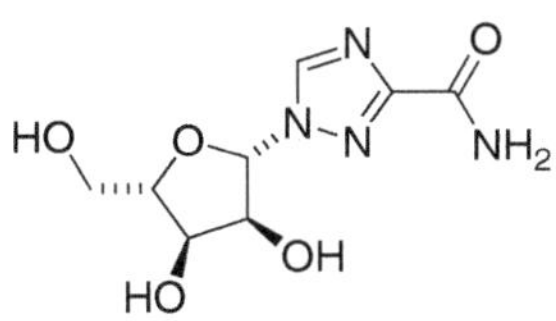	$t^{1}/_{2}$	3.5 h (iv)*	V	0.7*	LEVOVIRIN
	F	10%	CL	2.6*	*monkey
	pb	–	Mwt	244.2	Antiviral (hepatitis C)
	ur	21%*	PSA	144 Å^2	
			log P	−2.26	
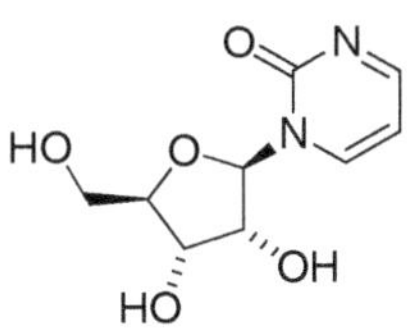	$t^{1}/_{2}$	1–2 h (iv)*	V	0.4–0.7*	ZEBULARINE
	F	<1%*	CL	4–11*	*monkey
	pb	57%	Mwt	228.2	DNA methylation inhibitor,
	ur	–	PSA	103 Å^2	antineoplastic
			log P	−0.83	

(continued overleaf)

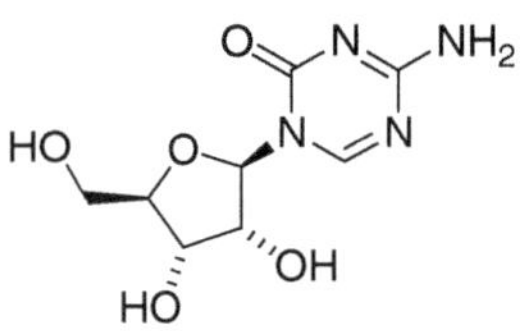	$t\frac{1}{2}$ 1–4 h (sc) *F* 18% pb – ur –	*V* 1.2±0.4 CL 43±13 Mwt 244.2 PSA 141 Å^2 log *P* −2.19	AZACYTIDINE Antineoplastic for sc administration Metabolism: phosphorylation
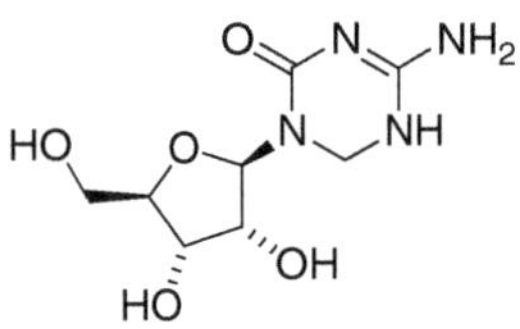	$t\frac{1}{2}$ 1–2 h (iv) *F* – pb – ur <20% (iv)	*V* – CL – Mwt 246.2 PSA 141 Å^2 log *P* −2.20	DIHYDROAZACYTIDINE Antineoplastic
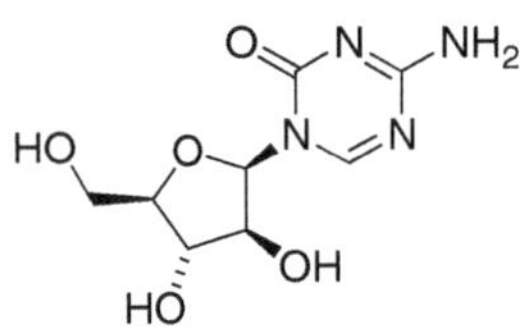	$t\frac{1}{2}$ 4 h *F* – pb – ur –	*V* 0.6* CL 18 Mwt 244.2 PSA 141 Å^2 log *P* −2.19	FAZARABINE, ARABINOSYL-5-AZACYTIDINE *monkey Antineoplastic, iv administration Metabolism: phosphorylation
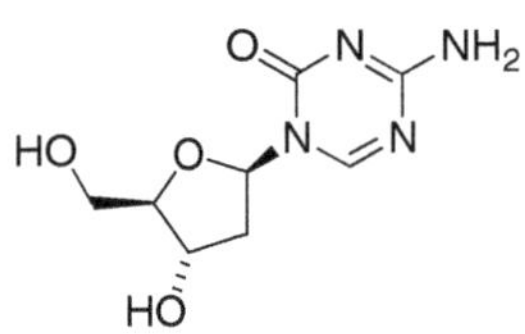	$t\frac{1}{2}$ 0.6 h (iv) *F* Low pb 0% ur –	*V* 0.2–4.6 CL 130 Mwt 228.2 PSA 121 Å^2 log *P* −1.87	DECITABINE Antineoplastic, crosses bbb; Metabolism: deamination, phosphorylation
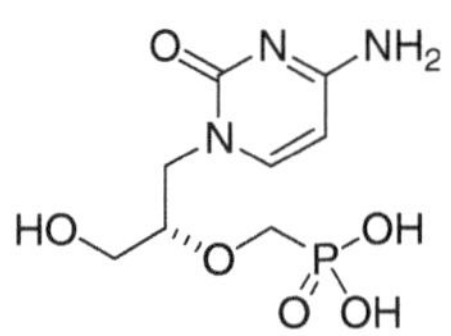	$t\frac{1}{2}$ 2–3 h *F* <5% pb 0–6% ur 80–100%*	*V* 0.49 CL 2.5 Mwt 279.2 PSA 155 Å^2 log *P* −1.47	CIDOFOVIR *iv Antiviral Metabolism: phosphorylation; prodrug: CMX001 (hexadecyloxypropyl phosphonate; $t\frac{1}{2}$ 1 h (monkey))
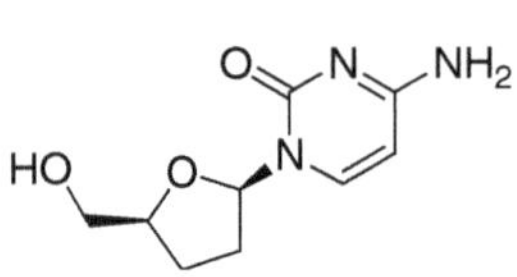	$t\frac{1}{2}$ 2±1 h *F* 88±17% pb <4% ur 65±17%	*V* 0.53±0.13 CL 4.1±1.2 Mwt 211.2 PSA 88.2 Å^2 log *P* −1.36	ZALCITABINE Antiviral Metabolism: phosphorylation
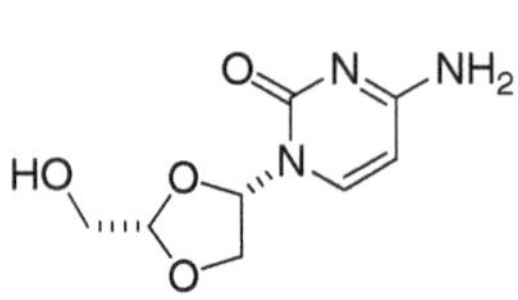	$t\frac{1}{2}$ 11±2 h *F* Low pb – ur 60% (iv)	*V* 1.0±0.7 CL 2.5±1.0 Mwt 213.2 PSA 97.4 Å^2 log *P* −0.78	TROXACITABINE, BCH-4556 Antineoplastic Metabolism: phosphorylation

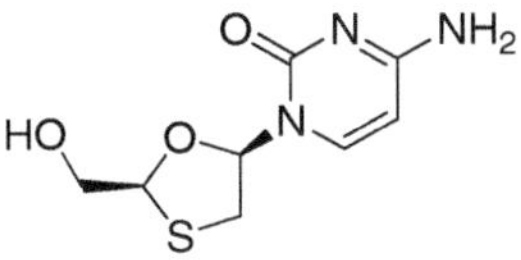	$t^{1}/_{2}$ 5–7 h *F* 86±16% pb 16–36% ur 55–85%	*V* 1.3 CL 4.8 Mwt 229.3 PSA 113 Å^2 log *P* −0.71	LAMIVUDINE Antiviral Metabolite: *trans*-sulfoxide
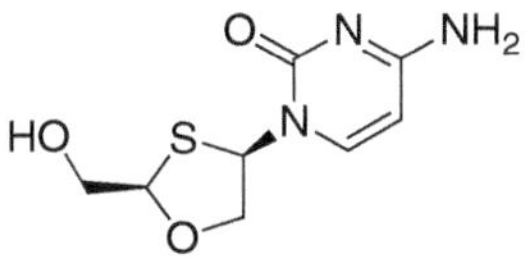	$t^{1}/_{2}$ 3–5 h *F* 65–80% pb <4% ur 65–80%	*V* 1 CL – Mwt 229.3 PSA 113 Å^2 log *P* −1.35	APRICITABINE Nucleoside reverse transcriptase inhibitor, antiviral (HIV) Metabolite: triphosphate
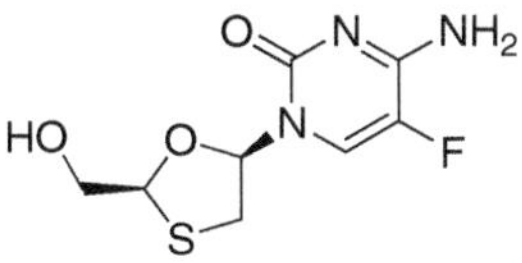	$t^{1}/_{2}$ 7–18 h *F* 75–93% pb <4% ur 75%	*V* 1.4±0.3 CL 4.7±1.5 Mwt 247.3 PSA 113 Å^2 log *P* −0.41	EMTRICITABINE Antiviral Metabolite: 3′-sulfoxide
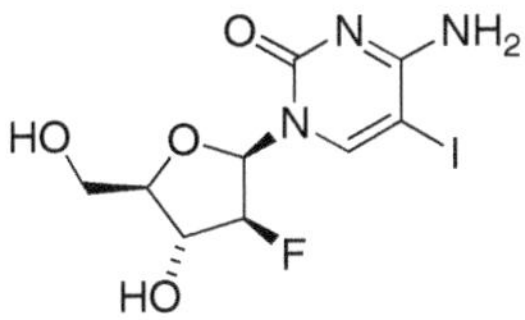	$t^{1}/_{2}$ 1.3 h (iv) *F* High pb – ur 12% (iv)	*V* – CL – Mwt 371.1 PSA 108 Å^2 log *P* −1.54	FIACITABINE Antiviral Metabolism: replacement of NH_2 by OH (to fialuridine)
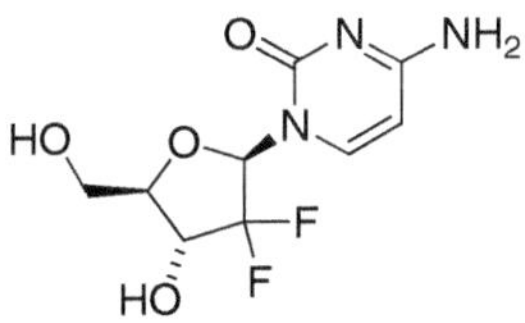	$t^{1}/_{2}$ 0.2–0.4 h *F* 10% pb 0% ur <10%	*V* 0.22–0.74 CL 17–54 Mwt 263.2 PSA 108 Å^2 log *P* −2.22	GEMCITABINE Antineoplastic Metabolites: triphosphate and difluorodeoxyuridine ($t^{1}/_{2}$ 89 h)
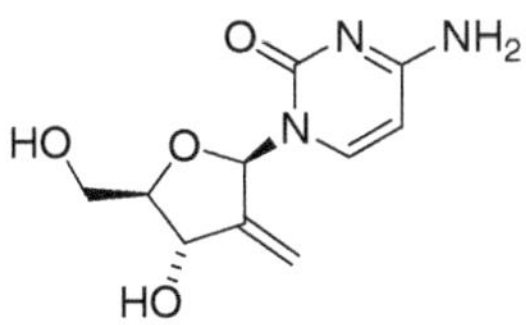	$t^{1}/_{2}$ 2–6 h *F* 10–80% pb – ur 30–50%	*V* 1.0–1.5 CL 4.1–6.2 Mwt 239.2 PSA 108 Å^2 log *P* −1.12	2-DEOXY-2-METHYLENECYTIDINE, DMDC Antineoplastic Metabolites: triphosphate and deoxymethyleneuridine
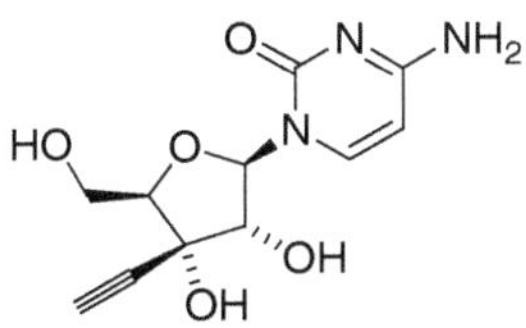	$t^{1}/_{2}$ 11±3 h* *F* – pb Low ur 62–71%*	*V* 1.4* CL 1.7±0.2* Mwt 267.2 PSA 129 Å^2 log *P* −2.57	3′-C-ETHINYLCYTIDINE, TAS-106 *iv Antineoplastic

(*continued overleaf*)

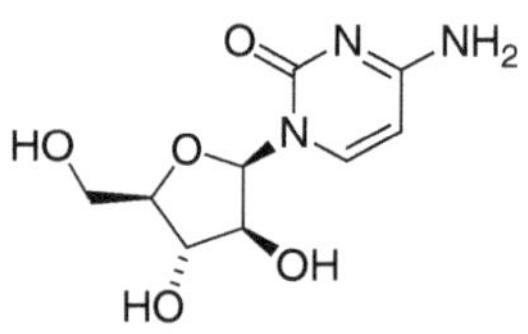	$t_{1/2}$ *F* pb ur	2.6±0.6 h 20% 13% 11±8%	*V* CL Mwt PSA log *P*	3.0±1.9 13±4 243.2 129 Å^2 −1.81	CYTARABINE Antineoplastic Metabolism: phosphorylation to triphosphate; 80–100% of the latter is excreted in urine
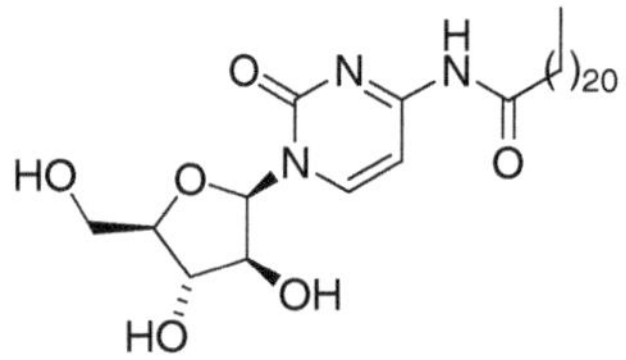	$t_{1/2}$ *F* pb ur	5.3 h (iv) – – –	*V* CL Mwt PSA log *P*	0.32 0.9 565.8 132 Å^2 8.62	ENOCITABINE Antineoplastic
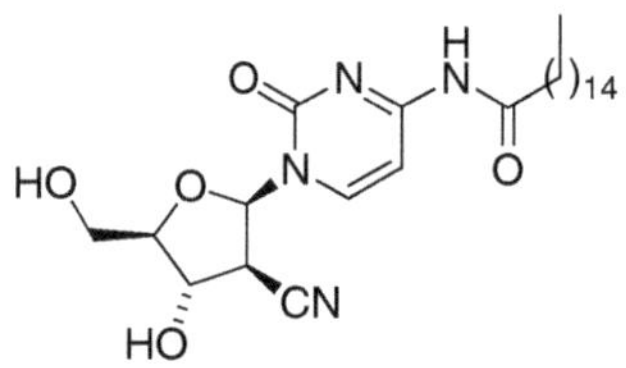	$t_{1/2}$ *F* pb ur	1–3 h – – –	*V* CL Mwt PSA log *P*	– – 490.6 135 Å^2 5.67	SAPACITABINE Prodrug of CNDAC Values: CNDAC on oral administration of prodrug
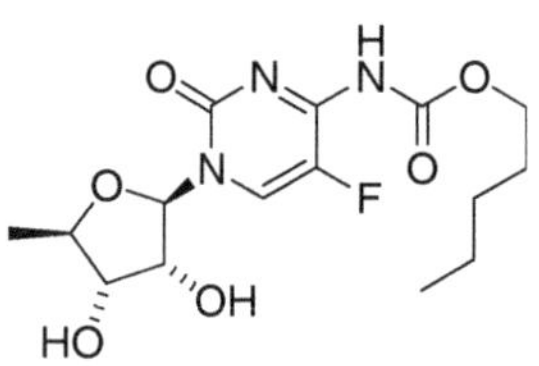	$t_{1/2}$ *F* pb ur	0.5–0.9 h 100% 54% 3%	*V* CL Mwt PSA log *P*	– 64 359.4 121 Å^2 0.97	CAPECITABINE Prodrug of 5-fluorouracil
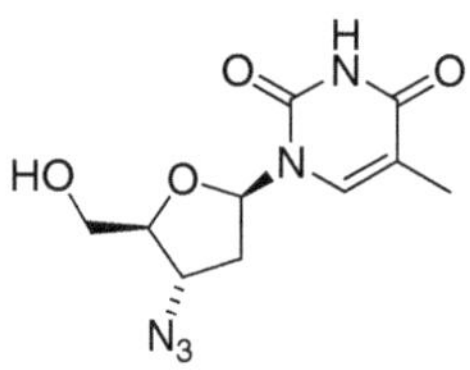	$t_{1/2}$ *F* pb ur	1.1±0.2 h 63±13% <25% 18±5%	*V* CL Mwt PSA log *P*	1.4±0.4 26±6 281.3 134.1 Å^2 −0.09	ZIDOVUDINE Antiviral Metabolism: glucuronidation, intracellular phosphorylation, reduction of azide to NH_2, causes granulocytopenia and anemia
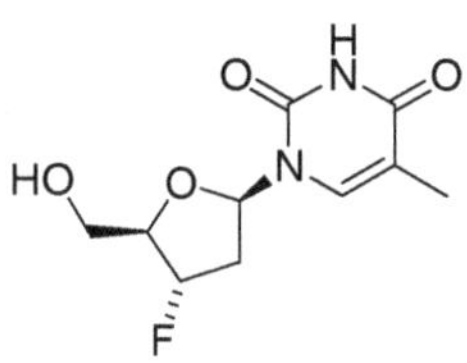	$t_{1/2}$ *F* pb ur	4–5 h High <10% –	*V* CL Mwt PSA log *P*	– – 244.2 78.9 Å^2 −0.20	ALOVUDINE Antiviral

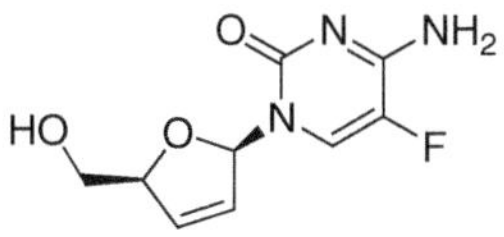	$t_{1/2}$ 20–100 h F 41% pb <1% ur 32%	V/F 23–40 CL/F 6–12 Mwt 227.2 PSA 88.2 Å^2 log P −1.28	ELVUCITABINE Antiviral (HIV) Metabolism: phosphorylation
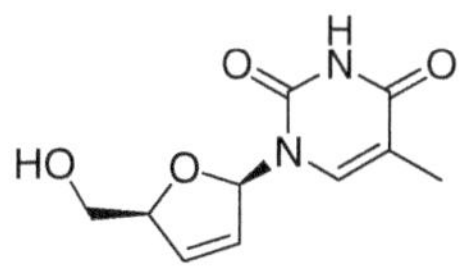	$t_{1/2}$ 1.3 h F 86±18% pb 0% ur 34±5%	V 0.7–1.0 CL 8.2 Mwt 224.2 PSA 78.9 Å^2 log P −0.65	STAVUDINE Antiviral Metabolism: hydroxylation, glucuronidation, epoxidation followed by reaction with nucleophiles
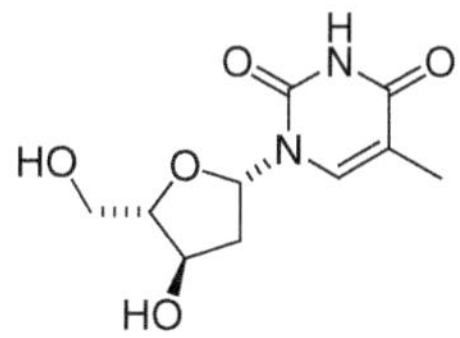	$t_{1/2}$ 40–49 h F 52% pb 3% ur 42%	V 8±4 CL 3 Mwt 242.2 PSA 99.1 Å^2 log P −0.84	TELBIVUDINE Antiviral No metabolites
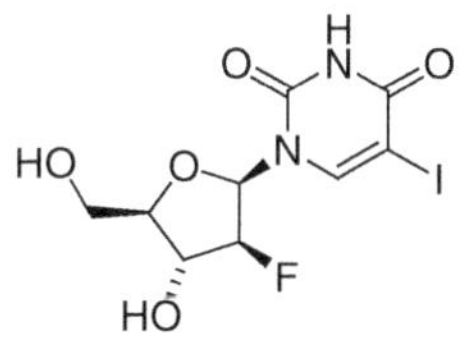	$t_{1/2}$ 4 h F – pb – ur –	V – CL – Mwt 372.1 PSA 99.1 Å^2 log P 0.12	FIALURIDINE Antiviral; discontinued because of excessive toxicity
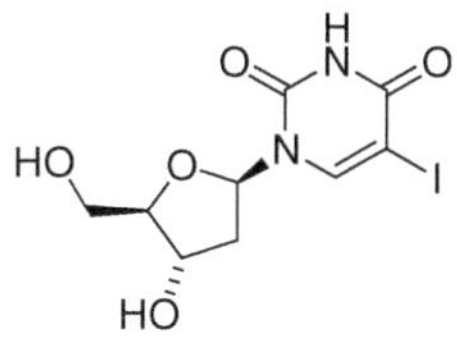	$t_{1/2}$ 0.5 h (iv) F – pb – ur –	V – CL – Mwt 354.1 PSA 99.1 Å^2 log P −0.59	IDOXURIDINE Antiviral Metabolites: idoxuracil, uracil
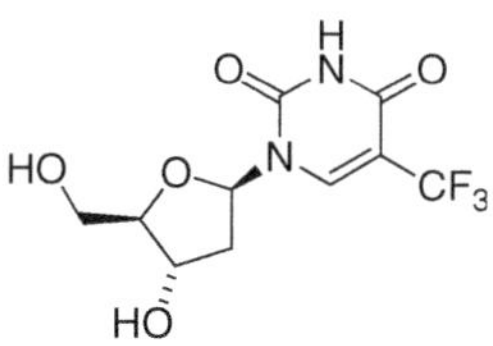	$t_{1/2}$ 18 min (iv) F – pb – ur –	V – CL – Mwt 296.2 PSA 99.1 Å^2 log P −0.02	TRIFLURIDINE, TRIFLUOROTHYMIDINE Antiviral; topical application only; too mutagenic for systemic application Metabolite: 5-carboxyuracil
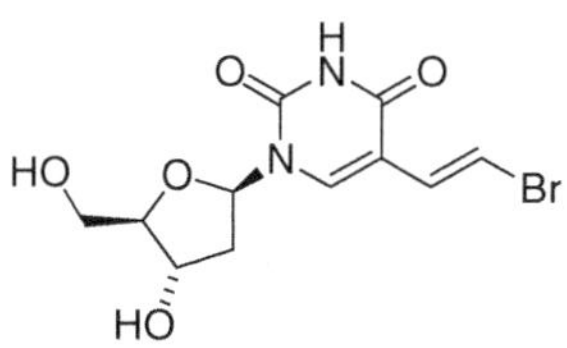	$t_{1/2}$ 16 h F 30% pb >95% ur <1%	V 1.2 CL 3.7 Mwt 333.1 PSA 99.1 Å^2 log P 0.29	BRIVUDINE Antiviral Inactive metabolites: bromovinyluracil, uracilacetic acid

(continued overleaf)

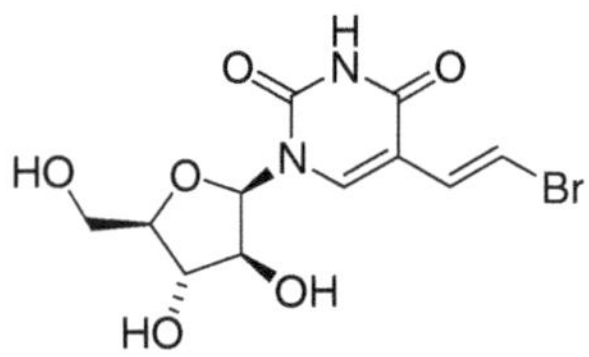	$t_{1/2}$ *F* pb ur	5–7 h 50–70% 98% 74%	*V* CL Mwt PSA log *P*	– – 349.1 119 Å^2 −0.03	SORIVUDINE Antiviral Metabolism: hydrolysis of aminal
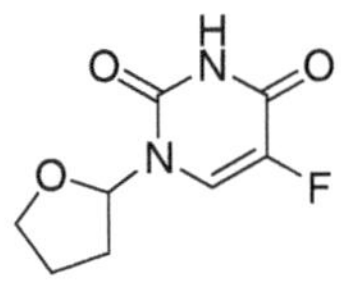	$t_{1/2}$ *F* pb ur	4–11 h High 52% 20%	*V* CL Mwt PSA log *P*	– – 200.2 58.6 Å^2 −0.60	TEGAFUR Antineoplastic Metabolism: hydrolysis of acetal to 5-fluorouracil
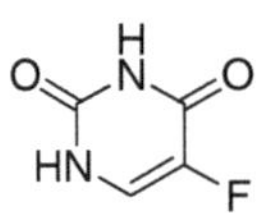	$t_{1/2}$ *F* pb ur	11±4 min 28% 10±2% <5%	*V* CL Mwt PSA log *P*	0.25±0.12 16±7 130.1 58.2 Å^2 −0.65	5-FLUOROURACIL Antineoplastic, metabolite of flucytosine Metabolite: 5,6-dihydro
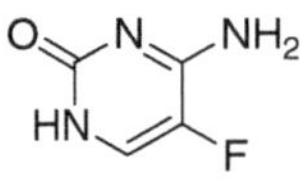	$t_{1/2}$ *F* pb ur	4.2±0.3 h 84±6% 4% >90%	*V* CL Mwt PSA log *P*	0.7–0.8 – 129.1 67.5 Å^2 −1.40	FLUCYTOSINE Antifungal Metabolites: 6-hydroxy, glucuronide, α-fluoro-β-alanine
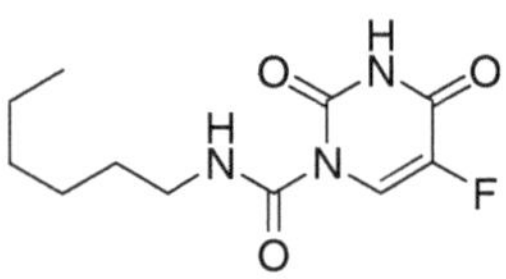	$t_{1/2}$ *F* pb ur	1.1 h	*V* CL Mwt PSA log *P*	– 257.3 78.5 Å^2 2.54	CARMOFUR Prodrug of 5-fluorouracil
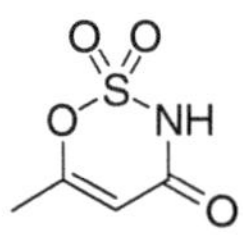	$t_{1/2}$ *F* pb ur	2.5 h 100% – >99%	*V* CL Mwt PSA log *P*	– – 163.2 80.8 Å^2 −0.88	ACESULFAME Artificial sweetener
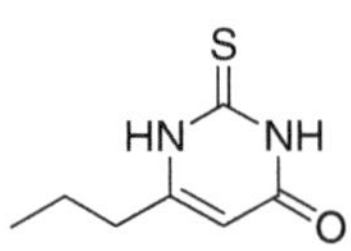	$t_{1/2}$ *F* pb ur	1–2 h 75% 80% <10%	*V* CL Mwt PSA log *P*	0.3–0.4 3.9 170.2 73.2 Å^2 1.15	PROPYLTHIOURACIL Antihyperthyroid, anabolic Metabolism: glucuronidation, sulfation, S-methylation

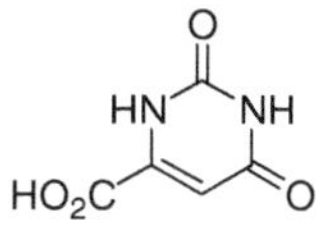	$t_{1/2}$ 1 h F <10% pb 0% ur –	V – CL – Mwt 156.1 PSA 95.5 Å^2 log P −0.84	OROTIC ACID Uricosuric
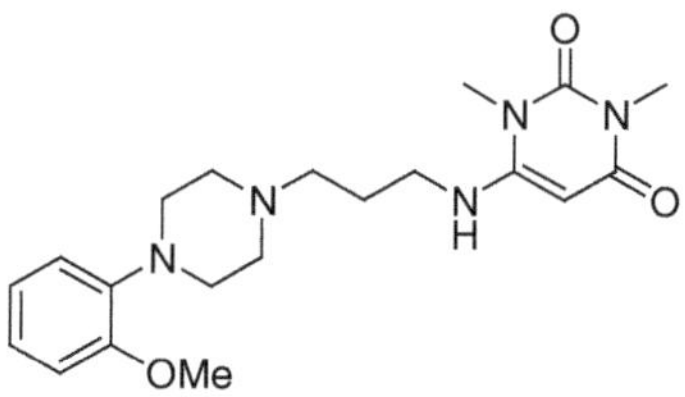	$t_{1/2}$ 2–5 h F 63–80% pb 75–80% ur 10–15%	V 0.4–0.8 CL 1.8–3.8 Mwt 387.5 PSA 68.4 Å^2 log P 1.64	URAPIDIL α_1-Adrenergic antagonist, antihypertensive Metabolites: 4-hydroxyaryl ($t_{1/2}$ 6–10 h), *N*-desmethyl ($t_{1/2}$ 10–16 h)
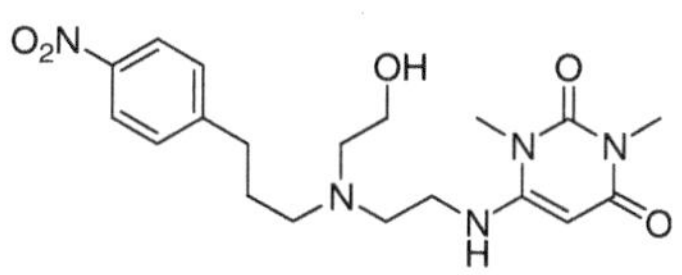	$t_{1/2}$ 1.0–1.5 h F – pb – ur 30%	V 0.14 CL – Mwt 405.5 PSA 122 Å^2 log P 2.48	NIFEKALANT Potassium channel blocker, antiarrhythmic Metabolism: glucuronidation
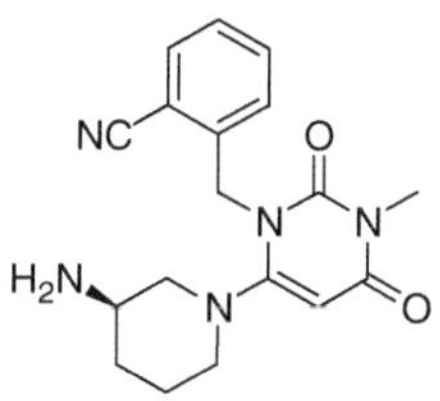	$t_{1/2}$ 21±9 h F 72–88%* pb – ur 60–71%	V 4.6±1.2 CL 2.7±0.7 Mwt 339.4 PSA 93.7 Å^2 log P 1.84	ALOGLIPTIN *monkey DPPIV inhibitor, antidiabetic Metabolites: *N*-desmethyl, *N*-acetyl

$t_{1/2}$, plasma half-life; *F*, oral bioavailability; pb, plasma protein binding; ur, excretion of unchanged drug in urine; V, volume of distribution; CL, clearance; Mwt, molecular weight; PSA, polar surface area; bbb, blood–brain barrier; CNDAC, 1-(2-C-cyano-2-deoxy-β-D-arabinopentofuranosyl)cytosine; DPPIV, dipeptidylpeptidase IV.

Table 61.2 Nucleoside analogs with bicyclic heteroarenes. *V* in l kg^{-1}; CL in ml min^{-1} kg^{-1}; Mwt in g mol^{-1}.

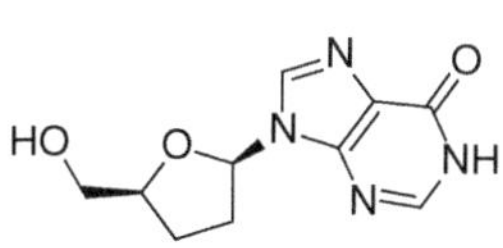	$t_{1/2}$ 1.4±0.3 h F 38±15% pb <5% ur 36±9%	V 1.0±0.2 CL 16±7 Mwt 236.2 PSA 83.8 Å^2 log P −1.33	DIDANOSINE Antiviral Metabolite: hypoxanthine
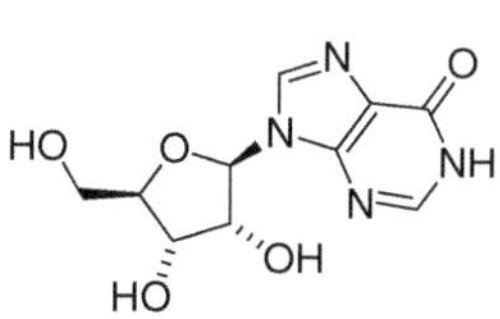	$t_{1/2}$ 15±2 h F – pb – ur –	V – CL – Mwt 268.2 PSA 129 Å^2 log P −1.97	INOSINE Natural nucleoside

(*continued overleaf*)

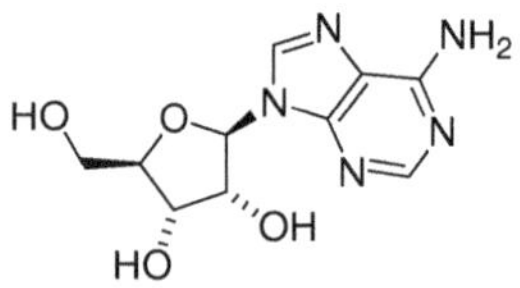	$t^1/_2$ <2 min (iv) F – pb 0% ur <5%	V – CL – Mwt 267.2 PSA 140 Å^2 log P −0.76	ADENOSINE Natural nucleoside
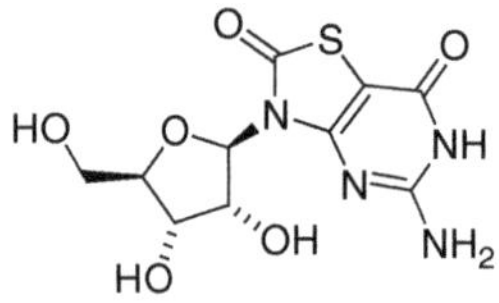	$t^1/_2$ 2 h F Low pb – ur –	V – CL – Mwt 316.3 PSA 183 Å^2 log P −2.17	ISATORIBINE Immunomodulator, antiviral (hepatitis)
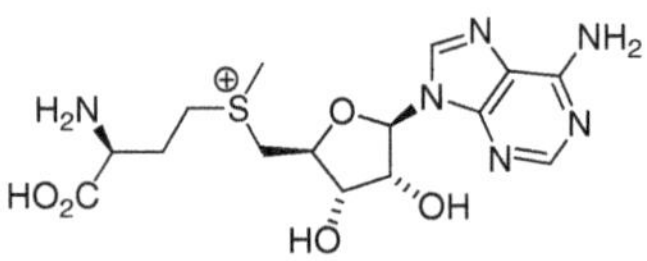	$t^1/_2$ 1.4 h F – pb 5% ur 34–40%	V 0.42 CL 3.4 Mwt 399.5 PSA 182.6 Å^2 log P −5.02	*S*-ADENOSYL-L-METHIONINE Natural electrophilic methylating agent
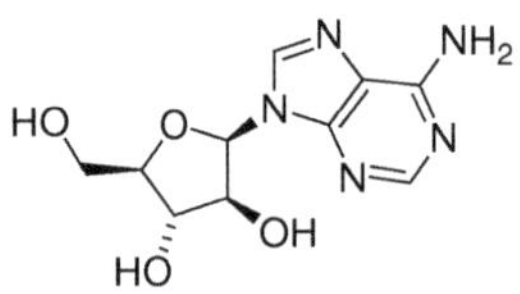	$t^1/_2$ 1 h F – pb 24–38% ur –	V – CL – Mwt 267.2 PSA 140 Å^2 log P −0.76	VIDARABINE Antiviral, crosses bbb, active metabolite: desamino (arabinosyl hypoxanthine, $t^1/_2$ 3.5 h)
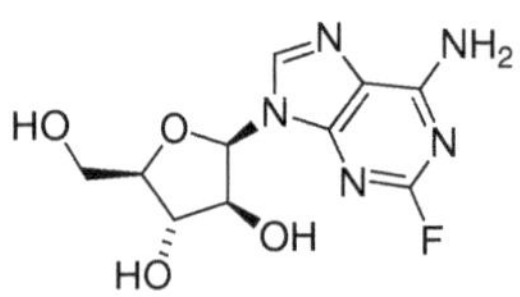	$t^1/_2$ 10–34 h* F 55–75% pb 19–29% ur 27–60%	V 0.3–2.7 CL 1.1–6.3 Mwt 285.2 PSA 140 Å^2 log P −0.80	FLUDARABINE *iv; antineoplastic; prodrug: fludarabine phosphate (F 55–75%, fludarabine on oral administration of prodrug) Metabolism: O-phosphorylation
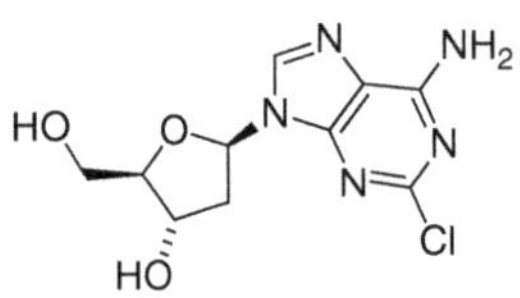	$t^1/_2$ 6–16 h F 37–51% pb 15–21% ur 30–50%*	V 1.3–9.2 CL 11–22 Mwt 285.7 PSA 119 Å^2 log P −0.98	CLADRIBINE *iv Antineoplastic, crosses bbb Metabolite: 2-chloroadenine
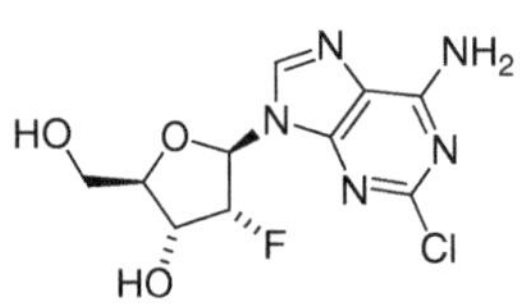	$t^1/_2$ 5 h* F Low** pb 47% ur 49–60%*	V 4.5 CL 1.6 Mwt 303.7 PSA 119 Å^2 log P 0.45	CLOFARABINE *iv; **rat: 50% Antineoplastic Metabolism: O-glucuronidation, O-sulfation, O-phosphorylation, aminal hydrolysis, replacement of NH_2 by OH, oxidation of CH_2OH to CO_2H

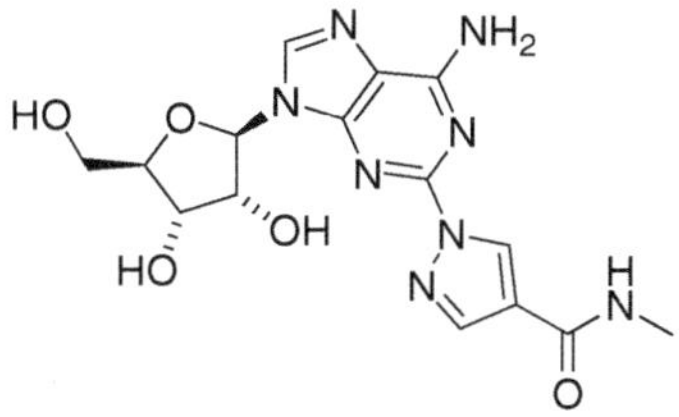

$t^1/_2$	2 h*	V	1.3
F	–	CL	10
pb	–	Mwt	390.4
ur	19–77%*	PSA	177 Å^2
		log P	−3.09

REGADENOSON
*iv
Adenosine A_{2A} agonist, vasodilator

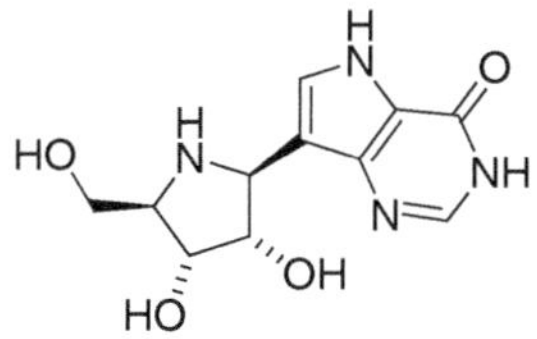

$t^1/_2$	12–20 h	V	1–2
F	40–50%	CL	1–2
pb	–	Mwt	266.3
ur	40–70% (iv)	PSA	130 Å^2
		log P	−3.82

FORODESINE
Antineoplastic

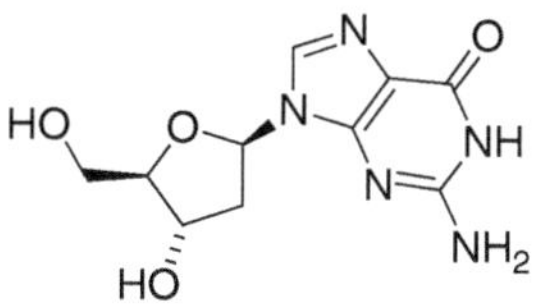

$t^1/_2$	1.3 h	V	–
F	–	CL	–
pb	–	Mwt	267.2
ur	–	PSA	135 Å^2
		log P	−2.01

2′-DEOXYGUANOSINE
Natural nucleoside

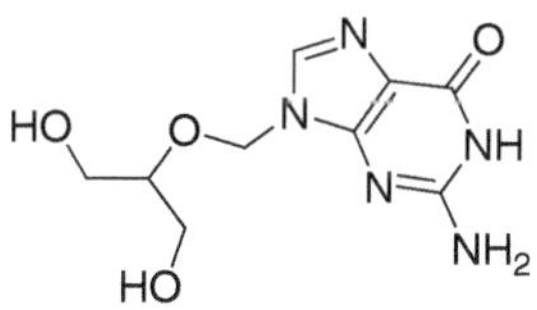

$t^1/_2$	4.3±1.6 h	V	1.1±0.3
F	3%	CL	4.6±1.8
pb	1–2%	Mwt	255.2
ur	73±31%	PSA	135 Å^2
		log P	−1.61

GANCICLOVIR
Antiviral
Metabolism: phosphorylation

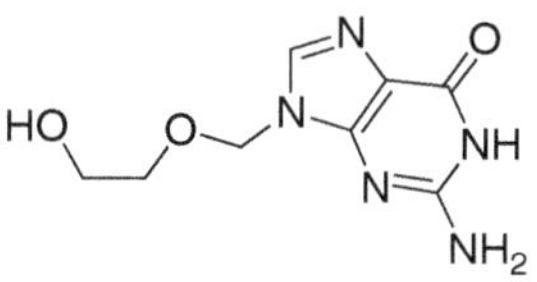

$t^1/_2$	2.4±0.7 h	V	0.7±0.2
F	15±5%	CL	3.4
pb	15±4%	Mwt	225.2
ur	75±10%	PSA	110 Å^2
		log P	−1.76

ACYCLOVIR
Antiviral; prodrug: ester with valine (valacyclovir); F (acyclovir on oral administration of valacyclovir) 54%
Metabolism: 2-hydroxylation of imidazole, oxidation to (carboxymethoxymethyl) guanine, phosphorylation

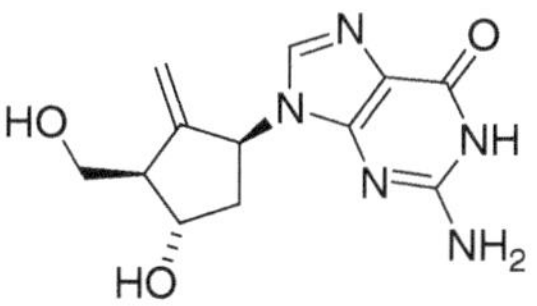

$t^1/_2$	128–149 h	V	1*
F	70%	CL	9
pb	13%	Mwt	277.3
ur	75%	PSA	121 Å^2
		log P	−0.96

ENTECAVIR
*monkey
Antiviral
Metabolites: phosphoryl ($t^1/_2$ 15 h), glucuronide

(*continued overleaf*)

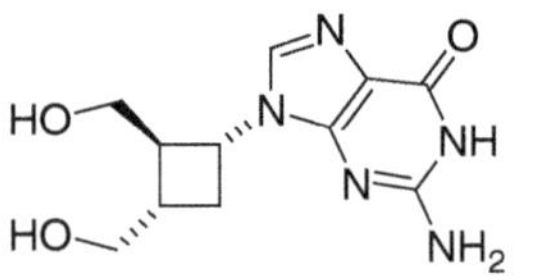	$t_{1/2}$ 10 h *F* 30–40% pb Low ur –	*V* – CL 11 Mwt 265.3 PSA 121 Å^2 log *P* −1.24	LOBUCAVIR Antiviral
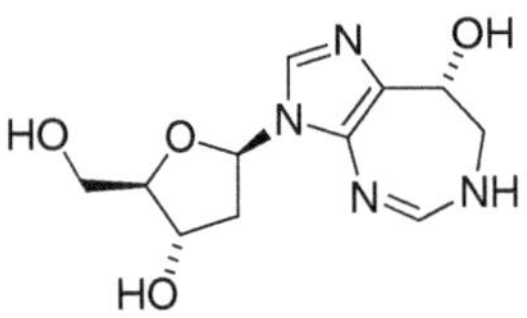	$t_{1/2}$ 5–10 h (iv) *F* Low pb 4% ur 96% (iv)	*V* 0.6 CL 2.8 Mwt 268.3 PSA 112 Å^2 log *P* −2.24	PENTOSTATIN Antineoplastic Crosses bbb
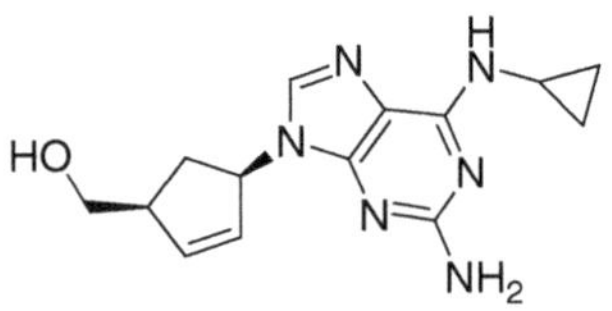	$t_{1/2}$ 1.5±0.6 h *F* 83% pb 50% ur 1.2%	*V* 0.84 CL 13 Mwt 286.3 PSA 97.0 Å^2 log *P* 0.72	ABACAVIR Antiviral Metabolism: glucuronidation, oxidation of alcohol to carboxylic acid
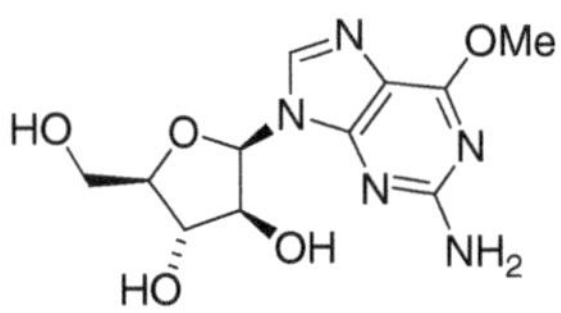	$t_{1/2}$ 0.5 h (iv) *F* <10% pb 20% ur –	*V* 4.9 CL 81 Mwt 297.3 PSA 149 Å^2 log *P* −0.26	NELARABINE Prodrug for arabinofuranosylguanine ($t_{1/2}$ 2–6 h) Metabolism: O-demethylation
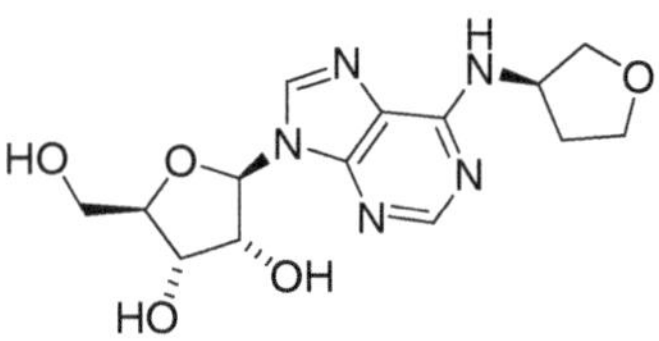	$t_{1/2}$ 0.5 h *F* – pb – ur –	*V* – CL – Mwt 337.3 PSA 135 Å^2 log *P* −1.38	TECADENOSON Adenosine A_1 agonist, antiarrhythmic
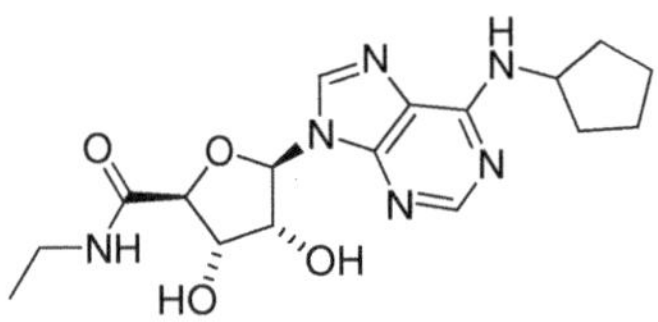	$t_{1/2}$ 2.5 h *F* – pb – ur –	*V* – CL – Mwt 376.4 PSA 134 Å^2 log *P* 2.29	SELODENOSON, DTI-0009 Adenosine A_1 agonist

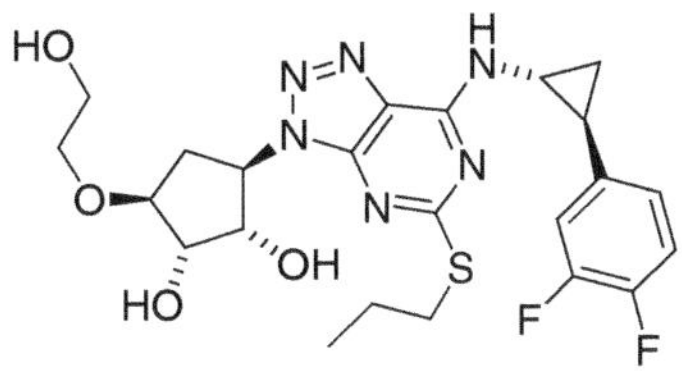	$t_{1/2}$ *F* pb ur	8±2 h Low <99.7% >0.05%	*V* CL Mwt PSA log *P*	1.3 – 522.6 164 Å^2 2.02	TICAGRELOR, AZD6140 Platelet aggregation inhibitor, antithrombotic Metabolism: O-de(hydroxyethylation), N-decyclopropylation
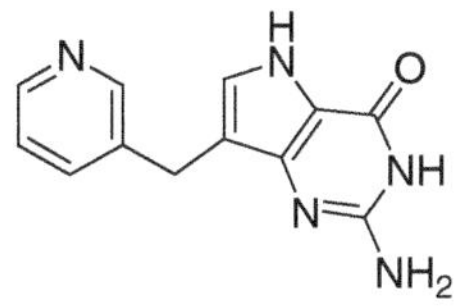	$t_{1/2}$ *F* pb ur	2–5 h 33–59% <14% 82%	*V* CL Mwt PSA log *P*	– 7.4 241.3 96.2 Å^2 −0.74	PELDESINE Nucleoside phosphorylase inhibitor, immunomodulator
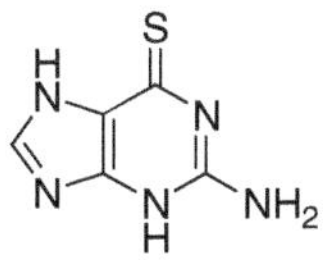	$t_{1/2}$ *F* pb ur	0.5–4.0 h 14–46% – Low	*V* CL Mwt PSA log *P*	0.15 9–15 167.2 111 Å^2 −0.54	THIOGUANINE Antineoplastic Metabolism: S-methylation, replacement of NH_2 by OH, C-hydroxylation of imidazole, conversion to thioguanosine-5′-phosphate
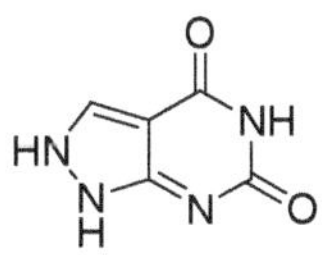	$t_{1/2}$ *F* pb ur	17–29 h – <5% –	*V* CL Mwt PSA log *P*	0.6–1.1 0.3–0.5 136.1 70.1 Å^2 0.55	OXYPURINOL Active metabolite of allopurinol
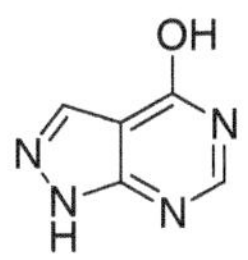	$t_{1/2}$ *F* pb ur	1–2 h 67–90% 3% <10%	*V* CL Mwt PSA log *P*	1.6 6–12 136.1 70.1 Å^2 −0.55	ALLOPURINOL Xanthine oxidase inhibitor, antiurolithic Active metabolite: oxypurinol (see above)
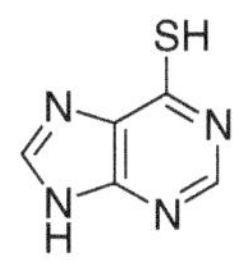	$t_{1/2}$ *F* pb ur	0.9±0.4 h 12±7% 19% 22±12%	*V* CL Mwt PSA log *P*	0.56±0.38 11±4 152.2 85.2 Å^2 −0.30	MERCAPTOPURINE Antineoplastic, immunosuppressant; metabolized intracellularly to 6-thioinosinate
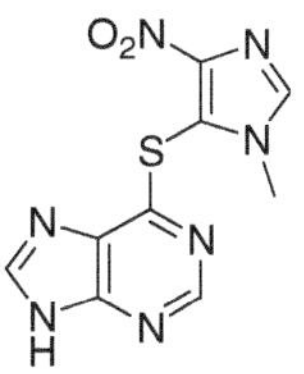	$t_{1/2}$ *F* pb ur	0.2±0.1 h 60±30% 30% <2%	*V* CL Mwt PSA log *P*	0.8±0.7 57±31 277.3 138 Å^2 −1.05	AZATHIOPRINE Prodrug of 6-mercaptopurine, immunosuppressant Metabolites: mercaptopurine, 6-thiouracil, 6-methylmercaptopurine

(*continued overleaf*)

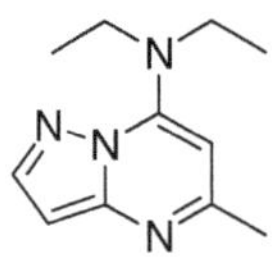	$t^1\!/\!_2$ 1.6–2.4 h *F* 100% pb – ur 2%	*V* 0.8±0.4 CL – Mwt 205.3 PSA 46.3 Å^2 log *P* 2.31	TRAPIDIL Vasodilator Metabolite: *N*-desethyl ($t^1\!/\!_2$ 4–5 h)
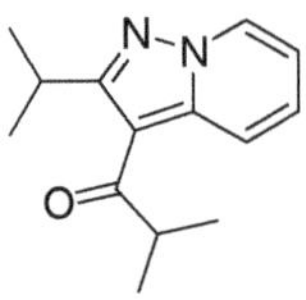	$t^1\!/\!_2$ 19 h *F* – pb >95% ur <0.1%	*V* – CL – Mwt 230.3 PSA 34.4 Å^2 log *P* 2.44	IBUDILAST Antiallergic, antiasthmatic, vasodilator, crosses bbb Metabolite: 6,7-dihydro
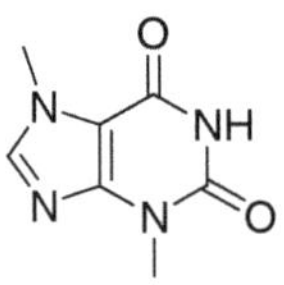	$t^1\!/\!_2$ 6–10 h *F* 80% pb 21% ur 10%	*V* 0.6 CL 1.0 Mwt 180.2 PSA 67.2 Å^2 log *P* −1.06	THEOBROMINE Antiasthmatic Metabolism: N-demethylation, hydroxylation of imidazole (to methylated uric acids)
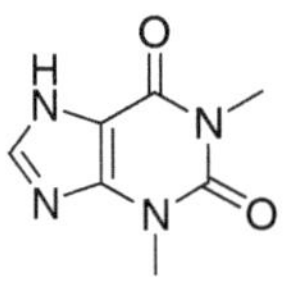	$t^1\!/\!_2$ 9±2 h *F* 96±8% pb 56±4% ur 18±3%	*V* 0.50±0.16 CL 0.6±0.2 Mwt 180.2 PSA 69.3 Å^2 log *P* −0.28	THEOPHYLLINE Antiasthmatic Metabolism: N-demethylation, hydroxylation of imidazole (to methylated uric acids)
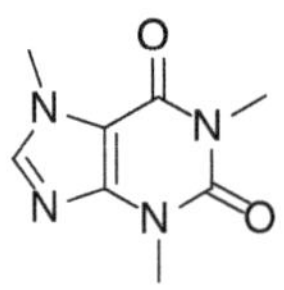	$t^1\!/\!_2$ 5±2 h *F* 100% pb 36±7% ur 1.1±0.5%	*V* 0.61±0.02 CL 1.4±0.5 Mwt 194.2 PSA 58.4 Å^2 log *P* −0.63	CAFFEINE CNS stimulant, smoking doubles clearance Metabolism: N-demethylation, oxidation to uric acid, imidazole ring opening
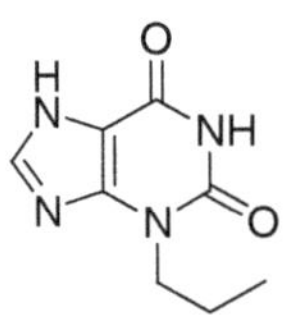	$t^1\!/\!_2$ 1–2 h *F* >85% pb 47% ur 85–91%	*V* 0.5–0.6 CL 3–4 Mwt 194.2 PSA 78.1 Å^2 log *P* 0.31	ENPROFYLLINE Bronchodilator
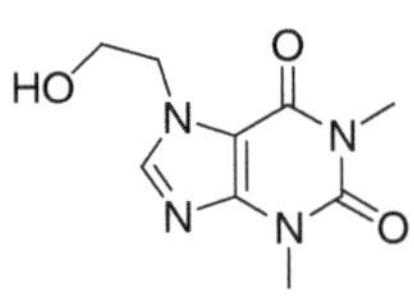	$t^1\!/\!_2$ 4 h *F* 80% pb – ur 16%	*V* 0.6 CL 1.8 Mwt 224.2 PSA 78.7 Å^2 log *P* −0.91	ETOFYLLINE Vasodilator Metabolism: hydroxylation

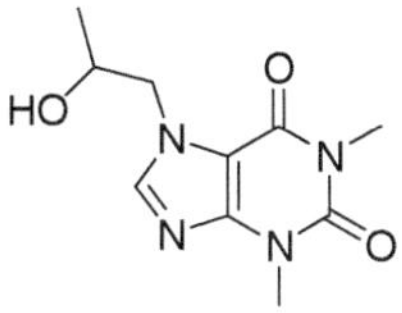	$t^1\!/_2$ 8–12 h *F* 100% pb – ur 18–29%	*V* 0.53–0.72 CL 0.78 Mwt 238.2 PSA 78.7 Å^2 log *P* −0.56	PROXYPHYLLINE Bronchodilator, vasodilator Metabolism: N3-demethylation
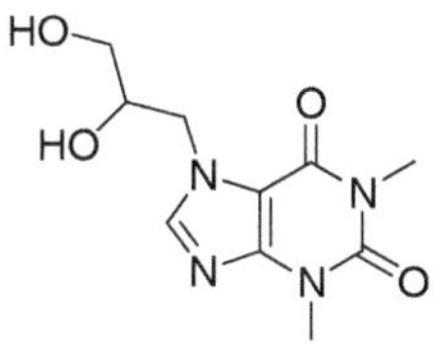	$t^1\!/_2$ 1.8–2.3 h *F* 68–82% pb <3% ur 85–88%	*V* 0.4–0.8 CL 4.8 Mwt 254.2 PSA 98.9 Å^2 log *P* −1.25	DYPHYLLINE, DIPROPHYLLINE Bronchodilator
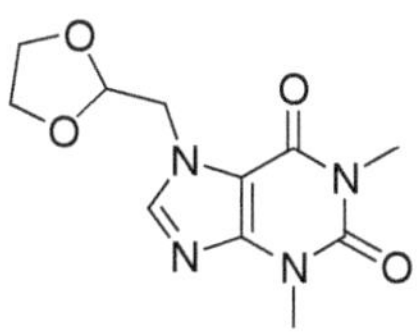	$t^1\!/_2$ 7±1 h *F* 63% pb 48% ur <4%	*V* 1 CL – Mwt 266.3 PSA 76.9 Å^2 log *P* −0.87	DOXOFYLLINE Bronchodilator Metabolite: (hydroxyethyl)theophylline
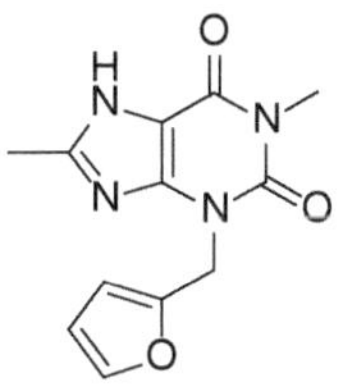	$t^1\!/_2$ 48±11 h *F* – pb – ur	*V* – CL – Mwt 260.3 PSA 82.4 Å^2 log *P* 0.85	FURAFYLLINE Antiasthmatic
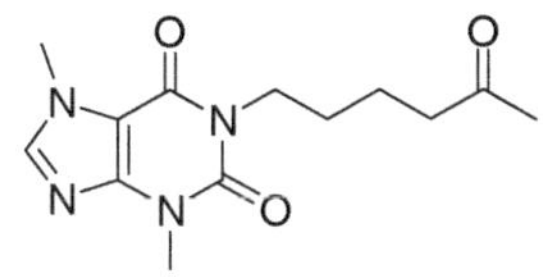	$t^1\!/_2$ 0.9±0.3 h *F* 33±13% pb 0% ur 0%	*V* 4.2±0.9 CL 60±13 Mwt 278.3 PSA 75.5 Å^2 log *P* −0.15	PENTOXIFYLLINE Hemorheologic agent Metabolism: reduction of ketone to alcohol, oxidation to 4-carboxybutyl and 3-carboxypropyl-3,7-dimethylxanthine
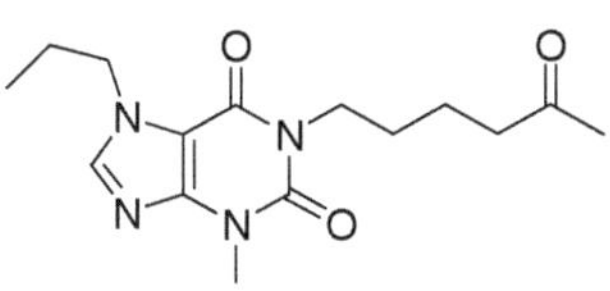	$t^1\!/_2$ 0.74 h *F* 4%* pb – ur –	*V* – CL – Mwt 306.4 PSA 75.5 Å^2 log *P* 0.87	PROPENTOPHYLLINE *rabbit Vasodilator Metabolism: reduction of ketone to alcohol
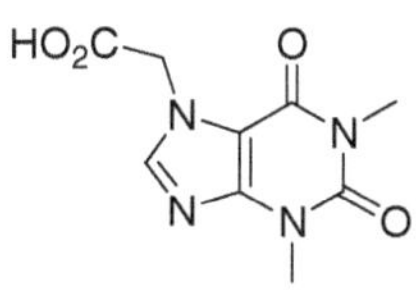	$t^1\!/_2$ 1–2 h *F* 1–2% pb – ur –	*V* 0.5 CL 7 Mwt 238.2 PSA 95.7 Å^2 log *P* −0.78	ACEFYLLINE Diuretic, cardiotonic, bronchodilator

(continued overleaf)

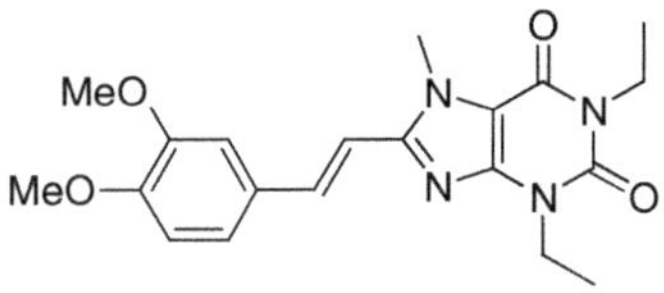	$t_{1/2}$ 47 h F – pb – ur –	V/F 3 CL/F 1.5 Mwt 384.4 PSA 76.9 Å^2 log P 2.63	ISTRADEFYLLINE, KW-6002 Adenosine A_{2A} antagonist for treatment of Parkinson's disease
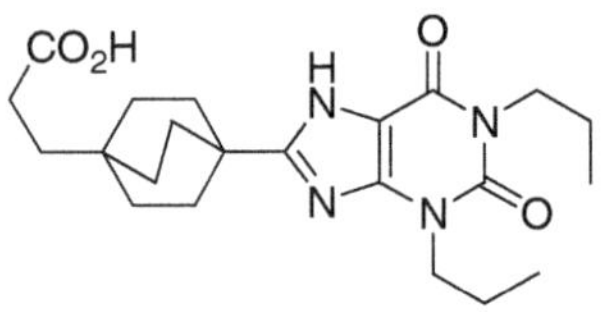	$t_{1/2}$ 14–25 h F 92%* pb – ur –	V 5.6 CL – Mwt 416.5 PSA 107 Å^2 log P 3.79	BG 9928 *monkey Adenosine A_1 antagonist, natriuresis enhancer
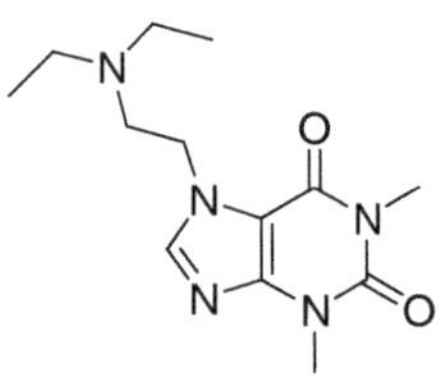	$t_{1/2}$ 1.6 h* F 84%* pb – ur 20%*	V – CL – Mwt 279.3 PSA 61.7 Å^2 log P 0.57	ETAMIPHYLLINE *rat Bronchodilator
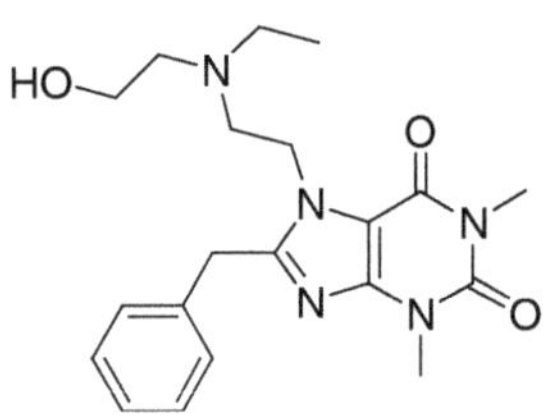	$t_{1/2}$ 1–2 h F – pb – ur –	V 18 CL – Mwt 385.5 PSA 81.9 Å^2 log P 1.08	BAMIFYLLINE Bronchodilator; active metabolite: des(hydroxyethyl) ($t_{1/2}$ 18 h)
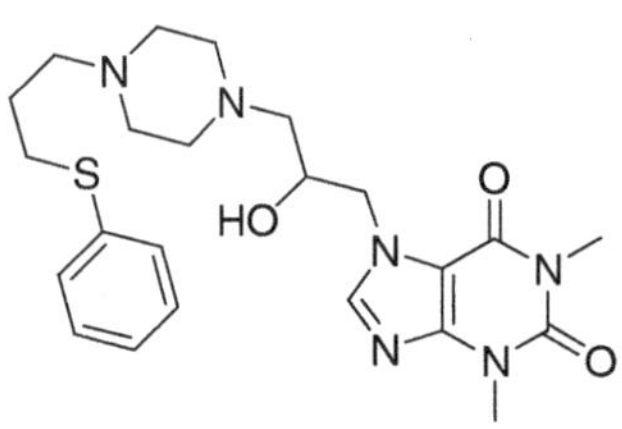	$t_{1/2}$ 16±8 h F – pb – ur –	V – CL – Mwt 472.6 PSA 110 Å^2 log P 1.91	TAZIFYLLINE Histamine H_1 antagonist
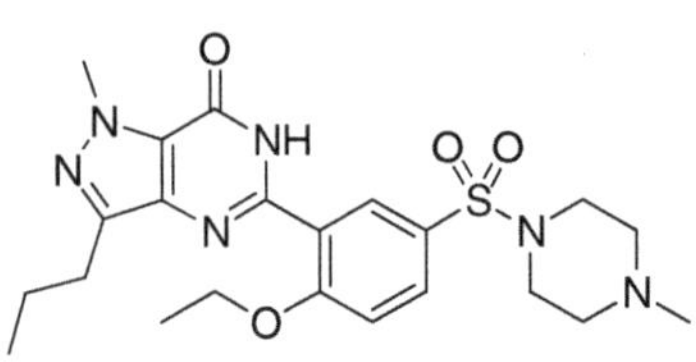	$t_{1/2}$ 3–5 h F 41% pb 96% ur <2%	V 1.5 CL 9.1 Mwt 474.6 PSA 113 Å^2 log P 2.28	SILDENAFIL Phosphodiesterase V inhibitor Metabolism: N-demethylation, oxidative degradation of piperazine

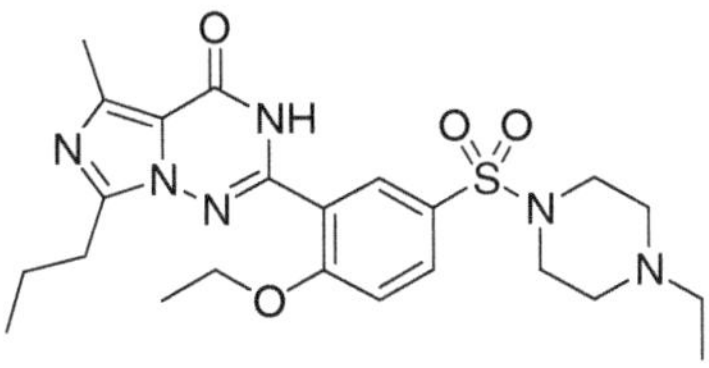	$t^{1/2}$ 4–5 h *F* 15% pb 95% ur 2–6%	*V* 3 CL 13 Mwt 488.6 PSA 118 Å^2 log *P* 3.03	VARDENAFIL Active metabolite: *N*-desethyl
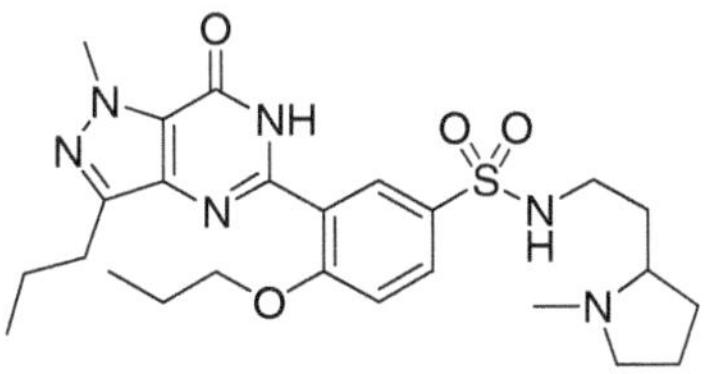	$t^{1/2}$ 7–12 h *F* 38–56%* pb 94% ur –	*V* – CL/*F* 7–23 Mwt 516.7 PSA 121 Å^2 log *P* 3.08	UDENAFIL, DA-8159 *rat Phosphodiesterase *V* inhibitor Metabolism: hydroxylation, N-demethylation, N-dealkylation of sulfonamide
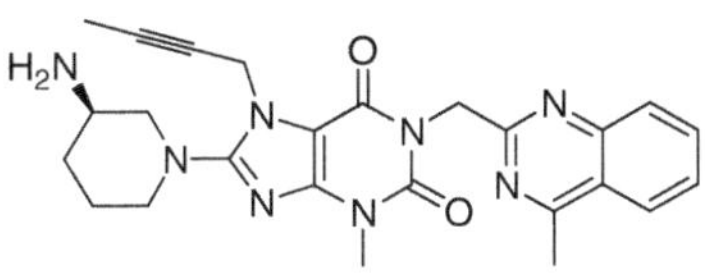	$t^{1/2}$ 113–130 h *F* 15–51%* pb 75–99% ur 2.4%	*V* 4–40 CL 0.2–1.1 Mwt 472.5 PSA 113 Å^2 log *P* 1.99	LINAGLIPTIN *rat DPPIV inhibitor, antidiabetic Metabolism: hydroxylation of propargylic CH_3, oxidation of aromatic CH_3 to CO_2H, acetylation of NH_2, oxidative deamination to piperidin-3-one, then reduction to alcohol
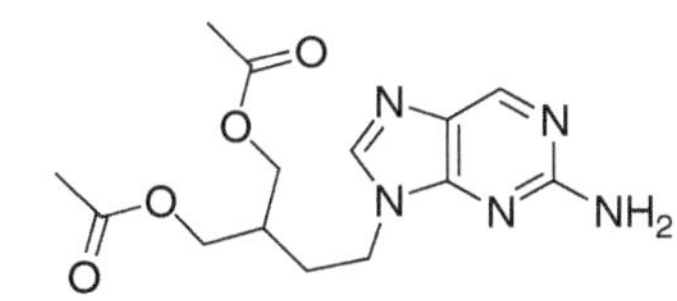	$t^{1/2}$ 2.3±0.4 h *F* 77±8% pb <20% ur 74±9%	*V* 0.98±0.13 CL 8.0±1.5 Mwt 321.3 PSA 122 Å^2 log *P* −0.12	FAMCICLOVIR Antiviral Prodrug of penciclovir (diol, *F* 5%); all data: diol on oral dosing of diacetate
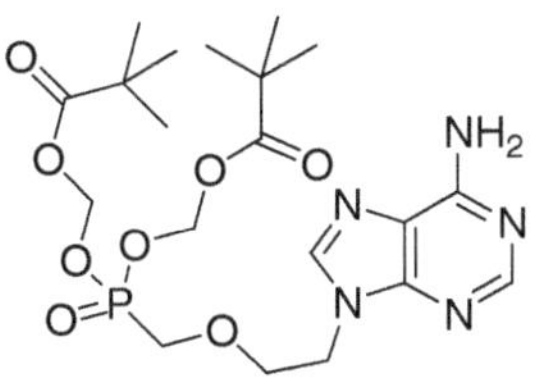	$t^{1/2}$ 5–11 h** *F* 60%* pb 4%* ur 45%*	*V* 0.42* CL 3.7* Mwt 501.5 PSA 172 Å^2 log *P* 2.45	ADEFOVIR DIPIVOXIL Prodrug of adefovir (phosphonic acid); *adefovir on oral dosing of prodrug **prodrug: 1.6 h (iv)
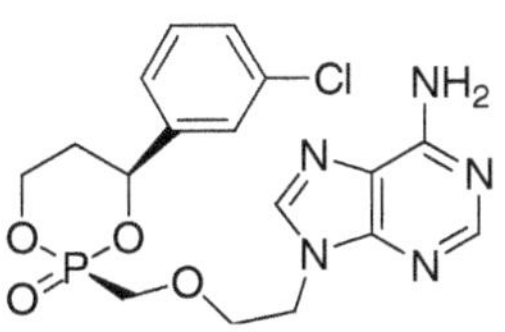	$t^{1/2}$ 4–14 h *F* 19%* pb – ur –	*V* – CL – Mwt 423.8 PSA 124 Å^2 log *P* 1.67	PRADEFOVIR *monkey Prodrug of adefovir (phosphonic acid)

(*continued overleaf*)

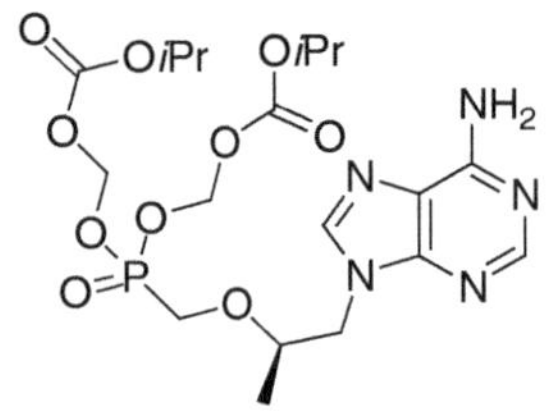

$t_{1/2}$	12–26 h*	V	0.83**
F	25%*	CL	3.8*
pb	<1%*	Mwt	519.4
ur	70–80%**	PSA	190 Å^2
		log P	2.04

TENOFOVIR DISOPROXIL
*tenofovir on oral dosing of prodrug
**tenofovir iv
Antiviral, prodrug of tenofovir (phosphonic acid)

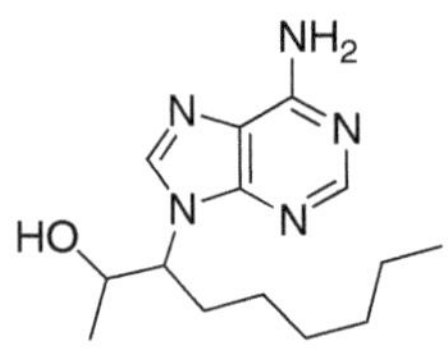

$t_{1/2}$	1–2 h	V	–
F	–	CL	–
pb	–	Mwt	277.4
ur	–	PSA	89.9 Å^2
		log P	2.61

ERYTHROHYDROXYNONYL-ADENINE (EHNA)
Adenosine deaminase inhibitor

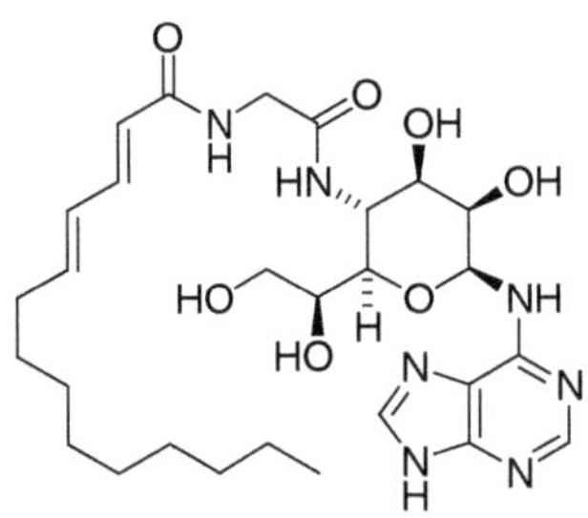

$t_{1/2}$	1.3 h (iv)	V	0.17
F	–	CL	2.7
pb	–	Mwt	589.7
ur	–	PSA	209 Å^2
		log P	1.60

KRN-5500
Antineoplastic

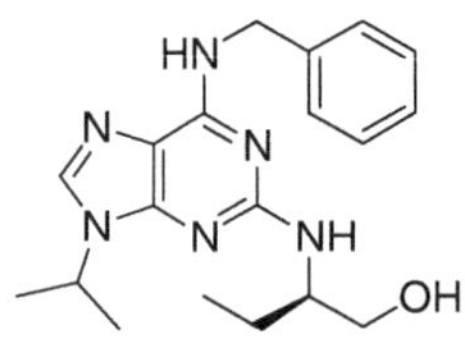

$t_{1/2}$	<1 h	V	–
F	–	CL	–
pb	90%	Mwt	354.5
ur	<3% (rat)	PSA	87.9 Å^2
		log P	1.68

ROSCOVITINE
CDK inhibitor; crosses bbb
Metabolism: oxidation of alcohol to carboxyl, N-dealkylation

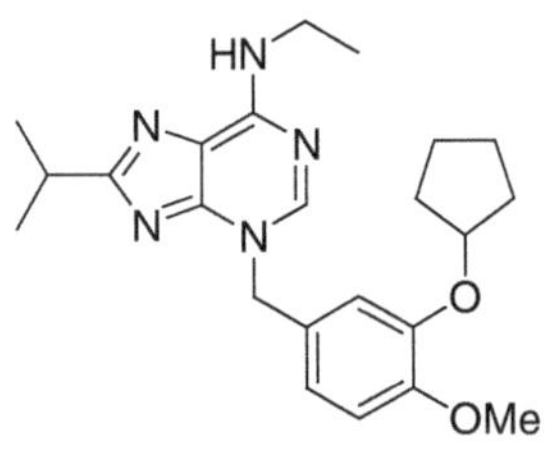

$t_{1/2}$	10±4 h	V	–
F	–	CL	–
pb	–	Mwt	409.5
ur	1%	PSA	74.1 Å^2
		log P	4.84

V-11294A
PDE4-inhibitor
Metabolite: *N*-desethyl ($t_{1/2}$ 10±2 h)

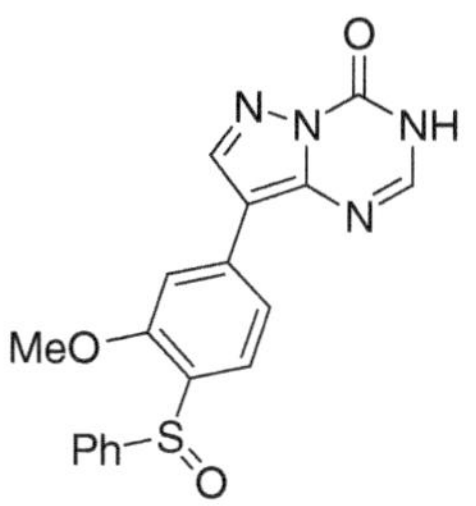	$t^{1}/_{2}$	1.7–1.9 h	*V*	–	BOF-4272 Xanthine oxidase inhibitor for treatment of hyperuricemia Metabolism: hydroxylation, reduction to sulfide, oxidation to sulfone
	F	–	CL	–	
	pb	–	Mwt	366.4	
	ur	1%	PSA	89.4 Å^2	
			log *P*	2.62	

$t^{1}/_{2}$, plasma half-life; *F*, oral bioavailability; pb, plasma protein binding; ur, excretion of unchanged drug in urine; V, volume of distribution; CL, clearance; Mwt, molecular weight; PSA, polar surface area; CDK, cyclin-dependent kinases; CNS, central nervous system; PDE, phosphodiesterase.

62
Dihydropyridines

Dihydropyridines are readily prepared from ammonia, acetoacetic esters, and aldehydes (Scheme 62.1). This one-step synthesis does not allow any structural modification of the 1,4-dihydropyridine substructure but only a limited variation of the substituents.

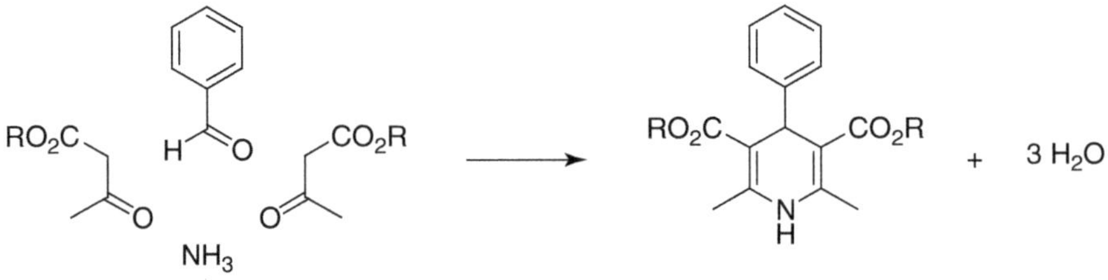

Scheme 62.1 Synthesis of dihydropyridines.

Dihydropyridines, discovered 1967 by Friedrich Bossert and Wulf Vater at Bayer, are one of the three main types of calcium channel blockers known (the other two are verapamil and diltiazem) and mainly act as vasodilators. Their target indications are angina and hypertension (Table 62.1).

Dihydropyridines such as nifedipine mainly undergo dehydrogenation to pyridines and ester hydrolysis *in vivo*. Some drugs of this type, though, have long half-lives, because the alkoxycarbonyl groups are in fact vinylogous carbamates, that is, the lone electron pair at the amino group stabilizes the ester against nucleophilic attack.

Lead Optimization for Medicinal Chemists: Pharmacokinetic Properties of Functional Groups and Organic Compounds, First Edition. Florencio Zaragoza Dörwald.

Table 62.1 Dihydropyridines (calcium channel blockers, antihypertensives). *V* in l kg^{-1}; CL in ml min^{-1} kg^{-1}; Mwt in g mol^{-1}.

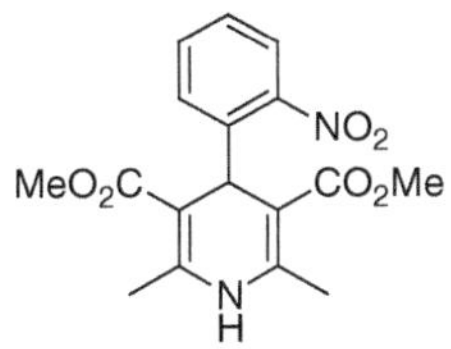	$t_{1/2}$ *F* pb ur	2.5±1.3 h 50±13% 96±1% <1%	*V* CL Mwt PSA log *P*	0.8±0.2 7±2 346.3 110 Å^2 3.58	NIFEDIPINE
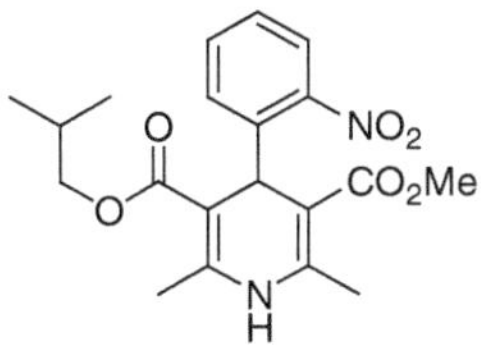	$t_{1/2}$ *F* pb ur	7–15 h 4–8% 99% 0%	*V* CL Mwt PSA log *P*	2–7 8–16 388.4 110 Å^2 4.38	NISOLDIPINE
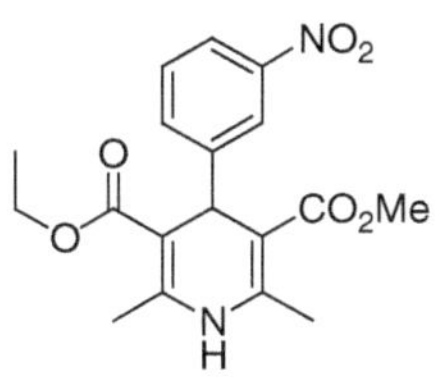	$t_{1/2}$ *F* pb ur	4±2 h 14–55% 98% <1%	*V* CL Mwt PSA log *P*	3.8±1.3 21±4 360.4 110 Å^2 3.81	NITRENDIPINE CYP3A4 substrate
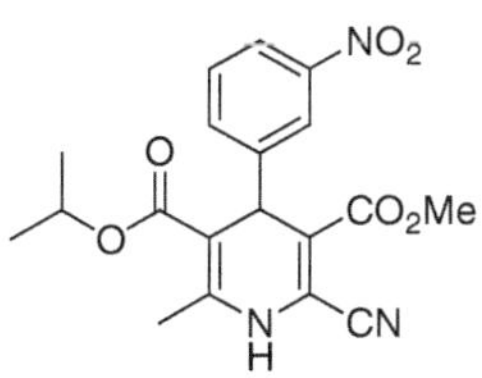	$t_{1/2}$ *F* pb ur	11±2 h 14–19% 98% 0%	*V* CL Mwt PSA log *P*	24 7–18 385.4 134 Å^2 3.38	NILVADIPINE
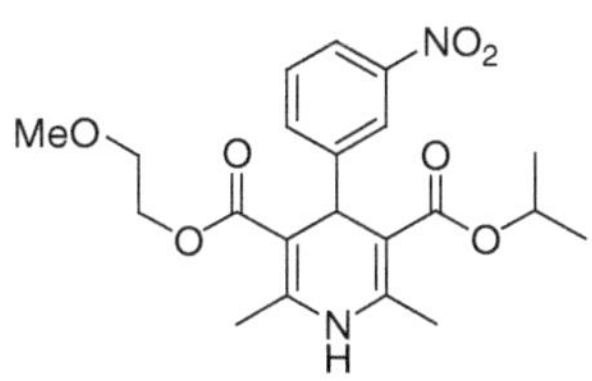	$t_{1/2}$ *F* pb ur	1–3 h 10±4% 98% <1%	*V* CL Mwt PSA log *P*	1.7±0.6 19±6 418.4 120 Å^2 4.05	NIMODIPINE CYP3A substrate
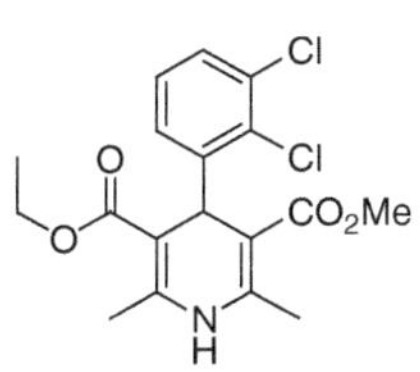	$t_{1/2}$ *F* pb ur	14±4 h 15±8% 99% <0.5%	*V* CL Mwt PSA log *P*	10±3 12±5 384.3 64.6 Å^2 4.76	FELODIPINE CYP3A substrate Metabolism: aromatization, ester hydrolysis, hydroxylation of methyl groups

(*continued overleaf*)

Structure					Name
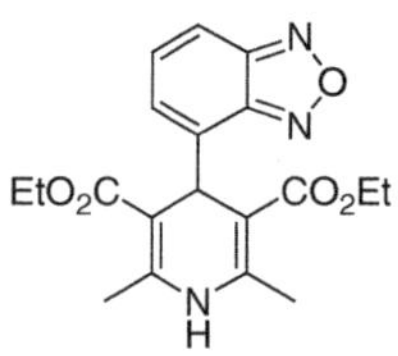	$t_{1/2}$ *F* pb ur	2–4 h – 97% –	*V* CL Mwt PSA log *P*	– 28 371.4 104 Å^2 3.88	DARODIPINE
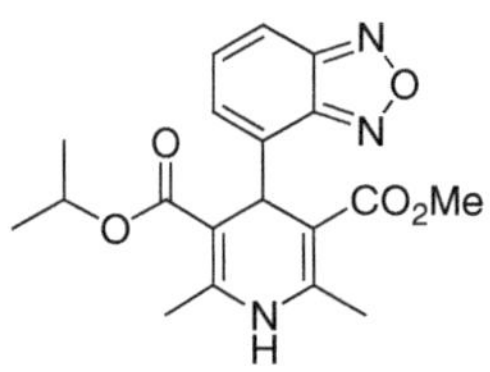	$t_{1/2}$ *F* pb ur	8±5 h 19±7% 97% 0%	*V* CL Mwt PSA log *P*	4±2 10±1 371.4 104 Å^2 3.73	ISRADIPINE
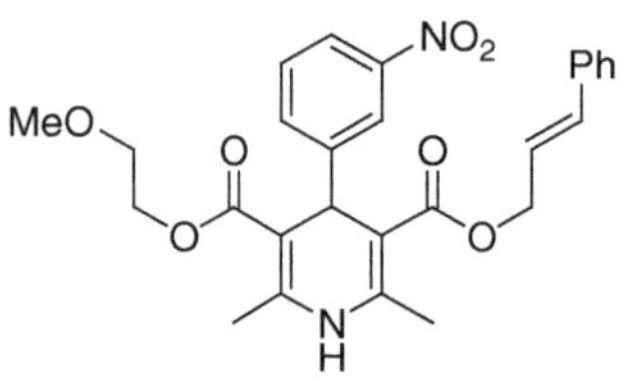	$t_{1/2}$ *F* pb ur	2–3 h – – –	*V* CL Mwt PSA log *P*	– – 492.5 120 Å^2 5.36	CILNIDIPINE, CINALDIPINE
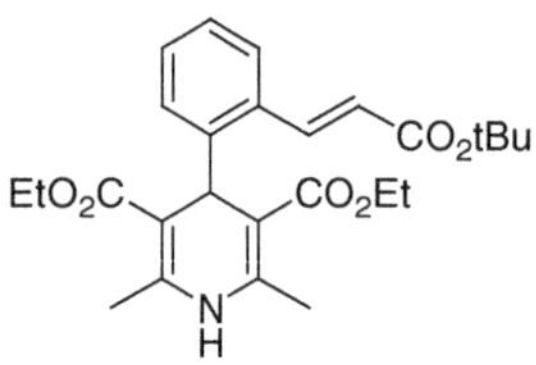	$t_{1/2}$ *F* pb ur	2–11 h 2–52% ≯95% 0%	*V* CL Mwt PSA log *P*	0.9–2.3 125 455.5 90.9 Å^2 5.99	LACIDIPINE
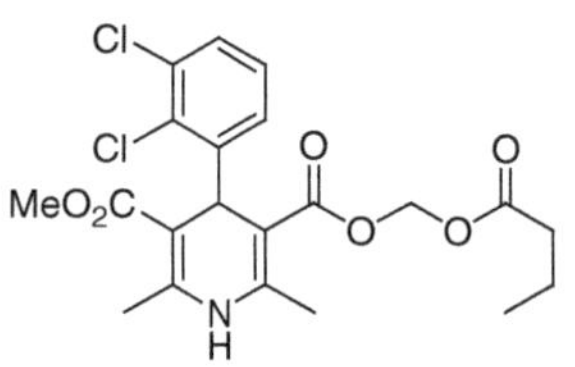	$t_{1/2}$ *F* pb ur	<10 min – >99% 0%	*V* CL Mwt PSA log *P*	0.6 142–170 456.3 90.9 Å^2 5.46	CLEVIDIPINE Antihypertensive for iv dosing
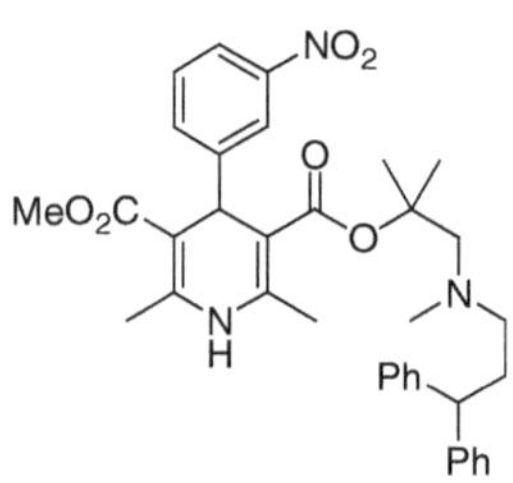	$t_{1/2}$ *F* pb ur	3–11 h 3–40% >98% 0%	*V* CL Mwt PSA log *P*	2.0–2.5 336 611.7 114 Å^2 7.56	LERCANIDIPINE CYP3A4 substrate

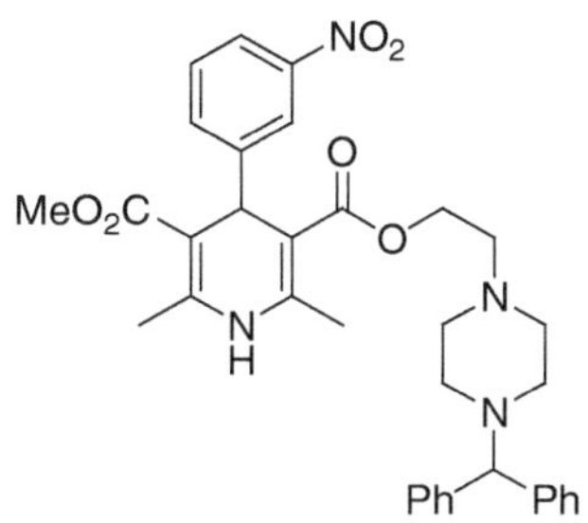

$t^{1}/_{2}$	4–8 h	*V*	–	MANIDIPINE
F	–	CL	–	
pb	99%	Mwt	610.7	
ur	0%	PSA	117 Å^2	
		log *P*	5.66	

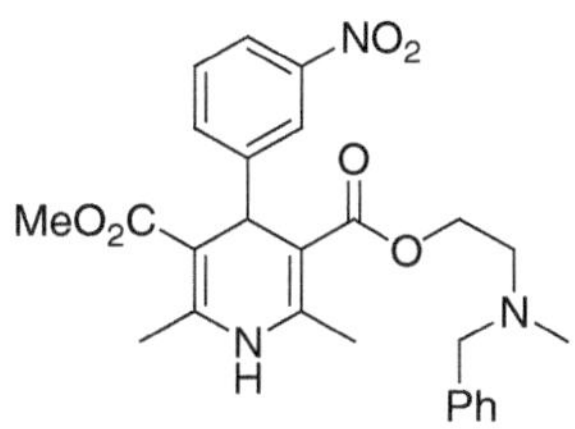

$t^{1}/_{2}$	1.3±0.5 h	*V*	1.1±0.3	NICARDIPINE
F	18±11%	CL	10.4±3.1	
pb	98–99%	Mwt	479.5	
ur	<1%	PSA	114 Å^2	
		log *P*	4.89	

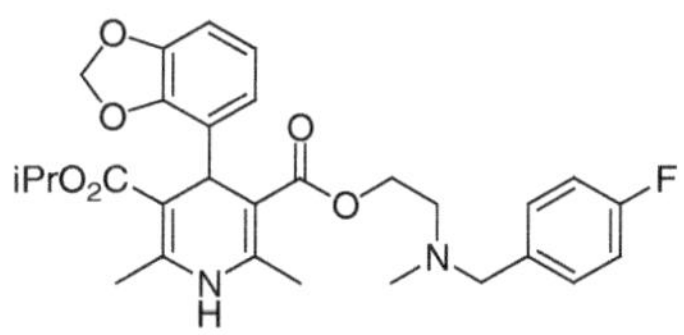

$t^{1}/_{2}$	10–24 h*	*V*	–	ELGODIPINE
F	–	CL	–	*iv
pb	–	Mwt	524.6	
ur	–	PSA	86.3 Å^2	
		log *P*	6.20	

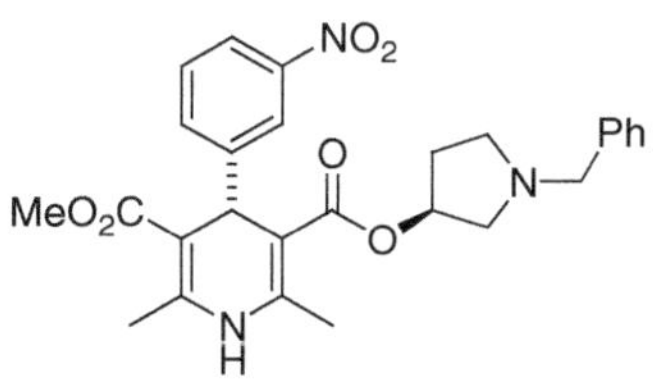

$t^{1}/_{2}$	8–20 h	*V*	–	BARNIDIPINE
F	3–4%	CL	9–10	CYP3A4 substrate
pb	92–99%	Mwt	491.5	Metabolism: ester hydrolysis,
ur	0%	PSA	114 Å^2	oxidation to pyridine, and
		log *P*	5.13	N-debenzylation

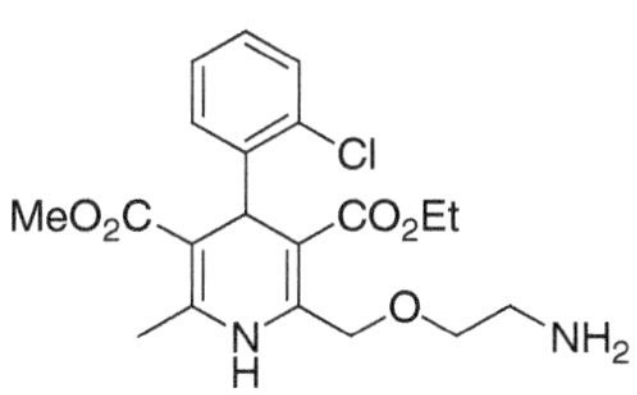

$t^{1}/_{2}$	39±8 h	*V*	16±4	AMLODIPINE
F	74±17%	CL	5.9±1.5	Metabolism: deamination
pb	93±1%	Mwt	408.9	
ur	5–10%	PSA	99.9 Å^2	
		log *P*	4.16	

$t^{1}/_{2}$, plasma half-life; *F*, oral bioavailability; pb, plasma protein binding; ur, excretion of unchanged drug in urine; V, volume of distribution; CL, clearance; Mwt, molecular weight; PSA, polar surface area.

63
Arenesulfonamides

The sulfonamide group may be either a replaceable structural element of drugs or an essential pharmacophore. Sulfonamides show exceptional chemical and metabolic stability and do not undergo hydrolysis *in vivo*. The half-life of these drugs mainly depends on their rate of renal excretion and metabolic degradation of other functional groups.

Mainly three types of drug owe their biological activity to a sulfonamide group: antibacterial 4-aminobenzenesulfonamides (inhibitors of folate biosynthesis by mimicking 4-aminobenzoic acid), hypoglycemic sulfonylureas (blockers of the ATP-sensitive potassium channel), and diuretics (carbonic anhydrase inhibitors) (Table 63.1).

63.1
Antibacterials

Prontosil, a prodrug of sulfanilamide, was patented in 1932 by the German chemists Josef Klarer and Fritz Mietzsch (I.G. Farben). The credit for the discovery of prontosils antibacterial properties went to Gerhard Domagk, a research director at I.G. Farben, who was awarded the 1938 Noble Prize in Medicine for his discovery. Prontosil was a true breakthrough in the treatment of bacterial infections, and numerous other improved sulfonamide antibacterials were developed later. As mimetics of 4-aminobenzoic acid, these compounds inhibit the bacterial biosynthesis of folate. The widespread use of these sulfonamides at high doses revealed some interesting side effects, which led to the development of the sulfonamide-based diuretics and the hypoglycemic sulfonylureas (Chapter 64).

63.2
Diuretics

Diuretics are used to treat several diseases, not all of them being directly related to renal function [1]. The adjustment of volume or composition of body fluids can be required for the treatment of heart failure, hypertension, cirrhosis, nephrotic syndrome, and renal failure.

Lead Optimization for Medicinal Chemists: Pharmacokinetic Properties of Functional Groups and Organic Compounds, First Edition. Florencio Zaragoza Dörwald.

The first diuretics used as such were plant extracts containing digitoxin, caffeine, or theobromine. In 1886, Ernst Jendrássik reported that Hg_2Cl_2 (calomel) had a strong diuretic effect. Extracts of digitalis to which small amounts of mercurous chloride had been added were used as diuretics until the 1950s. Two physicians, Alfred Vogl and Paul Saxl, at the First Medical University of Vienna discovered the diuretic properties of organomercurials. These were originally used only as antiseptics or to treat syphilis, but careful monitoring of the patients clearly revealed their potent diuretic effect. Compounds such as merbaphen, mersalyl, mercamphamide, mercaptomerin, and chlormerodrin were common diuretics until the 1950s (Scheme 63.1).

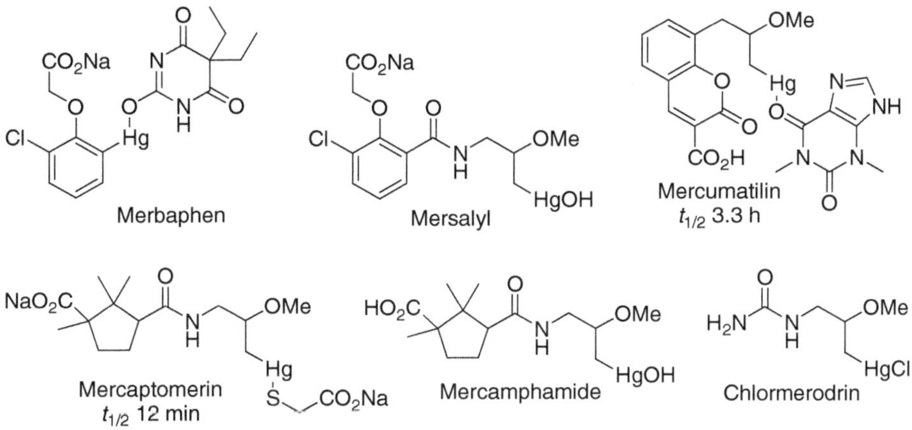

Scheme 63.1 Organomercurials as drugs.

Organomercurials act by binding to thiol groups in dehydrogenases located in renal tubules, thereby inhibiting these enzymes. Because of the (surprisingly low) toxicity of these organomercurials, other less toxic diuretics were sought. One of these was the Michael acceptor ethacrynic acid, launched in 1962, also an irreversible dehydrogenase inhibitor.

High doses of sulfonamide antibacterials, such as sulfanilamide (4-aminobenzenesulfonamide), also induce diuresis. In 1940, it was discovered that many sulfonamides with unsubstituted nitrogen (RSO_2NH_2) inhibit carbonic anhydrase, the enzyme that catalyzes the addition of water to CO_2. Because the kidneys contain large amounts of carbonic anhydrase, it was correctly proposed in 1941 that the diuretic effect of sulfonamides was due to inhibition of renal carbonic anhydrase. The search for more potent and selective sulfonamide-based carbonic anhydrase inhibitors led to the discovery of a large number of useful diuretics, the first being acetazolamide (1952) and chlorothiazide (1957).

Reference

1. Sneader, W. (1997) Diuretics. *Drug News Perspect.*, **10**, 364–369.

Table 63.1 Sulfonamides and related compounds. V in l kg^{-1}; CL in ml min^{-1} kg^{-1}; Mwt in g mol^{-1}.

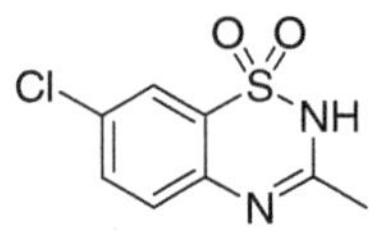	$t^1\!/_2$ F pb ur	48±12 h 86–96% 94±14% 35±15%	V CL Mwt PSA log P	0.21±0.02 0.06±0.02 230.7 66.9 Å^2 1.19	DIAZOXIDE Antihypoglycemic, antihypertensive Metabolism: hydroxylation of CH_3, then oxidation to CO_2H
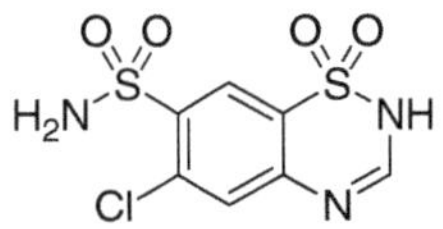	$t^1\!/_2$ F pb ur	1.5±0.2 h 33–56% 95±1% 92±5%	V CL Mwt PSA log P	0.20±0.08 4.5±1.7 295.7 135 Å^2 −0.15	CHLOROTHIAZIDE Diuretic, antihypertensive
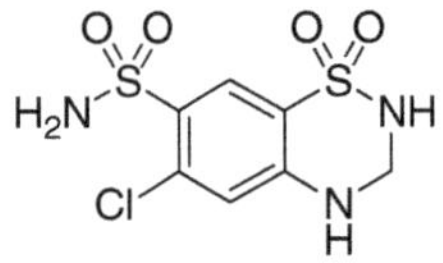	$t^1\!/_2$ F pb ur	7–15 h 71±15% 58±17% >95%	V CL Mwt PSA log P	0.8±0.3 4.9±1.1 297.7 135 Å^2 −0.02	HYDROCHLOROTHIAZIDE Diuretic Metabolism: hydrolysis of aminal (4% in urine)
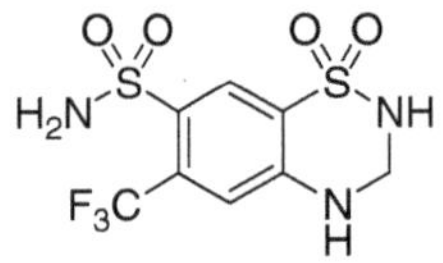	$t^1\!/_2$ F pb ur	10±7 h 63±12% 74% 50–70%	V CL Mwt PSA log P	5 6.4 331.3 135 Å^2 0.08	HYDROFLUMETHIAZIDE Diuretic, antihypertensive Metabolism: hydrolysis of aminal
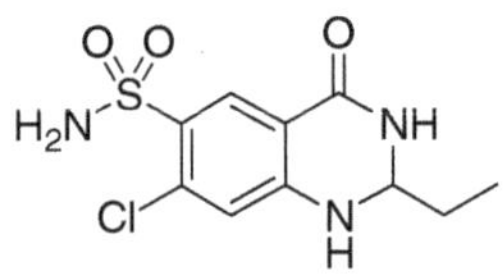	$t^1\!/_2$ F pb ur	6–15 h – – –	V CL Mwt PSA log P	– – 289.7 110 Å^2 3.59	QUINETHAZONE Diuretic, antihypertensive
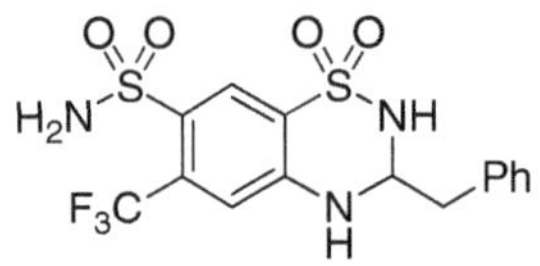	$t^1\!/_2$ F pb ur	3.8±1.2 h 60–90% 94% 30%	V CL Mwt PSA log P	1.0–1.5 4.3 421.4 135 Å^2 1.29	BENDROFLUMETHIAZIDE Diuretic, antihypertensive
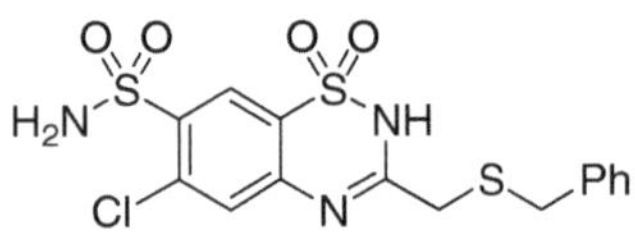	$t^1\!/_2$ F pb ur	5–15 h 60–70% 30% <1%	V CL Mwt PSA log P	– – 431.9 161 Å^2 2.58	BENZTHIAZIDE Diuretic, antihypertensive

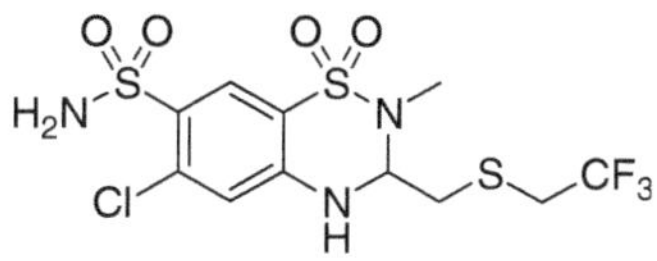	$t_{1/2}$ 26 h *F* High pb 80–85% ur 25%	*V* 4.1 CL – Mwt 439.9 PSA 152 Å^2 log *P* 2.05	POLYTHIAZIDE Diuretic, antihypertensive
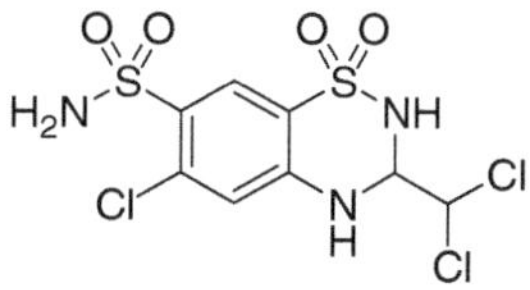	$t_{1/2}$ 2.3±1.9 h *F* 60% pb – ur 62–70%	*V* 0.4* CL 3.4 Mwt 380.7 PSA 135 Å^2 log *P* 0.13	TRICHLORMETHIAZIDE *calves Diuretic, antihypertensive
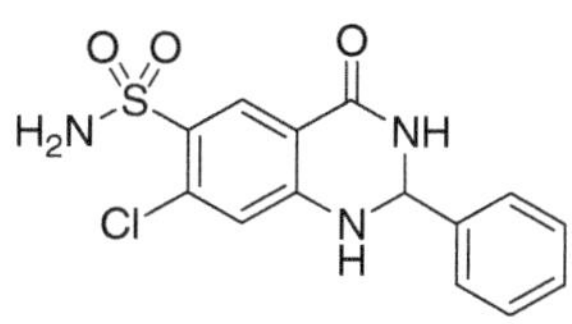	$t_{1/2}$ 17 h *F* >53% pb – ur –	*V* 10.6 CL – Mwt 337.8 PSA 110 Å^2 log *P* 4.17	FENQUIZONE Diuretic
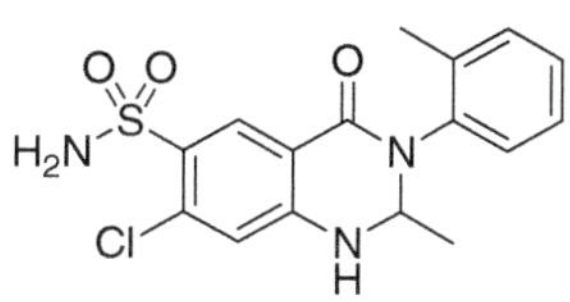	$t_{1/2}$ 14±8 h *F* 40–65% pb 95% ur 80–95%	*V* 1.6 CL 1.4 Mwt 365.8 PSA 101 Å^2 log *P* 3.16	METOLAZONE Diuretic, antihypertensive
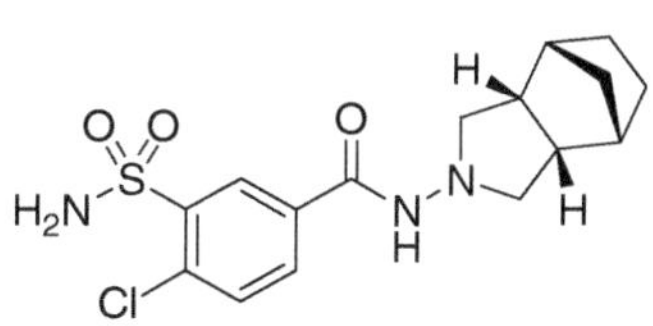	$t_{1/2}$ 10 h *F* >80% pb – ur <1.5%	*V* – CL – Mwt 369.9 PSA 101 Å^2 log *P* 2.21	TRIPAMIDE Diuretic, antihypertensive Metabolism: hydrazide hydrolysis, hydrazide N-acetylation, hydroxylation of bicycyloalkyl group
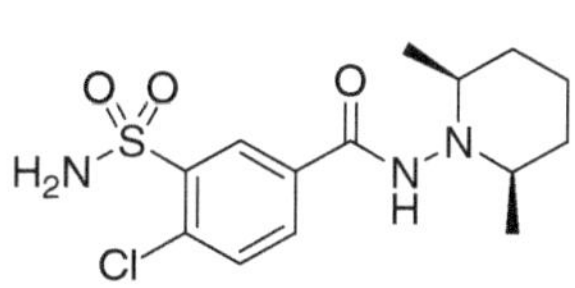	$t_{1/2}$ 9±3 h *F* 85% pb 46% ur 33%	*V* 1.5 CL 2.5 Mwt 345.8 PSA 101 Å^2 log *P* 2.08	CLOPAMIDE Antihypertensive, diuretic
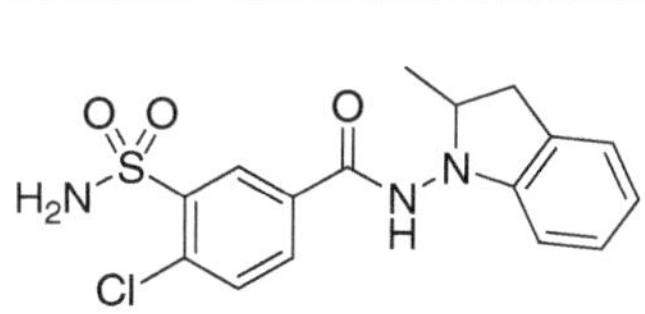	$t_{1/2}$ 18±6 h *F* 93% pb 79% ur 5–7%	*V* 0.4–1.6 CL – Mwt 365.8 PSA 101 Å^2 log *P* 1.96	INDAPAMIDE Antihypertensive, diuretic Metabolism: indoline 5-hydroxylation, dehydrogenation to indole, then oxidation to indol-3-one

(*continued overleaf*)

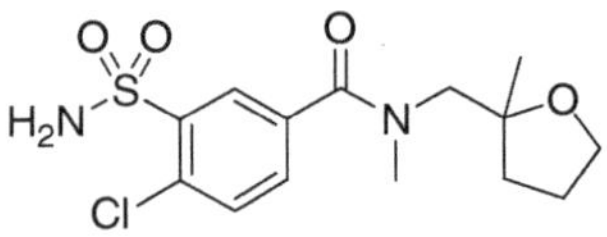	$t_{1/2}$ *F* pb ur	8±5 h >70% 64% 0.5%	*V* CL Mwt PSA log *P*	4.8–8.0 5.8–33 382.9 124 Å^2 1.79	MEFRUSIDE Diuretic Metabolism: oxidation to γ-lactone, then hydrolysis to hydroxy carboxylic acid
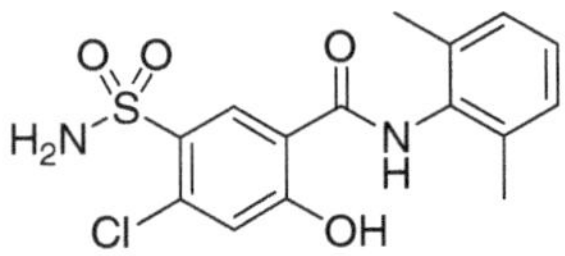	$t_{1/2}$ *F* pb ur	7±1 h 73% 99% 40%	*V* CL Mwt PSA log *P*	0.3 0.54 354.8 118 Å^2 2.19	XIPAMIDE Diuretic, antihypertensive Metabolism: O-glucuronidation
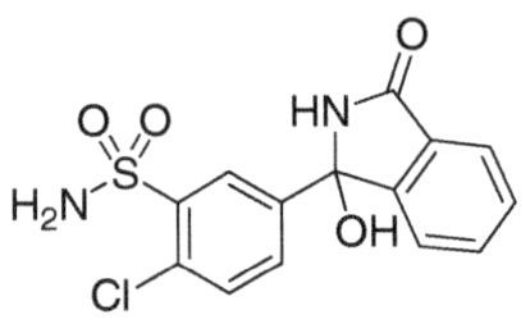	$t_{1/2}$ *F* pb ur	47±22 h 64±10% 75±1% 65±9%	*V* CL Mwt PSA log *P*	0.10±0.04 0.04±0.01 338.8 118 Å^2 0.70	CHLORTHALIDONE Diuretic, antihypertensive Metabolism: reversible ring opening to benzophenone
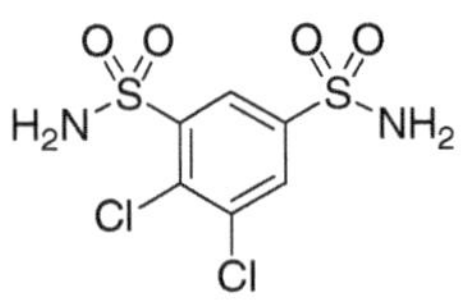	$t_{1/2}$ *F* pb ur	– – 55% –	*V* CL Mwt PSA log *P*	– – 305.2 137 Å^2 0.34	DICHLORPHENAMIDE Carbonic anhydrase inhibitor for treatment of glaucoma Duration of action: 6–12 h
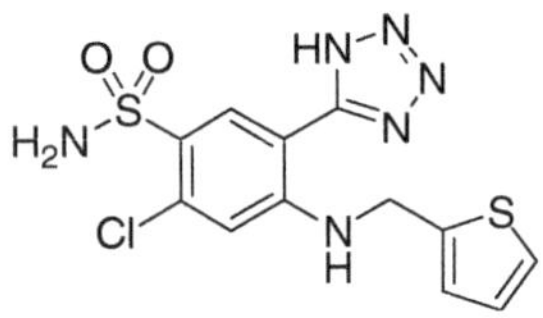	$t_{1/2}$ *F* pb ur	2–3 h 15–20% >95% 27%	*V* CL Mwt PSA log *P*	0.26 1.7 370.8 163 Å^2 1.40	AZOSEMIDE Diuretic Metabolism: N-dealkylation
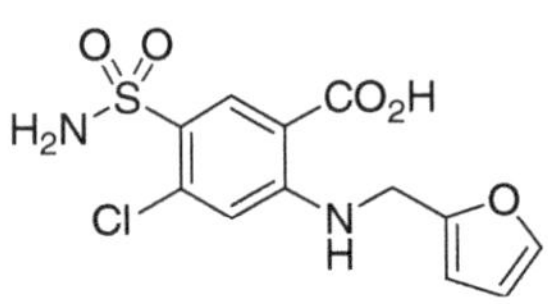	$t_{1/2}$ *F* pb ur	0.3–3.4 h 61±17% 98.8±0.2% 66±7%	*V* CL Mwt PSA log *P*	0.11±0.02 2.0±0.4 330.7 131 Å^2 2.30	FUROSEMIDE Diuretic Metabolism: acyl glucuronidation, N-dealkylation
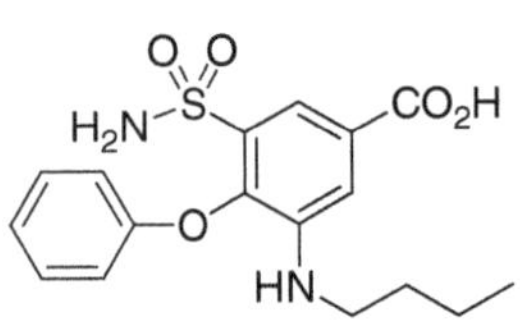	$t_{1/2}$ *F* pb ur	0.8±0.2 h 59–89% 99% 62±20%	*V* CL Mwt PSA log *P*	0.13±0.03 2.6±0.5 364.4 127 Å^2 2.88	BUMETANIDE Diuretic Metabolism: glucuronidation, β-, γ-, and δ-hydroxylation of butyl, N-debutylation, oxidation of CH_3 to CO_2H

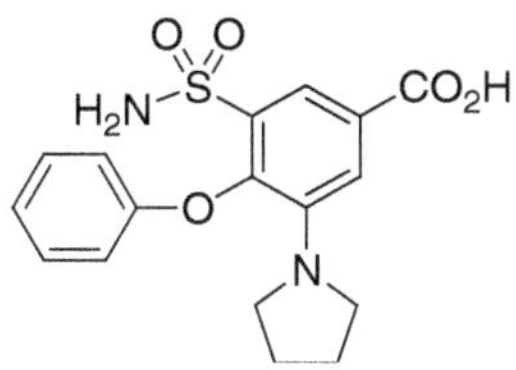	$t^1/_2$ F pb ur	1.0±0.5 h 80% 94% 50%	V CL Mwt PSA log P	0.25–0.30 3.8±0.8 (iv) 362.4 118 Å^2 2.30	PIRETANIDE Diuretic Metabolism: cleavage of pyrrolidine to aminobutanol
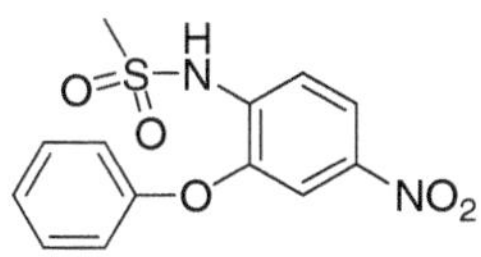	$t^1/_2$ F pb ur	2–5 h 50–70% 96–99% 1–3%	V CL Mwt PSA log P	0.16–0.39 0.5–1.8 308.3 110 Å^2 2.70	NIMESULIDE Antiinflammatory, hepatotoxic Main metabolite: 4-hydroxyphenoxy ($t^1/_2$ 6±3 h) Further metabolism: reduction of NO_2 to NH_2, then oxidation to quinoneimine, followed by reaction with nucleophiles
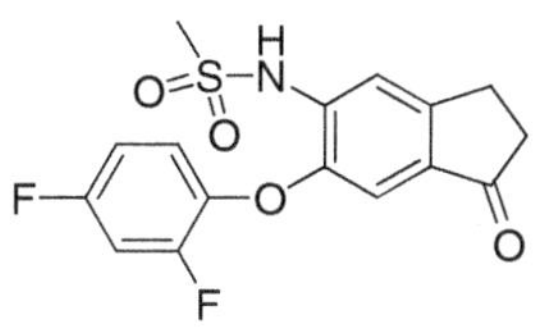	$t^1/_2$ F pb ur	5 h* (iv) 63%* – –	V CL Mwt PSA log P	– – 353.3 80.8 Å^2 2.05	FLOSULIDE *monkey COX-2 inhibitor
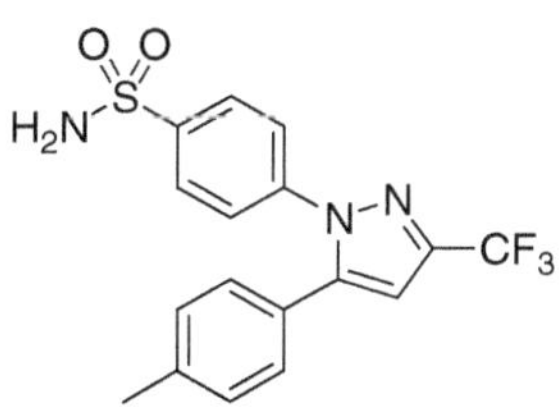	$t^1/_2$ F pb ur	11–16 h 40% >97% <2%	V CL Mwt PSA log P	5.7–7.1 6–8 381.4 81.4 Å^2 4.21	CELECOXIB Antiinflammatory, analgesic Metabolism (by CYP2C9): hydroxylation of CH_3, then oxidation to CO_2H and glucuronidation
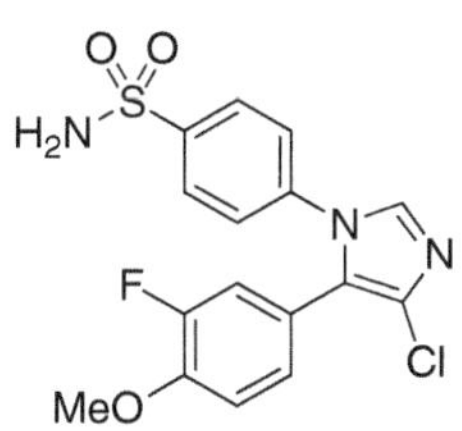	$t^1/_2$ F pb ur	17–26 h 45%* 90% <0.2%	V CL Mwt PSA log P	– – 381.8 90.7 Å^2 2.11	CIMICOXIB, UR-8880 *dog COX-2 inhibitor, antiinflammatory, antidepressant Metabolism: O-demethylation, then glucuronidation

(*continued overleaf*)

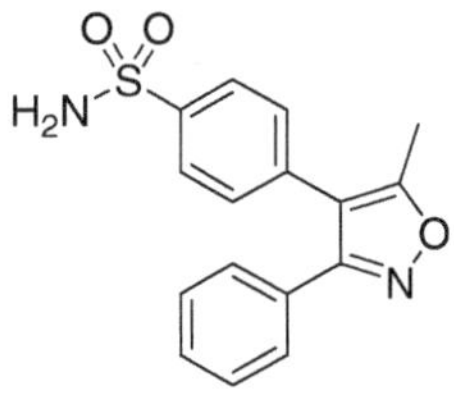

$t_{1/2}$	9±2 h	V	1.3
F	83%	CL	1.7
pb	98%	Mwt	314.4
ur	<5%	PSA	94.6 Å^2
		log P	1.71

VALDECOXIB
Antiinflammatory, analgesic
Metabolism: glucuronidation, reductive cleavage of N–O bond, 4-hydroxylation of Ph, hydroxylation of CH_3, then oxidation to CO_2H; withdrawn in 2005 for adverse skin reactions and increasing risk of myocardial infarction

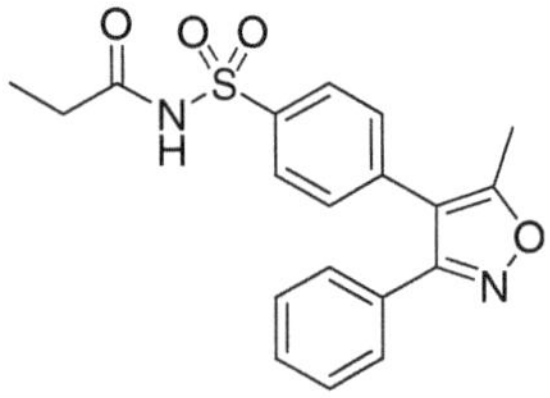

$t_{1/2}$	0.3–1.0 h	V	–
F	–	CL	–
pb	–	Mwt	370.4
ur	–	PSA	97.7 Å^2
		log P	1.72

PARECOXIB
Water soluble, injectable prodrug of valdecoxib
Metabolism: N-depropionylation

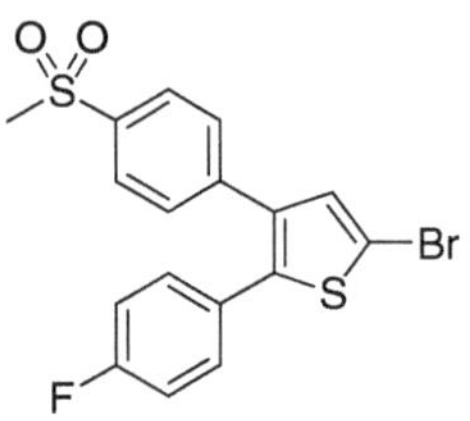

$t_{1/2}$	>200 h	V	–
F	–	CL	–
pb	–	Mwt	411.3
ur	–	PSA	70.8 Å^2
		log P	4.24

DuP 697
COX-2 inhibitor
Long half-life due to enterohepatic recirculation; development discontinued

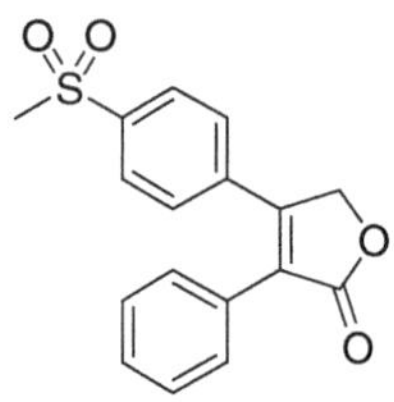

$t_{1/2}$	17 h	V	1.4
F	93%	CL	1.9
pb	87%	Mwt	314.4
ur	<1%	PSA	68.8 Å^2
		log P	1.34

ROFECOXIB
Antiinflammatory, analgesic
Metabolism: 4-hydroxylation and 3,4-dihydroxylation of Ph, hydrogenation of olefin, hydroxylation at CH_2O, reversible lactone hydrolysis; withdrawn in 2004 because of risk of thrombosis and myocardial infarction

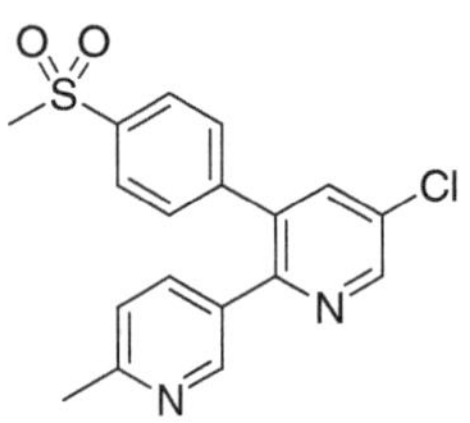

$t_{1/2}$	21–27 h	V	1.6–1.7
F	100%	CL	0.7–0.8
pb	92%	Mwt	358.8
ur	<1%	PSA	68.3 Å^2
		log P	2.21

ETORICOXIB
Antiinflammatory, analgesic
Metabolism: hydroxylation of CH_3, then oxidation to CO_2H and glucuronidation

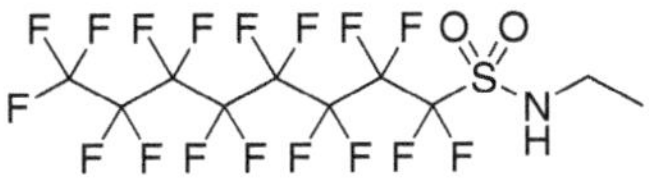	$t_{1/2}$ 16–20 h F – pb – ur –	V – CL – Mwt 527.2 PSA 54.6 Å^2 log P 7.79	SULFLURAMID Insecticide Metabolism: N-deethylation
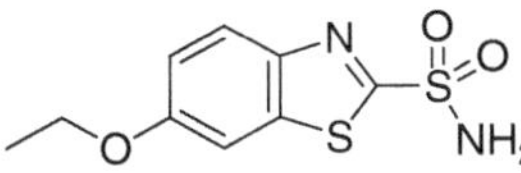	$t_{1/2}$ 4.0±1.5 h F 65% pb >95% ur 40%	V – CL – Mwt 258.3 PSA 119 Å^2 log P 2.14	ETHOXZOLAMIDE Carbonic anhydrase inhibitor, diuretic, antiglaucoma agent
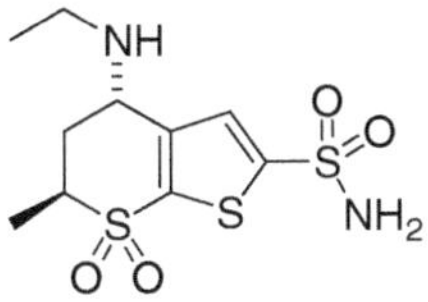	$t_{1/2}$ 133 d F – pb 33% ur 30–40%	V – CL – Mwt 324.4 PSA 151 Å^2 log P 0.10	DORZOLAMIDE Carbonic anhydrase inhibitor; antiglaucoma agent Metabolism (CYP2E1 and CYP3A2): N-deethylation
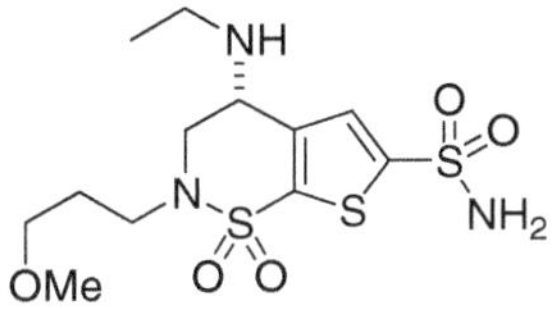	$t_{1/2}$ 111 d F – pb 60% ur High	V – CL – Mwt 383.5 PSA 164 Å^2 log P 0.28	BRINZOLAMIDE Carbonic anhydrase inhibitor; antiglaucoma agent Metabolism: N-deethylation, O-demethylation, N-demethoxypropylation
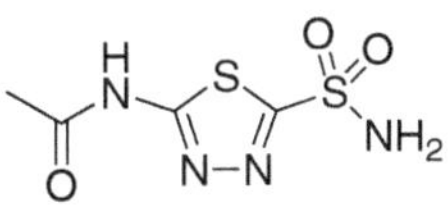	$t_{1/2}$ 7±5 h F 100% pb 85–95% ur 100%	V 0.2–0.4 CL 0.65 Mwt 222.3 PSA 152 Å^2 log P −0.26	ACETAZOLAMIDE Carbonic anhydrase inhibitor, antiglaucoma agent, diuretic
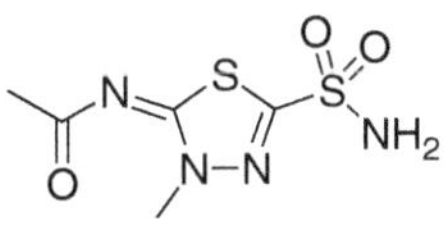	$t_{1/2}$ 14 h F 100% pb 72% ur 25–60%	V 0.07 CL 0.01 Mwt 236.3 PSA 139 Å^2 log P 0.13	METHAZOLAMIDE Carbonic anhydrase inhibitor; diuretic Metabolism: conjugation with cysteine (by nucleophilic displacement of SO_2NH_2 by thiol)
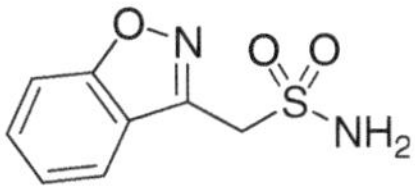	$t_{1/2}$ 59±9 h F 38–40% pb 50–60% ur 15–35%	V 1.2–1.8 CL 0.26–0.32 Mwt 212.2 PSA 94.6 Å^2 log P 0.72	ZONISAMIDE Antiepileptic Metabolism: N-acetylation and N-hydroxylation of sulfonamide, hydroxylation of CH_2, reductive N–O bond cleavage, then reduction of imine to amine and N-acetylation

(*continued overleaf*)

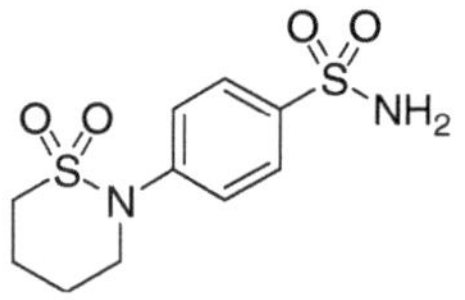	$t_{1/2}$ 9±3 h *F* >90% pb 29% ur 17–69%	*V* – CL – Mwt 290.4 PSA 114 Å^2 log *P* 0.56	SULTHIAME Anticonvulsant, carbonic anhydrase inhibitor Metabolism: hydroxylation
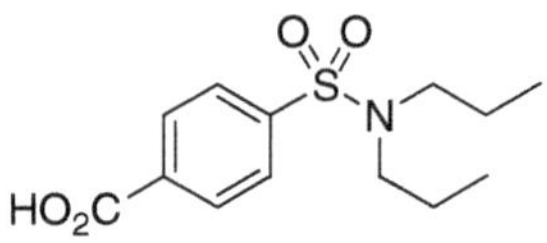	$t_{1/2}$ 4–12 h *F* 100% pb 90±5% ur 1.2±0.2%	*V* 0.17±0.03 CL 0.4 Mwt 285.4 PSA 83.1 Å^2 log *P* 2.51	PROBENECID Uricosuric Metabolism: acyl glucuronidation, hydroxylation at CH_2CH_3 and CH_3, N-depropylation, oxidation of CH_3 to CO_2H
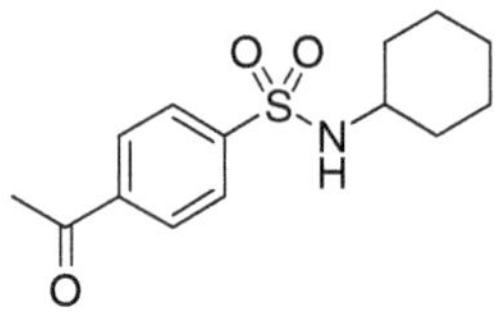	$t_{1/2}$ 1.0–1.3 h *F* – pb 90% ur Low	*V* – CL – Mwt 324.4 PSA 101 Å^2 log *P* 2.44	ACETOHEXAMIDE Antidiabetic Active metabolite: alcohol ($t_{1/2}$ 6.5±1.5 h) formed by reduction of the acetyl group
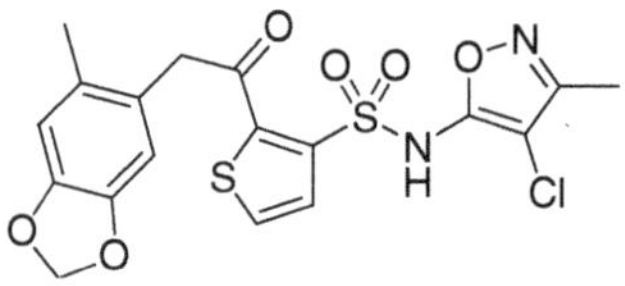	$t_{1/2}$ 7–13 h *F* 70–100% pb >99% ur <1%	*V/F* 1 CL/*F* 1.3–1.5 Mwt 454.9 PSA 144 Å^2 log *P* 2.77	SITAXENTAN Endothelin antagonist, treatment of pulmonary hypertension; crosses bbb Metabolism: benzylic hydroxylation, then oxidation to 1,2-diketone, hydrolysis of acetal, then O-methylation
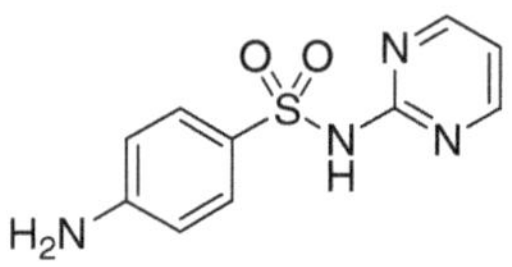	$t_{1/2}$ 9.9±4.3 h *F* 100% pb 54±4% ur 57±14%	*V* 0.29±0.04 CL 0.55±0.17 Mwt 250.3 PSA 106 Å^2 log *P* −0.07	SULFADIAZINE Antibacterial Metabolism: aniline N-acetylation and N-hydroxylation, pyrimidine 4-hydroxylation
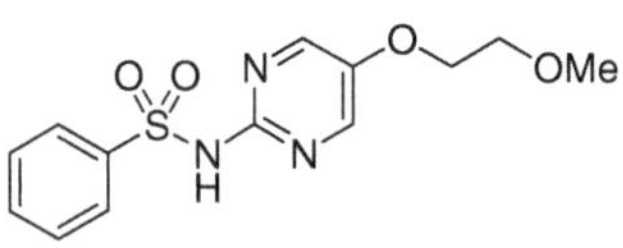	$t_{1/2}$ 4±2 h *F* – pb 80–90% ur 1%	*V* – CL – Mwt 309.3 PSA 98.8 Å^2 log *P* −0.07	GLYMIDINE Antidiabetic Metabolism: O-demethylation

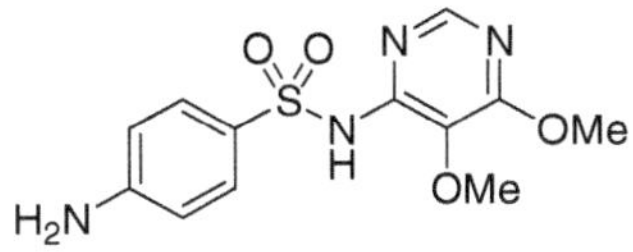	$t^1\!/_2$ *F* pb ur	100–230 h High 90% 30–60%	*V* CL Mwt PSA log *P*	0.37–0.41 – 310.33 125 Å^2 0.46	SULFADOXINE Antibacterial Metabolism: N-acetylation
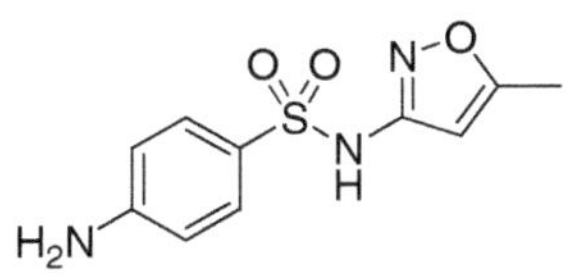	$t^1\!/_2$ *F* pb ur	10±5 h 100% 62±5% 14±2%	*V* CL Mwt PSA log *P*	0.21±0.02 0.32±0.04 253.3 107 Å^2 0.66	SULFAMETHOXAZOLE Antibacterial Metabolism: aniline N-acetylation and N-hydroxylation, N-glucuronidation of sulfonamide
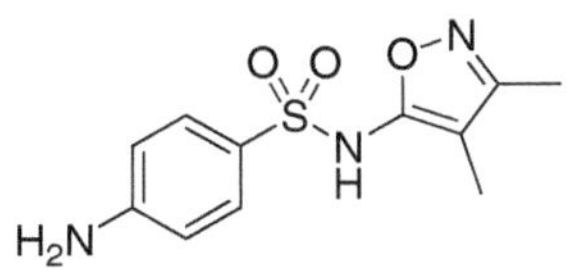	$t^1\!/_2$ *F* pb ur	6.6±0.7 h 96±14% 91±1% 49±8%	*V* CL Mwt PSA log *P*	0.15±0.02 0.33±0.01 267.3 107 Å^2 1.01	SULFISOXAZOLE Antibacterial Metabolism: aniline N-acetylation
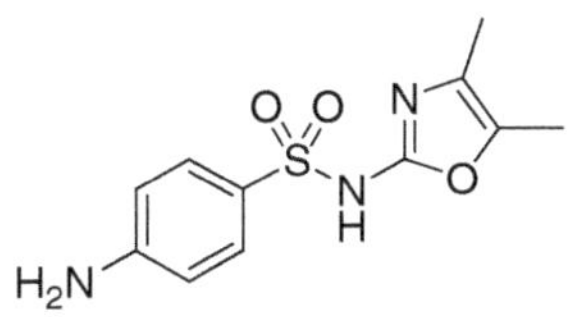	$t^1\!/_2$ *F* pb ur	4–11 h – 25% 65–71%	*V* CL Mwt PSA log *P*	0.1 – 267.3 107 Å^2 0.68	SULFAMOXOLE Antibacterial Metabolism: N-acetylation
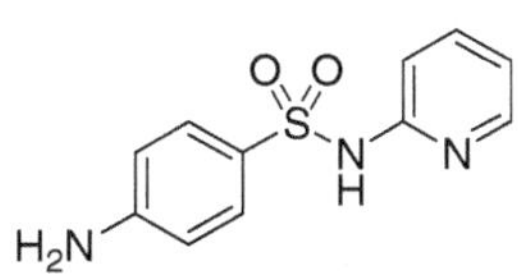	$t^1\!/_2$ *F* pb ur	9±4 h 20% 50% 10–20 %	*V* CL Mwt PSA log *P*	0.4–1.2 0.6–2.1 249.3 93.5 Å^2 0.47	SULFAPYRIDINE Metabolite of sulfasalazine; all values: sulfapyridine on oral dosing of sulfasalazine Metabolism: N-acetylation
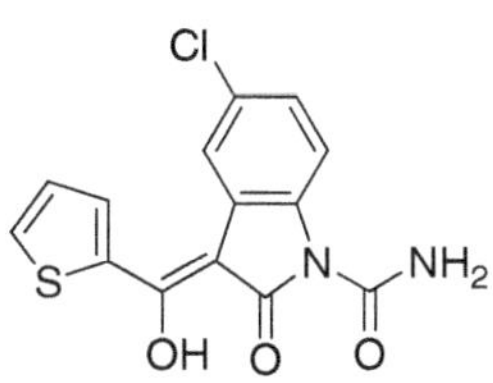	$t^1\!/_2$ *F* pb ur	23.5 h 89% >99% <1%	*V* CL Mwt PSA log *P*	0.11 0.06 320.8 112 Å^2 1.59	TENIDAP Antiinflammatory Metabolism: benzene and thiophene hydroxylation; hepatotoxic
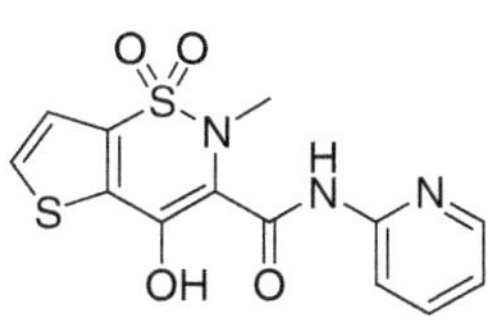	$t^1\!/_2$ *F* pb ur	68±8 h 100% >99% Negligible	*V* CL Mwt PSA log *P*	0.12–0.15 0.02–0.06 337.4 136 Å^2 0.01	TENOXICAM Antiinflammatory Metabolism: pyridine 5′-hydroxylation, thiophene 2-hydroxylation, glucuronidation

(*continued overleaf*)

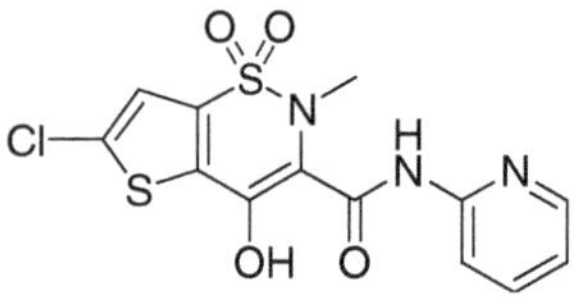	$t_{1/2}$ 4±1 h *F* 100% pb 99.7% ur Negligible	*V* 0.1–0.2 CL 0.36–0.81 Mwt 371.8 PSA 136 Å^2 log *P* 0.77	LORNOXICAM Antiinflammatory, analgesic; pyridyl: pK_a 5.5 Metabolism: pyridine 5′-hydroxylation, glucuronidation; 5′-hydroxylornoxicam does not undergo enterohepatic recirculation
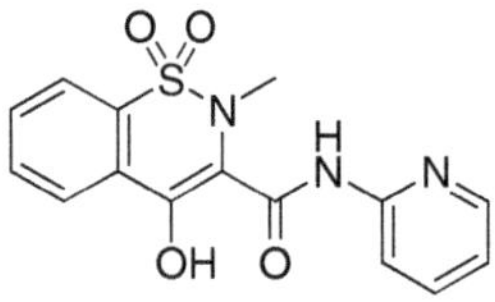	$t_{1/2}$ 48±8 h *F* 100% pb 99% ur <5%	*V/F* 0.15±0.03 CL/*F* 0.04 Mwt 331.4 PSA 108 Å^2 log *P* 0.59	PIROXICAM Antiinflammatory Metabolism: pyridine 5′-hydroxylation, then glucuronidation, hydrolysis of amide, then decarboxylation, cyclodehydration
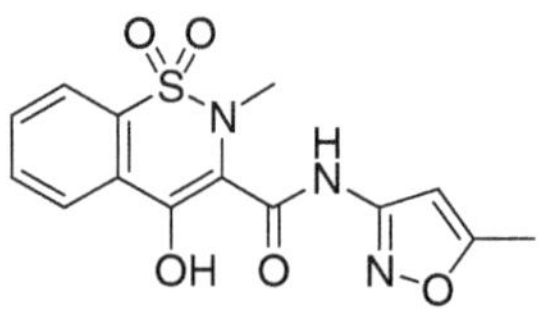	$t_{1/2}$ 22–45 h *F* 97% pb 98% ur 2%	*V* 0.2 CL 0.08 Mwt 335.3 PSA 121 Å^2 log *P* 0.90	ISOXICAM Antiinflammatory, does not cross bbb Metabolism: hydroxylation of CCH_3, withdrawn in 1985 because of fatal skin reactions
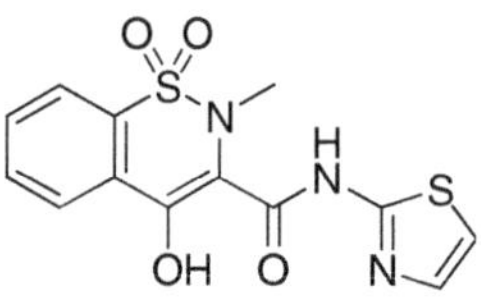	$t_{1/2}$ 24–96 h *F* High pb – ur –	*V* 0.2–0.3 CL – Mwt 337.4 PSA 136 Å^2 log *P* 2.36	SUDOXICAM Antiinflammatory, hepatotoxic due to hydrolysis and N-dealkylation to acylthiourea and reactive aldehyde; development stopped in phase III
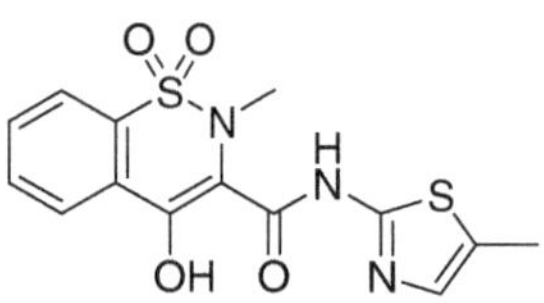	$t_{1/2}$ 19±5 h *F* 89–97% pb >99% ur <0.3%	*V* 0.14–0.21 CL 0.10–0.16 Mwt 351.4 PSA 136 Å^2 log *P* 2.66	MELOXICAM Nonhepatotoxic, antiinflammatory Metabolism (by CYP2C9): hydroxylation of CCH_3, then oxidation to thiazole carboxylic acid
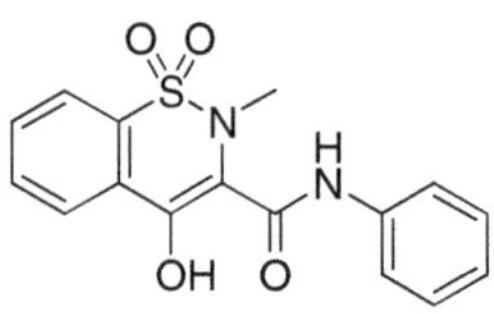	$t_{1/2}$ 21 h *F* – pb – ur	*V* – CL – Mwt 330.4 PSA 95.1 Å^2 log *P* 0.89	CP 14304 Metabolism: 4-hydroxylation of Ph

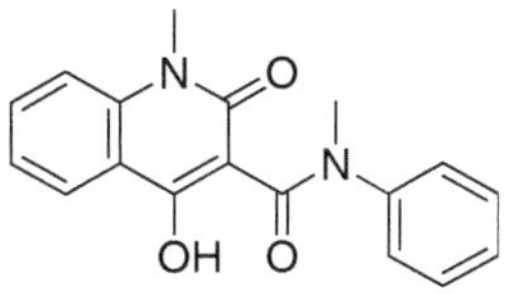	$t^{1/2}$	26–42 h	*V*	0.22	ROQUINIMEX Antineoplastic Metabolism: hydroxylation, N-demethylation, glucuronidation
	F	100%	CL	0.1	
	pb	–	Mwt	308.3	
	ur	1%	PSA	60.9 Å^2	
			log *P*	1.03	
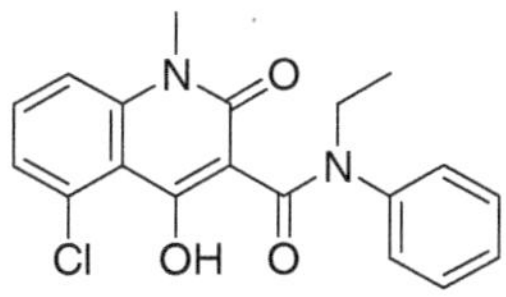	$t^{1/2}$	80 h	*V*	0.2	LAQUINIMOD Treatment of multiple sclerosis Metabolism: N-deethylation, N-demethylation, aniline and quinolone aromatic hydroxylation
	F	82–95%	CL	–	
	pb	99%	Mwt	356.8	
	ur	10%	PSA	60.9 Å^2	
			log *P*	2.69	

$t^{1/2}$, plasma half-life; *F*, oral bioavailability; pb, plasma protein binding; ur, excretion of unchanged drug in urine; V, volume of distribution; CL, clearance; Mwt, molecular weight; PSA, polar surface area; bbb, blood–brain barrier; COX, cyclooxygenase.

64
Sulfonylureas

The antidiabetic effect of sulfonylureas was discovered two times. In France, physicians noticed in 1942 that some sulfonamides used for the treatment of typhoid fever caused hypoglycemia in patients and experimental animals. Despite recognizing its importance, this finding remained unexploited during the war, as the lack of food also lowered the number of diabetics. In 1954, Hans Franke and Joachim Fuchs rediscovered the hypoglycemic properties of the sulfonylurea carbutamide by testing it on themselves. This compound was the first sulfonylurea used to treat type II diabetes, but had to be withdrawn because of adverse effects on bone marrow. It was replaced by tolbutamide (1956), and since then a large number of sulfonylureas with improved potency and pharmacokinetics (PK) properties were developed (Table 64.1). As the potency of sulfonylureas increased, toxicity became less of an issue.

Sulfonylureas block the ATP-sensitive potassium channel in pancreatic β-cells, and thereby stimulate the release of insulin. Hence, these drugs can only be used to treat type II diabetes, where the pancreas is still capable of producing insulin.

Table 64.1 Sulfonylureas. V in l kg^{-1}; CL in ml min^{-1} kg^{-1}; Mwt in g mol^{-1}.

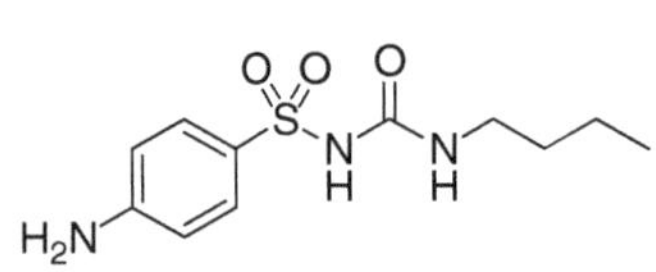	$t^1/_2$ F pb ur	30±6 h – – –	V CL Mwt PSA log P	– – 271.3 110 Å^2 1.01	CARBUTAMIDE Antidiabetic with high bone marrow toxicity Metabolism: N-acetylation
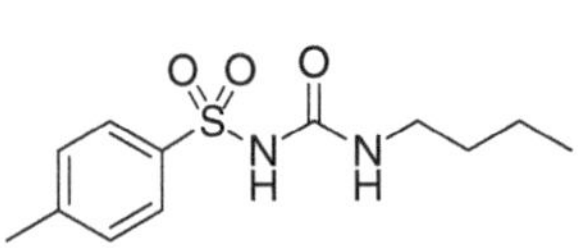	$t^1/_2$ F pb ur	5.9±1.4 h 93±10% 96±1% 0%	V CL Mwt PSA log P	0.10±0.02 0.24±0.04 270.4 83.7 Å^2 2.36	TOLBUTAMIDE Antidiabetic; CYP2C9 substrate Metabolism: benzylic hydroxylation, then oxidation to benzoic acid

Lead Optimization for Medicinal Chemists: Pharmacokinetic Properties of Functional Groups and Organic Compounds, First Edition. Florencio Zaragoza Dörwald.

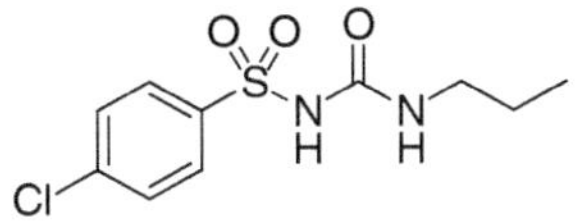	$t^{1/2}$ 33±6 h *F* >90% pb 96±0.6% ur 20±18%	*V/F* 0.1±0.01 CL/*F* 0.03±0.005 Mwt 276.7 PSA 83.7 Å^2 log *P* 2.30	CHLORPROPAMIDE Antidiabetic Metabolism: propyl 2- and 3-hydroxylation, N-depropylation, hydrolysis to 4-chlorobenzenesulfonamide
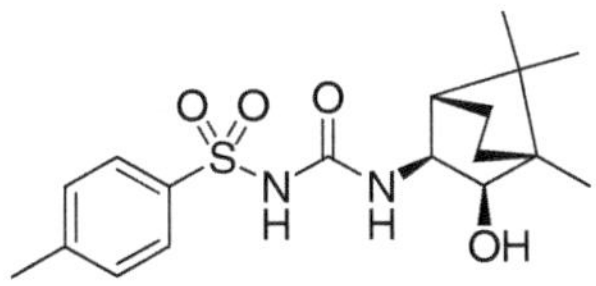	$t^{1/2}$ 8±3 h *F* 91% pb 95% ur –	*V* 0.25 CL – Mwt 366.5 PSA 104 Å^2 log *P* 2.69	GLIBORNURIDE Antidiabetic Metabolism: benzylic hydroxylation, then oxidation to benzoic acid
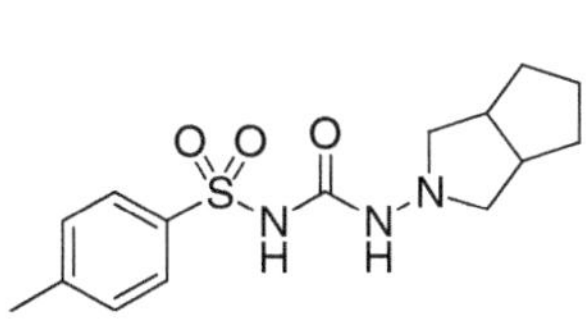	$t^{1/2}$ 14±6 h *F* 79–100% pb 85–95% ur <1%	*V* 0.18–0.34 CL 0.19 Mwt 323.4 PSA 86.9 Å^2 log *P* 1.61	GLICLAZIDE Antidiabetic Metabolism: glucuronidation, benzylic hydroxylation, then oxidation to benzoic acid, methylene hydroxylation at cyclopentane
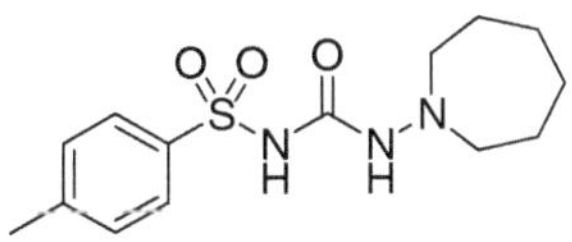	$t^{1/2}$ 6±2 h *F* – pb 94% ur 4–8%	*V* – CL – Mwt 311.4 PSA 86.9 Å^2 log *P* 1.54	TOLAZAMIDE Antidiabetic Metabolism: benzylic hydroxylation, then oxidation to benzoic acid, 4-hydroxylation of azepine, hydrolysis to toluenesulfonamide
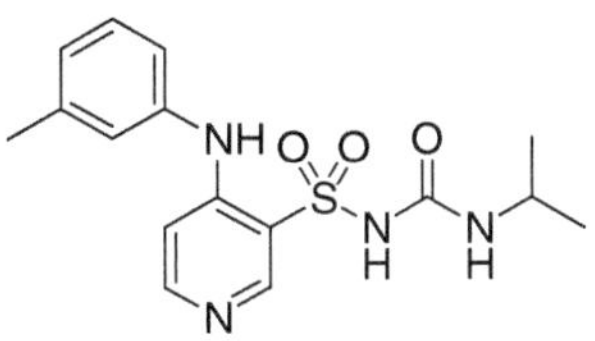	$t^{1/2}$ 4±1 h *F* 75–89% pb >99% ur 20–25%	*V* 0.19–0.23 CL 0.5–0.6 Mwt 348.4 PSA 109 Å^2 log *P* 1.96	TORSEMIDE, TORASEMIDE Diuretic Metabolism: benzylic hydroxylation and oxidation to carboxylic acid, hydroxylation of $CHCH_3$, aromatic 4-hydroxylation
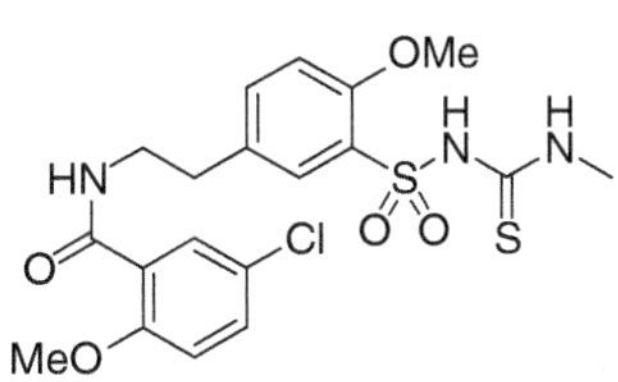	$t^{1/2}$ 0.6–0.7 h *F* 86% pb High ur –	*V* – CL 4–5 Mwt 472.0 PSA 146 Å^2 log *P* 1.54	CLAMIKALANT Potassium channel blocker, antiarrhythmic

(*continued overleaf*)

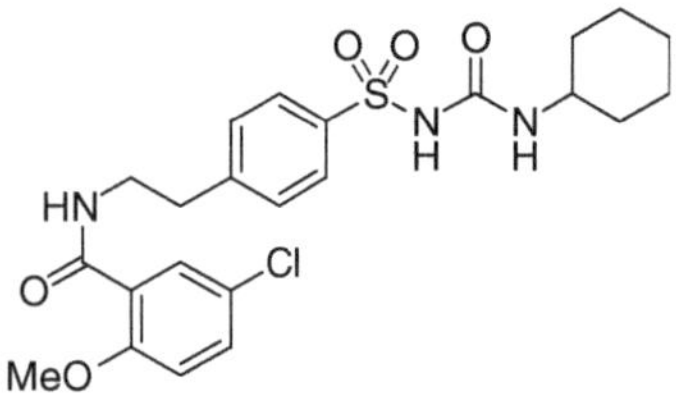

$t_{1/2}$	17±3 h	*V*	0.2±0.1	GLYBURIDE, GLIBENCLAMIDE
F	64–100%	CL	1.3±0.5	Antidiabetic
pb	99.8%	Mwt	494.0	Metabolism: 4-trans and 3-cis
ur	Negligible	PSA	122 Å²	hydroxylation of cyclohexyl
		log *P*	3.08	

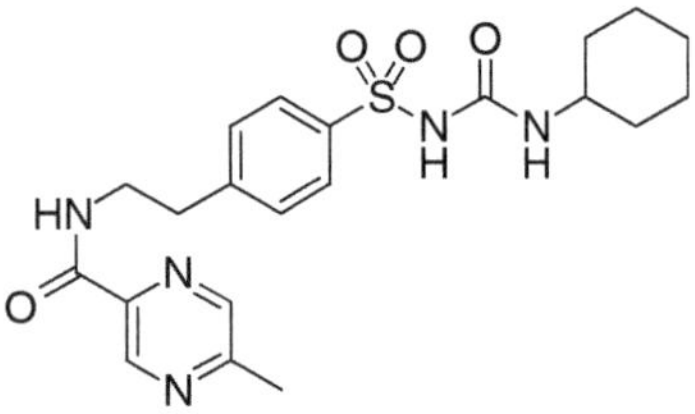

$t_{1/2}$	3.4±0.7 h	*V*	0.17±0.02	GLIPIZIDE
F	95%	CL	0.52±0.18	Antidiabetic
pb	98%	Mwt	445.5	Metabolism: cyclohexyl 3- and
ur	<5%	PSA	139 Å²	4-hydroxylation, hydroxylation of
		log *P*	1.88	CH_3

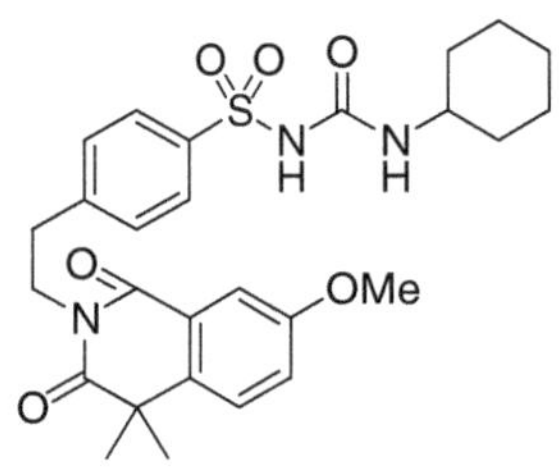

$t_{1/2}$	15±9 h	*V*	–	GLIQUIDONE
F	–	CL	–	Antidiabetic
pb	–	Mwt	527.6	Poorly water soluble
ur	Negligible	PSA	130 Å²	Metabolism: cyclohexyl
		log *P*	3.91	4-hydroxylation

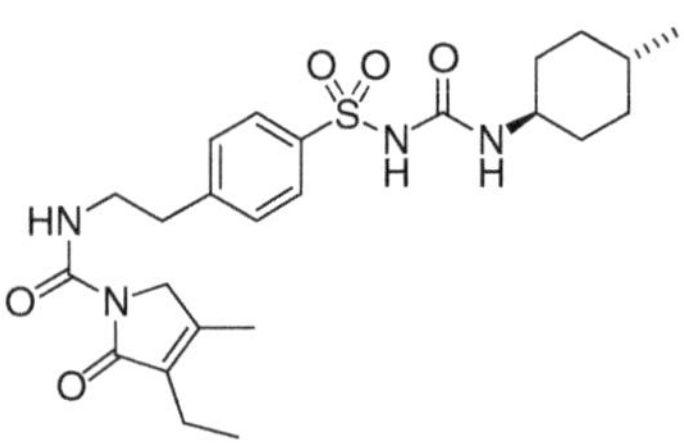

$t_{1/2}$	6±3 h	*V*	0.12	GLIMEPIRIDE
F	100%	CL	0.6–0.8	Antidiabetic
pb	99.4%	Mwt	490.6	Metabolism: hydroxylation of
ur	0%	PSA	133 Å²	cyclohexyl-bound CH_3 (to
		log *P*	3.45	hydroxyglimepiride; $t_{1/2}$ 3.1±0.4 h)

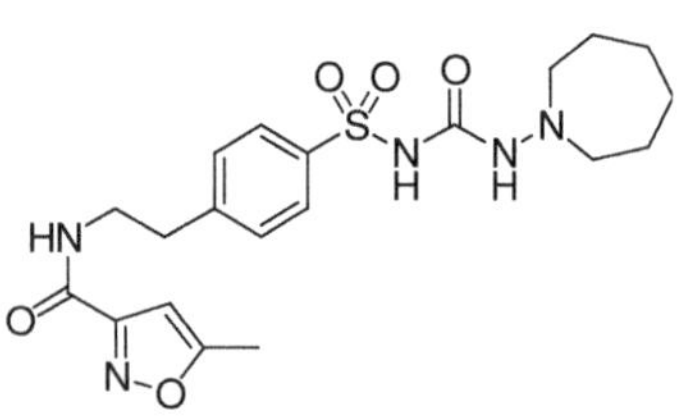

$t_{1/2}$	2 h	*V*	0.4	GLISOXEPID
F	95%	CL	0.9	Antidiabetic
pb	93%	Mwt	449.5	
ur	–	PSA	142 Å²	
		log *P*	0.81	

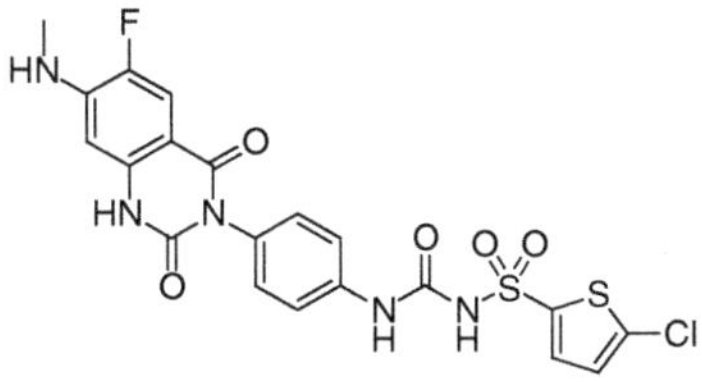

$t^1/_2$	11–12 h	V	–	ELINOGREL
F	50%	CL	–	Antiplatelet agent
pb	–	Mwt	524.0	Metabolism: N-demethylation
ur	–	PSA	173 Å^2	
		log P	3.00	

$t^1/_2$, plasma half-life; F, oral bioavailability; pb, plasma protein binding; ur, excretion of unchanged drug in urine; V, volume of distribution; CL, clearance; Mwt, molecular weight; PSA, polar surface area.

65
Benzodiazepines

Benzodiazepines belong to the few types of drug that originated from a purely chemical approach, without seeking inspiration in natural products. During the mid-1950s, Leo Sternbach at Hoffmann-La Roche Inc., Nutley, was trying to find a new type of tranquilizer and prepared a series of compounds related to his postdoctoral work [1]. Initially, these were believed to be heptoxdiazines, **A**, but turned out to be quinazoline-*N*-oxides, **B** (Scheme 65.1).

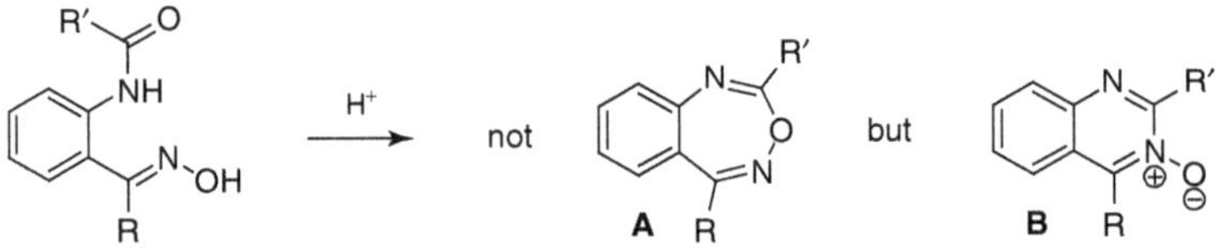

Scheme 65.1 Synthesis of quinazoline-*N*-oxides.

Because all the compounds tested were inactive, and other, more urgent work had to be done, the project was halted in the second half of 1955. Two years later, while cleaning up the laboratory, the crystalline product of the reaction of methylamine with quinazoline-*N*-oxide, **C** (Scheme 65.2) was found and submitted for direct testing in animals (something unthinkable today). This product turned out to be a potent tranquilizer of low toxicity and had good pharmacokinetic (PK) properties. Again, a wrong structure (**D**) had initially been assigned to this product, which turned out to be the benzodiazepine chlordiazepoxide (**E**). This compound was launched 1960 under the trademark Librium.

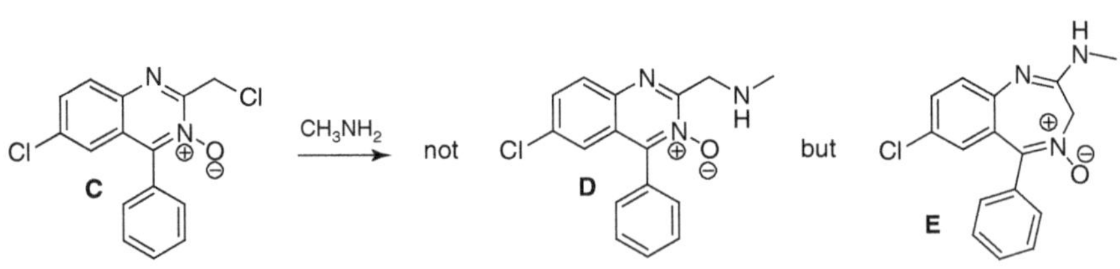

Chlordiazepoxide (librium)
$t_{1/2}$ 10 h, *F* 100%

Scheme 65.2 Synthesis of chlordiazepoxide.

Lead Optimization for Medicinal Chemists: Pharmacokinetic Properties of Functional Groups and Organic Compounds, First Edition. Florencio Zaragoza Dörwald.

Since then thousands of benzodiazepines were prepared and evaluated, and many of them reached the market (Table 65.1). These compounds are mainly used as anxiolytics and hypnotics.

The main metabolic pathways of representative benzodiazepines are sketched in Scheme 65.3. Often, the metabolites of benzodiazepines are biologically active, and the dosing must be adjusted accordingly. All benzodiazepines cross the blood–brain barrier (bbb).

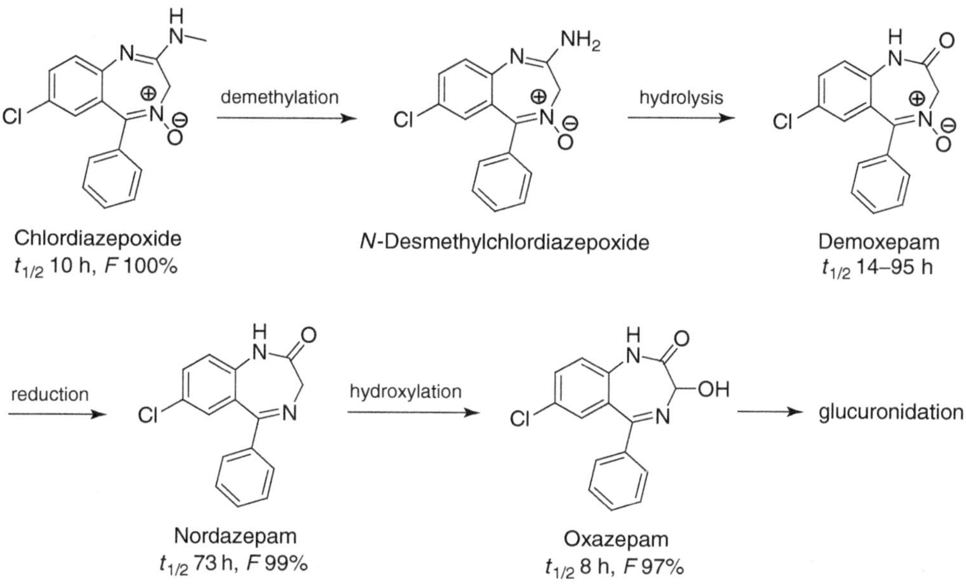

Scheme 65.3 Metabolism of benzodiazepines.

Reference

1. Sternbach, L.H. (1979) The benzodiazepine story. *J. Med. Chem.*, **22**, 1–7.

Table 65.1 Benzodiazepines and related compounds. V in l kg^{-1}, CL in ml min^{-1} kg^{-1}, and Mwt in g mol^{-1}.

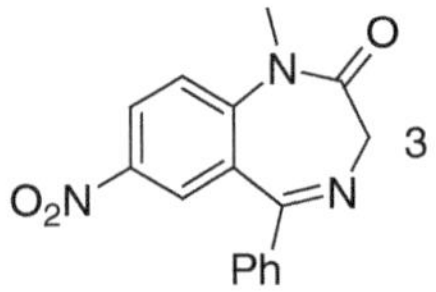	$t^{1/2}$ F pb ur	14–30 h 95% – –	V CL Mwt PSA log P	– – 295.3 78.5 Å^2 2.31	NIMETAZEPAM Anticonvulsant, muscle relaxant; Metabolism: hydroxylation (C-3), reduction of NO_2 to NH_2, then N-acetylation, and N-demethylation (slowest reaction)
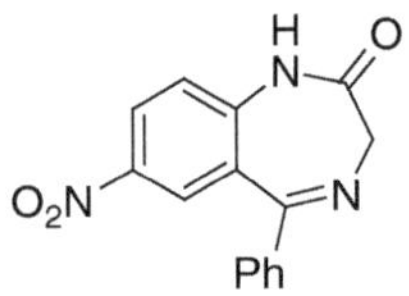	$t^{1/2}$ F pb ur	26±3 h 78±16% 87±1% <1%	V CL Mwt PSA log P	1.9±0.3 0.86±0.12 281.3 87.3 Å^2 2.1	NITRAZEPAM Anticonvulsant, hypnotic; Metabolism: reduction of NO_2 to NH_2, then N-acetylation
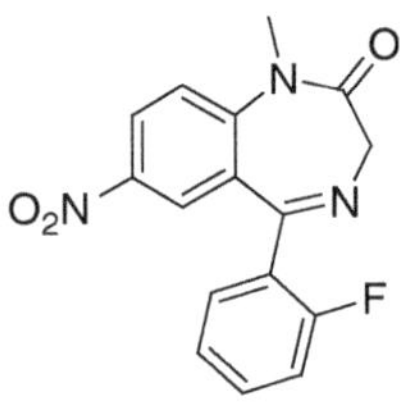	$t^{1/2}$ F pb ur	15±5 h 85% 77–79% <1%	V CL/F Mwt PSA log P	3.3±0.6 3.5±0.4 313.3 78.5 Å^2 2.1	FLUNITRAZEPAM Hypnotic Metabolism: reduction of NO_2 to NH_2, N-demethylation, hydroxylation of CH_2
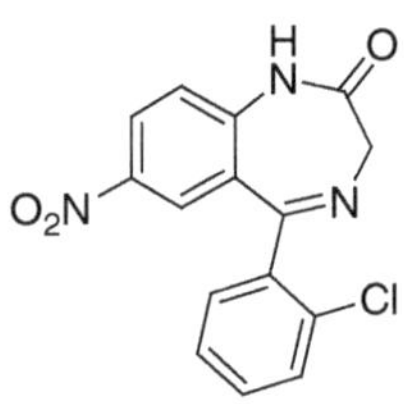	$t^{1/2}$ F pb ur	23±5 h 98±31% 86±0.5% <1%	V CL/F Mwt PSA log P	3.2±1.1 1.55±0.28 315.7 87.3 Å^2 2.8	CLONAZEPAM Anticonvulsant Metabolism: reduction of NO_2 to NH_2, then N-acetylation, hydroxylation of CH_2
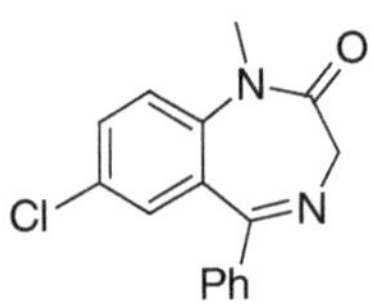	$t^{1/2}$ F pb ur	43±12 h 100±14% 99% <1%	V CL Mwt PSA log P	1.1±0.3 0.38±0.06 284.8 32.7 Å^2 2.7	DIAZEPAM (VALIUM) Anxiolytic, muscle relaxant Metabolism: N-demethylation, hydroxylation of CH_2 group, then glucuronidation
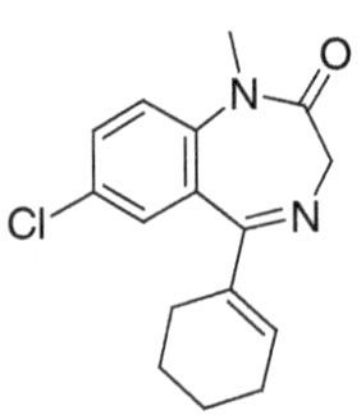	$t^{1/2}$ F pb ur	15±4 h – 70–90% –	V CL Mwt PSA log P	6.7±2.1 – 288.8 32.7 Å^2 2.63	TETRAZEPAM Muscle relaxant Metabolism: N-demethylation, allylic hydroxylation, then aromatization of cyclohexenyl to diazepam and nordazepam

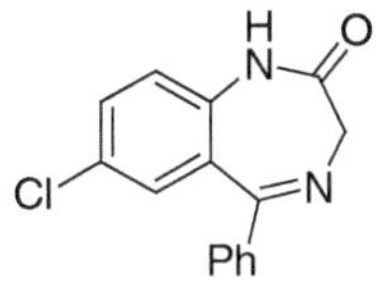	$t^{1}/_{2}$ *F* pb ur	73±33 h 99±6% 97.5% <1%	*V* CL Mwt PSA log *P*	0.78±0.12 0.14±0.05 270.7 41.5 Å^2 2.9	NORDAZEPAM Anxiolytic Metabolite of diazepam
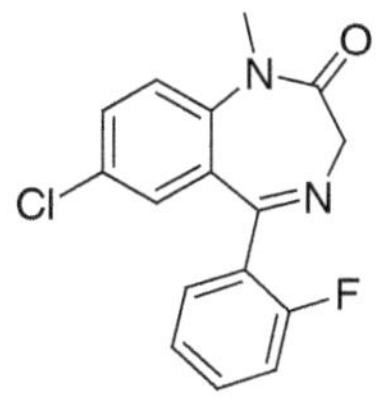	$t^{1}/_{2}$ *F* pb ur	23–48 h – – –	*V* CL Mwt PSA log *P*	– – 302.7 32.7 Å^2 2.9	FLUDIAZEPAM Anxiolytic Metabolism: N-demethylation
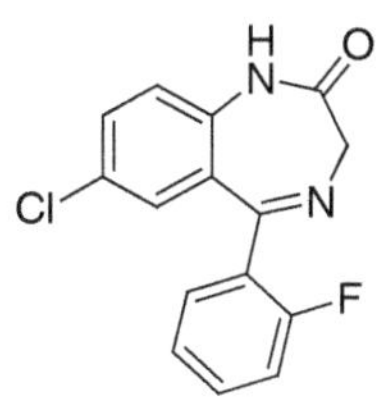	$t^{1}/_{2}$ *F* pb ur	74±24 h – 96.6% <1%	*V* CL/*F* Mwt PSA log *P*	22±7 4.5±2.3 288.7 41.5 Å^2 2.60	*N*-DESALKYLFLURAZEPAM Main metabolite of flurazepam, fludiazepam, and flutoprazepam
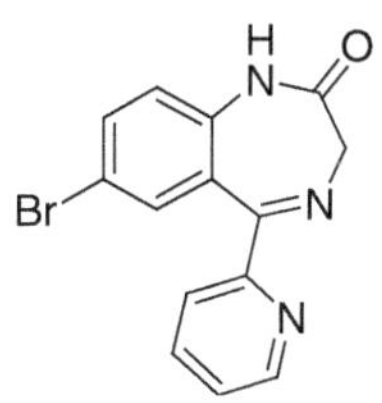	$t^{1}/_{2}$ *F* pb ur	15±5 h 60% 70% 2.3%	*V* CL Mwt PSA log *P*	1.1±0.3 0.7±0.1 316.2 54.4 Å^2 2.4	BROMAZEPAM Anxiolytic Metabolism: hydroxylation of CH_2, then O-glucuronidation, amide and imine hydrolysis, aromatic hydroxylation
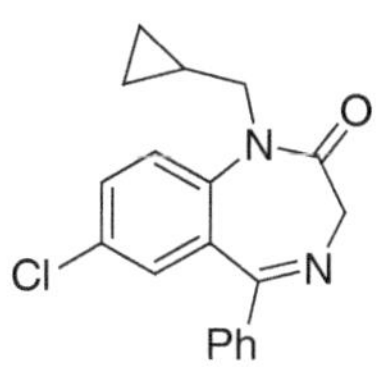	$t^{1}/_{2}$ *F* pb ur	1.3±0.7 h – – 0%	*V* CL/*F* Mwt PSA log *P*	14.4±5.1 140±100 324.8 32.7 Å^2 3.6	PRAZEPAM Anxiolytic Prodrug of nordazepam
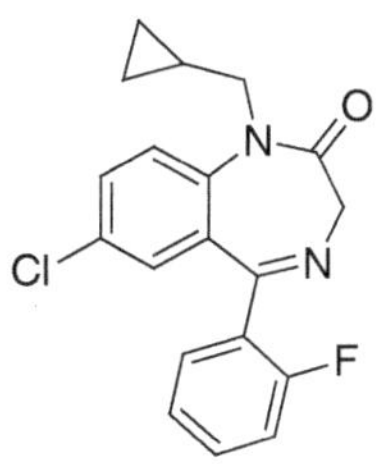	$t^{1}/_{2}$ *F* pb ur	<6 h 85±5% – –	*V* CL Mwt PSA log *P*	– – 342.8 32.7 Å^2 3.7	FLUTOPRAZEPAM Anxiolytic Metabolism: N-decyclopropylmethylation

(*continued overleaf*)

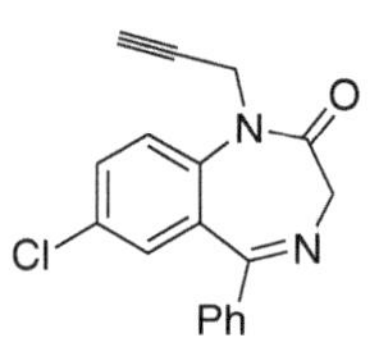	$t_{1/2}$ F pb ur	1–5 h 40–50% 94–97% –	V CL Mwt PSA log P	0.8–1.7 4–9 308.8 32.7 Å^2 2.9	PINAZEPAM Anxiolytic Metabolism: hydroxylation of CH_2CO; in rat liver microsomes depropargylation is eight times faster than demethylation of diazepam
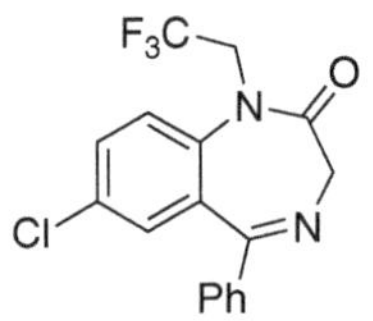	$t_{1/2}$ F pb ur	15±1 h – – –	V CL Mwt PSA log P	1 6.2 352.7 32.7 Å^2 3.54	HALAZEPAM Anxiolytic Metabolism: N-dealkylation to nordazepam
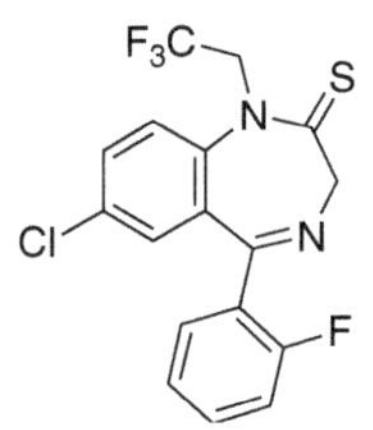	$t_{1/2}$ F pb ur	14–36 h 29–35% >95% –	V CL Mwt PSA log P	8.6 13.7 386.8 47.7 Å^2 4.03	QUAZEPAM Sedative, hypnotic Metabolites: *N*-desalkylflurazepam, 2-oxoquazepam ($t_{1/2}$ 40 h), and *N*-desalkylquazepam ($t_{1/2}$ 70–75 h)
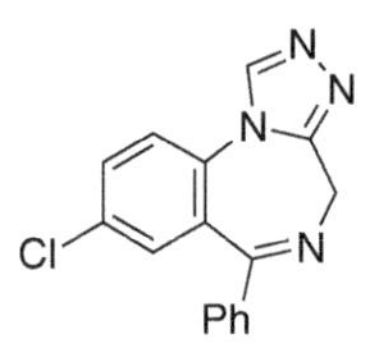	$t_{1/2}$ F pb ur	16±6 h 98%* 93% <5%	V CL Mwt PSA log P	3.1* 18* 294.7 43.1 Å^2 2.52	ESTAZOLAM *mice Sedative, hypnotic Metabolism: hydroxylation of triazole and CH_2, aromatic 4-hydroxylation, then glucuronidation, hydrolysis of imine
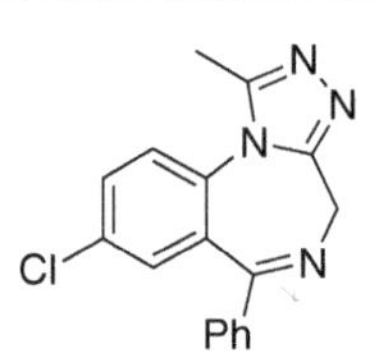	$t_{1/2}$ F pb ur	13±2 h 88±16% 71±3% 20%	V CL Mwt PSA log P	0.72±0.12 0.74±0.14 308.8 43.1 Å^2 2.3	ALPRAZOLAM Anxiolytic metabolism: hydroxylation of CH_3, and CH_2, imine hydrolysis
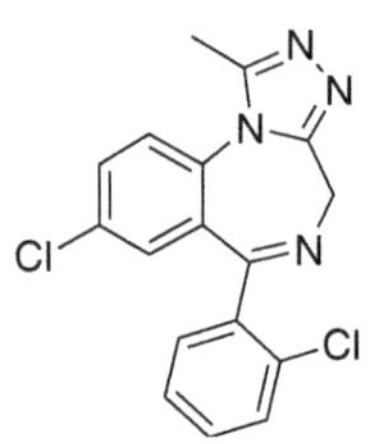	$t_{1/2}$ F pb ur	3±1 h 44% 90±2% 2%	V/F CL/F Mwt PSA log P	1.1±0.4 5.6±2.0 343.2 43.1 Å^2 2.9	TRIAZOLAM Sedative, hypnotic; CYP3A substrate Metabolism: hydroxylation of CH_3 and CH_2; withdrawn in 1991 in the United Kingdom

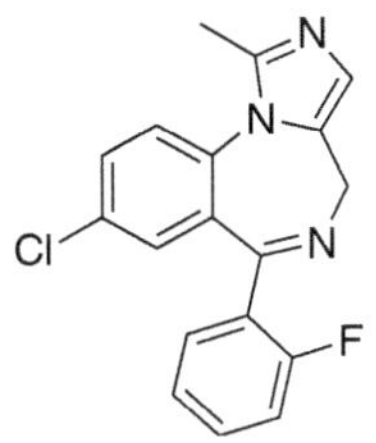	$t_{1/2}$ *F* pb ur	1.9±0.6 h 44±17% 95±2% 56±26%	*V* CL Mwt PSA log *P*	1.1±6 6.6±1.8 325.8 30.2 Å^2 3.1	MIDAZOLAM Anesthetic for iv dosing Metabolism: hydroxylation of CH_3 and CH_2
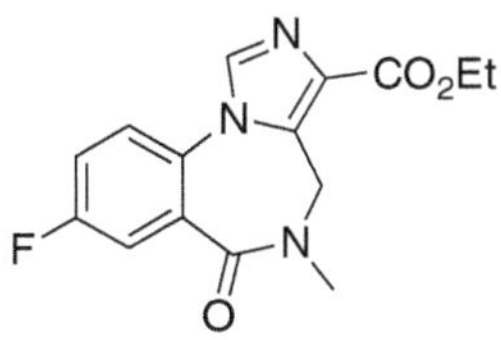	$t_{1/2}$ *F* pb ur	0.9±0.2 h 20% 42% <1%	*V* CL Mwt PSA log *P*	1.0±0.2 17±3 303.3 64.4 Å^2 2.15	FLUMAZENIL Benzodiazepine antagonist Metabolism: ester hydrolysis, then glucuronidation
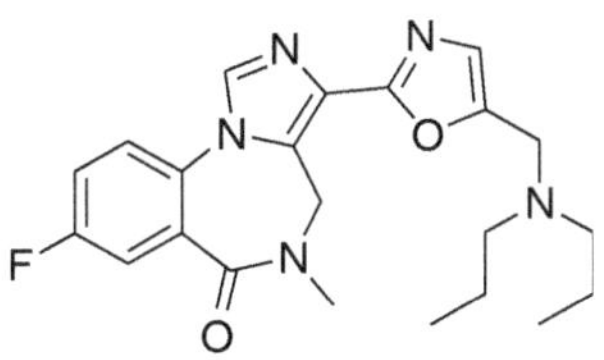	$t_{1/2}$ *F* pb ur	2 h – – –	*V* CL Mwt PSA log *P*	4 28 411.5 62.5 Å^2 2.39	RO 48-8684 Sedative Metabolism: N-depropylation
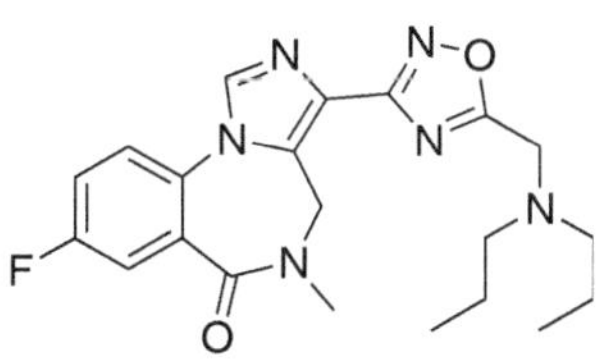	$t_{1/2}$ *F* pb ur	0.5 h – – –	*V* CL Mwt PSA log *P*	3 27 412.5 80.3 Å^2 2.53	RO 48-6791 Sedative Metabolism: N-depropylation
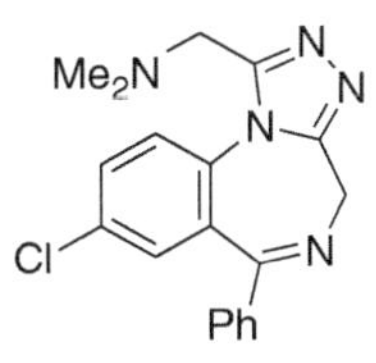	$t_{1/2}$ *F* pb ur	2.9±0.3 h 33–45% 64–75% <2%	*V* CL Mwt PSA log *P*	1.6 6.8 351.8 46.3 Å^2 1.27	ADINAZOLAM Antidepressant; Main metabolite: *N*-desmethyl ($t_{1/2}$ 2–8 h)
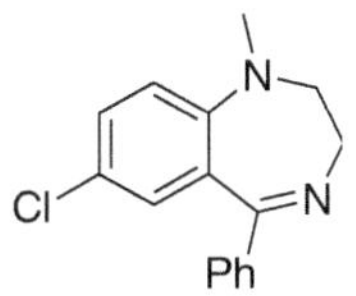	$t_{1/2}$ *F* pb ur	3±2 h – >95% –	*V* CL Mwt PSA log *P*	– – 270.8 15.6 Å^2 3.4	MEDAZEPAM Anxiolytic Active metabolite: diazepam and *N*-desmethylmedazepam ($t_{1/2}$ 40–80 h)

(*continued overleaf*)

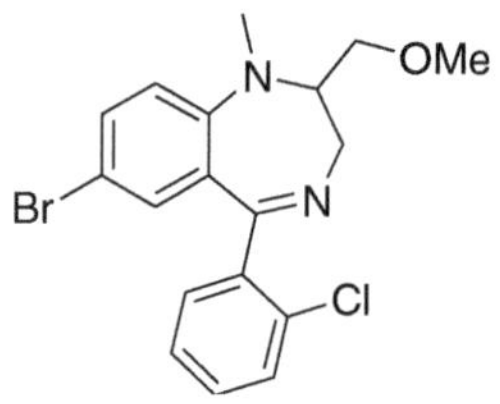	$t^1/_2$ *F* pb ur	11–23 h – – –	*V* CL Mwt PSA log *P*	– – 393.7 24.8 Å^2 3.86	METACLAZEPAM Anxiolytic Metabolism: N- and O-demethylation, aromatic hydroxylation, imine hydrolysis
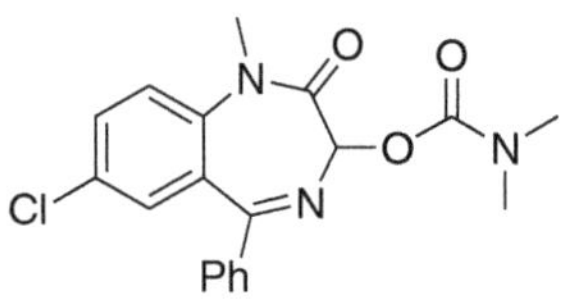	$t^1/_2$ *F* pb ur	6–11 h >90% – –	*V* CL Mwt PSA log *P*	– – 371.8 62.2 Å^2 2.6	CAMAZEPAM Anxiolytic Metabolism: hydrolysis of carbamate, N-demethylation
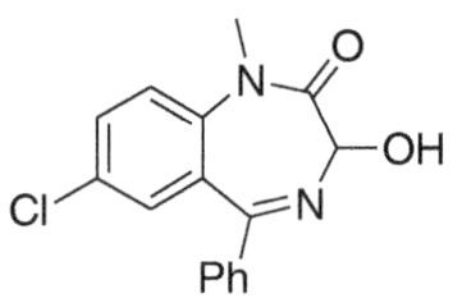	$t^1/_2$ *F* pb ur	11±6 h 91% 97.6% <1%	*V* CL Mwt PSA log *P*	0.95±0.34 1.0±0.3 300.8 52.9 Å^2 2.1	TEMAZEPAM Sedative, hypnotic Metabolism: N-demethylation, O-glucuronidation; withdrawn in 1999 in Sweden and Norway for risk of abuse
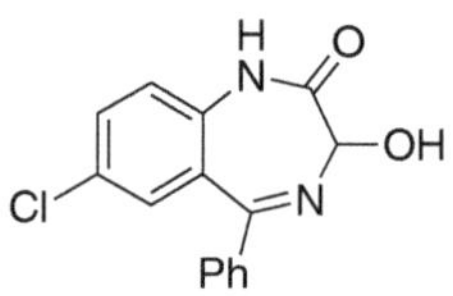	$t^1/_2$ *F* pb ur	8.0±2.4 h 97±11% 99±2% <1%	*V* CL Mwt PSA log *P*	0.6±0.2 1.05±0.36 286.7 61.7 Å^2 1.8	OXAZEPAM Anxiolytic Metabolism: glucuronidation
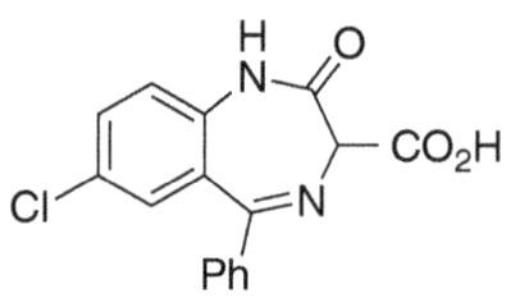	$t^1/_2$ *F* pb ur	2.0±0.9 h 5% – <1%	*V/F* CL/*F* Mwt PSA log *P*	0.33±0.17 1.8±0.2 314.7 78.8 Å^2 2.54	CLORAZEPIC ACID Anxiolytic; prodrug of nordazepam
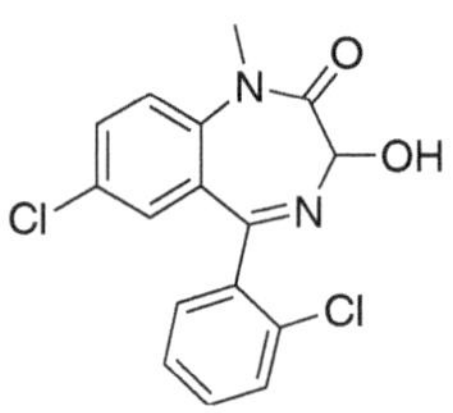	$t^1/_2$ *F* pb ur	10 h 80% 85% –	*V* CL Mwt PSA log *P*	4.6 4 335.2 52.9 Å^2 2.7	LORMETAZEPAM Sedative, hypnotic Metabolism: O-glucuronidation, N-demethylation

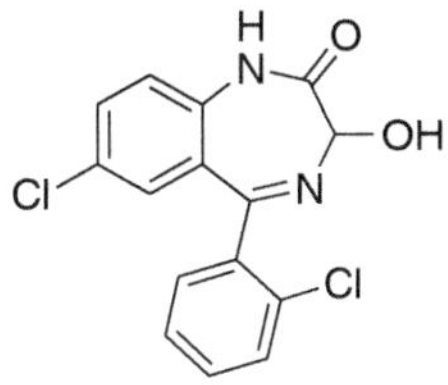

$t^1/_2$	14±5 h	V	1.3±0.2	LORAZEPAM
F	93±10%	CL	1.1±0.4	Anxiolytic
pb	91±2%	Mwt	321.2	Metabolism: glucuronidation
ur	<1%	PSA	61.7 Å^2	
		log P	2.5	

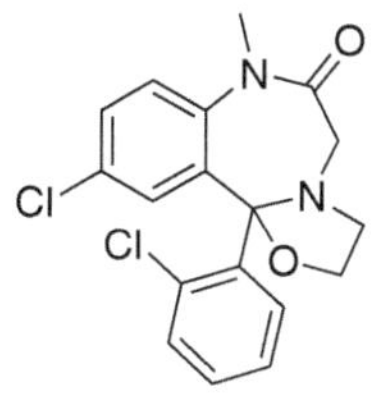

$t^1/_2$	66±11 h	V	–	CLOXAZOLAM
F	–	CL	–	Anxiolytic
pb	–	Mwt	363.2	Prodrug of
ur	–	PSA	32.8 Å^2	chlorodesmethyldiazepam
		log P	3.9	Metabolism: twofold N-dealkylation

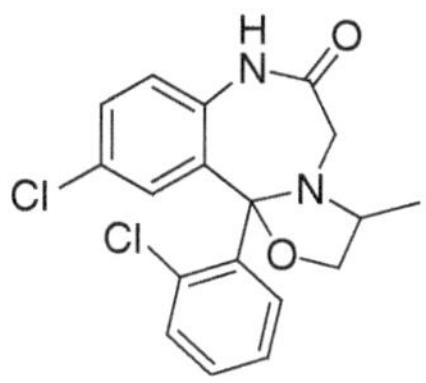

$t^1/_2$	70 h	V	–	MEXAZOLAM
F	–	CL	–	Anxiolytic
pb	–	Mwt	363.2	Metabolism: N-dealkylation to
ur	–	PSA	41.6 Å^2	chlorodesmethyldiazepam
		log P	4.4	

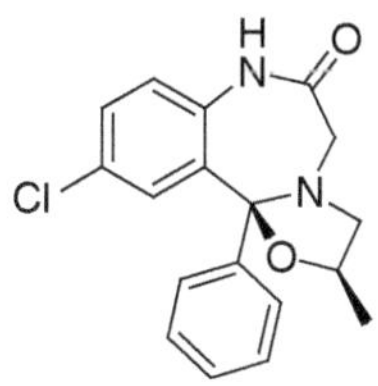

$t^1/_2$	10–20 h	V	–	OXAZOLAM
F	–	CL	–	Sedative, anxiolytic, anticonvulsant
pb	–	Mwt	328.8	Metabolism: N-dealkylation to
ur	–	PSA	41.6 Å^2	desmethyldiazepam
		log P	3.8	

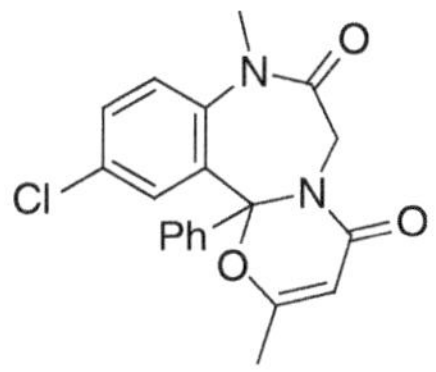

$t^1/_2$	2 h	V	–	KETAZOLAM
F	–	CL	–	Sedative, anxiolytic, anticonvulsant
pb	93%	Mwt	368.8	Active metabolite: *N*-desmethyl
ur	–	PSA	49.9 Å^2	($t^1/_2$ 52 h)
		log P	0.51	

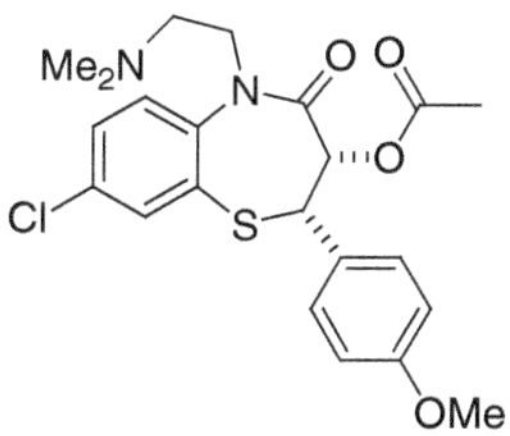

$t^1/_2$	14–16 h	V	9.8	CLENTIAZEM
F	35–50%	CL	13.8	Antihypertensive
pb	–	Mwt	449.0	Metabolism: deacetylation,
ur	–	PSA	84.4 Å^2	N-demethylation, O-demethylation,
		log P	4.46	aromatic hydroxylation (ortho to methoxy)

(*continued overleaf*)

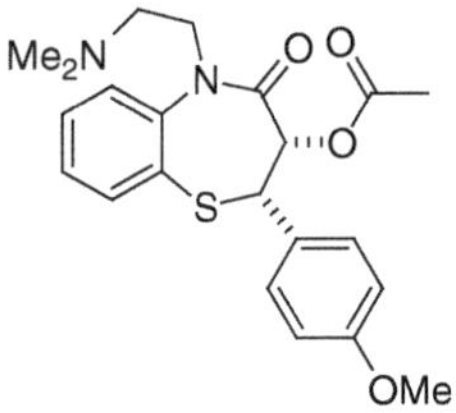

$t^{1}\!/_{2}$	5–6 h	V	3.1–1.2
F	44±10%	CL	12±4
pb	78±3%	Mwt	414.5
ur	0.2–4.0%	PSA	84.4 Å^2
		log P	4.73

DILTIAZEM
Antianginal, antihypertensive, antiarrhythmic
Metabolites: desacetyl ($t^{1}\!/_{2}$ 9±2 h), *N*-monodesmethyl ($t^{1}\!/_{2}$ 7.5±1.0 h)

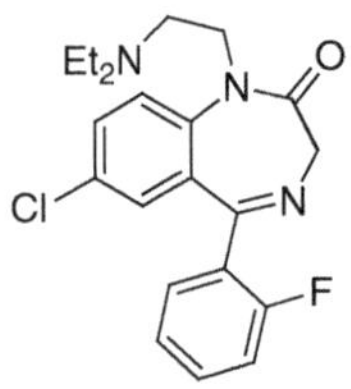

$t^{1}\!/_{2}$	2 h	V	3.4
F	83%	CL	4.5±2.3
pb	75%	Mwt	387.9
ur	<1%	PSA	35.9 Å^2
		log P	3.6

FLURAZEPAM
Sedative, hypnotic
Metabolism: N-dealkylation to 2-hydroxyethyl and *N*-desalkylflurazepam, hydroxylation of CH_2 to doxefazepam

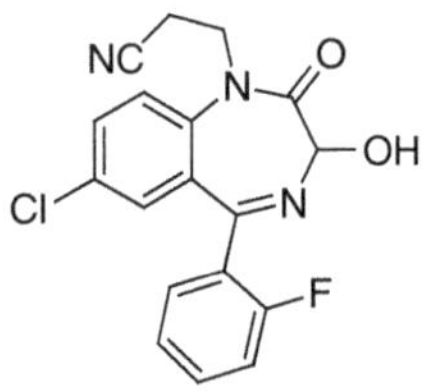

$t^{1}\!/_{2}$	9 h	V	–
F	–	CL	–
pb	–	Mwt	357.8
ur	–	PSA	76.7 Å^2
		log P	1.7

CINOLAZEPAM
Sedative, anxiolytic, anticonvulsant
Metabolism: N-dealkylation, glucuronidation

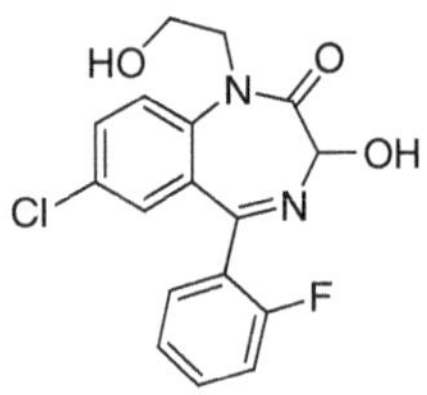

$t^{1}\!/_{2}$	3–4 h	V	–
F	–	CL	–
pb	–	Mwt	348.8
ur	–	PSA	73.1 Å^2
		log P	1.6

DOXEFAZEPAM
Sedative, anxiolytic
Metabolism: glucuronidation, N-dealkylation, oxidation of CH_2OH to CO_2H, and ring contraction to quinazolinone

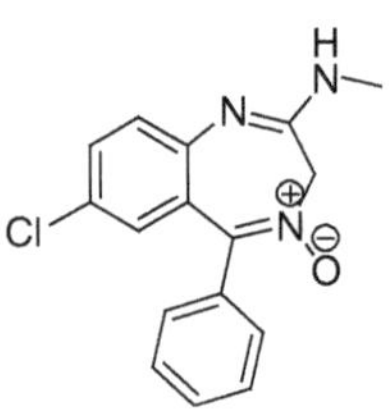

$t^{1}\!/_{2}$	10±3 h	V	0.30±0.03
F	100%	CL	0.54±0.49
pb	97±2%	Mwt	299.8
ur	<1%	PSA	53.1 Å^2
		log P	2.1

CHLORDIAZEPOXIDE
Sedative, hypnotic
Metabolism: N-demethylation, hydrolysis of amidine to amide (demoxepam; $t^{1}\!/_{2}$ 14–95 h), hydrolysis and reduction to desmethyldiazepam

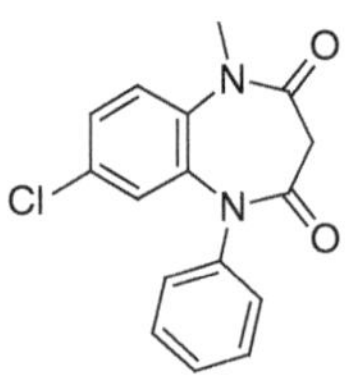

$t^{1}\!/_{2}$	20±10 h	V	0.9–1.5
F	90%	CL	0.35–0.65
pb	85–90%	Mwt	300.7
ur	<1%	PSA	40.6 Å^2
		log P	1.25

CLOBAZAM
Anxiolytic, anticonvulsant
Metabolism: N-demethylation (to *N*-desmethyl, $t^{1}\!/_{2}$ 36–46 h), aromatic 4-hydroxylation

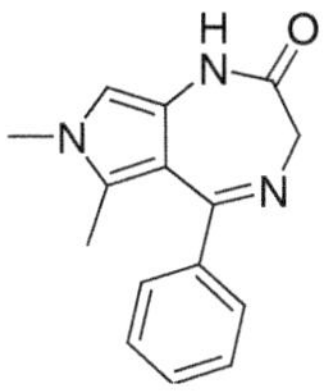	$t^1/_2$ 7.9±1.2 h *F* – pb – ur 63%	*V* – CL – Mwt 253.3 PSA 46.4 Å^2 log *P* 1.31	PREMAZEPAM Anxiolytic, sedative Metabolism: N-demethylation, hydroxylation of CH_3
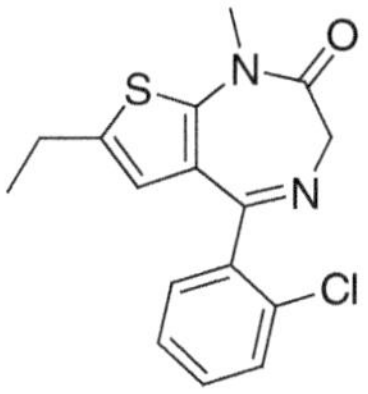	$t^1/_2$ 13±5 h *F* 2.5% pb 99% ur –	*V* 3–4 CL 9–10 Mwt 318.8 PSA 60.9 Å^2 log *P* 3.95	CLOTIAZEPAM Anxiolytic, sedative, hypnotic Metabolism: N-demethylation, hydroxylation
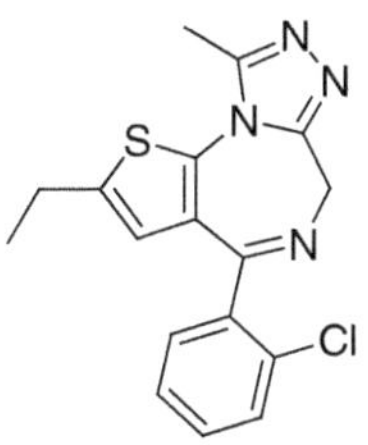	$t^1/_2$ 3–15 h *F* 87% pb 93% ur <3%	*V* – CL – Mwt 342.9 PSA 71.3 Å^2 log *P* 2.20	ETIZOLAM Anxiolytic, sedative, hypnotic Metabolism: hydroxylation of ethyl to 1-hydroxyethyl and of CH_3, then glucuronidation
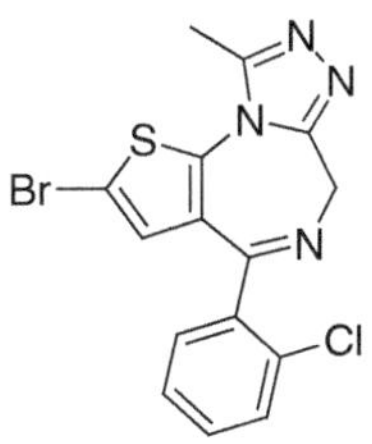	$t^1/_2$ 6±2 h *F* 70±22% pb 90% ur <1%	*V* 0.7 CL 1.7±0.4 Mwt 393.7 PSA 71.3 Å^2 log *P* 2.46	BROTIZOLAM Anxiolytic, sedative, hypnotic Metabolism: hydroxylation of CH_3, and CH_2
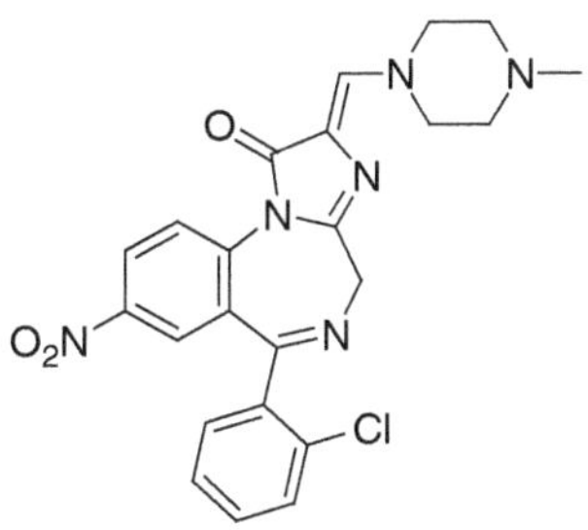	$t^1/_2$ 7±2 h *F* – pb – ur –	*V* – CL – Mwt 464.9 PSA 97.3 Å^2 log *P* 0.60	LOPRAZOLAM Anxiolytic, sedative Metabolism: formation of piperazine *N*-oxide ($t^1/_2$ 12–16 h), reduction of NO_2 to NH_2

(*continued overleaf*)

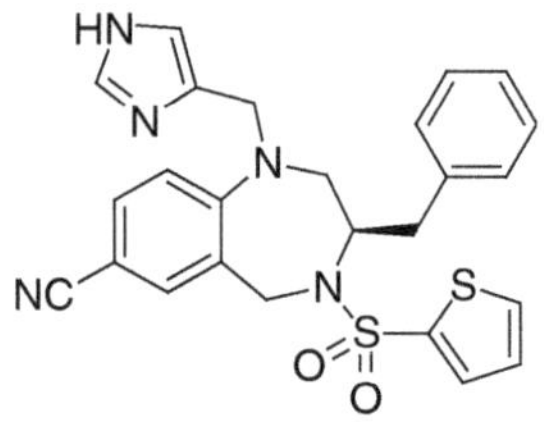	$t^{1}/_{2}$ 1.6 h (iv) *F* – pb – ur –	*V* 0.84 CL 9.6 Mwt 489.6 PSA 130 Å^2 log *P* 3.78	BMS-214662 Farnesyltransferase inhibitor, antineoplastic Metabolism: hydroxylation, glucuronidation
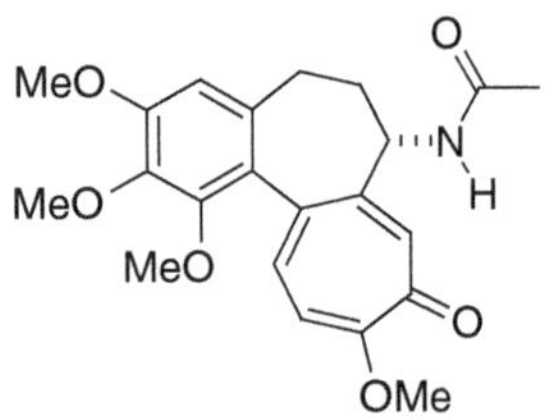	$t^{1}/_{2}$ 54–92 h *F* 37–44% pb 32% ur 10–25%	*V* 4.2 CL 1.9 Mwt 399.4 PSA 83.1 Å^2 log *P* 1.07	COLCHICINE Gout suppressant
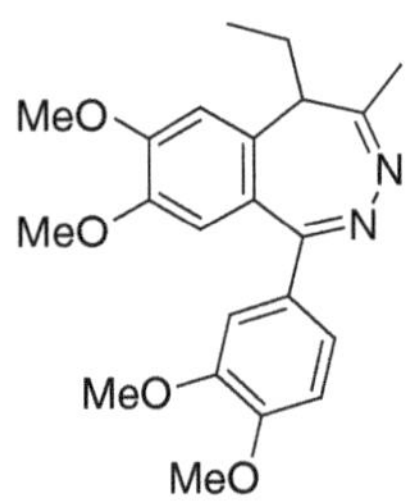	$t^{1}/_{2}$ 2.7–3.5 h *F* – pb – ur –	*V* 8–79 CL – Mwt 382.5 PSA 61.6 Å^2 log *P* 3.64	TOFISOPAM Anxiolytic Metabolism: sequential O-demethylation
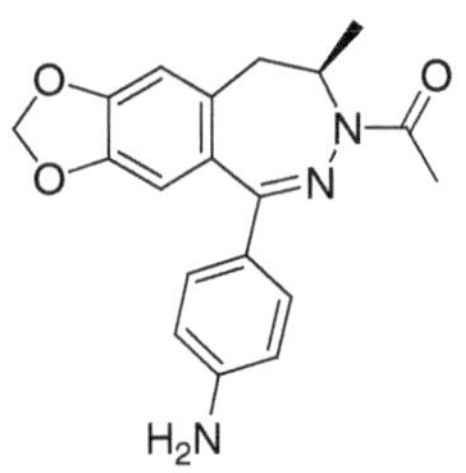	$t^{1}/_{2}$ 4–7 h *F* – pb 67–88% ur –	*V* – CL 6.7 Mwt 337.4 PSA 77.2 Å^2 log *P* 1.30	TALAMPANEL Anticonvulsant Metabolism: N-acetylation, reductive cleavage of acetal to 7-*O*-methylcatechol, glucuronidation

$t^{1}/_{2}$, plasma half-life; *F*, oral bioavailability; pb, plasma protein binding; ur, excretion of unchanged drug in urine; V, volume of distribution; CL, clearance; Mwt, molecular weight; PSA, polar surface area.

66
Steroids

Natural steroids are endogenous hormones of short plasma half-life, low oral bioavailability, and usually, little therapeutic value [1] (Scheme 66.1). Although these lipophilic compounds are quickly and completely absorbed from the gastrointestinal (GI) tract, first-pass metabolism is too extensive for oral dosing. Chemically modified steroids (agonists or antagonists), however, can show much improved pharmacokinetic properties (Table 66.1).

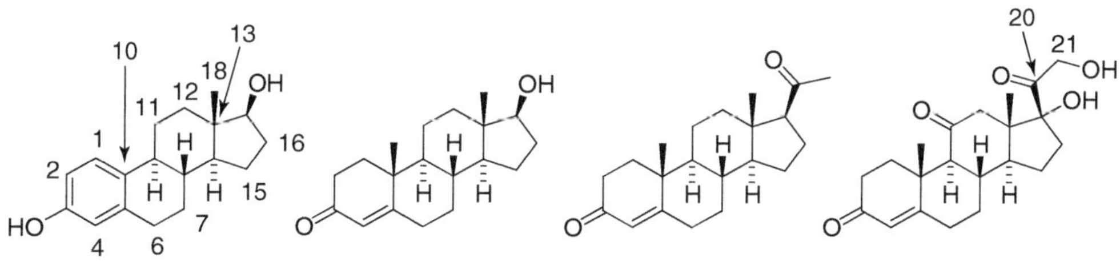

Estradiol (an estrogen, induces feminization) Testosterone (an androgen, induces virilization) Progesterone (a progestin, maintenance of pregnancy) Cortisone (a corticosteroid, an antiinflammatory)

Scheme 66.1 Biological activities of steroids.

Estrogens and progestins are mainly used as contraceptives for the treatment of female infertility and in hormone replacement therapy, for example, for the treatment of menopausal symptoms. Androgens are used as anabolics or in hormone replacement therapy. The growth of hormone-responsive neoplasms can be suppressed by antiestrogens or antiandrogens. Corticosteroids (e.g., cortisone) are mainly used as systemic or topical antiinflammatories [2]. A number of nonsteroids may also interact with steroid receptors as agonists or antagonists.

The main phase I metabolic transformations of steroids include the reduction/oxidation of ketones/alcohols, aromatic and allylic hydroxylation, hydrogenation of alkenes (mainly enones), and hydrolysis of esters. Phase II conjugation reactions include glucuronidation and sulfation of hydroxyl groups.

Lead Optimization for Medicinal Chemists: Pharmacokinetic Properties of Functional Groups and Organic Compounds, First Edition. Florencio Zaragoza Dörwald.

References

1. Sneader, W. (2000) The discovery of oestrogenic hormones. *J. Br. Menopause Soc.*, 129–133.
2. Khan, M.O.F. and Lee, H.J. (2008) Synthesis and pharmacology of anti-inflammatory steroidal antedrugs. *Chem. Rev.*, **108**, 5131–5145.

Table 66.1 Steroids. V is in l kg^{-1}, CL in ml min^{-1} kg^{-1} and Mwt in g mol^{-1}.

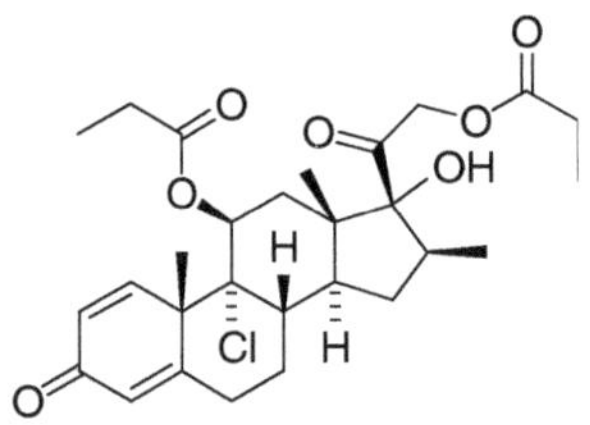	$t\frac{1}{2}$ *F* pb ur	0.5 h (iv) 15–20%* 87% <10%	*V* CL Mwt PSA log *P*	0.29 36 521.0 107 Å^2 4.07	BECLOMETHASONE DIPROPIONATE Antiallergic, antiasthmatic, topical antiinflammatory Metabolism: ester hydrolysis; *17-monopropionate ($t\frac{1}{2}$ 2.7 h) after oral administration of dipropionate
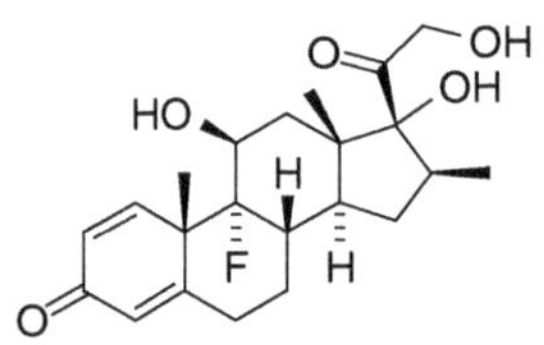	$t\frac{1}{2}$ *F* pb ur	5.6 h (iv) 80% 64% 5%	*V* CL Mwt PSA log *P*	1.3 2.8 392.5 94.8 Å^2 2.03	BETAMETHASONE Glucocorticoid Metabolism: glucuronidation
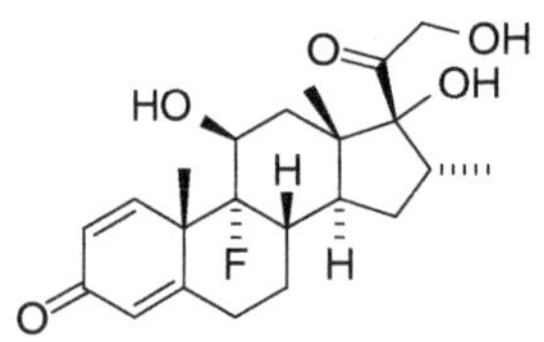	$t\frac{1}{2}$ *F* pb ur	3.0±0.8 h 78±14% 68±3% 8%	*V* CL Mwt PSA log *P*	0.8±0.2 3.7±0.9 392.5 94.8 Å^2 2.03	DEXAMETHASONE Antiinflammatory Metabolism: allylic hydroxylation, reduction of ketone to alcohol, glucuronidation, sulfation
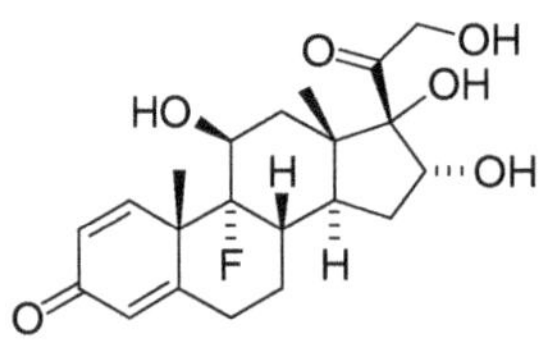	$t\frac{1}{2}$ *F* pb ur	1.4–3.6 h 23% 71–79% Low	*V* CL Mwt PSA log *P*	1.4–2.1 9–17 394.4 115 Å^2 0.53	TRIAMCINOLONE Glucocorticoid Metabolism: oxidation of 11-alcohol to ketone, hydrogenation of 4,5-alkene
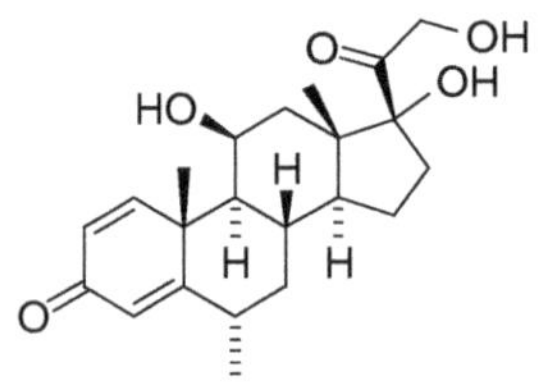	$t\frac{1}{2}$ *F* pb ur	2.3–4.0 h 80% 77% <10%	*V* CL Mwt PSA log *P*	1.7±0.3 6.1 374.5 94.8 Å^2 2.17	METHYLPREDNISOLONE Glucocorticoid, crosses bbb Metabolites: 20-β-hydroxy, oxidation of 11-hydroxy to ketone (methylprednisone)

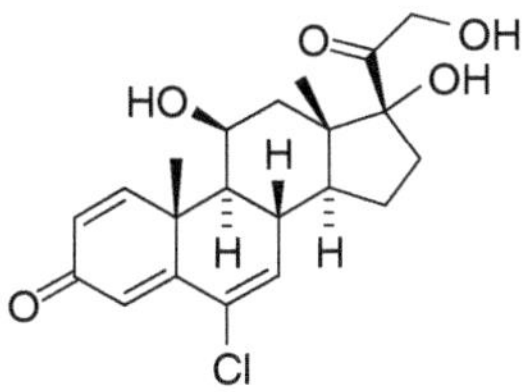

$t_{1/2}$	1.9±0.4 h	V	0.8±0.1	CLOPREDNOL
F	–	CL	4.1±0.7	Glucocorticoid
pb	74–94%	Mwt	392.9	Metabolism: reduction of
ur	–	PSA	94.8 Å^2	noncyclic ketone to alcohol
		log P	1.34	

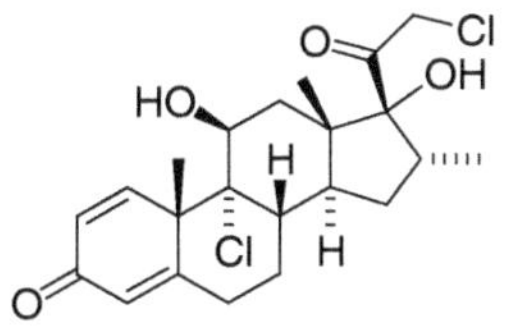

$t_{1/2}$	4.4–5.8 h*	V	4.4–4.7	MOMETASONE
F	<1%	CL	12	*iv
pb	98–99%	Mwt	427.4	Furoate: F <1%
ur	Low	PSA	74.6 Å^2	Antiinflammatory
		log P	2.68	Metabolism: reduction of ketone, formation of epoxide from 2-chloroethanol, allylic hydroxylation

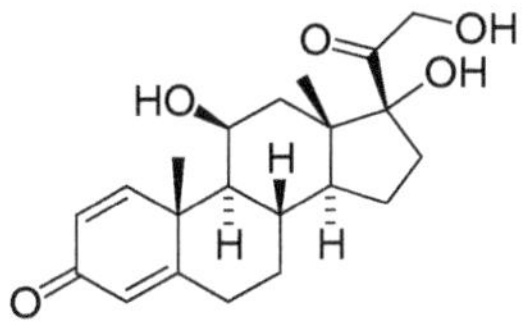

$t_{1/2}$	3.6±0.4 h	V	0.97±0.11	PREDNISONE
F	80±11%	CL	3.6±0.8	(CYCLOHEXANONE) or
pb	75±2%	Mwt	360.44*	PREDNISOLONE*
ur	3±2%	PSA	94.8 Å^{2}*	(CYCLOHEXANOL,
		log P	1.64*	DRAWN)
				Glucocorticoid
				Prednisone and prednisolone are interconverted *in vivo*

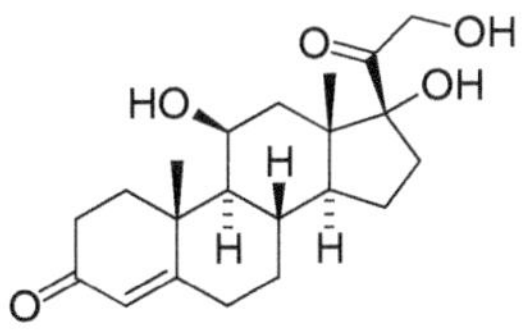

$t_{1/2}$	1.3–1.9 h	V	0.3–0.6	HYDROCORTISONE,
F	75–100%	CL	3.1–4.3	CORTISOL
pb	90%	Mwt	362.5	Glucocorticoid, crosses bbb
ur	<1%	PSA	94.8 Å^2	
		log P	1.76	

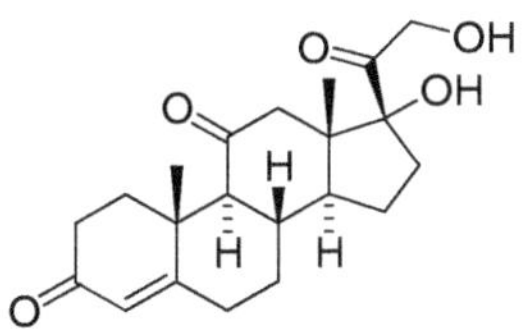

$t_{1/2}$	1.2 h	V	0.3*	CORTISONE
F	–	CL	–	*acetate
pb	95%*	Mwt	360.4	Glucocorticoid,
ur	–	PSA	91.7 Å^2	antiinflammatory
		log P	1.43	

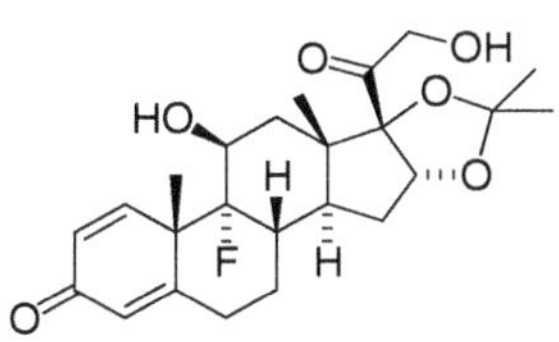

$t_{1/2}$	2.0±0.7 h	V	1.3±0.7	TRIAMCINOLONE
F	23±10%	CL	7.7±2.6	ACETONIDE
pb	40%	Mwt	434.5	Glucocorticoid
ur	1.0±0.6%	PSA	93.1 Å^2	Metabolism:
		log P	2.43	6-hydroxylation, oxidation to 21-carboxy, no acetal hydrolysis

(*continued overleaf*)

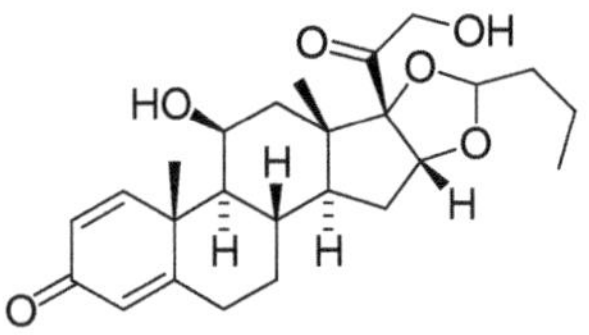

$t_{1/2}$	2–3 h (iv)	*V*	2.9±1.5
F	12±3%	CL	17±4
pb	88%	Mwt	430.5
ur	2%	PSA	93.1 Å^2
		log *P*	3.20

BUDESONIDE
Antiinflammatory
Metabolism: allylic hydroxylation, cyclopentane hydroxylation (at *C*H–O)

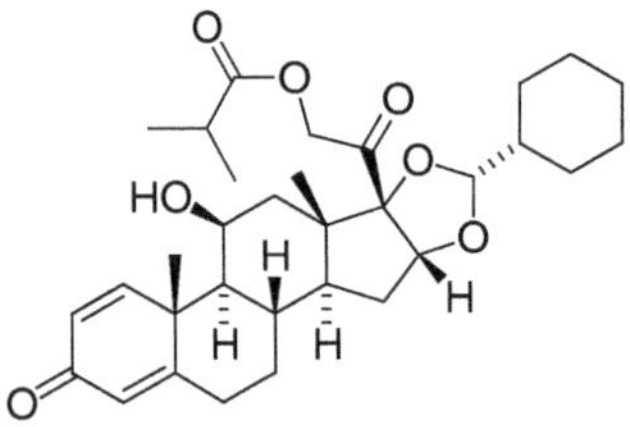

$t_{1/2}$	0.7 h	*V*	2.9
F	<0.5%	CL	39
pb	99%	Mwt	540.7
ur	–	PSA	99.1 Å^2
		log *P*	6.13

CICLESONIDE
Antiinflammatory
High first-pass metabolism (ester hydrolysis)

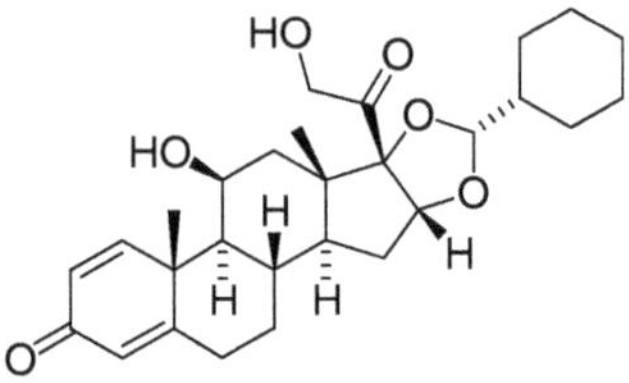

$t_{1/2}$	6–7 h	*V*	12.1
F	1%	CL	59
pb	99%	Mwt	470.6
ur	<20%	PSA	93.1 Å^2
		log *P*	4.14

DES-CICLESONIDE
Main active metabolite of ciclesonide

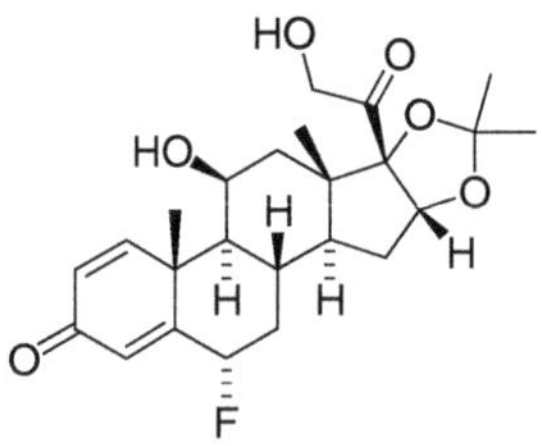

$t_{1/2}$	1.4–1.7 h*	*V*	1.4–1.9
F	6–10%	CL	13.7–14.3
pb	80%	Mwt	434.5
ur	–	PSA	93.1 Å^2
		log *P*	2.04

FLUNISOLIDE
*iv
Antiasthmatic
Metabolism: displacement of allylic fluoride by OH

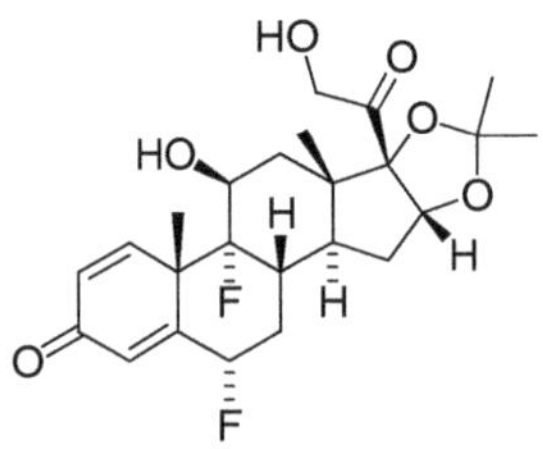

$t_{1/2}$	1.3–1.7 h	*V*	–
F	Low	CL	–
pb	–	Mwt	452.5
ur	–	PSA	93.1 Å^2
		log *P*	2.23

FLUOCINOLONE ACETONIDE
Topical antiinflammatory
Acetonide: improves absorption by skin

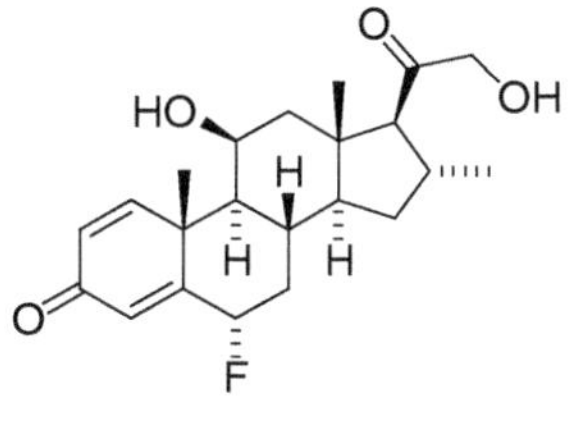

$t_{1/2}$	1–2 h	V	1
F	>90%	CL	6.5±2.1
pb	83–95%	Mwt	376.5
ur	Low	PSA	74.6 Å^2
		log P	1.83

FLUOCORTOLONE
Glucocorticoid, immunosuppressant
Metabolism: elimination of fluoride, substitution of fluoride by OH, glucuronidation, sulfation, oxidation of secondary alcohol to ketone

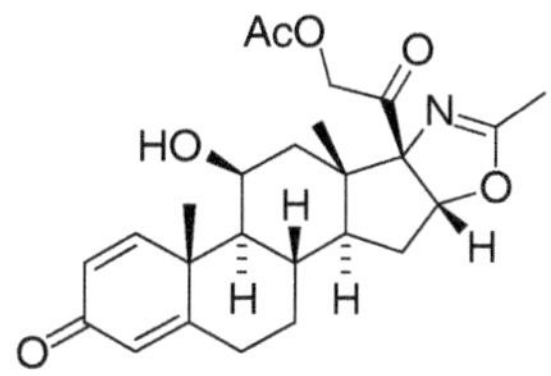

$t_{1/2}$	1.1–1.9 h	V	1.2–1.5
F	68%	CL	–
pb	40%	Mwt	441.5
ur	18%	PSA	102 Å^2
		log P	2.05

DEFLAZACORT
Antiinflammatory
Prodrug of desacetyldeflazacort; all values for metabolite

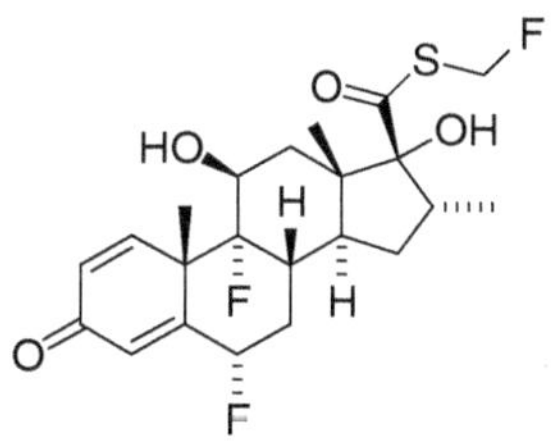

$t_{1/2}$	7.8 h (iv)	V	4.2–11.1
F	<1%	CL	12.8–22.7
pb	90%	Mwt	444.5
ur	<0.02%	PSA	99.9 Å^2
		log P	2.07

FLUTICASONE
Propionate: F <2%
Furoate: F 1.6%
Antiallergic, antiinflammatory
Metabolism: hydrolysis of ester, displacement of allylic fluoride by OH

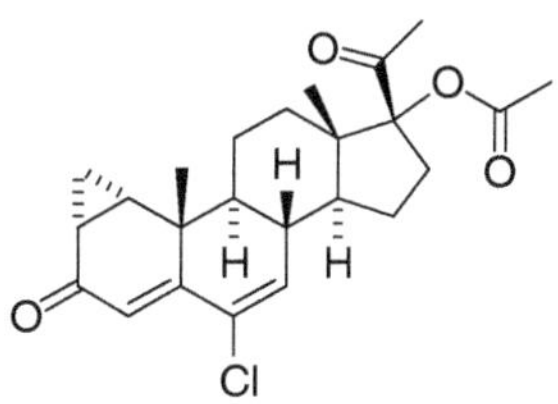

$t_{1/2}$	30–40 h	V	20
F	88–100%	CL	3.6
pb	95%	Mwt	416.9
ur	33%	PSA	60.4 Å^2
		log P	4.22

CYPROTERONE ACETATE
Antiandrogen, treatment of acne
Metabolite: 15-β-hydroxy- and 3-α-hydroxycyproterone

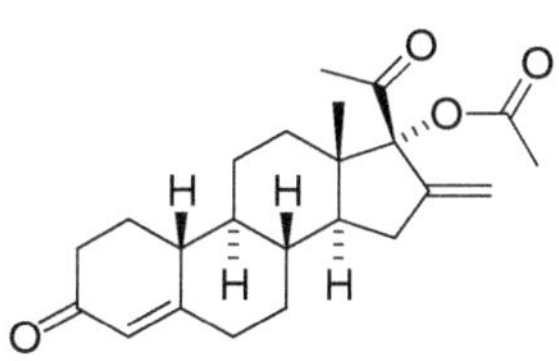

$t_{1/2}$	1–2 h	V	4
F	10%	CL	–
pb	–	Mwt	370.5
ur	–	PSA	60.4 Å^2
		log P	3.00

ELCOMETRINE
Contraceptive

(*continued overleaf*)

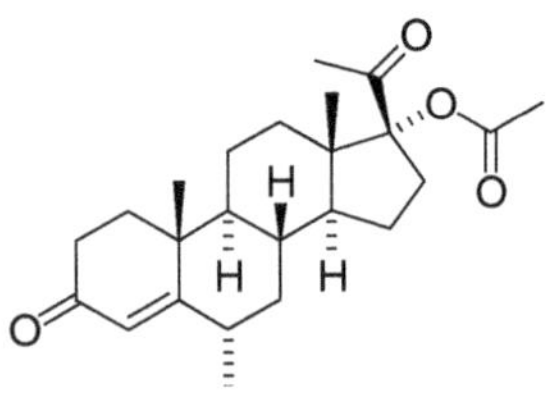

$t_{1/2}$	40–60 h*	*V*	3.7	MEDROXYPROGESTERONE ACETATE
F	<10%	CL	–	*im: 2 wk
pb	86–95%	Mwt	386.5	Crosses bbb
ur	1%	PSA	60.4 Å^2	Metabolism: reduction of enone to enol, then glucuronidation
		log *P*	4.17	

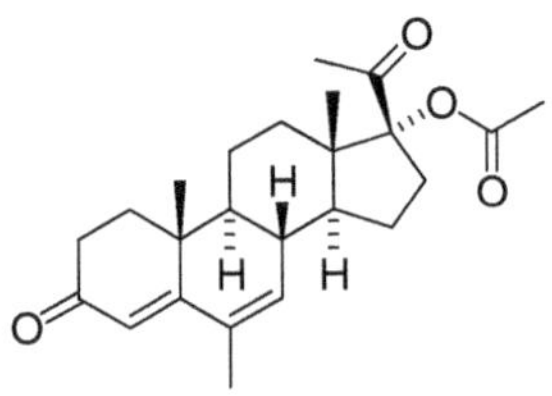

$t_{1/2}$	15–20 h	*V*	–	MEGESTROL ACETATE
F	–	CL	–	Progestogen, appetite stimulant
pb	–	Mwt	384.5	Metabolism: hydroxylation of cyclopentyl-CH_3
ur	50–70%	PSA	60.4 Å^2	
		log *P*	3.75	

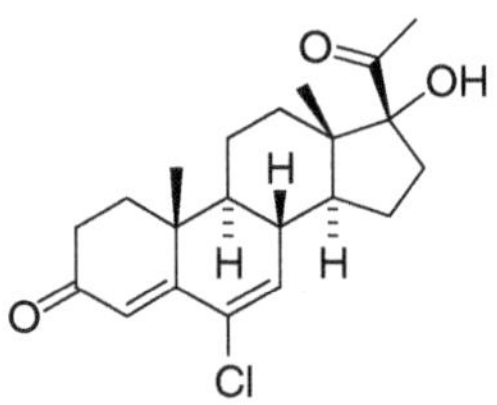

$t_{1/2}$	25–39 h	*V*	2–3	CHLORMADINONE
F	24%	CL	–	Contraceptive
pb	>95%	Mwt	362.9	Metabolism: reduction of cyclohexenone to cyclohexenol, 2-hydroxylation; withdrawn in 1972 because of carcinogenicity
ur	–	PSA	54.4 Å^2	
		log *P*	3.27	

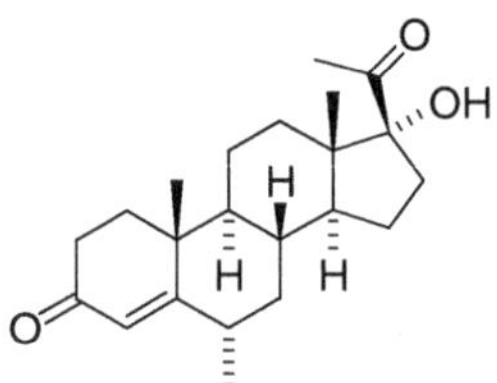

$t_{1/2}$	12–17 h	*V*	–	MEDROXYPROGESTERONE
F	–	CL	–	Progestogen
pb	86–90%	Mwt	344.5	
ur	–	PSA	54.4 Å^2	
		log *P*	3.58	

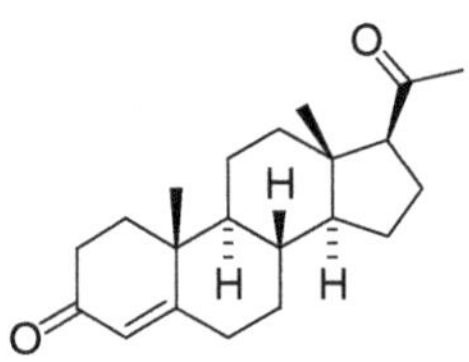

$t_{1/2}$	20 min	*V*	0.3–0.5	PROGESTERONE
F	<20%	CL	–	Progestogen, crosses bbb
pb	95–98%	Mwt	314.5	Metabolism: hydrogenation of alkene and of enone to enol
ur	–	PSA	34.1 Å^2	
		log *P*	3.83	

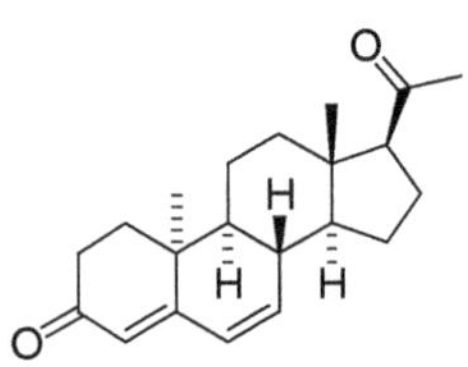

$t_{1/2}$	5–7 h	*V*	–	DYDROGESTERONE
F	28%	CL	–	Progestogen
pb	97%	Mwt	312.5	Active metabolite: 20α-dihydrodydrogesterone ($t_{1/2}$ 14–17 h)
ur	–	PSA	34.1 Å^2	
		log *P*	3.55	

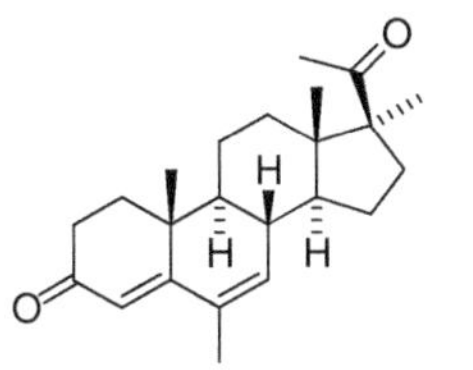

$t_{1/2}$	36 h	V	–	MEDROGESTONE
F	100%	CL	–	Progestogen
pb	–	Mwt	340.5	
ur	–	PSA	34.1 Å^2	
		log P	4.35	

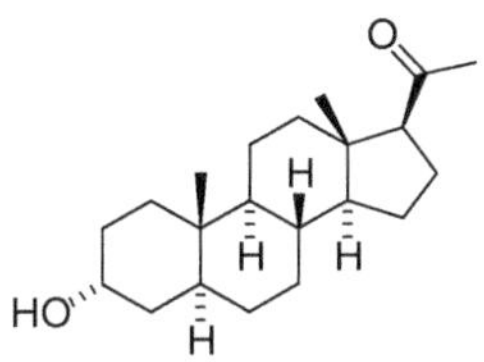

$t_{1/2}$	4.7 h (iv)	V	2	ELTANOLONE, PREGNANOLONE
F	Low	CL	22	Local anesthetic
pb	99%	Mwt	318.5	Metabolism:
ur	Negligible	PSA	37.3 Å^2	glucuronidation, reduction
		log P	5.04	of ketone to alcohol

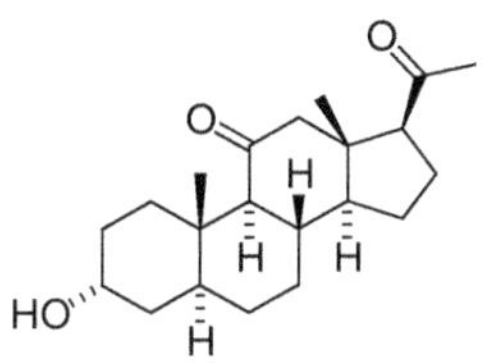

$t_{1/2}$	0.1–0.5 h	V	–	ALPHAXALONE
F	–	CL	–	Sedative, anesthetic,
pb	40%	Mwt	332.5	anticonvulsant
ur	–	PSA	54.4 Å^2	Metabolism:
		log P	3.28	glucuronidation

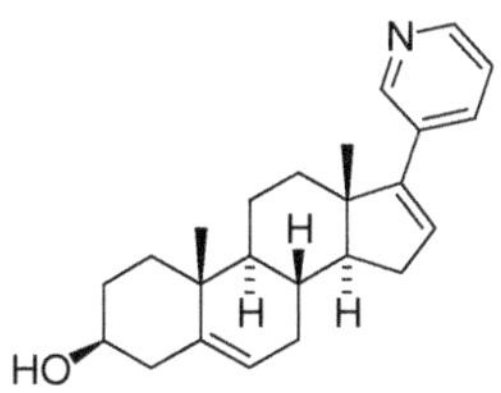

$t_{1/2}$	5–14 h	V	–	ABIRATERONE
F	Low	CL	79	CYP17 inhibitor, oncolytic;
pb	>99%	Mwt	349.5	prodrug: acetate
ur	–	PSA	33.1 Å^2	Metabolism: N-oxidation,
		log P	5.70	O-sulfation

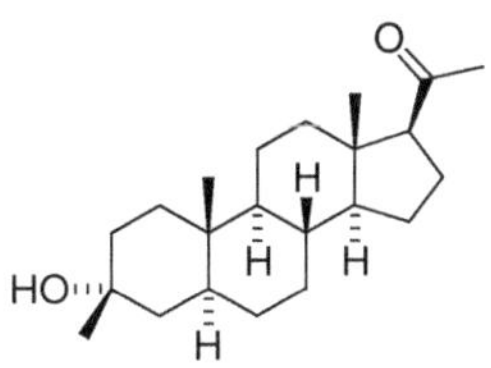

$t_{1/2}$	37–70 h	V/F	3.2*	GANAXOLONE
F	–	CL	–	*rat
pb	–	Mwt	332.5	GABA receptor modulator,
ur	–	PSA	37.3 Å^2	sedative, antiepileptic
		log P	5.42	

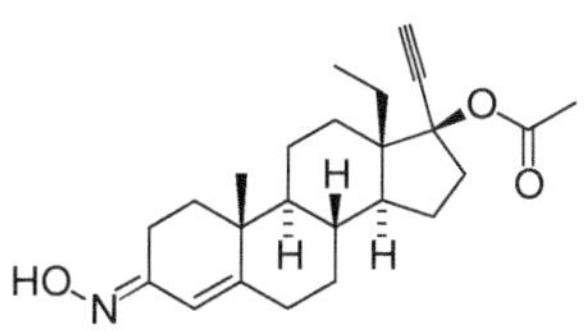

$t_{1/2}$	25 h*	V	–	NORGESTIMATE
F	22%**	CL	–	*desacetyl metabolite
pb	–	Mwt	369.5	**levonorgestrel on oral
ur	–	PSA	58.9 Å^2	dosing of norgestimate
		log P	5.13	Not detectable in plasma after oral dosing
				Metabolism: O-deacetylation, hydrolysis of oxime (to levonorgestrel)

(continued overleaf)

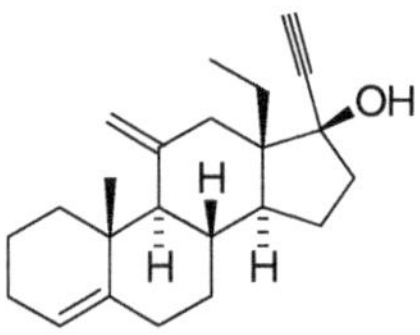	$t^1/_2$ – F 62–75%* pb – ur –	V – CL 2.4 Mwt 310.5 PSA 20.2 Å^2 log P 5.35	DESOGESTREL Oral contraceptive; rapid oxidation by CYP2C9 to etonogestrel; *etonogestrel on oral administration of desogestrel
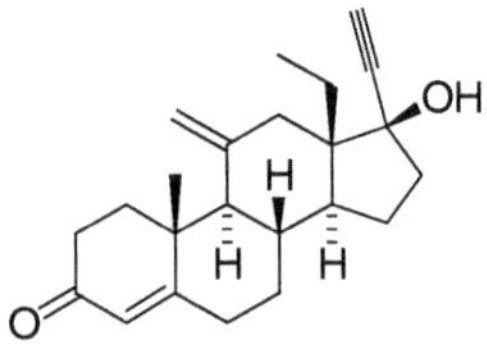	$t^1/_2$ 24–30 h F 82% pb 96–99% ur –	V 7.2 CL 2.6 Mwt 324.5 PSA 37.3 Å^2 log P 3.32	ETONOGESTREL, 3-KETODESOGESTREL Active metabolite of desogestrel
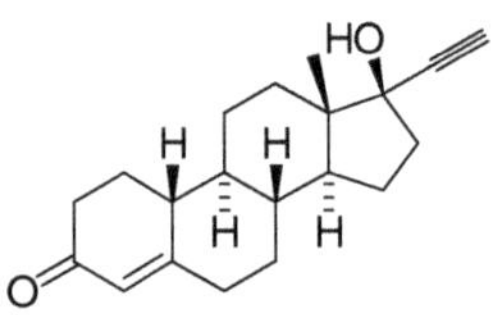	$t^1/_2$ 8.5±2.6 h F 64±14% pb 94±3% ur <4%	V 3.6±1.7 CL 5.9±1.1 Mwt 298.4 PSA 37.3 Å^2 log P 2.86	NORETHINDRONE, NORETHISTERONE Progestogen, oral contraceptive Metabolism: reduction of ketone to alcohol, hydrogenation of alkene
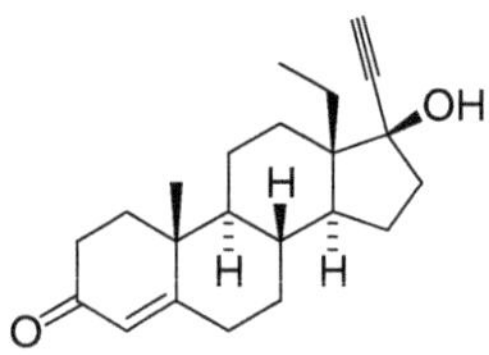	$t^1/_2$ 15±3 h F 94% pb 37±7% ur 52±8%	V 1.7 CL 1.5±0.6 Mwt 312.5 PSA 37.3 Å^2 log P 3.37	LEVONORGESTREL Progestogen, oral contraceptive Metabolism: hydrogenation of ketone and alkene, 16-hydroxylation
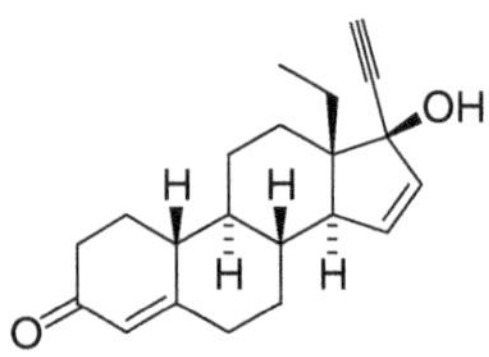	$t^1/_2$ 12–18 h F 100% pb 98–99% ur <1%	V 0.24–1.2 CL 0.8–0.9 Mwt 310.4 PSA 37.3 Å^2 log P 2.02	GESTODENE Oral contraceptive, progestogen Metabolism: reduction of ketone to alcohol
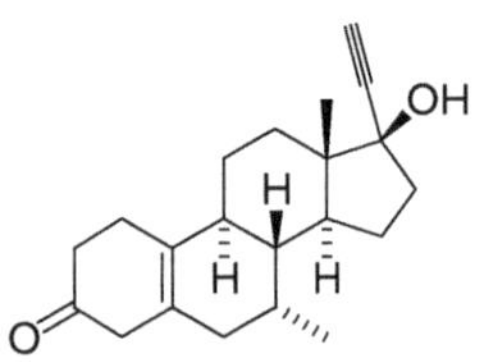	$t^1/_2$ 5.7–6.4 h F – pb – ur –	V – CL – Mwt 312.5 PSA 37.3 Å^2 log P 3.52	TIBOLONE Synthetic steroid for hormone replacement therapy Metabolism: reduction of ketone to active 3-α ($t^1/_2$ 27 h, monkey) and 3-β ($t^1/_2$ 17 h, monkey) hydroxy metabolites

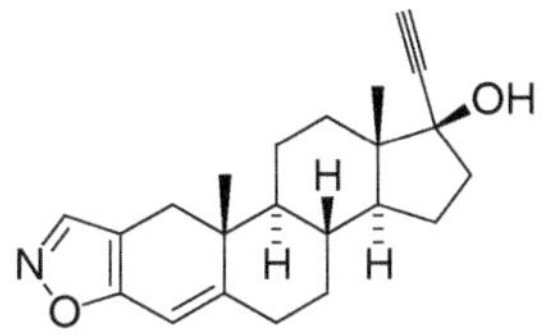	$t_{1/2}$ *F* pb ur	24 h Low – –	*V* CL Mwt PSA log *P*	3.4 – 337.5 46.3 Å^2 4.94	DANAZOL Antigonadotropin Metabolism: reductive cleavage of isoxazole to 2-hydroxymethylethisterone and ethisterone
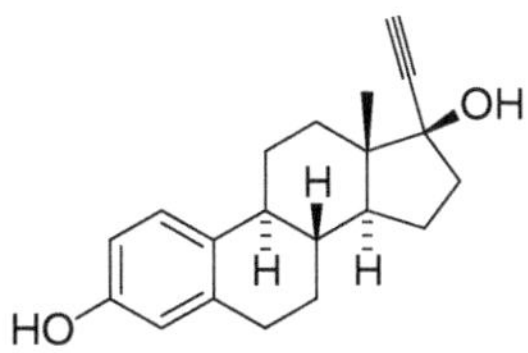	$t_{1/2}$ *F* pb ur	10±6 h 51±9% 95–98% 1–5%	*V* CL Mwt PSA log *P*	3.5±1.0 5.4±2.1 296.4 40.5 Å^2 4.11	ETHINYL ESTRADIOL Estrogen, oral contraceptive Metabolism: aromatic 2-hydroxylation, then O-methylation
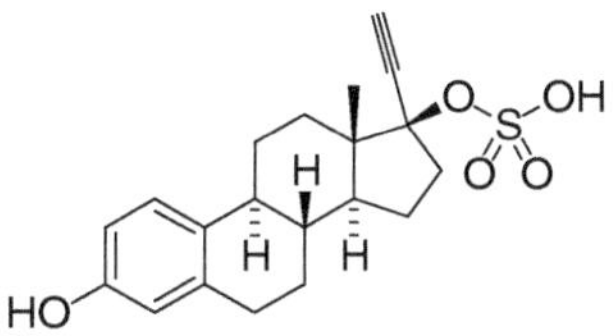	$t_{1/2}$ *F* pb ur	9.3 h (iv) – – –	*V* CL Mwt PSA log *P*	0.61 1.3 376.5 83.8 Å^2 1.33	ETHINYL ESTRADIOL-17-SULFATE Metabolite of ethinyl estradiol
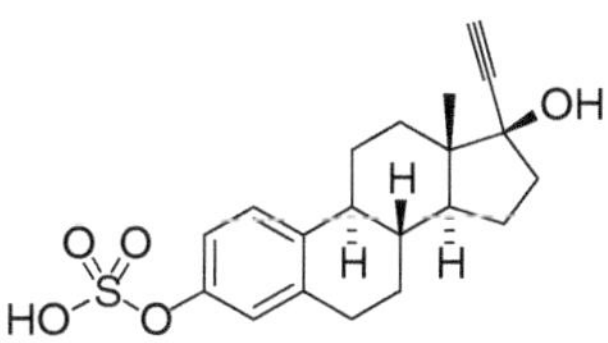	$t_{1/2}$ *F* pb ur	8.4 h (iv) – 99% –	*V* CL Mwt PSA log *P*	2.3 4.7 376.5 92.2 Å^2 3.29	ETHINYL ESTRADIOL-3-SULFATE Metabolite of ethinyl estradiol
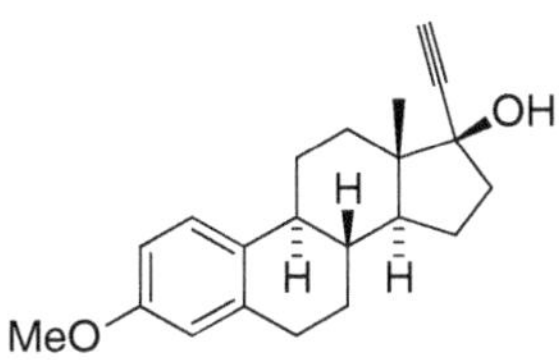	$t_{1/2}$ *F* pb ur	0.8–1.0 h – – –	*V* CL Mwt PSA log *P*	2 – 310.4 29.5 Å^2 4.94	MESTRANOL Prodrug of ethinyl estradiol, contraceptive Metabolism: O-demethylation
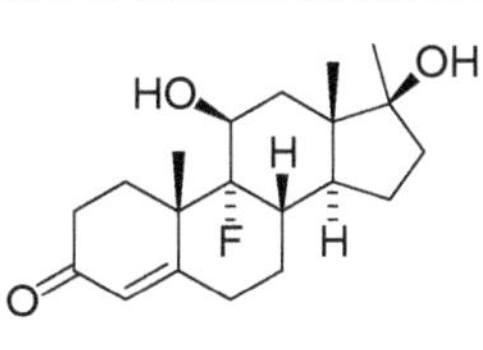	$t_{1/2}$ *F* pb ur	2–10 h <44% 99% –	*V* CL Mwt PSA log *P*	– – 336.4 57.5 Å^2 2.27	FLUOXYMESTERONE Androgen Metabolism: hydrogenation of ketone and C=C double bond, hydroxylation

(*continued overleaf*)

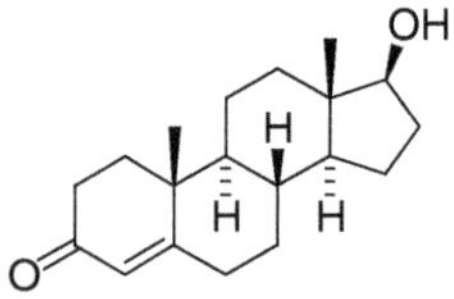

$t_{1/2}$	1–2 h	*V*	1	TESTOSTERONE
F	4.3%	CL	12.9	Natural androgen, crosses
pb	98%	Mwt	288.4	bbb
ur	–	PSA	37.3 Å^2	Metabolism: hydrogenation
		log *P*	3.18	of alkene

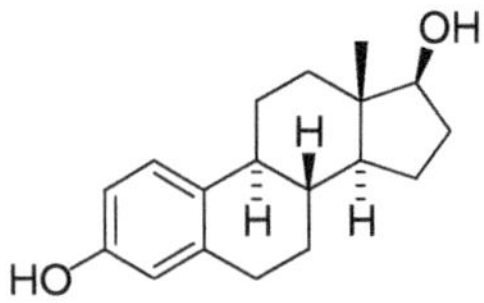

$t_{1/2}$	2–4 h	*V*	1.0	ESTRADIOL
F	3–5%	CL	30	Natural estrogen, crosses
pb	97–99%	Mwt	272.4	bbb; active metabolites:
ur	–	PSA	40.5 Å^2	estrone, estriol; prodrug
		log *P*	4.15	estradiol valerate: same oral
				bioavailability as estradiole

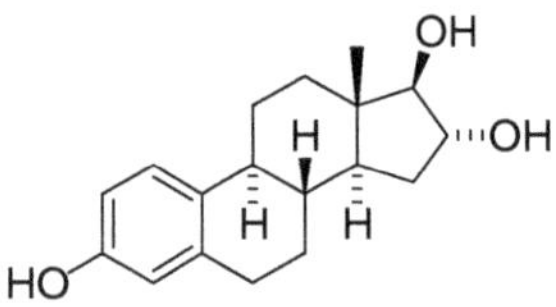

$t_{1/2}$	9–10 h	*V*	–	ESTRIOL
F	–	CL	–	Natural estrogen
pb	–	Mwt	288.4	Metabolites:
ur	–	PSA	60.7 Å^2	16-hydroxyestrone,
		log *P*	2.53	glucuronides

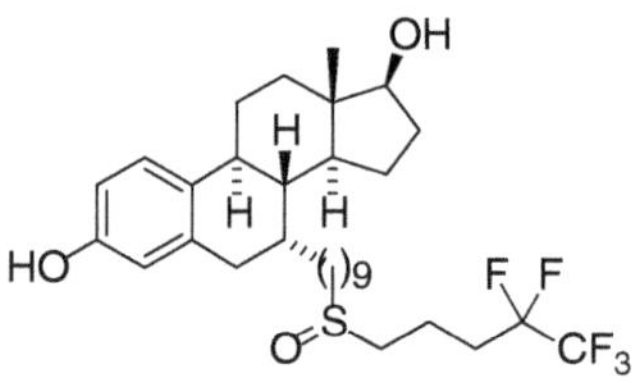

$t_{1/2}$	40–50 d*	*V*	3–5	FULVESTRANT
F	Low	CL		*im
pb	99%	Mwt	606.8	Antineoplastic
ur	<1%	PSA	76.7 Å^2	Metabolism: aromatic
		log *P*	7.92	hydroxylation,
				glucuronidation, sulfation

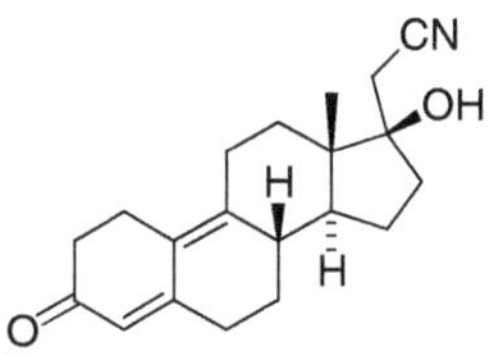

$t_{1/2}$	8–9 h	*V*	0.6	DIENOGEST
F	91%	CL	0.7	Progestogen, oral
pb	90%	Mwt	311.4	contraceptive
ur	<1%	PSA	61.1 Å^2	Metabolism: hydroxylation
		log *P*	2.58	

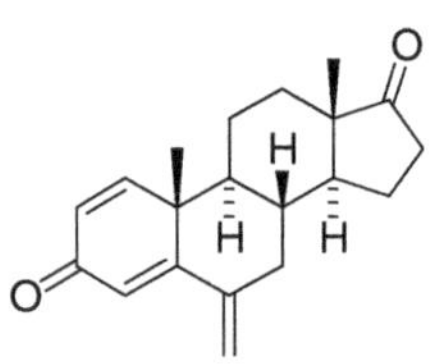

$t_{1/2}$	24–27 h	*V*	1.8*	EXEMESTANE
F	5–60%	CL/*F*	147	*dog
pb	90%	Mwt	296.4	Reductase inhibitor,
ur	<1%	PSA	34.1 Å^2	antineoplastic
		log *P*	2.43	Metabolism: reduction of
				cyclopentanone to
				cyclopentanol

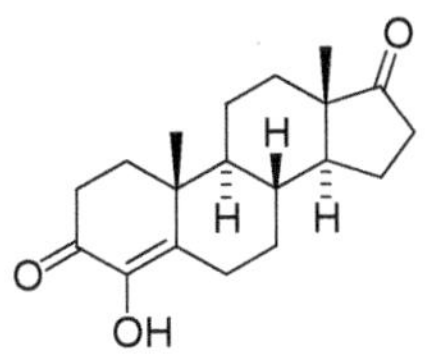

$t_{1/2}$	18 min (iv)	V	1.8	FORMESTANE
F	20–25%*	CL	2–3	*im
pb	82–86%	Mwt	302.4	Antineoplastic
ur	Negligible	PSA	54.4 Å^2	Metabolism:
		log P	1.79	glucuronidation of enol, reduction of cyclohexenone to cyclohexenol, then conjugation

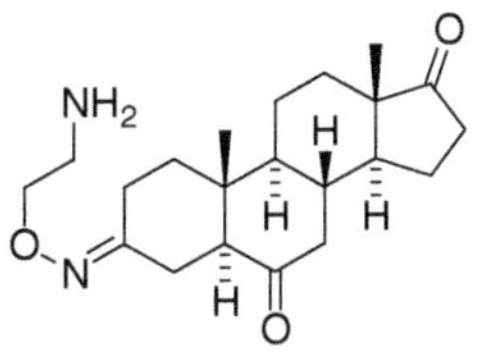

$t_{1/2}$	<1 h	V	2	ISTAROXIME
F	–	CL	60	Inotropic, lusitropic
pb	–	Mwt	360.5	
ur	–	PSA	81.8 Å^2	
		log P	1.89	

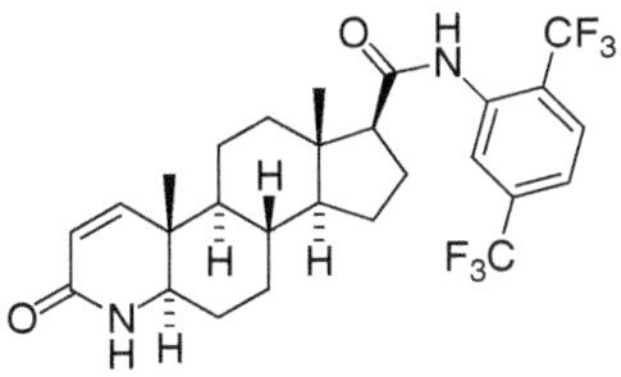

$t_{1/2}$	>240 h	V	4.6–7.7	DUTASTERIDE
F	40–94%	CL	0.15	5α-Reductase inhibitor
pb	>99%	Mwt	528.5	Metabolism: hydroxylation
ur	<1%	PSA	58.2 Å^2	
		log P	5.61	

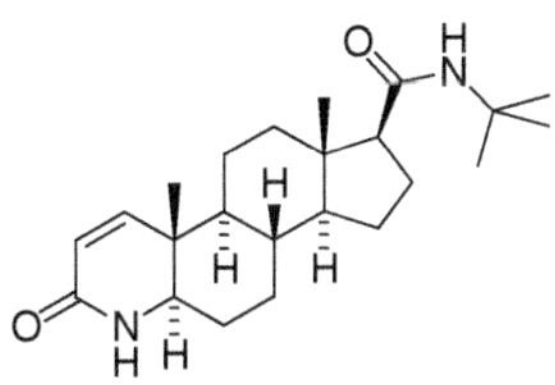

$t_{1/2}$	7.9±2.5 h	V	1.1±0.2	FINASTERIDE
F	63±21%	CL	2.3±0.8	5α-Reductase inhibitor,
pb	93%	Mwt	372.5	CYP3A4 substrate
ur	<1%	PSA	58.2 Å^2	Metabolism:
		log P	3.83	CH_3-hydroxylation, then oxidation of CH_3 to CO_2H

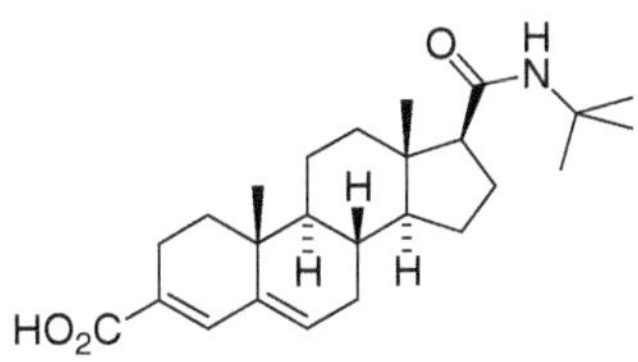

$t_{1/2}$	26 h	V	0.54	EPRISTERIDE
F	93%	CL	0.33	5α-Reductase inhibitor for
pb	97%	Mwt	399.6	treatment of benign
ur	–	PSA	66.4 Å^2	prostatic hypertrophy
		log P	5.34	

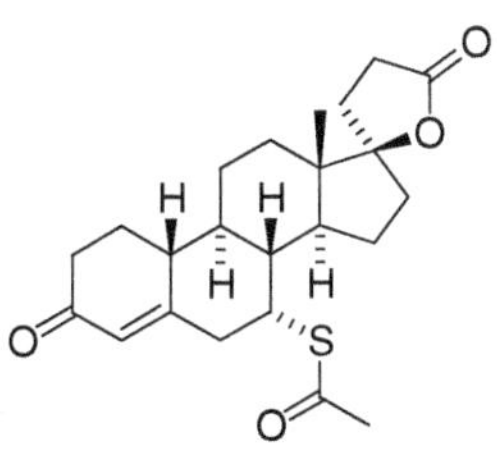

$t_{1/2}$	1.4 h	V	1.8	SPIRONOLACTONE
F	>90%	CL	4.2±1.7	Diuretic
pb	>90%	Mwt	416.6	Metabolism: deacetylation,
ur	<1%	PSA	85.7 Å^2	then S-methylation,
		log P	3.15	elimination of thioacetyl (to yield diene = canrenone)

(continued overleaf)

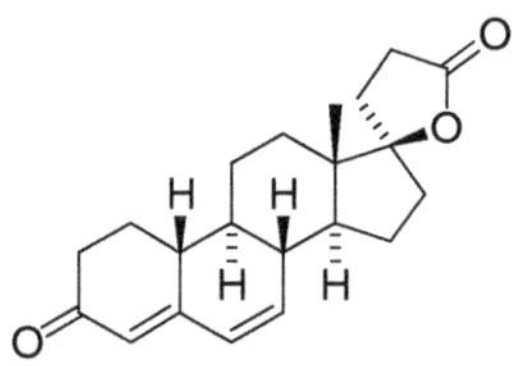

$t^1/_2$	3.9±1.2 h	V	–	CANRENONE
F	25±9%	CL	4.2±1.7	Diuretic
pb	–	Mwt	340.5	Metabolite of spironolactone
ur	–	PSA	43.4 Å^2	
		log P	3.51	

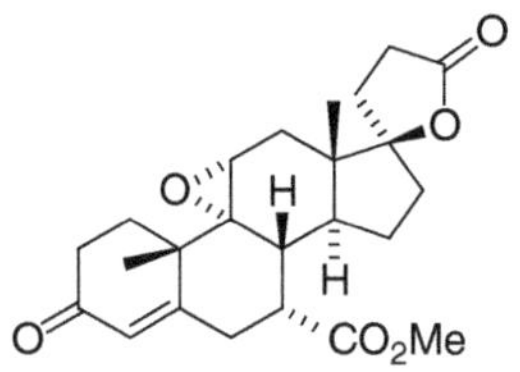

$t^1/_2$	4–6 h	V	0.6–1.3	EPLERENONE
F	–	CL	2.6	Antihypertensive
pb	49%	Mwt	414.5	Metabolism: lactone
ur	<2%	PSA	82.2 Å^2	hydrolysis, ester hydrolysis,
		log P	1.05	6β- and 21-hydroxylation

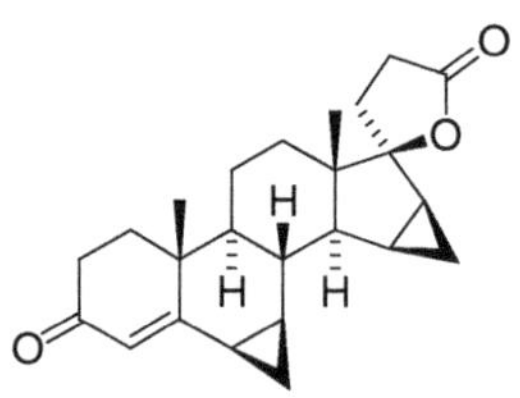

$t^1/_2$	35–39 h	V	3.7±4.2	DROSPIRENONE
F	76–85%	CL	1.2–1.5	Progestogen
pb	95–98%	Mwt	366.5	Metabolism: ring-opening
ur	Negligible	PSA	43.4 Å^2	of lactone, hydrogenation of
		log P	3.39	alkene, sulfation

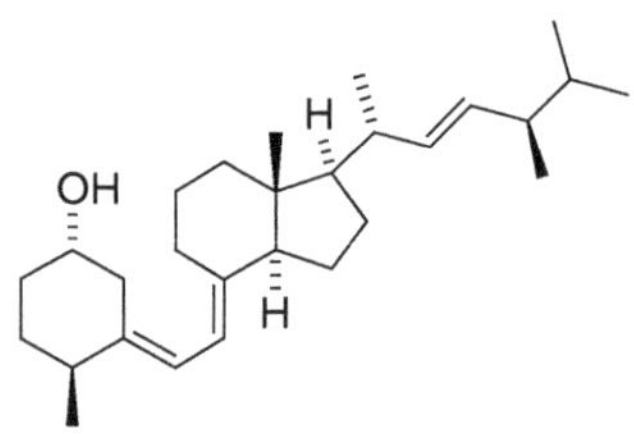

$t^1/_2$	11–13 h	V	–	DIHYDROTACHYSTEROL
F	–	CL	–	Vitamin D analog
pb	>99%	Mwt	398.7	Metabolism: dihydroxylation
ur	–	PSA	20.2 Å^2	
		log P	9.66	

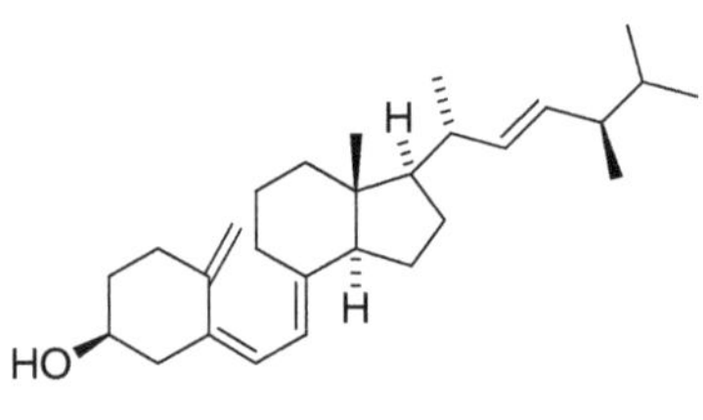

$t^1/_2$	19–48 h	V	–	ERGOCALCIFEROL,
F	55–99%	CL	–	Vitamin D_2
pb	>99.8%	Mwt	396.7	
ur	–	PSA	20.2 Å^2	
		log P	9.15	

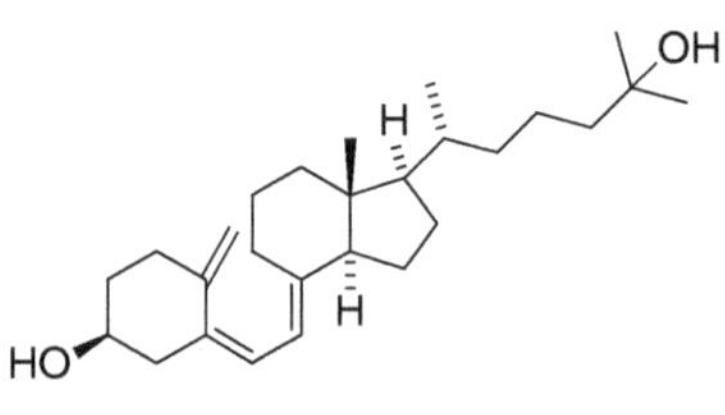

$t^1/_2$	10–22 d	V	–	CALCIFEDIOL
F	–	CL	–	Calcium regulator
pb	98%	Mwt	400.6	Metabolite: calcitriol
ur	–	PSA	40.5 Å^2	
		log P	7.08	

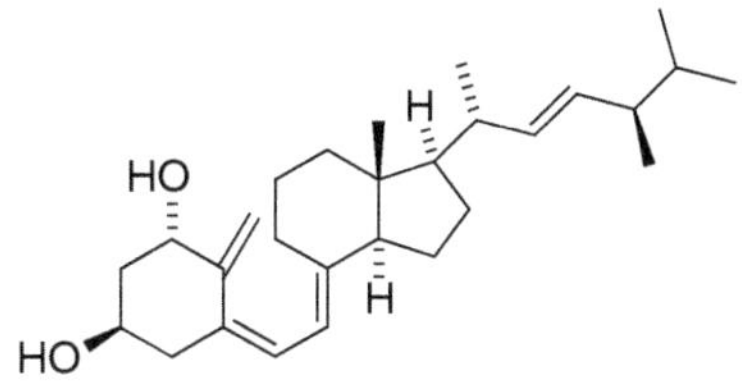	$t^1/_2$ 32–96 h *F* 42% pb >99% ur –	*V* – CL – Mwt 412.7 PSA 40.5 Å^2 log *P* 7.70	DOXERCALCIFEROL, 1α-HYDROXYVITAMIN D_2 Vitamin D_2 analog Metabolite: 1,25-dihydroxyvitamin D_2
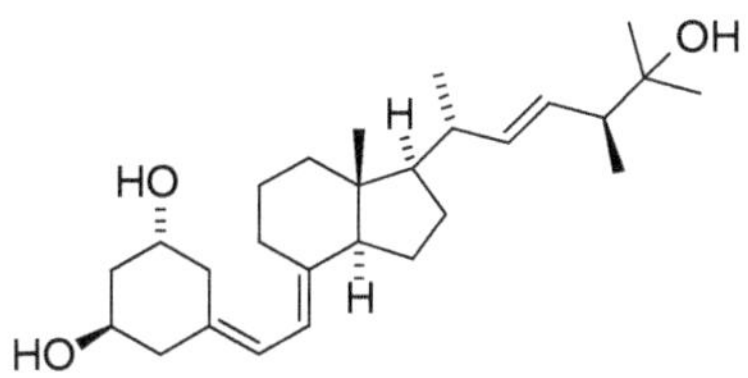	$t^1/_2$ 5–15 h *F* 72% pb >99% ur 0%	*V* 0.4 CL 0.9 Mwt 416.6 PSA 60.7 Å^2 log *P* 5.83	PARICALCITOL Vitamin D analog Metabolism: hydroxylation, dihydroxylation of olefins, glucuronidation
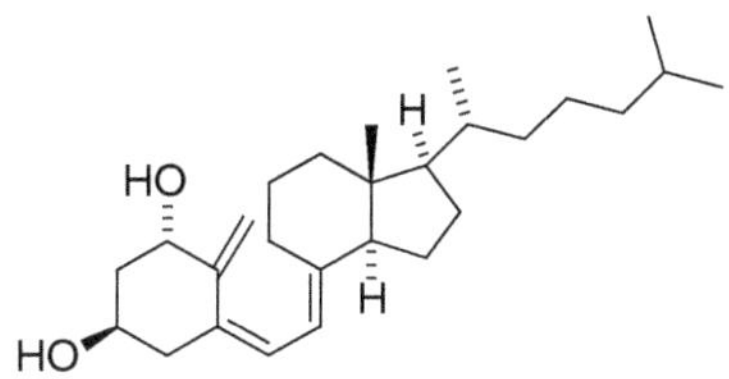	$t^1/_2$ 3 h *F* 71% pb >99% ur –	*V* – CL – Mwt 400.6 PSA 40.5 Å^2 log *P* 7.64	ALFACALCIDOL, 1-HYDROXYCHOLECALCIFEROL Vitamin D analog Metabolism: hydroxylation to calcitriol
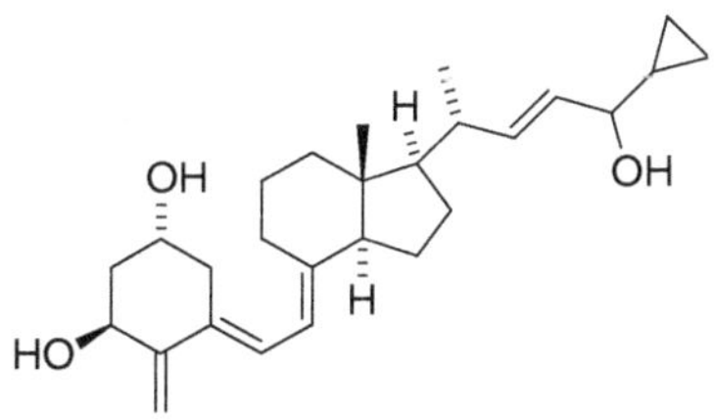	$t^1/_2$ <2 h *F* 5–6% pb – ur –	*V* – CL – Mwt 412.6 PSA 60.7 Å^2 log *P* 5.43	CALCIPOTRIOL Vitamin D analog for treatment of psoriasis Metabolism: oxidation of cyclopropylmethanol to ketone, then hydrogenation of enone to saturated ketone
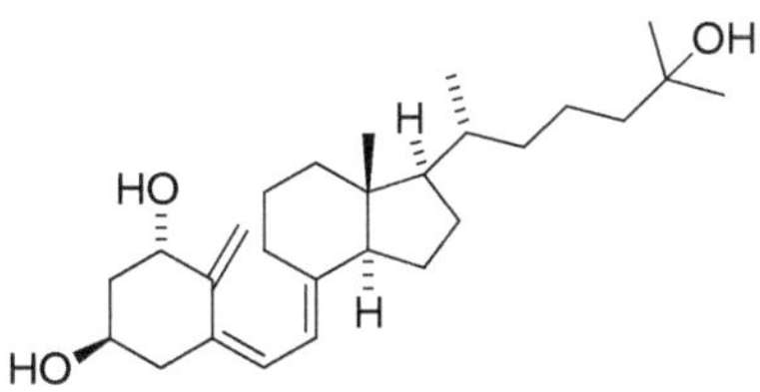	$t^1/_2$ 3–6 h *F* 62% pb >99% ur –	*V* 0.2–0.4* CL 6–21* Mwt 416.6 PSA 60.7 Å^2 log *P* 5.63	CALCITRIOL, VITAMIN D_3 *dog
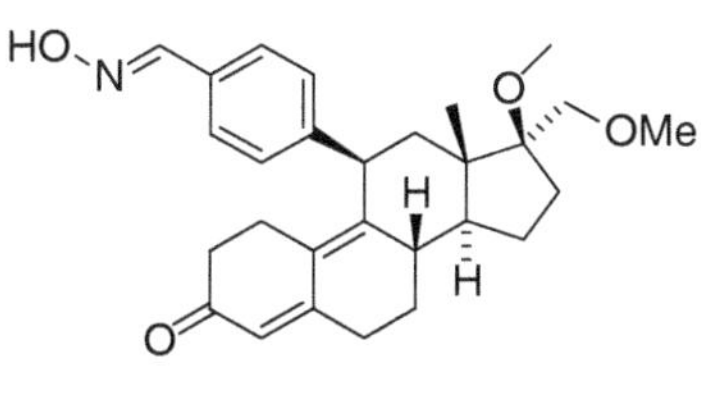	$t^1/_2$ 3.0–4.9 h *F* – pb – ur <0.2%	*V* – CL – Mwt 449.6 PSA 68.1 Å^2 log *P* 4.11	ASOPRISNIL Progesterone receptor modulator; active metabolite: $t^1/_2$: 4–5 h; further metabolism: O-demethylation

(*continued overleaf*)

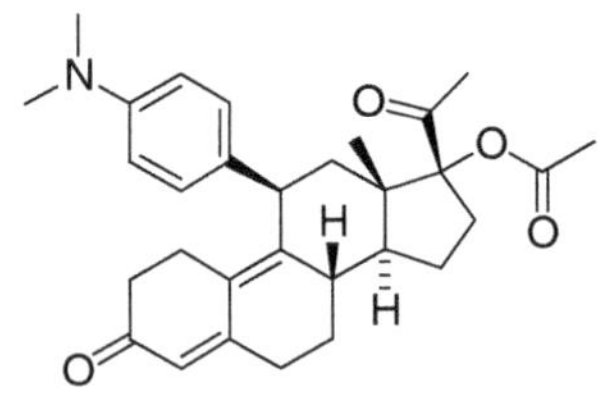

$t^{1}/_{2}$	32±6 h*	V	–	ULIPRISTAL ACETATE
F	56%	CL	20	*ulipristal
pb	97–99%	Mwt	475.6	Contraceptive
ur	–	PSA	63.7 Å^2	Metabolism: deacetylation to ulipristal, N-demethylation, hydroxylation
		log P	4.48	

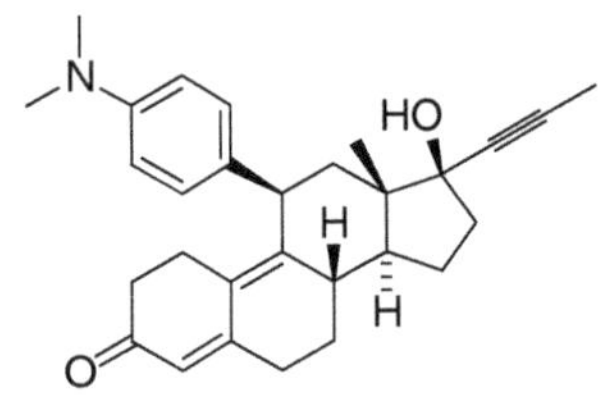

$t^{1}/_{2}$	15–48 h	V	0.4–0.7	MIFEPRISTONE
F	40–70%	CL	0.7	Abortifacient
pb	94–99%	Mwt	429.6	Metabolism:
ur	Negligible	PSA	40.5 Å^2	N-demethylation, hydroxylation of propargylic CH_3, then oxidation to aldehyde
		log P	6.19	

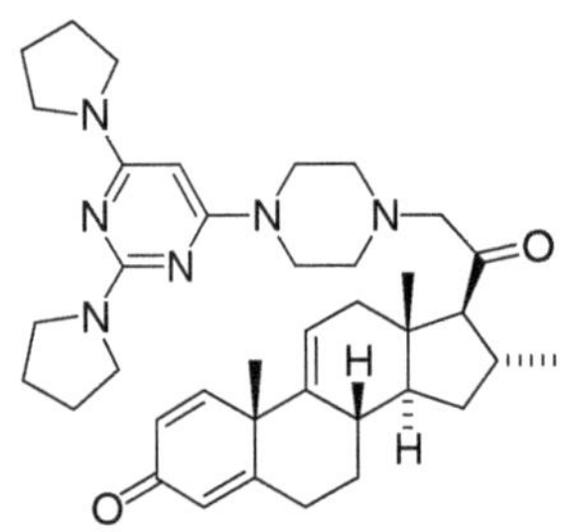

$t^{1}/_{2}$	16–52 h	V	4.7–6.8	TIRILAZAD
F	<12%	CL	5.6–10.5	Lipid peroxidation inhibitor
pb	>99%	Mwt	624.9	Metabolism: hydrogenation of 4–5 alkene (enone), hydroxylation
ur	–	PSA	72.9 Å^2	
		log P	4.89	

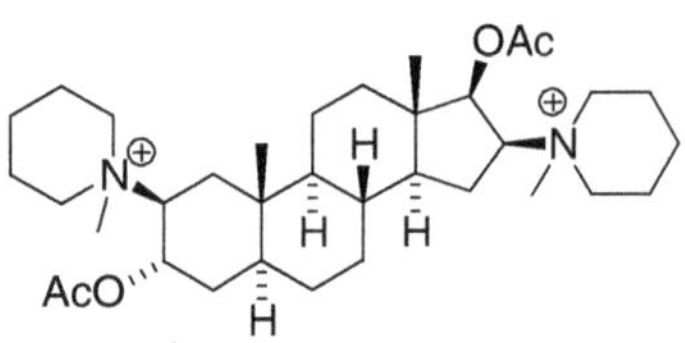

$t^{1}/_{2}$	2.3±0.4 h	V	0.26±0.07	PANCURONIUM
F	–	CL	1.8±0.4	Neuromuscular blocking agent
pb	7±2%	Mwt	572.9	Metabolism: deacetylation
ur	67±18%	PSA	52.6 Å^2	
		log P	−1.62	

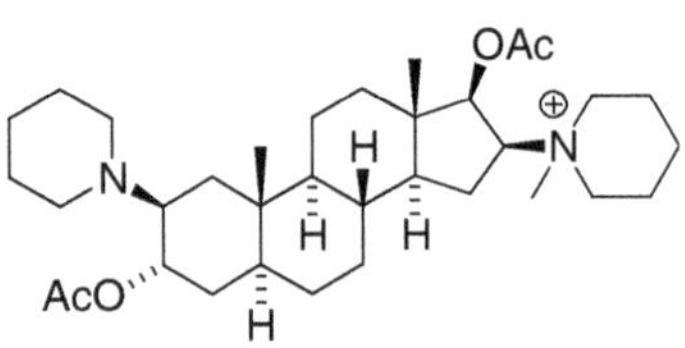

$t^{1}/_{2}$	1.5±0.7 h	V	0.21±0.08	VECURONIUM
F	–	CL	3.0±0.1	Neuromuscular blocking agent
pb	30±9%	Mwt	557.8	Metabolism: O-deacetylation
ur	18±5%	PSA	55.9 Å^2	
		log P	2.35	

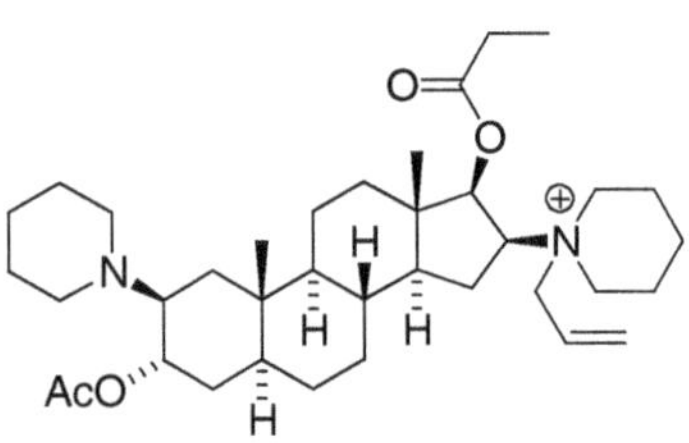

$t^{1}/_{2}$	2 h	V	0.3–0.4	RAPACURONIUM
F	–	CL	5–9	Neuromuscular blocking agent
pb	50–88%	Mwt	597.9	Metabolism:
ur	8%	PSA	55.8 Å^2	O-deacetylation; withdrawn in 2001 for causing fatal bronchospasm
		log P	2.32	

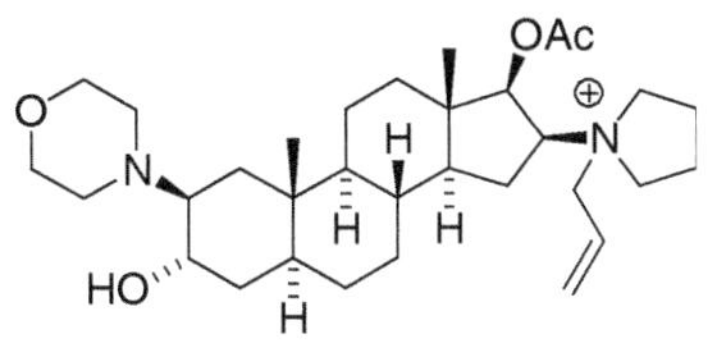

$t_{1/2}$	2.2–3.4 h	*V*	0.18–0.28	ROCURONIUM
F	–	CL	2.8–4.0	Neuromuscular blocking agent
pb	25%	Mwt	529.8	Metabolism: deacetylation
ur	9–22%	PSA	59.0 Å^2	
		log *P*	0.73	

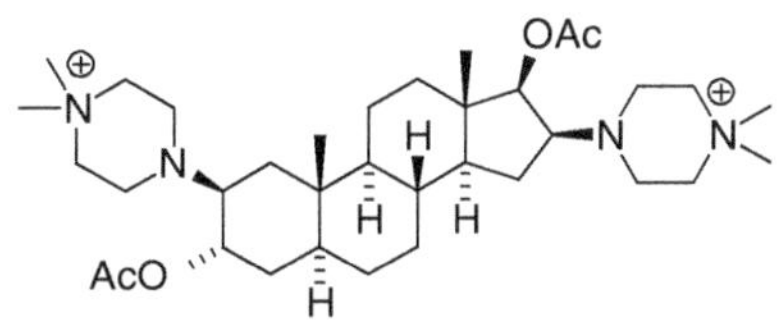

$t_{1/2}$	2.5±1.5 h	*V*	0.35±0.08	PIPECURONIUM, PIPECURIUM
F	–	CL	2.1±0.5	Neuromuscular blocking agent
pb	–	Mwt	602.9	Metabolism: deacetylation
ur	37–41%	PSA	59.1 Å^2	
		log *P*	−3.66	

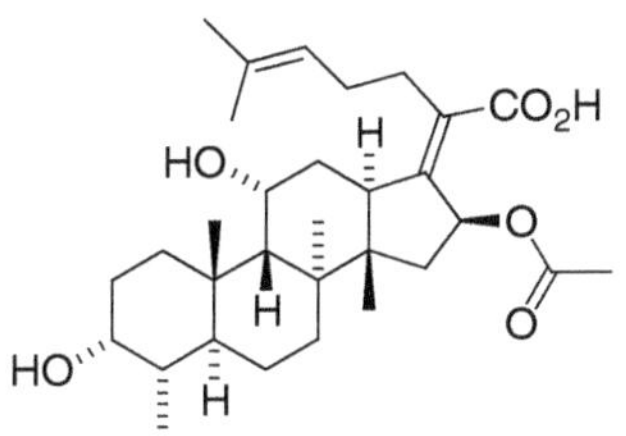

$t_{1/2}$	4–15 h	*V*	0.3	FUSIDIC ACID
F	91%	CL	–	Antibiotic
pb	98–99%	Mwt	516.7	Metabolism: oxidation of allylic CH_3 groups, glucuronidation
ur	<0.1%	PSA	104 Å^2	
		log *P*	5.76	

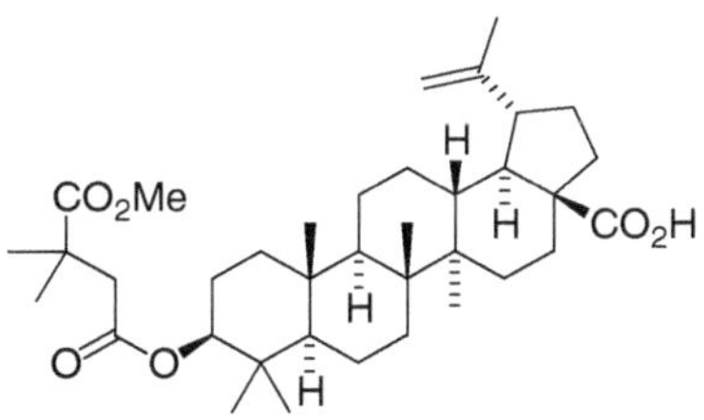

$t_{1/2}$	56–70 h	*V*	0.2	BEVIRIMAT
F	22–24%*	CL	0.05	*rat
pb	>99.5%	Mwt	584.8	HIV-1 maturation inhibitor
ur	<1%	PSA	101 Å^2	Metabolism: glucuronidation
		log *P*	10.2	

$t_{1/2}$, plasma half-life; *F*, oral bioavailability; pb, plasma protein binding; ur, excretion of unchanged drug in urine; V, volume of distribution; CL, clearance; Mwt, molecular weight; PSA, polar surface area; bbb, blood–brain barrier; GABA, γ-aminobutyric acid.

67
Anthracyclines

Tetracycline antibiotics, produced by *Streptomyces aureofaciens* and *Streptomyces rimosus*, are active against in a broad spectrum of bacteria. The first tetracycline, chlortetracycline, was introduced in 1948. These compounds inhibit bacterial protein synthesis by binding to the bacterial 30S ribosome. Being highly oxidized and heteroatom rich, tetracyclines undergo few metabolic transformations.

The anthracycline antibiotics daunorubicin and doxorubicin are also produced by a fungus (*Streptomyces peucetius*). These DNA intercalators have become valuable antitumor agents. However, as DNA/RNA-damaging agents, anthracyclines are mutagenic and carcinogenic as well. Because these natural products also show high cardiotoxicity, numerous analogs have been prepared, some with improved therapeutic index (Table 67.1).

Table 67.1 Anthracyclines and related compounds. *V* in l kg^{-1}; CL in ml min^{-1} kg^{-1}; Mwt in g mol^{-1}.

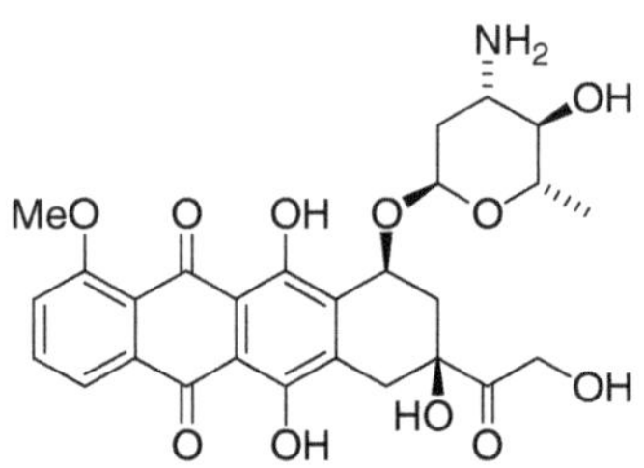	$t^1\!/_2$ 15–45 h *F* Low pb 75% ur 6–7%	*V* 13–52 CL 7.1–23.8 Mwt 543.5 PSA 206 Å^2 log *P* 0.24	EPIRUBICIN Antineoplastic Metabolism: glucuronidation, reduction of nonaromatic ketone to alcohol
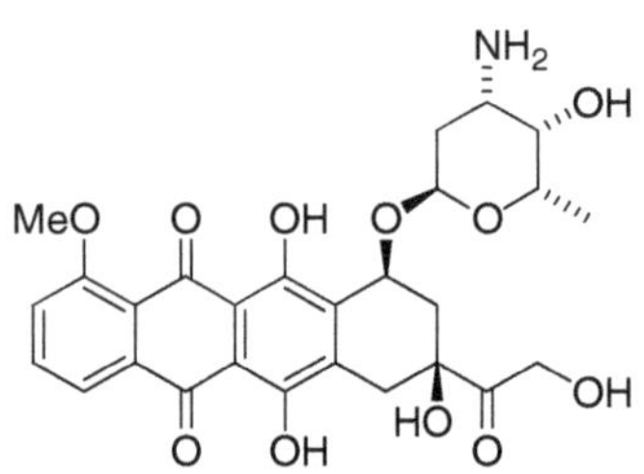	$t^1\!/_2$ 26±17 h *F* 5% pb 70–85% ur <7%	*V* 17±11 CL 17±8 Mwt 543.5 PSA 206 Å^2 log *P* 0.24	DOXORUBICIN Antineoplastic Prodrug: *N*-L-leucyl-doxorubicin

Lead Optimization for Medicinal Chemists: Pharmacokinetic Properties of Functional Groups and Organic Compounds, First Edition. Florencio Zaragoza Dörwald.

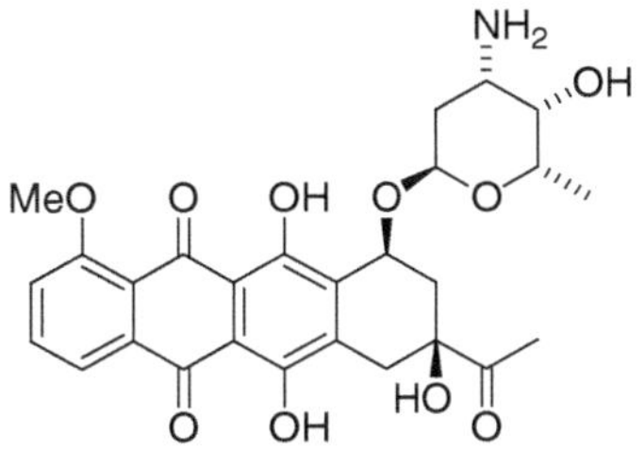

$t^1/_2$	20–30 h	*V*	39	DAUNORUBICIN
F	Low	CL	15–30	Antineoplasitc (leukemia); does not cross bbb
pb	50–60%	Mwt	527.5	Metabolism: reduction of ketone, O-demethylation, glycosidic cleavage, O-sulfation, O-glucuronidation
ur	5–18% (iv)	PSA	186 Å^2	
		log *P*	0.54	

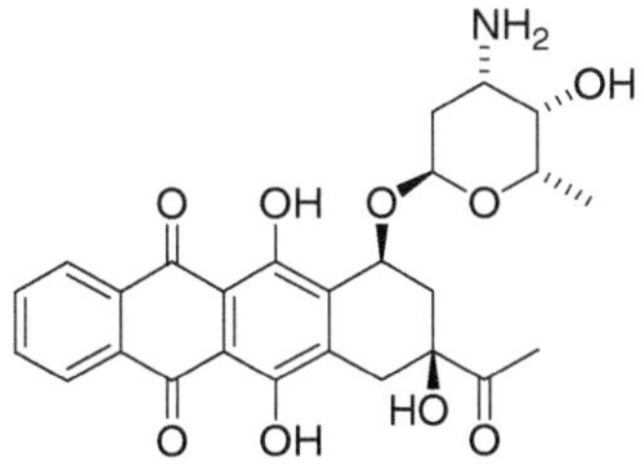

$t^1/_2$	6–35 h	*V*	37–64	IDARUBICIN, 4-DEMETHOXYDAUNORUBICIN
F	18–39%	CL	17–24	Antineoplastic, crosses bbb
pb	97%	Mwt	497.5	Metabolite: 13-dihydro (idarubicinol, $t^1/_2$ 81 h; pb 94%; ur < 5%)
ur	<5%	PSA	177 Å^2	
		log *P*	1.44	

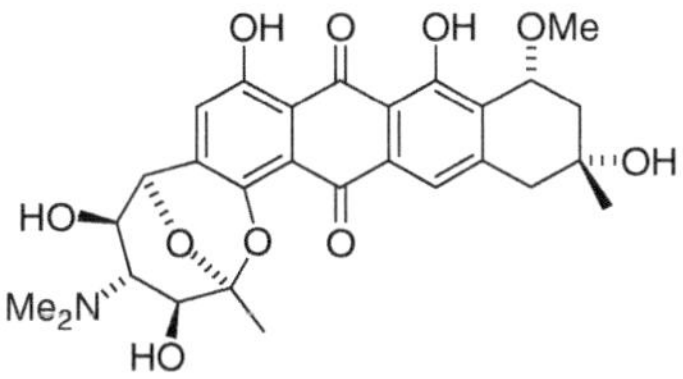

$t^1/_2$	11±6 h	*V*	30±9	MENOGARIL
F	32±12%	CL	17±7	Antineoplastic
pb	–	Mwt	541.6	Metabolism: N-demethylation
ur	17% (iv)	PSA	166 Å^2	
		log *P*	3.16	

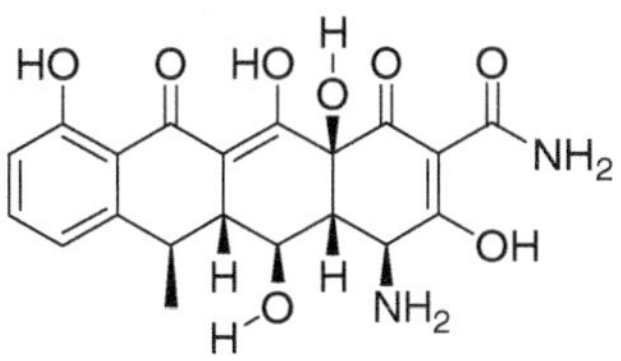

$t^1/_2$	16±6 h	*V*	0.75±0.32	DOXYCYCLINE
F	93%	CL	0.53±0.18	Antibacterial
pb	88±5%	Mwt	444.4	
ur	41±19%	PSA	182 Å^2	
		log *P*	1.78	

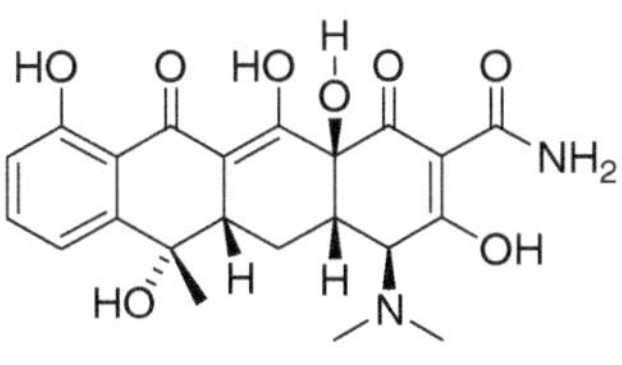

$t^1/_2$	10.6±1.5 h	*V*	1.5±0.1	TETRACYCLINE
F	77%	CL	1.67±0.24	Antibacterial
pb	65±3%	Mwt	444.4	
ur	58±8%	PSA	182 Å^2	
		log *P*	0.62	

(continued overleaf)

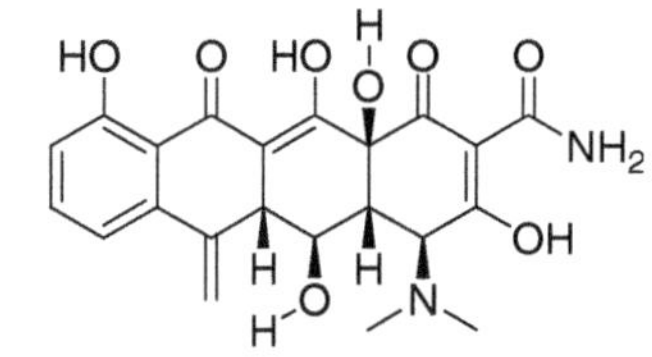

$t^1/_2$	14 h	V	–
F	<60%	CL	–
pb	78%	Mwt	442.4
ur	25%	PSA	182 Å^2
	·	log P	1.35

METHACYCLINE
Antibacterial

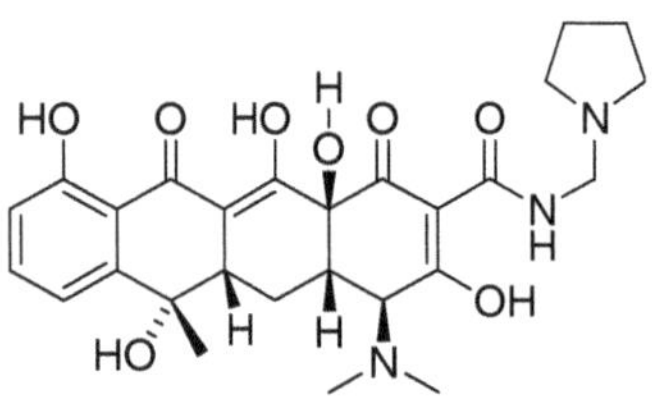

$t^1/_2$	5–9 h (iv)	V	0.54
F	Low	CL	0.97
pb	50%	Mwt	527.6
ur	>50% (iv)	PSA	171 Å^2
		log P	1.66

ROLITETRACYCLINE
Antibacterial, prodrug of tetracycline

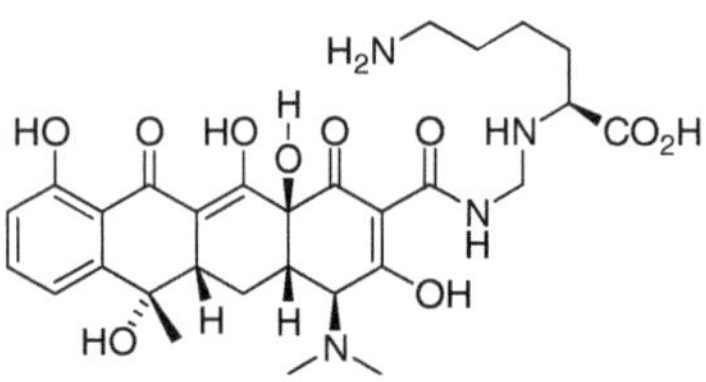

$t^1/_2$	9–12 h*	V	1.3–1.6*
F	70–80%*	CL	–
pb	55–64%*	Mwt	602.6
ur	25%*	PSA	243 Å^2
		log P	0.75

LYMECYCLINE
Antibacterial
Prodrug of tetracyclin; prodrug can not be detected in plasma after oral dosing
*tetracycline upon oral dosing of prodrug

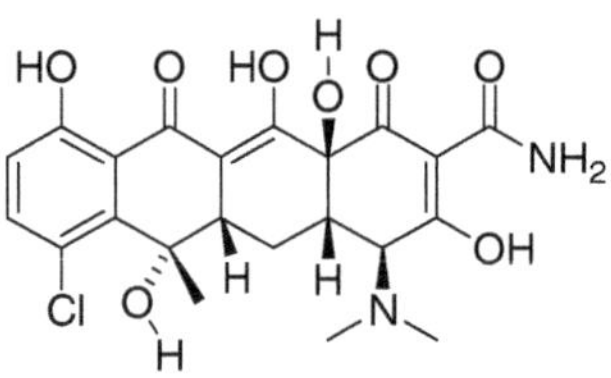

$t^1/_2$	7 h (iv)	V	0.9
F	–	CL	2
pb	48%	Mwt	478.9
ur	–	PSA	182 Å^2
		log P	1.32

CHLORTETRACYCLINE
Metabolism: epimerization, N-demethylation

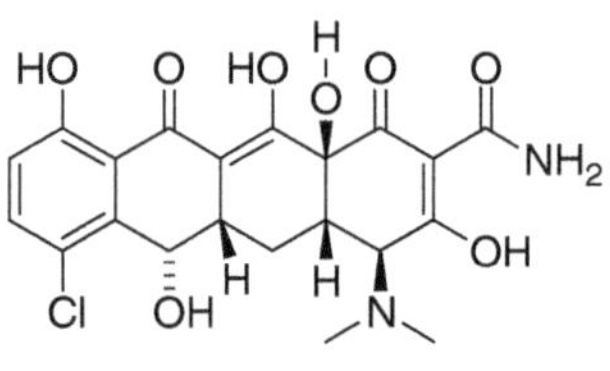

$t^1/_2$	14 h	V	1.3
F	60–80%	CL	1.1
pb	41–50%	Mwt	464.9
ur	30–50%	PSA	182 Å^2
		log P	0.94

DEMECLOCYCLINE
Antibacterial

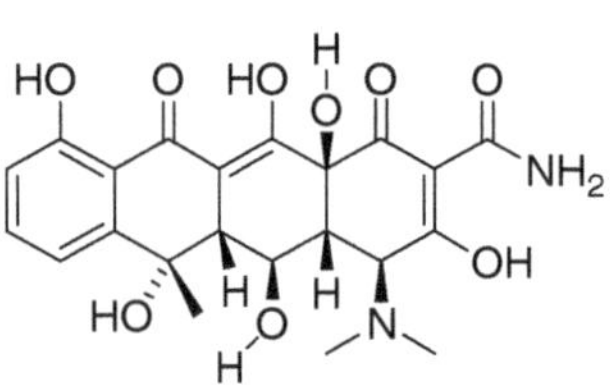

$t^1/_2$	10 h (iv)	V	1.7
F	3% (pig)	CL	2.0
pb	10%	Mwt	460.4
ur	–	PSA	202 Å^2
		log P	0.48

OXYTETRACYCLINE
Antibacterial
Metabolism: N-demethylation, epimerization

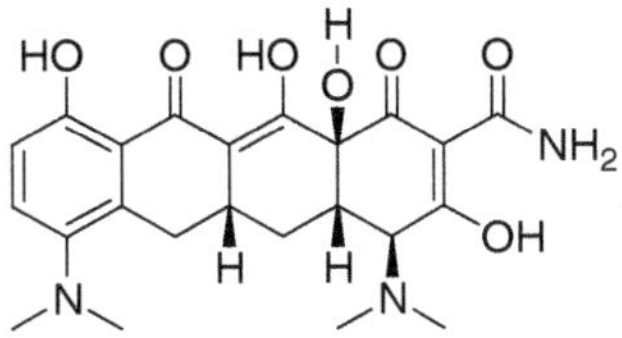	$t^{1}/_{2}$ 16±2 h *F* 95–100% pb 76% ur 11±2%	*V* 1.3±0.2 CL 1.0±0.3 Mwt 457.5 PSA 165 Å^2 log *P* 2.12	MINOCYCLINE Antibacterial Metabolism: 9-hydroxylation, N-demethylation
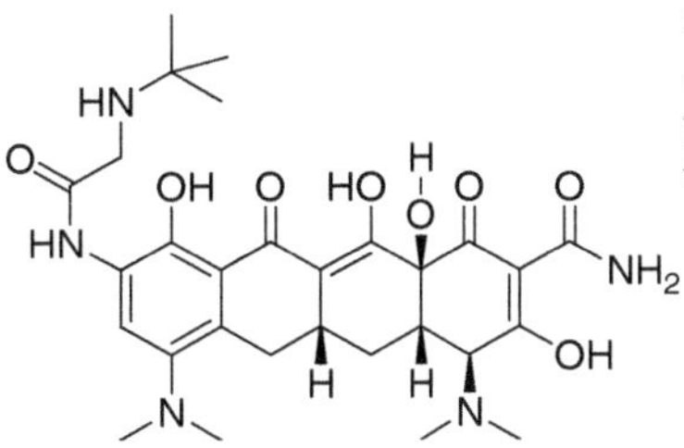	$t^{1}/_{2}$ 37–67 h* *F* Low pb 71–89% ur 15% (iv)	*V* 7–9 CL 3.3–5.0 Mwt 585.7 PSA 206 Å^2 log *P* −0.85	TIGECYCLINE *iv Antibacterial Metabolism: glucuronidation, glycinamide hydrolysis, then N-acetylation
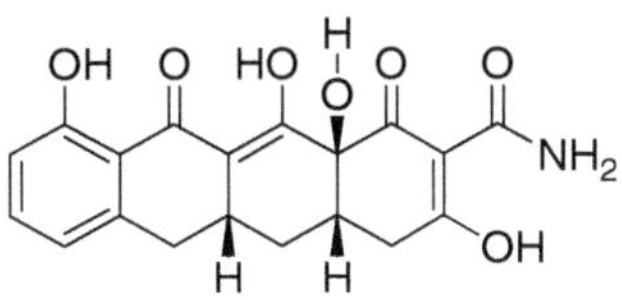	$t^{1}/_{2}$ 57 h *F* – pb 95% ur	*V/F* 0.63 CL/*F* 0.13 Mwt 371.3 PSA 158 Å^2 log *P* 0.84	METASTAT, COL-3 Matrix metalloproteinase inhibitor, antineoplastic
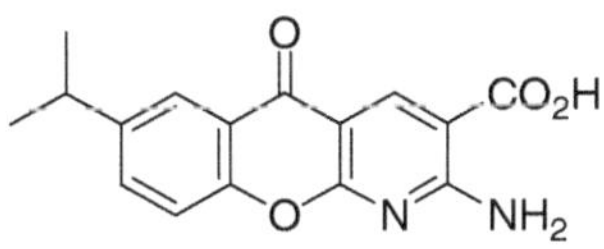	$t^{1}/_{2}$ 3.5±1.1 h *F* – pb – ur 17%	*V* – CL – Mwt 298.3 PSA 103 Å^2 log *P* 3.74	AMLEXANOX Antiallergic, antiasthmatic Metabolism: hydroxylation of *CH*$(CH_3)_2$, then conjugation
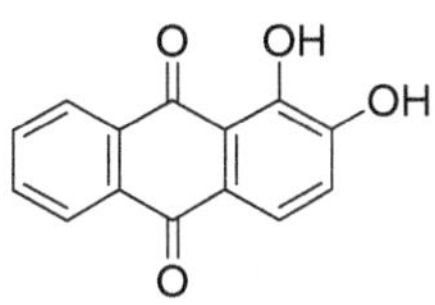	$t^{1}/_{2}$ 12 h *F* – pb – ur 18–36%	*V* – CL – Mwt 240.2 PSA 74.6 Å^2 log *P* 2.91	ALIZARIN Biological stain Metabolism: O-glucuronidation
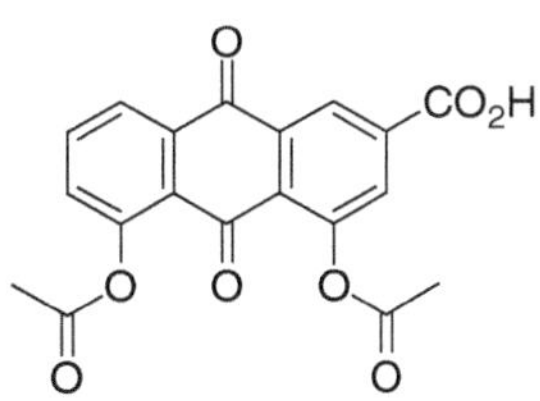	$t^{1}/_{2}$ 4–8 h* *F* 35%* pb 99%* ur 20%*	*V* – CL – Mwt 368.3 PSA 124 Å^2 log *P* 3.13	DIACEREIN, DIACETYLRHEIN *rhein upon oral dosing of diacerein Antiarthritic Metabolism: deacetylation to rhein, then glucuronidation, sulfation

(*continued overleaf*)

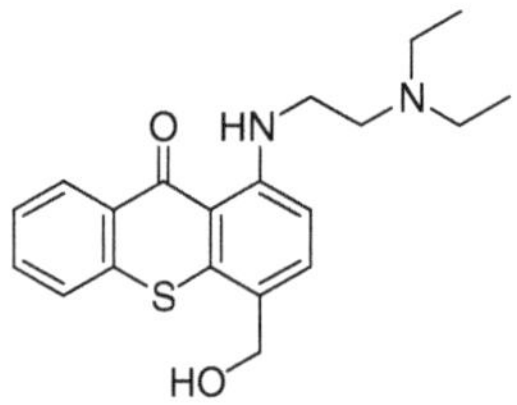

$t^1\!/_2$	3–5 h (iv)	*V*	–
F	Low	CL	–
pb	–	Mwt	356.5
ur	–	PSA	77.9 Å^2
		log *P*	3.81

HYCANTHONE
Hydroxylated metabolite of lucanthone, anthelmintic
Metabolism (monkey): N-deethylation

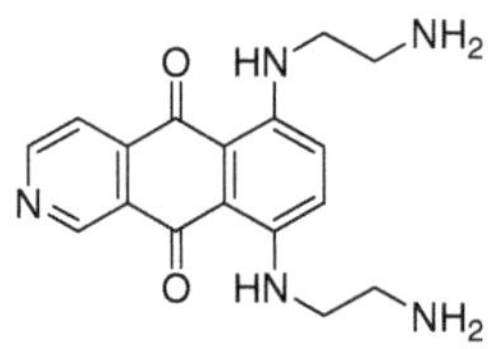

$t^1\!/_2$	12–32 h	*V*	10–30
F	Low	CL	13–22
pb	57%	Mwt	325.4
ur	<10%	PSA	123 Å^2
		log *P*	1.82

PIXANTRONE
Intercalator, topoisomerase II inhibitor, antineoplastic, immunosuppressant

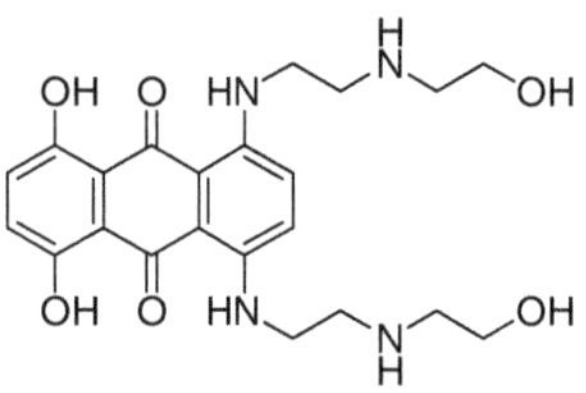

$t^1\!/_2$	10–18 d	*V*	15–32
F	–	CL	4.7–14
pb	75%	Mwt	444.5
ur	4%	PSA	163 Å^2
		log *P*	1.55

MITOXANTRONE
Antineoplastic, does not cross bbb
Metabolism: oxidation to dicarboxylic acid

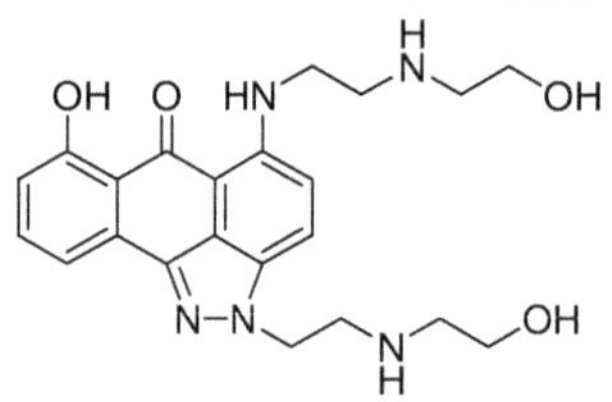

$t^1\!/_2$	24–51 h*	*V*	21–39
F	–	CL	12–20
pb	92–97%	Mwt	425.5
ur	<8%	PSA	132 Å^2
		log *P*	1.99

LOSOXANTRONE, CI 941
*iv
Antineoplastic
Metabolism: 2- and 4-hydroxylation of phenol

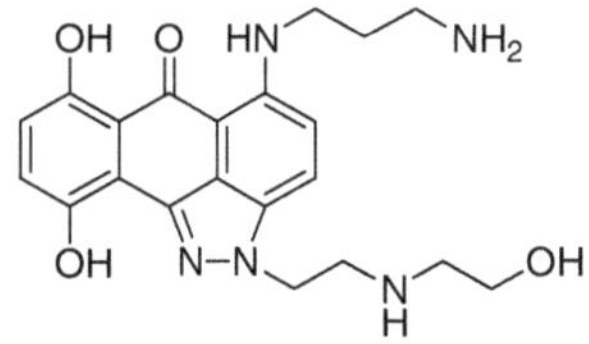

$t^1\!/_2$	0.5 h (iv)	*V*	0.8
F	–	CL	34
pb	–	Mwt	411.5
ur	5% (iv)	PSA	141 Å^2
		log *P*	1.96

PIROZANTRONE, OXANTRAZOLE
Antineoplastic

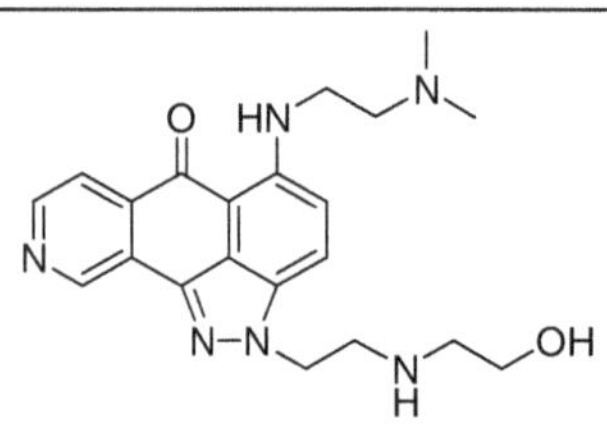

$t^1\!/_2$	45 h (iv)	*V*	57
F	–	CL	27
pb	–	Mwt	394.5
ur	<10% (iv)	PSA	90.4 Å^2
		log *P*	1.31

TOPIXANTRONE
Antineoplastic

$t_{1/2}$, plasma half-life; *F*, oral bioavailability; pb, plasma protein binding; ur, excretion of unchanged drug in urine; V, volume of distribution; CL, clearance; Mwt, molecular weight; PSA, polar surface area; bbb, blood–brain barrier.

68
Arylacetic, Benzoic, and Related Carboxylic Acids (NSAIDS)

Aryl- and heteroarylacetic acids and the closely related 2-arylpropionic acids are an important class of nonsteroidal antiinflammatory drugs (NSAIDs). A large number of drugs of this type have been developed (Table 68.1), but most of them show similar biological properties and no significant advantages if compared to the rest.

One of the first examples of this class of NSAIDs was ibuprofen, launched in 1964, which resulted from the screening of more than 600 alkanoic acids for antiinflammatory activity. Although the goal was a nonulcerogenic analgesic or antipyretic, ibuprofen was again burdened with this side effect, which seems to be inherent to all cyclooxygenase-1-inhibiting NSAIDs. Prostaglandines have a strong gastroprotective effect, and the suppression of their biosynthesis by inhibition of COX-1 (cyclooxygenase) inevitably increases the risk of gastric lesions.

Most of the drugs listed below undergo acyl glucuronidation *in vivo* and are then excreted in the urine and bile.

68.1
Salicylates

The use of willow bark extracts for the treatment of pain and fever was reported as early as 1554. Salicin (Scheme 68.1), the active glycoside of the willow bark, was isolated in pure form in 1829 and used for the treatment of rheuma until the 1950s.

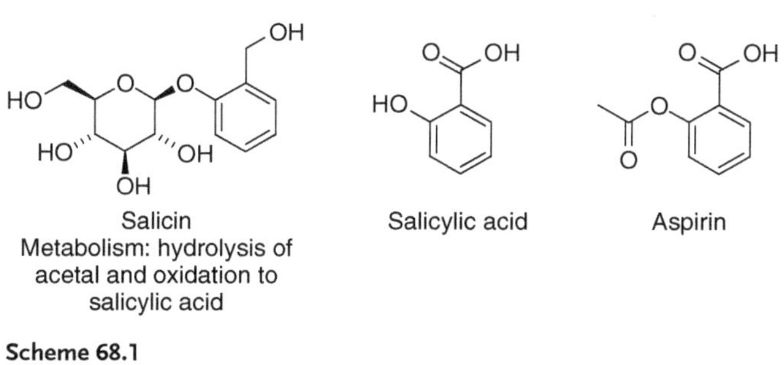

Scheme 68.1

Adolph Kolbe synthesized salicylic acid in 1874 and arranged to have Carl Emil Buss (St. Gallen, Switzerland) test it in typhoid patients as an alternative to the

Lead Optimization for Medicinal Chemists: Pharmacokinetic Properties of Functional Groups and Organic Compounds, First Edition. Florencio Zaragoza Dörwald.

antiseptic phenol. Buss found that salicylic acid, despite its phenol substructure, was not antiseptic, but lowered fever instead. This discovery led to the large-scale production and marketing of salicylic acid as antipyretic.

Unfortunately, salicylic acid causes stomach irritation and has such a bad taste that it may induce vomiting. With the aim of finding a better-tasting compound, the chemist Felix Hoffmann prepared (the known) acetylsalicylic acid in 1897 at Bayer. This compound, later named aspirin, turned out to be superior to salicylic acid and became the most widely used antipyretic [1].

Salicylates and other nonsteroidal antiinflammatory drugs (NSAIDs; 4-aminophenols, aryl/heteroarylacetic acids, and oxicams) suppress the biosynthesis of prostaglandins by inhibiting the cyclooxygenases (COX-1 and COX-2). One common side effect of all COX-1 inhibitors is their tendency to cause gastric damage (ulcers and erosions), as prostaglandins play an important role in the maintainance of the gastric mucosa. This does not seem to be the case for COX-2 inhibitors, which, however, seem to cause heart damage by increasing the viscosity of blood [2].

Numerous derivatives and analogs of salicylic acid have been marketed as antipyretics (Table 68.2), but none excelled aspirin. Anthranilic acid (2-aminobenzoic acid) is devoid of antipyretic effect, but 2-(arylamino)benzoic acids are potent analgetics and antiinflammatories.

References

1. (a) Sneader, W. (1995) Salicylates and their analogues. *Drug News Perspect.*, **8**, 644–648; (b) Sneader, W. (1996) Antipyretic analgesics. *Drug News Perspect.*, **9**, 61–64; (c) Sneader, W. (1996) Nonsteroidal antiinflammatory drugs. *Drug News Perspect.*, **9**, 181–185.
2. Blobaum, A.L. and Marnett, L.J. (2007) Structural and functional basis of cyclooxygenase inhibition. *J. Med. Chem.*, **50**, 1425–1441.

Table 68.1 Arylacetic acids, arenecarboxylic acids, and related compounds. V in l kg^{-1}, CL in ml min^{-1} kg^{-1}, Mwt in g mol^{-1}.

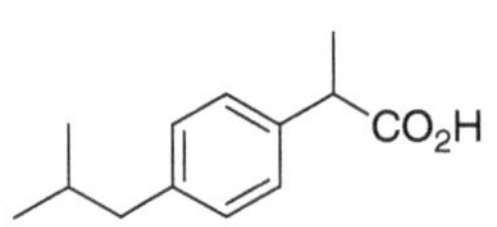	$t^1\!/_2$	2.0±0.5 h	V/F	0.15±0.02	IBUPROFEN Antiinflammatory, racemate Metabolism: glucuronidation, conversion to CoA-thioester, hydroxylation of isobutyl group at all positions, oxidation of $(CH_3)_2CH$ to $HO_2C(CH_3)CH$; prodrug: hydroxamic acid (ibuproxam)
	F	>80%	CL/F	0.75±0.20	
	pb	>99%	Mwt	206.3	
	ur	<1%	PSA	37.3 Å^2	
			log P	3.50	
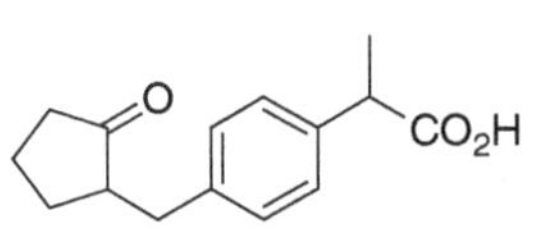	$t^1\!/_2$	–	V	–	LOXOPROFEN Antiinflammatory Metabolism: reduction of ketone to alcohol ($t^1\!/_2$ 1.3 h)
	F	–	CL	–	
	pb	97%	Mwt	246.3	
	ur	–	PSA	54.4 Å^2	
			log P	1.87	

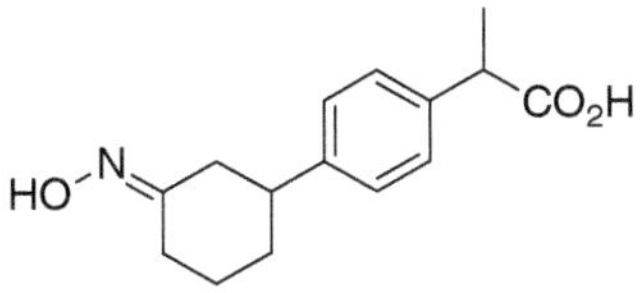	$t_{1/2}$ 1.6–3.8 h *F* 98% pb 93% ur 25%	*V* 0.26 CL 1.6 Mwt 261.3 PSA 69.9 Å^2 log *P* 2.31	XIMOPROFEN Antiinflammatory Metabolism: hydrolysis of oxime to ketone, then reduction to alcohol
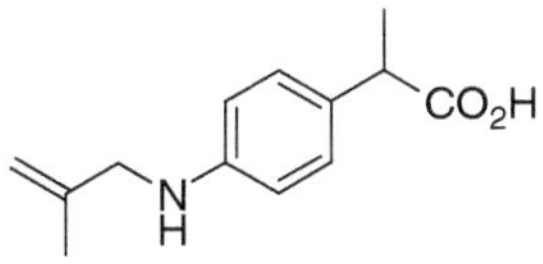	$t_{1/2}$ 2–4 h *F* – pb – ur –	*V* 0.17 CL 0.5–1.0 Mwt 219.3 PSA 49.3 Å^2 log *P* 2.51	ALMINOPROFEN Antiinflammatory
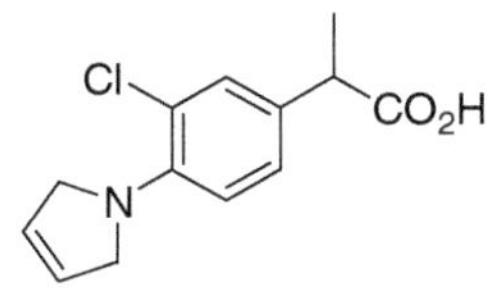	$t_{1/2}$ 6±1 h *F* – pb >99.5% ur <5%	*V* 0.11–0.17 CL 0.28 Mwt 251.7 PSA 40.5 Å^2 log *P* 2.71	PIRPROFEN Antiinflammatory Metabolism: dehydrogenation to pyrrole, epoxidation and trans-dihydroxylation of alkene, N-dealkylation, then N-acetylation
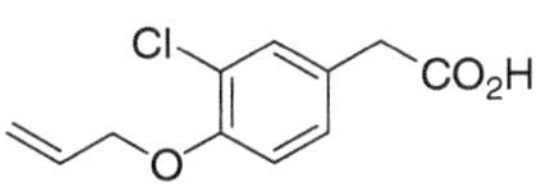	$t_{1/2}$ 2.0±0.5 h *F* – pb >99% ur 10–50%	*V* 0.1 CL – Mwt 226.7 PSA 46.5 Å^2 log *P* 2.77	ALCLOFENAC Analgesic, antipyretic, antiinflammatory Metabolism: epoxidation and dihydroxylation of alkene
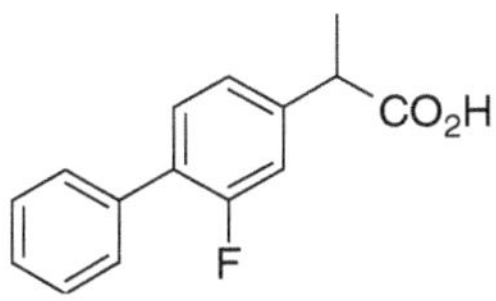	$t_{1/2}$ 5.5±1.4 h *F* 92% pb >99.5% ur 2±1%	*V/F* 0.15±0.02 CL/*F* 0.35±0.09 Mwt 244.3 PSA 37.3 Å^2 log *P* 3.66	FLURBIPROFEN Antiinflammatory, racemate Metabolism: 3- and 4-hydroxylation of Ph, then conjugation
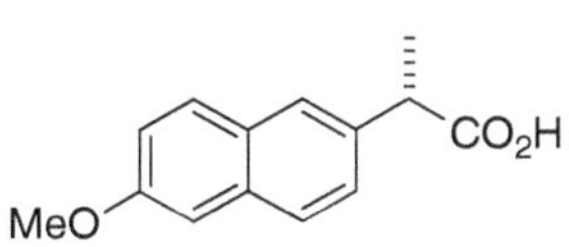	$t_{1/2}$ 14±1 h *F* 99% pb 99.7±0.1% ur <0.1%	*V/F* 0.16±0.02 CL/*F* 0.13±0.02 Mwt 230.3 PSA 46.5 Å^2 log *P* 2.88	NAPROXEN Antiinflammatory Metabolites: glucuronide, *O*-desmethyl
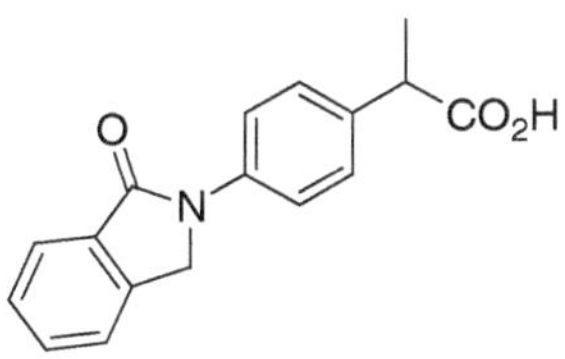	$t_{1/2}$ 3±1 h *F* 100% pb 99% ur –	*V* 0.11–0.17 CL – Mwt 281.3 PSA 57.6 Å^2 log *P* 2.82	INDOPROFEN Antiinflammatory Metabolism: glucuronidation

(*continued overleaf*)

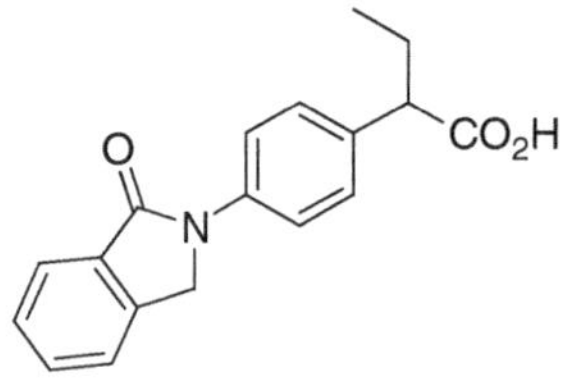	$t_{1/2}$	6±2 h	*V*	0.18–0.21	INDOBUFEN
	F	>75%	CL	0.32	Antithrombotic
	pb	>99%	Mwt	295.3	Metabolism (rat): aromatic 5- and 6-hydroxylation of indolone,
	ur	13%	PSA	57.6 Å^2	then conjugation
			log *P*	3.30	
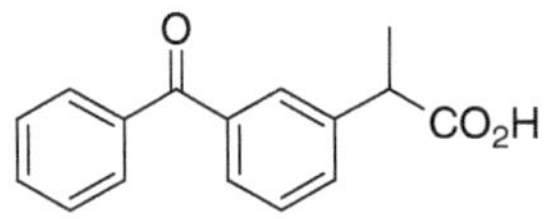	$t_{1/2}$	1.8±0.3 h	*V*	0.15±0.03	KETOPROFEN
	F	100%	CL	1.2±0.3	Antiinflammatory, analgesic; racemate
	pb	99%	Mwt	254.3	Metabolism: 3- and 4-hydroxylation of Ph, reduction
	ur	<1%	PSA	54.4 Å^2	of ketone to alcohol,
			log *P*	2.91	glucuronidation
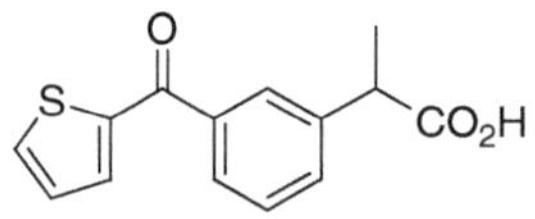	$t_{1/2}$	2.1 h	*V*	0.04	SUPROFEN
	F	92%	CL	0.76	Antiinflammatory
	pb	>99%	Mwt	260.3	Metabolism: reduction of ketone to alcohol, hydroxylation of
	ur	–	PSA	82.6 Å^2	thiophene, conjugation with glucuronic acid or taurine, oxidative degradation of
			log *P*	2.18	thiophene to 2-(4-carboxyphenyl)-propionic acid; withdrawn 1987 for causing flank pain
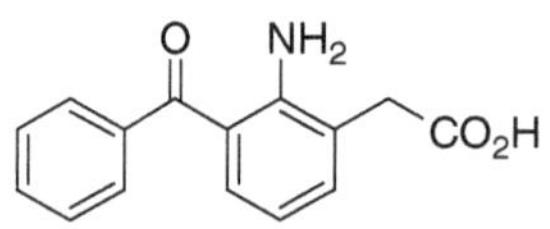	$t_{1/2}$	<1 h*	*V*	–	AMFENAC
	F	–	CL	–	*monkey
	pb	–	Mwt	255.3	Antiinflammatory
	ur	–	PSA	80.4 Å^2	Metabolism: lactam formation,
			log *P*	2.23	glucuronidation, hydroxylation
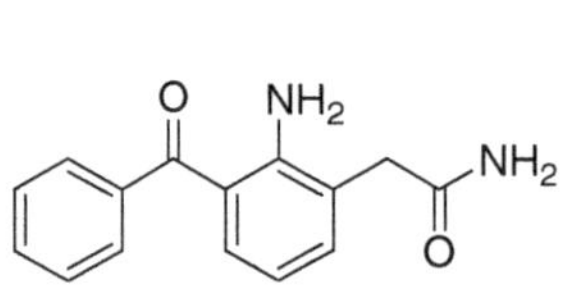	$t_{1/2}$	<1 h*	*V*	–	NEPAFENAC
	F	6% (rat)	CL	–	*monkey
	pb	99%	Mwt	254.3	Prodrug of amfenac (acid),
	ur	0%	PSA	86.2 Å^2	ophthalmic antiinflammatory
			log *P*	1.04	Metabolism: hydrolysis of amide, para-hydroxylation of aniline
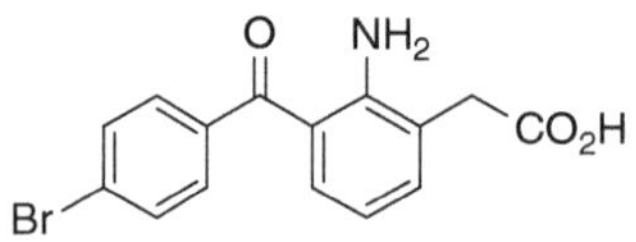	$t_{1/2}$	0.5–4.0 h	*V*	0.1	BROMFENAC
	F	67±20%	CL	1–2	Antiinflammatory
	pb	>99.8%	Mwt	334.2	Metabolism: N-glucosidation,
	ur	Low	PSA	80.4 Å^2	acylglucuronidation, lactam
			log *P*	2.02	formation; withdrawn in 1998 because of hepatotoxicity

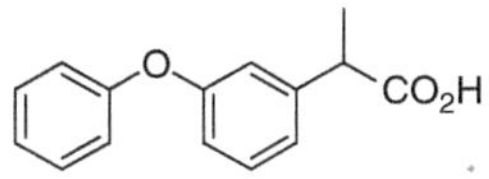	$t\frac{1}{2}$ 2.5±0.5 h *F* 85% pb >99% ur 30%	*V* 0.1 CL 0.13–0.60 Mwt 242.3 PSA 46.5 Å^2 log *P* 3.72	FENOPROFEN Antiinflammatory Metabolites: glucuronide, 4-hydroxyphenoxy
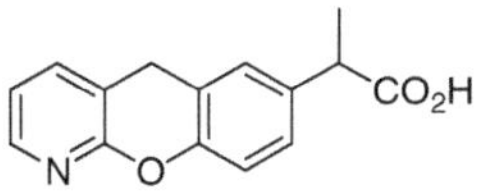	$t\frac{1}{2}$ 5 h *F* – pb – ur 1.3%	*V* – CL – Mwt 255.3 PSA 59.4 Å^2 log *P* 1.34	PRANOPROFEN Antiinflammatory Metabolism: glucuronidation
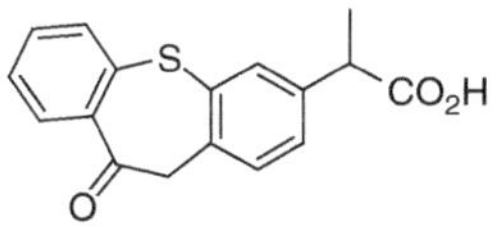	$t\frac{1}{2}$ 9.6 h* *F* 84%* pb 99.6% ur 3%	*V* 1.2* CL 1.4* Mwt 298.4 PSA 79.7 Å^2 log *P* 4.25	ZALTOPROFEN *rat Antiinflammatory Metabolism: glucuronidation, S-oxidation to sulfoxide
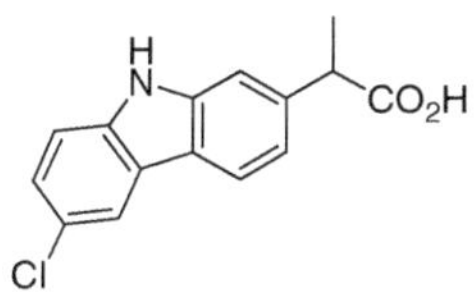	$t\frac{1}{2}$ 21±6 h *F* 90% pb 99% ur 5–12%	*V* Small CL – Mwt 273.7 PSA 53.1 Å^2 log *P* 3.84	CARPROFEN Antiinflammatory Metabolism: glucuronidation
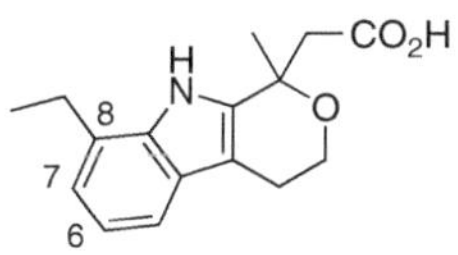	$t\frac{1}{2}$ 7±4 h *F* >80% pb 99% ur <1%	*V/F* 0.36±0.13 CL/*F* 0.78±0.27 Mwt 287.4 PSA 62.3 Å^2 log *P* 3.42	ETODOLAC Antiinflammatory Racemate Metabolites: glucuronide; 6- and 7-hydroxy, 1′-hydroxyethyl
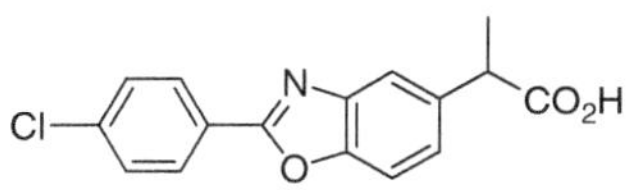	$t\frac{1}{2}$ 27–35 h *F* >95% pb >99.5% ur –	*V* 0.15 CL – Mwt 301.7 PSA 63.3 Å^2 log *P* 3.72	BENOXAPROFEN Antiinflammatory Metabolism: glucuronidation (to acylglucuronide, $t\frac{1}{2}$ 100 h); half-life is longer in elderly patients
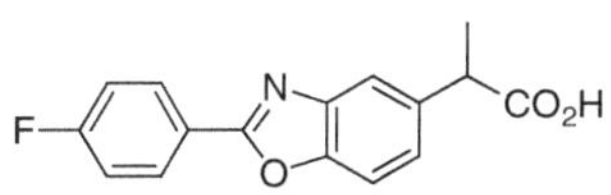	$t\frac{1}{2}$ 8±2 h *F* – pb – ur 10%	*V* 0.13–0.18* CL 0.4 Mwt 285.3 PSA 63.3 Å^2 log *P* 3.18	FLUNOXAPROFEN *monkey Antiinflammatory Metabolism: acyl glucuronidation
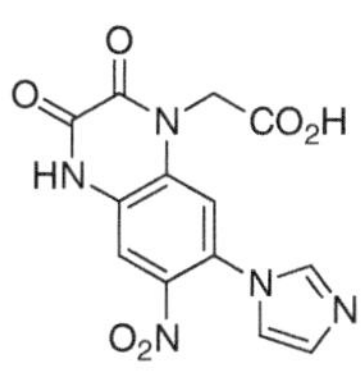	$t\frac{1}{2}$ 0.8 h (iv) *F* – pb – ur 95% (iv)	*V* 0.19 CL 5.3 Mwt 331.2 PSA 145 Å^2 log *P* 0.12	ZONAMPANEL NMDA antagonist Metabolism: reduction of NO_2 to hydroxylamine, then to amine

(continued overleaf)

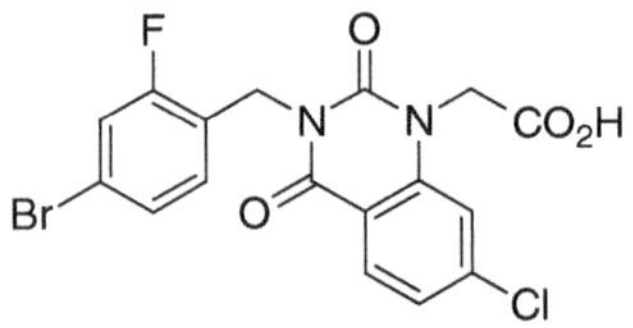	$t^1/_2$ 8 h F 65% (dog) pb >95% ur 27%	V – CL – Mwt 441.6 PSA 77.9 Å^2 log P 4.36	ZENARESTAT Aldose reductase inhibitor Metabolism: glucuronidation
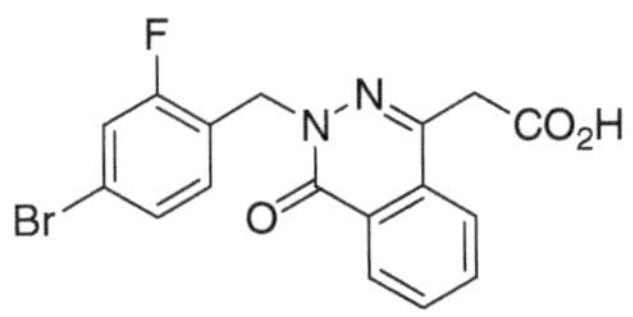	$t^1/_2$ 13 h F 90% pb – ur –	V – CL – Mwt 391.2 PSA 70.0 Å^2 log P 2.70	PONALRESTAT Aldose reductase inhibitor Metabolism: glucuronidation
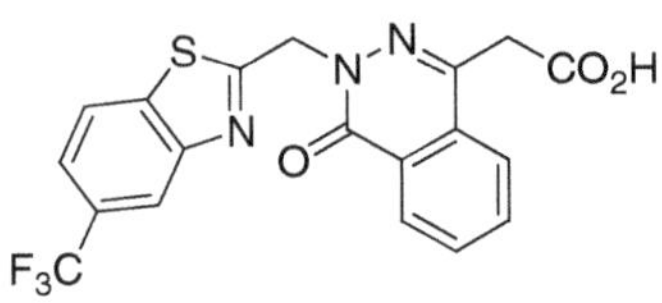	$t^1/_2$ 27–30 h F >45% pb >99% ur 35–45%	V/F 0.2 CL 0.08 Mwt 419.4 PSA 111 Å^2 log P 3.01	ZOPOLRESTAT Aldose reductase inhibitor for treatment of diabetic complications Metabolism: glucuronidation, N-dealkylation (to benzothiazole-2-carboxylic acid), aromatic hydroxylation, and dihydroxylation
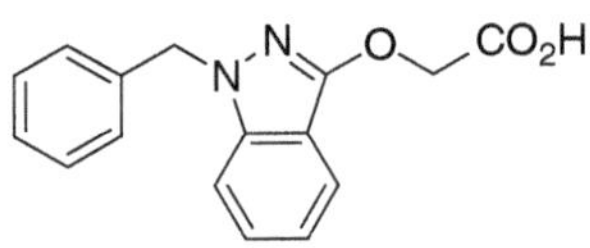	$t^1/_2$ 3–4 h F – pb >99% ur <10%	V 0.16 CL 0.6 Mwt 282.3 PSA 64.3 Å^2 log P 2.83	BENDAZAC Antiinflammatory Metabolites: 5-hydroxy, glucuronides
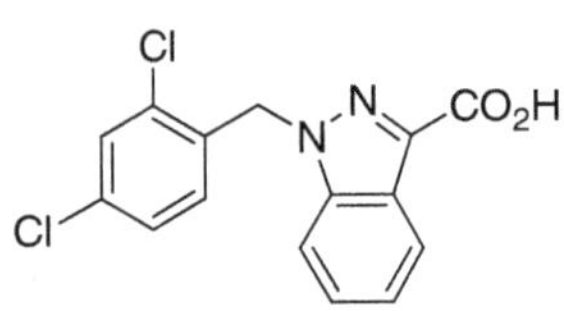	$t^1/_2$ 3–12 h F 50%* pb – ur High	V – CL – Mwt 321.2 PSA 55.1 Å^2 log P 4.63	LONIDAMINE *monkey Antineoplastic Metabolite: acyl glucuronide
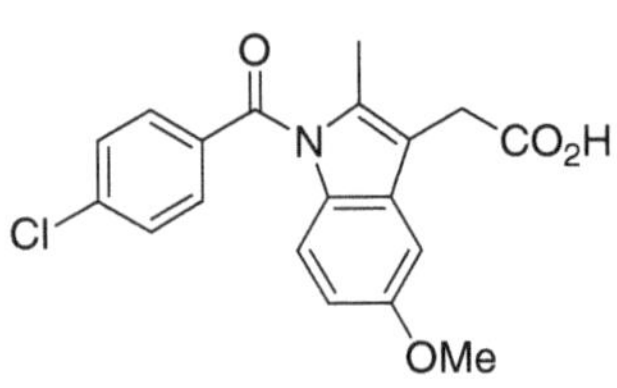	$t^1/_2$ 2.4±0.4 h F 98±21% pb 90% ur 15±8%	V 0.29±0.04 CL 1.4±0.2 Mwt 357.8 PSA 68.5 Å^2 log P 4.25	INDOMETHACIN Antiinflammatory Metabolism: O-demethylation, N-deacylation, glucuronidation; hydroxamic acid of indomethacin (oxametacine): $t^1/_2$ 11±1 h Metabolism: O-demethylation, glucuronidation, reductive N–O bond cleavage, debenzoylation; no hydrolysis to indomethacin

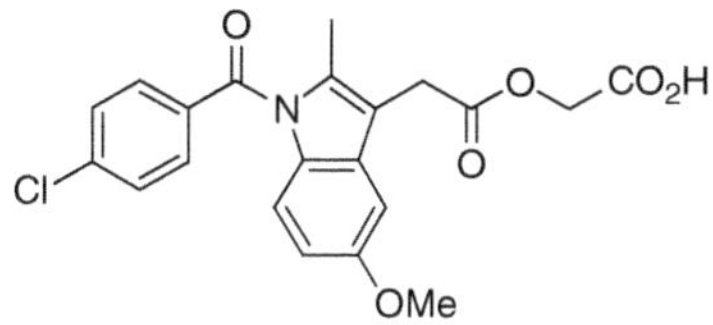	$t_{1/2}$ 4.5±2.8 h F 66–100% pb 88% ur –	V 1 CL – Mwt 415.8 PSA 94.8 Å^2 log P 4.49	ACEMETACIN Antiinflammatory Metabolism: glucuronidation, ester hydrolysis to indomethacin, O-demethylation
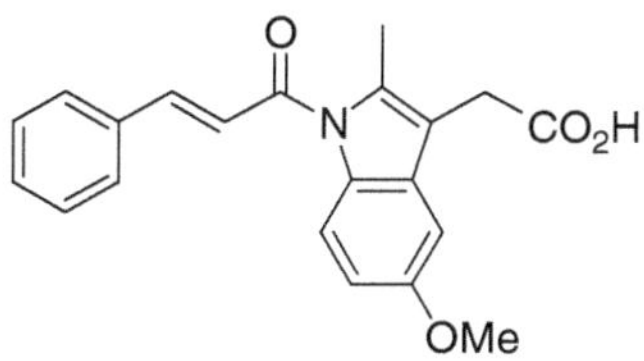	$t_{1/2}$ 3.8±0.2 h F – pb – ur –	V 0.28±0.03 CL 0.85±0.08 Mwt 349.4 PSA 68.5 Å^2 log P 3.91	CINMETACIN Antiinflammatory Metabolism: hydrogenation of olefin, O-demethylation, N-deacylation
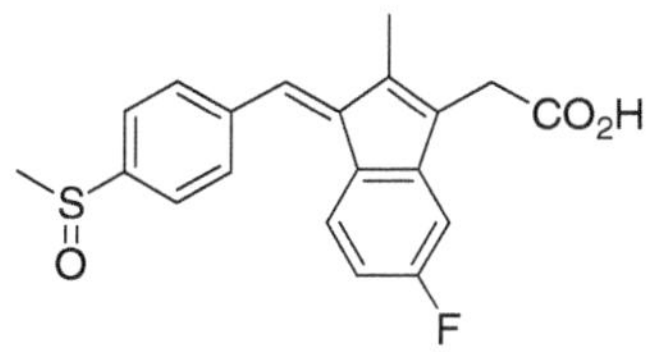	$t_{1/2}$ 15±4 h* F 90%* pb 94±1%* ur Negligible	V/F 2 CL 1.5±0.9* Mwt 356.4 PSA 73.6 Å^2 log P 2.55	SULINDAC *sulfide on oral dosing of sulindac Antiinflammatory Metabolism: oxidation to sulfone, reduction to sulfide
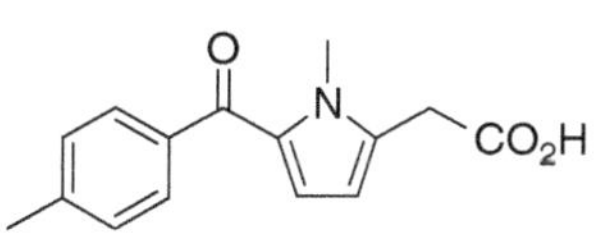	$t_{1/2}$ 4.9±0.3 h F >90% pb >99% ur 7±3%	V/F 0.54±0.07 CL/F 1.3±0.3 Mwt 257.3 PSA 59.3 Å^2 log P 2.68	TOLMETIN Antiinflammatory Metabolism: glucuronidation
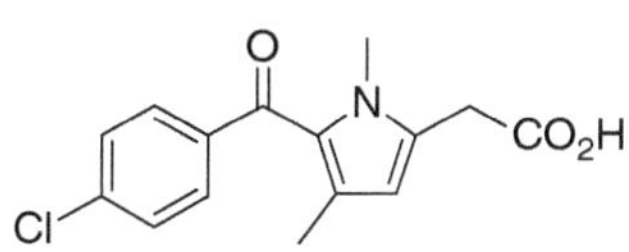	$t_{1/2}$ 4–10 h F 100% pb 98–99% ur 5–20%	V – CL 3.7–4.6 Mwt 291.7 PSA 59.3 Å^2 log P 3.36	ZOMEPIRAC Antiinflammatory Metabolism: glucuronidation, hydroxylation of CCH_3, cleavage to 4-chlorobenzoic acid; withdrawn in 1983 for causing anaphylaxis
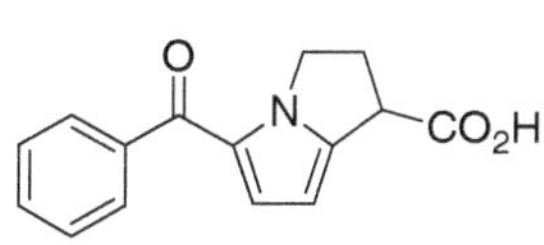	$t_{1/2}$ 5.3±1.2 h F 90±10% pb 99% ur 5–10%	V 0.21±0.04 CL 0.50±0.15 Mwt 255.3 PSA 59.3 Å^2 log P 2.68	KETOROLAC Antiinflammatory, analgesic Metabolism: glucuronidation, 4-hydroxylation of Ph
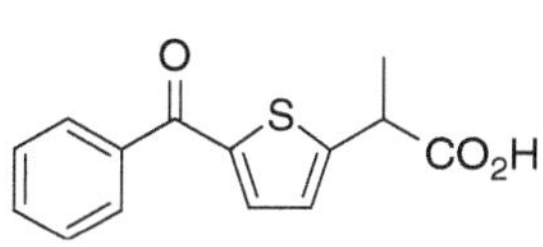	$t_{1/2}$ 2±1 h F 95±5% pb 98–99% ur <5%	V 0.04–0.17 CL 0.5–1.4 Mwt 260.3 PSA 82.6 Å^2 log P 2.57	TIAPROFENIC ACID Antiinflammatory Metabolism: reduction of ketone to alcohol, 4-hydroxylation of Ph

(*continued overleaf*)

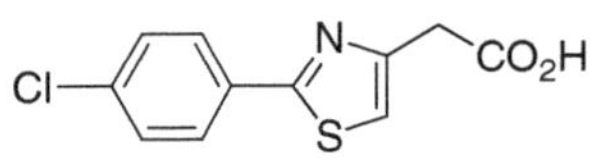	*t*½ 26–31 h *F* – pb – ur –	*V* – CL – Mwt 253.7 PSA 78.4 Å^2 log *P* 3.11	FENCLOZIC ACID, MYALEX Antiinflammatory, hepatotoxic
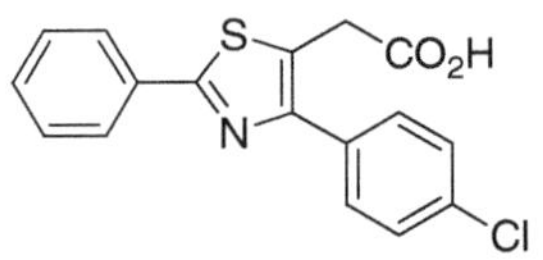	*t*½ 4 h *F* – pb 99.1% ur –	*V* – CL 8.6 Mwt 329.8 PSA 78.4 Å^2 log *P* 5.05	FENTIAZAC Antiinflammatory Metabolism: 4-hydroxylation of Ph
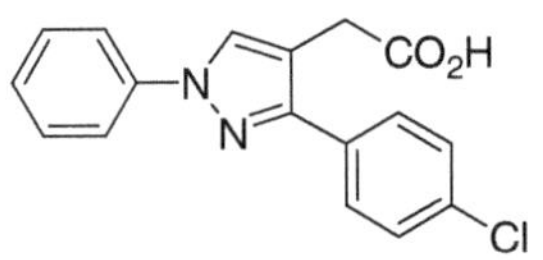	*t*½ 6–12 h *F* – pb 99% ur 2%	*V* – CL – Mwt 312.8 PSA 50.2 Å^2 log *P* 4.13	LONAZOLAC Antiinflammatory Metabolism: aromatic 4-hydroxylation, then sulfation
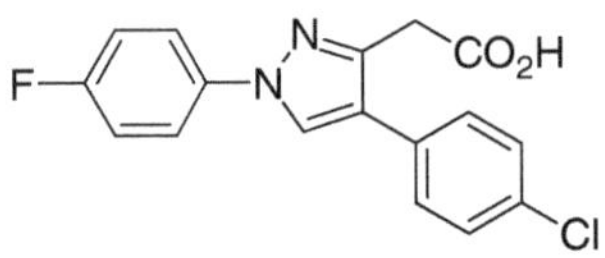	*t*½ 17±1 h *F* 100% pb 99% ur 10%	*V* – CL 0.25±0.32 Mwt 330.7 PSA 55.1 Å^2 log *P* 3.84	PIRAZOLAC Antiinflammatory Metabolism: glucuronidation
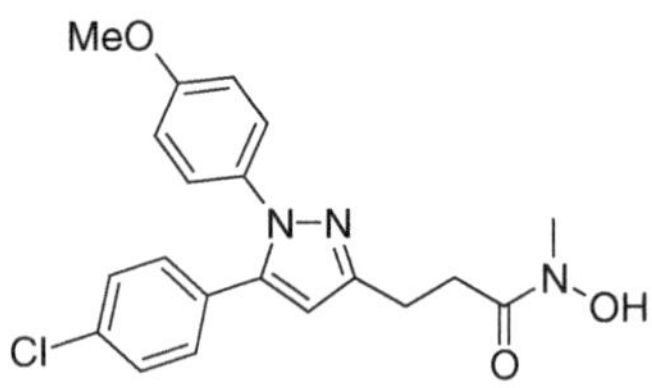	*t*½ 3.3 h *F* pb >98% ur –	*V* – CL – Mwt 385.8 PSA 67.6 Å^2 log *P* 3.16	TEPOXALIN Antiinflammatory Metabolite: carboxylic acid (*t*½ 12–14 h)
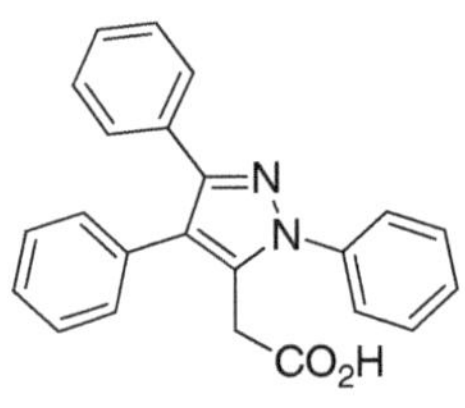	*t*½ 2.8±0.8 h *F* – pb 99% ur 14±5%	*V* – CL – Mwt 354.4 PSA 55.1 Å^2 log *P* 3.56	ISOFEZOLAC Antiinflammatory
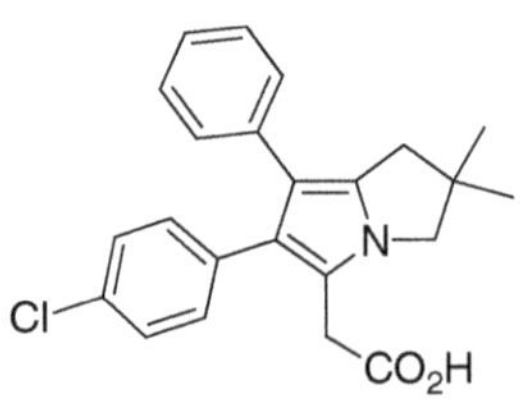	*t*½ 9–11 h *F* High pb – ur –	*V* – CL – Mwt 379.9 PSA 37.3 Å^2 log *P* 6.73	LICOFELONE Antiinflammatory Metabolism: methyl and phenyl hydroxylation, glucuronidation

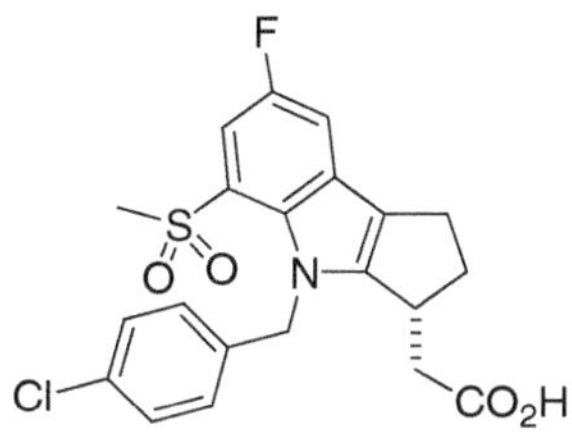	$t_{1/2}$ 17 h F 71% pb >99% ur 5%	V 1.1 CL – Mwt 435.9 PSA 84.8 Å^2 log P 3.82	LAROPIPRANT Prostaglandin (DP_1) antagonist, vasoconstrictor, in combination with niacin as antilipemic Metabolism: glucuronidation
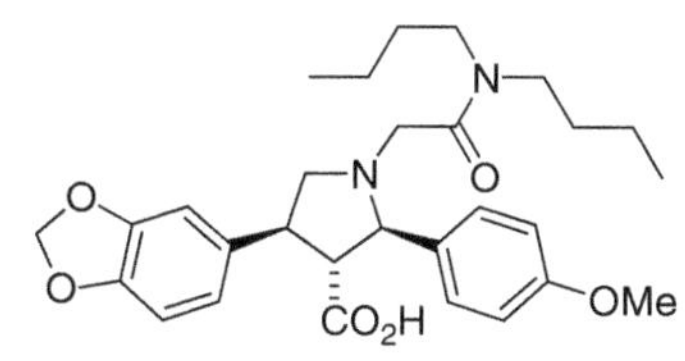	$t_{1/2}$ 20–25 h F – pb 99% ur <0.1%	V 6 CL 3–7 Mwt 510.6 PSA 88.5 Å^2 log P 5.80	ATRASENTAN Antineoplastic Metabolism: glucuronidation, N-debutylation, O-demethylation
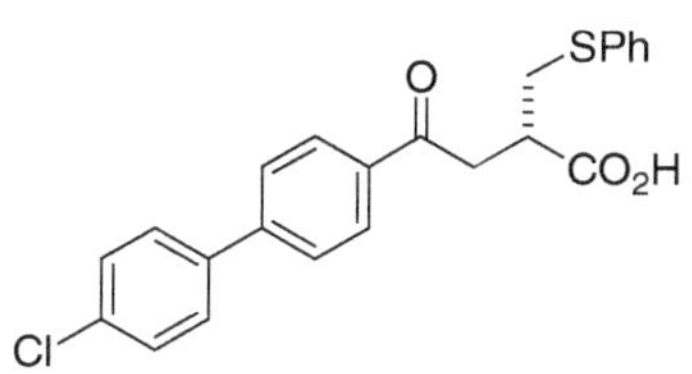	$t_{1/2}$ 61–90 h F – pb 99.99% ur –	V – CL – Mwt 410.9 PSA 79.7 Å^2 log P 6.51	TANOMASTAT, BAY-129566 Matrix metalloproteinase inhibitor, antineoplastic
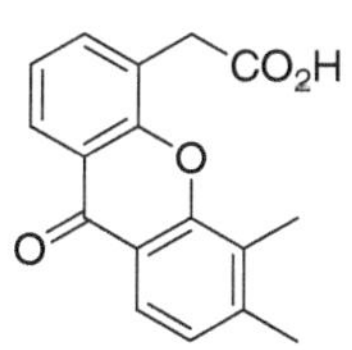	$t_{1/2}$ 8.1±4.3 h F 73%* pb 99% ur –	V – CL – Mwt 282.3 PSA 63.6 Å^2 log P 3.60	VADIMEZAN, ASA404, DMXAA *mice Antineoplastic Metabolism: glucuronidation
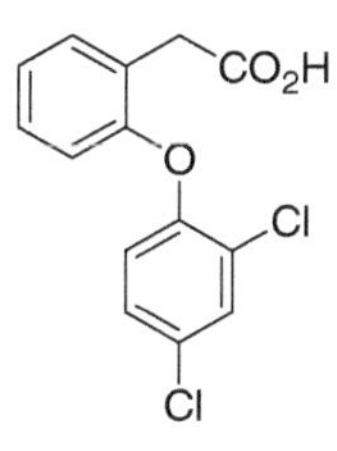	$t_{1/2}$ 29±9 h F 100% pb – ur –	V 0.20–0.25 CL – Mwt 297.1 PSA 46.5 Å^2 log P 4.59	FENCLOFENAC Antiinflammatory
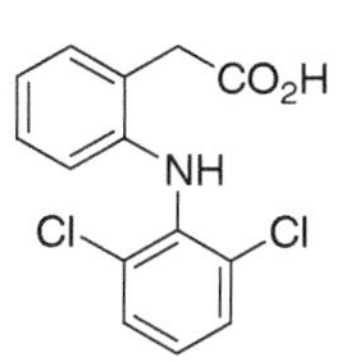	$t_{1/2}$ 1.1±0.2 h F 54±2% pb >99.5% ur <1%	V 0.17±0.11 CL 4.2±0.9 Mwt 296.2 PSA 49.3 Å^2 log P 4.55	DICLOFENAC Antiinflammatory Metabolized by CYP2C9 to 4′-hydroxy ($t_{1/2}$ 4.3 h); further metabolism: aromatic 5- and 3′-hydroxylation, then 4′-O-methylation

(*continued overleaf*)

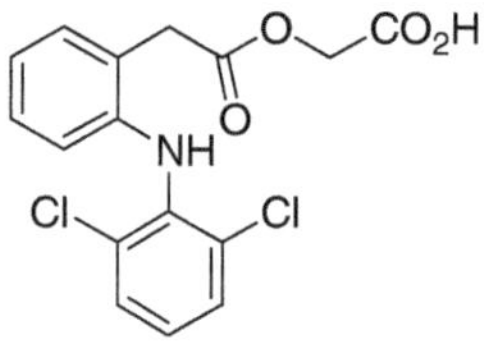	$t_{1/2}$ 4.9±1.4 h *F* – pb >99% ur <1%	*V* 0.36 CL – Mwt 354.2 PSA 75.6 Å^2 log *P* 3.81	ACECLOFENAC Antiinflammatory Metabolism: ester hydrolysis, aromatic 4′-hydroxylation of dichlorophenyl, then conjugation
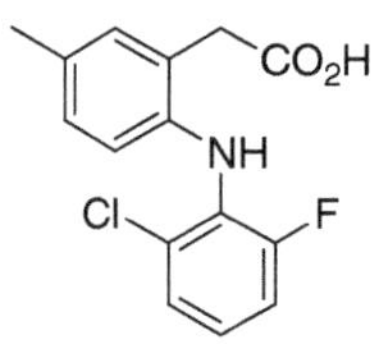	$t_{1/2}$ 4–7 h *F* 74% pb 98% ur 3.3%	*V* 0.14 CL 2.1 Mwt 293.7 PSA 49.3 Å^2 log *P* 3.90	LUMIRACOXIB Antiinflammatory Metabolism: oxidation of CH_3 to CO_2H, 4′-hydroxylation of dihaloarene, then glucuronidation; withdrawn in 2007 because of hepatotoxicity
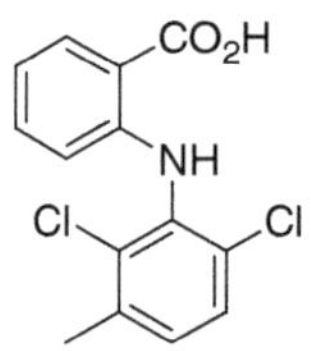	$t_{1/2}$ 3±1 h *F* 65±3%* pb 99.8% ur 3±1%	*V* 0.14–0.67 CL 3.2 Mwt 296.2 PSA 49.3 Å^2 log *P* 5.44	MECLOFENAMIC ACID *sheep Antiinflammatory Metabolism: glucuronidation, oxidation of CH_3 to CH_2OH, then to CO_2H
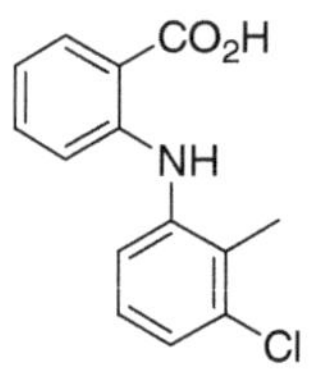	$t_{1/2}$ 2.5±0.6 h *F* 60% pb 99.7% ur <8%	*V* 0.16 CL 2.4±0.2 Mwt 261.7 PSA 49.3 Å^2 log *P* 5.47	TOLFENAMIC ACID Antiinflammatory Metabolism, aromatic hydroxylation (ortho to Cl), oxidation of CH_3 to CHO
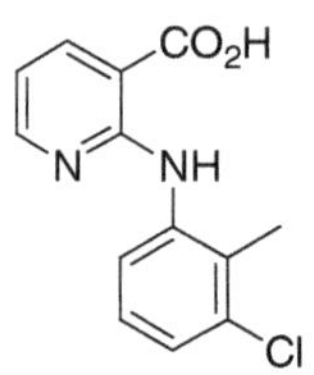	$t_{1/2}$ 1.5±0.1 h *F* – pb – ur –	*V* 0.11 CL – Mwt 262.7 PSA 62.2 Å^2 log *P* 4.71	CLONIXIN Analgesic Metabolism: aromatic 4′-(ortho to Cl)- and 5-hydroxylation, hydroxylation of CH_3
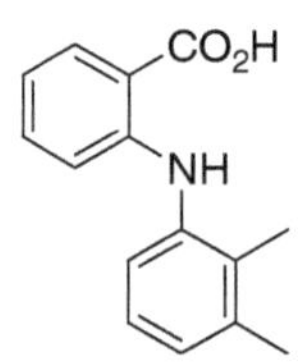	$t_{1/2}$ 3±1 h *F* Low pb 99% ur <6%	*V* 1.1 CL 5.4 Mwt 241.3 PSA 49.3 Å^2 log *P* 4.83	MEFENAMIC ACID Antiinflammatory Metabolism (by CYP2C9) to 3-hydroxymethyl mefenamic acid, then to 3-carboxymefenamic acid

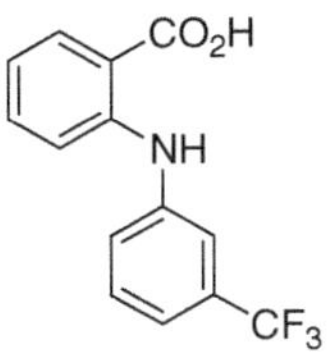	$t\frac{1}{2}$ 9 h *F* – pb >90% ur <1%	*V* – CL 1.2–2.3 Mwt 281.2 PSA 49.3 Å^2 log *P* 5.22	FLUFENAMIC ACID Antiinflammatory Prodrug: etofenamate (monoester of diethyleneglycol) for topical application
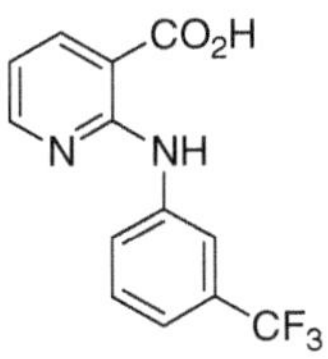	$t\frac{1}{2}$ 2 h *F* – pb 90–97% ur 5%	*V* 0.12 CL 0.7 Mwt 282.2 PSA 62.2 Å^2 log *P* 4.46	NIFLUMIC ACID Antiinflammatory Metabolism: glucuronidation, aromatic 4′- and 5-hydroxylation, then sulfation and glucuronidation
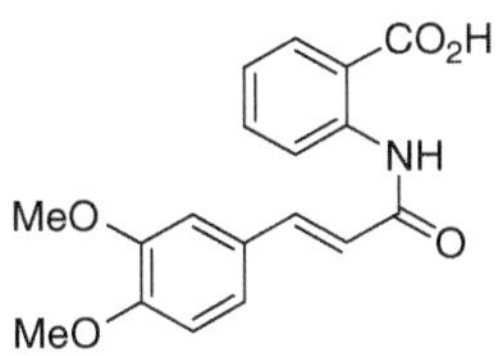	$t\frac{1}{2}$ 4–9 h *F* 0.2%* pb – ur –	*V*/*F* 0.09 CL/*F* 0.13 Mwt 327.3 PSA 84.9 Å^2 log *P* 3.17	TRANILAST *transdermal Antiallergic Metabolism: glucuronidation, 4-O-demethylation
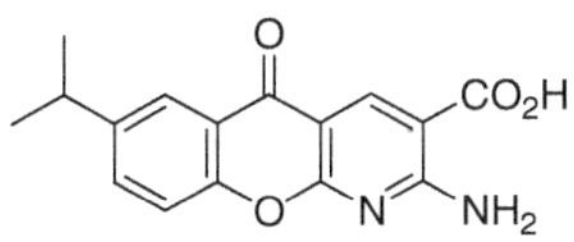	$t\frac{1}{2}$ 3.5±1.1 h *F* – pb – ur 17%	*V* – CL – Mwt 298.3 PSA 103 Å^2 log *P* 3.74	AMLEXANOX Antiallergic, antiasthmatic Metabolism: hydroxylation of *CH*Me$_2$, then conjugation
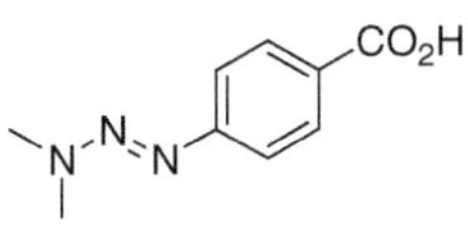	$t\frac{1}{2}$ 2.5 h (iv) *F* – pb – ur –	*V* 0.25 CL 1.3 Mwt 193.2 PSA 65.3 Å^2 log *P* 2.59	CB 10-277 Antineoplastic
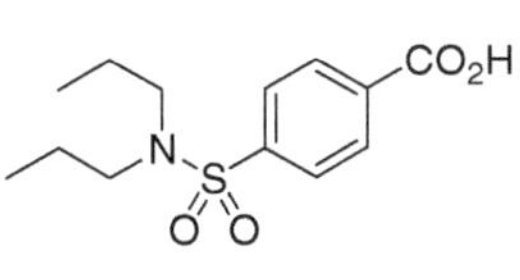	$t\frac{1}{2}$ 4–12 h *F* 100% pb 90±5% ur 1.2±0.2%	*V* 0.17±0.03 CL 0.4 Mwt 285.4 PSA 83.1 Å^2 log *P* 2.51	PROBENECID Uricosuric Metabolism: acyl glucuronidation, hydroxylation at *CH*$_2$Me and Me, N-depropylation, oxidation of CH_3 to CO_2H
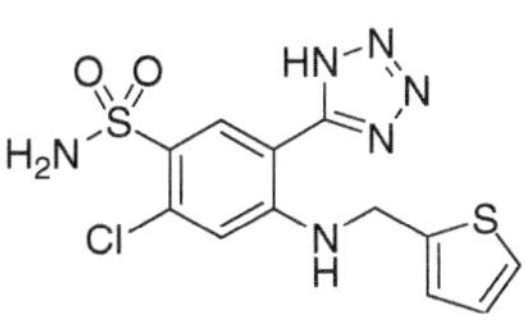	$t\frac{1}{2}$ 2–3 h *F* 15–20% pb >95% ur 27%	*V* 0.26 CL 1.7 Mwt 370.8 PSA 163 Å^2 log *P* 1.40	AZOSEMIDE Diuretic Metabolism: N-dealkylation

(continued overleaf)

Structure					Drug
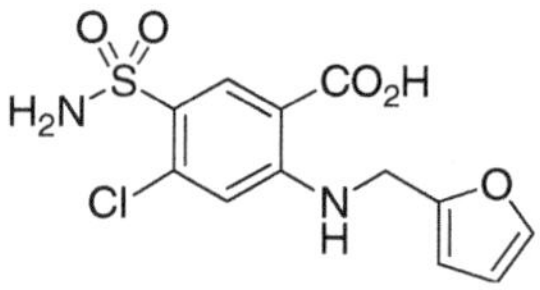	$t_{1/2}$	0.3–3.4 h	V	0.11±0.02	FUROSEMIDE Diuretic Metabolism: acyl glucuronidation, N-dealkylation
	F	61±17%	CL	2.0±0.4	
	pb	98.8±0.2%	Mwt	330.7	
	ur	66±7%	PSA	131 Å^2	
			log P	2.30	
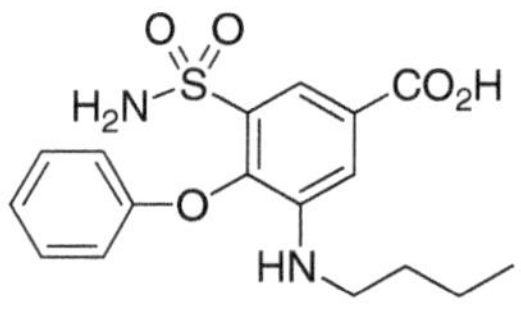	$t_{1/2}$	0.8±0.2 h	V	0.13±0.03	BUMETANIDE Diuretic Metabolism: glucuronidation, β-, γ-, and δ-hydroxylation of butyl, N-debutylation, oxidation of CH_3 to CO_2H
	F	59–89%	CL	2.6±0.5	
	pb	99%	Mwt	364.4	
	ur	62±20%	PSA	127 Å^2	
			log P	2.88	
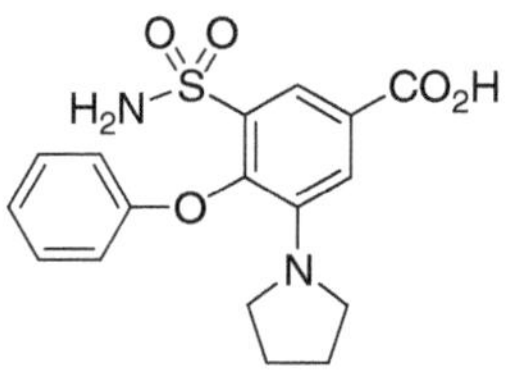	$t_{1/2}$	1.0±0.5 h	V	0.25–0.30	PIRETANIDE Diuretic Metabolism: cleavage of pyrrolidine to aminobutanol
	F	80%	CL	3.8±0.8 (iv)	
	pb	94%	Mwt	362.4	
	ur	50%	PSA	118 Å^2	
			log P	2.30	
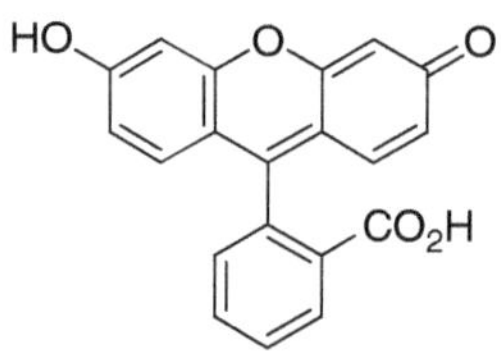	$t_{1/2}$	1.6 h (iv)	V	0.14	FLUORESCEIN Diagnostic aid Metabolism: glucuronidation
	F	–	CL	1.3	
	pb	85–90%	Mwt	332.3	
	ur	–	PSA	76.0 Å^2	
			log P	2.68	
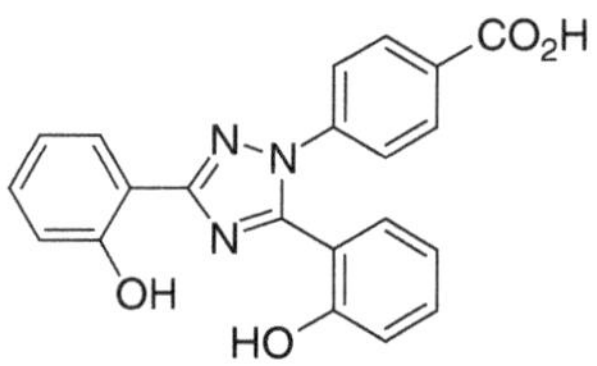	$t_{1/2}$	8–16 h	V	0.2	DEFERASIROX, ICL 670 Iron chelator Metabolism: glucuronidation
	F	74±13%	CL	0.9±0.2	
	pb	99%	Mwt	373.4	
	ur	2%	PSA	108 Å^2	
			log P	5.15	
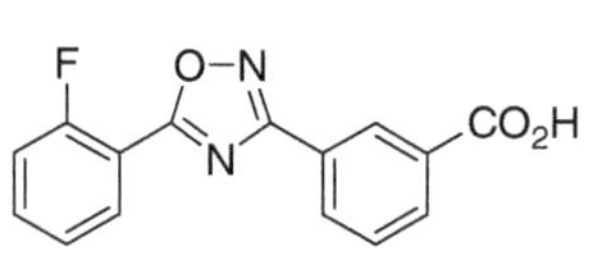	$t_{1/2}$	3–6 h	V	–	ATALUREN Nonsense mutation suppressor for treatment of cystic fibrosis and muscular dystrophy
	F	High	CL	–	
	pb	–	Mwt	284.2	
	ur	–	PSA	76.2 Å^2	
			log P	3.73	
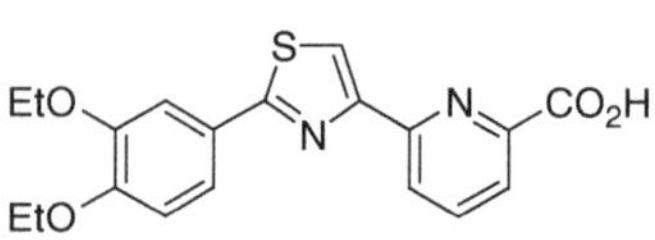	$t_{1/2}$	24 h	V	0.18	TETOMILAST, OPC-6535 PDE_4 inhibitor
	F	100%	CL	–	
	pb	>99.5%	Mwt	370.4	
	ur	–	PSA	110 Å^2	
			log P	3.48	

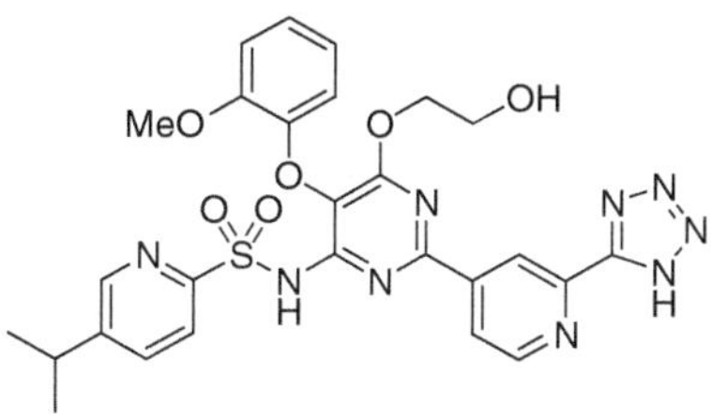

$t\frac{1}{2}$	3.2 h (iv)	V	0.3
F	–	CL	8–9
pb	>99%	Mwt	605.6
ur	2% (iv)	PSA	208 Å^2
		log P	0.88

TEZOSENTAN
Endothelin antagonist for treatment of heart failure
Metabolism: hydroxylation at $CHMe_2$

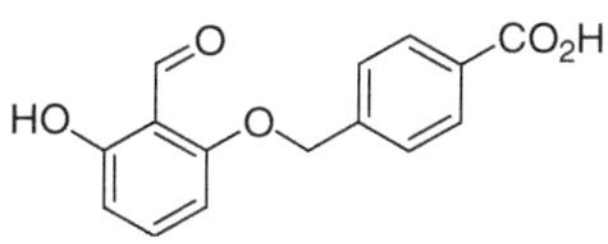

$t\frac{1}{2}$	195±95 h	V	0.3
F	–	CL	0.03
pb	>98.5%	Mwt	272.3
ur	–	PSA	83.8 Å^2
		log P	2.75

TUCARESOL
Antisickling agent, increases oxygen affinity of hemoglobin

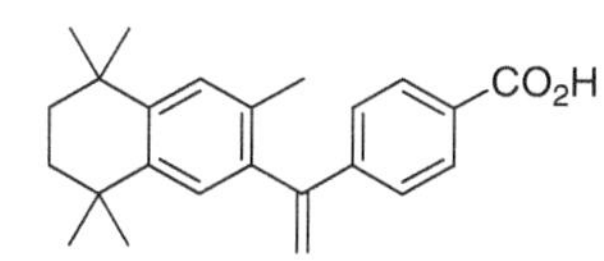

$t\frac{1}{2}$	7 h	V	1
F	8% (dog)	CL	–
pb	>99%	Mwt	348.5
ur	<1%	PSA	37.3 Å^2
		log P	6.90

BEXAROTENE
Antineoplastic
Metabolism: hydroxylation of CH_2CH_2, then oxidation of alcohol to ketone

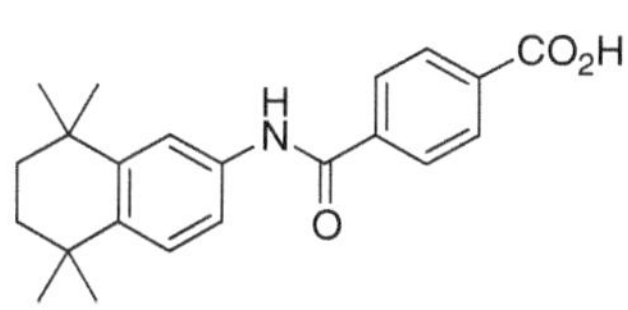

$t\frac{1}{2}$	3–5 h	V	–
F	–	CL	–
pb	>99%	Mwt	351.4
ur	–	PSA	66.4 Å^2
		log P	5.52

TAMIBAROTENE
Retinoic acid-α agonist, antileukemic
Metabolism: oxidation of CH_2 to ketone

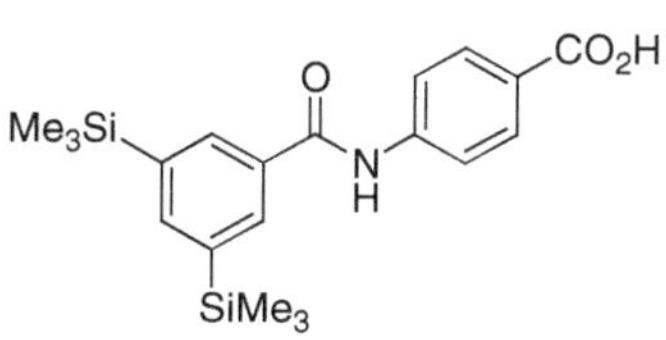

$t\frac{1}{2}$	6–8 h	V	0.7–0.9
F	62–94%*	CL	1.3
pb	–	Mwt	385.6
ur	–	PSA	66.4 Å^2
		log P	7.73

TAC-101
*rat, dog
Retinoic acid analog, antineoplastic

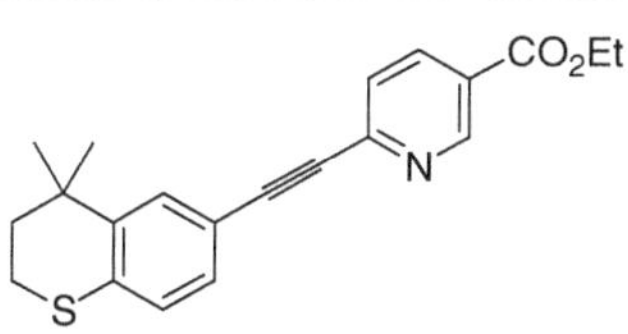

$t\frac{1}{2}$	14–20 h*	V	2*
F	1–5%**	CL	1.6*
pb	99%	Mwt	351.5
ur	0%	PSA	64.5 Å^2
		log P	5.99

TAZAROTENE
*acid on oral dosing of ester
**from topical dosing
Topical antipsoriatic
Metabolism: ester hydrolysis, oxidation to sulfoxide

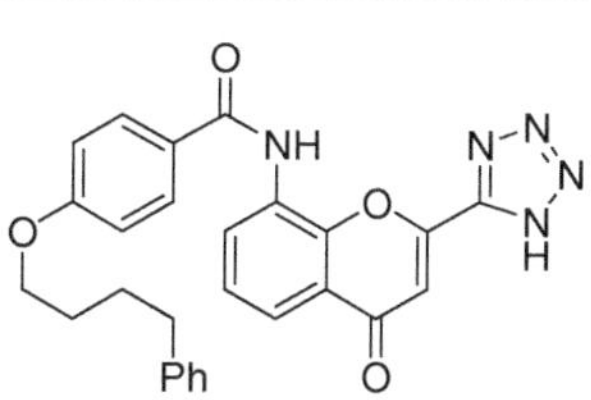

$t\frac{1}{2}$	1.7–9.3 h	V/F	1.53
F	<20%	CL/F	19
pb	>99%	Mwt	481.5
ur	Low	PSA	119 Å^2
		log P	4.20

PRANLUKAST
Antiasthmatic
Metabolism: phenyl 4-hydroxylation, then O-sulfation, benzylic hydroxylation, then oxidation to ketone

(*continued overleaf*)

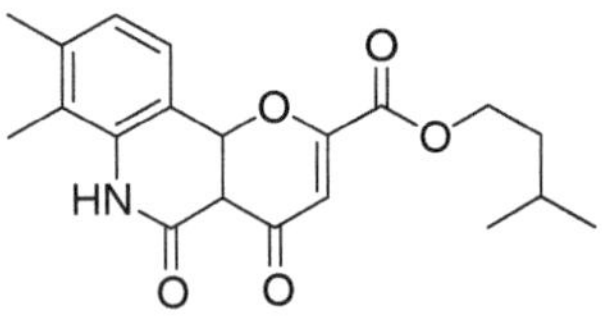	$t^{1/2}$ 6–10 h* F 5–80%* pb 97%* ur 7.4%*	V – CL – Mwt 355.4 PSA 81.7 Å^2 log P 4.89	REPIRINAST *acid on oral dosing of ester Antiallergic Ester not detectable in plasma on oral dosing
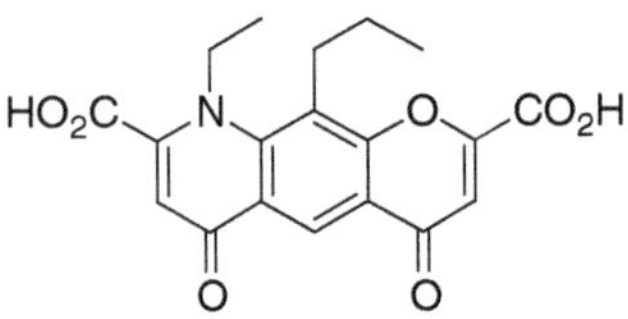	$t^{1/2}$ 1.1–3.3 h F 2–3%* pb 89% ur 64–81%	V 0.43 CL 10.2 Mwt 371.3 PSA 121 Å^2 log P 1.30	NEDOCROMIL *pulmonal: 6% Antiallergic, antiasthmatic
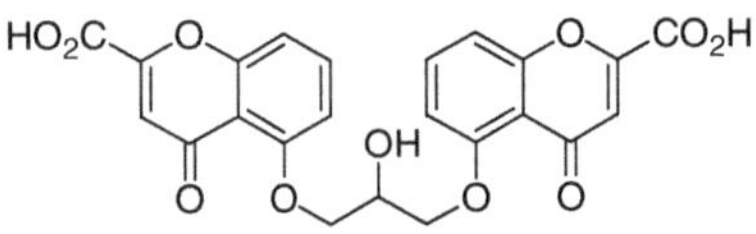	$t^{1/2}$ 1.0–1.5 h F 1–2% pb 57–69% ur 50%	V 0.32 CL 8 Mwt 468.4 PSA 166 Å^2 log P 2.00	CROMOGLICIC ACID, CROMOLYN Antiasthmatic, antiallergic
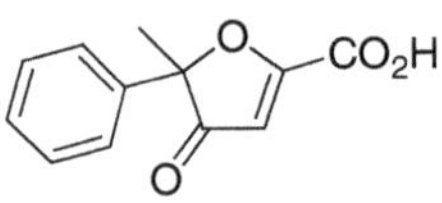	$t^{1/2}$ 1.4–1.6 h F – pb 88% ur 78–95%	V 0.5 CL 2.6 Mwt 218.2 PSA 63.6 Å^2 log P 2.04	ACIFRAN Antihyperlipoproteinemic
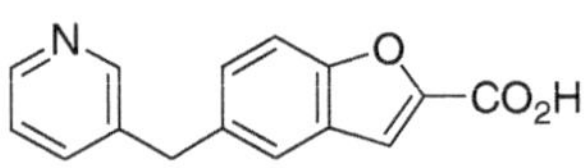	$t^{1/2}$ 2 h F 75–80%* pb – ur 62–78%	V – CL – Mwt 253.3 PSA 63.3 Å^2 log P 2.76	FUREGRELATE *dog Thromboxane A_2 synthase inhibitor
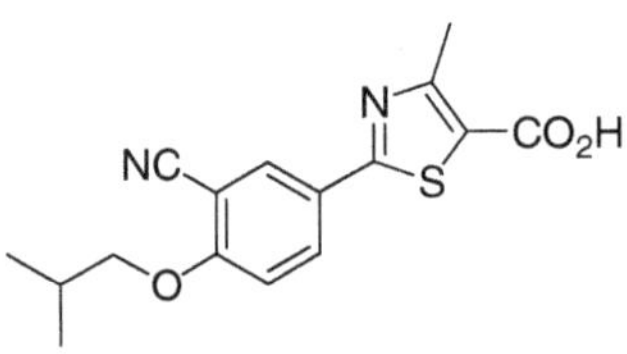	$t^{1/2}$ 5–8 h F 84% pb 99.2% ur 3%	V 0.8 CL 2.6–3.1 Mwt 316.4 PSA 111 Å^2 log P 4.87	FEBUXOSTAT Xanthine oxidase inhibitor for treatment of gout Metabolism: oxidation of isobutyl, hydroxylation, glucuronidation
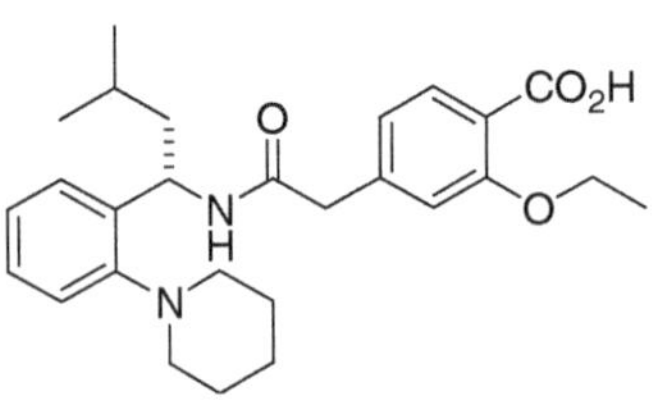	$t^{1/2}$ 0.5–1.0 h F 56–63% pb >98% ur <2%	V 0.3–0.9 CL 7.8 Mwt 452.6 PSA 78.9 Å^2 log P 4.69	REPAGLINIDE Antidiabetic Metabolism: glucuronidation, piperidine 3-hydroxylation, O-deethylation, 2-hydroxylation of isopropyl, piperidine cleavage to 5-aminovaleric acid, then aniline N-dealkylation

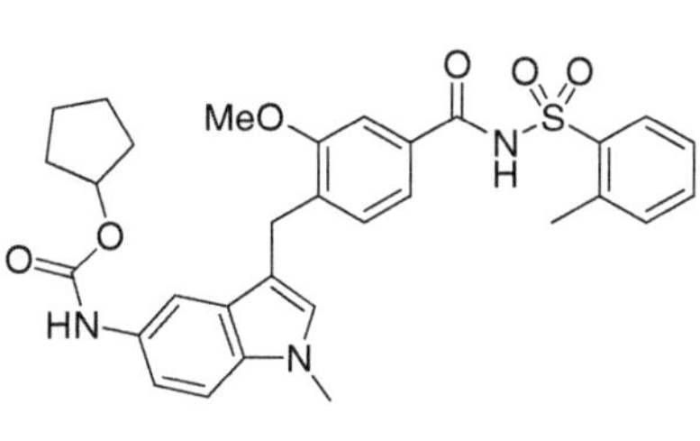	$t_{1/2}$ 8–16 h F – pb >99% ur <10%	V/F 1.1 CL/F 5.1 Mwt 575.7 PSA 124 Å^2 log P 5.04	ZAFIRLUKAST Antiasthmatic, leukotriene D_4 and E_4 antagonist, CYP2C9 substrate, and inhibitor Metabolism (animals): N-demethylation, hydrolysis of carbamate, then N-acetylation of aniline, hydroxylation of cyclopentyl group

$t_{1/2}$, plasma half-life; F, oral bioavailability; pb, plasma protein binding; ur, excretion of unchanged drug in urine; V, volume of distribution; CL, clearance; Mwt, molecular weight; PSA, polar surface area; NMDA, N-methyl D-aspartic acid; PDE, phosphodiesterase.

Table 68.2 Salicylic acid analogs. V in l kg^{-1}; CL in ml min^{-1} kg^{-1}; Mwt in g mol^{-1}.

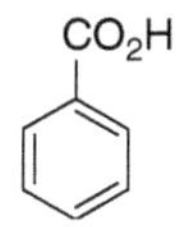	$t_{1/2}$ <6 h F 100% pb – ur –	V – CL – Mwt 122.1 PSA 37.3 Å^2 log P 1.56	BENZOIC ACID Metabolite: hippuric acid (N-benzoylglycine), benzoyl glucuronide
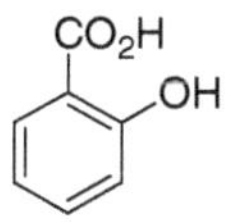	$t_{1/2}$ 2.4 h* F 100% pb 80–95% ur 2–30%	V 0.17±0.03 CL 0.18–0.88 Mwt 138.1 PSA 57.5 Å^2 log P 2.01	SALICYLIC ACID *300 mg Metabolism: aromatic 5-hydroxylation (to gentisic acid), conjugation with glucuronic acid and glycine
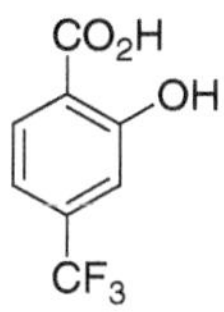	$t_{1/2}$ 34 h F – pb >99% ur –	V 0.13 CL 0.4 Mwt 206.1 PSA 57.5 Å^2 log P 4.25	2-HYDROXY-4-TRIFLUOROMETHYL BENZOIC ACID Main metabolite of triflusal
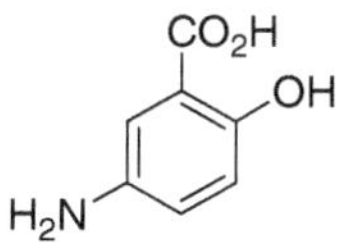	$t_{1/2}$ 1.5±1.0 h F 15–25% pb 43–60% ur <1%	V 0.26–0.33 CL 4–9 Mwt 153.1 PSA 83.6 Å^2 log P 0.74	MESALAZINE 5-AMINOSALICYLIC ACID Metabolism: N-acetylation

(continued overleaf)

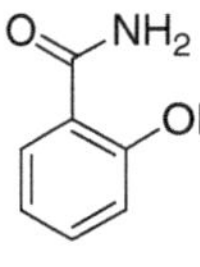	*t*½ 1.0–1.5 h *F* <10% pb – ur <2%	*V* – CL – Mwt 137.1 PSA 63.3 Å^2 log *P* 1.35	SALICYLAMIDE Main metabolites: *O*-glucuronide, *O*-sulfate, gentisamide (2,5-dihydroxybenzamide); no hydrolysis to salicylic acid
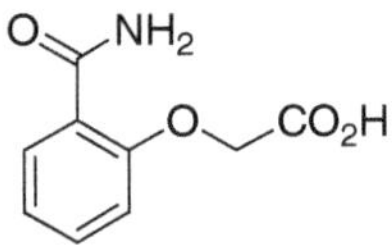	*t*½ 0.9±0.2 h* *F* – pb – ur –	*V* – CL – Mwt 195.2 PSA 89.6 Å^2 log *P* −0.61	SALICYLAMIDE *O*-ACETIC ACID; (2-CARBAMOYL-PHENOXY) ACETIC ACID *im Analgesic Metabolite: salicylamide
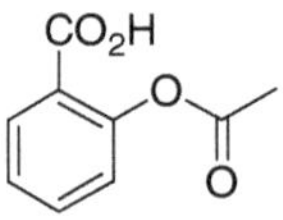	*t*½ 0.25 h *F* 68±3% pb 49% ur 1.4±1.2%	*V* 0.15±0.03 CL 9.3±1.1 Mwt 180.2 PSA 63.6 Å^2 log *P* 1.40	ACETYLSALICYLIC ACID, ASPIRIN Deacetylation: *t*½ 2 h Deacetylation of esters of acetylsalicylic acid: *t*½ 1–3 min
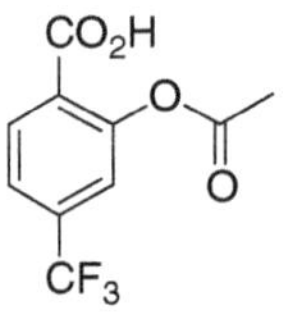	*t*½ 0.5 h *F* – pb – ur –	*V* 0.5 CL 10.8 Mwt 248.2 PSA 63.6 Å^2 log *P* 2.90	TRIFLUSAL Metabolism: deacetylation to 2-hydroxy-4-trifluoromethyl-benzoic acid (see above)
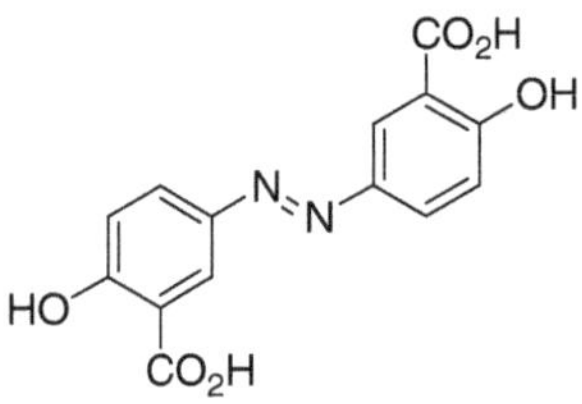	*t*½ 1 h *F* <5% pb 99% ur <1%	*V* 0.07 CL 1.2 Mwt 302.2 PSA 140 Å^2 log *P* 3.57	OLSALAZINE, MORDANT YELLOW Metabolite: 5-Aminosalicylic acid (*t*½ 1.5±1.0 h)
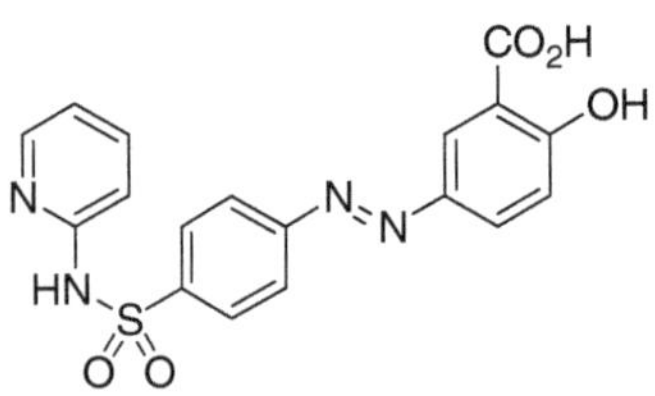	*t*½ 1.4–14 h *F* <20% pb 95–99% ur 10–15%	*V* 7.5 CL 0.26 Mwt 398.4 PSA 150 Å^2 log *P* 3.05	SULFASALAZINE Cleaved in the colon by gut bacteria to sulfapyridine and 5-aminosalicylic acid
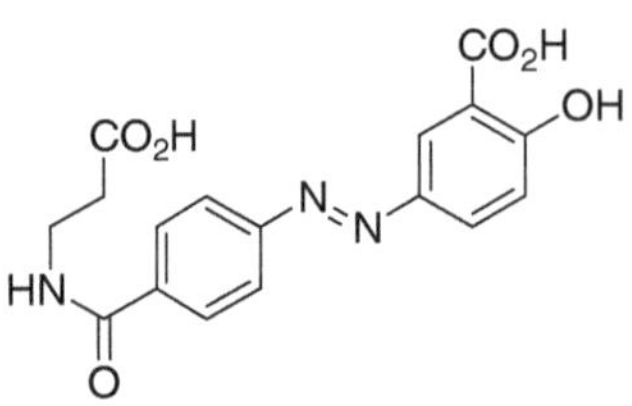	*t*½ 2–3 h *F* <1% pb >99% ur <1%*	*V* – CL 1.2 Mwt 357.3 PSA 149 Å^2 log *P* 3.11	BALSALAZIDE *10% *N*-acetyl-5-aminosalicylic acid; cleaved by gut bacteria to 5-aminosalicylic acid

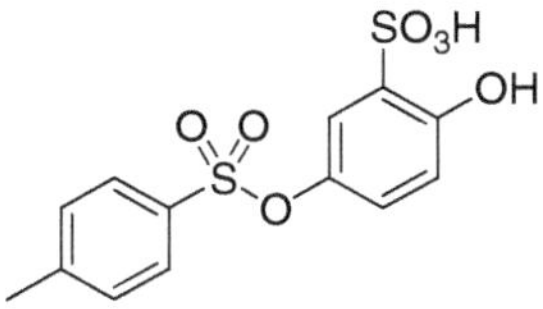	$t\frac{1}{2}$ 18 h *F* >60% pb – ur 28%	*V* – CL – Mwt 344.4 PSA 135 Å^2 log *P* 0.75	SULTOSILIC ACID Antihyperlipidemic Metabolism: hydroxylation and oxidation of CH_3 to CO_2H
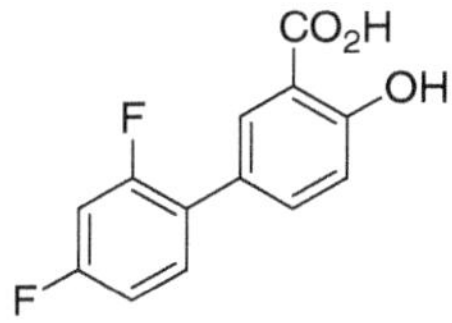	$t\frac{1}{2}$ 11±2 h *F* 90% pb 99.9% ur 6±3%	*V* 0.10±0.01 CL 0.10±0.02 Mwt 250.2 PSA 57.5 Å^2 log *P* 3.65	DIFLUNISAL Analgesic, antiinflammatory Metabolism: sulfation, phenolic and acyl glucuronidation (acyl glucuronide $t\frac{1}{2}$ 12–30 h)
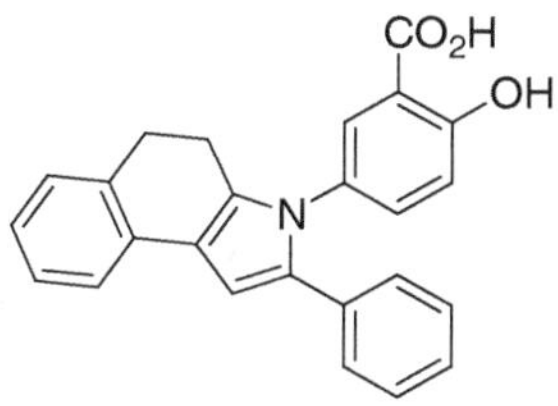	$t\frac{1}{2}$ 2 h *F* – pb – ur –	*V* – CL – Mwt 381.4 PSA 62.5 Å^2 log *P* 5.04	FENDOSAL Analgesic, antiinflammatory Metabolism: hydroxylation
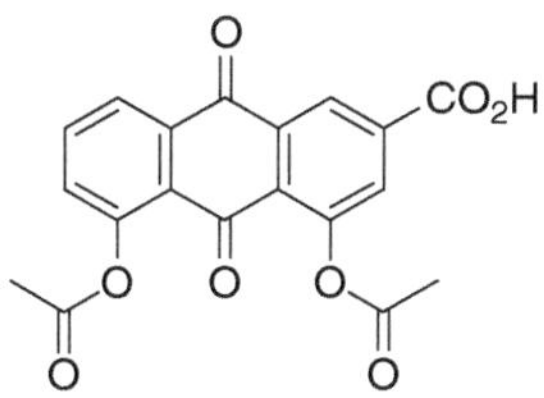	$t\frac{1}{2}$ 4–8 h* *F* 35%* pb 99%* ur 20%*	*V* – CL – Mwt 368.3 PSA 124 Å^2 log *P* 3.13	DIACEREIN, DIACETYLRHEIN *Rhein on oral dosing of diacerein Antiarthritic Metabolism: deacetylation to rhein, then glucuronidation, sulfation

69
Quinolonecarboxylic Acids (Gyrase Inhibitors)

Nalidixic acid was introduced in 1963 as an antibacterial for the treatment of urinary tract infections. Despite belonging to a new class of antibacterial agents, nalidixic acid was no great success, mainly because of its low potency, its side effect profile, and the quick development of resistance. As reflected by its low volume of distribution and high affinity to plasma proteins, nalidixic acid does not reach high concentrations in most organs and is therefore of little therapeutic value. These issues were addressed by Japanese and German researchers, who introduced a basic, positively charged group to enhance the solubility and volume of distribution, while lowering the affinity to plasma proteins. Thereby, the drug concentration in potentially infected tissues was increased and the interindividual variability of pharmacokinetic (PK) parameters reduced. These efforts led to compounds such as pipemidic acid (1974), norfloxacin (1978), pefloxacin (1979), ofloxacin (1982), and ciprofloxacin (1983) (Scheme 69.1).

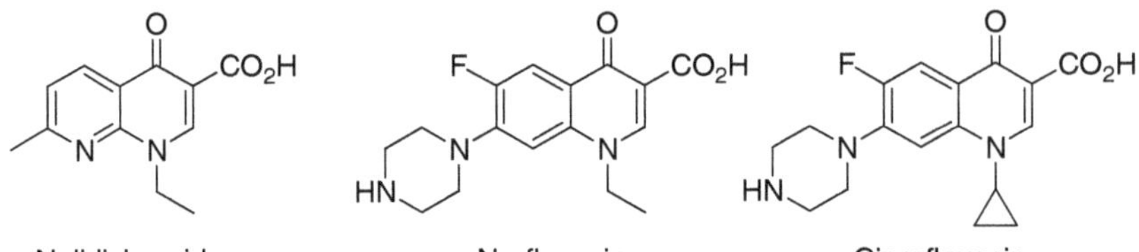

Nalidixic acid
$t_{1/2}$ 1–8 h, *F* 96%
pb 93–97%, *V* 0.5–0.6 l kg^{-1}

Norfloxacin
$t_{1/2}$ 5 h, *F* 35%
pb 15–20%, *V* 3.2 l kg^{-1}

Ciprofloxacin
$t_{1/2}$ 4 h, *F* 60%
pb 40%, *V* 1.8 l kg^{-1}

Scheme 69.1 Quinolone antibacterials.

Quinolone antibiotics are potent inhibitors of bacterial DNA gyrase, but much less potent inhibitors of eukaryotic type II topoisomerase. DNA gyrase is an enzyme capable of temporarily breaking DNA and then reassembling it. This enzyme belongs to the topoisomerases and is essential for the replication, transcription, recombination, and reparation of DNA, as well as for the compact packing of DNA within cells.

The newer quinolone antibiotics (Table 69.1) are mainly metabolized at the alkylamine substructures by oxidative N-dealkylation and conjugation.

Lead Optimization for Medicinal Chemists: Pharmacokinetic Properties of Functional Groups and Organic Compounds, First Edition. Florencio Zaragoza Dörwald.

Table 69.1 Quinolonecarboxylic acids and related compounds. V in l kg^{-1}; CL in ml min^{-1} kg^{-1}; Mwt in g mol^{-1}.

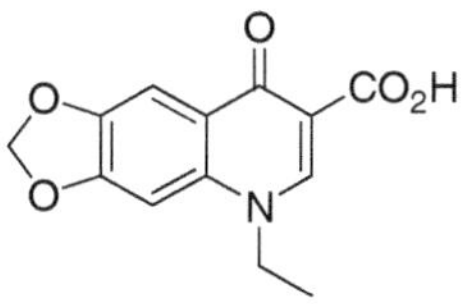	$t_{1/2}$ F pb ur	6–7 h 50% 77–85% 5%	V CL Mwt PSA log P	– – 261.2 76.1 Å^2 −0.38	OXOLINIC ACID Antibacterial Metabolism: acetal hydrolysis, then O-methylation by COMT
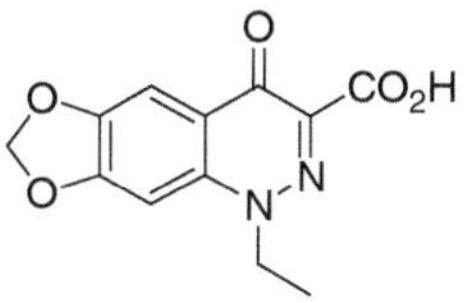	$t_{1/2}$ F pb ur	1–2 h 60–85% 43–63% 60–85%	V CL Mwt PSA log P	0.33 2.5 262.2 88.4 Å^2 0.49	CINOXACIN Antibacterial
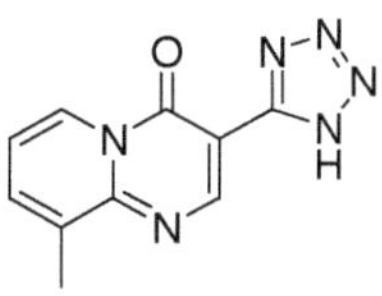	$t_{1/2}$ F pb ur	4.5 h – – 10–15%	V CL Mwt PSA log P	– – 228.2 87.1 Å^2 0.22	PEMIROLAST Antiallergic
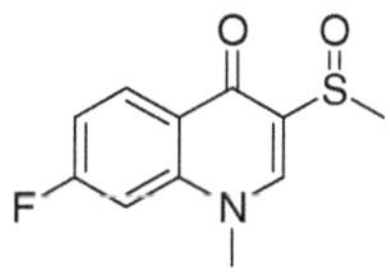	$t_{1/2}$ F pb ur	1.7±0.5 h 72% – 1%	V CL Mwt PSA log P	0.8 8.2 239.3 56.6 Å^2 −1.43	FLOSEQUINAN Antihypertensive Metabolite: sulfone ($t_{1/2}$ 30–40 h); withdrawn in 1993 because of excessive toxicity
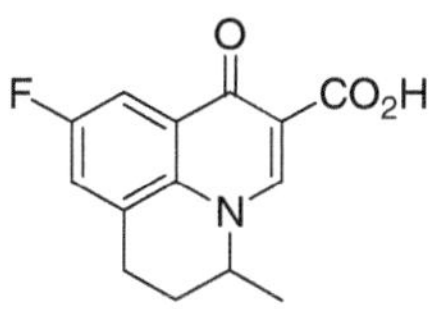	$t_{1/2}$ F pb ur	7 h 56–93%* 75%* –	V CL Mwt PSA log P	1.4* – 261.3 57.6 Å^2 0.88	FLUMEQUINE *calves Antibacterial Metabolism: benzylic hydroxylation, glucuronidation
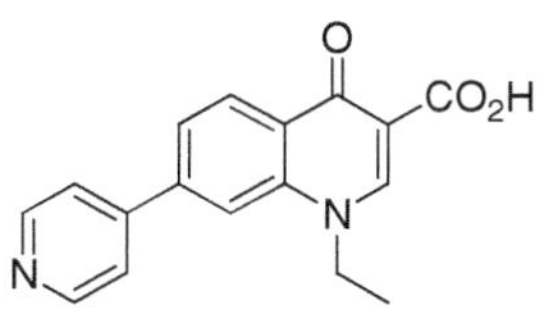	$t_{1/2}$ F pb ur	5±1 h >64% 70% 4–5%	V CL Mwt PSA log P	0.64±0.05 2.1±0.2 294.3 70.5 Å^2 0.57	ROSOXACIN Antibacterial Metabolism: pyridine N-oxidation, glucuronidation
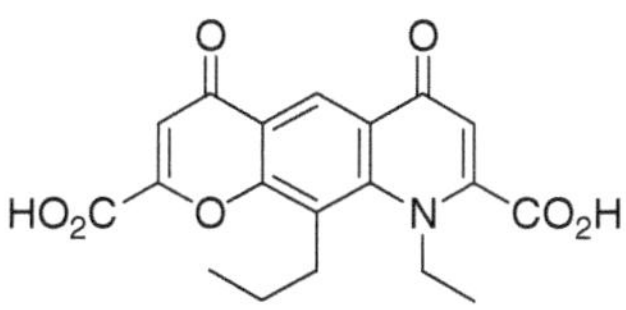	$t_{1/2}$ F pb ur	1.1–3.3 h 2–3%* 89% 64–81%	V CL Mwt PSA log P	0.43 10.2 371.3 121 Å^2 1.30	NEDOCROMIL *pulmonal: 6% Antiallergic, antiasthmatic

(continued overleaf)

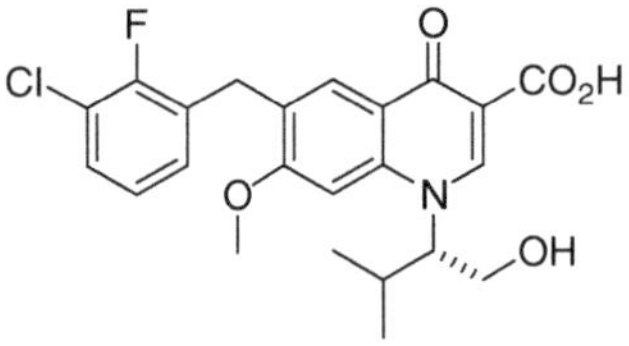

$t^{1}/_{2}$	4–9 h	V	–	ELVITEGRAVIR
F	–	CL	–	HIV-1 integrase inhibitor, antiviral
pb	–	Mwt	447.9	Metabolism: glucuronidation
ur	–	PSA	87.1 Å^2	
		log P	3.26	

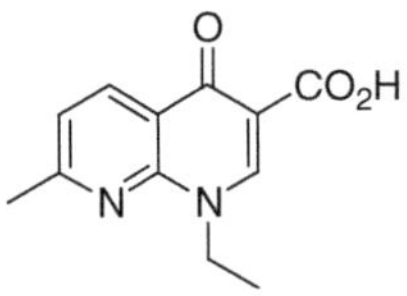

$t^{1}/_{2}$	1–8 h	V	0.5–0.6	NALIDIXIC ACID
F	96%	CL	–	Antibacterial
pb	93–97%	Mwt	232.2	Metabolites: acyl glucuronide,
ur	11–33%	PSA	70.5 Å^2	7-hydroxymethyl ($t^{1}/_{2}$ 3–6 h, pb
		log P	0.03	63%), 7-carboxy

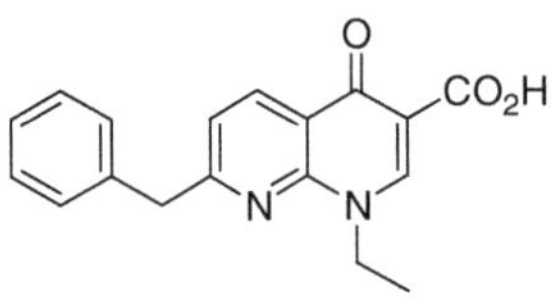

$t^{1}/_{2}$	12 h	V	–	AMFONELIC ACID
F	–	CL	–	Dopamine reuptake inhibitor,
pb	–	Mwt	308.3	locomotor stimulant
ur	–	PSA	70.5 Å^2	
		log P	1.38	

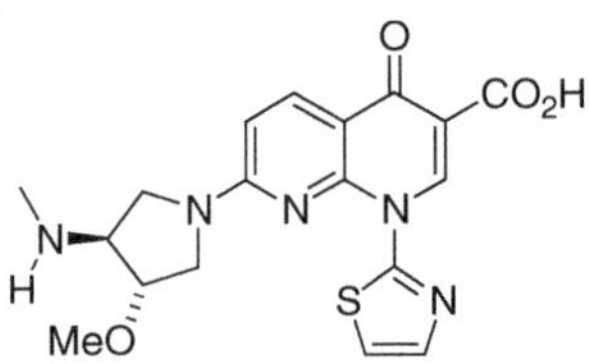

$t^{1}/_{2}$	6–24 h	V	4	VORELOXIN
F	–	CL	10	Antineoplastic, topoisomerase II
pb	50–75%	Mwt	401.4	inhibitor
ur	–	PSA	136 Å^2	Metabolism: O-demethylation,
		log P	−0.86	glucuronidation, N-dealkylation

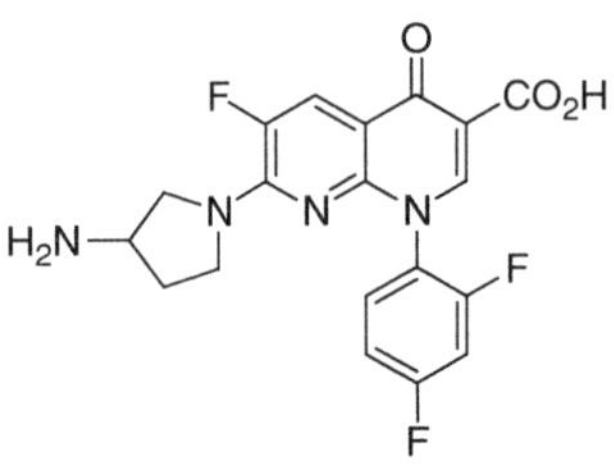

$t^{1}/_{2}$	3.5±0.2 h	V	1.5–7.3	TOSUFLOXACIN
F	–	CL	6.4–39	Antibacterial, racemate
pb	37%	Mwt	404.3	Metabolism: glucuronidation,
ur	34±9%	PSA	99.8 Å^2	replacement of NH_2 by OH
		log P	1.01	

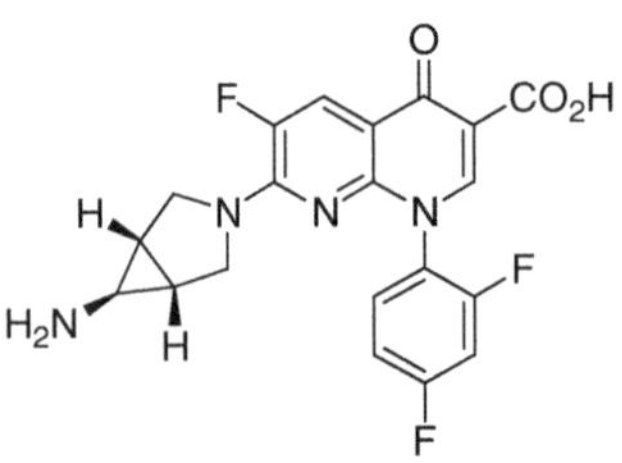

$t^{1}/_{2}$	11±1 h	V	1.3	TROVAFLOXACIN
F	90%	CL	1.4	Metabolism: acyl glucuronidation,
pb	76–88%	Mwt	416.4	N-sulfation, N-acetylation
ur	9.3±2.5%	PSA	99.8 Å^2	Prodrug: alatrofloxacin
		log P	1.08	(Chapter 33); withdrawn in 2001
				because of hepatotoxicity

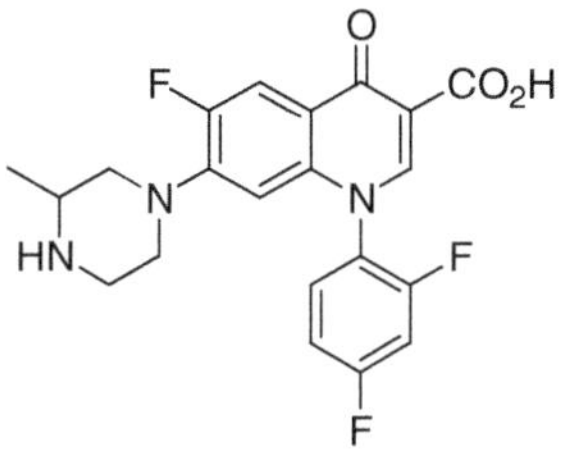	$t^1/_2$ 8.0±0.5 h *F* 90% pb 26% ur 74±18%	*V* 1.5 CL 3.4 Mwt 417.4 PSA 72.9 Å^2 log *P* 1.28	TEMAFLOXACIN Metabolism: oxidative degradation of piperazine to ethylenediamines; withdrawn in 1992 because of excessive toxicity (hemolytic uremic anemia)
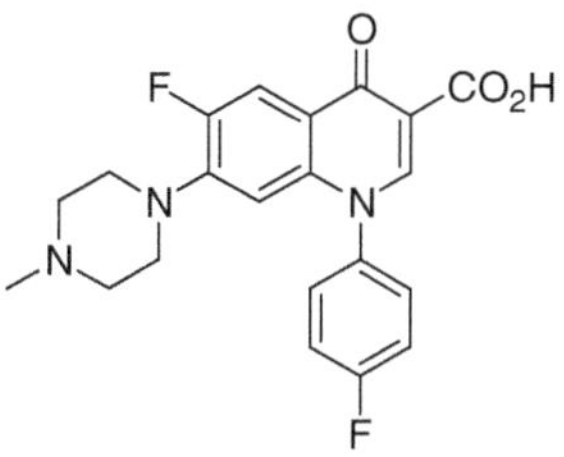	$t^1/_2$ 21–29 h *F* – pb – ur 10%	*V* 1.4–1.5 CL – Mwt 399.4 PSA 64.1 Å^2 log *P* 0.84	DIFLOXACIN Antibacterial, does not cross bbb Metabolism: N-demethylation, N-oxidation, oxidative degradation of piperazine to ethylenediamines
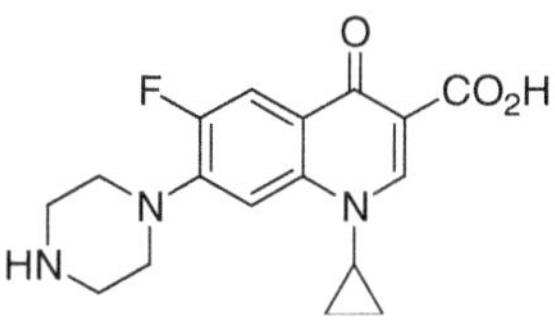	$t^1/_2$ 4.1±0.9 h *F* 60±12% pb 40% ur 65±12%	*V* 1.8±0.4 CL 6.0±1.2 Mwt 331.3 PSA 72.9 Å^2 log *P* 1.31	CIPROFLOXACIN Antibacterial, does not cross bbb Metabolism: piperazine 3′-hydroxylation, N-sulfation
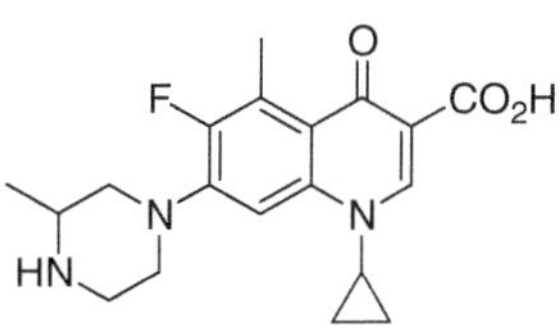	$t^1/_2$ 11±1 h *F* 72% pb 50% ur 9±3%	*V* 5–8 CL 3.9±0.9 Mwt 359.4 PSA 72.9 Å^2 log *P* 2.27	GREPAFLOXACIN Metabolism: glucuronidation, N-sulfation, oxidative degradation of piperazine; withdrawn in 1999 because of QT interval prolongation
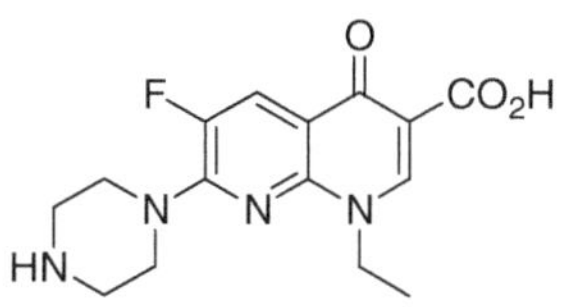	$t^1/_2$ 4.5±1.0 h *F* 87±16% pb 40% ur 45±11%	*V* 1.6±0.4 CL 5.0±1.2 Mwt 320.3 PSA 85.8 Å^2 log *P* 1.21	ENOXACIN Antibacterial, does not cross bbb Metabolism: N-acetylation, 3′-oxidation of piperazine to lactam, oxidative degradation of piperazine
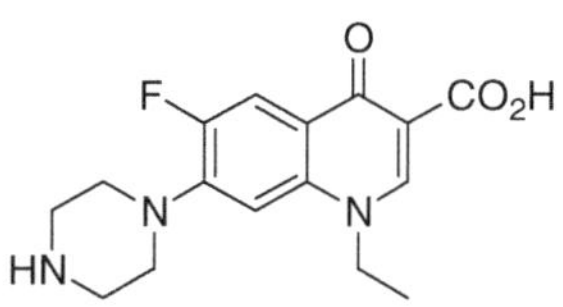	$t^1/_2$ 5.0±0.7 h *F* 35±5% pb 15–20% ur 29±3%	*V/F* 3.2±1.4 CL/*F* 7.2±3.0 Mwt 319.3 PSA 72.9 Å^2 log *P* 1.74	NORFLOXACIN Antibacterial Metabolism: N-formylation, N-acetylation, 3′-oxidation of piperazine to lactam, cleavage of piperazine to *N*-2-(aminoethyl)aniline and an unsubstituted aromatic amine

(*continued overleaf*)

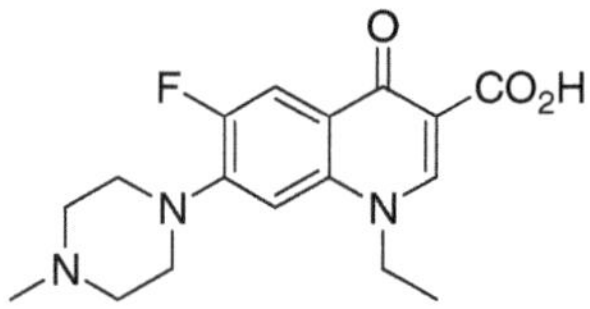

$t^{1}/_{2}$	10±1 h	V	1.5
F	100%	CL	2.1
pb	20–30%	Mwt	333.4
ur	42%	PSA	64.1 Å^2
		log P	1.92

PEFLOXACIN
Metabolism: glucuronidation, N-demethylation, 3′-oxidation to lactam, N-oxidation

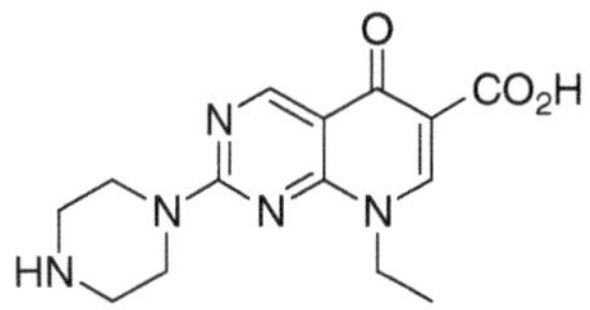

$t^{1}/_{2}$	3.4±0.2 h	V	1.9±0.2
F	93±11%	CL	6.3±0.5
pb	10–30%	Mwt	303.3
ur	90%	PSA	98.7 Å^2
		log P	−0.19

PIPEMIDIC ACID
Antibacterial
Metabolism: N-acetylation, N-formylation, 3′-oxidation of piperazine to lactam

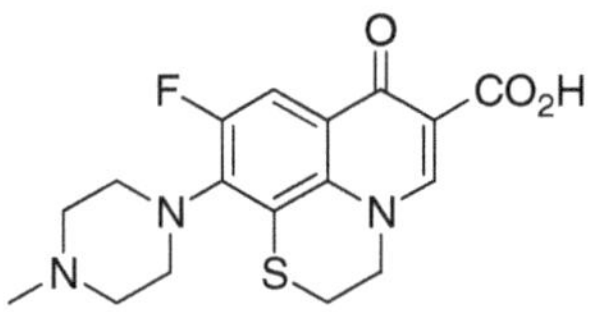

$t^{1}/_{2}$	38±3 h	V	2.0–2.1
F	70%*	CL	0.62
pb	60%	Mwt	363.4
ur	30–50%	PSA	89.4 Å^2
		log P	1.85

RUFLOXACIN
*monkey
Antibacterial
Metabolism: N-demethylation, oxidation to sulfoxide

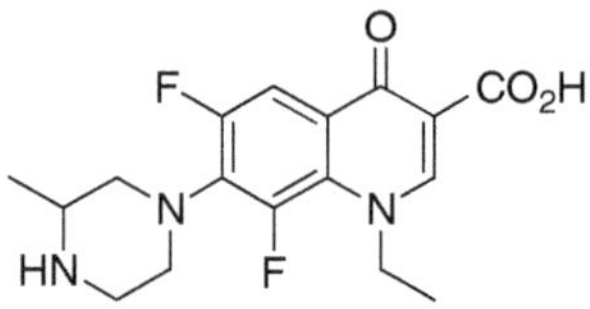

$t^{1}/_{2}$	8.0±1.4 h	V	2.3±0.3
F	97±2%	CL	3.3±0.5
pb	10–20%	Mwt	351.4
ur	65±9%	PSA	72.9 Å^2
		log P	2.46

LOMEFLOXACIN
Antibacterial, racemate
Metabolism: glucuronidation

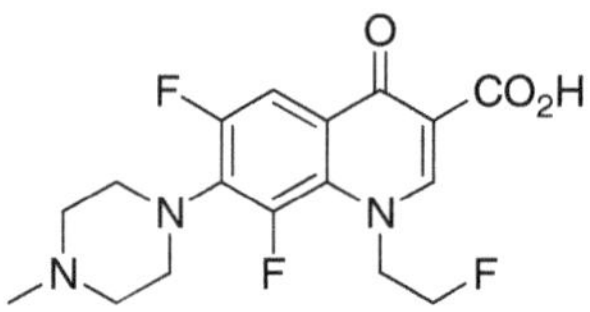

$t^{1}/_{2}$	11±2 h	V	1.5±0.3
F	>98%	CL	2.4
pb	25±5%	Mwt	369.3
ur	60–70%	PSA	64.1 Å^2
		log P	1.84

FLEROXACIN
Antibacterial, does not cross bbb
Metabolism: N-demethylation, N-oxidation

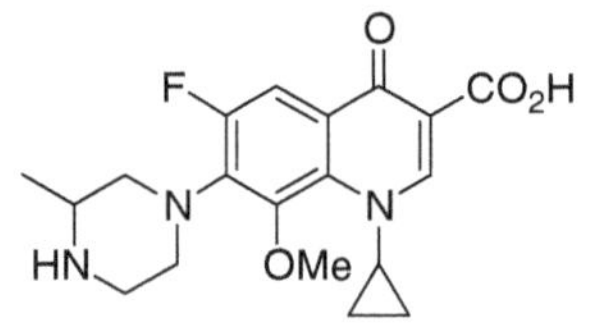

$t^{1}/_{2}$	6.5±1.5 h	V	2.2±0.2
F	96%	CL	3.3±0.5
pb	20%	Mwt	375.4
ur	82–88%	PSA	82.1 Å^2
		log P	2.10

GATIFLOXACIN
Antibacterial
Metabolism: oxidative degradation of piperazine

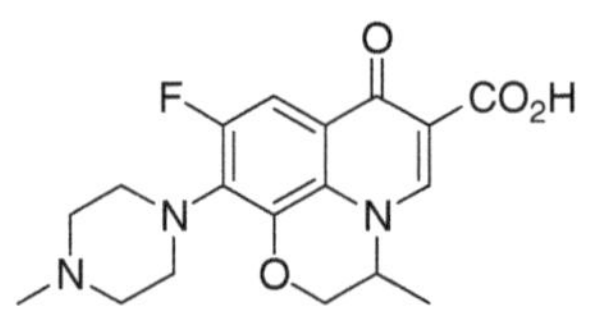

$t^{1}/_{2}$	5.7±1.0 h	V	1.8±0.3
F	100%	CL	3.5±0.7
pb	25±6%	Mwt	361.4
ur	64±16%	PSA	73.3 Å^2
		log P	1.86

OFLOXACIN
Antibacterial
Metabolism: N-demethylation, N-oxidation

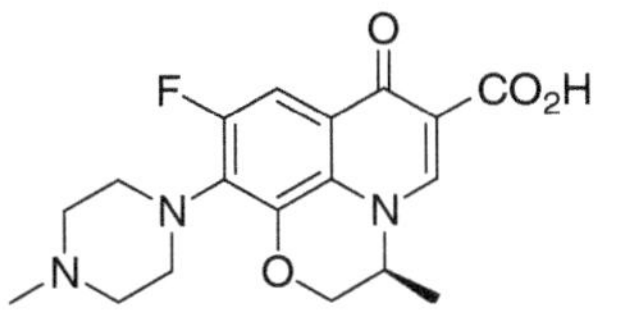	$t_{1/2}$ 7±1 h *F* 99% pb 24–52% ur >85%	*V* 1.3 CL – Mwt 361.4 PSA 73.3 Å^2 log *P* 1.86	LEVOFLOXACIN Antibacterial Does not racemize *in vivo* Metabolites: *N*-desmethyl, *N*-oxide
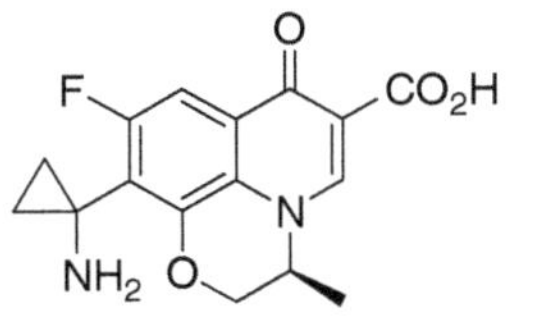	$t_{1/2}$ 4±2 h *F* – pb – ur 70%	*V* 0.9 CL 6.8 Mwt 318.3 PSA 92.9 Å^2 log *P* 0.35	PAZUFLOXACIN Parenteral antibacterial Metabolism: glucuronidation, oxidative degradation of aminocyclopropyl group (to 1-hydroxyethyl, 3-hydroxypropanoyl)
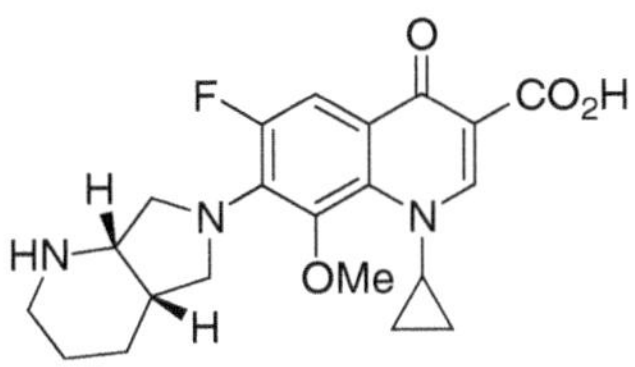	$t_{1/2}$ 12 h *F* 90% pb 39–52% ur 20±5%	*V* 2.0–3.6 CL 3.3±0.5 Mwt 401.4 PSA 82.1 Å^2 log *P* 1.90	MOXIFLOXACIN Antibacterial Metabolism: acyl glucuronidation, N-sulfation
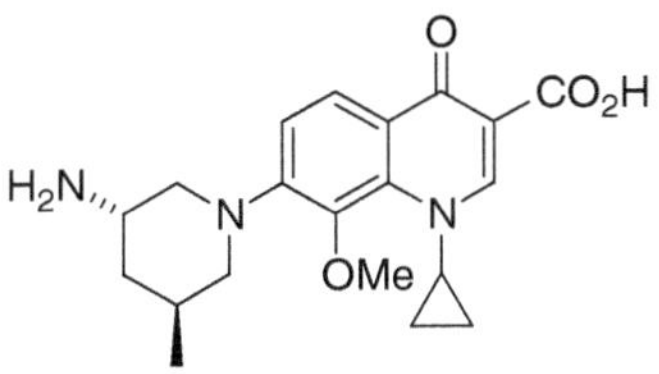	$t_{1/2}$ 9–16 h *F* – pb 16% ur 34–56%	*V* 3.4–5.3 CL 3.7 Mwt 371.4 PSA 96.1 Å^2 log *P* 0.94	NEMONOXACIN Antibacterial
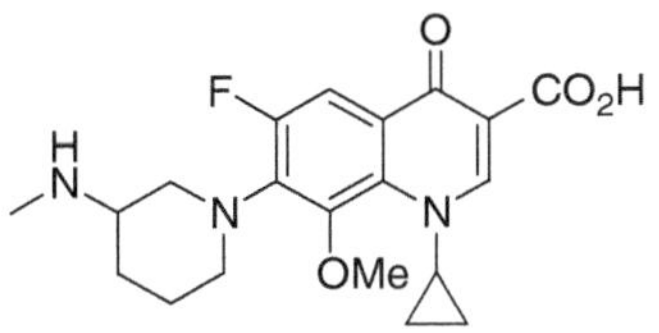	$t_{1/2}$ 11±4 h *F* 88%* pb – ur 23%*	*V/F* 0.7 CL 3.0±0.5 Mwt 389.4 PSA 82.1 Å^2 log *P* 1.54	BALOFLOXACIN *dog Antibacterial Metabolism: acyl glucuronidation, N-demethylation
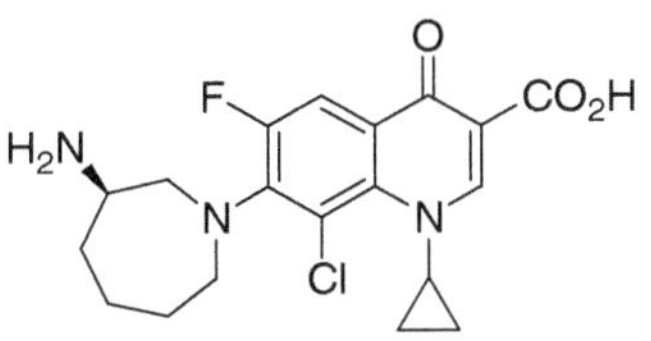	$t_{1/2}$ 7 h *F* – pb 39–44% ur 23%*	*V* – CL – Mwt 393.8 PSA 86.9 Å^2 log *P* 2.57	BESIFLOXACIN *animals Topical and ophthalmic antibacterial

(*continued overleaf*)

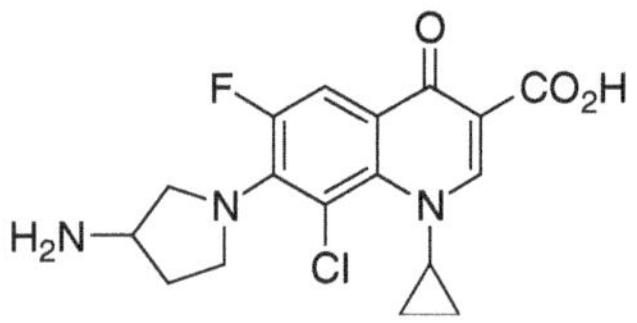

$t_{1/2}$	5.7±0.7 h	*V*	2.6±0.3	CLINAFLOXACIN
F	90%	CL	5.3±0.6	Antibacterial
pb	50–60%	Mwt	365.8	
ur	58±18%	PSA	86.9 Å^2	
		log *P*	2.45	

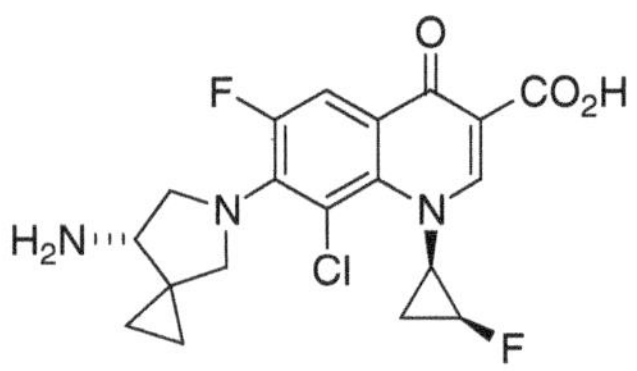

$t_{1/2}$	5.9±0.7 h	*V*	2.8	SITAFLOXACIN
F	89%	CL	4.6	Antibacterial
pb	49%	Mwt	409.8	Metabolism: N-acetylation, replacement of NH_2 by OH, then oxidation to ketone, ring scission of pyrrolidine to aminobutanol, then oxidation to aminobutyric acid
ur	26–86%	PSA	86.9 Å^2	
		log *P*	1.53	

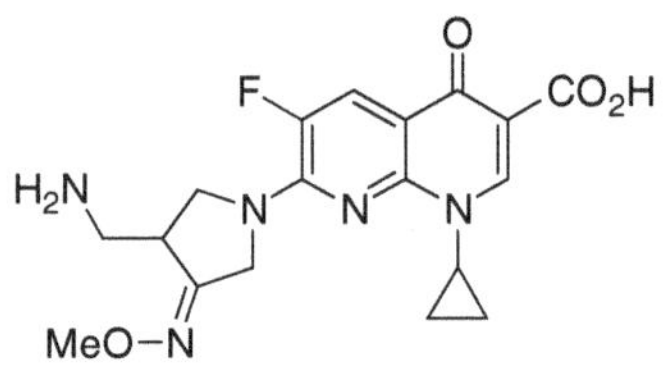

$t_{1/2}$	7.3±2.5 h	*V*	1.5–3.5	GEMIFLOXACIN
F	71%	CL	12.5±5.6	Antibacterial
pb	60–70%	Mwt	389.4	Metabolism: glucuronidation, N-acetylation, isomerization of oxime
ur	18±4%	PSA	121 Å^2	
		log *P*	1.04	

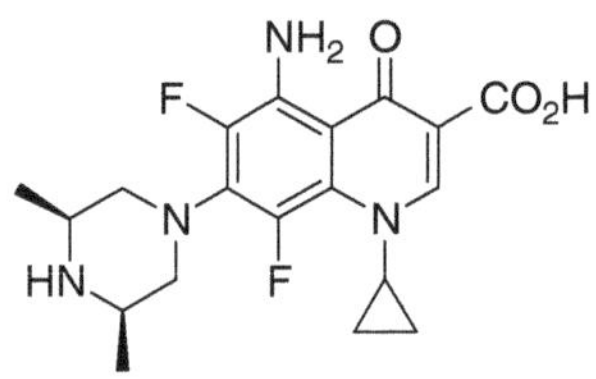

$t_{1/2}$	18±2 h	*V*	4–6	SPARFLOXACIN
F	90%	CL	2.7	Antibacterial
pb	45%	Mwt	392.4	Metabolism: acyl glucuronidation
ur	9–10%	PSA	98.9 Å^2	
		log *P*	2.60	

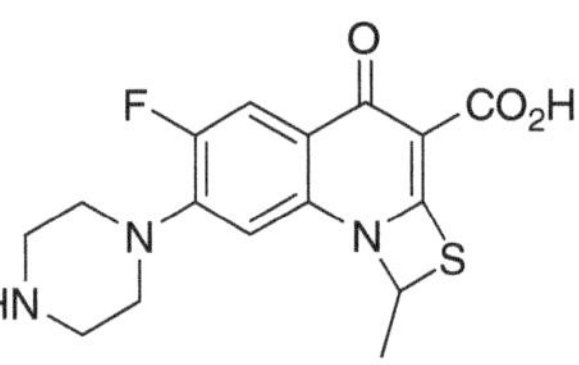

$t_{1/2}$	10–12 h*	*V*	19	ULIFLOXACIN
F	–	CL	–	*ulifloxacin on oral dosing of prulifloxacin (Chapter 23)
pb	41–59%	Mwt	349.4	Metabolism: acyl glucuronidation, oxidative cleavage of piperazine
ur	–	PSA	98.2 Å^2	Prodrug: prulifloxacin, which cannot be detected in plasma after oral dosing
		log *P*	2.53	

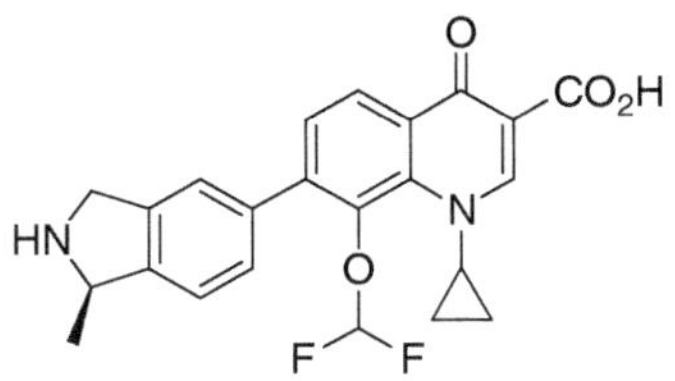	$t^1/_2$ 13–18 h *F* 76%* pb 87% ur 30–50%	*V* 1.0–1.5* CL – Mwt 426.4 PSA 78.9 Å^2 log *P* 1.62	GARENOXACIN *monkey Antibacterial Metabolism: glucuronidation, sulfation, probably forms reactive metabolite (isoindole) by oxidation of isoindoline		
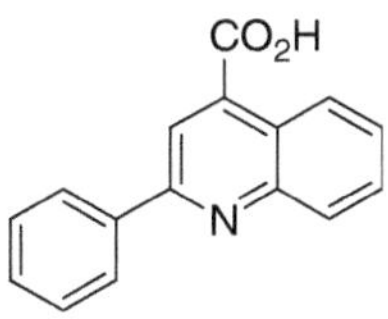	$t^1/_2$ 4 h *F* High pb – ur <5%	*V* – CL – Mwt 249.3 PSA 50.2 Å^2 log *P* 4.18	CINCHOPHEN Analgesic, induces ulcers, hepatotoxic Metabolism: aromatic hydroxylation		
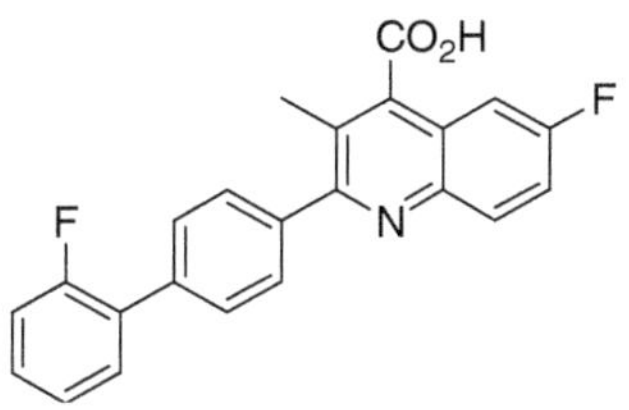	$t^1/_2$ 13–18 h *F* High pb >98% ur 0.1–6%*	*V* 0.24±0.08 CL 0.5±0.2 Mwt 375.4 PSA 50.2 Å^2 log *P* 6.69	BREQUINAR *iv Immunosuppressant		

$t_{1/2}$, plasma half-life; *F*, oral bioavailability; pb, plasma protein binding; ur, excretion of unchanged drug in urine; V, volume of distribution; CL, clearance; Mwt, molecular weight; PSA, polar surface area; bbb, blood–brain barrier; COMT, catechol-*O*-methyltransferase.

70
β-Lactams

In 1928, Alexander Fleming, a microbiologist working at St. Mary's Hospital in London, discovered a mold (*Penicillium notatum*), which inhibited bacterial growth in its proximity. He called the antibiotic penicillin and explored its usefulness as topical antiseptic, but failed to test its efficacy after parenteral dosing *in vivo*. More than a decade later, Florey, Chain, and Abraham at Oxford University administered penicillin broth parenterally to mice infected with Streptococci, and later to patients, with remarkable success. The production of penicillin broth, containing about 10% penicillin, was scaled up significantly in the United Kingdom and United States to meet the requirements of the Allied Forces during the war.

Pure, crystalline penicillin was obtained by Oskar Wintersteiner in 1943, who found that several different penicillins were produced by the mold, including 2-pentenylpenicillin (penicillin F) and benzylpenicillin (penicillin G) (Scheme 70.1). X-ray structural analysis by Hodkin in 1945 confirmed the β-lactam structure, and the total synthesis of phenoxymethylpenicillin was accomplished in 1957 by John C. Sheehan. This chemical synthesis was, however, more tedious and expensive than the fermentative route.

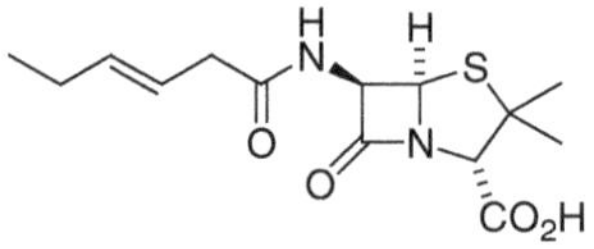

2-Pentenylpenicillin
penicillin F

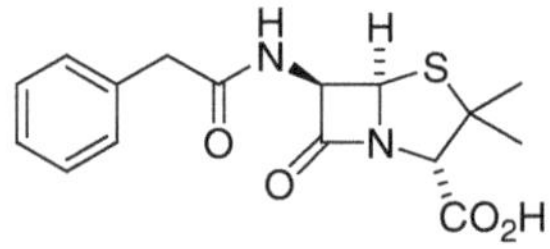

Benzylpenicillin
penicillin G
$t_{1/2}$ 30–40 min, F < 30%

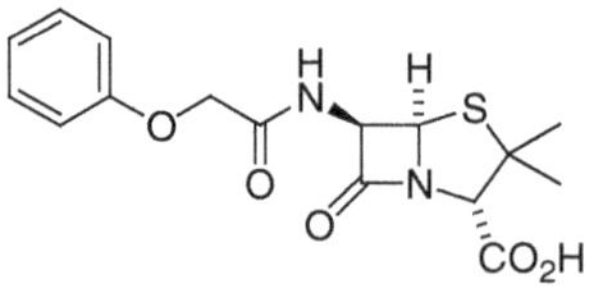

Phenoxymethylpenicillin
penicillin V
$t_{1/2}$ 0.5–1.0 h, F 40–60%

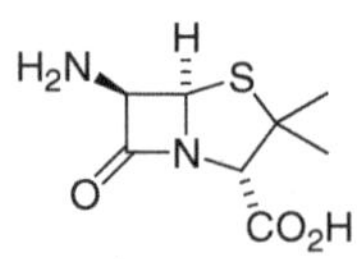

6-Aminopenicillamic acid

Scheme 70.1 Natural and synthetic penicillins.

Lead Optimization for Medicinal Chemists: Pharmacokinetic Properties of Functional Groups and Organic Compounds, First Edition. Florencio Zaragoza Dörwald.

Oral bioavailability of natural penicillins is poor, but side-chain variations quickly led to semisynthetic penicillins with improved pharmacokinetics (PK) (Table 70.1). While growing *Penicillium notatum*, the broth always contains 6-aminopenicillamic acid, which can be isolated and acylated with various acids to prepare other, unnatural penicillin analogs.

The most remarkable property of β-lactam antibiotics is their low toxicity to humans. Even intravenous doses of several grams are tolerated. Because of their ability to acylate plasma proteins (e.g., albumin), however, β-lactam allergies occasionally result.

Penicillins kill bacteria by acylating a bacterial transpeptidase involved in the formation of cell walls. Because of the massive use of β-lactam antibiotics in the past decades, resistance is emerging and posing a real threat to our ability to fight bacterial infections [1].

Most penicillins undergo β-lactam hydrolysis *in vivo* to lose their antibiotic activity. Because of their hydrophilicity, β-lactams are not oxidized to sulfones but mainly excreted renally, either unchanged or as hydrolyzed β-lactam.

A number of other β-lactams have been isolated, mainly from fungi, and used as leads for the development of antibiotics. These include the carbapenems, cephalosporins, and various β-lactamase inhibitors.

70.1
Cephalosporins

The biochemist Giuseppe Brotzu discovered the fungus *Cephalosporium acremonium* in 1945 in a sewer outlet near the Sardinian coast. This mold produces a number of β-lactam antibiotics that are weaker than penicillins but more resistant to bacterial penicillases. Of these, only cephalosporin C attained some therapeutic significance (Scheme 70.2).

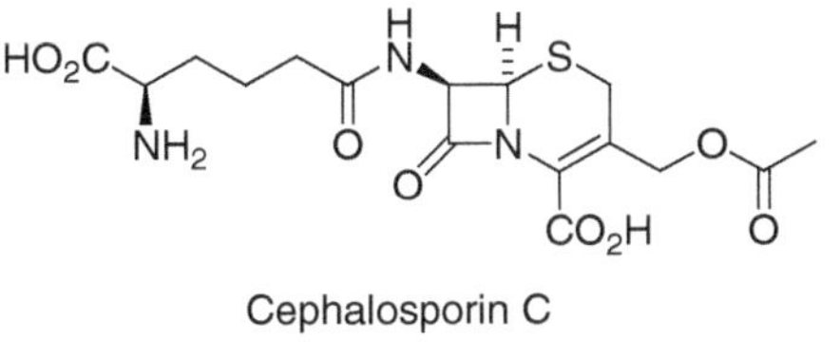

Scheme 70.2

As in the case of penicillins, many synthetic cephalosporin analogs with improved PK and potency have been prepared by side-chain modification of the natural cephalosporins (Table 70.2).

Reference

1. (a) Oelschlaeger, P., Ai, N., DuPrez, K.T., Welsh, W.J., and Toney, J.H. (2010) Evolving carbapenemases: can medicinal chemists advance one step ahead of the coming storm?

J. Med. Chem., **53**, 3013–3027; (b) Taubes, G. (2008) The bacteria fight back. *Science*, **321**, 356–361; (c) Walsh, C. and Wright, G. (2005) Antibiotic resistance. *Chem. Rev.*, **105**, 391–393; (d) Fisher, J.F., Meroueh, S.O., and Mobashery, S. (2005) Bacterial resistance to β-lactam antibiotics: compelling opportunism, compelling opportunity. *Chem. Rev.*, **105**, 395–424.

Table 70.1 Penicillins, carbapenems, and related compounds. *V* in l kg^{-1}; CL in ml min^{-1} kg^{-1}; Mwt in g mol^{-1}.

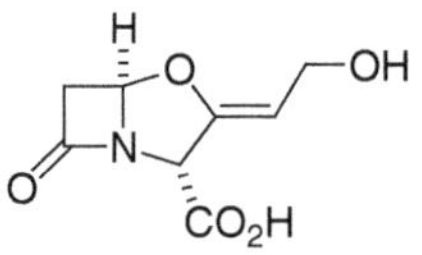	$t½$ F pb ur	0.9±0.1 h 75±21% 22% 43±14%	V CL Mwt PSA log P	0.21±0.05 3.6±1.0 199.2 87.1 Å^2 −0.32	CLAVULANIC ACID β-Lactamase inhibitor
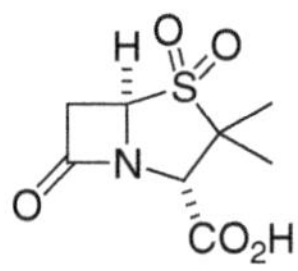	$t½$ F pb ur	1.0–1.7 h – 38% 75–85%	V CL Mwt PSA log P	0.32 3.7–5.1 233.2 100 Å^2 0.39	SULBACTAM β-Lactamase inhibitor
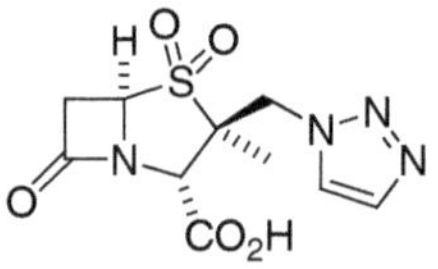	$t½$ F pb ur	0.7–1.2 h* 0% 20–23% 64%	V CL Mwt PSA log P	0.2–0.5 3.1 300.3 131 Å^2 0.60	TAZOBACTAM *iv β-Lactamase inhibitor
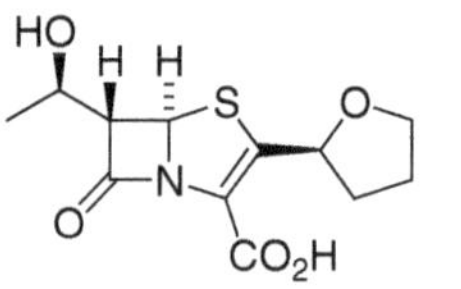	$t½$ F pb ur	0.9–1.3 h 20–30% 90–95% 14–20%	V CL Mwt PSA log P	– – 285.3 112 Å^2 −1.09	FAROPENEM
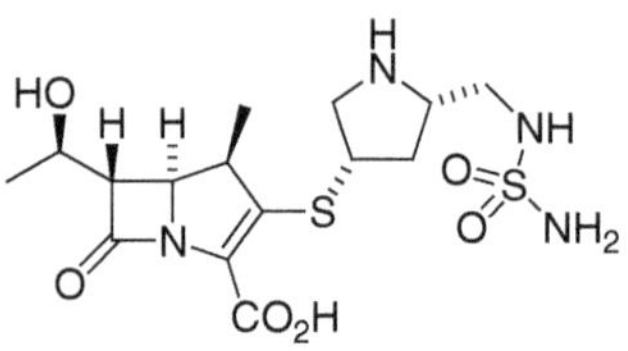	$t½$ F pb ur	1.1 h* Low 8% 79±6%*	V CL Mwt PSA log P	0.26 4.1 420.5 196 Å^2 −3.65	DORIPENEM *iv Metabolism: β-lactam hydrolysis, then conjugation with glycine and taurine
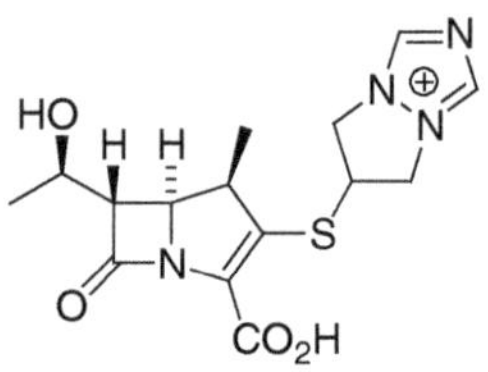	$t½$ F pb ur	1.1 h* – 8% 47–53%*	V CL Mwt PSA log P	0.2 2.5 351.4 99.6 Å^2 −4.12	BIAPENEM *iv

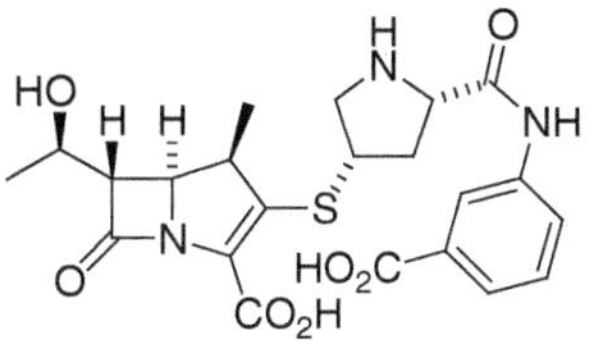	$t^1\!/_2$ *F* pb ur	3.8 h (iv) Low 85–95% 70–78%	*V* CL Mwt PSA log *P*	0.12 0.45 475.5 182 Å^2 −1.07	ERTAPENEM Metabolism: β-lactam hydrolysis
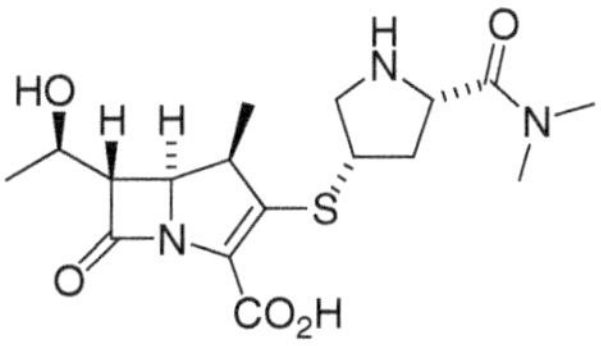	$t^1\!/_2$ *F* pb ur	1 h (iv) Low 13% 65% (iv)	*V* CL Mwt PSA log *P*	0.3 3.9 383.5 135 Å^2 −1.23	MEROPENEM
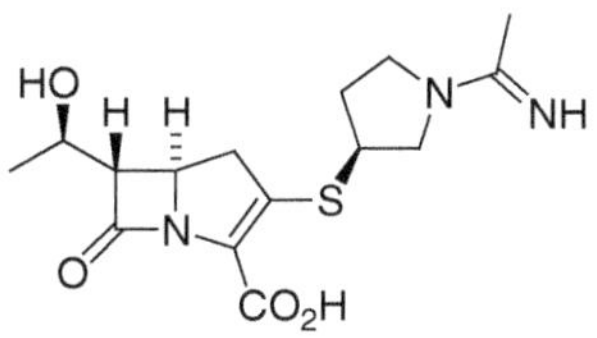	$t^1\!/_2$ *F* pb ur	1 h (iv) – – 30%	*V* CL Mwt PSA log *P*	0.19 2.6 339.4 130 Å^2 −2.85	PANIPENEM
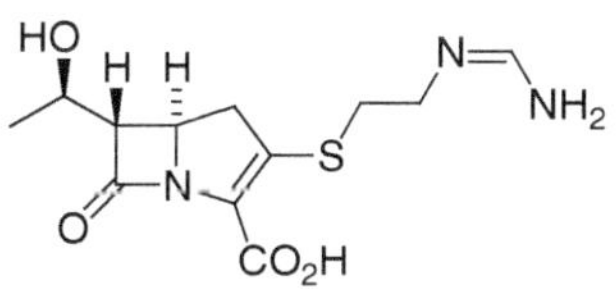	$t^1\!/_2$ *F* pb ur	0.9±0.1 h Low <20% 69±15%	*V* CL Mwt PSA log *P*	0.23±0.05 2.9±0.3 2994. 139 Å^2 −2.78	IMIPENEM Values for simultaneous administration with cilastatin
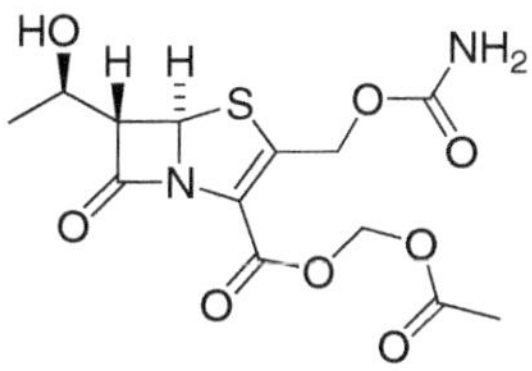	$t^1\!/_2$ *F* pb ur	0.7 h* 37–42%* – 16%*	*V* CL Mwt PSA log *P*	0.25* – 288.3 155 Å^2 −1.44	RITIPENEM *acid (ritipenem) on oral dosing of ester; no ester can be detected in plasma
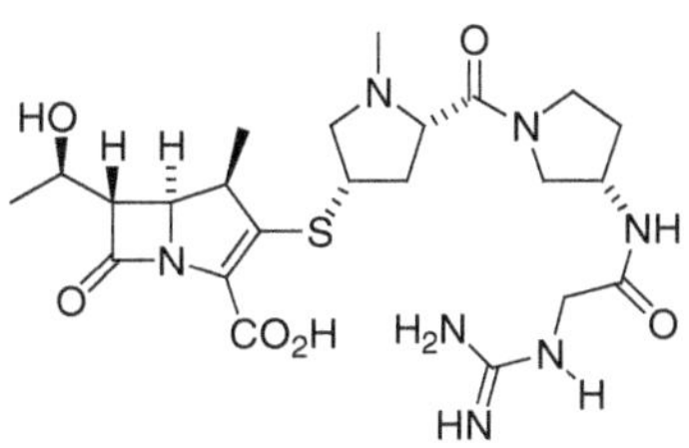	$t^1\!/_2$ *F* pb ur	1.7 h (iv) – 9% 70% (iv)	*V* CL Mwt PSA log *P*	0.23 1.9 537.6 218 Å^2 −5.00	TOMOPENEM

(continued overleaf)

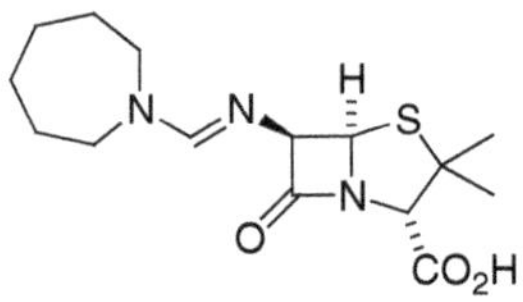

$t_{1/2}$	1 h	V	0.3–0.4	AMDINOCILLIN
F	Low	CL	5–6	Prodrug: amdinocillin pivoxil
pb	–	Mwt	325.4	
ur	63%	PSA	98.5 Å^2	
		log P	2.33	

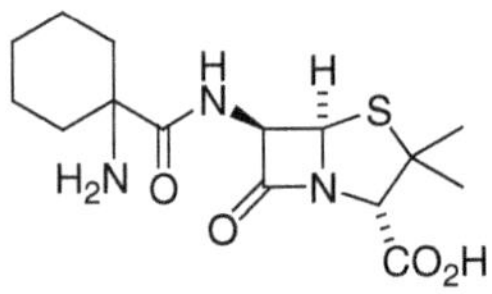

$t_{1/2}$	0.5–0.8 h	V	–	CYCLACILLIN
F	High	CL	7	
pb	18–30%	Mwt	341.4	
ur	24–91%	PSA	138 Å^2	
		log P	1.21	

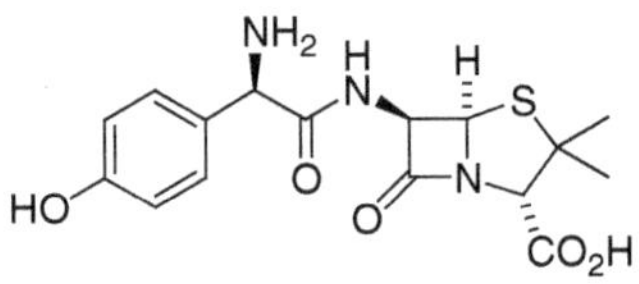

$t_{1/2}$	1.7±0.3 h	V	0.21±0.03	AMOXICILLIN
F	93±10%	CL	2.6±0.4	Metabolism: lactam and
pb	18%	Mwt	365.4	thioaminal hydrolysis,
ur	86±8%	PSA	158 Å^2	formation of diketopiperazine
		log P	0.88	

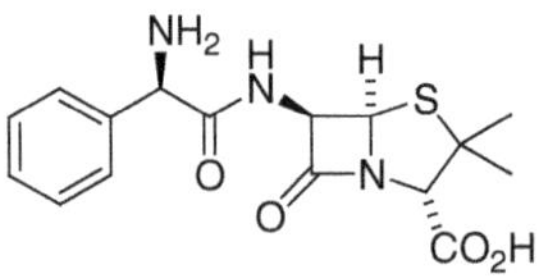

$t_{1/2}$	1.3±0.2 h	V	0.28±0.07	AMPICILLIN
F	62±17%	CL	1.7	Prodrug esters: lenampicillin
pb	18±2%	Mwt	349.4	(about 20% higher F),
ur	82±10%	PSA	113 Å^2	pivampicillin, talampicillin
		log P	−0.87	

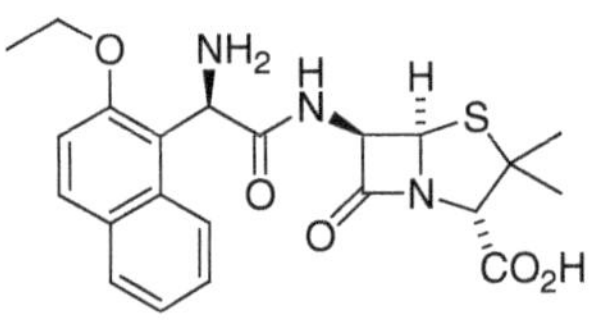

$t_{1/2}$	1.0±0.2 h	V	0.35±0.09	NAFCILLIN
F	36%	CL	7.5±1.9	
pb	89–90%	Mwt	414.5	
ur	27±5%	PSA	121 Å^2	
		log P	2.60	

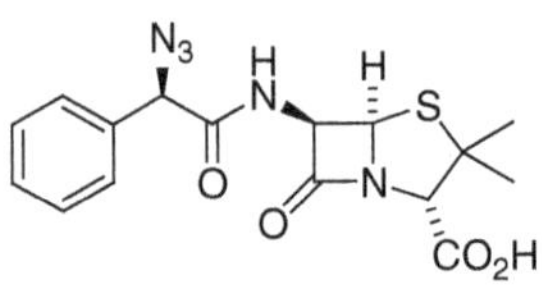

$t_{1/2}$	0.6–1.1 h	V	0.4–0.5	AZIDOCILLIN
F	57–64%	CL	–	Metabolism in animals:
pb	–	Mwt	375.4	reduction of azide to amine
ur	37–40%	PSA	136.5 Å^2	(minor extent), β-lactam
		log P	2.30	hydrolysis

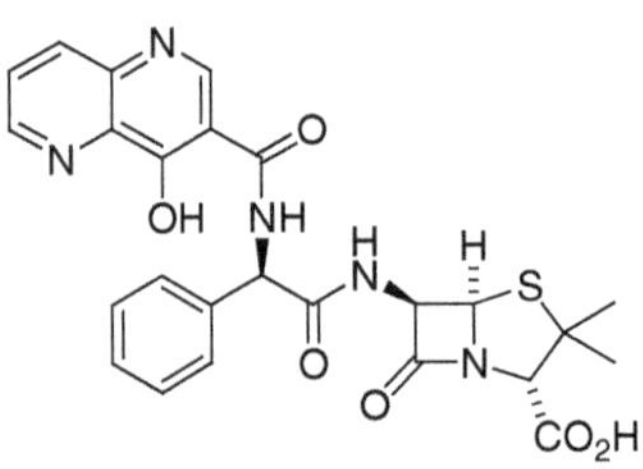

$t_{1/2}$	1.2±0.2 h*	V	0.19±0.03	APALCILLIN
F	–	CL	2.0±0.5	*iv
pb	38%	Mwt	521.5	
ur	18±3%	PSA	187 Å^2	
		log P	3.67	

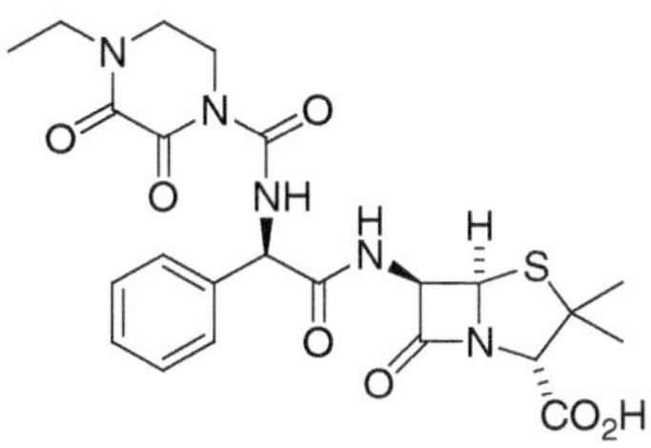	$t_{1/2}$ *F* pb ur	0.9±0.1 h Low 16–22% 71±14%	*V* CL Mwt PSA log *P*	0.18±0.03 2.6±0.7 517.6 182 Å^2 1.00	PIPERACILLIN
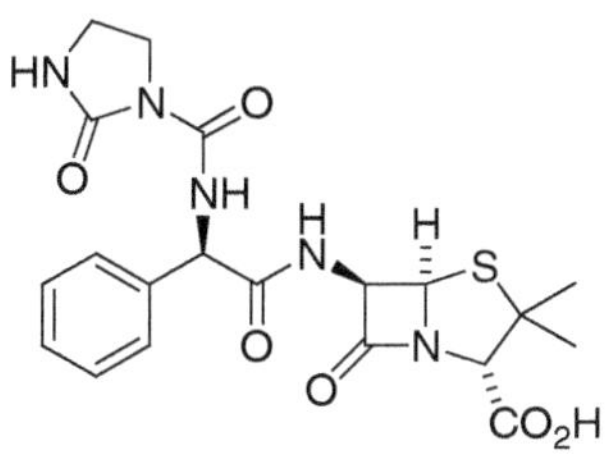	$t_{1/2}$ *F* pb ur	0.8–1.5 h Low 28±6% 65±9%	*V* CL Mwt PSA log *P*	0.26 2.5 461.5 173 Å^2 0.83	AZLOCILLIN
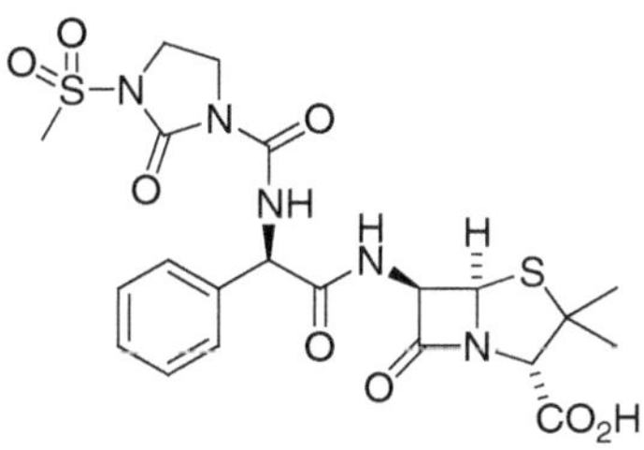	$t_{1/2}$ *F* pb ur	1.3±0.4 h 0% 29±13% 45±6%	*V* CL Mwt PSA log *P*	0.20±0.06 1.44 539.6 207 Å^2 0.33	MEZLOCILLIN
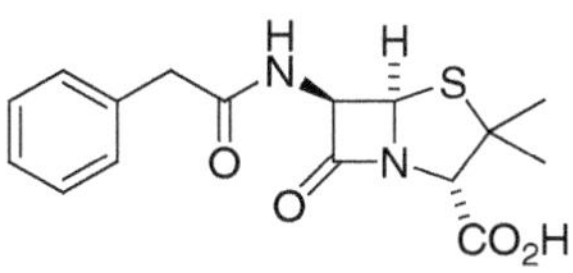	$t_{1/2}$ *F* pb ur	30–40 min <30% 45–65% –	*V* CL Mwt PSA log *P*	0.3–0.4 6.9 334.4 112 Å^2 1.92	BENZYLPENICILLIN, PENICILLIN G Diethylaminoethyl ester of benzylpenicillin: penethamate
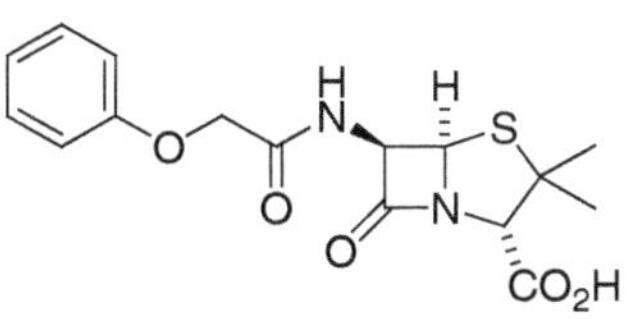	$t_{1/2}$ *F* pb ur	0.5–1.0 h 40–60% 55–78% 20–40%	*V* CL Mwt PSA log *P*	0.73 350.4 121 Å^2 1.94	PHENOXYMETHYL- PENICILLIN, PENICILLIN V
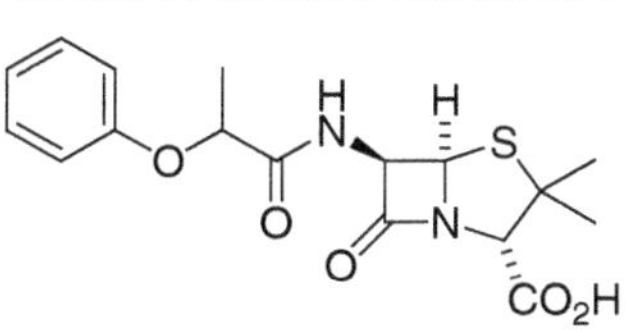	$t_{1/2}$ *F* pb ur	0.9 h (iv) 86% 75% 48%	*V* CL Mwt PSA log *P*	0.30 4.2 364.4 121 Å^2 2.29	PHENETHICILLIN

(continued overleaf)

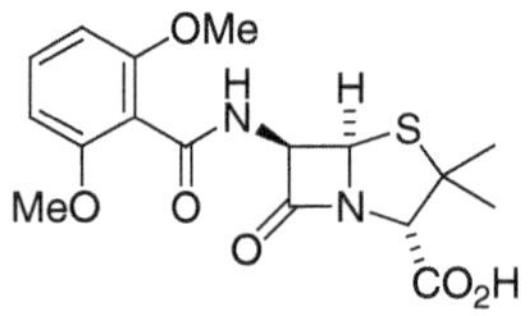

$t_{1/2}$	0.6–0.9 h*	*V*	0.3–0.4	METHICILLIN
F	–	CL	6.1±1.3	*iv
pb	57%	Mwt	380.4	
ur	88±17%	PSA	130 Å^2	
		log *P*	1.02	

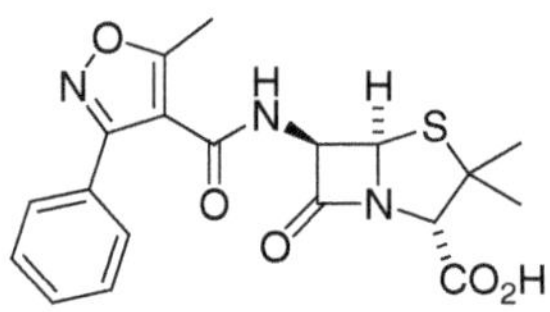

$t_{1/2}$	0.6±0.2 h	*V*	0.33±0.09	OXACILLIN
F	33%	CL	6.1±1.7	
pb	92.2±0.6%	Mwt	401.4	
ur	46±4%	PSA	138 Å^2	
		log *P*	2.55	

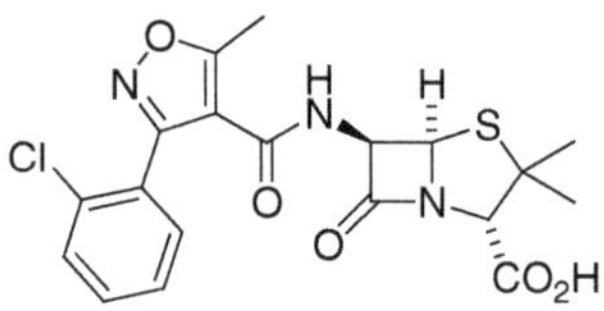

$t_{1/2}$	0.6±0.1 h	*V*	0.09±0.02	CLOXACILLIN
F	43±16%	CL	2.2±0.5	
pb	94.6±0.6%	Mwt	435.9	
ur	75±14%	PSA	138 Å^2	
		log *P*	2.38	

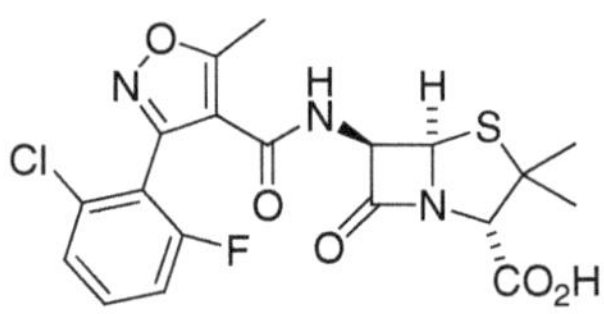

$t_{1/2}$	53 min	*V*	0.19	FLUCLOXACILLIN,
F	55%	CL	2.4	FLOXACILLIN
pb	95%	Mwt	453.9	
ur	35–76%	PSA	138 Å^2	
		log *P*	2.89	

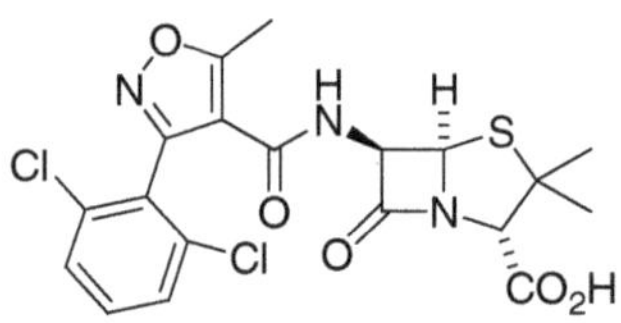

$t_{1/2}$	0.7±0.1 h	*V*	0.09±0.02	DICLOXACILLIN
F	68±18%	CL	1.6±0.3	
pb	96%	Mwt	470.3	
ur	60±7%	PSA	138 Å^2	
		log *P*	3.10	

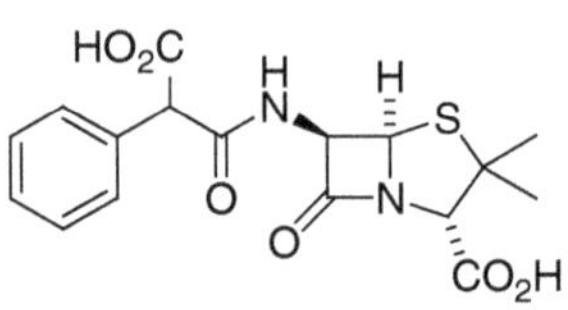

$t_{1/2}$	1.0–1.5 h	*V*	0.18	CARBENICILLIN
F	30–40%	CL	0.68–1.9	
pb	48%	Mwt	378.4	
ur	80–99%	PSA	149 Å^2	
		log *P*	1.13	

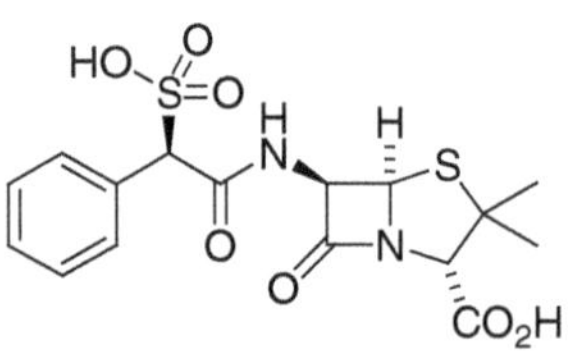

$t_{1/2}$	1.2±0.2 h*	*V*	0.15	SULBENICILLIN
F	Low	CL	1.6	*iv
pb	50%	Mwt	414.5	
ur	80%	PSA	175 Å^2	
		log *P*	−0.51	

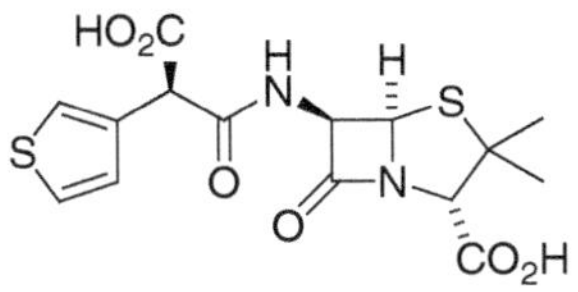	$t^1/_2$	1.2±0.2 h	*V*	0.21±0.04	TICARCILLIN
	F	Low	CL	1.6±0.3	
	pb	55–65%	Mwt	384.4	
	ur	77±12%	PSA	178 Å^2	
			log *P*	1.24	
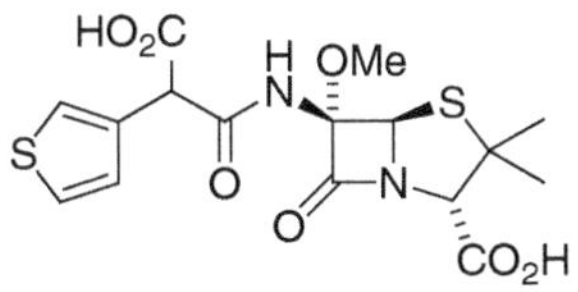	$t^1/_2$	4–6 h (iv)	*V*	0.27–0.30	TEMOCILLIN
	F	–	CL	0.4–0.7	
	pb	75–88%	Mwt	414.5	
	ur	67–72%	PSA	187 Å^2	
			log *P*	7.19	
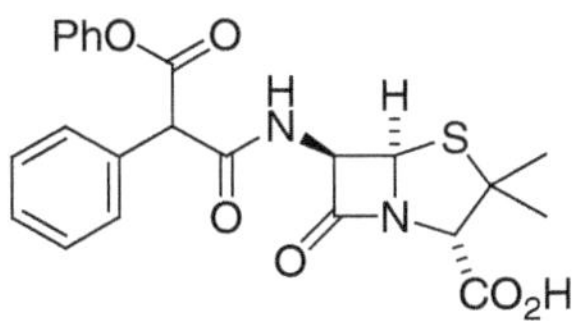	$t^1/_2$	1.2 h	*V*	–	CARFECILLIN
	F	43%*	CL	–	*acid (carbenicillin) on oral
	pb	–	Mwt	454.5	dosing of carfecillin
	ur	–	PSA	138 Å^2	
			log *P*	2.85	
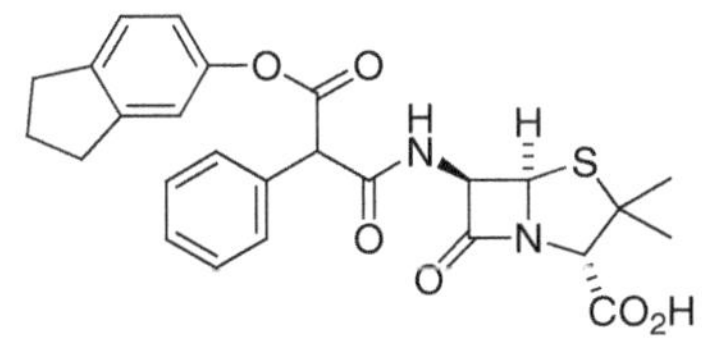	$t^1/_2$	1.2 h	*V*	–	CARINDACILLIN
	F	69%*	CL	–	*acid (carbenicillin) on oral
	pb	–	Mwt	494.6	dosing of carindacillin
	ur	–	PSA	138 Å^2	
			log *P*	3.58	

$t_{1/2}$, plasma half-life; *F*, oral bioavailability; pb, plasma protein binding; ur, excretion of unchanged drug in urine; V, volume of distribution; CL, clearance; Mwt, molecular weight; PSA, polar surface area.

Table 70.2 Cephalosporins and related compounds. *V* in l kg^{-1}; CL in ml min^{-1} kg^{-1}; Mwt in g mol^{-1}.

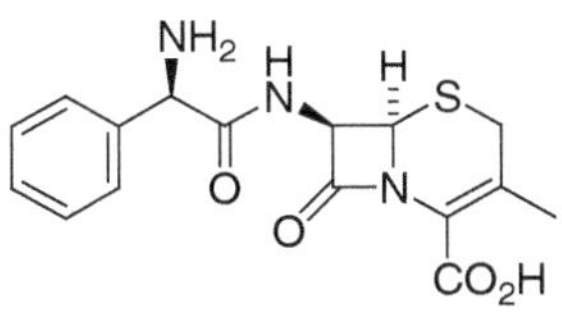	$t^1/_2$	0.9±0.2 h	*V*	0.26±0.03	CEPHALEXIN
	F	90±9%	CL	4.3±1.1	
	pb	14±3%	Mwt	347.4	
	ur	91±18%	PSA	138 Å^2	
			log *P*	0.35	
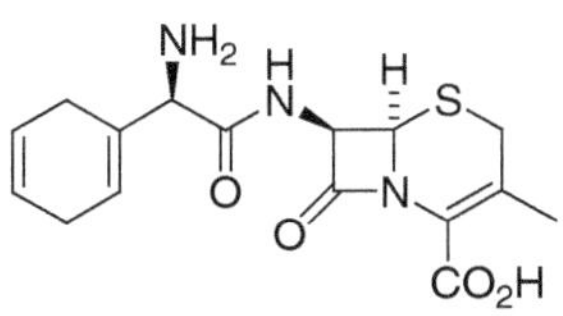	$t^1/_2$	0.9±0.3 h	*V*	0.46±0.07	CEPHRADINE
	F	94±12%	CL	4.8±0.6	
	pb	14±3%	Mwt	349.4	
	ur	86±10%	PSA	138 Å^2	
			log *P*	0.48	

(continued overleaf)

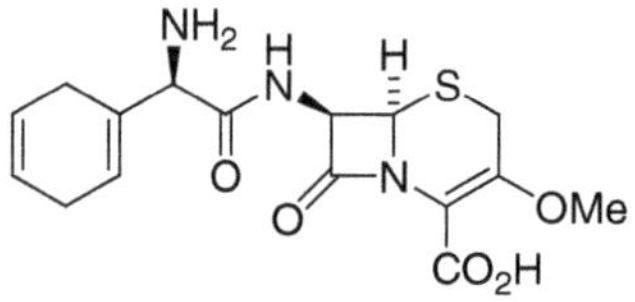

$t^1/_2$	1.0 h	V	0.25	CEFROXADINE
F	100%	CL	3.1	
pb	10%	Mwt	365.4	
ur	71%	PSA	147 Å^2	
		log P	1.46	

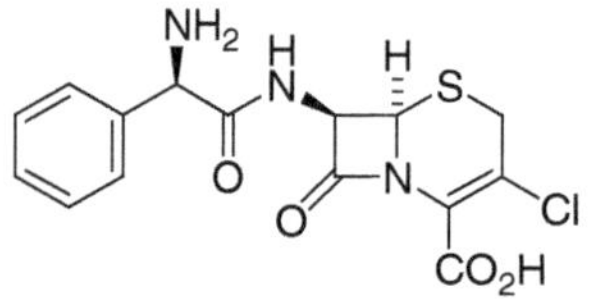

$t^1/_2$	0.7±0.3 h	V	0.36	CEFACLOR
F	52±17%	CL	6.1±1.5	
pb	25%	Mwt	367.8	
ur	52±17%	PSA	138 Å^2	
		log P	0.14	

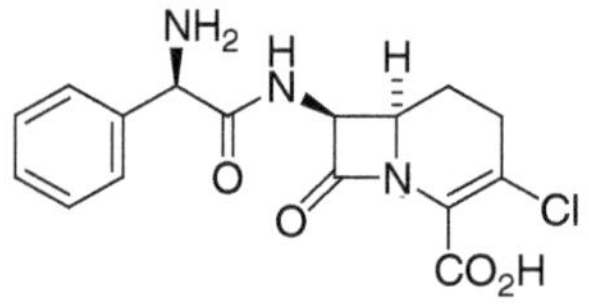

$t^1/_2$	1.2±0.3 h	V	0.32±0.10	LORACARBEF
F	94±16%	CL	1.7	
pb	25%	Mwt	349.8	
ur	94±16%	PSA	113 Å^2	
		log P	−1.60	

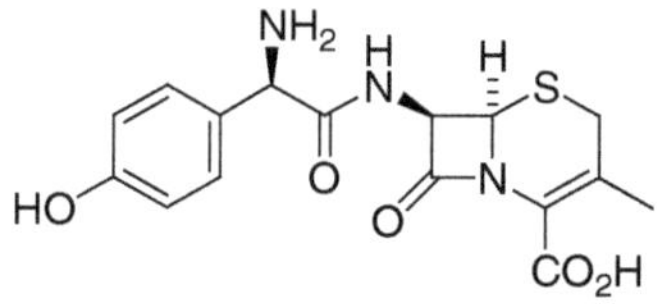

$t^1/_2$	1.2±0.2 h	V	0.24±0.05	CEFADROXIL
F	100%	CL	2.9±0.7	
pb	20%	Mwt	363.4	
ur	93±4%	PSA	158 Å^2	
		log P	−0.25	

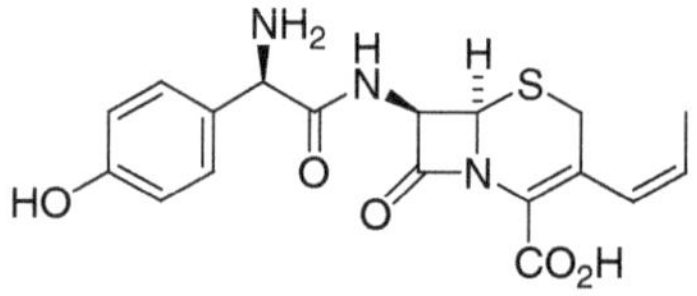

$t^1/_2$	1.5±0.4 h	V	0.22±0.02	CEFPROZIL
F	90±10%	CL	3.0±0.4	cis/trans 9 : 1
pb	40±4%	Mwt	389.4	
ur	73±7%	PSA	158 Å^2	
		log P	−0.60	

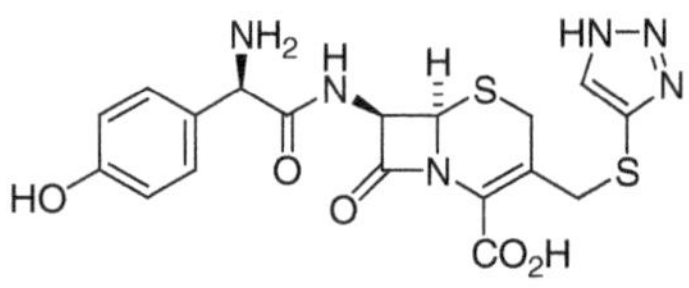

$t^1/_2$	1.2–1.4 h	V	0.22	CEFATRIZINE
F	50–75%	CL	3.3	
pb	60%	Mwt	462.5	
ur	42–60%	PSA	225 Å^2	
		log P	−1.07	

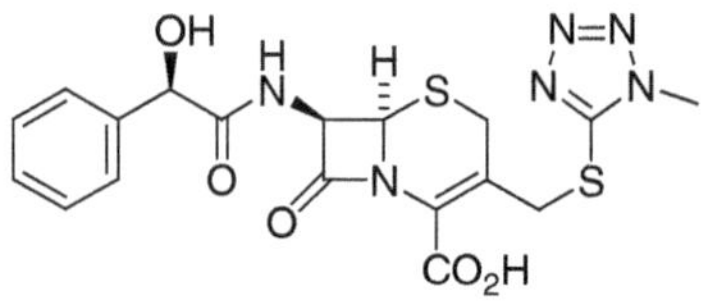

$t^1/_2$	0.8±0.1 h	V	0.16±0.05	CEFAMANDOLE
F	96±3%	CL	2.8±1.0	
pb	74%	Mwt	462.5	
ur	96±3%	PSA	201 Å^2	
		log P	−0.04	

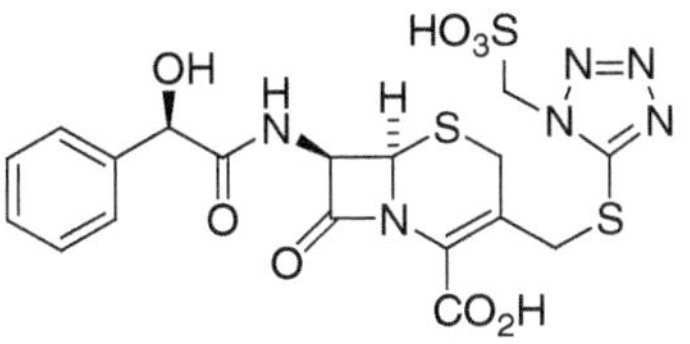

$t_{1/2}$	4.4±0.8 h	*V*	0.11±0.01	CEFONICID
F	–	CL	0.32±0.06	
pb	98%	Mwt	542.6	
ur	88±6%	PSA	264 Å^2	
		log *P*	−2.29	

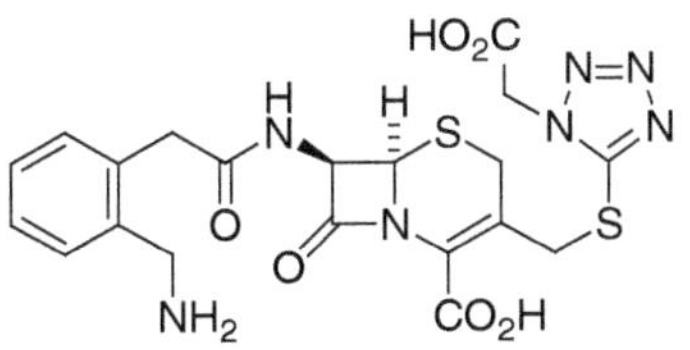

$t_{1/2}$	2.6±0.5 h	*V*	0.14±0.04	CEFORANIDE
F	–	CL	0.26–0.66	
pb	80–82%	Mwt	519.6	
ur	84±3%	PSA	244 Å^2	
		log *P*	−0.36	

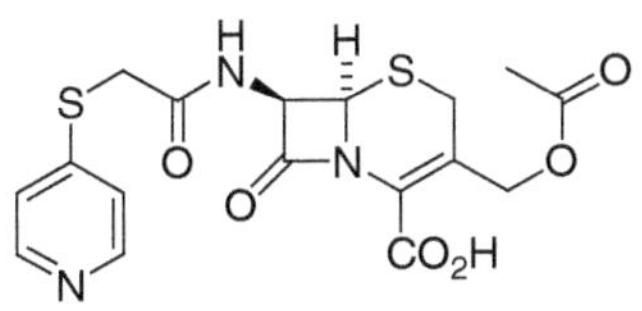

$t_{1/2}$	0.7±0.2 h	*V*	0.21±0.06	CEPHAPIRIN
F	–	CL	6.9±2.0	Metabolism: O-deacetylation,
pb	62±4%	Mwt	423.5	then lactonization
ur	48±7%	PSA	177 Å^2	
		log *P*	−0.41	

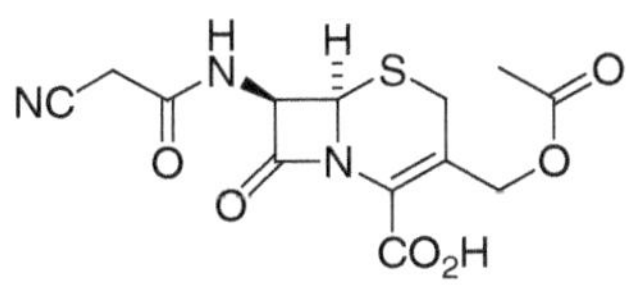

$t_{1/2}$	0.9 h (iv)	*V*	0.41	CEPHACETRILE
F	Low	CL	6.3	*rabbit
pb	27%*	Mwt	339.3	
ur	–	PSA	162 Å^2	
		log *P*	−1.60	

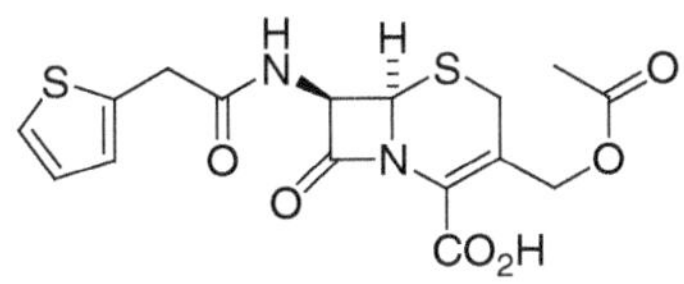

$t_{1/2}$	0.6±0.3 h	*V*	0.26±0.11	CEPHALOTHIN
F	–	CL	6.7±1.7	
pb	71±3%	Mwt	396.4	
ur	52%	PSA	167 Å^2	
		log *P*	0.09	

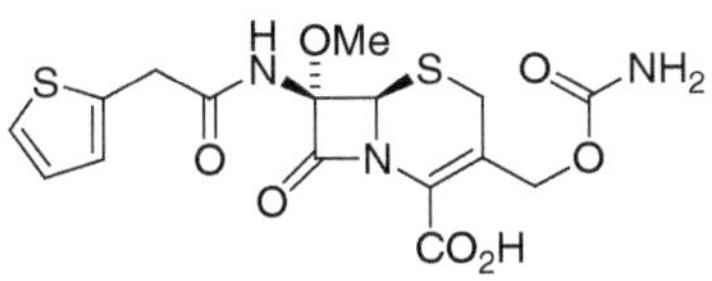

$t_{1/2}$	0.8±0.1 h	*V*	0.25±0.10	CEFOXITIN
F	Low	CL	3.3	
pb	73%	Mwt	427.5	
ur	79±5%	PSA	202 Å^2	
		log *P*	0.00	

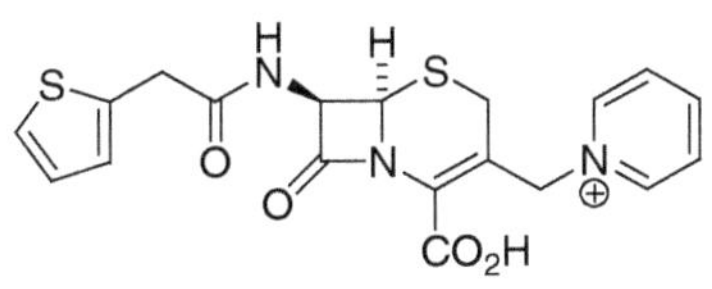

$t_{1/2}$	1.5 h (iv)	*V*	0.46	CEPHALORIDINE
F	Low	CL	3.3	
pb	20%	Mwt	416.5	
ur	–	PSA	90.6 Å^2	
		log *P*	−4.26	

(*continued overleaf*)

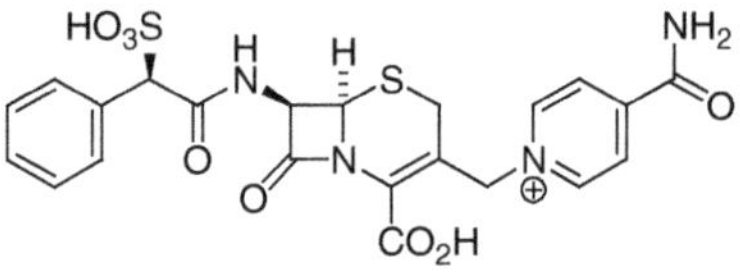

$t\frac{1}{2}$	1.8 h	*V*	0.28	CEFSULODIN
F	–	CL	1.85	
pb	15%	Mwt	533.6	
ur	–	PSA	188.1 Å^2	
		log *P*	−5.72	

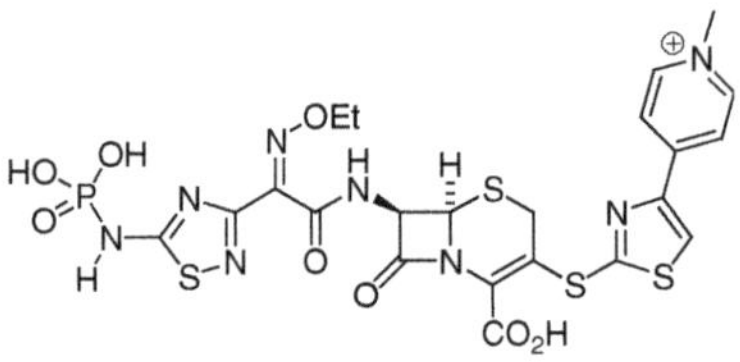

$t\frac{1}{2}$	2–3 h*	*V*	0.37*	CEFTAROLINE FOSAMIL
F	Low	CL	–	*dephosphorylated metabolite (ceftaroline) on iv dosing of prodrug
pb	<20%*	Mwt	684.7	
ur	–	PSA	207.5 Å^2	
		log *P*	−4.79	

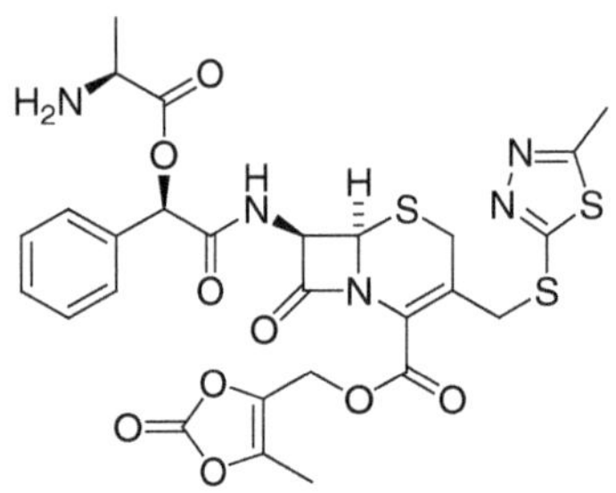

$t\frac{1}{2}$	1 h*	*V*	0.13	CEFCANEL DALOXATE
F	40%*	CL	2.6	*acid (cefcanel) on oral dosing of ester
pb	–	Mwt	661.7	Diester prodrug of cefcanel (hydroxyacid)
ur	<70%*	PSA	268 Å^2	
		log *P*	2.85	

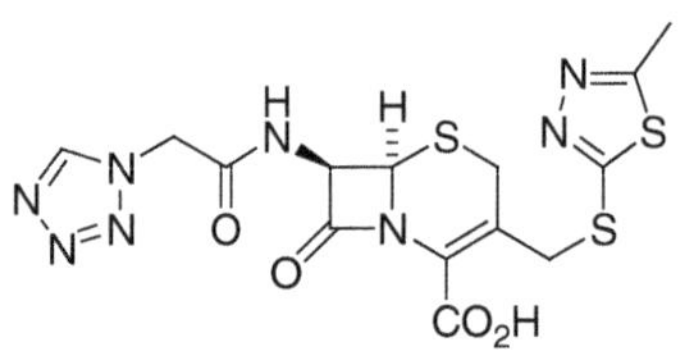

$t\frac{1}{2}$	1.8±0.4 h	*V*	0.14±0.04	CEFAZOLIN
F	>90%	CL	0.95±0.17	
pb	89±2%	Mwt	454.5	
ur	80±16%	PSA	235 Å^2	
		log *P*	−0.70	

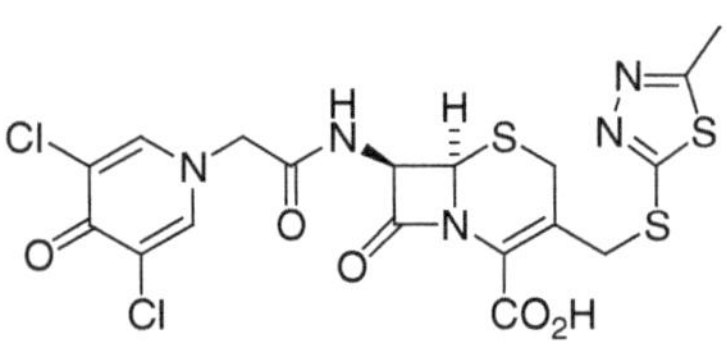

$t\frac{1}{2}$	1.6±0.2 h	*V*	0.15	CEFAZEDONE
F	–	CL	–	Similar PK as cefazolin
pb	93%	Mwt	548.4	
ur	–	PSA	212 Å^2	
		log *P*	−0.42	

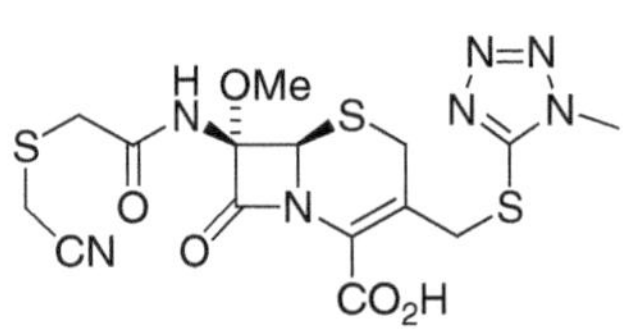

$t\frac{1}{2}$	1.5±0.3 h	*V*	0.18±0.04	CEFMETAZOLE
F	80±20%	CL	1.45±0.10	
pb	70%	Mwt	471.5	
ur	80±13%	PSA	239 Å^2	
		log *P*	−0.62	

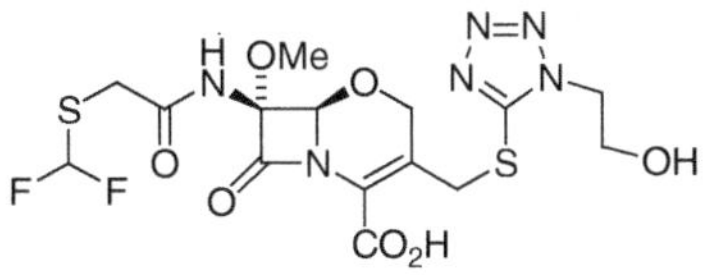	$t^1/_2$	50 min	V	0.2	FLOMOXEF
	F	–	CL	4.1	
	pb	29%	Mwt	496.5	
	ur	47%	PSA	220 Å^2	
			log P	−2.43	
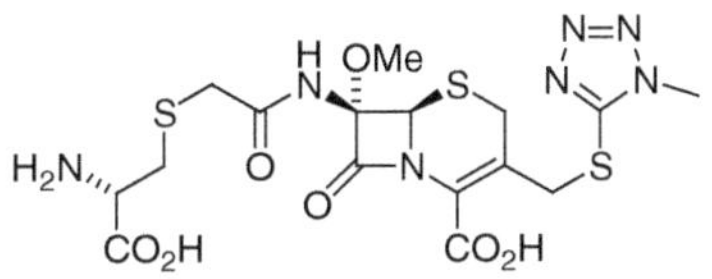	$t^1/_2$	2.4 h	V	–	CEFMINOX
	F	Low	CL	–	
	pb	–	Mwt	519.6	
	ur	–	PSA	279 Å^2	
			log P	−0.68	
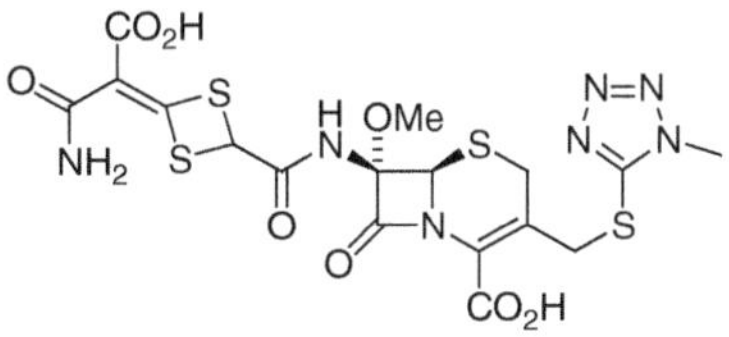	$t^1/_2$	3.6±1.0 h	V	0.14±0.03	CEFOTETAN
	F	–	CL	0.23	Metabolism: isomerization to
	pb	85±4%	Mwt	575.6	1,2-thiazole (isothiazole;
	ur	67±11%	PSA	316 Å^2	nucleophilic attack of
			log P	2.34	amide-NH at S)
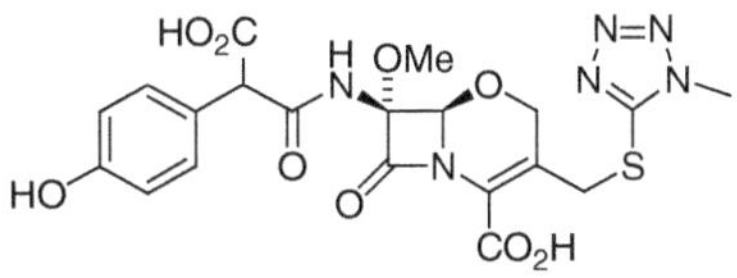	$t^1/_2$	2.1±0.7 h	V	0.25±0.08	MOXALACTAM,
	F	3%	CL	1.0	LATAMOXEF
	pb	53–67%	Mwt	520.5	Mixture of diastereomers
	ur	76±12%	PSA	232 Å^2	
			log P	−2.56	
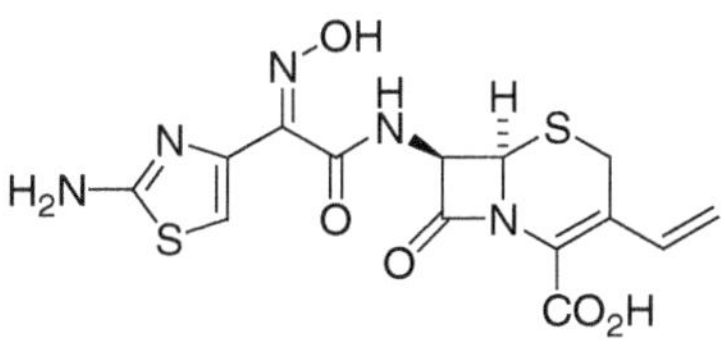	$t^1/_2$	1.7±0.6 h	V	0.27±0.08	CEFDINIR
	F	16–21%	CL	0.5±0.2	
	pb	60–70%	Mwt	395.4	
	ur	12–18%	PSA	212 Å^2	
			log P	−1.78	
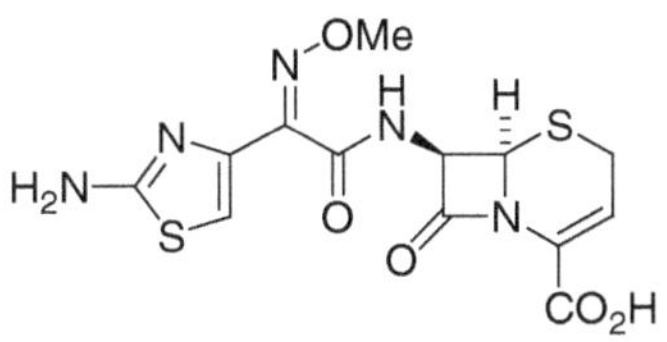	$t^1/_2$	1.8±0.7 h	V	0.36±0.19	CEFTIZOXIME
	F	Low	CL	1.1	
	pb	28±5%	Mwt	383.4	
	ur	93±8%	PSA	201 Å^2	
			log P	0.59	
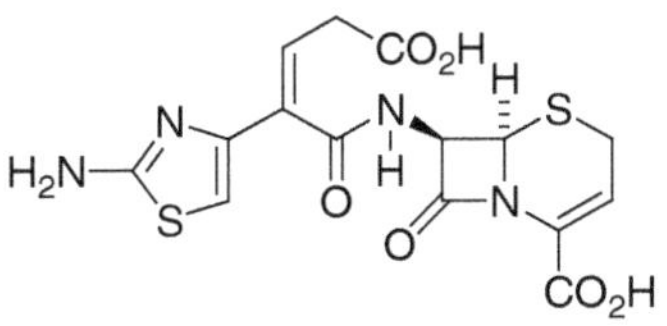	$t^1/_2$	2.5 h	V	0.21±0.03	CEFTIBUTEN
	F	100%	CL/F	1.3	
	pb	62–64%	Mwt	410.4	
	ur	7–26%	PSA	216 Å^2	
			log P	−3.19	

(*continued overleaf*)

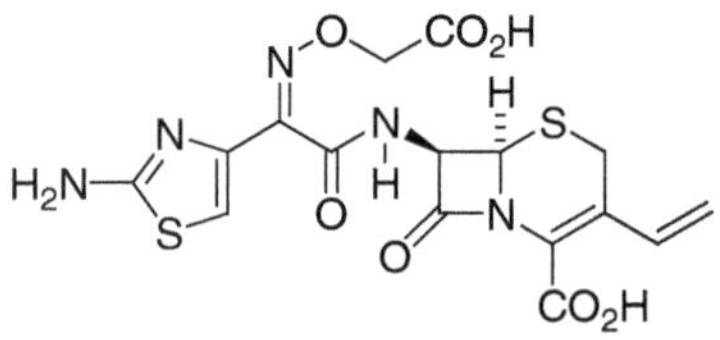

$t^1/_2$	3.0±0.4 h	V	0.30±0.03	CEFIXIME
F	47±15%	CL	1.3±0.2	
pb	67±1%	Mwt	453.5	
ur	41±7%	PSA	238 Å^2	
		log P	1.00	

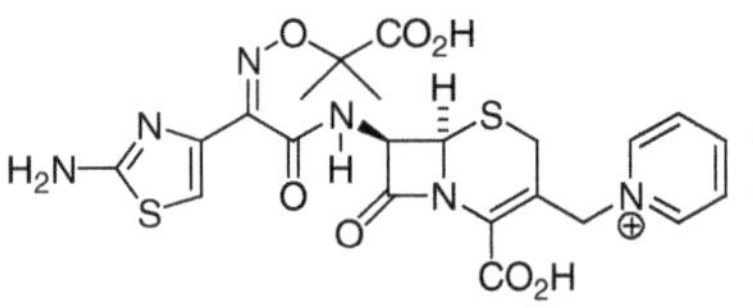

$t^1/_2$	1.6±0.1 h	V	0.23±0.02	CEFTAZIDIME
F	Low	CL	1.05	
pb	21±6%	Mwt	547.6	
ur	84±4%	PSA	188.4 Å^2	
		log P	−4.98	

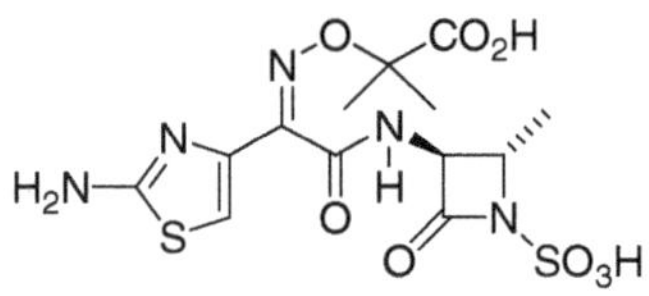

$t^1/_2$	1.7±0.2 h	V	0.16±0.02	AZTREONAM
F	<1%	CL	1.3±0.1	No metabolites identified
pb	56%	Mwt	435.4	
ur	68±8%	PSA	238 Å^2	
		log P	−0.66	

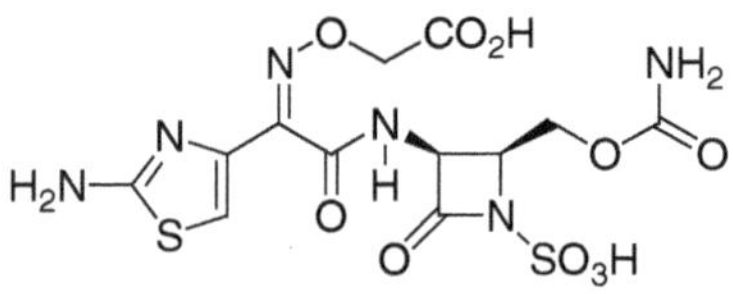

$t^1/_2$	1.6–1.8 h*	V	0.18	CARUMONAM
F	–	CL	1.5	*iv
pb	28%	Mwt	466.4	
ur	75–81%	PSA	291 Å^2	
		log P	−2.26	

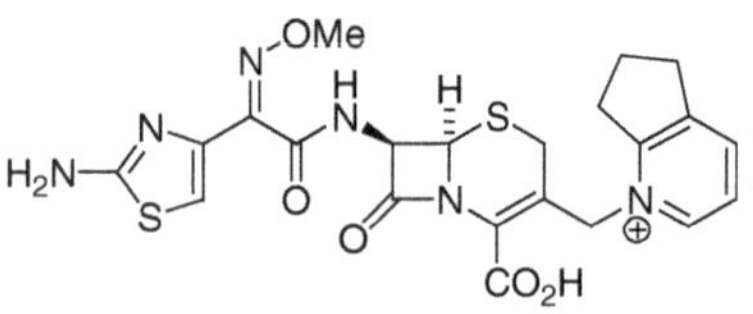

$t^1/_2$	1.7–2.2 h*	V	0.24	CEFPIROME
F	–	CL	1.9	*iv
pb	10%	Mwt	515.6	
ur	80–90%	PSA	151.1 Å^2	
		log P	−4.08	

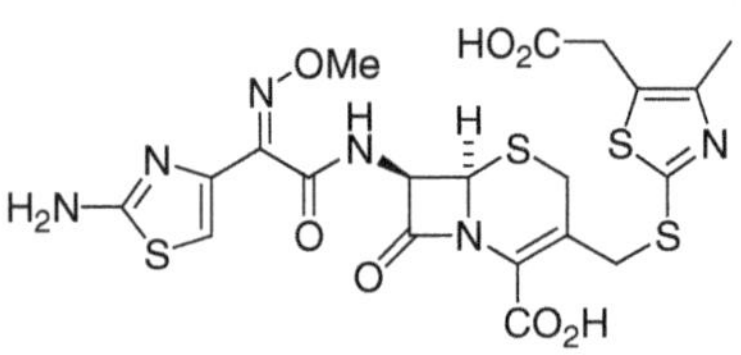

$t^1/_2$	2.7–4.2 h*	V	0.054	CEFODIZIME
F	–	CL	0.61	*iv
pb	82%	Mwt	584.7	
ur	80%	PSA	304 Å^2	
		log P	2.55	

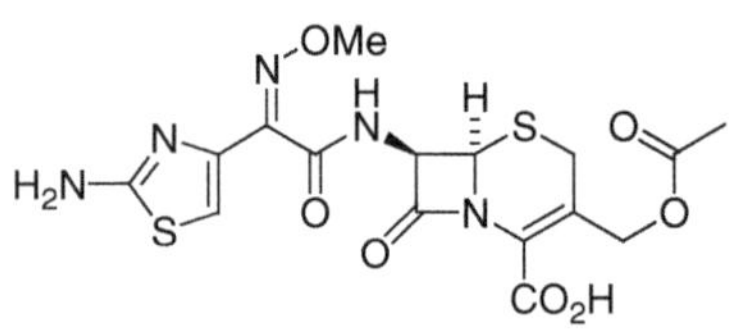

$t^1/_2$	1.1±0.3 h	V	0.23±0.06	CEFOTAXIME
F	Low	CL	3.7±0.6	Active metabolite: desacetyl
pb	36±3%	Mwt	455.5	($t^1/_2$ 1.5 h)
ur	55±10%	PSA	227 Å^2	
		log P	1.20	

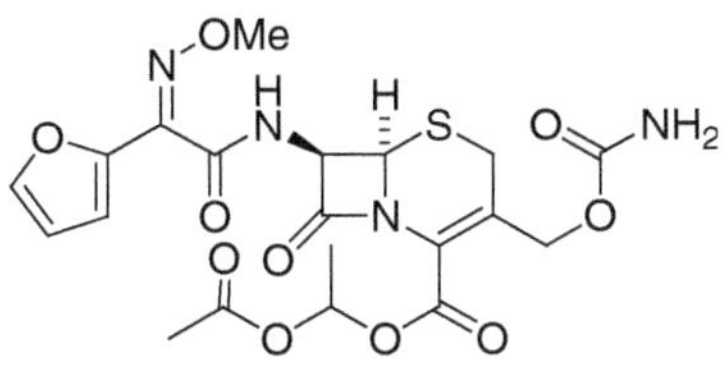

$t^1/_2$	1.7±0.6 h*	V	0.20±0.04
F	68%*	CL	0.94*
pb	33±6%*	Mwt	510.5
ur	96±10%*	PSA	214 Å^2
		log P	0.85

CEFUROXIME AXETIL
*acid (cefuroxime) on oral dosing of ester
Metabolism: ester hydrolysis, O-demethylation

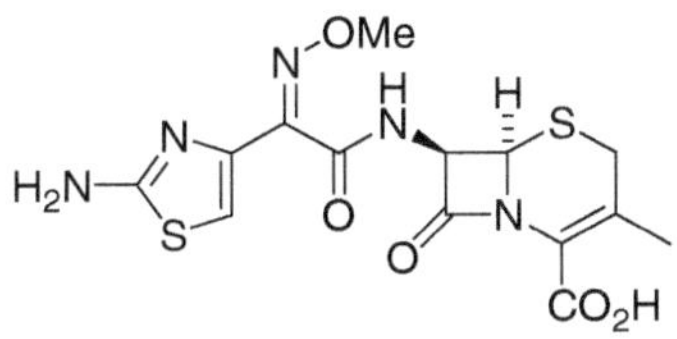

$t^1/_2$	2.1–2.9 h*	V	0.3–0.4
F	–	CL	1.7–2.3
pb	22%	Mwt	397.4
ur	63–99%*	PSA	201 Å^2
		log P	−0.24

CEFETAMET
*iv

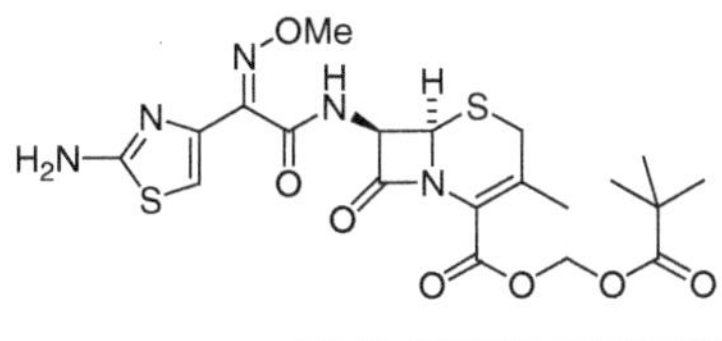

$t^1/_2$	2.5±0.3 h*	V	0.3*
F	45±13%*	CL	2.9–6.4*
pb	–	Mwt	511.6
ur	27–55%*	PSA	216 Å^2
		log P	2.44

CEFETAMET PIVOXIL
*acid (cefetamet) on oral dosing of ester
Prodrug of cefetamet (acid), no prodrug can be detected in plasma after po dosing

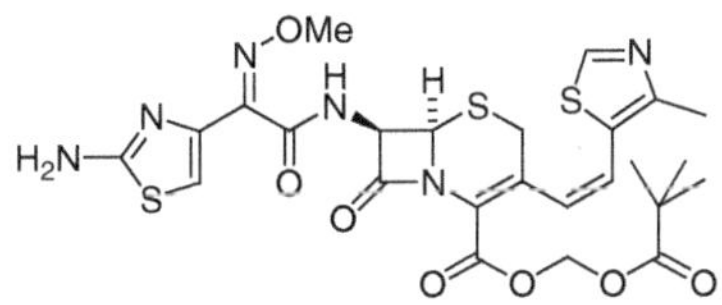

$t^1/_2$	1.6±0.4 h*	V	0.14±0.03*
F	14%*	CL	0.1–0.3*
pb	88%*	Mwt	620.7
ur	–	PSA	257 Å^2
		log P	3.35

CEFDITOREN PIVOXIL
*acid (cefditoren) on oral dosing of ester

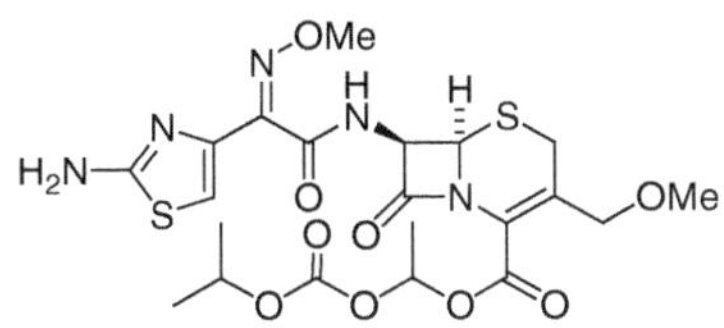

$t^1/_2$	2.3±0.3 h*	V	0.46±0.03*
F	52±5%*	CL	2.4±0.1*
pb	27±4%*	Mwt	557.6
ur	81±5%*	PSA	235 Å^2
		log P	2.17

CEFPODOXIME PROXETIL
*acid (cef) on oral dosing of ester

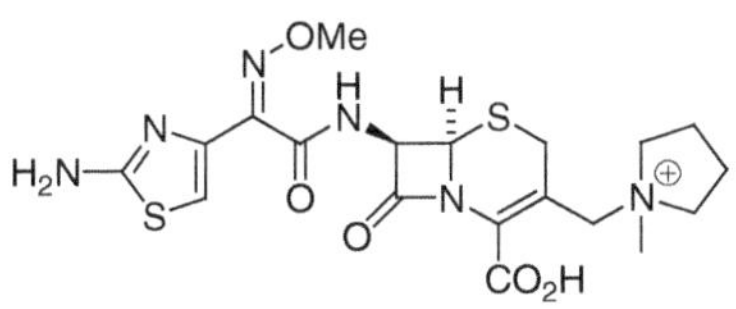

$t^1/_2$	1.9–2.5 h*	V	0.28
F	–	CL	2.2
pb	16–22%	Mwt	481.6
ur	76–85%*	PSA	147.2 Å^2
		log P	−4.40

CEFEPIME
*iv
Metabolism: epimerization, N-dealkylation (to *N*-methylpyrrolidine)

(continued overleaf)

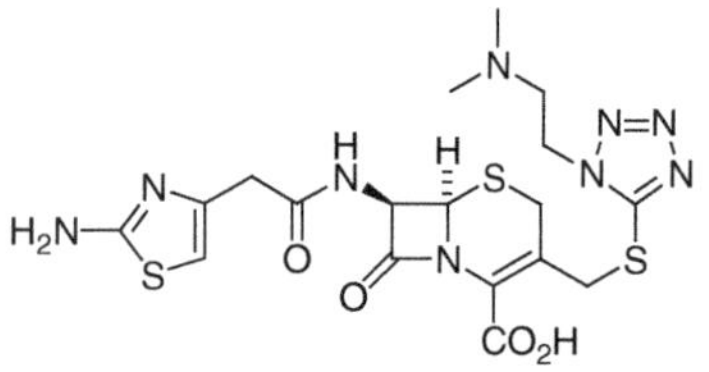

$t\frac{1}{2}$	0.6–1.1 h	*V*	–	CEFOTIAM
F	–	CL	6.7±1.5	Prodrug: cefotiam hexetil; *F*: 46% of cefotiam after oral dosing of prodrug
pb	40%	Mwt	525.6	
ur	82%	PSA	251 Å^2	
		log *P*	0.23	

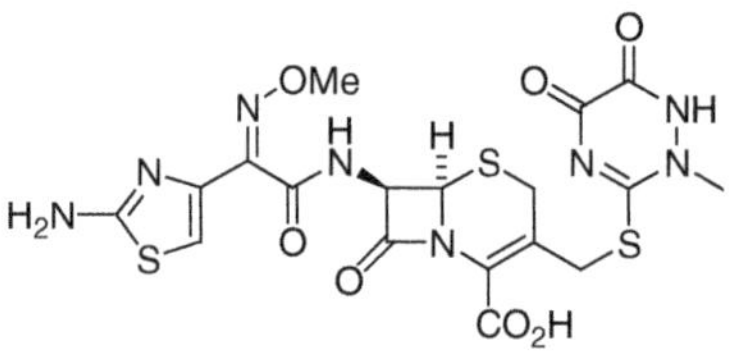

$t\frac{1}{2}$	7.3±1.6 h	*V*	0.16±0.03	CEFTRIAXONE
F	Low	CL	0.24±0.06	
pb	90–95%	Mwt	554.6	
ur	49±13%	PSA	288 Å^2	
		log *P*	−0.25	

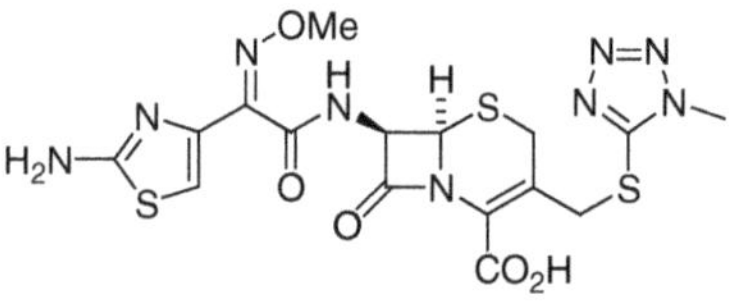

$t\frac{1}{2}$	1.1–1.6 h*	*V*	0.23	CEFMENOXIME
F	–	CL	3.3	*iv
pb	50–60%	Mwt	511.6	
ur	81%*	PSA	270 Å^2	
		log *P*	−0.91	

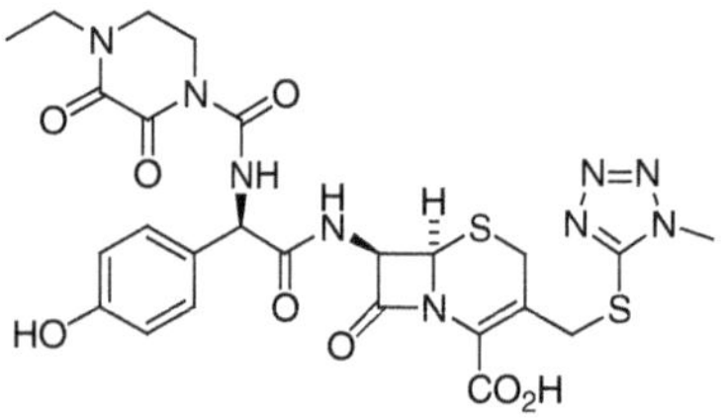

$t\frac{1}{2}$	2.2±0.3 h	*V*	0.14±0.03	CEFOPERAZONE
F	–	CL	1.2±0.2	Metabolism: hydrolysis of oxalamide (*CON*–CONH)
pb	89–93%	Mwt	645.7	
ur	29±4%	PSA	266 Å^2	
		log *P*	1.43	

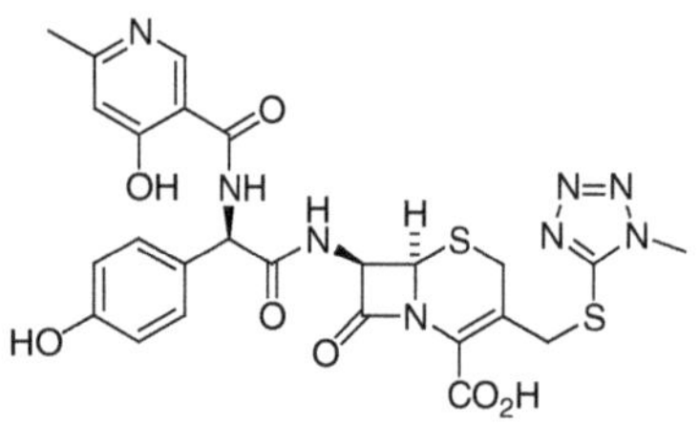

$t\frac{1}{2}$	5.4±1.4 h*	*V*	0.15±0.03	CEFPIRAMIDE
F	–	CL	0.43±0.08	*iv
pb	95–99%	Mwt	612.6	
ur	17–22%*	PSA	263 Å^2	
		log *P*	0.17	

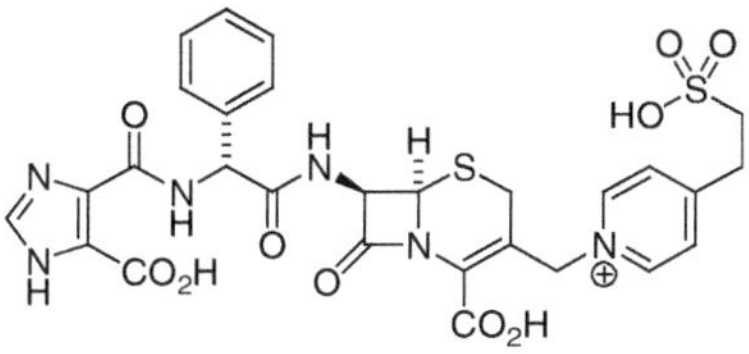	$t^{1}/_{2}$	1.9 h (iv)	V	0.30±0.05	CEFPIMIZOLE
	F	–	CL	1.7±0.3	
	pb	–	Mwt	671.7	
	ur	80%	PSA	240.0 Å^2	
			log P	−5.68	
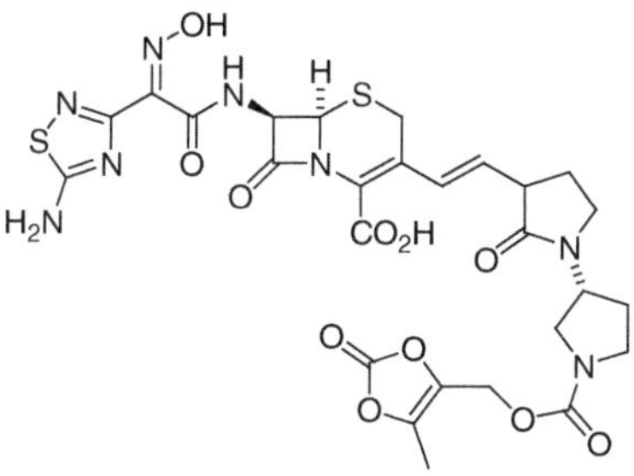	$t^{1}/_{2}$	3.3 h*	V	0.27	CEFTOBIPROLE MEDOCARIL
	F	–	CL	1.5	*amine (ceftobiprole) on iv dosing of carbamate
	pb	16–38%	Mwt	690.7	
	ur	70%	PSA	310 Å^2	
			log P	−1.02	

$t_{1/2}$, plasma half-life; F, oral bioavailability; pb, plasma protein binding; ur, excretion of unchanged drug in urine; V, volume of distribution; CL, clearance; Mwt, molecular weight; PSA, polar surface area.

71
Prostaglandin Analogs

Arachidonic acid is the biochemical precursor of the prostaglandins, prostacyclins, thromboxanes, and leukotrienes (Scheme 71.1). These hormones are carboxylic acids with 20 carbon atoms, also called *eicosanoids*. In 1930, the American gynecologists Raphael Kurzrok and Charles Lieb discovered that semen contained these substances, which cause the contraction or relaxation of uterine muscle. Since then, eicosanoids were found to be involved in a broad spectrum of biological activities.

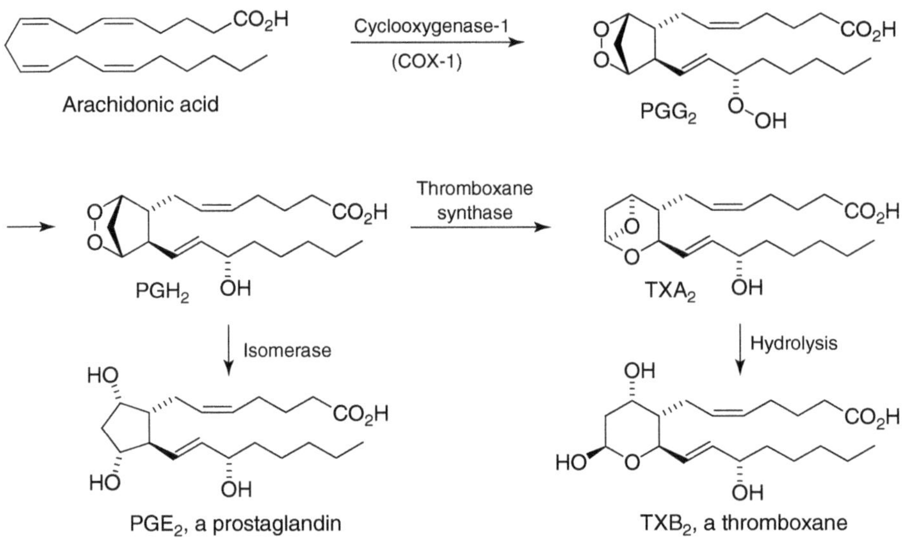

Scheme 71.1 Biosynthesis of eicosanoids.

Like the steroid hormones, eicosanoids are extremely potent. Oral doses of misoprostol, an analog of prostaglandin E_2 (PGE_2) suitable for the treatment of gastric ulcers, range from only 0.4 to 0.8 mg daily.

Because natural prostaglandins have low oral bioavailabilities and short half-lives, synthetic analogs with improved pharmacokinetics (PK) have been prepared (Table 71.1). These are used as abortifacients, gastroprotectants, and antithrombotics (e.g., for treatment of acute myocardial infarction).

Lead Optimization for Medicinal Chemists: Pharmacokinetic Properties of Functional Groups and Organic Compounds, First Edition. Florencio Zaragoza Dörwald.

Typical metabolic transformations include oxidation of hydroxyl groups, reduction of ketones, hydroxylation, and oxidative degradation. As for most nonhalogenated alkanes, these transformations proceed quickly, and most prostaglandin analogs undergo extensive first-pass metabolism.

Table 71.1 Prostaglandin analogs. V in l kg^{-1}; CL in ml min^{-1} kg^{-1}; Mwt in g mol^{-1}.

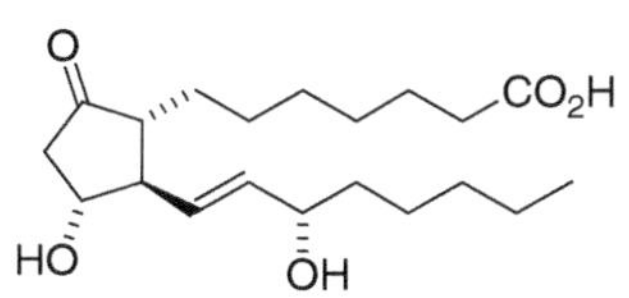	$t^1\!/_2$ 5–10 min F 14% pb 93% ur 0%	V 0.8 CL 20 Mwt 354.5 PSA 94.8 Å^2 log P 2.75	ALPROSTADIL, PGE$_1$ Treatment of erectile dysfunction Metabolism: oxidation of allylic OH to ketone
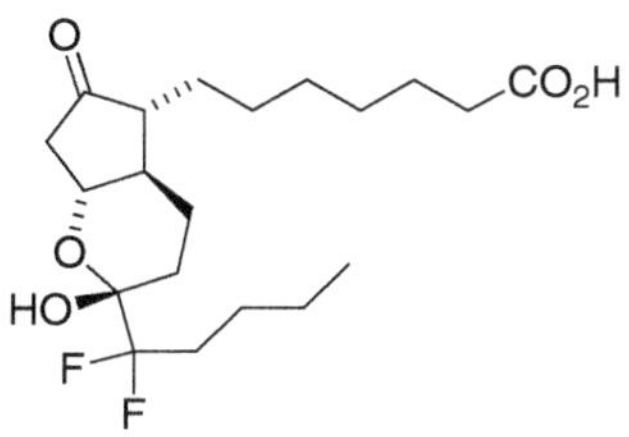	$t^1\!/_2$ 1–2 h* F Low pb 94% ur 0%	V – CL – Mwt 390.5 PSA 83.8 Å^2 log P 2.85	LUBIPROSTONE *reduced difluoroketone Laxative Metabolism: reduction of difluoroketone to alcohol
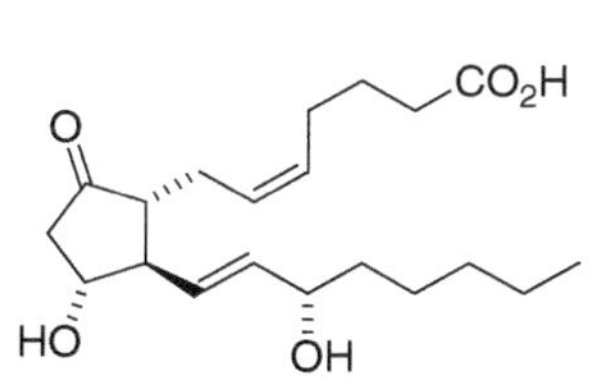	$t^1\!/_2$ 1–5 min F – pb 73% ur Low	V – CL – Mwt 352.5 PSA 94.8 Å^2 log P 2.97	DINOPROSTONE Abortifacient Metabolism: hydrogenation of double bonds, oxidation of exocyclic alcohol to ketone, oxidation of methyl to CO_2H, oxidative degradation of carboxyalkyl group
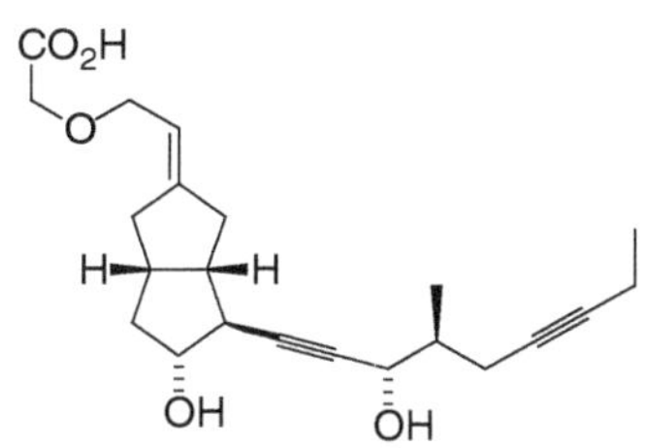	$t^1\!/_2$ 2.0±0.5 h F 100% pb – ur 60%	V 0.08 CL 3.8 Mwt 374.5 PSA 87 Å^2 log P 1.94	CICAPROST Inhibitor of platelet aggregation, vasodilator

(continued overleaf)

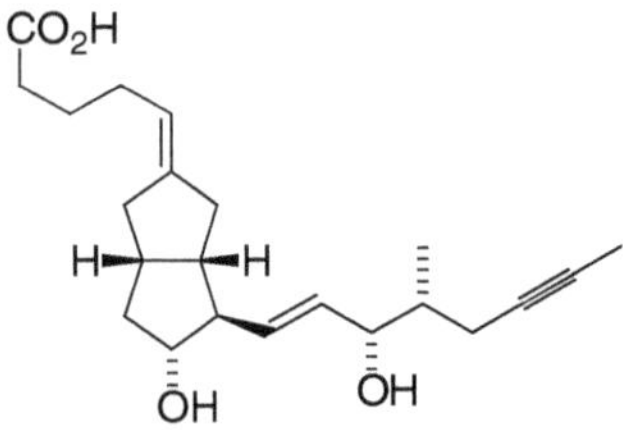	$t^1/_2$	0.3–0.5 h	*V*	0.6–0.8	ILOPROST
	F	15–20%	CL	16–20	Inhibitor of platelet aggregation, vasodilator; mixture of diastereomers
	pb	60%	Mwt	360.5	Metabolism: oxidation and hydroxylation of alkyl chains, oxidation of allylic OH to ketone
	ur	0%	PSA	77.8 Å^2	
			log *P*	3.28	
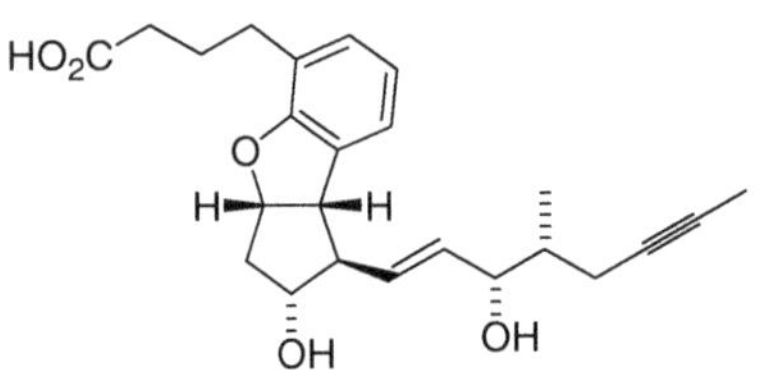	$t^1/_2$	1 h	*V*	–	BERAPROST
	F	–	CL	–	Antithrombotic, vasodilator
	pb	–	Mwt	398.5	Metabolism: oxidative degradation of arylbutyric acid to arylacetic acid, propargylic hydroxylation, isomerization of allylic alcohol to saturated ketone
	ur	0%	PSA	87.0 Å^2	
			log *P*	2.87	
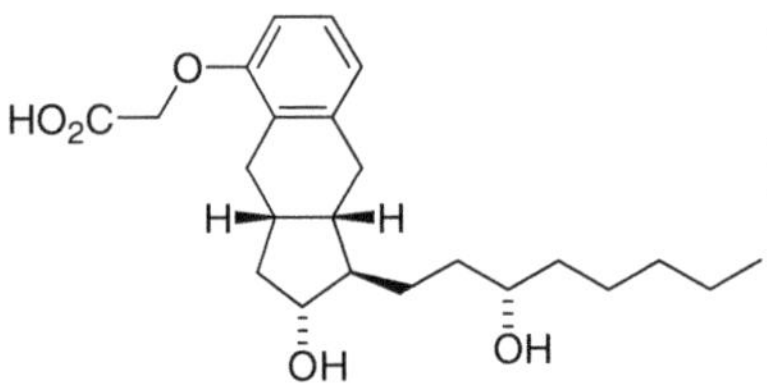	$t^1/_2$	0.9 h (iv)	*V*	0.8–1.1	TREPROSTINIL
	F	–	CL	11	Antihypertensive
	pb	91%	Mwt	390.5	Metabolism: glucuronidation, oxidation of octyl chain
	ur	4% (iv)	PSA	87.0 Å^2	
			log *P*	3.50	
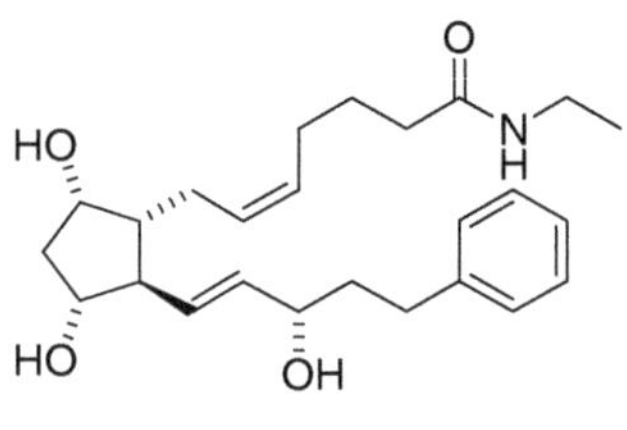	$t^1/_2$	45 min (iv)	*V*	0.67	BIMATOPROST
	F	–	CL	25	Lowers intraocular pressure, antiglaucoma agent
	pb	88%	Mwt	415.6	Metabolism: N-deethylation, glucuronidation, hydrolysis of amide
	ur	–	PSA	89.8 Å^2	
			log *P*	1.98	
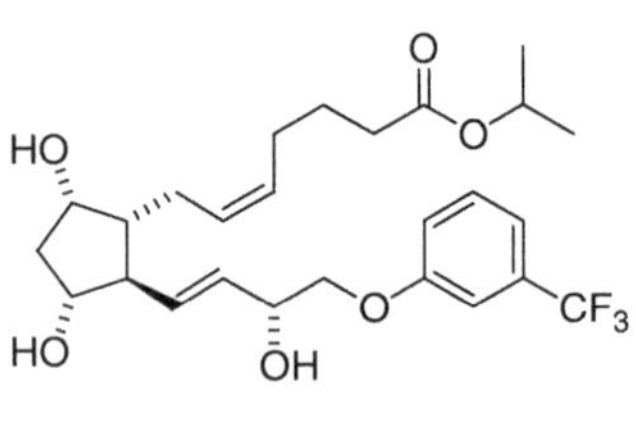	$t^1/_2$	17–86 min*	*V*	2.6* (rat)	TRAVOPROST
	F	Negligible	CL	–	*acid on oral dosing of ester
	pb	–	Mwt	500.6	Lowers intraocular pressure, antiglaucoma agent
	ur	<2%*	PSA	96.2 Å^2	Metabolism of acid: hydrogenation of double bond, oxidation to 15-ketone, oxidative degradation of upper side chain
			log *P*	3.98	

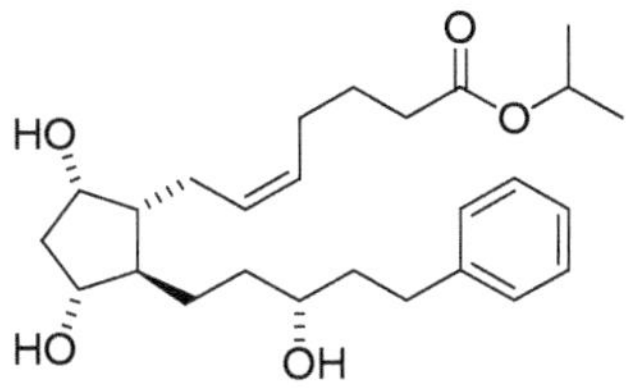	$t_{1/2}$ 17 min* *F* – pb 87%* ur –	*V* 0.16 CL – Mwt 432.6 PSA 87.0 Å^2 log *P* 3.65	LATANOPROST *acid Lowers intraocular pressure, antiglaucoma agent
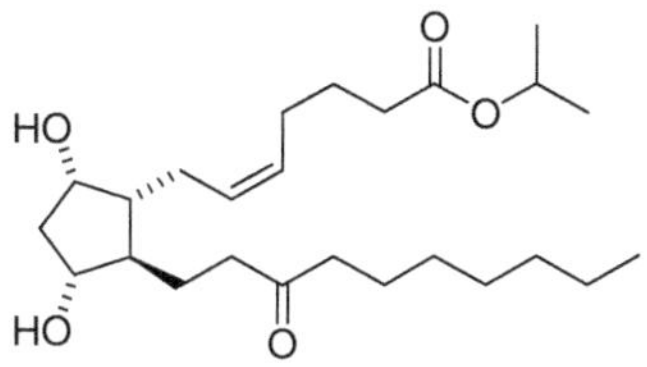	$t_{1/2}$ 14 min* *F* – pb – ur –	*V* – CL – Mwt 424.6 PSA 83.8 Å^2 log *P* 4.63	UNOPROSTONE ISOPROPYL *acid Lowers intraocular pressure, antiglaucoma agent
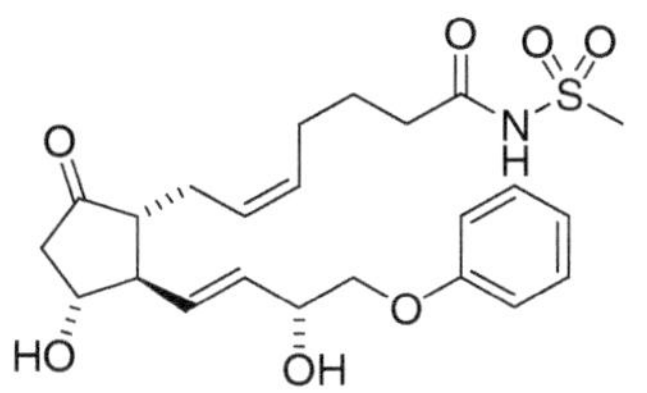	$t_{1/2}$ 2 h *F* – pb 20–30% ur Traces	*V* 17 CL – Mwt 465.6 PSA 138 Å^2 log *P* 1.52	SULPROSTONE Abortifacient Metabolism: elimination of water (to yield cyclopentenone), hydrogenation of *trans*-olefin and cyclodehydration (to yield pyran), hydrolysis of acylsulfonamide to CO_2H
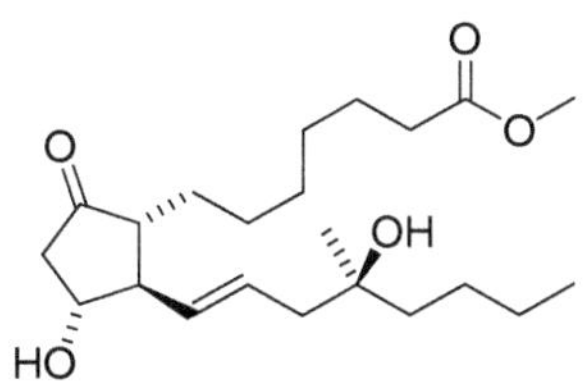	$t_{1/2}$ 0.5±0.4 h* *F* >80%* pb 81–89%* ur 1–4%*	*V* 14±8* CL 240±100* Mwt 382.5 PSA 83.8 Å^2 log *P* 3.80	MISOPROSTOL *acid on oral dosing of ester Antiulcerative; prodrug, not detectable in plasma after oral dose; mixture of two diastereomers Metabolism: oxidative degradation of alkyl chains
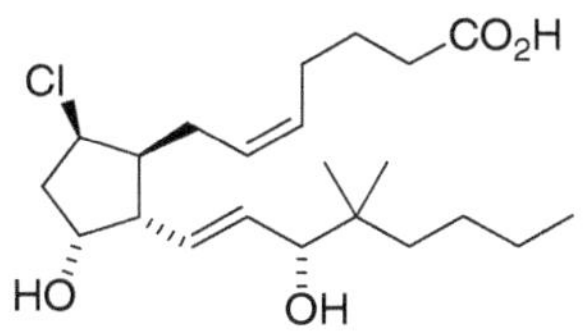	$t_{1/2}$ 4–49 min *F* 2% pb – ur –	*V* 0.16 CL 13.2 Mwt 401.0 PSA 77.8 Å^2 log *P* 4.28	NOCLOPROST Gastroprotective

(*continued overleaf*)

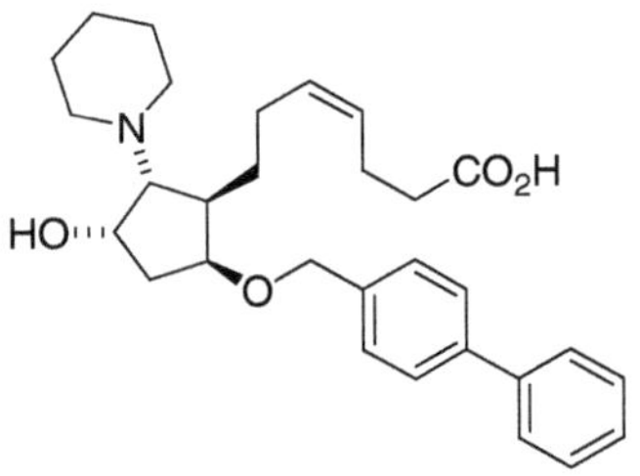

$t½$	1.1–1.4 h*	V	0.8–0.9	VAPIPROST
F	–	CL	7.3–10.3	*iv
pb	–	Mwt	477.6	Thromboxane antagonist
ur	<1%	PSA	70.0 Å^2	
		log P	5.68	

$t_{1/2}$, plasma half-life; F, oral bioavailability; pb, plasma protein binding; ur, excretion of unchanged drug in urine; V, volume of distribution; CL, clearance; Mwt, molecular weight; PSA, polar surface area; PGE$_1$, prostaglandin E$_1$.

72
Sartans

One of the strategies to treat hypertension ("the silent killer") is the administration of an angiotensin II antagonist (Chapter 53). Since the 1970s, a number of peptidomimetics were prepared for this purpose, but all the antagonists found (e.g., saralasin) were also partial agonists and useless because of their lack of oral bioavailability.

The first breakthrough in the development of a small-molecule angiotensin II antagonist was achieved in 1982 by Furakawa at Takeda, Japan [1]. His group patented a series of imidazole acetic acids, including the compounds S-8307 and S-8308 (Scheme 72.1), which were weak but selective angiotensin II antagonists devoid of agonistic activity. These imidazole derivatives were used as leads by David J. Carini and John V. Duncia at DuPont, who in 1983 started to optimize their potency and pharmacokinetic (PK) properties to finally arrive at losartan in March 1986. The main reason for DuPont's endeavor and final success was that the newly hired, unexperienced pharmacologist Pancras C. Wong tested S-8307 in rats at an unusually high concentration (100 mg kg^{-1}). This experiment clearly proved that S-8307 was indeed a selective, competitive angiotensin II antagonist, which worked *in vivo*. Similar experiments at lower concentrations, which had probably been conducted at Takeda and other companies, would not have revealed any antihypertensive activity of S-8307.

Lead Optimization for Medicinal Chemists: Pharmacokinetic Properties of Functional Groups and Organic Compounds, First Edition. Florencio Zaragoza Dörwald.

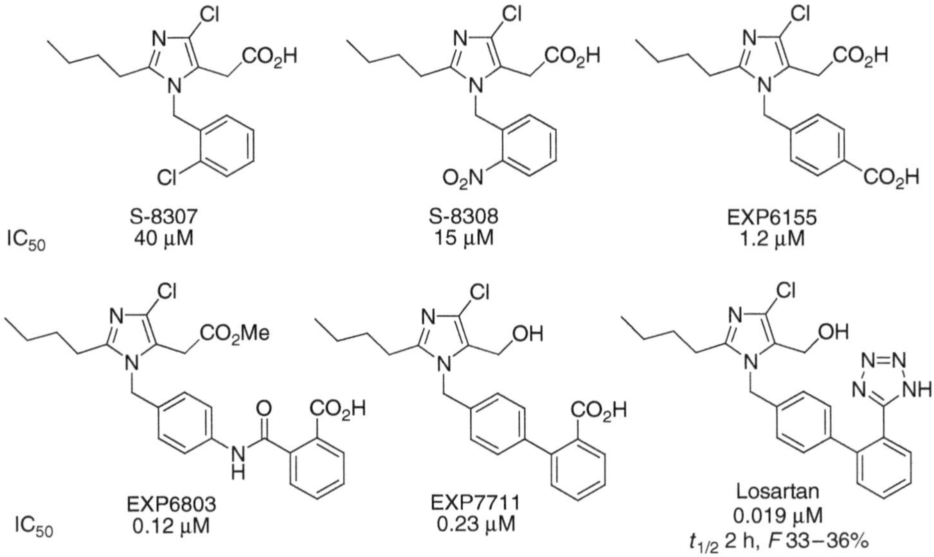

Scheme 72.1 Development of losartan.

Because the pharma team at DuPont was small and unexperienced, they opted for a collaboration with Merck & Co. to push forward the clinical development and marketing of the new drug. Losartan was approved by the FDA in 1995 and generated sales of more than $3 billion by 2005. Since then, a number of other angiotensin II antagonists have been developed as antihypertensive drugs (Table 72.1). These usually show fewer side effects than the angiotensin-converting enzyme (ACE) inhibitors.

Most sartans have low oral bioavailability, mainly because of poor absorption from the GI tract. These molecules are too large and too hydrophilic (large polar surface area (PSA)) to be efficiently absorbed. To enhance their lipophilicity, esters have been occasionally used as prodrugs.

Reference

1. (a) Bhardwaj, G. (2006) How the anti-hypertensive losartan was discovered. *Expert Opin. Drug Discov.*, **1**, 609–618; (b) Timmermans, P.B., Wong, P.C., Chiu, A.T., Herblin, W.F., Benfield, P., Carini, D.J., Lee, R.J., Wexler, R.R., Saye, J.A.M., and Smith, R.D. (1993) Angiotensin II receptors and angiotensin II receptor antagonists. *Pharmacol. Rev.*, **45**, 205–251.

Table 72.1 Antihypertensive angiotensin II receptor antagonists (sartans). V in l kg^{-1}; CL in ml min^{-1} kg^{-1}; Mwt in g mol^{-1}.

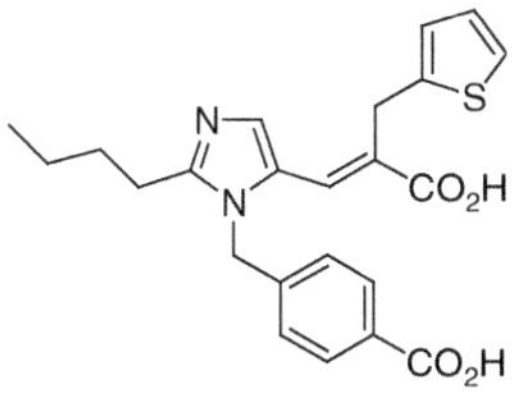	$t^1/_2$ F pb ur	6.8±2.3 h 13±7%* 98% 2–3%	V CL Mwt PSA log P	0.2 1.7–2.0 424.5 116 Å^2 4.96	EPROSARTAN *owing to poor absorbtion from GI tract Metabolism: acyl glucuronidation
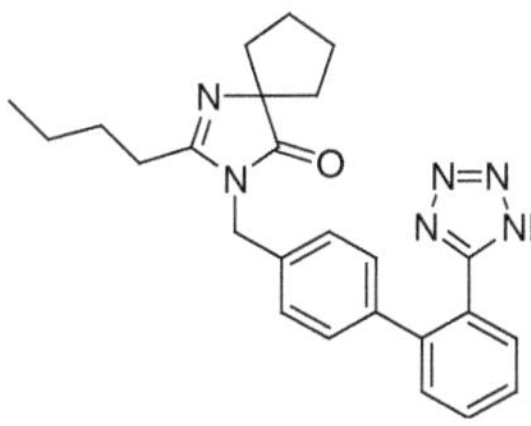	$t^1/_2$ F pb ur	11–15 h 60–80% 90–96% 1–2%	V CL Mwt PSA log P	0.8–1.3 2.4–2.7 428.5 87.1 Å^2 4.50	IRBESARTAN Metabolism: hydroxylation at butyl and cyclopentane, glucuronidation at tetrazole
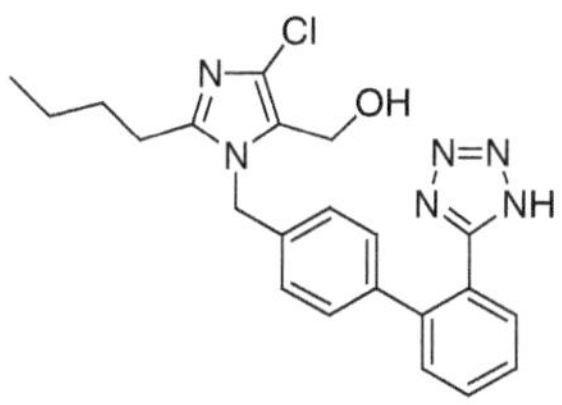	$t^1/_2$ F pb ur	2.0±0.5 h 33–36% 98.7% 9–12%	V CL Mwt PSA log P	0.4–0.5 6.8–8.5 422.9 92.5 Å^2 3.46	LOSARTAN Metabolism: oxidation of alcohol to carboxylic acid, hydroxylation of butyl, glucuronidation of tetrazole
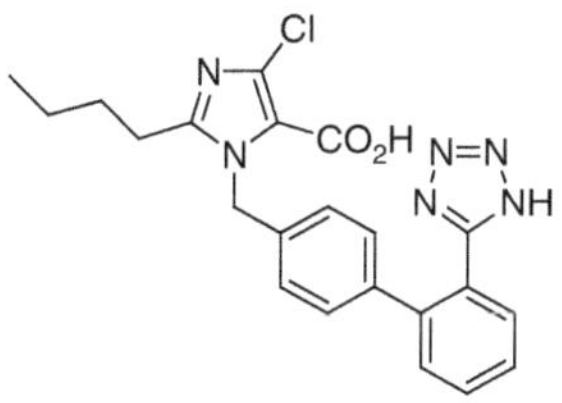	$t^1/_2$ F pb ur	7±3 h – 99.8% 55%	V CL Mwt PSA log P	0.13–0.17 0.6 436.9 105 Å^2 4.79	EXP 3174 Main active metabolite of losartan
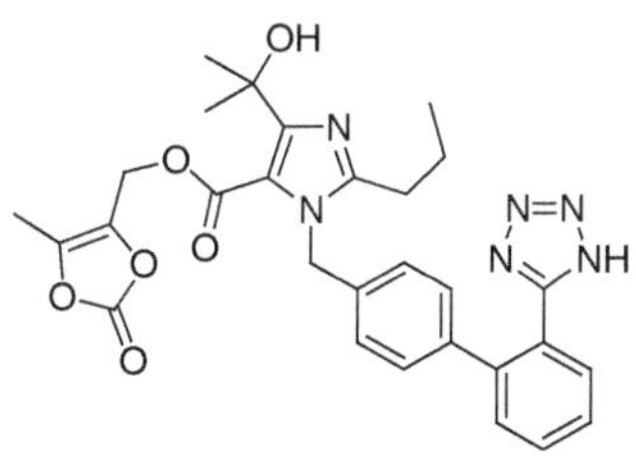	$t^1/_2$ F pb ur	10–15 h* 26%* 99%* 35–50%*	V CL Mwt PSA log P	0.21–0.28* – 558.6 154 Å^2 5.23	OLMESARTAN MEDOXOMIL Prodrug of olmesartan (carboxylic acid) *acid after oral administration of ester

(*continued overleaf*)

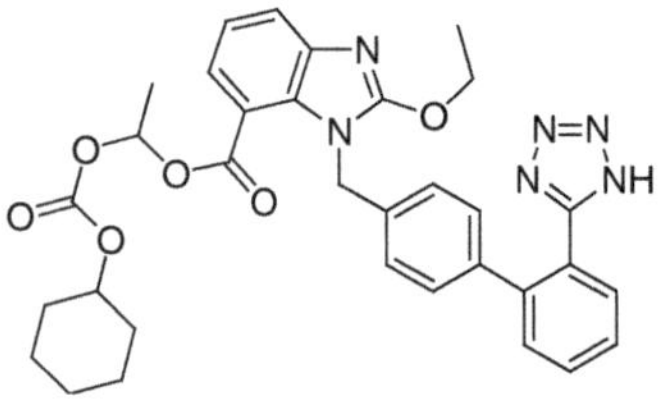

$t_{1/2}$	10±3 h*	V	0.13*	CANDESARTAN CILEXETIL
F	15–40%*	CL	0.4–3.2*	*acid (candesartan) on oral dosing of ester
pb	>99%*	Mwt	610.7	Further metabolism: O-deethylation
ur	33–50%*	PSA	143 Å^2	
		log P	7.79	

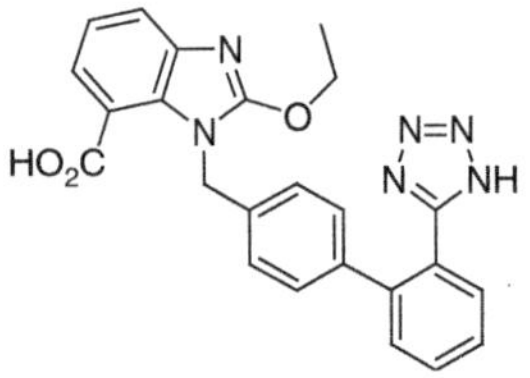

$t_{1/2}$	9–13 h	V	0.13	CANDESARTAN
F	15–40%	CL	0.4–3.2	
pb	>99%	Mwt	440.5	
ur	33–50%	PSA	114 Å^2	
		log P	5.01	

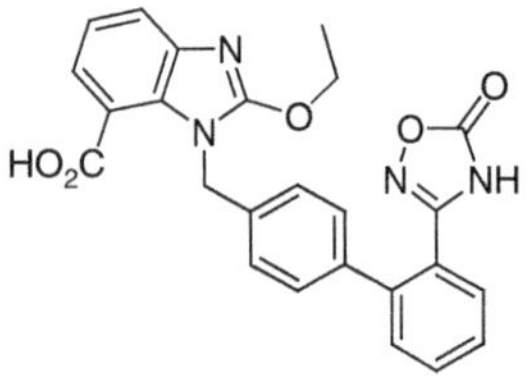

$t_{1/2}$	11 h*	V	0.25	AZILSARTAN
F	60%*	CL	–	*drug on oral dosing of prodrug azilsartan medoxomil
pb	>99%	Mwt	456.5	Metabolism: O-deethylation, decarboxylation
ur	15%*	PSA	115 Å^2	
		log P	5.63	

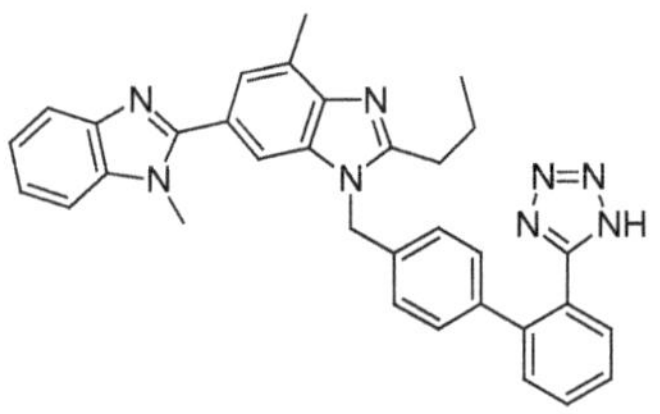

$t_{1/2}$	24 h	V	6.6–7.3	TELMISARTAN
F	42–57%	CL	12.3	Metabolism: glucuronidation
pb	99.5%	Mwt	514.6	
ur	<1%	PSA	72.9 Å^2	
		log P	7.73	

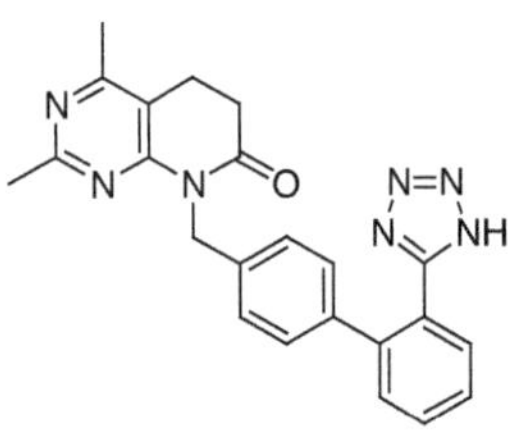

$t_{1/2}$	5.3±0.6 h	V	–	TASOSARTAN
F	60–79%	CL	–	Metabolism: 4-hydroxylation of piperidin-2-one, then elimination of water, hydroxylation of 7-CH_3, conversion to enoltasosartan (see below)
pb	99.8%	Mwt	411.5	
ur	–	PSA	101 Å^2	
		log P	3.42	

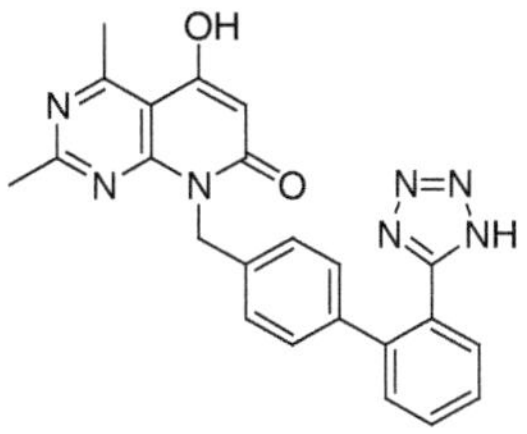	$t_{1/2}$ *F* pb ur	41±2 h – – –	*V* CL Mwt PSA log *P*	– – 425.4 121 Å^2 2.96	ENOLTASOSARTAN Main active metabolite of tasosartan
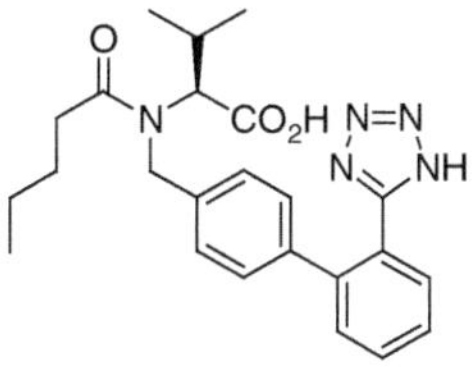	$t_{1/2}$ *F* pb ur	7.3±2.2 h 23–39% 85–99% 7–30%	*V* CL Mwt PSA log *P*	0.24 0.51 435.5 112 Å^2 4.74	VALSARTAN Metabolism: hydroxylation at CH_2CH_3

$t_{1/2}$, plasma half-life; *F*, oral bioavailability; pb, plasma protein binding; ur, excretion of unchanged drug in urine; V, volume of distribution; CL, clearance; Mwt, molecular weight; PSA, polar surface area.

73
Statins

Numerous studies in animals and humans have proven that high plasma levels of cholesterol (hypercholesterolemia) strongly enhance the mortality from coronary artery disease. It has also been shown that drug therapy for hypercholesterolemia is highly effective in reducing risk for cardiac, vascular, and cerebrovascular events.

Hypercholesterolemia may be controlled by diet or by treatment with niacin, fibrates, or 3-hydroxy-3-methylglutaryl coenzyme A (HMG-CoA) reductase inhibitors. HMG-CoA reductase catalyzes the rate-determining step of the biosynthesis of cholesterol (Scheme 73.1).

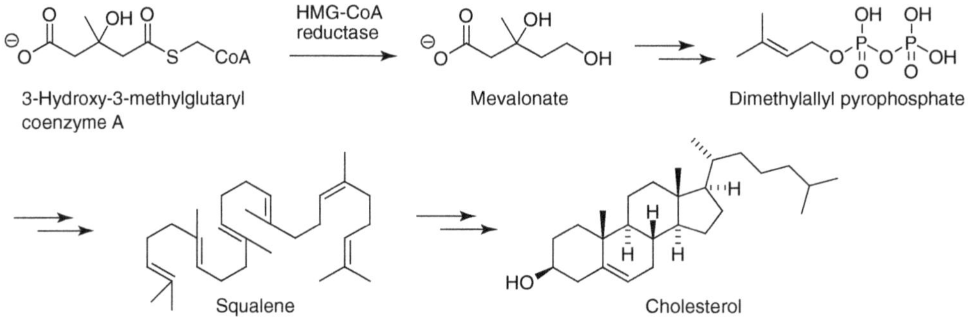

Scheme 73.1 Biosynthesis of cholesterol.

The first natural HMG-CoA reductase inhibitor (mevastatin or compactin) was isolated from a *Penicillium* mold in 1976. Two years later, a second inhibitor, closely related to mevastatin (lovastatin or mevinolin), was isolated from cultures of *Aspergillus*. Fortunately, it was soon discovered that the lipophilic, synthetically demanding decalin substructure of mevastatin and lovastatin allowed ample structural modifications without loss of inhibitory activity. This led to the development of numerous HMG-CoA reductase inhibitors devoid of decalin substructure, which have become bestselling drugs (Table 73.1).

If compared to the natural products, the newer HMG-CoA reductase inhibitors are metabolically more stable and show higher oral bioavailabilities and longer half-lives. Their synthesis is also much easier.

Lead Optimization for Medicinal Chemists: Pharmacokinetic Properties of Functional Groups and Organic Compounds, First Edition. Florencio Zaragoza Dörwald.

Table 73.1 Statins (HMG-CoA reductase inhibitors). V in l kg^{-1}; CL in ml min^{-1} kg^{-1}; Mwt in g mol^{-1}.

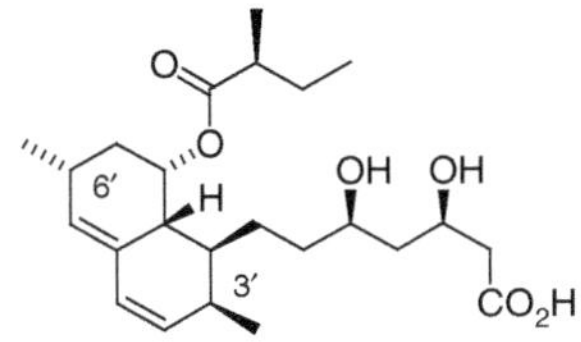	$t^1/_2$ F pb ur	1.4±0.3 h <5% 95% Negligible	V CL Mwt PSA log P	0.87 11±7 404.5 72.8 Å^2 4.31	LOVASTATIN Antilipemic, CYP3A4 substrate Metabolism: 3′- and 6′-hydroxylations, 3′-methyl hydroxylation, oxidation of alcohol to ketone, glucuronides
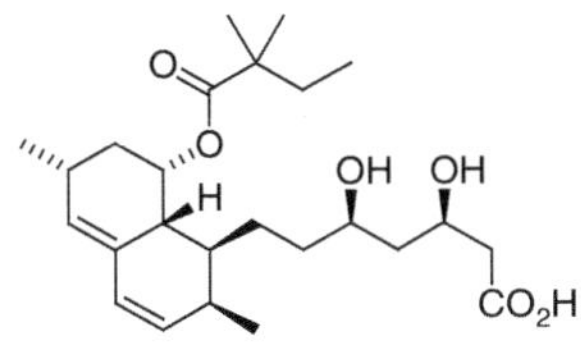	$t^1/_2$ F pb ur	1.9 h <5% 94% <5%	V CL Mwt PSA log P	2.3 (dog) 7.6 418.6 72.8 Å^2 4.72	SIMVASTATIN Antilipemic, CYP3A4 substrate
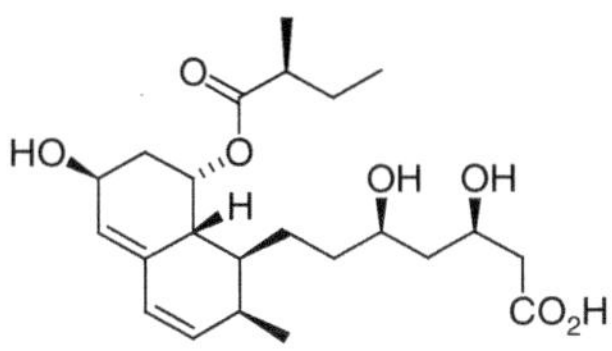	$t^1/_2$ F pb ur	1.8±0.8 h 18±8% 43–48% 47±7%	V CL Mwt PSA log P	0.46±0.04 3.5±2.4 424.5 124 Å^2 2.21	PRAVASTATIN Antilipemic Metabolism: reversible lactone hydrolysis, hydroxylation at MeCH_2, 1,4-dihydroxylation of diene, epimerization, and hydroxy group 1,5-migration of pentadienol
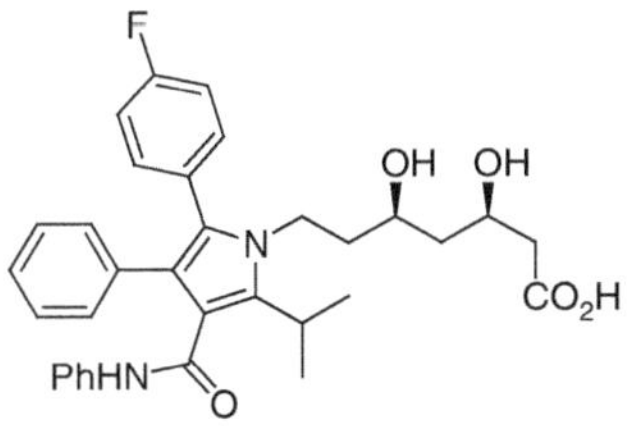	$t^1/_2$ F pb ur	15±7 h 14% >98% <2%	V CL Mwt PSA log P	5.9 31 558.6 107 Å^2 4.13	ATORVASTATIN Antilipemic Metabolism: 2- and 4-hydroxylations of aminophenyl
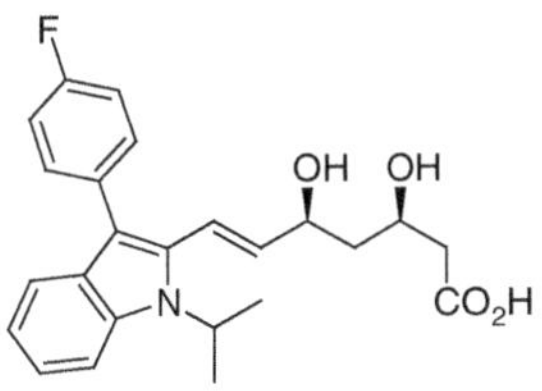	$t^1/_2$ F pb ur	2.7±1.3 h 9–50% >99% <1%	V CL Mwt PSA log P	0.15–0.42 7.9–16.2 411.5 82.7 Å^2 4.57	FLUVASTATIN Substrate and inhibitor of CYP2C9 Metabolism: 5- and 6-hydroxylations of indole, N-deisopropylation

(*continued overleaf*)

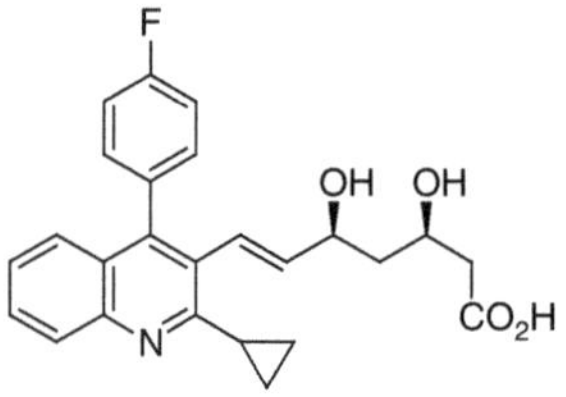	$t_{1/2}$ *F* pb ur	11 h 51–60% >99% <5%	*V* CL Mwt PSA log *P*	2.3 – 421.5 90.7 Å^2 3.45	PITAVASTATIN Antilipemic Metabolism: reversible lactonization
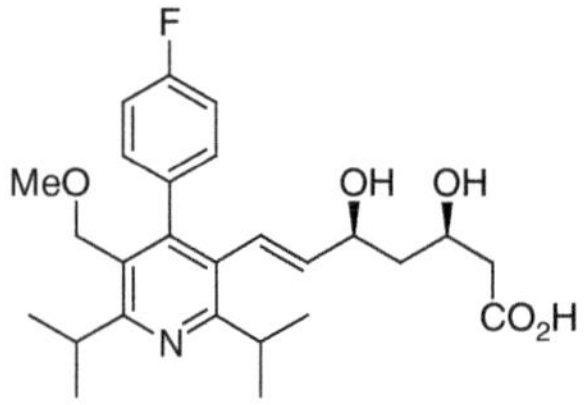	$t_{1/2}$ *F* pb ur	2–4 h 39–100% >99% 0%	*V* CL Mwt PSA log *P*	0.33 2.9–3.1 459.6 99.9 Å^2 3.71	CERIVASTATIN Metabolism: O-demethylation, hydroxylation of CCH_3; withdrawn in 2001 because of toxicity in combination with gemfibrozil
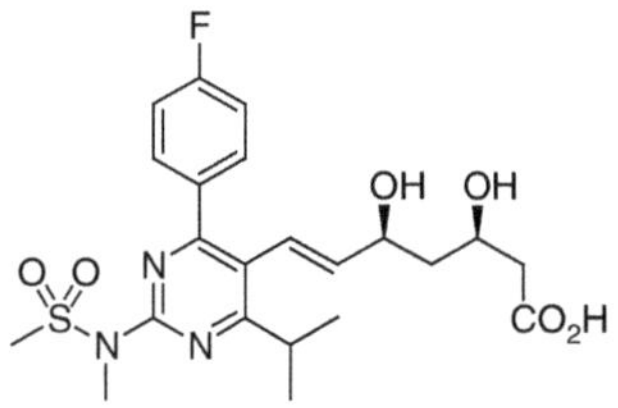	$t_{1/2}$ *F* pb ur	18–25 h 20% 88% 10%	*V* CL Mwt PSA log *P*	1.7 10.5 481.5 149 Å^2 0.42	ROSUVASTATIN Antilipemic Metabolism: N-demethylation

$t_{1/2}$, plasma half-life; *F*, oral bioavailability; pb, plasma protein binding; ur, excretion of unchanged drug in urine; V, volume of distribution; CL, clearance; Mwt, molecular weight; PSA, polar surface area.

74
Folic Acid Analogs (Antifolates)

Folic acid (vitamin B_9) is involved in the biosynthesis of nucleosides. A dietary deficiency of folic acid causes anemia and a decline in the number of blood-generating bone marrow cells. While investigating the effect of folic acid analogs on the growth of cancer cells, Sidney Farber found that some analogs promoted cell growth while others effectively suppressed it. During 1947, aminopterin and methotrexate were shown to induce temporary remissions in 30% of children with leukemia. In 1963, Roy Hertz discovered that antifolates consistently cured choriocarcinoma, the first instance of a cure for a solid tumor. Methotrexate can also be used to treat psoriasis and as an immunosuppressant.

Antifolates (Table 74.1) are inhibitors of dihydrofolate reductase, and thereby block the biosynthesis of thymine and purine. The main metabolic pathways include conjugation with glutamic acid and heteroaromatic hydroxylation.

Table 74.1 Folic acid analogs. V in l kg^{-1}; CL in ml min^{-1} kg^{-1}; Mwt in g mol^{-1}.

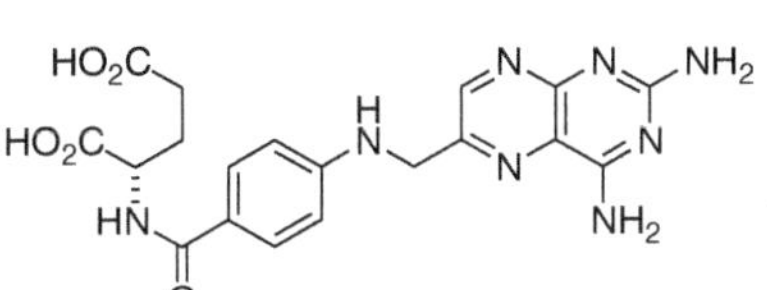	$t^1/_2$	3.6±0.3 h	V	0.6	AMINOPTERIN
	F	84±8%	CL	–	Antineoplastic
	pb	–	Mwt	440.4	
	ur	–	PSA	219 Å^2	
			log P	−0.93	
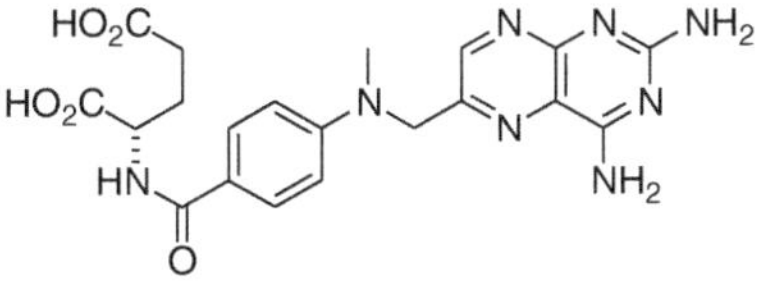	$t^1/_2$	7.2±2.1 h	V	0.55±0.19	METHOTREXATE
	F	70±27%	CL	2.1±0.8	Antineoplastic, antirheumatic
	pb	46±11%	Mwt	454.4	Metabolite: 7-hydroxy
	ur	81±9%	PSA	211 Å^2	
			log P	−0.45	

(*continued overleaf*)

Lead Optimization for Medicinal Chemists: Pharmacokinetic Properties of Functional Groups and Organic Compounds, First Edition. Florencio Zaragoza Dörwald.

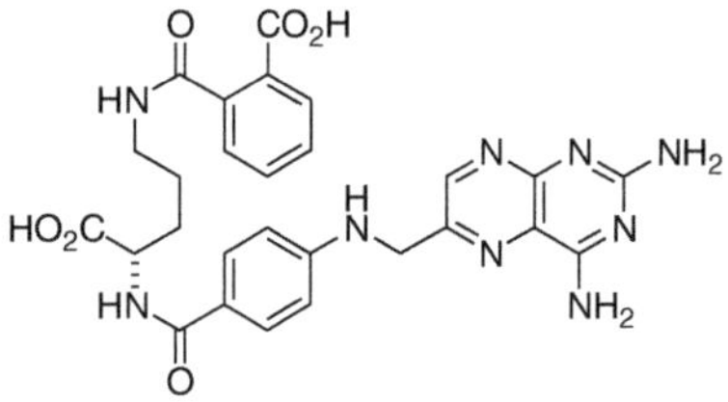

$t^{1}/_{2}$	5.3±1.2 h	V	0.2	PT-523
F	–	CL	0.5	Antineoplastic
pb	–	Mwt	573.6	
ur	40±15%	PSA	248 Å^2	
		log P	−0.46	

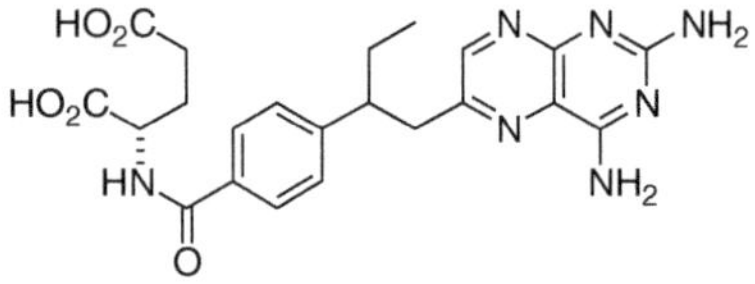

$t^{1}/_{2}$	7–12 h	V	–	EDATREXATE
F	–	CL	6.4±3.6	Antineoplastic
pb	30% (rat)	Mwt	467.5	Metabolite: 7-hydroxy
ur	13–55%	PSA	207 Å^2	
		log P	1.54	

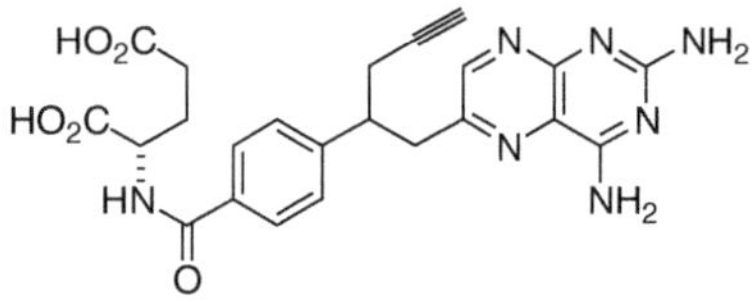

$t^{1}/_{2}$	12–18 h*	V	0.6–1.6	PRALATREXATE
F	–	CL	2.9–6.4	*iv
pb	67%	Mwt	477.5	Antineoplastic
ur	31–38%*	PSA	207 Å^2	Metabolism: glutamylation
		log P	0.23	

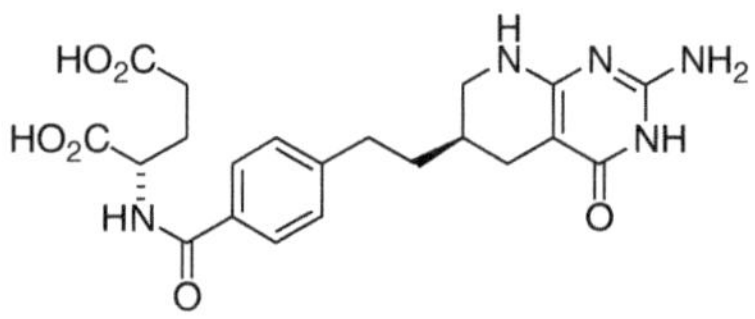

$t^{1}/_{2}$	43±28 h	V	–	LOMETREXOL
F	–	CL	0.8±0.2	Antineoplastic
pb	78%	Mwt	443.5	Metabolism: conjugation with glutamic acid to higher glutamates
ur	–	PSA	183 Å^2	
		log P	1.23	

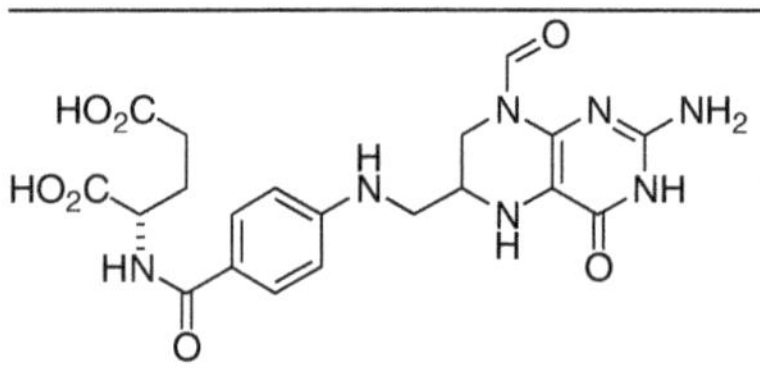

$t^{1}/_{2}$	9.3±2.8 h	V	3.2±1.2	LEUCOVORIN, FOLINIC ACID
F	40–100%	CL	3.9±0.8	Prodrug of tetrahydrofolic acid; mixture of stereoisomers
pb	40±5%	Mwt	473.4	
ur	10%	PSA	216 Å^2	
		log P	−1.43	

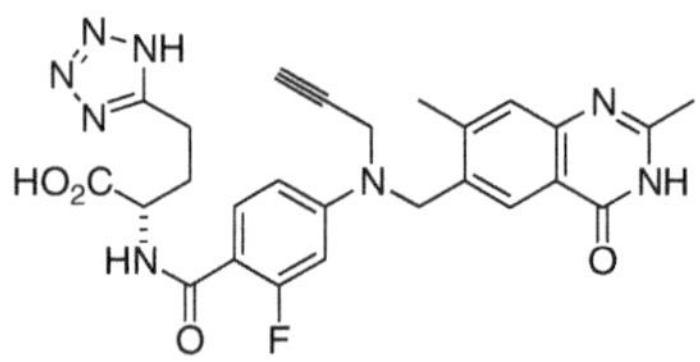

$t^{1}/_{2}$	2–6 d	V	0.3–1.0	ZD 9331
F	80% (dog)	CL	0.05–0.31	Antineoplastic
pb	98%	Mwt	532.5	
ur	–	PSA	166 Å^2	
		log P	1.48	

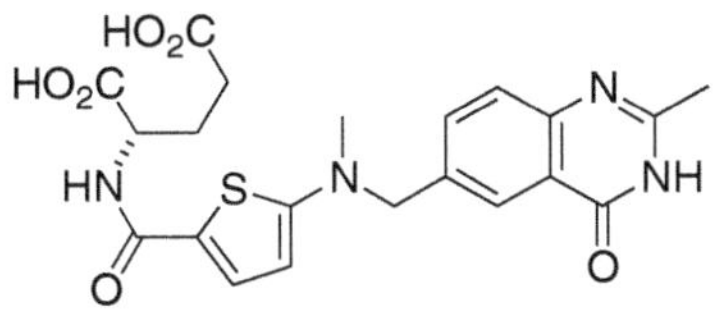	$t_{1/2}$	150–380 h	V	8	RALTITREXED, TOMUDEX
	F	10–20%*	CL	0.6–1.2	*rat
	pb	95%	Mwt	458.5	Antineoplastic
	ur	20–50%	PSA	177 $Å^2$	Metabolism: polyglutamation
			log P	0.86	
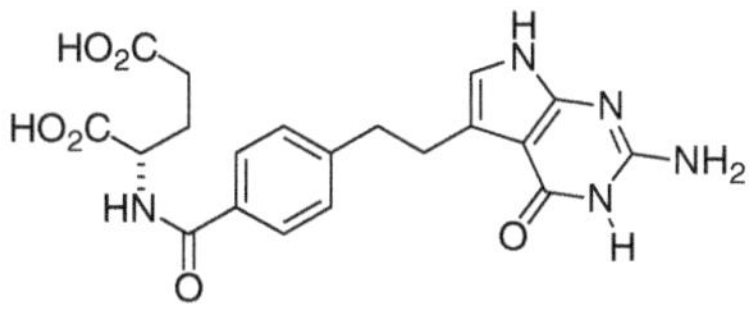	$t_{1/2}$	3.5 h*	V	0.3	PEMETREXED
	F	13%**	CL	1.4	*iv
	pb	81%	Mwt	427.4	**mice
	ur	70–90%*	PSA	187 $Å^2$	Antineoplastic
			log P	−0.70	

$t_{1/2}$, plasma half-life; F, oral bioavailability; pb, plasma protein binding; ur, excretion of unchanged drug in urine; V, volume of distribution; CL, clearance; Mwt, molecular weight; PSA, polar surface area.

75
Taxanes

In the early 1960s, large collections of natural products were screened at the National Cancer Institute (NCI, US). In 1962, one extract showed remarkable activity against several types of tumor. The active principle, paclitaxel (Taxol), was isolated from the bark of the Western yew tree (*Taxus brevifolia*) and its structure determined in 1971 by single-crystal X-ray crystallography by Mansukh C. Wani and coworkers. Paclitaxel was approved for treatment of ovarian cancer in 1992.

In contrast to the vinca alkaloids, which inhibit microtubule formation, paclitaxel acts by promoting microtubule formation, thereby preventing mitosis. Because of its poor solubility, it has to be solubilized with a detergent, Cremophor EL. Still, this formulation is far from optimal, and further, more soluble taxanes or prodrugs thereof are being developed (Table 75.1) [1].

Reference

1. Skwarczynski, M., Hayashi, Y., and Kiso, Y. (2006) Paclitaxel prodrugs: toward smarter delivery of anticancer agents. *J. Med. Chem.*, **49**, 7253–7269.

Table 75.1 Taxanes. *V* in l kg^{-1}; CL in ml min^{-1} kg^{-1}; Mwt in g mol^{-1}.

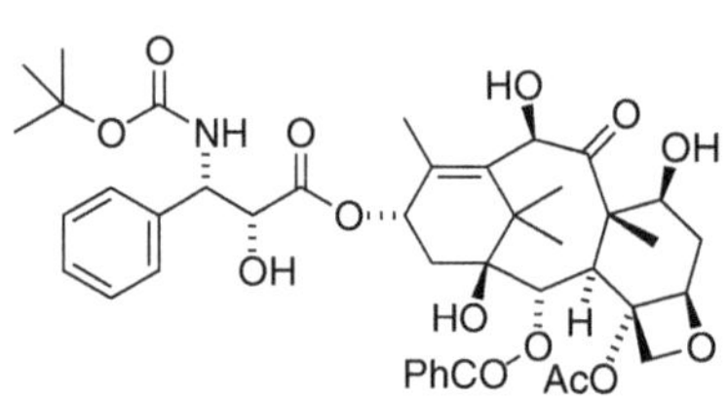	$t_{1/2}$ 11–18 h	*V*	1.7–3.7	DOCETAXEL	
	F –	CL	7.0–9.1	Antineoplastic; Metabolism:	
	pb 93–94%	Mwt	807.9	hydroxylation of *tert*-butyl	
	ur <4%	PSA	224 Å^2	and phenyl	
		log *P*	2.46		

Lead Optimization for Medicinal Chemists: Pharmacokinetic Properties of Functional Groups and Organic Compounds, First Edition. Florencio Zaragoza Dörwald.

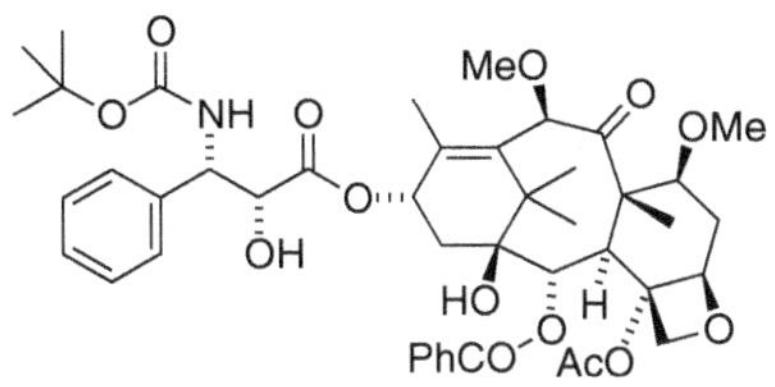	$t_{1/2}$ 62 h *F* – pb 89–92% ur 2.3% (iv)	*V* 71 CL 12 Mwt 835.9 PSA 202 Å^2 log *P* 3.30	CABAZITAXEL Antineoplastic; Metabolism: O-demethylation
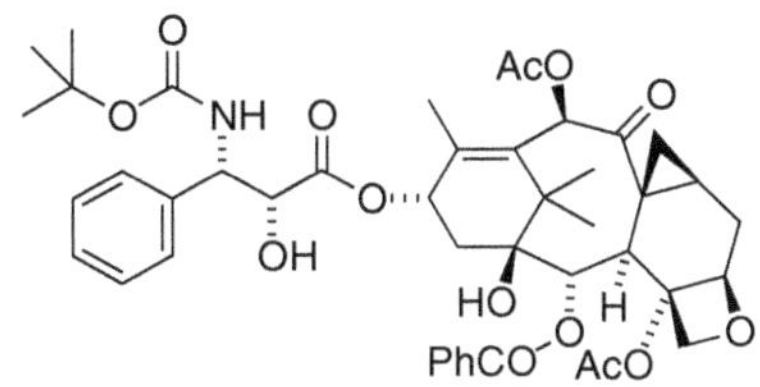	$t_{1/2}$ 20–24 h *F* – pb – ur 1%	*V* 25 CL 19 Mwt 831.9 PSA 210 Å^2 log *P* 5.66	RPR-109881A Antineoplastic
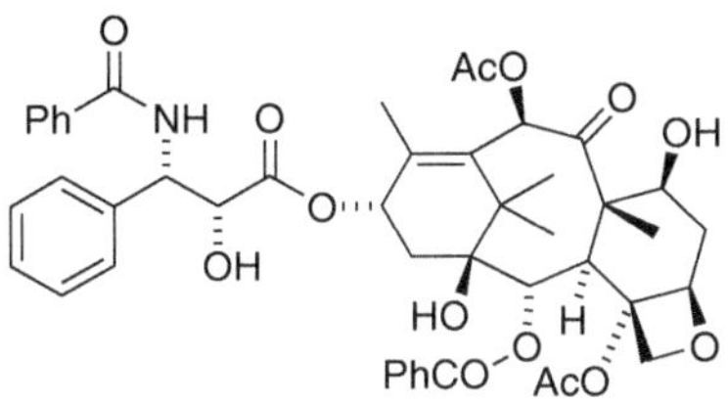	$t_{1/2}$ 3±1 h *F* Negligible pb 95–98% ur 5±2%	*V* 2.0±1.2 CL 5.5±3.5 Mwt 853.9 PSA 221 Å^2 log *P* 3.95	PACLITAXEL, TAXOL Antineoplastic; Metabolism: hydroxylation of phenyl group and methylene group of cyclohexane
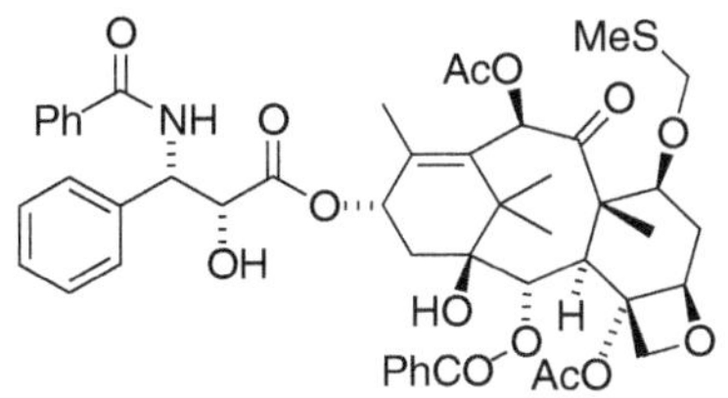	$t_{1/2}$ 34±6 h *F* – pb – ur –	*V* 10±2 CL 5.1±0.7 Mwt 914.0 PSA 236 Å^2 log *P* 5.00	BMS-184476 Antineoplastic; Metabolism: oxidation to sulfoxide
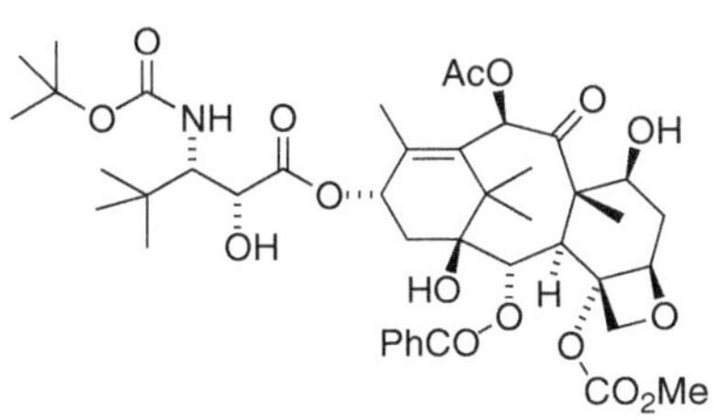	$t_{1/2}$ 22 h *F* 24% pb – ur <3.5%*	*V* – CL – Mwt 845.9 PSA 240 Å^2 log *P* 3.49	BMS-275183 *animals Antineoplastic; Metabolism: ester hydrolysis, hydroxylation of *tert*-butyl

(*continued overleaf*)

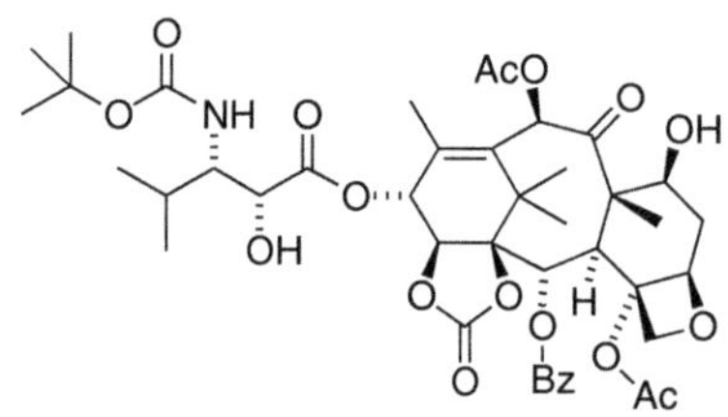	$t_{1/2}$ 70 h *F* 19–31% pb – ur –		*V* – CL – Mwt 871.9 PSA 246 Å^2 log *P* 7.56		ORTATAXEL, BAY 59-8862 Antineoplastic
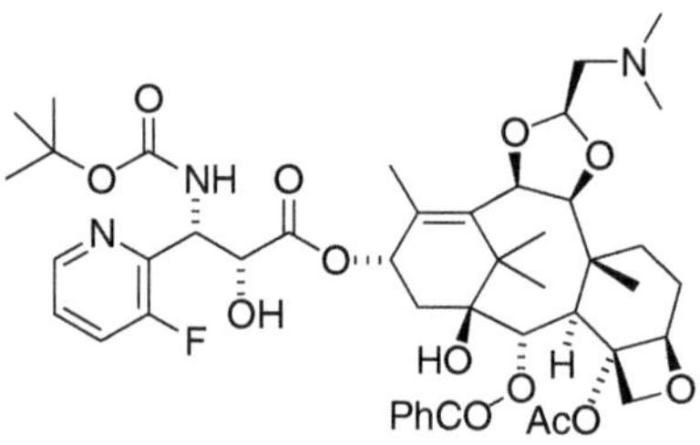	$t_{1/2}$ 167±77 h *F* 63%* pb 92–98% ur –		*V/F* 72±47 CL 6±3 Mwt 882.0 PSA 202 Å^2 log *P* 5.73		DJ-927 *monkey Antineoplastic

$t_{1/2}$, plasma half-life; *F*, oral bioavailability; pb, plasma protein binding; ur, excretion of unchanged drug in urine; V, volume of distribution; CL, clearance; Mwt, molecular weight; PSA, polar surface area.

76
Macrocyclic Compounds

A number of bioactive, macrocyclic natural products are being used as drugs or as lead structures. Despite being highly flexible, and thus having an entropic disadvantage when compared to smaller, less flexible molecules, some of these compounds bind to proteins with high affinity and selectivity [1]. Because of the entropic disadvantage of large, flexible compounds, it is unlikely that selective ligands will be found by screening large, oligomeric molecules, prepared randomly. For this reason, most of the compounds listed below originated from natural products. Cyclic peptides are listed in Table 55.1.

Rapamycin and tacrolimus were isolated in 1965 and 1975, respectively, from soil bacteria of Easter Island. These compounds bind selectively to a protein involved in the immune response (FK-binding protein 12) and are potent immunosuppressants, used to prevent transplant rejection.

The antibiotic erythromycin was also isolated from soil bacteria (1952, J. M. McGuire). Macrolide antibiotics such as erythromycin are bacteriostatic. These lactones bind reversibly to the bacterial 50S ribosomal subunit and thereby inhibit protein biosynthesis. A number of analogs with improved pharmacokinetic (PK) properties have been developed (Table 76.1). Macrolides are important leads for new antibiotics, as antibiotic-resistant bacterial strains continue to emerge (see Ref. [1] in Chapter 70).

Reference

1. Marsault, E. and Peterson, M.L. (2011) Macrocycles are great cycles: applications, opportunities, and challenges of synthetic macrocycles in drug discovery. *J. Med. Chem.*, **54**, 1961–2004.

Lead Optimization for Medicinal Chemists: Pharmacokinetic Properties of Functional Groups and Organic Compounds, First Edition. Florencio Zaragoza Dörwald.

Table 76.1 Macrocyclic compounds. *V* in l kg^{-1}, CL in ml min^{-1} kg^{-1}, Mwt in g mol^{-1}.

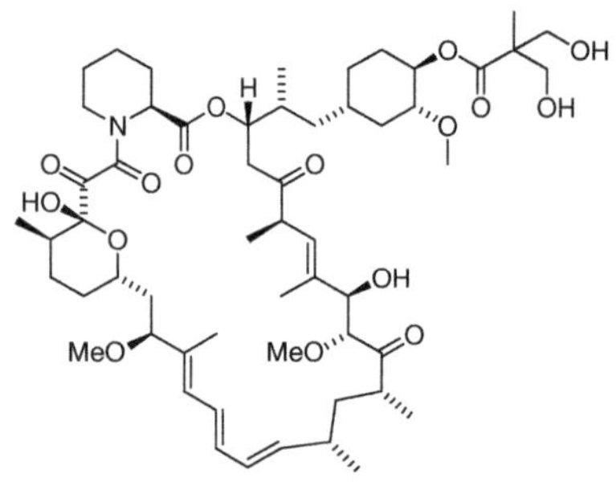	$t\frac{1}{2}$	18 h (iv)	*V*	2.7	TEMSIROLIMUS *monkey Antineoplastic Metabolism: exocyclic ester hydrolysis to sirolimus
	F	22%*	CL	7.7	
	pb	86%	Mwt	1030.3	
	ur	–	PSA	242 Å^2	
			log *P*	3.00	
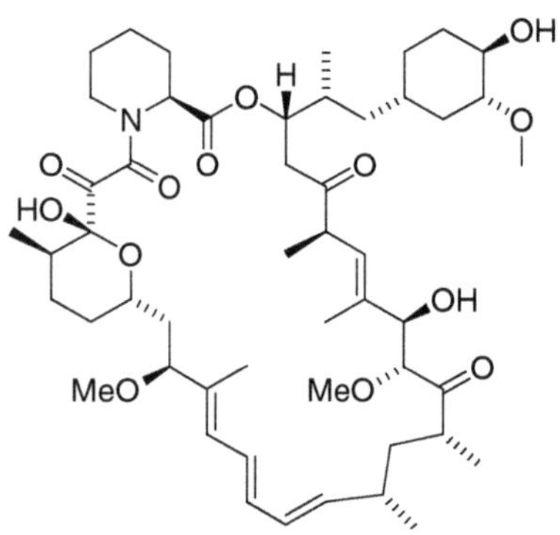	$t\frac{1}{2}$	40–86 h	*V*	23±10	RAPAMYCIN, SIROLIMUS Immunosuppressant Metabolism: O-demethylation, tetrahydropyran 4-hydroxylation
	F	15%	CL	4.6±2.0	
	pb	92–98%	Mwt	914.2	
	ur	<2%	PSA	195 Å^2	
			log *P*	4.26	
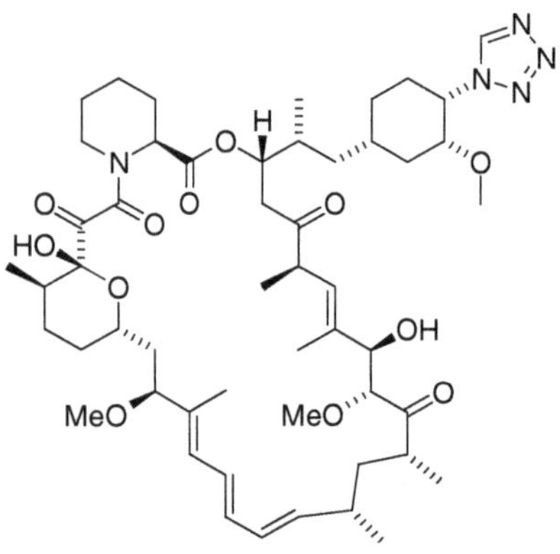	$t\frac{1}{2}$	35 h (iv)	*V*	1.3	ZOTAROLIMUS Immunosuppressant
	F	–	CL	0.7	
	pb	99%	Mwt	966.2	
	ur	0.1%	PSA	219 Å^2	
			log *P*	3.59	

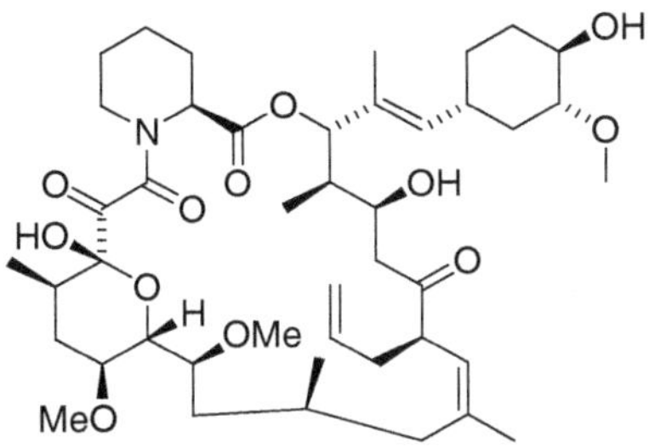

$t^1/_2$	15±7 h	V	0.88±0.31	TACROLIMUS, FK506
F	16±7%	CL	0.70±0.27	Immunosuppressant
pb	75–99%	Mwt	804.0	CYP3A4 substrate
ur	<1%	PSA	178 Å^2	Metabolism:
		log P	4.79	O-demethylation

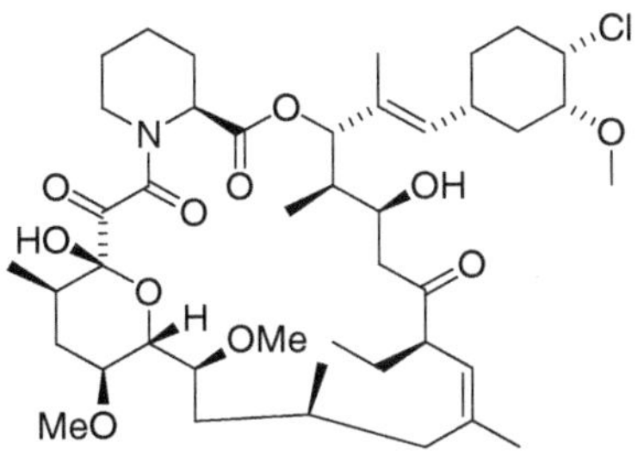

$t^1/_2$	35–92 h	V/F	40–138	PIMECROLIMUS
F	–	CL/F	13–21	Immunomodulator for
pb	74–87%	Mwt	810.5	topical application
ur	0%	PSA	158 Å^2	Metabolism:
		log P	5.76	O-demethylation

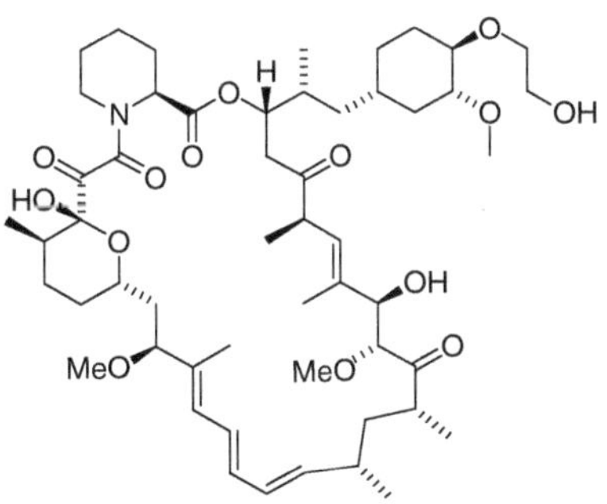

$t^1/_2$	30–43 h	V	4.7	EVEROLIMUS
F	16%*	CL	1.5–2.7	*rat
pb	74%	Mwt	958.2	Antineoplastic
ur	Negligible	PSA	205 Å^2	Metabolism:
		log P	3.39	hydroxylation

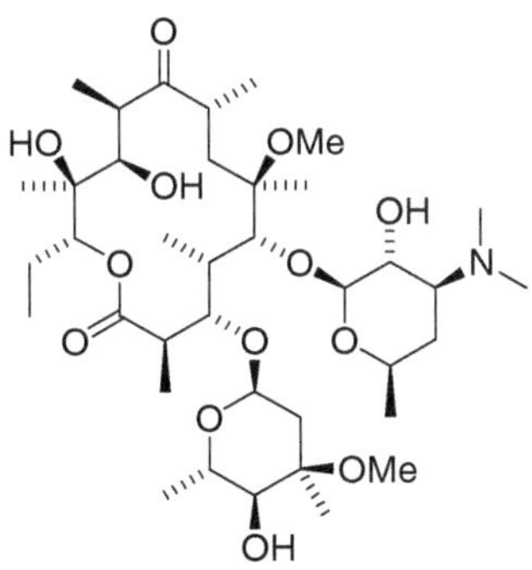

$t^1/_2$	3.3±0.5 h	V	2.6±0.5	CLARITHROMYCIN
F	55±8%	CL	7.3±1.9	Antibacterial
pb	46±4%	Mwt	748.0	
ur	36±7%	PSA	183 Å^2	
		log P	3.16	

(*continued overleaf*)

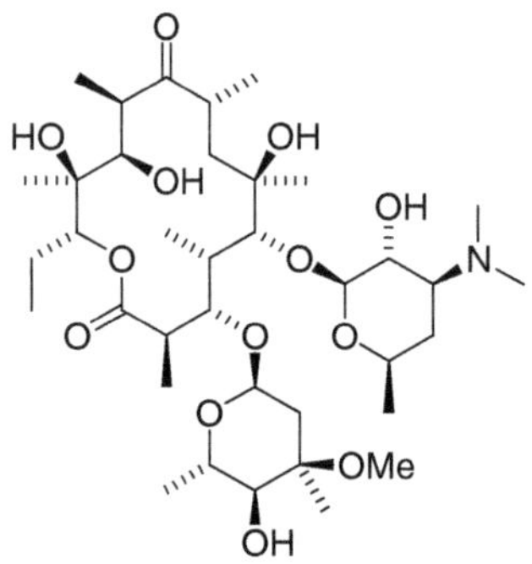

$t_{1/2}$	1.6±0.7 h	V	0.78±0.44	ERYTHROMYCIN
F	35±25%	CL	9.1±4.1	Antibacterial
pb	84±3%	Mwt	733.9	CYP3A substrate
ur	12±7%	PSA	194 Å^2	
		log P	1.91	

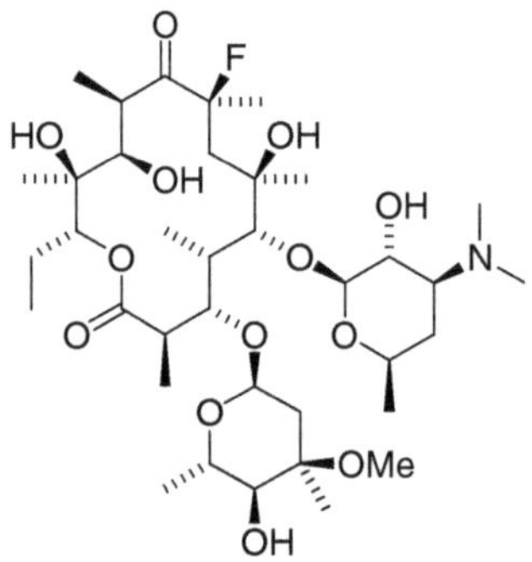

$t_{1/2}$	8–10 h	V/F	5.5	FLURITHROMYCIN
F	Good	CL/F	8	Antibacterial
pb	–	Mwt	751.9	
ur	–	PSA	194 Å^2	
		log P	1.89	

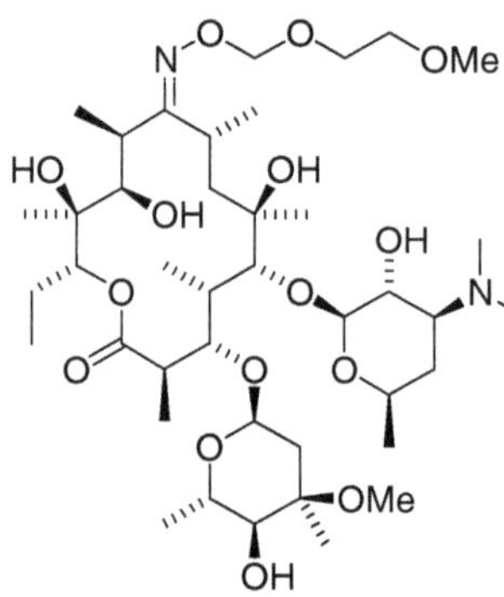

$t_{1/2}$	12 h	V	0.44	ROXITHROMYCIN
F	60%	CL	–	Metabolism:
pb	96%	Mwt	837.1	deglycosylation, N- and
ur	<10%	PSA	217 Å^2	O-demethylation,
		log P	2.84	formaldehyde acetal hydrolysis

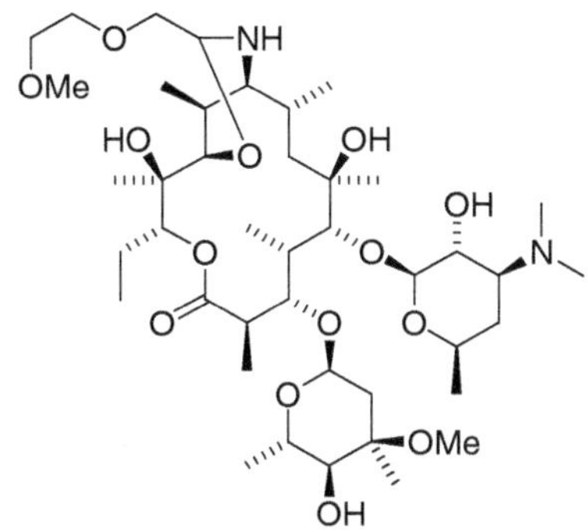

$t_{1/2}$	16–55 h*	V	8–16*	DIRITHROMYCIN
F	10%*	CL	–	Prodrug of erythromyclylamine (hydrolysis of hemiaminal)
pb	15–30%*	Mwt	835.1	*erythromyclylamine on oral administration of dirithromycin
ur	2%	PSA	196 Å^2	
		log P	3.38	

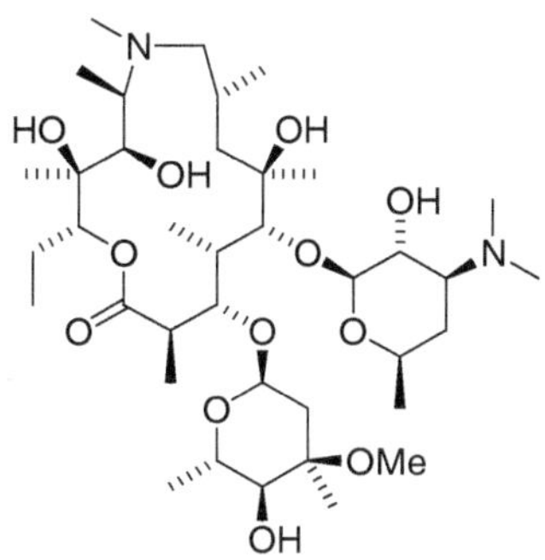

$t_{1/2}$	40 h	V	31	AZITHROMYCIN
F	37%	CL	9	Antibacterial
pb	7–50%	Mwt	749.0	Metabolism: N- and O-demethylation, hydroxylation, hydrolysis of glycosidic bonds
ur	12%	PSA	180 Å^2	
		log P	3.33	

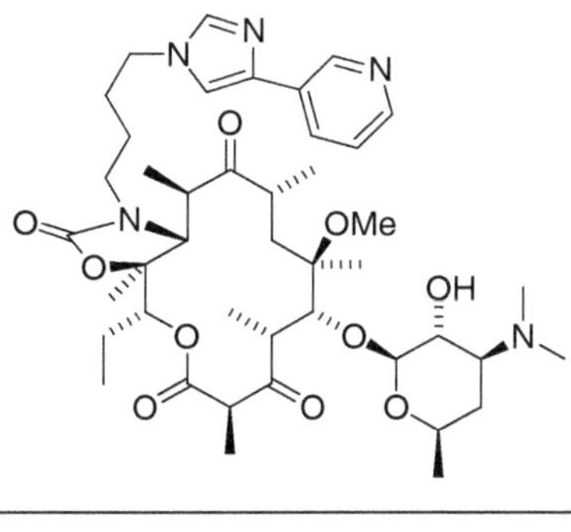

$t_{1/2}$	10–14 h	V	2.0–3.2	TELITHROMYCIN
F	57%	CL	–	Antibacterial
pb	60–70%	Mwt	812.0	Metabolism: N-dealkylation of imidazole to primary alcohol
ur	20–23% (iv)	PSA	172 Å^2	
		log P	5.22	

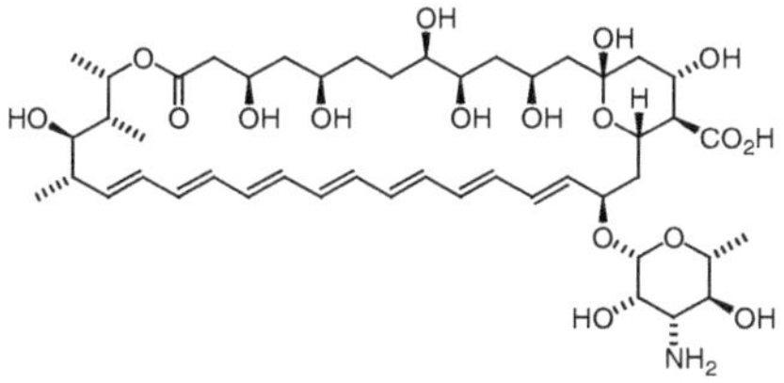

$t_{1/2}$	18±7 h	V	0.76±0.52	AMPHOTERICIN B
F	–	CL	0.46±0.20	Antifungal
pb	>90%	Mwt	924.1	
ur	2–5%	PSA	327 Å^2	
		log P	0.06	

(*continued overleaf*)

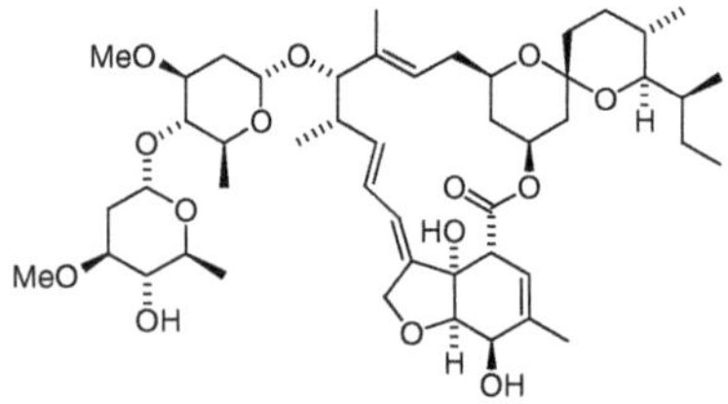

$t_{1/2}$	27 h	V	0.5	IVERMECTIN
F	–	CL	0.3	Antiparasitic (nematodes, arthropods)
pb	93%	Mwt	875.1	Metabolism:
ur	0%	PSA	170 Å^2	hydroxylation,
		log P	5.97	O-demethylation

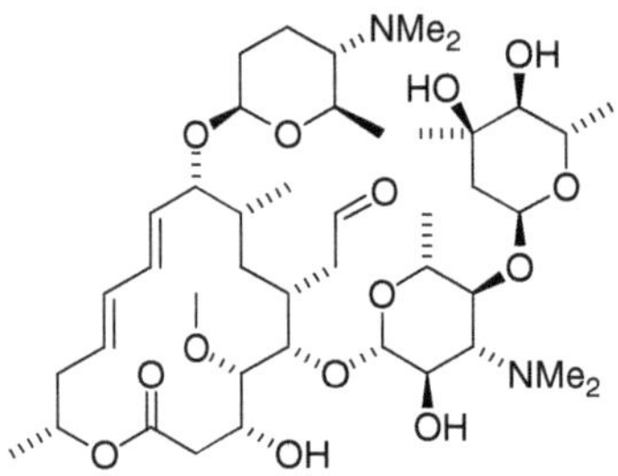

$t_{1/2}$	4–8 h	V	6	SPIRAMYCIN
F	30–40%	CL	8–22	Antibacterial
pb	18%	Mwt	843.1	Metabolism:
ur	14%	PSA	195.4 Å^2	deglycosylation of
		log P	2.30	disaccharide

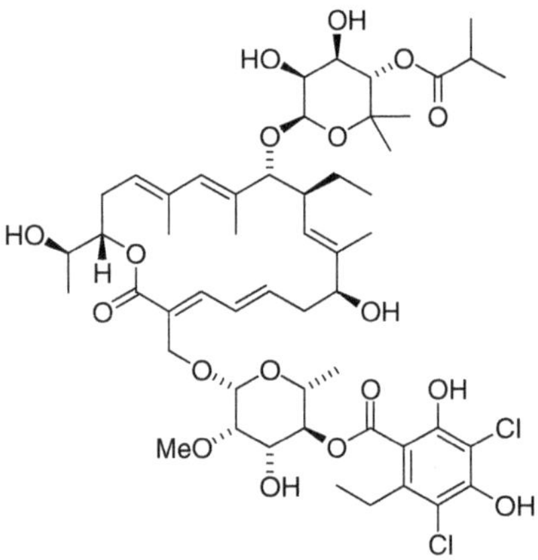

$t_{1/2}$	1–3 h	V	–	FIDAXOMICIN
F	Low	CL	–	Antibacterial, RNA
pb	–	Mwt	1058.0	polymerase inhibitor
ur	0%	PSA	267 Å^2	
		log P	11.0	

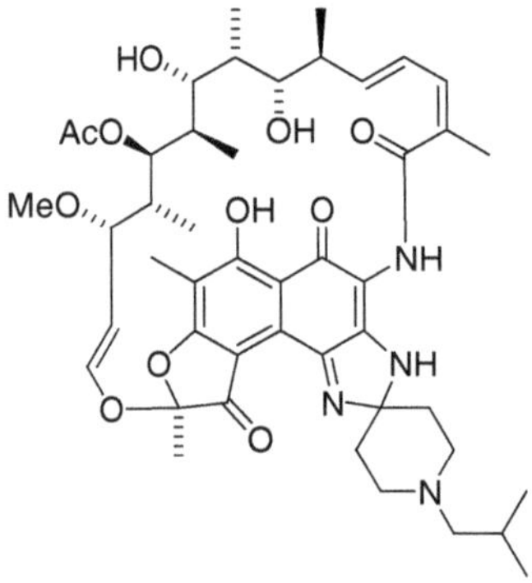

$t_{1/2}$	47±12 h	V/F	40±15	RIFABUTIN
F	20±7%	CL/F	12±3	Antibacterial
pb	85±2%	Mwt	847.0	Metabolism:
ur	7±3%	PSA	206 Å^2	O-deacetylation
		log P	4.49	

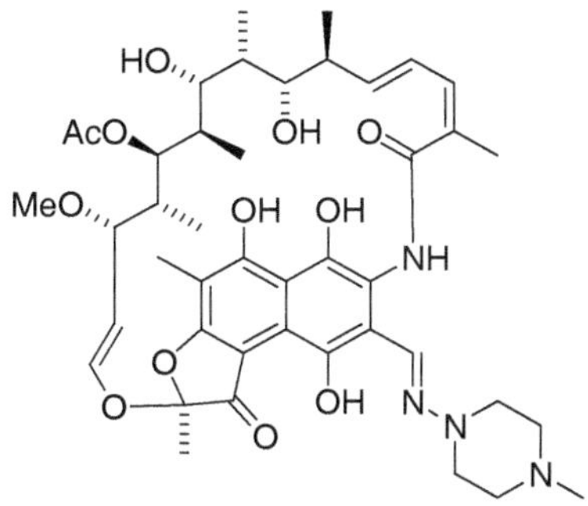

$t_{1/2}$	3.5±0.8 h	*V*	0.97±0.36	RIFAMPIN
F	90–100%	CL	3.5±1.6	Antibacterial
pb	89±1%	Mwt	822.9	Metabolism:
ur	7±3%	PSA	220 Å^2	O-deacetylation
		log *P*	2.05	

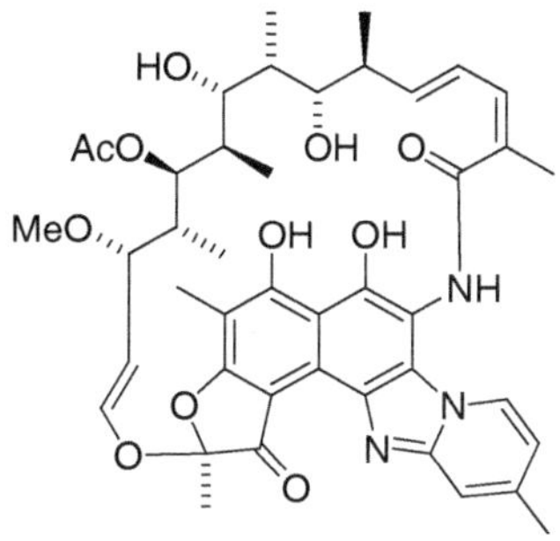

$t_{1/2}$	6 h	*V*	–	RIFAXIMIN
F	<0.4%	CL	–	Antibacterial
pb	68%	Mwt	785.9	Metabolism: deacetylation
ur	<0.1%	PSA	198 Å^2	
		log *P*	3.35	

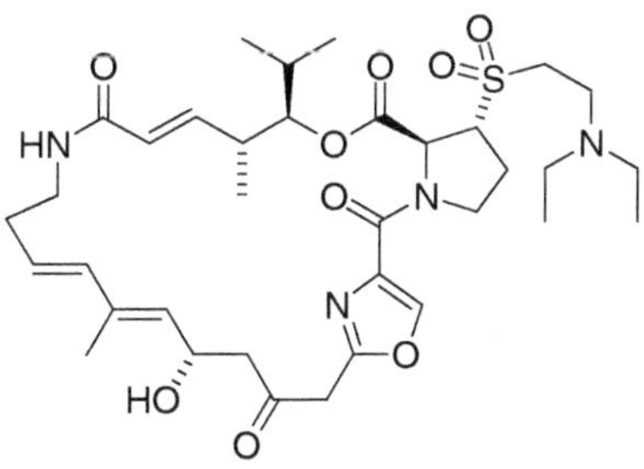

$t_{1/2}$	0.7 h (iv)	*V*	0.39	DALFOPRISTIN
F	–	CL	16	Antibacterial
pb	18%	Mwt	690.9	
ur	0%	PSA	185 Å^2	
		log *P*	−0.94	

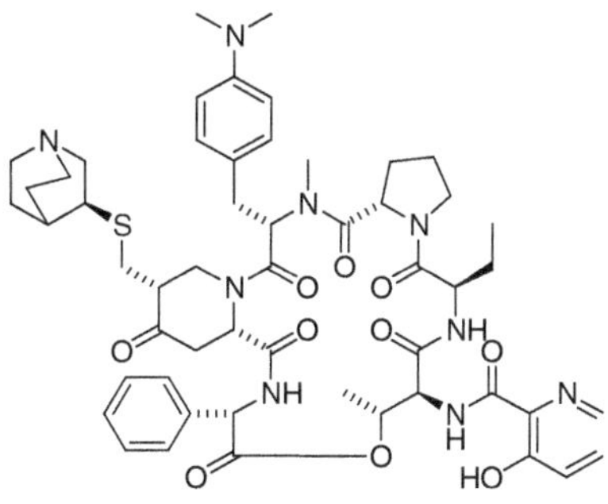

$t_{1/2}$	1–3 h (iv)	*V*	0.7	QUINUPRISTIN
F	–	CL	16	Antibacterial
pb	66%	Mwt	1022.2	
ur	–	PSA	231.2 Å^2	
		log *P*	1.97	

(*continued overleaf*)

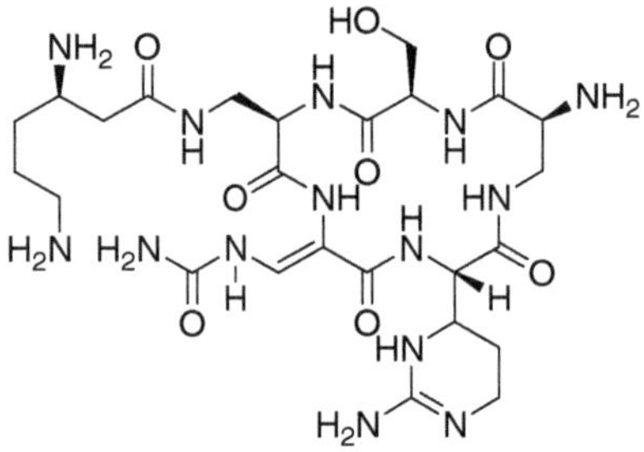

$t^1\!/_2$	5.2±1.9 h	*V*	0.4±0.1	CAPREOMYCIN 1A
F	Low	CL	0.61	Antibacterial
pb	–	Mwt	668.7	
ur	57±19%	PSA	378.4 Å^2	
		log *P*	−8.65	

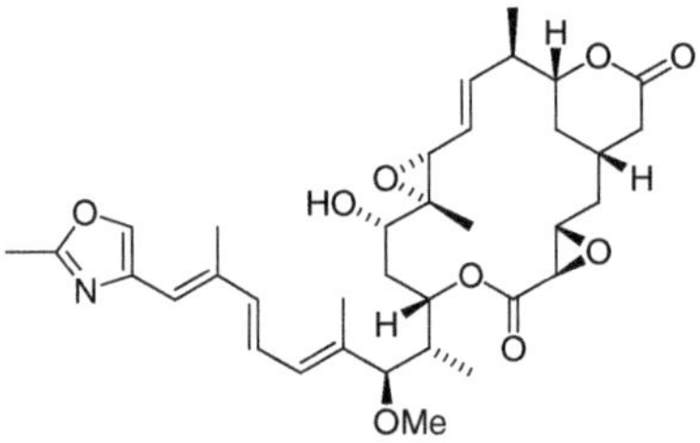

$t^1\!/_2$	4–19 min*	*V*	0.7–4.2	RHIZOXIN
F	–	CL	35–434	*iv
pb	–	Mwt	625.8	Antineoplastic
ur	–	PSA	133 Å^2	
		log *P*	4.30	

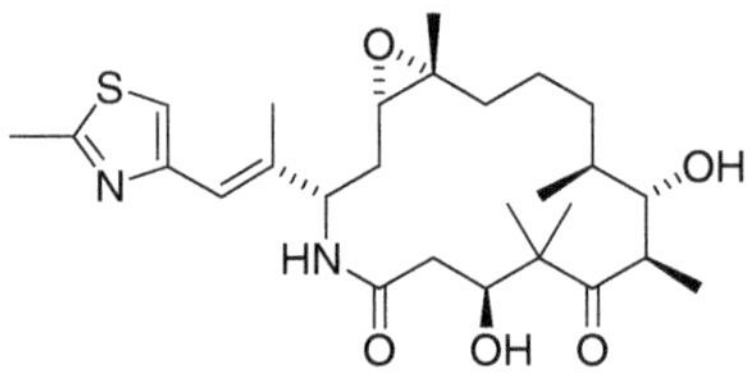

$t^1\!/_2$	17–52 h (iv)	*V*	11	IXABEPILONE
F	–	CL	10	Analog of epothilone B
pb	67–77%	Mwt	506.7	(replacement of lactone
ur	6% (iv)	PSA	140 Å^2	by lactam),
		log *P*	1.77	microtubule-stabilizing
				agent, antineoplastic

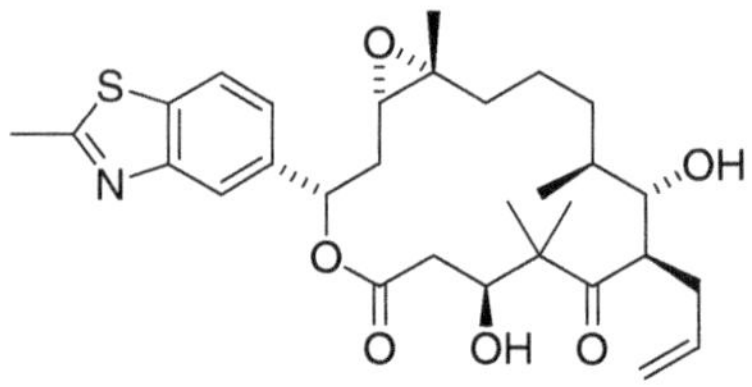

$t^1\!/_2$	53–83 h*	*V*	36–58	SAGOPILONE
F	–	CL	10–19	*β; α: 3 min
pb	–	Mwt	543.7	Microtubule-stabilizing
ur	–	PSA	134 Å^2	agent, antineoplastic
		log *P*	3.19	

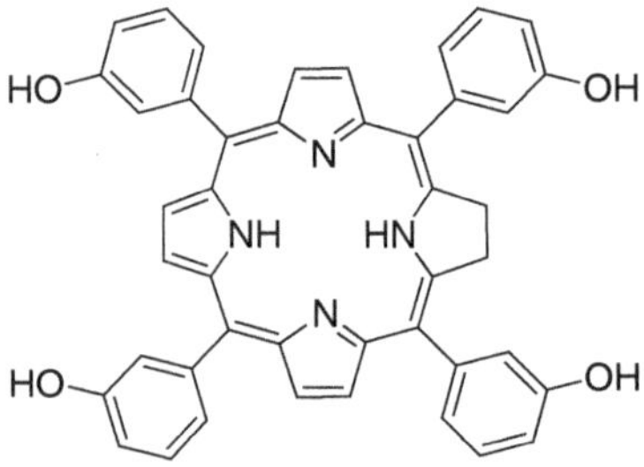

$t^1\!/_2$	45–65 h*	*V*	0.5	TEMOPORFIN
F	–	CL	0.07	*iv
pb	85–88%	Mwt	680.8	Photosensitizer, no
ur	Negligible	PSA	138 Å^2	metabolism
		log *P*	9.60	

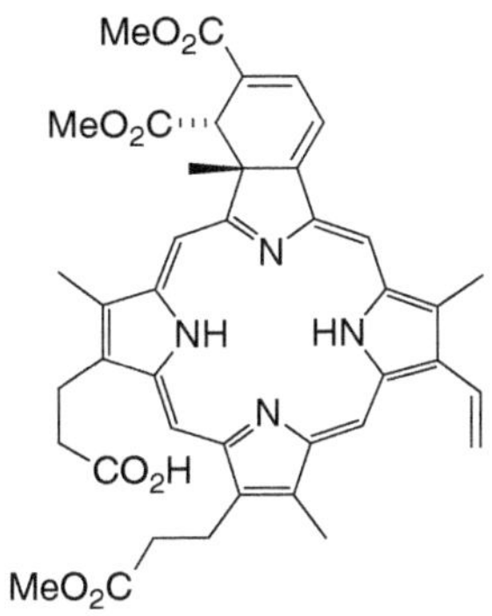	$t_{1/2}$ 5–6 h (iv) *F* – pb 91% ur <0.01%	*V* 0.6 CL 1.7–2.0 Mwt 794.8 PSA 184.6 Å² log *P* 7.38	VERTEPORFIN Photosensitizer
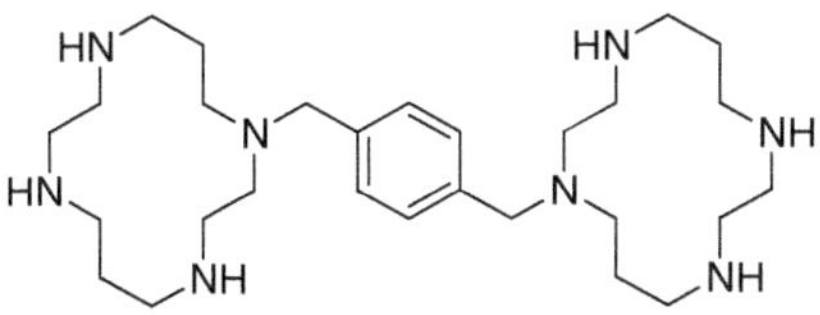	$t_{1/2}$ 3–5 h *F* 3% pb 58% ur 70%	*V* 0.3 CL 1.3 Mwt 502.8 PSA 78.7 Å² log *P* −0.82	PLERIXAFOR α-Chemokine antagonist, immunostimulant; no metabolism

$t_{1/2}$, plasma half-life; *F*, oral bioavailability; pb, plasma protein binding; ur, excretion of unchanged drug in urine; V, volume of distribution; CL, clearance; Mwt, molecular weight; PSA, polar surface area.

Index

Bold page numbers = tabulated PK data

Lead Optimization for Medicinal Chemists: Pharmacokinetic Properties of Functional Groups and Organic Compounds, First Edition. Florencio Zaragoza Dörwald.

b

d

f

g

h

m

o

t

w

x